MAMMAL SPECIES OF THE WORLD

Compiled for the
Parties to the Convention on International Trade in Endangered Species
of Wild Fauna and Flora to serve as a
standard reference to mammalian nomenclature

with scientific advice from members of the
American Society of Mammalogists
and its Checklist Committee

Robert S. Hoffmann, Coordinator

A. L. Gardner, Chairman
R. L. Brownell
R. S. Hoffmann
K. F. Koopman
G. G. Musser
D. A. Schlitter

MAMMAL SPECIES
OF THE WORLD

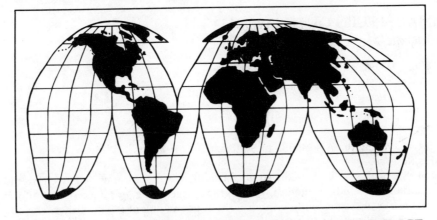

A TAXONOMIC AND GEOGRAPHIC REFERENCE

Edited by

**James H. Honacki
Kenneth E. Kinman
and
James W. Koeppl**

Published as a Joint Venture of
Allen Press, Inc.
and
The Association of Systematics Collections
Lawrence, Kansas, U.S.A.

MAMMAL SPECIES OF THE WORLD

A TAXONOMIC AND GEOGRAPHIC REFERENCE

Published June 21, 1982

Library of Congress Cataloging in Publication Data

Mammal species of the world.

1. Mammals—Classification. 2. Mammals—Geographical distribution. I. Honacki, James H.
II. Kinman, Kenneth E. III. Koeppl, James W. IV. Association of Systematics Collections.
QL708.M35   1982   599'.0012   82-8751
ISBN 0-89327-235-3   AACR2

Copies of **Mammal Species of the World: A Taxonomic and Geographic Reference**
can be ordered from:

JVA-Mammals
P.O. Box 368
Lawrence, Kansas 66044
U.S.A.
(913) 864-4867

ISBN 0-942924-00-2

Preface

A checklist of species is an invaluable tool for both researchers and the interested public. A well-constructed list provides an authoritative reference to the scientific name of a species and may also be a starting point for rewarding research efforts. In recent years, checklist preparation has been viewed as a byproduct of research activity and therefore has not received adequate professional or financial support. Now, when research in the biological sciences is attracting considerable attention in areas ranging from environmental and species conservation to genetic engineering, a reference as basic as a checklist should be an essential element in fostering accurate communications.

Taxonomic checklists are ordinarily prepared by a small number of specialists; the present volume is unique in that 189 professional mammalogists from 23 countries (especially members of the American Society of Mammalogists) have cooperated over the past three years in its production with the Association of Systematics Collections and Allen Press, Inc., both of Lawrence, Kansas, U.S.A.

The project was initiated in response to a request made in 1979 to the Association of Systematics Collections by the Parties to the Convention on International Trade in Endangered Species of Wild Fauna and Flora (CITES) for an authoritative, up-to-date checklist of the species of mammals of the world. One of us (SRE), then contacted the other (RSH, at that time President of the American Society of Mammalogists) seeking his advice. Our discussions led to the conclusion that the optimal objective would be development of an information management system that would involve the cooperation of numerous specialists in various mammalian taxa in concert with computer technology and data base management strategies, so that it would be easier to alter and correct the content of the list as new information became available. Hoffmann volunteered, on behalf of the American Society of Mammalogists, to serve as coordinator of the project, if it could be initiated. Fortunately, the U.S. Department of Energy, desiring a similar listing, provided funding to develop the initial list of species. Other world checklists available at that time were surveyed. It was decided to begin with a hybrid of two: the mammal portion of *Vertebrates of the World*, a computer-based list prepared by the Gainesville Field Station of the U.S. Fish and Wildlife Service, and of particular importance, an unpublished manuscript that had been compiled by Kenneth E. Kinman, one of the editors of this volume. The first draft of the present list was compiled by the editors from these two principal sources, and the categories of information that would be carried in the list were defined.

All members of the American Society of Mammalogists were then sent a letter by the project coordinator, outlining the nature of the project and soliciting the assistance of individuals willing to contribute to the project by serving as reviewers. The response was gratifying, with offers to review anything from one species to entire orders of mammals. It was also encouraging to find that while of the 255 members responding, two thirds were North American residents, the remainder were from other parts of the world, clearly supporting our desire that the project should be international in its scope with respect to the reviewers. The collaboration of the Association of Systematics Collections with the American Society of Mammalogists was subsequently ratified by the Boards of Directors of the two groups, in May and June 1980, respectively.

During this phase of the project, we also adopted two fundamental principles that have added significantly to the quality of the checklist and, it is hoped, will provide a precedent for future efforts of this type. These are: 1) that the systematic and nomenclatorial decisions would be the province of the professional research community (mammalogists in this instance); and 2) that the necessary computer and administrative, clerical, and information management skills would be provided by an organization with experience and credibility in these areas (in this case the Association of Systematics Collections). A further corollary to this first principle was one of non-exclusion; any mammalogist who wished to collaborate in the project was encouraged to do so. This has naturally led to a variety of conflicting opinions being expressed concerning certain taxonomic problems. The editors and the coordinator have attempted to assess these different opinions fairly in the light of the published scientific literature. However, all opinions have been included in the form of "Comments" when germane to the problem. Given the variety of sources in which systematic mammalogists publish, we have undoubtedly overlooked certain critical papers. We have attempted to survey all of the primary literature through 1980, and also have included such information published in 1981 as has come to our attention. In this, we have been greatly assisted by the *ad hoc* committee appointed by the American Society of Mammalogists to review the later drafts of the checklist. These individuals have invested a great deal of effort to improve the accuracy of the list, and without their help, it would be more imperfect than it is. For any omissions, and for other errors involving systematics that may have been committed, the coordinator and editors take full responsibility. Likewise, technical errors arising from the entry and management of the information, including standardization of citations, and index compilation, are the responsibility of the Association of Systematics Collections.

Systematic biology is a dynamic field in which scientific names change, or the higher taxonomic relationships of species are altered as a result of enhanced understanding of the evolutionary relationships of species. We estimate that in the past six years new taxonomic work has affected the status of about 10% of the species of Mammalia. Therefore, this edition should not be viewed as the final or definitive statement on the mammal species of the world, but rather as an effort to represent the "state of the art" of mammalian taxonomy. Data base management strategies and computer technology have been used from the outset in the development of the list. It is our hope that it will be a continuously maintained and improved reference system on mammalian systematics, of which this first edition is only one possible outcome. It is the intention of the Association of Systematics Collections, with the cooperation of the American Society of Mammalogists, to correct and revise the data base continuously. You can assist in this endeavor by informing us of errors in the present text. Moreover, if you have an expertise in a particular taxonomic group, we would like to encourage you to share your knowledge by serving as a reviewer/contributor. We welcome your comments as well as any references to pertinent literature that may have been overlooked. Your interest and cooperation will improve the breadth and quality of future editions.

ROBERT S. HOFFMANN,
Project Coordinator
Past President,
American Society of
Mammalogists

STEPHEN R. EDWARDS,
Executive Director,
Association of Systematics
Collections

Table of Contents

Introduction

The dynamic, rapidly evolving nature of mammalian systematics, documented by an enormous literature, has long hindered the compilation of a detailed, comprehensive world checklist. Information about the taxonomy of mammals is distributed throughout a wide variety of publications. Since complete evaluation of this literature is now a task too great for a single individual, the work in producing this book was divided among 189 editors, reviewers, and consultants from 23 countries who were specialists in the various taxa. The reviewers contributing to this volume have searched the literature in mammalogy, from the earliest publications to the most recent, and have provided references published in many languages. In a broader sense, however, the information about the 4,170 species of mammals considered in this book is the product of countless past students united by a common interest in the diversity of mammals. Perhaps the most important feature of the present volume is that it identifies those gaps in our understanding in greatest need of additional work, and will thus serve as a starting point for those who will follow.

While the decision to consult many professional mammalogists was, we think, theoretically sound, it presented a practical information management problem. The breadth of information and the difficulty in compiling many diverse, often conflicting views has undoubtedly resulted in inadequacies in the present volume. Also, many taxa will benefit from future review by other mammalogists with special expertise. For these reasons, the present volume is conceived as a first step in establishing an ongoing taxonomic data base for mammals.

The volume has been assembled and printed using computer technology capable of rapid text modification and sorting. This permitted automated cross-indexing of the many names employed, and the standardization of literature references and spelling. Even so, complete uniformity of citation, punctuation, format, and of spelling of place names, has not been achieved in this edition due to constraints imposed by our publishing schedule. Moreover, diacritical marks and orthographies employed in languages other than English have had to be omitted in this volume because of computer limitations. Nevertheless, the computerized data base provides the potential for future production of revised editions with minimal investment of time and money.

Within the last hundred years, the first comprehensive assessment of all mammals of the world was undertaken by Trouessart (1898–1899, 1904–1905). More recently, E. P. Walker and colleagues brought out, in 1964, the first edition of *Mammals of the World*. This three volume work, now in its third edition (1975), was systematically organized (but included much additional biological information) to the generic level, and frequently listed the species recognized in a genus; a photograph or other illustration was provided for each genus. Walker's *Mammals of the World* was the basis for the three volume work by V. E. Sokolov, *Systematics of Mammals*, published in the Russian language between 1973 and 1977; within each generic account there was a list of species recognized, together with a brief description of geographic distribution. Finally, Corbet and Hill (1980) listed the species of the world, their abbreviated distributions and common names, and provided citations to major regional works used in compiling their list, as well as citations to literature used in modifying presentations in the major regional works. It is illuminating to compare the present volume with that of Corbet and Hill to evaluate the differences in treatment of some taxa. These differences reflect, in many cases, variation in interpretation and emphasis of literature, and in some cases, the inclusion of recent literature not available to Corbet and Hill.

The Process of Compilation

It may be useful to recount the process of compiling and editing the information contained in *Mammal Species of the World* so that the reader will understand better how the final form of the book was achieved. A preliminary checklist of species names was developed in 1979, based on a manuscript list developed by Kenneth E. Kinman in 1976–1977, and the mammal section of *Vertebrates of the World* (Gainesville Field Station, U.S. Fish and Wildlife Service), which was in turn based on the International Species Inventory System (ISIS) list (Seal and Makey, 1974). This preliminary list was compared with available regional checklists, including page proofs for Hall (1981) which the author generously made available to us. Where differences were detected, the most recent primary literature was consulted to resolve questions, and explanatory comments and literature citations were included to reflect alternative treatments. We realized that many museums contain specimens bearing obsolete names (often correctly employed when the specimen was catalogued) as does the scientific and popular literature. We therefore decided to associate these synonyms with currently used names, in order to enhance the utility of the checklist. This stage of development resulted in a draft checklist for review by professional mammalogists.

Members of the American Society of Mammalogists and other professional mammalogists were invited to contribute, and were asked which taxa they wished to review. Responses were received from 255 individuals and, early in 1980, all were sent appropriate portions of the draft checklist. Reviewers were asked to provide, for each species, the following categories of information: 1) author of the scientific name of the species, and citation; 2) type locality; 3) brief verbal description of distribution; 4) citations of revisions or reviews, important synonyms, and, when necessary, explanatory comments. The number of completed reviews received was 150, with about 85 percent of the species of mammals in the draft checklist covered by at least one reviewer. Subsequently, additional reviewers were found for the remainder of the species. During 1980 and early 1981, the reviews were compiled by the editors and coordinator, discrepancies were checked in the literature, additional references were incorporated and the second draft compilation was returned to reviewers for comment, together with questions about discrepancies and any other issues that had arisen in the course of compiling the draft. The reviewers' corrections and additions were then incorporated into a third draft. At this point, it was evident that several families had received inadequate review with respect to the volume of recent literature available and still more reviewers were solicited to update those sections.

The resulting new information was incorporated into a penultimate draft that was forwarded to the Checklist Committee of the American Society of

Mammalogists, whose members were Alfred L. Gardner (Chairman), Robert L. Brownell, Robert S. Hoffmann, Karl F. Koopman, Guy G. Musser, and Duane A. Schlitter. Members of this committee provided the final review of the text, detected many errors, contributed to the resolution of many of the remaining taxonomic and nomenclatorial problems, and provided much additional useful information previously overlooked, derived from recently published literature. The importance of their contributions to this effort cannot be overemphasized. Unfortunately, a few of the problems pointed out by the committee remain unresolved in this volume.

The process of compiling successive drafts involved the comparison of various contributions, and the inclusion of all information for which there was common agreement. Often, however, there was not agreement. In such cases, we endeavored to verify the information with original published sources, but it has not been possible to check all original sources, and because of this some errors may remain. Of particular concern was the status of the names of taxa. When reviewers disagreed, the editors and coordinator consulted the primary literature and the most recent work was generally followed, with positions taken by other reviewers noted, and pertinent literature cited. Reviewers were then requested to make any further changes they deemed necessary; this process established the relationships of contending names in this list.

Two categories of information provided in this volume were not included in the review process. The Protected Status of various taxa was derived from the Federal Register (U.S. Gov. Print. Off., Washington, D.C.) and is current to January 1, 1982. The International Species Inventory System (ISIS) numbers are used in a computerized monitoring system for zoo animals in order to locate captive live specimens, and to apply correct names to zoo animals, were derived from Seal and Makey (1974). The reader will note that not all species recognized here were included in the ISIS list, and that in both ISIS numbers and Protected Status names, a number of misspelled names occur; in such cases they are individually listed.

Organization of the Book

Species names included in this volume apply to living or recently extinct species (possibly alive during the preceding 500 years); in cases where the continued existence of a species is in doubt, this has been noted in the Comment section.

Taxonomic Arrangement—Many authors have revised the higher taxonomy of mammals since Simpson (1945) published his classical arrangement, among them Ride (1964), Anderson and Jones (1967), and McKenna (1975). These treatments were modified by Corbet and Hill (1980, *A world list of mammalian species*), and their arrangement of the orders and families has been generally followed in this volume (see also Eisenberg, 1981, for a somewhat different arrangement). When in certain cases it has been necessary to augment or modify this arrangement, this has been noted and documented by citation. Generic and specific names are ordered alphabetically within families to facilitate their location. Other taxonomic categories (subfamily, subgenus) have been employed when it was deemed particularly useful.

The scientific names employed for domesticated mammals, and for the wild ancestors of domesticated forms, are in all cases the earliest available valid names, as called for by the Code of the International Commission on Zoological Nomenclature. A different approach has been used by Corbet (1978) and Corbet and Hill (1980), in which names based on domesticated forms were not employed, and the first valid name based on a wild individual was used instead (except for *Camelus dromedarius*, with no known wild ancestor).

Scientific Name and Authority—The non-systematist may wonder why the name(s) of the author(s) of a scientific name, followed by the year in which the species was first described, *i.e.*, *Mus musculus* Linnaeus, 1758, is sometimes enclosed in parentheses, *i.e.*, *Vulpes vulpes* (Linnaeus, 1758). Species that were originally named by their author in a genus other than the one in which they are now placed have parentheses surrounding the author and date; thus, Linnaeus, when he first named the red fox *vulpes*, placed it in the genus *Canis*, instead of *Vulpes*. Should it be replaced in *Canis*, the parentheses would be removed, *i.e.*, *Canis vulpes* Linnaeus, 1758. Following the species name, authority, and date is the citation of the work in which the original description appeared. Usually, only the first page of the work on which the species name appears follows the title citation. However, some systematists follow the tradition of citing more than the first page, and may include references to figures or plates; sometimes the full pagination is listed instead. We have attempted to cite the first page on which the name appears, but verification has not been possible in all cases, and sometimes citations appear as they were received from reviewers.

Type Locality—The type locality is the geographical location at which the type material of a species was collected. This information has been arranged where possible with the current country name (sometimes abbreviated) followed by state, province, or district, and specific locality. Overseas possessions are listed by local name, with the abbreviated country name in parentheses, *i.e.*, Falkland Isl. (U.K.), if this is not otherwise apparent. Altitude above sea level has been included when provided, as have global coordinates in some cases. Metric equivalents of U.S. units of measurement have also been inserted to assist the user. When appropriate, restrictions of the type locality made by revisers have been included in the Comment section. We have followed *The International Atlas*, Rand-McNally, 1979, and some recent political changes may not be included. In particular, several Pacific island groups, such as Vanuatu (formerly, New Hebrides) which have recently become independent are not listed as such.

Distribution—The geographical range of each species is summarized using contemporary political units or, in some cases, geographical names for brevity. For example, island names such as Borneo are used, rather than the several political units on Borneo. However, these geographical names are usually used only when the entire area is included in the range of the species. An attempt has been made to standardize usage and spelling, but some inconsistencies may remain. In particular, we have made no attempt to employ the new Pinyin system of transliteration for Chinese; although a few reviewers used it, most transliterations of Chinese place-names are in the traditional form. Most Russian names are expressed in the transliteration system of the U.S. Board of Geographic Names. Place-names from other non-Latin alphabets are given as received from reviewers. Distribution records resulting from human introduction are noted. Maps of the distributions of Palearctic and North American mammals are provided in Corbet (1978) and Hall (1981) respectively. Other maps and detailed distribution information can be located in the regional works listed in the Bibliography.

Comments—Considerable effort has been expended to provide a taxonomic list consistent with recent literature. Commonly used synonyms are included here, together with citation of the authority who determined their status. These are usually either names of later origin than the species name used (junior synonyms) or names that are systematically invalid; they are equivalent to the primary name, or are applied to a subset of the species, as in subspecies names. Relationships among names are indicated by phrases such as "includes," "included in," "conspecific with," or their negatives. Such phrases refer to genus or species categories, unless otherwise specified. In a few cases, synonyms are indicated by an "equals" sign. Synonyms resulting from recent changes in status have been included, as well as names considered by the editors to be potential sources of confusion. For example, *Citellus* (widely regarded as invalid) is included in the Comment section under *Spermophilus* since the former is widely used in the Old World, and the latter throughout North America. Taxonomic and nomenclatorial alternatives are explained by appropriate documentation, including opinions of the International Commission on Zoological Nomenclature (ICZN); revisions and additional literature sources are also listed. Often, in the interest of brevity, secondary reference sources are cited to document taxonomic evidence, and reference to the primary sources should be looked for therein. Personal opinions or unpublished information are identified by the reviewer's initials in parentheses. When appropriate, other data such as type locality restriction, occurrence of hybridization, and of species known only from the type locality, from a single specimen, or from a few specimens, are included in the Comment section. Subgenera or subfamilies are included in some large groups to indicate relationships, and to make the list easier to compare with other publications. Absence of a Comment section under any species may mean that either it is systematically non-controversial, or too poorly known to engender comment.

Protected Status—Mammal species governed by the regulations for the U.S. Endangered Species Act (U.S. ESA) and those listed in the appendices of the Convention on International Trade in Endangered Species of Wild Fauna and Flora (CITES) have been annotated in the text in this category. U.S. ESA endangered or threatened categories and Appendices I, II, and III of CITES are included. Where entire families or orders have been protected, they are noted under those categories as well as under each species affected. The Federal Register (for U.S. ESA) and the CITES appendices should be consulted for details. This information was current January 1, 1982.

ISIS Numbers—Numbers assigned to species of mammals in the International Species Inventory System have been included in this category to facilitate comparisons with data bases arranged in the ISIS format. Number listings under the species names in the text contain all names which apply to the usage in this volume, but which may have been listed under different names in ISIS (Seal and Makey, 1974).

Attribution—The contributors who supplied information and reviewed portions of *Mammal Species of the World* have been listed in the Reviewed By category. Ordinarily, each reviewer was listed once, in the highest appropriate taxonomic category up to family, and not under each species reviewed; *i.e.,* individuals listed at the genus level reviewed all or nearly all species in that genus. Those individuals who reviewed an entire order have been listed under all its included families so that the reader can easily find the names of reviewers. Comments attributable to individual reviewers are identified by their initials in parentheses.

Bibliographic Treatment—The Bibliography contains the works consulted in the compilation of this text, and literature that has been cited in brief form. It also includes citations of recent regional works that should be consulted for more detailed distributional information (see also Hickman, 1981). The text itself contains citations to many papers and books that bear on specific problems discussed in the comments, or that contain the original species description, but does not contain titles for reasons of space economy. Abbreviations of journals have been standardized using BIOSIS whenever possible; journals not listed in BIOSIS were generally listed as cited by the reviewers. It has not been possible to verify every citation supplied by the reviewers, and some inconsistencies and errors undoubtedly occur in the text, especially among transliterated and translated sources.

Index—The index contains all taxonomic names included in this volume including synonyms, *nomina nuda, nomina oblita,* and other invalid names. Page references to all recognized generic and species names employed in this volume are in boldface type. Secondary listings, such as synonyms or references to other taxa, are in italics. Species names are individually listed alphabetically, and are also associated with the genus in the alphabetical listing.

A final word to the user who is not a systematic biologist—do not be intimidated or discouraged by the amount of controversy surrounding definition of species limits in many groups of mammals. Disagreements have been emphasized in this volume in order to highlight areas that call for further systematic work. Mammals are no worse off in this respect than other groups of animals, and mammals are better known than many. This project has been designed to summarize our present understanding, so as to extend it even further.

R. S. HOFFMANN, J. H. HONACKI, K. E. KINMAN, AND J. W. KOEPPL

Acknowledgments

When a work as complex as *Mammal Species of the World* is completed, it is virtually impossible to recognize each of the numerous contributions worthy of mention that have made the work possible. We depended upon the hundreds of professional mammalogists who have devoted their careers to the systematic study of the Mammalia. Each has made a contribution to this synthesis.

The 189 individuals listed at the end of this introductory section, by focusing on specific taxonomic groups or geographic regions, have accomplished, in a cooperative effort, the first modern discipline-wide review of a major taxonomic group and provided the substance and quality of this edition. The fact that these people devoted considerable personal time and effort exemplifies their commitment to advancing the field of mammalogy, and also insured that their knowledge would be shared with others who rely on this information. To each, we express our profound gratitude for contributing to the successful culmination of this initial checklist.

Within the list of reviewers, it is appropriate to recognize the exceptional contributions of certain individuals. The Checklist Committee, formed by the American Society of Mammalogists, deserves special recognition; A. L. Gardner, as Chairman, and his committee comprised of R. L. Brownell, R. S. Hoffmann, K. F. Koopman, G. G. Musser, and D. A. Schlitter worked intensively to complete their final review of the manuscript. Gardner and Koopman collated all committee comments and Koopman in addition was a primary reviewer of all species of Chiroptera, and read the entire manuscript, a herculean effort. Olga L. Rossolimo reviewed all extant species of the Soviet Union; Wang Sung (S. Wang) of the People's Republic of China provided information on distributions of pertinent species that has significantly enhanced our understanding of this poorly-known region. Richard G. Van Gelder reviewed more than half of the taxa. J. Ramirez-Pulido (Mexico), and A. Langguth (South America), by providing extensive contributions for the mammals of these areas, have also added significantly to the completion of this work.

The greatest single contribution toward the completion of the work was provided by Robert S. Hoffmann, who in addition to making significant contributions to the content of the volume, has served as the primary liaison with the professional community. His personal and professional dedication to the development of this checklisting system has been extensive. His arbitration of sometimes strong differences of opinion and his continued commitment in conjunction with numerous other responsibilities have unquestionably assured our success. It is through his efforts that the basic pattern of our checklisting series has been established and will eventually be duplicated in other disciplines. The Association is profoundly grateful for this commitment and these contributions.

Special attribution to the editors of this volume is required, in particular to James H. Honacki. Normally, employed editors are exempt from special recognition, beyond the bibliographic honors afforded the role. However, in this instance, his service as the senior editor of this volume far exceeded any normal terms of employment; completion of this task has resulted from a deeper dedication. Through all these months, Jim Honacki in particular has consistently

7

pursued the objectives of the program. The Association, and those who find this volume useful, clearly owe him and his editorial colleagues a debt of gratitude.

Special thanks are extended to A. Michael Neuner and Lee Hubbell, who invested considerable personal time and effort toward completion of this work. We are grateful to E. Raymond Hall, for his advice, use of his excellent library, and the opportunity to review page-proofs of *The Mammals of North America*, which have enhanced the scope afforded this compilation. N. J. Dippenaar, P. Dollinger, R. Jachowski, G. Pilleri, and P. Swanepoel, although not reviewers of record, provided significant assistance in other ways. B. H. Blake and M. Green made available translations of important papers.

Finally, we would like to express our gratitude to the U.S. Department of Energy for making funds available (Grant No. DE-AC02-79EV10258) for the initial development phase of this project. The Wildlife Permit Office of the Fish and Wildlife Service has also contributed financial support for the project. Numerous other government agencies and private organizations have provided moral support and endorsements of the project for which we are grateful. In addition, the Parties to the Convention on International Trade in Endangered Species of Wild Fauna and Flora have continued to endorse the development of this listing system for the purpose of serving as their standard reference to mammalian nomenclature.

STEPHEN R. EDWARDS, for the
Association of Systematics Collections

List of Contributors

Jean-Pierre Airoldi
University of Bern
Bern, Switzerland

F. R. Allison
South Brisbane, Australia

Milos Andera
Charles University
Praha, Czechoslovakia

Elaine Anderson
Denver, Colorado, U.S.A.

Karen L. Anderson
University of Connecticut
Storrs, Connecticut, U.S.A.

Stephen Anderson
University of Southern California
Los Angeles, California, U.S.A.

David E. Babb
Bishop, California, U.S.A.

Ruben Barquez
Tucuman, Argentina

Ronald E. Barry, Jr.
Unity College
Unity, Maine, U.S.A.

Marc Bekoff
University of Colorado
Boulder, Colorado, U.S.A.

Deb K. Bennett
University of Kansas
Lawrence, Kansas, U.S.A.

Troy L. Best
Eastern New Mexico University
Portales, New Mexico, U.S.A.

Brad R. Blood
University of Southern California
Los Angeles, California, U.S.A.

Mark S. Boyce
University of Wyoming
Laramie, Wyoming, U.S.A.

Janet K. Braun
University of Oklahoma
Norman, Oklahoma, U.S.A.

Fred J. Brenner
Grove City College
Grove City, Pennsylvania, U.S.A.

Harold E. Broadbooks
Southern Illinois University
Edwardsville, Illinois, U.S.A.

Robert L. Brownell, Jr., U.S. Fish
  and Wildlife Service
San Simeon, California, U.S.A.

John E. Bucher
University of Kansas
Lawrence, Kansas, U.S.A.

William Caire
Central State University
Edmond, Oklahoma, U.S.A.

Joan R. Callahan
University of Georgia
Athens, Georgia, U.S.A.

Douglas Candland
Bucknell University
Lewisburg, Pennsylvania, U.S.A.

Raymond P. Canham
Southern Methodist University
Dallas, Texas, U.S.A.

Domenico Capolongo
Roma, Italy

Michael D. Carleton
National Museum of Natural
  History
Washington, D.C., U.S.A.

Leslie N. Carraway
Oregon State University
Corvallis, Oregon, U.S.A.

Jean Chaline
Institut des Sciences de la Terre
Dijon, France

Joseph A. Chapman
Frostburg State College
Frostburg, Maryland, U.S.A.

Jerry R. Choate
Fort Hays State University
Hays, Kansas, U.S.A.

Garrett C. Clough
Nasson College
Springvale, Maine, U.S.A.

E. Lendell Cockrum
University of Arizona
Tucson, Arizona, U.S.A.

Harold J. Coolidge
Beverly, Massachusetts, U.S.A.

Gordon B. Corbet
British Museum (Natural History)
London, England

Earl of Cranbrook
Suffolk, England

Kai Curry-Lindahl
United Nations Environmental
   Program
Nairobi, Kenya

W. B. Davis
Texas A&M University
College Station, Texas, U.S.A.

Ron DeBry
University of Kansas
Lawrence, Kansas, U.S.A.

Eric Delson
American Museum of Natural
   History
New York, New York, U.S.A.

Wendell E. Dodge
University of Massachusetts
Amherst, Massachusetts, U.S.A.

Daryl P. Domning
Howard University
Washington, D.C., U.S.A.

Nicole Duplaix
TRAFFIC USA
Washington, D.C., U.S.A.

Conrad R. Durst
Larchmont, New York, U.S.A.

Harold J. Egoscue
Grantsville, Utah, U.S.A.

John F. Eisenberg
National Zoological Park
Washington, D.C., U.S.A.

Charles L. Elliott
Brigham Young University
Provo, Utah, U.S.A.

Dieter Ernst
Hamburg
Federal Republic of Germany

Howard E. Evans
Cornell University
Ithaca, New York, U.S.A.

Mark Fitzsimmons
Albany County Planning Board
Albany, New York, U.S.A.

John E. C. Flux
D.S.I.R. Ecology Division
Lower Hutt, New Zealand

Roy Fontaine
Bucknell University
Lewisburg, Pennsylvania, U.S.A.

Jack Fooden
Field Museum of Natural History
Chicago, Illinois, U.S.A.

Jiri Gaisler
J. E. Purkyne University
Brno, Czechoslovakia

Udo Ganslosser
Zool. Inst., abt. Wirbeltiermorph.
Heidelberg
Federal Republic of Germany

Alfred L. Gardner, U.S. Fish and
   Wildlife Laboratory
Washington, D.C., U.S.A.

R. K. Ghose
C.S.I.R.O.
Canberra, A.C.T.
Australia

Gregory E. Glass
University of Kansas
Lawrence, Kansas, U.S.A.

David C. Gordon
Victoria, New York, U.S.A.

Edwin Gould
Johns Hopkins University
Baltimore, Maryland, U.S.A.

Cynthia L. Gray
San Diego, California, U.S.A.

Thomas A. Griffiths
Illinois Wesleyan University
Bloomington, Illinois, U.S.A.

Colin P. Groves
Australian National University
Canberra, A.C.T.
Australia

Peter Grubb
London, England

Richard J. Guenzel
University of Wyoming
Laramie, Wyoming, U.S.A.

David J. Hafner
University of New Mexico
Albuquerque, New Mexico, U.S.A.

James G. Hallett
University of California
Irvine, California, U.S.A.

David L. Harrison
Harrison Zoological Museum
Kent, England

E. Blake Hart
Northern State College
Aberdeen, South Dakota, U.S.A.

Lawrence R. Heaney
University of Michigan
Ann Arbor, Michigan, U.S.A.

Clyde A. Hill
San Diego Zoological Park
San Diego, California, U.S.A.

Hendrik N. Hoeck
Charles Darwin Research Station
Santa Cruz, Galapagos
Ecuador

Robert S. Hoffmann
University of Kansas
Lawrence, Kansas, U.S.A.

Thor Holmes
University of Kansas
Lawrence, Kansas, U.S.A.

James H. Honacki
University of Kansas
Lawrence, Kansas, U.S.A.

David G. Huckaby
California State University
Long Beach, California, U.S.A.

Rainer Hutterer
Zool. Forschungsinst.
Museum A. Koenig, Bonn
Federal Republic of Germany

Stephen H. Jenkins
University of Nevada
Reno, Nevada, U.S.A.

Murray L. Johnson
University of Puget Sound
Tacoma, Washington, U.S.A.

Gwilym S. Jones
Northeastern University
Boston, Massachusetts, U.S.A.

J. Knox Jones, Jr.
Texas Tech University
Lubbock, Texas, U.S.A.

Jeffrey P. Jorgenson
Arlington, Virginia, U.S.A.

Jane Ann Junge
University of Kansas
Lawrence, Kansas, U.S.A.

Toshio Kasuya
Ocean Research Institute
Tokyo, Japan

C. William Kilpatrick
University of Vermont
Burlington, Vermont, U.S.A.

Judith E. King
University of South Wales
Kensington, N.S.W.
Australia

Kenneth Kinman
Hoisington, Kansas, U.S.A.

John A. W. Kirsch
Harvard University
Cambridge, Massachusetts, U.S.A.

Wighart Von Koenigswald
Hessicher Landesmuseum
Darmstadt
Federal Republic of Germany

James W. Koeppl
University of Kansas
Lawrence, Kansas, U.S.A.

Ilse Kohler
Ober-Ramstadt-Wembach
Federal Republic of Germany

Karl F. Koopman
American Museum of Natural
   History
New York, New York, U.S.A.

Sheila Kortlucke
University of Kansas
Lawrence, Kansas, U.S.A.

Karl R. Kranz
National Zoological Park
Washington, D.C., U.S.A.

David W. Kuehn
Grand Rapids, Minnesota, U.S.A.

Thomas E. Lacher, Jr.
University of Pittsburgh
Pittsburgh, Pennsylvania, U.S.A.

Alfredo Langguth
Universidade Federal de Paraiba
Joao Pessoa, Paraiba
Brazil

Nigel Langham
University of the South Pacific
Suva, Fiji

Timothy E. Lawlor
Humboldt State University
Arcata, California, U.S.A.

Douglas M. Lay
University of North Carolina
Chapel Hill, North Carolina, U.S.A.

Sanford R. Leffler
Camarillo, California, U.S.A.

Howard Levenson
University of Kansas
Lawrence, Kansas, U.S.A.

Stephen L. Lindsay
Walla Walla College
College Place, Washington, U.S.A.

Alicia V. Linzey
Virginia Polytechnic Institute and
    State University
Blacksburg, Virginia, U.S.A.

Donald W. Linzey
Blacksburg, Virginia, U.S.A.

Irma Lira
Mexico City, D.F.
Mexico

Ricardo Lopez-Wilchis
Mexico City, D.F.
Mexico

Thomas R. Loughlin
National Marine Mammal
    Laboratory
Seattle, Washington, U.S.A.

Michael A. Mares
University of Oklahoma
Norman, Oklahoma, U.S.A.

Basil J. Marlow
Australian Museum
Sydney, Australia

Joe T. Marshall, U.S. Fish and
    Wildlife Laboratory
Washington, D.C., U.S.A.

Larry D. Martin
University of Kansas
Lawrence, Kansas, U.S.A.

John J. Mayer
University of Connecticut
Storrs, Connecticut, U.S.A.

W. Bruce McGillivray
University of Kansas
Lawrence, Kansas, U.S.A.

Gary McGrath
University of Kansas
Lawrence, Kansas, U.S.A.

Thomas J. McIntyre
Rockville, Maryland, U.S.A.

George M. McKay
Macquarie University
North Ryde, N.S.W.
Australia

Charles A. McLaughlin
San Diego Zoological Park
San Diego, California, U.S.A.

James G. Mead
National Museum of Natural
    History
Washington, D.C., U.S.A.

J. Meester
University of Natal
Durban, South Africa

J. I. Menzies
Devon, England

Richard M. Mitchell, U.S. Fish and
    Wildlife Service
Washington, D.C., U.S.A.

Jesus Molinari
Universidad de los Andes
Merida, Venezuela

Gary S. Morgan
National Museum of Natural
    History
Washington, D.C., U.S.A.

Alan E. Muchlinski
California State University
Los Angeles, California, U.S.A.

Carolina Mudespacher
Mexico City, D.F.
Mexico

Guy G. Musser
American Museum of Natural
    History
New York, New York, U.S.A.

Kenneth Myers
C.S.I.R.O.
Canberra, A.C.T.
Australia

Nancy A. Neff
American Museum of Natural
    History
New York, New York, U.S.A.

Masaharu Nishiwaki
University of the Ryukyus
Tokyo, Japan

Margaret A. O'Connell
Texas Tech University
Lubbock, Texas, U.S.A.

Ricardo A. Ojeda
University of Pittsburgh
Pittsburgh, Pennsylvania, U.S.A.

Richard D. Orr
Pittsburgh, Pennsylvania, U.S.A.

R. D. Owen
University of Oklahoma
Norman, Oklahoma, U.S.A.

Aida Parkinson
University of New Mexico
Albuquerque, New Mexico, U.S.A.

R. B. Patten
San Diego, California, U.S.A.

Oliver P. Pearson
University of California
Berkeley, California, U.S.A.

Thomas W. Pearson
University of Kansas
Lawrence, Kansas, U.S.A.

W. F. Perrin
Southwest Fisheries Center
La Jolla, California, U.S.A.

R. H. Pine
George Williams College
Downers Grove, Illinois, U.S.A.

John J. Pizzimenti
University of Illinois, Chicago
    Circle
Chicago, Illinois, U.S.A.

Roger A. Powell
North Carolina State University
Raleigh, North Carolina, U.S.A.

Jose Ramirez-Pulido
Universidad Autonoma
    Metropolitano-Iztapalapa
Mexico City, D.F.
Mexico

Osvaldo A. Reig
University of Caracas
Caracas, Venezuela

Charles A. Repenning, U.S.
    Geological Survey
Menlo Park, California, U.S.A.

Dale W. Rice
National Marine Mammal
    Laboratory
Seattle, Washington, U.S.A.

Mark S. Rich
San Diego Zoological Park
San Diego, California, U.S.A.

Ingo Rieger
Zurich, Switzerland

C. Brian Robbins
National Museum of Natural
    History
Washington, D.C., U.S.A.

Peter Roben
Universitat Heidelberg
Heidelberg
Federal Republic of Germany

T. J. Robinson
University of Pretoria
Pretoria
South Africa

Aryan I. Roest
California Polytechnic State
    University
San Luis Obispo, California, U.S.A.

Olga L. Rossolimo
Moscow State University
Moscow, U.S.S.R.

Lothar Schlawe
Berlin
Federal Republic of Germany

Duane A. Schlitter
Carnegie Museum of Natural
    History
Pittsburgh, Pennsylvania, U.S.A.

William D. Severinghaus
Champaign, Illinois, U.S.A.

Scott D. Shalaway
Oklahoma State University
Stillwater, Oklahoma, U.S.A.

Alan H. Shoemaker
Riverbanks Zoological Park
Columbia, South Carolina, U.S.A.

Jeheskel Shoshani
Wayne State University
Detroit, Michigan, U.S.A.

George A. Sidorowicz
Toronto, Ontario
Canada

C. David Simpson
Texas Tech University
Lubbock, Texas, U.S.A.

Andrew T. Smith
Arizona State University
Tempe, Arizona, U.S.A.

Dana P. Snyder
Amherst College
Amherst, Massachusetts, U.S.A.

Friederike Spitzenberger
Natural History Museum
Vienna, Austria

Howard J. Stains
Southern Illinois University
Carbondale, Illinois, U.S.A.

Barbara R. Stein
University of Kansas
Lawrence, Kansas, U.S.A.

Gerhard Storch
Forschungs-Institut Senckenberg
Frankfurt am Mein
Federal Republic of Germany

Nicholas Sullivan
Philadelphia, Pennsylvania, U.S.A.

Dallas A. Sutton
California State University
Chico, California, U.S.A.

Valdir Antonio Taddei
Instituto de Biociencias, UNESP
Sao Jose do Rio Preto
Brazil

Robert H. Tamarin
Boston University
Boston, Massachusetts, U.S.A.

James R. Tamsitt
Royal Ontario Museum
Toronto, Ontario
Canada

Carol J. Terry
University of Kansas
Lawrence, Kansas, U.S.A.

Guillermina Urbano-Villa
Universidad Nacional Autonoma
   Mexico
Mexico City, D.F.
Mexico

Erik Van der Straeten
Rijksuniversitair Centrum
   Antwerpen
Groenenborgerlaan
Antwerpen, Belgium

Richard G. Van Gelder
American Museum of Natural
   History
New York, New York, U.S.A.

Dirk Van Vuren
Napa, California, U.S.A.

Eduardo Vernier
Club Alpino Italiano
Padua, Italy

Bernardo Villa-Ramirez
Instituto de Biologia
Universidad Nacional Autonoma
   Mexico
Mexico City, D.F.
Mexico

Robert S. Voss
University of Michigan
Ann Arbor, Michigan, U.S.A.

S. Wang (Wang Sung)
Academia Sinica
Beijing (Peking)
China

C. H. S. Watts
Inst. Med. Vet. Science
Adelaide, S. Australia
Australia

W. David Webster
University of North Carolina
Wilmington, North Carolina, U.S.A.

Marla L. Weston
University of British Columbia
Vancouver, B.C.
Canada

Ralph M. Wetzel
University of Connecticut
Storrs, Connecticut, U.S.A.

Kenneth T. Wilkins
University of Florida
Gainesville, Florida, U.S.A.

Daniel F. Williams
California State College, Stanislaus
Turlock, California, U.S.A.

Michael R. Willig
University of Pittsburgh
Pittsburgh, Pennsylvania, U.S.A.

Charles A. Woods
Florida State Museum
Gainesville, Florida, U.S.A.

W. Chris Wozencraft
University of Kansas
Lawrence, Kansas, U.S.A.

Alan C. Ziegler
Bishop Museum
Honolulu, Hawaii, U.S.A.

CHECKLIST OF
MAMMAL SPECIES OF THE WORLD

ORDER MONOTREMATA
ISIS NUMBER: 5301401000000000000.

Family Tachyglossidae
REVIEWED BY: B. J. Marlow (BJM); R. G. Van Gelder (RGVG); A. C. Ziegler (ACZ).
ISIS NUMBER: 5301401001000000000.

Tachyglossus Illiger, 1811. Prodr. Syst. Mammal. et Avium., p. 114.
ISIS NUMBER: 5301401001001000000.

Tachyglossus aculeatus (Shaw and Nodder, 1792). Nat. Misc., 3, pl. 109.
TYPE LOCALITY: Australia, New South Wales, Sydney (=New Holland).
DISTRIBUTION: S. and E. New Guinea; Australian mainland, Kangaroo Isl. (off South Australia), Tasmania.
COMMENT: Includes *lawesi, typicus* and *setosus;* see Ride, 1970:231.
ISIS NUMBER: 5301401001001001001.

Zaglossus Gill, 1877. Ann. Rec. Sci. Indus., May:p. 171.
PROTECTED STATUS: CITES - Appendix II as *Zaglossus* spp.
ISIS NUMBER: 5301401001002000000.

Zaglossus bruijni (Peters and Doria, 1876). Ann. Mus. Stor. Nat. Genova, 9:183.
TYPE LOCALITY: Indonesia, Irian Jaya, Vogelkop, Manokwari Div., Arfak Mtns.
DISTRIBUTION: Interior New Guinea; Salawatti Isl.
COMMENT: Includes *bartoni* and *bubuensis;* see Van Deusen and George, 1969, Am. Mus. Novit., 2383:22.
PROTECTED STATUS: CITES - Appendix II as *Zaglossus* spp.
ISIS NUMBER: 5301401001002002001 as *Z. bruijni.*
5301401001002001001 as *Z. bartoni.*

Family Ornithorhynchidae
REVIEWED BY: B. J. Marlow (BJM); R. G. Van Gelder (RGVG).
ISIS NUMBER: 5301401002000000000.

Ornithorhynchus Blumenbach, 1800. Gotting. Gelehrt. Anz., 1:609-610.
ISIS NUMBER: 5301401002001000000.

Ornithorhynchus anatinus (Shaw and Nodder, 1799). Nat. Misc., 10, pl. 385-386.
TYPE LOCALITY: Australia, New South Wales, Sydney (=New Holland).
DISTRIBUTION: Queensland, New South Wales, S.E. South Australia, Victoria, Tasmania.
ISIS NUMBER: 5301401002001001001.

ORDER MARSUPIALIA

REVIEWED BY: J. A. W. Kirsch (JAWK); R. G. Van Gelder (RGVG).

COMMENT: Includes Marsupicarnivora, Paucituberculata, Peramelina, and Diprotodonta of Ride, 1964. Includes Polyprotodontia, Paucituberculata and Peramelina; see Kirsch, 1976, *in* Hunsaker, ed., The Biology of the Marsupials. See Kirsch and Calaby, 1977:10–12, for a comparison of higher classifications of marsupials.

ISIS NUMBER: 5301402000000000000.

Family Didelphidae

REVIEWED BY: J. A. W. Kirsch (JAWK); R. H. Pine (RHP); R. G. Van Gelder (RGVG).

COMMENT: Placed in the order Polyprotodontia by Kirsch, 1976, *in* Hunsaker, ed., The Biology of the Marsupials; but also see Kirsch and Calaby, 1977. Does not include *Dromiciops;* see Kirsch and Calaby, 1977:11.

ISIS NUMBER: 5301402001000000000.

Caluromys J. A. Allen, 1900. Bull. Am. Mus. Nat. Hist., 13:189.

ISIS NUMBER: 5301402001001000000.

Caluromys derbianus (Waterhouse, 1841). Jardine's Natur. Libr., 11:94.

REVIEWED BY: J. Ramirez-Pulido (JRP).

TYPE LOCALITY: Colombia, Cauca, Cauca Valley.

DISTRIBUTION: W. Ecuador to Veracruz, Mexico.

COMMENT: Reviewed by Bucher and Hoffmann, 1980, Mamm. Species, 140:1–4.

ISIS NUMBER: 5301402001001001001.

Caluromys lanatus (Illiger, 1815). Abh. Preuss. Akad. Wiss., 1811:107.

TYPE LOCALITY: Paraguay, Caazapa.

DISTRIBUTION: E. Colombia; W. Venezuela; Brazil; Bolivia; Paraguay; E. Ecuador; E. Peru.

ISIS NUMBER: 5301402001001002001.

Caluromys philander (Linnaeus, 1758). Syst. Nat., 10th ed., 1:54.

TYPE LOCALITY: Surinam.

DISTRIBUTION: Brazil; Venezuela, including Margarita Isl.; French Guiana; Guyana; Surinam; Trinidad.

ISIS NUMBER: 5301402001001003001.

Caluromysiops Sanborn, 1951. Fieldiana Zool., 31:474.

COMMENT: Generic distinctness from *Caluromys* dubious (RHP).

ISIS NUMBER: 5301402001002000000.

Caluromysiops irrupta Sanborn, 1951. Fieldiana Zool., 31:474.

TYPE LOCALITY: Peru, Cuzco, Quincemil.

DISTRIBUTION: S.E. Peru.

ISIS NUMBER: 5301402001005001001.

Chironectes Illiger, 1811. Prodr. Syst. Mammal. et Avium., p. 76.

REVIEWED BY: J. Ramirez-Pulido (JRP).

ISIS NUMBER: 5301402001003000000.

Chironectes minimus (Zimmermann, 1780). Geogr. Gesch. Mensch. Vierf. Thiere, 2:317.

TYPE LOCALITY: French Guiana, Cayenne.

DISTRIBUTION: Oaxaca and Tabasco (Mexico) to Peru, N.E. Argentina, S. and N.E. Brazil, French Guiana, Guyana, Surinam, and Trinidad.

COMMENT: Reviewed by Marshall, 1978, Mamm. Species, 109:1–6.

ISIS NUMBER: 5301402001003001001.

Didelphis Linnaeus, 1758. Syst. Nat., 10th ed., 1:54.

REVIEWED BY: M. A. O'Connell (MAO); J. Ramirez-Pulido (JRP).

ISIS NUMBER: 5301402001004000000.

Didelphis albiventris Lund, 1840. Kongl. Dansk. Vid. Naturv. Math. Selsk. Afhandl., p. 20.
TYPE LOCALITY: Brazil, Minas Gerais, Lagoa Santa.
DISTRIBUTION: Ecuador; Colombia; W. Venezuela; E. Brazil; Uruguay; Paraguay; W.
Bolivia; Peru; N. and C. Argentina.
COMMENT: Formerly known under the name *azarae*; see Hershkovitz, 1969, Quart.
Rev. Biol., 44:1–70.

Didelphis marsupialis Linnaeus, 1758. Syst. Nat., 10th ed., 1:54.
TYPE LOCALITY: Surinam.
DISTRIBUTION: Tamaulipas (Mexico) to Peru, Bolivia, Paraguay, N.E. Argentina,
Trinidad, and Lesser Antilles.
COMMENT: *D. azarae* is a junior synonym; see Hershkovitz, 1969, Quart. Rev. Biol.,
44:1–70. Revised by Gardner, 1973, Spec. Publ. Mus. Texas Tech Univ., 4:1–81.
Type locality discussed by Thomas, 1911, Proc. Zool. Soc. Lond., 1911:143.
ISIS NUMBER: 5301402001004002001 as *D. marsupialis*..
5301402001004001001 as *D. azarae*.

Didelphis virginiana Kerr, 1792. Anim. Kingdom, p. 193.
TYPE LOCALITY: U.S.A., Virginia.
DISTRIBUTION: N. Costa Rica to E. and C. United States, and S.E. Ontario (Canada).
Introduced on Pacific coast (U.S.A.).
COMMENT: Revised by Gardner, 1973, Spec. Publ. Mus. Texas Tech. Univ., 4:1–81.
Reviewed by McManus, 1974, Mamm. Species, 40:1–6.
ISIS NUMBER: 5301402001004003001.

Glironia Thomas, 1912. Ann. Mag. Nat. Hist., ser. 8, 9:239.
ISIS NUMBER: 5301402001006000000.

Glironia venusta Thomas, 1912. Ann. Mag. Nat. Hist., ser. 8, 9:240.
TYPE LOCALITY: Peru, Huanaco, Pozuzo.
DISTRIBUTION: N. and Amazonian Peru; E. Ecuador; N. Bolivia.
COMMENT: Includes *criniger*; see Marshall, 1978, Mamm. Species, 107:1–3.
ISIS NUMBER: 5301402001006002001 as *G. venusta*.
5301402001006001001 as *G. criniger*.

Lestodelphys Tate, 1934. J. Mammal., 15:154.
ISIS NUMBER: 5301402001007000000.

Lestodelphys halli (Thomas, 1921). Ann. Mag. Nat. Hist., ser. 9, 8:137.
TYPE LOCALITY: Argentina, Santa Cruz, Cabo Tres Puntas.
DISTRIBUTION: S. Patagonia (Argentina).
COMMENT: Reviewed by Marshall, 1977, Mamm. Species, 81:1–3.
ISIS NUMBER: 5301402001007001001.

Lutreolina Thomas, 1910. Ann. Mag. Nat. Hist., ser. 8, 5:247.
ISIS NUMBER: 5301402001008000000.

Lutreolina crassicaudata (Desmarest, 1804). Nouv. Dict. Hist. Nat., 24:19.
TYPE LOCALITY: Paraguay, Asuncion.
DISTRIBUTION: Venezuela; Guyana; S. Brazil; Uruguay; Paraguay; Bolivia; N. Argentina.
COMMENT: Reviewed by Marshall, 1978, Mamm. Species, 91:1–4.
ISIS NUMBER: 5301402001008001001.

Marmosa Gray, 1821. Lond. Med. Repos., 15:308.
REVIEWED BY: C. A. Hill (CAH).
COMMENT: *Marmosa* spp. are conventionally distributed among three subgenera,
Marmosa, *Stegomarmosa*, and *Thylamys*; see Kirsch and Calaby, 1977:14. Revised by
Tate, 1933, Bull. Am. Mus. Nat. Hist., 66:6–236.
ISIS NUMBER: 5301402001009000000.

Marmosa aceramarcae Tate, 1931. Am. Mus. Novit., 493:12.
TYPE LOCALITY: Bolivia, La Paz, Rio Aceramarca, 10,800 ft. (3292 m).
DISTRIBUTION: Bolivia.
ISIS NUMBER: 5301402001009001001.

Marmosa agilis (Burmeister, 1854). Syst. Uebers. Thiere Bras., 1:139.
TYPE LOCALITY: Brazil, Minas Gerais, Lagoa Santa.
DISTRIBUTION: Brazil; Paraguay; N. Argentina; Uruguay; Bolivia; E. Peru.
ISIS NUMBER: 5301402001009002001.

Marmosa agricolai Moojen, 1943. Bol. Mus. Nac. Rio de J., Zool., 1:2.
TYPE LOCALITY: Brazil, Ceara, Crato.
DISTRIBUTION: Known only from the type locality.
ISIS NUMBER: 5301402001009003001.

Marmosa alstoni (J. A. Allen, 1900). Bull. Am. Mus. Nat. Hist., 13:189.
TYPE LOCALITY: Costa Rica, Cartago, Tres Rios.
DISTRIBUTION: W. Colombia; Central America north to Belize and Honduras.
COMMENT: Included in *cinerea* by Pine, 1973, *in* Collins, but treated as a separate
species by Kirsch and Calaby, 1977:14.

Marmosa andersoni Pine, 1972. J. Mammal., 53:279.
TYPE LOCALITY: Peru, Cuzco, Hda. Villa Carmen (Cosnipata).
DISTRIBUTION: Known only from the type locality.

Marmosa canescens (J. A. Allen, 1893). Bull. Am. Mus. Nat. Hist., 5:235.
REVIEWED BY: J. Ramirez-Pulido (JRP).
TYPE LOCALITY: Mexico, Oaxaca, Santo Domingo de Guzman.
DISTRIBUTION: S. Sonora to Oaxaca, Yucatan, and Tres Marias Isls. (Mexico).
ISIS NUMBER: 5301402001009004001.

Marmosa cinerea (Temminck, 1824). Monogr. Mamm., 1:46.
TYPE LOCALITY: Brazil, Bahia, Rio Mucuri.
DISTRIBUTION: E. Colombia; Venezuela; French Guiana; Guyana; Surinam; E. and S.
Brazil; Paraguay; Missiones Prov., Argentina.
ISIS NUMBER: 5301402001009005001.

Marmosa constantiae Thomas, 1904. Proc. Zool. Soc. Lond., 1903 (2):243.
TYPE LOCALITY: Brazil, Mato Grosso, Chapada.
DISTRIBUTION: Mato Grosso, Brazil; W. Bolivia; N. W. Argentina.
ISIS NUMBER: 5301402001009006001.

Marmosa cracens Handley and Gordon, 1979. *In* Eisenberg, ed., Vertebrate ecology in the
Northern Neotropics, Smithson. Inst., p. 66.
TYPE LOCALITY: Venezuela, Falcon, 14 km E.N.E. Mirimire, near La Pastora (11° 12′ N.,
68° 37′ W.), 150 m.
DISTRIBUTION: Known only from vicinity of type locality.

Marmosa domina Thomas, 1920. Ann. Mag. Nat. Hist., ser. 9, 6:280.
TYPE LOCALITY: Brazil, Para, Villa Braga, above the Rio Tapajos.
DISTRIBUTION: Amazonian Brazil.
ISIS NUMBER: 5301402001009007001.

Marmosa dryas Thomas, 1898. Ann. Mag. Nat. Hist., ser. 7, 1:456.
TYPE LOCALITY: Venezuela, Merida, Culata.
DISTRIBUTION: W. Venezuela; E. Colombia.
ISIS NUMBER: 5301402001009008001.

Marmosa elegans (Waterhouse, 1839). Zool. H.M.S. "Beagle," Mammalia, p. 95.
TYPE LOCALITY: Chile, Coquimbo, Valparaiso.
DISTRIBUTION: N. and C. Chile; S. and S.W. Bolivia; N.W. Argentina; S. Peru.
ISIS NUMBER: 5301402001009009001.

Marmosa emiliae Thomas, 1909. Ann. Mag. Nat. Hist., ser. 8, 3:379.
TYPE LOCALITY: Brazil, Para.
DISTRIBUTION: N.E. Brazil; Surinam.

COMMENT: May include *juninensis* and *parvidens;* see Kirsch and Calaby, 1977:18; but see also Pine, 1981, Mammalia, 45:55–70.
ISIS NUMBER: 5301402001009010001.

Marmosa formosa Shamel, 1930. J. Mammal., 11:311.
TYPE LOCALITY: Argentina, Formosa, Riacho Pilaga, 16 km N.W. of Km 182.
DISTRIBUTION: Formosa Prov., N. Argentina.
COMMENT: Formerly regarded as a subspecies of *velutina* by Cabrera, 1958:33; but see Kirsch and Calaby, 1977:14.
ISIS NUMBER: 5301402001009011001.

Marmosa fuscata Thomas, 1896. Ann. Mag. Nat. Hist., ser. 6, 18:13.
TYPE LOCALITY: Venezuela, Merida, Rio Albarregas.
DISTRIBUTION: Venezuela; N. and C. Colombia; Trinidad.
COMMENT: Includes *carri;* see Handley and Gordon, 1979, *in* Eisenberg, ed., Vertebrate ecology in the Northern Neotropics, Smithson. Inst., p. 66.
ISIS NUMBER: 5301402001009012001.

Marmosa germana Thomas, 1904. Ann. Mag. Nat. Hist., ser. 7, 13:143.
TYPE LOCALITY: Ecuador, Napo-Pastaza, Sarayacu (Rio Bobonanza).
DISTRIBUTION: E. Ecuador; E. Peru; perhaps W. Brazil and W. Colombia.
ISIS NUMBER: 5301402001009013001.

Marmosa grisea (Desmarest, 1827). Dict. Sci. Nat., 67:398.
TYPE LOCALITY: Paraguay, Tapua (N.E. of Asuncion).
DISTRIBUTION: Paraguay.
ISIS NUMBER: 5301402001009014001.

Marmosa handleyi Pine, 1981. Mammalia, 45:67.
TYPE LOCALITY: Colombia, Antioquia, 9 km S. Valdivia, 1400 m. (7° ll' N., 75° 27' W.).
DISTRIBUTION: Known only from the vicinity of the type locality.
COMMENT: Known from only two specimens.

Marmosa impavida (Tschudi, 1844). Fau. Peru, p. 148, pl. 9.
TYPE LOCALITY: Peru, Junin, Montana de Vitoc (near Chanchamayo).
DISTRIBUTION: Mountains of Panama to Venezuela, W. Brazil, and S. Peru.
ISIS NUMBER: 5301402001009015001.

Marmosa incana (Lund, 1840). Kongl. Dansk. Vid. Selsk. Naturv. Math. Afhandl., p. 21.
TYPE LOCALITY: Brazil, Minas Gerais, Lagoa Santa.
DISTRIBUTION: E. Brazil.
ISIS NUMBER: 5301402001009016001.

Marmosa invicta Goldman, 1912. Smithson. Misc. Collect., 60(2):3.
TYPE LOCALITY: Panama, Darien, Cana.
DISTRIBUTION: Panama.
COMMENT: Reviewed by Pine, 1981, Mammalia, 45:55–70, who thought it closest to *parvidens.*
ISIS NUMBER: 5301402001009017001.

Marmosa juninensis Tate, 1931. Am. Mus. Novit., 493:13.
TYPE LOCALITY: Peru, Junin, Utcuyaco, between Tarma and Chanchamayo, 4800 ft. (1463 m).
DISTRIBUTION: C. Peru.
COMMENT: Possibly a synonym of *emiliae;* see Kirsch and Calaby, 1977:18. Considered a subspecies of *parvidens* by Pine, 1981, Mammalia, 45:55–70.

Marmosa karimii Petter, 1968. Mammalia, 32:315.
TYPE LOCALITY: Brazil, Pernambuco, Exu.
DISTRIBUTION: N.E. and C. Brazil.
ISIS NUMBER: 5301402001009018001.

Marmosa lepida (Thomas, 1888). Ann. Mag. Nat. Hist., ser. 6, 1:158.
TYPE LOCALITY: Peru, Loreto, Santa Cruz, near Yurimaguas.
DISTRIBUTION: Surinam to Bolivia, E. Peru, and Ecuador. Also Colombia (RHP).
ISIS NUMBER: 5301402001009019001.

Marmosa leucastra Thomas, 1927. Ann. Mag. Nat. Hist., ser. 9, 20:607.
    TYPE LOCALITY: Peru, Amazonas, Tambo Carrizal (S. of Chachapoyas).
    DISTRIBUTION: N. Peru.
    ISIS NUMBER: 5301402001009020001.

Marmosa mapiriensis Tate, 1931. Am. Mus. Novit., 493:3.
    TYPE LOCALITY: Bolivia, La Paz, Ticunhuaya (Rio Tipuani), 4800 ft. (1463 m).
    DISTRIBUTION: N.W. Bolivia and adjacent S. Peru.
    ISIS NUMBER: 5301402001009021001.

Marmosa marica Thomas, 1898. Ann. Mag. Nat. Hist., ser. 7, 1:455.
    TYPE LOCALITY: Venezuela, Merida, Rio Albarregas.
    DISTRIBUTION: W. Venezuela.
    ISIS NUMBER: 5301402001009022001.

Marmosa mexicana Merriam, 1897. Proc. Biol. Soc. Wash., 11:44.
    REVIEWED BY: J. Ramirez-Pulido (JRP).
    TYPE LOCALITY: Mexico, Oaxaca, Juquila.
    DISTRIBUTION: Tamaulipas (Mexico) to W. Panama.
    ISIS NUMBER: 5301402001009023001.

Marmosa microtarsus (Wagner, 1842). Arch. Naturgesch., 8(1):359.
    TYPE LOCALITY: Brazil, Sao Paulo, Ipanema.
    DISTRIBUTION: E. Brazil.
    ISIS NUMBER: 5301402001009024001.

Marmosa murina (Linnaeus, 1758). Syst. Nat., 10th ed., 1:56.
    TYPE LOCALITY: Surinam.
    DISTRIBUTION: French Guiana; Guyana; Surinam; Amazonian and N.E. Brazil;
        Venezuela; Ecuador; N. Peru; Tobago.
    ISIS NUMBER: 5301402001009025001.

Marmosa noctivaga (Tschudi, 1844). Fau. Peru, p. 148.
    TYPE LOCALITY: Peru, Junin, Montana de Vitoc (near Chanchamayo).
    DISTRIBUTION: Amazonian Brazil to E. Ecuador; W. Bolivia; C. and E. Peru.
    ISIS NUMBER: 5301402001009026001.

Marmosa ocellata Tate, 1931. Am. Mus. Novit., 493:7.
    TYPE LOCALITY: Bolivia, Santa Cruz, Buena Vista, 500 m.
    DISTRIBUTION: C. Bolivia.
    ISIS NUMBER: 5301402001009027001.

Marmosa parvidens Tate, 1931. Am. Mus. Novit., 493:13.
    TYPE LOCALITY: Guyana, Hyde Park (Rio Demerara), 20 ft. (6 m).
    DISTRIBUTION: Guyana; Venezuela. Also in Surinam, Amazonian Brazil, Colombia, and
        Peru (RHP).
    COMMENT: Possibly a synonym of *emiliae*; see Kirsch and Calaby, 1977:18; but see also
        Pine, 1981, Mammalia, 45:55–70.
    ISIS NUMBER: 5301402001009028001.

Marmosa phaea Thomas, 1899. Ann. Mag. Nat. Hist., ser. 7, 3:44.
    TYPE LOCALITY: Ecuador, San Pablo.
    DISTRIBUTION: Ecuador; W. Colombia.
    COMMENT: A form "resembling" *phaea* occurs in Panama; see Handley, 1966, *in*
        Wenzel and Tipton, ed., Ectoparasites of Panama, Field Mus. Nat. Hist., pp. 1–861.
    ISIS NUMBER: 5301402001009029001.

Marmosa pusilla (Desmarest, 1804). Nouv. Dict. Hist. Nat., 24:19.
    TYPE LOCALITY: Paraguay, San Ignacio.
    DISTRIBUTION: N. Argentina; S.W. Bolivia; Paraguay.
    ISIS NUMBER: 5301402001009030001.

Marmosa quichua Thomas, 1899. Ann. Mag. Nat. Hist., ser. 7, 3:43.
    TYPE LOCALITY: Peru, Cuzco, Ocabamba.
    DISTRIBUTION: E. Peru.
    ISIS NUMBER: 5301402001009031001.

Marmosa rapposa Thomas, 1899. Ann. Mag. Nat. Hist., ser. 7, 3:42.
TYPE LOCALITY: Peru, Cuzco, Rio Vilcanota, N. of Cuzco.
DISTRIBUTION: N.W. Bolivia; S.E. Peru.
ISIS NUMBER: 5301402001009032001.

Marmosa regina Thomas, 1898. Ann. Mag. Nat. Hist., ser. 7, 2:274.
TYPE LOCALITY: Colombia, Cundinamarca, Bogota area.
DISTRIBUTION: Dept. Cundinamarca, Colombia.
ISIS NUMBER: 5301402001009033001.

Marmosa robinsoni Bangs, 1898. Proc. Biol. Soc. Wash., 12:95.
TYPE LOCALITY: Venezuela, Isla Margarita, Nueva Esparta.
DISTRIBUTION: N.W. Peru; Ecuador, including Puna Isl.; W. and N. Colombia; N.
    Venezuela including Margarita Isl.; Belize; Honduras, including Roatan Isl.;
    Panama including San Miguel and Saboga Isls.; Trinidad; Tobago; Grenada
    (Lesser Antilles).
COMMENT: Includes *mitis*; see Cabrera, 1958:25.
ISIS NUMBER: 5301402001009034001.

Marmosa rubra Tate, 1931. Am. Mus. Novit., 493:6.
TYPE LOCALITY: Peru, Loreto, mouth of Rio Curaray.
DISTRIBUTION: E. Ecuador; N.E. and S. Peru.
ISIS NUMBER: 5301402001009035001.

Marmosa scapulata (Burmeister, 1856). Erlaut. Fau. Brasil, p. 79.
TYPE LOCALITY: Brazil, Minas Gerais.
DISTRIBUTION: Known only from type locality.
ISIS NUMBER: 5301402001009036001.

Marmosa tatei Handley, 1956. J. Wash. Acad. Sci., 46:402.
TYPE LOCALITY: Peru, Ancash, Chasquitambo.
DISTRIBUTION: Known only from type locality.
ISIS NUMBER: 5301402001009037001.

Marmosa tyleriana Tate, 1931. Am. Mus. Novit., 493:6.
TYPE LOCALITY: Venezuela, Amazonas, Mt. Duida plateau, upper Orinoco River, 4800
    ft. (1463 m).
DISTRIBUTION: Bolivar and Amazonas, S. Venezuela.
ISIS NUMBER: 5301402001009038001.

Marmosa unduaviensis Tate, 1931. Am. Mus. Novit., 493:11.
TYPE LOCALITY: Bolivia, La Paz, Pitiguaya (Rio Unduavi), 5800 ft. (1768 m).
DISTRIBUTION: W. Bolivia.
ISIS NUMBER: 5301402001009039001.

Marmosa velutina (Wagner, 1842). Arch. Naturgesch., 8(1):360.
TYPE LOCALITY: Brazil, Sao Paulo, Ipanema.
DISTRIBUTION: E. Brazil.
COMMENT: *M. velutina* reported from Argentina is assignable to *formosa*; see Pine,
    1975, Mammalia, 39:321–322.
ISIS NUMBER: 5301402001009040001.

Marmosa xerophila Handley and Gordon, 1979. *In* Eisenberg, ed., Vertebrate ecology in
    the Northern Neotropics, Smithson. Inst., p. 68.
TYPE LOCALITY: Colombia, Guajira, 37 km N.N.E. Paraguaipoa, near Cojoro, La Isla, 15
    m.
DISTRIBUTION: Extreme N.E. Colombia; N.W. Venezuela.

Marmosa yungasensis Tate, 1931. Am. Mus. Novit., 493:7.
TYPE LOCALITY: Bolivia, La Paz, Pitiguaya (Rio Unduavi), 5600 ft. (1707 m).
DISTRIBUTION: N.W. Brazil to Ecuador; W. Bolivia.
ISIS NUMBER: 5301402001009041001.

Metachirus Burmeister, 1854. Syst. Uebers. Thiere Bras., 1:135.
COMMENT: Controversy exists regarding the use of *Metachirus*; see Pine, 1973, Proc.
Biol. Soc. Wash., 86:391–402. Its usage here follows Tate, 1939, Bull. Am. Mus.
Nat. Hist., 76:161–162; Haltenorth, 1958:17; Husson, 1978:24–28; and Corbet and
Hill, 1980:9. Hall, 1981:17, employed the name *Philander* for this genus; but see
comments under *Philander*.
ISIS NUMBER: 5301402001010000000.

Metachirus nudicaudatus (E. Geoffroy, 1803). Cat. Mammal. Mus. d'Hist. Nat., Paris, p. 142.
TYPE LOCALITY: French Guiana, Cayenne.
DISTRIBUTION: Nicaragua to Peru, E. Bolivia, Paraguay, N.E. Argentina, and S. Brazil;
French Guiana; Guyana; Surinam.
ISIS NUMBER: 5301402001010001001.

Monodelphis Burnett, 1830. Quart. J. Sci. Lit. and Art., 28:351.
REVIEWED BY: C. A. Hill (CAH).
ISIS NUMBER: 5301402001011000000.

Monodelphis adusta (Thomas, 1897). Ann. Mag. Nat. Hist., ser. 6, 20:219.
TYPE LOCALITY: Colombia, Depto. Cundinamarca.
DISTRIBUTION: Panama; C. Colombia; Ecuador; E. Peru; W. Bolivia.
COMMENT: Includes *osgoodi*; see Cabrera, 1958:6; *osgoodi* may be a valid species
according to Kirsch and Calaby, 1977:18. Includes *melanops*; see Handley, 1966, *in*
Wenzel and Tipton, ed., Ectoparasites of Panama, Field Mus. Nat. Hist., p. 754;
and Kirsch and Calaby, 1977:18.
ISIS NUMBER: 5301402001011001001 as *M. adusta*.
                5301402001011008001 as *M. osgoodi*.

Monodelphis americana (Muller, 1776). Natursyst. Suppl., 7:36.
TYPE LOCALITY: Brazil, Pernambuco.
DISTRIBUTION: S., E., and C. Brazil to French Guiana, Guyana, and Surinam.
COMMENT: Pine, 1976, J. Mammal., 57:785–787, and Pine, 1977, Mammalia, 41:235,
considered *umbristriata* and *iheringi* to be distinct species, not subspecies of
americana; see also Miranda-Ribeiro, 1936, Rev. Museu Paulista, 20:422, and Kirsch
and Calaby, 1977:18, who also consider *theresa* as a possibly valid species (RGVG).
ISIS NUMBER: 5301402001011002001.

Monodelphis brevicaudata (Erxleben, 1777). Syst. Regn. Anim., p. 80.
TYPE LOCALITY: Surinam.
DISTRIBUTION: Amazonian Brazil; French Guiana; Guyana; Surinam; W. Venezuela;
adjacent Colombia.
COMMENT: *M. touan* is probably a junior synonym of *brevicaudata* (RHP).
ISIS NUMBER: 5301402001011003001.

Monodelphis dimidiata (Wagner, 1847). Abh. Akad. Wiss. Munchen, 5(1):151, note.
TYPE LOCALITY: Uruguay, Maldonado.
DISTRIBUTION: C. Brazil; Uruguay; Pampean Argentina.
COMMENT: Includes *fosteri*; see Reig, 1964, Publs. Mus. Cienc. Nat. de Mar del Plata,
1:203–224; Kirsch and Calaby, 1977.
ISIS NUMBER: 5301402001011004001 as *M. dimidiata*.
                5301402001011006001 as *M. fosteri*.

Monodelphis domestica (Wagner, 1842). Arch. Naturgesch., 8:359.
TYPE LOCALITY: Brazil, Mato Grosso, Cuiaba.
DISTRIBUTION: E. and C. Brazil; Bolivia; Paraguay.
COMMENT: Cabrera, 1958:8 included *maraxina* in this species; but see Pine, 1980 (1979),
Mammalia, 43:497.
ISIS NUMBER: 5301402001011005001.

Monodelphis henseli (Thomas, 1888). Ann. Mag. Nat. Hist., ser. 6, 1:159.
TYPE LOCALITY: Brazil, Rio Grande do Sul, Taquara.
DISTRIBUTION: S. Brazil; S.E. Paraguay; N.E. Argentina.

COMMENT: Probably a junior synonym of *M. sorex*; see Pine, 1980 (1979), Mammalia, 43:495.

ISIS NUMBER: 5301402001011007001.

Monodelphis iheringi Thomas, 1888. Ann. Mag. Nat. Hist., ser. 6, 1:159.
TYPE LOCALITY: Brazil, Rio Grande do Sul, Taquara.
DISTRIBUTION: S. Brazil.
COMMENT: Formerly included in *americana*; see Pine, 1977, Mammalia, 41:235.

Monodelphis kunsi Pine, 1975. Mammalia, 39:321.
TYPE LOCALITY: Bolivia, Beni, La Granja (Rio Itonamas).
DISTRIBUTION: Known only from the type locality.

Monodelphis maraxina Thomas, 1923. Ann. Mag. Nat. Hist., ser. 9, 12:157.
TYPE LOCALITY: Brazil, Para, Caldeirao.
DISTRIBUTION: Marajo Isl. (and vicinity?), Brazil.
COMMENT: Formerly included in *domestica* by Cabrera, 1958:8; but see Pine, 1980 (1979), Mammalia, 43:497, who considered *maraxina* a distinct species.

Monodelphis orinoci (Thomas, 1899). Ann. Mag. Nat. Hist., ser. 7, 3:154.
TYPE LOCALITY: Venezuela, Bolivar, Caicara.
DISTRIBUTION: Orinoco Basin in S. Venezuela; W. Guyana; N. Brazil.
COMMENT: Formerly included in *brevicaudata* by Cabrera, 1958; separated by Reig *et al.*, 1977, Biol. J. Linn. Soc., 9:191-216.

Monodelphis scalops (Thomas, 1888). Ann. Mag. Nat. Hist., ser. 6, 1:158.
TYPE LOCALITY: Brazil, Rio de Janeiro, Theresopolis.
DISTRIBUTION: E. Brazil.
ISIS NUMBER: 5301402001011009001.

Monodelphis sorex (Hensel, 1872). Abh. Preuss. Akad. Wiss., p. 122.
TYPE LOCALITY: Brazil, Rio Grande do Sul, Taquara.
DISTRIBUTION: S. Brazil.
COMMENT: Probably includes *henseli*; see Pine, 1980 (1979) Mammalia, 43:495.
ISIS NUMBER: 5301402001011010001.

Monodelphis touan (Shaw, 1800). Gen. Zool., 1(2):432.
TYPE LOCALITY: French Guiana, Cayenne.
DISTRIBUTION: Brazil; Paraguay; N. Argentina; French Guiana; Guyana; Surinam.
COMMENT: Includes *emiliae*; see Cabrera, 1958:9. *M. emiliae* is considered a valid species according to Pine and Handley, in prep (RHP) and *touan* is probably a junior synonym of *brevicaudata* (RHP).

Monodelphis umbristriata Miranda-Ribeiro, 1936. Rev. Museu Paulista, 20:422.
TYPE LOCALITY: Brazil, E. Goiaz.
DISTRIBUTION: E. Brazil.
COMMENT: Formerly included in *americana*; see Pine, 1976, J. Mammal., 57:785.

Monodelphis unistriata (Wagner, 1842). Arch. Naturgesch., 8:360.
TYPE LOCALITY: Brazil, Sao Paulo, Itarara.
DISTRIBUTION: Sao Paulo, Brazil.
ISIS NUMBER: 5301402001011011001.

Philander Tiedemann, 1808. Allgemeine Zoologie, Mensch und Saugthiere, Landshut, Vol. 1, p. 426.
COMMENT: Pine, 1973, Proc. Biol. Soc. Wash., 86:391-402, presented arguments for using the name *Metachirops* Matschie, 1916 (Sitzb. Ges. Naturf. Fr. Berlin, p. 262) for this genus, as did Husson, 1978:24-28; they have been followed by Corbet and Hill, 1980:9, and Hall, 1981:8. However, evidence supports the usage followed here; see Hershkovitz, 1981, Proc. Biol. Soc. Wash., 93:943-946; Gardner, 1981, J. Mammal., 62:447.
ISIS NUMBER: 5301402001012000000.

Philander mcilhennyi Gardner and Patton, 1972. Occas. Pap. Mus. Zool. La. St. Univ., 43:2.
TYPE LOCALITY: Peru, Loreto, Balta (Rio Curanja), 300 m.
DISTRIBUTION: Known only from the type locality.
ISIS NUMBER: 5301402001012001001.

Philander opossum (Linnaeus, 1758). Syst. Nat., 10th ed., 1:55.
REVIEWED BY: J. Ramirez-Pulido (JRP).
TYPE LOCALITY: Surinam, Paramaribo.
DISTRIBUTION: Tamaulipas (Mexico) to E. Peru, W. Bolivia, Paraguay, N.E. Argentina, and Brazil.
ISIS NUMBER: 5301402001012002001.

Family Microbiotheriidae
REVIEWED BY: C. A. Hill (CAH); J. A. W. Kirsch (JAWK); R. G. Van Gelder (RGVG).
COMMENT: *Dromiciops* was placed in this otherwise fossil family by Kirsch, 1977, *in* Hunsaker, ed., The Biology of the Marsupials, pp. 1–50. Also see Marshall, 1978, Mamm. Species, 99:1, who included the Microbiotheriinae as a subfamily of the Didelphidae.

Dromiciops Thomas, 1894. Ann. Mag. Nat. Hist., ser. 6, 14:186.
ISIS NUMBER: 5301402001005000000.

Dromiciops australis (Philippi, 1894). An. Univ. Chile, 86:31 (plate).
TYPE LOCALITY: Chile, Valdivia Province, Union.
DISTRIBUTION: S. and C. Chile including Chiloe Isl.; adjacent Argentina.
COMMENT: Reviewed by Marshall, 1978, Mamm. Species, 99:1–5, who considered this species assignable to Didelphidae, subfamily Microbiotheriinae.

Family Caenolestidae
REVIEWED BY: C. A. Hill (CAH); J. A. W. Kirsch (JAWK); R. G. Van Gelder (RGVG).
COMMENT: Reviewed by Marshall, 1980, Fieldiana Geol., N.S., 5:1–145.
ISIS NUMBER: 5301402005000000000.

Caenolestes Thomas, 1895. Ann. Mag. Nat. Hist., ser. 6, 16:367.
ISIS NUMBER: 5301402005001000000.

Caenolestes caniventer Anthony, 1921. Am. Mus. Novit., 20:6.
TYPE LOCALITY: Ecuador, Prov. Oro, El Chiral, 1600 m.
DISTRIBUTION: S. Ecuador, west of the Andes.
ISIS NUMBER: 5301402005001001001.

Caenolestes convelatus Anthony, 1924. Am. Mus. Novit., 120:1.
TYPE LOCALITY: Ecuador, Prov. Pichincha, Las Maquinas, 2100 m.
DISTRIBUTION: N. Ecuador, W. part of Andes.
ISIS NUMBER: 5301402005001002001.

Caenolestes fuliginosus (Tomes, 1863). Proc. Zool. Soc. Lond., 1863:51.
TYPE LOCALITY: Ecuador, Prov. Pichincha, paramos of Chimborazo or Pichincha.
DISTRIBUTION: N. Andean Ecuador; N. Peru.
COMMENT: May include *obscurus*; see Kirsch and Calaby, 1977:19.
ISIS NUMBER: 5301402005001003001.

Caenolestes obscurus Thomas, 1895. Ann. Mag. Nat. Hist., ser. 6, 16:367.
TYPE LOCALITY: Colombia, Cundinamarca, Estancia La Selva (Bogota).
DISTRIBUTION: Andean Colombia; adjacent extreme W. Venezuela.
COMMENT: Probably conspecific with *fuliginosus*; see Kirsch and Calaby, 1977:19.
ISIS NUMBER: 5301402005001004001.

Caenolestes tatei Anthony, 1923. Am. Mus. Novit., 55:1.
TYPE LOCALITY: Ecuador, Prov. Azuay, Molleturo, 2280 m.
DISTRIBUTION: S. Ecuador (in W. Andes).
ISIS NUMBER: 5301402005001005001.

Lestoros Oehser, 1934. J. Mammal., 15:240.
ISIS NUMBER: 5301402005002000000.

Lestoros inca (Thomas, 1917). Smithson. Misc. Collect., 68(4):3.
TYPE LOCALITY: Peru, Cuzco, Torontoy.
DISTRIBUTION: S. Andean Peru.
ISIS NUMBER: 5301402005002001001.

Rhyncholestes Osgood, 1924. Field Mus. Nat. Hist. Publ. Zool. Ser., 14:170.
ISIS NUMBER: 5301402005003000000.

Rhyncholestes raphanurus Osgood, 1924. Field Mus. Nat. Hist. Publ. Zool. Ser., 14:170.
TYPE LOCALITY: Chile, Prov. Chiloe, mouth of Rio Inio.
DISTRIBUTION: S.C. Chile, Chiloe Island and adjacent mainland of Chile (Mt. Osorno).
ISIS NUMBER: 5301402005003001001.

Family Dasyuridae

REVIEWED BY: C. A. Hill (CAH); J. A. W. Kirsch (JAWK); B. J. Marlow (BJM); R. G. Van
Gelder (RGVG); A. C. Ziegler (ACZ) (New Guinea).
COMMENT: Some authors include Thylacinidae and Myrmecobiidae in this family; see
Vaughan, 1978:39; but also see Ride, 1964, J. Proc. Roy. Soc. West Aust.,
47:97–131, Archer and Kirsch, 1977, Proc. Linn. Soc. N.S.W., 102:18–25, and
Kirsch and Calaby, 1977:15, who retained these families. Revised by Tate, 1947,
Bull. Am. Mus. Nat. Hist., 88:97–156.
ISIS NUMBER: 5301402002000000000.

Antechinus Macleay, 1841. Ann. Mag. Nat. Hist., 8:242.
COMMENT: Includes *Parantechinus* and *Pseudantechinus;* see Haltenorth, 1958:18; Ride,
1964, W. Aust. Nat., 9:58–65.
ISIS NUMBER: 5301402002010000000.

Antechinus apicalis (Gray, 1842). Ann. Mag. Nat. Hist., 9:518.
TYPE LOCALITY: Australia, S.W. Western Australia.
DISTRIBUTION: Inland periphery of S.W. Western Australia.
PROTECTED STATUS: U.S. ESA - Endangered.
ISIS NUMBER: 5301402002010001001.

Antechinus bellus (Thomas, 1904). Nov. Zool., 11:229.
TYPE LOCALITY: Australia, Northern Territory, South Alligator River.
DISTRIBUTION: N. Northern Territory (Australia).
ISIS NUMBER: 5301402002010002001.

Antechinus bilarni Johnson, 1954. Proc. Biol. Soc. Wash., 67:77.
TYPE LOCALITY: Australia, Northern Territory, Oenpelli (12° 20′ S. and 133° 3′ E.).
DISTRIBUTION: Northern Territory (Australia).
COMMENT: Included in *macdonnellensis* by Ride, 1970:116, but see Kirsch and
Calaby, 1977:15.

Antechinus flavipes (Waterhouse, 1838). Proc. Zool. Soc. Lond., 1837:75.
TYPE LOCALITY: Australia, New South Wales, north of Hunter River.
DISTRIBUTION: Cape York Peninsula (Queensland) to Victoria and S.E. South Australia;
S.W. Western Australia.
COMMENT: Does not include *godmani;* see Kirsch and Calaby, 1977:15.
ISIS NUMBER: 5301402002010003001.

Antechinus godmani (Thomas, 1923). Ann. Mag. Nat. Hist., ser. 9, 11:174.
TYPE LOCALITY: Australia, Queensland, Ravenshoe.
DISTRIBUTION: N.E. Queensland (Australia).
COMMENT: Formerly included in *flavipes* by Haltenorth, 1958:18; but also see
Kirsch and Calaby, 1977:15.
ISIS NUMBER: 5301402002010004001.

Antechinus leo Van Dyck, 1980. Aust. Mammal., 3:1.
  TYPE LOCALITY: Australia, Queensland, Cape York Peninsula, Nesbit River, Buthen
    Buthen (13° 21' S, 143° 28' E).
  DISTRIBUTION: Cape York Peninsula from the Iron Range to the southern limit of the
    McIlwraith Range (Queensland, Australia).

Antechinus macdonnellensis (Spencer, 1896). Rept. Horn Sci. Exped. Cent. Aust., Zool.,
    2:27.
  TYPE LOCALITY: Australia, Northern Territory, south of Alice Springs.
  DISTRIBUTION: N. Western Australia; Northern Territory; central deserts.
  ISIS NUMBER: 5301402002010005001.

Antechinus melanurus (Thomas, 1899). Ann. Mus. Stor. Nat. Genova, 20:191.
  TYPE LOCALITY: Papua New Guinea, Central Prov., Astrolabe Range, Moroka, 1300 m.
  DISTRIBUTION: New Guinea.
  ISIS NUMBER: 5301402002010007001.

Antechinus minimus (E. Geoffroy, 1803). Bull. Sci. Soc. Philom. Paris, 81:159.
  TYPE LOCALITY: Australia, Tasmania.
  DISTRIBUTION: Coastal S.E. South Australia to Tasmania.
  ISIS NUMBER: 5301402002010008001.

Antechinus naso (Jentink, 1911). Notes Leyden Mus., 33:236.
  TYPE LOCALITY: Indonesia, Irian Jaya, Djajawidjaja Div., Helwig Mtns., south of Mt.
    Wilhelmina, about 2000 m.
  DISTRIBUTION: Interior New Guinea.
  COMMENT: Possibly includes *Murexia longicaudata parva* (ACZ).
  ISIS NUMBER: 5301402002010009001.

Antechinus rosamondae Ride, 1964. W. Aust. Nat., 9:58.
  TYPE LOCALITY: Australia, Western Australia, Woodstock Station (via Marble Bar).
  DISTRIBUTION: N.W. Western Australia.
  ISIS NUMBER: 5301402002010010001.

Antechinus stuartii Macleay, 1841. Ann. Mag. Nat. Hist., 8:242.
  TYPE LOCALITY: Australia, New South Wales, Manly (Spring Cove, Sydney Harbour).
  DISTRIBUTION: E. Queensland; E. New South Wales; Victoria (Australia).
  ISIS NUMBER: 5301402002010011001.

Antechinus swainsonii (Waterhouse, 1840). Mag. Nat. Hist. (Charlesworth), 4:299.
  TYPE LOCALITY: Australia, Tasmania.
  DISTRIBUTION: S.E. Queensland, E. New South Wales, E. and S.E. Victoria, coastal S.E.
    South Australia, Tasmania.
  ISIS NUMBER: 5301402002010012001.

Antechinus wilhelmina Tate, 1947. Bull. Am. Mus. Nat. Hist., 88:130.
  TYPE LOCALITY: Indonesia, Irian Jaya, Djajawidjaja Div., 9 km. N. Lake Habbema, north
    of Mt. Wilhelmina, 2800 m.
  DISTRIBUTION: C. New Guinea.
  ISIS NUMBER: 5301402002010013001.

Dasycercus Peters, 1875. Sitzb. Ges. Naturf. Fr. Berlin, p. 73.
  ISIS NUMBER: 5301402002001000000.

Dasycercus cristicauda (Krefft, 1867). Proc. Zool. Soc. Lond., 1866:435.
  TYPE LOCALITY: Australia, South Australia, probably Lake Alexandrina.
  DISTRIBUTION: Arid Australia from N.W. Western Australia to S.W. Queensland.
  ISIS NUMBER: 5301402002001001001.

Dasyuroides Spencer, 1896. Rept. Horn Sci. Exped. Cent. Aust., Zool., 2:36.
  ISIS NUMBER: 5301402002002000000.

Dasyuroides byrnei Spencer, 1896. Rept. Horn Sci. Exped. Cent. Aust., Zool., 2:36.
  TYPE LOCALITY: Australia, Northern Territory, Charlotte Waters.
  DISTRIBUTION: Junction of Northern Territory, South Australia and Queensland (C.
    Australia).
  ISIS NUMBER: 5301402002002001001.

Dasyurus E. Geoffroy, 1796. Mag. Encyclop., ser. 2, 3:469.
  COMMENT: Includes *Dasyurops, Dasyurinus* and *Satanellus;* see Haltenorth, 1958:20.
  ISIS NUMBER: 5301402002003000000.

Dasyurus albopunctatus Schlegel, 1880. Notes Leyden Mus., 2:51.
  TYPE LOCALITY: Indonesia, Irian Jaya, Vogelkop, Manokwari Div., Arfak Mts., Sapoea.
  DISTRIBUTION: New Guinea.
  ISIS NUMBER: 5301402002003001001.

Dasyurus geoffroii Gould, 1841. Proc. Zool. Soc. Lond., 1840:151.
  TYPE LOCALITY: Australia, New South Wales, Liverpool Plains.
  DISTRIBUTION: Western Australia, South Australia, Northern Territory, S. Queensland,
    W. New South Wales, N.W. Victoria. S.C. New Guinea.
  COMMENT: See Archer, 1979, *in* Tyler, ed., The Status of Endangered Australasian
    Wildlife, and Waithman, 1979, Aust. Zool., 20(2):313–326 for a discussion of
    distribution. Australian populations extant only in Western Australia (BJM); New
    Guinea populations stable (ACZ).
  ISIS NUMBER: 5301402002003002001.

Dasyurus hallucatus Gould, 1842. Proc. Zool. Soc. Lond., 1842:41.
  TYPE LOCALITY: Australia, Northern Territory, Port Essington.
  DISTRIBUTION: N. Northern Territory, N. and N.E. Queensland, N. Western Australia.
  ISIS NUMBER: 5301402002003003001.

Dasyurus maculatus (Kerr, 1792). Anim. Kingdom, p. 170.
  TYPE LOCALITY: Australia, New South Wales, Port Jackson.
  DISTRIBUTION: E. Queensland, E. New South Wales, E. and S. Victoria, S.E. South
    Australia, Tasmania.
  ISIS NUMBER: 5301402002003004001.

Dasyurus viverrinus (Shaw, 1800). Gen. Zool., 1(2), Mammalia, p. 491.
  TYPE LOCALITY: Australia, New South Wales, Sydney.
  DISTRIBUTION: Coast of E. New South Wales, E. and S. Victoria, and S.E. South
    Australia. Kangaroo Isl., King Isl., and Tasmania (BJM).
  COMMENT: *D. quoll* Zimmermann, 1777, is invalid; this work was rejected by
    ICZN Opinion 257.
  PROTECTED STATUS: U.S. ESA - Endangered.
  ISIS NUMBER: 5301402002003005001.

Murexia Tate and Archbold, 1937. Bull. Am. Mus. Nat. Hist., 73:335 (footnote), 339.
  ISIS NUMBER: 5301402002011000000.

Murexia longicaudata (Schlegel, 1866). Ned. Tijdschr. Dierk., 3:356.
  TYPE LOCALITY: Indonesia, Aru Islands, Wonoumbai.
  DISTRIBUTION: New Guinea; Aru Islands.
  COMMENT: The subspecies *parva* is probably conspecific with *Antechinus naso* (ACZ).
  ISIS NUMBER: 5301402002011001001.

Murexia rothschildi (Tate, 1938). Nov. Zool., 41:58.
  TYPE LOCALITY: Papua New Guinea, Central Prov., head of Aroa River, about 1220 m.
  DISTRIBUTION: S.E. New Guinea.
  ISIS NUMBER: 5301402002011002001.

Myoictis Gray, 1858. Proc. Zool. Soc. Lond., 1858:112.
ISIS NUMBER: 5301402002004000000.

Myoictis melas (Muller, 1840). *In* Temminck, Verh. Nat. Ges. Ned. Overz. Bezitt. Land-en
Volkenkunde, Zool., p. 20.
TYPE LOCALITY: Indonesia, Irian Jaya, Fakfak Div., Lobo Dist., near Triton Bay, Mt.
Lamantsjieri.
DISTRIBUTION: New Guinea; Salawatti Isl.; Aru Isls.
ISIS NUMBER: 5301402002004001001.

Neophascogale Stein, 1933. Z. Saugetierk., 8:87.
ISIS NUMBER: 5301402002005000000.

Neophascogale lorentzi (Jentink, 1911). Notes Leyden Mus., 33:234.
TYPE LOCALITY: Indonesia, Irian Jaya, Djajawidjaja Div., Helwig Mtns., south of Mt.
Wilhelmina, 2600 m.
DISTRIBUTION: C. New Guinea.
ISIS NUMBER: 5301402002005001001 as *N. lorentzii (sic)*.

Ningaui Archer, 1975. Mem. Queensl. Mus., 17(2):239.
COMMENT: An undescribed species of *Ningaui* occurs in Northern Territory (Australia);
see Johnson and Roff, 1980, Aust. Mammal., 3:127–129.

Ningaui ridei Archer, 1975. Mem. Queensl. Mus., 17(2):246.
TYPE LOCALITY: Australia, Western Australia, 38.6 km E.N.E. Laverton (28° 30' S. and
122° 47' E.).
DISTRIBUTION: Western Australia (deserts).

Ningaui timealeyi Archer, 1975. Mem. Queensl. Mus., 17(2):244.
TYPE LOCALITY: Australia, Western Australia, 32.2 km S.E. Mt. Robinson.
DISTRIBUTION: N.W. Western Australia.

Phascogale Temminck, 1824. Monogr. Mamm., 1:23, 56.
ISIS NUMBER: 5301402002012000000.

Phascogale calura Gould, 1844. Proc. Zool. Soc. Lond., 1844:104.
TYPE LOCALITY: Australia, Western Australia, Williams River.
DISTRIBUTION: Inland S.W. Western Australia, Northern Territory, South Australia,
N.W. Victoria, S.W. New South Wales.
ISIS NUMBER: 5301402002012001001.

Phascogale tapoatafa (Meyer, 1793). Zool. Entdeck., p. 28.
TYPE LOCALITY: Australia, New South Wales, Sydney.
DISTRIBUTION: Western Australia, S.E. South Australia, S. Victoria, E. New South
Wales, E. Queensland, Northern Territory.
ISIS NUMBER: 5301402002012002001.

Phascolosorex Matschie, 1916. Mitt. Zool. Mus. Berlin, 8:263.
ISIS NUMBER: 5301402002006000000.

Phascolosorex doriae (Thomas, 1886). Ann. Mus. Stor. Nat. Genova, 4:208.
TYPE LOCALITY: Indonesia, Irian Jaya, Vogelkop, Manokwari Div., Arfak Mtns., Mori.
DISTRIBUTION: W. interior New Guinea.
ISIS NUMBER: 5301402002006001001.

Phascolosorex dorsalis (Peters and Doria, 1876). Ann. Mus. Stor. Nat. Genova, 8:335.
TYPE LOCALITY: Indonesia, Irian Jaya, Vogelkop, Manokwari Div., Arfak Mtns., Hatam.
DISTRIBUTION: Interior New Guinea.
ISIS NUMBER: 5301402002006002001.

Planigale Troughton, 1928. Rec. Aust. Mus., 16:282.
COMMENT: Revised by Archer, 1976, Mem. Queensl. Mus., 17(3):341–365.
ISIS NUMBER: 5301402002013000000.

Planigale gilesi Aitken, 1972. Rec. S. Aust. Mus., 16(10):1.
TYPE LOCALITY: Australia, South Australia, Ann Creek Station (No. 3 bore) (28° 18′ S. and 136° 29′ 40″ E.).
DISTRIBUTION: N. South Australia, N.W. New South Wales. Queensland (RGVG).

Planigale ingrami (Thomas, 1906). Abstr. Proc. Zool. Soc. Lond., 32:6.
TYPE LOCALITY: Australia, Northern Territory, Alexandria.
DISTRIBUTION: N. and E. Queensland, Northern Territory, N.E. and C. Western Australia.
COMMENT: Includes *subtilissima*; see Archer, 1976, Mem. Queensl. Mus., 17(3):351.
PROTECTED STATUS: U.S. ESA - Endangered as *P. subtilissima* only.
ISIS NUMBER: 5301402002013001001 as *P. ingrami*.
5301402002013003001 as *P. subtilissima*.

Planigale maculata (Gould, 1851). Mammal. Aust., 1, pl. 44.
TYPE LOCALITY: Australia, New South Wales, Clarence River.
DISTRIBUTION: Queensland; New South Wales; Western Australia; South Australia.
COMMENT: Transferred to *Planigale* from *Antechinus* by Archer, 1976, Mem. Queensl. Mus., 17(3):346.
ISIS NUMBER: 5301402002010006001 as *Antechinus maculatus*.

Planigale novaeguineae Tate and Archbold, 1941. Am. Mus. Novit., 1101:7.
TYPE LOCALITY: Papua New Guinea, Central Prov., Rona Falls, Laloki River (vicinity Port Moresby), 250 m.
DISTRIBUTION: S. New Guinea
ISIS NUMBER: 5301402002013002001.

Planigale tenuirostris Troughton, 1928. Rec. Aust. Mus., 16:285.
TYPE LOCALITY: Australia, New South Wales, Bourke or Wilcannia.
DISTRIBUTION: N.C. New South Wales; S.C. Queensland; C. Western Australia.
PROTECTED STATUS: U.S. ESA - Endangered.
ISIS NUMBER: 5301402002013004001.

Sarcophilus I. Geoffroy and F. Cuvier, 1837. Hist. Nat. Mamm., 4(60):6.
ISIS NUMBER: 5301402002007000000.

Sarcophilus harrisii (Boitard, 1841). Jardin des Plantes, p. 290.
TYPE LOCALITY: Australia, Tasmania.
DISTRIBUTION: Tasmania; perhaps S. Victoria.
ISIS NUMBER: 5301402002007001001.

Sminthopsis Thomas, 1887. Ann. Mus. Stor. Nat. Genova, ser. 2, 4:503.
COMMENT: Includes *Antechinomys* as a subgenus; formerly considered a valid genus by Archer, 1977, Mem. Queensl. Mus., 18:17–29, but considered a subgenus by Archer, 1979, Aust. Zool., 20(2):329; see also Kirsch and Calaby, 1977:15. An additional undescribed species has been reported from S. New Guinea and N. Australia; see Archer, 1979, *in* Tyler, ed., The Status of Endangered Australasian Wildlife and Waithman, 1979, Aust. Zool., 20(2):313–326. Revised by Archer, 1981.
ISIS NUMBER: 5301402002014000000 as *Sminthopsis*.
5301402002009000000 as *Antechinomys*.

Sminthopsis butleri Archer, 1979. Aust. Zool., 20(2):329.
TYPE LOCALITY: Australia, Western Australia, Kalumburu (14° 15′ S, 126° 40′ E).
DISTRIBUTION: Known only from the type locality.

Sminthopsis crassicaudata (Gould, 1844). Proc. Zool. Soc. Lond., 1844:105.
TYPE LOCALITY: Australia, Western Australia, Williams River.
DISTRIBUTION: South Australia; S.W. Queensland; S. Northern Territory; S. Western Australia; New South Wales; W. Victoria.
COMMENT: Includes *centralis* and *ferruginea*; see Archer, 1979, Aust. Zool., 20(2):329; 1981:176.
ISIS NUMBER: 5301402002014001001.

Sminthopsis douglasi Archer, 1979. Aust. Zool., 20(2):337.
TYPE LOCALITY: Australia, Queensland, Cloncurry River watershed, Julia Creek (20° 40′ S, 141° 40′ E).
DISTRIBUTION: Known only from type locality and Richmond in Cloncurry River Watershed, Queensland, Australia.

Sminthopsis granulipes Troughton, 1932. Rec. Aust. Mus., 18:350.
TYPE LOCALITY: Australia, Western Australia, King George's Sound (Albany).
DISTRIBUTION: S.W. Western Australia.
ISIS NUMBER: 5301402002014003001.

Sminthopsis hirtipes Thomas, 1898. Nov. Zool., 5:3.
TYPE LOCALITY: Australia, Northern Territory, Charlotte Waters.
DISTRIBUTION: Central deserts in Northern Territory and Western Australia (also coastal scrub 500 km N. of Perth).
ISIS NUMBER: 5301402002014004001.

Sminthopsis laniger (Gould, 1856). Mammal. Aust., 1, pl. 33.
TYPE LOCALITY: Australia, interior New South Wales.
DISTRIBUTION: Western Australia; Northern Territory; N. Victoria; W. New South Wales; Queensland; N.W. South Australia.
COMMENT: Subgenus *Antechinomys*; see Archer, 1979, Aust. Zool., 20(2):329; 1981:187. Includes *spenceri*; see Archer, 1977, Mem. Queensl. Mus., 18:19.
PROTECTED STATUS: U.S. ESA - Endangered as *Antechinomys laniger*.
ISIS NUMBER: 5301402002009001001 as *Antechinomys laniger*.
5301402002009002001 as *Antechinomys spenceri*.

Sminthopsis leucopus (Gray, 1842). Ann. Mag. Nat. Hist., 10:261.
TYPE LOCALITY: Australia, Tasmania.
DISTRIBUTION: S. and S.E. Victoria; Tasmania.
COMMENT: Includes *ferruginifrons, mitchelli,* and *leucogenys*; see Archer, 1979, Aust. Zool., 20(2):329; 1981:102.
ISIS NUMBER: 5301402002014005001 as *S. leucopis (sic)*.

Sminthopsis longicaudata Spencer, 1909. Proc. Roy. Soc. Victoria, (new series):449.
TYPE LOCALITY: Australia, Western Australia.
DISTRIBUTION: Western Australia.
COMMENT: Known from only four specimens; see Ride, 1970:201.
PROTECTED STATUS: CITES - Appendix I and U.S. ESA - Endangered.
ISIS NUMBER: 5301402002014006001.

Sminthopsis macroura (Gould, 1845). Proc. Zool. Soc. Lond., 1845:79.
TYPE LOCALITY: Australia, Queensland, Darling Downs.
DISTRIBUTION: N. New South Wales; Queensland; Northern Territory; N. South Australia; Western Australia.
COMMENT: Includes *froggatti, larapinta, stalkeri,* and *monticola*; see Archer, 1979, Aust. Zool., 20(2):329; 1981:148.
ISIS NUMBER: 5301402002014007001 as *S. macroura*.
5301402002014002001 as *S. froggatti*.

Sminthopsis murina (Waterhouse, 1838). Proc. Zool. Soc. Lond., 1837:76.
TYPE LOCALITY: Australia, New South Wales, N. of Hunter's River.
DISTRIBUTION: S.W. Western Australia; S.E. South Australia; Victoria; New South Wales; E. Queensland.
COMMENT: Includes *albipes, fuliginosa,* and *tatei*; see Archer, 1979, Aust. Zool., 20(2):329; 1981:94–99.
ISIS NUMBER: 5301402002014008001.

Sminthopsis ooldea Troughton, 1965. Proc. Linn. Soc. N.S.W., 1964, 89:316.
TYPE LOCALITY: Australia, South Australia, Ooldea.
DISTRIBUTION: Edge of Nullarbor Plain (South Australia); Western Australia; S. Northern Territory.
COMMENT: Originally described as a subspecies of *murina*, but considered a distinct species by Kirsch and Calaby, 1977:15; Archer, 1975, Mem. Queensl. Mus., 17(2):243.

Sminthopsis psammophila Spencer, 1895. Proc. Roy. Soc. Victoria, 7 (new series):223.
TYPE LOCALITY: Australia, Northern Territory, Lake Amadeus.
DISTRIBUTION: S.W. Northern Territory and Eyre Peninsula (South Australia).
COMMENT: Known from five specimens; see Archer, 1981:215.
PROTECTED STATUS: CITES - Appendix I and U.S. ESA - Endangered.
ISIS NUMBER: 5301402002014010001.

Sminthopsis virginiae (Tarragon, 1847). Rev. Zool. Paris, p. 177.
TYPE LOCALITY: Australia, Queensland, Herbert Vale.
DISTRIBUTION: N. Queensland, N. Northern Territory, Australia; Aru Isls., Indonesia;
S. New Guinea.
COMMENT: Includes *nitela, rufigenis, lumholtzi,* and *rona;* see Archer, 1979, Aust. Zool.,
20(2):329; 1981:132. In reference to *rufigenis,* see also Kirsch and Calaby, 1977, *in*
Stonehouse and Gilmore, p. 15. According to ACZ, De Tarragon's 1847
description of *S. virginiae* did not specify a type locality; Collett, 1887, Proc. Zool.
Soc. Lond., 1886:548, apparently named a different *S. virginiae* in 1887 from
Herbert Vale and it was subsequently renamed *S. lumholtzi;* see Archer, 1981:136.
ISIS NUMBER: 5301402002014009001 as *S. nitela.*
5301402002014011001 as *S. rufigenis.*

Family Myrmecobiidae
REVIEWED BY: C. A. Hill (CAH); J. A. W. Kirsch (JAWK); B. J. Marlow (BJM); R. G. Van
Gelder (RGVG).
COMMENT: Some authors include this family in the Dasyuridae; see Vaughan, 1978:39;
but also see Ride, 1964, J. Proc. Roy. Soc. West Aust., 47:97–131; Archer and
Kirsch, 1977, Proc. Linn. Soc. N.S.W., 102:18–25; and Kirsch and Calaby, 1977:15,
who retained this family.

Myrmecobius Waterhouse, 1836. Proc. Zool. Soc. Lond., 1836:69.
COMMENT: See comment under the family.
ISIS NUMBER: 5301402002008000000.

Myrmecobius fasciatus Waterhouse, 1836. Proc. Zool. Soc. Lond., 1836:69.
TYPE LOCALITY: Australia, Western Australia, 90 mi. (145 km) S.E. mouth of Swan
River.
DISTRIBUTION: S. and E. Western Australia; N.W. South Australia; formerly in S.W.
New South Wales.
PROTECTED STATUS: U.S. ESA - Endangered.
ISIS NUMBER: 5301402002008001001.

Family Thylacinidae
REVIEWED BY: C. A. Hill (CAH); J. A. W. Kirsch (JAWK); R. G. Van Gelder (RGVG).
COMMENT: Some authors include this family in the Dasyuridae; see Vaughan, 1978:39;
but also see Ride, 1964, J. Proc. Roy. Soc. West Aust., 47:97–131; Archer and
Kirsch, 1977, Proc. Linn. Soc. N.S.W., 102:18–25; and Kirsch and Calaby, 1977:15,
who retained this family.

Thylacinus Temminck, 1824. Monogr. Mamm., 1:23, 60, 267.
ISIS NUMBER: 5301402002015000000.

Thylacinus cynocephalus (Harris, 1808). Trans. Linn. Soc. Lond., 9:174.
TYPE LOCALITY: Australia, Tasmania.
DISTRIBUTION: Tasmania (Australia).
COMMENT: Possibly extinct (RGVG); tracks and sightings continue to be reported; see
Ride, 1970:201.
PROTECTED STATUS: CITES - Appendix I and U.S. ESA - Endangered.
ISIS NUMBER: 5301402002015001001.

Family Notoryctidae
REVIEWED BY: C. A. Hill (CAH); J. A. W. Kirsch (JAWK); B. J. Marlow (BJM); R. G. Van
Gelder (RGVG).
ISIS NUMBER: 5301402003000000000.

Notoryctes Stirling, 1891. Trans. R. Soc. S. Aust., 14:154.
COMMENT: Includes *Psammoryctes;* see Iredale and Troughton, 1934, Mem. Aust. Mus., 6:16.
ISIS NUMBER: 5301402003001000000.

Notoryctes typhlops (Stirling, 1889). Trans. R. Soc. S. Aust., 12:158.
TYPE LOCALITY: Australia, Northern Territory, Indracowrie, 100 mi. (161 km) from Charlotte Waters.
DISTRIBUTION: Western deserts from Ooldea (South Australia) to Charlotte Waters and N.W. Western Australia; Northern Territory.
ISIS NUMBER: 5301402003001001001.

Family Peramelidae
REVIEWED BY: C. A. Hill (CAH); J. A. W. Kirsch (JAWK); B. J. Marlow (BJM); R. G. Van Gelder (RGVG); A. C. Ziegler (ACZ) (New Guinea).
COMMENT: Some authors include the family Thylacomyidae in this family; see Vaughan, 1978:39; but also see Archer and Kirsch, 1977, Proc. Linn. Soc. N.S.W., 102:18–25. Revised by Tate, 1948, Bull. Am. Mus. Nat. Hist., 92:313–346.
ISIS NUMBER: 5301402004000000000.

Chaeropus Ogilby, 1838. Proc. Zool. Soc. Lond., 1838:26.
ISIS NUMBER: 5301402004001000000.

Chaeropus ecaudatus (Ogilby, 1838). Proc. Zool. Soc. Lond., 1838:25.
TYPE LOCALITY: Australia, New South Wales, banks of Murray River.
DISTRIBUTION: S.W. New South Wales; Victoria; S. Northern Territory; N. South Australia; Western Australia.
COMMENT: Possibly extinct, last taken in 1907; see Ride, 1970:200.
PROTECTED STATUS: CITES - Appendix I and U.S. ESA - Endangered.
ISIS NUMBER: 5301402004001001001.

Echymipera Lesson, 1842. Nouv. Tabl. Regn. Anim. Mammal., p. 192.
ISIS NUMBER: 5301402004002000000.

Echymipera clara Stein, 1932. Z. Saugetierk., 7:256.
TYPE LOCALITY: Indonesia, Irian Jaya, Tjenderawasih Div., Japen Isl.
DISTRIBUTION: N.C. New Guinea.
ISIS NUMBER: 5301402004002001001.

Echymipera kalubu (Fischer, 1829). Synopsis Mammal., p. 274.
TYPE LOCALITY: Indonesia, Irian Jaya, Sorong Div., Waigeo Island.
DISTRIBUTION: New Guinea and adjacent small islands; Bismarck Arch.; Mysol and Salawatti Isls.
COMMENT: The name *kalubu* has been attributed to Lesson, 1828, Dict. Class. Hist. Nat., 13:200; but see Husson, 1955, Nova Guinea, 6:290.
ISIS NUMBER: 5301402004002002001.

Echymipera rufescens (Peters and Doria, 1875). Ann. Mus. Stor. Nat. Genova, 7:541.
TYPE LOCALITY: Indonesia, Kei Islands.
DISTRIBUTION: Cape York Peninsula (Queensland); New Guinea and certain small islands off S.E. coast; Kei and Aru Isls.
ISIS NUMBER: 5301402004002003001.

Isoodon Desmarest, 1817. Nouv. Dict. Hist. Nat., nouv. ed., 16:409.
COMMENT: Includes *Thylacis* of Haltenorth, 1958, which was an incorrect usage; see Van Deusen and Jones, 1967, *in* Anderson and Jones, p. 74; and Lidicker and Follett, 1968, Proc. Biol. Soc. Wash., 81:251–256.
ISIS NUMBER: 5301402004003000000.

Isoodon auratus (Ramsay, 1887). Proc. Linn. Soc. N.S.W., ser. 2, 2:551.
TYPE LOCALITY: Australia, Western Australia, Derby.
DISTRIBUTION: N. and E. Western Australia; Barrow Island; Northern Territory; inland South Australia.

COMMENT: Includes *barrowensis*; see Ride, 1970:96.
ISIS NUMBER: 5301402004003001001.

Isoodon macrourus (Gould, 1842). Proc. Zool. Soc. Lond., 1842:41.
TYPE LOCALITY: Australia, Northern Territory, Port Essington.
DISTRIBUTION: N.E. Western Australia; N. Northern Territory, E. Queensland; New
    South Wales; S. and E. New Guinea.
ISIS NUMBER: 5301402004003002001.

Isoodon obesulus (Shaw, 1797). Nat. Misc., 8:298.
TYPE LOCALITY: Australia, New South Wales, Sydney.
DISTRIBUTION: E. New South Wales; S. Victoria; S.E. South Australia; N. Queensland;
    S.W. Western Australia; Nuyts Arch.; Tasmania.
ISIS NUMBER: 5301402004003003001.

Microperoryctes Stein, 1932. Z. Saugetierk., 7:256.
ISIS NUMBER: 5301402004005000000.

Microperoryctes murina Stein, 1933. Z. Saugetierk., 7:257.
TYPE LOCALITY: Indonesia, Irian Jaya, Paniai Div., Weyland Mtns., Sumuri Mtn., 2500
    m.
DISTRIBUTION: W. interior New Guinea.
ISIS NUMBER: 5301402004005001001.

Perameles E. Geoffroy, 1804. Ann. Mus. Hist. Nat. Paris, 4:56.
COMMENT: This name was also used by Geoffroy, 1804, Bull. Sci. Soc. Philom. Paris,
    3(80):249.
ISIS NUMBER: 5301402004006000000.

Perameles bougainville Quoy and Gaimard, 1824. Voy. "Uranie", Zool., p. 56.
TYPE LOCALITY: Australia, Western Australia, Shark's Bay.
DISTRIBUTION: S. South Australia; N.W. Victoria; W. New South Wales; S. Western
    Australia; Bernier and Dorre Isls.
COMMENT: Includes *fasciata*; see Ride, 1970:100. Extirpated throughout most of its
    former mainland Australian range (BJM).
PROTECTED STATUS: CITES - Appendix I and U.S. ESA - Endangered.
ISIS NUMBER: 5301402004006001001.

Perameles eremiana Spencer, 1897. Proc. Roy. Soc. Victoria, 9 (new series):9.
TYPE LOCALITY: Australia, Northern Territory, Burt Plain (N. of Alice Springs).
DISTRIBUTION: N. South Australia; S. Northern Territory; Great Victoria Desert
    (Western Australia).
COMMENT: Possibly extinct; see Ride, 1970:200.
PROTECTED STATUS: U.S. ESA - Endangered.
ISIS NUMBER: 5301402004006002001.

Perameles gunnii Gray, 1838. Ann. Nat. Hist., 1:107.
TYPE LOCALITY: Australia, Tasmania.
DISTRIBUTION: S. Victoria; Tasmania.
ISIS NUMBER: 5301402004006003001.

Perameles nasuta E. Geoffroy, 1804. Ann. Mus. Hist. Nat. Paris, 4:62.
TYPE LOCALITY: Australia, New South Wales, Sydney.
DISTRIBUTION: E. Queensland; E. New South Wales; E. Victoria.
ISIS NUMBER: 5301402004006004001.

Peroryctes Thomas, 1906. Proc. Zool. Soc. Lond., 1906:476.
ISIS NUMBER: 5301402004007000000.

Peroryctes broadbenti (Ramsay, 1879). Proc. Linn. Soc. N.S.W., 3:402, pl. 27.
TYPE LOCALITY: Papua New Guinea, Central Prov., banks of Goldie River (a tributary
    of the Laloki River) inland from Port Moresby.
DISTRIBUTION: S.E. New Guinea.

COMMENT: Included in *raffrayanus* by Laurie and Hill, 1954:10, but considered a distinct species by Van Deusen and Jones, 1967, *in* Anderson and Jones, p. 74.

Peroryctes longicauda (Peters and Doria, 1876). Ann. Mus. Stor. Nat. Genova, 8:335.
TYPE LOCALITY: Indonesia, Irian Jaya, Vogelkop, Manokwari Div., Arfak Mtns., Hatam, 1520 m.
DISTRIBUTION: Interior New Guinea.
COMMENT: Possibly includes *papuensis* (ACZ).
ISIS NUMBER: 5301402004007001001.

Peroryctes papuensis Laurie, 1952. Bull. Br. Mus. (Nat. Hist.), Zool., 1:291.
TYPE LOCALITY: Papua New Guinea, Milne Bay Prov., Mt. Mura, (30 mi (48 km) N.W. Mt. Simpson), Boneno, ca. 1220–1525 m.
DISTRIBUTION: S.E. interior New Guinea.
COMMENT: Does not include *broadbenti* (ACZ), contrary to Collins, 1973, who suggested that they may be conspecific. Possibly should be included in *longicauda* (ACZ).
ISIS NUMBER: 5301402004007002001.

Peroryctes raffrayanus (Milne-Edwards, 1878). Ann. Sci. Nat. Paris, 7(Article 11):1–2.
TYPE LOCALITY: Indonesia, Irian Jaya, Vogelkop, Manokwari (?) Div., Amberbaki.
DISTRIBUTION: New Guinea.
COMMENT: Laurie and Hill, 1954:10, included *broadbenti* in this species; but see Van Deusen and Jones, 1967, *in* Anderson and Jones, p. 74.
ISIS NUMBER: 5301402004007003001.

Rhynchomeles Thomas, 1920. Ann. Mag. Nat. Hist., 6:429–430.
ISIS NUMBER: 5301402004008000000.

Rhynchomeles prattorum Thomas, 1920. Ann. Mag. Nat. Hist., 6:429–430.
TYPE LOCALITY: Indonesia, Seram Isl., Mt. Manusela, 1800 m.
DISTRIBUTION: Seram Isl. (Indonesia).
ISIS NUMBER: 5301402004008001001.

Family Thylacomyidae
REVIEWED BY: C. A. Hill (CAH); J. A. W. Kirsch (JAWK); B. J. Marlow (BJM); R. G. Van Gelder (RGVG).
COMMENT: Separated from Peramelidae by Archer and Kirsch, 1977, Proc. Linn. Soc. N.S.W., 102:18–25. First use of group name (as Thalacomyinae) by Bensley, 1903 (JAWK).

Macrotis Reid, 1837. Proc. Zool. Soc. Lond., 1836:131.
COMMENT: Not preoccupied by *Macrotis* Dejean, 1833, a *nomen nudum*; see Troughton, 1932, Aust. Zool., 7(3):219–236. Archer and Kirsch placed *Thylacomys* (a junior synonym of *Macrotis*) in a separate family, Thylacomyidae, rather than Peramelidae; see Archer and Kirsch, 1977, Proc. Linn. Soc. N.S.W., 102:18–25.
ISIS NUMBER: 5301402004004000000.

Macrotis lagotis (Reid, 1837). Proc. Zool. Soc. Lond., 1836:129.
TYPE LOCALITY: Australia, Western Australia, Swan River.
DISTRIBUTION: Western Australia; South Australia; Northern Territory; W. New South Wales; S.W. Queensland.
PROTECTED STATUS: CITES - Appendix I and U.S. ESA - Endangered.
ISIS NUMBER: 5301402004004001001.

Macrotis leucura (Thomas, 1887). Ann. Mag. Nat. Hist., ser. 5, 19:397.
TYPE LOCALITY: Australia, South Australia, Adelaide (uncertain).
DISTRIBUTION: C. Australia.
COMMENT: Possibly extinct; see Ride, 1970:200.
PROTECTED STATUS: CITES - Appendix I and U.S. ESA - Endangered.
ISIS NUMBER: 5301402004004002001.

Family Phalangeridae
REVIEWED BY: C. P. Groves (CPG); C. A. Hill (CAH); J. A. W. Kirsch (JAWK); B. J.
Marlow (BJM); R. G. Van Gelder (RGVG); A. C. Ziegler (ACZ)(New Guinea).
COMMENT: Does not include Phascolarctidae; see Ride, 1970:22.
ISIS NUMBER: 5301402006000000000.

Phalanger Storr, 1780. Prodr. Meth. Mammal., p. 38.
COMMENT: Includes *Spilocuscus;* see Ride, 1970:248. Revised by Tate, 1945, Am. Mus.
Novit., 1283:1–41; Feiler, 1978, Zool. Abh. Mus. Tierk. Dresden, 34:385–395;
35:1–30, 161–168; and George, 1979 *in* Tyler, ed., The Status of Endangered
Australasian Wildlife.
ISIS NUMBER: 5301402006001000000.

Phalanger carmelitae Thomas, 1898. Ann. Mus. Stor. Nat. Genova, 19:5.
TYPE LOCALITY: Papua New Guinea, Central Prov., upper Vanapa River.
DISTRIBUTION: Interior New Guinea.
COMMENT: Formerly included in *vestitus;* see George, 1979 *in* Tyler, ed., The Status of
Endangered Australasian Wildlife, p. 94.

Phalanger celebensis (Gray, 1858). Proc. Zool. Soc. Lond., 1858:105.
TYPE LOCALITY: Indonesia, Sulawesi, Sulawesi Selatan, Ujung Pandang (=Macassar).
DISTRIBUTION: Sulawesi; Peleng Isl.; Sanghir Isls.; Taliabu (=Sula) Isls.; Obi Isl.
COMMENT: *P. rothschildi* from Obi Isl. (included here) may be a distinct species (CPG).
ISIS NUMBER: 5301402006001002001.

Phalanger gymnotis (Peters and Doria, 1875). Ann. Mus. Stor. Nat. Genova, 7:543.
TYPE LOCALITY: Indonesia, Aru Islands, Gialnhegen Island.
DISTRIBUTION: New Guinea; Aru Isls, Wetar Isl, Timor Isl, and other small Indonesian
Isls.
COMMENT: Distribution poorly known. Type locality restricted by Van der Feen, 1962,
Ann. Mus. Civ. Stor. Nat. Genova, 73:40. May include *leucippus* (ACZ).
ISIS NUMBER: 5301402006001003001.

Phalanger interpositus Stein, 1933. Z. Saugetierk., 8:90.
TYPE LOCALITY: Indonesia, Irian Jaya, Paniai Div., Weyland Mtns., Mt. Kunupi,
1500–2000 m.
DISTRIBUTION: N. interior New Guinea.
COMMENT: Formerly included in *orientalis;* see George, 1979, *in* Tyler, ed., The Status
of Endangered Australasian Wildlife, p. 98.

Phalanger leucippus Thomas, 1898. Ann. Mus. Stor. Nat. Genova, 19:7.
TYPE LOCALITY: Papua New Guinea, Central Prov., upper Vanapa River.
DISTRIBUTION: New Guinea (except Vogelkop).
COMMENT: Removed from *gymnotis* by Feiler, 1978, Zool. Abh. Mus. Tierk. Dresden,
34:391–392, and Van der Feen, 1962, Ann. Mus. Civ. Stor. Nat. Genova, 73:44.
May be conspecific with *gymnotis* (ACZ).

Phalanger lullulae Thomas, 1896. Novit. Zool., 3:528.
TYPE LOCALITY: Papua New Guinea, Milne Bay Prov., Woodlark Isl.
DISTRIBUTION: Woodlark Isl. (New Guinea).
COMMENT: Formerly included in *orientalis;* see George, 1979, *in* Tyler, ed., The Status
of Endangered Australasian Wildlife, p. 97.

Phalanger maculatus (E. Geoffroy, 1803). Cat. Mamm. Mus. Hist. Nat., Paris, p. 149.
TYPE LOCALITY: Indonesia, Irian Jaya, Vogelkop, Manokwari Div., Manokwari.
DISTRIBUTION: New Guinea and adjacent small islands; Aru and Kei isls., Seram,
Ambon and Saleyer isls.; Cape York Peninsula (Queensland).
COMMENT: Laurie and Hill, 1954:14, regarded Geoffroy, 1803, as unavailable,
preferring Desmarest, 1818, Nouv. Dict. Hist. Nat., 25:472. Feiler, 1978, Zool.
Abh. Mus. Tierk. Dresden, 35:3–13, included *atrimaculatus* in this species, but
George, 1979, *in* Tyler, ed., The Status of Endangered Australasian Wildlife, p. 98,
placed it in *rufoniger*. All three may be conspecific (ACZ).
PROTECTED STATUS: CITES - Appendix II.
ISIS NUMBER: 5301402006001004001.

Phalanger orientalis (Pallas, 1766). Misc. Zool., p. 61.
　　TYPE LOCALITY: Indonesia, Amboina Isl., Maluku.
　　DISTRIBUTION: Timor and Seram Isls. to New Guinea, and adjacent small Isls.;
　　　　Bismarck Arch.; Solomon Islands; and E. Cape York (Queensland,
　　　　Australia).
　　COMMENT: Formerly included *interpositus* and *lullulae*; see George, 1979, *in* Tyler, ed.,
　　　　The Status of Endangered Australasian Wildlife, pp. 97–98.
　　PROTECTED STATUS: CITES - Appendix II.
　　ISIS NUMBER: 5301402006001005001.

Phalanger rufoniger Zimara, 1937. Anz. Acad. Wiss. Wien, 74:35.
　　TYPE LOCALITY: Papua New Guinea, Morobe Prov., Sattelberg.
　　DISTRIBUTION: N. New Guinea.
　　COMMENT: Includes *atrimaculatus*; see George, 1979, *in* Tyler, ed., The Status of
　　　　Endangered Australasian Wildlife, p. 98; but also see Feiler, 1978, Zool. Abh. St.
　　　　Mus. Tierk, Dresden, 35:3–13, who placed it in *maculatus*.
　　ISIS NUMBER: 5301402006001001001 as *P. atrimaculatus*.

Phalanger ursinus (Temminck, 1824). Monogr. Mamm., 1:10.
　　TYPE LOCALITY: Indonesia, Sulawesi, Sulawesi Utara, Minahasa, Manado.
　　DISTRIBUTION: Sulawesi; Peleng Isl.; Talaut Isls.; Togian Isl.; Muna Isl.; Butung Isl.;
　　　　Lembeh Isl.
　　ISIS NUMBER: 5301402006001006001.

Phalanger vestitus (Milne-Edwards, 1877). C. R. Acad. Sci. Paris, 85:1080.
　　TYPE LOCALITY: Indonesia, Irian Jaya, Vogelkop, Sorong Div., Tamrau Range, Karons
　　　　Mtns.
　　DISTRIBUTION: Interior New Guinea.
　　COMMENT: Formerly included *carmelitae*; see George, 1979, *in* Tyler, ed., The Status of
　　　　Endangered Australasian Wildlife, p. 94. They are sympatric east of 144° E. and
　　　　between 300 and 2000 m. (CPG), or 141° E. and 900 and 2000 m. (ACZ).
　　ISIS NUMBER: 5301402006001007001.

Trichosurus Lesson, 1828. Dict. Class. Hist. Nat., 13:333.
　　ISIS NUMBER: 5301402006002000000.

Trichosurus arnhemensis Collett, 1897. Proc. Zool. Soc. Lond., 1897:328.
　　TYPE LOCALITY: Australia, Northern Territory, Daly River.
　　DISTRIBUTION: N. Northern Territory; N.E. Western Australia; Barrow Island.
　　ISIS NUMBER: 5301402006002001001.

Trichosurus caninus (Ogilby, 1836). Proc. Zool. Soc. Lond., 1835:191.
　　TYPE LOCALITY: Australia, New South Wales, Hunter River.
　　DISTRIBUTION: S.E. Queensland; E. New South Wales; E. Victoria.
　　ISIS NUMBER: 5301402006002002001.

Trichosurus vulpecula (Kerr, 1792). Anim. Kingdom, 1:198.
　　TYPE LOCALITY: Australia, New South Wales, Sydney.
　　DISTRIBUTION: Australia; except N. Northern Territory, N.W. Western Australia, and
　　　　Barrow Island; Tasmania.
　　COMMENT: Introduced into New Zealand; see Wodzicki, 1950, Bull. Dept. Sci. Ind.
　　　　Res., 98:1–255.
　　ISIS NUMBER: 5301402006002003001.

Wyulda Alexander, 1918. J. R. Soc. West. Aust. (1917–1918), 4:31.
　　ISIS NUMBER: 5301402006003000000.

Wyulda squamicaudata Alexander, 1918. J. R. Soc. West. Aust. (1917–1918), 4:31.
　　TYPE LOCALITY: Australia, Western Australia, Wyndham.
　　DISTRIBUTION: N.E. Western Australia.
　　PROTECTED STATUS: U.S. ESA - Endangered.
　　ISIS NUMBER: 5301402006003001001.

Family Burramyidae
REVIEWED BY: J. A. W. Kirsch (JAWK); B. J. Marlow (BJM); R. G. Van Gelder (RGVG);
A. C. Ziegler (ACZ)(New Guinea).
ISIS NUMBER: 5301402008000000000.

Acrobates Desmarest, 1818. Nouv. Dict. Hist. Nat., nouv. ed., 25:405.
ISIS NUMBER: 5301402008001000000.

Acrobates pygmaeus (Shaw, 1793). Zool. New Holland, 1:5.
TYPE LOCALITY: Australia, New South Wales, Sydney.
DISTRIBUTION: E. Queensland to S.E. South Australia, inland to Deniliquin, New South
Wales.
COMMENT: Tate, 1938, Novit. Zool., 41:60, believed the single specimen (of *A.
pulchellus* which is considered a synonym of *pygmaeus*) obtained in N.W. New
Guinea was probably an introduction as a pet (ACZ).
ISIS NUMBER: 5301402008001001001.

Burramys Broom, 1896. Proc. Linn. Soc. N.S.W., 10:564.
ISIS NUMBER: 5301402008002000000.

Burramys parvus Broom, 1896. Proc. Linn. Soc. N.S.W., 10:564.
TYPE LOCALITY: Australia, New South Wales, Taralga (fossil).
DISTRIBUTION: Mountains of E. Victoria; perhaps New South Wales.
COMMENT: Figured in Proc. Linn. Soc. N.S.W., 10, plate 25, page 273. Known from
single living specimen; see Ride, 1970:14–17.
PROTECTED STATUS: CITES - Appendix II and U.S. ESA - Endangered.
ISIS NUMBER: 5301402008002001001.

Cercartetus Gloger, 1841. Hand. Hilfsb. Nat., 1:85.
COMMENT: Includes *Eudromicia*; see Kirsch and Calaby, 1977:16.
ISIS NUMBER: 5301402008003000000.

Cercartetus caudatus (Milne-Edwards, 1877). C. R. Acad. Sci. Paris, 85:1079.
TYPE LOCALITY: Indonesia, Irian Jaya, Vogelkop, Manokwari Div., Arfak Mtns.
DISTRIBUTION: Interior New Guinea; Fergusson Isl. (Papua New Guinea); N.E.
Queensland, Australia.
COMMENT: Includes *macrura*; see Ride, 1970:224.
ISIS NUMBER: 5301402008003001001.

Cercartetus concinnus (Gould, 1845). Proc. Zool. Soc. Lond., 1845:2.
TYPE LOCALITY: Australia, Western Australia, Swan River.
DISTRIBUTION: S.W. Western Australia, S. and S.E. South Australia, W. Victoria, S.W.
New South Wales.
ISIS NUMBER: 5301402008003002001.

Cercartetus lepidus (Thomas, 1888). Cat. Marsup. and Monotr. Br. Mus., p. 142.
TYPE LOCALITY: Australia, Tasmania.
DISTRIBUTION: Tasmania, adjacent Australia, and Kangaroo Island (South Australia).
ISIS NUMBER: 5301402008003003001.

Cercartetus nanus (Desmarest, 1818). Nouv. Dict. Hist. Nat., nouv. ed., 25:477.
TYPE LOCALITY: Australia, Tasmania, Ile Maria.
DISTRIBUTION: S.E. South Australia, E. New South Wales, Victoria, and Tasmania.
ISIS NUMBER: 5301402008003004001.

Distoechurus Peters, 1874. Ann. Mus. Stor. Nat. Genova, 6:303.
ISIS NUMBER: 5301402007003000000.

Distoechurus pennatus (Peters, 1874). Ann. Mus. Stor. Nat. Genova, 6:303.
TYPE LOCALITY: Indonesia, Irian Jaya, Vogelkop, Manokwari Div., "Andai." (Probably
= Arfak Mtns., Hatam, 1520 m).
DISTRIBUTION: New Guinea.
COMMENT: Type locality discussed by Van der Feen, 1962, Ann. Mus. Civ. Stor. Nat.
Genova, 73:52.
ISIS NUMBER: 5301402007003001001.

Family Petauridae
REVIEWED BY: C. A. Hill (CAH); J. A. W. Kirsch (JAWK); B. J. Marlow (BJM); G. M. McKay (GMM); R. G. Van Gelder (RGVG); A. C. Ziegler (ACZ)(New Guinea).
ISIS NUMBER: 5301402007000000000.

Dactylopsila Gray, 1858. Proc. Zool. Soc. Lond., 1858:109.
COMMENT: Includes *Dactylonax*; see Haltenorth, 1958:28.
ISIS NUMBER: 5301402007002000000 as *Dactylopsila*.
5301402007001000000 as *Dactylonax*.

Dactylopsila megalura Rothschild and Dollman, 1932. Abstr. Proc. Zool. Soc. Lond., 353:14.
TYPE LOCALITY: Indonesia, Irian Jaya, Paniai Div., Weyland Range, Gebroeders Mtns.
DISTRIBUTION: Interior New Guinea.
COMMENT: Considered a subspecies of *trivirgata* by Ziegler, 1977, *in* Stonehouse and Gilmore, p. 131.
ISIS NUMBER: 5301402007002001001.

Dactylopsila palpator Milne-Edwards, 1888. Mem. Cent. Soc. Philom. Paris, p. 174.
TYPE LOCALITY: "South coast of New Guinea."
DISTRIBUTION: Interior New Guinea.
ISIS NUMBER: 5301402007001001001 as *Dactylonax palpator*.

Dactylopsila tatei Laurie, 1952. Bull. Br. Mus. (Nat. Hist.), Zool., 1:278.
TYPE LOCALITY: Papua New Guinea, Milne Bay Prov., Fergusson Isl., Faralulu Dist., mtns. above Taibutu Village, 610–915 m.
DISTRIBUTION: Fergusson Isl., Papua New Guinea.
COMMENT: Considered a subspecies of *trivirgata* by Ziegler, 1977, *in* Stonehouse and Gilmore, p. 131; considered a distinct species by George, 1979, *in* Tyler, ed., The Status of Endangered Australasian Wildlife, p. 94.
ISIS NUMBER: 5301402007002002001.

Dactylopsila trivirgata Gray, 1858. Proc. Zool. Soc. Lond., 1858:111.
TYPE LOCALITY: Indonesia, Aru Islands.
DISTRIBUTION: New Guinea and adjacent small isls.; Aru Isls; N.E. Queensland (Australia).
ISIS NUMBER: 5301402007002003001.

Gymnobelideus McCoy, 1867. Ann. Mag. Nat. Hist., ser. 3, 20:287.
ISIS NUMBER: 5301402007004000000.

Gymnobelideus leadbeateri McCoy, 1867. Ann. Mag. Nat. Hist., ser. 3, 20:287.
TYPE LOCALITY: Australia, Victoria, Bass River.
DISTRIBUTION: S. and S.E. Victoria.
ISIS NUMBER: 5301402007004001001.

Petaurus Shaw, 1791. Nat. Misc., 2, pl. 60.
ISIS NUMBER: 5301402007006000000.

Petaurus abidi Ziegler, 1981. Austr. Mammal., 4:81.
TYPE LOCALITY: Papua New Guinea, West Sepik Prov., Mt. Somero.
DISTRIBUTION: N.C. New Guinea.

Petaurus australis Shaw, 1791. Nat. Misc., 2, pl. 60.
TYPE LOCALITY: Australia, New South Wales, Sydney.
DISTRIBUTION: Coastal Queensland, New South Wales, and Victoria.
ISIS NUMBER: 5301402007006001001.

Petaurus breviceps Waterhouse, 1839. Proc. Zool. Soc. Lond., 1838:152.
TYPE LOCALITY: Australia, New South Wales.
DISTRIBUTION: S.E. South Australia to Cape York Peninsula (Queensland); Tasmania (introduction); N. Northern Territory; N.E. Western Australia; New Guinea and adjacent small islands; Bismarck Arch.; Aru Isls.; N. Moluccas.
COMMENT: Reviewed by Smith, 1973, Mamm. Species, 30:1–5.
ISIS NUMBER: 5301402007006002001.

Petaurus norfolcensis (Kerr, 1792). Anim. Kingdom, 1:270.
    TYPE LOCALITY: Australia, New South Wales, Sydney.
    DISTRIBUTION: E. Queensland; E. New South Wales; E. Victoria.
    ISIS NUMBER: 5301402007006003001.

Pseudocheirus Ogilby, 1837. Mag. Nat. Hist. (Charlesworth), 1:457.
    COMMENT: Includes *Petropseudes* and *Hemibelideus;* see Haltenorth, 1958:30. Includes
        Pseudochirops; see Kirsch and Calaby, 1977:16. Revised by Tate, 1945, Am. Mus.
        Novit., 1287:1–30. Ride, 1970, reinstated *Petropseudes* and *Hemibelideus* to generic
        level, but Kirsch and Calaby, 1977, reduced them again. *Petropseudes* appears to
        belong in *Pseudochirops* but the affinities of *Hemibelideus lemuroides* are much closer
        to *Schoinobates (=Petauroides); Hemibelideus* probably deserves generic rank (GMM).
    ISIS NUMBER: 5301402007008000000 as *Pseudocheirus.*
                5301402007005000000 as *Hemibelideus.*
                5301402007007000000 as *Petropseudes.*

Pseudocheirus albertisi (Peters, 1874). Ann. Mus. Stor. Nat. Genova, 6:303.
    TYPE LOCALITY: Indonesia, Irian Jaya, Vogelkop, Manokwari Div., Arfak Mtns., Hatam,
        1520 m.
    DISTRIBUTION: N. and W. New Guinea.
    COMMENT: Formerly included in *Pseudochirops;* see Kirsch and Calaby, 1977:16.
    ISIS NUMBER: 5301402007008001001.

Pseudocheirus archeri (Collett, 1884). Proc. Zool. Soc. Lond., 1884:381.
    TYPE LOCALITY: Australia, North Queensland, Herbert River District.
    DISTRIBUTION: N.E. Queensland (Australia).
    COMMENT: Formerly included in *Pseudochirops;* see Kirsch and Calaby, 1977:16.
    ISIS NUMBER: 5301402007008002001.

Pseudocheirus canescens (Waterhouse, 1846). Nat. Hist. Mammal., 1:306.
    TYPE LOCALITY: Indonesia, Irian Jaya, Fakfak Div., Triton Bay.
    DISTRIBUTION: New Guinea; Salawatti Isl.
    ISIS NUMBER: 5301402007008003001.

Pseudocheirus caroli Thomas, 1921. Ann. Mag. Nat. Hist., 8:357.
    TYPE LOCALITY: Indonesia, Irian Jaya, Paniai Div., Weyland Range, Menoo Valley, Mt.
        Kunupi, 1830 m.
    DISTRIBUTION: W.C. New Guinea.
    ISIS NUMBER: 5301402007008004001.

Pseudocheirus corinnae Thomas, 1897. Ann. Mus. Stor. Nat. Genova, 18:142.
    TYPE LOCALITY: Papua New Guinea, Central Prov., upper Vanapa River.
    DISTRIBUTION: Interior New Guinea.
    COMMENT: Formerly included in *Pseudochirops;* see Kirsch and Calaby, 1977:16.
    ISIS NUMBER: 5301402007008005001.

Pseudocheirus cupreus Thomas, 1897. Ann. Mus. Stor. Nat. Genova, 18:145.
    TYPE LOCALITY: Papua New Guinea, Owen Stanley Range.
    DISTRIBUTION: Interior New Guinea.
    COMMENT: Formerly included in *Pseudochirops;* see Kirsch and Calaby, 1977:16.
    ISIS NUMBER: 5301402007008006001.

Pseudocheirus dahli Collett, 1895. Zool. Anz., 18(490):464.
    TYPE LOCALITY: Australia, Northern Territory, Mary River.
    DISTRIBUTION: Northern Territory; N.W. Western Australia.
    COMMENT: Formerly included in *Petropseudes;* see Kirsch and Calaby, 1977:16, and
        comments under genus.
    ISIS NUMBER: 5301402007007001001 as *Petropseudes dahli.*

Pseudocheirus forbesi Thomas, 1887. Ann. Mag. Nat. Hist., 19:146.
    TYPE LOCALITY: Papua New Guinea, Central Prov., Astrolabe Range, near Port
        Moresby, Sogeri, 458 m.
    DISTRIBUTION: Interior New Guinea.
    ISIS NUMBER: 5301402007008007001.

Pseudocheirus herbertensis (Collett, 1884). Proc. Zool. Soc. Lond., 1884:383.
TYPE LOCALITY: Australia, Queensland, Herbert Vale.
DISTRIBUTION: N.E. Queensland.
ISIS NUMBER: 5301402007008008001.

Pseudocheirus lemuroides (Collett, 1884). Proc. Zool. Soc. Lond., 1884:385.
TYPE LOCALITY: Australia, North Queensland.
DISTRIBUTION: N.E. Queensland.
COMMENT: Formerly included in *Hemibelideus*; see Haltenorth, 1958:30; Kirsch and
Calaby, 1977:16.
ISIS NUMBER: 5301402007005001001 as *Hemibelideus lemuroides*.

Pseudocheirus mayeri Rothschild and Dollman, 1932. Abstr. Proc. Zool. Soc. Lond., 353:15.
TYPE LOCALITY: Indonesia, Irian Jaya, Paniai Div., Weyland Range, Gebroeders Mtns.,
1830 m.
DISTRIBUTION: C. interior New Guinea.
COMMENT: Includes *pygmaeus*; see Laurie and Hill, 1954:21.
ISIS NUMBER: 5301402007008009001 as *P. mayeri*.
5301402007008011001 as *P. pygmaeus*.

Pseudocheirus peregrinus (Boddaert, 1785). Elench. Anim., p. 78.
TYPE LOCALITY: Australia, Queensland, Endeavour River.
DISTRIBUTION: Cape York Peninsula (Queensland) to S.E. South Australia and S.W.
Western Australia; Tasmania; Isls. of the Bass Straits.
COMMENT: Includes *laniginosus, cooki (=convolutor), victoriae, rubidus,* and *occidentalis;* see
Ride, 1970:246.
ISIS NUMBER: 5301402007008010001.

Pseudocheirus schlegeli Jentink, 1884. Notes Leyden Mus., 6:110.
TYPE LOCALITY: Indonesia, Irian Jaya, Vogelkop, Manokwari Div., Arfak Mtns.
DISTRIBUTION: Extreme N.W. New Guinea.
ISIS NUMBER: 5301402007008012001.

Schoinobates Lesson, 1842. Nouv. Tabl. Regn. Anim. Mammal., p. 190.
COMMENT: This name was used by Lesson only for *Petaurista leucogenys* Temminck,
1823, the giant flying squirrel. McKay (in press) argues for a return to *Petauroides*
Thomas, 1888.
ISIS NUMBER: 5301402007009000000.

Schoinobates volans (Kerr, 1792). Anim. Kingdom, 1:199.
TYPE LOCALITY: Australia, New South Wales, Sydney.
DISTRIBUTION: E. Australia, from Dandenong Ranges (Victoria) to Rockhampton
(Queensland).
ISIS NUMBER: 5301402007009001001.

Family Macropodidae
REVIEWED BY: C. P. Groves (CPG); C. A. Hill (CAH); J. A. W. Kirsch (JAWK); B. J.
Marlow (BJM); R. G. Van Gelder (RGVG); A. C. Ziegler (ACZ)(New Guinea).
COMMENT: Revised by Tate, 1948, Bull. Am. Mus. Nat. Hist., 91:233–352.
ISIS NUMBER: 5301402012000000000.

Aepyprymnus Garrod, 1875. Proc. Zool. Soc. Lond., 1875:59.
ISIS NUMBER: 5301402012001000000.

Aepyprymnus rufescens (Gray, 1837). Mag. Nat. Hist. (Charlesworth), 1:584.
TYPE LOCALITY: Australia, New South Wales.
DISTRIBUTION: N.E. Victoria, E. New South Wales, and E. Queensland.
ISIS NUMBER: 5301402012001001001.

Bettongia Gray, 1837. Mag. Nat. Hist. (Charlesworth), 1:584.
PROTECTED STATUS: CITES - Appendix I as *Bettongia* spp.
ISIS NUMBER: 5301402012002000000.

Bettongia gaimardi (Desmarest, 1822). Tabl. Encycl. Meth. Mammal., 2:542.
TYPE LOCALITY: Australia, New South Wales, Port Jackson.
DISTRIBUTION: Coastal S.E. Queensland and N. New South Wales, south to S.W.
Victoria. Tasmania.
PROTECTED STATUS: CITES - Appendix I as *Bettongia* spp. U.S. ESA - Endangered as *B.
gaimardi.*
ISIS NUMBER: 5301402012002001001.

Bettongia lesueuri (Quoy and Gaimard, 1824). Voy. "Uranie," Zool., p. 64.
TYPE LOCALITY: Australia, Western Australia, Dirk Hartog's Island (Shark's Bay).
DISTRIBUTION: Dampier Land (Western Australia); South Australia; Dirk Hartog's Isl.;
Barrow Isl.; Bernier and Dorre Isls.; Northern Territory; S.W. New South Wales.
PROTECTED STATUS: CITES - Appendix I as *Bettongia* spp. and U.S. ESA -
Endangered as *B. lesueur (sic).*
ISIS NUMBER: 5301402012002002001 as *B. lesueur (sic).*

Bettongia penicillata Gray, 1837. Mag. Nat. Hist. (Charlesworth), 1:584.
TYPE LOCALITY: Australia, New South Wales.
DISTRIBUTION: S.W. Western Australia; S. South Australia including St. Francis Isl.;
N.W. Victoria; C. New South Wales; E. Queensland.
COMMENT: Includes *tropica;* see Sharman *et al.,* 1980, Aust. J. Zool., 28; this form may
be extinct; see Ride, 1970:199.
PROTECTED STATUS: CITES - Appendix I as *Bettongia* spp. and U.S. ESA -
Endangered as *B. penicillata.*
U.S. ESA - Endangered as *B. tropica.*
ISIS NUMBER: 5301402012002003001 as *B. penicillata.*
5301402012002004001 as *B. tropica.*

Caloprymnus Thomas, 1888. Cat. Marsup. and Monotr. Br. Mus., p. 114.
ISIS NUMBER: 5301402012003000000.

Caloprymnus campestris (Gould, 1843). Proc. Zool. Soc. Lond., 1843:81.
TYPE LOCALITY: Australia, South Australia.
DISTRIBUTION: Extreme N.E. South Australia; S.W. Queensland.
COMMENT: Possibly extinct; see Ride, 1970:198.
PROTECTED STATUS: CITES - Appendix I and U.S. ESA - Endangered.
ISIS NUMBER: 5301402012003001001.

Dendrolagus Muller, 1840. *In* Temminck, Verhandl. Nat. Gesch. Nederland Overz. Bezitt.,
Zool., p. 30.
REVIEWED BY: U. Ganslosser (UG).
ISIS NUMBER: 5301402012006000000.

Dendrolagus bennettianus De Vis, 1887. Proc. Roy. Soc. Queensl., 3(1886):11.
TYPE LOCALITY: Australia, Queensland, Daintree River.
DISTRIBUTION: N.E. Queensland.
COMMENT: Considered a subspecies of *dorianus* by Haltenorth, 1958; but see Ride,
1970:223, and Kirsch and Calaby, 1977:17.
PROTECTED STATUS: CITES - Appendix II.
ISIS NUMBER: 5301402012006001001.

Dendrolagus dorianus Ramsay, 1883. Proc. Linn. Soc. N.S.W., 8:17.
TYPE LOCALITY: Papua New Guinea, "ranges behind Mt. Astrolabe."
DISTRIBUTION: Interior New Guinea.
COMMENT: Does not include *bennettianus;* see Ride, 1970:223; Kirsch and Calaby,
1977:17.
ISIS NUMBER: 5301402012006002001.

Dendrolagus goodfellowi Thomas, 1908. Ann. Mag. Nat. Hist., ser. 8, 2:452.
TYPE LOCALITY: Papua New Guinea, Owen Stanley Range, vic. Mt. Obree, 8000 ft.
(2438 m).
DISTRIBUTION: Interior E. New Guinea.

COMMENT: Includes *spadix*; see Kirsch and Calaby, 1977:21; Lidicker and Ziegler, 1968, Univ. Calif., Publ. Zool., 87:23. Collins, 1973, regarded this species as a subspecies of the earlier named *matschiei*; but see also Ganslosser, 1980, Zool. Anz., 205:43–66.
ISIS NUMBER: 5301402012006003001.

Dendrolagus inustus Muller, 1840. *In* Temminck, Verhandl. Nat. Gesch. Nederland Overz. Bezitt., Zool., p. 20.
TYPE LOCALITY: Indonesia, Irian Jaya, Fakfak Div., Lobo Dist., near Triton Bay, Mt. Lamantsjieri.
DISTRIBUTION: N. and extreme W. New Guinea, possibly Mysol Isl. and W. Schouten Isls.
COMMENT: Considered a subspecies of *ursinus* by Haltenorth, 1958; but see Kirsch and Calaby, 1977:17. A highly distinct species (CPG).
PROTECTED STATUS: CITES - Appendix II.
ISIS NUMBER: 5301402012006004001.

Dendrolagus lumholtzi Collett, 1884. Proc. Zool. Soc. Lond., 1884:387.
TYPE LOCALITY: Australia, Queensland, Herbert Vale.
DISTRIBUTION: N.E. Queensland.
PROTECTED STATUS: CITES - Appendix II.
ISIS NUMBER: 5301402012006005001.

Dendrolagus matschiei Forster and Rothschild, 1907. Nov. Zool., 14:506.
TYPE LOCALITY: Papua New Guinea, Morobe Prov., Rawlinson Mtns.
DISTRIBUTION: Extreme N.E. interior New Guinea; Umboi Isl. (Introduced?).
COMMENT: Includes *deltae*; see Kirsch and Calaby, 1977:21; Lidicker and Ziegler, 1968, Univ. Calif., Publ. Zool., 87:23. See also comments under *goodfellowi*.
ISIS NUMBER: 5301402012006006001.

Dendrolagus ursinus Temminck, 1836. Discours preliminaire destine a servir d'introduction al faune du Japon, 6(2).
TYPE LOCALITY: Indonesia, Irian Jaya, Fakfak Div., Lobo Dist., near Triton Bay, Mt. Lamantsjieri.
DISTRIBUTION: Extreme N.W. New Guinea.
COMMENT: Does not include *inustus*; see Kirsch and Calaby, 1977:17. Correct original citation presented by Husson, 1955, Nova Guinea, 6:302.
PROTECTED STATUS: CITES - Appendix II.

Dorcopsis Schlegel and Muller, 1845. *In* Temminck, Verhandl. Nat. Gesch. Nederland Overz. Bezitt., Zool., p. 130.
REVIEWED BY: U. Ganslosser (UG).
COMMENT: Includes *Dorcopsulus*; see Ziegler, 1977:134; Kirsch and Calaby, 1977:21. This group is in need of revision (CAH).
ISIS NUMBER: 5301402012007000000 as *Dorcopsis*.
                5301402012008000000 as *Dorcopsulus*.

Dorcopsis atrata Van Deusen, 1957. Am. Mus. Novit., 1826:5.
TYPE LOCALITY: Papua New Guinea, Milne Bay Prov., Goodenough Island, eastern slopes, near "Top Camp", about 1600 m.
DISTRIBUTION: Goodenough Isl.
ISIS NUMBER: 5301402012007001001.

Dorcopsis hageni Heller, 1897. Abh. Zool. Anthrop.-Ethnology Mus. Dresden, 6(8):7.
TYPE LOCALITY: Papua New Guinea, Madang Prov., near Astrolabe Bay, Stefansort.
DISTRIBUTION: N.C. New Guinea.
ISIS NUMBER: 5301402012007002001.

Dorcopsis macleayi Miklouho-Maclay, 1885. Proc. Linn. Soc. N.S.W., 10:145, 149.
TYPE LOCALITY: Papua New Guinea, Central Prov., "inland from Port Moresby."
DISTRIBUTION: Extreme S.E. New Guinea.
COMMENT: Transferred from *Dorcopsulus* by Ziegler, 1977:134.
ISIS NUMBER: 5301402012008001001 as *Dorcopsulus macleayi*.

Dorcopsis vanheurni Thomas, 1922. Ann. Mag. Nat. Hist., 9:264.
　　TYPE LOCALITY: Indonesia, Irian Jaya, Djajawidjaja Div., Doormanpad-bivak (3° 30′ S.
　　　　and 138° 30′ E.), 1410 m.
　　DISTRIBUTION: Interior New Guinea.
　　COMMENT: Probably conspecific with *macleayi*; see Kirsch and Calaby, 1977:21.
　　　　Transferred from *Dorcopsulus* by Ziegler, 1977:134.
　　ISIS NUMBER: 5301402012008002001 as *Dorcopsulus vanheurni.*

Dorcopsis veterum (Lesson, 1827). Voy. autour du Monde, sur.... "la Coquille," Zool.,
　　1:164.
　　TYPE LOCALITY: Indonesia, Irian Jaya, Vogelkop, Manokwari Div., Dorei
　　　　(=Manokwari), Lobo Bay.
　　DISTRIBUTION: W., S., and E. New Guinea; Mysol and Salawatti Isls; Aru Isls.
　　COMMENT: *D. muelleri* is a junior synonym; see Kirsch and Calaby, 1977:21;
　　　　Husson, 1955, Nova Guinea, 6:299. George and Schuerer, 1978, Int. Zoo
　　　　Yearb., 18:152–156 rejected *veterum* as based on *Dendrolagus,* probably
　　　　inustus, and employed *muelleri.*
　　ISIS NUMBER: 5301402012007003001 as *D. muelleri.*

Hypsiprymnodon Ramsay, 1876. Proc. Linn. Soc. N.S.W., 1:33.
　　ISIS NUMBER: 5301402012004000000.

Hypsiprymnodon moschatus Ramsay, 1876. Proc. Linn. Soc. N.S.W., 1:34.
　　TYPE LOCALITY: Australia, Queensland, Rockingham Bay.
　　DISTRIBUTION: N.E. Queensland (Australia).
　　ISIS NUMBER: 5301402012004001001.

Lagorchestes Gould, 1841. Monogr. Macrop., 1, pl. 12 (text).
　　REVIEWED BY: U. Ganslosser (UG).
　　ISIS NUMBER: 5301402012009000000.

Lagorchestes asomatus Finlayson, 1943. Trans. Roy. Soc. South Aust., 67:319.
　　TYPE LOCALITY: Australia, Northern Territory, between Mt. Farewell and Lake Mackay.
　　DISTRIBUTION: Known only from the type locality.
　　COMMENT: Known from a single unsexed skull; see Kirsch and Calaby, 1977:22.
　　ISIS NUMBER: 5301402012009001001.

Lagorchestes conspicillatus Gould, 1842. Proc. Zool. Soc. Lond., 1841:82.
　　TYPE LOCALITY: Australia, Western Australia, Barrow Island.
　　DISTRIBUTION: N. Western Australia and adjacent isls.; N. Northern Territory; N. and
　　　　W. Queensland.
　　ISIS NUMBER: 5301402012009002001.

Lagorchestes hirsutus Gould, 1844. Proc. Zool. Soc. Lond., 1844:32.
　　TYPE LOCALITY: Australia, Western Australia, York district.
　　DISTRIBUTION: C. Western Australia; C. Australia; Dorre Isl. and Bernier Isl. (Western
　　　　Australia).
　　PROTECTED STATUS: CITES - Appendix I and U.S. ESA - Endangered.
　　ISIS NUMBER: 5301402012009003001.

Lagorchestes leporides Gould, 1841. Proc. Zool. Soc. Lond., 1840:93.
　　TYPE LOCALITY: Australia, interior New South Wales.
　　DISTRIBUTION: W. New South Wales; E. South Australia; N.W. Victoria.
　　COMMENT: Possibly extinct; not recorded for more than a century; see Kirsch and
　　　　Calaby, 1977:22.
　　ISIS NUMBER: 5301402012009004001.

Lagostrophus Thomas, 1887. Proc. Zool. Soc. Lond., 1886:544.
　　ISIS NUMBER: 5301402012010000000.

Lagostrophus fasciatus (Peron and Lesueur, 1807). Voy. Terres. Austral., Atlas, pl. 27, 1:114.
　　TYPE LOCALITY: Australia, Western Australia, Bernier Island (Shark's Bay).
　　DISTRIBUTION: Bernier Isl. and Dorre Isl. (Western Australia); S.W. Western Australia;
　　　　perhaps South Australia.
　　PROTECTED STATUS: CITES - Appendix I and U.S. ESA - Endangered.
　　ISIS NUMBER: 5301402012010001001.

Macropus Shaw, 1790. Nat. Misc., 1, pl. 23 (text).
COMMENT: Includes *Megaleia* and *Protemnodon* (*sensu* Haltenorth, 1958); see Kirsch and
Calaby, 1977:17, 22. Rationale for present usage of *Macropus* given in Calaby,
1966, CSIRO Wildl. Res., 10:1–55. Ride, 1962, Aust. J. Sci., 24:367–372, discussed
generic nomenclature for all Macropodinae. Van Gelder, 1977, Am. Mus. Novit.,
2635:1–25, included *Thylogale* and *Wallabia* in this genus; but see Kirsch and
Calaby, 1977:17; Corbet and Hill, 1980:17–18.
ISIS NUMBER: 5301402012011000000 as *Macropus*.
5301402012012000000 as *Magaleia* (*sic*).

Macropus agilis (Gould, 1842). Proc. Zool. Soc. Lond., 1841:81.
TYPE LOCALITY: Australia, Northern Territory, Port Essington.
DISTRIBUTION: N.E. Western Australia; Northern Territory; Queensland; S. New
Guinea; Kiriwina Isls. and other Isls. off the S.E. coast of New Guinea.
ISIS NUMBER: 5301402012011001001.

Macropus antilopinus (Gould, 1842). Proc. Zool. Soc. Lond., 1841:80.
TYPE LOCALITY: Australia, Northern Territory, Port Essington.
DISTRIBUTION: N. Queensland; Northern Territory; N.E. Western Australia.
ISIS NUMBER: 5301402012011002001.

Macropus bernardus Rothschild, 1904. Nov. Zool., 10:543.
TYPE LOCALITY: Australia, Northern Territory, head of South Alligator River.
DISTRIBUTION: Interior of N. Northern Territory.
COMMENT: Possibly extinct; see Ride, 1970:198.
ISIS NUMBER: 5301402012011003001.

Macropus dorsalis (Gray, 1837). Mag. Nat. Hist. (Charlesworth), 1:583.
TYPE LOCALITY: Australia, New South Wales, prob. interior (Namoi Hills).
DISTRIBUTION: E. Queensland; E. New South Wales.
ISIS NUMBER: 5301402012011004001.

Macropus eugenii (Desmarest, 1817). Nouv. Dict. Hist. Nat., nouv. ed., 17:38.
TYPE LOCALITY: Australia, South Australia, Nuyt's Arch., St. Peter's Island.
DISTRIBUTION: S.W. Western Australia; South Australia; Kangaroo Isl.; Wallaby Isl. and
other islands.
ISIS NUMBER: 5301402012011005001.

Macropus fuliginosus (Desmarest, 1817). Nouv. Dict. Hist. Nat., nouv. ed., 17:35.
TYPE LOCALITY: Australia, South Australia, Kangaroo Island.
DISTRIBUTION: S.W. New South Wales; Victoria; South Australia; S.W. Western
Australia; Tasmania; King Isl.; Kangaroo Isl.
COMMENT: See Kirsch and Poole, 1972, Aust. J. Zool., 20:315–339, for discussion of
species limits and subspecies included in this taxon and in *giganteus*.
PROTECTED STATUS: U.S. ESA - Threatened.
ISIS NUMBER: 5301402012011006001.

Macropus giganteus Shaw, 1790. Nat. Misc., 1, pl. 33 (text).
TYPE LOCALITY: Australia, Queensland, Cooktown (="New Holland").
DISTRIBUTION: E. and C. Queensland; Victoria; New South Wales; S. South Australia;
Tasmania.
COMMENT: Opinion 760 of the ICZN placed this name on the Official List of Specific
Names in Zoology; see Anon., 1966, Bull. Zool. Nomencl., 22:292–295; and
Calaby, Mack, and Ride, 1963, Bull. Zool. Nomencl., 20:376–379 for discussion.
Revised by Kirsch and Poole, 1972, Aust. J. Zool., 20:315–319.
PROTECTED STATUS: U.S. ESA - Threatened, all subspecies except *M. g. tasmaniensis*.
U.S. ESA - Endangered as *M. g. tasmaniensis* subspecies only.
ISIS NUMBER: 5301402012011007001.

Macropus greyi Waterhouse, 1846. Nat. Hist. Mammal., 1:122.
TYPE LOCALITY: Australia, South Australia, Coorong.
DISTRIBUTION: S.E. South Australia and adjacent Victoria.
COMMENT: Probably extinct; see Ride, 1970:47; Kirsch and Calaby, 1977:22. Probably
conspecific with *irma* (CPG).
ISIS NUMBER: 5301402012011008001.

Macropus irma (Jourdan, 1837). C. R. Acad. Sci. Paris, 5:523.
TYPE LOCALITY: Australia, Western Australia, Swan River.
DISTRIBUTION: S.W. Western Australia.
ISIS NUMBER: 5301402012011009001.

Macropus parma Waterhouse, 1846. Nat. Hist. Mammal., 1:149.
TYPE LOCALITY: Australia, New South Wales, Sydney.
DISTRIBUTION: E. New South Wales.
COMMENT: Introduced to New Zealand (including Kawau Isl.); see Wodzicki and Flux, 1967, Aust. J. Sci., 29:429–430.
PROTECTED STATUS: U.S. ESA - Endangered.
ISIS NUMBER: 5301402012011010001.

Macropus parryi Bennett, 1835. Proc. Zool. Soc. Lond., 1834:151.
TYPE LOCALITY: Australia, New South Wales, Stroud (near Port Stephens).
DISTRIBUTION: E. Queensland; N.E. New South Wales.
COMMENT: Formerly included in *Protemnodon;* see Haltenorth, 1958:39; but also see Kirsch and Calaby, 1977.
ISIS NUMBER: 5301402012011011001.

Macropus robustus Gould, 1841. Proc. Zool. Soc. Lond., 1840:92.
TYPE LOCALITY: Australia, New South Wales, interior (summit of mtns.).
DISTRIBUTION: Western Australia; South Australia; S. Northern Territory; Queensland; New South Wales; Barrow Isl.
COMMENT: Includes *cervinus* and *erubescens;* see Richardson and Sharman, 1976, J. Zool. Lond., 179:499–513.
ISIS NUMBER: 5301402012011012001.

Macropus rufogriseus (Desmarest, 1817). Nouv. Dict. Hist. Nat., nouv. ed., 17:36.
TYPE LOCALITY: Australia, Tasmania, King Island.
DISTRIBUTION: S.E. South Australia; Victoria; E. Queensland; E. New South Wales; Tasmania; King Isl. and adjacent islands.
ISIS NUMBER: 5301402012011013001.

Macropus rufus (Desmarest, 1822). Tabl. Encycl. Meth. Mammal., 2:541.
TYPE LOCALITY: Australia, New South Wales, Blue Mtns.
DISTRIBUTION: Mainland, mid-latitude Australia.
PROTECTED STATUS: U.S. ESA - Threatened as *Megaleia rufa.*
ISIS NUMBER: 5301402012012001001 as *Megaleia rufa.*

Onychogalea Gray, 1841. J. Two Exped. Aust., 2:402.
ISIS NUMBER: 5301402012013000000.

Onychogalea fraenata (Gould, 1841). Proc. Zool. Soc. Lond., 1840:92.
TYPE LOCALITY: Australia, New South Wales, interior.
DISTRIBUTION: S. Queensland; interior New South Wales.
COMMENT: Extinct throughout most of its former range (CPG).
PROTECTED STATUS: CITES - Appendix I as *O. frenata (sic).* U.S. ESA - Endangered as *O. frenata (sic).*
ISIS NUMBER: 5301402012013001001.

Onychogalea lunata (Gould, 1841). Proc. Zool. Soc. Lond., 1840:93.
TYPE LOCALITY: Australia, Western Australia, coast.
DISTRIBUTION: S.C. and S.W. Western Australia; perhaps Victoria and New South Wales.
COMMENT: Extinct throughout most of its former range (CPG).
PROTECTED STATUS: CITES - Appendix I and U.S. ESA - Endangered.
ISIS NUMBER: 5301402012013002001.

Onychogalea unguifer (Gould, 1841). Proc. Zool. Soc. Lond., 1840:93.
TYPE LOCALITY: Australia, Western Australia, Derby (Kings sound).
DISTRIBUTION: N. Australia (Western Australia; Northern Territory; Queensland).
ISIS NUMBER: 5301402012013003001.

Peradorcas Thomas, 1904. Nov. Zool., 11:226.
COMMENT: Probably congeneric with *Petrogale*; see Kitchener and Sanson, 1978, Rec. W. Aust. Mus., 6:269–285.
ISIS NUMBER: 5301402012014000000.

Peradorcas concinna (Gould, 1842). Proc. Zool. Soc. Lond., 1842:57.
TYPE LOCALITY: Australia, Western Australia, Wyndham.
DISTRIBUTION: N. Northern Territory; N.E. Western Australia.
ISIS NUMBER: 5301402012014001001.

Petrogale Gray, 1837. Mag. Nat. Hist. (Charlesworth), 1:583.
COMMENT: Revision of this genus is underway by Sharman, *et al.*; a preliminary account of their arrangement is provided by Poole, 1979, *in* Tyler, ed., The Status of Endangered Australasian Wildlife, pp. 19–22. Kitchener and Sanson, 1978, Rec. W. Aust. Mus., 6:269–285 considered this genus as probably congeneric with *Peradorcas*.
ISIS NUMBER: 5301402012015000000.

Petrogale brachyotis Gould, 1841. Proc. Zool. Soc. Lond., 1840:128.
TYPE LOCALITY: Australia, Western Australia, Hanover Bay.
DISTRIBUTION: Coast of N.W. Australia; N. Northern Territory.
ISIS NUMBER: 5301402012015001001.

Petrogale burbidgei Kitchener and Sanson, 1978. Rec. W. Aust. Mus., 6:269–285.
TYPE LOCALITY: Australia, Western Australia, Mitchell Plateau, Crystal Creek (14° 30′ S. and 125° 47′20″ E.).
DISTRIBUTION: Kimberly (Western Australia), Bonaparte Arch, and adjacent isls.

Petrogale godmani Thomas, 1923. Abstr. Proc. Zool. Soc. Lond., 235:13.
TYPE LOCALITY: Australia, Queensland, Cooktown (Black Mt.).
DISTRIBUTION: Cape York Peninsula, N. Queensland (Australia).
COMMENT: Revision of the genus is underway by Sharman *et al.*, Included in *penicillata* in a preliminary account of their arrangement by Poole, 1979, *in* Tyler, ed., The Status of Endangered Australasian Wildlife, p. 21.
ISIS NUMBER: 5301402012015002001.

Petrogale penicillata (Gray, 1827). Anim. Kingdom (Cuvier), Mamm., 3, plate only.
TYPE LOCALITY: Australia, New South Wales, Sydney.
DISTRIBUTION: Mainland Australia; Mondraine Isl. and adjacent Isls.; Groote Eylandt (off Northern Territory); Pearson Isls. (S. Australia).
COMMENT: Includes *inornata*; see Ride, 1970:223. Revision of the genus is underway by Sharman *et al.*; a preliminary account of their arrangement is provided by Poole, 1979, *in* Tyler, ed., The Status of Endangered Australasian Wildlife, pp. 19–22; *inornata, godmani* and *purpureicollis* are included in *penicillata*.
ISIS NUMBER: 5301402012015003001.

Petrogale purpureicollis Le Souef, 1924. Aust. Zool., 3:274.
TYPE LOCALITY: Australia, Queensland, Dajarra.
DISTRIBUTION: N.E. Queensland (Australia).
COMMENT: Revision of the genus is underway by Sharman *et al.*, Included in *penicillata* in a preliminary account of their arrangement by Poole, 1979, *in* Tyler, ed., The Status of Endangered Australasian Wildlife, p. 21.
ISIS NUMBER: 5301402012015004001.

Petrogale rothschildi Thomas, 1904. Nov. Zool., 11:366.
TYPE LOCALITY: Australia, Western Australia, Cossack River.
DISTRIBUTION: C. to W. Western Australia.
ISIS NUMBER: 5301402012015005001.

Petrogale xanthopus Gray, 1855. Proc. Zool. Soc. Lond., 1854:259.
TYPE LOCALITY: Australia, South Australia, Strange.
DISTRIBUTION: S.W. Queensland; South Australia; N.W. New South Wales; Victoria.
PROTECTED STATUS: U.S. ESA - Endangered.
ISIS NUMBER: 5301402012015006001.

Potorous Desmarest, 1804. Nouv. Dict. Hist. Nat., 1st ed., 24, Tabl. Meth. Mamm., p. 20.
ISIS NUMBER: 5301402012005000000.

Potorous longipes Seebeck and Johnston, 1980. Aust. J. Zool., 28:121.
TYPE LOCALITY: Australia, Victoria, Bellbird Creek, 32 km E. Orbost.
DISTRIBUTION: Victoria (Australia).
COMMENT: Known only from four specimens; first collected in 1968.

Potorous platyops (Gould, 1844). Proc. Zool. Soc. Lond., 1844:103.
TYPE LOCALITY: Australia, Western Australia, Swan River.
DISTRIBUTION: S.W. Western Australia; formerly on Kangaroo Isl.
COMMENT: Possibly extinct; see Ride, 1970:199.
ISIS NUMBER: 5301402012005002001.

Potorous tridactylus (Kerr, 1792). Anim. Kingdom, 1:198.
TYPE LOCALITY: Australia, New South Wales, Sydney.
DISTRIBUTION: S.E. Queensland; coastal New South Wales; N.E. Victoria; S.E. South
    Australia; S.W. Western Australia; Tasmania; King Island.
COMMENT: Includes *apicalis*; see Kirsch and Calaby, 1977:21.
ISIS NUMBER: 5301402012005003001 as *P. tridactylus*.
    5301402012005001001 as *P. apicalis*.

Setonix Lesson, 1842. Nouv. Tabl. Regn. Anim. Mammal., p. 194.
ISIS NUMBER: 5301402012016000000.

Setonix brachyurus (Quoy and Gaimard, 1830). Voy. Astrol., Zool., 1:114.
TYPE LOCALITY: Australia, Western Australia, King George's Sound (Albany).
DISTRIBUTION: S.W. Western Australia; Rottnest Isl.; Bald Isl.
PROTECTED STATUS: U.S. ESA - Endangered.
ISIS NUMBER: 5301402012016001001.

Thylogale Gray, 1837. Mag. Nat. Hist. (Charlesworth), 1:583.
COMMENT: Included in *Macropus* by Van Gelder, 1977, Am. Mus. Novit., 2635:1–25;
    but see Kirsch and Calaby, 1977:17.
ISIS NUMBER: 5301402012017000000.

Thylogale billardieri (Desmarest, 1822). Tabl. Encycl. Meth. Mammal., 2:542.
TYPE LOCALITY: Australia, Tasmania.
DISTRIBUTION: S.E. South Australia; Victoria; Tasmania; Isls. in Bass Strait.
ISIS NUMBER: 5301402012017001001 as *T. billardierii (sic)*.

Thylogale brunii (Schreber, 1778). Saugethiere, 3:551.
TYPE LOCALITY: Indonesia, Aru Islands.
DISTRIBUTION: C. and E. New Guinea and adjacent small islands; Bismarck
    Arch.; Aru Isls.
COMMENT: *T. bruijni* is a later spelling; see Haltenorth, 1958:38.
ISIS NUMBER: 5301402012017002001 as *T. bruijni (sic)*.

Thylogale stigmatica (Gould, 1860). Mammal. Aust., 2, Part 12, pl. 33–34.
TYPE LOCALITY: Australia, Queensland, Point Cooper (N. of Rockingham Bay).
DISTRIBUTION: E. Queensland; E. New South Wales; S.C. New Guinea.
COMMENT: Citation for original description given as Gould, 1860, Proc. Zool. Soc.
    Lond., 1860:375, by some authors, but this is dated Nov. 13, while Mammal.
    Aust., Part 12 was published Nov. 1.
ISIS NUMBER: 5301402012017003001.

Thylogale thetis (Lesson, 1828). Monogr. Mamm., p. 229.
TYPE LOCALITY: Australia, New South Wales, Sydney.
DISTRIBUTION: E. Queensland; E. New South Wales.
ISIS NUMBER: 5301402012017004001.

Wallabia Trouessart, 1905. Cat. Mamm. Viv. Foss., Suppl. fasc., 4:834.
COMMENT: Included in *Macropus* by Van Gelder, 1977, Am. Mus. Novit., 2635:1–25;
    but see Kirsch and Calaby, 1977:17, 22.
ISIS NUMBER: 5301402012018000000.

Wallabia bicolor (Desmarest, 1804). Nouv. Dict. Hist. Nat., 1st ed., p. 357.
TYPE LOCALITY: Australia, New South Wales.
DISTRIBUTION: E. Queensland; E. New South Wales; Victoria; S.E. South Australia;
Stradbroke Island.
ISIS NUMBER: 5301402012018001001.

Family Phascolarctidae
REVIEWED BY: C. A. Hill (CAH); J. A. W. Kirsch (JAWK); B. J. Marlow (BJM); R. G. Van
Gelder (RGVG).
COMMENT: Formerly included in the Phalangeridae; see Ride, 1970:225.
ISIS NUMBER: 5301402010000000000.

Phascolarctos Blainville, 1816. Nouv. Bull. Sci. Soc. Philom. (Paris), p. 108.
ISIS NUMBER: 5301402010001000000.

Phascolarctos cinereus (Goldfuss, 1817). Saugethiere (Schreber), part 65, pl. 155, Aa, Ac.
TYPE LOCALITY: Australia, New South Wales.
DISTRIBUTION: S.E. Queensland, E. New South Wales, S.E. South Australia, and
Victoria. Introduced on Kangaroo Isl., South Australia (BJM).
ISIS NUMBER: 5301402010001001001.

Family Vombatidae
REVIEWED BY: C. A. Hill (CAH); J. A. W. Kirsch (JAWK); B. J. Marlow (BJM); R. G. Van
Gelder (RGVG).
COMMENT: Phascolomyidae is based on *Phascolomys*, a junior synonym; see Haltenorth,
1958:32.
ISIS NUMBER: 5301402011000000000.

Lasiorhinus Gray, 1863. Ann. Mag. Nat. Hist., ser. 3, 11:458.
ISIS NUMBER: 5301402011001000000.

Lasiorhinus krefftii (Owen, 1873). Philos. Trans. R. Soc. Lond., 162:178, pl. 17, 20.
TYPE LOCALITY: Australia, Queensland, Moonie River.
DISTRIBUTION: S.E. Queensland.
COMMENT: Includes *gillespiei* and *barnardi*; see Kirsch and Calaby, 1977:23, who stated
that only a single remnant population of *krefftii* remained at the type locality of
barnardi.
PROTECTED STATUS: CITES - Appendix I as *L. krefftii*. U.S. ESA - Endangered as *L.
barnardi*. U.S. ESA - Endangered as *L. gillespiei*.
ISIS NUMBER: 5301402011001001001 as *L. barnardi*.
5301402011001002001 as *L. gillespiei*.

Lasiorhinus latifrons (Owen, 1845). Proc. Zool. Soc. Lond., 1845:82.
TYPE LOCALITY: Australia, South Australia.
DISTRIBUTION: E. Queensland, S. South Australia, S.E. Western Australia.
ISIS NUMBER: 5301402011001003001.

Vombatus E. Geoffroy, 1803. Bull. Sci. Soc. Philom. Paris, 72:185.
COMMENT: *Phascolomis* E. Geoffroy, 1803 is a junior synonym; see Haltenorth, 1958:32.
ISIS NUMBER: 5301402011002000000.

Vombatus ursinus (Shaw, 1800). Gen. Zool., 1, Mammalia, p. 504.
TYPE LOCALITY: Australia, Tasmania, Bass Strait, Cape Barren Island.
DISTRIBUTION: E. New South Wales, S. Victoria, S.E. South Australia, Tasmania, Isls. in
the Bass Strait, and S.E. Queensland.
ISIS NUMBER: 5301402011002001001.

Family Tarsipedidae
REVIEWED BY: C. A. Hill (CAH); J. A. W. Kirsch (JAWK); B. J. Marlow (BJM); R. G. Van
Gelder (RGVG).
ISIS NUMBER: 5301402009000000000.

Tarsipes Gray, 1842. Ann. Mag. Nat. Hist., 9:40.
ISIS NUMBER: 5301402009001000000.

Tarsipes spenserae Gray, 1842. Ann. Mag. Nat. Hist., 9:40.
TYPE LOCALITY: Australia, Western Australia, King George's Sound (Albany).
DISTRIBUTION: S.W. Western Australia.
COMMENT: *T. spenserae* is considered a misspelling by JAWK because the
name was presented as a patronym for Spencer. Gray, 1842, Ann. Mag.
Nat. Hist., 9:40 misspelled the name as *spenserae* in the original
description and therefore it must be retained (BJM). Mahoney, 1981,
Aust. Mammal., 4:135–138, presented evidence that *Tarsipes rostratus*
Gervais and Verreaus, 1842, predates *T. spenserae* Gray, 1842.
ISIS NUMBER: 5301402009001001001.

ORDER EDENTATA (=XENARTHRA)

COMMENT: Reviewed by Montgomery, ed., 1982, The evolution and ecology of sloths, anteaters and armadillos (Mammalia, Xenarthra=Edentata), Smithson. Inst. Press, Washington, D.C.

ISIS NUMBER: 5301407000000000000.

Family Myrmecophagidae

REVIEWED BY: R. M. Wetzel (RMW).

ISIS NUMBER: 5301407001000000000.

Cyclopes Gray, 1821. Lond. Med. Repos., 15:305.

REVIEWED BY: J. Ramirez-Pulido (JRP).

ISIS NUMBER: 5301407001001000000.

Cyclopes didactylus (Linnaeus, 1758). Syst. Nat., 10th ed., 1:35.

TYPE LOCALITY: Surinam.

DISTRIBUTION: S.E. Veracruz and S.E. Oaxaca (Mexico) to Colombia; west of Andes possibly to N.W. Peru; east of Andes, Colombia and Venezuela, south to Bolivia (Santa Cruz) and Brazil (Acre, Amazonas, Para, Maranhhao, Pernambuco and Alagoas); Trinidad. Type locality restricted by Thomas, 1911, Proc. Zool. Soc. Lond., 1911:132.

ISIS NUMBER: 5301407001001001001.

Myrmecophaga Linnaeus, 1758. Syst. Nat., 10th ed., 1:35.

ISIS NUMBER: 5301407001002000000.

Myrmecophaga tridactyla Linnaeus, 1758. Syst. Nat., 10th ed., 1:35.

TYPE LOCALITY: Brazil, Pernambuco.

DISTRIBUTION: S. Belize and Guatemala to South America; west of Andes to N.W. Ecuador; east of Andes, Colombia and Venezuela, south to the Gran Chaco (Bolivia, Paraguay, Argentina) on the west and Uruguay on the east. Type locality restricted by Thomas, 1911, Proc. Zool. Soc. Lond., 1911:132.

PROTECTED STATUS: CITES - Appendix II.

ISIS NUMBER: 5301407001002001001.

Tamandua Gray, 1825. Ann. Philos., 10:343.

ISIS NUMBER: 5301407001003000000.

Tamandua mexicana (Saussure, 1860). Rev. Mag. Zool. Paris, ser. 2, 12:9.

REVIEWED BY: J. Ramirez-Pulido (JRP).

TYPE LOCALITY: Mexico, Tabasco.

DISTRIBUTION: San Luis Potosi and Tamaulipas (Mexico) to Depto. Lambayeque (N.W. Peru) and N.W. Venezuela; Puna Isl. (Ecuador).

COMMENT: Separated from *tetradactyla* by Wetzel, 1975, Proc. Biol. Soc. Wash., 88:104.

Tamandua tetradactyla (Linnaeus, 1758). Syst. Nat., 10th ed., 1:35.

TYPE LOCALITY: Brazil, Pernambuco.

DISTRIBUTION: E. of Andes, from N. South America, south to Paraguay, N. Argentina, and Uruguay; Trinidad.

COMMENT: Includes *longicaudata*; see Wetzel, 1975, Proc. Biol. Soc. Wash., 88:105. *T. longicaudata mexianae* Hagmann, 1908 (see Cabrera, 1958:203), is a *nomen nudum* according to Wetzel, 1975, Proc. Biol. Soc. Wash., 88:106. Type locality restricted by Thomas, 1911, Proc. Zool. Soc. Lond., 1911:133.

PROTECTED STATUS: CITES - Appendix III as *T. tetradactyla* (Guatemala). CITES - Appendix II as *T. t. chapadensis* subspecies only.

ISIS NUMBER: 5301407001003002001 as *T. tetradactyla*.
                    5301407001003001001 as *T. longicaudata*.

Family Bradypodidae

REVIEWED BY: R. M. Wetzel (RMW).

ISIS NUMBER: 5301407002000000000.

Bradypus Linnaeus, 1758. Syst. Nat., 10th ed., 1:34.
  ISIS NUMBER: 5301407002001000000.

Bradypus torquatus Desmarest, 1816. Nouv. Dict. Hist. Nat. Paris, Applique au Arts and
    Agr., 2nd ed., 4:353.
  TYPE LOCALITY: Brazil.
  DISTRIBUTION: Coastal forests of S.E. Brazil, probably now restricted to Atlantic
    drainage of Bahia, Espirito Santo, and Rio de Janeiro; formerly north to Rio
    Grande del Norte. See Wetzel and Avila-Pires, 1980, Rev. Brasil Biol., 40:834.
  COMMENT: *B. torquatus* Illiger, 1811, Prodr. Syst. Mamm. et Avium., p. 108, is a *nomen
    nudum*; see Thomas, 1917, Ann. Mag. Nat. Hist., ser. 8, 19:352.
  PROTECTED STATUS: U.S. ESA - Endangered.
  ISIS NUMBER: 5301407002001003001.

Bradypus tridactylus Linnaeus, 1758. Syst. Nat., 10th ed., 1:34.
  TYPE LOCALITY: Surinam.
  DISTRIBUTION: Bolivar and T.F. Amazonas (S. Venezuela); forests of Guianas, south to
    the Amazon and the Rio Negro rivers.
  COMMENT: Type locality restricted by Thomas, 1911, Proc. Zool. Soc. Lond., 1911:132.
    Reviewed by Wetzel and Kock, 1973, Proc. Biol. Soc. Wash., 86:29–30.
  ISIS NUMBER: 5301407002001004001.

Bradypus variegatus Schinz, 1825. Das Thierreich, 1:510.
  TYPE LOCALITY: Brazil, possibly Bahia.
  DISTRIBUTION: E. Honduras to S.E. Brazil, Prov. Misiones (N.E. Argentina), west to S.E.
    Bolivia and N.W. Argentina, and E. Peru, Ecuador and Colombia.
  COMMENT: Includes *boliviensis*, *griseus*, and *infuscatus*; see Wetzel and Kock, 1973, Proc.
    Biol. Soc. Wash., 86:27; Wetzel and Avila-Pires, 1980, Rev. Brasil. Biol., 40:833;
    Wetzel, 1982, *in* Montgomery, ed., The evolution and ecology of sloths, anteaters
    and armadillos (Mammalia, Xenarthra = Edentata), Smithson. Inst. Press,
    Washington, D.C. Type locality discussed by Mertens, 1925,
    Senckenberg. Biol., 7:23.
  PROTECTED STATUS: CITES - Appendix II as *B. boliviensis*.
    CITES - Appendix III as *B. griseus* (Costa Rica).
  ISIS NUMBER: 5301407002001001001 as *B. boliviensis* only.
    5301407002001002001 as *B. infuscatus* only.

Family Choloepidae
  REVIEWED BY: R. M. Wetzel (RMW).
  COMMENT: Formerly included in Bradypodidae; see Patterson and Pascual, 1968, Q.
    Rev. Biol., 43:409–451; Hoffstetter, 1969, Bull. Mus. Nat. Hist. Nat. Paris, ser. 2,
    41:91–103. Webb, 1982, *in* Montgomery, ed., The evolution and ecology of sloths,
    anteaters and armadillos (Mammalia, Xenarthra = Edentata), Smithson. Inst.
    Press, Washington, D.C., places Choloepidae in Megalonychidae.

Choloepus Illiger, 1811. Prodr. Syst. Mamm. et Avium., p. 108.
  ISIS NUMBER: 5301407002002000000.

Choloepus didactylus (Linnaeus, 1758). Syst. Nat., 10th ed., 1:35.
  TYPE LOCALITY: Surinam.
  DISTRIBUTION: Guianas; Delta of Rio Orinoco and south of Rio Orinoco (Venezuela);
    south to Maranhao on east, and west along both banks of Rio Amazonas (Brazil)
    to Amazon Basin of Colombia, Ecuador and Peru.
  COMMENT: Reviewed by Wetzel and Avila-Pires, 1980, Rev. Brasil. Biol., 40:834. Type
    locality restricted by Thomas, 1911, Proc. Zool. Soc. Lond., 1911:132.
  ISIS NUMBER: 5301407002002001001.

Choloepus hoffmanni Peters, 1859. Monatsb. Preuss. Akad. Wiss. Berlin, 1858:128.
  TYPE LOCALITY: Costa Rica, Heredia, Volcan Barbara.
  DISTRIBUTION: N. Nicaragua to South America; west of Andes to Esmeraldas (Ecuador);
    east of Andes to Rio Solimoes (Peru) and N. Mato Grosso (Brazil).

COMMENT: See Wetzel and Avila-Pires, 1980, Rev. Brasil. Biol., 40:834–835, for
discussion of type locality.
PROTECTED STATUS: CITES - Appendix III (Costa Rica).
ISIS NUMBER: 5301407002002002001.

Family Megalonychidae
COMMENT: Known only from sub-Recent fossils from the Greater Antilles; see Hall,
1981:274–275. See also comment under Choloepidae.

Family Dasypodidae
REVIEWED BY: R. M. Wetzel (RMW).
ISIS NUMBER: 5301407003000000000.

Cabassous McMurtrie, 1831. Anim. Kingdom, 1:164.
COMMENT: Revised by Wetzel, 1980, Ann. Carnegie Mus., 49:323–357.
ISIS NUMBER: 5301407003002000000.

Cabassous centralis (Miller, 1899). Proc. Biol. Soc. Wash., 13:7.
TYPE LOCALITY: Honduras, Cortes, Chamelecon.
DISTRIBUTION: Honduras and probably Guatemala to Venezuela west of the Cordillera
Oriental.
PROTECTED STATUS: CITES - Appendix III (Costa Rica).
ISIS NUMBER: 5301407003002001001.

Cabassous chacoensis Wetzel, 1980. Ann. Carnegie Mus., 49(2):335.
TYPE LOCALITY: Paraguay, Depto. Presidente Hayes, 5–7 km W. Estancia Juan de
Zalazar.
DISTRIBUTION: Gran Chaco (N.W. Argentina, W. Paraguay, S.E. Bolivia); probably Mato
Grosso do Sul (Brazil).

Cabassous tatouay (Desmarest, 1804). Tabl. Meth. Mamm., 24:28.
TYPE LOCALITY: S.E. Paraguay at 27° S.
DISTRIBUTION: S. Brazil, south to Rio Grande do Sul; Uruguay, N.E. Argentina and
Paraguay west to Rio Paraguay.
COMMENT: Type locality discussed by Cabrera, 1958:219.
PROTECTED STATUS: CITES - Appendix III as *C. gymnurus (tatouay)* (Uruguay).
ISIS NUMBER: 5301407003002004001.

Cabassous unicinctus (Linnaeus, 1758). Syst. Nat., 10th ed., 1:50.
TYPE LOCALITY: Surinam.
DISTRIBUTION: South America E. of Andes, Venezuela south to Mato Grosso do Sul,
Minas Gerais, Goias, and Marananhao (Brazil).
COMMENT: Includes *gymnurus, hispidus, loricatus, lugubris,* and *latirostris;* see Wetzel,
1980, Ann. Carnegie Mus., 49:344. Type locality corrected by Thomas, 1911, Proc.
Zool. Soc. Lond., 1911:141.
ISIS NUMBER: 5301407003002005001 as *C. unicinctus.*
             5301407003002002001 as *C. hispidus.*
             5301407003002003001 as *C. loricatus.*

Chaetophractus Fitzinger, 1871. Sitzb. Math-Nat. K. Akad. Wiss., 64(1):268.
ISIS NUMBER: 5301407003003000000.

Chaetophractus nationi (Thomas, 1894). Ann. Mag. Nat. Hist., ser. 6, 13:71.
TYPE LOCALITY: Bolivia, Oruro.
DISTRIBUTION: Cochabamba, Orura, and La Paz (Bolivia) and possibly puna of
Argentina and Chile.
COMMENT: Distribution and status uncertain, may be a subspecies of *vellerosus;* see
Wetzel, 1982, *in* Montgomery, ed., The evolution and ecology of sloths, anteaters,
and armadillos (Mammalia, Xenarthra = Edentata), Smithson. Inst. Press,
Washington, D.C.
ISIS NUMBER: 5301407003003001001.

Chaetophractus vellerosus (Gray, 1865). Proc. Zool. Soc. Lond., 1865:376.
  TYPE LOCALITY: Bolivia, Santa Cruz, Santa Cruz (de la Sierra).
  DISTRIBUTION: S.E. Bolivia through W. Paraguay and Chaco of Argentina to Mendoza
    and C. Buenos Aires Prov.
  COMMENT: Reviewed by Wetzel, 1982, *in* Montgomery, ed., Smithson. Inst. Press,
    Washington, D.C.
  ISIS NUMBER: 5301407003003002001.

Chaetophractus villosus (Desmarest, 1804). Tabl. Meth. Mamm., 24:28.
  TYPE LOCALITY: Argentina, Buenos Aires Province, between 35° S. and 36° S.
  DISTRIBUTION: Chaco of Paraguay; Uruguay, Mendoza, Cordoba, Santa Fe, south to
    Santa Cruz (Argentina) and Chikchico (Chile); probably Chaco of Argentina and
    Bolivia.
  COMMENT: Type locality discussed by Cabrera, 1958:214.
  ISIS NUMBER: 5301407003003003001.

Chlamyphorus Harlan, 1825. Ann. Lyc. Nat. Hist., 1:235.
  COMMENT: Includes *Burmeisteria* and *Calyptophractus*. *Burmeisteria* Gray is antedated by
    Burmeisteria Salter, a trilobite, with *Calyptophractus* Fitzinger, 1871, the next
    available name; see Wetzel, 1982, *in* Montgomery, ed., Smithson. Inst. Press,
    Washington, D.C.
  ISIS NUMBER: 5301407003004000000 as *Chlamyphorus*.
      5301407003001000000 as *Burmeisteria*.

Chlamyphorus retusus (Burmeister, 1863). Abh. Ges. Naturf. Halle, 7:167.
  TYPE LOCALITY: Bolivia, Santa Cruz, Santa Cruz (de la Sierra).
  DISTRIBUTION: Gran Chaco, from provinces of Formosa, Chaco, and Salta of Argentina
    through W. Paraguay to C. Santa Cruz, Bolivia.
  COMMENT: Reviewed by Wetzel, 1982, *in* Montgomery, ed., Smithson. Inst. Press,
    Washington, D.C.
  ISIS NUMBER: 5301407003001001001 as *Burmeisteria retusa*.

Chlamyphorus truncatus Harlan, 1825. Ann. Lyc. Nat. Hist., 1:235.
  TYPE LOCALITY: Argentina, Mendoza Prov., 33° 25′ S. and 69° 45′ E.
  DISTRIBUTION: Argentina.
  PROTECTED STATUS: U.S. ESA - Endangered.
  ISIS NUMBER: 5301407003004001001.

Dasypus Linnaeus, 1758. Syst. Nat., 10th ed., 1:50.
  COMMENT: Reviewed by Wetzel and Mondolfi, 1979, *in* Eisenberg, J.F. (ed.), Vertebrate
    ecology in the northern Neotropics, Smithson. Inst. Press, Washington, D.C., pp.
    43–63.
  ISIS NUMBER: 5301407003005000000.

Dasypus hybridus (Desmarest, 1804). Tabl. Meth. Mamm., 24:28.
  TYPE LOCALITY: Paraguay, Depto. Misiones, San Ignacio.
  DISTRIBUTION: Rio Grande do Sul (S. Brazil); Uruguay; Paraguay; south in Argentina to
    Mendoza and Rio Negro.
  COMMENT: Partially includes *mazzai* Yepes (a composite); see Wetzel and Mondolfi,
    1979:52–53.
  ISIS NUMBER: 5301407003005001001 as *D. hybridus*.
      5301407003005003001 as *D. mazzai*.

Dasypus kappleri Krauss, 1862. Arch. Naturgesch., 28(1):20.
  TYPE LOCALITY: Surinam, Marowijne River.
  DISTRIBUTION: Colombia, east of Andes; Venezuela, south of Rio Orinoco; Guianas,
    south through the Amazon Basin of Brazil, Ecuador and Peru.
  COMMENT: Reviewed by Wetzel and Mondolfi, 1979:56.
  ISIS NUMBER: 5301407003005002001.

Dasypus novemcinctus Linnaeus, 1758. Syst. Nat., 10th ed., 1:51.
REVIEWED BY: J. Ramirez-Pulido (JRP).
TYPE LOCALITY: Brazil, Pernambuco.
DISTRIBUTION: N. Argentina, to Guianas and Venezuela and north to Morelos (Mexico), thence to N. Sinaloa (Mexico), N.E. Kansas and South Carolina (U.S.A.); Grenada (Lesser Antilles); Tobago, Trinidad, Margarita Isl. (Venezuela), Mexiana Isl. (Brazil).
COMMENT: Partially includes *mazzai* Yepes (a composite); see Wetzel and Mondolfi, 1979:50, 52. Type locality discussed by Cabrera, 1958:225.
ISIS NUMBER: 5301407003005004001.

Dasypus pilosus (Fitzinger, 1856). Versamml. Deutsch. Nat. Arzte, Wien, 6:123.
TYPE LOCALITY: Montane Peru.
DISTRIBUTION: Known only from four localities in montane Peru.
COMMENT: Reviewed by Wetzel and Mondolfi, 1979:56.
ISIS NUMBER: 5301407003005005001.

Dasypus sabanicola Mondolfi, 1968. Mem. Soc. Cienc. Nat. La Salle, 27:151.
TYPE LOCALITY: Venezuela, Apure, Achaguas, Hato Macanillal.
DISTRIBUTION: Llanos of Venezuela and E. Colombia.

Dasypus septemcinctus Linnaeus, 1758. Syst. Nat., 10th ed., 1:51.
TYPE LOCALITY: Brazil, Pernambuco.
DISTRIBUTION: Lower Amazon Basin (Brazil) to Chaco, Formosa, and Salta (Argentina).
COMMENT: Reviewed by Wetzel and Mondolfi, 1979:53. Type locality discussed by Cabrera, 1958:226.
ISIS NUMBER: 5301407003005006001.

Euphractus Wagler, 1830. Naturliches Syst. Amphibien, p. 36.
COMMENT: Moeller, 1968, Zool. Jahrb. Abt. Syst. Oekol. Geogr. Tiere, 85:411–528, included *Chaetophractus* and *Zaedyus* in this genus, but see Wetzel, 1982, *in* Montgomery, ed., Smithson. Inst. Press, Washington, D.C., who considered them to be distinct genera.
ISIS NUMBER: 5301407003006000000.

Euphractus sexcinctus (Linnaeus, 1758). Syst. Nat., 10th ed., 1:51.
TYPE LOCALITY: Brazil, Para.
DISTRIBUTION: S. Surinam, Brazil; Uruguay; N. Argentina; Paraguay; S.E. Bolivia.
COMMENT: Reviewed by Wetzel, 1982, *in* Montgomery, ed., Smithson. Inst. Press, Washington, D.C. Type locality restricted by Thomas, 1911, Proc. Zool. Soc. Lond., 1911:141.
ISIS NUMBER: 5301407003006001001.

Priodontes F. Cuvier, 1825. Des Dentes des Mammiferes., p. 257.
ISIS NUMBER: 5301407003007000000.

Priodontes maximus Kerr, 1792. Anim. Kingdom, p. 112.
TYPE LOCALITY: French Guiana, Cayenne.
DISTRIBUTION: South America east of the Andes, from Venezuela to Guianas and Colombia, south to Salta, Formosa, Misiones and Chaco (N. Argentina).
COMMENT: Includes *giganteus*; see Thomas, 1880, Proc. Zool. Soc. Lond., 1880:402; Allen, 1895, Bull. Am. Mus. Nat. Hist., 7:187; Hershkovitz, 1959, J. Mammal., 40:350; Wetzel and Lovett, 1974, Univ. Conn. Occ. Pap. Biol. Sci. Ser., 2:209; Handley, 1976, Brigham Young Univ. Sci. Bull., 20(5):44.
PROTECTED STATUS: CITES - Appendix I as *P. giganteus* (=*maximus*). U.S. ESA - Endangered as *P. maximus* (=*giganteus*).
ISIS NUMBER: 5301407003007001001 as *P. giganteus*.

Tolypeutes Illiger, 1811. Prodr. Syst. Mamm. et Avium., p. 111.
ISIS NUMBER: 5301407003008000000.

Tolypeutes matacus (Desmarest, 1804). Tabl. Meth. Mamm., p. 28.
  TYPE LOCALITY: Argentina, Tucuman, Tucuman.
  DISTRIBUTION: Santa Cruz (S.E. Bolivia), east to Mato Grosso (S.W. Brazil) south
    through the Gran Chaco to Santa Cruz (S. Argentina).
  COMMENT: Sanborn, 1930, J. Mammal., 11:66, reduced *matacus* to a subspecies of
    tricinctus, but it was retained as a species by Wetzel, 1982, *in* Montgomery, ed.,
    Smithson. Inst. Press, Washington, D.C. Type locality was restricted by Sanborn,
    1930, J. Mammal., 11:62.
  ISIS NUMBER: 5301407003008001001.

Tolypeutes tricinctus (Linnaeus, 1758). Syst. Nat., 10th ed., 1:56.
  TYPE LOCALITY: Brazil, Pernambuco.
  DISTRIBUTION: N.E. Brazil south of the Amazon River.
  COMMENT: Reviewed by Wetzel, 1982, *in* Montgomery, ed., Smithson. Inst. Press,
    Washington, D.C. Type locality restricted by Sanborn, 1930, J. Mammal., 11:62.
  ISIS NUMBER: 5301407003008002001.

Zaedyus Ameghino, 1889. Acta Acad. Nac. Cienc. Cordoba, 6:867.
  ISIS NUMBER: 5301407003009000000.

Zaedyus pichiy (Desmarest, 1804). Tabl. Meth. Mamm., 24:28.
  TYPE LOCALITY: Argentina, Buenos Aires, Bahia Blanca.
  DISTRIBUTION: Mendoza, San Luis, and Buenos Aires, south to Rio Santa Cruz
    (Argentina); Andean grasslands in Chile, south to Strait of Magellan.
  COMMENT: Reviewed by Wetzel, 1982, *in* Montgomery, ed., Smithson. Inst. Press,
    Washington, D.C.
  ISIS NUMBER: 5301407003009001001.

ORDER INSECTIVORA
ISIS NUMBER: 5301403000000000000

Family Solenodontidae
REVIEWED BY: G. S. Morgan (GSM); R. G. Van Gelder (RGVG); C. A. Woods (CAW).
ISIS NUMBER: 5301403001000000000.

Solenodon Brandt, 1833. Mem. Acad. Imp. Sci., St. Petersbourg, ser. 6, Sci. Math. Phys. Nat., 2:459.
COMMENT: Includes *Atopogale;* see Varona, 1974:6.
ISIS NUMBER: 5301403001001000000.

Solenodon cubanus Peters, 1861. Monatsb. Preuss. Akad. Wiss. Berlin, p. 169.
TYPE LOCALITY: Cuba, Oriente Prov., Bayamo.
DISTRIBUTION: Oriente Prov. (Cuba).
COMMENT: Sometimes placed in a distinct genus *(Atopogale);* see Hall and Kelson, 1959:22; Hall, 1981:22.
PROTECTED STATUS: U.S. ESA - Endangered as *Atopogale cubana.*
ISIS NUMBER: 5301403001001001001.

Solenodon paradoxus Brandt, 1833. Mem. Acad. Imp. Sci., St. Petersbourg, ser. 6, Sci. Math. Phys. Nat., 2:459.
TYPE LOCALITY: Dominican Republic, "Hispaniola".
DISTRIBUTION: Haiti, Dominican Republic (Hispaniola).
COMMENT: A variable species in color and morphology (CAW).
PROTECTED STATUS: U.S. ESA - Endangered.
ISIS NUMBER: 5301403001001002001.

Family Nesophontidae
COMMENT: Known only from sub-Recent fossils from the Greater Antilles; see Hall, 1981:22-24.

Family Tenrecidae
REVIEWED BY: J. F. Eisenberg (JFE); E. Gould (EG); R. G. Van Gelder (RGVG).
COMMENT: Includes Potamogalidae; see Corbet, 1974, Part 1.2:1.
ISIS NUMBER: 5301403002000000000 as Tenrecidae.
          5301403003000000000 as Potamogalidae.

Dasogale G. Grandidier, 1928. Bull. Acad. Malgache, 11:85.
ISIS NUMBER: 5301403002001000000.

Dasogale fontoynonti G. Grandidier, 1928. Bull. Acad. Malgache, 11:85.
TYPE LOCALITY: Madagascar; E. forests.
DISTRIBUTION: E. Madagascar.
COMMENT: Only two specimens exists in poorly preserved condition; these specimens may be young *Setifer setosus* with aberrant or retained early teeth (EG).
ISIS NUMBER: 5301403002001001001.

Echinops Martin, 1838. Proc. Zool. Soc. Lond., 1838:17.
ISIS NUMBER: 5301403002002000000.

Echinops telfairi Martin, 1838. Proc. Zool. Soc. Lond., 1838:17.
TYPE LOCALITY: Madagascar.
DISTRIBUTION: S. Madagascar.
ISIS NUMBER: 5301403002002001001.

Geogale Milne-Edwards and G. Grandidier, 1872. Ann. Sci. Nat. Zool., 15(art. 19):1.
ISIS NUMBER: 5301403002006000000.

Geogale aurita Milne-Edwards and G. Grandidier, 1872. Ann. Sci. Nat. Zool., 15(art. 19):1.
TYPE LOCALITY: Madagascar; Murundava.
DISTRIBUTION: N.E. and S.W. Madagascar, in Lamboharana, Tulear, and Fenerive.
COMMENT: Includes *Cryptogale*; see Genest and Petter, 1971, Part 1.1:3.
ISIS NUMBER: 5301403002006001001.

Hemicentetes Mivart, 1871. Proc. Zool. Soc. Lond., 1871:72.
ISIS NUMBER: 5301403002003000000.

Hemicentetes semispinosus (G. Cuvier, 1798). Tabl. Elem. Hist. Nat. Anim., 1798 Reg.
Anim., 1st ed., p. 136.
TYPE LOCALITY: Madagascar.
DISTRIBUTION: Madagascar, in E. forests.
COMMENT: Includes *nigriceps*; see Genest and Petter, 1971, Part 1.1:2. Eisenberg and
Gould, 1970, Smithson. Contrib. Zool., p. 27 believe *nigriceps* is distinct from
semispinosus.
ISIS NUMBER: 5301403002003002001 as *H. semispinosus*.
5301403002003001001 as *H. nigriceps*.

Leptogale Thomas, 1918. Ann. Mag. Nat. Hist., ser. 9, 1:302.
COMMENT: May be congeneric with *Microgale*; see Corbet and Hill, 1980:22.

Leptogale gracilis (Major, 1896). Ann. Mag. Nat. Hist., ser. 6, 18:318.
TYPE LOCALITY: Madagascar, Ambohimitombo Forest.
DISTRIBUTION: S. Madagascar.
COMMENT: A transitional form between the genus *Microgale* and the genus *Oryzorictes*;
see Heim de Balsac, 1972.
ISIS NUMBER: 5301403002008007001 as *Microgale gracilis*.

Limnogale Major, 1896. Ann. Mag. Nat. Hist., ser. 6, 18:318.
ISIS NUMBER: 5301403002007000000.

Limnogale mergulus Major, 1896. Ann. Mag. Nat. Hist., ser. 6, 18:318.
TYPE LOCALITY: Madagascar, Imasindrary, N.E. Betsileo.
DISTRIBUTION: E. Madagascar, freshwater streams (including the Vohitra,
Andranotobaka, Amborompotsy, and Antsampandrono rivers).
ISIS NUMBER: 5301403002007001001.

Microgale Thomas, 1882. J. Linn. Soc. Lond., Zool., 16:319.
COMMENT: Includes *Nesogale* and *Paramicrogale*; does not include *Leptogale*; see Heim
de Balsac, 1972; Eisenberg and Gould, 1970, Smithson. Contrib. Zool., p. 27.
ISIS NUMBER: 5301403002008000000.

Microgale brevicaudata G. Grandidier, 1899. Bull. Mus. Hist. Nat. Paris, 5:349.
TYPE LOCALITY: Madagascar, Mahanara, 75 km S. of Vohemar.
DISTRIBUTION: N.E. Madagascar.
ISIS NUMBER: 5301403002008001001.

Microgale cowani Thomas, 1883. J. Linn. Soc. Lond., Zool., 16:319.
TYPE LOCALITY: Madagascar, E. Betsileo.
DISTRIBUTION: Ankafana forest (E. Madagascar).
ISIS NUMBER: 5301403002008002001.

Microgale crassipes Milne-Edwards, 1893. Ann. Sci. Nat. Zool., 15:98.
TYPE LOCALITY: Madagascar, near Tananarive.
DISTRIBUTION: Known only from the type locality in C. Madagascar.
ISIS NUMBER: 5301403002008003001.

Microgale decaryi G. Grandidier, 1928. Bull. Mus. Hist. Nat. Paris, 34:69.
TYPE LOCALITY: Madagascar, Andragomana Grotto, south of Fort Dauphin.
DISTRIBUTION: S. Madagascar.
COMMENT: Formerly included in *Paramicrogale*; see Heim de Balsac, 1972.
ISIS NUMBER: 5301403002008004001.

Microgale dobsoni Thomas, 1884. Ann. Mag. Nat. Hist., 14:337.
   TYPE LOCALITY: Madagascar, C. Betsileo, Nandesen Forest.
   DISTRIBUTION: C. Madagascar, central part of E. escarpment of C. Plateau, and
      Manohilaky, west of Ambatandrazaka.
   COMMENT: Formerly included in *Nesogale*; see Heim de Balsac, 1972.
   ISIS NUMBER: 5301403002008005001.

Microgale drouhardi G. Grandidier, 1934. Bull. Mus. Hist. Nat. Paris, 6:474.
   TYPE LOCALITY: Madagascar, near Diego Suarez.
   DISTRIBUTION: N. and S.E. Madagascar.
   ISIS NUMBER: 5301403002008006001.

Microgale longicaudata Thomas, 1883. J. Linn. Soc. Lond., Zool., 16:319.
   TYPE LOCALITY: Madagascar, E. Betsileo, Ankafana Forest.
   DISTRIBUTION: Ankafana forest (E. Madagascar).
   ISIS NUMBER: 5301403002008008001.

Microgale longirostris Major, 1896. Ann. Mag. Nat. Hist., ser. 6, 18:318.
   TYPE LOCALITY: Madagascar, Ampitambe.
   DISTRIBUTION: S.E. Madagascar, central part of E. escarpment of E. plateau.
   ISIS NUMBER: 5301403002008009001.

Microgale majori Thomas, 1918. Ann. Mag. Nat. Hist., ser. 9, 1:302.
   TYPE LOCALITY: Madagascar, Ankafana Forest.
   DISTRIBUTION: E. Betsileo (E. Madagascar).
   ISIS NUMBER: 5301403002008010001.

Microgale melanorrhachis Morrison-Scott, 1948. Proc. Zool. Soc. Lond., 1948:817.
   TYPE LOCALITY: Madagascar, Perinet, near Moramanga, 3000 ft. (914 m).
   DISTRIBUTION: E. Madagascar; type locality, and 6 mi. (10 km) E. Ivohibe, 5000 ft. (1524
      m), 400 mi. (644 km) S. of type locality.

Microgale occidentalis (G. Grandidier and Petit, 1931). Bull. Soc. Zool. Fr., 56(2):126.
   TYPE LOCALITY: W. Madagascar, Maintirano.
   DISTRIBUTION: W. Madagascar.
   COMMENT: Formerly included in *Paramicrogale*; see Heim de Balsac, 1972.
   ISIS NUMBER: 5301403002008011001.

Microgale parvula G. Grandidier, 1934. Bull. Mus. Hist. Nat. Paris, 6:476.
   TYPE LOCALITY: Madagascar, near Diego Suarez.
   DISTRIBUTION: N. Madagascar.
   ISIS NUMBER: 5301403002008012001.

Microgale principula Thomas, 1926. Ann. Mag. Nat. Hist., ser. 9, 17:250.
   TYPE LOCALITY: Madagascar, Midongy-Du-Sud.
   DISTRIBUTION: S.E. Madagascar.
   ISIS NUMBER: 5301403002008013001.

Microgale prolixicaudata G. Grandidier, 1937. Bull. Mus. Hist. Nat. Paris, 9:347.
   TYPE LOCALITY: Madagascar, region of Diego Suarez.
   DISTRIBUTION: N. Madagascar.
   COMMENT: May be a subspecies of *longicaudata* (JFE).
   ISIS NUMBER: 5301403002008014001.

Microgale pusilla Major, 1896. Ann. Mag. Nat. Hist., ser. 6, 18:461.
   TYPE LOCALITY: Madagascar, Forest of Ikongo, near Vinanitelo (Fort Carnot).
   DISTRIBUTION: Known only from the type locality.
   ISIS NUMBER: 5301403002008015001.

Microgale sorella Thomas, 1926. Ann. Mag. Nat. Hist., ser. 9, 17:250.
   TYPE LOCALITY: Madagascar, Forest of Beforona, inland from Andevorante.
   DISTRIBUTION: N. Madagascar.
   ISIS NUMBER: 5301403002008016001.

Microgale taiva Major, 1896. Ann. Mag. Nat. Hist., ser. 6, 18:461.
TYPE LOCALITY: Madagascar, Tanala of E. Betsileo, Forest of Ambohimitombo.
DISTRIBUTION: S.E. Madagascar.
ISIS NUMBER: 5301403002008017001.

Microgale talazaci Major, 1896. Ann. Mag. Nat. Hist., ser. 6, 18:318.
TYPE LOCALITY: Madagascar, Forest of Independent Tanala of Ikongo, Betsileo.
DISTRIBUTION: E. Madagascar, north to Perinet on the C. Plateau.
COMMENT: Formerly included in *Nesogale;* see Heim de Balsac, 1972.
ISIS NUMBER: 5301403002008018001.

Microgale thomasi Major, 1896. Ann. Mag. Nat. Hist., ser. 6, 18:318.
TYPE LOCALITY: Madagascar, N.E. Betsileo, Forest of Ampitambe.
DISTRIBUTION: S.E. Madagascar.
ISIS NUMBER: 5301403002008019001.

Micropotamogale Heim de Balsac, 1954. C. R. Acad. Sci. Paris, 239:102.
COMMENT: Includes *Mesopotamogale;* see Corbet, 1974, Part 1.2:2.
ISIS NUMBER: 5301403003001000000.

Micropotamogale lamottei Heim de Balsac, 1954. C. R. Acad. Sci. Paris, 239:102.
TYPE LOCALITY: Guinea, Ziela, Mt. Nimba.
DISTRIBUTION: S. Guinea; adjacent Liberia; Ivory Coast.
ISIS NUMBER: 5301403003001001001.

Micropotamogale ruwenzorii (de Witte and Frechkop, 1955). Bull. Inst. Roy. Sci. Nat. Belg., 31(84):1.
TYPE LOCALITY: Zaire, W. slopes of Mt. Ruwenzorii.
DISTRIBUTION: Ruwenzori region (Uganda); E.C. Zaire, west of Lake Edward and Lake Kivu.
COMMENT: Heim de Balsac, 1956, C. R. Acad. Sci. Paris, 242:2257, proposed a new genus, *Mesopotamogale,* for this species; regarded as a subgenus by Corbet, 1974, Part 1.2:2.
ISIS NUMBER: 5301403003001002001.

Oryzorictes A. Grandidier, 1870. Rev. Mag. Zool. Paris, 22:49.
COMMENT: *Nesoryctes* considered a subgenus of *Oryzorictes;* see Heim de Balsac, 1972.
ISIS NUMBER: 5301403002009000000.

Oryzorictes hova A. Grandidier, 1870. Rev. Mag. Zool. Paris, 22:49.
TYPE LOCALITY: Madagascar, near rice fields of Ankay and Antsihanaka.
DISTRIBUTION: C. Madagascar, central plateau.
ISIS NUMBER: 5301403002009001001.

Oryzorictes talpoides G. Grandidier and Petit, 1930. Bull. Mus. Hist. Nat. Paris, ser. 2, 2(5):498.
TYPE LOCALITY: Madagascar, Majunga Prov., coastal plain of Marovoay.
DISTRIBUTION: N.W. Madagascar.
ISIS NUMBER: 5301403002009002001.

Oryzorictes tetradactylus Milne-Edwards and G. Grandidier, 1882. Le Naturaliste, 4:55.
TYPE LOCALITY: Madagascar, Ampitambe, Sirabe, Imerina.
DISTRIBUTION: C. Madagascar.
COMMENT: *O. niger* Major, 1896, Ann. Mag. Nat. Hist., ser. 6, 18:318 is considered a melanistic form of *tetradactylus;* see Thomas, 1918, Ann. Mag. Nat. Hist., ser. 9, 1:302. Formerly included in *Nesoryctes;* see Heim de Balsac, 1972.
ISIS NUMBER: 5301403002009003001.

Potamogale du Chaillu, 1860. Proc. Boston Soc. Nat. Hist., 7:363.
ISIS NUMBER: 5301403003002000000.

Potamogale velox (du Chaillu, 1860). Proc. Boston Soc. Nat. Hist., 7:363.
TYPE LOCALITY: Gabon, Equatorial Africa.
DISTRIBUTION: C. Africa, from Nigeria to Angola, east to the Rift Valley.
ISIS NUMBER: 5301403003002001001.

Setifer Froriep, 1806. *In* Dumeril, Analit. Zool. Zusatzen, p. 15.
ISIS NUMBER: 5301403002004000000.

Setifer setosus (Schreber, 1777). Saugethiere, 3:584.
TYPE LOCALITY: Madagascar.
DISTRIBUTION: Madagascar, in rainforest on C. plateau region.
ISIS NUMBER: 5301403002004001001.

Tenrec Lacepede, 1799. Tabl. Meth. Mamm., *in* Buffon's, Hist. Nat., 14:156.
ISIS NUMBER: 5301403002005000000.

Tenrec ecaudatus (Schreber, 1777). Saugethiere, 3:584.
TYPE LOCALITY: Madagascar.
DISTRIBUTION: Madagascar; Comoro Isls.; introduced on Reunion, Mauritius, and the
Seychelles (JFE).
ISIS NUMBER: 5301403002005001001.

Family Chrysochloridae
REVIEWED BY: J. Meester (JM); R. G. Van Gelder (RGVG).
COMMENT: For widely divergent treatments see Simonetta, 1968, Monitore Zool. Ital.,
n.s., 2(suppl.):27–55; Meester, 1974, Part 1.3; and Petter, 1981, Mammalia,
45:49–53.
ISIS NUMBER: 5301403004000000000.

Amblysomus Pomel, 1848. Arch. Sci. Phys. Nat. Geneve, 9:244.
COMMENT: Includes *Neamblysomus*; see Ellerman *et al.*, 1953:34.
ISIS NUMBER: 5301403004001000000.

Amblysomus gunningi (Broom, 1908). Ann. Transvaal Mus., 1:14.
TYPE LOCALITY: South Africa, E. Transvaal, Woodbush Hill, Zoutpansberg.
DISTRIBUTION: Transvaal (South Africa).
COMMENT: Formerly in monotypic genus *Neamblysomus* Roberts, 1924; see Meester,
1974, Part 1.3:6.
ISIS NUMBER: 5301403004001001001.

Amblysomus hottentotus (A. Smith, 1829). Zool. J., 4:436.
TYPE LOCALITY: South Africa, "Interior parts of South Africa," E. Cape Province,
Grahamstown.
DISTRIBUTION: Natal, Lesotho, and Transvaal to S. Cape Prov. (South Africa).
COMMENT: Includes *devilliersi* and *marleyi* as subspecies and *pondoliae, longiceps,
albifrons, garneri, drakensbergensis, natalensis,* and *orangensis* as synonyms; see
Meester, 1974, Part 1.3:6.
ISIS NUMBER: 5301403004001002001.

Amblysomus iris Thomas and Schwann, 1905. Abstr. Proc. Zool. Soc. Lond., 18:23.
TYPE LOCALITY: South Africa, Natal, Umfolosi Station (Zululand).
DISTRIBUTION: S. Cape Prov., Natal and E. Transvaal (South Africa).
COMMENT: Includes *corriae* and *septentrionalis* as subspecies and *littoralis* as synonym;
see Meester, 1974, Part 1.3:6.

Amblysomus julianae Meester, 1972. Ann. Transvaal Mus., 28(4):35.
TYPE LOCALITY: South Africa, Transvaal, Pretoria, Willows.
DISTRIBUTION: Transvaal; type locality and Kruger Nat. Park (South Africa).

Calcochloris Mivart, 1868. J. Anat. Phys., ser. 2, 1:133.
COMMENT: Includes *Crysotricha*; see Ellerman *et al.*, 1953:33. Placed in *Amblysomus* by
Petter, 1981, Mammalia, 45:49–53.

Calcochloris obtusirostris (Peters, 1851). Monatsb. Preuss. Akad. Wiss. Berlin, p. 467.
TYPE LOCALITY: Mozambique, Inhambane, 24° S.
DISTRIBUTION: S. Mozambique; S.E. Zimbabwe; N. Natal, N.E. Transvaal, Zululand,
(South Africa).

COMMENT: Includes *chrysillus* and *limpopoensis* as subspecies; see Roberts, 1951, The Mammals of South Africa, p. 114–115.

ISIS NUMBER: 5301403004001004001 as *Amblysomus obtusirostris*.

Chlorotalpa Roberts, 1924. Ann. Transvaal Mus., 10:64.
COMMENT: Included in *Amblysomus* by Ellerman *et al.*, 1953:34, and Petter, 1981, Mammalia, 45:49–53. Includes *Carpitalpa* and *Kilimitalpa*; see Meester, 1974, Part 1.3:4; *Carpitalpa* was regarded by Simonetta, 1968, Monitore Zool. Ital., 2(suppl.):27–55, as a valid genus.

Chlorotalpa arendsi Lundholm, 1955. Ann. Transvaal Mus., 22:285.
TYPE LOCALITY: Zimbabwe, Inyanga.
DISTRIBUTION: E. Zimbabwe.
COMMENT: Formerly included in *Carpitalpa* by Simonetta, 1968, Monitore Zool. Ital., 2(suppl.):27–55; but see Meester, 1974, Part 1.3:4.

Chlorotalpa duthiae (Broom, 1907). Trans. S. Afr. Philos. Soc., 18:292.
TYPE LOCALITY: South Africa, Cape Province, Knysna.
DISTRIBUTION: S. Cape Prov. (South Africa).

Chlorotalpa leucorhina (Huet, 1885). Nouv. Arch. Mus. Hist. Nat. Paris, Bull., 8:8.
TYPE LOCALITY: Africa, "Gulf of Guinea Coast."
DISTRIBUTION: N. Angola; S. Zaire; Cameroun; Central African Repulic.
COMMENT: Includes *congicus* and *luluanus* as synonyms and *cahni* as a subspecies; see Meester, 1974, Part 1.3:5, who placed this species in *Chlorotalpa*. Included in *Chrysochloris* by Allen, 1939, Checklist of African Mammals, Bull. Mus. Comp. Zool., 83:1–763; included in *Amblysomus* by Simonetta, 1968, Monitore Zool. Ital., 2(suppl):27–55, and Petter, 1981, Mammalia, 45:49–53.
ISIS NUMBER: 5301403004001003001 as *Amblysomus leucorhinus*.
5301403004002002001 as *Chrysochloris congicus*.

Chlorotalpa sclateri (Broom, 1907). Ann. Mag. Nat. Hist., ser. 7, 19:263.
TYPE LOCALITY: South Africa, Cape Prov., Beaufort West.
DISTRIBUTION: Cape Prov., E. Orange Free State, S. Transvaal (South Africa); Lesotho.
COMMENT: Includes *guillarmodi* and *shortridgei* as synonyms and *montana* as a subspecies; see Meester, 1974, Part 1.3:4. Included in *Amblysomus* by Petter, 1981, Mammalia, 45:49–53.
ISIS NUMBER: 5301403004001005001 as *Amblysomus sclateri*.

Chlorotalpa tytonis (Simonetta, 1968). Monitore Zool. Ital., n.s., 2(suppl.):31.
TYPE LOCALITY: Somalia, Giohar (=Villaggio Duca Degli Abruzzi).
DISTRIBUTION: Known only from the type locality.
COMMENT: Assigned to *Amblysomus* by Simonetta, 1968, Monitore Zool. Ital., n.s., 2(suppl):31, and Petter, 1981, Mammalia, 45:49–53; Meester, 1974, Part 1.3:5, placed this species in *Chlorotalpa*.
ISIS NUMBER: 5301403004001007001 as *Amblysomus tytonis*.

Chrysochloris Lacepede, 1799. Tabl. Mamm., p. 7.
ISIS NUMBER: 5301403004002000000.

Chrysochloris asiatica (Linnaeus, 1758). Syst. Nat., 10th ed., 1:53.
TYPE LOCALITY: South Africa, "Sibiria." Usually taken as South Africa, Cape of Good Hope.
DISTRIBUTION: W. Cape Province and Robben Isl. (South Africa); perhaps Namibia, Damaraland.
COMMENT: Includes *damarensis, namaquensis, minor, bayoni, concolor, taylori*, and probably *visagiei*; see Meester, 1974, Part 1.3:3. See Ellerman *et al.*, 1953:40 for discussion of type locality.
ISIS NUMBER: 5301403004002001001.

Chrysochloris stuhlmanni Matschie, 1894. Sitzb. Ges. Naturf. Fr. Berlin, p. 123.
TYPE LOCALITY: Uganda, Ruwenzori region, Ukonjo and Kinyawanga.
DISTRIBUTION: Uganda; Kenya; N. Zaire; Tanzania.

COMMENT: Includes *tropicalis, vermiculus, fosteri*; see Meester, 1974, Part 1.3:3, who placed *stuhlmanni* in *Chrysochloris*. Lundholm, 1955, Ann. Transvaal Mus., 22:279–303, proposed the name *Chlorotalpa (Kilimitalpa)* for this species. Simonetta, 1968, Monitore Zool. Ital., n.s., 2(suppl.):31, regarded it as a synonym of *Carpitalpa*. He placed *arendsi, stuhlmanni* and *fosteri* in *Carpitalpa* and *tropicalis* in *Chlorotalpa*.

ISIS NUMBER: 5301403004002004001 as *C. stuhlmanni*.
5301403004002003001 as *C. fosteri*.
5301403004002005001 as *C. vermiculus*.
5301403004001006001 as *Amblysomus tropicalis*.

Chrysochloris visagiei Broom, 1950. Ann. Transvaal Mus., 21:238.
TYPE LOCALITY: South Africa, Cape Prov., 54 mi. (87 km) E. Calvinia, Gouna.
DISTRIBUTION: Known only from the type locality.
COMMENT: Probably an aberrant *asiatica*; see Meester, 1974, Part 1.3:3. Simonetta, 1968, Monitore Zool. Ital., n.s., 2(suppl.):31, included it in *asiatica* as a subspecies.

Chrysospalax Gill, 1883. Standard Nat. Hist., 5 (Mamm.):137.
COMMENT: Includes *Bematiscus*; see Ellerman *et al.*, 1953:41.
ISIS NUMBER: 5301403004003000000.

Chrysospalax trevelyani (Gunther, 1875). Proc. Zool. Soc. Lond., 1875:311.
TYPE LOCALITY: South Africa, Cape Prov., Pirie Forest, near King William's Town.
DISTRIBUTION: Cape Prov. (South Africa).
ISIS NUMBER: 5301403004003001001.

Chrysospalax villosus (A. Smith, 1833). S. Afr. J., 2:81.
TYPE LOCALITY: South Africa, Durban, "Towards Natal."
DISTRIBUTION: Transvaal, Natal, E. Cape Prov. (South Africa).
COMMENT: Includes *dobsoni, leschae, rufopallidus, rufus,* and *transvaalensis*; see Meester, 1974, Part 1.3:2. See Roberts, 1951, The Mammals of South Africa, p. 121 for discussion of type locality.
ISIS NUMBER: 5301403004003002001.

Cryptochloris Shortridge and Carter, 1938. Ann. S. Afr. Mus., 32:284.
COMMENT: A synonym of *Chrysochloris* according to Simonetta, 1968, Monitore Zool. Ital., n.s., 2(suppl.):31; but see Meester, 1974, Part 1.3:2.
ISIS NUMBER: 5301403004004000000.

Cryptochloris wintoni (Broom, 1907). Ann. Mag. Nat. Hist., ser. 7, 19:264.
TYPE LOCALITY: South Africa, Little Namaqualand, Nolloth Prov.
DISTRIBUTION: S.W. Cape Prov., Little Namaqualand (South Africa).
ISIS NUMBER: 5301403004004001001.

Cryptochloris zyli Shortridge and Carter, 1938. Ann. S. Afr. Mus., 32:284.
TYPE LOCALITY: South Africa, West Cape Prov., Caompagnies Drift, 10 mi. (16 km) inland from Lamberts Bay.
DISTRIBUTION: Known only from the type locality.
COMMENT: Considered a subspecies of *wintoni* by Ellerman *et al.*, 1953:39. Considered morphologically separable; see Meester, 1974, Part 1.3:3.

Eremitalpa Roberts, 1924. Ann. Transvaal Mus., 10:63.
ISIS NUMBER: 5301403004005000000.

Eremitalpa granti (Broom, 1907). Ann. Mag. Nat. Hist., ser. 7, 19:265.
TYPE LOCALITY: South Africa, Little Namaqualand, Garies.
DISTRIBUTION: S.W. Cape Prov., Little Namaqualand (South Africa); Namib Desert (Namibia).
COMMENT: Includes *cana* as synonym and *namibensis* as a subspecies; see Meester, 1974, Part 1.3:4.
ISIS NUMBER: 5301403004005001001.

Family Erinaceidae
REVIEWED BY: G. B. Corbet (GBC); O. L. Rossolimo (OLR) (U.S.S.R.); R. G. Van Gelder (RGVG); S. Wang (SW) (China).
COMMENT: Reviewed by Corbet, 1974, Part 1.4:1–3.
ISIS NUMBER: 5301403005000000000.

Echinosorex Blainville, 1838. C. R. Acad. Sci. Paris, 6:742.
COMMENT: *Gymnura* Lesson, 1827, is preoccupied; see Ellerman and Morrison-Scott, 1951:17; Medway, 1977:15.
ISIS NUMBER: 5301403005001000000.

Echinosorex gymnurus (Raffles, 1821). Trans. Linn. Soc. Lond., 13:272.
TYPE LOCALITY: Indonesia, Sumatra, Bencoolen.
DISTRIBUTION: S., E., and N.C. Borneo; Labuan Isl.; Peninsular Thailand; Malay Peninsula; Sumatra.
ISIS NUMBER: 5301403005001001001.

Erinaceus Linnaeus, 1758. Syst. Nat., 10th ed., 1:52.
REVIEWED BY: J. Meester (JM).
COMMENT: Includes *Atelerix* and *Aethechinus*; see Corbet, 1974, Part 1.4:2.
ISIS NUMBER: 5301403005008000000 as *Erinaceus*.
5301403005006000000 as *Aethechinus*.
5301403005007000000 as *Atelerix*.

Erinaceus albiventris Wagner, 1841. Schreber's Saugethiere, Suppl. 2:22.
TYPE LOCALITY: Unknown, Senegambia?
DISTRIBUTION: Senegal to S. Somalia and the Zambezi River.
COMMENT: Includes *pruneri, spiculus, Atelerix faradjius, A. langi*, and *A. spinifex*; see Corbet, 1974, Part 1.4:2, and Ansell, 1974, Puku, suppl., 1:1–48. Gureev, 1979:167, listed this as a subspecies of *frontalis*, without comment. See Allen, 1939:20 for discussion of type locality.
ISIS NUMBER: 5301403005007001001 as *Atelerix albiventris*.

Erinaceus algirus Lereboullet, 1842. Mem. Soc. Hist. Nat. Strasbourg., 3(2), art. QQ:4.
TYPE LOCALITY: Algeria, Oran.
DISTRIBUTION: Mauritania; Rio de Oro; S.W. Morocco to Libya; Canary Isls.; Balearic Isls.; Malta; coastal France and Spain (probably introduced).
COMMENT: Gureev, 1979:167, listed this species as a subspecies of *frontalis*, without comment.
ISIS NUMBER: 5301403005008001001.

Erinaceus concolor Martin, 1838. Proc. Zool. Soc. Lond., 1837:103.
TYPE LOCALITY: Turkey, near Trebizond.
DISTRIBUTION: E. Europe; S. Russia; Asia Minor to Israel.
COMMENT: Formerly included in *europaeus*; but see Kratochvil, 1975, Zool. Listy, 24:297–312; Kral, 1967, Zool. Listy, 16:239–252; Orlov, 1969, *in* II All-Union Theriol. Conf., p. 6, Novosibirsk. OLR includes this species in *europaeus*, as did Gureev, 1979:164, and Gromov and Baranova, 1981:9. Includes *roumanicus*; see Corbet, 1978:14.

Erinaceus europaeus Linnaeus, 1758. Syst. Nat., 10th ed., 1:52.
TYPE LOCALITY: Sweden, S. Gothland Isl., Wamlingbo.
DISTRIBUTION: Europe and W. Siberia; Shensi (China) east to W. and S. Korea, Manchuria, and adjacent Siberia.
COMMENT: Includes *amurensis*; see Corbet, 1978:14; Gromov and Baranova, 1981:9. Also see Corbet and Hill, 1980:24, who treat it as a species; GBC now believes *amurensis* is best treated as a separate species until relationships with *europaeus* and *concolor* are resolved.
ISIS NUMBER: 5301403005008002001.

Erinaceus frontalis A. Smith, 1831. S. Afr. Quart. J., 2:10,29.
TYPE LOCALITY: South Africa, "Cape Colony"; N. Graaff Reinet Dist.
DISTRIBUTION: South Africa; Zimbabwe; Namibia; S.W. Angola.

COMMENT: Includes *angolae*; see Corbet, 1974, Part 1.4:2. Type locality restricted by
Ellerman *et al.*, 1953:18.
PROTECTED STATUS: CITES - Appendix II.
ISIS NUMBER: 5301403005006001001 as *Aethechinus angolae*.

Erinaceus sclateri Anderson, 1895. Proc. Zool. Soc. Lond., 1895:415.
TYPE LOCALITY: "Somaliland."
DISTRIBUTION: N. Somalia.
COMMENT: Provisionally distinct from *albiventris*; see Corbet, 1974, Part 1.4:3.

Hemiechinus Fitzinger, 1866. Sitzb. Math-Nat. K. Akad. Wiss., 1:565.
REVIEWED BY: J. Meester (JM).
COMMENT: May be a subgenus of *Erinaceus*; see Gureev, 1979:168; Gromov and
Baranova, 1981:9. Corbet, 1978:15 considered *Hemiechinus* a distinct genus.
ISIS NUMBER: 5301403005009000000.

Hemiechinus auritus (Gmelin, 1770). Nova Comm. Acad. Sci. Petrop., 14:519.
TYPE LOCALITY: U.S.S.R., Astrakhansk. Obl., near Astrakhan.
DISTRIBUTION: Coast of Libya and Egypt; Cyprus; Transcaucasia and Israel to N.W.
India and Gobi Desert; N. China (SW).
COMMENT: Includes *megalotis*; see Corbet, 1978:15. Includes *aegyptius*, *libycus*, and
metwallyi, possibly as subspecies; see Corbet, 1974, Part 1.4:1.
ISIS NUMBER: 5301403005009001001 as *H. auritus*.
5301403005009002001 as *H. megalotis*.

Hemiechinus dauuricus (Sundevall, 1842). K. Sven. Vetenskapsakad Akad. Handl.,
1841:237.
TYPE LOCALITY: U.S.S.R., Transbaikalia, Dauuria.
DISTRIBUTION: S. Lake Baikal to Ningsiahiu (China), and east to N.E. China (SW).
COMMENT: Includes *sibiricus*; see Corbet, 1978:15 Bull. Zool. Nomencl., 35:123–124;
Russian authors usually spell this name *dauricus*.

Hemiechinus sylvaticus Ma, 1964. Acta Zootax. Sin., 1:35.
TYPE LOCALITY: China, Shansi Prov., N. slope of Li-shan Mtn., Qin-shui Dist.
DISTRIBUTION: Known only from type locality.
COMMENT: *Erinaceus hughi* Thomas, 1908 is probably conspecific (GBC), but was
included in *E. europeaus* by Ellerman and Morrison-Scott, 1951:21, and in *H.
dauuricus* by Corbet, 1978:15.

Hylomys Muller, 1839. *In* Temminck, Verhandl. Nat. Gesch. Nederland Overz. Bezitt.,
Zool., 50.
ISIS NUMBER: 5301403005002000000.

Hylomys suillus Muller, 1839. *In* Temminck, Verhandl. Nat. Gesch. Nederland Overz.
Bezitt., Zool., 25, 50.
TYPE LOCALITY: Indonesia, Java.
DISTRIBUTION: S. Yunnan (China); E. Burma; Thailand; Laos; N. and C. Vietnam;
Malaya; Sumatra; Borneo; Java.
ISIS NUMBER: 5301403005002001001.

Neohylomys Shaw and Wong, 1959. Acta Zool. Sin., 11:422.
ISIS NUMBER: 5301403005003000000.

Neohylomys hainanensis Shaw and Wong, 1959. Acta Zool. Sin., 11:422.
TYPE LOCALITY: China, Hainan Isl., Pai-sa Hsian.
DISTRIBUTION: Hainan (China).
COMMENT: The name Wong is also spelled Wang in other publications (R.
S. Hoffmann).
ISIS NUMBER: 5301403005003001001.

Neotetracus Trouessart, 1909. Ann. Mag. Nat. Hist., 4:389.
ISIS NUMBER: 5301403005004000000.

Neotetracus sinensis Trouessart, 1909. Ann. Mag. Nat. Hist., 4:390.
TYPE LOCALITY: China, Szechwan, Tatsienlu, 2545 m.
DISTRIBUTION: N.E. Burma; N. Vietnam; Yunnan; Szechwan; Kweichow (China) (SW).
ISIS NUMBER: 5301403005004001001.

Paraechinus Trouessart, 1879. Rev. Mag. Zool. Paris, ser. 3, 7:242.
REVIEWED BY: J. Meester (JM).
COMMENT: May be a subgenus of *Erinaceus*; see Gureev, 1979:176. Corbet and Hill, 1980:25 and Corbet, 1978:16, considered *Paraechinus* a distinct genus.
ISIS NUMBER: 5301403005010000000.

Paraechinus aethiopicus (Ehrenberg, 1833). Symb. Phys. Mamm., 2, sig. K.
TYPE LOCALITY: Sudan (Anglo-Egyptian Sudan), deserts of Dongola.
DISTRIBUTION: Morocco and Mauritania to N. Somalia, Egypt through Arabia to Iraq.
COMMENT: Includes *blancalis*, *deserti* (syn. *wassifi*), *dorsalis* and *pectoralis* as subspecies; see Corbet, 1974, Part 1.4:2; but see also Osborn and Helmy, 1980, Fieldiana Zool., n.s., 5:64,70, who listed *deserti* and *dorsalis* as distinct species.
ISIS NUMBER: 5301403005010001001.

Paraechinus hypomelas (Brandt, 1836). Bull. Sci. St. Petersb., 1:32.
TYPE LOCALITY: "N. Persia," probably N. Iran, Khorassan.
DISTRIBUTION: Arabia; Tanb Isl. (Persian Gulf); Iran; Afghanistan; Turkmenia, Tadzhikistan, and W. Kazakhstan (S. U.S.S.R.); N.W. India.
COMMENT: Gureev, 1979:176; Gromov and Baranova, 1981:10 listed this as a subspecies of *aethiopicus*; see Corbet, 1978:16, who considered *hypomelas* a distinct species. See Ognev, 1927, Zool. Anz., 69:210–212 for discussion of type locality.
ISIS NUMBER: 5301403005010002001.

Paraechinus micropus (Blyth, 1846). J. Asiat. Soc. Bengal, 15:170.
TYPE LOCALITY: Pakistan, Bhawalpur.
DISTRIBUTION: Pakistan and N.W. India; S. India.
COMMENT: Includes *nudiventris* and *intermedius*; see Corbet, 1978:16–17. Gureev, 1979:176, listed this as a subspecies of *aethiopicus* without comment.
ISIS NUMBER: 5301403005010003001.

Podogymnura Mearns, 1905. Proc. U.S. Nat. Mus., 28:436.
ISIS NUMBER: 5301403005005000000.

Podogymnura truei Mearns, 1905. Proc. U.S. Nat. Mus., 28:437
TYPE LOCALITY: Philippine Isls., Mindanao, Davao, Mount Apo.
DISTRIBUTION: Mindanao.
ISIS NUMBER: 5301403005005001001.

Family Soricidae
REVIEWED BY: R. Hutterer (RH); G. S. Jones (GSJ); O. L. Rossolimo (OLR)(U.S.S.R.); R. G. Van Gelder (RGVG); S. Wang (SW)(China).
COMMENT: Revised by Gureev, 1971, [Shrew fauna of the World], Leningrad; Gureev, 1979.
ISIS NUMBER: 5301403007000000000.

Anourosorex Milne-Edwards, 1870. C. R. Acad. Sci. Paris, 70:341.
COMMENT: Subfamily Soricinae, tribe Neomyini; see Repenning, 1967:61.
ISIS NUMBER: 5301403007001000000.

Anourosorex squamipes Milne-Edwards, 1872. Rech. Mamm., p. 264.
TYPE LOCALITY: China, Szechwan Prov., probably Moupin.
DISTRIBUTION: Shensi and Hupeh, south to Yunnan (China); Taiwan; N. and W. Burma; Assam and Bhutan (India); N. Vietnam; Thailand.
COMMENT: Includes *assamensis*; see Ellerman and Morrison-Scott, 1951:87; Repenning, 1967:53.
ISIS NUMBER: 5301403007001001001.

Blarina Gray, 1838. Proc. Zool. Soc. Lond., 1837:124.
    COMMENT: Subfamily Soricinae, tribe Blarinini; see Repenning, 1967:37.
    ISIS NUMBER: 5301403007002000000.

Blarina brevicauda (Say, 1823). *In* Long, Account of an Exped. from Pittsburgh to the
    Rocky Mtns., Vol. 1:164.
    TYPE LOCALITY: U.S.A., Nebraska, Washington Co., near Blair (formerly Engineer
    Cantonment), W. bank of Missouri River.
    DISTRIBUTION: S. Canada west to C. Saskatchewan and east to S.E. Canada, south to S.
    Nebraska and N. Virginia (U.S.A.).
    COMMENT: McGaugh and Genoways, 1976, Museology, 2:81; and Handley, 1979, *in*
    Kirk, The Great Dismal Swamp, Univ. Press, 311 pp. included *telmalestes* in this
    species; but also see Gureev, 1979:425, Tate *et al.*, 1980, Proc. Biol. Soc. Wash.,
    93:50–60, and Hall, 1981:57, who listed *telmalestes* as a distinct species.
    ISIS NUMBER: 5301403007002001001.

Blarina carolinensis (Bachman, 1837). J. Acad. Nat. Sci. Phila., 7:366.
    REVIEWED BY: J. R. Choate (JRC).
    TYPE LOCALITY: U.S.A., E. South Carolina.
    DISTRIBUTION: S. Illinois east to N. Virginia, and south through E. Texas and Florida
    (U.S.A.).
    COMMENT: For specific status see Handley, 1979; Tate *et al.*, 1980, Proc. Biol. Soc.
    Wash., 93:50–60. Hall, 1981:54, listed *carolinensis* as a subspecies of *brevicauda*.
    ISIS NUMBER: 5301403007002002001.

Blarina hylophaga Elliot, 1899. Field Columb. Mus. Publ., Zool. Ser., 1:287.
    TYPE LOCALITY: U.S.A., Oklahoma, Murray Co., Dougherty.
    DISTRIBUTION: S. Nebraska and S.W. Iowa south to S. Texas; east to Missouri and N.W.
    Arkansas (U.S.A.). Perhaps N.E. Louisiana.
    COMMENT: Formerly included in *carolinensis*, but separated as a distinct species by
    George, *et al.*, 1981, Ann. Carnegie Mus., 50(21):493–513.

Blarina telmalestes Merriam, 1895. N. Am. Fauna, 10:15.
    TYPE LOCALITY: U.S.A., Virginia, Norfolk Co., Dismal Swamp.
    DISTRIBUTION: Dismal Swamp, and adjacent E. Virginia and N.E. North Carolina
    (U.S.A.).
    COMMENT: Formerly included in *brevicauda* by McGaugh and Genoways, 1976,
    Museology, 2:81; and Handley, 1979. Tate *et al.*, 1980, Proc. Biol. Soc. Wash.,
    93:50–60, and Hall, 1981:57, listed *telmalestes* as a distinct species.
    ISIS NUMBER: 5301403007002003001.

Blarinella Thomas, 1911. Proc. Zool. Soc. Lond., 1911:166.
    COMMENT: Subfamily Soricinae; tribe Soricini; see Repenning, 1967:61.
    ISIS NUMBER: 5301403007003000000.

Blarinella quadraticauda (Milne-Edwards, 1872). Rech. Mamm., p. 261.
    TYPE LOCALITY: China, Szechwan Prov., Moupin.
    DISTRIBUTION: Shensi, Kansu, Szechwan, Yunnan (China); N. Burma.
    COMMENT: Includes *griselda* and *wardi*; see Ellerman and Morrison-Scott, 1951:56.
    ISIS NUMBER: 5301403007003001001.

Chimarrogale Anderson, 1877. J. Asiat. Soc. Bengal, 46:262.
    COMMENT: Includes *Crossogale*; see Harrison, 1958, Ann. Mag. Nat. Hist., 13:282.
    Misspelled in Ellerman and Morrison-Scott, 1951:87, and Corbet, 1978:31 as
    "*Chimmarogale.*" Subfamily Soricinae; tribe Neomyini; see Repenning, 1967:61.
    ISIS NUMBER: 5301403007004000000.

Chimarrogale himalayica (Gray, 1842). Ann. Mag. Nat. Hist., 10:261.
    TYPE LOCALITY: India, Punjab, Chamba.
    DISTRIBUTION: Kashmir through S.E. Asia to Sumatra and Borneo; C. and S. China;
    Japan; Taiwan.

COMMENT: Includes *leander, platycephala, varennei,* and probably *hantu;* see Corbet, 1978:31. Gureev, 1979:458, listed *leander, hantu, platycephala,* and *varennei* as distinct species without comment. Species reviewed by Jones and Mumford, 1971, J. Mammal., 52:228–232. Medway, 1977:21 considered *phaeura* as a subspecies of *himalayica;* but Corbet, 1978:32, maintained *styani* and *phaeura* as separate species. ISIS NUMBER: 5301403007004002001 as *C. platycephala.*

Chimarrogale phaeura Thomas, 1898. Ann. Mag. Nat. Hist., ser. 7, 2:246.
TYPE LOCALITY: Malaysia, Sabah, Mount Kinabalu, Saiap.
DISTRIBUTION: Borneo, Sumatra.
COMMENT: Formerly included in *himalayica;* see Medway, 1977:21; but also see Corbet, 1978:32, who considered it a separate species. Ellerman and Morrison-Scott, 1966:87, included *sumatrana* in this species, but see Gureev, 1979:458, who listed it as a distinct species.
ISIS NUMBER: 5301403007004001001.

Chimarrogale styani De Winton, 1899. Proc. Zool. Soc. Lond., 1899:574.
TYPE LOCALITY: China, N.W. Szechwan, Yangliupa.
DISTRIBUTION: Shensi, Szechwan (China); N. Burma.
COMMENT: Regarded as a distinct species by Jones and Mumford, 1971, J. Mammal., 52:229; Corbet, 1978:32.

Crocidura Wagler, 1832. Isis, p. 275.
REVIEWED BY: F. Spitzenberger (FS)(Palearctic).
COMMENT: Includes *Praesorex, Rhinomus, Leucodon, Paurodus* and *Heliosorex;* see Allen, 1939, Bull. Mus. Comp. Zool. Harv. Univ., 83:29; Heim de Balsac and Meester, 1977, Part 1:21. Subfamily Crocidurinae; see Repenning, 1967:71. Revised by Jenkins, 1976:271–309. Gureev, 1979:388, listed *Praesorex* as a distinct genus.
ISIS NUMBER: 5301403007005000000.

Crocidura aequicauda Robinson and Kloss, 1918. J. Fed. Malay St. Mus., 8:22.
TYPE LOCALITY: Indonesia, Sumatra; Korinchi Peak, Sungai Kring, 7200 ft. (2194 m).
DISTRIBUTION: Sumatra; Peninsular Malaya.
ISIS NUMBER: 5301403007005002001.

Crocidura agadiri Vesmanis and Vesmanis, 1979. Zool. Abh. Mus. Tierk. Dresden, 36(2):67.
TYPE LOCALITY: Morocco, Agadir Prov., 15 km N. Tiznit, 29° 49′ N., 9° 39′ W.
DISTRIBUTION: Known from a single specimen from the type locality.
COMMENT: RH expresses doubt about the validity of this species.

Crocidura aleksandrisi Vesmanis, 1977. Bonn. Zool. Beitr., 28:3.
TYPE LOCALITY: Libya, Cyrenaica, 5 km W. Tocra.
DISTRIBUTION: Cyrenaica, Apollonia (Libya).

Crocidura allex Osgood, 1910. Field Mus. Nat. Hist. Zool. Publ. Ser., 10(3):20.
TYPE LOCALITY: Kenya, Naivasha.
DISTRIBUTION: Mt. Kilimanjaro and highlands of Kenya; N. Tanzania.
COMMENT: Includes *alpina* and *zinki* as synonyms; see Heim de Balsac and Meester, 1977, Part 1:9. Gureev, 1979:413, listed *alpina* as a distinct species without comment.
ISIS NUMBER: 5301403007005004001.

Crocidura andamanensis Miller, 1902. Proc. U.S. Nat. Mus., 24:777.
TYPE LOCALITY: India, Andaman Isls., South Andaman Isl., MacPherson Strait.
DISTRIBUTION: Andaman Isls., Bay of Bengal.
COMMENT: A species of *Suncus,* according to Krumbiegel, 1978, Saugetierk. Mitt., 26:71.
ISIS NUMBER: 5301403007005005001.

Crocidura anthonyi Heim de Balsac, 1940. Bull. Mus. Hist. Nat. Paris, ser. 2, 12:382.
TYPE LOCALITY: Tunisia, Gafsa.
DISTRIBUTION: Gafsa (Tunisia).
COMMENT: Unidentifiable. Holotype has been lost and possibly not from Tunisia at all; see Heim de Balsac and Meester, 1977, Part 1:9, and Heim de Balsac, 1968, Bonn. Zool. Beitr., 19:181–188.

Crocidura arethusa Dollman, 1915. Ann. Mag. Nat. Hist., ser. 8, 16:144.
TYPE LOCALITY: Nigeria, Bauchi Prov., Kabwir.
DISTRIBUTION: Kabwir, Maiduguri (N. Nigeria).
COMMENT: Probably a subspecies of *sericea* (RH).
ISIS NUMBER: 5301403007005006001.

Crocidura ariadne Pieper, 1979. Bonn. Zool. Beitr., 29:282.
REVIEWED BY: M. Andera (MA).
TYPE LOCALITY: Greece, Crete, Iraklion Prov., Agio Pnevma.
DISTRIBUTION: Known only from the type locality.

Crocidura armenica Gureev, 1963. Mammal Fauna of the U.S.S.R., 1:118.
TYPE LOCALITY: U.S.S.R., Armyansk S.S.R., 14 km down river from Garni.
DISTRIBUTION: Armenia (U.S.S.R.).
COMMENT: Revised by Gureev, 1979:401, who considered *armenica* distinct from
pergrisea; but see Dolgov and Yudin, 1975, Trans. Biol. Inst. Novosibirsk, 23:5-40,
and Corbet, 1978:29, who considered it a subspecies; Gromov and Baranova,
1981:25 listed it as a distinct species.

Crocidura attenuata Milne-Edwards, 1872. Rech. Mamm., p. 263.
TYPE LOCALITY: China, Szechwan Prov., Moupin.
DISTRIBUTION: S. and C. China; Hainan Isl.; Taiwan; Kashmir; Punjab, Kumaon, Nepal,
Bhutan, and Assam (India); Burma; Christmas Isl. (Indian Ocean).
COMMENT: Includes *trichura*; see Jenkins, 1976:297.
ISIS NUMBER: 5301403007005007001 as *C. attenuata*.
5301403007005127001 as *C. trichura*.

Crocidura baileyi Osgood, 1936. Field Mus. Nat. Hist. Zool. Publ. Ser., 20:225.
TYPE LOCALITY: Ethiopia, Simien Mtns., Ras Dashan (= Mt. Geech).
DISTRIBUTION: Ethiopian Highlands, west of the Rift Valley.
ISIS NUMBER: 5301403007005008001.

Crocidura bartelsii Jentink, 1910. Notes Leyden Mus., 32:197.
TYPE LOCALITY: Indonesia, Java, Mount Pangerango.
DISTRIBUTION: Java.
COMMENT: Probably conspecific with *monticola*; see Jenkins, 1976:303.
ISIS NUMBER: 5301403007005009001 as *C. bartelsi (sic)*.

Crocidura beatus Miller, 1910. Proc. U.S. Nat. Mus., 38:392.
TYPE LOCALITY: Philippines, Mindanao, summit of Mt. Bliss, 1461 m.
DISTRIBUTION: Known only from the type locality.
ISIS NUMBER: 5301403007005011001.

Crocidura beccarii Dobson, 1886. Ann. Mus. Civ. Stor. Nat. Genova, ser. 2, 4:556.
TYPE LOCALITY: Indonesia, Sumatra, Mt. Singalang.
DISTRIBUTION: Sumatra.
ISIS NUMBER: 5301403007005012001.

Crocidura bicolor Bocage, 1889. J. Sci. Math. Phys. Nat. Lisboa, ser. 2, 1:29.
TYPE LOCALITY: Angola, Mossamedes Dist., Gambos.
DISTRIBUTION: N. South Africa to Angola and Sudan.
COMMENT: Includes *marita, cuninghamei, hendersoni, tephrogaster,* and *woosnami;* see
Heim de Balsac and Meester, 1977, Part 1:10.
ISIS NUMBER: 5301403007005014001 as *C. bicolor*.
5301403007005078001 as *C. marita*.

Crocidura bloyeti Dekeyser, 1943. Bull. Mus. Hist. Nat. Paris, ser. 2, 15:155.
TYPE LOCALITY: Tanzania, Irangi, Kondoa.
DISTRIBUTION: Known only from the type locality.
ISIS NUMBER: 5301403007005015001.

Crocidura bolivari Morales Agacino, 1934. Bol. Soc. Esp. Nat. Hist., 34:93.
TYPE LOCALITY: Western Sahara, Villa Cisneros.
DISTRIBUTION: Chad; Senegal; Guinea-Bissau; Western Sahara; Morocco.

COMMENT: Formerly included in *sericea* by Heim de Balsac and Meester, 1977, Part 1:24; Vesmanis and Vesmanis, 1979, Zool. Abh. Mus. Tierk. Dresden, 36(2):39–40, listed *bolivari* as a distinct species but considered its status uncertain (perhaps in *sericea*). RH expresses doubt about the validity of this species.
ISIS NUMBER: 5301403007005016001.

Crocidura bottegi Thomas, 1898. Ann. Mus. Civ. Stor. Nat. Genova, ser. 2, 18:677.
TYPE LOCALITY: Ethiopia, between Badditu and Dime, N.E. Lake Rudolf (= L. Turkana).
DISTRIBUTION: Guinea to S. Ethiopia and N. Kenya.
COMMENT: Includes *eburnea* and *obscurior;* see Heim de Balsac and Meester, 1977, Part 1:10.
ISIS NUMBER: 5301403007005017001.

Crocidura bovei Dobson, 1887. Ann. Mus. Civ. Stor. Nat. Genova, ser. 2, 5:425.
TYPE LOCALITY: Zaire, Vivi.
DISTRIBUTION: Known only from the type locality.
ISIS NUMBER: 5301403007005018001.

Crocidura brevicauda Jentink, 1890. *In* Weber, Zool. Ergebnisse einer Reise in Niederl. Ost-Indien, 1:124.
TYPE LOCALITY: Indonesia, Java, Tijibodas, near Sindanglaja.
DISTRIBUTION: Java.
COMMENT: Probably a subspecies of *fuliginosa;* see Jenkins, 1976.
ISIS NUMBER: 5301403007005020001.

Crocidura brunnea Jentink, 1888. Notes Leyden Mus., 10:164.
TYPE LOCALITY: Indonesia, Java.
DISTRIBUTION: Java; Sumatra; Moluccas; Sulawesi.
COMMENT: May include *doriae;* Laurie and Hill, 1954:30, doubted the occurrence of *doriae* east of Wallace's line. Jenkins, 1976, considered *brunnea* as "probably *fuliginosa*." See also comments under *fuliginosa*.
ISIS NUMBER: 5301403007005021001.

Crocidura butleri Thomas, 1911. Ann. Mag. Nat. Hist., ser. 8, 8:375.
TYPE LOCALITY: Sudan, Bahr-el-Ghazal, between Chakchak and Dem Zubeir.
DISTRIBUTION: Sudan; S. Somalia; Kenya.
COMMENT: Includes *percivali* and *aridula;* see Heim de Balsac and Meester, 1977, Part 1:10. Gureev, 1979:408, listed *percivali* as a distinct species without comment.
ISIS NUMBER: 5301403007005023001 as *C. butleri.*
5301403007005108001 as *C. percivali.*

Crocidura buttikoferi Jentink, 1888. Notes Leyden Mus., 10:47.
TYPE LOCALITY: Liberia, Robertsport.
DISTRIBUTION: Guinea-Bissau to Liberia; S.E. Nigeria; S. Cameroun.
COMMENT: Includes *attila;* see Heim de Balsac and Meester, 1977, Part 1:11.
ISIS NUMBER: 5301403007005022001.

Crocidura caliginea Hollister, 1916. Bull. Am. Mus. Nat. Hist., 35:664.
TYPE LOCALITY: Zaire, Medje.
DISTRIBUTION: Known only from the type locality.
ISIS NUMBER: 5301403007005024001.

Crocidura chaouianensis Vesmanis and Vesmanis, 1979. Zool. Abh. Mus. Tierk. Dresden, 36(2):64.
TYPE LOCALITY: Morocco, Settat Prov., 3 km N. Settat, 33° 02' N., 7° 46' W.
DISTRIBUTION: Settat Prov., Morocco.
COMMENT: RH expresses doubt about the validity of this species.

Crocidura cinderella Thomas, 1911. Ann. Mag. Nat. Hist., ser. 8, 8:119.
TYPE LOCALITY: Gambia, Gemenjulla.
DISTRIBUTION: Known with certainty only from the type locality; perhaps also Mali.
ISIS NUMBER: 5301403007005026001.

Crocidura congobelgica Hollister, 1916. Bull. Am. Mus. Nat. Hist., 35:670.
TYPE LOCALITY: Zaire, Lubila, near Bafwasende.
DISTRIBUTION: N.E. Zaire.
ISIS NUMBER: 5301403007005027001.

Crocidura crenata Brosset, DuBost and Heim de Balsac, 1965. Mammalia, 29:268.
TYPE LOCALITY: Gabon, Belinga.
DISTRIBUTION: Type locality and Makokou (Gabon).

Crocidura crossei Thomas, 1895. Ann. Mag. Nat. Hist., ser. 6, 16:53.
TYPE LOCALITY: Nigeria, Asaba (150 mi. (241 km) up the Niger River).
DISTRIBUTION: Lowland forest from Nigeria to Ivory Coast.
COMMENT: Includes *ebriensis, ingoldbyi* and *jouvenetae;* see Heim de Balsac and Meester, 1977, Part 1:12.
ISIS NUMBER: 5301403007005028001 as *C. crossei.*
        5301403007005059001 as *C. ingoldbyu (sic).*
        5301403007005061001 as *C. jouvenetae.*

Crocidura cyanea (Duvernoy, 1838). Mem. Soc. Hist. Nat. Strasbourg., 2:2.
TYPE LOCALITY: Africa, "La Riviere des Elephants, au sud de l'Afrique" = Cistrasdal.
DISTRIBUTION: South Africa to Angola, N.E. Zaire, S. Sudan, Ethiopia and Somalia.
COMMENT: Includes *boydi, lutrella, sacralis, parvipes* (see Heim de Balsac and Meester, 1977, Part 1:12), *argentata* and *martensi* (see Ellerman *et al.,* 1953:27). Heim de Balsac and Meester, 1977, Part 1:12, included *smithii* in this species; but see Hutterer, 1981, Bonn. Zool. Beitr., 31:237, who considered *smithii* a distinct species. Gureev, 1979:411–412, listed *argentata, boydi, lutrella, martensi, parvipes,* and *sacralis* as distinct species without comment.
ISIS NUMBER: 5301403007005029001 as *C. cyanea.*
        5301403007005019001 as *C. boydi.*
        5301403007005106001 as *C. parvipes.*

Crocidura cypria Bate, 1904. Proc. Zool. Soc. Lond., 1903:344.
TYPE LOCALITY: Cyprus.
DISTRIBUTION: Cyprus.
COMMENT: Considered a distinct species by Spitzenberger, 1978, Ann. Nat. Hist. Mus. Wien, 81:409; but see also Richter, 1970, Zool. Abh. Mus. Tierk. Dresden, 31(17):293–304, who treated it as a subspecies of *gueldenstaedti,* and Corbet, 1978:28, who treated it as a subspecies of *russula.*

Crocidura denti Dollman, 1915. Ann. Mag. Nat. Hist., ser. 8, 16:377.
TYPE LOCALITY: Zaire, Ituri Forest, between Mawambi and Avakubi.
DISTRIBUTION: N.E. Zaire and adjacent Uganda; S. Cameroun; perhaps N. Angola.
COMMENT: Considered a distinct species by Heim de Balsac, 1959, Bonn. Zool. Beitr., 10:198.

Crocidura dolichura Peters, 1876. Monatsb. Preuss. Akad. Wiss. Berlin, 1876:475.
TYPE LOCALITY: Cameroun, Bonjongo.
DISTRIBUTION: S. Cameroun; Bioko; Gabon; Cent. Afr. Rep.; Congo Rep.; Zaire and adjacent Uganda; Liberia; Guinea; Ivory Coast; Ghana.
COMMENT: Includes *ludia, muricauda,* and *polia;* see Heim de Balsac and Meester, 1977, Part 1:13.
ISIS NUMBER: 5301403007005030001 as *C. dolichura.*
        5301403007005070001 as *C. ludia.*
        5301403007005088001 as *C. muricauda.*
        5301403007005112001 as *C. polia.*

Crocidura douceti Heim de Balsac, 1958. Mem. Inst. Fr. Afr. Noire, 53:329.
TYPE LOCALITY: Ivory Coast, Adiopodoume.
DISTRIBUTION: Ivory Coast; Guinea at Seredou.
COMMENT: Probably closely related to *suaveolens;* see Heim de Balsac and Meester, 1977, Part 1:13–14.
ISIS NUMBER: 5301403007005031001.

Crocidura dracula Thomas, 1912. Ann. Mag. Nat. Hist., 9:686.
TYPE LOCALITY: China, Yunnan, "probably near" Mongtze (Mengtze).
DISTRIBUTION: S. China; N. Thailand; Vietnam; N. Borneo.
COMMENT: Medway, 1977:18–20, included *dracula* in *fuliginosa,* but Lekagul and McNeely, 1977:28, considered *dracula* a distinct species.

Crocidura dsinezumi (Temminck, 1844). *In* Siebold, Fauna Japonica, Part 2:26.
TYPE LOCALITY: Japan, Kyushu.
DISTRIBUTION: Japan; Ryukyu Isls.; Quelpart Isl. (Korea).
COMMENT: The spelling of the name was clarified by Corbet, 1978:28, and in Bull. Zool. Nomencl., 35:125, but misprinted in the same paper. Includes *chisai* and *quelpartis;* see Corbet, 1978:28; but also see Jameson and Jones, 1977, Proc. Biol. Soc. Wash., 90:459–482, who placed these taxa in *russula.*

Crocidura edwardsiana Trouessart, 1880. Le Naturaliste, 1:330.
TYPE LOCALITY: Philippines, Sulu Isl., Jolo.
DISTRIBUTION: Known only from the type locality.
ISIS NUMBER: 5301403007005032001.

Crocidura eisentrauti Heim de Balsac, 1957. Zool. Jahrb. Abt. Syst. Oekol. Geogr. Tiere, 85:616.
TYPE LOCALITY: Cameroun, Mount Cameroun, Johannes-Albrecht Hut, above de Buea.
DISTRIBUTION: Mt. Cameroun (Cameroun).
COMMENT: May be conspecific with *vulcani;* see Heim de Balsac and Meester, 1977, Part 1:13.
ISIS NUMBER: 5301403007005033001.

Crocidura elgonius Osgood, 1910. Ann. Mag. Nat. Hist., ser. 8, 5:369.
TYPE LOCALITY: Kenya, Mt. Elgon, Kirui.
DISTRIBUTION: Mt. Elgon (W. Kenya); N.E. Tanzania.
COMMENT: Regarded as a distinct species related to *suaveolens;* see Heim de Balsac and Meester, 1977, Part 1:13.

Crocidura elongata Miller and Hollister, 1921. Proc. Biol. Soc. Wash., 34:101.
TYPE LOCALITY: Indonesia, Sulawesi, Temboan (S.W. from Tondano Lake).
DISTRIBUTION: Known only from the type locality.
ISIS NUMBER: 5301403007005034001.

Crocidura erica Dollman, 1915. Ann. Mag. Nat. Hist., ser. 8, 15:145.
TYPE LOCALITY: Angola, Pungo Andongo.
DISTRIBUTION: W. Angola.
COMMENT: Resembles *hirta* in cranial dimensions; see Heim de Balsac and Meester, 1977, Part 1:13.

Crocidura essaouiranensis Vesmanis and Vesmanis, 1979. Zool. Abh. Mus. Tierk. Dresden, 36(2):61.
TYPE LOCALITY: Morocco, Safi Prov., 5 km N. Essaouira, 31° 31′ N., 9° 46′ W.
DISTRIBUTION: Known from a single specimen from the type locality.
COMMENT: RH questions the validity of this species.

Crocidura ferruginea Heuglin, 1865. Nova Acta Acad. Caes. Leop.-Carol., Halle, 32:36.
TYPE LOCALITY: Sudan, Bahr-el-Ghazal.
DISTRIBUTION: Sudan.
COMMENT: Unidentifiable. Probably a synonym of *flavescens;* see Heim de Balsac and Meester, 1977, Part 1:13. Type locality taken from Heuglin's papers (RH).
ISIS NUMBER: 5301403007005035001.

Crocidura fischeri Pagenstecher, 1885. Jb. Hamb. Wiss. Anst., 2:34.
TYPE LOCALITY: Tanzania, Nguruman = "just N. of Lake Natran and at 2° S., hence in Kenya."
DISTRIBUTION: S. Ethiopia; Kenya; N. Tanzania; S. and W. Zaire.
COMMENT: Includes *voi.* Possibly related to *flavescens;* see Heim de Balsac and Meester, 1977, Part 1:14. Gureev, 1979:409, listed *voi* as a distinct species without comment.

See Moreau, *et al.*, 1946, Proc. Zool. Soc. Lond., 1945:395 for discussion of type locality.

ISIS NUMBER: 5301403007005036001 as *C. fischeri*.
5301403007005132001 as *C. voi*.

Crocidura flavescens (I. Geoffroy, 1827). Dict. Class. Hist. Nat., 11:324; Mem. Mus. Hist. Nat. Paris, 15:126.
TYPE LOCALITY: South Africa, "La Cafrerie et le pays des Hottentots" = King William's Town.
DISTRIBUTION: Egypt; Senegal to Ethiopia to South Africa; Gabon; Rio Muni.
COMMENT: Includes *anchietae, doriana, fuscosa, hedenborgiana, manni, martiensseni, nyansae, occidentalis* and *sururae;* see Heim de Balsac and Meester, 1977, Part 1:14. Also includes *olivieri;* see Corbet, 1978:30. Gureev, 1979:406–407, listed these forms as distinct species without comment. Dieterlen and Heim de Balsac, 1979, Saugetierk. Mitt., 27:241–287, listed *occidentalis* as a distinct species. Revised by Heim de Balsac and Barloy, 1966, Mammalia, 30:601–633. *C. odorata guineensis* Heim de Balsac, 1968, Mammalia, 32:384–385, is a junior homonym of *C. doriana guineensis* Cabrera, 1903, Mem. R. Soc. Hist. Nat. Madrid, 1:22.
ISIS NUMBER: 5301403007005037001 as *C. flavescens*.
5301403007005043001 as *C. fuscosa*.
5301403007005052001 as *C. hedenborgiana*.
5301403007005075001 as *C. manni*.
5301403007005079001 as *C. martiensseni*.
5301403007005097001 as *C. nyaansae (sic)*.
5301403007005098001 as *C. occidentalis*.
5301403007005100001 as *C. olivieri*.

Crocidura floweri Dollman, 1915. Ann. Mag. Nat. Hist., ser. 8, 15:515; 17:192.
TYPE LOCALITY: Egypt, Gizeh.
DISTRIBUTION: Egypt.
COMMENT: The affinities of this species are unknown; see Heim de Balsac and Meester, 1977, Part 1:15. Gureev, 1979:177 referred to this species as *C. floveri* Dollman, 1916.
ISIS NUMBER: 5301403007005038001.

Crocidura foxi Dollman, 1915. Ann. Mag. Nat. Hist., ser. 8, 15:514; 16:143.
TYPE LOCALITY: Nigeria, Panyam.
DISTRIBUTION: N. Nigeria to Ghana.
COMMENT: A member of the *poensis* group and probably an earlier name for *theresae* (RH).
ISIS NUMBER: 5301403007005039001.

Crocidura fuliginosa (Blyth, 1855). J. Asiat. Soc. Bengal, 24:362.
TYPE LOCALITY: Burma, Schwegyin, near Pegu.
DISTRIBUTION: Burma; adjacent China; N. India; Java; Borneo; Sumatra; Malay Peninsula and adjacent isls.; Sulawesi.
COMMENT: Includes *aagaardi, aoris, balvensis, doriae, foetida, gravida, grisescens, klossi, malayana, mansumensis, maporensis, negligens, praedix,* and *tionis;* see Jenkins, 1976:284. Medway, 1977:18–21, also included *dracula* in this species, but Lekagul and McNeely, 1977:28 considered *dracula* a distinct species. Gureev, 1979:405, listed *doriae* as a distinct species without comment, but Medway, 1977:18–21, regarded it as a synonym of *foetida*.
ISIS NUMBER: 5301403007005040001 as *C. fuliginosa*.
5301403007005001001 as *C. aagaardi*.
5301403007005076001 as *C. maporensis*.
5301403007005089001 as *C. negligens*.

Crocidura fulvastra (Sundevall, 1843). K. Sven. Vetenskapsakad Akad. Handl., 1842:172.
TYPE LOCALITY: Sudan, Bahr-el-Abiad.
DISTRIBUTION: Known only from the type locality.
COMMENT: Includes *strauchii;* see Heim de Balsac and Meester, 1977, Part 1:15.
ISIS NUMBER: 5301403007005041001.

Crocidura fumosa Thomas, 1904. Ann. Mag. Nat. Hist., ser. 7, 14:238.
TYPE LOCALITY: Kenya, Western slope of Mt. Kenya, 2600 m.
DISTRIBUTION: N. Tanzania; Kenya; Uganda; in mountains.
COMMENT: Includes *alchemillae* and *montis;* see Heim de Balsac and Meester, 1977, Part 1:15.
ISIS NUMBER: 5301403007005042001.

Crocidura fuscomurina (Heuglin, 1865). Leopoldina, 5, *in* Nouv. Acta Acad. Caes. Leop.-Carol., 32:36.
TYPE LOCALITY: Sudan, Bahr-el-Ghazal, Meshra-el-Req.
DISTRIBUTION: Known only from type locality.
COMMENT: Regarded as unidentifiable by Heim de Balsac and Meester, 1977, Part 1:16. RH considers it to be conspecific with *bicolor.*

Crocidura glassi Heim de Balsac, 1966. Mammalia, 30:448.
TYPE LOCALITY: Ethiopia, Harar, Gara Mulata Mtns.
DISTRIBUTION: Ethiopian highlands east of Rift Valley.

Crocidura gouliminensis Vesmanis and Vesmanis, 1979. Zool. Abh. Mus. Tierk. Dresden, 36(2):52.
TYPE LOCALITY: Morocco, Agadir Prov., 28 km S.W. Goulimine, 28° 46′ N., 10° 14′ W.
DISTRIBUTION: Known only from the type locality.
COMMENT: RH questions the validity of this species.

Crocidura gracilipes Peters, 1870. Monatsb. Preuss. Akad. Wiss. Berlin, p. 584.
TYPE LOCALITY: Tanzania, Mt. Kilamanjaro, "auf der Reise nach dem Kilimandscharo"; somewhat uncertain.
DISTRIBUTION: South Africa to S. Sudan, S. Ethiopia, W. Somalia; perhaps Guinea and Guinea-Bissau; Cameroun.
COMMENT: Includes *altae, hildegardeae, holobrunneus, ibeana, lutreola, maanjae, phaios, procera, rubecula, silacea* and *virgata;* see Heim de Balsac and Meester, 1977, Part 1:16. Gureev, 1979:410–412, listed *hildegardeae, ibeana, lutreola, maanjae,* and *silacea* as distinct species without comment. Dieterlen and Heim de Balsac, 1979, Saugetierk. Mitt., 27:241–287, listed *hildegardeae* as a distinct species. For comment on type locality, see Moreau, *et al.,* 1946, Proc. Zool. Soc. Lond., 1945:395.
ISIS NUMBER: 5301403007005048001 as *C. gracilipes.*
5301403007005053001 as *C. hildegardeae.*
5301403007005121001 as *C. silacea.*

Crocidura grandis Miller, 1911. Proc. U.S. Nat. Mus., 38:393.
TYPE LOCALITY: Philippines, Mindanao, Grand Malindang Mtn.
DISTRIBUTION: Known only from the type locality.
ISIS NUMBER: 5301403007005049001.

Crocidura grassei Brosset, DuBost and Heim de Balsac, 1965. Biologia Gabon., 1:165.
TYPE LOCALITY: Gabon, Belinga.
DISTRIBUTION: Belinga (Gabon); Boukoko (Central African Republic); Yaounde (Cameroun).

Crocidura grayi Dobson, 1890. Ann. Mag. Nat. Hist., ser. 6, 6:494.
TYPE LOCALITY: Philippines, Luzon.
DISTRIBUTION: Luzon (Philippine Isls.).
ISIS NUMBER: 5301403007005050001.

Crocidura greenwoodi Heim de Balsac, 1966. Monitore Zool. Ital., 74(suppl.):215.
TYPE LOCALITY: Somalia, Gelib.
DISTRIBUTION: S. Somalia.

Crocidura gueldenstaedti Pallas, 1811. Zoogr. Rosso-Asiat., 1:132.
TYPE LOCALITY: U.S.S.R., Georgian S.S.R., near Dushet (N. of Tbilisi).
DISTRIBUTION: Transcaucasia (U.S.S.R.); Asia Minor.
COMMENT: Revised by Richter, 1970, Zool. Abh. Mus. Tierk. Dresden, 31(17):294–304. May be conspecific with *russula;* see Corbet, 1978:28.

Crocidura halconus Miller, 1910. Proc. U.S. Nat. Mus., 38:391.
TYPE LOCALITY: Philippines, Mindoro, Mount Halcon.
DISTRIBUTION: Mindoro (Philippine Isls.).
ISIS NUMBER: 5301403007005051001.

Crocidura heljanensis Vesmanis, 1975. Senckenberg. Biol., 56:12.
TYPE LOCALITY: Algeria, Oran, Heljani, near St. Eugene.
DISTRIBUTION: Algeria; Morocco.
COMMENT: May be conspecific with *russula*; see Jenkins, 1976, Mammalia, 40:166; but
also see Vesmanis and Vesmanis, 1979, Zool. Abh. Mus. Tierk. Dresden, 36(2):54,
who considered *heljanensis* a distinct species.

Crocidura hirta Peters, 1852. Reise nach Mossambique, Saugethiere, p. 78.
TYPE LOCALITY: Mozambique, Tette, 17° S.
DISTRIBUTION: Zaire; N. and S.E. Tanzania; South Africa; Mozambique; Zimbabwe;
N.E. Namibia; Zambia; Uganda; perhaps Kenya and S. Somalia.
COMMENT: Includes *annellata, beirae, canescens, deserti, langi, luimbalensis,* and *velutina;*
see Heim de Balsac and Meester, 1977, Part 1:17. Gureev, 1979:407–408, listed
beirae and *deserti* as distinct species without comment.
ISIS NUMBER: 5301403007005055001 as *C. hirta.*
5301403007005062001 as *C. langi.*
5301403007005129001 as *C. velutina.*

Crocidura hispida Thomas, 1913. Ann. Mag. Nat. Hist., ser. 8, 11:468.
TYPE LOCALITY: India, Andaman Isls., Middle Andaman Isl. (northern end).
DISTRIBUTION: Middle Andaman Isl. (Andaman Isls.).
ISIS NUMBER: 5301403007005056001.

Crocidura horsfieldi (Tomes, 1856). Ann. Mag. Nat. Hist., 17:23.
TYPE LOCALITY: Sri Lanka.
DISTRIBUTION: Sri Lanka; N. Thailand to C. Vietnam; Nepal, Mysore, and Ladak
(India); Yunnan, Fukien, Hainan Isl.(China); Taiwan; Ryukyu Isls.
COMMENT: Includes *indochinensis*; see Jenkins, 1976:285; Jameson and Jones, 1977, Proc.
Biol. Soc. Wash., 90:461–465.
ISIS NUMBER: 5301403007005057001 as *C. horsfieldi.*
5301403007005058001 as *C. indochinensis.*

Crocidura jacksoni Thomas, 1904. Ann. Mag. Nat. Hist., ser. 7, 14:238.
TYPE LOCALITY: Kenya, Ravine Station.
DISTRIBUTION: S. Kenya; N. Tanzania; perhaps Uganda.
COMMENT: Includes *amalae*; see Heim de Balsac and Meester, 1977, Part 1:17.
ISIS NUMBER: 5301403007005060001.

Crocidura kivuana Heim de Balsac, 1968. Biologia Gabon., 4:319.
TYPE LOCALITY: Zaire, Kivu, Tschibati.
DISTRIBUTION: Mt. Kivu, and Mt. Kahusi (Zaire).

Crocidura lamottei Heim de Balsac, 1968. Mammalia, 32:386.
TYPE LOCALITY: Ivory Coast, Lamto (Savanne).
DISTRIBUTION: Ivory Coast; Togo; perhaps Senegal.
COMMENT: See Heim de Balsac and Meester, 1977, Part 1:18, for discussion of
distribution.

Crocidura lanosa Heim de Balsac, 1968. Biologia Gabon., 4:309.
TYPE LOCALITY: Zaire, Kivu, Lemera.
DISTRIBUTION: Uinka (Rwanda); Kivu, Irangi (E. Zaire).

Crocidura lasiura Dobson, 1890. Ann. Mag. Nat. Hist., 5:31.
TYPE LOCALITY: N.E. China, (Manchuria), Ussuri River.
DISTRIBUTION: Ussuri Region (U.S.S.R.) and N.E. China to Korea; Kiangsu (China)
(SW).
COMMENT: Includes *campuslincolnensis*; see Corbet, 1978:29.
ISIS NUMBER: 5301403007005063001.

Crocidura latona Hollister, 1916. Bull. Am. Mus. Nat. Hist., 35:667.
TYPE LOCALITY: Zaire, Medje.
DISTRIBUTION: Medje, Avakubi, Tshuapa, Ubembo, and Ikela (Zaire).
ISIS NUMBER: 5301403007005064001.

Crocidura lea Miller and Hollister, 1921. Proc. Biol. Soc. Wash., 34:102.
TYPE LOCALITY: Indonesia, Sulawesi, Temboan.
DISTRIBUTION: Known only from the type locality.
ISIS NUMBER: 5301403007005065001.

Crocidura lepidura Lyon, 1908. Proc. U.S. Nat. Mus., 34:662.
TYPE LOCALITY: Indonesia, Sumatra, Kateman River.
DISTRIBUTION: Sumatra.
ISIS NUMBER: 5301403007005066001.

Crocidura leucodon (Hermann, 1780). *In* Zimmermann, Geogr. Gesch., 2:382.
REVIEWED BY: M. Andera (MA).
TYPE LOCALITY: France, Bas Rhin, vicinity of Strasbourg.
DISTRIBUTION: France to the Volga and Caucasus; Elburz Mtns.; Asia Minor; Israel;
    Lebanon; Lesbos Isl. (Aegean Sea).
COMMENT: Reviewed by Richter, 1970, Zool. Abh. Mus. Tierk. Dresden,
    31(17):293–304, and Gureev, 1979. Includes *persica*; see Dolgov, 1979, Arch. Zool.
    Mus. Moscow St. Univ., 18:257–263. Includes *lasia*; see Gureev, 1979:396, and
    Jenkins, 1976:301; but also see Corbet, 1978:30. Includes *caspica* from Iran; see
    Jenkins, 1976:299. Gureev, 1979:394; Gromov and Baranova, 1981:124, listed
    persica as a distinct species without comment.
ISIS NUMBER: 5301403007005067001.

Crocidura levicula Miller and Hollister, 1921. Proc. Biol. Soc. Wash., 34:103.
TYPE LOCALITY: Indonesia, Sulawesi, Pinedapa.
DISTRIBUTION: Known only from the type locality.
ISIS NUMBER: 5301403007005068001.

Crocidura littoralis Heller, 1910. Smithson. Misc. Coll., 56(15):5.
TYPE LOCALITY: Uganda, Butiaba.
DISTRIBUTION: Uganda and Zaire to Cameroun.
COMMENT: This species has been included in *monax*, but is now regarded as a distinct
    species by Dieterlen and Heim de Balsac, 1979, Saugetierk. Mitt., 27:255.
ISIS NUMBER: 5301403007005069001.

Crocidura lucina Dippenaar, 1980. Ann. Transvaal Mus., 32:134–138.
TYPE LOCALITY: Ethiopia, Web River, near Dinshu.
DISTRIBUTION: Web River, Dinshu, Mt. Albasso, E. of Rift Valley, 3,000 m (Ethiopia).

Crocidura luluae Matschie, 1926. Z. Saugetierk., 1:111.
TYPE LOCALITY: Zaire, Kananga, Luluabourg, "Kapeleskse".
DISTRIBUTION: Known only from the type locality.
ISIS NUMBER: 5301403007005071001.

Crocidura luna Dollman, 1910. Ann. Mag. Nat. Hist., ser. 8, 5:175.
TYPE LOCALITY: Zaire, Shaba Prov., Bunkeya River, Katanga.
DISTRIBUTION: S. Zimbabwe and N. Mozambique to N. Angola, N.E. Zaire and C.
    Ethiopia.
COMMENT: Includes *electa, garambae, inyangai, johnstoni, macmillani, raineyi, schistacea,*
    selina, and *umbrosa*; see Heim de Balsac and Meester, 1977, Part 1:18. Gureev,
    1979:411, listed *electa* and *raineyi* as distinct species without comment.
ISIS NUMBER: 5301403007005072001 as *C. luna.*
    5301403007005113001 as *C. raineyi.*

Crocidura lusitania Dollman, 1915. Ann. Mag. Nat. Hist., ser. 8, 17:198.
TYPE LOCALITY: Mauritania, Trarza country.
DISTRIBUTION: Mauritania; Senegal; Morocco.
ISIS NUMBER: 5301403007005073001.

Crocidura macarthuri St. Leger, 1934. Ann. Mag. Nat. Hist., ser. 10, 13:559.
  TYPE LOCALITY: Kenya, Tana River, Merifano (20 mi. (32 km) from mouth of Tana
    River).
  DISTRIBUTION: Merifano, Moyale, and Ijara (Kenya).
  ISIS NUMBER: 5301403007005074001.

Crocidura macowi Dollman, 1915. Ann. Mag. Nat. Hist., ser. 8, 16:378.
  TYPE LOCALITY: Kenya, Mt. Nyiro, S. of Lake Rudolf (=L. Turkana).
  DISTRIBUTION: Known only from the type locality.

Crocidura maquassiensis Roberts, 1946. Ann. Transvaal Mus., 20:312.
  TYPE LOCALITY: South Africa, W. Transvaal, Maquassi, Klipkuil.
  DISTRIBUTION: Transvaal (South Africa); Inyanga (Zimbabwe).
  COMMENT: Includes *malani;* see Heim de Balsac and Meester, 1977, Part 1:19.

Crocidura mariquensis (A. Smith, 1849). Illustr. Zool. S. Afr. Mamm., pl. 44, f. 1.
  TYPE LOCALITY: South Africa, "Wood near the tropic of Capricorn" = W. Transvaal,
    Morico River, near its junction with Limpopo.
  DISTRIBUTION: N.E. Namibia; South Africa; Mozambique; W. Zimbabwe; N. Botswana;
    Zambia; S.C. Angola; perhaps S.E. Zaire.
  COMMENT: Includes *neavei, pilosa* and *sylvia;* see Heim de Balsac and Meester, 1977,
    Part 1:19. Gureev, 1979:411, 415, listed these forms as distinct species without
    comment. Reviewed by Dippenaar, 1979, Ann. Transvaal Mus., 32:1–34.
  ISIS NUMBER: 5301403007005077001 as *C. mariquensis.*
            5301403007005110001 as *C. pilosa.*

Crocidura maurisca Thomas, 1904. Ann. Mag. Nat. Hist., ser. 7, 14:239.
  TYPE LOCALITY: Uganda, Entebbe.
  DISTRIBUTION: Entebbe, Echuya Swamp (Uganda); Kaimosi (Kenya); Tanzania.
  ISIS NUMBER: 5301403007005080001.

Crocidura maxi Sody, 1936. Natuurk. Tijdschr. Ned.-Ind., 96:53.
  TYPE LOCALITY: Indonesia, Java, East Besoeki.
  DISTRIBUTION: Java.
  COMMENT: Possibly conspecific with *monticola;* see Jenkins, 1976:303.
  ISIS NUMBER: 5301403007005081001.

Crocidura melanorhyncha Jentink, 1910. Notes Leyden Mus., 32:198.
  TYPE LOCALITY: Indonesia, Java, Mt. Pangerango, 3000 ft. (914 m).
  DISTRIBUTION: Java.
  COMMENT: Possibly conspecific with *fuliginosa;* see Jenkins, 1976:303.
  ISIS NUMBER: 5301403007005082001.

Crocidura mesatanensis Vesmanis and Vesmanis, 1979. Zool. Abh. Mus. Tierk. Dresden,
    36(2):67.
  TYPE LOCALITY: Morocco, Khowribga Prov., 10 km S. Oued Zem, 31° 31′ N., 9° 46′ W.
  DISTRIBUTION: Known from three specimens from Safi Prov. and Khowribga Prov.
    (Morocco).
  COMMENT: RH questions the validity of this species.

Crocidura mindorus Miller, 1910. Proc. U.S. Nat. Mus., 38:392.
  TYPE LOCALITY: Philippines, Mindoro, Mt. Halcon, 1938 m.
  DISTRIBUTION: Known only from the type locality.
  ISIS NUMBER: 5301403007005083001.

Crocidura minuta Otten, 1917. Med. Burgerl. Geneesk. Dienst. Ned. Ind., 6:103.
  TYPE LOCALITY: Indonesia, Java, E. Java.
  DISTRIBUTION: Java.
  COMMENT: May be conspecific with *monticola;* see Jenkins, 1976:303.
  ISIS NUMBER: 5301403007005084001.

Crocidura miya Phillips, 1929. Spolia Zeylan., 15:113.
  TYPE LOCALITY: Sri Lanka, Kandyan Hills, Nilambe Dist., Moolgama, 3000 ft. (914 m).
  DISTRIBUTION: Sri Lanka.
  ISIS NUMBER: 5301403007005085001.

Crocidura monax Thomas, 1910. Ann. Mag. Nat. Hist., ser. 8, 6:310.
TYPE LOCALITY: Tanzania, Mt. Kilimanjaro, Rombo, 6000 ft. (1828 m).
DISTRIBUTION: N. Tanazania; W. Kenya; N.E. Zaire.
COMMENT: Belongs to the *littoralis* group; see Dieterlen and Heim de Balsac, 1979,
Saugetierk. Mitt., 27:255. Includes *oritis* and *ultima;* see Heim de Balsac and
Meester, 1977, Part 1:19–20. Gureev, 1979:414–415, listed *ultima* as a distinct
species without comment.
ISIS NUMBER: 5301403007005086001 as *C. monax.*
             5301403007005102001 as *C. oritis.*
             5301403007005128001 as *C. ultima.*

Crocidura monticola Peters, 1870. Monatsb. Preuss. Akad. Wiss. Berlin, p. 584.
TYPE LOCALITY: Indonesia, Java, Mount Lawu, near Surakarta.
DISTRIBUTION: Borneo; Java to Flores and probably Timor.
ISIS NUMBER: 5301403007005087001.

Crocidura mutesae Heller, 1910. Smithson. Misc. Coll., 56(15):3.
TYPE LOCALITY: Uganda, Kampala.
DISTRIBUTION: Uganda.
COMMENT: May be related to *suahelae* (RH).

Crocidura nana Dobson, 1890. Ann. Mag. Nat. Hist., ser. 6, 5:225.
TYPE LOCALITY: Somalia, Dollo.
DISTRIBUTION: Somalia; Ethiopia; Nile delta, possibly Nile valley (Egypt).
COMMENT: See Heim de Balsac and Meester, 1977, Part 1:20, for a discussion of
possible relationship with *religiosa;* but also see Corbet, 1978:27.

Crocidura nanilla Thomas, 1909. Ann. Mag. Nat. Hist., ser. 8, 4:99.
TYPE LOCALITY: Uganda, probably Entebbe.
DISTRIBUTION: Uganda and Kenya to Guinea.
COMMENT: Includes *rudolfi;* see Heim de Balsac and Meester, 1977, Part 1:20. For
discussion of "small *Crocidura,*" see Heim de Balsac, 1968, Mammalia, 32:379.

Crocidura neglecta Jentink, 1888. Notes Leyden Mus., 10:165.
TYPE LOCALITY: Indonesia, Sumatra.
DISTRIBUTION: Sumatra.

Crocidura negrina Rabor, 1952. Chicago Acad. Sci. Nat. Hist. Misc., 96:6.
TYPE LOCALITY: Philippines, Negros Isl., Cuernos de Negros Mtn., Dayongan, 1300 m.
DISTRIBUTION: Negros Isl. (Philippine Isls.).
ISIS NUMBER: 5301403007005090001.

Crocidura nicobarica Miller, 1902. Proc. U.S. Nat. Mus., 24:776.
TYPE LOCALITY: India, Nicobar Isl., Great Nicobar Isl.
DISTRIBUTION: Great Nicobar Isl. (Nicobar Isls.).
COMMENT: This is a species of *Suncus,* according to Krumbiegel, 1978, Saugetierk.
Mitt., 26:71.
ISIS NUMBER: 5301403007005091001.

Crocidura nigricans Bocage, 1889. J. Sci. Math. Phys. Nat. Lisboa, ser. 2, 1:28.
TYPE LOCALITY: Angola, Benguela Dist., Quindumbo.
DISTRIBUTION: Angola.
COMMENT: Provisionally unidentifiable; see Heim de Balsac and Meester, 1977, Part
1:20.
ISIS NUMBER: 5301403007005092001.

Crocidura nigripes Miller and Hollister, 1921. Proc. Biol. Soc. Wash., 34:101.
TYPE LOCALITY: Indonesia, Sulawesi, Temboan, S.W. from Tondano Lake.
DISTRIBUTION: N.E. and C. Sulawesi.
ISIS NUMBER: 5301403007005093001.

Crocidura nigrofusca Matschie, 1895. Saugeth. Deutsch-Ost-Afrikas, p. 33.
TYPE LOCALITY: Zaire, W. of Semliki, Kinyawanga, Wukalala, "between old Beni and
Lesse" in Semliki Valley.
DISTRIBUTION: Zaire; perhaps Angola.
COMMENT: Status uncertain; see Heim de Balsac and Meester, 1977, Part 1:20.
ISIS NUMBER: 5301403007005094001.

Crocidura nimbae Heim de Balsac, 1956. Mammalia, 20:131.
TYPE LOCALITY: Ivory Coast, Mt. Nimba, Zouguepo.
DISTRIBUTION: Nimba, Zouguepo (Ivory Coast); Lapie, Deaple (Liberia); Guinea.
COMMENT: May be related to *wimmeri;* see Heim de Balsac and Meester, 1977, Part
   1:20.
ISIS NUMBER: 5301403007005095001.

Crocidura niobe Thomas, 1906. Ann. Mag. Nat. Hist., ser. 7, 18:138.
TYPE LOCALITY: Uganda, Ruwenzori East, = Mubukee Valley, 6000 ft. (1828 m).
DISTRIBUTION: Uganda; S.W. Ethiopia. Zaire (RH).
COMMENT: Ethiopian records of Corbet and Yalden, 1972, Bull. Br. Mus. (Nat. Hist.)
   Zool., 22:224, and Yalden *et al.,* 1976, Monitore Zool. Ital., n.s., suppl., 8(1):12.
ISIS NUMBER: 5301403007005096001.

Crocidura odorata (Le Conte, 1857). Proc. Acad. Nat. Sci. Phila., p. 11.
TYPE LOCALITY: Gabon, Cette Cama.
DISTRIBUTION: Guinea to Gabon.
COMMENT: Includes *giffardi* and *goliath;* see Heim de Balsac and Meester, 1977, Part
   1:21. Gureev, 1979:388, 405, listed *giffardi* as a distinct species of *Crocidura* and
   goliath as a distinct species of *Praesorex* without comment.
ISIS NUMBER: 5301403007005099001 as *C. odorata.*
            5301403007005045001 as *C. giffardi.*
            5301403007005047001 as *C. goliath.*

Crocidura orientalis Jentink, 1890. *In* Weber, Zool. Ergebnisse einer Reise in Niederl. Ost-
   Indien, 1:124.
TYPE LOCALITY: Indonesia, Java, Tjibodas.
DISTRIBUTION: Java.
COMMENT: Probably conspecific with *fuliginosa;* see Jenkins, 1976.
ISIS NUMBER: 5301403007005101001.

Crocidura palawanensis Taylor, 1934. Monogr. Bur. Sci. Manila, 30:88.
TYPE LOCALITY: Philippines, Palawan, Sir J. Brooke Point.
DISTRIBUTION: Known only from the type locality.
ISIS NUMBER: 5301403007005103001.

Crocidura paradoxura Dobson, 1886. Ann. Mus. Civ. Stor. Nat. Genova, 4:566.
TYPE LOCALITY: Indonesia, Sumatra, Mt. Singalang, 2000 m.
DISTRIBUTION: Sumatra.
ISIS NUMBER: 5301403007005104001.

Crocidura parvacauda Taylor, 1934. Monogr. Bur. Sci. Manila, 30:83.
TYPE LOCALITY: Philippines, Mindanao, Cotabato, Saub.
DISTRIBUTION: Known only from the type locality.
ISIS NUMBER: 5301403007005105001.

Crocidura pasha Dollman, 1915. Ann. Mag. Nat. Hist., ser. 8, 15:517; 17:195–196.
TYPE LOCALITY: Sudan, Atbara River.
DISTRIBUTION: S. Sudan; N.E. Zaire; N. Nigeria; Niger; Upper Volta; Mali; Ivory Coast.
   Uganda (GSJ).
COMMENT: Includes *glebula;* see Heim de Balsac and Meester, 1977, Part 1:21. Gureev,
   1979:414, listed *glebula* as a distinct species without comment.
ISIS NUMBER: 5301403007005107001 as *C. pasha.*
            5301403007005046001 as *C. glebula.*

Crocidura pergrisea Miller, 1913. Proc. Biol. Soc. Wash., 26:113.
TYPE LOCALITY: Pakistan, Baltistan, Shigar, Skozo Loomba, 9500 ft. (2896 m).
DISTRIBUTION: Mountains from S.W. Asia Minor; Kopet Dag to Afghanistan; Kashmir;
   Tien Shan; Nepal.
COMMENT: Formerly included *armenica;* see Gureev, 1979:401, and comments under
   armenica.
ISIS NUMBER: 5301403007005109001.

Crocidura phaeura Osgood, 1936. Field Mus. Nat. Hist. Publ. Ser. Zool., 20:228.
TYPE LOCALITY: Ethiopia, Sidamo, N.E. of Allata, west base of Mt. Guramba.
DISTRIBUTION: Known only from the type locality.

Crocidura picea Sanderson, 1940. Trans. Zool. Soc. Lond., 24:682.
TYPE LOCALITY: Cameroun, Mamfe Div., Assumbo, Tinta.
DISTRIBUTION: Known only from the type locality.

Crocidura pitmani Barclay, 1932. Ann. Mag. Nat. Hist., ser. 10, 10:440.
TYPE LOCALITY: Zambia, Maluwe-Serenje Dist.
DISTRIBUTION: C. and N. Zambia.

Crocidura planiceps Heller, 1910. Smithson. Misc. Coll., 56(15):5.
TYPE LOCALITY: Uganda, Lado Enclave, Rhino Camp.
DISTRIBUTION: Uganda; Sudan; Zaire.
COMMENT: May be related to *suaveolens;* see Heim de Balsac, 1968, Mammalia, 32:411, and Heim de Balsac and Meester, 1977, Part 1:22.

Crocidura poensis Fraser, 1843. Proc. Zool. Soc. Lond., 1842:200.
TYPE LOCALITY: Equatorial Guinea, Bioko, Clarence.
DISTRIBUTION: Bioko; Gabon; Zaire; N. Angola; Cameroun to Liberia; San Tome Isl.
COMMENT: Includes *batesi, calabarensis, nigeriae, pamela, schweitzeri, stampflii* and *thomensis;* see Heim de Balsac and Meester, 1977, Part 1:22. Gureev, 1979:406, 409, listed *batesi* and *schweitzeri* as distinct species without comment.
ISIS NUMBER: 5301403007005111001 as *C. poensis.*
5301403007005010001 as *C. batesi.*
5301403007005119001 as *C. schweitzeri.*
5301403007005126001 as *C. thomensis.*

Crocidura religiosa I. Geoffroy, 1827. Mem. Mus. Hist. Nat. Paris, 15:128.
TYPE LOCALITY: Egypt, Thebes.
DISTRIBUTION: Nile Valley (Egypt).
COMMENT: Described from embalmed specimens from tombs at Thebes; see Corbet, 1978:27, who selected a neotype.
ISIS NUMBER: 5301403007005114001.

Crocidura rhoditis Miller and Hollister, 1921. Proc. Biol. Soc. Wash., 34:102.
TYPE LOCALITY: Indonesia, Sulawesi, Temboan.
DISTRIBUTION: S.E. Sulawesi.
ISIS NUMBER: 5301403007005115001.

Crocidura roosevelti Heller, 1910. Smithson. Misc. Coll., 56(15):6.
TYPE LOCALITY: Uganda, Lado Enclave, Rhino Camp.
DISTRIBUTION: Muita (Angola); Ouaddah (Central Afr. Rep.); Lado (Uganda): Garamba Natl. Park (Zaire).
ISIS NUMBER: 5301403007005116001.

Crocidura russula (Hermann, 1780). *In* Zimmermann, Geogr. Gesch., 2:382.
REVIEWED BY: M. Andera (MA).
TYPE LOCALITY: France, Bas Rhin, near Strasbourg.
DISTRIBUTION: S. and C. Europe; Morocco to Tunisia; large isls. of Mediterranean; Atlantic isls. off France; Israel to Asia Minor; N. Iraq; Kashmir; Pamirs; S. China; Hainan Isl.(SW); Taiwan.
COMMENT: Includes *caudata* and *sicula;* see Corbet, 1978:28, but Corbet and Hill, 1980:30, listed *sicula* as a distinct species without comment; RGVG agrees. Also includes *hosletti, pamirensis, pullata, rapax,* and *vorax* (see Ellerman and Morrison-Scott, 1966:81; Jameson and Jones, 1977, Proc. Biol. Soc. Wash., 90:459–482), and *foucauldi* (see Vesmanis and Vesmanis, 1979, Zool. Abh. Mus. Tierk. Dresden, 36(2):32). Heim de Balsac, 1968, Bonn. Zool. Beitr., 19:181–188, and Heim de Balsac and Meester, 1977, Part 1:15, considered *foucauldi* a distinct species. Gureev, 1979:396, 403, listed *caudata, pamirensis, rapax,* and *vorax* as distinct species, but listed *foucauldi* as a subspecies of *russula.* Gromov and Baranova,

1981:25, listed *pamirensis* as a distinct species. See also comments under *cypria* and *dsinezumi*. Reviewed by Richter, 1970, Zool. Abh. Mus. Tierk. Dresden, 31(17):293–304.

ISIS NUMBER: 5301403007005117001 as *C. russula*.
5301403007005025001 as *C. caudata*.

Crocidura sansibarica Neumann, 1900. Zool. Jahrb. Abt. Syst. Oekol. Geogr. Tiere, 13:54.
TYPE LOCALITY: Zanzibar, Mojoni.
DISTRIBUTION: Zanzibar; Pemba Isl.
ISIS NUMBER: 5301403007005118001.

Crocidura sericea (Sundevall, 1843). K. Sven. Vetenskapsakad Akad. Handl., 1842:173, 177.
TYPE LOCALITY: Sudan, near Bahr-el-Abiad.
DISTRIBUTION: Kenya and W. Ethiopia to Guinea-Bissau and Senegal.
COMMENT: Includes *beta, diana, hindei, macrodon* and *marrensis*; see Heim de Balsac and Meester, 1977, Part 1:23 who also included *bolivari* in *sericea*; Vesmanis and Vesmanis, 1979, Zool. Abh. Mus. Tierk. Dresden, 36(2):39, regarded *bolivari* as a distinct species; RH considers *bolivari* a subspecies of *sericea*. Gureev, 1979:408, listed *beta* and *hindei* as distinct species without comment.
ISIS NUMBER: 5301403007005120001 as *C. sericea*.
5301403007005013001 as *C. beta*.
5301403007005054001 as *C. hindei*.

Crocidura sibirica Dukelski, 1930. Zool. Anz., 88(1–4):75.
TYPE LOCALITY: U.S.S.R., S. Krasnoyarsky Krai, upper Yenisei River, 96 km S. of Minusinsk, Oznatchenoie.
DISTRIBUTION: C. Asia from Lake Issyk Kul to Upper Ob River; Lake Baikal; Sinkiang (China) (SW); perhaps Mongolia.
COMMENT: Includes *ognevi*; see Yudin, 1971, [Insectivorous Mammals of Siberia], Nauka, Novosibirsk, p. 146. OLR reported that this species does not occur in Mongolia; but see also Corbet, 1978:29; Sokolov and Orlov, 1980:50.

Crocidura smithii Thomas, 1895. Ann. Mag. Nat. Hist., ser. 6, 15:51.
TYPE LOCALITY: Ethiopia, Webi Shebeli, near Finik.
DISTRIBUTION: Arid regions of Somalia and Ethiopia. Senegal (RH).
COMMENT: This species is omitted from several modern lists or included in *cyanea* (see Heim de Balsac and Meester, 1977, Part 1:12), but has been regarded as a distinct species by Heim de Balsac, 1966, Monitore Zool. Ital., 74:205; Yalden, *et al.*, 1976, Monitore Zool. Ital., 8:14; Hutterer, 1981, Bonn. Zool. Beitr., 31:237; and Hutterer, 1981, Mammalia, 45:388–391.
ISIS NUMBER: 5301403007005122001 as *C. smith (sic)*.

Crocidura somalica Thomas, 1895. Ann. Mag. Nat. Hist., ser. 6, 16:52.
TYPE LOCALITY: Somalia, Webi Shebeli (about 44° E. near Geladi).
DISTRIBUTION: Webi Shebeli, Belet Amin, Afgoi (Somalia). Ethiopia (RH).
ISIS NUMBER: 5301403007005123001.

Crocidura suahelae Heller, 1912. Smithson. Misc. Coll., 60(12):6.
TYPE LOCALITY: Kenya, Mazeras.
DISTRIBUTION: Mazeras (Kenya); Bagamoya (Tanzania); perhaps Bangui (Central African Republic).
COMMENT: This species has been included in *hirta*; see Allen, 1939:38; but is considered a species by Heim de Balsac and Meester, 1977, Part 1:24.

Crocidura suaveolens (Pallas, 1811). Zoogr. Rosso-Asiat., 1:133.
REVIEWED BY: M. Andera (MA).
TYPE LOCALITY: U.S.S.R., Crimea, Kherssones, near Sevastopol.
DISTRIBUTION: S.W. and C. Europe; N. Africa discontinuously to Korea, China, and Taiwan.
COMMENT: Includes *dinniki, hyrcania* and *matruhensis*; see Corbet, 1978:27. Dolgov (in litt.) includes *pamirensis* in this species (OLR); but also see Ellerman and Morrison-Scott, 1951:81, who included it in *russula*. Gureev, 1979:393, 402 and

Gromov and Baranova, 1981:24, 26, listed *hyrcania* and *dinniki* as distinct species without comment.
ISIS NUMBER: 5301403007005124001.

Crocidura susiana Redding and Lay, 1978. Z. Saugetierk., 43:307.
TYPE LOCALITY: Iran, Khuzistan Province, 8 km S.S.W. of Dezful (32° 19' N. and 48° 21' E.).
DISTRIBUTION: Known only from vicinity of Dezful (S.W. Iran).

Crocidura tarfayensis Vesmanis and Vesmanis, 1979. Zool. Abh. Mus. Tierk. Dresden, 36(2):47.
TYPE LOCALITY: Morocco, Agadir Prov., 8 km S. Tarfaya, 27° 50' N., 12° 30' W.
DISTRIBUTION: Known only from the type locality from a single specimen.
COMMENT: RH questions the validity of this species.

Crocidura tenuis (Muller, 1839). *In* Temminck, Verhandl. Nat. Gesch. Nederland Overz. Bezitt., Zool., 1:26, 50.
TYPE LOCALITY: Indonesia, Timor.
DISTRIBUTION: Timor.
ISIS NUMBER: 5301403007005125001.

Crocidura tephra Setzer, 1956. Proc. U.S. Nat. Mus., 106:466.
TYPE LOCALITY: Sudan, Equatoria Prov., Torit.
DISTRIBUTION: Known only from the type locality.

Crocidura thalia Dippenaar, 1980. Ann. Transvaal Mus., 32:138-147.
TYPE LOCALITY: Ethiopia, N.W. Bale Prov., Gedeb Mts., S.E. Dodola, 2600 m.
DISTRIBUTION: Ethiopian highlands.
COMMENT: Member of the *luna--fumosa* species complex.

Crocidura theresae Heim de Balsac, 1968. Mammalia, 32:398.
TYPE LOCALITY: Upper Guinea, Nzerekore.
DISTRIBUTION: Ghana to Guinea.
COMMENT: Probably a synonym of *foxi* (RH).

Crocidura tizuitensis Vesmanis and Vesmanis, 1979. Zool. Abh. Mus. Tierk. Dresden, 36(2):54.
TYPE LOCALITY: Morocco, Agadir Prov., 15 km N. Tiznit, 29° 49' N., 9° 39' W.
DISTRIBUTION: Agadir Prov., Morocco.
COMMENT: RH questions the validity of this species.

Crocidura turba Dollman, 1910. Ann. Mag. Nat. Hist., ser. 8, 5:176.
TYPE LOCALITY: Zambia, Lake Bangweolo, Chilui (=Chilubi) Isl.
DISTRIBUTION: N. Angola; Zambia; S. Zaire.
COMMENT: Includes *angolae*; see Heim de Balsac and Meester, 1977, Part 1:24-25. Range not exactly known, due to confusion with *zaodon* (RH).

Crocidura usambarae Dippenaar, 1980. Ann. Transvaal Mus., 32:138-147.
TYPE LOCALITY: Tanzania, W. Usambara Mts., Shume, 16 mi N. Lushoto.
DISTRIBUTION: Magamba, Shume (N.E. Tanzania), perhaps also Ngozi Crater, S.W. Tanzania.

Crocidura viaria (I. Geoffroy, 1834). *In* Zool. Voy. de Belanger Indes-Orient., p. 127.
TYPE LOCALITY: Senegal.
DISTRIBUTION: Senegal.
COMMENT: May be an earlier name for *sericea*; see Heim de Balsac, 1968:394, and Heim de Balsac and Meester, 1977, Part 1:25.
ISIS NUMBER: 5301403007005130001.

Crocidura villosa Robinson and Kloss, 1918. J. Fed. Malay St. Mus., 8:21.
TYPE LOCALITY: Indonesia, Sumatra, Korinchi, Sungai Kumbang, 4700 ft. (1433 m).
DISTRIBUTION: Sumatra.
COMMENT: Probably conspecific with *fuliginosa*; see Jenkins, 1976:302.
ISIS NUMBER: 5301403007005131001.

Crocidura vosmaeri Jentink, 1888. Notes Leyden Mus., 10:165.
TYPE LOCALITY: Indonesia, Banka Isl.
DISTRIBUTION: Banka Isl. (W. Indonesia).
ISIS NUMBER: 5301403007005133001.

Crocidura vulcani Heim de Balsac, 1956. Mammalia, 20:134.
TYPE LOCALITY: Cameroun, Mt. Cameroun, Bibundi Crater.
DISTRIBUTION: Known only from the type locality.
COMMENT: May be an aberrant form of *eisentrauti;* see Heim de Balsac and Meester,
1977, Part 1:25.
ISIS NUMBER: 5301403007005134001.

Crocidura weberi Jentink, 1890. *In* Weber Zool. Ergebnisse Reise Med. Ost.-Indien, 1:124.
TYPE LOCALITY: Indonesia, Sumatra, Singbarak.
DISTRIBUTION: Sumatra.
ISIS NUMBER: 5301403007005135001.

Crocidura whitakeri De Winton, 1898. Proc. Zool. Soc. Lond., 1897:954.
TYPE LOCALITY: Morocco, Sierzet, between Morocco City and Mogador.
DISTRIBUTION: Mauritania; Rio de Oro; Morocco; Algeria; Tunisia.

Crocidura wimmeri Heim de Balsac and Aellen, 1958. Rev. Suisse Zool., 65:952.
TYPE LOCALITY: Ivory Coast, Adiopodoume.
DISTRIBUTION: Gabon; Cameroun; Ivory Coast; Liberia.
COMMENT: May be related to *nimbae;* see Heim de Balsac and Meester, 1977, Part 1:25.
Kuhn, 1965, Senckenberg. Biol., 46(5):321–340, considered these forms possibly
conspecific.
ISIS NUMBER: 5301403007005136001.

Crocidura xantippe Osgood, 1910. Field Mus. Nat. Hist. Publ. Ser. Zool., 10:19.
TYPE LOCALITY: Kenya, Voi.
DISTRIBUTION: Nyiru, Voi, Tsavo (S.E. Kenya); Usambara Mtns. (Tanzania).
ISIS NUMBER: 5301403007005137001.

Crocidura yankariensis Hutterer and Jenkins, 1980. Bull. Br. Mus. (Nat. Hist.) Zool., 39:305.
TYPE LOCALITY: Nigeria, Bauchi State, 16 km E. of Yankari Game Reserve boundary,
Futuk (9° 50′ N., 10° 55′ E.).
DISTRIBUTION: Futuk and Wikki (N. Nigeria).

Crocidura zaianensis Vesmanis and Vesmanis, 1979. Zool. Abh. Mus. Tierk. Dresden,
36(2):62.
TYPE LOCALITY: Morocco, Settat Prov., 5 km N. Ben-Silmane, 33° 39′ N., 7° 08′ W.
DISTRIBUTION: Known only from the type locality from a single specimen.
COMMENT: RH questions the validity of this species.

Crocidura zaodon Osgood, 1910. Field Mus. Nat. Hist. Publ. Ser. Zool., 10:21.
TYPE LOCALITY: Kenya, Nairobi.
DISTRIBUTION: Ethiopia throughout E. Africa to Zambia and Angola, and W.
Cameroun.
COMMENT: Includes *ansorgei, cabrerai, kempi, lakiunde, nilotica, nyikae, provocax, soricoides,
tarella,* and *zena;* see Heim de Balsac and Meester, 1977, Part 1:25–26. Gureev,
1979:409, listed *ansorgei, nilotica,* and *zena* as distinct species without comment.

Crocidura zaphiri Dollman, 1915. Ann. Mag. Nat. Hist., ser. 8, 15:509; 16:66.
TYPE LOCALITY: Ethiopia, Kaffa, Charada Forest.
DISTRIBUTION: Kaffa, (S. Ethiopia); Kaimosi, Kisumu (Kenya).
COMMENT: Includes *simiolus;* see Osgood, 1936, Field Mus. Nat. Hist. Publ. Ser. Zool.,
20:224; Heim de Balsac and Meester, 1977, Part 1:26.
ISIS NUMBER: 5301403007005138001.

Crocidura zarudnyi Ognev, 1928. Mammals of Eastern Europe and Northern Asia, 1:341.
TYPE LOCALITY: Iran, Baluchistan (border).
DISTRIBUTION: S.E. Iran; W. Pakistan; Afghanistan.

Crocidura zimmeri Osgood, 1936. Field Mus. Nat. Hist. Publ. Ser. Zool., 20:223.
TYPE LOCALITY: Zaire, Katanga Prov., Lualaba River, Katobwe, near Bukoma.
DISTRIBUTION: Known only from the type locality.
ISIS NUMBER: 5301403007005139001.

Cryptotis Pomel, 1848. Arch. Sci. Phys. Nat. Geneve, 9:249.
REVIEWED BY: J. R. Choate (JRC).
COMMENT: Revised in part by Choate, 1970, Univ. Kans. Publ. Mus. Nat. Hist.,
19:195–317. Subfamily Soricinae, tribe Blarinini; see Repenning, 1967:37.
Formerly included *C. surinamensis* which was transferred to *Sorex araneus* by
Husson, 1963, Studies on the Fauna of Surinam and other Guyanas, 13:35–37.
Gureev, 1979:433–437, did not cite Choate *(op. cit.)*, and recognized those species
Choate considered synonyms. A key to the genus was published by Choate and
Fleharty, 1974, Mamm. Species, 44:1–3.
ISIS NUMBER: 5301403007006000000.

Cryptotis avia G. M. Allen, 1923. Proc. N. Engl. Zool. Club, 8:37
TYPE LOCALITY: Colombia, El Jerjon, east of Bogota.
DISTRIBUTION: E. Cordillera of Colombia.
ISIS NUMBER: 5301403007006001001 as *C. avius (sic)*.

Cryptotis endersi Setzer, 1950. J. Wash. Acad. Sci., 40:300.
TYPE LOCALITY: Panama, Chiriqui Bocas del Toro, Cylindro.
DISTRIBUTION: Known only from the type locality.
ISIS NUMBER: 5301403007006003001.

Cryptotis goldmani (Merriam, 1895). N. Am. Fauna, 10:25.
REVIEWED BY: J. Ramirez-Pulido (JRP).
TYPE LOCALITY: Mexico, Guerrero, Mountains near Chilpancingo, 10,000 ft. (3048 m).
DISTRIBUTION: Guatemala to Jalisco (Mexico).
COMMENT: Includes *alticola, euryrhynchis, fossor, frontalis, griseoventris*, and *guerrerensis;*
see Choate, 1970, Univ. Kans. Publ. Mus. Nat. Hist., 19:247; Hutterer, 1980,
Mammalia, 44:413.
ISIS NUMBER: 5301403007006006001 as *C. goldmani.*
5301403007006004001 as *C. fossor.*
5301403007006005001 as *C. frontalis.*
5301403007006009001 as *C. griseoventris.*
5301403007006010001 as *C. guerrerensis.*

Cryptotis goodwini Jackson, 1933. Proc. Biol. Soc. Wash., 46:81.
TYPE LOCALITY: Guatemala, Quezaltenango, Calel, 10,200 ft. (3109 m).
DISTRIBUTION: S. Guatemala; W. El Salvador; S. Mexico.
COMMENT: Reviewed by Choate and Fleharty, 1974, Mamm. Species, 44:1–3; Hutterer,
1980, Mammalia, 44:413.
ISIS NUMBER: 5301403007006007001.

Cryptotis gracilis Miller, 1911. Proc. Biol. Soc. Wash., 24:221.
TYPE LOCALITY: Costa Rica, Talamanca (=Limon), near base of Pico Blanco, head of
Lari River, 6000 ft. (1828 m).
DISTRIBUTION: S.C. Honduras to W. Panama.
COMMENT: Includes *jacksoni;* see Choate, 1970, Univ. Kans. Publ. Mus. Nat. Hist.,
19:281.
ISIS NUMBER: 5301403007006008001 as *C. gracilis.*
5301403007006011001 as *C. jacksoni.*

Cryptotis magna (Merriam, 1895). N. Am. Fauna, 10:28.
REVIEWED BY: J. Ramirez-Pulido (JRP).
TYPE LOCALITY: Mexico, Oaxaca, Totontepec, 6800 ft. (2073 m).
DISTRIBUTION: N.C. Oaxaca (Mexico).
COMMENT: Reviewed by Robertson and Rickart, 1975, Mamm. Species, 61:1–2.
ISIS NUMBER: 5301403007006012001.

Cryptotis mexicana (Coues, 1877). Bull. U.S. Geol. Geogr. Surv. Terr., 3:652.
  REVIEWED BY: J. Ramirez-Pulido (JRP).
  TYPE LOCALITY: Mexico, Veracruz, Jalapa (=Xalapa).
  DISTRIBUTION: Tamaulipas to Chiapas (Mexico).
  COMMENT: Includes *Notiosorex (Xenosorex) phillipsii* Schaldach, 1966, Saugetierk. Mitt.,
    14:289; see Choate, 1969, Proc. Biol. Soc. Wash., 82:469–476 (GSJ). Reviewed by
    Choate, 1973, Mamm. Species, 28:1–3.
  ISIS NUMBER: 5301403007006014001.

Cryptotis montivaga (Anthony, 1921). Am. Mus. Novit., 20:5.
  TYPE LOCALITY: Ecuador, Azuay Prov., Bestion, 3000 m.
  DISTRIBUTION: Andean zone of Ecuador.
  ISIS NUMBER: 5301403007006016001 as *C. montivagus* (sic).

Cryptotis nigrescens (J. A. Allen, 1895). Bull. Am. Mus. Nat. Hist., 7:339.
  REVIEWED BY: J. Ramirez-Pulido (JRP).
  TYPE LOCALITY: Costa Rica, "San Isidro (San Jose)."
  DISTRIBUTION: Guerrero and Yucatan (Mexico) to Panama.
  COMMENT: Includes *mera, micrura, zeteki,* and *tersus;* see Choate, 1970, Univ. Kans.
    Publ. Mus. Nat. Hist., 19:279.
  ISIS NUMBER: 5301403007006017001 as *C. nigrescens.*
                5301403007006013001 as *C. mera.*
                5301403007006015001 as *C. micrura.*
                5301403007006023001 as *C. tersus.*
                5301403007006025001 as *C. zeteki.*

Cryptotis parva (Say, 1823). *In* Long, Account of an Exped. from Pittsburgh to the Rocky
    Mtns., Vol. 1:163.
  REVIEWED BY: J. Ramirez-Pulido (JRP).
  TYPE LOCALITY: U.S.A., Nebraska, Washington Co., W. bank Missouri River, near Blair.
  DISTRIBUTION: W. Connecticut (U.S.A.) and S.W. tip of Ontario (Canada) to N.E. corner
    of Mexico south to Panama.
  COMMENT: Includes *micrura, olivacea, pergracilis,* and *celatus;* see Choate, 1970, Univ.
    Kans. Publ. Mus. Nat. Hist., 19:260–268. Reviewed by Whitaker, 1974, Mamm.
    Species, 43:1–8.
  ISIS NUMBER: 5301403007006019001 as *C. parva.*
                5301403007006002001 as *C. celatus.*
                5301403007006018001 as *C. olivacea.*
                5301403007006020001 as *C. pergracilis.*

Cryptotis squamipes (J. A. Allen, 1912). Bull. Am. Mus. Nat. Hist., 31:93.
  TYPE LOCALITY: Colombia, Paramo, 65 km W. Popayan, 3100 m.
  DISTRIBUTION: S. Cordillera Occidental of Colombia.
  ISIS NUMBER: 5301403007006021001.

Cryptotis thomasi (Merriam, 1897). Proc. Biol. Soc. Wash., 11:227.
  TYPE LOCALITY: Colombia, Savanna of Bogota, near Bogota, 3000 m.
  DISTRIBUTION: Cordillera Oriental of Colombia, Ecuador, S.W. Venezuela.
  ISIS NUMBER: 5301403007006024001.

Diplomesodon Brandt, 1852. *In* Baer and Helmersen, Beitr. Russ. Reichs., 17:299.
  COMMENT: Subfamily Crocidurinae; see Repenning, 1967:15.
  ISIS NUMBER: 5301403007007000000.

  Diplomesodon pulchellum (Lichtenstein, 1823). *In* Eversmann, Reise von Orenburg nach
    Bokhara, p. 124.
    TYPE LOCALITY: U.S.S.R., W. Kazakhstan, Sands on E. Bank of Ural River.
    DISTRIBUTION: W. and S. Kazakstan, Uzbekistan, Turkmenia (U.S.S.R.).
    ISIS NUMBER: 5301403007007001001.

Feroculus Kelaart, 1852. Prodr. Faun. Zeylanica, p. 31.
  COMMENT: Subfamily Crocidurinae; see Repenning, 1967:15.
  ISIS NUMBER: 5301403007008000000.

Feroculus feroculus (Kelaart, 1850). J. Ceylon Branch Asiat. Soc., 2(5):211.
TYPE LOCALITY: Sri Lanka, Nuwara Eliya.
DISTRIBUTION: C. Highlands of Sri Lanka.
ISIS NUMBER: 5301403007008001001.

Megasorex Hibbard, 1950. Contrib. Mus. Paleontol. Univ. Mich., 8:129.
REVIEWED BY: J. Ramirez-Pulido (JRP).
COMMENT: Subfamily Soricinae, tribe Neomyinae.

Megasorex gigas (Merriam, 1897). Proc. Biol. Soc. Wash., 11:227.
TYPE LOCALITY: Mexico, Jalisco, near San Sebastian, Mtns. near Milpillas.
DISTRIBUTION: Nayarit to Oaxaca (Mexico).
COMMENT: Formerly included in *Notiosorex*; see Repenning, 1967:56 and review by
Armstrong and Jones, 1972, Mamm. Species, 16:12, who considered *Megasorex* a
distinct genus; see also Hall, 1981:65–66, who included *gigas* in *Notiosorex* without
comment.
ISIS NUMBER: 5301403007013002001 as *Notiosorex gigas*.

Myosorex Gray, 1838. Proc. Zool. Soc. Lond., 1837:124.
COMMENT: Includes *Surdisorex* and *Congosorex*; see Heim de Balsac and Meester, 1977,
Part 1:2, and Heim de Balsac and Lamotte, 1951, Mammalia, 20:140–167.
Subfamily Crocidurinae; see Repenning, 1967:15. Gureev, 1979:375, listed
Surdisorex as a genus without comment.
ISIS NUMBER: 5301403007010000000 as *Myosorex*.
5301403007021000000 as *Surdisorex*.

Myosorex babaulti Heim de Balsac and Lamotte, 1956. Mammalia, 20:150.
TYPE LOCALITY: Zaire, Kivu.
DISTRIBUTION: Mt. Kivu (Zaire).
COMMENT: Formerly included in *blarina*; see Dieterlen and Heim de Balsac, 1979,
Saugetierk. Mitt., 27:264.
ISIS NUMBER: 5301403007010001001.

Myosorex blarina Thomas, 1906. Ann. Mag. Nat. Hist., ser. 7, 18:139.
TYPE LOCALITY: Uganda, Ruwenzori East, Mubuku Valley, 10000 ft. (3048 m).
DISTRIBUTION: Uganda; Zaire; Tanzania.
COMMENT: Formerly included *babaulti*; see Dieterlen and Heim de Balsac, 1979,
Saugetierk. Mitt., 27:264. Includes *zinki*; see Heim de Balsac and Meester, 1977,
Part 1:4.
ISIS NUMBER: 5301403007010002001.

Myosorex cafer (Sundevall, 1846). Ofv. Kongl. Svenska Vet.-Akad. Forhandl. Stockholm,
3:119.
TYPE LOCALITY: South Africa, "E Caffraria interiore et Port-Natal".
DISTRIBUTION: E. South Africa; W. Mozambique; E. Zimbabwe.
COMMENT: Includes *affinis*, *sclateri*, *swinnyi*, *talpinus*, and *tenuis*; see Heim de Balsac and
Meester, 1977, Part 1:4. Revised by Meester, 1958, J. Mammal., 39:325.
ISIS NUMBER: 5301403007010003001.

Myosorex eisentrauti Heim de Balsac, 1968. Bonn. Zool. Beitr., 19:20.
TYPE LOCALITY: Equatorial Guinea, Bioko, Pic Santa Isabel, 2400 m.
DISTRIBUTION: Lake Manenguba, Rumpi Hills, Mt. Lefo (Cameroun) and Lake Oku
(Bioko).
COMMENT: Includes *okuensis* and *rumpii*; see Heim de Balsac and Meester, 1977, Part
1:4–5.
ISIS NUMBER: 5301403007010004001.

Myosorex geata (Allen and Loveridge, 1927). Proc. Boston Soc. Nat. Hist., 38:417.
TYPE LOCALITY: Tanzania, Uluguru Mtns., Nyingwa.
DISTRIBUTION: Tanzania.
COMMENT: Formerly in *Crocidura*; see Heim de Balsac, 1967, Mammalia, 31:610.

Myosorex longicaudatus Meester and Dippenaar, 1978. Ann. Transvaal Mus., 31:30.
TYPE LOCALITY: South Africa, Cape Prov., 14 km N.N.E. Knysna, Diepwalle State
Forest Station, 33° 57′ S., 23° 10′ E.
DISTRIBUTION: S. Cape Prov. (South Africa).

Myosorex norae (Thomas, 1906). Ann. Mag. Nat. Hist., ser. 7, 18:223.
TYPE LOCALITY: Kenya, E. side of Aberdare Range, near Nyeri.
DISTRIBUTION: Aberdare Range (Kenya).
COMMENT: Formerly in *Surdisorex;* see Heim de Balsac and Meester, 1977, Part 1:2.
ISIS NUMBER: 5301403007021001001 as *Surdisorex norae.*

Myosorex polli Heim de Balsac and Lamotte, 1956. Mammalia, 20:155.
TYPE LOCALITY: Zaire, Kasai, Lubondai.
DISTRIBUTION: Known only from the type locality.
ISIS NUMBER: 5301403007010005001.

Myosorex polulus (Hollister, 1916). Smithson. Misc. Coll., 66(1):1.
TYPE LOCALITY: Kenya, West side of Mt. Kenya, 10,700 ft. (3261 m).
DISTRIBUTION: Mt. Kenya (Kenya).
COMMENT: May be a subspecies of *norae;* see Heim de Balsac and Meester, 1977, Part
1:3.
ISIS NUMBER: 5301403007021002001 as *Surdisorex polulus.*

Myosorex preussi Matschie, 1893. Sitzb. Ges. Naturf. Fr. Berlin, p. 177.
TYPE LOCALITY: Cameroun, Mt. Cameroun, Buea.
DISTRIBUTION: Known only from the type locality.
COMMENT: Formerly in *Sylvisorex;* see Heim de Balsac and Meester, 1977, Part 1:5.
ISIS NUMBER: 5301403007022007001 as *Sylvisorex preussi.*

Myosorex schalleri Heim de Balsac, 1967. Mammalia, 31:610.
TYPE LOCALITY: Zaire, Mwenga Dist., Albert Natl. Park, Nzombe.
DISTRIBUTION: Known only from the type locality.
COMMENT: Probably a young *M. blarina;* see Heim de Balsac and Meester, 1977, Part
1:3.
ISIS NUMBER: 5301403007010006001.

Myosorex varius (Smuts, 1832). Enumer. Mamm. Cap., p. 108.
TYPE LOCALITY: South Africa, Cape of Good Hope, Algoa Bay (Port Elizabeth).
DISTRIBUTION: South Africa; Lesotho.
COMMENT: Includes *capensis, herpestes, pondoensis* and *transvaalensis;* see Heim de Balsac
and Meester, 1977, Part 1:4. Revised by Meester, 1958, J. Mammal., 39:325.
ISIS NUMBER: 5301403007010007001.

Nectogale Milne-Edwards, 1870. C. R. Acad. Sci. Paris, 70:341.
COMMENT: Subfamily Soricinae, tribe Neomyini; see Repenning, 1967:45.
ISIS NUMBER: 5301403007011000000.

Nectogale elegans Milne-Edwards, 1870. C. R. Acad. Sci. Paris, 70:341.
TYPE LOCALITY: China, Szechwan, Moupin.
DISTRIBUTION: Szechwan, Yunnan, Kansu, Shensi, and Tibet (China); N. Burma; Nepal;
Sikkim (India); Bhutan.
ISIS NUMBER: 5301403007011001001.

Neomys Kaup, 1829. Skizz. Europ. Thierwelt, 1:117.
REVIEWED BY: M. Andera (MA).
COMMENT: Subfamily Soricinae, tribe Neomyini; see Repenning, 1967:45.
ISIS NUMBER: 5301403007012000000.

Neomys anomalus Cabrera, 1907. Ann. Mag. Nat. Hist., ser. 7, 20:214.
TYPE LOCALITY: Spain, Madrid, San Martin de la Vega.
DISTRIBUTION: Montane woodlands of W. Europe; lowlands of E. Europe; perhaps W.
Asia Minor.
ISIS NUMBER: 5301403007012001001.

Neomys fodiens (Pennant, 1771). Synopsis Quadrupeds, p. 308.
TYPE LOCALITY: Germany, Berlin.
DISTRIBUTION: Europe to N.W. Mongolia, Lake Baikal, Yenesi River, and Tien Shan (China); disjunct in Sakhalin Isl. and adjacent Siberia; Kirin (China) (SW).
ISIS NUMBER: 5301403007012002001.

Neomys schelkovnikovi Satunin, 1913. Trud. Obshch. Izuch. Chernomorsk. Poberezh., 2:24.
TYPE LOCALITY: U.S.S.R., Georgian S.S.R., Svanetiya, Ushkul.
DISTRIBUTION: Armenia, Georgia (U.S.S.R.).

Notiosorex Coues, 1877. Bull. U.S. Geol. Geogr. Surv. Terr., 3:646.
REVIEWED BY: J. Ramirez-Pulido (JRP).
COMMENT: Subfamily Soricinae, tribe Neomyini; see Repenning, 1967:45. Formerly included *Megasorex*; see Repenning, 1967:56 and Armstrong and Jones, 1972, Mamm. Species, 16:1-2, who considered it a distinct genus; but also see Hall, 1981:65–66. *Notiosorex (Xenosorex) phillipsii* is a synonym of *Cryptotis mexicana*; see Choate, 1969, Proc. Biol. Soc. Wash., 82:475.
ISIS NUMBER: 5301403007013000000.

Notiosorex crawfordi (Coues, 1877). Bull. U.S. Geol. Geogr. Surv. Terr., 3:631.
TYPE LOCALITY: U.S.A., Texas, El Paso Co., Fort Bliss, 2 mi. (3 km) above El Paso.
DISTRIBUTION: S.W. and S.C. United States to Baja California and N. Michoacan (Mexico).
COMMENT: Includes *evotis*; see Armstrong and Jones, 1971, J. Mammal., 52:747–757. Reviewed by Armstrong and Jones, 1972, Mamm. Species, 17:1–5.
ISIS NUMBER: 5301403007013001001.

Paracrocidura Heim de Balsac, 1956. Rev. Zool. Bot. Afr., 54:137.
COMMENT: Subfamily Crocidurinae; see Repenning, 1967:15.
ISIS NUMBER: 5301403007014000000.

Paracrocidura schoutedeni Heim de Balsac, 1956. Rev. Zool. Bot. Afr., 54:137.
TYPE LOCALITY: Zaire, Kasai, Lubondaie (75 km S. of Luluabourg), Tshimbulu (Dibaya).
DISTRIBUTION: Cameroun; Gabon; Congo Rep.; Zaire; Central African Republic.
ISIS NUMBER: 5301403007014001001.

Podihik Deraniyagala, 1958. Admin. Rep. Dir. Natl. Mus. Ceylon, 1957, Part 4, Ed., Sci., Art., (E):5.
COMMENT: Questionably subfamily Soricinae; see Repenning, 1967:57.
ISIS NUMBER: 5301403007015000000.

Podihik kura Deraniyagala, 1958. Admin. Rep. Dir. Natl. Mus. Ceylon, 1957, Part 4, Ed., Sci., Art., (E):5.
TYPE LOCALITY: Sri Lanka, North-Central Prov., Madirigiriya, Kandulu River.
DISTRIBUTION: N.C. Sri Lanka.
COMMENT: J. E. Hill, 1975, *in* Walker *et al.*, p. 146, believes that the paratype (1 of 2 specimens known) is *Suncus etruscus*, placing the validity of the genus in doubt; see also Repenning, 1967:57. Probably a *Suncus* (RH).
ISIS NUMBER: 5301403007015001001.

Scutisorex Thomas, 1913. Ann. Mag. Nat. Hist., ser. 8, 11:321.
COMMENT: Subfamily Crocidurinae; see Repenning, 1967:15.
ISIS NUMBER: 5301403007016000000.

Scutisorex somereni (Thomas, 1910). Ann. Mag. Nat. Hist., ser. 8, 6:113.
TYPE LOCALITY: Uganda, near Kampala, Kyetume.
DISTRIBUTION: N. Zaire; Uganda; Rwanda.
COMMENT: Includes *congicus*; see Heim de Balsac and Meester, 1977, Part 1:7. Gureev, 1979:386, listed *congicus* as a distinct species without comment.
ISIS NUMBER: 5301403007016002001 as *S. somereni*.
5301403007016001001 as *S. congicus*.

Solisorex Thomas, 1924. Spolia Zeylan., 13:94.
  COMMENT: Subfamily Crocidurinae; see Repenning, 1967:15.
  ISIS NUMBER: 5301403007017000000.

Solisorex pearsoni Thomas, 1924. Spolia Zeylan., 13:94.
  TYPE LOCALITY: Sri Lanka, Hakgala, near Nuwara Eliya.
  DISTRIBUTION: C. Highlands of Sri Lanka.
  ISIS NUMBER: 5301403007017001001.

Sorex Linnaeus, 1758. Syst. Nat., 10th ed., 1:53.
  REVIEWED BY: R. S. Hoffmann (RSH); J. A. Junge (JAJ); D. W. Linzey (DWL)(N.
    America); J. Ramirez-Pulido (JRP) (Mexico).
  COMMENT: Includes *Microsorex* as a subgenus (closely allied to subgenus *Otisorex*); see
    Diersing, 1980, J. Mammal., 61:76. Includes *Ognevia* as a subgenus; see Corbet,
    1978:17. N. American species reviewed by Junge and Hoffmann, 1981:1–48.
    Subfamily Soricinae, tribe Soricini; see Repenning, 1967:29.
  ISIS NUMBER: 5301403007018000000 as *Sorex.*
            5301403007009000000 as *Microsorex.*

Sorex alaskanus Merriam, 1900. Proc. Wash. Acad. Sci., 2:18.
  TYPE LOCALITY: U.S.A., Alaska, Glacier Bay, Point Gustavus.
  DISTRIBUTION: Known only from the type locality.
  COMMENT: Perhaps a subspecies of *palustris*; see Hall, 1981:43.
  ISIS NUMBER: 5301403007018001001.

Sorex alpinus Schinz, 1837. Neue Denkschr. Allgem. Schweiz. Gesell. Naturwiss.
    Neuchatel, 1:13.
  REVIEWED BY: M. Andera (MA).
  TYPE LOCALITY: Switzerland, Uri Canton, St. Gothard Pass.
  DISTRIBUTION: Montane forests of central Europe; including Pyrenees, Balkans,
    Carpathians, Tatra, Sudeten, Harz, and Jura Mtns.
  ISIS NUMBER: 5301403007018002001.

Sorex araneus Linnaeus, 1758. Syst. Nat., 10th ed., 1:53.
  REVIEWED BY: M. Andera (MA).
  TYPE LOCALITY: Sweden, Uppsala.
  DISTRIBUTION: W. Europe (excluding Iberia and Ireland) to Lake Baikal and Yenesei
    River (U.S.S.R.).
  COMMENT: Referred to as "chromosome type B", or "NF 40 karyospecies"; see Meylan
    and Hausser, 1978, Mammalia, 42:115–122. See also comments under *coronatus*
    and *granarius*. Does not include *gemellus*; see Hausser, 1978, Mammalia,
    42:329–341, who included *gemellus* in *coronatus*. Includes *Blarina pyrrhonota*
    Jentink, 1910, which was assigned to *Cryptotis surinamensis* by Cabrera, 1958:47;
    however, Husson, 1963, Studies on the Fauna of Surinam and other Guyanas,
    13:35–37, showed that the locality information was incorrect and that it is *Sorex
    araneus.*
  ISIS NUMBER: 5301403007018003001 as *S. araneus.*
            5301403007006022001 as *Cryptotis surinamensis.*

Sorex arcticus Kerr, 1792. Anim. Kingdom, p. 206.
  TYPE LOCALITY: Canada, Ontario, settlement at mouth of Severn River (now Fort
    Severn), Hudson Bay.
  DISTRIBUTION: W.C. Yukon and S.W. Northwest Terr. to S. Quebec, Nova Scotia, and S.
    New Brunswick (Canada); N. North Dakota, South Dakota, Minnesota, and
    Wisconsin (U.S.A.).
  COMMENT: Palearctic species currently referred to *arcticus* (Gromov and Baranova,
    1981:18) have been referred to *tundrensis* by Junge *et al.,* 1981, Abst. Tech. Pap.,
    61st Ann. Meet., Am. Soc. Mamm., No. 45; see also Sokolov and Orlov, 1980:47,
    Junge and Hoffmann, 1981:42.
  ISIS NUMBER: 5301403007018004001.

Sorex arizonae Diersing and Hoffmeister, 1977. J. Mammal., 58:329.
  TYPE LOCALITY: U.S.A., Arizona, Cochise Co., upper end Miller Canyon, 10 mi. (16 km) S., 4.75 mi. (7.6 km) E. Fort Huachuca, near spring at lower edge of Douglas Fir zone, Huachuca Mtns.
  DISTRIBUTION: Mtns. of S.E. Arizona, and S.W. New Mexico (U.S.A.); Chihuahua (Mexico)
  COMMENT: See Conway and Schmitt, 1978, J. Mammal., 59:631 for discussion of U.S. distribution and Caire, *et al.*, 1978, Southwest. Nat., 23:532–533 for discussion of Mexican distribution.

Sorex asper Thomas, 1914. Ann. Mag. Nat. Hist., ser. 8, 13:565.
  TYPE LOCALITY: China or U.S.S.R., Tien Shan Mtns., Tekes Valley.
  DISTRIBUTION: Tien Shan Mtns. in Kirgizia (U.S.S.R.) and adjacent Sinkiang (China).
  COMMENT: May include *excelsus*; see Corbet, 1978:21.

Sorex bedfordiae Thomas, 1911. Abstr. Proc. Zool. Soc. Lond., 90:30.
  TYPE LOCALITY: China, Szechwan, Omisan, 9500 ft. (2896 m).
  DISTRIBUTION: Montane forests of S. Kansu and W. Shensi to Yunnan (China); adjacent Burma and Nepal.
  COMMENT: Includes *gomphus, nepalensis,* and *wardii;* formerly considered a subspecies of *cylindricauda;* see Corbet, 1978:24.

Sorex bendirii (Merriam, 1884). Trans. Linn. Soc. New York, 2:217.
  TYPE LOCALITY: U.S.A., Oregon, Klamath Co., 1 mi. (1.6 km) from Williamson River, 18 mi. (29 km) S.E. Fort Klamath.
  DISTRIBUTION: Coastal areas of N.W. California to Washington (U.S.A.); Port Moody, Brit. Col. (Canada).
  COMMENT: Reviewed by Pattie, 1973, Mamm. Species, 27:1–2.
  ISIS NUMBER: 5301403007018005001.

Sorex buchariensis Ognev, 1921. Ann. Mus. Zool. Acad. Sci. St. Petersb., 22:320.
  TYPE LOCALITY: U.S.S.R., River Davan-su, N.W. Russian Pamir Mtns., Peter the Great Ridge.
  DISTRIBUTION: Pamir Mts.; Tadzhikistan. Probably N. Afghanistan and N. Pakistan (OLR).
  COMMENT: Considered a subspecies of *thibetanus* by Dolgov and Hoffmann, 1977, Zool. Zh., 56:1687–1692. Retained by Hutterer, 1979, Z. Saugetierk., 44:65–80, who considered *thibetanus* a *nomen dubium.*
  ISIS NUMBER: 5301403007018006001.

Sorex caecutiens Laxmann, 1788. Nova Acta Acad. Sci. Petrop., 1785, 3:285.
  REVIEWED BY: M. Andera (MA).
  TYPE LOCALITY: U.S.S.R., Siberia, Lake Baikal (neighborhood of Irkutsk).
  DISTRIBUTION: Taiga and tundra zones from E. Europe to E. Siberia, south to C. Ukraine, N. Kazakhstan, Altai Mts., C. Mongolia, Kansu (China) and N.E. China, to Korea; Sakhalin; Hokkaido, N. Honshu, and Shikoku (Japan).
  COMMENT: Includes *cansulus, macropygmaeus, buxtoni,* and *centralis;* see Corbet, 1978:20. However *cansulus* may not be conspecific (JAJ and RSH). Corbet, 1978:20 included *shinto* in this species (see comment under *shinto).*
  ISIS NUMBER: 5301403007018007001.

Sorex caucasicus Satunin, 1913. Trud. Obshch. Izuch. Chernomorsk. Poberezh., 2:24.
  TYPE LOCALITY: U.S.S.R., Georgian S.S.R., Tbilisi, Bakuryani.
  DISTRIBUTION: Caucasus; N. Turkey.

Sorex cinereus Kerr, 1792. Anim. Kingdom, p. 206.
  TYPE LOCALITY: Canada, Ontario, Ft. Severn, at mouth of Severn River.
  DISTRIBUTION: Anadyr to Kamchatka and Kurile Isls. (Paramushir) (N.E. Siberia); N. America throughout Alaska and Canada and southward along the Rocky and Appalachian Mtns. to 45 degrees.
  COMMENT: Includes *beringianus;* see Okhotina, 1977, Acta Theriol. 22:191–206, and Corbet, 1978:22. Does not include *haydeni, fontinalis,* and *jacksoni;* see Junge and

Hoffmann, 1981, Occ. Pap. Mus. Nat. Hist. Univ. Kansas 94:1–48, and references cited therein. Gureev, 1979:349, listed *beringianus* and *jacksoni* as distinct species without comment.

ISIS NUMBER: 5301403007018008001.

Sorex coronatus Miller, 1828. Faune de Maine-et-Loire. I. Rosier, Paris, p. 18.
TYPE LOCALITY: France, Maine-et-Loire, Blou.
DISTRIBUTION: N. Spain to W. West Germany, W. Austria, Liechtenstein, N. Switzerland; Jersey (Channel Isls.).
COMMENT: Considered a distinct species by Meylan and Hausser, 1978, Mammalia, 42:115. Includes *gemellus*; referred to as "chromosome type A" or "NF 44 karyospecies"; see Hausser, 1978, Mammalia, 42:329–341.

Sorex cylindricauda Milne-Edwards, 1872. Nouv. Arch. Mus. Hist. Nat. Paris, Bull., 7:92.
TYPE LOCALITY: China, W. Szechwan, Moupin.
DISTRIBUTION: Montane forests of N. Szechwan.
COMMENT: Formerly included *bedfordiae*; see Corbet, 1978:24.
ISIS NUMBER: 5301403007018009001.

Sorex daphaenodon Thomas, 1907. Proc. Zool. Soc. Lond., 1907:407.
TYPE LOCALITY: U.S.S.R., Sakhalin Isl., Darine, 25 mi. (40 km) N.W. Korsakoff.
DISTRIBUTION: Urals to the Kolyma River (Siberia); Sakhalin; Paramushir Isl. (N. Kuriles); N. Mongolia; N.E. China.
ISIS NUMBER: 5301403007018010001.

Sorex dispar Batchelder, 1911. Proc. Biol. Soc. Wash., 24:97.
TYPE LOCALITY: U.S.A., New York, Essex Co., Beedes (Keene Hts.).
DISTRIBUTION: Appalachian Mtns. from North Carolina to Maine; S. New Brunswick (Canada).
COMMENT: For comparison with *gaspensis*, see Kirkland and Van Deusen, 1979, Am. Mus. Novit., 2765:1–21. Reviewed by Kirkland, 1981, Mamm. Species, 155:1–4.
ISIS NUMBER: 5301403007018011001.

Sorex emarginatus Jackson, 1925. Proc. Biol. Soc. Wash., 38:129.
TYPE LOCALITY: Mexico, Jalisco, Sierra Madre, near Bolanos, 7,600 ft. (2316 m).
DISTRIBUTION: Durango, Zacatecas, and Jalisco (Mexico).
COMMENT: Findley, 1955, Univ. Kans. Publ. Mus. Nat. Hist., 9:1–68 considered this a subspecies of *oreopolus*. However, it is assignable to the subgenus *Sorex* whereas *oreopolus* belongs to the subgenus *Otisorex*; see Diersing and Hoffmeister, 1977, J. Mammal., 58:313. See comments under *oreopolus* and *ventralis*.

Sorex excelsus G. M. Allen, 1923. Am. Mus. Novit., 100:4.
TYPE LOCALITY: China, W. Yunnan, summit of Ho Shan, Peitei, 30 mi. (48 km) S. Chungtien, 13,000 ft. (3962 m).
DISTRIBUTION: Yunnan, (China).
COMMENT: Possibly a subspecies of *asper*; see Corbet, 1978:21.

Sorex fontinalis Hollister, 1911. Proc. U.S. Nat. Mus., 40:378.
TYPE LOCALITY: U.S.A. Maryland, Prince Georges Co., Cold Spring Swamp.
DISTRIBUTION: S. Pennsylvania, Delaware, Maryland, N.E. Virginia (U.S.A.).
COMMENT: Separated from *cinereus* by Junge and Hoffmann, 1981:38; see also Kirkland, 1977, Proc. Penn. Acad. Sci., 51:43–46.

Sorex fumeus Miller, 1895. N. Am. Fauna, 10:50.
TYPE LOCALITY: U.S.A., New York, Madison Co., Peterboro.
DISTRIBUTION: S. Ontario, S. Quebec, New Brunswick, Nova Scotia (Canada); all of New England and Appalachian Mtns. and adjacent areas to N.E. Georgia (U.S.A.).
ISIS NUMBER: 5301403007018012001.

Sorex gaspensis Anthony and Goodwin, 1924. Am. Mus. Novit., 109:1.
TYPE LOCALITY: Canada, Quebec, Gaspe Peninsula, Mt. Albert, 2000 ft. (610 m).
DISTRIBUTION: Gaspe Peninsula, New Brunswick, Nova Scotia, Cape Breton Isl. (Canada).

COMMENT: For comparison with *dispar*, see Kirkland and Van Deusen, 1979, Am. Mus.
Novit., 2765:1–21. Reviewed by Kirkland, 1981, Mamm. Species, 155:1–4.
ISIS NUMBER: 5301403007018013001.

Sorex gracillimus Thomas, 1907. Proc. Zool. Soc. Lond., 1907:408.
TYPE LOCALITY: U.S.S.R., Siberia, Sakhalin Isl., Darine, 25 mi. (40 km) N.W. Korsakoff.
DISTRIBUTION: Southern shore of the Sea of Okhotsk to N. Korea; Sakhalin Isl.;
Hokkaido (Japan); Kurile Isls. (U.S.S.R.); N.E. China.

Sorex granarius Miller, 1910. Ann. Mag. Nat. Hist., ser. 8, 6:458.
TYPE LOCALITY: Spain, Segovia, La Granja.
DISTRIBUTION: Pyrenees; N.W. half of Iberian Peninsula.
COMMENT: Afforded specific rank by Hausser, *et al.*, 1975, Bull. Soc. Vaudoise. Sci.
Nat., 72(348):241–252.

Sorex haydeni Baird, 1858. Mammals, *in* Repts. Expl. Surv...., 8(1):29.
TYPE LOCALITY: U.S.A., "Nebraska," Fort Union (later Fort Buford, now Mondak,
Montana, near Buford, Williams Co., North Dakota.)
DISTRIBUTION: S.E. Alberta, S. Saskatchewan, S.W. Manitoba (Canada); N.W. Montana
southeast to N.C. Kansas, east to western and southern Minnesota (U.S.A.).
COMMENT: Separated from *cinereus* by Junge and Hoffmann, 1981:36; see also Van Zyll
de Jong, 1980, J. Mammal., 61:66–75.

Sorex hosonoi Imaizumi, 1954. Bull. Nat. Sci. Mus. Tokyo, 35:94.
TYPE LOCALITY: Japan, Honshu, Nagano Pref., Tokiwa Mura, Kita-Azumi Gun.
DISTRIBUTION: Montane forests of C. Honshu (Japan).

Sorex hoyi Baird, 1858. Mammals, *in* Repts. Expl. Surv...., 8(1):32.
TYPE LOCALITY: U.S.A., Wisconsin, Racine.
DISTRIBUTION: Alaska; Canada (except Keewatin and Newfoundland) to C. Colorado in
Rocky Mtns.; E. U.S.A. to N. Georgia.
COMMENT: Formerly included in *Microsorex*; includes *thompsoni*; see Diersing, 1980, J.
Mammal., 61:76. Reviewed by Long, 1974, Mamm. Species, 33:1–4.
ISIS NUMBER: 5301403007009001001 as *Microsorex hoyi*.
5301403007009002001 as *Microsorex thompsoni*.

Sorex hydrodromus Dobson, 1889. Ann. Mag. Nat. Hist., ser. 6, 4:373.
TYPE LOCALITY: U.S.A., Alaska, "Unalaska Isl."
DISTRIBUTION: Known only from St. Paul Isl., Pribilof Isls.
COMMENT: Includes *pribilofensis*; see Hoffmann and Peterson, 1967, Syst. Zool., 16:131;
Hall, 1981:30; Junge and Hoffmann, 1981:36. Gureev, 1979:365, listed *pribilofensis*
as a distinct species without comment.
ISIS NUMBER: 5301403007018015001 as *S. hydrodromus*.
5301403007018031001 as *S. pribilofensis*.

Sorex isodon Turov, 1924. C. R. Acad. Sci. Paris, p. 111.
TYPE LOCALITY: U.S.S.R., Buryat-Mongolsk A.S.S.R., Barguzin Ridge, N.E. corner of
Lake Baikal, Sosnovka.
DISTRIBUTION: S.E. Norway and Finland through Siberia to the Pacific coast;
Kamchatka; Sakhalin Isl.; Kurile Isls.; N.E. China (SW).
COMMENT: Possibly conspecific with *sinalis*; see Corbet, 1978:22, who included *gravesi*
and *ruthenus* with *sinalis*. Dolgov, 1964, Zool. Zh., 43:893–903; Siivonen, 1969,
Aquilo Ser. Zool., 7:42–49; and Kozlovskii and Orlov, 1971, Zool. Zh.,
50:1056–1062 provided evidence of specific distinctness and validity of the name
isodon.

Sorex jacksoni Hall and Gilmore, 1932. Univ. Calif. Publ. Zool., 38:392.
TYPE LOCALITY: U.S.A., St. Lawrence Isl., Sevoonga, 2 mi. E. North cape.
DISTRIBUTION: Known only from St. Lawrence Isl. (Bering Sea).
COMMENT: Separated from *cinereus* by Junge and Hoffmann, 1981:36. Placed in *arcticus*
species group by Hall and Gilmore, 1932, Univ. Calif. Publ. Zool., 38:392; in
cinereus species group by Hoffmann and Peterson, 1967, Syst. Zool., 16:127–136.
ISIS NUMBER: 5301403007018016001.

Sorex juncensis Nelson and Goldman, 1909. Proc. Biol. Soc. Wash., 22:27.
TYPE LOCALITY: Mexico, Baja California, Socorro, 15 mi. (24 km) S. San Quintin.
DISTRIBUTION: Known only from the type locality.
COMMENT: Very similar to, and probably conspecific with, *ornatus;* see Junge and
Hoffmann, 1981:34.
ISIS NUMBER: 5301403007018017001.

Sorex kozlovi Stroganov, 1952. Byull. Mosk. Ova. Ispyt. Prir. Otd. Biol., 57:21.
TYPE LOCALITY: China, E. Tibet, Dze-Chyu River.
DISTRIBUTION: Known only from the type locality.
COMMENT: Known from a single specimen. Regarded as conspecific with *thibetanus* by
Dolgov and Hoffmann, 1977, Zool. Zh., 56:1687–1692, and with *buchariensis* by
Corbet, 1978:23, but its specific status retained by Hutterer, 1979, Z. Saugetierk.,
44:65–80, and Gureev, 1979:361.

Sorex longirostris Bachman, 1837. J. Acad. Nat. Sci. Phila., 7:370.
TYPE LOCALITY: U.S.A., South Carolina, Hume Plantation, swamps of Santee River (=
Cat Isl., mouth of Santee River).
DISTRIBUTION: S.E. U.S.A. (except S. Florida) west to C. Louisiana, N.W. Arkansas, S.W.
and N.E. Missouri, S. and E. Illinois, and W. and S. Indiana.
COMMENT: Reviewed by French, 1980, Mamm. Species, 143:1–3.
ISIS NUMBER: 5301403007018018001.

Sorex lyelli Merriam, 1902. Proc. Biol. Soc. Wash., 15:75.
TYPE LOCALITY: U.S.A., California, Tuolumne Co., Mt. Lyell.
DISTRIBUTION: C. Sierra Nevada, California (U.S.A.).
ISIS NUMBER: 5301403007018019001.

Sorex macrodon Merriam, 1895. N. Am. Fauna, 10:82.
TYPE LOCALITY: Mexico, Veracruz, Orizaba, 4200 ft. (1280 m).
DISTRIBUTION: Veracruz, in mountains from 4000–9500 ft. (1676–2896 m) and Puebla
(Mexico); see Heaney and Birney, 1977 (1976), Southwest. Nat., 21:543–545.
COMMENT: Similar to, and probably conspecific with, *veraepacis;* see Junge and
Hoffmann, 1981:43. See Heaney and Birney, 1977, Southwest. Nat., 21:543–545 for
discussion of distribution.
ISIS NUMBER: 5301403007018020001.

Sorex merriami Dobson, 1890. A Monogr. of the Insectivora - Systematic and Anatomical.
Part 3 (Soricidae), Fasc. l, pl. 23.
TYPE LOCALITY: U.S.A. Montana, Bighorn Co., Little Bighorn River, ca. l mi. (1.6 km)
above Ft. Custer(=Hardin).
DISTRIBUTION: E.C. Washington to N. and E. California to S. and C. Arizona,
northeastward to W. Nebraska, E. Wyoming and Montana; perhaps W. North
Dakota (U.S.A.).
COMMENT: Reviewed by Armstrong and Jones, 1971, Mamm. Species, 2:1–2.
ISIS NUMBER: 5301403007018021001.

Sorex milleri Jackson, 1947. Proc. Biol. Soc. Wash., 60:131.
TYPE LOCALITY: Mexico, Coahuila, Sierra del Carmen, Madera Camp, 8000 ft. (2438 m).
DISTRIBUTION: Coahuila, Nuevo Leon (Mexico).
ISIS NUMBER: 5301403007018022001.   ·

Sorex minutissimus Zimmermann, 1780. Geogr. Gesch. Mensch. Vierf. Thiere, 2:385.
REVIEWED BY: M. Andera (MA).
TYPE LOCALITY: U.S.S.R., Kemerovsk. Obl., Mariinsk(=Kiiskoe), bank of Kiia River
(near Yenesei R.).
DISTRIBUTION: Taiga zone from Norway, Sweden and Estonia to E. Siberia; Sakhalin;
Hokkaido (Japan); Mongolia; Szechwan (China); perhaps Honshu (Japan).
COMMENT: Includes *hawkeri;* see Corbet, 1978:19.
ISIS NUMBER: 5301403007018014001 as *S. hawkeri.*

Sorex minutus Linnaeus, 1766. Syst. Nat., 12th ed., 1:73.
REVIEWED BY: M. Andera (MA).
TYPE LOCALITY: U.S.S.R., Altaisky Krai, near Barnaul.
DISTRIBUTION: W. Europe to Yenesei River and Lake Baikal, south through Altai and
Tien Shan Mtns.; west to Nepal, N. China (SW).
COMMENT: An isolated population in the Tien Shan Mtns., *heptopotamicus*, is
considered a distinct species by some authors (JAJ and RSH). Formerly included
gracillimus, which is regarded as specifically distinct by all Russian authors; see
Corbet, 1978:19–20, who also included *planiceps* and *thibetanus;* but see Dolgov
and Hoffmann, 1977, Zool. Zh., 56:16871692, and Hutterer, 1979, Z. Saugetierk.,
44:65–80. *S. volnuchini* may represent a distinct species based on its karyotype; see
Orlov and Kozlovsky, 1971, Vestn. Moscow Univ., Biol., 2:12–16.
ISIS NUMBER: 5301403007018023001.

Sorex mirabilis Ognev, 1937. Byull. Mosk. Ova. Ispyt. Prir. Otd. Biol., 46(5):268.
TYPE LOCALITY: U.S.S.R., Primorsky Krai, Kishinka River.
DISTRIBUTION: N. Korea, N.E. China, Ussuri region (U.S.S.R.).
COMMENT: Placed in monotypic subgenus *Ognevia* by Heptner and Dolgov, 1967, Zool.
Zh., 56:1419–1422. Not conspecific with *pacificus;* see Hoffmann, 1971, Z.
Saugetierk., 36:193–200.

Sorex monticolus Merriam, 1890. N. Am. Fauna, 3:43.
TYPE LOCALITY: U.S.A., Arizona, Coconino Co., San Francisco Mtn., 11500 ft. (3505 m).
DISTRIBUTION: Montane boreal and coastal coniferous forest and alpine areas from
Alaska to California and New Mexico, east to Montana, Wyoming, and Colorado
(U.S.A.) and to W. Manitoba (Canada); Chihuahua, Durango (Mexico).
COMMENT: Includes *obscurus* and *durangae*, which were previously included in *vagrans*
and *saussurei* respectively; see Hennings and Hoffmann, 1977, Occas. Pap. Mus.
Nat. Hist. Univ. Kansas, 68:1–35.
ISIS NUMBER: 5301403007018025001 as *S. obscurus.*

Sorex nanus Merriam, 1895. N. Am. Fauna, 10:81.
TYPE LOCALITY: U.S.A., Colorado, Larimer Co., Estes Park.
DISTRIBUTION: Rocky Mtns. from Montana to New Mexico; South Dakota; Arizona
(U.S.A.).
COMMENT: Very similar to, and perhaps conspecific with, *tenellus;* see review by
Hoffmann and Owen, 1980, Mamm. Species, 131:1–4.
ISIS NUMBER: 5301403007018024001.

Sorex oreopolus Merriam, 1892. Proc. Biol. Soc. Wash., 7:173.
TYPE LOCALITY: Mexico, Jalisco, N. slope Sierra Nevada de Colima, ca. 10000 ft. (3048
m).
DISTRIBUTION: Jalisco (Mexico); if including *S. vagrans orizabae*, then east to Puebla and
Veracruz (Mexico) (JAJ and RSH).
COMMENT: Contrary to Findley, 1955, Univ. Kans. Publ. Mus. Nat. Hist., 7:613–618,
this species does not include *emarginatus* or *ventralis*, which are in the subgenus
Sorex (Diersing and Hoffmeister, 1977, J. Mammal., 58:331). Subgenera
characterized by Diersing, 1980, J. Mammal., 61:76. *S. vagrans orizabae* may prove
to be a junior synonym of *oreopolus;* see Junge and Hoffmann, 1981:43.
ISIS NUMBER: 5301403007018026001.

Sorex ornatus Merriam, 1895. N. Am. Fauna, 10:79.
TYPE LOCALITY: U.S.A., California, Kern Co., San Emigdio Canyon, Mt. Pinos.
DISTRIBUTION: California coastal ranges from north of San Francisco Bay to N. Baja
California; tip of Baja California.
COMMENT: Includes *willeti;* see Von Bloeker, 1967, Proc. Symp. on the Biol. of the
Calif. Isls., p. 247. Probably includes *juncensis* and *sinuosus;* see Hall, 1981:40;
Junge and Hoffmann, 1981:34.
ISIS NUMBER: 5301403007018027001.

Sorex pacificus Coues, 1877. Bull. U.S. Geol. Geogr. Surv. Terr., 3(3):650.
TYPE LOCALITY: U.S.A., Oregon, Douglas Co., Ft. Umpqua.
DISTRIBUTION: Forests of coastal Oregon and California, south to San Francisco Bay
    (U.S.A.).
COMMENT: Not conspecific with *mirabilis*; see Hoffmann, 1971, Z. Saugetierk.,
    36:193–200. May prove to be conspecific with *monticolus*; see Findley, 1955, Univ.
    Kansas Publ. Mus. Nat. Hist., 9:1–68; Junge and Hoffmann, 1981:28–31.
ISIS NUMBER: 5301403007018028001.

Sorex palustris Richardson, 1828. Zool. J., 3:517.
TYPE LOCALITY: Canada, "from marshy places from Hudson's Bay to the Rocky Mtns."
DISTRIBUTION: Montane and boreal areas of North America below the tree line from
    Alaska to the Sierra Nevada, Rocky, and Appalachian mtns.
COMMENT: Hall, 1981:43 noted that *alaskanus* may be a subspecies of *palustris*.
ISIS NUMBER: 5301403007018029001.

Sorex planiceps Miller, 1911. Proc. Biol. Soc. Wash., 24:242.
TYPE LOCALITY: India, Kashmir, Kuistwar, Dachin, 9000 ft. (2743 m).
DISTRIBUTION: Kashmir and N. Pakistan.
COMMENT: Considered a subspecies of *thibetanus* by Dolgov and Hoffmann, 1977, Zool.
    Zh., 56:1687–1692. Retained by Hutterer, 1979, Z. Saugetierk., 44:65–80, who
    considered *thibetanus* a *nomen dubium*.

Sorex preblei Jackson, 1922. J. Wash. Acad. Sci., 12:263.
TYPE LOCALITY: U.S.A., Oregon, Malheur Co., Jordan Valley, 4200 ft. (1280 m).
DISTRIBUTION: Columbia Plateau of Washington, Oregon and Nevada to W. Great
    Plains of Montana (U.S.A.).
ISIS NUMBER: 5301403007018030001.

Sorex raddei Satunin, 1895. Arch. Naturgesch., 1:109.
TYPE LOCALITY: U.S.S.R., Georgian S.S.R., near Kutais.
DISTRIBUTION: Transcaucasia; N. Turkey.
COMMENT: Includes *batis*; see Corbet, 1978:21.

Sorex roboratus Hollister, 1913. Smithson. Misc. Coll., 60(24):2.
TYPE LOCALITY: U.S.S.R., Siberia, Altai Mtns., Tapucha.
DISTRIBUTION: Altai Mtns.
COMMENT: If this species is conspecific with *vir*, the name *roboratus* apparently has
    priority; see Yudin, 1971, [Insectivorous Mammals of Siberia], Nauka,
    Novosibirsk, p. 105. Considered a subspecies of *araneus* by Gromov and Baranova,
    1981:17.

Sorex samniticus Altobello, 1926. Bol. Inst. Zool. Univ. Roma, 3:102.
TYPE LOCALITY: Italy, Campobasso Prov., Molise, 700 m.
DISTRIBUTION: Italy.
COMMENT: Formerly included in *araneus*; considered a distinct species by Graf *et al.*,
    1979, Bonn. Zool. Beitr., 30:14.

Sorex saussurei Merriam, 1892. Proc. Biol. Soc. Wash., 7:173.
TYPE LOCALITY: Mexico, Jalisco, N. slope Sierra Nevada de Colima, ca. 8000 ft. (2438
    m).
DISTRIBUTION: Coahuila and Durango to Chiapas (Mexico); Guatemala.
ISIS NUMBER: 5301403007018032001.

Sorex sclateri Merriam, 1897. Proc. Biol. Soc. Wash., 11:228.
TYPE LOCALITY: Mexico, Chiapas, Tumbala, 5000 ft. (1524 m).
DISTRIBUTION: Known only from the type locality.
ISIS NUMBER: 5301403007018033001.

Sorex shinto Thomas, 1906. Proc. Zool. Soc. Lond., 1905:338.
TYPE LOCALITY: Japan, N. Hondo, Aomori Ken, Makado.
DISTRIBUTION: Honshu, Shikoku, Hokkaido (Japan); Kunashir Isl., Sakhalin (U.S.S.R.).
COMMENT: Included in *caecutiens* by Corbet, 1978:20, but Imaizumi, 1970, The
    Handbook of Japanese Land Mammals, Vol. 1, treated *shinto* as a separate species.

Sorex sinalis Thomas, 1912. Ann. Mag. Nat. Hist., 10:398.
REVIEWED BY: M. Andera (MA).
TYPE LOCALITY: China, Shensi, 45 mi. (72 km) S.E. Feng-hsiang-fu, 10,500 ft. (3200 m).
DISTRIBUTION: C. China.
COMMENT: May be conspecific with *isodon*; see Corbet, 1978:22, for discussion.

Sorex sinuosus Grinnell, 1913. Univ. Calif. Publ. Zool., 10:187.
TYPE LOCALITY: U.S.A., California, Solano Co., Grizzly Isl. (near Suisun).
DISTRIBUTION: Grizzly Isl. and adjacent tidal marshes of north San Francisco Bay.
COMMENT: May prove to be a subspecies of *ornatus*; see Hall, 1981:40; Junge and
Hoffmann, 1981:34.
ISIS NUMBER: 5301403007018034001.

Sorex stizodon Merriam, 1895. N. Am. Fauna, 10:98.
TYPE LOCALITY: Mexico, Chiapas, San Cristobal, 9000 ft. (2743 m).
DISTRIBUTION: Known only from the type locality.
COMMENT: Similar to *ventralis*; see Junge and Hoffmann, 1981:45.
ISIS NUMBER: 5301403007018035001.

Sorex tenellus Merriam, 1895. N. Am. Fauna, 10:81.
TYPE LOCALITY: U.S.A., California, Inyo Co., along Lone Pine Creek at upper edge of
Alabama Hills at ca. 5000 ft. (1524 m) (near Lone Pine, Owens Valley).
DISTRIBUTION: Mountains of W.C. Nevada and E.C. California (U.S.A.).
COMMENT: Similar to, and perhaps conspecific with, *nanus*; see review by Hoffmann
and Owen, 1980, Mamm. Species, 131:1–4.
ISIS NUMBER: 5301403007018036001.

Sorex trowbridgii Baird, 1858. Mammals, *in* Repts. Expl. Surv...., 8(1):13.
TYPE LOCALITY: U.S.A., Oregon, Clatsop Co., Astoria, mouth of the Columbia River.
DISTRIBUTION: Coastal ranges from Washington to California (U.S.A.); S.W. British
Columbia (Canada).
ISIS NUMBER: 5301403007018038001.

Sorex tundrensis Merriam, 1900. Proc. Wash. Acad. Sci., 2:16.
TYPE LOCALITY: U.S.A., Alaska, St. Michael.
DISTRIBUTION: Siberia, from the Pechora R. to Chukotka, south to the Altai Mtns.;
Mongolia and N.E. China; Alaska (U.S.A.); Yukon, N.W. Northwest Territories
(Canada).
COMMENT: Youngman, 1975, Mammals of the Yukon Terr., Nat. Mus. Can. Publ. Zool.,
10, provided evidence that *tundrensis* is specifically distinct from *arcticus*.
Palearctic populations formerly referred to *arcticus* were referred to *tundrensis* by
Junge *et al.*, 1981, Abst. Tech. Pap., 61st Ann. Meet., Am. Soc. Mamm., No. 45; see
comments under *arcticus*. Kozlovskii, 1976, Zool. Zh., 50:756–762 found *irkutensis*
and *sibiriensis* to be karyotypically distinct.

Sorex unguiculatus Dobson, 1890. Ann. Mag. Nat. Hist., 5:115.
TYPE LOCALITY: U.S.S.R., Sakhalin.
DISTRIBUTION: U.S.S.R. Pacific coast, from S. Khabarovsky Krai to Primorsky Krai;
Sakhalin; Kurile Isls. (U.S.S.R.); N.E. China; N. Korea; Hokkaido (Japan).

Sorex vagrans Baird, 1858. Mammals, *in* Repts. Expl. Surv...., 8(1):15.
TYPE LOCALITY: U.S.A., Washington, Pacific Co., Shoalwater Bay.
DISTRIBUTION: Riparian and montane areas of the northern Great Basin and Columbia
Plateau, north to S. British Columbia and Vancouver Isl. (Canada); east to W.
Montana, W. Wyoming, Wasatch Mtns. (N. Utah); C. Nevada to Sierra Nevadas,
(Calif.) and south to San Francisco Bay (U.S.A.); possibly Transvolcanic Belt
(=Transverse Volcanic Belt) in Mexico.
COMMENT: Includes *trigonirostris*; see Hennings and Hoffmann, 1977, Occas. Pap. Mus.
Nat. Hist. Univ. Kansas, 68:8. *S. v. orizabae* of Mexico may be conspecific with *S.
oreopolus*; see comments under *oreopolus*, and Junge and Hoffmann, 1981:43.
ISIS NUMBER: 5301403007018039001 as *S. vagrans*.
5301403007018037001 as *S. trigonirostris*.

Sorex ventralis Merriam, 1895. N. Am. Fauna, 10:75.
TYPE LOCALITY: Mexico, Oaxaca, Cerro San Felipe, 1,000 ft. (305 m).
DISTRIBUTION: N.W. Puebla to Oaxaca (Mexico).
COMMENT: Considered a subspecies of *oreopolus* by Findley, 1955, Misc. Publ. Univ.
Kans. Mus. Nat. Hist., 7:613–618, but *ventralis* is in the subgenus *Sorex* and
oreopolus is in the subgenus *Otisorex;* see Diersing and Hoffmeister, 1977, J.
Mammal., 58:331. Similar to *saussurei* but smaller; see Junge and Hoffmann,
1981:44. See also comments under *oreopolus* and *emarginatus.*

Sorex veraepacis Alston, 1877. Proc. Zool. Soc. Lond., 1877:445.
TYPE LOCALITY: Guatemala, Verapaz, Alta Verapaz, Coban.
DISTRIBUTION: Chiapas, Oaxaca, Guerrero (Mexico); Guatemala.
ISIS NUMBER: 5301403007018040001.

Sorex vir G. M. Allen, 1914. Proc. N. Engl. Zool. Club, 5:52.
TYPE LOCALITY: U.S.S.R., N.E. Yakutia, Nizhne Kolymsk (near mouth of Kolyma
River).
DISTRIBUTION: Siberia, west to Ural, perhaps south to the Altai; N. and E. Mongolia.
COMMENT: If *roboratus* and *vir* are conspecific, *vir* is apparently a junior synonym; see
Yudin, 1971, [Insectivorous Mammals of Siberia], Nauka, Novosibirsk, p. 105; but
see also Corbet, 1978:22.

Soriculus Blyth, 1854. J. Asiat. Soc. Bengal, 23:733.
REVIEWED BY: R. M. Mitchell (RMM).
COMMENT: Includes *Chodsigoa* and *Episoriculus;* see Corbet, 1978:24; but also see
Repenning, 1967:52, and Jameson and Jones, 1977, Proc. Biol. Soc. Wash.,
90:474–475, who considered *Episoriculus* a distinct genus. Subfamily Soricinae,
tribe Neomyini; see Repenning, 1967:45. Gureev, 1979:450–452, listed *Chodsigoa* as
a subgenus of *Notiosorex.*
ISIS NUMBER: 5301403007019000000.

Soriculus baileyi Thomas, 1941. J. Bombay Nat. Hist. Soc., 22:683.
TYPE LOCALITY: India, Assam, Hishmi Hills, Tsu River, 7500 ft.
DISTRIBUTION: Himalayas in Tibet (SW), Nepal; Assam (India); probably Bhutan.
COMMENT: Considered a subspecies of *caudatus* by Ellerman and Morrison-Scott,
1951:59; but also see Abe, 1971, J. Fac. Agric. Hokkaido Univ., 56:367–423, who
found *baileyi* and *caudatus* in sympatry.

Soriculus caudatus (Horsfield, 1851). Cat. Mamm. Mus. E. India Co., p. 135.
TYPE LOCALITY: India, W. Bengal, Darjeeling.
DISTRIBUTION: N. India to N. Vietnam, Szechwan and Yunnan (China).
COMMENT: Includes *soluensis;* see Gruber, 1969, *in* Hellmich, ed., Khumbu Himal.,
3(2):197–312. Subgenus *Episoriculus;* see Ellerman and Morrison-Scott, 1951:59; but
also see Repenning, 1967:47.
ISIS NUMBER: 5301403007019001001.

Soriculus fumidus Thomas, 1913. Ann. Mag. Nat. Hist., 11:216.
TYPE LOCALITY: Taiwan, Mt. Arisan, 8000 ft. (2438 m).
DISTRIBUTION: Taiwan.
COMMENT: Formerly included in *caudatus* by Ellerman and Morrison-Scott, 1951:59,
but see Jameson and Jones, 1977, Proc. Biol. Soc. Wash., 90:474, who included
sodalis in *fumidus.* Subgenus *Episoriculus;* see Ellerman and Morrison-Scott, 1951:59,
but also see Repenning, 1967:47.

Soriculus gruberi Weigel, 1969. Ergeb. Forsch. Unternehmens Nepal Himalaya, 3(2):170.
TYPE LOCALITY: Nepal, Junbesi (S. of Mt. Everest).
DISTRIBUTION: Solukhumbu region (N.E. Nepal).
COMMENT: Probably a synonym of *baileyi* (K. F. Koopman).
ISIS NUMBER: 5301403007019002001.

Soriculus hypsibius De Winton, 1899. Proc. Zool. Soc. Lond., 1899:574.
TYPE LOCALITY: China, Szechwan, Yangliupa.
DISTRIBUTION: Yunnan; Szechwan, Shensi, Kansu, Hopeh, Fukien (China).

COMMENT: Includes *parva;* see Corbet, 1978:24. Subgenus *Chodsigoa;* see Ellerman and Morrison-Scott, 1951:59; but also see Repenning, 1967:48.
ISIS NUMBER: 5301403007019003001 as *S. hypsibius.*
5301403007019007001 as *S. parva.*

Soriculus leucops (Horsfield, 1855). Ann. Mag. Nat. Hist., 16:111.
TYPE LOCALITY: Nepal.
DISTRIBUTION: N. India; Nepal; Sikkim; N. Burma; Szechwan, Yunnan (China).
COMMENT: Subgenus *Episoriculus;* see Ellerman and Morrison-Scott, 1951:59; but also see Repenning, 1967:47.
ISIS NUMBER: 5301403007019004001.

Soriculus lowei Osgood, 1932. Field Mus. Nat. Hist. Publ. Ser. Zool., 18:249.
TYPE LOCALITY: Vietnam, Tonkin, Chapa.
DISTRIBUTION: N. Vietnam.
COMMENT: Subgenus *Chodsigoa;* see Ellerman and Morrison-Scott, 1951:59; but also see Repenning, 1967:48.
ISIS NUMBER: 5301403007019005001.

Soriculus nigrescens (Gray, 1842). Ann. Mag. Nat. Hist., 10:261.
TYPE LOCALITY: India, W. Bengal, Darjeeling.
DISTRIBUTION: N. India; Sikkim; Bhutan; Nepal; Tibet (SW); N. Burma.
COMMENT: Subgenus *Soriculus;* see Ellerman and Morrison-Scott, 1951:59; Repenning, 1967:52.
ISIS NUMBER: 5301403007019006001.

Soriculus salenskii Kastschenko, 1907. Ann. Mus. Zool. Acad. Sci. St. Petersb., 10:253.
TYPE LOCALITY: China, Szechwan, Linganfu.
DISTRIBUTION: N. Burma; Yunnan, Szechwan (China); N. Thailand.
COMMENT: Subgenus *Chodsigoa;* see Ellerman and Morrison-Scott, 1951:59, but also see Repenning, 1967:48.
ISIS NUMBER: 5301403007019008001.

Soriculus smithii (Thomas, 1911). Abstr. Proc. Zool. Soc. Lond., 90:4.
TYPE LOCALITY: China, Szechwan, Tatsienlu.
DISTRIBUTION: Shensi, Szechwan (China).
COMMENT: Regarded as a separate species by Corbet, 1978:24, but as a subspecies of *salenskii* by Ellerman and Morrison-Scott, 1951:60. Subgenus *Chodsigoa;* see Ellerman and Morrison-Scott, 1951:59, but also see Repenning, 1967:48.

Suncus Ehrenberg, 1832. *In* Hemprich and Ehrenberg, Symb. Phys. Mamm., 2:k.
COMMENT: This genus has been regarded as part of *Crocidura* by Lekagul and McNeely, 1977:35, but it was retained by Corbet, 1978:30. Includes *Pachyura, Paradoxodon,* and *Plerodus;* see Meester and Lambrechts, 1971, Ann. Transvaal Mus., 27(1):1–14, who revise the South African species. Subfamily Crocidurinae; see Repenning, 1967:15.
ISIS NUMBER: 5301403007020000000.

Suncus ater Medway, 1965. J. Malay. Branch R. Asiat. Soc., 36:38.
TYPE LOCALITY: Malaysia, Sabah, Gunong (=Mt.) Kinabalu, Lumu-Lumu, 5500 ft. (1676 m).
DISTRIBUTION: Known only from the type locality.
COMMENT: Reviewed by Medway, 1977:16–17.
ISIS NUMBER: 5301403007020001001.

Suncus dayi Dobson, 1888. Ann. Mag. Nat. Hist., 1:428.
TYPE LOCALITY: India, Cochin, Trichur.
DISTRIBUTION: S. India.
ISIS NUMBER: 5301403007020002001.

Suncus etruscus (Savi, 1822). Nuovo Giorn. de Letterati, Pisa, 1:60.
  REVIEWED BY: M. Andera (MA).
  TYPE LOCALITY: Italy, Pisa.
  DISTRIBUTION: S. Europe; Caucasus; Turkmenistan; W. Asia Minor; Afghanistan; Iraq;
    Yemen; Israel; Nile Delta; Tunisia to Morocco; Ethiopia; N. Nigeria; Guinea;
    Madagascar; Comoro Isls.; India; Sri Lanka; Burma; Thailand; Yunnan
    (China)(SW); Malaya; Borneo. The West African records are doubtful (K. F.
    Koopman).
  COMMENT: Includes *coquerelii* and *madagascariensis;* see Heim de Balsac and Meester,
    1977, Part 1:6. Includes *hosei* and *malayanus;* see Medway, 1977:17. Gureev,
    1979:383, listed *coquerelii* and *madagascariensis* as distinct species without comment.
  ISIS NUMBER: 5301403007020003001 as *S. etruscus.*
                5301403007020008001 as *S. madagascariensis.*

Suncus infinitesimus (Heller, 1912). Smithson. Misc. Coll., 60(12):5.
  TYPE LOCALITY: Kenya, Laikipia Plateau, Rumruti, 7000 ft. (2134 m).
  DISTRIBUTION: South Africa to Kenya; Central African Republic; perhaps Nigeria.
  COMMENT: Includes *chriseos* and *ubanguiensis;* see Heim de Balsac and Meester, 1977,
    Part 1:6. Gureev, 1979:383, listed *chriseos* as a distinct species without comment.
  ISIS NUMBER: 5301403007020004001.

Suncus lixus (Thomas, 1898). Proc. Zool. Soc. Lond., 1897:930.
  TYPE LOCALITY: Malawi, Nyika Plateau (between 10 and 11° S. and between 33° 40' and
    34° 10' E.).
  DISTRIBUTION: Kenya; Zaire; Angola; Transvaal (South Africa); N. Botswana; Malawi;
    Zambia; Tanzania.
  COMMENT: Includes *aequatorialis* and *gratulus;* see Heim de Balsac and Meester, 1977,
    Part 1:6. Gureev, 1979:383, listed *gratulus* as a distinct species without comment.
  ISIS NUMBER: 5301403007020006001.

Suncus luzoniensis (Peters, 1870). Monatsb. Preuss. Akad. Wiss. Berlin, p. 595.
  TYPE LOCALITY: Philippine Isls., Luzon, Daraga, Albay.
  DISTRIBUTION: Luzon, Marinduque, and Cebu (Philippine Isls.).
  ISIS NUMBER: 5301403007020007001.

Suncus mertensis Kock, 1974. Senckenberg. Biol., 55:198.
  TYPE LOCALITY: Indonesia, Flores Isl., Rana Mese.
  DISTRIBUTION: Flores Isl.

Suncus murinus (Linnaeus, 1766). Syst. Nat., 12th ed., 1:74.
  TYPE LOCALITY: Indonesia, Java.
  DISTRIBUTION: Sri Lanka; India; Kashmir; Nepal; Sikkim; Java; Sulawesi and
    surrounding islands; Guam; Japan; S. China; Taiwan; Israel; Egypt to Tanzania;
    Arabia; Sinai; Iraq; Madagascar; Zanzibar; Comoro Isls.
  COMMENT: Includes *albicauda, auriculata, crassicaudus, duvernoyi, leucura, mauritiana, sacer,*
    and *geoffroyi;* see Heim de Balsac and Meester, 1977, Part 1:5. Gureev, 1979:383,
    listed *leucura* as a distinct species without comment. Much of the present
    distribution is the result of human agency (K. F. Koopman).
  ISIS NUMBER: 5301403007020009001 as *S. murinus.*
                5301403007005003001 as *Crocidura albicauda.*
                5301403007005044001 as *Crocidura geoffroyi.*
                5301403007020005001 as *S. leucura.*

Suncus occultidens (Hollister, 1913). Proc. U.S. Nat. Mus., 46:303.
  TYPE LOCALITY: Philippine Isls., Panay, Iloilo.
  DISTRIBUTION: Mindanao, Panay Isl., Negros Isl. (Philippine Isls.).
  ISIS NUMBER: 5301403007020010001.

Suncus palawanensis (Taylor, 1934). Philippine Monogr. Bur. Sci. Manila, 30:78.
  TYPE LOCALITY: Philippine Isls., Palawan, Taytay.
  DISTRIBUTION: Palawan (Philippine Isls.).
  ISIS NUMBER: 5301403007020012001.

Suncus remyi Brosset, DuBost and Heim de Balsac, 1965. Biologia Gabon., 1:170.
TYPE LOCALITY: Gabon, Makokou.
DISTRIBUTION: Belinga and Makokou (Gabon).
ISIS NUMBER: 5301403007020013001.

Suncus stoliczkanus (Anderson, 1877). J. Asiat. Soc. Bengal, 46:270.
TYPE LOCALITY: India, Bombay.
DISTRIBUTION: Pakistan; India; Nepal. Bangladesh (RGVG).
ISIS NUMBER: 5301403007020015001.

Suncus varilla (Thomas, 1895). Ann. Mag. Nat. Hist., ser. 6, 16:54.
TYPE LOCALITY: South Africa, Cape Prov., East London.
DISTRIBUTION: South Africa; N.E. Zimbabwe; Tanzania; E. Zaire; Malawi; Zambia.
COMMENT: Includes *minor, natalensis, orangiae, tulbaghensis,* and *warreni;* see Heim de
    Balsac and Meester, 1977, Part 1:6. Gureev, 1979:383, listed *orangiae* and *warreni* as
    distinct species without comment.
ISIS NUMBER: 5301403007020016001.

Sylvisorex Thomas, 1904. Abstr. Proc. Zool. Soc. Lond., 10:12.
COMMENT: This genus has been regarded as part of *Suncus* by Smithers and Tello,
    1976, Mus. Mem. Salisbury, 8:1–184, but was retained by Ansell, 1978:13.
    Subfamily Crocidurinae; see Repenning, 1967:15.
ISIS NUMBER: 5301403007022000000.

Sylvisorex granti Thomas, 1907. Ann. Mag. Nat. Hist., ser. 7, 19:118.
TYPE LOCALITY: Uganda, Ruwenzori East, Mubuku Valley, 10,000 ft. (3048 m).
DISTRIBUTION: Cameroun; Uganda; Rwanda; E. Zaire; Kenya; Tanzania.
COMMENT: Includes *camerounensis* and *inundus;* see Heim de Balsac and Meester, 1977,
    Part 1:8.
ISIS NUMBER: 5301403007022001001.

Sylvisorex johnstoni (Dobson, 1888). Proc. Zool. Soc. Lond., 1887:577.
TYPE LOCALITY: Cameroun, Rio del Rey.
DISTRIBUTION: S.W. Cameroun; Gabon; Boiko; Zaire.
ISIS NUMBER: 5301403007022002001.

Sylvisorex lunaris Thomas, 1906. Ann. Mag. Nat. Hist., ser. 7, 18:139.
TYPE LOCALITY: Uganda, Ruwenzori East, Mubuku Valley, 12,500 ft. (3810 m).
DISTRIBUTION: Medje (N.E. Zaire) to Ruwenzori East (Uganda) and Birunga Volcanoes
    (Kivu, Zaire).
COMMENT: Includes *oriundus* and *ruandae;* see Heim de Balsac and Meester, 1977, Part
    1:8. Gureev, 1979:380–381, listed *oriundus* and *ruandae* as distinct species.
ISIS NUMBER: 5301403007022003001 as *S. lunaris.*
    5301403007022006001 as *S. oriundus.*
    5301403007020014001 as *Suncus ruandae.*

Sylvisorex megalura (Jentink, 1888). Notes Leyden Mus., 10:48.
TYPE LOCALITY: Liberia, Junk River, Schieffelinsville.
DISTRIBUTION: Upper Guinea to Ethiopia; Angola and Kenya to W. Mozambique and E.
    Zimbabwe.
COMMENT: Includes *angolensis, gemmeus, infuscus, irene, phaeopus, sheppardi, sorella,* and
    sorelloides. See Heim de Balsac and Meester, 1977, Part 1:7–8. Gureev, 1979:381,
    listed *sorella* as a distinct species without comment.
ISIS NUMBER: 5301403007022004001.

Sylvisorex morio (Gray, 1862). Proc. Zool. Soc. Lond., 1862:180.
TYPE LOCALITY: Cameroun, Mt. Cameroun.
DISTRIBUTION: Cameroun; Bioko, perhaps Nigeria.
COMMENT: Includes *isabellae;* see Heim de Balsac and Meester, 1977, Part 1:8.
ISIS NUMBER: 5301403007022005001.

Sylvisorex ollula Thomas, 1913. Ann. Mag. Nat. Hist., ser. 8, 11:321.
TYPE LOCALITY: Cameroun, Bitye, Ja River, 2000 ft. (610 m).
DISTRIBUTION: S. Cameroun; W. Zaire; Gabon.

COMMENT: Possibly only subspecifically distinct from *suncoides;* see Heim de Balsac and Meester, 1977, Part 1:8.
ISIS NUMBER: 5301403007020011001 as *Suncus ollula.*

Sylvisorex suncoides Osgood, 1936. Field Mus. Nat. Hist. Publ. Ser. Zool., 20:217.
TYPE LOCALITY: Zaire, W. slope Mt. Ruwenzori, Kalongi.
DISTRIBUTION: E. Zaire; Uganda; perhaps Rwanda.
COMMENT: Probably a subspecies of *ollula;* see Heim de Balsac and Meester, 1977, Part 1:8.
ISIS NUMBER: 5301403007022008001.

Family Talpidae
REVIEWED BY: O. L. Rossolimo (OLR)(U.S.S.R.); R. G. Van Gelder (RGVG); S. Wang (SW)(China).
COMMENT: Revised by Gureev, 1979. *Desmana* and *Galemys* are sometimes placed in a separate family, Desmanidae; see Bobrinskii, *et al.,* 1965:43; Flint *et al.,* 1965:27.
ISIS NUMBER: 5301403006000000000.

Condylura Illiger, 1811. Prodr. Syst. Mamm. et Avium., p. 125.
ISIS NUMBER: 5301403006001000000.

Condylura cristata (Linnaeus, 1758). Syst. Nat., 10th ed., 1:53.
TYPE LOCALITY: U.S.A., Pennsylvania.
DISTRIBUTION: Georgia and N.W. South Carolina (U.S.A.) to Nova Scotia and Labrador (Canada); Great Lakes region to S.E. Manitoba.
COMMENT: Reviewed by Peterson and Yates, 1980, Mamm. Species, 129:1–4.
ISIS NUMBER: 5301403006001001001.

Desmana Guldenstaedt, 1777. Besehaft. Berlin Ges. Naturforsch Fr., 3:108.
ISIS NUMBER: 5301403006002000000.

Desmana moschata (Linnaeus, 1758). Syst. Nat., 10th ed., 1:59.
TYPE LOCALITY: U.S.S.R., "Russia aquosis."
DISTRIBUTION: U.S.S.R.; Don, Volga, and S. Ural rivers and their tributaries; introduced into Tachan and Tartas rivers (Ob basin) and Dnepr River.
ISIS NUMBER: 5301403006002001001.

Galemys Kaup, 1829. Skizz. Europ. Thierwelt, 1:119.
ISIS NUMBER: 5301403006004000000.

Galemys pyrenaicus (E. Geoffroy, 1811). Ann. Mus. Hist. Nat. Paris, 17:193.
TYPE LOCALITY: France, Hautes-Pyrenees, near Tarbes.
DISTRIBUTION: Rivers of Pyrenees Mtns. of Spain and France; N.W. Spain and Portugal.
ISIS NUMBER: 5301403006004001001.

Mogera Pomel, 1848. Arch. Sci. Phys. Nat. Geneve, 9:246.
COMMENT: Formerly included in *Talpa* by Corbet, 1978:35; but see Imaizumi, 1970, The Handbook of Japanese Land Mammals, Vol. 1; Gureev, 1979:276.

Mogera latouchei Thomas, 1907. Proc. Zool. Soc. Lond., 1907:463.
TYPE LOCALITY: China, Fukien, Kuatun, 3500 ft. (1067 m).
DISTRIBUTION: S.E. China, Hainan Isl.
COMMENT: Formerly included in *Talpa micrura* by Ellerman and Morrison-Scott, 1951:40; but see Corbet, 1978:33; Gureev, 1979:279.

Mogera robusta Nehring, 1891. Sitzb. Ges. Naturf. Fr. Berlin, 6:95.
TYPE LOCALITY: U.S.S.R., Primorsky Krai, near Vladivostok.
DISTRIBUTION: Korea to N. E. China and adjacent Siberia; Japan (S. Honshu, Shikoku, Kyushu, Sado Isl.).
COMMENT: Includes *coreana;* see Corbet, 1978:36. Includes *kobeae,* which Imaizumi, 1970, The Handbook of Japanese Land Mammals, Vol. 1, regarded as a separate species. Formerly included in *Talpa;* but see Imaizumi, 1970, *op cit.;* Gureev, 1979; Gromov and Baranova, 1981:16. See also comment under *wogura.*

Mogera wogura (Temminck, 1842). Siebold's Fauna Japonica, Mamm., 1:19.
  TYPE LOCALITY: Japan, Kyushu, Nagasaki.
  DISTRIBUTION: Japan (Honshu, Kyushu, Tane, Amakusa, Tsushima and other isls.);
    Taiwan.
  COMMENT: Includes *insularis;* but see Corbet, 1978:35. Formerly included in *Talpa;* see
    Imaizumi, 1970, The Handbook of Japanese Land Mammals, Vol. 1; Gureev, 1979;
    Gromov and Baranova, 1981:15. According to SW, *coreana* is a subspecies of
    wogura.

Neurotrichus Gunther, 1880. Proc. Zool. Soc. Lond., 1880:441.
  ISIS NUMBER: 5301403006005000000.

Neurotrichus gibbsii (Baird, 1858). Mammals, *in* Repts. Expl. Surv....., 8(1):76.
  TYPE LOCALITY: U.S.A., Washington, Pierce Co., Naches Pass, 4500 ft. (1372 m).
  DISTRIBUTION: S.W. British Columbia (Canada) to W.C. California (U.S.A.).
  ISIS NUMBER: 5301403006005001001.

Parascalops True, 1894. Diagnoses New N. Am. Mamm., p. 2. (preprint of Proc. U.S. Nat.
    Mus., 17:242).
  ISIS NUMBER: 5301403006006000000.

Parascalops breweri (Bachman, 1842). Boston J. Nat. Hist., 4:32.
  TYPE LOCALITY: Eastern N. America.
  DISTRIBUTION: S.E. Canada; N.E. U.S.A.
  COMMENT: Reviewed by Hallett, 1978, Mamm. Species, 98:1–4.
  ISIS NUMBER: 5301403006006001001.

Parascaptor Gill, 1875. Bull. U.S. Geol. Geogr. Surv. Terr., I, 2:110.
  COMMENT: Formerly included in *Talpa;* see Gureev, 1979; but also see Corbet and Hill,
    1980:33, and Corbet, 1978:32.

Parascaptor leucura (Blyth, 1850). J. Asiat. Soc. Bengal, 19:215, pl. 4.
  TYPE LOCALITY: India, Assam, Khasi Hills, Cherrapunji.
  DISTRIBUTION: Assam (India); Yunnan and Szechwan (China) (SW).
  COMMENT: Formerly included in *T. micrura;* see Ellerman and Morrison-Scott, 1951:40;
    but also see Corbet, 1978:33. Also see comment under *Talpa parvidens.*

Scalopus E. Geoffroy, 1803. Cat. Mamm. Mus. Nat. Hist. Nat. Paris, p. 77.
  REVIEWED BY: J. Ramirez-Pulido (JRP).
  ISIS NUMBER: 5301403006007000000.

Scalopus aquaticus (Linnaeus, 1758). Syst. Nat., 10th ed., 1:53.
  TYPE LOCALITY: U.S.A., Pennsylvania, Philadelphia.
  DISTRIBUTION: N. Tamaulipas and N. Coahuila (Mexico) through E. U.S.A. to
    Massachusetts and Minnesota.
  COMMENT: Includes *inflatus* and *montanus;* see Yates and Schmidly, 1977, Occas. Pap.
    Mus. Texas Tech Univ., 45:1–36. Reviewed by Yates and Schmidly, 1978, Mamm.
    Species, 105:1–4. Gureev, 1979:254 listed *aereus* and *inflatus* as distinct species
    without comment. Hall, 1981:72 included *aereus* and *inflatus* in *aquaticus.*
  ISIS NUMBER: 5301403006007001001 as *S. aquaticus.*
               5301403006007002001 as *S. inflatus.*
               5301403006007003001 as *S. montanus.*

Scapanulus Thomas, 1912. Ann. Mag. Nat. Hist., 10:396.
  ISIS NUMBER: 5301403006008000000.

Scapanulus oweni Thomas, 1912. Ann. Mag. Nat. Hist., 10:397.
  TYPE LOCALITY: China, Kansu, 23 mi. (37 km) S.E. Taochou, 9000 ft. (2743 m).
  DISTRIBUTION: C. China; Shensi, Kansu, Szechwan and Tsinghai.
  ISIS NUMBER: 5301403006008001001.

Scapanus Pomel, 1848. Arch. Sci. Phys. Nat. Geneve, 9:247.
   ISIS NUMBER: 5301403006009000000.

   Scapanus latimanus (Bachman, 1842). Boston J. Nat. Hist., 4:34.
      REVIEWED BY: J. Ramirez-Pulido (JRP).
      TYPE LOCALITY: U.S.A., California, Santa Clara Co., Santa Clara.
      DISTRIBUTION: S.C. Oregon (U.S.A.) to N. Baja California (Mexico).
      ISIS NUMBER: 5301403006009001001.

   Scapanus orarius True, 1896. Proc. U.S. Nat. Mus., 19:52.
      TYPE LOCALITY: U.S.A., Washington, Pacific Co., Willapa Bay.
      DISTRIBUTION: S.W. British Columbia (Canada) to N.W. California; W.C. Idaho; N.
         Oregon; C. and S.E. Washington (U.S.A.).
      ISIS NUMBER: 5301403006009002001.

   Scapanus townsendii (Bachman, 1839). J. Acad. Nat. Sci. Phila., 8:58.
      TYPE LOCALITY: U.S.A., Washington, Clark Co., vicinity of Vancouver.
      DISTRIBUTION: S.W. British Columbia (Canada) to N.W. California (U.S.A.).
      ISIS NUMBER: 5301403006009003001.

Scaptochirus Milne-Edwards, 1867. Ann. Sci. Nat. Zool., 7:375.
   COMMENT: Formerly included in *Talpa*; see Gureev, 1979; but also see Corbet, 1978:36,
      and Corbet and Hill, 1980:33.

   Scaptochirus moschatus Milne-Edwards, 1867. Ann. Sci. Nat. Zool., 7:375.
      TYPE LOCALITY: China, Swanhwafu, 100 mi. (161 km) N.W. of Peking.
      DISTRIBUTION: N. China.

Scaptonyx Milne-Edwards, 1872. *In* David, Nouv. Arch. Mus. Hist. Nat. Paris, 7:Bull., p. 92.
   ISIS NUMBER: 5301403006010000000.

   Scaptonyx fusicaudatus Milne-Edwards, 1872. *In* David, Nouv. Arch. Mus. Hist. Nat. Paris,
      7:Bull., p. 92.
      TYPE LOCALITY: China, vicinity of Kukunor (Lake).
      DISTRIBUTION: Tsinghai, Shensi, Szechwan and Yunnan (China); N. Burma.
      ISIS NUMBER: 5301403006010001001 as *S. fusicaudus (sic)*.

Talpa Linnaeus, 1758. Syst. Nat., 10th ed., 1:52.
   REVIEWED BY: D. Capolongo (DC); F. Spitzenberger (FS).
   COMMENT: Includes *Euroscaptor* and *Asioscalops*; formerly included *Parascaptor, Mogera,*
      and *Scaptochirus*; see Gureev, 1979; but also see Corbet and Hill, 1980:33.
      Imaizumi, 1970, The Handbook of Japanese Land Mammals, Vol. 1, considered
      Euroscaptor a monotypic genus containing *mizura*. This entire group needs
      revision (OLR).
   ISIS NUMBER: 5301403006011000000.

   Talpa altaica Nikolsky, 1883. Trans. Soc. Nat. Petersburg, 14:165.
      TYPE LOCALITY: U.S.S.R., R.S.F.S.R., Altai Mtns., valley of Tourak.
      DISTRIBUTION: S. Transbaikalia; Siberia from 70° N. (Yenesei R.) to N. Mongolia, and
         between the Ob and Lena rivers.
      COMMENT: Placed by Yudin, 1971, [Insectivorous Mammals of Siberia], Nauka,
         Novosibirsk, p. 52, in genus *Asioscalops*; but see Corbet, 1978:33; and Gureev,
         1979:265.

   Talpa caucasica Satunin, 1908. Mitt. Kaukas. Mus., 4:5.
      TYPE LOCALITY: U.S.S.R., Stavropolsky Krai, Stavropol.
      DISTRIBUTION: Stavropolsky Krai, Krasnodarsky Krai, and Daghestan (S.W. U.S.S.R.),
         N. of the Caucasus Mtns.

   Talpa caeca Savi, 1822. Nuovo Giorn. de Letterati, Pisa, 1:265.
      TYPE LOCALITY: Italy, Pisa.
      DISTRIBUTION: Iberian peninsula; Apennines and Alps (Italy); and Switzerland to the
         Caucasus (U.S.S.R.).
      ISIS NUMBER: 5301403006011001001.

Talpa europaea Linnaeus, 1758. Syst. Nat., 10th ed., 1:52.
TYPE LOCALITY: Sweden, Kristianstad, Engelholm.
DISTRIBUTION: France and Britain to the Ob and Irtysh rivers (U.S.S.R.), north to S.
Sweden and south to N. Italy and the Sea of Azov.
ISIS NUMBER: 5301403006011002001.

Talpa grandis Miller, 1940. J. Mammal., 21:444.
TYPE LOCALITY: China, Szechwan, Omei-Shan.
DISTRIBUTION: N. and S. Bakbo and Cha-pa (Vietnam); S. China.
COMMENT: In *Euroscaptor* group of *Talpa* (OLR); see Gureev, 1979:272. Regarded as a
synonym of *longirostris* by Ellerman and Morrison-Scott, 1966:40.

Talpa klossi Thomas, 1929. Ann. Mag. Nat. Hist., 38:206.
TYPE LOCALITY: Thailand, 10 mi. (16 km) N.W. Raheng-Tak, Hue Nya Pla.
DISTRIBUTION: N. and W. Thailand to Cha-pa and Bakpo (Vietnam) and S. China.
COMMENT: Formerly included in *micrura*; see Gureev, 1979:272.

Talpa longirostris Milne-Edwards, 1870. C. R. Acad. Sci. Paris, 70:341.
TYPE LOCALITY: China, Szechwan, Moupin.
DISTRIBUTION: C. and S. China (SW); possibly Tibet and Kashmir.
COMMENT: Formerly included in *micrura* by Ellerman and Morrison-Scott, 1966:40 and
Corbet, 1978:35. In the *Euroscaptor* group of *Talpa*; see Gureev, 1979:272.

Talpa micrura Hodgson, 1841. Calcutta J. Nat. Hist., 2:221.
TYPE LOCALITY: Nepal, central and northern hills.
DISTRIBUTION: Nepal; Sikkim, Assam (India).
COMMENT: Formerly included *klossi* and *longirostris*; see Corbet, 1978:33.
ISIS NUMBER: 5301403006011003001.

Talpa mizura Gunther, 1880. Proc. Zool. Soc. Lond., 1880:441.
TYPE LOCALITY: Japan, Honshu, near Yokohama.
DISTRIBUTION: Honshu (Japan).
COMMENT: Imaizumi, 1970, The Handbook of Japanese Land Mammals, Vol. 1,
included this species in the genus *Euroscaptor*. Corbet, 1978:35 placed this species
in *Talpa*.
ISIS NUMBER: 5301403006011004001.

Talpa parvidens Miller, 1940. J. Mammal., 21:203.
TYPE LOCALITY: Vietnam, Di Linh, Blao Forest Station.
DISTRIBUTION: Known from type locality and Rakho on the Chinese border.
COMMENT: Ellerman and Morrison-Scott, 1966:40, included this species in *T. micrura
leucura*. Corbet, 1978:33, mentioned *leucura* as a species, but Gureev, 1979:274 also
listed *parvidens* in the *Euroscaptor* group of *Talpa*. See comment under *P. leucura*.

Talpa romana Thomas, 1902. Ann. Mag. Nat. Hist., 10:516.
TYPE LOCALITY: Italy, Rome, Ostia.
DISTRIBUTION: C. and S. Italy; Sicily; Corfu and Macedonia (Greece and Yugoslavia);
Transcaucasus (U.S.S.R.).

Talpa streeti Lay, 1965. Fieldiana Zool., 44:227.
TYPE LOCALITY: Iran, Kurdistan, Hezar Darreh.
DISTRIBUTION: Iran.
ISIS NUMBER: 5301403006011005001 as *T. streetorum (sic)*.

Uropsilus Milne-Edwards, 1872. *In* David, Nouv. Arch. Mus. Hist. Nat. Paris, 7:Bull., p. 92.
COMMENT: Includes *Nasillus* and *Rhynchonax*; see Ellerman and Morrison-Scott,
1966:31; Corbet and Hill, 1980:33; but also see Gureev, 1979:201-204, who listed
Rhynchonax and *Nasillus* as distinct genera.
ISIS NUMBER: 5301403006012000000.

Uropsilus soricipes Milne-Edwards, 1872. *In* David, Nouv. Arch. Mus. Hist. Nat. Paris,
7:Bull., p. 92.
TYPE LOCALITY: China, Szechwan, Moupin.
DISTRIBUTION: Shensi, Szechwan, Yunnan (China); N. Burma.

COMMENT: Includes *andersoni, gracilis,* and *investigator* according to Ellerman and
Morrison-Scott, 1966:32; but see Gureev, 1979:201–202, who listed these as
distinct species without comment.
ISIS NUMBER: 5301403006012001001.

Urotrichus Temminck, 1841. Het. Instit. K. Ned. Inst., p. 212.
COMMENT: Includes *Dymecodon;* see Corbet, 1978:37, and Ellerman and Morrison-Scott,
1951:33–34. See also Imaizumi, 1970, The Handbook of Japanese Land Mammals,
Vol. 1, p. 123, who considered *Dymecodon* a distinct genus.
ISIS NUMBER: 5301403006013000000 as *Urotrichus.*
            5301403006003000000 as *Dymecodon.*

Urotrichus pilirostris (True, 1886). Proc. U.S. Nat. Mus., 9:97.
TYPE LOCALITY: Japan, Honshu, Bay of Yeddo, Enoshima (Yenosima).
DISTRIBUTION: Montane forests of Honshu, Shikoku, Kyushu (Japan).
COMMENT: Formerly included in *Dymecodon;* see Corbet, 1978:37.
ISIS NUMBER: 5301403006003001001 as *Dymecodon pilirostris.*

Urotrichus talpoides Temminck, 1841. Het. Instit. K. Ned. Inst., p. 215.
TYPE LOCALITY: Japan, Kyushu, Nagasaki.
DISTRIBUTION: Grassland and forest of Honshu, Shikoku, Kyushu (Japan); Dogo Isl., N.
Tsushima Isl. (Japan).
ISIS NUMBER: 5301403006013001001.

ORDER SCANDENTIA

COMMENT: Often included in Insectivora; but see McKenna, 1975:41.

Family Tupaiidae

REVIEWED BY: Earl of Cranbrook (Cranbrook); N. Langham (NL).

COMMENT: The ordinal placing of this family is controversial, but all evidence shows that it is a coherent natural group; see Luckett, ed., 1980, Comparative Biology and Evolutionary Relationships of Tree Shrews, New York, Plenum, 314 pp.; Campbell, 1974, Mammal Rev., 4:125–143; Dene *et al.*, 1978, J. Mammal., 59:697–706. Le Gros Clark, 1971, The Antecedents of Man, 3rd ed., and Van Valen, 1965, Evolution, 19:137–151, discussed affinities of this family; also see Campbell, 1966, Evolution, 20:276–281; Campbell, 1966, Science, 153:436. A bibliography, 1780–1969, was prepared by Elliott, 1971.

ISIS NUMBER: 5301403009000000000.

Anathana Lyon, 1913. Proc. U.S. Nat. Mus., 45:120.

ISIS NUMBER: 5301403009001000000.

Anathana ellioti (Waterhouse, 1850). Proc. Zool. Soc. Lond., 1849:107.

TYPE LOCALITY: India, Andhra Pradesh, "hills between Cuddapah and Nellox (sic)," (=Velikanda Range).

DISTRIBUTION: India, in dry and moist forests of the peninsula, south of Ganges River.

ISIS NUMBER: 5301403009001001001.

Dendrogale Gray, 1848. Proc. Zool. Soc. Lond., 1848:23.

ISIS NUMBER: 5301403009002000000.

Dendrogale melanura (Thomas, 1892). Ann. Mag. Nat. Hist., ser. 6, 9:252.

TYPE LOCALITY: Malaysia, Sarawak, Mt. Dulit, 5000 ft. (1524 m).

DISTRIBUTION: Mountains of N.E. Sarawak and Kinabalu and Trus Madi, Sabah, nowhere below 3000 ft. (914 m).

ISIS NUMBER: 5301403009002001001.

Dendrogale murina (Schlegel and Mueller, 1845). *In* Temminck, Verhandl. Nat. Gesch. Nederland Bezitt., Zool., p. 167, pls. 26, 27.

TYPE LOCALITY: Indonesia, Kalimantan Barat Prov., Pontianak (Probably erroneous).

DISTRIBUTION: From eastern Thailand, Chatraburi and Trat Provinces, through Cambodia (Kampuchea) to Vietnam.

COMMENT: See Lyon, 1913, Proc. U.S. Nat. Mus., 45:1–186 for discussion of type locality.

ISIS NUMBER: 5301403009002002001.

Ptilocercus Gray, 1848. Proc. Zool. Soc. Lond., 1848:24.

COMMENT: Subfamily Ptilocercinae; see Campbell, 1974, Mammal. Rev., 4:125–143.

ISIS NUMBER: 5301403009003000000.

Ptilocercus lowii Gray, 1848. Proc. Zool. Soc. Lond., 1848:24.

TYPE LOCALITY: Malaysia, Sarawak, "caught in the Rajah's house", *i.e.*, Kuching.

DISTRIBUTION: Southern peninsular Thailand, peninsular Malaysia, Sumatra, including Riau Isls. and Batu Isls., Banka, Serasan Isl. and Borneo, including Labuan Isl.

ISIS NUMBER: 5301403009003001001.

Tupaia Raffles, 1821. Trans. Linn. Soc. Lond., 13:256.

ISIS NUMBER: 5301403009004000000.

Tupaia dorsalis Schlegel, 1857. Handl. Beoef. Dierk., 1:59, 447, pl. 3.

TYPE LOCALITY: Borneo.

DISTRIBUTION: The mainland of Borneo at low to moderate elevations (except S.E.).

COMMENT: United with *tana* in the subgenus *Tana* by Lyon, 1913, Proc. U.S. Nat. Mus., 45:1–186, but this is not supported by immunological evidence; see Dene *et al.*, 1978, J. Mammal., 59:697–706.

ISIS NUMBER: 5301403009004001001.

Tupaia glis (Diard, 1820). Asiat. J. Mon. Reg., 10:478.
 TYPE LOCALITY: Malaysia, Penang Isl.
 DISTRIBUTION: S.E. Asia from Sikkim and S. China, including Hainan Isl., to Java and
  Borneo and various surrounding islands.
 COMMENT: Includes *longipes, salatana;* see Chasen, 1940, Bull. Raffles Mus., 15:4;
  Sorenson and Conaway, 1964, Sabah Soc. J., 2:77–91. Includes *chinensis;* see
  Warkentin and Conaway, 1969, J. Mammal., 50:817–818; includes *natunae* and
  lucida which have been placed with *splendidula;* see Lyon, 1913, Proc. U.S. Nat.
  Mus., 45:1–186, but also see Medway, 1961, Treubia, 25:265–272. Includes *belangeri*
  and *ferruginea;* see Chasen, 1940, Bull. Raffles Mus., 15:1; Napier and Napier,
  1967:456; Lekagul and McNeely, 1977:5; Dene *et al.,* 1978, J. Mammal., 59:697–706.
 ISIS NUMBER: 5301403009004002001.

Tupaia gracilis Thomas, 1893. Ann. Mag. Nat. Hist., ser. 6, 12:53.
 TYPE LOCALITY: Malaysia, Sarawak, Baram Dist., Apoh River at base of Mt. Batu Song.
 DISTRIBUTION: Borneo (except S.E.), west to Karimata Isl., Belitung Isl., and Banka Isl.;
  north to Banggi Isl.
 ISIS NUMBER: 5301403009004003001.

Tupaia javanica Horsfield, 1822. Zool. Res. Java, 3 (pages unnumbered).
 TYPE LOCALITY: Indonesia, Java, Jawa Timur Prov., perhaps near Banjuwangi.
 DISTRIBUTION: Indonesian Isls. of Nias, Sumatra, Java, and Bali.
 COMMENT: For type locality clarification, see Lyon, 1913, Proc. U.S. Nat. Mus.,
  45:1–186.
 ISIS NUMBER: 5301403009004004001.

Tupaia minor Guenther, 1876. Proc. Zool. Soc. Lond., 1876:426.
 TYPE LOCALITY: Malaysia, Sabah, mainland "opposite the island of Labuan."
 DISTRIBUTION: Southern peninsular Thailand, peninsular Malayasia, Sumatra, Lingga
  Archipelago, Borneo, and offshore islands of Laut, Banggi and Balambangan.
 ISIS NUMBER: 5301403009004005001.

Tupaia montana Thomas, 1892. Ann. Mag. Nat. Hist., ser. 6, 9:252.
 TYPE LOCALITY: Malaysia, Sarawak, Mt. Dulit, 5000 ft. (1524 m).
 DISTRIBUTION: Mountains of Sarawak and W. Sabah; recorded from 1200 to 10400 ft.
  (366–3170 m) on Mt. Kinabalu.
 COMMENT: Distribution in the central highlands of Borneo on the Indonesian side of
  the border is unknown (Cranbrook).
 ISIS NUMBER: 5301403009004006001.

Tupaia nicobarica (Zelebor, 1869). Reise "Novara", Zool. Theil., 1:17, pl. 1.
 TYPE LOCALITY: India, Nicobar Isls., Great Nicobar Isl.
 DISTRIBUTION: Great and Little Nicobar Isls.
 COMMENT: A distinctive species of uncertain affinity (Cranbrook).
 ISIS NUMBER: 5301403009004007001.

Tupaia palawanensis Thomas, 1894. Ann. Mag. Nat. Hist., ser. 6, 13:367.
 TYPE LOCALITY: Philippine Isls., Palawan Isl.
 DISTRIBUTION: Philippines, islands of Palawan, Busuanga, Cuyo, and Culion.
 COMMENT: The taxon *palawanensis* was considered by Thomas, 1894, Ann. Mag. Nat.
  Hist., ser. 6, 13:367, to be a geographical representative of *Tupaia glis* when he
  described it, but see Corbet and Hill, 1980:34.
 ISIS NUMBER: 5301403009004008001.

Tupaia picta Thomas, 1892. Ann. Mag. Nat. Hist., ser. 6, 9:251.
 TYPE LOCALITY: Malaysia, Sarawak, Baram Dist.
 DISTRIBUTION: Borneo, in N. Sarawak and parts of E. Kalimantan.
 COMMENT: The species' distribution is poorly known (Cranbrook).
 ISIS NUMBER: 5301403009004009001.

Tupaia splendidula Gray, 1865. Proc. Zool. Soc. Lond., 1865:322, pl. 12.

TYPE LOCALITY: Borneo.

DISTRIBUTION: Borneo, south of about 1° N.; Karimata Isl.; Laut Isl.; Bunguran Isl. (North Natuna Isls.).

COMMENT: The status of this species and its synonyms was discussed by Medway, 1961, Treubia, 25:265-272. *T. glis natunae* and *T. glis lucida* were associated with *splendidula* by Lyon, 1913, Proc. U.S. Nat. Mus., 45:1-186; but also see Chasen, 1940, Bull. Raffles Mus., 15:4; Medway, 1961, Treubia, 25:265-272. Includes *carimatae* and *muelleri*; see Medway, 1961, *loc. cit.*

ISIS NUMBER: 5301403009004010000.

Tupaia tana Raffles, 1821. Trans. Linn. Soc. New York, 13:257.

TYPE LOCALITY: Indonesia, west coast of Sumatra, by inference in Sumatra, Selatan Prov., in the neighborhood of Bengkulu (="Bencoolen")(Cranbrook).

DISTRIBUTION: Sumatra, including offshore islands of Tuanku, Batu group, Lingga group, Banga, Belitung, Tambelon, and Serasan groups; Borneo, Banggi Isl.

COMMENT: *T. tana* is the type of *Tana* Lyon = *Lyonogale* Conisbee, but immunological studies (Dene *et al.*, 1978, J. Mammal., 59:697-706) do not support the generic or subgeneric standing of this taxon.

ISIS NUMBER: 5301403009004011001.

Urogale Mearns, 1905. Proc. U.S. Nat. Mus., 28:435.

ISIS NUMBER: 5301403009005000000.

Urogale everetti (Thomas, 1892). Ann. Mag. Nat. Hist., ser. 6, 9:250.

TYPE LOCALITY: Philippine Isls., Mindanao, Zamboanga.

DISTRIBUTION: Mindanao (Philippines).

COMMENT: The species was considered by Lyon, 1913, Proc. U.S. Nat. Mus., 45:1-186, to be closest to *Tupaia tana*.

ISIS NUMBER: 5301403009005001001.

ORDER DERMOPTERA
ISIS NUMBER: 5301404000000000000.

Family Cynocephalidae
REVIEWED BY: J. H. Honacki (JH).
ISIS NUMBER: 5301404001000000000.

Cynocephalus Boddaert, 1768. Dierk. Meng., 2:8.
COMMENT: Includes *Galeopterus;* see Ellerman and Morrison-Scott, 1951:89.
ISIS NUMBER: 5301404001001000000.

Cynocephalus variegatus (Audebert, 1799). Hist. Nat. Singes Makis, sig. Rr. Java.
TYPE LOCALITY: Indonesia, Java.
DISTRIBUTION: Indochina to Java and Borneo.
ISIS NUMBER: 5301404001001001001.

Cynocephalus volans (Linnaeus, 1758). Syst. Nat., 10th ed., 1:30.
TYPE LOCALITY: Philippine Isls., S. Luzon, Pampanga Prov.
DISTRIBUTION: Philippine Isls.
ISIS NUMBER: 5301404001001002001.

ORDER CHIROPTERA
ISIS NUMBER: 5301405000000000000.

Family Pteropodidae
REVIEWED BY: K. F. Koopman (KFK).
ISIS NUMBER: 5301405001000000000.

Acerodon Jourdan, 1837. L'Echo du Monde Savant, 4, No. 275, p. 156.
COMMENT: Very closely related to and probably congeneric with *Pteropus* (KFK).
ISIS NUMBER: 5301405001001000000.

Acerodon celebensis (Peters, 1867). Monatsb. Preuss. Akad. Wiss. Berlin, p. 333.
TYPE LOCALITY: Indonesia, Sulawesi.
DISTRIBUTION: Sulawesi, Saleyer Isl., and Sula Mangoli, Sula Isls.
ISIS NUMBER: 5301405001001001001.

Acerodon humilis K. Andersen, 1909. Ann. Mag. Nat. Hist., ser. 7, 3:24–25.
TYPE LOCALITY: Indonesia, Talaud Isls., Lirong.
DISTRIBUTION: Talaud Isls.
ISIS NUMBER: 5301405001001002001.

Acerodon jubatus (Eschscholtz, 1831). Zool. Atl., Part 4:1.
TYPE LOCALITY: Philippines, Luzon, Manila.
DISTRIBUTION: Philippines.
ISIS NUMBER: 5301405001001003001.

Acerodon lucifer Elliot, 1896. Field Columb. Mus. Publ., Zool. Ser., 1:78.
TYPE LOCALITY: Philippines, Panay, Concepcion.
DISTRIBUTION: Philippines.
ISIS NUMBER: 5301405001001004001.

Acerodon mackloti (Temminck, 1837). Monogr. Mamm., 2:69.
TYPE LOCALITY: Indonesia, Timor.
DISTRIBUTION: Lombok; Sumbawa; Flores; Alor Isl.; Sumba; Timor.
ISIS NUMBER: 5301405001001005001.

Aethalops Thomas, 1923. Proc. Zool. Soc. Lond., 1923:178.
COMMENT: Replacement name for *Aethalodes* (preoccupied) (KFK).
ISIS NUMBER: 5301405001002000000.

Aethalops alecto (Thomas, 1923). Ann. Mag. Nat. Hist., ser. 9, 11:251.
TYPE LOCALITY: Indonesia, Sumatra, Indrapura Peak, 7300 ft. (2225 m).
DISTRIBUTION: Malaya; Sumatra; Borneo; Java.
COMMENT: Includes *aequalis*; see Hill, 1961, Proc. Zool. Soc. Lond., 1961:639.
ISIS NUMBER: 5301405001002001001 as *A. alecto*.
5301405001024002001 as *Pteropus aequalis*.

Alionycteris Kock, 1969. Senckenberg. Biol., 50:319.

Alionycteris paucidentata Kock, 1969. Senckenberg. Biol., 50:322.
TYPE LOCALITY: Philippines, Mindanao, Bukidion Prov., Mt. Katanglad.
DISTRIBUTION: Mindanao (Philippines).

Aproteles Menzies, 1977. Aust. J. Zool., 25:330.
REVIEWED BY: A. C. Ziegler (ACZ).

Aproteles bulmerae Menzies, 1977. Aust. J. Zool., 25:331.
TYPE LOCALITY: Papua New Guinea, Chimbu Prov., 2 km S.E Chuave Govt. Sta., 1530 m.
DISTRIBUTION: New Guinea.
COMMENT: Originally described from fossil material, but since found living; see Hyndman and Menzies, 1980, J. Mammal., 61:159.

Balionycteris Matschie, 1899. Flederm. Berliner Mus. Naturk., p. 80.
ISIS NUMBER: 5301405001003000000.

Balionycteris maculata (Thomas, 1893). Ann. Mag. Nat. Hist., ser. 6, 11:341.
TYPE LOCALITY: Malaysia, Sarawak.
DISTRIBUTION: Thailand; Malaya; Borneo; Durian and Galang Isls. (Rhio Arch.).
ISIS NUMBER: 5301405001003001001.

Boneia Jentink, 1879. Notes Leyden Mus., 1:117.
ISIS NUMBER: 5301405001004000000.

Boneia bidens Jentink, 1879. Notes Leyden Mus., 1:117.
TYPE LOCALITY: Indonesia, N. Sulawesi, Bone (near Gerontalo).
DISTRIBUTION: N. Sulawesi.
ISIS NUMBER: 5301405001004001001.

Casinycteris Thomas, 1910. Ann. Mag. Nat. Hist., ser. 8, 6:111.
ISIS NUMBER: 5301405001005000000.

Casinycteris argynnis Thomas, 1910. Ann. Mag. Nat. Hist., ser. 8, 6:111.
TYPE LOCALITY: Cameroun, Ja River, Bitye.
DISTRIBUTION: Cameroun to E. Zaire.
ISIS NUMBER: 5301405001005001001.

Chironax K. Andersen, 1912. Cat. Chiroptera Br. Mus., 2nd ed., p. 658.
ISIS NUMBER: 5301405001006000000.

Chironax melanocephalus (Temminck, 1825). Monogr. Mamm., 1:190.
TYPE LOCALITY: Indonesia, W. Java, Bantam.
DISTRIBUTION: Thailand; Malaya; Sumatra; Java; Nias Isl.; Sulawesi.
ISIS NUMBER: 5301405001006001001.

Cynopterus F. Cuvier, 1824. Des Dentes des Mammiferes, p. 248.
ISIS NUMBER: 5301405001007000000.

Cynopterus archipelagus Taylor, 1934. Monogr. Bur. Sci. Manila, p.182.
TYPE LOCALITY: Philippines, Polillo Isl.
DISTRIBUTION: Polillo Isl. (Philippines).
COMMENT: Based on a single immature specimen. Possibly a variant of *brachyotis*
(KFK).
ISIS NUMBER: 5301405001007001001.

Cynopterus brachyotis (Muller, 1838). Tijdschr. Nat. Gesch. Physiol., 5:146.
TYPE LOCALITY: Borneo, Dewei River.
DISTRIBUTION: Sri Lanka; S.E. Asia; Malaysia; Philippines; Nicobar Isls.; Andaman
Islands; Sumatra; Borneo; Sulawesi; Talaud Isls. and adjacent small islands.
COMMENT: Does not include *angulatus*, which was transferred to *sphinx*; see Hill and
Thonglongya, 1972, Bull. Br. Mus. (Nat. Hist.) Zool., 22:175.
ISIS NUMBER: 5301405001007002001.

Cynopterus horsfieldi Gray, 1843. List Mamm. Br. Mus., p.38.
TYPE LOCALITY: Indonesia, Java
DISTRIBUTION: Thailand; Malaya; Borneo; Java; Sumatra; adjacent small islands.
COMMENT: Includes *harpax*; see Hill, 1961, Proc. Zool. Soc. Lond., 1961:632–634.
ISIS NUMBER: 5301405001007003001.

Cynopterus sphinx (Vahl, 1797). Skr. Nat. Selsk. Copenhagen, 4(1):123.
REVIEWED BY: S. Wang (SW).
TYPE LOCALITY: India, Madras, Tranquebar.
DISTRIBUTION: Sri Lanka; India; S. China; S.E. Asia; Malaya; Sumatra; Java; Timor;
adjacent small islands.
COMMENT: Includes *angulatus*; see Hill and Thonglongya, 1972, Bull. Br. Mus. (Nat.
Hist.) Zool., 22:175.
ISIS NUMBER: 5301405001007004001.

Dobsonia Palmer, 1898. Proc. Biol. Soc. Wash., 12:114.
REVIEWED BY: A. C. Ziegler (ACZ).
ISIS NUMBER: 5301405001008000000.

Dobsonia beauforti Bergmans, 1975. Beaufortia, 23(295):3.
TYPE LOCALITY: Indonesia, Irian Jaya, So Rong Div., Waigeo Isl., Njanjef.
DISTRIBUTION: Waigeo Isl. (off west end of New Guinea).
COMMENT: Closely related to *viridis*; see Bergmans, 1975, Beaufortia, 23(295):11.

Dobsonia crenulata K. Andersen, 1909. Ann. Mag. Nat. Hist., ser. 7, 4:532.
TYPE LOCALITY: Indonesia, Molucca Isls., Ternate.
DISTRIBUTION: Rau, Morotai, Gililo, Ternate and Batchian Isls. (N. Moluccas).
COMMENT: Considered a subspecies of *viridis*; see Rabor, 1952, Chicago Acad. Sci., Nat. Hist. Misc., 96, but see also Bergmans, 1975, Beaufortia, 295(23).
ISIS NUMBER: 5301405001008001001.

Dobsonia exoleta K. Andersen, 1909. Ann. Mag. Nat. Hist., ser. 8, 4:533.
TYPE LOCALITY: Indonesia, Sulawesi, Minahassa, Tomohon.
DISTRIBUTION: Sulawesi and adjacent small islands.
ISIS NUMBER: 5301405001008002001.

Dobsonia inermis K. Andersen, 1909. Ann. Mag. Nat. Hist., ser. 8, 4:532.
TYPE LOCALITY: Solomon Isls., San Cristobal Isl.
DISTRIBUTION: Solomon Isls.
ISIS NUMBER: 5301405001008003001.

Dobsonia minor (Dobson, 1879). Proc. Zool. Soc. Lond., 1878:875.
TYPE LOCALITY: Indonesia, Irian Jaya, Manokwari Div., Amberbaki.
DISTRIBUTION: C. and W. New Guinea and adjacent small islands.
ISIS NUMBER: 5301405001008004001.

Dobsonia moluccensis (Quoy and Gaimard, 1830). *In* d'Urville, Voy. "Astrolabe," Zool., 1:86.
TYPE LOCALITY: Indonesia, Molucca Isls., Amboina Isl.
DISTRIBUTION: Bismarck Arch.; New Guinea; Aru Isls.; Batanta, Mysol, and Waigeo Isls. (off W. New Guinea); Molucca Isls.; N. Queensland.
COMMENT: Does not include *pannietensis*; see Bergmans, 1979, Beaufortia, 29(355):201–207. Includes *anderseni*; see Koopman, 1979, Am. Mus. Novit., 2690:6.
ISIS NUMBER: 5301405001008005001 as *D. moluccense (sic)*.

Dobsonia pannietensis (De Vis, 1905). Ann. Queensl. Mus., 6:36.
TYPE LOCALITY: Papua New Guinea, Louisiade Arch., Panniet Isl.
DISTRIBUTION: Louisiade Arch.; D'Entrecasteaux Isls.; Trobriand Isls.
COMMENT: Considered a subspecies of *moluccensis* by Laurie and Hill, 1954:42, but as a separate species by Bergmans, 1979, Beaufortia, 29:201–207.

Dobsonia peroni (E. Geoffroy, 1810). Ann. Mus. Hist. Nat. Paris, 15:104.
TYPE LOCALITY: Indonesia, Lesser Sunda Isls., Timor.
DISTRIBUTION: Sumba; Timor; Flores; Sumbawa; Nusa Penida (near Bali); Komodo; Alor; Wetar; Babar.
ISIS NUMBER: 5301405001008006001.

Dobsonia praedatrix K. Andersen, 1909. Ann. Mag. Nat. Hist., ser. 8, 4:532.
TYPE LOCALITY: Papua New Guinea, Bismarck Arch., "Duke of York group".
DISTRIBUTION: Bismarck Arch.
ISIS NUMBER: 5301405001008007001.

Dobsonia remota Cabrera, 1920. Bol. Soc. Esp. Nat. Hist., 20:107.
TYPE LOCALITY: Papua New Guinea, Trobriand Isls., Kiriwina Isl. (=Trobriand Isl.).
DISTRIBUTION: Trobriand Isls.
COMMENT: Almost certainly a subspecies of *pannietensis* (KFK). A record from Bougainville Isl. is based on a misidentified *inermis*; see Bergmans, 1979, Beaufortia, 29(355).
ISIS NUMBER: 5301405001008008001.

Dobsonia viridis (Heude, 1896). Mem. Hist. Nat. Emp. Chin., 3:176.
TYPE LOCALITY: Indonesia, Molucca Isls., Kei Isls.
DISTRIBUTION: Kei Isls.; Amboina; Buru; Seram; Banda Isls.; Negros Isl. (Philippines).
ISIS NUMBER: 5301405001008009001.

Dyacopterus K. Andersen, 1912. Cat. Chiroptera Br. Mus., 1:651.
ISIS NUMBER: 5301405001009000000.

Dyacopterus spadiceus (Thomas, 1890). Ann. Mag. Nat. Hist., ser. 6, 5:235.
TYPE LOCALITY: Malaysia, Sarawak, Baram.
DISTRIBUTION: Sumatra; Borneo; Malaya.
COMMENT: Includes *brooksi;* see Hill, 1961, Proc. Zool. Soc. Lond., 1961:637; but also
see Peterson, 1969, Life Sci. Occ. Pap. R. Ont. Mus., 13:1–4.
ISIS NUMBER: 5301405001009001001.

Eidolon Rafinesque, 1815. Analyse de la Nature, p. 54.
ISIS NUMBER: 5301405001010000000.

Eidolon helvum (Kerr, 1792). Anim. Kingdom, 1(1):xvii,91.
TYPE LOCALITY: Senegal.
DISTRIBUTION: Senegal to Ethiopia to South Africa; Madagascar; Yemen; Aden; islands
in Guinea Gulf and off eastern Africa.
COMMENT: Includes *dupreanum* and *sabaeum;* see Hayman and Hill, 1971, Part 2:11.
ISIS NUMBER: 5301405001010001001.

Eonycteris Dobson, 1873. Proc. Asiat. Soc. Bengal, p. 148.
ISIS NUMBER: 5301405001030000000.

Eonycteris major K. Andersen, 1910. Ann. Mag. Nat. Hist., ser. 8, 6:625.
TYPE LOCALITY: Malaysia, Sarawak, Mt. Dulit.
DISTRIBUTION: Borneo; Philippines; Mentawi Isls.
COMMENT: Includes *robusta* and *longicauda;* see Tate, 1942, Bull. Am. Mus. Nat. Hist.,
80:344. Corbet and Hill, 1980:42, listed *robusta* as a distinct species without
comment.
ISIS NUMBER: 5301405001030001001 as *E. major.*
5301405001030002001 as *E. robusta.*

Eonycteris rosenbergi (Jentink, 1899). Notes Leyden Mus., 11:210.
TYPE LOCALITY: Indonesia, N. Sulawesi, Gorontalo.
DISTRIBUTION: N. Sulawesi.
COMMENT: Almost certainly a subspecies of *spelaea* (KFK).
ISIS NUMBER: 5301405001030003001.

Eonycteris spelaea (Dobson, 1871). Proc. Asiat. Soc. Bengal, p. 105, 106.
REVIEWED BY: S. Wang (SW).
TYPE LOCALITY: Burma, Tenasserim, Moulmein.
DISTRIBUTION: N. India; Burma; S. China; Thailand; Malaya; Sumatra; Java; Borneo;
Sumba; Timor; Philippines; Andaman Isls.; perhaps Sulawesi.
COMMENT: Almost certainly includes *rosenbergi* (KFK).
ISIS NUMBER: 5301405001030004001.

Epomophorus Bennett, 1836. Proc. Zool. Soc. Lond., 1835:149.
ISIS NUMBER: 5301405001011000000.

Epomophorus angolensis Gray, 1870. Cat. Monkeys, Lemurs, and Fruit-eating Bats Br. Mus.,
p. 125.
TYPE LOCALITY: Angola, Benguella.
DISTRIBUTION: W. Angola; N.W. Namibia.
ISIS NUMBER: 5301405001011001001.

Epomophorus crypturus Peters, 1852. Reise nach Mossambique, Saugethiere, p. 26.
TYPE LOCALITY: Mozambique, Tete.
DISTRIBUTION: E. Zaire; S. Tanzania; E. Angola; Zambia; Malawi; N. Botswana;
Zimbabwe; Mozambique; E. South Africa.
COMMENT: Includes *parvus* (KFK).
ISIS NUMBER: 5301405001011003001.

Epomophorus gambianus (Ogilby, 1835). Proc. Zool. Soc. Lond., 1835:100.
TYPE LOCALITY: Gambia.
DISTRIBUTION: Senegal to S. Ethiopia.
COMMENT: Does not include *parvus* (KFK).
ISIS NUMBER: 5301405001011004001.

Epomophorus labiatus (Temminck, 1837). Monogr. Mamm., 2:83.
TYPE LOCALITY: Sudan, Blue Nile Prov., Sennar.
DISTRIBUTION: Senegal; Nigeria to Ethiopia; Sudan; Kenya; Uganda; Tanzania; E. Zaire;
Congo Republic.
COMMENT: Includes *anurus;* see Koopman, 1975, Bull. Am. Mus. Nat. Hist., 154:364.
ISIS NUMBER: 5301405001011005001 as *E. labiatus.*
5301405001011002001 as *E. anurus.*

Epomophorus minor Dobson, 1880. Proc. Zool. Soc. Lond., 1879:715.
TYPE LOCALITY: Tanzania, Zanzibar.
DISTRIBUTION: Ethiopia to Zaire, Zambia, and Malawi; Zanzibar.

Epomophorus pousarguesi Trouessart, 1904. Cat. Mamm. Viv. Foss., 5th suppl., p. 55.
TYPE LOCALITY: Central African Republic, between Makorou and Yabanda.
DISTRIBUTION: Central African Republic.
COMMENT: Concerning type locality, see Bergmans, 1978, J. Nat. Hist., 12:684.
ISIS NUMBER: 5301405001011006001.

Epomophorus reii Aellen, 1950. Rev. Suisse Zool., 57:559.
TYPE LOCALITY: Cameroun, Garva region, Rei Bouba.
DISTRIBUTION: Cameroun.
ISIS NUMBER: 5301405001011007001.

Epomophorus wahlbergi (Sundevall, 1846). Ofv. Kongl. Svenska Vet.-Akad. Forhandl.
Stockholm, 3(4):118.
TYPE LOCALITY: South Africa, Natal, near Durban.
DISTRIBUTION: Cameroun to Somalia, south to Angola and South Africa (Liberian
record probably erroneous); Pemba Isl.
ISIS NUMBER: 5301405001011008001.

Epomops Gray, 1870. Cat. Monkeys, Lemurs, and Fruit-eating Bats Br. Mus., p. 126.
ISIS NUMBER: 5301405001012000000.

Epomops buettikoferi (Matschie, 1899). Megachiroptera Berlin Mus., p. 45.
TYPE LOCALITY: Liberia, Junk River, Schlieffelinsville.
DISTRIBUTION: Guinea to Ghana; perhaps Nigeria.
ISIS NUMBER: 5301405001012001001.

Epomops dobsoni (Bocage, 1899). J. Sci. Math. Phys. Nat. Lisboa, ser. 2, 1:1.
TYPE LOCALITY: Angola, Benguela, Quindumbo.
DISTRIBUTION: Angola to Rwanda, Tanzania, Zambia, and N. Botswana.
ISIS NUMBER: 5301405001012002001.

Epomops franqueti (Tomes, 1860). Proc. Zool. Soc. Lond., 1860:54.
TYPE LOCALITY: Gabon.
DISTRIBUTION: Ivory Coast to Sudan, Uganda, and N.W. Tanzania to Zambia and
Angola.
ISIS NUMBER: 5301405001012003001.

Haplonycteris Lawrence, 1939. Bull. Mus. Comp. Zool., 86:31.
  ISIS NUMBER: 5301405001013000000.

  Haplonycteris fischeri Lawrence, 1939. Bull. Mus. Comp. Zool., 86:33.
    TYPE LOCALITY: Philippines, Mindoro, Mt. Halcyon.
    DISTRIBUTION: Mindoro and Luzon (Philippines).
    ISIS NUMBER: 5301405001013001001.

Harpyionycteris Thomas, 1896. Ann. Mag. Nat. Hist., ser. 6, 18:243.
  ISIS NUMBER: 5301405001038000000.

  Harpyionycteris celebensis Miller and Hollister, 1921. Proc. Biol. Soc. Wash., 34:99.
    TYPE LOCALITY: Indonesia, Sulawesi, middle Sulawesi, Gimpoe.
    DISTRIBUTION: Sulawesi.
    COMMENT: Considered a subspecies of *whiteheadi* by Laurie and Hill, 1954:48, but as a
      separate species by Peterson and Fenton, 1970, R. Ont. Mus. Life Sci. Occ. Pap.,
      17:13–14.

  Harpyionycteris whiteheadi Thomas, 1896. Ann. Mag. Nat. Hist., ser. 6, 18:244.
    TYPE LOCALITY: Philippines, Mindoro Isl., 5,000 ft. (1524 m).
    DISTRIBUTION: Negros Isl., Mindanao, Mindoro (Philippines).
    ISIS NUMBER: 5301405001038001001.

Hypsignathus H. Allen, 1861. Proc. Acad. Nat. Sci. Phila., p. 156.
  ISIS NUMBER: 5301405001014000000.

  Hypsignathus monstrosus H. Allen, 1861. Proc. Acad. Nat. Sci. Phila., p. 157.
    TYPE LOCALITY: Gabon.
    DISTRIBUTION: Gambia to Ethiopia, south to Zambia and Angola; Bioko.
    ISIS NUMBER: 5301405001014001001.

Latidens Thonglongya, 1972. J. Bombay Nat. Hist. Soc., 69:151.

  Latidens salimalii Thonglongya, 1972. J. Bombay Nat. Hist. Soc., 69:153.
    TYPE LOCALITY: India, Madras, Madurai Dist., High Wavy Mts., 2500 ft. (762 m).
    DISTRIBUTION: S. India.

Macroglossus F. Cuvier, 1824. Des Dentes des Mammiferes, p. 248.
  REVIEWED BY: A. C. Ziegler (ACZ)(New Guinea).
  ISIS NUMBER: 5301405001031000000.

  Macroglossus fructivorus Taylor, 1934. Monogr. Bur. Sci. Manila, p. 125.
    TYPE LOCALITY: Philippines, Mindanao, Cotabato, Tatayan.
    DISTRIBUTION: Mindanao (Philippines).
    COMMENT: Probably a synonym of *lagochilus* (KFK).
    ISIS NUMBER: 5301405001031001001.

  Macroglossus minimus (E. Geoffroy, 1810). Ann. Mus. Hist. Nat. Paris, 15:97.
    TYPE LOCALITY: Indonesia, Java.
    DISTRIBUTION: Thailand to Philippines, New Guinea, Bismarck Arch., Solomon Isls.,
      and N. Australia.
    COMMENT: Lekagul and McNeely, 1977:50, listed *lagochilus* as a synonym of this
      species without comment.
    ISIS NUMBER: 5301405001031003001 as *M. minimus.*
              5301405001031002001 as *M. lagochilus.*

  Macroglossus sobrinus K. Andersen, 1911. Ann. Mag. Nat. Hist., ser. 8, 3:642.
    TYPE LOCALITY: Malaysia, Perak, Gunong Igari, 2,000 ft. (610 m).
    DISTRIBUTION: S.E. Asia; Sumatra; Java; adjacent small islands.
    COMMENT: This is the species previously called *minimus;* see Lekagul and McNeely,
      1977:50.

Megaerops Peters, 1865. Monatsb. Preuss. Akad. Wiss. Berlin, p. 256.
ISIS NUMBER: 5301405001015000000.

Megaerops ecaudatus (Temminck, 1837). Monogr. Mamm., 2:94.
TYPE LOCALITY: Indonesia, Sumatra, Padang.
DISTRIBUTION: N. and W. Borneo; W. Sumatra; Malaya; Thailand; Vietnam.
ISIS NUMBER: 5301405001015001001.

Megaerops kusnotoi Hill and Boedi, 1978. Mammalia, 42:427.
TYPE LOCALITY: Indonesia, Java, Sukabumi, Lengkong, Hanjuang Ciletuh, 700 m.
DISTRIBUTION: Java.

Megaerops wetmorei Taylor, 1934. Monogr. Bur. Sci. Manila, p. 191.
TYPE LOCALITY: Philippines, Mindanao, Cotabato, Tatayan.
DISTRIBUTION: Philippines.
ISIS NUMBER: 5301405001015002001.

Megaloglossus Pagenstecher, 1885. Zool. Anz., 8:245.
ISIS NUMBER: 5301405001032000000.

Megaloglossus woermanni Pagenstecher, 1885. Zool. Anz., 8:245.
TYPE LOCALITY: Gabon, Sibange farm.
DISTRIBUTION: Guinea to E. and S. Zaire; Uganda; Bioko; Gabon; N. Angola.
ISIS NUMBER: 5301405001032001001.

Melonycteris Dobson, 1877. Proc. Zool. Soc. Lond., 1877:119.
REVIEWED BY: A. C. Ziegler (ACZ).
COMMENT: Includes *Nesonycteris*; see Phillips, 1968, Univ. Kans. Publ. Mus. Nat. Hist., 16:814.
ISIS NUMBER: 5301405001033000000.

Melonycteris aurantius Phillips, 1966. J. Mammal., 47:24.
TYPE LOCALITY: Solomon Isls., Florida Isl., Haleta, 10 m.
DISTRIBUTION: Florida Isl., Choiseul Isl. (Solomon Isls.).

Melonycteris melanops Dobson, 1877. Proc. Zool. Soc. Lond., 1877:119.
TYPE LOCALITY: Papua New Guinea, Bismarck Arch., "Duke of York Isl."
DISTRIBUTION: Bismarck Arch.; E. New Guinea.
COMMENT: The New Guinea record is highly questionable (KFK and ACZ). Type locality given in K. Andersen, 1912, Cat. Chiroptera Br. Mus., 2nd ed. p. 790, is "New Ireland, coast adjacent to Duke of York Isl." (ACZ).
ISIS NUMBER: 5301405001033001001.

Melonycteris woodfordi (Thomas, 1887). Ann. Mag. Nat. Hist., 19:147.
TYPE LOCALITY: Solomon Isls., Alu Isl. (near Shortland island).
DISTRIBUTION: Solomon Isls.; Bougainville Isl.
ISIS NUMBER: 5301405001033002001.

Micropteropus Matschie, 1899. Megachiroptera Berlin Mus., p. 36, 57.
ISIS NUMBER: 5301405001016000000.

Micropteropus grandis Sanborn, 1950. Publ. Cult. Comp. Diamantes Angola, 10:55.
TYPE LOCALITY: Angola, Lunda, Dundo.
DISTRIBUTION: N. Angola; Congo Republic.
ISIS NUMBER: 5301405001016001001.

Micropteropus intermedius Hayman, 1963. Publ. Cult. Comp. Diamantes Angola, 66:100.
TYPE LOCALITY: Angola, Lunda, Dundo.
DISTRIBUTION: N. Angola; S.E. Zaire.
ISIS NUMBER: 5301405001016002001.

Micropteropus pusillus (Peters, 1867). Monatsb. Preuss. Akad. Wiss. Berlin, p. 870.
TYPE LOCALITY: Gambia.
DISTRIBUTION: Gambia to Ethiopia, south to Angola, Burundi, and Zambia.

Myonycteris Matschie, 1899. Megachiroptera Berlin Mus., p. 61, 63.
ISIS NUMBER: 5301405001017000000.

Myonycteris brachycephala (Bocage, 1889). J. Sci. Math. Phys. Nat. Lisboa, ser. 2, 1:198.
TYPE LOCALITY: Sao Tome and Principe, Sao Tome Isl.
DISTRIBUTION: Sao Tome Isl. (Gulf of Guinea).
ISIS NUMBER: 5301405001017001001.

Myonycteris relicta Bergmans, 1980. Zool. Meded. Rijks. Mus. Nat. Hist. Leiden, 14:126.
TYPE LOCALITY: Kenya, Coast prov., Shimba Hills, Lukore area, Mukanda river.
DISTRIBUTION: Shimba Hills, Kenya; Nguru and Usambara Mts., Tanzania.

Myonycteris torquata (Dobson, 1878). Cat. Chiroptera Br. Mus., p. 71, 76.
TYPE LOCALITY: N. Angola.
DISTRIBUTION: Sierra Leone to Uganda, south to Angola and Zambia; Bioko.
COMMENT: Includes *leptodon* and *wroughtoni*; see Hayman and Hill, 1971, Part 2:12–13.
ISIS NUMBER: 5301405001017002001.

Nanonycteris Matschie, 1899. Megachiroptera Berlin Mus., p. 36, 58.
ISIS NUMBER: 5301405001018000000.

Nanonycteris veldkampi (Jentink, 1888). Notes Leyden Mus., 10:51.
TYPE LOCALITY: Liberia, Fisherman Lake, Buluma.
DISTRIBUTION: Guinea to Congo Republic.
ISIS NUMBER: 5301405001018001001.

Neopteryx Hayman, 1946. Ann. Mag. Nat. Hist., ser. 11, 12:569.
ISIS NUMBER: 5301405001019000000.

Neopteryx frosti Hayman, 1946. Ann. Mag. Nat. Hist., ser. 11, 12:571.
TYPE LOCALITY: Indonesia, W. Sulawesi, Tamalanti, 3300 ft. (1006 m).
DISTRIBUTION: W. Sulawesi.
ISIS NUMBER: 5301405001019001001.

Notopteris Gray, 1859. Proc. Zool. Soc. Lond., 1859:36.
ISIS NUMBER: 5301405001034000000.

Notopteris macdonaldi Gray, 1859. Proc. Zool. Soc. Lond., 1859:38.
TYPE LOCALITY: Fiji Isls., Viti Levu.
DISTRIBUTION: New Hebrides; New Caledonia; Fiji Isls.; Caroline Isls.
COMMENT: Includes *neocaledonica*; see Sanborn, 1950, Fieldiana Zool., 31:329, 330.
ISIS NUMBER: 5301405001034001001.

Nyctimene Borkhausen, 1797. Deutsche Fauna, 1:86.
REVIEWED BY: A. C. Ziegler (ACZ)(New Guinea).
ISIS NUMBER: 5301405001036000000.

Nyctimene aello (Thomas, 1900). Ann. Mag. Nat. Hist., ser. 7, 5:216.
TYPE LOCALITY: Papua New Guinea, Milne Bay Prov., Milne Bay.
DISTRIBUTION: New Guinea; Molucca Isls.
ISIS NUMBER: 5301405001036001001.

Nyctimene albiventer (Gray, 1863). Proc. Zool. Soc. Lond., 1862:262.
TYPE LOCALITY: Indonesia, Molucca Isls., Morotai Isl.
DISTRIBUTION: New Guinea; Molucca Isls; Solomon Isls.; Kei Isls.; N. Queensland;
    Bismarck Arch.
COMMENT: Formerly included *draconilla*; see Greig-Smith, 1975, Sci. In New Guinea,
    3:119–120.
ISIS NUMBER: 5301405001036002001.

Nyctimene cephalotes (Pallas, 1767). Spicil. Zool., 3:10.
TYPE LOCALITY: Indonesia, Molucca Isls., Amboina.
DISTRIBUTION: Sulawesi; Timor; Molucca Isls., Numfoor Isl., Umboi Isl. (off N. coast
    New Guinea).

COMMENT: See K. Andersen, 1912, Cat. Chiroptera Br. Mus., 2nd ed., p. 707, for discussion of type locality.
ISIS NUMBER: 5301405001036003001.

Nyctimene cyclotis K. Andersen, 1910. Ann. Mag. Nat. Hist., ser. 7, 6:623.
TYPE LOCALITY: Indonesia, Irian Jaya, Manokwari Div., Arfak Mtns.
DISTRIBUTION: New Guinea.
ISIS NUMBER: 5301405001036004001.

Nyctimene draconilla Thomas, 1922. Nova Guinea, 13:725.
TYPE LOCALITY: Indonesia, Irian Jaya, Southern Div., Lorentz River, Bivak Isl.
DISTRIBUTION: New Guinea.
COMMENT: Considered a subspecies of *albiventer* by Laurie and Hill, 1954:46, but also see Greig-Smith, 1975, Sci. In New Guinea, 3:119-120.

Nyctimene major (Dobson, 1877). Proc. Zool. Soc. Lond., 1877:117.
TYPE LOCALITY: Papua New Guinea, Bismarck Arch., Duke of York Isl.
DISTRIBUTION: Fergusson Isl.; Kiriwana Isl.; D'Entrecasteaux, Trobriand Isls., Bismarck and Louisiade Archs.; Solomon Isls.; Heath Isl.; Goodenough Isl. (off east end of New Guinea).
COMMENT: A New Guinea record is almost certainly erroneous; see Koopman, 1979, Am. Mus. Novit., 2690:6.
ISIS NUMBER: 5301405001036005001.

Nyctimene malaitensis Phillips, 1968. Univ. Kans. Publ. Mus. Nat. Hist., 16:822.
TYPE LOCALITY: Solomon Isls., Malaita Isl.
DISTRIBUTION: Malaita Isl. (Solomon Isls.).

Nyctimene minutus K. Andersen, 1910. Ann. Mag. Nat. Hist., ser. 7, 6:622.
TYPE LOCALITY: Indonesia, Sulawesi, Minahassa, Tondano.
DISTRIBUTION: Sulawesi; Amboina Isls.
ISIS NUMBER: 5301405001036006001.

Nyctimene robinsoni Thomas, 1904. Ann. Mag. Nat. Hist., ser. 7, 14:196.
TYPE LOCALITY: Australia, Queensland, Cooktown.
DISTRIBUTION: E. Queensland.
ISIS NUMBER: 5301405001036007001.

Nyctimene sanctacrucis Troughton, 1931. Proc. Linn. Soc. N.S.W., 56:206.
TYPE LOCALITY: Solomon Isls., Santa Cruz Isls.
DISTRIBUTION: Santa Cruz Isls.

Otopteropus Kock, 1969. Senckenberg. Biol., 50:329.

Otopteropus cartilagonodus Kock, 1969. Senckenberg. Biol., 50:333.
TYPE LOCALITY: Philippines, Luzon, Mountain Prov., Sitio Pactil.
DISTRIBUTION: Luzon (Philippines).

Paranyctimene Tate, 1942. Am. Mus. Novit., 1204:1.
REVIEWED BY: A. C. Ziegler (ACZ).
ISIS NUMBER: 5301405001037000000.

Paranyctimene raptor Tate, 1942. Am. Mus. Novit., 1204:1.
TYPE LOCALITY: Papua New Guinea, Western Prov., Fly River, Oroville Camp. *(ca.* 4 mi. (6 km) below Elavala River mouth).
DISTRIBUTION: New Guinea.
ISIS NUMBER: 5301405001037001001.

Penthetor K. Andersen, 1912. Cat. Chiroptera Br. Mus., p. 665.
ISIS NUMBER: 5301405001020000000.

Penthetor lucasi (Dobson, 1880). Ann. Mag. Nat. Hist., ser. 5, 6:163.
TYPE LOCALITY: Malaysia, Sarawak.
DISTRIBUTION: Malaysia; Borneo; Rhio Arch.
ISIS NUMBER: 5301405001020001001.

Plerotes K. Andersen, 1910. Ann. Mag. Nat. Hist., ser. 8, 5:97.
ISIS NUMBER: 5301405001021000000.

Plerotes anchietai (Seabra, 1900). J. Sci. Math. Phys. Nat. Lisboa, ser. 2, 6:116.
TYPE LOCALITY: Angola, Benguela, Galanga.
DISTRIBUTION: Angola; Zambia; S. Zaire.
ISIS NUMBER: 5301405001021001001.

Ptenochirus Peters, 1861. Monatsb. Preuss. Akad. Wiss. Berlin, p. 707.
ISIS NUMBER: 5301405001022000000.

Ptenochirus jagori (Peters, 1861). Monatsb. Preuss. Akad. Wiss. Berlin, p. 707.
TYPE LOCALITY: Philippines, Luzon, Albay, Daraga.
DISTRIBUTION: Mindanao, Luzon, Mindoro, Tablas and Negros Isls. (Philippines).
ISIS NUMBER: 5301405001022001001.

Ptenochirus minor Yoshiyuki, 1979. Bull. Nat. Sci. Mus. Tokyo, ser. A (Zool.), 5:75.
TYPE LOCALITY: Philippines, Mindanao, Davao, Mt. Talomo, Baracatan.
DISTRIBUTION: Mindanao and Palawan (Philippines).

Pteralopex Thomas, 1888. Ann. Mag. Nat. Hist., ser. 6, 1:155.
REVIEWED BY: A. C. Ziegler (ACZ)(New Guinea).
ISIS NUMBER: 5301405001023000000.

Pteralopex acrodonta Hill and Beckon, 1978. Bull. Br. Mus. (Nat. Hist.) Zool., 34:68.
TYPE LOCALITY: Fiji Isls., Taveuni Isl., Des Voeux Peak, *ca.* 3840 ft. (1170 m).
DISTRIBUTION: Fiji Isls.

Pteralopex anceps K. Andersen, 1909. Ann. Mag. Nat. Hist., ser. 7, 3:266.
TYPE LOCALITY: Papua New Guinea, Bougainville Isl.
DISTRIBUTION: Bougainville and Choiseul Isls., (Solomon Isls.).

Pteralopex atrata Thomas, 1888. Ann. Mag. Nat. Hist., ser. 6, 1:155.
TYPE LOCALITY: Solomon Isls., Guadalcanal, Aola.
DISTRIBUTION: Ysabel and Guadalcanal (Solomon Isls.).
COMMENT: Does not include *anceps;* see Hill and Beckon, 1978, Bull. Br. Mus. (Nat. Hist.) Zool., 34:67, 68.
ISIS NUMBER: 5301405001023001001.

Pteropus Erxleben, 1777. Syst. Anim., p. 130.
REVIEWED BY: N. Sullivan (NS); S. Wang (SW)(China); A. C. Ziegler (ACZ)(New Guinea).
COMMENT: Originally named *Pteropus* by Brisson, 1762, *in* Regnum Animale, which is not available (ACZ).
ISIS NUMBER: 5301405001024000000.

Pteropus admiralitatum Thomas, 1894. Ann. Mag. Nat. Hist., ser. 6, 13:293.
TYPE LOCALITY: Papua New Guinea, Bismarck Arch., Admiralty Isls.
DISTRIBUTION: Solomon Isls.; Admiralty Isls.; New Britain; Tabar Isls.
COMMENT: Includes *solomonis, colonus,* and *goweri;* see Laurie and Hill, 1954:33.
ISIS NUMBER: 5301405001024001001.

Pteropus alecto Temminck, 1837. Monogr. Mamm., 2:75.
TYPE LOCALITY: Indonesia, Sulawesi, Manado (=Menado).
DISTRIBUTION: Sulawesi; N. and E. Australia; Saleyer Isl.; Lombok; Bawean Isl.; Kangean Isls.; Sumba Isl.; Savu Isl.; S. New Guinea.
COMMENT: Includes *gouldi;* see Tate, 1942, Bull. Am. Mus. Nat. Hist., 80:336, 337.
ISIS NUMBER: 5301405001024004001.

Pteropus anetianus Gray, 1870. Cat. Monkeys, Lemurs, and Fruit-eating Bats Br. Mus., p. 101.
TYPE LOCALITY: New Hebrides, Aneiteum (France - U.K.).
DISTRIBUTION: New Hebrides; Banks Isls.
COMMENT: Includes *eotinus, bakeri,* and *banksiana;* see Felten and Kock, 1972, Senckenberg. Biol., 53:182–185.

Pteropus argentatus Gray, 1844. Voy. "Sulphur", Zool., 1:30.
TYPE LOCALITY: Indonesia, Moluccas Isls., Amboina Isl. (uncertain).
DISTRIBUTION: Amboina; perhaps Sulawesi.
COMMENT: Sulawesi specimens probably referable to *Acerodon celebensis* (KFK).
ISIS NUMBER: 5301405001024005001.

Pteropus arquatus Miller and Hollister, 1921. Proc. Biol. Soc. Wash., 34:100.
TYPE LOCALITY: Indonesia, Sulawesi, Koelawi.
DISTRIBUTION: C. Sulawesi.
COMMENT: Probably should be transferred to *Acerodon* (KFK).
ISIS NUMBER: 5301405001024006001.

Pteropus balutus Hollister, 1913. Proc. Biol. Soc. Wash., 26:111.
TYPE LOCALITY: Philippines, Sarangani Isls., Balut Isl.
DISTRIBUTION: Sarangani Isls. (S. Philippines).
ISIS NUMBER: 5301405001024007001.

Pteropus caniceps Gray, 1870. Cat. Monkeys, Lemurs, and Fruit-eating Bats Br. Mus., p. 107.
TYPE LOCALITY: Indonesia, Molucca Isls., Halmahera Isls., Batjan.
DISTRIBUTION: Halmahera Isls.; Sangihe Isls.; Sulawesi; Sula Isls.
COMMENT: Includes *dobsoni*; see Laurie and Hill, 1954:34.
ISIS NUMBER: 5301405001024008001 as *P. caniceps*.
5301405001024014001 as *P. dobsoni*.

Pteropus chrysoproctus Temminck, 1837. Monogr. Mamm., 2:67.
TYPE LOCALITY: Indonesia, Molucca Isls., Amboina.
DISTRIBUTION: Amboina; Buru; Seram; Sangihe Isls.; small islands east of Seram.
ISIS NUMBER: 5301405001024009001.

Pteropus conspicillatus Gould, 1850. Proc. Zool. Soc. Lond., 1849:109.
TYPE LOCALITY: Australia, Queensland, Fitzroy Isl.
DISTRIBUTION: Halmahera Isls.; New Guinea and adjacent islands; N.E. Queensland.
ISIS NUMBER: 5301405001024012001.

Pteropus dasymallus Temminck, 1825. Monogr. Mamm., 1:180.
TYPE LOCALITY: Japan, Ryukyu Isls., Kuchinoerabu Isl.
DISTRIBUTION: Ryukyu Isls.; Taiwan; extreme S. Kyushu (Japan).
COMMENT: Includes *daitoensis*; see Kuroda, 1933, J. Mammal., 14:314.
ISIS NUMBER: 5301405001024013001.

Pteropus faunulus Miller, 1902. Proc. U.S. Nat. Mus., 24:785.
TYPE LOCALITY: India, Nicobar Isls., Car Nicobar Isl.
DISTRIBUTION: Nicobar Isls. (Bay of Bengal).

Pteropus fundatus Felten and Kock, 1972. Senckenberg. Biol., 53:186.
TYPE LOCALITY: New Hebrides, Banks Isls., Mota Isl. (France - U.K.).
DISTRIBUTION: Banks Isls. (N. New Hebrides).

Pteropus giganteus (Brunnich, 1782). Dyrenes Historie, 1:45.
TYPE LOCALITY: India, Bengal.
DISTRIBUTION: Maldive Isls.; India; Sri Lanka; Pakistan; Burma; Andaman Isls. Tsinghai (China)(SW).
COMMENT: Includes *ariel*; see Hill, 1958, J. Bombay Nat. Hist. Soc., 55:5, 6.
ISIS NUMBER: 5301405001024015001.

Pteropus gilliardi Van Deusen, 1969. Am. Mus. Novit., 2371:5.
TYPE LOCALITY: Papua New Guinea, Bismarck Arch., New Britain, Whiteman Mts., Wild Dog Ridge, *ca.* 1600 m.
DISTRIBUTION: New Britain Isl. (Bismarck Arch.).
ISIS NUMBER: 5301405001024016001.

Pteropus griseus (E. Geoffroy, 1810). Ann. Mus. Hist. Nat. Paris, 15:94.
TYPE LOCALITY: Indonesia, Lesser Sunda Isls., Timor.
DISTRIBUTION: Timor; Samao Isl.; Dyampea Isl.; Bonerato Isl.; Saleyer Isl.; Sulawesi; Banda Isls.; perhaps S. Luzon (Philippines).

COMMENT: Includes *mimus;* see Laurie and Hill, 1954:33.
ISIS NUMBER: 5301405001024017001 as *P. griseus.*
5301405001024029001 as *P. mimus.*

Pteropus howensis Troughton, 1931. Proc. Linn. Soc. N.S.W., 56:204.
TYPE LOCALITY: Solomon Isls., Ontong Java Isl.
DISTRIBUTION: Ontong Java Isl. (Solomon Isls.).

Pteropus hypomelanus Temminck, 1853. Esquisses Zool. sur la Cote de Guine, p. 61.
TYPE LOCALITY: Indonesia, Molucca Isls., Ternate Isl.
DISTRIBUTION: Maldive Isls.; New Guinea through Indonesia to Vietnam and Thailand,
and adjacent islands; Solomon Isls.; N.E. Australia (accidental).
COMMENT: Includes *brunneus;* see Ride, 1970:180.
ISIS NUMBER: 5301405001024018001.

Pteropus insularis Hombron and Jacquinot, 1842. Voy. Pole Sud. Mamm., p. 24.
TYPE LOCALITY: Caroline Isl., Truk Isl., Hogoleu (Pac. Isls. Trust Terr., U.S.).
DISTRIBUTION: Truk Isls. (C. Caroline Isls.).

Pteropus leucopterus Temminck, 1853. Esquisses Zool. sur la Cote de Guine, p. 60.
TYPE LOCALITY: Philippines.
DISTRIBUTION: Luzon (Philippines).
ISIS NUMBER: 5301405001024019001.

Pteropus leucotis Sanborn, 1950. Proc. Biol. Soc. Wash., 63:189.
TYPE LOCALITY: Philippines, Calamian Isls., Busuanga Isl., Singay.
DISTRIBUTION: Palawan, Busuanga Isl. (Philippines).
COMMENT: Probably should be transferred to *Acerodon* (KFK).

Pteropus livingstonei Gray, 1866. Proc. Zool. Soc. Lond., 1866:66.
TYPE LOCALITY: Comoro Isls., Anjouan Isl.
DISTRIBUTION: Comoro Isls.
ISIS NUMBER: 5301405001024020001.

Pteropus lombocensis Dobson, 1878. Cat. Chiroptera Br. Mus., p. 34.
TYPE LOCALITY: Indonesia, Lesser Sunda Isls., Lombok Isl.
DISTRIBUTION: Alor Isl. (near Timor); Lombok; Flores.
ISIS NUMBER: 5301405001024021001.

Pteropus lylei K. Andersen, 1908. Ann. Mag. Nat. Hist., ser. 8, 2:367.
TYPE LOCALITY: Thailand, Bangkok.
DISTRIBUTION: Thailand; S. Vietnam.
ISIS NUMBER: 5301405001024022001.

Pteropus macrotis Peters, 1867. Monatsb. Preuss. Akad. Wiss. Berlin, p. 327.
TYPE LOCALITY: Indonesia, Aru Isls., Wokam Isl.
DISTRIBUTION: New Guinea; Aru Isls.
ISIS NUMBER: 5301405001024023001.

Pteropus mahaganus Sanborn, 1931. Field Mus. Nat. Hist. Publ. Ser. Zool., 2:19.
TYPE LOCALITY: Solomon Isls., Ysabel Isl., Tunnibul.
DISTRIBUTION: Bougainville Isl.; Ysabel Isl. (Solomon Isls.).
ISIS NUMBER: 5301405001024024001.

Pteropus mariannus Desmarest, 1822. Mammalogie, Part 2, Suppl., p. 547.
TYPE LOCALITY: Marianna Isls., Guam (U.S.).
DISTRIBUTION: Palau Isls.; Mariana Isls.; Caroline Isls.; Ryukyu Isls.
COMMENT: Includes *loochooensis, pelewensis, ualanus,* and *yapensis;* see Kuroda, 1938, A
List of the Japanese Mammals, 122 pp.; Corbet and Hill, 1980:37–38, listed these
forms as distinct species without comment.
ISIS NUMBER: 5301405001024025001.

Pteropus mearnsi Hollister, 1913. Proc. Biol. Soc. Wash., 26:112.
TYPE LOCALITY: Philippines, Basilan Isl., Isabella.
DISTRIBUTION: Mindanao, Basilan Isl. (Philippines).
ISIS NUMBER: 5301405001024026001.

Pteropus melanopogon Peters, 1867. Monatsb. Preuss. Akad. Wiss. Berlin, p. 330.
TYPE LOCALITY: Indonesia, Molucca Isls., Amboina.
DISTRIBUTION: Aru Isls.; Kei Isls.; Amboina; Buru; Seram; Banda Isls.; Timor Laut;
Sangihe Isls.; adjacent islands.
COMMENT: Does not include *sepikensis;* see Koopman, 1979, Am. Mus. Novit., 2690:5.
See comment under *neohibernicus.*
ISIS NUMBER: 5301405001024027001.

Pteropus melanotus Blyth, 1863. Cat. Mamm. Mus. Asiat. Soc. Calcutta, p. 20.
TYPE LOCALITY: India, Nicobar Isls.
DISTRIBUTION: Nicobar Isls.; Engano Isl. and Nias Isl. (W. Sumatra); Christmas Isl.;
Andaman Isls.
COMMENT: Includes *satyrus;* see Hill, 1971, J. Bombay Nat. Hist. Soc., 68:6, 7.
ISIS NUMBER: 5301405001024028001.

Pteropus molossinus Temminck, 1853. Esquisses Zool. sur la Cote de Guine, p. 62.
TYPE LOCALITY: Caroline Isls., Ponape (Pac. Isls. Trust Terr., U.S.).
DISTRIBUTION: Mortlock Isl., Ponape Isl. (Caroline Isls.).

Pteropus neohibernicus Peters, 1876. Monatsb. Preuss. Akad. Wiss. Berlin, p. 317.
TYPE LOCALITY: Papua New Guinea, Bismarck Arch., New Ireland Isl.
DISTRIBUTION: Bismarck Arch.; Admiralty Isls.; New Guinea; Mysol Isl.; Ghebi Isl.
COMMENT: Includes *sepikensis;* see Koopman, 1979, Am. Mus. Novit., 2690:5.
ISIS NUMBER: 5301405001024030001.

Pteropus niger (Kerr, 1792). Anim. Kingdom, 1:90.
TYPE LOCALITY: Mascarene Isls., Reunion Isl. (France).
DISTRIBUTION: Reunion; Mauritius (Mascarene Isls.).
ISIS NUMBER: 5301405001024031001.

Pteropus nitendiensis Sanborn, 1930. Am. Mus. Novit., 435:2.
TYPE LOCALITY: Solomon Isls., Santa Cruz Isls., Ndeni Isl.
DISTRIBUTION: Ndeni (Santa Cruz Isls.).

Pteropus ocularis Peters, 1867. Monatsb. Preuss. Akad. Wiss. Berlin, p. 326.
TYPE LOCALITY: Indonesia, Molucca Isls., Seram Isl.
DISTRIBUTION: Seram; Buru (Moluccas).
ISIS NUMBER: 5301405001024032001.

Pteropus ornatus Gray, 1870. Cat. Monkeys, Lemurs, and Fruit-eating Bats Br. Mus., p. 105.
TYPE LOCALITY: New Caledonia, Noumea (France).
DISTRIBUTION: Loyalty Isls.; New Caledonia.
COMMENT: Includes *auratus;* see Felten, 1964, Senckenberg. Biol., 45:678–680.

Pteropus personatus Temminck, 1825. Monogr. Mamm., 1:189.
TYPE LOCALITY: Indonesia, Molucca Isls., Ternate.
DISTRIBUTION: Halmahera Isls.; N. Sulawesi.
ISIS NUMBER: 5301405001024033001.

Pteropus phaeocephalus Thomas, 1882. Proc. Zool. Soc. Lond., 1881:756.
TYPE LOCALITY: Caroline Isls., Mortlock Isl. (Pac. Isls. Trust Terr., U.S.).
DISTRIBUTION: Mortlock Isl. (C. Caroline Isls.).

Pteropus pilosus K. Andersen, 1908. Ann. Mag. Nat. Hist., ser. 8, 2:369.
TYPE LOCALITY: Caroline Isls., Palau Isls. (Pac. Isls. Trust Terr., U.S.).
DISTRIBUTION: Palau Isls.

Pteropus pohlei Stein, 1933. Z. Saugetierk., 8:93.
TYPE LOCALITY: Indonesia, Irian Jaya, Tjenderawasih Div., Japen Isl.
DISTRIBUTION: Japen Isl. (off N.W. New Guinea).
ISIS NUMBER: 5301405001024034001.

Pteropus poliocephalus Temminck, 1825. Monogr. Mamm., 1:179.
TYPE LOCALITY: Australia.
DISTRIBUTION: E. Australia, from Cape York to Victoria.
ISIS NUMBER: 5301405001024035001.

Pteropus pselaphon Lay, 1829. Zool. J., 4:457.
TYPE LOCALITY: Japan, Bonin Isls.
DISTRIBUTION: Bonin Isls.; Volcano Isls.

Pteropus pumilus Miller, 1911. Proc. U.S. Nat. Mus., 38:394.
TYPE LOCALITY: Indonesia, Miangas Isl.
DISTRIBUTION: Miangas Isl. (between Talaud Isls. and Mindanao, Philippines).
ISIS NUMBER: 5301405001024036001.

Pteropus rayneri Gray, 1870. Cat. Monkeys, Lemurs, and Fruit-eating Bats Br. Mus., p. 108.
TYPE LOCALITY: Solomon Isls., Guadalcanal Isl.
DISTRIBUTION: Solomon Isls.
COMMENT: Includes *cognatus*; see Hill, 1962, *in* Wolff, The Nat. Hist. of Rennel Isl. and
Brit. Solomon Isls., 4:9. Corbet and Hill, 1980:36, listed *cognatus* as a distinct
species without comment.
ISIS NUMBER: 5301405001024037001 as *P. rayneri*.
5301405001024010001 as *P. cognatus*.

Pteropus rodricensis Dobson, 1878. Cat. Chiroptera Br. Mus., p. 36.
TYPE LOCALITY: Mascarene Isls., Rodrigues (U.K.).
DISTRIBUTION: Rodrigues Isl. (Mascarene Isls.).
ISIS NUMBER: 5301405001024038001.

Pteropus rufus E. Geoffroy, 1803. Cat. Mamm. Mus. Nat. Hist. Nat. Paris, p. 47.
TYPE LOCALITY: Madagascar.
DISTRIBUTION: Madagascar.
ISIS NUMBER: 5301405001024039001.

Pteropus samoensis Peale, 1848. Mammalia and Ornithology *in* U.S. Expl. Exped., 8:20.
TYPE LOCALITY: Samoan Isls., Tutuila Isl. (Amer. Samoa).
DISTRIBUTION: Fiji Isls.; Samoan Isls.
COMMENT: Includes *nawaiensis*; see Hill and Beckon, 1978, Bull. Br. Mus. (Nat. Hist.)
Zool., 34:65.

Pteropus sanctacrucis Troughton, 1930. Rec. Aust. Mus., 18:3.
TYPE LOCALITY: Solomon Isls., Santa Cruz Isls., Ndeni Isl.
DISTRIBUTION: Santa Cruz Isls.
COMMENT: Listed as a form of *nitendiensis* by Corbet and Hill, 1980:37; the reference
given does not substantiate this (KFK).

Pteropus scapulatus Peters, 1862. Monatsb. Preuss. Akad. Wiss. Berlin, p. 574.
TYPE LOCALITY: Australia, Queensland, Cape York.
DISTRIBUTION: Australia; S. New Guinea; adjacent small islands.
ISIS NUMBER: 5301405001024040001.

Pteropus seychellensis Milne-Edwards, 1877. Bull. Sci. Soc. Philom. Paris, ser. 7, 2:221.
TYPE LOCALITY: Seychelle Isls., Mahe Isl.
DISTRIBUTION: Seychelle Isls.; Comoro Isls.; Aldabra Isl.; Mafia Isl. (off Tanzania).
COMMENT: Includes *aldabrabensis* and *comorensis*; see Hill, 1971, Philos. Trans. R. Soc.
Lond., B260:574.
ISIS NUMBER: 5301405001024041001 as *P. seychellensis*.
5301405001024003001 as *P. aldabrensis (sic)*.
5301405001024011001 as *P. comorensis*.

Pteropus speciosus K. Andersen, 1908. Ann. Mag. Nat. Hist., ser. 8, 2:364.
TYPE LOCALITY: Philippines, Malanipa Isl. (off west end of Mindanao).
DISTRIBUTION: Sibutu Isl., Malanipa Isl., Sulu Arch., Mindanao, Negros Isl.
(Philippines); Solombo Besar and Mata Siri (Java Sea).
ISIS NUMBER: 5301405001024042001.

Pteropus subniger (Kerr, 1792). Anim. Kingdom, p. 91.
TYPE LOCALITY: Mascarene Isls., Reunion Isl. (France).
DISTRIBUTION: Reunion, Mauritius (Mascarene Isls.).
COMMENT: Probably extinct; see Cheke and Dahl, 1981, Mammalia, 45:205–238.
ISIS NUMBER: 5301405001024043001.

Pteropus tablasi Taylor, 1934. Monogr. Bur. Sci. Manila, 3:169.
TYPE LOCALITY: Philippines, Tablas Isl., Odiongan.
DISTRIBUTION: Philippines.
ISIS NUMBER: 5301405001024044001.

Pteropus temmincki Peters, 1867. Monatsb. Preuss. Akad. Wiss. Berlin, p. 331.
TYPE LOCALITY: Indonesia, Molucca Isls., Amboina Isl.
DISTRIBUTION: Buru; Amboina; Seram; Bismarck Arch.; nearby small islands; perhaps
    Timor Isl.
COMMENT: See K. Andersen, 1912, Cat. Chiroptera Br. Mus., p. 318, for clarification of
    the type locality.
ISIS NUMBER: 5301405001024045001.

Pteropus tokudae Tate, 1934. Am. Mus. Novit., 713:1.
TYPE LOCALITY: Mariana Isls., Guam (U.S.).
DISTRIBUTION: Mariana Isls.
COMMENT: Possibly extinct on Guam (NS).

Pteropus tonganus Quoy and Gaimard, 1830. In d'Urville, Voy. "Astrolabe", Zool., 1:74.
TYPE LOCALITY: Tonga Isls., Tongatapu Isl.
DISTRIBUTION: Karkar Isl. (off N.E. New Guinea) and Rennell Isl. (Solomon Isls.),
    south to New Caledonia, east to Cook Isls.
COMMENT: Includes *vanikorensis* and *geddiei*; see Sanborn, 1931, Field Mus. Nat. Hist.
    Publ. Ser. Zool., 18:14; Felten and Kock, 1972, Senckenberg. Biol., 53:180–181.
ISIS NUMBER: 5301405001024046001.

Pteropus tuberculatus Peters, 1869. Monatsb. Preuss. Akad. Wiss. Berlin, p. 393.
TYPE LOCALITY: Solomon Isls., Santa Cruz Isls., Vanikoro Isl.
DISTRIBUTION: Vanikoro Isl. (Santa Cruz Isls.).

Pteropus vampyrus (Linnaeus, 1758). Syst. Nat., 10th ed., 1:31.
TYPE LOCALITY: Indonesia, Java.
DISTRIBUTION: Indochina; Malay Peninsula; Sumatra; Borneo; Java; Philippines; Lesser
    Sunda Isls.; adjacent small islands.
COMMENT: Includes *intermedius*; see Lekagul and McNeely, 1977:77. Corbet and Hill,
    1980:37, listed *intermedius* as a distinct species without comment.
ISIS NUMBER: 5301405001024047001.

Pteropus vetulus Jouan, 1863. Mem. Soc. Imp. Sci. Nat. Cherbourg, 9:90.
TYPE LOCALITY: New Caledonia (France).
DISTRIBUTION: New Caledonia.
COMMENT: Includes *macmillani*; see Felten, 1964, Senckenberg. Biol., 45:671.

Pteropus voeltzkowi Matschie, 1909. Sitzb. Ges. Naturf. Fr. Berlin, p. 486.
TYPE LOCALITY: Tanzania, Pemba Isl., Fufuni.
DISTRIBUTION: Pemba Isl. (off coast of Tanzania).
ISIS NUMBER: 5301405001024048001.

Pteropus woodfordi Thomas, 1888. Ann. Mag. Nat. Hist., ser. 6, 1:156.
TYPE LOCALITY: Solomon Isls., Guadalcanal Isl., Aola.
DISTRIBUTION: Fauro Isl. to Guadalcanal Isl. (Solomon Isls.).
ISIS NUMBER: 5301405001024049001.

Rousettus Gray, 1821. Lond. Med. Repos., 15:299.
REVIEWED BY: A. C. Ziegler (ACZ)(New Guinea).
COMMENT: Includes *Lissonycteris*; see Koopman, 1975, Bull. Am. Mus. Nat. Hist.,
    154:361–362. Corbet and Hill, 1980:36, listed *Lissonycteris* as a distinct genus
    without comment.
ISIS NUMBER: 5301405001025000000.

Rousettus aegyptiacus (E. Geoffroy, 1810). Ann. Mus. Hist. Nat. Paris, 15:96.
TYPE LOCALITY: Egypt, Giza.
DISTRIBUTION: Senegal, Egypt, Cyprus, and Turkey, south to South Africa; Pakistan to
    Aden (S. Yemen); Oman; adjacent small islands.

COMMENT: Includes *leachi* and *arabicus;* see Hayman and Hill, 1971, Part 2:11; Corbet, 1978:38.
ISIS NUMBER: 5301405001025001001.

Rousettus amplexicaudatus (E. Geoffroy, 1810). Ann. Mus. Hist. Nat. Paris, 15:96.
TYPE LOCALITY: Indonesia, Lesser Sunda Isls., Timor Isl.
DISTRIBUTION: Kampuchea; Thailand; Malay Peninsula through Indonesia to New Guinea, Bismarck Arch., and Solomon Isls.; Philippines.
COMMENT: Includes *stresemanni;* see Koopman, 1979, Am. Mus. Novit., 2960:4.
ISIS NUMBER: 5301405001025002001 as *R. amplexicaudatus.*
5301405001025008001 as *R. stresemanni.*

Rousettus angolensis (Bocage, 1898). J. Sci. Math. Phys. Nat. Lisboa, ser. 2, 5:133.
TYPE LOCALITY: Angola, Quibula, Cahata, Pungo Andongo.
DISTRIBUTION: Senegal and Angola to Ethiopia and Mozambique; Bioko.
COMMENT: Formerly included in *Lissonycteris;* see Koopman, 1975, Bull. Am. Mus. Nat. Hist., 154:361–362.
ISIS NUMBER: 5301405001025003001.

Rousettus celebensis K. Andersen, 1907. Ann. Mag. Nat. Hist., ser. 7, 19:503, 509.
TYPE LOCALITY: Indonesia, Sulawesi, Mt. Masarang, 3500 ft. (1067 m).
DISTRIBUTION: Sulawesi; Sangihe Isls.
ISIS NUMBER: 5301405001025004001.

Rousettus lanosus Thomas, 1906. Ann. Mag. Nat. Hist., ser. 7, 18:137.
TYPE LOCALITY: Uganda, Ruwenzori East, Mubuku Valley, 13,000 ft. (3962 m).
DISTRIBUTION: Uganda; Kenya; Tanzania; S. Ethiopia; E. Zaire.
ISIS NUMBER: 5301405001025005001.

Rousettus leschenaulti (Desmarest, 1820). Encyclop. Method. Mamm., 1:110.
TYPE LOCALITY: India, Pondicherry.
DISTRIBUTION: Sri Lanka; Pakistan to Vietnam and S. China; Java; Bali.
COMMENT: Includes *seminudus;* see Sinha, 1970, Mammalia, 34:82.
ISIS NUMBER: 5301405001025006001 as *R. leschenaulti.*
5301405001025007001 as *R. seminudus* (sic).

Rousettus madagascariensis G. Grandidier, 1928. Bull. Acad. Malgache, ll:91.
TYPE LOCALITY: Madagascar, Beforona (between Tananarive and Andevoranto).
DISTRIBUTION: Madagascar.
COMMENT: Considered a subspecies of *lanosus* by Hayman and Hill, 1971, Part 2:12; but see Bergmans, 1977, Mammalia, 41:67–74.

Rousettus obliviosus Kock, 1978. Proc. 4th Int. Bat Res. Conf. Nairobi, p. 208.
TYPE LOCALITY: Comoro Isls., Grand Comoro, near Boboni, 640 m.
DISTRIBUTION: Comoro Isls.

Rousettus spinalatus Bergmans and Hill, 1980. Bull. Br. Mus. (Nat. Hist.) Zool., 38:95.
TYPE LOCALITY: Indonesia, N. Sumatra, near Madan or near Prapat.
DISTRIBUTION: Sumatra; Borneo.

Scotonycteris Matschie, 1894. Sitzb. Ges. Naturf. Fr. Berlin, p. 200.
ISIS NUMBER: 5301405001026000000.

Scotonycteris ophiodon Pohle, 1943. Sitzb. Ges. Naturf. Fr. Berlin, p. 76.
TYPE LOCALITY: Cameroun, Bipindi.
DISTRIBUTION: Liberia to Congo Republic.
ISIS NUMBER: 5301405001026001001.

Scotonycteris zenkeri Matschie, 1894. Sitzb. Ges. Naturf. Fr. Berlin, p. 202.
TYPE LOCALITY: Cameroun, Yaunde.
DISTRIBUTION: Liberia to Gabon to E. Zaire; Bioko.
ISIS NUMBER: 5301405001026002001.

Sphaerias Miller, 1906. Proc. Biol. Soc. Wash., 19:83.
ISIS NUMBER: 5301405001027000000.

  Sphaerias blanfordi (Thomas, 1891). Ann. Mus. Civ. Stor. Nat. Genova, ser. 2, 10:884, 921, 922.
  REVIEWED BY: S. Wang (SW).
  TYPE LOCALITY: Burma, Karin Hills.
  DISTRIBUTION: N. India; Bhutan; Tibet (SW); Burma; N. Thailand; S.W. China.
  ISIS NUMBER: 5301405001027001001.

Styloctenium Matschie, 1899. Megachiroptera Berlin Mus., p. 33.
ISIS NUMBER: 5301405001028000000.

  Styloctenium wallacei (Gray, 1866). Proc. Zool. Soc. Lond., 1866:65.
  TYPE LOCALITY: Indonesia, Sulawesi, Macassar.
  DISTRIBUTION: Sulawesi.
  ISIS NUMBER: 5301405001028001001.

Syconycteris Matschie, 1899. Megachiroptera Berlin Mus., p. 94, 95, 98.
REVIEWED BY: A. C. Ziegler (ACZ).
ISIS NUMBER: 5301405001035000000.

  Syconycteris australis (Peters, 1867). Monatsb. Preuss. Akad. Wiss. Berlin, p. 13.
  TYPE LOCALITY: Australia, Queensland, Rockhampton.
  DISTRIBUTION: E. Queensland; New South Wales; New Guinea; Louisiade Arch.;
      D'Entrecasteaux Isls.; Trobriand Isls.; Molucca Isls.; Bismarck Arch.; various
      adjacent small islands.
  COMMENT: Includes *naias* and *crassa*; see Lidicker and Ziegler, 1968, Univ. Calif. Publ.
      Zool., 87:34, and Koopman, 1979, Am. Mus. Novit., 2960:8.
  ISIS NUMBER: 5301405001035001001 as *S. australis*.
                    5301405001035002001 as *S. crassa*.
                    5301405001035003001 as *S. naias*.

Thoopterus Matschie, 1899. Megachiroptera Berlin Mus., p. 72, 73, 77.
ISIS NUMBER: 5301405001029000000.

  Thoopterus nigrescens (Gray, 1870). Cat. Monkeys, Lemurs, and Fruit-eating Bats Br. Mus.,
  p. 123.
  TYPE LOCALITY: Indonesia, Molucca Isls., Morotai.
  DISTRIBUTION: Molucca Isls.; N. Sulawesi.
  ISIS NUMBER: 5301405001029001001.

Family Rhinopomatidae
  REVIEWED BY: K. F. Koopman (KFK).
  ISIS NUMBER: 5301405002000000000.

Rhinopoma E. Geoffroy, 1818. Descrip. de L'Egypte, 2:113.
  COMMENT: Revised by Hill, 1977, Bull. Br. Mus. (Nat. Hist.) Zool., 32:29–43.
  ISIS NUMBER: 5301405002001000000.

  Rhinopoma hardwickei Gray, 1831. Zool. Misc., 1:37.
  TYPE LOCALITY: India.
  DISTRIBUTION: India to Israel and Arabia to Morocco, Mauritania, Nigeria, Sudan, and
      Kenya; Socotra Isl.
  ISIS NUMBER: 5301405002001001001.

  Rhinopoma microphyllum (Brunnich, 1782). Dyrenes Historie, 1:50.
  TYPE LOCALITY: Egypt, Giza.
  DISTRIBUTION: Morocco and Senegal to Thailand; Sumatra.
  ISIS NUMBER: 5301405002001002001.

  Rhinopoma muscatellum Thomas, 1903. Ann. Mag. Nat. Hist., ser. 7, 11:498.
  TYPE LOCALITY: Oman, Muscat, Wadi Bani Ruha.
  DISTRIBUTION: Oman; W. and S. Iran; S. Afghanistan.

Family Emballonuridae
REVIEWED BY: K. F. Koopman (KFK); J. Ramirez-Pulido (JRP)(Mexico).
ISIS NUMBER: 5301405003000000000.

Balantiopteryx Peters, 1867. Monatsb. Preuss. Akad. Wiss. Berlin, p. 476.
REVIEWED BY: B. Villa-Ramirez (BVR).
ISIS NUMBER: 5301405003001000000.

Balantiopteryx infusca (Thomas, 1897). Ann. Mag. Nat. Hist., ser. 6, 20:546.
TYPE LOCALITY: Ecuador, Esmeraldas, Cachabi.
DISTRIBUTION: W. Ecuador.
ISIS NUMBER: 5301405003001001001.

Balantiopteryx io Thomas, 1904. Ann. Mag. Nat. Hist., ser. 7, 13:252.
TYPE LOCALITY: Guatemala, Alta Verapaz, Rio Dolores (near Coban).
DISTRIBUTION: S. Veracruz (Mexico) to E.C. Guatemala.
ISIS NUMBER: 5301405003001002001.

Balantiopteryx plicata Peters, 1867. Monatsb. Preuss. Akad. Wiss. Berlin, p. 476.
TYPE LOCALITY: Costa Rica, Puntarenas.
DISTRIBUTION: Costa Rica to C. Sonora, S. Baja California (Mexico).
ISIS NUMBER: 5301405003001003001.

Centronycteris Gray, 1838. Mag. Zool. Bot., 2:499.
REVIEWED BY: B. Villa-Ramirez (BVR).
ISIS NUMBER: 5301405003002000000.

Centronycteris maximiliani (Fischer, 1829). Synopsis Mamm., p. 122.
TYPE LOCALITY: Brazil, Espirito Santo, Rio Jucy, Fazenda do Coroaba.
DISTRIBUTION: S. Veracruz (Mexico) to Peru, Brazil, and Guianas.
ISIS NUMBER: 5301405003002001001.

Coleura Peters, 1867. Monatsb. Preuss. Akad. Wiss. Berlin, p. 479.
ISIS NUMBER: 5301405003003000000.

Coleura afra (Peters, 1852). Reise nach Mossambique, Saugethiere, p. 51.
TYPE LOCALITY: Mozambique, Tete.
DISTRIBUTION: Guinea-Bissau to Somalia, south to Angola, Zaire, and Mozambique;
    Aden.
ISIS NUMBER: 5301405003003001001.

Coleura seychellensis Peters, 1869. Monatsb. Preuss. Akad. Wiss. Berlin for 1868, p. 367.
TYPE LOCALITY: Seychelle Isls., Mahe Isl.
DISTRIBUTION: Seychelle Isls.; possibly Zanzibar.
COMMENT: The Zanzibar record is extremely dubious (KFK).
ISIS NUMBER: 5301405003003002001.

Cormura Peters, 1867. Monatsb. Preuss. Akad. Wiss. Berlin, p. 475.
ISIS NUMBER: 5301405003004000000.

Cormura brevirostris (Wagner, 1843). Arch. Naturgesch., ser. 9, 1:367.
TYPE LOCALITY: Brazil, Amazonas, Rio Negro, Marabitanas.
DISTRIBUTION: Nicaragua to Peru, Amazonian Brazil, and Guianas.
ISIS NUMBER: 5301405003004001001.

Cyttarops Thomas, 1913. Ann. Mag. Nat. Hist., ser. 8, 11:134.
ISIS NUMBER: 5301405003010000000.

Cyttarops alecto Thomas, 1913. Ann. Mag. Nat. Hist., ser. 8, 11:135.
TYPE LOCALITY: Brazil, Para, Mocajatuba.
DISTRIBUTION: Nicaragua; Costa Rica; Guyana; Amazonian Brazil.
COMMENT: Reviewed by Starrett, 1972, Mamm. Species, 13:1–2.
ISIS NUMBER: 5301405003010001001.

Diclidurus Wied, 1820. Isis von Oken for 1819, p. 1629.
COMMENT: Includes *Depanycteris;* see Ojasti and Linares, 1976, Acta Biol. Venez.,
7:421–441. Corbet and Hill, 1980:46, listed *Depanycteris* as a distinct genus without
comment.
ISIS NUMBER: 5301405003012000000 as *Diclidurus.*
5301405003011000000 as *Depanycteris.*

Diclidurus albus Wied, 1820. Isis von Oken for 1819, p. 1630.
REVIEWED BY: B. Villa-Ramirez (BVR).
TYPE LOCALITY: Brazil, Bahia, Rio Pardo, Canavieiras.
DISTRIBUTION: Nayarit (Mexico) to E. Brazil and Trinidad.
COMMENT: Includes *virgo;* see Goodwin, 1969, Bull. Am. Mus. Nat. Hist., 141:48, 49.
Corbet and Hill, 1980:46, listed *virgo* as a distinct species without comment.
ISIS NUMBER: 5301405003012001001.

Diclidurus ingens Hernandez-Camacho, 1955. Caldasia, 7:87.
TYPE LOCALITY: Colombia, Caqueta, Rio Putumayo, Puerto Leguizamo.
DISTRIBUTION: Venezuela; S.E. Colombia; Guyana; N.W. Brazil.
ISIS NUMBER: 5301405003012002001.

Diclidurus isabellus (Thomas, 1920). Ann. Mag. Nat. Hist., ser. 9, 6:271.
TYPE LOCALITY: Brazil, Amazonas, Manacapuru (lower Solimoes River).
DISTRIBUTION: N.W. Brazil; Venezuela.
COMMENT: Formerly included in *Depanycteris;* see Ojasti and Linares, 1976, Acta Biol.
Venez., 7:421–441.
ISIS NUMBER: 5301405003011001001 as *Depanycteris isabella.*

Diclidurus scutatus Peters, 1869. Monatsb. Preuss. Akad. Wiss. Berlin, p. 400.
TYPE LOCALITY: Brazil, Para, Belem.
DISTRIBUTION: Amazonian Brazil; Venezuela; Peru; Guyana; Surinam.
ISIS NUMBER: 5301405003012003001.

Emballonura Temminck, 1838. Tijdschr. Nat. Gesch. Physiol., 5:22.
REVIEWED BY: A. C. Ziegler (ACZ)(New Guinea).
ISIS NUMBER: 5301405003005000000.

Emballonura alecto (Eydoux and Gervais, 1836). Mag. Zool. Paris, 6:7.
TYPE LOCALITY: Philippines, Luzon, Manila.
DISTRIBUTION: Philippines; Borneo; N. Sulawesi and adjacent small islands.
COMMENT: Includes *rivalis;* see Medway, 1977:44.
ISIS NUMBER: 5301405003005001001 as *E. alecto.*
5301405003005009001 as *E. rivalis.*

Emballonura atrata Peters, 1874. Monatsb. Preuss. Akad. Wiss. Berlin, p. 693.
TYPE LOCALITY: Madagascar.
DISTRIBUTION: E. and C. Madagascar.
ISIS NUMBER: 5301405003005002001.

Emballonura beccarii Peters and Doria, 1881. Ann. Mus. Civ. Stor. Nat. Genova, 16:693.
TYPE LOCALITY: Indonesia, Irian Jaya, Tjenderawasih Div., Japen Isl., Ansus.
DISTRIBUTION: New Guinea; Kei Isls.; Trobriand Isls.
ISIS NUMBER: 5301405003005003001.

Emballonura dianae Hill, 1956. *In* Wolff, The Nat. Hist. Rennell Isl., Br. Solomon Isls.,
1:74.
TYPE LOCALITY: Solomon Isls., Rennell Isl., near Tigoa, Te-Abagua Cave, about 35 m.
DISTRIBUTION: Rennell and Malaita Isls. (Solomon Isls.).

Emballonura furax Thomas, 1911. Ann. Mag. Nat. Hist., ser. 8, 7:384.
TYPE LOCALITY: Indonesia, Irian Jaya, Kapare River, Whitewater Camp., 400 ft. (122 m).
DISTRIBUTION: New Guinea.
ISIS NUMBER: 5301405003005004001.

Emballonura monticola Temminck, 1838. Tijdschr. Nat. Gesch. Physiol., 5:25.
TYPE LOCALITY: Indonesia, Java, Mt. Munara.
DISTRIBUTION: Thailand to Malaya; Sumatra; Rhio Arch.; Banka; Billiton; Engano; Babi
Isls.; Anamba Isls.; Batu Isls.; Nias Isl.; Mentawi Isls.; Borneo; Java; Sulawesi;
Karimata Isl.
ISIS NUMBER: 5301405003005005001.

Emballonura nigrescens (Gray, 1843). Ann. Mag. Nat. Hist., ser. 1, 11:117.
TYPE LOCALITY: Indonesia, Molucca Isls., Amboina Isl.
DISTRIBUTION: New Guinea; Kei Isls.; Halmahera Isls.; Schouten Isls.; Sulawesi;
Amboina; Buru; Bismarck Archipelago; Solomon Isls.; adjacent small islands.
COMMENT: Includes *papuana*; see Laurie and Hill, 1954:49. Includes *solomonis*,
considered a distinct species by McKean, 1972, Res. Tech. Pap., Comm. Sci. Res.
Org., 26:35.
ISIS NUMBER: 5301405003005006001 as *E. nigrescens*.
5301405003005007001 as *E. papuana*.

Emballonura raffrayana Dobson, 1879. Proc. Zool. Soc. Lond., 1878:876.
TYPE LOCALITY: Indonesia, Irian Jaya, Tjenderawasih Div., Numfoor Isl.
DISTRIBUTION: Seram Isl.; Kei Isls.; New Guinea; Choiseul, Ysabel and Malaita Isls.
(Solomon Isls.); Bismarck Arch.; Sulawesi.
COMMENT: For clarification of type locality, see Thomas, 1914, Ann. Mag. Nat. Hist.,
ser. 8, 13:442.
ISIS NUMBER: 5301405003005008001.

Emballonura semicaudata (Peale, 1848). Mammalia and Ornithology, *in* U.S. Expl. Exped.,
8:23.
TYPE LOCALITY: Samoa.
DISTRIBUTION: Mariana Isls.; Palau Isls.; New Hebrides; Fijis; Samoa.
COMMENT: Includes *rotensis*, which may belong with *sulcata* (KFK).

Emballonura sulcata Miller, 1911. Proc. Biol. Soc. Wash., 24:161.
TYPE LOCALITY: Caroline Isls., Truk Isls., Uola Isl. (Pac. Isls. Trust Terr., U.S.).
DISTRIBUTION: Truk and Ponape Isls. (Caroline Isls.).

Peropteryx Peters, 1867. Monatsb. Preuss. Akad. Wiss. Berlin, p. 472.
REVIEWED BY: B. Villa-Ramirez (BVR).
COMMENT: Includes *Peronymus*; see Cabrera, 1958:52. Corbet and Hill, 1980:45, listed
Peronymus as a distinct genus following Sanborn, 1937, Field Mus. Nat. Hist.
Publ. Zool. Ser., 20:321–354.
ISIS NUMBER: 5301405003006000000.

Peropteryx kappleri Peters, 1867. Monatsb. Preuss. Akad. Wiss. Berlin, p. 473.
TYPE LOCALITY: Surinam.
DISTRIBUTION: S. Veracruz and Oaxaca (Mexico) to Surinam, S. and E. Brazil, and Peru.
ISIS NUMBER: 5301405003006001001.

Peropteryx leucoptera Peters, 1867. Monatsb. Preuss. Akad. Wiss. Berlin, p. 474.
TYPE LOCALITY: Surinam
DISTRIBUTION: Peru; N. and E. Brazil; Venezuela; French Guiana; Guyana; and
Surinam.
COMMENT: Formerly included in *Peronymus*; see Cabrera, 1958:52.
ISIS NUMBER: 5301405003006002001 as *P. leucopterus (sic)*.

Peropteryx macrotis (Wagner, 1843). Arch. Naturgesch., ser. 9, 1:367.
TYPE LOCALITY: Brazil, Mato Grosso.
DISTRIBUTION: Oaxaca, Guerrero, and Yucatan (Mexico) to Peru, Paraguay, and S. and
E. Brazil; Tobago; Margarita Isl.; Aruba Isl.; Trinidad; Grenada (Lesser Antilles).
COMMENT: Includes *trinitatis*; see Goodwin, 1961, Bull. Am. Mus. Nat. Hist., 122:216.
Corbet and Hill, 1980:45, listed *trinitatis* as a distinct species without comment.
ISIS NUMBER: 5301405003006003001.

Rhynchonycteris Peters, 1867. Monatsb. Preuss. Akad. Wiss. Berlin, p. 477.
REVIEWED BY: B. Villa-Ramirez (BVR).
ISIS NUMBER: 5301405003007000000.

Rhynchonycteris naso (Wied-Neuwied, 1820). Reise nach Brasilien, 1:251.
TYPE LOCALITY: Brazil, Bahia, Rio Mucuri, near Morro d'Arara.
DISTRIBUTION: E. Oaxaca and C. Veracruz (Mexico) to C. and E. Brazil, Peru, Bolivia,
French Guiana, Guyana, and Surinam; Trinidad.
COMMENT: See Avila-Pires, 1965, Am. Mus. Novit., 2209:9, for clarification of the type
locality.
ISIS NUMBER: 5301405003007001001.

Saccolaimus Temminck, 1838. Tijdschr. Nat. Gesch. Physiol., 5:14.
REVIEWED BY: F. R. Allison (FRA)(Australia).
COMMENT: Considered a subgenus of *Taphozous* by Ellerman and Morrison-Scott,
1951:104, and Corbet and Hill, 1980:45, but see Barghorn, 1977, Am. Mus. Novit.,
2618:5.

Saccolaimus flaviventris Peters, 1867. Proc. Zool. Soc. Lond., 1866:430.
TYPE LOCALITY: Australia.
DISTRIBUTION: Australia (except Tasmania).
COMMENT: Includes *hargravei* and *affinis*; see Troughton, 1925, Rec. Aust. Mus.,
14:313–341.
ISIS NUMBER: 5301405003009002001 as *Taphozous flaviventris*.

Saccolaimus mixtus Troughton, 1925. Rec. Aust. Mus., 14:322.
REVIEWED BY: A. C. Ziegler (ACZ).
TYPE LOCALITY: Papua New Guinea, Central Prov., Port Moresby.
DISTRIBUTION: S.E. New Guinea; N.E. Queensland (Australia).
ISIS NUMBER: 5301405003009010001 as *Taphozous mixtus*.

Saccolaimus peli (Temminck, 1853). Esquisses Zool. sur la Cote de Guine, p. 82.
TYPE LOCALITY: Ghana, Boutry River.
DISTRIBUTION: Liberia to Zaire to W. Kenya; Gabon; N.E. Angola.
ISIS NUMBER: 5301405003009013001 as *Taphozous peli*.

Saccolaimus pluto (Miller, 1911). Proc. U.S. Nat. Mus., 38:396.
TYPE LOCALITY: Philippines, Mindanao, near Zamboanga.
DISTRIBUTION: Philippines.
COMMENT: Includes *capito*; see Lawrence, 1939, Bull. Mus. Comp. Zool., 86:42. Corbet
and Hill, 1980:45, listed *capito* as a distinct species without comment. Probably a
subspecies of *saccolaimus* (KFK).

Saccolaimus saccolaimus (Temminck, 1838). Tijdschr. Nat. Gesch. Physiol., 5:14.
TYPE LOCALITY: Indonesia, Java.
DISTRIBUTION: India and Sri Lanka through S.E. Asia to Sumatra, Borneo, Java, and
Timor; New Guinea; N.E. Queensland (Australia); Guadalcanal Isl. (Solomon
Isls.).
COMMENT: Includes *affinis*, *flavomaculatus*, and *nudicluniatus*; see Medway, 1977:45;
Goodwin, 1979, Bull. Am. Mus. Nat. Hist., 163:102. Corbet and Hill, 1980:45,
listed *nudicluniatus* as a distinct species without comment.
ISIS NUMBER: 5301405003009016001 as *Taphozous saccolaimus*.
5301405003009011001 as *Taphozous nudicluniatus*.

Saccopteryx Illiger, 1811. Prodr. Syst. Mamm. et Avium., p. 121.
REVIEWED BY: B. Villa-Ramirez (BVR)(C. America).
ISIS NUMBER: 5301405003008000000.

Saccopteryx bilineata (Temminck, 1838). Tijdschr. Nat. Gesch. Physiol., 5:33.
TYPE LOCALITY: Surinam.
DISTRIBUTION: Colima to Oaxaca and Veracruz (Mexico) to Bolivia, Peru, French
Guiana, Guyana, Surinam, and E. Brazil south to Rio de Janiero; Trinidad.
ISIS NUMBER: 5301405003008001001.

Saccopteryx canescens Thomas, 1901. Ann. Mag. Nat. Hist., ser. 7, 7:366.
TYPE LOCALITY: Brazil, Para, Obidos.
DISTRIBUTION: Colombia; Venezuela; Guianas; N. Brazil; Peru.
COMMENT: Includes *pumila*; see Husson, 1962, Zool. Meded. Rijks. Mus. Nat. Hist.
Leiden, 58:46.
ISIS NUMBER: 5301405003008002001 as *S. canescens*.
5301405003008005001 as *S. pumilis (sic)*.

Saccopteryx gymnura Thomas, 1901. Ann. Mag. Nat. Hist., ser. 7, 7:367.
TYPE LOCALITY: Brazil, Para, Santarem.
DISTRIBUTION: Amazonian Brazil; perhaps Venezuela.
ISIS NUMBER: 5301405003008003001.

Saccopteryx leptura (Schreber, 1774). Die Saugethiere, 1(8):57.
TYPE LOCALITY: Surinam.
DISTRIBUTION: Chiapas and Tabasco (Mexico) to E. Brazil and Peru; Guianas; Margarita
Isl.; Trinidad and Tobago.
ISIS NUMBER: 5301405003008004001.

Taphozous E. Geoffroy, 1818. Descrip. de L'Egypte, 2:113.
REVIEWED BY: F. R. Allison (FRA)(Australia); A. C. Ziegler (ACZ).
COMMENT: Includes *Liponycteris* but not *Saccolaimus*; see Hayman and Hill, 1971, Part
2:15, and Barghorn, 1977, Am. Mus. Novit., 2618:5.
ISIS NUMBER: 5301405003009000000.

Taphozous australis Gould, 1854. Mamm. Aust., p. 3.
TYPE LOCALITY: Australia, Queensland, Albany Isl. (off Cape York).
DISTRIBUTION: N. Queensland; Torres Strait islands; S.E. New Guinea (perhaps
accidental).
COMMENT: Includes *fumosus*; see Troughton, 1925, Rec. Aust. Mus., 14:332. Tate, 1952,
Bull. Am. Mus. Nat. Hist., 98:607, included *georgianus* in this species, but also see
McKean and Price, 1967, Mammalia, 31:101–119.
ISIS NUMBER: 5301405003009001001.

Taphozous georgianus Thomas, 1915. J. Bombay Nat. Hist. Soc., 24:62.
TYPE LOCALITY: Australia, Western Australia, King Georges Sound.
DISTRIBUTION: Australia.
COMMENT: Includes *troughtoni*; see McKean and Price, 1967, Mammalia, 31:101–119.
ISIS NUMBER: 5301405003009003001.

Taphozous hamiltoni Thomas, 1920. Ann. Mag. Nat. Hist., ser. 9, 5:142.
TYPE LOCALITY: Sudan, Equatoria, Mongalla.
DISTRIBUTION: S. Sudan; Chad; Kenya.
ISIS NUMBER: 5301405003009004001.

Taphozous hildegardeae Thomas, 1909. Ann. Mag. Nat. Hist., ser. 8, 4:98.
TYPE LOCALITY: Kenya, Coast Province, Rabai (near Mombasa).
DISTRIBUTION: Kenya; N.E. Tanzania; Zanzibar.
ISIS NUMBER: 5301405003009005001.

Taphozous hilli Kitchener, 1980. Rec. W. Aust. Mus., 8:162.
TYPE LOCALITY: Australia, Western Australia, Hamersley range, near Mt. Bruce.
DISTRIBUTION: Western Australia, Northern Territory.
COMMENT: From the description of of this species, it appears to to be close to and
possibly conspecific with *kapalgensis* (KFK).

Taphozous kapalgensis McKean and Friend, 1979. Vict. Nat., 96:239.
TYPE LOCALITY: Australia, Northern Territory, S. Alligator River, near Rookery Point,
Kapalga.
DISTRIBUTION: Northern Territory (Australia).

Taphozous longimanus Hardwicke, 1825. Trans. Linn. Soc. Lond., 14:525.
TYPE LOCALITY: India, Bengal, Calcutta.
DISTRIBUTION: Sri Lanka; India to Kampuchea; Malay Peninsula; Sumatra; Borneo;
Java; Flores; Bali.
ISIS NUMBER: 5301405003009007001.

Taphozous mauritianus E. Geoffroy, 1818. Descrip. de L'Egypte, 2:127.
TYPE LOCALITY: Mauritius.
DISTRIBUTION: South Africa to Sudan and Somalia to Senegal; Mauritius; Madagascar;
Reunion; Assumption Isl.; Aldabra Isl.
ISIS NUMBER: 5301405003009008001.

Taphozous melanopogon Temminck, 1841. Monogr. Mamm., 2:287.
REVIEWED BY: S. Wang (SW).
TYPE LOCALITY: Indonesia, W. Java; Bantam.
DISTRIBUTION: Sri Lanka; India; Burma; Laos; Vietnam; S. China; Malay Peninsula and
adjacent islands; Sumatra; Java; Borneo; Savu Isl.; Sumbawa; Timor; Kei Isls.;
perhaps Lombok.
ISIS NUMBER: 5301405003009009001.

Taphozous nudiventris Cretzschmar, 1830. *In* Ruppell, Atlas Reise Nordl. Afr., Saugeth., p.
70.
TYPE LOCALITY: Egypt, Giza.
DISTRIBUTION: Mauritania, Senegal, and Guinea-Bissau, south to Tanzania and east to
Burma.
COMMENT: Includes *kachensis*; see Felten, 1962, Senckenberg. Biol., 43:175. Formerly
included in *Liponycteris*; see Hayman and Hill, 1971, Part 2:15; and Barghorn,
1977, Am. Mus. Novit., 2618:5.
ISIS NUMBER: 5301405003009012001 as *T. nudiventris.*
5301405003009006001 as *T. kachhensis (sic).*

Taphozous perforatus E. Geoffroy, 1818. Descrip. de L'Egypte, 2:126.
TYPE LOCALITY: Egypt, Kom Ombo.
DISTRIBUTION: Senegal to Zaire to Botswana and Mozambique, Tanzania and Kenya;
Somalia to Sudan and Egypt; S. Arabia; S. Iran; Pakistan; N.W. India.
COMMENT: Includes *senegalensis* and *sudani*; see Hayman and Hill, 1971, Part 2:16.
ISIS NUMBER: 5301405003009014001.

Taphozous philippinensis Waterhouse, 1845. Proc. Zool. Soc. Lond., 1845:9.
TYPE LOCALITY: Philippines.
DISTRIBUTION: Philippines.
COMMENT: Includes *solifer*; see Ellerman and Morrison-Scott, 1951:105. Probably a
subspecies of *melanopogon* (KFK). Corbet and Hill, 1980:45, listed *solifer* as a
distinct species without comment.
ISIS NUMBER: 5301405003009015001.

Taphozous theobaldi Dobson, 1872. Proc. Asiat. Soc. Bengal, p. 152.
TYPE LOCALITY: Burma, Tenasserim.
DISTRIBUTION: C. India to Indochina; Java.
COMMENT: A record from Malaya appears to be in error; see Medway, 1969, The Wild
Mammals of Malaya, p. 18.
ISIS NUMBER: 5301405003009017001.

Family Craseonycteridae
REVIEWED BY: K. F. Koopman (KFK).

Craseonycteris Hill, 1974. Bull. Br. Mus. (Nat. Hist.) Zool., 27:304.

Craseonycteris thonglongyai Hill, 1974. Bull. Br. Mus. (Nat. Hist.) Zool., 27:305.
TYPE LOCALITY: Thailand, Kanchanaburi, Ban Sai Yoke.
DISTRIBUTION: Thailand; known only from the type locality.
COMMENT: Reviewed by Hill and Smith, 1981, Mamm. Species, 160:1–4.

Family Nycteridae
REVIEWED BY: K. F. Koopman (KFK).
ISIS NUMBER: 5301405005000000000.

Nycteris G. Cuvier and E. Geoffroy, 1795. Mag. Encyclop., 2:186.
ISIS NUMBER: 5301405005001000000.

Nycteris arge Thomas, 1903. Ann. Mag. Nat. Hist., ser. 7, 12:633.
TYPE LOCALITY: Cameroun, Efulen.
DISTRIBUTION: Sierra Leone to S. and E. Zaire; W. Kenya; S.W. Sudan; N.E. Angola; Bioko.
COMMENT: Includes *intermedia;* see Hayman and Hill, 1971, Part 2:19.
ISIS NUMBER: 5301405005001002001 as *N. arge.*
5301405005001006001 as *N. intermedia.*

Nycteris gambiensis (K. Andersen, 1912). Ann. Mag. Nat. Hist., ser. 8, 10:548.
TYPE LOCALITY: Senegal, Dialakoto.
DISTRIBUTION: Senegal; Guinea; Sierra Leone; Ghana; Gambia; Togo; Upper Volta; Benin.
ISIS NUMBER: 5301405005001003001.

Nycteris grandis Peters, 1865. Monatsb. Preuss. Akad. Wiss. Berlin, p. 358.
TYPE LOCALITY: "Guinea".
DISTRIBUTION: Senegal to Zaire to Kenya, Tanzania and Uganda to Zambia, Malawi, Zimbabwe, and Mozambique; Zanzibar; Pemba. Possibly Namibia.
ISIS NUMBER: 5301405005001004001.

Nycteris hispida (Schreber, 1775). Saugethiere, 1:169, 188.
TYPE LOCALITY: Senegal.
DISTRIBUTION: Senegal to Somalia to Angola and South Africa; Zanzibar; Bioko.
COMMENT: Includes *aurita;* see Koopman, 1975, Bull. Am. Mus. Nat. Hist., 154:377, 378. Includes *pallida;* see Hayman and Hill, 1971, Part 2:18.
ISIS NUMBER: 5301405005001005001.

Nycteris javanica E. Geoffroy, 1813. Ann. Mus. Hist. Nat. Paris, 20:20.
TYPE LOCALITY: Indonesia, Java.
DISTRIBUTION: Java; Bali.
COMMENT: Does not include *tragata;* see Ellerman and Morrison-Scott, 1955, Suppl. to Chasen, 1940, Handlist of Malaysian Mamm., p. 9.
ISIS NUMBER: 5301405005001007001 as *N. javanicus (sic).*

Nycteris macrotis Dobson, 1876. Monogr. Asiat. Chiroptera, p. 80.
TYPE LOCALITY: Sierra Leone.
DISTRIBUTION: Senegal to Ethiopia, south to Zimbabwe and Malawi; Zanzibar.
COMMENT: Includes *aethiopica, luteola,* and *major* (of J. A. Allen, 1917, not K. Andersen, 1912); see Koopman, 1975, Bull. Am. Mus. Nat. Hist., 154:378.
ISIS NUMBER: 5301405005001008001 as *N. macrotis.*
5301405005001001001 as *N. aethiopica.*

Nycteris madagascariensis G. Grandidier, 1937. Bull. Mus. Hist. Nat. Paris, ser. 2, 9:353.
TYPE LOCALITY: Madagascar, Valley of the Rodo.
DISTRIBUTION: Madagascar.

Nycteris major (K. Andersen, 1912). Ann. Mag. Nat. Hist., ser. 8, 10:547.
TYPE LOCALITY: Cameroun, Ja River.
DISTRIBUTION: Benin; Cameroun; Congo Republic; S. and E. Zaire; Gabon.
COMMENT: Includes *avakubia;* see Koopman, 1965, Am. Mus. Novit., 2219:6.
ISIS NUMBER: 5301405005001009001.

Nycteris nana (K. Andersen, 1912). Ann. Mag. Nat. Hist., ser. 8, 10:547.
TYPE LOCALITY: Equatorial Guinea, Rio Muni, Benito River.
DISTRIBUTION: Ghana to N.E. Angola and S. and E. Zaire; W. Kenya; S.W. Sudan; Uganda; W. Tanzania.
COMMENT: Includes *tristis;* see Koopman, 1975, Bull. Am. Mus. Nat. Hist., 154:376.
ISIS NUMBER: 5301405005001010001.

Nycteris parisii (De Beaux, 1924). Atti. Soc. Ital. Sci. Nat. Mus. Civ. Stor. Nat. Milano, 62:254.
TYPE LOCALITY: Somalia, Upper Juba, Ballei.
DISTRIBUTION: N. Cameroun; S. Somalia; Ethiopia; perhaps S. Tanzania.
COMMENT: A poorly known species of dubious validity (KFK).
ISIS NUMBER: 5301405005001011001.

Nycteris thebaica E. Geoffroy, 1818. Descrip. de L'Egypte, 2:119.
TYPE LOCALITY: Egypt, Thebes (near Luxor).
DISTRIBUTION: S.W. Arabia; Israel; Sinai; Egypt to Morocco, Senegal, Benin, Somalia and Kenya, thence south to South Africa in open country; Zanzibar; Pemba.
ISIS NUMBER: 5301405005001012001.

Nycteris tragata (K. Andersen, 1912). Ann. Mag. Nat. Hist., ser. 8, 10:546.
TYPE LOCALITY: Malaysia, Sarawak, Bidi caves.
DISTRIBUTION: Burma; Thailand; Malaya; Sumatra; Borneo.
COMMENT: Clearly distinct from *javanica* (KFK).

Nycteris vinsoni Dalquest, 1965. J. Mammal., 46:256.
TYPE LOCALITY: Mozambique, Save River (212 km S.S.W. Beira).
DISTRIBUTION: Mozambique.
COMMENT: A poorly known species apparently closest to *thebaica* (KFK).

Nycteris woodi K. Andersen, 1914. Ann. Mag. Nat. Hist., ser. 8, 13:563.
TYPE LOCALITY: Zambia, Chilanga.
DISTRIBUTION: S.E. Zimbabwe; C. and E. Zambia; S.W. Tanzania.
ISIS NUMBER: 5301405005001013001.

Family Megadermatidae
REVIEWED BY: K. F. Koopman (KFK).
ISIS NUMBER: 5301405006000000000.

Cardioderma Peters, 1873. Monatsb. Preuss. Akad. Wiss. Berlin, p. 488.
ISIS NUMBER: 5301405006001000000.

Cardioderma cor (Peters, 1872). Monatsb. Preuss. Akad. Wiss. Berlin, p. 194.
TYPE LOCALITY: Ethiopia.
DISTRIBUTION: Ethiopia; Somalia; Kenya; Uganda; E. Sudan; Tanzania; Zanzibar.
ISIS NUMBER: 5301405006001001001.

Lavia Gray, 1838. Mag. Zool. Bot., 2:490.
ISIS NUMBER: 5301405006002000000.

Lavia frons (E. Geoffroy, 1810). Ann. Mus. Hist. Nat. Paris, 15:192.
TYPE LOCALITY: Senegal.
DISTRIBUTION: Senegal to Somalia, south to Zambia; Zanzibar.
ISIS NUMBER: 5301405006002001001.

Macroderma Miller, 1906. Proc. Biol. Soc. Wash., 19:84.
REVIEWED BY: F. R. Allison (FRA).
ISIS NUMBER: 5301405006003000000.

Macroderma gigas (Dobson, 1880). Proc. Zool. Soc. Lond., 1880:461.
TYPE LOCALITY: Australia, Queensland, Wilson's River, Mt. Margaret.
DISTRIBUTION: N. and C. Australia.
COMMENT: Includes *saturata*; see Douglas, 1962, W. Aust. Nat., 8:59–61.
ISIS NUMBER: 5301405006003001001.

Megaderma E. Geoffroy, 1810. Ann. Mus. Hist. Nat. Paris, 15:197.
ISIS NUMBER: 5301405006004000000.

Megaderma lyra E. Geoffroy, 1810. Ann. Mus. Hist. Nat. Paris, 15:190.
TYPE LOCALITY: India, Madras.
DISTRIBUTION: Afghanistan to S. China, south to Sri Lanka and Malaya.
ISIS NUMBER: 5301405006004001001.

Megaderma spasma (Linnaeus, 1758). Syst. Nat., 10th ed., 1:32.
TYPE LOCALITY: Indonesia, Molucca Islands, Ternate.
DISTRIBUTION: Sri Lanka and India through S.E. Asia to Java, the Philippines and Molucca Isls; various adjacent islands.
ISIS NUMBER: 5301405006004002001.

Family Rhinolophidae
REVIEWED BY: K. F. Koopman (KFK).
COMMENT: Includes Hipposideridae, see Vaughan, 1978:39; Koopman and Jones, 1970,
in Slaughter and Walton, S. Methodist Univ. Press, pp. 24–25; but also see
Swanepoel, *et al.*, 1980, Ann. Transvaal Mus., 32(7):157.
ISIS NUMBER: 5301405007000000000.

Anthops Thomas, 1888. Ann. Mag. Nat. Hist., ser. 6, 1:156.
REVIEWED BY: A. C. Ziegler (ACZ).
ISIS NUMBER: 5301405007001000000.

Anthops ornatus Thomas, 1888. Ann. Mag. Nat. Hist., ser. 6, 1:156.
TYPE LOCALITY: Solomon Isls., Guadalcanal Isl., Aola.
DISTRIBUTION: Solomon Isls.
ISIS NUMBER: 5301405007001001001.

Asellia Gray, 1838. Mag. Zool. Bot., 2:493.
ISIS NUMBER: 5301405007002000000.

Asellia patrizii De Beaux, 1931. Ann. Mus. Civ. Stor. Nat. Genova, 55:186.
TYPE LOCALITY: Ethiopia, Dancalia, Gaare.
DISTRIBUTION: N. Ethiopia; and islands in the Red Sea.
ISIS NUMBER: 5301405007002001001.

Asellia tridens (E. Geoffroy, 1813). Ann. Mus. Hist. Nat. Paris, 20:265.
TYPE LOCALITY: Egypt, Qena, near Luxor.
DISTRIBUTION: Pakistan to Arabia, Sinai and Israel; Egypt to Morocco, Senegal, Chad,
Sudan and S. Somalia; Socotra; perhaps Zanzibar.
ISIS NUMBER: 5301405007002002001.

Aselliscus Tate, 1941. Am. Mus. Novit., 1140:2.
REVIEWED BY: A. C. Ziegler (ACZ).
ISIS NUMBER: 5301405007005000000.

Aselliscus stoliczkanus (Dobson, 1871). Proc. Asiat. Soc. Bengal, p. 106.
TYPE LOCALITY: Malaysia, Penang.
DISTRIBUTION: Malaya; Indochina; Burma; S. China; Thailand.
COMMENT: Includes *trifidus* and *wheeleri*; see Sanborn, 1952, Chicago Acad. Sci. Nat.
Hist. Misc., 97:3.
ISIS NUMBER: 5301405007005001001.

Aselliscus tricuspidatus (Temminck, 1835). Monogr. Mamm., 2:20.
TYPE LOCALITY: Indonesia, Molucca Isls., Amboina.
DISTRIBUTION: Moluccas; New Guinea; Bismarck Arch.; Solomon Isls.; Santa Cruz Isls.;
New Hebrides; adjacent small islands.
ISIS NUMBER: 5301405007005002001.

Cloeotis Thomas, 1901. Ann. Mag. Nat. Hist., ser. 7, 8:28.
ISIS NUMBER: 5301405007006000000.

Cloeotis percivali Thomas, 1901. Ann. Mag. Nat. Hist., ser. 7, 8:28.
TYPE LOCALITY: Kenya, Coast Prov., Takaungu.
DISTRIBUTION: Kenya; Tanzania; S. Zaire; Mozambique; Zambia; Zimbabwe; S.E.
Botswana; Swaziland; Transvaal (South Africa).
ISIS NUMBER: 5301405007006001001.

Coelops Blyth, 1848. J. Asiat. Soc. Bengal, 17:251.
COMMENT: Includes *Chilophylla*; see Ellerman and Morrison-Scott, 1951:131.
ISIS NUMBER: 5301405007007000000.

Coelops frithi Blyth, 1848. J. Asiat. Soc. Bengal, 17:251.
TYPE LOCALITY: India, Bengal, Sunderbans.
DISTRIBUTION: N.E. India to S. China and Vietnam, south to Malaya, Taiwan, Java, and
Bali.
ISIS NUMBER: 5301405007007001001.

Coelops hirsutus (Miller, 1911). Proc. U.S. Nat. Mus., 38:395.
TYPE LOCALITY: Philippines, Mindoro Isl., Alag River.
DISTRIBUTION: Mindoro (Philippines).
COMMENT: Probably a subspecies of *robinsoni;* see Hill, 1972, Bull. Br. Mus. (Nat. Hist.)
Zool., 23:30.
ISIS NUMBER: 5301405007007002001.

Coelops robinsoni Bonhote, 1908. J. Fed. Malay St. Mus., 3:4.
TYPE LOCALITY: Malaysia, Pahang, foot of Mt. Tahan.
DISTRIBUTION: Malay Peninsula (including S. Thailand).
ISIS NUMBER: 5301405007007003001.

Hipposideros Gray, 1831. Zool. Misc., 1:37.
REVIEWED BY: F. R. Allison (FRA)(Australia); A. C. Ziegler (ACZ)(New Guinea).
COMMENT: Revised by Hill, 1963, Bull. Br. Mus. (Nat. Hist.) Zool., 11:1–129.
ISIS NUMBER: 5301405007008000000.

Hipposideros abae J. A. Allen, 1917. Bull. Am. Mus. Nat. Hist., 37:432.
TYPE LOCALITY: Zaire, Oriental, Aba.
DISTRIBUTION: Guinea-Bissau to S.W. Sudan and Uganda.
ISIS NUMBER: 5301405007008001001.

Hipposideros armiger (Hodgson, 1835). J. Asiat. Soc. Bengal, 4:699.
TYPE LOCALITY: Nepal.
DISTRIBUTION: N. India; Nepal; Burma; S. China; Vietnam; Laos; Thailand; Malay
Peninsula; Taiwan.
ISIS NUMBER: 5301405007008002001.

Hipposideros ater Templeton, 1848. J. Asiat. Soc. Bengal, 17:252.
TYPE LOCALITY: Sri Lanka, Western Prov., Colombo.
DISTRIBUTION: Sri Lanka; India to Malaya, through East Indies, Philippines, and New
Guinea to N. Queensland, N. Northern Territory and Kimberley (N. Western
Australia).
COMMENT: Formerly included in *bicolor,* but see Hill, 1963, Bull. Br. Mus. (Nat. Hist.)
Zool., 11:30.
ISIS NUMBER: 5301405007008003001.

Hipposideros beatus K. Andersen, 1906. Ann. Mag. Nat. Hist., ser. 7, 17:279.
TYPE LOCALITY: Equatorial Guinea, Rio Muni, 15 mi. (24 km) from Benito River.
DISTRIBUTION: Sierra Leone; Liberia; Ghana; Nigeria; Cameroun; Rio Muni; Gabon; N.
Zaire.
ISIS NUMBER: 5301405007008004001.

Hipposideros bicolor (Temminck, 1834). Tijdschr. Nat. Gesch. Physiol., 1:19.
REVIEWED BY: S. Wang (SW).
TYPE LOCALITY: Indonesia, Java, Anger coast.
DISTRIBUTION: India to S. China, Taiwan, and Vietnam, through Malaysia to the
Philippines, Sulawesi, Timor, and adjacent small islands.
COMMENT: Includes *pomona, erigens,* and *gentilis;* see Hill, 1963, Bull. Br. Mus. (Nat.
Hist.) Zool., 11:27.
ISIS NUMBER: 5301405007008005001 as *H. bicolor.*
5301405007008023001 as *H. gentilis.*
5301405007008036001 as *H. pomona.*

Hipposideros breviceps Tate, 1941. Bull. Am. Mus. Nat. Hist., 78:358.
TYPE LOCALITY: Indonesia, Sumatra, Mentawai Isls., N. Pagi Isl.
DISTRIBUTION: Mentawai Isls.

Hipposideros caffer (Sundevall, 1846). Ofv. Kongl. Svenska Vet.-Akad. Forhandl.
Stockholm, 3(4):118.
TYPE LOCALITY: South Africa, Natal, near Durban.
DISTRIBUTION: Most of subsaharan Africa except the central forested region; Morocco;
Yemen; Zanzibar; Pemba.
COMMENT: Includes *braima, nanus,* and *auriantiaca;* see Hayman and Hill, 1971, Part
2:29.
ISIS NUMBER: 5301405007008006001.

Hipposideros calcaratus (Dobson, 1877). Proc. Zool. Soc. Lond., 1877:122.
  TYPE LOCALITY: Papua New Guinea, Bismarck Archipelago, Duke of York Isl.
  DISTRIBUTION: New Guinea; Bismarck Arch.; Solomon Isls.; adjacent small islands.
  COMMENT: Includes *cupidus*; see Smith and Hill, 1981, Los Ang. Cty. Mus. Contrib.
    Sci., 331:8.
  ISIS NUMBER: 5301405007008007001 as *H. calcaratus.*
               5301405007008013001 as *H. cupidus.*

Hipposideros camerunensis Eisentraut, 1956. Zool. Jahrb. Abt. Syst. Oekol. Geogr. Tiere,
    84:526.
  TYPE LOCALITY: Cameroun, near Buea.
  DISTRIBUTION: Cameroun; E. Zaire.
  ISIS NUMBER: 5301405007008008001.

Hipposideros cineraceus Blyth, 1853. J. Asiat. Soc. Bengal, 22:410.
  TYPE LOCALITY: Pakistan, Punjab, Salt Range, near Pind Dadan Khan.
  DISTRIBUTION: Pakistan; India to Vietnam and Borneo; adjacent small islands.
  ISIS NUMBER: 5301405007008010001.

Hipposideros commersoni (E. Geoffroy, 1813). Ann. Mus. Hist. Nat. Paris, 20:263.
  TYPE LOCALITY: Madagascar, Fort Dauphin.
  DISTRIBUTION: Gambia to Ethiopia, south to Namibia, Botswana, Zimbabwe, and
    Mozambique; Madagascar; adjacent small islands.
  ISIS NUMBER: 5301405007008011001.

Hipposideros coronatus (Peters, 1871). Monatsb. Preuss. Akad. Wiss. Berlin, p. 327.
  TYPE LOCALITY: Philippines, Mindanao, Surigao, Mainit.
  DISTRIBUTION: N.E. Mindanao (Philippines).

Hipposideros coxi Shelford, 1901. Ann. Mag. Nat. Hist., ser. 7, 8:113.
  TYPE LOCALITY: Malaysia, Sarawak, Mt. Penrisen, 4200 ft. (1280 m).
  DISTRIBUTION: Sarawak (Borneo).
  ISIS NUMBER: 5301405007008012001.

Hipposideros crumeniferus Lesueur and Petit, 1807. *In* Peron, Voyage Decouv. Terres
    Australes, Atlas, pl. 35.
  TYPE LOCALITY: Indonesia, Timor.
  DISTRIBUTION: Timor.
  COMMENT: Based on plate only; not certainly determinable; see Laurie and Hill,
    1954:56; Hill, 1963, Bull. Br. Mus. (Nat. Hist.) Zool., 11:23.
  ISIS NUMBER: 5301405007008015001.

Hipposideros curtus G. M. Allen, 1921. Rev. Zool. Afr., 9:194.
  TYPE LOCALITY: Cameroun, Sakbayeme.
  DISTRIBUTION: Cameroun; Bioko; perhaps Nigeria.
  COMMENT: Includes *sandersoni*; see Hill, 1963, Bull. Br. Mus. (Nat. Hist.) Zool., 11:60.
  ISIS NUMBER: 5301405007008014001.

Hipposideros cyclops (Temminck, 1853). Esquisses Zool. sur la Cote de Guine, p. 75.
  TYPE LOCALITY: Ghana, Boutry River.
  DISTRIBUTION: Kenya and S. Sudan to Senegal and Guinea-Bissau; Bioko.
  ISIS NUMBER: 5301405007008016001.

Hipposideros diadema (E. Geoffroy, 1813). Ann. Mus. Hist. Nat. Paris, 20:263.
  TYPE LOCALITY: Indonesia, Lesser Sunda Isls., Timor Isl.
  DISTRIBUTION: Burma and Indochina through Malaya and East Indies to New Guinea,
    Bismarck Arch., Solomon Isls. and N.E. and N.C. Australia; Philippines.
  ISIS NUMBER: 5301405007008017001.

Hipposideros dinops K. Andersen, 1905. Ann. Mag. Nat. Hist., ser. 7, 16:502.
  TYPE LOCALITY: Solomon Isls., New Georgia Group, Rubiana Isl.
  DISTRIBUTION: Rubiana, Ysabel, Malaita (Solomon Isls.), and Bougainville Isl.; Peleng
    Isl.; Sulawesi.
  COMMENT: Includes *pelingensis*; see Hill, 1963, Bull. Br. Mus. (Nat. Hist.) Zool., 11:113.
  ISIS NUMBER: 5301405007008018001 as *H. dinops.*
               5301405007008035001 as *H. pelingensis.*

Hipposideros doriae (Peters, 1871). Monatsb. Preuss. Akad. Wiss. Berlin, p. 326.
TYPE LOCALITY: Malaysia, Sarawak.
DISTRIBUTION: Borneo.
COMMENT: May be an earlier name for *sabanus*, which it antedates; see Hill, 1963, Bull. Br. Mus. (Nat. Hist.) Zool., 11:24, 46, 47.

Hipposideros dyacorum Thomas, 1902. Ann. Mag. Nat. Hist., ser. 7, 9:271.
TYPE LOCALITY: Malaysia, Sarawak, Baram, Mt. Mulu.
DISTRIBUTION: Borneo.
ISIS NUMBER: 5301405007008019001.

Hipposideros fuliginosus (Temminck, 1853). Esquisses Zool. sur la Cote de Guine, p. 77.
TYPE LOCALITY: "Guinea coast".
DISTRIBUTION: Guinea to Zaire.
ISIS NUMBER: 5301405007008020001.

Hipposideros fulvus Gray, 1838. Mag. Zool. Bot., 2:492.
TYPE LOCALITY: India, Mysore, Dharwar.
DISTRIBUTION: Pakistan to Vietnam, south to Sri Lanka.
ISIS NUMBER: 5301405007008021001.

Hipposideros galeritus Cantor, 1846. J. Asiat. Soc. Bengal, 15:183.
TYPE LOCALITY: Malaysia, Penang. Probably Pinang State (ACZ).
DISTRIBUTION: Sri Lanka and India through S.E. Asia and East Indies to Philippines, New Guinea, Solomon Isls. and N. Queensland (Australia); Santa Cruz Isls.; New Hebrides.
COMMENT: Includes *cervinus* and *longicauda*; see Hill, 1963, Bull. Br. Mus. (Nat. Hist.) Zool., 11:56.
ISIS NUMBER: 5301405007008022001 as *H. galeritus*.
5301405007008009001 as *H. cervinus*.
5301405007008028001 as *H. longicauda*.

Hipposideros inexpectatus Laurie and Hill, 1954. List of land mammals of New Guinea, Celebes, and adjacent islands, Br. Mus. (Nat. Hist.), p. 60.
TYPE LOCALITY: Indonesia, Sulawesi, Poso(=Posso).
DISTRIBUTION: N. Sulawesi.
ISIS NUMBER: 5301405007008024001.

Hipposideros jonesi Hayman, 1947. Ann. Mag. Nat. Hist., ser. 11, 14:71.
TYPE LOCALITY: Sierra Leone, Makeni.
DISTRIBUTION: Sierra Leone and Guinea to Mali, Ghana and Nigeria.
ISIS NUMBER: 5301405007008025001.

Hipposideros lankadiva Kelaart, 1850. J. Sri Lanka Branch Asiat. Soc., 2(2):216.
TYPE LOCALITY: Sri Lanka, Kandy.
DISTRIBUTION: Sri Lanka; S. and C. India.
ISIS NUMBER: 5301405007008026001.

Hipposideros larvatus (Horsfield, 1823). Zool. Res. Java, 6, pl. 9.
TYPE LOCALITY: Indonesia, Java.
DISTRIBUTION: Bangladesh to Indochina; Yunnan, Kwangsi, Hainan (China)(SW); and through Malaya to Sumatra, Java, Borneo, Sumba, and adjacent small islands.
ISIS NUMBER: 5301405007008027001.

Hipposideros lekaguli Thonglongya and Hill, 1974. Mammalia, 38:286.
TYPE LOCALITY: Thailand, Saraburi, Kaeng Khoi, Phu Nam Tok Tak Kwang.
DISTRIBUTION: Thailand.

Hipposideros lylei Thomas, 1913. Ann. Mag. Nat. Hist., ser. 8, 12:88.
TYPE LOCALITY: Thailand, 50 mi. (80 km) N. Chiengmai, Chiengdao Cave.
DISTRIBUTION: Burma; Thailand; Malaya.
ISIS NUMBER: 5301405007008029001.

Hipposideros maggietaylorae Smith and Hill, 1981. Los Ang. Cty. Mus. Contrib. Sci., 331:9.
TYPE LOCALITY: Papua New Guinea, Bismarck Arch., New Ireland, 1.3 km S., 3 km E.,
Lakuramau Plantation.
DISTRIBUTION: New Guinea; Bismarck Arch.
COMMENT: Formerly confused with *H. calcaratus* (KFK).

Hipposideros marisae Aellen, 1954. Rev. Suisse Zool., 61:474.
TYPE LOCALITY: Ivory Coast, Duekoue, White Leopard Rock.
DISTRIBUTION: Ivory Coast; Liberia; Guinea.
ISIS NUMBER: 5301405007008030001.

Hipposideros megalotis (Heuglin, 1862). Nova Acta Acad. Caes. Leop.-Carol., Halle, 29(8):4,
8.
TYPE LOCALITY: Ethiopia, Eritrea, Bogos Land, Keren.
DISTRIBUTION: Ethiopia; Kenya.

Hipposideros muscinus (Thomas and Doria, 1886). Ann. Mus. Civ. Stor. Nat. Genova, 4:201.
TYPE LOCALITY: Papua New Guinea, Western Prov., Fly River.
DISTRIBUTION: E. New Guinea.
ISIS NUMBER: 5301405007008031001.

Hipposideros nequam K. Andersen, 1918. Ann. Mag. Nat. Hist., ser. 9, 2:380.
TYPE LOCALITY: Malaysia, Selangor, Klang.
DISTRIBUTION: Malaya.
COMMENT: Known only from the type locality, see Hill, 1963, Bull. Br. Mus. (Nat.
Hist.) Zool., 11:36.
ISIS NUMBER: 5301405007008032001.

Hipposideros obscurus (Peters, 1861). Monatsb. Preuss. Akad. Wiss. Berlin, p. 707.
TYPE LOCALITY: Philippines, Luzon, Camarines, Paracale.
DISTRIBUTION: Philippines.
ISIS NUMBER: 5301405007008033001.

Hipposideros papua (Thomas and Doria, 1886). Ann. Mus. Civ. Stor. Nat. Genova, 4:204.
TYPE LOCALITY: Indonesia, Irian Jaya, Tjenderawasih Div., Misori Isl. (=Biak Isl.),
Korido.
DISTRIBUTION: Biak Isl.
ISIS NUMBER: 5301405007008034001.

Hipposideros pratti Thomas, 1891. Ann. Mag. Nat. Hist., ser. 6, 7:527.
REVIEWED BY: S. Wang (SW).
TYPE LOCALITY: China, Szechwan, Kiatingfu.
DISTRIBUTION: S. China; Indochina.
ISIS NUMBER: 5301405007008037001.

Hipposideros pygmaeus (Waterhouse, 1843). Proc. Zool. Soc. Lond., 1843:67.
TYPE LOCALITY: Philippines.
DISTRIBUTION: Luzon, Negros Isl. (Philippines).

Hipposideros ridleyi Robinson and Kloss, 1911. J. Fed. Malay St. Mus., 4:241.
TYPE LOCALITY: Malaysia, Singapore, Botanic Gardens.
DISTRIBUTION: Malaya.
COMMENT: Known only from the type locality.
ISIS NUMBER: 5301405007008038001.

Hipposideros ruber (Noack, 1893). Zool. Jahrb. Abt. Syst. Oekol. Geogr. Tiere, 7:586.
TYPE LOCALITY: Tanzania, Eastern Province, Ngerengere River.
DISTRIBUTION: Senegal to Ethiopia, south to Angola, Zambia, Zimbabwe, Malawi, and
Tanzania; Bioko; Principe and Sao Tome Isls.
COMMENT: Closely related to and probably conspecific with *caffer*; see Koopman, 1975,
Bull. Am. Mus. Nat. Hist., 154:393.
ISIS NUMBER: 5301405007008039001.

Hipposideros sabanus Thomas, 1898. Ann. Mag. Nat. Hist., ser. 7, 1:243.
TYPE LOCALITY: Malaysia, Sarawak, Lawas.
DISTRIBUTION: Borneo; Sumatra; Malaya.
COMMENT: Possibly a synonym of *doriae*; see Hill, 1963, Bull. Br. Mus. (Nat. Hist.) Zool., 11:47.
ISIS NUMBER: 5301405007008040001.

Hipposideros schistaceus K. Andersen, 1918. Ann. Mag. Nat. Hist., ser. 9, 2:382.
TYPE LOCALITY: India, Mysore, Bellary.
DISTRIBUTION: S. India.

Hipposideros semoni Matschie, 1903. Denks. Med. Nat. Ges. Jena (Semon Zool. Forsch. Austr.), 8:774 (Heft 6:132).
TYPE LOCALITY: Australia, Queensland, Cooktown.
DISTRIBUTION: N. Queensland (Australia); E. New Guinea.
ISIS NUMBER: 5301405007008041001.

Hipposideros speoris (Schneider, 1800). *In* Schreber, Die Saugethiere, pl. 59b.
TYPE LOCALITY: India, Madras, Tranquebar.
DISTRIBUTION: Peninsular India; Sri Lanka.
ISIS NUMBER: 5301405007008042001.

Hipposideros stenotis Thomas, 1913. Ann. Mag. Nat. Hist., ser. 8, 12:206.
TYPE LOCALITY: Australia, Northern Territory, Mary River.
DISTRIBUTION: Northern Territory, N. Western Australia, N. Queensland (Australia); perhaps New Guinea.
COMMENT: The New Guinea record is probably erroneous; see Hill, 1963, Bull. Br. Mus. (Nat. Hist.) Zool., 11:87.
ISIS NUMBER: 5301405007008043001.

Hipposideros turpis Bangs, 1901. Am. Nat., 35:561.
TYPE LOCALITY: Japan, Ryukyu Isls., Sakishima Isls., Ishigaki Isl.
DISTRIBUTION: Peninsular Thailand; Ishigaki Isl, Yonakuni Isl., Iriomote Isl. (S. Ryukyu Isls.).

Hipposideros wollastoni Thomas, 1913. Ann. Mag. Nat. Hist., ser. 8, 12:205.
TYPE LOCALITY: Indonesia, Irian Jaya, Utakwa River, 2500 ft. (762 m).
DISTRIBUTION: W. New Guinea.
ISIS NUMBER: 5301405007008044001.

Paracoelops Dorst, 1947. Bull. Mus. Hist. Nat. Paris, ser. 2, 19:436.
ISIS NUMBER: 5301405007009000000.

Paracoelops megalotis Dorst, 1947. Bull. Mus. Hist. Nat. Paris, ser. 2, 19:436.
TYPE LOCALITY: Vietnam, Annam, Vinh.
DISTRIBUTION: C. Vietnam.
ISIS NUMBER: 5301405007009001001.

Rhinolophus Lacepede, 1799. Tabl. Mamm., p. 15.
REVIEWED BY: F. R. Allison (FRA)(Australia); J. Gaisler (JG) (Palearctic); O. L. Rossolimo (OLR)(U.S.S.R.); E. Vernier (EV)(Europe); A. C. Ziegler (ACZ)(New Guinea).
COMMENT: Includes *Rhinomegalophus*; see Thonglongya, 1973, Mammalia, 37:587.
ISIS NUMBER: 5301405007003000000 as *Rhinolophus*.
5301405007004000000 as *Rhinomegalophus*.

Rhinolophus acuminatus Peters, 1871. Monatsb. Preuss. Akad. Wiss. Berlin, p. 308.
TYPE LOCALITY: Indonesia, Java.
DISTRIBUTION: Sumatra; Borneo; Java; Lombok; Bali; Palawan (Philippines); Nias Isl., Engano Isl. (Sumatra); Thailand; Laos; Kampuchea.
ISIS NUMBER: 5301405007003001001.

Rhinolophus adami Aellen and Brosset, 1968. Rev. Suisse Zool., 75:443.
TYPE LOCALITY: Congo Republic, Kouilou.
DISTRIBUTION: Congo Republic.
ISIS NUMBER: 5301405007003002001.

Rhinolophus affinis Horsfield, 1823. Zool. Res. Java, 6, pl. figs. a, b.
REVIEWED BY: S. Wang (SW).
TYPE LOCALITY: Indonesia, Java.
DISTRIBUTION: India to S. China through Malaysia to Sumba; Andaman Isls.; perhaps
    Sri Lanka.
COMMENT: Includes *andamanensis*; see Sinha, 1973, Mammalia, 37:612–613.
ISIS NUMBER: 5301405007003003001.

Rhinolophus alcyone Temminck, 1852. Esquisses Zool. sur la Cote de Guine, p. 80.
TYPE LOCALITY: Ghana, Boutry River.
DISTRIBUTION: Senegal to Uganda, S.W. Sudan, N. Zaire, and Gabon; Bioko.
ISIS NUMBER: 5301405007003004001.

Rhinolophus anderseni Cabrera, 1909. Bol. Soc. Esp. Nat. Hist., p. 305.
TYPE LOCALITY: Philippines, Luzon (uncertain).
DISTRIBUTION: Palawan, Luzon (Philippines).
ISIS NUMBER: 5301405007003005001.

Rhinolophus arcuatus Peters, 1871. Monatsb. Preuss. Akad. Wiss. Berlin, p. 305.
TYPE LOCALITY: Philippines, Luzon.
DISTRIBUTION: Wetter Isl. (Flores Sea); Philippines; Sumatra; Sarawak (Borneo).
ISIS NUMBER: 5301405007003006001.

Rhinolophus blasii Peters, 1867. Monatsb. Preuss. Akad. Wiss. Berlin for 1866, p. 17.
TYPE LOCALITY: Italy and Yugoslavia.
DISTRIBUTION: Transvaal (South Africa) to S. Zaire; Ethiopia; Somalia; Morocco;
    Algeria; Tunisia; Turkey; Yemen; Israel; Jordan; Syria; Iran; Yugoslavia; Albania;
    Bulgaria; Rumania; Transcaucasia; Turkmenia; Afghanistan; Pakistan; Italy;
    Greece; Cyprus.
COMMENT: Includes *brockmani*; see Koopman, 1975, Bull. Am. Mus. Nat. Hist., 154:383.
    Includes *clivosus* Blasius, 1857 (not Cretzschmar, 1828); see Ellerman and
    Morrison-Scott, 1951:120.
ISIS NUMBER: 5301405007003008001.

Rhinolophus borneensis Peters, 1861. Monatsb. Preuss. Akad. Wiss. Berlin, p. 709.
TYPE LOCALITY: Malaysia, Sabah, Labuan Isl.
DISTRIBUTION: N. Borneo; Labuan Isl.; Banguey Isl.; Karimata Isls.; South Natuna Isls;
    Kampuchea; Vietnam; Java; Bali; Timor; Sulawesi.
COMMENT: Includes *javanicus, celebensis,* and *chaseni*; see Goodwin, 1979, Bull. Amer.
    Mus. Nat. Hist., 163:104; Hill and Thonglongya, 1972, Bull Br. Mus. (Nat. Hist.),
    Zool., 22:187. Corbet and Hill, 1980:48, 49, listed *javanicus* and *celebensis* as distinct
    species without comment.
ISIS NUMBER: 5301405007003010001 as *R. borneensis.*
              5301405007003013001 as *R. celebensis.*
              5301405007003031001 as *R. javanicus.*

Rhinolophus capensis Lichtenstein, 1823. Verz. Doblet. Mus. Univ. Berlin, p. 4.
TYPE LOCALITY: South Africa, Cape Province, Cape of Good Hope.
DISTRIBUTION: Cape Prov., Natal (South Africa); Zambia; Zimbabwe; Mozambique.
ISIS NUMBER: 5301405007003012001.

Rhinolophus clivosus Cretzschmar, 1828. *In* Ruppell, Atlas Reise Nordl. Afr., Saugeth., p.
    47.
TYPE LOCALITY: Saudi Arabia, Muwaylih(=Mohila).
DISTRIBUTION: Transcaucasia to Afghanistan; Arabia to Algeria; subsaharan Africa to
    Cameroun and South Africa.
COMMENT: Includes *bocharicus*; see Aellen, 1959, Rev. Suisse Zool., 66:362–366. *R.
    bocharicus* is considered a species by OLR and by DeBlase, 1980, Fieldiana Zool.,
    N.S., 4:94–97. This species does not include *deckenii* or *silvestris*; see Koopman,
    1975, Bull. Am. Mus. Nat. Hist., 154:386.
ISIS NUMBER: 5301405007003014001.

Rhinolophus coelophyllus Peters, 1867. Proc. Zool. Soc. Lond., 1866:426.
TYPE LOCALITY: Burma, Salaween River.
DISTRIBUTION: Malaya; Thailand; Burma.
COMMENT: Does not include *shameli;* see Hill and Thonglongya, 1972, Bull. Br. Mus.
(Nat. Hist.) Zool., 22:183–186.
ISIS NUMBER: 5301405007003015001.

Rhinolophus cognatus K. Andersen, 1906. Ann. Mus. Civ. Stor. Nat. Genova, ser. 3, 2:181.
TYPE LOCALITY: India, Andaman Isls., S. Andaman Isl., Port Blair.
DISTRIBUTION: Andaman Isls.

Rhinolophus cornutus Temminck, 1835. Monogr. Mamm., 2:37.
REVIEWED BY: S. Wang (SW).
TYPE LOCALITY: Japan.
DISTRIBUTION: Ryukyu Isls. and Japan; perhaps SE China.
COMMENT: Does not include *blythi;* see Hill and Yoshiyuki, 1980, Bull. Nat. Sci. Mus.
Tokyo, ser. A (Zool.), 6:186; but also see Corbet, 1978:43. See also comments
under *pusillus.*
ISIS NUMBER: 5301405007003016001.

Rhinolophus creaghi Thomas, 1896. Ann. Mag. Nat. Hist., ser. 6, 18:244.
TYPE LOCALITY: Malaysia, Sabah, Sandakan.
DISTRIBUTION: Borneo; Madura Isl.; Java; Timor.
COMMENT: Includes *canuti* and *pilosus;* see Hill, 1958, Sarawak Mus. J., 8:470–475.
ISIS NUMBER: 5301405007003017001 as *R. creaghi.*
5301405007003011001 as *R. canuti.*

Rhinolophus darlingi K. Andersen, 1905. Ann. Mag. Nat. Hist., ser. 7, 15:70.
TYPE LOCALITY: Zimbabwe, Mazoe.
DISTRIBUTION: Transvaal (South Africa); Namibia; S. Angola; N. and W. Botswana;
Zimbabwe; Malawi; Mozambique; Tanzania.
COMMENT: Includes *barbertonensis;* see Hayman and Hill, 1971, Part 2:23.
ISIS NUMBER: 5301405007003018001.

Rhinolophus deckenii Peters, 1868. Monatsb. Preuss. Akad. Wiss. Berlin for 1867, p. 705.
TYPE LOCALITY: Tanzania, "Zanzibar coast" (mainland opposite Zanzibar).
DISTRIBUTION: Uganda; Kenya; Tanzania; Pemba; Zanzibar.
COMMENT: Treated as a subspecies of *clivosus* by Hayman and Hill, 1971, Part 2:23; but
see Koopman, 1975, Bull. Am. Mus. Nat. Hist., 154:386.

Rhinolophus denti Thomas, 1904. Ann. Mag. Nat. Hist., ser. 7, 13:386.
TYPE LOCALITY: South Africa, Cape Province, Kuruman.
DISTRIBUTION: N. Cape Prov. (South Africa); Namibia; Botswana; Zimbabwe;
Mozambique; Guinea.
ISIS NUMBER: 5301405007003019001.

Rhinolophus eloquens K. Andersen, 1905. Ann. Mag. Nat. Hist., ser. 7, 15:74.
TYPE LOCALITY: Uganda, Entebbe.
DISTRIBUTION: Uganda; S. Somalia; S. Sudan; N.E. Zaire; Kenya; Rwanda; N. Tanzania;
Pemba; Zanzibar.
COMMENT: Includes *perauritus;* see Koopman, 1975, Bull. Am. Mus. Nat. Hist., 154:389.

Rhinolophus euryale Blasius, 1853. Arch. Naturgesch., 19(1):49.
TYPE LOCALITY: Italy, Milan.
DISTRIBUTION: Transcaucasia to Israel and S. Europe; Turkmenia; Iran; Algeria;
Morocco; Tunisia; adjacent islands; perhaps Egypt.
COMMENT: Revised by DeBlase, 1972, Isr. J. Zool., 21:1–12.
ISIS NUMBER: 5301405007003022001.

Rhinolophus euryotis Temminck, 1835. Monogr. Mamm., 2:26.
TYPE LOCALITY: Indonesia, Molucca Isls., Amboina Isl.
DISTRIBUTION: Aru Isls.; Buru, Amboina, Seram, and Timor Laut Isls.; Kei Isls.;
Batchian Isl.; New Guinea; Bismarck Arch.; Sulawesi; adjacent small islands.
ISIS NUMBER: 5301405007003023001.

Rhinolophus feae K. Andersen, 1905. Ann. Mus. Civ. Stor. Nat. Genova, 3:474.
TYPE LOCALITY: Burma, Karin hills, Biapo.
DISTRIBUTION: Burma; Thailand.

Rhinolophus ferrumequinum (Schreber, 1774). Saugethiere, 1:174, pl. 62.
TYPE LOCALITY: France.
DISTRIBUTION: S. England to Caucasus to N. Japan, south to Morocco (but not Egypt)
    through Iran and Himalayas to China; adjacent small islands.
COMMENT: Revised by Strelkov *et al.*, 1978, Proc. Zool. Inst. Acad. Sci. U.S.S.R.,
    79:3–71.
ISIS NUMBER: 5301405007003024001.

Rhinolophus fumigatus Ruppell, 1842. Mus. Senckenbergianum, 3:132, 155.
TYPE LOCALITY: Ethiopia, Shoa.
DISTRIBUTION: Somalia; Ethiopia; Sudan; Kenya; Tanzania; Rwanda; Burundi; Zaire;
    Nigeria; Sierra Leone; Togo; Benin; Senegal; Gambia; Guinea; Upper Volta;
    Ghana; Niger; Nigeria; Cameroun; Central African Republic; Zambia; Zimbabwe;
    Mozambique; Angola; Namibia; S. Africa.
COMMENT: Does not include *eloquens* or *perauritus*, but does include *aethiops*; see
    Koopman, 1975, Bull. Am. Mus. Nat. Hist., 154:389–390.
ISIS NUMBER: 5301405007003025001.

Rhinolophus hildebrandti Peters, 1878. Monatsb. Preuss. Akad. Wiss. Berlin, p. 195.
TYPE LOCALITY: Kenya, Taita, Ndi.
DISTRIBUTION: Transvaal (South Africa) and Mozambique to Ethiopia, S. Sudan, and
    N.E. Zaire.
ISIS NUMBER: 5301405007003026001.

Rhinolophus hipposideros (Bechstein, 1800). *In* Pennant, Uebers Vierf. Thiere, 2:629.
TYPE LOCALITY: France.
DISTRIBUTION: Ireland, Iberia and Morocco through S. Europe and N. Africa to
    Kirghizia and Kashmir; Arabia; Sudan; Ethiopia.
COMMENT: Revised by Felten *et al.*, 1977, Senckenberg. Biol., 58:1–44.
ISIS NUMBER: 5301405007003027001.

Rhinolophus imaizumii Hill and Yoshiyuki, 1980. Bull. Nat. Sci. Mus. Tokyo, ser. A (Zool.),
    6:180.
TYPE LOCALITY: Japan, Ryukyu Isls., Yayeyama Isls., Iriomote Isl., Otomi-do cave.
DISTRIBUTION: Iriomote Isl. (S. Japan).

Rhinolophus importunus Chasen, 1939. Treubia, 17:188.
TYPE LOCALITY: Indonesia, Java, Wijnkoops Bay.
DISTRIBUTION: Java.
ISIS NUMBER: 5301405007003029001.

Rhinolophus inops K. Andersen, 1905. Ann. Mag. Nat. Hist., ser. 7, 16:284, 651.
TYPE LOCALITY: Philippines, Mindanao, Davao, Mt. Apo, Todaya, 1325 m.
DISTRIBUTION: Mindanao (Philippines).
ISIS NUMBER: 5301405007003030001.

Rhinolophus keyensis Peters, 1871. Monatsb. Preuss. Akad. Wiss. Berlin, p. 371.
TYPE LOCALITY: Indonesia, Molucca Isls., Kei Isl.
DISTRIBUTION: Batchian Isl; (Halmahera Isls.); Seram; Goram Isl. (S.E. of Seram); Kei
    Isls.; Wetter Isl. (Flores Sea).
ISIS NUMBER: 5301405007003032001.

Rhinolophus landeri Martin, 1838. Proc. Zool. Soc. Lond., 1837:101.
TYPE LOCALITY: Equatorial Guinea, Bioko.
DISTRIBUTION: Senegal to Ethiopia, Somalia and Kenya, south to South Africa;
    Namibia; W. Angola; Bioko; Zaire; Zanzibar.
COMMENT: Includes *angolensis* and *guineensis*, but not *brockmani*; see Hayman and Hill,
    1971, Part 2; but also see Bohme and Hutterer, 1979, Bonn. Zool. Beitr., 29:306,

307; and Koopman, 1975, Bull. Am. Mus. Nat. Hist., 154:388, who also included *dobsoni.*
ISIS NUMBER: 5301405007003033001.

Rhinolophus lepidus Blyth, 1844. J. Asiat. Soc. Bengal, 13:486.
TYPE LOCALITY: India, Bengal, Calcutta (uncertain).
DISTRIBUTION: Afghanistan; N. India; Burma; Thailand; Szechwan, Yunnan (China); Malaya; Sumatra.
COMMENT: Includes *monticola* and *refulgens;* see Hill and Yoshiyuki, 1980, Bull. Nat. Sci. Mus. Tokyo, ser. A (Zool.), 6:186; but also see Sinha, 1973, Mammalia, 37:620-621, who considered *monticola* a distinct species.
ISIS NUMBER: 5301405007003035001 as *R. lepidus.*
5301405007003050001 as *R. refulgens.*

Rhinolophus luctus Temminck, 1835. Monogr. Mamm., 2:24.
REVIEWED BY: S. Wang (SW).
TYPE LOCALITY: Indonesia, Java.
DISTRIBUTION: India; Nepal; Sikkim; Burma; Sri Lanka; S. China; Taiwan; Vietnam; Laos; Thailand; Malay Peninsula; Sumatra; Java; Borneo; Bali.
COMMENT: Includes *lanosus;* see Ellerman and Morrison-Scott, 1951:121.
ISIS NUMBER: 5301405007003036001 as *R. luctus.*
5301405007003034001 as *R. lanosus.*

Rhinolophus maclaudi Pousargues, 1897. Bull. Mus. Hist. Nat. Paris, 3:358.
TYPE LOCALITY: Guinea, Conakry.
DISTRIBUTION: Guinea; E. Zaire; W. Uganda; Rwanda.
COMMENT: Includes *ruwenzorii* and *hilli;* see Smith and Hood, 1980, Proc. 5th Int. Bat Res. Conf., p. 170.
ISIS NUMBER: 5301405007003037001 as *R. maclaudi.*
5301405007003054001 as *R. ruwenzorii.*

Rhinolophus macrotis Blyth, 1844. J. Asiat. Soc. Bengal, 13:485.
TYPE LOCALITY: Nepal.
DISTRIBUTION: N. India to S. China, Vietnam, and Malaya; Sumatra; Guimaras Isl. (Philippines).
COMMENT: Includes *episcopus* and *hirsutus;* see Ellerman and Morrison-Scott, 1951:122; Tate, 1943, Am. Mus. Novit., 1519:2. Corbet and Hill, 1980:48, listed *hirsutus* as a distinct species without comment.
ISIS NUMBER: 5301405007003038001 as *R. macrotis.*
5301405007003020001 as *R. episcopus.*
5301405007003028001 as *R. hirsutus.*

Rhinolophus madurensis K. Andersen, 1918. Ann. Mag. Nat. Hist., ser. 9, 2:375.
TYPE LOCALITY: Indonesia, Madura Isl. (off Java).
DISTRIBUTION: Madura Isl.
ISIS NUMBER: 5301405007003039001.

Rhinolophus malayanus Bonhote, 1903. Fasc. Malayenses Zool., 1:15.
TYPE LOCALITY: Thailand, Patani, Biserat.
DISTRIBUTION: Thailand; Laos; Vietnam.
ISIS NUMBER: 5301405007003040001.

Rhinolophus marshalli Thonglongya, 1973. Mammalia, 37:590.
TYPE LOCALITY: Thailand, Chantaburi, Amphoe Pong Nam Ron, foothills of Khao Soi Dao Thai.
DISTRIBUTION: Thailand.

Rhinolophus megaphyllus Gray, 1834. Proc. Zool. Soc. Lond., 1834:52.
TYPE LOCALITY: Australia, New South Wales, Murrumbidgee River.
DISTRIBUTION: E. New Guinea; Misima Isl. (Louisiade Arch.); Goodenough Isl. (D'Entrecasteaux Isls.); Bismarck Arch.; Queensland; New South Wales; S.E. Victoria.
ISIS NUMBER: 5301405007003041001.

Rhinolophus mehelyi Matschie, 1901. Sitzb. Ges. Naturf. Fr. Berlin, p. 225.
    TYPE LOCALITY: Rumania, Bucharest.
    DISTRIBUTION: Portugal; Spain; France; Rumania; Yugoslavia; Bulgaria; Greece;
        Transcaucasia; Morocco to Cyrenaica (Libya); Mediterranean islands; Iran;
        Afghanistan; Asia Minor; Israel; Egypt.
    COMMENT: Revised by DeBlase, 1972, Isr. J. Zool., 21:1–12.
    ISIS NUMBER: 5301405007003042001.

Rhinolophus mitratus Blyth, 1844. J. Asiat. Soc. Bengal, 13:483.
    TYPE LOCALITY: India, Orissa, Chaibassa.
    DISTRIBUTION: N. India.

Rhinolophus monoceros K. Andersen, 1905. Proc. Zool. Soc. Lond., 1905:131.
    TYPE LOCALITY: Taiwan, Baksa.
    DISTRIBUTION: Taiwan.
    COMMENT: Probably a subspecies of *cornutus* (KFK).
    ISIS NUMBER: 5301405007003045001.

Rhinolophus nereis K. Andersen, 1905. Proc. Zool. Soc. Lond., 1905:90.
    TYPE LOCALITY: Indonesia, Anamba Isls., Siantan Isl.
    DISTRIBUTION: Siantan Isl. (Anamba Isls.); Bungaran Isl. (North Natuna Isls.).
    ISIS NUMBER: 5301405007003046001.

Rhinolophus osgoodi Sanborn, 1939. Field Mus. Nat. Hist. Publ. Ser. Zool., 24:40.
    TYPE LOCALITY: China, Yunnan, N. of Likiang, Nguluko.
    DISTRIBUTION: Yunnan (China).

Rhinolophus paradoxolophus (Bourret, 1951). Bull. Mus. Hist. Nat. Paris, ser. 2, 33:607.
    TYPE LOCALITY: Vietnam, Tonkin, Lao Key, near Chapa, 1700 m.
    DISTRIBUTION: N. Vietnam; Thailand.
    ISIS NUMBER: 5301405007004001001 as *Rhinomegalophus paradoxolophus.*

Rhinolophus pearsoni Horsfield, 1851. Cat. Mamm. Mus. E. India Co., p. 33.
    REVIEWED BY: S. Wang (SW).
    TYPE LOCALITY: India, W. Bengal, Darjeeling.
    DISTRIBUTION: N. India; Burma; Szechwan, Anwei, and Fukien (China) to N. Vietnam;
        Thailand; Malaya.
    ISIS NUMBER: 5301405007003047001.

Rhinolophus philippinensis Waterhouse, 1843. Proc. Zool. Soc. Lond., 1843:68.
    TYPE LOCALITY: Philippines, Luzon.
    DISTRIBUTION: Mindoro, Luzon, Mindanao and Negros (Philippines); Kei Isls.; S.
        Sulawesi; Timor; N. Borneo; N.E. Queensland.
    ISIS NUMBER: 5301405007003048001.

Rhinolophus pusillus Temminck, 1834. Tijdschr. Nat. Gesch. Physiol., 1:29.
    TYPE LOCALITY: Indonesia, Java.
    DISTRIBUTION: India; Thailand; Malaya; Mentawi Isls.; Java; small adjacent islands.
    COMMENT: Includes *blythi, minutillus,* and *pagi;* see Hill and Yoshiyuki, 1980, Bull. Nat.
        Sci. Mus. Tokyo, ser. A (Zool.) 6:186. Corbet and Hill, 1980:49, listed *minutillus* as
        a distinct species without comment.
    ISIS NUMBER: 5301405007003049001 as *R. pusillus.*
                5301405007003009001 as *R. blythi.*
                5301405007003044001 as *R. minutillus.*

Rhinolophus rex G. M. Allen, 1923. Am. Mus. Novit., 85:3.
    REVIEWED BY: S. Wang (SW).
    TYPE LOCALITY: China, Szechwan, Wanhsien.
    DISTRIBUTION: S.W. China.
    ISIS NUMBER: 5301405007003051001.

Rhinolophus robinsoni K. Andersen, 1918. Ann. Mag. Nat. Hist., ser. 9, 2:375.
    TYPE LOCALITY: Thailand, Surat Thani, Bandon.
    DISTRIBUTION: Malaya; Thailand; adjacent small islands.
    COMMENT: Includes *klossi;* see Medway, 1969:24.
    ISIS NUMBER: 5301405007003052001.

Rhinolophus rouxi Temminck, 1835. Monogr. Mamm., 2:306.
TYPE LOCALITY: India, Pondicherry and Calcutta.
DISTRIBUTION: Sri Lanka and India to S. China and Vietnam.
COMMENT: Includes *petersi;* see Sinha, 1973, Mammalia, 37:614,615.
ISIS NUMBER: 5301405007003053001.

Rhinolophus rufus Eydoux and Gervais, 1836. Zool. Voy. "Favorite", p. 9.
TYPE LOCALITY: Philippines, Luzon, Manila.
DISTRIBUTION: Philippines.
COMMENT: Name revived by Lawrence, 1939, Bull. Mus. Comp. Zool., 86:47–50.

Rhinolophus sedulus K. Andersen, 1905. Ann. Mag. Nat. Hist., ser. 7, 16:247.
TYPE LOCALITY: Malaysia, Sarawak.
DISTRIBUTION: Malaya; Borneo.
ISIS NUMBER: 5301405007003055001.

Rhinolophus shameli Tate, 1943. Am. Mus. Novit., 1219:3.
TYPE LOCALITY: Thailand, Koh Chang Isl.
DISTRIBUTION: Burma; Thailand; Kampuchea; Malaya.
COMMENT: Described as a subspecies of *coelophyllus,* but see Hill and Thonglongya,
    1972, Bull. Br. Mus. (Nat. Hist.) Zool., 22:183–186.
ISIS NUMBER: 5301405007003056001.

Rhinolophus silvestris Aellen, 1959. Arch. Sci. Phys. Nat. Geneve, 42:228.
TYPE LOCALITY: Gabon, Latoursville, N'Dumbu Cave.
DISTRIBUTION: Gabon; Congo Republic.
COMMENT: Considered a subspecies of *clivosus* by Hayman and Hill, 1971, Part 2:23;
    but see Koopman, 1975, Bull. Am. Mus. Nat. Hist., 154:386.

Rhinolophus simplex K. Andersen, 1905. Proc. Zool. Soc. Lond., 1905:76.
TYPE LOCALITY: Indonesia, Lesser Sunda Isls., Lombok, 2500 ft. (762 m).
DISTRIBUTION: Lombok, Sumbawa, Komodo Isl. (Lesser Sunda Isls.).
ISIS NUMBER: 5301405007003057001.

Rhinolophus simulator K. Andersen, 1904. Ann. Mag. Nat. Hist., ser. 7, 14:384.
TYPE LOCALITY: Zimbabwe, Mazoe.
DISTRIBUTION: South Africa to S. Sudan and Ethiopia; Cameroun; N. Nigeria.
COMMENT: Includes *alticolus* and *bembanicus;* see Koopman, 1975, Bull. Am. Mus. Nat.
    Hist., 154:387; Hayman and Hill, 1971, Part 2:25.
ISIS NUMBER: 5301405007003058001.

Rhinolophus stheno K. Andersen, 1905. Proc. Zool. Soc. Lond., 1905:91.
TYPE LOCALITY: Malaysia, Selangor.
DISTRIBUTION: Malaya; Thailand; Sumatra; Java.
ISIS NUMBER: 5301405007003059001.

Rhinolophus subbadius Blyth, 1844. J. Asiat. Soc. Bengal, 13:486.
TYPE LOCALITY: Nepal.
DISTRIBUTION: Assam (India); Nepal; N. Vietnam; Burma.
ISIS NUMBER: 5301405007003060001.

Rhinolophus subrufus K. Andersen, 1905. Ann. Mag. Nat. Hist., ser. 7, 16:283.
TYPE LOCALITY: Philippines, Luzon, Manila.
DISTRIBUTION: Mindanao, Luzon, Mindoro, and Negros Isl. (Philippines).
COMMENT: Includes *bunkeri;* see Lawrence, 1939, Bull. Mus. Comp. Zool., 86:52,53.
ISIS NUMBER: 5301405007003061001.

Rhinolophus swinnyi Gough, 1908. Ann. Transvaal Mus., 1:72.
TYPE LOCALITY: South Africa, Cape Province, Pondoland, Ngqeleni Dist.
DISTRIBUTION: South Africa to S. Zaire and Zanzibar.
COMMENT: Probably a subspecies of *denti* (KFK).
ISIS NUMBER: 5301405007003062001.

Rhinolophus thomasi K. Andersen, 1905. Proc. Zool. Soc. Lond., 1905:100.
TYPE LOCALITY: Burma, Karin Hills.
DISTRIBUTION: S.E. Burma; N. Vietnam; Thailand; Yunnan (China).
ISIS NUMBER: 5301405007003063001.

Rhinolophus toxopeusi Hinton, 1925. Ann. Mag. Nat. Hist., ser. 9, 16:256.
TYPE LOCALITY: Indonesia, Molucca Isls., Buru, 1400 m.
DISTRIBUTION: Buru.
ISIS NUMBER: 5301405007003064001.

Rhinolophus trifoliatus Temminck, 1834. Tijdschr. Nat. Gesch. Physiol., 1:24.
TYPE LOCALITY: Indonesia, Java.
DISTRIBUTION: Malay Peninsula; Sumatra; Rhio Archipelago; Banguey Isl.; Borneo; Java; S.W. Thailand; Burma; N.E. India; Banka Isl.; Nias Isl. (W. Sumatra).
ISIS NUMBER: 5301405007003065001.

Rhinolophus virgo K. Andersen, 1905. Proc. Zool. Soc. Lond., 1905:88.
TYPE LOCALITY: Philippines, Luzon, Camarines Sur, Pasacao.
DISTRIBUTION: Philippines.
ISIS NUMBER: 5301405007003066001.

Rhinolophus yunanensis Dobson, 1872. J. Asiat. Soc. Bengal, 41:336.
TYPE LOCALITY: China, Yunnan, Hotha.
DISTRIBUTION: Yunnan (China); Thailand; N.E. India.
COMMENT: See Lekagul and McNeely, 1977:152, 154 for distinction of this species from *pearsoni*.

Rhinonicteris Gray, 1847. Proc. Zool. Soc. Lond., 1847:16.
REVIEWED BY: F. R. Allison (FRA).
COMMENT: *Rhinonicteris* is the correct spelling if original orthography is adhered to. *Rhinonycteris* Gray, 1866, Proc. Zool. Soc. Lond., 1866:81, is sometimes used (FRA).
ISIS NUMBER: 5301405007010000000.

Rhinonicteris aurantia (Gray, 1845). *In* Eyre, Central Australia, 1:405.
TYPE LOCALITY: Australia, Northern Territory, Port Essington.
DISTRIBUTION: N. Western Australia, Northern Territory, N.W. Queensland (Australia).
ISIS NUMBER: 5301405007010001001.

Triaenops Dobson, 1871. J. Asiat. Soc. Bengal, 40:455.
ISIS NUMBER: 5301405007011000000.

Triaenops furculus Trouessart, 1906. Bull. Mus. Hist. Nat. Paris, 1906, 7:446.
TYPE LOCALITY: Madagascar, near Tulear, St. Augustine Bay.
DISTRIBUTION: N. and W. Madagascar; Aldabra Isl.
COMMENT: Includes *aurita*; see Hayman and Hill, 1971, Part 2:30.
ISIS NUMBER: 5301405007011001001.

Triaenops humbloti Milne-Edwards, 1881. C. R. Acad. Sci. Paris, 91:1035.
TYPE LOCALITY: Madagascar (E. coast).
DISTRIBUTION: E. Madagascar.
COMMENT: Probably only a color variant of *rufus*; see Hayman and Hill, 1971, Part 2:30.
ISIS NUMBER: 5301405007011002001.

Triaenops persicus Dobson, 1871. J. Asiat. Soc. Bengal, 40:455.
TYPE LOCALITY: Iran, Shiraz, 4750 ft. (1448 m).
DISTRIBUTION: Somalia; Ethiopia; Kenya; Tanzania; Uganda; Zanzibar; Mozambique; Yemen; Oman; Congo Republic; Iran; perhaps Egypt.
COMMENT: Includes *afer* and *majusculus*; see Hayman and Hill, 1971, Part 2:30.
ISIS NUMBER: 5301405007011003001.

Triaenops rufus Milne-Edwards, 1881. C. R. Acad. Sci. Paris, 91:1035.
TYPE LOCALITY: Madagascar (E. coast).
DISTRIBUTION: E. Madagascar.
ISIS NUMBER: 5301405007011004001.

Family Noctilionidae
REVIEWED BY: W. B. Davis (WBD); K. F. Koopman (KFK); J. Ramirez-Pulido (JRP).
ISIS NUMBER: 5301405004000000000.

Noctilio Linnaeus, 1766. Syst. Nat., 12th ed., 1:88.
  ISIS NUMBER: 5301405004001000000.

Noctilio albiventris Desmarest, 1818. Nouv. Dict. Hist. Nat. Paris, 23:15.
  TYPE LOCALITY: Brazil, Bahia, Rio Sao Francisco.
  DISTRIBUTION: Honduras to Guianas, E. Brazil, N. Argentina, and Peru.
  COMMENT: Formerly referred to as *labialis*; see Davis, 1976, J. Mammal., 57:687–707.
  ISIS NUMBER: 5301405004001001001 as *N. labialis.*

Noctilio leporinus (Linnaeus, 1758). Syst. Nat., 10th ed., 1:32.
  REVIEWED BY: B. Villa-Ramirez (BVR).
  TYPE LOCALITY: Surinam.
  DISTRIBUTION: Sinaloa (Mexico) to Guianas, S. Brazil, Paraguay, N. Argentina, and Peru; Trinidad; Greater and Lesser Antilles; S. Bahamas.
  ISIS NUMBER: 5301405004001002001.

Family Mormoopidae
  REVIEWED BY: J. K. Jones, Jr. (JKJ); K. F. Koopman (KFK); J. Ramirez-Pulido (JRP)(Mexico).
  COMMENT: Revised by Smith, 1972, Univ. Kansas Mus. Nat. Hist. Misc. Publ., 56:55.
  ISIS NUMBER: 5301405009000000000.

Mormoops Leach, 1821. Trans. Linn. Soc. Lond., 13:76.
  COMMENT: This name is used instead of *Aello* following opinion 462 of the ICZN.
  ISIS NUMBER: 5301405009001000000.

Mormoops blainvillii Leach, 1821. Trans. Linn. Soc. Lond., 13:77.
  TYPE LOCALITY: Jamaica.
  DISTRIBUTION: Greater Antilles; adjacent small islands.
  ISIS NUMBER: 5301405009001001001.

Mormoops megalophylla (Peters, 1864). Monatsb. Preuss. Akad. Wiss. Berlin, p. 381.
  REVIEWED BY: B. Villa-Ramirez (BVR).
  TYPE LOCALITY: Mexico, Coahuila, Parras.
  DISTRIBUTION: S. Texas, S. Arizona (U.S.A.), and Baja California (Mexico) to N. Ecuador, N. Colombia, and N. Venezuela; Aruba, Curacao, Bonaire (Dutch West Indies); Trinidad; Margarita Isl.
  ISIS NUMBER: 5301405009001002001.

Pteronotus Gray, 1838. Mag. Zool. Bot., 2:500.
  REVIEWED BY: B. Villa-Ramirez (BVR).
  COMMENT: Includes *Chilonycteris*; see Smith, 1972, Misc. Publ. Univ. Kans. Mus. Nat. Hist., 56:55.
  ISIS NUMBER: 5301405009002000000.

Pteronotus davyi Gray, 1838. Mag. Zool. Bot., 2:500.
  TYPE LOCALITY: Trinidad and Tobago, Trinidad.
  DISTRIBUTION: N.W. Peru and N.E. Brazil to Puebla thence to S. Sonora and S. Tamaulipas (Mexico); Trinidad; S. Lesser Antilles.
  ISIS NUMBER: 5301405009002001001.

Pteronotus fuliginosus (Gray, 1843). Proc. Zool. Soc. Lond., 1843:20.
  TYPE LOCALITY: Haiti, Port au Prince.
  DISTRIBUTION: Cuba; Jamaica; Hispaniola; Puerto Rico.
  COMMENT: Includes *torrei*; *P. quadridens* may be the correct name for this species; see Silva, 1976, Poeyana, 153:7.
  ISIS NUMBER: 5301405009002002001.

Pteronotus gymnonotus Natterer, 1843. *In* Wagner, Wiegmann's Archive Naturgesch., 9:367.
  TYPE LOCALITY: Brazil, Mato Grosso, Cuiaba.
  DISTRIBUTION: S. Veracruz (Mexico) to Peru, S.W. Brazil, and Guyana.
  COMMENT: Includes *suapurensis*; see Smith, 1977, J. Mammal., 58:245.
  ISIS NUMBER: 5301405009002006001 as *P. suapurensis.*

Pteronotus macleayii (Gray, 1839). Ann. Nat. Hist., 4:5.
TYPE LOCALITY: Cuba, Habana, Guanabacoa.
DISTRIBUTION: Cuba; Jamaica.
ISIS NUMBER: 5301405009002003001.

Pteronotus parnellii (Gray, 1843). Proc. Zool. Soc. Lond., 1843:50.
TYPE LOCALITY: Jamaica.
DISTRIBUTION: Peru, Brazil, Guianas, and Venezuela to Tlaxcala to S. Sonora and S.
    Tamaulipas (Mexico); Cuba; Jamaica; Puerto Rico; Hispaniola; Trinidad; Margarita
    Isl.; La Gonave Isl. (Haiti).
ISIS NUMBER: 5301405009002004001.

Pteronotus personatus (Wagner, 1843). Arch. Naturg., 9:367.
TYPE LOCALITY: Brazil, Mato Grosso, Sao Vicente.
DISTRIBUTION: N. Colombia, Peru, Brazil, and Surinam to S. Sonora and S. Tamaulipas
    (Mexico); Trinidad.
COMMENT: Includes *psilotis;* see Smith, 1972, Misc. Publ. Univ. Kans. Mus. Nat. Hist.,
    56:92.   ·
ISIS NUMBER: 5301405009002005001.

Family Phyllostomidae

REVIEWED BY: J. K. Jones, Jr. (JKJ); K. F. Koopman (KFK); J. Ramirez-Pulido
    (JRP)(Mexico); J. R. Tamsitt (JRT).
COMMENT: Includes Desmodontidae; see Jones and Carter, 1976:7. For use of this
    familial name rather than Phyllostomatidae, see Handley *et al.*, 1980, Proc. Fifth
    Int. Bat Res. Conf., p. 10.
ISIS NUMBER: 5301405008000000000.

Ametrida Gray, 1847. Proc. Zool. Soc. Lond., p. 15.
REVIEWED BY: R. Barquez (RB); J. Molinari (JML).
ISIS NUMBER: 5301405008023000000.

Ametrida centurio Gray, 1847. Proc. Zool. Soc. Lond., 1847:15.
TYPE LOCALITY: Brazil, Para, Belem.
DISTRIBUTION: Amazonian Brazil; Guyana; Surinam; Venezuela; Trinidad; Bonaire Isl.
COMMENT: Includes *minor;* see Jones and Carter, 1976:29.
ISIS NUMBER: 5301405008023001001.

Anoura Gray, 1838. Mag. Zool. Bot., 2:490.
REVIEWED BY: R. Barquez (RB); T. A. Griffiths (TAG); J. Molinari (JML).
COMMENT: Includes *Lonchoglossa;* see Cabrera, 1958:74.
ISIS NUMBER: 5301405008015000000.

Anoura caudifer (E. Geoffroy, 1818). Mem. Mus. Hist. Nat. Paris, 4:418.
TYPE LOCALITY: Brazil, Rio de Janeiro.
DISTRIBUTION: Colombia; Venezuela; Guianas; Brazil; Ecuador; Peru; Bolivia.
ISIS NUMBER: 5301405008015002001.

Anoura cultrata Handley, 1960. Proc. U.S. Nat. Mus., 112:463.
TYPE LOCALITY: Panama, Darien, Rio Pucro, Tacarcuna Village, 3200 ft. (975 m).
DISTRIBUTION: Costa Rica; Panama; Venezuela; Colombia; Peru.
COMMENT: Includes *brevirostrum* and *werckleae;* see Nagorsen and Tamsitt, 1981, J.
    Mammal., 62:82–100.
ISIS NUMBER: 5301405008015003001.

Anoura geoffroyi Gray, 1838. Mag. Zool. Bot., 2:490.
REVIEWED BY: B. Villa-Ramirez (BVR).
TYPE LOCALITY: Brazil, Rio de Janeiro.
DISTRIBUTION: Peru, Bolivia, N.W. Argentina, S.E. Brazil, French Guiana and Surinam
    to S. Tamaulipas and N. Sinaloa (Mexico); Trinidad; Grenada (Lesser Antilles).
COMMENT: Includes *apollinari;* see Sanborn, 1933, Field Mus. Nat. Hist. Publ. Ser.
    Zool., 20:26.
ISIS NUMBER: 5301405008015004001.

Ardops Miller, 1906. Proc. Biol. Soc. Wash., 19:84.
REVIEWED BY: G. S. Morgan (GSM).
COMMENT: Revised by Jones and Schwartz, 1967, Proc. U.S. Nat. Mus., 124:1–13.
Included under *Stenoderma* by Varona, 1974:24–26, and by Simpson, 1945:58; but
see Jones and Carter, 1976:28.
ISIS NUMBER: 5301405008024000000.

Ardops nichollsi (Thomas, 1891). Ann. Mag. Nat. Hist., ser. 6, 7:529.
TYPE LOCALITY: Dominica (Lesser Antilles).
DISTRIBUTION: Lesser Antilles, from St. Eustatius to St. Vincent.
COMMENT: Includes *montserratensis, luciae,* and *annectens;* see Jones and Schwartz, 1967,
Proc. U.S. Nat. Mus., 124:1–13. Reviewed by Jones and Genoways, 1973, Mamm.
Species, 24:1–2.
ISIS NUMBER: 5301405008024001001.

Ariteus Gray, 1838. Mag. Zool. Bot., 2:491.
REVIEWED BY: G. S. Morgan (GSM).
COMMENT: Included as a subgenus of *Stenoderma* by Varona, 1974:24, and Simpson,
1945:58; but see Jones and Carter, 1976:29.
ISIS NUMBER: 5301405008025000000.

Ariteus flavescens (Gray, 1831). Zool. Misc., 1:37.
TYPE LOCALITY: Not designated in original publication.
DISTRIBUTION: Jamaica.
ISIS NUMBER: 5301405008025001001.

Artibeus Leach, 1821. Trans. Linn. Soc. Lond., 13:75.
REVIEWED BY: R. Barquez (RB)(S. America); J. Molinari (JML)(S. America); B. Villa-
Ramirez (BVR)(Mexico).
COMMENT: Includes *Enchisthenes;* see Jones and Carter, 1979:8.
ISIS NUMBER: 5301405008026000000.

Artibeus anderseni Osgood, 1916. Field Mus. Nat. Hist. Publ. Ser. Zool., 10:212.
TYPE LOCALITY: Brazil, Rondonia, Porto Velho.
DISTRIBUTION: W. Brazil; Bolivia; Ecuador; Peru; French Guiana.
COMMENT: Previously considered a subspecies of *cinereus;* but see Koopman, 1978:14.

Artibeus aztecus K. Andersen, 1906. Ann. Mag. Nat. Hist., ser. 7, 18:422.
TYPE LOCALITY: Mexico, Morelos, Tetela del Volcan.
DISTRIBUTION: W. Panama; Costa Rica; Honduras, El Salvador, Guatemala; and Chiapas;
Guerrero, to S. Nuevo Leon and C. Sinaloa (Mexico).
COMMENT: Not a subspecies of *cinereus;* see Jones and Carter, 1976:27. Revised by
Davis, 1969, Southwest. Nat., 14:15–29.
ISIS NUMBER: 5301405008026001001.

Artibeus cinereus (Gervais, 1856). Exped. Castelnau Zool., p. 36.
TYPE LOCALITY: Brazil, Para, Belem.
DISTRIBUTION: Veracruz (Mexico) to Guianas, N.E. Brazil, Bolivia, and Amazonian
Peru; Trinidad; Tobago; Grenada.
COMMENT: Includes *quadrivittatus* and *rosenbergi;* see Cabrera, 1958:87–88. Includes
glaucus, watsoni and *bogotensis;* see Jones and Carter, 1979:8, and Jones and Carter,
1976:26. Includes *pumilio;* see Koopman, 1978:13–14.
ISIS NUMBER: 5301405008026002001.

Artibeus concolor Peters, 1865. Monatsb. Preuss. Akad. Wiss. Berlin, p. 357.
TYPE LOCALITY: Surinam, Paramaribo.
DISTRIBUTION: Guianas; Venezuela; Colombia; N.Brazil; Peru.

Artibeus fraterculus Anthony, 1924. Am. Mus. Novit., 114:5.
TYPE LOCALITY: Ecuador, El Oro, Portovelo, 2000 ft. (610 m).
DISTRIBUTION: Ecuador; Peru.
COMMENT: Considered a subspecies of *jamaicensis* by Jones and Carter, 1976:2; but see
Koopman, 1978:14.

Artibeus fuliginosus Gray, 1838. Mag. Zool. Bot., 2:487.
  TYPE LOCALITY: "South America".
  DISTRIBUTION: Colombia; Venezuela; Guyana; Ecuador; Peru; Bolivia; Amazonian
    Brazil.
  COMMENT: Probably only a subspecies of *jamaicensis;* see Koopman, 1978:15; but also
    see Handley, 1976, Brigham Young Univ. Sci. Bull., 20(5):31.

Artibeus hartii Thomas, 1892. Ann. Mag. Nat. Hist., ser. 6, 10:409.
  TYPE LOCALITY: Trinidad and Tobago, Trinidad, Port of Spain.
  DISTRIBUTION: Bolivia and Venezuela to Jalisco and Tamaulipas (Mexico); Arizona
    (USA); Trinidad.
  COMMENT: Formerly included in *Enchisthenes;* see Jones and Carter, 1979:8. Webster
    and Jones, 1980, Occas. Pap. Mus. Texas Tech Univ., 68:1–6, reviewed the Arizona
    distribution.

Artibeus hirsutus K. Andersen, 1906. Ann. Mag. Nat. Hist., ser. 7, 18:420.
  TYPE LOCALITY: Mexico, Michoacan, La Salada.
  DISTRIBUTION: C. Sonora to Guerrero (Mexico).
  ISIS NUMBER: 5301405008026004001.

Artibeus inopinatus Davis and Carter, 1964. Proc. Biol. Soc. Wash., 77:119.
  TYPE LOCALITY: Honduras, Choluteca, Choluteca.
  DISTRIBUTION: El Salvador; Honduras; Nicaragua.
  COMMENT: Possibly a subspecies of *hirsutus* (JKJ).
  ISIS NUMBER: 5301405008026005001.

Artibeus jamaicensis Leach, 1821. Trans. Linn. Soc. Lond., 13:75.
  TYPE LOCALITY: Jamaica.
  DISTRIBUTION: Sinaloa and Tamaulipas (Mexico) to Guianas, Trinidad; Tobago; Greater
    Antilles; Lesser Antilles. Amazonian Brazil (JRT; KFK disagrees).
  COMMENT: Includes *trinitatis* and *preceps* (KFK). Bats often treated as *jamaicensis* most
    probably belong to several additional species; they are *fraterculus* and *planirostris*
    (Koopman, 1978:14–16) and *fuliginosus* (Handley, 1976, Brigham Young Univ. Sci.
    Bull., 20(5):31) and are so treated here.
  ISIS NUMBER: 5301405008026006001.

Artibeus lituratus (Olfers, 1818). *In* Eschwege, Neue Bibl. Reisenb., p. 224.
  TYPE LOCALITY: Paraguay, Asuncion.
  DISTRIBUTION: Sinaloa and Tamaulipas (Mexico) to S. Brazil, N. Argentina, and Bolivia;
    Trinidad; Tobago; S. Lesser Antilles.
  COMMENT: Includes *palmarum* but not *fallax, hercules,* or *preceps* (KFK). Includes
    intermedius; see Jones and Carter, 1976:28. See also comments under *jamaicensis.*
  ISIS NUMBER: 5301405008026007001.

Artibeus phaeotis (Miller, 1902). Proc. Acad. Nat. Sci. Phila., 54:405.
  TYPE LOCALITY: Mexico, Yucatan, Chichen-Itza.
  DISTRIBUTION: Veracruz and Sinaloa (Mexico) to Ecuador and Guyana.
  COMMENT: Includes *nanus* and *turpis;* see Jones and Lawlor, 1965, Univ. Kans. Publ.
    Mus. Nat. Hist., 16:412, and Davis, 1970, Southwest. Nat., 14:389–402. Includes
    ravus (KFK).
  ISIS NUMBER: 5301405008026008001.

Artibeus planirostris (Spix, 1823). Sim. Vespert. Brasil., p. 66.
  TYPE LOCALITY: Brazil, Bahia, Salvador.
  DISTRIBUTION: Colombia and Venezuela, south to N. Argentina and east to E. Brazil.
  COMMENT: Includes *hercules* and *fallax;* see Koopman, 1978:15. It is possible that *fallax*
    is the correct name of this biological species (KFK). See also comments under
    lituratus.

Artibeus toltecus (Saussure, 1860). Rev. Mag. Zool. Paris, ser. 2, 12:427.
  TYPE LOCALITY: Mexico, Veracruz, Mirador.
  DISTRIBUTION: Panama to Nuevo Leon and Sinaloa (Mexico).
  COMMENT: Not a subspecies of *cinereus;* see Jones and Carter, 1976:27. Revised by
    Davis, 1969, Southwest. Nat., 14:15–29. Does not include *ravus* (KFK).
  ISIS NUMBER: 5301405008026011001.

Brachyphylla Gray, 1834. Proc. Zool. Soc. Lond., 1833:122.
   REVIEWED BY: T. A. Griffiths (TAG); G. S. Morgan (GSM).
   COMMENT: Revised by Swanepoel and Genoways, 1978, Bull. Carnegie Mus. Nat. Hist.,
      12:1–53.
   ISIS NUMBER: 5301405008027000000.

Brachyphylla cavernarum Gray, 1834. Proc. Zool. Soc. Lond., 1833:123.
   TYPE LOCALITY: St. Vincent (Lesser Antilles) (U.K.).
   DISTRIBUTION: Puerto Rico, Virgin Isls. and throughout Lesser Antilles south to St.
      Vincent and Barbados.
   COMMENT: Includes *minor*; see Swanepoel and Genoways, 1978, Bull. Carnegie Mus.
      Nat. Hist., 12:39, and Varona, 1974:27.
   ISIS NUMBER: 5301405008027001001.

Brachyphylla nana Miller, 1902. Proc. Acad. Nat. Sci. Phila., 54:409.
   TYPE LOCALITY: Cuba, Pinar del Rio, El Guama.
   DISTRIBUTION: Cuba; Hispaniola (Haiti and Dominican Republic); Jamaica (extinct);
      Grand Cayman; Middle Caicos (S.E. Bahamas).
   COMMENT: Includes *pumila*; see Jones and Carter, 1976:30, and Swanepoel and
      Genoways, 1978, Bull. Carnegie Mus. Nat. Hist., 12:1–53. Considered a subspecies
      of *cavernarum* by Buden, 1977, J. Mammal., 58:221–225.
   ISIS NUMBER: 5301405008027002001.

Carollia Gray, 1838. Mag. Zool. Bot., 2:488.
   REVIEWED BY: R. Barquez (RB); J. Molinari (JML); B. Villa-Ramirez (BVR)(Mexico).
   COMMENT: Revised by Pine, 1972, Tech. Monogr. Texas Agric. Exp. Sta. Texas A and
      M Univ., 8:1–125.
   ISIS NUMBER: 5301405008021000000.

Carollia brevicauda (Schinz, 1821). Das Thierreich, 1:164.
   TYPE LOCALITY: Brazil, Espirito Santo, Jucu River, Fazenda de Coroaba.
   DISTRIBUTION: San Luis Potosi (Mexico) to Peru, Bolivia, and E. Brazil.
   COMMENT: Long confused with *perspicillata* or *subrufa*; see Pine, 1972, Tech. Monogr.
      Texas Agric. Exp. Sta. Texas A and M Univ., 8:1–125.
   ISIS NUMBER: 5301405008021001001.

Carollia castanea H. Allen, 1890. Proc. Am. Philos. Soc., 28:19.
   TYPE LOCALITY: Costa Rica, Angostura.
   DISTRIBUTION: Honduras to Peru, Bolivia, Venezuela and Guianas.
   ISIS NUMBER: 5301405008021002001.

Carollia perspicillata (Linnaeus, 1758). Syst. Nat., 10th ed., 1:31.
   TYPE LOCALITY: Surinam.
   DISTRIBUTION: Veracruz and Yucatan Peninsula (Mexico) to Peru, Bolivia, Paraguay,
      S.E. Brazil and Guianas; Trinidad; Tobago; Grenada (Lesser Antilles); perhaps
      Jamaica, N. Lesser Antilles. Paraguay (JRT).
   COMMENT: Includes *tricolor*; see Pine, 1972, Tech. Monogr. Texas Agric. Exp. Sta. Texas
      A and M Univ., 8:1–125.
   ISIS NUMBER: 5301405008021003001.

Carollia subrufa (Hahn, 1905). Proc. Biol. Soc. Wash., 18:247.
   TYPE LOCALITY: Mexico, Oaxaca, N.W. Tapanatepec, Sta. Efigenia.
   DISTRIBUTION: Jalisco (Mexico) to N.W. Nicaragua.
   COMMENT: Not a subspecies of *castanea*; see Pine, 1972, Tech. Monogr. Texas Agric.
      Exp. Sta. Texas A and M Univ., 8:1–125.
   ISIS NUMBER: 5301405008021004001.

Centurio Gray, 1842. Ann. Mag. Nat. Hist., ser. 1, 10:259.
   REVIEWED BY: B. Villa-Ramirez (BVR).
   ISIS NUMBER: 5301405008028000000.

Centurio senex Gray, 1842. Ann. Mag. Nat. Hist., ser. 1, 10:259.
   TYPE LOCALITY: Nicaragua, Chinandega, Realejo.
   DISTRIBUTION: Venezuela to Tamaulipas and Sinaloa (Mexico); Trinidad; Tobago.

COMMENT: Reviewed by Paradiso, 1967, Mammalia, 31:595–604; Snow, *et al.*, 1980, Mamm. Species, 138:1–3.
ISIS NUMBER: 5301405008028001001.

Chiroderma Peters, 1860. Monatsb. Preuss. Akad. Wiss. Berlin, p. 747.
REVIEWED BY: R. Barquez (RB); J. Molinari (JML); V. A. Taddei (VAT); B. Villa-Ramirez (BVR).
COMMENT: Reviewed by Goodwin, 1958, Am. Mus. Novit., 1877:2–3.
ISIS NUMBER: 5301405008029000000.

Chiroderma doriae Thomas, 1891. Ann. Mus. Civ. Stor. Nat. Genova, ser. 2, 10:881.
TYPE LOCALITY: Brazil, Minas Gerais.
DISTRIBUTION: Minas Gerais and Sao Paulo (S.E. Brazil).
ISIS NUMBER: 5301405008029001001.

Chiroderma improvisum Baker and Genoways, 1976. Occas. Pap. Mus. Texas Tech Univ., 39:2.
TYPE LOCALITY: Guadeloupe (Lesser Antilles), Basse Terre, 2 km S. and 2 km E. Baie-Mahault (France).
DISTRIBUTION: Guadeloupe (Lesser Antilles). Montserrat (Lesser Antilles) (VAT).
COMMENT: Reviewed by Jones and Baker, 1980, Mamm. Species, 134:1–2.

Chiroderma salvini Dobson, 1878. Cat. Chiroptera Br. Mus., p. 532.
TYPE LOCALITY: Costa Rica.
DISTRIBUTION: Bolivia, Colombia, Ecuador and Venezuela to Veracruz and Chihuahua (Mexico).
ISIS NUMBER: 5301405008029002001.

Chiroderma trinitatum Goodwin, 1958. Am. Mus. Novit., 1877:1.
TYPE LOCALITY: Trinidad and Tobago, Trinidad, Cumaca, 1000 ft. (305 m).
DISTRIBUTION: Panama to Venezuela, Amazonian Brazil, Bolivia and Peru; Trinidad. Guyana; Surinam (VAT).
COMMENT: Includes *gorgasi*; see Jones and Carter, 1976:25.
ISIS NUMBER: 5301405008029003001.

Chiroderma villosum Peters, 1860. Monatsb. Preuss. Akad. Wiss. Berlin, p. 748.
TYPE LOCALITY: Brazil.
DISTRIBUTION: Veracruz (Mexico) to Surinam, C. Brazil, Bolivia and Peru; Trinidad and Tobago.
COMMENT: Includes *jesupi* and *isthmicum*; see Handley, 1960, Proc. U.S. Nat. Mus., 112:466. See Carter and Dolan, 1978, Spec. Publ. Mus. Texas Tech Univ., 15:59, for type locality.
ISIS NUMBER: 5301405008029004001.

Choeroniscus Thomas, 1928. Ann. Mag. Nat. Hist., ser. 10, 1:122.
REVIEWED BY: R. Barquez (RB); T. A. Griffiths (TAG); J. Molinari (JML).
ISIS NUMBER: 5301405008030000000.

Choeroniscus godmani (Thomas, 1903). Ann. Mag. Nat. Hist., ser. 7, 11:288.
REVIEWED BY: B. Villa-Ramirez (BVR).
TYPE LOCALITY: Guatemala.
DISTRIBUTION: Sinaloa (Mexico) to Venezuela, Guyana (JRT), and Surinam.
ISIS NUMBER: 5301405008030001001.

Choeroniscus intermedius (J. A. Allen and Chapman, 1893). Bull. Am. Mus. Nat. Hist., 5:207.
TYPE LOCALITY: Trinidad and Tobago, Trinidad, Princestown.
DISTRIBUTION: Trinidad; Peru; Guyana; Surinam; Amazonian Brazil.
COMMENT: Distribution poorly known (TAG). Closely related to *minor* and *inca* (RB).
ISIS NUMBER: 5301405008030003001.

Choeroniscus minor (Peters, 1868). Monatsb. Preuss. Akad. Wiss. Berlin, p. 366.
TYPE LOCALITY: Surinam.
DISTRIBUTION: Guyana; Surinam; N. Brazil; N.C. Colombia; Ecuador; Peru; Bolivia;
Venezuela.
COMMENT: Includes *inca;* see Koopman, 1978, Am. Mus. Novit., 2651:8; and Jones and
Carter, 1979:8.
ISIS NUMBER: 5301405008030004001.

Choeroniscus periosus Handley, 1966. Proc. Biol. Soc. Wash., 79:84.
TYPE LOCALITY: Colombia, Valle, 27 km S. Buenaventura, Rio Raposo.
DISTRIBUTION: W. Colombia.
ISIS NUMBER: 5301405008030005001.

Choeronycteris Tschudi, 1844. Untersuchungen uber die Fauna Peruana, p. 70.
REVIEWED BY: T. A. Griffiths (TAG); B. Villa-Ramirez (BVR).
ISIS NUMBER: 5301405008031000000.

Choeronycteris mexicana Tschudi, 1844. Untersuchungen uber die Fauna Peruana, p. 72.
TYPE LOCALITY: Mexico.
DISTRIBUTION: Honduras to S. California, Arizona, and New Mexico (USA); perhaps
Venezuela.
COMMENT: It is doubtful that *ponsi* (described from N.W. Venezuela) is referrable to
this genus; see Jones and Carter, 1976:18.
ISIS NUMBER: 5301405008031002001.

Chrotopterus Peters, 1865. Monatsb. Preuss. Akad. Wiss. Berlin, p. 505.
REVIEWED BY: R. Barquez (RB); J. Molinari (JML); B. Villa-Ramirez (BVR).
ISIS NUMBER: 5301405008004000000.

Chrotopterus auritus (Peters, 1856). Monatsb. Preuss. Akad. Wiss. Berlin, p. 415.
TYPE LOCALITY: Brazil, Santa Catarina.
DISTRIBUTION: C. Veracruz (Mexico) to the Guianas, S. Brazil, Bolivia, Peru, Paraguay,
and N. Argentina.
COMMENT: See Carter and Dolan, 1978, Spec. Publ. Mus. Texas Tech Univ., 15:37, for
corrected type locality.
ISIS NUMBER: 5301405008004001001.

Desmodus Wied-Neuweid, 1826. Beitr. Naturgesch. Brasil, 2:231.
REVIEWED BY: J. Molinari (JML); B. Villa-Ramirez (BVR).
COMMENT: Includes *Diaemus;* see Jones and Carter, 1979:9. Formerly included in
Desmodontidae; see Jones and Carter, 1976:31.
ISIS NUMBER: 5301405008001000000.

Desmodus rotundus (E. Geoffroy, 1810). Ann. Mus. Hist. Nat. Paris, 15:181.
TYPE LOCALITY: Paraguay, Asuncion.
DISTRIBUTION: Uruguay, N. Argentina, and N. Chile to Sonora and Tamaulipas
(Mexico); Margarita Isl.; Trinidad.
ISIS NUMBER: 5301405008001001001.

Desmodus youngi Jentink, 1893. Notes Leyden Mus., 15:282.
TYPE LOCALITY: Guyana, Berbice River, upper Canje Creek.
DISTRIBUTION: Tamaulipas (Mexico) to N. Argentina and E. Brazil; Trinidad; Margarita
Isl.
COMMENT: May represent a distinct genus, *Diaemus* (JKJ).

Diphylla Spix, 1823. Sim. Vespert. Brasil., p. 68.
REVIEWED BY: J. Molinari (JML); B. Villa-Ramirez (BVR).
COMMENT: Formerly included in Desmodontidae; see Jones and Carter, 1976:31.
ISIS NUMBER: 5301405008003000000.

Diphylla ecaudata Spix, 1823. Sim. Vespert. Brasil., p. 68.
TYPE LOCALITY: Brazil, Bahia, San Francisco River.
DISTRIBUTION: S. Texas (U.S.A.) to Peru, E. Brazil, and Venezuela.
ISIS NUMBER: 5301405008003001001.

Ectophylla H. Allen, 1892. Proc. U.S. Nat. Mus., 15:441.
REVIEWED BY: R. Barquez (RB); J. Molinari (JML); W. D. Webster (WDW).
COMMENT: Includes *Mesophylla;* see Goodwin and Greenhall, 1962, Am. Mus. Novit.,
2080:2–10; and Jones and Carter, 1976:25.
ISIS NUMBER: 5301405008032000000.

Ectophylla alba H. Allen, 1892. Proc. U.S. Nat. Mus., 15:442.
TYPE LOCALITY: Nicaragua, Comarca de El Cabo(=Rio Segovia).
DISTRIBUTION: Nicaragua to W. Panama.
ISIS NUMBER: 5301405008032001001.

Ectophylla macconnelli (Thomas, 1901). Ann. Mag. Nat. Hist., ser. 7, 8:143.
TYPE LOCALITY: Guyana, Essequibo Dist., Kunuku Mts.
DISTRIBUTION: Costa Rica to Peru, Bolivia and Amazonian Brazil; Trinidad.
COMMENT: Often placed in a separate genus *(Mesophylla);* see Greenbaum *et al.,* 1975,
Bull. South. Calif. Acad. Sci., 74:156, 158, 159.
ISIS NUMBER: 5301405008032002001.

Erophylla Miller, 1906. Proc. Biol. Soc. Wash., 19:84.
REVIEWED BY: T. A. Griffiths (TAG); G. S. Morgan (GSM).
COMMENT: Revised by Buden, 1976, Proc. Biol. Soc. Wash., 89:1–16. Included as a
subgenus of *Phyllonycteris* by Varona, 1974:29.
ISIS NUMBER: 5301405008047000000.

Erophylla sezekorni (Gundlach, 1861). Monatsb. Preuss. Akad. Wiss. Berlin for 1860, p.
818.
TYPE LOCALITY: Cuba, Pinar del Rio, Santa Cruz de los Pinos, Rangel.
DISTRIBUTION: Cuba, Jamaica, Hispaniola, Puerto Rico, Bahamas, and Cayman Isls.
COMMENT: Includes *bombifrons;* see Buden, 1976, Proc. Biol. Soc. Wash., 89:14; reviewed
by Baker *et al.,* 1978, Mamm. Species, 115:1–5. Based on differences in size of ears,
shape of rostrum, inflation of braincase, and certain dental characters, *E.*
bombifrons (Hispaniola and Puerto Rico) should probably be regarded as a distinct
species. Although recognition of *Erophylla bombifrons* represents my unpublished
opinion, this species has been universally recognized up until the last several
years; see, Varona, 1974:29 and Hall, 1981:170 (GSM). *E. bombifrons* is probably a
distinct species (KFK).
ISIS NUMBER: 5301405008047002001.

Glossophaga E. Geoffroy, 1818. Mem. Mus. Hist. Nat. Paris, 4:418.
REVIEWED BY: R. Barquez (RB)(S. America); T. A. Griffiths (TAG); J. Molinari (JML); B.
Villa-Ramirez (BVR); W. D. Webster (WDW).
COMMENT: Revised by Miller, 1913. Proc. U.S. Nat. Mus., 46:413–429.
ISIS NUMBER: 5301405008016000000.

Glossophaga commissarisi Gardner, 1962. Los Ang. Cty. Mus. Contrib. Sci., 54:1.
TYPE LOCALITY: Mexico, Chiapas, 10 km S.E. Tonala.
DISTRIBUTION: N. Sinaloa (Mexico) to Panama. Perhaps N. South America to N. Peru
(WDW and TAG)
ISIS NUMBER: 5301405008016001001.

Glossophaga leachii Gray, 1844. Mammalia, *in* Zool. Voy. H. M. S. "Sulfur," 1:18.
TYPE LOCALITY: Nicaragua, Chinandega, Realejo.
DISTRIBUTION: Costa Rica to Guerrero, Morelos, and Tlaxcala (Mexico). Colima, Jalisco
(Mexico)(WDW).
COMMENT: Originally considered a subspecies of *soricina;* see Jones and Carter,
1976:14. Includes *morenoi;* see Villa, 1966, Inst. Biol. Univ. Nac. Auton. Mexico,
491 pp. Includes *alticola* Davis, 1944, by synonymy; see Webster and Jones, 1980,
Occas. Pap. Mus. Texas Tech Univ., 71:4.

Glossophaga longirostris Miller, 1898. Proc. Acad. Nat. Sci. Phila., p. 330.
TYPE LOCALITY: Colombia, Magdalena, Sierra Nevada de Santa Marta.
DISTRIBUTION: N. Colombia; Venezuela; Guyana; Margarita Isl.; Trinidad; Tobago;
Grenada; Dominica; Curacao; Bonaire; Aruba.

COMMENT: Includes *elongata;* see Jones and Carter, 1976:14; and Koopman, 1958, Evolution, 12:437.
ISIS NUMBER: 5301405008016003001.

Glossophaga mexicana Webster and Jones, 1980. Occas. Pap. Mus. Texas Tech Univ., 71:6.
TYPE LOCALITY: Mexico, Oaxaca, Rio Guamol, 34 mi. (55 km) S. (by Hwy.190) La Ventosa Jct.
DISTRIBUTION: Chiapas and Oaxaca (Mexico).

Glossophaga soricina (Pallas, 1766). Misc. Zool., p. 48.
TYPE LOCALITY: Surinam.
DISTRIBUTION: N. Sinaloa, Tres Marias Isls.(Mexico) to Guianas, S.E. Brazil, N. Argentina, Ecuador and Peru; Margarita Isl.; Trinidad; Jamaica; perhaps Bahama Isls.
ISIS NUMBER: 5301405008016004001.

Hylonycteris Thomas, 1903. Ann. Mag. Nat. Hist., ser. 7, 11:286.
REVIEWED BY: T. A. Griffiths (TAG); B. Villa-Ramirez (BVR); W. D. Webster (WDW).
ISIS NUMBER: 5301405008034000000.

Hylonycteris underwoodi Thomas, 1903. Ann. Mag. Nat. Hist., ser. 7, 11:287.
TYPE LOCALITY: Costa Rica, San Jose, Rancho Redondo.
DISTRIBUTION: W. Panama to Oaxaca to Jalisco and C. Veracruz (Mexico).
COMMENT: Reviewed by Jones and Homan, 1974, Mamm. Species, 32:1-2.
ISIS NUMBER: 5301405008034001001.

Leptonycteris Lydekker, 1891. *In* Flower and Lydekker, Intro. Mamm. Living Extinct, p. 674.
REVIEWED BY: T. A. Griffiths (TAG); W. D. Webster (WDW).
COMMENT: Revised by Hoffmeister, 1957, J. Mammal., 38:454-461. Reviewed by Davis and Carter, 1962, Proc. Biol. Soc. Wash., 75:193-198.
ISIS NUMBER: 5301405008035000000.

Leptonycteris curasoae Miller, 1900. Proc. Biol. Soc. Wash., 13:126.
REVIEWED BY: J. Molinari (JML).
TYPE LOCALITY: Curacao, Willemstad (Netherlands).
DISTRIBUTION: Curacao; Bonaire; Aruba; N.E. Colombia; N. Venezuela; Margarita Isl.
ISIS NUMBER: 5301405008035001001.

Leptonycteris nivalis (Saussure, 1860). Rev. Mag. Zool. Paris, ser. 2, 12:492.
REVIEWED BY: B. Villa-Ramirez (BVR).
TYPE LOCALITY: Mexico, Veracruz, Mt. Orizaba.
DISTRIBUTION: S.E. Arizona and W. Texas (U.S.A.) to S. Guatemala.
COMMENT: Includes *yerbabuenae;* see Watkins, *et al.,* 1972, Spec. Publ. Mus. Texas Tech Univ., 1:16.
ISIS NUMBER: 5301405008035002001.

Leptonycteris sanborni Hoffmeister, 1957. J. Mammal., 38:456.
REVIEWED BY: B. Villa-Ramirez (BVR).
TYPE LOCALITY: U.S.A., Arizona, Cochise Co., 10 mi. (16 km) S.S.E. Ft. Huachuca, mouth of Miller canyon.
DISTRIBUTION: S. Arizona and New Mexico (U.S.A.) to El Salvador.
COMMENT: *L. yerbabuenae* may be the correct name for this species; see Ramirez-Pulido and Alvarez, 1972, Southwest. Nat., 16:249-259; but also see Watkins, *et al.,* 1972, Spec. Publ. Mus. Texas Tech Univ., 1:16.
ISIS NUMBER: 5301405008035003001.

Lichonycteris Thomas, 1895. Ann. Mag. Nat. Hist., ser. 6, 16:55.
REVIEWED BY: R. Barquez (RB); T. A. Griffiths (TAG); J. Molinari (JML).
ISIS NUMBER: 5301405008036000000.

Lichonycteris degener Miller, 1931. J. Mammal., 12:411.
TYPE LOCALITY: Brazil, Para, Belem (=Para).
DISTRIBUTION: Amazonian Brazil; Venezuela; Bolivia.
ISIS NUMBER: 5301405008036001001.

Lichonycteris obscura Thomas, 1895. Ann. Mag. Nat. Hist., ser. 6, 16:55.
TYPE LOCALITY: Nicaragua, Managua, Managua.
DISTRIBUTION: Guatemala to Surinam and Peru.
ISIS NUMBER: 5301405008036002001.

Lionycteris Thomas, 1913. Ann. Mag. Nat. Hist., ser. 8, 12:270.
REVIEWED BY: R. Barquez (RB); T. A. Griffiths (TAG); J. Molinari (JML); V. A. Taddei (VAT).
ISIS NUMBER: 5301405008017000000.

Lionycteris spurrelli Thomas, 1913. Ann. Mag. Nat. Hist., ser. 8, 12:271.
TYPE LOCALITY: Colombia, Choco, Condoto.
DISTRIBUTION: E. Panama; Guyana; Surinam; Amazonian Brazil; Venezuela; Colombia; Amazonian Peru.
COMMENT: A junior synonym is *Lancophylla* sp. (sic) (part), Piccinini, 1974, Bol. Mus. Para. Emilio Goeldi, Belem, 77:13; see Taddei *et al.*, 1978, Bol. Mus. Para. Emilio Goeldi, Belem, 92:1. Williams and Genoways, 1980, Ann. Carnegie Mus., 49, extended the range to Surinam.
ISIS NUMBER: 5301405008017001001.

Lonchophylla Thomas, 1903. Ann. Mag. Nat. Hist., ser. 7, 12:458.
REVIEWED BY: R. Barquez (RB); T. A. Griffiths (TAG); J. Molinari (JML); V. A. Taddei (VAT).
ISIS NUMBER: 5301405008018000000.

Lonchophylla bokermanni Sazima *et al.*, 1978. Rev. Brasil. Biol., 38:82.
TYPE LOCALITY: Brazil, Minas Gerais, Jaboticatubas, Serra do Cipo.
DISTRIBUTION: S.E. Brazil.
COMMENT: Apparently related to *hesperia* (VAT).

Lonchophylla handleyi Hill, 1980. Bull. Br. Mus. (Nat. Hist.) Zool., 38:233.
TYPE LOCALITY: Ecuador, Morona, Santiago, Los Tayos (03° 07' S, 18° 12' W.).
DISTRIBUTION: Ecuador; Peru.
COMMENT: Formerly confused with *robusta*; see Hill, 1980, *op. cit.*

Lonchophylla hesperia G. M. Allen, 1908. Bull. Mus. Comp. Zool. Harv. Univ., 52:35.
TYPE LOCALITY: Peru, Tumbes, Zorritos.
DISTRIBUTION: N. Peru; Ecuador.
COMMENT: Known only from five specimens; see Gardner, 1976, Occas. Pap. Mus. Zool. La. St. Univ., 48:5.
ISIS NUMBER: 5301405008018001001.

Lonchophylla mordax Thomas, 1903. Ann. Mag. Nat. Hist., ser. 7, 12:459.
TYPE LOCALITY: Brazil, Bahia, Lamarao.
DISTRIBUTION: Costa Rica to Ecuador; E. Brazil; perhaps Peru and Bolivia.
COMMENT: Includes *concava*; see Handley, 1966, *in* Wenzel and Tipton, eds, Ectoparasites of Panama, p. 763; but also see Jones and Carter, 1976:16, who provisionally recognized it as a species.
ISIS NUMBER: 5301405008018002001.

Lonchophylla robusta Miller, 1912. Proc. U.S. Nat. Mus., 42:23.
TYPE LOCALITY: Panama, Canal Zone, Rio Chilibrillo, near Alahuela.
DISTRIBUTION: Nicaragua to Venezuela and Ecuador.
COMMENT: Although specimens from Panama are larger than those from Costa Rica, the species is still regarded as monotypic; see Walton, 1963, Tulane Stud. Zool., 10(2):87–90.
ISIS NUMBER: 5301405008018003001.

Lonchophylla thomasi J. A. Allen, 1904. Bull. Am. Mus. Nat. Hist., 20:230.
TYPE LOCALITY: Venezuela, Bolivar, Ciudad Bolivar.
DISTRIBUTION: E. Panama; Colombia; Venezuela; Guianas; Amazonian Brazil; Peru; Bolivia.

COMMENT: Specimens of this species have frequently been confused with *concava*, *mordax*, and *Lionycteris spurrelli*; see Taddei *et al.*, 1978, Bol. Mus. Para. Emilio Goeldi, Belem, 92:1, 2; and Koopman, 1978, Am. Mus. Novit., 2651:1–33.
ISIS NUMBER: 5301405008018004001.

Lonchorhina Tomes, 1863. Proc. Zool. Soc. Lond., 1863:81.
REVIEWED BY: R. Barquez (RB); J. Molinari (JML); W. D. Webster (WDW).
COMMENT: Reviewed by Hernandez-Camacho and Cadena-G., 1978, Caldasia, 12:200–251.
ISIS NUMBER: 5301405008005000000.

Lonchorhina aurita Tomes, 1863. Proc. Zool. Soc. Lond., p. 83.
REVIEWED BY: B. Villa-Ramirez (BVR).
TYPE LOCALITY: Trinidad and Tobago, Trinidad.
DISTRIBUTION: Oaxaca (Mexico) to S.E. Brazil; Venezuela; Bolivia; Peru; Ecuador; Trinidad; perhaps New Providence Isl. (Bahama Isls.).
COMMENT: Includes *occidentalis*; see Jones and Carter, 1976:10, who also discussed the questionable Bahamian record.
ISIS NUMBER: 5301405008005001001.

Lonchorhina marinkellei Hernandez-Camacho and Cadena-G., 1978. Caldasia, 12:229.
TYPE LOCALITY: Colombia, Vaupes, near Mitu, Durania.
DISTRIBUTION: E. Colombia.

Lonchorhina orinocensis Linares and Ojasti, 1971. Novid. Cient. Contrib. Occas. Mus. Hist. Nat. La Salle, Ser. Zool., 36:2.
TYPE LOCALITY: Venezuela, Bolivar, 50 km N.E. Puerto Paez, Boca de Villacoa.
DISTRIBUTION: Venezuela; Colombia.

Macrophyllum Gray, 1838. Mag. Zool. Bot., 2:489.
REVIEWED BY: R. Barquez (RB); J. Molinari (JML); B. Villa-Ramirez (BVR); W. D. Webster (WDW).
ISIS NUMBER: 5301405008006000000.

Macrophyllum macrophyllum (Schinz, 1821). Das Thierreich, 1:163.
TYPE LOCALITY: Brazil, Bahia, Rio Mucuri.
DISTRIBUTION: Tabasco (Mexico) to Venezuela, Guianas, Peru, S.E. Brazil, and N.E. Argentina.
COMMENT: Reviewed by Harrison, 1975, Mamm. Species, 62:1–3.
ISIS NUMBER: 5301405008006001001.

Macrotus Gray, 1843. Proc. Zool. Soc. Lond., 1843:21.
REVIEWED BY: B. Villa-Ramirez (BVR); W. D. Webster (WDW).
COMMENT: Revised by Anderson and Nelson, 1965, Am. Mus. Novit., 2212:1–39.
ISIS NUMBER: 5301405008007000000.

Macrotus californicus Baird, 1858. Proc. Acad. Nat. Sci. Phila., 10:116.
TYPE LOCALITY: U.S.A., California, Imperial Co., Old Fort Yuma.
DISTRIBUTION: N. Sinaloa and S.W. Chihuahua (Mexico) to S. Nevada, S. California (U.S.A.); Baja California and Tamaulipas (Mexico).
COMMENT: For a comparison with *waterhousii*, see Davis and Baker, 1974, Syst. Zool., 23:26, 34.

Macrotus waterhousii Gray, 1843. Proc. Zool. Soc. Lond., 1843:21.
TYPE LOCALITY: Haiti.
DISTRIBUTION: Sonora to Hidalgo and Colima (Mexico) to Guatemala; Bahama Isls.; Jamaica; Cuba; Isle of Pines; Hispaniola; Beata Isl.
COMMENT: Includes *mexicanus*; see Anderson and Nelson, 1965, Am. Mus. Novit., 2212:25. Reviewed by Anderson, 1969, Mamm. Species, 1:1–4.
ISIS NUMBER: 5301405008007001001.

Micronycteris Gray, 1866. Proc. Zool. Soc. Lond., 1866:113.
    REVIEWED BY: R. Barquez (RB); J. Molinari (JML); B. Villa-Ramirez (BVR); W. D.
        Webster (WDW).
    COMMENT: Includes *Barticonycteris;* see Koopman, 1978, Am. Mus. Novit., 2651:4.
        Corbet and Hill, 1980:54, listed *Barticonycteris* as a distinct genus without
        comment. Revised by Sanborn, 1949, Fieldiana Zool., 31:215-233.
    ISIS NUMBER: 5301405008008000000.

Micronycteris behni (Peters, 1865). Monatsb. Preuss. Akad. Wiss. Berlin, p. 505.
    TYPE LOCALITY: Brazil, Mato Grosso, Cuiaba(=Cuyaba).
    DISTRIBUTION: C. Brazil; S. Peru.
    ISIS NUMBER: 5301405008008001001.

Micronycteris brachyotis (Dobson, 1879). Proc. Zool. Soc. Lond., 1878:880.
    TYPE LOCALITY: French Guiana, Cayenne.
    DISTRIBUTION: Oaxaca (Mexico) to French Guiana and Amazonian Brazil; Trinidad.
    COMMENT: Includes *platyceps;* see Jones and Carter, 1976:9.
    ISIS NUMBER: 5301405008008002001.

Micronycteris daviesi (Hill, 1964). Mammalia, 28:557.
    TYPE LOCALITY: Guyana, Essequibo Prov., Potaro road, 24 mi. (39 km) from Bartica.
    DISTRIBUTION: Costa Rica; Guyana; Surinam; Peru.
    COMMENT: Formerly included in *Barticonycteris;* see Koopman, 1978, Am. Mus. Novit.,
        2651:4.

Micronycteris hirsuta (Peters, 1869). Monatsb. Preuss. Akad. Wiss. Berlin, p. 397.
    TYPE LOCALITY: Costa Rica, Guanacaste, Pozo Azul.
    DISTRIBUTION: Honduras to Guyana, Trinidad, Amazonian Brazil, and Peru.
    ISIS NUMBER: 5301405008008003001.

Micronycteris megalotis (Gray, 1842). Ann. Mag. Nat. Hist., 10:257.
    TYPE LOCALITY: Brazil, Sao Paulo, Pereque.
    DISTRIBUTION: S. Tamaulipas and S. Jalisco (Mexico) to Peru; Guyana; French Guiana;
        Surinam; S.E. Brazil; Trinidad; Tobago; Margarita Isl.; Grenada.
    COMMENT: Includes *microtis;* see Gardner *et al.*, 1970, J. Mammal., 51:715; and Jones *et*
        al., 1977, Occ. Pap. Mus. Texas Tech Univ., 47:6.
    ISIS NUMBER: 5301405008008004001.

Micronycteris minuta (Gervais, 1856). Exped. Castelnau Zool., p. 50.
    TYPE LOCALITY: Brazil, Bahia, Capela Nova.
    DISTRIBUTION: Nicaragua to S. Brazil; Peru; Venezuela; Trinidad. Bolivia (JRT).
    ISIS NUMBER: 5301405008008005001.

Micronycteris nicefori Sanborn, 1949. Fieldiana Zool., 31:230.
    TYPE LOCALITY: Colombia, Norte de Santander, Cucuta.
    DISTRIBUTION: Nicaragua to N. Colombia, Venezuela, Amazonian Brazil, and Peru;
        Trinidad. S. Mexico; Guyana (JRT).
    ISIS NUMBER: 5301405008008006001.

Micronycteris pusilla Sanborn, 1949. Fieldiana Zool., 31:228.
    TYPE LOCALITY: Brazil, Amazonas, Tahuapunta (Vaupes River).
    DISTRIBUTION: N.W. Brazil; E. Colombia.
    ISIS NUMBER: 5301405008008008001.

Micronycteris schmidtorum Sanborn, 1935. Field Mus. Nat. Hist. Publ. Ser. Zool., 20:81.
    TYPE LOCALITY: Guatemala, Izabal, Bobos.
    DISTRIBUTION: S. Mexico to Venezuela.
    ISIS NUMBER: 5301405008008009001.

Micronycteris sylvestris (Thomas, 1896). Ann. Mag. Nat. Hist., ser. 6, 18:302.
    TYPE LOCALITY: Costa Rica, Guanacaste, Hda. Miravalles, between 1400 and 2000 ft.
        (427-610 m).
    DISTRIBUTION: Peru and S.E.Brazil to Nayarit and Veracruz (Mexico); Trinidad. Guyana
        (JRT).
    ISIS NUMBER: 5301405008008010001.

Mimon Gray, 1847. Proc. Zool. Soc. Lond., 1847:14.
REVIEWED BY: R. Barquez (RB); J. Molinari (JML); B. Villa-Ramirez (BVR)(Mexico); W. D. Webster (WDW).
COMMENT: Includes *Anthorhina*; see Handley, 1960, Proc. U.S. Nat. Mus., 112:459–479.
ISIS NUMBER: 5301405008009000000.

Mimon bennettii (Gray, 1838). Mag. Zool. Bot., 2:483.
TYPE LOCALITY: Brazil, Sao Paulo, Ipanema.
DISTRIBUTION: Guyana; Surinam; S.E. Brazil.
COMMENT: Formerly included *cozumelae*; see Schaldach, 1965, An. Inst. Biol. Univ. Nac. Auton. Mex., 35:132; and Villa, 1966, Inst. Biol. Univ. Nac. Auton. Mexico, p. 216; but also see Jones and Carter, 1976:12.
ISIS NUMBER: 5301405008009001001.

Mimon cozumelae Goldman, 1914. Proc. Biol. Soc. Wash., 27:75.
TYPE LOCALITY: Mexico, Quintana Roo, Cozumel Isl.
DISTRIBUTION: C. Veracruz and Oaxaca (Mexico) to N. Colombia.
COMMENT: Considered a subspecies of *bennettii* by Schaldach, 1965, An. Inst. Biol. Univ. Nac. Auton. Mex., 35:132 and Villa, 1966, Inst. Biol. Univ. Nac. Auton. Mexico, p. 216; but see Jones and Carter, 1976:12. Almost certainly a subspecies of *bennettii* (KFK,JKJ). Closely related to *bennettii* but monotypic (BVR).

Mimon crenulatum (E. Geoffroy, 1810). Ann. Mus. Hist. Nat. Paris, 15:193.
TYPE LOCALITY: Brazil, Bahia.
DISTRIBUTION: Campeche (Mexico) to Guianas, Brazil, Bolivia, Ecuador and E. Peru; Trinidad.
COMMENT: Includes *koepckeae*; see Jones and Carter, 1979:8 and Koopman, 1978:5. Also includes *longifolium, peruanum,* and *picatum*; see Handley, 1960, Proc. U.S. Nat. Mus., 112:462,463, who also discussed the type locality.
ISIS NUMBER: 5301405008009002001.

Monophyllus Leach, 1821. Trans. Linn. Soc. Lond., 13:75.
REVIEWED BY: T. A. Griffiths (TAG); W. D. Webster (WDW).
COMMENT: Reviewed by Schwartz and Jones, 1967, Proc. U.S. Nat. Mus., 124:1–20. A key to the genus was published by Homan and Jones, 1975, Mamm. Species, 57:1–3.
ISIS NUMBER: 5301405008019000000.

Monophyllus plethodon Miller, 1900. Proc. Wash. Acad. Sci., 2:35.
TYPE LOCALITY: Barbados (Lesser Antilles), St. Michael parish.
DISTRIBUTION: Lesser Antilles from Anguilla to Barbados. Subfossils known from Puerto Rico (JRT and WDW).
COMMENT: Includes *luciae*; see Schwartz and Jones, 1967, Proc. U.S. Nat. Mus., 124:13. Reviewed by Homan and Jones, 1975, Mamm. Species, 58:1–2.
ISIS NUMBER: 5301405008019001001.

Monophyllus redmani Leach, 1821. Trans. Linn. Soc. Lond., 13:76.
TYPE LOCALITY: Jamaica.
DISTRIBUTION: Cuba; Hispaniola; Puerto Rico; Jamaica; S. Bahama Isls.
COMMENT: Includes *cubanus, portoricensis,* and *clinedaphus*; see Schwartz and Jones, 1967, Proc. U.S. Nat. Mus., 124:13. Reviewed by Homan and Jones, 1975, Mamm. Species, 57:1–3.

Musonycteris Schaldach and McLaughlin, 1960. Los Ang. Cty. Mus. Contrib. Sci., 37:2.
REVIEWED BY: T. A. Griffiths (TAG); B. Villa-Ramirez (BVR).
COMMENT: Included in *Choeronycteris* by Handley, 1966, Proc. Biol. Soc. Wash., 79:85, 86; but see Phillips, 1971, Misc. Publ. Univ. Kans. Mus. Nat. Hist., 54:99.

Musonycteris harrisoni Schaldach and McLaughlin, 1960. Los Ang. Cty. Mus. Contrib. Sci., 37:3.
TYPE LOCALITY: Mexico, Colima, 2 km S.E. Pueblo Juarez.
DISTRIBUTION: Colima, S. Michoacan, C. Guerrero (Mexico).

Phylloderma Peters, 1865. Monatsb. Preuss. Akad. Wiss. Berlin, p. 513.
REVIEWED BY: R. Barquez (RB).
ISIS NUMBER: 5301405008010000000.

Phylloderma stenops Peters, 1865. Monatsb. Preuss. Akad. Wiss. Berlin, p. 513.
TYPE LOCALITY: French Guiana, Cayenne.
DISTRIBUTION: S. Mexico to Brazil and Bolivia. Peru (JRT, RB).
COMMENT: Includes *septentrionalis;* see Jones and Carter, 1976:13. Bolivian form
reviewed by Barquez and Ojeda, 1979, Neotropica, 25(73).
ISIS NUMBER: 5301405008010001001.

Phyllonycteris Gundlach, 1861. Monatsb. Preuss. Akad. Wiss. Berlin for 1860, p. 817.
REVIEWED BY: T. A. Griffiths (TAG); G. S. Morgan (GSM).
COMMENT: Includes the fossil species *major* from Puerto Rico (KFK, TAG).
ISIS NUMBER: 5301405008048000000.

Phyllonycteris aphylla (Miller, 1898). Proc. Acad. Nat. Sci. Phila., 50:334.
TYPE LOCALITY: Jamaica.
DISTRIBUTION: Jamaica.
ISIS NUMBER: 5301405008048001001.

Phyllonycteris poeyi Gundlach, 1861. Monatsb. Preuss. Akad. Wiss. Berlin for 1860, p. 817.
TYPE LOCALITY: Cuba, Matanzas, Canimar (cafetal "San Antonio el Fundador").
DISTRIBUTION: Cuba and Hispaniola (Haiti and Dominican Republic). Isle of Pines
(JRT).
COMMENT: Includes *obtusa;* see Jones and Carter, 1976:30; from Hispaniola this
subspecies was described from fossil remains and was only recently discovered as
a living animal; see Klingener, *et al.,* 1978, Ann. Carnegie Mus., 47:90–92. Corbet
and Hill, 1980:61, listed *obtusa* as a distinct species without comment.
ISIS NUMBER: 5301405008048004001.

Phyllops Peters, 1865. Monatsb. Preuss. Akad. Wiss. Berlin, p. 356.
REVIEWED BY: G. S. Morgan (GSM).
COMMENT: Included in *Stenoderma* by Varona, 1974:24; Simpson, 1945:58, and Silva-
Taboada, 1979, Los Murcielagos de Cuba, p. 199; but see Jones and Carter,
1976:28; and Corbet and Hill, 1980:60.
ISIS NUMBER: 5301405008037000000.

Phyllops falcatus (Gray, 1839). Ann. Nat. Hist., ser. 1, 4:1.
TYPE LOCALITY: Cuba, Habana, Guanabacoa.
DISTRIBUTION: Cuba; as fossil, Isle of Pines (JRT).
ISIS NUMBER: 5301405008037001001.

Phyllops haitiensis (J. A. Allen, 1908). Bull. Am. Mus. Nat. Hist., 24:581.
TYPE LOCALITY: Dominican Republic, Cana Honda.
DISTRIBUTION: Hispaniola (Haiti and Dominican Republic).
COMMENT: Probably a subspecies of *falcatus;* see Jones and Carter, 1976:29. Almost
certainly a subspecies of *falcatus* (KFK).
ISIS NUMBER: 5301405008037002001.

Phyllostomus Lacepede, 1799. Tabl. Mamm., p. 16.
REVIEWED BY: R. Barquez (RB).
ISIS NUMBER: 5301405008011000000.

Phyllostomus discolor Wagner, 1843. Arch. Naturgesch., 9(1):366.
REVIEWED BY: B. Villa-Ramirez (BVR).
TYPE LOCALITY: Brazil, Mato Grosso, Cuiaba.
DISTRIBUTION: C. Veracruz (Mexico) to Guianas, S.E. Brazil and Peru; Trinidad;
Margarita Isl. Bolivia (JRT). N. Argentina (RB).
ISIS NUMBER: 5301405008011001001.

Phyllostomus elongatus (E. Geoffroy, 1810). Ann. Mus. Hist. Nat. Paris, 15:182.
TYPE LOCALITY: Brazil, Mato Grosso, Rio Branco.
DISTRIBUTION: Bolivia, E. Peru, Ecuador, and Colombia to Surinam and E. Brazil.
ISIS NUMBER: 5301405008011002001.

Phyllostomus hastatus (Pallas, 1767). Spicil. Zool., 3:7.
TYPE LOCALITY: Surinam.
DISTRIBUTION: Honduras to N. and E. Brazil and Guianas; Peru; Trinidad; Tobago; Margarita Isl.; Bolivia.
ISIS NUMBER: 5301405008011003001.

Phyllostomus latifolius (Thomas, 1901). Ann. Mag. Nat. Hist., ser. 7, 8:142.
TYPE LOCALITY: Guyana, Essequibo Prov., Mt. Kanuku.
DISTRIBUTION: Guyana; Surinam; S.E. Colombia; Amazonian Brazil.
ISIS NUMBER: 5301405008011004001.

Platalina Thomas, 1928. Ann. Mag. Nat. Hist., ser. 10, 8:120.
REVIEWED BY: R. Barquez (RB); T. A. Griffiths (TAG).
ISIS NUMBER: 5301405008020000000.

Platalina genovensium Thomas, 1928. Ann. Mag. Nat. Hist., ser. 10, 8:121.
TYPE LOCALITY: Peru, near Lima.
DISTRIBUTION: Peru.
ISIS NUMBER: 5301405008020001001.

Pygoderma Peters, 1863. Monatsb. Preuss. Akad. Wiss. Berlin, p. 83.
REVIEWED BY: R. A. Ojeda (RAO).
ISIS NUMBER: 5301405008038000000.

Pygoderma bilabiatum (Wagner, 1843). Arch. Naturgesch., 1:366.
TYPE LOCALITY: Brazil, Sao Paulo, Ipanema.
DISTRIBUTION: Surinam; Bolivia; S.C. Brazil; Paraguay; N. Argentina.
COMMENT: The reported North American occurrence is erroneous; see Jones and Carter, 1976:29.
ISIS NUMBER: 5301405008038001001.

Rhinophylla Peters, 1865. Monatsb. Preuss. Akad. Wiss. Berlin, p. 355.
REVIEWED BY: R. Barquez (RB)(S. America).
ISIS NUMBER: 5301405008022000000.

Rhinophylla alethina Handley, 1966. Proc. Biol. Soc. Wash., 79:86.
TYPE LOCALITY: Colombia, Valle, 27 km S. Buenaventura, Raposo River.
DISTRIBUTION: W. Colombia.
ISIS NUMBER: 5301405008022001001.

Rhinophylla fischerae Carter, 1966. Proc. Biol. Soc. Wash., 79:235.
TYPE LOCALITY: Peru, Loreto, 61 mi. (98 km) S.E. Pucallpa.
DISTRIBUTION: Peru; Ecuador; S.E. Colombia; Amazonian Brazil.
ISIS NUMBER: 5301405008022002001.

Rhinophylla pumilio Peters, 1865. Monatsb. Preuss. Akad. Wiss. Berlin, p. 355.
TYPE LOCALITY: Brazil, Bahia.
DISTRIBUTION: Surinam to Colombia, Ecuador, Peru and Bolivia; N. and E. Brazil.
ISIS NUMBER: 5301405008022003001.

Scleronycteris Thomas, 1912. Ann. Mag. Nat. Hist., ser. 8, 10:404.
REVIEWED BY: R. Barquez (RB); T. A. Griffiths (TAG).
ISIS NUMBER: 5301405008039000000.

Scleronycteris ega Thomas, 1912. Ann. Mag. Nat. Hist., ser. 8, 10:405.
TYPE LOCALITY: Brazil, Amazonas, Ega.
DISTRIBUTION: Amazonian Brazil; S. Venezuela.
ISIS NUMBER: 5301405008039001001.

Sphaeronycteris Peters, 1882. Sitzb. Preuss. Akad. Wiss., 45:988.
REVIEWED BY: R. Barquez (RB).
ISIS NUMBER: 5301405008040000000.

Sphaeronycteris toxophyllum Peters, 1882. Sitzb. Preuss. Akad. Wiss., 45:989.
TYPE LOCALITY: Peru, Loreto, Pebas.
DISTRIBUTION: Colombia to Venezuela, Peru and Bolivia; Amazonian Brazil.
ISIS NUMBER: 5301405008040001001.

Stenoderma E. Geoffroy, 1818. Descrip. de L'Egypte, 2:114.
REVIEWED BY: G. S. Morgan (GSM).
COMMENT: Some authors include *Ardops, Phyllops,* and *Ariteus* in *Stenoderma;* see Varona, 1974:23–26, and Simpson, 1945:58, but most mammalogists follow the arrangement presented here (GSM); see also Jones and Carter, 1976:28–29.
ISIS NUMBER: 5301405008041000000.

Stenoderma rufum Desmarest, 1820. Mammalogie, p. 117.
TYPE LOCALITY: Not designated in original publication (probably Virgin Isls.).
DISTRIBUTION: Puerto Rico, Virgin Isls. (St. John and St. Thomas).
COMMENT: Reviewed by Genoways and Baker, 1972, Mamm. Species, 18:1–4.
ISIS NUMBER: 5301405008041001001.

Sturnira Gray, 1842. Ann. Mag. Nat. Hist., ser. 1, 10:257.
REVIEWED BY: R. Barquez (RB); B. Villa-Ramirez (BVR).
COMMENT: Includes *Corvira* and *Sturnirops;* see Jones and Carter, 1976:21. Davis, 1980, Occas. Pap. Mus. Texas Tech Univ., 70:4,5, gave a key to all but one of the species recognized here.
ISIS NUMBER: 5301405008042000000.

Sturnira aratathomasi Peterson and Tamsitt, 1968. R. Ont. Mus. Life Sci. Occ. Pap., 12:1.
TYPE LOCALITY: Colombia, Valle, 2 km S. Pance (ca. 20 km S.W. Cali), 1650 m.
DISTRIBUTION: S.W. Colombia; W. Ecuador.
ISIS NUMBER: 5301405008042001001.

Sturnira bidens Thomas, 1915. Ann. Mag. Nat. Hist., ser. 8, 16:310.
TYPE LOCALITY: Ecuador, Napo, Baeza, Upper Coca River, 6500 ft. (1981 m).
DISTRIBUTION: Ecuador; E. Peru; S. Colombia; Venezuela.
COMMENT: Formerly included in *Corvira;* see Gardner and O'Neill, 1969, Occas. Pap. Mus. Zool., Louisiana State Univ., 38:1–8; Jones and Carter, 1976:21.
ISIS NUMBER: 5301405008042002001.

Sturnira bogotensis Shamel, 1927. Proc. Biol. Soc. Wash., 40:129.
TYPE LOCALITY: Colombia, Cundinamarca, Bogota.
DISTRIBUTION: Venezuela; Colombia; Ecuador; Peru; Bolivia.
COMMENT: Usually confused with *ludovici,* but recognized by Handley, 1976, Brigham Young Univ. Sci. Bull., 20(5):25. See also comment under *ludovici.*

Sturnira erythromos (Tschudi, 1844). Fauna Peruana, p. 64.
TYPE LOCALITY: Peru.
DISTRIBUTION: Peru; Venezuela; Bolivia. E. slopes of Andes in Colombia (JRT).
ISIS NUMBER: 5301405008042003001.

Sturnira lilium (E. Geoffroy, 1810). Ann. Mus. Hist. Nat. Paris, 15:181.
TYPE LOCALITY: Paraguay, Asuncion.
DISTRIBUTION: Lesser Antilles; S. Sonora and S. Tamaulipas (Mexico) to N. Argentina, Uruguay, and E. Brazil; Trinidad; perhaps Jamaica. Peru (JRT).
COMMENT: Includes *angeli* and *paulsoni;* see Jones and Carter, 1976:20.
ISIS NUMBER: 5301405008042004001.

Sturnira ludovici Anthony, 1924. Am. Mus. Novit., 139:8.
TYPE LOCALITY: Ecuador, Pichincha, near Gualea.
DISTRIBUTION: Bolivia; Peru; Ecuador; Venezuela to S. Sinaloa, Durango and S. Tamaulipas (Mexico).

COMMENT: Includes *hondurensis* and *oporophilum*; see Jones and Carter, 1976:22. Middle American and Bolivian records may well pertain to *bogotensis*; Peruvian ones definitely do (KFK).

ISIS NUMBER: 5301405008042005001.

Sturnira luisi Davis, 1980. Occas. Pap. Mus. Texas Tech Univ., 70:1.
TYPE LOCALITY: Costa Rica, Alajuela, 11 mi. (18 km) N.E. Naranjo, Cariblanco, 3000 ft. (914 m).
DISTRIBUTION: Costa Rica, to Ecuador and Peru.
COMMENT: The presence of this species in Panama and Colombia has not been verified; previously confused with *Sturnira ludovici* (JRT).

Sturnira magna de la Torre, 1966. Proc. Biol. Soc. Wash., 79:267.
TYPE LOCALITY: Peru, Loreto, Iquitos, Rio Maniti, Santa Cecilia.
DISTRIBUTION: N.E. and C. Peru; Bolivia; Colombia; Ecuador.
ISIS NUMBER: 5301405008042006001.

Sturnira mordax (Goodwin, 1938). Am. Mus. Novit., 976:1.
TYPE LOCALITY: Costa Rica, Cartago, El Sauce Peralta.
DISTRIBUTION: Costa Rica.
COMMENT: Possibly the true Middle American representative of *ludovici* (KFK). Formerly included in *Sturnirops*; see Jones and Carter, 1976:21.
ISIS NUMBER: 5301405008042007001.

Sturnira nana Gardner and O'Neill, 1971. Occas. Pap. Mus. Zool. La. St. Univ., 42:1.
TYPE LOCALITY: Peru, Ayacucho, Huanhuachayo, 1660 m.
DISTRIBUTION: S. Peru.
ISIS NUMBER: 5301405008042008001.

Sturnira thomasi de la Torre and Schwartz, 1966. Proc. Biol. Soc. Wash., 79:299.
TYPE LOCALITY: Guadeloupe (Lesser Antilles), Sofaia, 1200 ft. (366 m) (France).
DISTRIBUTION: Guadeloupe (Lesser Antilles).
COMMENT: Reviewed by Jones and Genoways, 1975, Mamm. Species, 68:1–2. May be a subspecies of *lilium* (RB), but see Jones and Phillips, 1976, Occas. Pap. Mus. Texas Tech Univ., 40:1–16.
ISIS NUMBER: 5301405008042009001.

Sturnira tildae de la Torre, 1959. Chicago Acad. Sci. Nat. Hist. Misc., 166:1.
TYPE LOCALITY: Trinidad and Tobago, Trinidad, Arima Valley.
DISTRIBUTION: Brazil; Surinam; Guyana; Venezuela; Trinidad; Ecuador; Peru; Bolivia. E. Colombia (JRT).
ISIS NUMBER: 5301405008042010001.

Tonatia Gray, 1827. *In* Griffith, Cuvier's Anim. Kingd., 5:71.
REVIEWED BY: R. Barquez (RB).
ISIS NUMBER: 5301405008012000000.

Tonatia bidens (Spix, 1823). Sim. Vespert. Brasil., p. 65.
TYPE LOCALITY: Brazil, Bahia, Rio Sao Francisco.
DISTRIBUTION: Guatemala and Honduras to Peru, Venezuela, and Guianas; Paraguay; Brazil; Trinidad.
ISIS NUMBER: 5301405008012001001.

Tonatia brasiliense (Peters, 1866). Monatsb. Preuss. Akad. Wiss. Berlin, p. 674.
REVIEWED BY: B. Villa-Ramirez (BVR).
TYPE LOCALITY: Brazil, Bahia.
DISTRIBUTION: S. Veracruz (Mexico) to Peru, Brazil and Trinidad.
COMMENT: Includes *minuta, nicaraguae*, and *venezuelae*; see Jones and Carter, 1979:7; but also see Gardner, 1978, Occas. Pap. Mus. Zool. La. St. Univ., 48:3.
ISIS NUMBER: 5301405008012002001.

Tonatia carrikeri (J. A. Allen, 1910). Bull. Am. Mus. Nat. Hist., 28:147.
TYPE LOCALITY: Venezuela, Bolivar, Rio Mocho.
DISTRIBUTION: Colombia; Venezuela; Guyana; Surinam; Bolivia; Peru.
ISIS NUMBER: 5301405008012003001.

Tonatia evotis Davis and Carter, 1978. Occas. Pap. Mus. Texas Tech Univ., 53:8.
TYPE LOCALITY: Guatemala, Izabal, 25 km S.S.W. Puerto Barrios.
DISTRIBUTION: S. Mexico; Guatemala; Honduras.

Tonatia schulzi Genoways and Williams, 1980. Ann. Carnegie Mus., 49:205.
TYPE LOCALITY: Surinam, Brokopondo, 3 km S.W. Rudi Koppelvliegveld.
DISTRIBUTION: Surinam.

Tonatia silvicola (d'Orbigny, 1836). Voy. Amer. Merid. Atlas Zool., 4:11, pl. 7.
TYPE LOCALITY: Bolivia, Yungas between Secure and Isiboro rivers.
DISTRIBUTION: Campeche (Mexico) to Bolivia, N. Argentina, Paraguay, Guianas, N. and
E. Brazil.
COMMENT: Includes *laephotis* and *amblyotis*; see Davis and Carter, 1978, Occas. Pap.
Mus. Texas Tech Univ., 53:1–12.
ISIS NUMBER: 5301405008012005001.

Trachops Gray, 1847. Proc. Zool. Soc. Lond., 1847:14.
REVIEWED BY: R. Barquez (RB); B. Villa-Ramirez (BVR).
ISIS NUMBER: 5301405008013000000.

Trachops cirrhosus (Spix, 1823). Sim. Vespert. Brasil., p. 64.
TYPE LOCALITY: Brazil, Pernambuco.
DISTRIBUTION: Oaxaca (Mexico) to Guianas, Brazil, Bolivia and Peru; Trinidad.
ISIS NUMBER: 5301405008013001001.

Uroderma Peters, 1865. Monatsb. Preuss. Akad. Wiss. Berlin, p. 588.
REVIEWED BY: R. Barquez (RB); B. Villa-Ramirez (BVR); W. B. Davis (WBD).
COMMENT: Revised by Davis, 1968, J. Mammal., 49:676–698.
ISIS NUMBER: 5301405008043000000.

Uroderma bilobatum Peters, 1866. Monatsb. Preuss. Akad. Wiss. Berlin, p. 392.
TYPE LOCALITY: Brazil, Sao Paulo.
DISTRIBUTION: Veracruz and Oaxaca (Mexico) to Peru, Bolivia and Brazil; Trinidad.
ISIS NUMBER: 5301405008043001001.

Uroderma magnirostrum Davis, 1968. J. Mammal., 49:679.
TYPE LOCALITY: Honduras, Valle, 10 km E. San Lorenzo.
DISTRIBUTION: Oaxaca (Mexico) to Venezuela, Peru, and Bolivia; Amazonian Brazil.
Guerrero (Mexico) (BVR).
ISIS NUMBER: 5301405008043002001.

Vampyressa Thomas, 1900. Ann. Mag. Nat. Hist., ser. 7, 5:270.
REVIEWED BY: R. Barquez (RB).
COMMENT: Includes *Vampyriscus*; see Jones and Carter, 1976:25.
ISIS NUMBER: 5301405008044000000.

Vampyressa bidens (Dobson, 1878). Cat. Chiroptera Br. Mus., p. 535.
TYPE LOCALITY: Peru, Loreto, Santa Cruz (Rio Huallaga).
DISTRIBUTION: Surinam to Colombia to Peru; Amazonian Brazil.
COMMENT: Formerly included in *Vampyriscus*; see Jones and Carter, 1976:25.
ISIS NUMBER: 5301405008044001001.

Vampyressa brocki Peterson, 1968. Life Sci. Contrib. R. Ont. Mus., 73:1.
TYPE LOCALITY: Guyana, Rupununi, *ca.* 40 mi. (64 km) E. Dadanawa, at Ow-wi-dy-wau
(Oshi Wau head, near Marara Waunowa), Kuitaro River.
DISTRIBUTION: Guyana; Surinam; S.E. Colombia; Amazonian Brazil.

Vampyressa melissa Thomas, 1926. Ann. Mag. Nat. Hist., ser. 9, 18:157.
TYPE LOCALITY: Peru, Amazonas, Chachapoyas, Puca Tambo, 1480 m.
DISTRIBUTION: Peru.
ISIS NUMBER: 5301405008044002001.

Vampyressa nymphaea Thomas, 1909. Ann. Mag. Nat. Hist., ser. 8, 4:230.
TYPE LOCALITY: Colombia, Choco, Novita (San Juan River).
DISTRIBUTION: W. Colombia to Nicaragua.
ISIS NUMBER: 5301405008044003001.

Vampyressa pusilla (Wagner, 1843). Abh. Munch. Akad. Wiss., 5:173.
REVIEWED BY: B. Villa-Ramirez (BVR).
TYPE LOCALITY: Brazil, Rio de Janiero, Sapitiba.
DISTRIBUTION: Veracruz (Mexico) to Peru, Venezuela, and S.E. Brazil.
COMMENT: Includes *thyone, venilla,* and *nattereri;* see Jones and Carter, 1976:24.
ISIS NUMBER: 5301405008044004001.

Vampyrodes Thomas, 1900. Ann. Mag. Nat. Hist., ser. 7, 5:270.
REVIEWED BY: R. Barquez (RB); B. Villa-Ramirez (BVR).
ISIS NUMBER: 5301405008045000000.

Vampyrodes caraccioli (Thomas, 1889). Ann. Mag. Nat. Hist., ser. 6, 4:167.
TYPE LOCALITY: Trinidad and Tobago, Trinidad.
DISTRIBUTION: Oaxaca (Mexico) to Peru, Bolivia (KFK) and N. Brazil; Trinidad; Tobago.
COMMENT: Includes *major;* see Jones and Carter, 1976:24.
ISIS NUMBER: 5301405008045001001.

Vampyrops Peters, 1865. Monatsb. Preuss. Akad. Wiss. Berlin, p. 356.
REVIEWED BY: R. Barquez (RB).
COMMENT: A synonym is *Platyrrhinus* Saussure, 1860, *nec Platyrrhinus* Fabricius, 1801, an emendation of *Platyrhinus* Clairville, 1798 (a beetle); see de la Torre and Starrett, 1959, Chicago Acad. Sci. Nat. Hist. Misc., 167:1. Hall, 1981:144, did not consider *Platyrrhinus* Saussure, 1860, to be preoccupied, maintaining that Fabricius' name was an incorrect spelling of Clairville's name, which is not a homonym.
ISIS NUMBER: 5301405008046000000.

Vampyrops aurarius Handley and Ferris, 1972. Proc. Biol. Soc. Wash., 84:522.
TYPE LOCALITY: Venezuela, Bolivar, 85 km S.S.E. El Dorado, 1000 m.
DISTRIBUTION: S. Venezuela.
COMMENT: May be a synonym of *dorsalis;* see Jones and Carter, 1976:23.

Vampyrops brachycephalus Rouk and Carter, 1972. Occas. Pap. Mus. Texas Tech Univ., 1:1.
TYPE LOCALITY: Peru, Huanuco, 3 mi. (5 km) S. Tingo Maria, 2400 ft. (732 m).
DISTRIBUTION: N. Brazil; Colombia to Surinam; Ecuador; Peru.
COMMENT: Includes *latus;* see Jones and Carter, 1976:23.
ISIS NUMBER: 5301405008046001001.

Vampyrops dorsalis (Thomas, 1900). Ann. Mag. Nat. Hist., ser. 7, 5:269.
TYPE LOCALITY: Ecuador, Paramba, 1100 m.
DISTRIBUTION: Panama to Peru and Bolivia. Costa Rica (RB).
COMMENT: Although the named forms *umbratus, oratus,* and *aquilus* were regarded as synonyms of *dorsalis* by Carter and Rouk, 1973, J. Mammal., 54:976, see also Handley, 1976, Brigham Young Univ. Sci. Bull., Biol. Ser., 20(5):28, who considered them a distinct species for which the oldest name is *umbratus.* See also comments under *umbratus.*
ISIS NUMBER: 5301405008046002001.

Vampyrops helleri (Peters, 1867). Monatsb. Preuss. Akad. Wiss. Berlin, p. 392.
REVIEWED BY: B. Villa-Ramirez (BVR).
TYPE LOCALITY: Mexico.
DISTRIBUTION: Veracruz (Mexico) to Peru, Bolivia, Paraguay, and Amazonian Brazil; Trinidad.
COMMENT: Includes *zarhinus;* see Jones and Carter, 1976:23; and Gardner and Carter, 1972, J. Mammal., 53:78.
ISIS NUMBER: 5301405008046003001.

Vampyrops infuscus Peters, 1881. Monatsb. Preuss. Akad. Wiss. Berlin for 1880, p. 259.
TYPE LOCALITY: Peru, Cajamarca, Hualgayoc, Hac. Ninabamba.
DISTRIBUTION: Colombia to Peru, Bolivia, and N.W. Brazil.
COMMENT: Includes *intermedius* and *fumosus;* see Gardner and Carter, 1972, J. Mammal., 53:72, who designated a neotype.

Vampyrops lineatus (E. Geoffroy, 1810). Ann. Mus. Hist. Nat. Paris, 15:180.
TYPE LOCALITY: Paraguay, Asuncion.
DISTRIBUTION: Colombia to Peru, Bolivia, Uruguay, S. and E. Brazil, and N. Argentina.
COMMENT: Includes *nigellus;* see Jones and Carter, 1979:8.
PROTECTED STATUS: CITES - Appendix III (Uruguay).
ISIS NUMBER: 5301405008046005001.

Vampyrops recifinus Thomas, 1901. Ann. Mag. Nat. Hist., ser. 7, 8:192.
TYPE LOCALITY: Brazil, Pernambuco, Recife.
DISTRIBUTION: E. Brazil; Guyana.
COMMENT: May be conspecific with *lineatus* (RB).

Vampyrops umbratus Lyon, 1902. Proc. Biol. Soc. Wash., 15:151.
TYPE LOCALITY: Colombia, Magdalena, San Miguel (Macotama River).
DISTRIBUTION: Panama; N. and W. Colombia; N. Venezuela.
COMMENT: Formerly included in *dorsalis* by Carter and Rouk, 1973, J. Mammal., 54:976; but also see Handley, 1976, Brigham Young Univ. Sci. Bull., Biol. ser., 20(5):28, who by inference included *oratus* and *aquilus* in this species. Hall, 1981:146, included this species in *dorsalis.*

Vampyrops vittatus (Peters, 1860). Monatsb. Preuss. Akad. Wiss. Berlin, p. 225.
TYPE LOCALITY: Venezuela, Carabobo, Puerto Cabello.
DISTRIBUTION: Costa Rica to Venezuela and Peru. Bolivia (RB).
ISIS NUMBER: 5301405008046007001.

Vampyrum Rafinesque, 1815. Analyse de la Nature, p. 54.
ISIS NUMBER: 5301405008014000000.

Vampyrum spectrum (Linnaeus, 1758). Syst. Nat., 10th ed., 1:31.
TYPE LOCALITY: Surinam.
DISTRIBUTION: S. Veracruz (Mexico) to Ecuador and Peru, S.W. Brazil, and Guianas; Trinidad; perhaps Jamaica.
ISIS NUMBER: 5301405008014001001.

Family Natalidae
REVIEWED BY: K. F. Koopman (KFK); J. Ramirez-Pulido (JRP) (Mexico).
ISIS NUMBER: 5301405010000000000.

Natalus Gray, 1838. Mag. Zool. Bot., 2:496.
ISIS NUMBER: 5301405010001000000.

Natalus lepidus (Gervais, 1837). L'Inst. Paris, 5(218):253.
TYPE LOCALITY: Cuba.
DISTRIBUTION: Cuba; Bahama Isls.
ISIS NUMBER: 5301405010001001001.

Natalus micropus Dobson, 1880. Proc. Zool. Soc. Lond., 1880:443.
TYPE LOCALITY: Jamaica, Kingston.
DISTRIBUTION: Cuba; Jamaica; Hispaniola; Bahama Isls.; Old Providence Isl. (Colombia).
COMMENT: Includes *brevimanus, macer,* and *tumidifrons;* see Varona, 1974:32. *N. tumidifrons,* however, may be a distinct species (KFK). Reviewed by Kerridge and Baker, 1978, Mamm. Species, 114:1–3.
ISIS NUMBER: 5301405010001003001.

Natalus stramineus Gray, 1838. Mag. Zool. Bot., 2:496.
REVIEWED BY: B. Villa-Ramirez (BVR).
TYPE LOCALITY: Brazil, Minas Gerais, Lagoa Santa.
DISTRIBUTION: S. Baja California, Nuevo Leon, and Sonora (Mexico) to Brazil; Lesser Antilles.
COMMENT: Includes *major* and *primus;* see Varona, 1974:32. Includes *mexicanus* and *dominicensis;* see Goodwin, 1959, Am. Mus. Novit., 1977(2):6.
ISIS NUMBER: 5301405010001004001 as *N. stramineus.*
5301405010001002001 as *N. major.*

Natalus tumidirostris Miller, 1900. Proc. Biol. Soc. Wash., 13:160.
TYPE LOCALITY: Curacao, Hatto (Netherlands).
DISTRIBUTION: Venezuela; Colombia; Trinidad; Curacao; Bonaire.
ISIS NUMBER: 5301405010001005001.

Family Furipteridae
REVIEWED BY: K. F. Koopman (KFK).
ISIS NUMBER: 5301405011000000000.

Amorphochilus Peters, 1877. Monatsb. Preuss. Akad. Wiss. Berlin, p. 185.
ISIS NUMBER: 5301405011001000000.

Amorphochilus schnablii Peters, 1877. Monatsb. Preuss. Akad. Wiss. Berlin, p. 185.
TYPE LOCALITY: Peru, Tumbes, Tumbes.
DISTRIBUTION: W. Peru; W. Ecuador; Puna Isl.; N. Chile.
ISIS NUMBER: 5301405011001001001.

Furipterus Bonaparte, 1837. Iconogr. Fauna Ital., 1, fasc. 21.
ISIS NUMBER: 5301405011002000000.

Furipterus horrens (F. Cuvier, 1828). Mem. Mus. Hist. Nat. Paris, 16:150.
TYPE LOCALITY: French Guiana, Mana River.
DISTRIBUTION: Costa Rica to Peru and E. Brazil; Trinidad.
ISIS NUMBER: 5301405011002001001.

Family Thyropteridae
REVIEWED BY: J. K. Jones, Jr. (JKJ); K. F. Koopman (KFK).
ISIS NUMBER: 5301405012000000000.

Thyroptera Spix, 1823. Sim. Vespert. Brasil., p. 61.
ISIS NUMBER: 5301405012001000000.

Thyroptera discifera (Lichtenstein and Peters, 1855). Monatsb. Preuss. Akad. Wiss. Berlin, p. 335.
TYPE LOCALITY: Venezuela, Carabobo, Puerto Cabello.
DISTRIBUTION: Nicaragua to Guianas, Amazonian Brazil, and Peru.
COMMENT: Reviewed by Wilson, 1978, Mamm. Species, 104:1–3.
ISIS NUMBER: 5301405012001001001.

Thyroptera tricolor Spix, 1823. Sim. Vespert. Brasil., p. 61.
REVIEWED BY: B. Villa-Ramirez (BVR).
TYPE LOCALITY: Brazil, Amazon River.
DISTRIBUTION: Veracruz (Mexico) to Guianas, E. Brazil and Peru; Trinidad.
COMMENT: Reviewed by Wilson and Findley, 1977, Mamm. Species, 71:1–3.
ISIS NUMBER: 5301405012001002001.

Family Myzopodidae
REVIEWED BY: J. K. Jones, Jr. (JKJ); K. F. Koopman (KFK).
ISIS NUMBER: 5301405013000000000.

Myzopoda Milne-Edwards and A. Grandidier, 1878. Bull. Sci. Soc. Philom. Paris, ser. 7, 2:220.
ISIS NUMBER: 5301405013001000000.

Myzopoda aurita Milne-Edwards and A. Grandidier, 1878. Bull. Sci. Soc. Philom. Paris, ser. 7, 2:220.
TYPE LOCALITY: Madagascar.
DISTRIBUTION: Madagascar.
COMMENT: Reviewed by Schliemann and Maas, 1978, Mamm. Species, 116:1–2.
ISIS NUMBER: 5301405013001001001.

Family Vespertilionidae
REVIEWED BY: J. K. Jones, Jr. (JKJ); K. F. Koopman (KFK); N. Sullivan (NS); J. Ramirez-Pulido (JRP)(Mexico); O. L. Rossolimo (OLR)(U.S.S.R.); S. Wang (SW)(China).
ISIS NUMBER: 5301405014000000000.

Antrozous H. Allen, 1862. Proc. Acad. Nat. Sci. Phila., 14:248.
REVIEWED BY: B. Villa-Ramirez (BVR).
COMMENT: Includes *Bauerus*; see Pine *et al.*, 1971, J. Mammal., 52:663–669.
ISIS NUMBER: 5301405014030000000.

Antrozous dubiaquercus Van Gelder, 1959. Am. Mus. Novit., 1973:2.
TYPE LOCALITY: Mexico, Nayarit, Tres Marias Isls., Maria Magdalena Isl.
DISTRIBUTION: Maria Magdalena Isl. (Nayarit); Veracruz (Mexico); Honduras.
COMMENT: Includes *meyeri*; see Pine, 1967, Southwest. Nat., 12:484–485; Jones *et al.*, 1977, Occas. Pap. Mus. Texas Tech Univ., 47:27. Engstrom and Wilson, 1981, Ann. Carnegie Mus., 50:371–383, placed this species in the monotypic genus *Bauerus*.
ISIS NUMBER: 5301405014030001001.

Antrozous koopmani Orr and Silva, 1960. Proc. Biol. Soc. Wash., 73:84.
TYPE LOCALITY: Cuba, Pinar del Rio, San Juan y Martinez, Hoyo Garcia Cave.
DISTRIBUTION: Cuba.
COMMENT: Originally described from a single skull, but since known from several specimens; see Silva, 1979, Los Murcielagos de Cuba.
ISIS NUMBER: 5301405014030002001.

Antrozous pallidus (Le Conte, 1856). Proc. Acad. Nat. Sci. Phila., 7:437.
TYPE LOCALITY: U.S.A., Texas, El Paso Co., El Paso.
DISTRIBUTION: Queretaro (Mexico) to S.C. Kansas, through S. Nevada to S.C. British Columbia (Canada), south through California to Baja California (Mexico).
COMMENT: Includes *bunkeri*; see Morse and Glass, 1960, J. Mammal., 41:10–15.
ISIS NUMBER: 5301405014030003001.

Barbastella Gray, 1821. Lond. Med. Repos., 15:300.
REVIEWED BY: E. Vernier (EV).
ISIS NUMBER: 5301405014002000000.

Barbastella barbastellus (Schreber, 1774). Saugethiere, 1:168.
TYPE LOCALITY: France, Burgundy.
DISTRIBUTION: England and W. Europe to Caucasus; Turkey; Crimea; Morocco; larger Mediterranean islands; perhaps Senegal.
ISIS NUMBER: 5301405014002001001.

Barbastella leucomelas (Cretzschmar, 1826). *In* Ruppell, Atlas Reise Nordl. Afr., Saugeth., p. 73.
TYPE LOCALITY: Egypt, Sinai.
DISTRIBUTION: Caucasus to Pamirs, N. Iran, Afganistan, Himalayas, and W. China; Honshu, Hokkaido (Japan); Sinai; N. Ethiopia; perhaps Indo-China.
ISIS NUMBER: 5301405014002002001.

Chalinolobus Peters, 1866. Monatsb. Preuss. Akad. Wiss. Berlin, p. 679.
  REVIEWED BY: F. R. Allison (FRA)(Australia).
  COMMENT: Includes *Glauconycteris*; see Koopman, 1971, Am. Mus. Novit., 2451:1-4.
    Reviewed by Tate, 1942, Bull. Am. Mus. Nat. Hist., 80:221-227; Ryan, 1966, J.
    Mammal., 47:86-91; and Van Deusen and Koopman, 1971, Am. Mus. Novit.,
    2468:1-30.
  ISIS NUMBER: 5301405014003000000.

Chalinolobus alboguttatus (J. A. Allen, 1917). Bull. Am. Mus. Nat. Hist., 37:449.
  TYPE LOCALITY: Zaire, Oriental, Medje.
  DISTRIBUTION: Zaire; Cameroun.
  COMMENT: Transferred from *Glauconycteris*; see Koopman, 1971, Am. Mus. Novit.,
    2451:1-4.
  ISIS NUMBER: 5301405014003001001.

Chalinolobus argentatus (Dobson, 1875). Proc. Zool. Soc. Lond., 1875:385.
  TYPE LOCALITY: Cameroun, Western Province, Mt. Cameroun.
  DISTRIBUTION: Cameroun to Kenya, south to Angola and Tanzania; perhaps Uganda.
  COMMENT: Transferred from *Glauconycteris*; see Koopman, 1971, Am. Mus. Novit.,
    2451:1-4.
  ISIS NUMBER: 5301405014003002001 as *C. argentata (sic)*.

Chalinolobus beatrix (Thomas, 1901). Ann. Mag. Nat. Hist., ser. 7, 8:256.
  TYPE LOCALITY: Equatorial Guinea, Rio Muni, Benito River, 15 mi. (24 km) from
    mouth.
  DISTRIBUTION: Ivory Coast; Gabon; Rio Muni; Cameroun; Zaire; Uganda; Kenya; Congo
    Republic.
  COMMENT: Transferred from *Glauconycteris*; see Koopman, 1971, Am. Mus. Novit.,
    2451:1-4. Includes *humeralis*; see Koopman, 1971, Am. Mus. Novit., 2451:7, 8.
  ISIS NUMBER: 5301405014003003001.

Chalinolobus dwyeri Ryan, 1966. J. Mammal., 47:89.
  TYPE LOCALITY: Australia, New South Wales, 14 mi. (23 km) S. Inverell, Copeton.
  DISTRIBUTION: New South Wales. Adjacent part of Queensland (NS, FRA).
  ISIS NUMBER: 5301405014003004001.

Chalinolobus egeria (Thomas, 1913). Ann. Mag. Nat. Hist., ser. 8, 11:144.
  TYPE LOCALITY: Cameroun, Western Province, Bibundi.
  DISTRIBUTION: Cameroun; Uganda.
  COMMENT: Transferred from *Glauconycteris*; see Koopman, 1971, Am. Mus. Novit.,
    2451:1-4.
  ISIS NUMBER: 5301405014003005001.

Chalinolobus gleni (Peterson and Smith, 1973). R. Ont. Mus. Life Sci. Occ. Pap., 22:3.
  TYPE LOCALITY: Cameroun, near Lomie.
  DISTRIBUTION: Cameroun; Uganda.
  COMMENT: Transferred from *Glauconycteris*; see Koopman, 1971, Am. Mus. Novit.,
    2451:1-4.

Chalinolobus gouldii Gray, 1841. *In* Grey, J. Two Exped. Aust., 2:401, 405.
  TYPE LOCALITY: Australia, Tasmania, Launceston.
  DISTRIBUTION: Australia but not Cape York Penninsula N. of Cardwell (FRA);
    Tasmania; Norfolk Isl.; New Caledonia (KFK).
  COMMENT: Includes *neocaledonicus*; see Koopman, 1971, Am. Mus. Novit., 2451:5. Also
    includes *venatoris* (FRA).
  ISIS NUMBER: 5301405014003006001.

Chalinolobus morio Gray, 1841. *In* Grey, J. Two Exped. Aust., 2:400, 405.
  TYPE LOCALITY: Australia, Tasmania.
  DISTRIBUTION: S. Australia; Tasmania. Perhaps Western Australia.
  COMMENT: Includes *australis, microdon* and *signifer*; see Tate, 1942, Bull. Am. Mus. Nat.
    Hist., 80:221-297.
  ISIS NUMBER: 5301405014003007001.

Chalinolobus nigrogriseus Gould, 1852. Mamm. Aust., Part 4, Vol. 3, pl. 43.
REVIEWED BY: A. C. Ziegler (ACZ).
TYPE LOCALITY: Australia, Queensland, vic. of Moreton Bay.
DISTRIBUTION: N. and E. Australia; S.E. New Guinea. Adjacent small islands (KFK).
COMMENT: Includes *rogersi*; see Van Deusen and Koopman, 1971, Am. Mus. Novit.,
    2468:10–27; but also see Ride, 1970:176.
ISIS NUMBER: 5301405014003008001.

Chalinolobus picatus Gould, 1852. Mamm. Aust., ser. 4, Vol. 3, pl. 43.
TYPE LOCALITY: Australia, New South Wales, Capt. Sturt's Depot.
DISTRIBUTION: N.W. New South Wales; C. and S. Queensland; South Australia (FRA).
ISIS NUMBER: 5301405014003009001.

Chalinolobus poensis (Gray, 1842). Ann. Mag. Nat. Hist., ser. 1, 10:258.
TYPE LOCALITY: Nigeria, Abo (lower Niger River).
DISTRIBUTION: Senegal to Uganda; Bioko.
COMMENT: Transferred from *Glauconycteris*; see Koopman, 1971, Am. Mus. Novit.,
    2451:1–4.
ISIS NUMBER: 5301405014003010001.

Chalinolobus superbus (Hayman, 1939). Ann. Mag. Nat. Hist., ser. 2, 3:219–223.
TYPE LOCALITY: Zaire, Oriental, Ituri Dist., Pawa.
DISTRIBUTION: Ivory Coast; Ghana; N.E. Zaire.
COMMENT: Transferred from *Glauconycteris*; see Koopman, 1971, Am. Mus. Novit.,
    2451:1–4. Known only from two specimens (NS).
ISIS NUMBER: 5301405014003011001 as *C. superba (sic)*.

Chalinolobus tuberculatus Forster, 1844. Descrip. Animal. Itinere Maris Aust. Terras,
    1772–74:62.
TYPE LOCALITY: New Zealand.
DISTRIBUTION: New Zealand and adjacent small islands.
ISIS NUMBER: 5301405014003012001.

Chalinolobus variegatus (Tomes, 1861). Proc. Zool. Soc. Lond., 1861:36.
TYPE LOCALITY: Namibia, Otjoro.
DISTRIBUTION: Senegal to Somalia, south to South Africa.
COMMENT: Transferred from *Glauconycteris*; see Koopman, 1971, Am. Mus. Novit.,
    2451:1–4. Includes *machadoi*; see Koopman, 1971, Am. Mus. Novit., 2451:6.
ISIS NUMBER: 5301405014003013001.

Eptesicus Rafinesque, 1820. Ann. Nature, p. 2.
REVIEWED BY: F. R. Allison (FRA)(Australia); J. Gaisler (JG) (Palearctic).
COMMENT: Middle and South American species reviewed by Davis, 1965, J. Mammal.,
    46:229–240, and Davis, 1966, Southwest. Nat., 11:245–274. Australian species
    reviewed by McKean, *et al.*, 1978, Aust. J. Zool., 26:529–537, and Carpenter *et al.*,
    1978, Aust. J. Zool., 26:629–638.
ISIS NUMBER: 5301405014004000000.

Eptesicus bobrinskoi Kuzyakin, 1935. Bull. Soc. Nat. Moscow, 44:435.
TYPE LOCALITY: U.S.S.R., Kazakhstan, 65 km E. Aralsk, Tyulek Wells in Aral-Kara-Kum
    desert.
DISTRIBUTION: Kazakhstan, Uzbekistan, Turkmenia (U.S.S.R.); N.W. Iran.
COMMENT: Revised by Hanak and Gaisler, 1971, Vestn. Cs. Spol. Zool., 35(1):11–24.
ISIS NUMBER: 5301405014004002001.

Eptesicus bottae (Peters, 1869). Monatsb. Preuss. Akad. Wiss. Berlin, p. 406.
TYPE LOCALITY: Yemen.
DISTRIBUTION: Turkey, Egypt, and Yemen, east to Kazakstan and Iran.
COMMENT: Includes *innesi, hingstoni, ognevi,* and *anatolicus*; see Corbet, 1978:57; and
    DeBlase, 1971, Fieldiana Zool., 58:9–14. Revised by Hanak and Gaisler, 1971,
    Vestn. Cs. Spol. Zool., 35(1):11–24.
ISIS NUMBER: 5301405014004003001.

Eptesicus brasiliensis (Desmarest, 1819). Nouv. Dict. Hist. Nat. Paris, 2nd ed., 35:478.
 REVIEWED BY: W. B. Davis (WBD); B. Villa-Ramirez (BVR).
 TYPE LOCALITY: Brazil, Goias.
 DISTRIBUTION: San Luis Potosi (Mexico), south to N. Argentina and Uruguay; Trinidad; Tobago.
 COMMENT: Includes *chiriquinus, melanopterus,* and *andinus;* see Koopman, 1978:19.
 ISIS NUMBER: 5301405014004004001 as *E. brasiliensis.*
              5301405014004001001 as *E. andinus.*
              5301405014004007001 as *E. chiriquinus.*
              5301405014004017001 as *E. melanopterus.*

Eptesicus brunneus (Thomas, 1880). Ann. Mag. Nat. Hist., ser. 5, 6:165.
 TYPE LOCALITY: Nigeria, Eastern region, Calabar.
 DISTRIBUTION: Nigeria; Cameroun; Zaire; Ivory Coast.
 COMMENT: Known only from one specimen which is in poor condition (NS).
 ISIS NUMBER: 5301405014004005001.

Eptesicus capensis (A. Smith, 1829). Zool. J., 4:435.
 TYPE LOCALITY: South Africa, Cape Province, Grahamstown.
 DISTRIBUTION: Guinea to Ethiopia, south to South Africa; Madagascar.
 COMMENT: Includes *notius;* see Koopman, 1975, Bull. Am. Mus. Nat. Hist., 154:405.
 ISIS NUMBER: 5301405014004006001 as *E. capensis.*
              5301405014004022001 as *E. notius.*

Eptesicus demissus Thomas, 1916. J. Fed. Malay St. Mus., 7:1.
 TYPE LOCALITY: Thailand, Surat Thani, Khao Nong.
 DISTRIBUTION: Peninsular Thailand.

Eptesicus diminutus Osgood, 1915. Field Mus. Nat. Hist. Publ. Ser. Zool., 10:197.
 REVIEWED BY: W. B. Davis (WBD).
 TYPE LOCALITY: Brazil, Bahia, Rio Preto, Sao Marcello.
 DISTRIBUTION: Venezuela; Brazil; Paraguay; Uruguay; N. Argentina.
 COMMENT: Includes *fidelis;* see Williams, 1978, Ann. Carnegie Mus., 47:380–382, and also for use of *diminutus* instead of *dorianus.*
 ISIS NUMBER: 5301405014004008001 as *E. diminutus.*
              5301405014004009001 as *E. fidelis.*

Eptesicus douglasi Kitchener, 1976. Rec. W. Aust. Mus., 4:295, 296.
 TYPE LOCALITY: Australia, Western Australia, Kimberley, Napier Ranges, Tunnel Creek.
 DISTRIBUTION: Kimberley (N. Western Australia).

Eptesicus flavescens (Seabra, 1900). J. Sci. Math. Phys. Nat. Lisboa, ser. 2, 6:23.
 TYPE LOCALITY: Angola, Galanga.
 DISTRIBUTION: Angola; Burundi.

Eptesicus floweri (De Winton, 1901). Ann. Mag. Nat. Hist., ser. 7, 7:46.
 TYPE LOCALITY: Sudan, Khartoum, Wad Marium.
 DISTRIBUTION: Sudan; Mali.
 COMMENT: Includes *lowei;* see Braestrup, 1935, Vidensk. Medd. dansk naturk. Foren., 99:73–130; but also see Hayman and Hill, 1971, Part 2:45.
 ISIS NUMBER: 5301405014004010001.

Eptesicus furinalis (d'Orbigny, 1847). Voy. Amer. Merid. Atlas Zool., 4:13.
 REVIEWED BY: W. B. Davis (WBD); B. Villa-Ramirez (BVR).
 TYPE LOCALITY: Argentina, Corrientes.
 DISTRIBUTION: N. Argentina, Paraguay, Brazil, and Guianas to Jalisco and San Luis Potosi (Mexico).
 COMMENT: Includes *montosus;* see Koopman, 1978:19.
 ISIS NUMBER: 5301405014004011001 as *E. furinalis.*
              5301405014004019001 as *E. montosus.*

Eptesicus fuscus (Beauvois, 1796). Cat. Raisonne Mus. Peale Phila., p. 18.
REVIEWED BY: W. B. Davis (WBD); F. J. Brenner (FJB); B. Villa-Ramirez (BVR).
TYPE LOCALITY: U.S.A., Pennsylvania, Philadelphia.
DISTRIBUTION: S. Canada to Colombia and Venezuela; Greater Antilles; Bahamas;
    perhaps Alaska and Barbados.
COMMENT: Closely similar to *serotinus* with which it may be conspecific (KFK).
ISIS NUMBER: 5301405014004012001.

Eptesicus guadeloupensis Genoways and Baker, 1975. Occas. Pap. Mus. Texas Tech Univ.,
    34:1.
TYPE LOCALITY: Guadeloupe (Lesser Antilles), Basse Terre, 2 km S. and 2 km E. Baiae-
    Mahault (France).
DISTRIBUTION: Guadeloupe (Lesser Antilles).

Eptesicus guineensis (Bocage, 1889). J. Sci. Math. Phys. Nat. Lisboa, ser. 2, 1:6.
TYPE LOCALITY: Guinea-Bissau, Bissau.
DISTRIBUTION: Senegal and Guinea to Ethiopia and N.E. Zaire; perhaps Tanzania.
COMMENT: This species is called *pusillus* by Hayman and Hill, 1971, Part 2:44; but also
    see Koopman, 1975, Bull. Am. Mus. Nat. Hist., 154:406, 407.
ISIS NUMBER: 5301405014004013001.

Eptesicus hottentotus (A. Smith, 1833). S. Afr. J., 2:59.
TYPE LOCALITY: South Africa, Cape Province, Uitenhage.
DISTRIBUTION: Cape Prov. (South Africa); Namibia; Malawi; Zambia; Zimbabwe;
    Mozambique; Angola.
ISIS NUMBER: 5301405014004014001.

Eptesicus innoxius (Gervais, 1841). Voy. la Bonite Zool., lam. 2.
REVIEWED BY: W. B. Davis (WBD).
TYPE LOCALITY: Peru, Piura, Amotape.
DISTRIBUTION: N.W. Peru; W. Ecuador; Puna Isl. (Ecuador).
ISIS NUMBER: 5301405014004015001.

Eptesicus kobayashii Mori, 1928. Zool. Mag. (Tokyo), 40:292.
TYPE LOCALITY: Korea, Nando, Heian, Heijo.
DISTRIBUTION: Korea.
COMMENT: Status uncertain; see Corbet, 1978:58.

Eptesicus loveni Granvik, 1924. Acta Univ. Lund, ser. 2, 21(3):12.
TYPE LOCALITY: Kenya, Mt. Elgon, E. slopes, 8000 ft. (2438 m).
DISTRIBUTION: W. Kenya.
COMMENT: Known only from the type specimen (NS).
ISIS NUMBER: 5301405014004016001.

Eptesicus lynni Shamel, 1945. Proc. Biol. Soc. Wash., 58:107.
TYPE LOCALITY: Jamaica, 3 mi E. Montego Bay.
DISTRIBUTION: Jamaica.
COMMENT: Probably only a subspecies of *fuscus* (KFK). Arnold *et al.*, 1980, J. Mammal.,
    61:319–322, suggested that it is probably not conspecific with *fuscus*.

Eptesicus melckorum Roberts, 1919. Ann. Transvaal Mus., 6:113.
TYPE LOCALITY: South Africa, Cape Province, Berg River, Kersfontein.
DISTRIBUTION: Cape Prov. (South Africa); Zambia; Mozambique.
COMMENT: This species has not been clearly distinguished from *capensis* (KFK).
ISIS NUMBER: 5301405014004018001.

Eptesicus nasutus (Dobson, 1877). J. Asiat. Soc. Bengal, 46:311.
TYPE LOCALITY: Pakistan, Sind, Shikarpur.
DISTRIBUTION: Arabia; Iraq; Iran; Afghanistan; Pakistan.
COMMENT: Does not include *bobrinskoi*; see Corbet, 1978:57. Includes *walli*; see DeBlase,
    1980, Fieldiana Zool., N.S. 4:182–188. Revised by Gaisler, 1970, Acta Sci. Nat.
    Brno, 4(6):1–56.
ISIS NUMBER: 5301405014004020001 as *E. nasutus.*
    5301405014004031001 as *E. walli.*

Eptesicus nilssoni (Keyserling and Blasius, 1839). Arch. Naturgesch., 5(1):315.
REVIEWED BY: W. B. Davis (WBD); E. Vernier (EV).
TYPE LOCALITY: Sweden.
DISTRIBUTION: W. and E. Europe to E. Siberia, and N.W. China (SW); north beyond
Arctic Circle in Scandinavia, south to Iraq, the Elburz Mts., Pamirs and W. China
(not Tibet (SW)); Nepal; Honshu, Hokkaido (Japan); Sakhalin.
COMMENT: Includes *propinquus*; see Davis, 1965, J. Mammal., 46:230. Includes
japonensis; see Corbet, 1978:57. Revised by Wallin, 1969, Zool. Bidr. Upps.,
37:223–440.
ISIS NUMBER: 5301405014004021001.

Eptesicus pachyotis (Dobson, 1871). Proc. Asiat. Soc. Bengal, p. 211.
TYPE LOCALITY: India, Assam, Khasi Hills.
DISTRIBUTION: Assam (India); N. Burma; N. Thailand.
ISIS NUMBER: 5301405014004023001.

Eptesicus platyops (Thomas, 1901). Ann. Mag. Nat. Hist., ser. 7, 8:31.
TYPE LOCALITY: Nigeria, Western Region, Lagos.
DISTRIBUTION: Nigeria; Senegal.
COMMENT: Known from only from two specimens (NS).
ISIS NUMBER: 5301405014004024001.

Eptesicus pumilus Gray, 1841. *In* Grey, J. Two Exped. Austr., 2:406.
TYPE LOCALITY: Australia, New South Wales, Yarrundi.
DISTRIBUTION: N. Western Australia; Northern Territory; Queensland; New South
Wales; South Australia. Victoria (KFK).
COMMENT: Includes *darlingtoni* and *caurinus*; see McKean *et al.*, 1978, Aust. J. Zool.,
26:533.
ISIS NUMBER: 5301405014004025001.

Eptesicus regulus (Thomas, 1906). Proc. Zool. Soc. Lond., 1906:470, 471.
TYPE LOCALITY: Australia, Western Australia, King Georges Sound, King River (near
Albany).
DISTRIBUTION: S.W. and S.E. Australia.
COMMENT: See McKean *et al.*, 1978, Aust. J. Zool., 26:532, 534. Because of the
confusion about this species, *Pipistrellus* Kaup, 1829, *sensu* Thomas, 1906, and
Registrellus Troughton, 1943 should be included in the synonymy of *Eptesicus*; see
Hill, 1966, Mammalia, 30:302–307.

Eptesicus rendalli (Thomas, 1889). Ann. Mag. Nat. Hist., ser. 6, 3:362.
TYPE LOCALITY: Gambia, Bathurst.
DISTRIBUTION: Gambia to Somalia, south to Botswana, Malawi, and Mozambique.
ISIS NUMBER: 5301405014004026001.

Eptesicus sagittula McKean *et al.*, 1978. Aust. J. Zool., 26:535.
TYPE LOCALITY: Australia, New South Wales, 13 km N.W. Braidwood.
DISTRIBUTION: S.E. Australia; Lord Howe Isl. A single record from Tasmania (FRA).

Eptesicus serotinus Schreber, 1774. Saugethiere, 1:167.
REVIEWED BY: E. Vernier (EV).
TYPE LOCALITY: France.
DISTRIBUTION: W. Europe through S. Asiatic Russia to Himalayas, Thailand and China,
north to Korea; Taiwan; S. England; N. Africa; most islands in Mediterranean.
COMMENT: Includes *sodalis*; see Corbet, 1978:57. Includes *horikawai*; see Jones, 1975,
Quart. J. Taiwan Mus., 28:189. Revised by Gaisler, 1970, Acta Sci. Nat. Brno,
4(6):1–56. See comment under *fuscus*.
ISIS NUMBER: 5301405014004027001 as *E. serotinus*.
5301405014004028001 as *E. sodalis*.

Eptesicus somalicus (Thomas, 1901). Ann. Mag. Nat. Hist., ser. 7, 8:32.
TYPE LOCALITY: Somalia, Northwest Province, Hargeisa.
DISTRIBUTION: Guinea-Bissau to Somalia, south to Namibia and South Africa.
COMMENT: Includes *zuluensis*; see Koopman, 1975, Bull. Am. Mus. Nat. Hist., 154:404,
405.
ISIS NUMBER: 5301405014004029001 as *E. somalicus*.
5301405014004032001 as *E. zuluensis*.

Eptesicus tatei Ellerman and Morrison-Scott, 1951, p. 158.
TYPE LOCALITY: India, Darjeeling.
DISTRIBUTION: N.E. India.

Eptesicus tenuipinnis (Peters, 1872). Monatsb. Preuss. Akad. Wiss. Berlin, p. 263.
TYPE LOCALITY: "Guinea".
DISTRIBUTION: Senegal to Kenya, south to Angola and Zaire.
COMMENT: *E. bicolor* is tentatively included here by Hayman and Hill, 1971, Part 2:43, but may be an older name for *Pipistrellus anchietai;* see Koopman, 1975, Bull. Am. Mus. Nat. Hist., 154:404.
ISIS NUMBER: 5301405014004030001.

Eptesicus vulturnus Thomas, 1914. Ann. Mag. Nat. Hist., ser. 8, 13:440.
TYPE LOCALITY: Australia, Tasmania.
DISTRIBUTION: C. Queensland; S.E. Australia; Tasmania.
COMMENT: Includes *pygmaeus;* see McKean *et al.,* 1978, Aust. J. Zool., 26:532.

Euderma H. Allen, 1892. Proc. Acad. Nat. Sci. Phila., 43:467.
REVIEWED BY: B. Villa-Ramirez (BVR).
ISIS NUMBER: 5301405014005000000.

Euderma maculatum (J. A. Allen, 1891). Bull. Am. Mus. Nat. Hist., 3:195.
TYPE LOCALITY: U.S.A., California, Los Angeles Co., Santa Clara Valley, Castac Creek mouth.
DISTRIBUTION: S. W. Canada and W. United States to Queretaro (Mexico).
COMMENT: Reviewed by Watkins, 1977, Mamm. Species, 77:1–4.
ISIS NUMBER: 5301405014005001001.

Eudiscopus Conisbee, 1953. Genera Subgenera Rec. Mamm., p. 30.
COMMENT: Replacement name for *Discopus* Osgood, 1932, preoccupied by *Discopus* Thompson, 1864 (a beetle).
ISIS NUMBER: 5301405014006000000.

Eudiscopus denticulus (Osgood, 1932). Field Mus. Nat. Hist. Publ. Ser. Zool., 18:236.
TYPE LOCALITY: Laos, Phong Saly, 4000 ft. (1219 m).
DISTRIBUTION: Laos; C. Burma.
COMMENT: Reviewed by Koopman, 1972, Mamm. Species, 19:1–2.
ISIS NUMBER: 5301405014006001001.

Glischropus Dobson, 1875. Proc. Zool. Soc. Lond., 1875:472.
ISIS NUMBER: 5301405014007000000.

Glischropus javanus Chasen, 1939. Treubia, 17:189.
TYPE LOCALITY: Indonesia, Java, West Java, Mt. Pangrango.
DISTRIBUTION: Java.

Glischropus tylopus (Dobson, 1875). Proc. Zool. Soc. Lond., 1875:473.
TYPE LOCALITY: Malaysia, Sabah.
DISTRIBUTION: Burma; Thailand; Malaya; Sumatra; Borneo; S.W. Philippines; N. Molucca Isls.
ISIS NUMBER: 5301405014007001001.

Harpiocephalus Gray, 1842. Ann. Mag. Nat. Hist., ser. 1, 10:259.
ISIS NUMBER: 5301405014027000000.

Harpiocephalus harpia (Temminck, 1840). Monogr. Mamm., 2:219.
TYPE LOCALITY: Indonesia, Java, Mt. Gede.
DISTRIBUTION: India to Taiwan and Vietnam, south to Molucca Isls. and Java.
ISIS NUMBER: 5301405014027001001.

Hesperoptenus Peters, 1869. Monatsb. Preuss. Akad. Wiss. Berlin for 1868, p. 626.
ISIS NUMBER: 5301405014008000000.

Hesperoptenus blanfordi (Dobson, 1877). J. Asiat. Soc. Bengal, 46:312.
TYPE LOCALITY: Burma, Tenasserim.
DISTRIBUTION: Thailand; Malay Peninsula; Burma.
ISIS NUMBER: 5301405014008001001.

Hesperoptenus doriae (Peters, 1869). Monatsb. Preuss. Akad. Wiss. Berlin for 1868, p. 626.
TYPE LOCALITY: Malaysia, Sarawak.
DISTRIBUTION: N. Borneo; Malay Peninsula.
ISIS NUMBER: 5301405014008002001.

Hesperoptenus tickelli (Blyth, 1851). J. Asiat. Soc. Bengal, 20:157.
TYPE LOCALITY: India, Bihar, Chaibasa.
DISTRIBUTION: India; Sri Lanka; Burma; Thailand; Andaman Isls.; perhaps S.W. China.
ISIS NUMBER: 5301405014008003001.

Hesperoptenus tomesi Thomas, 1905. Ann. Mag. Nat. Hist., ser. 7, 16:575.
TYPE LOCALITY: Malaysia, Malacca.
DISTRIBUTION: Borneo; Malay Peninsula.
ISIS NUMBER: 5301405014008004001.

Histiotus Gervais, 1856. Exped. Castelnau Zool., p. 77.
ISIS NUMBER: 5301405014009000000.

Histiotus alienus Thomas, 1916. Ann. Mag. Nat. Hist., ser. 8, 17:276.
TYPE LOCALITY: Brazil, Santa Catarina, Joinville.
DISTRIBUTION: S.E. Brazil; Uruguay.
ISIS NUMBER: 5301405014009001001.

Histiotus macrotus (Poeppig, 1835). Reise Chile Peru Amaz., 1:451.
TYPE LOCALITY: Chile, Bio-Bio, Antuco.
DISTRIBUTION: Chile; S. Bolivia; N.W. Argentina; S. Peru.
COMMENT: Includes *laephotis*; see Cabrera, 1958:108.
ISIS NUMBER: 5301405014009002001.

Histiotus montanus (Philippi and Landbeck, 1861). Arch. Naturgesch., p. 289.
TYPE LOCALITY: Chile, Santiago Cordillera.
DISTRIBUTION: Chile; Argentina; Uruguay; S. Peru; Colombia; Ecuador; perhaps N.
    Peru, W. Bolivia, and S. Brazil; Venezuela.
ISIS NUMBER: 5301405014009003001.

Histiotus velatus (I. Geoffroy, 1824). Ann. Sci. Nat. Zool., 3:446.
TYPE LOCALITY: Brazil, Parana, Curitiba.
DISTRIBUTION: S. Brazil; Paraguay.
ISIS NUMBER: 5301405014009004001.

Ia Thomas, 1902. Ann. Mag. Nat. Hist., ser. 7, 10:163.
COMMENT: Considered a subgenus of *Pipistrellus* by Ellerman and Morrison-Scott,
    1951:162; but see Topal, 1970, Opusc. Zool. Budapest, 10:344, 345.

Ia io Thomas, 1902. Ann. Mag. Nat. Hist., ser. 7, 10:164.
TYPE LOCALITY: China, Hupeh, Chungyang.
DISTRIBUTION: S. China; Laos; Vietnam; Thailand; N.E. India.
COMMENT: Includes *beaulieui* and *longimana*; see Topal, 1970, Opusc. Zool. Budapest,
    10:342, 343.
ISIS NUMBER: 5301405014019021001 as *Pipistrellus io*.

Idionycteris Anthony, 1923. Am. Mus. Novit., 54:1.
REVIEWED BY: B. Villa-Ramirez (BVR).
COMMENT: KFK finds the case for considering *Idionycteris* a separate genus (Williams
    et al., 1970, J. Mammal., 51:602–606) unconvincing and is inclined to retain it in
    Plecotus; also see Handley, 1959, Proc. U.S. Nat. Mus., 112:459–479.

Idionycteris phyllotis (G. M. Allen, 1916). Bull. Mus. Comp. Zool., 60:352.
  TYPE LOCALITY: Mexico, San Luis Potosi.
  DISTRIBUTION: Distrito Federal (Mexico) to S. Utah and S. Nevada (U.S.A.).
  COMMENT: Formerly included in *Plecotus*; see Williams *et al.*, 1970, J. Mammal.,
    51:602–606.
  ISIS NUMBER: 5301405014020003001 as *Plecotus phyllotis*.

Kerivoula Gray, 1842. Ann. Mag. Nat. Hist., ser. 1, 10:258.
  COMMENT: Does not include *Phoniscus*; see Hill, 1965, Mammalia, 29:524–528; and
    Corbet and Hill, 1980:77; but also see Ryan, 1965, J. Mammal., 46:518.
  ISIS NUMBER: 5301405014029000000.

Kerivoula africana Dobson, 1878. Cat. Chiroptera Br. Mus., p. 335.
  TYPE LOCALITY: Tanzania, coast opposite Zanzibar Isl.
  DISTRIBUTION: Tanzania.
  COMMENT: Known only from the type specimen (NS).
  ISIS NUMBER: 5301405014029001001.

Kerivoula agnella Thomas, 1908. Ann. Mag. Nat. Hist., ser. 8, 2:372.
  REVIEWED BY: A. C. Ziegler (ACZ).
  TYPE LOCALITY: Papua New Guinea, Louisiade Archipelago, Misima Isl.
  DISTRIBUTION: Louisiade Archipelago; D'Entrecasteaux Isls.
  ISIS NUMBER: 5301405014029002001.

Kerivoula argentata Tomes, 1861. Proc. Zool. Soc. Lond., 1861:32.
  TYPE LOCALITY: Namibia, Otjoro.
  DISTRIBUTION: S. Kenya to Angola and Namibia; Natal (South Africa); Mozambique;
    Malawi; Zambia.
  ISIS NUMBER: 5301405014029003001.

Kerivoula cuprosa Thomas, 1912. Ann. Mag. Nat. Hist., ser. 8, 10:41.
  TYPE LOCALITY: Cameroun, Ja River, Bitye.
  DISTRIBUTION: S. Cameroun; Kenya; Zaire; Ghana.
  ISIS NUMBER: 5301405014029005001.

Kerivoula eriophora (Heuglin, 1877). Reise Nordost-Afrika, 2:34.
  TYPE LOCALITY: Ethiopia, Belegaz Valley, between Semien and Wogara.
  DISTRIBUTION: Ethiopia.
  COMMENT: Very poorly known; may be conspecific with *africana* which it antedates;
    see Hayman and Hill, 1971, Part 2:53.

Kerivoula hardwickei (Horsfield, 1824). Zool. Res. Java, Part 8:28.
  TYPE LOCALITY: Indonesia, Java.
  DISTRIBUTION: India; Sri Lanka; Burma; Indochina; Szechwan, Kwangsi, Fukien
    (China); Malaya; Borneo; Java; Sumatra; Engano; Mentawi Isls.; Sulawesi; Bali;
    Flores; Sumba; Sumbawa; Kangean Isl.; Peleng Isl.; Philippines.
  COMMENT: Includes *flora*; see Hill, 1965, Mammalia, 29:543.
  ISIS NUMBER: 5301405014029006001.

Kerivoula lanosa (A. Smith, 1847). Illustr. Zool. S. Afr. Mamm., pl. 50.
  TYPE LOCALITY: South Africa, Cape Province, 200 mi. (322 km) E. Capetown.
  DISTRIBUTION: Liberia to Ethiopia, south to South Africa.
  COMMENT: Includes *harrisoni* and *muscilla*; see Hill, 1977, Rev. Zool. Afr., 91:623–633.
  ISIS NUMBER: 5301405014029009001 as *K. lanosa*.
              5301405014029007001 as *K. harrisoni*.
              5301405014029011001 as *K. muscilla*.

Kerivoula minuta Miller, 1898. Proc. Acad. Nat. Sci. Phila., p. 321.
  TYPE LOCALITY: Thailand, Trang Province.
  DISTRIBUTION: Malaya; S. Thailand.
  ISIS NUMBER: 5301405014029010001.

Kerivoula muscina Tate, 1941. Bull. Am. Mus. Nat. Hist., 78:586.
REVIEWED BY: A. C. Ziegler (ACZ).
TYPE LOCALITY: Papua New Guinea, Western Province, Lake Daviumbu, *ca.* 20 m.
DISTRIBUTION: C. New Guinea.
ISIS NUMBER: 5301405014029012001.

Kerivoula myrella Thomas, 1914. Ann. Mag. Nat. Hist., ser. 8, 13:438.
REVIEWED BY: A. C. Ziegler (ACZ).
TYPE LOCALITY: Papua New Guinea, Bismarck Archipelago, Admiralty Isls., Manus Isl.
DISTRIBUTION: Bismarck Arch.
ISIS NUMBER: 5301405014029013001.

Kerivoula papillosa (Temminck, 1840). Monogr. Mamm., 2:220.
TYPE LOCALITY: Indonesia, Java, Bantam.
DISTRIBUTION: N.E. India; Indochina; Malaya; Java; Sumatra; Borneo.
ISIS NUMBER: 5301405014029014001.

Kerivoula pellucida (Waterhouse, 1845). Proc. Zool. Soc. Lond., 1845:6.
TYPE LOCALITY: Philippines.
DISTRIBUTION: Borneo; Philippines; Java; Sumatra; Malaya.
COMMENT: Includes *bombifrons;* see Hill, 1965, Mammalia, 29:539.
ISIS NUMBER: 5301405014029016001 as *K. pellucida.*
5301405014029004001 as *K. bombifrons.*

Kerivoula phalaena Thomas, 1912. Ann. Mag. Nat. Hist., ser. 8, 10:281.
TYPE LOCALITY: Ghana, Bibianaha.
DISTRIBUTION: Ghana; Cameroun; Liberia; Congo Republic; Zaire.
ISIS NUMBER: 5301405014029017001.

Kerivoula picta (Pallas, 1767). Spicil. Zool., 3:7.
TYPE LOCALITY: Indonesia, Molucca Isls., Ternate Isl.
DISTRIBUTION: Sri Lanka; India to Vietnam, Malaya, and Kwangtung, Hainan (China); Borneo; Sumatra; Bali; Java; Molucca Isls.; Lombok.
ISIS NUMBER: 5301405014029018001.

Kerivoula smithi Thomas, 1880. Ann. Mag. Nat. Hist., ser. 5, 6:166.
TYPE LOCALITY: Nigeria, Calabar.
DISTRIBUTION: Nigeria; Cameroun; N. and E. Zaire; Kenya.
ISIS NUMBER: 5301405014029021001.

Kerivoula whiteheadi Thomas, 1894. Ann. Mag. Nat. Hist., ser. 6, 14:460.
TYPE LOCALITY: Philippines, Luzon, Isabella.
DISTRIBUTION: Philippines; Borneo; S. Thailand; Malaya.
COMMENT: Includes *pusilla* and *bicolor;* see Hill, 1965, Mammalia, 29:532-533.
ISIS NUMBER: 5301405014029022001 as *K. whiteheadi.*
5301405014029019001 as *K. pusilla.*

Laephotis Thomas, 1901. Ann. Mag. Nat. Hist., ser. 7, 7:460.
COMMENT: Considered monotypic by Hayman and Hill, 1971, Part 2:49; but see Hill, 1974, Bull. Br. Mus. (Nat. Hist.) Zool., 27:73-82.
ISIS NUMBER: 5301405014010000000.

Laephotis angolensis Monard, 1935. Archos. Mus. Bocage, 6:45.
TYPE LOCALITY: Angola, Tyihumbwe, 15 km W. Dala.
DISTRIBUTION: Angola; Zaire.

Laephotis botswanae Setzer, 1971. Proc. Biol. Soc. Wash., 84:260, 263.
TYPE LOCALITY: Botswana, 50 mi. (80 km) W. and 12 mi. (19 km) S. Shakawe.
DISTRIBUTION: Zaire; Zambia; Botswana.

Laephotis namibensis Setzer, 1971. Proc. Biol. Soc. Wash., 84:259.
TYPE LOCALITY: Namibia, Gobabeb, Kuiseb River.
DISTRIBUTION: Namibia. Probably S.W. Cape Province (South Africa); reported as *L. wintoni* by Rautenbach and Nel, 1980, Ann. Transvaal Mus., 32:111.

Laephotis wintoni Thomas, 1901. Ann. Mag. Nat. Hist., ser. 7, 7:460.
TYPE LOCALITY: Kenya, Kitui, 1150 m.
DISTRIBUTION: Ethiopia; Kenya.
ISIS NUMBER: 5301405014010001001.

Lasionycteris Peters, 1866. Monatsb. Preuss. Akad. Wiss. Berlin, p. 648.
REVIEWED BY: B. Villa-Ramirez (BVR).
ISIS NUMBER: 5301405014011000000.

Lasionycteris noctivagans (Le Conte, 1831). *In* McMurtie, Animal Kingdom Cuvier,
1(App.):431.
TYPE LOCALITY: "Eastern United States".
DISTRIBUTION: S.E. Alaska; S. Canada; United States (except extreme southern parts);
N.E. Mexico; Bermuda.
ISIS NUMBER: 5301405014011001001.

Lasiurus Gray, 1831. Zool. Misc., 1:38.
REVIEWED BY: B. Villa-Ramirez (BVR).
COMMENT: Treated under the name *Nycteris* by Hall, 1981:219. Includes *Dasypterus*; see
Hall and Jones, 1961, Univ. Kans. Publ. Mus. Nat. Hist., 14:73–98.
ISIS NUMBER: 5301405014012000000.

Lasiurus borealis (Muller, 1776). Ritter Carl Linne Natursystem, Suppl., p. 20.
TYPE LOCALITY: U.S.A., New York.
DISTRIBUTION: Chile, Argentina, Uruguay and Brazil to Zacatecas (Mexico), thence to
S. Utah, S. Nevada and S.W. British Columbia and through E. United States to S.
Canada; Jamaica; Cuba; Hispaniola; Puerto Rico; Bermuda; Bahamas; Trinidad;
Galapagos.
COMMENT: Includes *degelidus, minor,* and *pfeifferi;* see Varona, 1974:36. Includes
brachyotis; see Niethammer, 1964, Mammalia, 28:595.
ISIS NUMBER: 5301405014012001001 as *L. borealis.*
5301405014012002001 as *L. brachyotis.*
5301405014012005001 as *L. degelidus.*

Lasiurus castaneus Handley, 1960. Proc. U.S. Nat. Mus., 112:468.
TYPE LOCALITY: Panama, Darien, Rio Pucro, Tacarcuna Village, 3200 ft. (975 m).
DISTRIBUTION: E. Panama.
ISIS NUMBER: 5301405014012003001.

Lasiurus cinereus (Beauvois, 1796). Cat. Raisonne Mus. Peale Phila., p. 18.
TYPE LOCALITY: U.S.A., Pennsylvania, Philadelphia.
DISTRIBUTION: Colombia and Venezuela to C. Chile, Uruguay, and C. Argentina;
Hawaii; N. and C. Mexico throughout the United States to S. British Columbia,
S.E. Mackensie, Hudson Bay and S. Quebec (Canada); Galapagos; Bermuda;
accidental on Cuba, Hispaniola, Iceland, and the Orkney Isls., Guatemala and S.
Mexico (BVR, KFK).
COMMENT: Includes *semotus;* see Sanborn and Crespo, 1957, Bol. Mus. Argentino
Cienc. Nat. Bernardino Rivadavia, 4:12.
PROTECTED STATUS: U.S. ESA - Endangered as *L. c. semotus* subspecies only.
ISIS NUMBER: 5301405014012004001.

Lasiurus ega (Gervais, 1856). Exped. Castelnau Zool., p. 73.
TYPE LOCALITY: Brazil, Amazonas, Ega.
DISTRIBUTION: S. California, Arizona and S. Texas (U.S.A.) to Argentina, Uruguay, and
Brazil; Trinidad.
COMMENT: Transferred from *Dasypterus;* see Hall and Jones, 1961, Univ. Kans. Publ.
Mus. Nat. Hist., 14:73–98.
ISIS NUMBER: 5301405014012006001.

Lasiurus egregius (Peters, 1871). Monatsb. Preuss. Akad. Wiss. Berlin for 1870, p. 275.
TYPE LOCALITY: Brazil, Santa Catarina.
DISTRIBUTION: S. Brazil; E. Panama.
COMMENT: Transferred from *Dasypterus;* see Hall and Jones, 1961, Univ. Kans. Publ.
Mus. Nat. Hist., 14:73–98.

Lasiurus intermedius H. Allen, 1862. Proc. Acad. Nat. Sci. Phila., 14:246.
   TYPE LOCALITY: Mexico, Tamaulipas, Matamoros.
   DISTRIBUTION: Honduras to S. Sinaloa and through Texas to New Jersey (U.S.A.);
      Cuba.
   COMMENT: Includes *floridanus* and *insularis*; see Hall and Jones, 1961, Univ. Kans. Publ.
      Mus. Nat. Hist., 14:84–87; but see also Silva, 1976, Poeyana, 153:10–15.
      Transferred from *Dasypterus*; see Hall and Jones, 1961, Univ. Kans. Publ. Mus.
      Nat. Hist., 14:73–98. Reviewed by Webster *et al.*, 1980, Mamm. Species, 132:1–3.
   ISIS NUMBER: 5301405014012007001.

Lasiurus seminolus (Rhoads, 1895). Proc. Acad. Nat. Sci. Phila., 47:32.
   TYPE LOCALITY: U.S.A., Florida, Pinellas Co., Tarpon Springs.
   DISTRIBUTION: C. Florida and E. Texas to S.E. Oklahoma and N. Virginia; S. Texas; S.E.
      Pennsylvania and S. New York (U.S.A.); Bermuda. N. Veracruz (Mexico) record
      unverified.
   COMMENT: May be only a localized color phase of *borealis*; see Koopman *et al.*, 1958, J.
      Mammal., 39:168; but also see Hall, 1981:224.
   ISIS NUMBER: 5301405014012008001.

Mimetillus Thomas, 1904. Abstr. Proc. Zool. Soc. Lond., 10:12, 22.
   ISIS NUMBER: 5301405014013000000.

Mimetillus moloneyi (Thomas, 1891). Ann. Mag. Nat. Hist., ser. 6, 7:528.
   TYPE LOCALITY: Nigeria, Western Region, Lagos.
   DISTRIBUTION: Sierra Leone to Gabon, Zaire, Ethiopia, and Kenya, south to Tanzania,
      N. Zambia, and Angola; Bioko.
   ISIS NUMBER: 5301405014013001001.

Miniopterus Bonaparte, 1837. Fauna Ital., 1, fasc. 20.
   REVIEWED BY: A. C. Ziegler (ACZ)(New Guinea).
   COMMENT: Reviewed by Peterson, 1981, Can. J. Zool., 59:828–843. See Goodwin, 1979,
      Bull. Am. Mus. Nat. Hist., 163:119.
   ISIS NUMBER: 5301405014026000000.

Miniopterus australis Tomes, 1858. Proc. Zool. Soc. Lond., 1858:125.
   REVIEWED BY: F. R. Allison (FRA).
   TYPE LOCALITY: New Caledonia, Loyalty Isls., Lifu (21° S. and 167.3° E.) (France).
   DISTRIBUTION: India to Hainan (China), southeast to Loyalty Isls. and N.E. Australia.
   COMMENT: Includes *pusillus, witkampi, tibialis,* and *paululus* (KFK from a variety of
      sources, some probably erroneous). Goodwin, 1979, Bull. Am. Mus. Nat. Hist.,
      163:118–120, recognized *pusillus* as a distinct species.
   ISIS NUMBER: 5301405014026001001.

Miniopterus fraterculus Thomas and Schwann, 1906. Proc. Zool. Soc. Lond., 1906:162.
   TYPE LOCALITY: South Africa, Cape Province, Knysna.
   DISTRIBUTION: Cape Province, Natal, E. Transvaal (South Africa); S. Malawi; Zambia;
      Angola; Mozambique.
   ISIS NUMBER: 5301405014026002001.

Miniopterus inflatus Thomas, 1903. Ann. Mag. Nat. Hist., ser. 7, 12:634.
   TYPE LOCALITY: Cameroun, Efulen.
   DISTRIBUTION: Ethiopia; Somalia; Kenya; Uganda; E. and S. Zaire; Cameroun; Gabon;
      Mozambique; Burundi.
   ISIS NUMBER: 5301405014026003001.

Miniopterus medius Thomas and Wroughton, 1909. Proc. Zool. Soc. Lond., 1909:382.
   TYPE LOCALITY: Indonesia, Java, Tji-Tandoei River, Kaliputjang.
   DISTRIBUTION: Thailand and Philippines to New Guinea, Bismarck Arch., Solomon and
      Loyalty Isls.
   COMMENT: Includes *macrocneme*; see Hill, 1971, J. Nat. Hist., 5:577, 578.
   ISIS NUMBER: 5301405014026004001.

Miniopterus minor Peters, 1867. Monatsb. Preuss. Akad. Wiss. Berlin for 1866, p. 885.
TYPE LOCALITY: Tanzania, coast opposite Zanzibar Isl.
DISTRIBUTION: Kenya; Tanzania; Zaire; Congo Republic; Madagascar; Sao Tome Isl;
Comoro Isls.
ISIS NUMBER: 5301405014026005001.

Miniopterus robustior Revilliod, 1914. *In* Sarasin and Roux, Nova Caledonia, A. Zool.,
1:359.
TYPE LOCALITY: New Caledonia, Loyalty Isls., Lifu Isl., Quepenee (France).
DISTRIBUTION: Loyalty Isls. (E. of New Caledonia).
COMMENT: See Hill, 1971, J. Nat. Hist., 5:579, for the status of this species.

Miniopterus schreibersi (Kuhl, 1819). Ann. Wetterau Ges. Naturk., 4(2):185.
REVIEWED BY: F. R. Allison (FRA); E. Vernier (EV).
TYPE LOCALITY: Rumania, Banat, near Coronini, Kolumbacs Cave.
DISTRIBUTION: S. Europe and Morocco through the Caucasus and Iran to most of China
and Japan; most of Oriental region; New Guinea; Solomon Isls.; Australia;
subsaharan Africa; Madagascar. Bismarck Arch. and Bougainville Isl. (ACZ).
COMMENT: Includes *fuliginosus, ravus, magnater* (KFK from a variety of sources, some
probably erroneous). Reviewed by Crucitti, 1976, Ann. Mus. Civ. Stor. Nat.
Genova, 81:131–138.
ISIS NUMBER: 5301405014026006001.

Miniopterus tristis (Waterhouse, 1845). Proc. Zool. Soc. Lond., 1845:3.
TYPE LOCALITY: Philippine Isls.
DISTRIBUTION: Philippines; Sulawesi; New Guinea; possibly Bismarck Arch.; Solomon
Isls.; New Hebrides.
ISIS NUMBER: 5301405014026007001.

Murina Gray, 1842. Ann. Mag. Nat. Hist., ser. 1, 10:258.
COMMENT: Includes *Harpiola*; see Corbet and Hill, 1980:75.
ISIS NUMBER: 5301405014028000000.

Murina aenea Hill, 1964. Fedn. Mus. J., Kuala Lumpur, N.S., 8:57.
TYPE LOCALITY: Malaysia, Pahang, Bentong Dist., near Janda Baik, Ulu Chemperoh.
DISTRIBUTION: Malaya.
ISIS NUMBER: 5301405014028001001.

Murina aurata Milne-Edwards, 1872. Rech. Hist. Nat. Mamm., p. 250.
TYPE LOCALITY: China, Szechwan, Moupin.
DISTRIBUTION: Nepal to S.W. China; Manchuria; S.E. Siberia; Sakhalin; Burma; Korea;
Japan.
COMMENT: Formerly included *ussuriensis;* see Maeda, 1980, Mammalia, 44:547.
ISIS NUMBER: 5301405014028002001.

Murina balstoni Thomas, 1908. Ann. Mag. Nat. Hist., ser. 8, 2:370.
TYPE LOCALITY: Indonesia, Java, Preanger, Tasimalaja.
DISTRIBUTION: Java.
ISIS NUMBER: 5301405014028003001.

Murina canescens Thomas, 1923. Ann. Mag. Nat. Hist., ser. 9, 11:254.
TYPE LOCALITY: Indonesia, Sumatra, Nias Isl.
DISTRIBUTION: Nias Isl., (off west coast of Sumatra).

Murina cyclotis Dobson, 1872. Proc. Asiat. Soc. Bengal, p. 210.
TYPE LOCALITY: India, Darjeeling.
DISTRIBUTION: Sri Lanka and India to Kwangtung (SW) and Hainan (China); Vietnam,
south to Malaya; perhaps Philippines.
ISIS NUMBER: 5301405014028004001.

Murina florium Thomas, 1908. Ann. Mag. Nat. Hist., ser. 8, 2:371.
REVIEWED BY: A. C. Ziegler (ACZ).
TYPE LOCALITY: Indonesia, Lesser Sunda Isls., Flores.
DISTRIBUTION: Flores, Sumbawa, Seram, Goram Isl., and Buru Isls.; Sulawesi; New
Guinea.
ISIS NUMBER: 5301405014028005001.

Murina fusca Sowerby, 1922. J. Mammal., 3:46.
TYPE LOCALITY: China, Manchuria, Kirin, Imienpo area.
DISTRIBUTION: Manchuria.
COMMENT: Listed as a subspecies of *leucogaster* by Ellerman and Morrison-Scott,
1951:185; and Corbet, 1978:62; but see Wallin, 1969, Zool. Bidr. Upps., 37:355.

Murina grisea Peters, 1872. Monatsb. Preuss. Akad. Wiss. Berlin, p. 258.
TYPE LOCALITY: India, Uttar Pradesh, Dehra Dun, Mussooree, Jeripanee, 5500 ft. (1676 m).
DISTRIBUTION: N.W. Himalayas.
COMMENT: Transferred from *Harpiola*; see Corbet and Hill, 1980:75.
ISIS NUMBER: 5301405014028006001.

Murina huttoni (Peters, 1872). Monatsb. Preuss. Akad. Wiss. Berlin, p. 257.
TYPE LOCALITY: India, Uttar Pradesh, Dehra Dun.
DISTRIBUTION: Tibet; N.W. India to Indochina; Fukien (China); Thailand; Malaya. SW
doubts Tibetan record.
ISIS NUMBER: 5301405014028007001.

Murina leucogaster Milne-Edwards, 1872. Rech. Hist. Nat. Mamm., p. 252.
TYPE LOCALITY: China, Szechwan, Moupin.
DISTRIBUTION: E. Himalayas; Shensi; Szechwan, Fukien (China); Upper Yenesei; Altai;
Inner Mongolia; Korea; Ussuri region; Sakhalin; Honshu, Kyushu (Japan).
ISIS NUMBER: 5301405014028008001.

Murina puta Kishida, 1924. Zool. Mag. (Tokyo), 36:130–133.
TYPE LOCALITY: Taiwan, Chang Hua, Erh-Shui.
DISTRIBUTION: Taiwan.
COMMENT: Status uncertain, but the only *Murina* known from Taiwan.

Murina suilla (Temminck, 1840). Monogr. Mamm., 2:224.
TYPE LOCALITY: Indonesia, Java.
DISTRIBUTION: Java; Borneo; Sumatra; Malaya.
ISIS NUMBER: 5301405014028009001.

Murina tenebrosa Yoshiyuki, 1970. Bull. Nat. Sci. Mus. Tokyo, 13:195.
TYPE LOCALITY: Japan, Tsushima Isls., Kamishima Isl., Sago.
DISTRIBUTION: Tsushima Isls. (between Japan and Korea).
COMMENT: Known only from the holotype (NS).
ISIS NUMBER: 5301405014028010001.

Murina tubinaris (Scully, 1881). Proc. Zool. Soc. Lond., 1881:200.
TYPE LOCALITY: Pakistan, Gilgit.
DISTRIBUTION: N. Pakistan; N.E. India; Burma; Laos; Vietnam.
COMMENT: Listed as a subspecies of *huttoni* by Ellerman and Morrison-Scott, 1951:186;
but also see Hill, 1964, Fedn. Mus. J., Kuala Lumpur, N.S., 8:49,50.
ISIS NUMBER: 5301405014028011001.

Murina ussuriensis Ognev, 1913. Ann. Mus. Zool. Acad. Sci. St. Petersb., 18:402.
TYPE LOCALITY: U.S.S.R., Primorsky Krai, Kreis Imansky.
DISTRIBUTION: Ussuri region, Kurile Isls., and Sakhalin (U.S.S.R.); Korea; Hokkaido,
Honshu, Tsushima and Yaku Isls. (Japan).
COMMENT: Formerly included in *aurata*; see Maeda, 1980, Mammalia, 44:547.

Myotis Kaup, 1829. Skizz. Europ. Thierwelt, 1:106.
REVIEWED BY: A. Parkinson (AP)(Western Hemisphere); E. Vernier (EV)(Europe); B.
Villa-Ramirez (BVR)(Mexico).
COMMENT: Includes *Pizonyx, Cistugo*, and *Anamygdon*; see Findley, 1972, Syst. Zool.,
21:43, Hayman and Hill, 1971, Part 2:33, and Phillips and Birney, 1968, Proc. Biol.
Soc. Wash., 81:495. Neotropical species revised by LaVal, 1973, Sci. Bull. Nat.
Hist. Mus. Los Ang. Cty., 15:1–54.
ISIS NUMBER: 5301405014014000000.

Myotis abei Yoshikura, 1944. Zool. Mag. (Tokyo), 56:6.
TYPE LOCALITY: U.S.S.R., Sakhalin, Shirutoru.
DISTRIBUTION: Known only from the type locality.

Myotis adversus (Horsfield, 1824). Zool. Res. Java, Part 8.
REVIEWED BY: F. R. Allison (FRA); A. C. Ziegler (ACZ).
TYPE LOCALITY: Indonesia, Java.
DISTRIBUTION: Taiwan and Malaya, south and east to New Guinea, Bismarck Arch.,
Solomon Isls. to New Hebrides and coastal Australia (except Western Australia);
perhaps Tibet (China).
COMMENT: Includes *taiwanensis*; see Ellerman and Morrison-Scott, 1951:149; but see
also Findley, 1972, Syst. Zool., 21:43. Includes *carimatae* and *abbotti*; see Tate, 1941,
Bull. Am. Mus. Nat. Hist., 78:551. Includes *Anamygdon solomonis*; see Phillips and
Birney, 1968, Proc. Biol. Soc. Wash., 81:495. Medway, 1969:35, listed this species
from Thailand, but Lekagul and McNeely, 1977:206, did not (FRA).
ISIS NUMBER: 5301405014014001001.

Myotis aelleni Baud, 1979. Rev. Suisse Zool., 86:268.
TYPE LOCALITY: Argentina, Chubut, El Hoyo de Epuyen.
DISTRIBUTION: S.W. Argentina.

Myotis albescens (E. Geoffroy, 1806). Ann. Mus. Hist. Nat. Paris, 8:204.
TYPE LOCALITY: Paraguay, Paraguari, Yaguaron (of neotype).
DISTRIBUTION: S. Veracruz (Mexico) to Uruguay, N. Argentina, Peru, and Surinam.
COMMENT: Includes *argentatus*; see LaVal, 1973, Bull. Los Ang. Cty. Mus. Nat. Hist. Sci.
Soc., 15:25.
ISIS NUMBER: 5301405014014002001 as *M. albescens.*
5301405014014005001 as *M. argentatus.*

Myotis altarium Thomas, 1911. Abstr. Proc. Zool. Soc. Lond., 90:3.
TYPE LOCALITY: China, Szechwan, Omi San.
DISTRIBUTION: Szechwan, Kweichow (China).
ISIS NUMBER: 5301405014014003001.

Myotis annectans (Dobson, 1871). Proc. Asiat. Soc. Bengal, p. 213.
TYPE LOCALITY: India, Assam, Naga Hills.
DISTRIBUTION: Assam (India) to Thailand.
COMMENT: Includes *primula*; see Topal, 1970, Ann. Hist.-Nat. Mus. Nat. Hungarici
Zool., 62:373–375.
ISIS NUMBER: 5301405014014004001.

Myotis atacamensis (Lataste, 1892). Actes Soc. Sci. Chile, 1:80.
TYPE LOCALITY: Chile, Antofogasta, San Pedro de Atacama.
DISTRIBUTION: S. Peru; N. Chile. Coastal desert (AP).
COMMENT: Listed as a subspecies of *chiloensis* by Cabrera, 1958:99. Includes *nicholsoni*;
see LaVal, 1973, Bull. Los Ang. Cty. Mus. Nat. Hist. Sci. Soc., 15:18–20.

Myotis auriculus Baker and Stains, 1955. Univ. Kans. Publ. Mus. Nat. Hist., 9:83.
TYPE LOCALITY: Mexico, Tamaulipas, Sierra de Tamaulipas, 10 mi. (16 km) W., 2 mi. (3
km) S. Piedra, 1200 ft. (366 m).
DISTRIBUTION: S.W. Arizona and S.E. and C. New Mexico (U.S.A.) to S. Jalisco; S.C.
Nuevo Leon to C. Veracruz (Mexico).
COMMENT: Listed as a subspecies of *evotis* by Hall and Kelson, 1959:169; but also see
Genoways and Jones, 1969, Southwest. Nat., 14:1–12; Hall, 1981:205.
ISIS NUMBER: 5301405014014006001.

Myotis australis (Dobson, 1878). Cat. Chiroptera Br. Mus., p. 317.
TYPE LOCALITY: Australia, New South Wales.
DISTRIBUTION: New South Wales.
COMMENT: Poorly known, the type and only specimen possibly being incorrectly
labelled (KFK), or a vagrant individual of *muricola* (FRA). See also Aust. Bat Res.
News, 9.
ISIS NUMBER: 5301405014014007001.

Myotis austroriparius (Rhoads, 1897). Proc. Acad. Nat. Sci. Phila., 49:227.
TYPE LOCALITY: U.S.A., Florida, Pinellas Co., Tarpon Springs.
DISTRIBUTION: S.E. United States, north to Indiana.
COMMENT: Reviewed by LaVal, 1970, J. Mammal., 51:542–552.
ISIS NUMBER: 5301405014014008001.

Myotis bechsteini (Kuhl, 1818). Ann. Wetterau Ges. Naturk., 4(1):30.
TYPE LOCALITY: Germany, Hessen, Hanau.
DISTRIBUTION: Europe to Caucasus and N.W. Transcaucasia; England; S. Sweden.
ISIS NUMBER: 5301405014014010001.

Myotis blythii (Tomes, 1857). Proc. Zool. Soc. Lond., 1857:53.
TYPE LOCALITY: India, Rajasthan, Nasirabad.
DISTRIBUTION: Mediterranean zone of Europe and N.W. Africa; Crimea, Caucasus, Asia Minor, Israel to Kirgizia, Afghanistan, and Himalayas; N.W. Altai; Inner Mongolia; Shensi (China).
COMMENT: Includes *omari, oxygnathus, ancilla,* and *risorius;* see Corbet, 1978:50, Strelkov, 1972, Acta Theriol., 17:355–380, Felten, 1977, Senckenberg. Biol., 58:1–44, and Bogan, 1978, Proc. 4th Int. Bat Res. Conf., p. 217–230.
ISIS NUMBER: 5301405014014011001.

Myotis bocagei (Peters, 1870). J. Sci. Math. Phys. Nat. Lisboa, ser. 1, 3:125.
TYPE LOCALITY: Angola, Duque de Braganca.
DISTRIBUTION: Liberia to S. Yemen, south to Angola, Zambia, and Transvaal (South Africa).
COMMENT: Includes *dogalensis;* see Corbet, 1978:49.
ISIS NUMBER: 5301405014014012001.

Myotis brandti (Eversmann, 1845). Bull. Soc. Nat. Moscow, 18:505.
TYPE LOCALITY: U.S.S.R., Orenburgsk. Obl., S. Ural, Sakmara River (OLR). "Foothills of the Ural Mountains."
DISTRIBUTION: Britain to Kazakhstan, S. Siberia (U.S.S.R.), Mongolia, south to Spain and Greece.
COMMENT: Listed as a subspecies of *mystacinus* by Ellerman and Morrison-Scott, 1951:139; but also see Corbet, 1978:48.

Myotis browni Taylor, 1934. Monogr. Bur. Sci. Manila, p. 288.
TYPE LOCALITY: Philippines, Mindanao, Cotabato, Saub.
DISTRIBUTION: Mindanao (Philippines).
COMMENT: Probably a subspecies of *muricola;* see Findley, 1972, Syst. Zool., 21:42.
ISIS NUMBER: 5301405014014013001.

Myotis californicus (Audubon and Bachman, 1842). J. Acad. Nat. Sci. Phila., ser. 1, 8:285.
TYPE LOCALITY: U.S.A., California, Monterey.
DISTRIBUTION: S. Alaska Panhandle (U.S.A.) to Baja California and Chiapas (Mexico).
COMMENT: See Miller and Allen, 1928, Bull. U.S. Nat. Mus., 144:153, for discussion of type.
ISIS NUMBER: 5301405014014014001.

Myotis capaccinii (Bonaparte, 1837). Fauna Ital., 1, fasc. 20.
TYPE LOCALITY: Italy, Sicily.
DISTRIBUTION: Mediterranean zone and islands of Europe and N.W. Africa; S. Asia Minor; Israel; S. Iraq; S. Iran; Uzbekistan (U.S.S.R.).
COMMENT: See comment under *macrodactylus.*
ISIS NUMBER: 5301405014014015001.

Myotis chiloensis (Waterhouse, 1840). Zool. Voy. H.M.S. "Beagle," Mammalia, p. 5.
TYPE LOCALITY: Chile, Chiloe Isl., Islets on eastern side.
DISTRIBUTION: C. and S. Chile.
COMMENT: See LaVal, 1973, Bull. Los Ang. Cty. Mus. Nat. Hist. Sci. Soc., 15:43, 44, for restriction of the scope of this species.
ISIS NUMBER: 5301405014014016001.

Myotis chinensis (Tomes, 1857). Proc. Zool. Soc. Lond., 1857:52.
TYPE LOCALITY: "Southern China".
DISTRIBUTION: Szechwan and Yunnan to Kiangsu (China); Hong Kong; N. Thailand.
COMMENT: Included in *myotis* by Ellerman and Morrison-Scott, 1951:144; but also see
    Lekagul and McNeely, 1977:206.

Myotis cobanensis Goodwin, 1955. Am. Mus. Novit., 1744:2.
TYPE LOCALITY: Guatemala, Alta Verapaz, Coban, 1305 m.
DISTRIBUTION: C. Guatemala.
COMMENT: Listed as a subspecies of *velifer* by Goodwin, 1955, Am. Mus. Novit., 1744:2;
    but see de la Torre, 1958, Proc. Biol. Soc. Wash., 71:167–170; Hall, 1981:197.
    Known only from the holotype.

Myotis dasycneme (Boie, 1825). Isis Jena, p. 1200.
TYPE LOCALITY: Denmark, Jutland, Dagbieg (near Wiborg).
DISTRIBUTION: N.E. Europe and W. Siberia, west to E. France and Switzerland, east to
    Yenesei River, south to Austria and N.W. Kazakhstan (U.S.S.R.); single record
    from Manchuria.
COMMENT: Probably includes *surinamensis*; see Carter and Dolan, 1978, Spec. Publ.
    Mus. Texas Tech Univ., 15:73.
ISIS NUMBER: 5301405014014017001.

Myotis daubentoni (Kuhl, 1819). Ann. Wetterau Ges. Naturk., 4(2):195.
TYPE LOCALITY: Germany, Hessen, Hanau.
DISTRIBUTION: Europe and S. Siberia, east to Kamtschatka, Vladivostok, Sakhalin
    (U.S.S.R.); Korea; Manchuria, E., S. China (SW); Britain and Ireland; S.
    Scandinavia; Assam (India).
COMMENT: Includes *laniger*; see Ellerman and Morrison-Scott, 1951:147.
ISIS NUMBER: 5301405014014018001 as *M. daubentoni*.
             5301405014014035001 as *M. laniger*.

Myotis dominicensis Miller, 1902. Proc. Biol. Soc. Wash., 15:243.
TYPE LOCALITY: Dominica (Lesser Antilles).
DISTRIBUTION: N. Lesser Antilles.
COMMENT: Listed as a subspecies of *nigricans* by Hall and Kelson, 1959:176, but see
    LaVal, 1973, Bull. Los Ang. Cty. Mus. Nat. Hist. Sci. Soc., 15:16, 17; Hall,
    1981:200.

Myotis dryas K. Andersen, 1907. Ann. Mus. Civ. Stor. Nat. Genova, 3:33.
TYPE LOCALITY: India, Andaman Isls., South Andaman Isl., Port Blair.
DISTRIBUTION: Andaman Isls.
COMMENT: Listed as a subspecies of *adversus* by Ellerman and Morrison-Scott,
    1951:149; but also see Hill, 1976, J. Bombay Nat. Hist. Soc., 73:436.

Myotis elegans Hall, 1962. Univ. Kans. Publ. Mus. Nat. Hist., 14:163–164.
TYPE LOCALITY: Mexico, Veracruz, 12.5 mi. (20 km) N. Tihuatlan.
DISTRIBUTION: San Luis Potosi (Mexico) to Costa Rica.
ISIS NUMBER: 5301405014014020001.

Myotis emarginatus (E. Geoffroy, 1806). Ann. Mus. Hist. Nat. Paris, 8:198.
TYPE LOCALITY: France, Ardennes, Givet, Charlemont.
DISTRIBUTION: S. Europe, north to Netherlands and S. Poland, Crimea, Caucasus,
    Kopet Dag, east to Uzbekistan (U.S.S.R.) and E. Iran; Israel; Morocco; Algeria;
    Tunisia; Lebanon; Afghanistan.
ISIS NUMBER: 5301405014014021001.

Myotis evotis (H. Allen, 1864). Smithson. Misc. Coll., 7:48.
TYPE LOCALITY: U.S.A., California, Monterey.
DISTRIBUTION: S. British Columbia, S. Alberta, S. Saskatchewan (Canada) to W. New
    Mexico and S. California (U.S.A.); Baja California (Mexico).
COMMENT: See Genoways and Jones, 1969, Southwest. Nat., 14:1. This species is in
    need of revision (BVR).
ISIS NUMBER: 5301405014014022001.

Myotis findleyi Bogan, 1978. J. Mammal., 59:524.
TYPE LOCALITY: Mexico, Nayarit, Tres Marias Isls., Maria Magdalena Isl.
DISTRIBUTION: Tres Marias Isls.
COMMENT: Very close to and possibly conspecific with *nigricans* (KFK).

Myotis formosus (Hodgson, 1835). J. Asiat. Soc. Bengal, 4:700.
TYPE LOCALITY: Nepal.
DISTRIBUTION: Afghanistan to Kweichow, Kwangsi, Kiangsu and Fukien (China)(SW);
Taiwan; Korea; Tsushima Isl. (Japan); Philippines; Sumatra; Java; Sulawesi; Bali.
COMMENT: Includes *bartelsi, flavus, hermani, rufopictus,* and *weberi;* see Findley, 1972,
Syst. Zool., 21:42.
ISIS NUMBER: 5301405014014023001 as *M. formosus.*
5301405014014009001 as *M. bartelsi.*
5301405014014029001 as *M. hermani.*
5301405014014054001 as *M. rufopictus.*
5301405014014066001 as *M. weberi.*

Myotis fortidens Miller and Allen, 1928. Bull. U.S. Nat. Mus., 144:54.
TYPE LOCALITY: Mexico, Tabasco, Teapa.
DISTRIBUTION: Sonora (Mexico) to Guatemala. W. Texas (U.S.A.) (NS).
ISIS NUMBER: 5301405014014024001.

Myotis frater G. M. Allen, 1923. Am. Mus. Novit., 85:6.
TYPE LOCALITY: China, Fukien, Yenping.
DISTRIBUTION: Afghanistan, Uzbekistan, S. Siberia (U.S.S.R.) to Korea, Heilungkiang
(China)(SW), S.E. China, and S.E. U.S.S.R.; Honshu (Japan).
COMMENT: Includes *longicaudatus;* see Corbet, 1978:49; but also see Wang, 1959, Acta
Zool. Sin., 11:344–352.
ISIS NUMBER: 5301405014014025001.

Myotis goudoti (A. Smith, 1834). S. Afr. Quart. J., 2:244.
TYPE LOCALITY: Madagascar.
DISTRIBUTION: Madagascar; Anjouan Isl. (Comoro Isls.).
ISIS NUMBER: 5301405014014026001.

Myotis grisescens A. H. Howell, 1909. Proc. Biol. Soc. Wash., 22:46.
TYPE LOCALITY: U.S.A., Tennessee, Marion Co., Nickajack Cave, near Shellmound.
DISTRIBUTION: Florida Panhandle to Kentucky to E. Kansas and N.E. Oklahoma.
PROTECTED STATUS: U.S. ESA - Endangered.
ISIS NUMBER: 5301405014014027001.

Myotis hasseltii (Temminck, 1840). Monogr. Mamm., 2:225.
TYPE LOCALITY: Indonesia, Java.
DISTRIBUTION: Thailand; Kampuchea; Vietnam; Malaya; Sumatra; Rhio Arch.; Borneo;
Java; Sri Lanka.
COMMENT: Includes *continentis;* see Hill and Thonglongya, 1972, Bull. Br. Mus. (Nat.
Hist.) Zool., 22:188.
ISIS NUMBER: 5301405014014028001.

Myotis herrei Taylor, 1934. Monogr. Bur. Sci. Manila, p. 290.
TYPE LOCALITY: Philippines, Luzon, Rizal Province, Montalban.
DISTRIBUTION: N. Philippines.
ISIS NUMBER: 5301405014014030001.

Myotis horsfieldii (Temminck, 1840). Monogr. Mamm., 2:226.
TYPE LOCALITY: Indonesia, Java.
DISTRIBUTION: S.E. China; Thailand; Malaya; Java; Bali; Borneo; Sulawesi.
COMMENT: Includes *lepidus* and *deignani;* see Hill, 1972, Bull. Br. Mus. (Nat. Hist.)
Zool., 23:31, Hill, 1974, Bull. Br. Mus. (Nat. Hist.) Zool., 27:132 respectively.
ISIS NUMBER: 5301405014014031001.

Myotis hosonoi Imaizumi, 1954. Bull. Nat. Sci. Mus. Tokyo, 1:44.
TYPE LOCALITY: Japan, Honshu, Nagano Pref., Tokiwa-Mura, Koumito.
DISTRIBUTION: Honshu (Japan).

Myotis ikonnikovi Ognev, 1912. Ann. Mus. Zool. Acad. Sci. St. Petersb., 16:477.
TYPE LOCALITY: U.S.S.R., Primorsk. Krai (= Ussuri Region), Iman Dist.
DISTRIBUTION: Ussuri region and N. Korea to Lake Baikal (U.S.S.R.), the Altai, and
Mongolia, N.E. China; Sakhalin and Hokkaido.
COMMENT: Probably a subspecies of *muricola*; see Corbet, 1978:48.

Myotis insularum (Dobson, 1878). Cat. Chiroptera Br. Mus., p. 313.
TYPE LOCALITY: Samoa.
DISTRIBUTION: Samoa.
COMMENT: Poorly known, the type and only specimen possibly being incorrectly
labelled (KFK).

Myotis jeannei Taylor, 1934. Monogr. Bur. Sci. Manila, p. 284.
TYPE LOCALITY: Philippines, Mindanao, Zamboanga, Caldera.
DISTRIBUTION: Mindanao (Philippines).
COMMENT: Probably a subspecies of *adversus*; see Tate, 1941, Bull. Am. Mus. Nat. Hist.,
78:551.
ISIS NUMBER: 5301405014014032001.

Myotis keaysi J. A. Allen, 1914. Bull. Am. Mus. Nat. Hist., 33:383.
TYPE LOCALITY: Peru, Puno, Inca Mines.
DISTRIBUTION: S. Tamaulipas (Mexico) to S.E. Peru and Venezuela; Trinidad.
COMMENT: Revised by LaVal, 1973, Bull. Los Ang. Cty. Mus. Nat. Hist. Sci. Soc., 15.
ISIS NUMBER: 5301405014014033001.

Myotis keenii (Merriam, 1895). Am. Nat., 29:860.
REVIEWED BY: F. J. Brenner (FJB).
TYPE LOCALITY: Canada, British Columbia, Queen Charlotte Isls., Graham Isl., Massett.
DISTRIBUTION: Alaska Panhandle to W. Washington (U.S.A.); Saskatchewan to Prince
Edward Isl. (Canada), south to C. Arkansas and Florida Panhandle (U.S.A.).
COMMENT: Includes *septentrionalis*, possibly a separate species; see Van Zyll de Jong,
1979, Can. J. Zool., 57. Reviewed by Fitch and Shump, 1979, Mamm. Species,
121:1-3.
ISIS NUMBER: 5301405014014034001.

Myotis leibii (Audubon and Bachman, 1842). J. Acad. Nat. Sci. Phila., ser. 1, 8:284.
TYPE LOCALITY: U.S.A., Pennsylvania, Erie Co.
DISTRIBUTION: N. Baja California and Zacatecas (Mexico) to S.C. British Columbia
(Canada), east through Iowa and Missouri to Maine (U.S.A.) and S. Quebec
(Canada).
COMMENT: Called *subulatus* in Hall and Kelson, 1959:175, and Hall, 1981:187; but see
Glass and Baker, 1968, Proc. Biol. Soc. Wash., 81:259.
ISIS NUMBER: 5301405014014036001.

Myotis lesueuri Roberts, 1919. Ann. Transvaal Mus., 6:112.
TYPE LOCALITY: South Africa, Cape Province, Paarl Dist., Lormarins.
DISTRIBUTION: S.W. Cape Province (South Africa).
COMMENT: Transferred from *Cistugo*; see Hayman and Hill, 1971, Part 2:33.
ISIS NUMBER: 5301405014014037001.

Myotis levis (I. Geoffroy, 1824). Ann. Sci. Nat. Zool., ser. 1, 3:444-445.
TYPE LOCALITY: "Southern Brazil."
DISTRIBUTION: N. Argentina; S.E. Brazil; Uruguay.
COMMENT: Included in *ruber* by Cabrera, 1958:102, but also see LaVal, 1973, Bull. Los
Ang. Cty. Mus. Nat. Hist. Sci. Soc., 15:36-40.

Myotis longipes (Dobson, 1873). Proc. Asiat. Soc. Bengal, p. 110.
TYPE LOCALITY: India, Kashmir, Bhima Devi Caves, 6000 ft. (1829 m).
DISTRIBUTION: Afghanistan; Kashmir.
COMMENT: Included in *capaccinii* by Ellerman and Morrison-Scott, 1951:148; but
considered a distinct species by Hanak and Gaisler, 1969, Zool. Listy, 18:195-206.
ISIS NUMBER: 5301405014014038001.

Myotis lucifugus (Le Conte, 1831). *In* McMurtie, Animal Kingdom Cuvier, 1(App.):431.
REVIEWED BY: F. J. Brenner (FJB).
TYPE LOCALITY: U.S.A., Georgia, Liberty Co., near Riceboro.
DISTRIBUTION: Alaska to Labrador, south to Distrito Federal (Mexico).
COMMENT: Includes *occultus*; see Findley and Jones, 1967, J. Mammal., 48:429-443.
Hybridizes with *yumanensis* in some areas; see Parkinson, 1979, J. Mamm.,
60:489-504. Reviewed by Fenton and Barclay, 1980, Mamm. Species, 142:1-8.
ISIS NUMBER: 5301405014014039001.

Myotis macrodactylus (Temminck, 1840). Monogr. Mamm., 2:231.
TYPE LOCALITY: Japan.
DISTRIBUTION: Japan; Kurile Isls.; S. China; S.E. U.S.S.R.
COMMENT: Includes *fimbriatus*, probably conspecific with *capaccinii*; see Wallin, 1969,
Zool. Bidr. Upps., 37:294-296; but see also Corbet, 1978:51; Findley, 1972, Syst.
Zool., 21:43.
ISIS NUMBER: 5301405014014040001.

Myotis macrotarsus (Waterhouse, 1845). Proc. Zool. Soc. Lond., 1845:5.
TYPE LOCALITY: Philippines.
DISTRIBUTION: Philippines; N. Borneo.
ISIS NUMBER: 5301405014014041001.

Myotis martiniquensis LaVal, 1973. Bull. Los Ang. Cty. Mus. Nat. Hist. Sci. Soc., 15:35.
TYPE LOCALITY: Martinique (Lesser Antilles), Tartane, 6 km E. La Trinite (France).
DISTRIBUTION: Martinique, Barbados (Lesser Antilles).
COMMENT: May occur on other islands in the Lesser Antilles (AP).

Myotis milleri Elliot, 1903. Field Columb. Mus. Publ., Zool. Ser., 3:172.
TYPE LOCALITY: Mexico, Baja California, Sierra San Pedro Martir, La Grulla.
DISTRIBUTION: N. Baja California (Mexico).
COMMENT: Revised by Miller and Allen, 1928, Bull. U.S. Nat. Mus., 144:118.

Myotis montivagus (Dobson, 1874). J. Asiat. Soc. Bengal, 43:237.
TYPE LOCALITY: China, Yunnan, Hotha.
DISTRIBUTION: Yunnan to Fukien and Chihli (China); India; Burma; Malaya.
COMMENT: Includes *peytoni*; see Hill, 1962, Proc. Zool. Soc. Lond., 139:126-130.
ISIS NUMBER: 5301405014014042001 as *M. montivagus*.
5301405014014051001 as *M. peytoni*.

Myotis morrisi Hill, 1971. Bull. Br. Mus. (Nat. Hist.) Zool., 21:43.
TYPE LOCALITY: Ethiopia, Walaga, Didessa River mouth.
DISTRIBUTION: Ethiopia.
ISIS NUMBER: 5301405014014043001.

Myotis muricola (Gray, 1846). Cat. Hodgson Coll. Br. Mus., p. 4.
TYPE LOCALITY: Nepal.
DISTRIBUTION: Afghanistan to Taiwan and New Guinea.
COMMENT: Includes *caliginosus, moupinensis, latirostris*, and *ater*; see Findley, 1972, Syst.
Zool., 21:42, 43.

Myotis myotis (Borkhausen, 1797). Deutsche Fauna, 1:80.
TYPE LOCALITY: Germany, Thuringia.
DISTRIBUTION: C. and S. Europe, east to Ukraine; S. England; most Mediterranean
islands; Asia Minor; Lebanon; Israel.
COMMENT: See Corbet, 1978:50, for content of this species.
ISIS NUMBER: 5301405014014044001.

Myotis mystacinus (Kuhl, 1819). Ann. Wetterau Ges. Naturk., 4(2):202.
TYPE LOCALITY: Germany.
DISTRIBUTION: Ireland and Scandinavia to Ussuri region (U.S.S.R.), Sakhalin, and
Japan, south to Morocco, N. Iran, and C., E. and S. China; Kurile Isls.
COMMENT: Includes *davidi*; see Corbet, 1978:47, 48, for content of this species.
ISIS NUMBER: 5301405014014045001 as *M. mystacinus*.
5301405014014019001 as *M. davidi*.

Myotis nathalinae Tupinier, 1977. Mammalia, 41:327.
TYPE LOCALITY: Spain, Ciudad Real, Cabezarrubias.
DISTRIBUTION: Spain; France; Switzerland; Poland.
COMMENT: Reviewed by Ruprecht, 1981, Acta Theriol., 26:349–357.

Myotis nattereri (Kuhl, 1818). Ann. Wetterau Ges. Naturk., 4(1):33.
TYPE LOCALITY: Germany, Hessen, Hanau.
DISTRIBUTION: Europe (except S.E. and Scandinavia); Morocco; Israel; Crimea and
    Caucasus to S.E. Siberia and Korea; Kirin (China)(SW); Honshu, Kyushu (Japan).
ISIS NUMBER: 5301405014014046001.

Myotis nesopolus Miller, 1900. Proc. Biol. Soc. Wash., 13:123.
TYPE LOCALITY: Curacao, Willemstad (Netherlands).
DISTRIBUTION: N.E. Venezuela, Curacao, Bonaire.
COMMENT: Includes *larensis*; see Genoways and Williams, 1979, Ann. Carnegie Mus.,
    48:316–320.

Myotis nigricans (Schinz, 1821). Thierreich, 1:179.
TYPE LOCALITY: Brazil, Espirito Santo, between Itapemirin and Iconha Rivers.
DISTRIBUTION: S.W. Nayarit and Tamaulipas (Mexico) to Peru, N. Argentina and S.
    Brazil; Guianas; Trinidad; Tobago; Grenada (Lesser Antilles).
COMMENT: Includes *carteri*; see Corbet and Hill, 1980:65, but see Bogan, 1978, J.
    Mammal., 59:520–524. Neotype designated by LaVal, 1973, Bull. Los Ang. Cty.
    Mus. Nat. Hist. Sci. Soc., 15. Reviewed by Wilson and LaVal, 1974, Mamm.
    Species, 39:1–3.
ISIS NUMBER: 5301405014014047001.

Myotis oreias (Temminck, 1840). Monogr. Mamm., 2:270.
TYPE LOCALITY: Singapore.
DISTRIBUTION: Malaya.
COMMENT: Known only from the holotype. May be a synonym of *mystacinus* (NS).
ISIS NUMBER: 5301405014014048001.

Myotis oxyotus (Peters, 1867). Monatsb. Preuss. Akad. Wiss. Berlin, p. 19.
TYPE LOCALITY: Ecuador, Mount Chimborazo, between 2743 and 3048 m.
DISTRIBUTION: Venezuela to Bolivia; Panama; Costa Rica.
COMMENT: Revised by LaVal, 1973, Bull. Los Ang. Cty. Mus. Nat. Hist. Sci. Soc., 15.

Myotis ozensis Imaizumi, 1954. Bull. Nat. Sci. Mus. Tokyo, 1:49.
TYPE LOCALITY: Japan, Honshu, Gunma Pref., Ozegahara, 1400 m.
DISTRIBUTION: Honshu (Japan).

Myotis patriciae Taylor, 1934. Monogr. Bur. Sci. Manila, p. 286.
TYPE LOCALITY: Philippines, Mindanao, Agusan Province.
DISTRIBUTION: Mindanao (Philippines).
COMMENT: Almost certainly a synonym of *browni* (KFK).
ISIS NUMBER: 5301405014014049001.

Myotis peninsularis Miller, 1898. Ann. Mag. Nat. Hist., ser. 7, 2:124.
TYPE LOCALITY: Mexico, Baja California, San Jose del Cabo.
DISTRIBUTION: S. Baja California (Mexico).
COMMENT: Listed as a subspecies of *velifer* by Hall and Kelson, 1959:166, but see
    Hayward, 1970, WRISCI (W. New Mexico Univ.), 1(4):5; Hall, 1981:196.

Myotis pequinius Thomas, 1908. Proc. Zool. Soc. Lond., 1908:637.
TYPE LOCALITY: China, Hopeh, 30 mi. (48 km) W. Peking.
DISTRIBUTION: Hopeh, Shantung, Honan and Kiangsu (China).
ISIS NUMBER: 5301405014014050001.

Myotis peshwa (Thomas, 1915). J. Bombay Nat. Hist. Soc., 23:611.
TYPE LOCALITY: India, Maharashtra, Poona.
DISTRIBUTION: India; perhaps Sri Lanka.
COMMENT: Listed as a subspecies of *adversus* by Ellerman and Morrison-Scott,
    1951:149; but see Hill, 1976, J. Bombay Nat. Hist. Soc., 73:433–436.

Myotis planiceps Baker, 1955. Proc. Biol. Soc. Wash., 68:165.
TYPE LOCALITY: Mexico, Coahuila, 7 mi. (11 km) S. and 4 mi. (6 km) E. Bella Union, 7200 ft. (2195 m).
DISTRIBUTION: S.E. Coahuila; W.C. Nuevo Leon; N.E. Zacatecas (Mexico).
COMMENT: Reviewed by Matson, 1975, Mamm. Species, 60:1–2.

Myotis pruinosus Yoshiyuki, 1971. Bull. Nat. Sci. Mus. Tokyo, 14:305.
TYPE LOCALITY: Japan, Honshu, Iwate Pref., Waga-Gun, Waga-Machi.
DISTRIBUTION: Honshu (Japan).

Myotis ricketti (Thomas, 1894). Ann. Mag. Nat. Hist., ser. 6, 14:300.
TYPE LOCALITY: China, Fukien, Foochow.
DISTRIBUTION: Fukien, Anhwei, Kiangsu, Shantung, Yunnan (China) and Hong Kong.
COMMENT: *M. pilosus* may be the oldest name for this species; see Ellerman and Morrison-Scott, 1951:150.
ISIS NUMBER: 5301405014014052001.

Myotis ridleyi Thomas, 1898. Ann. Mag. Nat. Hist., ser. 7, 1:361.
TYPE LOCALITY: Malaysia, Selangor.
DISTRIBUTION: Malaya; perhaps Sumatra.
COMMENT: Transferred from *Pipistrellus*; see Medway, 1978:35.
ISIS NUMBER: 5301405014019041001 as *Pipistrellus ridleyi*.

Myotis riparius Handley, 1960. Proc. U.S. Nat. Mus., 112:466–468.
TYPE LOCALITY: Panama, Darien, Rio Puero, Tacarcuna Village.
DISTRIBUTION: Honduras to Uruguay and E. Brazil; Trinidad.
COMMENT: Originally described as a subspecies of *simus*; *M. guaycuru* may be the oldest name for this species; see LaVal, 1973, Bull. Los Ang. Cty. Mus. Nat. Hist. Sci. Soc., 15:32–35.

Myotis rosseti (Oey, 1951). Beaufortia, 1(8):4.
TYPE LOCALITY: Kampuchea.
DISTRIBUTION: Kampuchea; Thailand.
COMMENT: Originally described as a species of *Glischropus*; see Hill and Topal, 1973, Bull. Br. Mus. (Nat. Hist.) Zool., 24:447–453.
ISIS NUMBER: 5301405014019042001 as *Pipistrellus rosseti*.

Myotis ruber (E. Geoffroy, 1806). Ann. Mus. Hist. Nat. Paris, 8:204.
TYPE LOCALITY: Paraguay, Neembucu, Sapucay (neotype locality).
DISTRIBUTION: S.E. Brazil; Paraguay; N.E. Argentina.
COMMENT: Does not include *levis*; revised by LaVal, 1973, Bull. Los Ang. Cty. Mus. Nat. Hist. Sci. Soc., 15, who with Miller and Allen, 1928, Bull. U.S. Nat. Mus., 144:199, discussed the type.
ISIS NUMBER: 5301405014014053001.

Myotis scotti Thomas, 1927. Ann. Mag. Nat. Hist., ser. 9, 19:554.
TYPE LOCALITY: Ethiopia, Shoa, Djem-djem Forest (ca. 40 mi. (64 km) W. Addis Ababa), 8000 ft. (2438 m).
DISTRIBUTION: Ethiopia.
ISIS NUMBER: 5301405014014055001.

Myotis seabrai Thomas, 1912. Ann. Mag. Nat. Hist., ser. 8, 10:205.
TYPE LOCALITY: Angola, Mossamedes.
DISTRIBUTION: N.W. Cape Prov. (South Africa); Namibia; S.W. Angola.
COMMENT: Transferred from *Cistugo*; see Hayman and Hill, 1971, Part 2:33.
ISIS NUMBER: 5301405014014056001.

Myotis sicarius Thomas, 1915. J. Bombay Nat. Hist. Soc., 23:608.
TYPE LOCALITY: India, N. Sikkim.
DISTRIBUTION: Sikkim (N.E. India).
ISIS NUMBER: 5301405014014057001.

Myotis siligorensis (Horsfield, 1855). Ann. Mag. Nat. Hist., ser. 2, 16:102.
TYPE LOCALITY: Nepal, Siligori.
DISTRIBUTION: N. India to S. China and Vietnam, south to Malaya.
ISIS NUMBER: 5301405014014058001.

Myotis simus Thomas, 1901. Ann. Mag. Nat. Hist., ser. 7, 7:541.
  TYPE LOCALITY: Peru, Loreto, Sarayacu (Ucayali River).
  DISTRIBUTION: Colombia; Ecuador; Peru; N. Brazil; Bolivia; Paraguay.
  COMMENT: Revised by LaVal, 1973, Bull. Los Ang. Cty. Mus. Nat. Hist. Sci. Soc., 15.
  ISIS NUMBER: 5301405014014059001.

Myotis sodalis Miller and Allen, 1928. Bull. U.S. Nat. Mus., 144:130.
  TYPE LOCALITY: U.S.A., Indiana, Crawford Co., Wyandotte Cave.
  DISTRIBUTION: Vermont and Massachusetts to Florida Panhandle, west to S.W.
    Wisconsin and N.E. Oklahoma.
  PROTECTED STATUS: U.S. ESA - Endangered.
  ISIS NUMBER: 5301405014014060001.

Myotis stalkeri Thomas, 1910. Ann. Mag. Nat. Hist., ser. 8, 5:384.
  TYPE LOCALITY: Indonesia, Molucca Isls., Kei Isl., Ara.
  DISTRIBUTION: Kei Isls. (Molucca Isls.).
  ISIS NUMBER: 5301405014014061001.

Myotis thysanodes Miller, 1897. N. Am. Fauna, 13:80.
  TYPE LOCALITY: U.S.A., California, Kern Co., Tehachapi Mountains, Old Fort Tejon.
  DISTRIBUTION: Chiapas (Mexico) to S.W. South Dakota (U.S.A.) and S.C. British
    Columbia (Canada).
  COMMENT: Revised by Miller and Allen, 1928, Bull. U.S. Nat. Mus., 144:122-129.
    Reviewed by O'Farrell and Studier, 1980, Mamm. Species, 137:1-5.
  ISIS NUMBER: 5301405014014062001.

Myotis tricolor (Temminck, 1832). *In* Smuts, Enum. Mamm. Capensium, p. 106.
  TYPE LOCALITY: South Africa, Cape Province, Capetown.
  DISTRIBUTION: Ethiopia and Zaire, south to South Africa.
  ISIS NUMBER: 5301405014014063001.

Myotis velifer (J. A. Allen, 1890). Bull. Am. Mus. Nat. Hist., 3:177.
  TYPE LOCALITY: Mexico, Jalisco, Guadalajara, Santa Cruz del Valle.
  DISTRIBUTION: Honduras to S.C. Kansas, C. Utah and S.E. California (U.S.A.).
  COMMENT: See Hayward, 1970, WRISCI (W. New Mexico Univ.), 1(1):4-5, for scope of
    this species. Reviewed by Fitch *et al.*, 1981, Mamm. Species, 149:1-5.
  ISIS NUMBER: 5301405014014064001.

Myotis vivesi Menegaux, 1901. Bull. Mus. Hist. Nat. Paris, 7:323.
  TYPE LOCALITY: Mexico, Baja California, Partida Isl.
  DISTRIBUTION: Coast of Sonora and Baja California (Mexico), chiefly on small islands.
  COMMENT: Transferred from *Pizonyx*; see Findley, 1972, Syst. Zool., 21:43. Needs
    revision due to difference from typical *Myotis* (BVR).

Myotis volans (H. Allen, 1866). Proc. Acad. Nat. Sci. Phila., 18:282.
  TYPE LOCALITY: Mexico, Baja California, Cabo San Lucas.
  DISTRIBUTION: W. Jalisco to C. Veracruz and Mexico (Mexico); Alaska Panhandle
    (U.S.A.) to Baja California (Mexico), east to N. Nuevo Leon (Mexico), W. South
    Dakota (U.S.A.), and C. Alberta (Canada).
  COMMENT: Revised by Miller and Allen, 1928, Bull. U.S. Nat. Mus., 144:135-147.
  ISIS NUMBER: 5301405014014065001.

Myotis welwitschii (Gray, 1866). Proc. Zool. Soc. Lond., 1866:211.
  TYPE LOCALITY: Angola.
  DISTRIBUTION: South Africa; Zimbabwe; Mozambique; Zambia; Malawi; Zaire;
    Tanzania; Kenya; Ethiopia; Angola.
  ISIS NUMBER: 5301405014014067001.

Myotis yumanensis (H. Allen, 1864). Smithson. Misc. Coll., 7:58.
  TYPE LOCALITY: U.S.A., California, Imperial Co., Old Fort Yuma.
  DISTRIBUTION: Morelos, Michoacan and Baja California (Mexico) to Colorado, Utah,
    and W. Nevada, north to Oregon, Washington and British Columbia, east to N.W.
    Wyoming and S.E. Montana (U.S.A.).
  COMMENT: See comment under *lucifugus*.

Nyctalus Bowdich, 1825. Excursions in Madeira and Porto Santo, p. 36.
REVIEWED BY: E. Vernier (EV).
ISIS NUMBER: 5301405014015000000.

Nyctalus aviator (Thomas, 1911). Ann. Mag. Nat. Hist., ser. 8, 8:380.
TYPE LOCALITY: Japan, Honshu, Tokyo.
DISTRIBUTION: Hokkaido, Shikoku, Kyushu, Tsushima, Iki (Japan); Korea; E. China.
COMMENT: Listed as a subspecies of *lasiopterus* by Ellerman and Morrison-Scott,
    1951:161; but also see Corbet, 1978:56.

Nyctalus lasiopterus (Schreber, 1780). *In* Zimmermann, Geogr. Gesch., 2:412.
TYPE LOCALITY: Northern Italy (uncertain).
DISTRIBUTION: W. Europe to Urals, Caucasus, Asia Minor, Iran and Ust-Urt Plateau
    (Kazakhstan, U.S.S.R.).
COMMENT: See Corbet, 1978:55–56, for content of this species.
ISIS NUMBER: 5301405014015003001.

Nyctalus leisleri (Kuhl, 1818). Ann. Wetterau Ges. Naturk., 4(1):46.
TYPE LOCALITY: Germany, Hessen, Hanau.
DISTRIBUTION: W. Europe to Urals and Caucasus; Britain; Ireland; Madeira; Azores; W.
    Himalayas; E. Afghanistan.
COMMENT: Includes *azoreum* and *verrucosus*; see Corbet, 1978:55.
ISIS NUMBER: 5301405014015004001 as *N. leisleri.*
    5301405014015001001 as *N. azoreum.*
    5301405014015008001 as *N. verrucosus.*

Nyctalus montanus (Barrett-Hamilton, 1906). Ann. Mag. Nat. Hist., ser. 7, 17:99.
TYPE LOCALITY: India, Uttar Pradesh, Dehra Dun, Mussooree.
DISTRIBUTION: Afghanistan; Pakistan; N. India; Nepal.
COMMENT: Listed as a subspecies of *leisleri* by Ellerman and Morrison-Scott, 1951:159;
    but also see Corbet, 1978:55.
ISIS NUMBER: 5301405014015005001.

Nyctalus noctula (Schreber, 1774). Saugethiere, 1:166.
TYPE LOCALITY: France.
DISTRIBUTION: Europe to Urals and Caucasus; Morocco; S.E. Asia Minor to Israel; W.
    Turkestan to S.W. Siberia, Himalayas, and China, south to Malaya; Taiwan;
    Honshu (Japan).
COMMENT: Includes *furvus* and *velutinus*; see Corbet, 1978:55.
ISIS NUMBER: 5301405014015006001.

Nycticeius Rafinesque, 1819. J. Phys. Chim. Hist. Nat. Arts Paris, 88:417.
REVIEWED BY: F. R. Allison (FRA)(Australia).
COMMENT: Formerly included *Scotoecus*; see Hill, 1974, Bull. Br. Mus. (Nat. Hist.)
    Zool., 27:169–171. For content of this genus; see Koopman, 1978, Proc. 4th Int. Bat
    Res. Conf. Nairobi, p. 165–171.
ISIS NUMBER: 5301405014016000000.

Nycticeius balstoni (Thomas, 1906). Abstr. Proc. Zool. Soc. Lond., 1906(2), 31:2.
REVIEWED BY: A. C. Ziegler (ACZ).
TYPE LOCALITY: Australia, Western Australia, Laverton, North Pool, 503 m.
DISTRIBUTION: Western Australia; South Australia; Northern Territory; Queensland;
    New South Wales; Victoria; E. New Guinea.
COMMENT: Includes *orion* and *sanborni*; see Koopman, 1978, Proc. 4th Int. Bat Res.
    Conf. Nairobi, p. 168, 170, 171; but also see Hall and Richards, 1979, Queensl.
    Mus., Booklet, 12, and comment under *influatus.*
ISIS NUMBER: 5301405014016010001 as *N. sanborni.*

Nycticeius greyii (Gray, 1842). Zool. Voy. "Erebus" and "Terror," pl. 20.
TYPE LOCALITY: Australia, Northern Territory, Port Essington.
DISTRIBUTION: Western Australia, excluding the southwest; Northern Territory,
    Queensland (Australia). Records from inland South Australia, Victoria, and New
    South Wales may refer to *balstoni.*
ISIS NUMBER: 5301405014016004001 as *N. greyi (sic).*

Nycticeius humeralis (Rafinesque, 1818). Am. Mon. Mag., 3(6):445.
REVIEWED BY: B. Villa-Ramirez (BVR).
TYPE LOCALITY: U.S.A., Kentucky.
DISTRIBUTION: N. Veracruz (Mexico) to E. Kansas, the Great Lakes, and Pennsylvania,
south to Florida and the Gulf coast (U.S.A.); Cuba.
COMMENT: Includes *cubanus*; see Varona, 1974:37; but also see Hall, 1981:227.
Reviewed by Watkins, 1972, Mamm. Species, 23:1–4.
ISIS NUMBER: 5301405014016006001 as *N. humeralis*.
5301405014016002001 as *N. cubanus*.

Nycticeius influatus (Thomas, 1924). Ann. Mag. Nat. Hist., ser. 9, 13:540.
TYPE LOCALITY: Australia, Queensland, Prairie.
DISTRIBUTION: N.C. Queensland.
COMMENT: There is some doubt about the distinctness of *influatus* and *balstoni*; see
Koopman, 1978, Proc. 4th Int. Bat Res. Conf. Nairobi, p. 165–171. However they
are quite distinct where sympatric (KFK).
ISIS NUMBER: 5301405014016007001.

Nycticeius rueppellii (Peters, 1866). Monatsb. Preuss. Akad. Wiss. Berlin, p. 21.
TYPE LOCALITY: Australia, New South Wales, Sydney.
DISTRIBUTION: E. Queensland; E. New South Wales.
ISIS NUMBER: 5301405014016009001.

Nycticeius schlieffeni (Peters, 1860). Monatsb. Preuss. Akad. Wiss. Berlin, p. 223.
TYPE LOCALITY: Egypt, Cairo.
DISTRIBUTION: S.W. Arabia; Egypt to Somalia to Mozambique, Botswana, South Africa,
and Namibia; Mauritania and Ghana to Sudan and Tanzania.
COMMENT: Includes *cinnamomeus*; see Koopman, 1975, Bull. Am. Mus. Nat. Hist.,
154:412–413.
ISIS NUMBER: 5301405014016011001.

Nyctophilus Leach, 1822. Trans. Linn. Soc. Lond., 13:78.
REVIEWED BY: F. R. Allison (FRA)(Australia); A. C. Ziegler (ACZ)(New Guinea).
COMMENT: Includes *Lamingtona*; Australian species reviewed by Hall and Richards,
1979, Queensl. Mus., Booklet, 12. This genus is in need of a complete revision
(FRA).
ISIS NUMBER: 5301405014032000000 as *Nyctophilus*.
5301405014031000000 as *Lamingtona*.

Nyctophilus arnhemensis Johnson, 1959. Proc. Biol. Soc. Wash., 72:184.
TYPE LOCALITY: Australia, Northern Territory, Cape Arnhem Peninsula, S. of Yirkala,
Rocky Bay. (12° 13′ S. and 136° 47′ E.).
DISTRIBUTION: Arnhem Land and Groote Eylandt (N. Australia).
ISIS NUMBER: 5301405014032001001.

Nyctophilus bifax Thomas, 1915. Ann. Mag. Nat. Hist., ser. 8, 15:496.
TYPE LOCALITY: Australia, Queensland, Herberton.
DISTRIBUTION: N. Northern Territory; N. Queensland; N. Western Australia.
COMMENT: Includes *daedalus*, which may prove to be a distinct species based on
differences in bacula (FRA). See also comment under *gouldi*.
ISIS NUMBER: 5301405014032002001.

Nyctophilus geoffroyi Leach, 1822. Trans. Linn. Soc. Lond., 13:78.
TYPE LOCALITY: Australia, Western Australia, King George Sound.
DISTRIBUTION: Northern and central Australia, including Western Australia, Northern
Territory and Queensland to latitude 18° S; Tasmania.
ISIS NUMBER: 5301405014032003001.

Nyctophilus gouldi Tomes, 1858. Proc. Zool. Soc. Lond., 1858:31.
TYPE LOCALITY: Australia, Queensland, Moreton Bay.
DISTRIBUTION: S.E. Queensland; E. New South Wales; S. Victoria; Tasmania.

COMMENT: May include Tasmanian records from Ride, 1970:164, of *timoriensis* and *sherrini* (FRA). If *gouldi* is separated from *timoriensis*, then it should probably include *bifax* (KFK). See also Hall and Richards, 1979, Bats of Eastern Australia, Queensl. Mus., Booklet, 12.

Nyctophilus lophorhina (McKean and Calaby, 1968). Mammalia, 32:373.
TYPE LOCALITY: Papua New Guinea, Northern Province, Mt. Lamington. (8° 55′ S. and 48° 10′ E.).
DISTRIBUTION: S.E. New Guinea.
COMMENT: Formerly placed in the genus *Lamingtona* by McKean and Calaby, 1968, Mammalia, 32:373. Very similar to *N. microtis*, of which it is almost certainly a synonym; see Hill and Koopman (in press), Bull. Br. Mus. (Nat. Hist.), Zool., 41.
ISIS NUMBER: 5301405014031001001 as *Lamingtona lophorhina*.

Nyctophilus microdon Laurie and Hill, 1954:78.
TYPE LOCALITY: Papua New Guinea, Western Highlands (?) Prov., Welya (W. of Hagen Range, 7000 ft. (2134 m)).
DISTRIBUTION: E. New Guinea. E.C. New Guinea (ACZ).
ISIS NUMBER: 5301405014032004001.

Nyctophilus microtis Thomas, 1888. Ann. Mag. Nat. Hist., ser. 6, 2:226.
TYPE LOCALITY: Papua New Guinea, Central Prov., Astrolabe Range, Sogeri.
DISTRIBUTION: E. New Guinea.
ISIS NUMBER: 5301405014032005001.

Nyctophilus timoriensis E. Geoffroy, 1806. Ann. Mus. Hist. Nat. Paris, 8:200.
TYPE LOCALITY: Indonesia, Timor (uncertain).
DISTRIBUTION: Queensland; New South Wales; Victoria, South Australia; Western Australia; Tasmania; Papua New Guinea; perhaps Timor.
COMMENT: This bat has been confused with the smaller *gouldi* in coastal S.E. Queensland (FRA). See Hall and Richards, 1979, Bats of Eastern Australia, Queensl. Mus., Booklet, 12.
ISIS NUMBER: 5301405014032006001.

Nyctophilus walkeri Thomas, 1892. Ann. Mag. Nat. Hist., ser. 6, 9:405.
TYPE LOCALITY: Australia, Northern Territory, Adelaide River.
DISTRIBUTION: Northern Territory; N. Western Australia.
ISIS NUMBER: 5301405014032007001.

Otonycteris Peters, 1859. Monatsb. Preuss. Akad. Wiss. Berlin, p. 223.
ISIS NUMBER: 5301405014017000000.

Otonycteris hemprichi Peters, 1859. Monatsb. Preuss. Akad. Wiss. Berlin, p. 223.
TYPE LOCALITY: Egypt, Nile Valley south of Assuan (Aswan).
DISTRIBUTION: The desert zone from Algeria through Egypt and Arabia to Tadzhikistan, Afghanistan, and Kashmir.
ISIS NUMBER: 5301405014017001001.

Pharotis Thomas, 1914. Ann. Mag. Nat. Hist., ser. 8, 14:381.
REVIEWED BY: A. C. Ziegler (ACZ).
ISIS NUMBER: 5301405014033000000.

Pharotis imogene Thomas, 1914. Ann. Mag. Nat. Hist., ser. 8, 14:382.
TYPE LOCALITY: Papua New Guinea, Central Prov., Lower Kemp Welch River, Kamali.
DISTRIBUTION: S.E. New Guinea.
ISIS NUMBER: 5301405014033001001.

Philetor Thomas, 1902. Ann. Mag. Nat. Hist., ser. 7, 9:220.
REVIEWED BY: A. C. Ziegler (ACZ).
ISIS NUMBER: 5301405014018000000.

Philetor brachypterus (Temminck, 1840). Monogr. Mamm., 2:215.
  TYPE LOCALITY: Indonesia, Sumatra, Padang Dist.
  DISTRIBUTION: Malaya; Sumatra; Java; Borneo; New Guinea; New Britain Isl.,
    Mindanao (Philippines). Perhaps Banka Isl.
  COMMENT: Includes *rohui* and *verecundus*; see Hill, 1971, Zool. Meded. Rijks. Mus. Nat.
    Hist. Leiden, 45:143.
  ISIS NUMBER: 5301405014019008001 as *Pipistrellus brachypterus.*
    5301405014018001001 as *Philetor rohui.*

Phoniscus Miller, 1905. Proc. Biol. Soc. Wash., 18:229.
  COMMENT: Probably a subgenus of *Kerivoula*; but see Hill, 1965, Mammalia,
    29:524–528.

Phoniscus aerosa (Tomes, 1858). Proc. Zool. Soc. Lond., 1858:333.
  TYPE LOCALITY: "Eastern coast of South Africa."
  DISTRIBUTION: Possibly Africa.
  COMMENT: See Hill, 1965, Mammalia, 29:552–554, for placement of this species.

Phoniscus atrox Miller, 1905. Proc. Biol. Soc. Wash., 18:230.
  TYPE LOCALITY: Indonesia, Sumatra, near Kateman River.
  DISTRIBUTION: S. Thailand; Malaya; Sumatra.

Phoniscus jagori (Peters, 1866). Monatsb. Preuss. Akad. Wiss. Berlin, p. 399.
  TYPE LOCALITY: Philippines, Samar.
  DISTRIBUTION: Borneo; Java; Bali; Sulawesi; Samar Isl. (Philippines).
  COMMENT: Includes *javanus* and *rapax*; see Hill, 1965, Mammalia, 29:549–550.
  ISIS NUMBER: 5301405014029008001 as *Kerivoula javana.*
    5301405014029020001 as *Kerivoula rapax.*

Phoniscus papuensis (Dobson, 1878). Cat. Chiroptera Br. Mus., p. 339.
  REVIEWED BY: F. R. Allison (FRA); A. C. Ziegler (ACZ).
  TYPE LOCALITY: Papua New Guinea, Central Prov., Port Moresby.
  DISTRIBUTION: S.E. New Guinea; Queensland.
  ISIS NUMBER: 5301405014029015001 as *Kerivoula papuensis.*

Pipistrellus Kaup, 1829. Skizz. Europ. Thierwelt, 1:98.
  REVIEWED BY: D. L. Harrison (DLH); E. Vernier (EV)(Europe).
  COMMENT: Includes *Scotozous*; see Ellerman and Morrison-Scott, 1951:162–163; but also
    see Hill, 1976, Bull. Br. Mus. (Nat. Hist.) Zool., 30:25.
  ISIS NUMBER: 5301405014019000000.

Pipistrellus aero Heller, 1912. Smithson. Misc. Coll., 60(12):3.
  TYPE LOCALITY: Kenya, Mathews Range, Mt. Gargues.
  DISTRIBUTION: N.W. Kenya. Perhaps Ethiopia (DLH).
  ISIS NUMBER: 5301405014019002001.

Pipistrellus affinis (Dobson, 1871). Proc. Asiat. Soc. Bengal, p. 213.
  TYPE LOCALITY: Burma, Bhamo.
  DISTRIBUTION: N.E. Burma; Yunnan (China); N. India.
  ISIS NUMBER: 5301405014019003001.

Pipistrellus anchietai (Seabra, 1900). J. Sci. Math. Phys. Nat. Lisboa, ser. 2, 6:26, 120.
  TYPE LOCALITY: Angola, Cahata.
  DISTRIBUTION: Angola; S. Zaire; Zambia.
  COMMENT: The oldest name for this species may be *bicolor*; see Koopman, 1975, Bull.
    Am. Mus. Nat. Hist., 154:404.
  ISIS NUMBER: 5301405014019004001.

Pipistrellus anthonyi Tate, 1942. Bull. Am. Mus. Nat. Hist., 80:252.
  TYPE LOCALITY: Burma, Changyinku, 7000 ft. (2134 m).
  DISTRIBUTION: N. Burma.

Pipistrellus arabicus Harrison, 1979. Mammalia, 43:575.
  TYPE LOCALITY: Oman, Wadi Sahtan (23° 22' N., 57° 18' E.).
  DISTRIBUTION: Oman.

Pipistrellus ariel Thomas, 1904. Ann. Mag. Nat. Hist., ser. 7, 14:157.
  TYPE LOCALITY: Sudan, Kassala Province, Wadi Alagi (22° N. and 35° E.), 2000 ft. (610 m).
  DISTRIBUTION: Egypt; N. Sudan.
  ISIS NUMBER: 5301405014019006001.

Pipistrellus babu Thomas, 1915. J. Bombay Nat. Hist. Soc., 24:30.
  TYPE LOCALITY: Pakistan, Punjab, Rawalpindi, Murree, 8000 ft. (2438 m).
  DISTRIBUTION: Afghanistan; Pakistan; N. India; Sikkim; Nepal; Bhutan; Burma; China.
  ISIS NUMBER: 5301405014019007001.

Pipistrellus bodenheimeri Harrison, 1960. Durban Mus. Novit., 5:261.
  TYPE LOCALITY: Israel, 40 km N. Eilat, Wadi Araba, Yotwata.
  DISTRIBUTION: Israel; S. Yemen; Oman; perhaps Socotra Isl.

Pipistrellus cadornae Thomas, 1916. J. Bombay Nat. Hist. Soc., 24:416.
  TYPE LOCALITY: India, Darjeeling, Pashok, 3500 ft. (1067 m).
  DISTRIBUTION: N.E. India; Burma; Thailand.
  COMMENT: Listed as a subspecies of *savii* by Ellerman and Morrison-Scott, 1951:170, but see Hill, 1962, Proc. Zool. Soc. Lond., 1962:133.
  ISIS NUMBER: 5301405014019009001.

Pipistrellus ceylonicus (Kelaart, 1852). Prodr. Faun. Zeylanica, p. 22.
  TYPE LOCALITY: Sri Lanka, Trincomalee.
  DISTRIBUTION: Pakistan; India; Sri Lanka; Burma; Kwangsi, Hainan (China); N. Vietnam; N. Borneo.
  ISIS NUMBER: 5301405014019010001.

Pipistrellus circumdatus (Temminck, 1840). Monogr. Mamm., 2:214.
  TYPE LOCALITY: Indonesia, Java, Tapos.
  DISTRIBUTION: Java; Malaya; Burma; N.E. India.
  ISIS NUMBER: 5301405014019011001.

Pipistrellus coromandra (Gray, 1838). Mag. Zool. Bot., 2:498.
  TYPE LOCALITY: India, Coromandel Coast, Pondicherry.
  DISTRIBUTION: Afghanistan; Pakistan; India; Sri Lanka; Nepal; Bhutan; Burma; S. China; Thailand; Vietnam; Nicobar Isls.
  COMMENT: Does not include *aladdin*; see Corbet, 1978:53. See comment under *P. pipistrellus.*
  ISIS NUMBER: 5301405014019012001.

Pipistrellus crassulus Thomas, 1904. Ann. Mag. Nat. Hist., ser. 7, 13:206.
  TYPE LOCALITY: Cameroun, Efulen.
  DISTRIBUTION: Cameroun; Zaire.
  ISIS NUMBER: 5301405014019013001.

Pipistrellus deserti Thomas, 1902. Proc. Zool. Soc. Lond., 1902(2):4.
  TYPE LOCALITY: Libya, Fezzan, Mursuk.
  DISTRIBUTION: Libya; Algeria; N. Sudan; Upper Volta; Egypt.
  ISIS NUMBER: 5301405014019014001.

Pipistrellus dormeri (Dobson, 1875). Proc. Zool. Soc. Lond., 1875:373.
  TYPE LOCALITY: India, Mysore, Bellary Hills.
  DISTRIBUTION: N.W., S. and E. India; Pakistan; perhaps Taiwan.
  COMMENT: Sometimes placed in a separate genus, *Scotozous*; see Tate, 1942, Bull. Amer. Mus. Nat. Hist., 80:221; Corbet and Hill, 1980:68; but also see Ellerman and Morrison-Scott, 1951:162–163; Corbet, 1978:51.
  ISIS NUMBER: 5301405014019015001.

Pipistrellus eisentrauti Hill, 1968. Bonn. Zool. Beitr., 19:45.
  TYPE LOCALITY: Cameroun, Western Province, Rumpi Highlands, Dikume-Balue.
  DISTRIBUTION: Cameroun; Ivory Coast; Kenya.
  ISIS NUMBER: 5301405014019016001.

Pipistrellus endoi Imaizumi, 1959. Bull. Nat. Sci. Mus. Tokyo, 4:363–371.
TYPE LOCALITY: Japan, Honshu, Iwate Pref., Ninohe-Gun, Ashiro-cho, Horobe.
DISTRIBUTION: Honshu (Japan).
ISIS NUMBER: 5301405014019017001.

Pipistrellus hesperus (H. Allen, 1864). Smithson. Misc. Coll., 7:43.
REVIEWED BY: B. Villa-Ramirez (BVR).
TYPE LOCALITY: U.S.A., California, Imperial Co., Old Fort Yuma.
DISTRIBUTION: Washington to S.W. Oklahoma (U.S.A.), and Baja California, south to
   Hidalgo and Guerrero (Mexico).
ISIS NUMBER: 5301405014019018001.

Pipistrellus imbricatus (Horsfield, 1824). Zool. Res. Java, Part 8.
TYPE LOCALITY: Indonesia, Java.
DISTRIBUTION: Java; Kangean Isl.; Bali; Palawan, Luzon, and Negros Isl. (Philippines)
   (KFK).
ISIS NUMBER: 5301405014019019001.

Pipistrellus inexspectatus Aellen, 1959. Arch. Sci. Phys. Nat. Geneve, 12:226.
TYPE LOCALITY: Cameroun, Upper Benoue Valley, Ngaaouyanga.
DISTRIBUTION: Cameroun; Zaire; Uganda (KFK). Perhaps Sudan (DLH, NS).
ISIS NUMBER: 5301405014019020001.

Pipistrellus javanicus (Gray, 1838). Mag. Zool. Bot., 2:498.
TYPE LOCALITY: Indonesia, Java.
DISTRIBUTION: S. and C. Japan; S. Ussuri region; Korea; China, through S.E. Asia to
   Java and the Philippines; perhaps Australia.
COMMENT: Includes *abramus*, "*tralatitius*," *camortae*, *meyeni*, and *irretitus*; see Laurie and
   Hill, 1954:67; Ellerman and Morrison-Scott, 1951:165; Hill, 1967, J. Bombay Nat.
   Hist. Soc., 64:7; Koopman, 1973, Period. Biol. Zagreb, 75:115.
ISIS NUMBER: 5301405014019023001 as *P. javanicus*.
   5301405014019001001 as *P. abramus*.
   5301405014019022001 as *P. irretitus*.
   5301405014019050001 as *P. tralatitius*.

Pipistrellus joffrei (Thomas, 1915). Ann. Mag. Nat. Hist., ser. 8, 15:225.
TYPE LOCALITY: Burma, Kachin Hills.
DISTRIBUTION: N. Burma.
COMMENT: Transferred from *Nyctalus*; see Hill, 1966, Bull. Br. Mus. (Nat. Hist.) Zool.,
   14:383.
ISIS NUMBER: 5301405014015002001 as *Nyctalus joffrei*.

Pipistrellus kitcheneri Thomas, 1916. Ann. Mag. Nat. Hist., ser. 8, 15:229.
TYPE LOCALITY: Malaysia, Kalimantan, Barito River.
DISTRIBUTION: Borneo.
COMMENT: Listed as a subspecies of *imbricatus* by Chasen, 1940, Bull. Raffles Mus., No.
   15; but also see Tate, 1942, Bull. Amer. Mus. Nat. Hist., 80:221; Medway, 1977:55.

Pipistrellus kuhlii (Natterer, 1819). *In* Kuhl, Ann. Wetterau Ges. Naturk., 4(2):199.
TYPE LOCALITY: Italy, Friuli-Venezia Giulia, Trieste.
DISTRIBUTION: S. Europe through the Caucasus to Kazakhstan and Pakistan; S.W. Asia;
   most of Africa.
ISIS NUMBER: 5301405014019024001 as *P. kuhli (sic)*.

Pipistrellus lophurus Thomas, 1915. J. Bombay Nat. Hist. Soc., 23:413.
TYPE LOCALITY: Burma, Tenasserim, Victoria Province, Maliwun.
DISTRIBUTION: Peninsular Burma.
ISIS NUMBER: 5301405014019025001.

Pipistrellus macrotis (Temminck, 1840). Monogr. Mamm., 2:218.
TYPE LOCALITY: Indonesia, Sumatra, Padang.
DISTRIBUTION: Malaya; Sumatra; Bali; adjacent small islands.
COMMENT: Listed as a subspecies of *imbricatus* by Medway, 1969:39; but see Tate, 1942,
   Bull. Am. Mus. Nat. Hist., 80:249; Corbet and Hill, 1980:67.

Pipistrellus maderensis (Dobson, 1878). Cat. Chiroptera Br. Mus., p. 231.
TYPE LOCALITY: Madeira Isls., Madeira Isl. (Portugal).
DISTRIBUTION: Madeira Isl.; Las Palmas, Canary Isls.
ISIS NUMBER: 5301405014019026001.

Pipistrellus mimus Wroughton, 1899. J. Bombay Nat. Hist. Soc., 12:722.
TYPE LOCALITY: India, Gujarat, Surat Dist., Dangs, Mheskatri.
DISTRIBUTION: Afghanistan; Pakistan; Sri Lanka; India; Nepal; Sikkim; Bhutan; Burma;
    Vietnam; Thailand.
ISIS NUMBER: 5301405014019027001.

Pipistrellus minahassae (Meyer, 1899). Abh. Zool. Anthrop.-Ethnology. Mus. Dresden,
    7(7):14.
TYPE LOCALITY: Indonesia, Sulawesi, Minahassa, Tomohon.
DISTRIBUTION: Sulawesi.
ISIS NUMBER: 5301405014019028001.

Pipistrellus mordax (Peters, 1866). Monatsb. Preuss. Akad. Wiss. Berlin, p. 402.
TYPE LOCALITY: Indonesia, Java.
DISTRIBUTION: Sri Lanka; India; Java.
ISIS NUMBER: 5301405014019029001.

Pipistrellus musciculus Thomas, 1913. Ann. Mag. Nat. Hist., ser. 8, 11:316.
TYPE LOCALITY: Cameroun, Ja River, Bitye, 2000 ft. (610 m).
DISTRIBUTION: Cameroun; Zaire; Gabon.
ISIS NUMBER: 5301405014019031001.

Pipistrellus nanulus Thomas, 1904. Ann. Mag. Nat. Hist., ser. 7, 14:198.
TYPE LOCALITY: Cameroun, Efulen.
DISTRIBUTION: Sierra Leone to Uganda; Bioko.
ISIS NUMBER: 5301405014019032001.

Pipistrellus nanus (Peters, 1852). Reise nach Mossambique, Saugethiere, p. 63.
TYPE LOCALITY: Mozambique, Inhambane.
DISTRIBUTION: South Africa to Sudan, Ethiopia, Somalia, Mali, and Sierra Leone;
    Madagascar; Pemba; Zanzibar.
COMMENT: The oldest name for this species is *africanus;* see Koopman, 1975, Bull. Am.
    Mus. Nat. Hist., 154:400.
ISIS NUMBER: 5301405014019033001.

Pipistrellus nathusii (Keyserling and Blasius, 1839). Arch. Naturgesch., 5(1):320.
TYPE LOCALITY: Germany, Berlin.
DISTRIBUTION: W. Europe to Urals, Caucasus, and W. Asia Minor; S. England.
ISIS NUMBER: 5301405014019034001.

Pipistrellus peguensis Sinha, 1969. Proc. Zool. Soc. Calcutta, 22:83.
TYPE LOCALITY: Burma, Pegu.
DISTRIBUTION: Burma.
ISIS NUMBER: 5301405014019036001.

Pipistrellus permixtus Aellen, 1957. Rev. Suisse Zool., 64:200.
TYPE LOCALITY: Tanzania, Dar-es-Salaam.
DISTRIBUTION: Tanzania.
ISIS NUMBER: 5301405014019037001.

Pipistrellus petersi (Meyer, 1899). Abh. Zool. Anthrop.-Ethnology. Mus. Dresden, 7(7):13.
TYPE LOCALITY: Indonesia, Sulawesi, North Sulawesi, Minahassa.
DISTRIBUTION: Sulawesi; Buru and Amboina (Molucca Isls.).
ISIS NUMBER: 5301405014019038001.

Pipistrellus pipistrellus (Schreber, 1774). Saugethiere, 1:167.
TYPE LOCALITY: France.
DISTRIBUTION: British Isles, S. Scandinavia, and W. Europe to the Volga and Caucasus;
    Morocco; Asia Minor and Israel to Kashmir, Kazakhstan, and Sinkiang (China).
    Perhaps Korea, Japan and Taiwan.
COMMENT: Includes *aladdin;* see Corbet, 1978:53.
ISIS NUMBER: 5301405014019039001.

Pipistrellus pulveratus (Peters, 1871). *In* Swinhoe, Proc. Zool. Soc. Lond., p. 618.
 TYPE LOCALITY: China, Fukien, Amoy.
 DISTRIBUTION: Szechwan, Yunnan, Hunan, Kiangsu, Fukien (China), Hong Kong;
   Thailand.
 ISIS NUMBER: 5301405014019040001.

Pipistrellus rueppelli (Fischer, 1829). Synops. Mamm., p. 109.
 TYPE LOCALITY: Sudan, Northern Province, Dongola.
 DISTRIBUTION: Senegal, Algeria, Egypt, and Iraq, south to Botswana and Zimbabwe;
   Zanzibar.
 ISIS NUMBER: 5301405014019043001.

Pipistrellus rusticus (Tomes, 1861). Proc. Zool. Soc. Lond., 1861:35.
 TYPE LOCALITY: Namibia, Damaraland, Olifants Vlei.
 DISTRIBUTION: Ghana and Ethiopia, south to South Africa.
 ISIS NUMBER: 5301405014019044001.

Pipistrellus savii (Bonaparte, 1837). Fauna Ital., 1, fasc. 20.
 TYPE LOCALITY: Italy, Pisa.
 DISTRIBUTION: Iberia, Morocco, and the Canary and Cape Verde Isls. through the
   Caucasus and Mongolia to Korea, N.E. China (SW) and Japan, southeastward
   through Iran and Afghanistan to N. India and Burma.
 ISIS NUMBER: 5301405014019045001.

Pipistrellus societatis Hill, 1972. Bull. Br. Mus. (Nat. Hist.) Zool., 23:34.
 TYPE LOCALITY: Malaysia, Pahang, Gunong Benom.
 DISTRIBUTION: Malaya.
 COMMENT: Known only from the holotype (NS).
 ISIS NUMBER: 5301405014019046001.

Pipistrellus stenopterus (Dobson, 1875). Proc. Zool. Soc. Lond., 1875:470.
 TYPE LOCALITY: Malaysia, Sarawak.
 DISTRIBUTION: Malaya; Sumatra; N. Borneo; Rhio Arch.; Mindanao (Philippines).
 COMMENT: Transferred from *Nyctalus*; see Medway, 1977:56.
 ISIS NUMBER: 5301405014015007001 as *Nyctalus stenopterus*.

Pipistrellus sturdeei Thomas, 1915. Ann. Mag. Nat. Hist., ser. 8, 15:230.
 TYPE LOCALITY: Japan, Bonin Isls., Hillsboro Isl.
 DISTRIBUTION: Bonin Isls.

Pipistrellus subflavus (F. Cuvier, 1832). Nouv. Ann. Mus. Hist. Nat. Paris, 1:17.
 REVIEWED BY: B. Villa-Ramirez (BVR).
 TYPE LOCALITY: U.S.A., Georgia.
 DISTRIBUTION: Nova Scotia, S. Quebec (Canada), and S.E. Minnesota, south to C.
   Florida (U.S.A.), Guatamala and Honduras.
 ISIS NUMBER: 5301405014019047001.

Pipistrellus tasmaniensis (Gould, 1858). Mamm. Aust., 3, pl. 48.
 REVIEWED BY: F. R. Allison (FRA).
 TYPE LOCALITY: Australia, Tasmania.
 DISTRIBUTION: S. Australia except South Australia; Tasmania.
 ISIS NUMBER: 5301405014019048001.

Pipistrellus tenuis (Temminck, 1840). Monogr. Mamm., 2:229.
 REVIEWED BY: F. R. Allison (FRA).
 TYPE LOCALITY: Indonesia, Sumatra.
 DISTRIBUTION: Thailand to New Guinea, Bismarck Arch., Solomon Isls., New Hebrides
   and N. Australia; Cocos Keeling Isl.; Christmas Isl.
 COMMENT: Includes *angulatus, collinus, papuanus, murrayi, subulidens,* and *sewelanus;* see
   Koopman, 1973, Period. Biol. Zagreb, 75:115. One of the first three above-named
   forms may prove to be a separate species (ACZ). See also McKean and Price, 1978,
   Mammalia, 42:346.
 ISIS NUMBER: 5301405014019049001 as *P. tenuis.*
              5301405014019005001 as *P. angulatus.*
              5301405014019030001 as *P. murrayi.*
              5301405014019035001 as *P. papuanus.*

Plecotus E. Geoffroy, 1818. Descrip. de L'Egypte, 2:112.
REVIEWED BY: E. Vernier (EV); B. Villa-Ramirez (BVR).
COMMENT: Includes *Corynorhinus*; see Anderson, 1972, Bull. Am. Mus. Nat. Hist., 148:255. Does not include *Idionycteris phyllotis*; see Williams *et al.*, 1970, J. Mammal., 51:602–606; but also see Handley, 1959, Proc. U.S. Nat. Mus., 110:95–246. A key to the genus was published by Jones, 1977, Mamm. Species, 69:1–4.
ISIS NUMBER: 5301405014020000000.

Plecotus auritus (Linnaeus, 1758). Syst. Nat., 10th ed., 1:32.
TYPE LOCALITY: Sweden.
DISTRIBUTION: Norway, Ireland, and Spain to Sakhalin, Japan, N. China and Nepal.
ISIS NUMBER: 5301405014020001001.

Plecotus austriacus (Fischer, 1829). Synopsis Mamm., p. 117.
TYPE LOCALITY: Austria, Vienna.
DISTRIBUTION: England, Spain, and Senegal to Mongolia and W. China; Canary and Cape Verde Isls.
COMMENT: Included in *auritus* by Ellerman and Morrison-Scott, 1951:181; but also see Corbet, 1978:61.
ISIS NUMBER: 5301405014020002001.

Plecotus mexicanus (G. M. Allen, 1916). Bull. Mus. Comp. Zool., 60:347.
TYPE LOCALITY: Mexico, Chihuahua, Pacheco.
DISTRIBUTION: N.W. Mexico to S.E. Mexico; Cozumel Isl.
COMMENT: Listed as a subspecies of *townsendii* by Hall and Kelson, 1959:200; but see Handley, 1959, Proc. U.S. Nat. Mus., 110:141–151; Hall, 1981:233. Formerly in genus *Corynorhinus*; see Anderson, 1972, Bull. Am. Mus. Nat. Hist., 148:255.

Plecotus rafinesquii Lesson, 1827. Mon. Mammal., p. 96.
TYPE LOCALITY: U.S.A., Illinois, Wabash Co., Mt. Carmel.
DISTRIBUTION: S.E. U.S.A. from Virginia to S. Missouri, south to Louisiana and C. Florida.
COMMENT: Formerly in genus *Corynorhinus*; see Anderson, 1972, Bull. Am. Mus. Nat. Hist., 148:255. Reviewed by Jones, 1977, Mamm. Species, 69:1–4.
ISIS NUMBER: 5301405014020004001.

Plecotus townsendii Cooper, 1837. Ann. Lyc. Nat. Hist., 4:73.
TYPE LOCALITY: U.S.A., Washington, Clark Co., Fort Vancouver.
DISTRIBUTION: S. British Columbia (Canada) through W. U.S.A. to Oaxaca (Mexico), east through Oklahoma to W. Virginia.
COMMENT: Formerly in genus *Corynorhinus*; see Anderson, 1972, Bull. Am. Mus. Nat. Hist., 148:255.
PROTECTED STATUS: U.S. ESA - Endangered as *P. t. ingens* subspecies only. U.S. ESA - Endangered as *P. t. virginianus* subspecies only.
ISIS NUMBER: 5301405014020005001.

Rhogeessa H. Allen, 1866. Proc. Acad. Nat. Sci. Phila., 18:285.
REVIEWED BY: B. Villa-Ramirez (BVR).
COMMENT: Includes *Baeodon*; see Jones *et al.*, 1977, Occas. Pap. Mus. Texas Tech Univ., 47:26. Revised by LaVal, 1973, Occas. Pap. Mus. Nat. Hist. Univ. Kansas, 19:1–47.
ISIS NUMBER: 5301405014021000000 as *Rhogeessa*.
5301405014001000000 as *Baeodon*.

Rhogeessa alleni Thomas, 1892. Ann. Mag. Nat. Hist., ser. 6, 10:477.
TYPE LOCALITY: Mexico, Jalisco, near Autlan, Santa Rosalia.
DISTRIBUTION: Oaxaca to S. Zacatecas (Mexico).
ISIS NUMBER: 5301405014001001001 as *Baeodon alleni*.

Rhogeessa gracilis Miller, 1897. N. Am. Fauna, 13:126.
TYPE LOCALITY: Mexico, Puebla, Piaxtla, 1100 m.
DISTRIBUTION: N. Jalisco to Oaxaca (Mexico).
COMMENT: Reviewed by Jones, 1977, Mamm. Species, 76:1–2.
ISIS NUMBER: 5301405014021001001.

Rhogeessa minutilla Miller, 1897. Proc. Biol. Soc. Wash., 11:139.
TYPE LOCALITY: Venezuela, Margarita Isl.
DISTRIBUTION: Colombia; Venezuela; Margarita Isl.
COMMENT: Listed as a subspecies of *parvula* by Cabrera, 1958:111; but see LaVal, 1973, Occas. Pap. Mus. Nat. Hist. Univ. Kansas, 19:37–38.
ISIS NUMBER: 5301405014021002001.

Rhogeessa mira LaVal, 1973. Occas. Pap. Mus. Nat. Hist. Univ. Kansas, 19:26.
TYPE LOCALITY: Mexico, Michoacan, 20 km N. El Infernillo.
DISTRIBUTION: S. Michoacan (Mexico).
ISIS NUMBER: 5301405014021003001.

Rhogeessa parvula H. Allen, 1866. Proc. Acad. Nat. Sci. Phila., 18:285.
TYPE LOCALITY: Mexico, Nayarit, Tres Marias Isls.
DISTRIBUTION: Oaxaca to Sonora (Mexico); Tres Marias Isls.
COMMENT: For scope of this species, see LaVal, 1973, Occas. Pap. Mus. Nat. Hist. Univ. Kansas, 19:21–26.
ISIS NUMBER: 5301405014021004001.

Rhogeessa tumida H. Allen, 1866. Proc. Acad. Nat. Sci. Phila., 18:286.
TYPE LOCALITY: Mexico, Veracruz, Mirador.
DISTRIBUTION: Tamaulipas (Mexico) to Ecuador and N.E. Brazil; Trinidad.
COMMENT: Listed as a subspecies of *parvula* by Hall and Kelson, 1959:196; but also see LaVal, 1973, Occas. Pap. Mus. Nat. Hist. Univ. Kansas, 19:21–26; and Hall, 1981:228, who included *bombyx* and *velilla*.
ISIS NUMBER: 5301405014021005001.

Scotoecus Thomas, 1901. Ann. Mag. Nat. Hist., ser. 7, 7:263.
COMMENT: Considered a subgenus of *Nycticeius* by Hayman and Hill, 1971, Part 2:36; but see Hill, 1974, Bull. Br. Mus. (Nat. Hist.) Zool., 27:169–171.

Scotoecus albofuscus Thomas, 1890. Ann. Mus. Civ. Stor. Nat. Genova, 29:84.
TYPE LOCALITY: Gambia, Bathurst.
DISTRIBUTION: Gambia to Kenya and Mozambique.
ISIS NUMBER: 5301405014016001001 as *Nycticeius albofuscus*.

Scotoecus hirundo De Winton, 1899. Ann. Mag. Nat. Hist., ser. 7, 4:355.
TYPE LOCALITY: Ghana, Gambaga.
DISTRIBUTION: Senegal to Ethiopia, south to Angola and Zambia.
COMMENT: Includes *hindei*; see Hayman and Hill, 1971, Part 2:36; Robbins, 1980, Mammalia, 44:84; but see also Hill, 1974, Bull. Br. Mus. (Nat. Hist.) Zool., 27:177–184.
ISIS NUMBER: 5301405014016005001 as *Nycticeius hirundo*.

Scotoecus pallidus Dobson, 1876. Monogr. Asiat. Chiroptera, App. D:186.
TYPE LOCALITY: Pakistan, Punjab, Lahore, Mian Mir.
DISTRIBUTION: Pakistan; N. India.
COMMENT: Included in *Nycticeius* by Ellerman and Morrison-Scott, 1951:177; but see Hill, 1974, Bull. Br. Mus. (Nat. Hist.) Zool., 27:171–174.
ISIS NUMBER: 5301405014016008001 as *Nycticeius pallidus*.

Scotomanes Dobson, 1875. Proc. Zool. Soc. Lond., 1875:371.
COMMENT: Includes *Scoteinus*; see Sinha and Chakraborty, 1971, Proc. Zool. Soc. Calcutta, 24:53–57.
ISIS NUMBER: 5301405014022000000.

Scotomanes emarginatus (Dobson, 1871). Proc. Asiat. Soc. Bengal, p. 211.
TYPE LOCALITY: "India."
DISTRIBUTION: India.
COMMENT: Included in *Nycticeius* by Ellerman and Morrison-Scott, 1951:177; but see Hill, 1974, Bull. Br. Mus. (Nat. Hist.), Zool., 27:171–174.
ISIS NUMBER: 5301405014016003001 as *Nycticeius emarginatus*.

Scotomanes ornatus (Blyth, 1851). J. Asiat. Soc. Bengal, 20:511.
TYPE LOCALITY: India, Assam, Khasi Hills, Cherrapunji.
DISTRIBUTION: N.E. India; Sikkim; Burma; S. China; Thailand; Vietnam.
ISIS NUMBER: 5301405014022001001.

Scotophilus Leach, 1821. Trans. Linn. Soc. Lond., 13:69, 71.
REVIEWED BY: C. B. Robbins (CBR).
COMMENT: Includes *Pachyotus;* see Walker, *et al.,* 1975:359.
ISIS NUMBER: 5301405014023000000.

Scotophilus borbonicus (E. Geoffroy, 1806). Ann. Mus. Hist. Nat. Paris, 8:201.
TYPE LOCALITY: Reunion Isl. (France).
DISTRIBUTION: Madagascar; Reunion Isl.
COMMENT: Included in *leucogaster* by Hayman and Hill, 1971, Part 2:50–51; but see
    Koopman, 1975, Bull. Amer. Mus. Nat. Hist., 154:414–416. Hill, 1980, Zool.
    Meded., Leiden, 55:287–295, considered African *viridis* and *damarensis,* and
    possibly *leucogaster* to be conspecific with *borbonicus,* but Robbins *et al.,* in press,
    rejected any affinity with African mainland species. May be extinct; see Cheke
    and Dahl, 1981, Mammalia, 45:217, who also pointed out that records from
    Mauritius are erroneus. Also see comment under *leucogaster.*

Scotophilus celebensis Sody, 1928. Natuurk. Tijdschr. Ned.-Ind., 88:90.
TYPE LOCALITY: Indonesia, Sulawesi, Toli Toli.
DISTRIBUTION: Sulawesi.
ISIS NUMBER: 5301405014023001001.

Scotophilus dinganii (A. Smith, 1833). S. Afr. Quart. J., 2:59.
TYPE LOCALITY: South Africa, Port Natal (= Durban).
DISTRIBUTION: Senegal and Sierra Leone east to Somalia and S. Yemen and south to
    South Africa and Namibia.
COMMENT: Includes *planirostris, colias, herero,* and *pondoensis* (see Robbins *et al.,* in
    press). Placed in *leucogaster* by Koopman, 1975, Bull. Amer. Mus. Nat. Hist.,
    154:414–416 (as *nigrita*) and Koopman *et al.,* 1978, Amer. Mus. Novit., 2643:4–5;
    but see also Schlitter, *et al.,* 1980, Ann. Transvaal Mus., 32:231–239.

Scotophilus heathi (Horsfield, 1831). Proc. Zool. Soc. Lond., 1831:113.
TYPE LOCALITY: India, Madras.
DISTRIBUTION: Afghanistan to S. China, Hainan, south to Sri Lanka and Vietnam.
ISIS NUMBER: 5301405014023003001.

Scotophilus kuhli Leach, 1822. Trans. Linn. Soc. Lond., 13:71.
TYPE LOCALITY: "India".
DISTRIBUTION: Pakistan to Taiwan, south to Sri Lanka and Malaya, southeast to
    Philippines and Aru Isls.; Nicobar Isls.; Hainan.
COMMENT: Generally called *temmincki;* but see Hill and Thonglongya, 1972, Bull. Br.
    Mus. (Nat. Hist.) Zool., 22:191–193.
ISIS NUMBER: 5301405014023004001 as *S. kuhlii (sic).*
                  5301405014023007001 as *S. temmincki.*

Scotophilus leucogaster (Cretzschmar, 1826). *In* Ruppell, Atlas Reise Nordl. Afr., Saugeth.,
    p. 71.
TYPE LOCALITY: Sudan, Kordofan Region, Brunnen Nedger (Nedger Well).
DISTRIBUTION: Mauritania and Senegal to N. Kenya and Ethiopia; south to South
    Africa.
COMMENT: Includes *damarensis.* For scope of this species see Koopman, 1975, Bull. Am.
    Mus. Nat. Hist., 154:414–416 (as *nigrita*) and Koopman *et al.,* 1978, Am. Mus.
    Novit., 2643:4–5. However, Robbins, 1978, J. Mammal., 59:212–213, has shown
    that the name *nigrita* was misapplied to another species, for which the next
    available name was *dinganii,* and which would be included in *leucogaster* if
    Koopman's (1975) reasoning were followed. However, Schlitter *et al.,* 1980, Ann.
    Transvaal Mus., 32:231–239, listed *dinganii* as distinct from *leucogaster,* as does
    CBR. Also see comment under *borbonicus.*
ISIS NUMBER: 5301405014023005001.

Scotophilus nigrita (Schreber, 1774). Saugethiere, 1:171.
    TYPE LOCALITY: Senegal.
    DISTRIBUTION: Senegal to Sudan and Kenya to Mozambique.
    COMMENTS: Includes *gigas* and *alvenslebeni*. For a review of this species see Robbins,
        1978, J. Mammal., 59:212-213. The identity of this species is clear and is the
        senior synonym of *gigas*. Specimens previously called *nigrita* should be called
        dinganii; see Robbins, 1978, J. Mammal., 59:212-213. KFK suggests that *nigrita*
        should be regarded as a *nomen dubium*, in which case the next available name
        would be *gigas*.
    ISIS NUMBER: 5301405014023006001 as *S. nigrita*.
                 5301405014023002001 as *S. gigas*.

Scotophilus nigritellus De Winton, 1899. Ann. Mag. Nat. Hist., ser. 7, 4:355.
    TYPE LOCALITY: Ghana, Gambaga.
    DISTRIBUTION: Senegal to Sudan.
    COMMENT: Listed as a subspecies of *leucogaster* by Hayman and Hill, 1971, Part 2:50,
        but see Koopman *et al.*, 1978, Am. Mus. Novit., 2643:4-5. According to CBR, a
        subspecies of *viridis*.

Scotophilus nux Thomas, 1904. Ann. Mag. Nat. Hist., ser. 7, 13:208.
    TYPE LOCALITY: Cameroon, Efulen.
    DISTRIBUTION: Sierra Leone to Kenya.
    COMMENT: Has been most often recognized as a subspecies of *nigrita* (see Allen, 1939,
        Bull. Mus. Comp. Zool., 83:1-763; Rosevear, 1965, The Bats of West Africa, Br.
        Mus. (Nat. Hist.), 418 pp.; Hayman and Hill, 1971, Part 2:50). Koopman *et al.*,
        1978, Am. Mus. Novit., 2643:4-5 listed it as a subspecies of *leucogaster*.

Scotophilus robustus Milne-Edwards, 1881. C. R. Acad. Sci. Paris, 91:1035.
    TYPE LOCALITY: Madagascar.
    DISTRIBUTION: Madagascar.
    COMMENTS: Recognized as a subspecies of *nigrita* (= *dinganii*) by Hayman and Hill,
        1971, Pt. 2:50. However, Robbins *et al.*, in press, considered it specifically distinct
        from *dinganii* and *borbonicus*.

Scotophilus viridis (Peters, 1852). Reise nach Mossambique, Saugethiere, p. 67.
    TYPE LOCALITY: Mozambique, Mozambique Isl., 15° S.
    DISTRIBUTION: Ethiopia south to Mozambique and South Africa.
    COMMENT: Included in *leucogaster* by Hayman and Hill, 1971, Part 2:50, but see
        Koopman, 1975, Bull. Am. Mus. Nat. Hist., 154:414-416; Schlitter, *et al.*, 1980,
        Ann. Transvaal Mus., 32:231-239. According to CBR, includes *nigritellus*.

Tomopeas Miller, 1900. Ann. Mag. Nat. Hist., ser. 7, 6:570.
    ISIS NUMBER: 5301405014034000000.

Tomopeas ravus Miller, 1900. Ann. Mag. Nat. Hist., ser. 7, 6:571.
    TYPE LOCALITY: Peru, Cajamarca, Yayan, 1000 m.
    DISTRIBUTION: W. Peru.
    ISIS NUMBER: 5301405014034001001.

Tylonycteris Peters, 1872. Monatsb. Preuss. Akad. Wiss. Berlin, p. 703.
    ISIS NUMBER: 5301405014024000000.

Tylonycteris pachypus (Temminck, 1840). Monogr. Mamm., 2:217.
    TYPE LOCALITY: Indonesia, Java, Bantam.
    DISTRIBUTION: India to S. China and Vietnam to Philippines and Lesser Sunda Isls.;
        Andaman Isls.
    ISIS NUMBER: 5301405014024001001.

Tylonycteris robustula Thomas, 1915. Ann. Mag. Nat. Hist., ser. 8, 15:227.
    TYPE LOCALITY: Malaysia, Sarawak, Upper Sarawak.
    DISTRIBUTION: S. China to Sulawesi and Lesser Sunda Isls.
    COMMENT: Includes *malayana*; see Medway, 1969:37.
    ISIS NUMBER: 5301405014024002001.

Vespertilio Linnaeus, 1758. Syst. Nat., 10th ed., 1:31.
ISIS NUMBER: 5301405014025000000.

Vespertilio murinus Linnaeus, 1758. Syst. Nat., 10th ed., 1:32.
REVIEWED BY: E. Vernier (EV).
TYPE LOCALITY: Sweden.
DISTRIBUTION: Norway and Britain to Ussuri region (U.S.S.R.) and Afghanistan; Sinkiang, Kansu (China); N.E. China (SW).
ISIS NUMBER: 5301405014025001001.

Vespertilio orientalis Wallin, 1969. Zool. Bidr. Upps., 37:307.
TYPE LOCALITY: Japan, Honshu, Iwate pref., Nikoke dist., Ashiro, Horobe.
DISTRIBUTION: Honshu (Japan); Shansi to Szechwan and Fukien (China); Taiwan.

Vespertilio superans Thomas, 1899. Proc. Zool. Soc. Lond., 1898:770.
TYPE LOCALITY: China, Hupeh, Ichang, Sesalin.
DISTRIBUTION: C. Yangtze to upper Amur, Ussuri region and Korea; Kyushu, Honshu (Japan).
COMMENT: Includes *namiyei, aurijunctus,* and *motoyoshii;* see Corbet, 1978:58.
ISIS NUMBER: 5301405014025002001.

Family Mystacinidae
REVIEWED BY: K. F. Koopman (KFK).
ISIS NUMBER: 5301405015000000000.

Mystacina Gray, 1843. Voy. "Sulphur," Zool., p. 23.
ISIS NUMBER: 5301405015001000000.

Mystacina tuberculata Gray, 1843. Voy. "Sulphur," Zool., p. 23.
TYPE LOCALITY: New Zealand.
DISTRIBUTION: New Zealand; Stewart Isl.
ISIS NUMBER: 5301405015001001001.

Family Molossidae
REVIEWED BY: K. F. Koopman (KFK); J. Ramirez-Pulido (JRP) (Mexico).
COMMENT: Revised by Freeman, 1981:1–171.
ISIS NUMBER: 5301405016000000000.

Chaerephon Dobson, 1874. J. Asiat. Soc. Bengal, 43:144.
REVIEWED BY: F. R. Allison (FRA)(Australia); D. L. Harrison (DLH); J. I. Menzies (JIM)(New Guinea); A. C. Ziegler (ACZ)(New Guinea).
COMMENT: Formerly included in *Tadarida;* see Freeman, 1981:60, 133, 150.

Chaerephon aloysiisabaudiae (Festa, 1907). Bol. Mus. Zool. Anat. Comp. Univ. Torino, 22(546):1.
TYPE LOCALITY: Uganda, Toro.
DISTRIBUTION: Ghana; Gabon; Zaire; Uganda; perhaps Ethiopia.
ISIS NUMBER: 5301405016011004001 as *Tadarida aloysiisabaudiae.*

Chaerephon ansorgei (Thomas, 1913). Ann. Mag. Nat. Hist., ser. 8, 11:318.
TYPE LOCALITY: Angola, Malange.
DISTRIBUTION: Cameroun to Ethiopia, south to Angola and Mozambique.
COMMENT: Distinct from *bivittata;* see Eger and Peterson, 1979, Can. J. Zool., 57:1889.
ISIS NUMBER: 5301405016011005001 as *Tadarida ansorgei.*

Chaerephon bemmeleni (Jentink, 1879). Notes Leyden Mus., 1:125.
TYPE LOCALITY: Liberia.
DISTRIBUTION: Liberia; Cameroun; Sudan; Zaire; Uganda; Kenya; Tanzania.
COMMENT: Includes *cistura;* see Koopman, 1975, Bull. Am. Mus. Nat. Hist., 154:425.
ISIS NUMBER: 5301405016011009001 as *Tadarida bemmeleni.*

Chaerephon bivittata (Heuglin, 1861). Nova Acta Acad. Caes. Leop.-Carol., Halle, 29(8):413.
TYPE LOCALITY: Ethiopia, Eritrea, Keren.
DISTRIBUTION: Sudan; Ethiopia; Uganda; Kenya; Tanzania; Zambia; Zimbabwe; Mozambique.
ISIS NUMBER: 5301405016011010001 as *Tadarida bivittata.*

Chaerephon chapini J. A. Allen, 1917. Bull. Am. Mus. Nat. Hist., 37:461.
TYPE LOCALITY: Zaire, Oriental, Faradje.
DISTRIBUTION: Ethiopia; Zaire; Uganda; Angola; Namibia; Botswana; Zimbabwe.
Zambia (DLH).
ISIS NUMBER: 5301405016011012001 as *Tadarida chapini.*

Chaerephon gallagheri (Harrison, 1975). Mammalia, 39:313.
TYPE LOCALITY: Zaire, Kivu, 30 km S.W. Kindu, Scierie Forest (3° 10′ S. and 25° 49′ E.).
DISTRIBUTION: Zaire.

Chaerephon jobensis (Miller, 1902). Proc. Biol. Soc. Wash., 15:246.
TYPE LOCALITY: Indonesia, Irian Jaya, Tjenderawasih Div., Japen Isl., Ansus.
DISTRIBUTION: New Guinea; N. and C. Australia; Solomon Isls.; New Hebrides; Fiji.
Perhaps Bismarck Arch. (ACZ).
COMMENT: Listed as a subspecies of *plicata* by Laurie and Hill, 1954:63; but see Hill,
1961, Mammalia, 25:54-55, who included *solomonis* in this species.
ISIS NUMBER: 5301405016011019001 as *Tadarida jobensis.*

Chaerephon johorensis (Dobson, 1873). Proc. Asiat. Soc. Bengal, p. 22.
TYPE LOCALITY: Malaysia, Johore.
DISTRIBUTION: Malaya; Sumatra.
ISIS NUMBER: 5301405016011020001 as *Tadarida johorensis.*

Chaerephon major (Trouessart, 1897). Cat. Mamm. Viv. Foss., 1st ed., Part 1, p. 146.
TYPE LOCALITY: N. Sudan, 5th Cataract of the Nile.
DISTRIBUTION: Mali; Upper Volta; Ghana; Togo; Nigeria; Niger; Sudan; N.E. Zaire;
Uganda; Tanzania.
ISIS NUMBER: 5301405016011027001 as *Tadarida major.*

Chaerephon nigeriae Thomas, 1913. Ann. Mag. Nat. Hist., ser. 8, 11:319.
TYPE LOCALITY: Nigeria, Northern Region, Zaria Province.
DISTRIBUTION: Ghana and Niger to Saudi Arabia and Ethiopia, south to Namibia and
Zimbabwe. Botswana (DLH).
COMMENT: Saudi Arabia record by Nader and Kock, 1980, Senckenberg. Biol.,
60(1979):131-135.
ISIS NUMBER: 5301405016011032001 as *Tadarida nigeriae.*

Chaerephon plicata (Buchanan, 1800). Trans. Linn. Soc. Lond., 5:261.
REVIEWED BY: S. Wang (SW).
TYPE LOCALITY: India, Bengal.
DISTRIBUTION: India and Sri Lanka to S. China and Vietnam, southeast to Philippines,
Borneo and Bali; Hainan; Cocos Keeling Isl.
COMMENT: Includes *luzonus*; see Hill, 1961, Mammalia, 25:52.
ISIS NUMBER: 5301405016011036001 as *Tadarida plicata.*

Chaerephon pumila (Cretzschmar, 1826). Ruppell Atlas Reise Nordl. Afr., p. 69.
TYPE LOCALITY: Ethiopia, Eritrea, Massawa.
DISTRIBUTION: Senegal to Yemen, south to South Africa; Bioko; Pemba; Zanzibar;
Aldabra Isl.; Madagascar.
COMMENT: Includes *pusillus*; see Hayman and Hill, 1971, Part 2:64.
ISIS NUMBER: 5301405016011037001 as *Tadarida pumila.*

Chaerephon russata J. A. Allen, 1917. Bull. Am. Mus. Nat. Hist., 37:458.
TYPE LOCALITY: Zaire, Oriental, Medje.
DISTRIBUTION: Ghana; Cameroun; Zaire.
COMMENT: Considered a distinct species by Freeman, 1981:60, 109, but also see p. 152.
ISIS NUMBER: 5301405016011038001 as *Tadarida russata.*

Cheiromeles Horsfield, 1824. Zool. Res. Java, Part 8.
REVIEWED BY: J. I. Menzies (JIM).
ISIS NUMBER: 5301405016001000000.

Cheiromeles parvidens Miller and Hollister, 1921. Proc. Biol. Soc. Wash., 34:100.
TYPE LOCALITY: Indonesia, Sulawesi, Pinedapa.
DISTRIBUTION: S.E. Philippines; Sulawesi.
COMMENT: Probably conspecific with *torquatus* (KFK).
ISIS NUMBER: 5301405016001001001.

Cheiromeles torquatus Horsfield, 1824. Zool. Res. Java, Part 8.
TYPE LOCALITY: Malaysia, Penang.
DISTRIBUTION: Malaya, Sumatra, Java, Borneo, S.W. Philippines, and nearby small
islands.
ISIS NUMBER: 5301405016001002001.

Eumops Miller, 1906. Proc. Biol. Soc. Wash., 19:85.
REVIEWED BY: B. Villa-Ramirez (BVR).
COMMENT: Revised by Eger, 1977, Life Sci. Contrib. R. Ont. Mus., 110:1–69.
ISIS NUMBER: 5301405016002000000.

Eumops auripendulus (Shaw, 1800). Gen. Zool. Syst. Nat. Hist., 1(1):137.
TYPE LOCALITY: French Guiana.
DISTRIBUTION: Oaxaca and Chiapas (Mexico) to Peru, N. Argentina, Bolivia, Paraguay,
E. Brazil, and Trinidad.
COMMENT: Called *abrasus* in Hall and Kelson, 1959:211; but see Husson, 1962, Zool.
Meded. Rijks. Mus. Nat. Hist. Leiden, 58:243–246; Hall, 1981:248.
ISIS NUMBER: 5301405016002003001 as *E. auripendulus*.
5301405016002001001 as *E. abrasus*.

Eumops bonariensis (Peters, 1874). Monatsb. Preuss. Akad. Wiss. Berlin, p. 232.
TYPE LOCALITY: Argentina, Buenos Aires.
DISTRIBUTION: S. Veracruz and Tabasco (Mexico) to Bolivia, Paraguay, N. Argentina,
Uruguay, Brazil, and Guianas.
ISIS NUMBER: 5301405016002004001.

Eumops dabbenei Thomas, 1914. Ann. Mag. Nat. Hist., ser. 8, 13:481.
TYPE LOCALITY: Argentina, Chaco.
DISTRIBUTION: Colombia; Venezuela; Paraguay; Argentina.
COMMENT: Includes *mederai* (KFK).

Eumops glaucinus (Wagner, 1843). Arch. Naturgesch., 9(1):368.
TYPE LOCALITY: Brazil, Mato Grosso, Cuiaba.
DISTRIBUTION: Colima, Michoacan and Distrito Federal (Mexico) to Peru, Bolivia,
Paraguay, N. Argentina, Brazil, and Surinam; Jamaica; Cuba; Florida (U.S.A.).
COMMENT: Includes *floridanus*; see Eger, 1977, R. Ont. Mus. Life Sci. Contrib., 110:43.
ISIS NUMBER: 5301405016002005001.

Eumops hansae Sanborn, 1932. J. Mammal., 13:356.
TYPE LOCALITY: Brazil, Santa Catarina, Joinville, Colonia Hansa.
DISTRIBUTION: Costa Rica; Panama; Venezuela; Guyana; Brazil.
COMMENT: Includes *amazonicus*; see Gardner, *et al.*, 1970, J. Mammal., 51:727; Eger,
1977, R. Ont. Mus. Life Sci. Contrib., 110:44.
ISIS NUMBER: 5301405016002006001 as *E. hansae*.
5301405016002002001 as *E. amazonicus*.

Eumops maurus (Thomas, 1901). Ann. Mag. Nat. Hist., ser. 8, 7:141.
TYPE LOCALITY: Guyana, Kanuku Mtns.
DISTRIBUTION: Guyana; Surinam.
COMMENT: Includes *geijskesi*; see Eger, 1977, R. Ont. Mus. Life Sci. Contrib., 110:46–48.
ISIS NUMBER: 5301405016002007001.

Eumops perotis (Schinz, 1821). Thierreich, 1:870.
TYPE LOCALITY: Brazil, Rio de Janiero, Campos do Goita Cazes, Villa Sao Salvador.
DISTRIBUTION: San Francisco Bay and Texas (U.S.A.) to Zacatecas (Mexico); Colombia to
N. Argentina and E. Brazil; Cuba; perhaps N. Chile.

COMMENT: Includes *trumbulli;* see Koopman, 1978, Am. Mus. Novit., 2651:22; but also see Eger, 1977, R. Ont. Mus. Life Sci. Contrib., 110:53.
ISIS NUMBER: 5301405016002008001 as *E. perotis.*
5301405016002009001 as *E. trumbulli.*

Eumops underwoodi Goodwin, 1940. Am. Mus. Novit., 1075:2.
TYPE LOCALITY: Honduras, La Paz, 6 km N. Chinacla.
DISTRIBUTION: S. Arizona (U.S.A.) to Nicaragua.
ISIS NUMBER: 5301405016002010001.

Molossops Peters, 1865. Monatsb. Preuss. Akad. Wiss. Berlin, p. 575.
COMMENT: Includes *Cynomops* and *Neoplatymops* as subgenera; see Jones *et al.,* 1977, Occas. Pap. Mus. Texas Tech Univ., 47:28, and Freeman, 1981:133, respectively.
ISIS NUMBER: 5301405016003000000 as *Molossops.*
5301405016006000000 as *Neoplatymops.*

Molossops abrasus (Temminck, 1827). Monogr. Mamm., 1:232.
TYPE LOCALITY: "Brazil."
DISTRIBUTION: N. Argentina; Venezuela; Guyana; Surinam; Peru; Brazil; Bolivia; Paraguay.
COMMENT: Called *brachymeles* by Cabrera, 1958:118-119, and Freeman, 1981:155 (who included it in the subgenus *Cynomops);* but see Carter and Dolan, 1978, Spec. Publ. Mus. Texas Tech Univ., 15:84.
ISIS NUMBER: 5301405016003002001 as *M. brachymeles.*

Molossops aequatorianus Cabrera, 1917. Trab. Mus. Nac. Cienc. Nat. Zool., 31:20.
TYPE LOCALITY: Ecuador, Los Rios, Babahoyo.
DISTRIBUTION: Ecuador.
ISIS NUMBER: 5301405016003001001.

Molossops greenhalli (Goodwin, 1958). Am. Mus. Novit., 1877:3.
REVIEWED BY: B. Villa-Ramirez (BVR).
TYPE LOCALITY: Trinidad and Tobago, Trinidad, Port of Spain, Botanic Gardens.
DISTRIBUTION: Nayarit; Jalisco, Guerrero, and Oaxaca (Mexico) to Honduras; Costa Rica to Ecuador and Venezuela; Trinidad; Surinam; N.E. Brazil.
COMMENT: In subgenus *Cynomops;* see Freeman, 1981:155.
ISIS NUMBER: 5301405016003003001.

Molossops mattogrossensis Vieira, 1942. Argent. Zool. Sao Paulo, 3:430.
REVIEWED BY: M. R. Willig (MRW).
TYPE LOCALITY: Brazil, Mato Grosso, Juruena River, Sao Simao.
DISTRIBUTION: Guyana; Venezuela; C. and N.E. Brazil.
COMMENT: Listed as a subspecies of *Molossops temminckii* by Cabrera, 1958:117; but see Peterson, 1965, R. Ont. Mus. Life Sci. Contrib., 64:3-5, who considered *Neoplatymops* a distinct genus. Subgenus *Neoplatymops,* according to Freeman, 1981:155.
ISIS NUMBER: 5301405016006001001 as *Neoplatymops mattogrossensis.*

Molossops neglectus Williams and Genoways, 1980. Ann. Carnegie Mus., 49(25):489.
TYPE LOCALITY: Surinam, Surinam, 1 km S., 2 km E. Powaka (5° 25′ N, 55° 3′ W).
DISTRIBUTION: Known only from the type locality.

Molossops planirostris (Peters, 1865). Monatsb. Preuss. Akad. Wiss. Berlin, p. 575.
TYPE LOCALITY: French Guiana, Cayenne.
DISTRIBUTION: Panama to Peru, N. Argentina, Paraguay, Brazil and Guianas.
COMMENT: Includes *milleri* and *paranus;* see Koopman, 1978, Am. Mus. Novit., 2651:20; but also see Handley, 1976, Brigham Young Univ. Sci. Bull., Biol. Ser., 20(5):1-89; and Williams and Genoways, 1980, Ann. Carnegie Mus., 49:487-498, who regarded *paranus* as a distinct species. Subgenus *Cynomops* according to Freeman, 1981:155.

ISIS NUMBER: 5301405016003005001 as *M. planirostris.*
5301405016003004001 as *M. milleri.*

Molossops temminckii (Burmeister, 1854). Syst. Uebers. Thiere Bras., p. 72.
TYPE LOCALITY: Brazil, Minas Gerais, Lagoa Santa.
DISTRIBUTION: Colombia; Peru; Bolivia; S. Brazil; Paraguay; N. Argentina; Uruguay.
ISIS NUMBER: 5301405016003006001.

Molossus E. Geoffroy, 1805. Ann. Mus. Hist. Nat. Paris, 6:151.
REVIEWED BY: B. Villa-Ramirez (BVR).
ISIS NUMBER: 5301405016004000000.

Molossus ater E. Geoffroy, 1805. Bull. Sci. Soc. Philom. Paris, 3(96):279.
TYPE LOCALITY: French Guiana, Cayenne.
DISTRIBUTION: E. Tamaulipas and W. Sinaloa (Mexico) to Peru, N. Argentina,
Paraguay, Brazil and Guianas; Trinidad.
COMMENT: Called *rufus* by Hall and Kelson, 1959:214, and Cabrera, 1958:132; but see
Husson, 1962, Zool. Meded. Rijks. Mus. Nat. Hist. Leiden, 58:259–264; Hall,
1981:252. Includes *malagai*; see Jones, 1965, Proc. Biol. Soc. Wash., 78:93. Includes
macdougalli; see Jones *et al.*, 1977, Occas. Pap. Mus. Texas Tech Univ., 47:31, but
also see Hall, 1981:253, who placed it in *pretiosus.*
ISIS NUMBER: 5301405016004001001.

Molossus bondae J. A. Allen, 1904. Bull. Am. Mus. Nat. Hist., 20:228.
TYPE LOCALITY: Colombia, Magdalena, Bonda.
DISTRIBUTION: Honduras to Ecuador and Venezuela; Cozumel Isl. (Mexico).
COMMENT: May include *coibensis*, according to Freeman, 1981:157.
ISIS NUMBER: 5301405016004003001.

Molossus molossus (Pallas, 1766). Misc. Zool., p. 49–50.
TYPE LOCALITY: Martinique (Lesser Antilles), (France).
DISTRIBUTION: N. Sinaloa and S. Tamaulipas (Mexico) to Peru, N. Argentina, Uruguay,
Brazil and Guianas; Greater and Lesser Antilles; Margarita Isl.; Curacao; Bonaire;
Trinidad; Tobago.
COMMENT: Includes *fortis, milleri, debilis,* and *tropidorhynchus*; see Varona, 1974:42.
Called *major* by Hall and Kelson, 1959:216, and Cabrera, 1958:130; but see
Husson, 1962, Zool. Meded. Rijks. Mus. Nat. Hist. Leiden, 58:251–259. Hall,
1981:255, included *coibensis* in *molossus.* Antillean populations reviewed by
Genoways, *et al.*, 1981, Ann. Carnegie Mus., 50:475:492. Includes *burnesi*; see
Freeman, 1981:158.
ISIS NUMBER: 5301405016004004001 as *M. molossus.*
5301405016004006001 as *M. tropidorhynchus.*

Molossus pretiosus Miller, 1902. Proc. Acad. Nat. Sci. Phila., p. 396.
TYPE LOCALITY: Venezuela, Caracas, LaGuaira.
DISTRIBUTION: Guerrero, Oaxaca (Mexico); Nicaragua to Venezuela and Guyana.
COMMENT: Listed as a synonym of *rufus* by Cabrera, 1958:132; but see Jones *et al.*,
1977, Occas. Pap. Mus. Texas Tech Univ., 47:31.

Molossus sinaloae J. A. Allen, 1906. Bull. Am. Mus. Nat. Hist., 22:236.
TYPE LOCALITY: Mexico, Sinaloa, Esquinapa.
DISTRIBUTION: S. Sinaloa (Mexico) to Costa Rica.
COMMENT: Formerly included *trinitatus*; see Freeman, 1981:158.
ISIS NUMBER: 5301405016004005001.

Molossus trinitatus Goodwin, 1959, Am. Mus. Novit., 1967:1.
TYPE LOCALITY: Trinidad and Tobago, Trinidad, Port of Spain, Belmont.
DISTRIBUTION: S. Costa Rica to Surinam; Trinidad.
COMMENT: Included in *sinaloae* by Ojasti and Linares, 1971, Acta Biol. Venez.,
7:421–441; Hall, 1981:254; but see Freeman, 1981:158.

Mops Lesson, 1842. Nouv. Tabl. Regn. Anim. Mammal., p. 18.
REVIEWED BY: D. L. Harrison (DLH); J. I. Menzies (JIM).

COMMENT: Formerly included in *Tadarida;* includes *Philippinopterus* and *Xiphonycteris;* see Freeman, 1981:158.
ISIS NUMBER: 5301405016012000000 as *Xiphonycteris.*

Mops brachypterus (Peters, 1852). Reise nach Mossambique, Saugethiere, p. 59.
TYPE LOCALITY: Mozambique, Mozambique Isl. (15° S. and 40° 42' E.).
DISTRIBUTION: Uganda; Kenya; Tanzania; Mozambique; Zanzibar.
COMMENT: May be conspecific with (and an older name than) *leonis* or *thersites* (KFK).

Mops condylurus (A. Smith, 1833). S. Afr. Quart. J., 1:54.
TYPE LOCALITY: South Africa, Natal, Durban.
DISTRIBUTION: Senegal to Somalia and South Africa; Madagascar.
ISIS NUMBER: 5301405016011013001 as *Tadarida condylura.*

Mops congicus J. A. Allen, 1917. Bull. Am. Mus. Nat. Hist., 37:467.
TYPE LOCALITY: Zaire, Oriental, Medje.
DISTRIBUTION: Ghana; Nigeria; Cameroun; Zaire; Uganda.
COMMENT: Does not include *trevori;* see Peterson, 1972, R. Ont. Mus. Life Sci. Contrib., 85:2–12.
ISIS NUMBER: 5301405016011014001 as *Tadarida congica.*

Mops demonstrator (Thomas, 1903). Ann. Mag. Nat. Hist., ser. 7, 12:504.
TYPE LOCALITY: Sudan, Equatoria, Mongalla.
DISTRIBUTION: Upper Volta; Sudan; Zaire; Uganda.
ISIS NUMBER: 5301405016011015001 as *Tadarida demonstrator.*

Mops leonis (Thomas, 1908). Ann. Mag. Nat. Hist., ser. 8, 2:373.
TYPE LOCALITY: Sierra Leone.
DISTRIBUTION: Sierra Leone to Bioko, Gabon, E. Zaire and Uganda.
ISIS NUMBER: 5301405016011023001 as *Tadarida leonis.*

Mops midas (Sundevall, 1843). Kongl. Svenska Vet.-Akad. Handl. Stockholm, for 1842, p. 207.
TYPE LOCALITY: Sudan, Blue Nile, White Nile River, West bank, Jebel el Funj.
DISTRIBUTION: Senegal to Yemen, south to South Africa; Madagascar.
ISIS NUMBER: 5301405016011028001 as *Tadarida midas.*

Mops mops (De Blainville, 1840). Osteogr. Vespertilio, p. 101.
TYPE LOCALITY: Indonesia, Sumatra.
DISTRIBUTION: Malaya; Sumatra; Borneo; perhaps Java.
ISIS NUMBER: 5301405016011030001 as *Tadarida mops.*

Mops nanulus J. A. Allen, 1917. Bull. Am. Mus. Nat. Hist., 37:477.
TYPE LOCALITY: Zaire, Oriental, Niangara.
DISTRIBUTION: Sierra Leone; Ghana; Nigeria; Ethiopia; Zaire; Uganda; Kenya.
COMMENT: Probably only a subspecies of *spurrelli* (KFK).
ISIS NUMBER: 5301405016011031001 as *Tadarida nanula.*

Mops niangarae J. A. Allen, 1917. Bull. Am. Mus. Nat. Hist., 37:468.
TYPE LOCALITY: Zaire, Niangara.
DISTRIBUTION: Known only from the holotype.
COMMENT: Peterson, 1972, R. Ont. Mus. Life Sci. Contrib., 85:4–12, included this species in *trevori;* Hayman and Hill, 1971, Part 2, listed it as a subspecies of *T. congica* (=*Mops congicus*). Freeman, 1981:111, 159, considered it a distinct species pending collection of additional specimens.

Mops niveiventer Cabrera and Ruxton, 1926. Ann. Mag. Nat. Hist., ser. 9, 17:594.
TYPE LOCALITY: Zaire, Kasai Occidental, Luluabourg.
DISTRIBUTION: Zaire; Rwanda; Burundi; Tanzania; Angola; Zambia; Mozambique.
COMMENT: Probably a subspecies of *demonstrator;* records from N. Botswana and Madagascar definitely *condylurus* (KFK).
ISIS NUMBER: 5301405016011033001 as *Tadarida niveiventer.*

Mops petersoni (El Rayah, 1981). R. Ont. Mus. Life Sci. Occas. Pap., 36:3.
TYPE LOCALITY: Cameroun, 15 km S. Kumba (4°39'N., 9°26'E.).
DISTRIBUTION: Cameroun and Ghana.
COMMENT: Described in *Tadarida* (*Xiphonycteris*), but see comments under *Tadarida* and *Mops.*

Mops sarasinorum (Meyer, 1899). Abh. Zool. Anthrop.-Ethnology. Mus. Dresden, 7(7):15.
TYPE LOCALITY: Indonesia, Sulawesi, Batulappa (North of Lake Tempe).
DISTRIBUTION: Sulawesi and adjacent small islands; Philippines.
COMMENT: Includes *lanei* (formerly included in *Philippinopterus*); see Freeman,
1981:160. Hill, 1961, Mammalia, 25:42–43, considered *lanei* a distinct species.
ISIS NUMBER: 5301405016011039001 as *Tadarida sarasinorum*.

Mops spurrelli (Dollman, 1911). Ann. Mag. Nat. Hist., ser. 8, 7:211.
TYPE LOCALITY: Ghana, Bibianaha.
DISTRIBUTION: Ivory Coast; Ghana; Togo; Benin; Rio Muni; Zaire; Bioko.
COMMENT: Transferred from *Xiphonycteris*; see Koopman, 1975, Bull. Am. Mus. Nat.
Hist., 154:419–421; Freeman, 1981:160.
ISIS NUMBER: 5301405016012001001 as *Xiphonycteris spurrelli*.

Mops thersites (Thomas, 1903). Ann. Mag. Nat. Hist., ser. 7, 12:634.
TYPE LOCALITY: Cameroun, Efulen.
DISTRIBUTION: Sierra Leone; Liberia; Ivory Coast; Ghana; Nigeria; Cameroun; Rio
Muni; Zaire; Rwanda. Bioko; perhaps Mozambique and Zanzibar (DLH).
ISIS NUMBER: 5301405016011042001 as *Tadarida thersites*.

Mops trevori J. A. Allen, 1917. Bull. Am. Mus. Nat. Hist., 37:468.
TYPE LOCALITY: Zaire, Oriental, Faradje.
DISTRIBUTION: N.W. Zaire; Uganda.
COMMENT: Formerly included *niangarae*; see Freeman, 1981:111, 159.
ISIS NUMBER: 5301405016011044001 as *Tadarida trevori*.

Mormopterus Peters, 1865. Monatsb. Preuss. Akad. Wiss. Berlin, p. 258.
REVIEWED BY: F. R. Allison (FRA)(Australia); D. L. Harrison (DLH); J. I. Menzies
(JIM)(New Guinea); A. C. Ziegler (ACZ)(New Guinea).
COMMENT: Formerly included in *Tadarida*; see Koopman, 1975, Bull. Am. Mus. Nat.
Hist., 154:419–421; but also see Freeman, 1981:133, 160, who included *Sauromys*
and *Platymops* as subgenera.
ISIS NUMBER: 5301405016008000000 as *Platymops*.
5301405016010000000 as *Sauromys*.

Mormopterus acetabulosus (Hermann, 1804). Observ. Zool., p. 19.
TYPE LOCALITY: Mauritius, Port Louis.
DISTRIBUTION: Ethiopia; Reunion; Mauritius; Madagascar; South Africa.
ISIS NUMBER: 5301405016011001001 as *Tadarida acetabulosus*.

Mormopterus beccarii Peters, 1881. Monatsb. Preuss. Akad. Wiss. Berlin, p. 484.
TYPE LOCALITY: Indonesia, Molucca Isls., Amboina Isl.
DISTRIBUTION: Molucca Isls.; New Guinea; adjacent small islands; N. Australia (FRA).
COMMENT: Includes *astrolabiensis*; see Freeman, 1981:160.
ISIS NUMBER: 5301405016011008001 as *Tadarida beccarii*.

Mormopterus doriae K. Andersen, 1907. Ann. Mus. Civ. Stor. Nat. Genova, 3(38):42.
TYPE LOCALITY: Indonesia, Sumatra, Deli, Soekaranda.
DISTRIBUTION: Sumatra.

Mormopterus jugularis (Peters, 1865). *In* Sclater, Proc. Zool. Soc. Lond., 1865:468.
TYPE LOCALITY: Madagascar, Antananarivo.
DISTRIBUTION: Madagascar.
ISIS NUMBER: 5301405016011021001 as *Tadarida jugularis*.

Mormopterus kalinowskii (Thomas, 1893). Proc. Zool. Soc. Lond., 1893:334.
TYPE LOCALITY: "Central Peru."
DISTRIBUTION: Peru; N. Chile.

Mormopterus loriae (Thomas, 1897). Ann. Mus. Civ. Stor. Nat. Genova, 18:609.
TYPE LOCALITY: Papua New Guinea, Central Prov., Kamali (Kemp Welch River mouth).
DISTRIBUTION: New Guinea; N. Australia.
COMMENT: Probably conspecific with *planiceps*; see Hill, 1961, Mammalia, 25:45–46; but
see also Felten, 1964, Senckenberg. Biol., 45:2–8; Freeman, 1981:161.
ISIS NUMBER: 5301405016011025001 as *Tadarida loriae*.

Mormopterus minutus (Miller, 1899). Bull. Am. Mus. Nat. Hist., 12:173.
TYPE LOCALITY: Cuba, Las Villas, Trinidad, San Pablo.
DISTRIBUTION: Cuba.

Mormopterus norfolkensis (Gray, 1839). Ann. Mag. Nat. Hist., ser. 11, 4:7.
TYPE LOCALITY: Norfolk Isl. (east of Australia), (uncertain).
DISTRIBUTION: Norfolk Isl.? S.E. Queensland, E. New South Wales, Australia (FRA).
COMMENT: There is considerable doubt as to the status of this species (KFK); see Hill,
    1961, Mammalia, 25:44.
ISIS NUMBER: 5301405016011034001 as *Tadarida norfolkensis*.

Mormopterus petrophilus (Roberts, 1917). Ann. Transvaal Mus., 6:4.
TYPE LOCALITY: South Africa, Transvaal, near Rustenburg, Bleskap.
DISTRIBUTION: South Africa; Namibia; Botswana; Zimbabwe; Mozambique.
COMMENT: Formerly included in *Sauromys* by Peterson, 1965, Life Sci. Contrib. R. Ont.
    Mus., 64:12; but also see Freeman, 1981:133, 161, who considered *Sauromys* a
    subgenus.
ISIS NUMBER: 5301405016010001001 as *Sauromys petrophilus*.

Mormopterus phrudus (Handley, 1956). Proc. Biol. Soc. Wash., 69:197.
TYPE LOCALITY: Peru, Cuzco, Machu Picchu, Urubamba River, San Miguel Bridge.
DISTRIBUTION: Peru.

Mormopterus planiceps (Peters, 1866). Monatsb. Preuss. Akad. Wiss. Berlin, p. 23.
TYPE LOCALITY: Australia. Probably New South Wales, Sydney.
DISTRIBUTION: S. and C. Australia.
COMMENT: See Iredale and Troughton, 1934, Mem. Aust. Mus., 6:101 for discussion of
    the type locality. May include *loriae*; see Hill, 1961, Mammalia, 25:45–46.
ISIS NUMBER: 5301405016011035001 as *Tadarida planiceps*.

Mormopterus setiger Peters, 1878. Monatsb. Preuss. Akad. Wiss. Berlin, p. 196.
TYPE LOCALITY: Kenya, Taita.
DISTRIBUTION: S. Sudan; Ethiopia; Kenya.
COMMENT: Formerly included in *Platymops* by Harrison and Fleetwood, 1960, Durban
    Mus. Novit., 15:277–278. In subgenus *Platymops*, see Freeman, 1981:133, 161.
ISIS NUMBER: 5301405016008001001 as *Platymops setiger*.

Myopterus E. Geoffroy, 1818. Descrip. de L'Egypte, 2:113.
COMMENT: Includes *Eomops*; see Hayman and Hill, 1971, Part 2:56.
ISIS NUMBER: 5301405016005000000.

Myopterus albatus Thomas, 1915. Ann. Mag. Nat. Hist., ser. 8, 16:469.
TYPE LOCALITY: Zaire, Oriental, Uele River.
DISTRIBUTION: Ivory Coast; N.E. Zaire.
COMMENT: Probably a subspecies of *daubentonii*; Freeman, 1981:162 listed *daubentonii*
    under *albatus*; but see Hill, 1969, Mammalia, 33:727–729.
ISIS NUMBER: 5301405016005001001.

Myopterus daubentonii Desmarest, 1820. Encyclop. Method. Mamm., 1:132.
TYPE LOCALITY: Senegal.
DISTRIBUTION: Senegal.
COMMENT: Holotype lost; see comments under *albatus*.
ISIS NUMBER: 5301405016005002001.

Myopterus whitleyi (Scharff, 1900). Ann. Mag. Nat. Hist., ser. 7, 6:569.
TYPE LOCALITY: Nigeria, Mid-Western Region, Benin City.
DISTRIBUTION: Ghana; Nigeria; Cameroun; Zaire; Uganda.
ISIS NUMBER: 5301405016005003001.

Nyctinomops Miller, 1902. Proc. Acad. Nat. Sci. Phila., 54:393.
REVIEWED BY: D. L. Harrison (DLH); B. Villa-Ramierz (BVR).
COMMENT: Formerly included in *Tadarida*; see Hall 1981:243; but also see Freeman,
    1981, Fieldiana Zool., N.S., 7:124.

Nyctinomops aurispinosus (Peale, 1848). Mammalia and Ornothology, *in* U.S. Expl. Exped., 8:21.
TYPE LOCALITY: Brazil, Rio Grande do Norte, 100 mi. (161 km) off Cape Sao Roque.
DISTRIBUTION: S. Sonora and S. Tamaulipas (Mexico) to Peru and Brazil.
COMMENT: Includes *similis*; see Aellen, 1970, Rev. Suisse Zool., 77:27.
ISIS NUMBER: 5301405016011006001 as *Tadarida aurispinosa.*
5301405016011040001 as *Tadarida similis.*

Nyctinomops femorosaccus (Merriam, 1889). N. Am. Fauna, 2:23.
TYPE LOCALITY: U.S.A., California, Riverside Co., Palm Springs.
DISTRIBUTION: Oaxaca and Guerrero (Mexico) to S. New Mexico, S. Arizona, S. California (U.S.A.) and Baja California (Mexico).
ISIS NUMBER: 5301405016011016001 as *Tadarida femorosacca.*

Nyctinomops laticaudatus (E. Geoffroy, 1805). Ann. Mus. Hist. Nat. Paris, 6:156.
TYPE LOCALITY: Paraguay, Asuncion.
DISTRIBUTION: Tamaulipas to E. Oaxaca and C. Guerrero (Mexico) to N.W. Peru, Paraguay, and Brazil; Trinidad; Cuba. N. Argentina (DLH).
COMMENT: Includes *yucatanicus, europs,* and *gracilis;* see Silva-Taboada and Koopman, 1964, Am. Mus. Novit., 2174:3, 4; Freeman, 1981:162.
ISIS NUMBER: 5301405016011022001 as *Tadarida laticaudata.*
5301405016011018001 as *Tadarida gracilis.*
5301405016011045001 as *Tadarida yucatanica (sic).*

Nyctinomops macrotis (Gray, 1839). Ann. Mag. Nat. Hist., ser. 11, 4:5.
TYPE LOCALITY: Cuba.
DISTRIBUTION: S.W. British Columbia; C. Iowa; S. California to S.W. Kansas (U.S.A.), south to Uruguay; Cuba; Jamaica; Hispaniola.
COMMENT: Called *Tadarida molossa* by Hall and Kelson, 1959:208, and *Tadarida macrotis* by Hall, 1981:245; but see Husson, 1962, Zool. Meded. Rijks. Mus. Nat. Hist. Leiden, 58:256–259. May be a monotypic species, but it needs systematic revision (BVR).
ISIS NUMBER: 5301405016011026001 as *Tadarida macrotis.*
5301405016011029001 as *Tadarida molossus (sic).*

Otomops Thomas, 1913. J. Bombay Nat. Hist. Soc., 22:91.
REVIEWED BY: J. I. Menzies (JIM)(New Guinea); A. C. Ziegler (ACZ)(New Guinea).
ISIS NUMBER: 5301405016007000000.

Otomops formosus Chasen, 1939. Treubia, 17:186.
TYPE LOCALITY: Indonesia, Java, Tjibadak.
DISTRIBUTION: Java.
ISIS NUMBER: 5301405016007001001.

Otomops martiensseni (Matschie, 1897). Arch. Naturgesch., 63(1):84.
TYPE LOCALITY: Tanzania, Tanga, Magrotto Plantation.
DISTRIBUTION: Djibouti to Angola and South Africa; Madagascar.
ISIS NUMBER: 5301405016007002001.

Otomops papuensis Lawrence, 1948. J. Mammal., 29:413.
TYPE LOCALITY: Papua New Guinea, Gulf Prov., Vailala River.
DISTRIBUTION: S.E. New Guinea.
ISIS NUMBER: 5301405016007003001.

Otomops secundus Hayman, 1952. *In* Laurie, Bull. Br. Mus. (Nat. Hist.), Zool., 1:314.
TYPE LOCALITY: Papua New Guinea, Madang Prov., Tapu.
DISTRIBUTION: N.E. New Guinea.
COMMENT: May be a subspecies of *papuensis,* according to the describer (ACZ).
ISIS NUMBER: 5301405016007004001.

Otomops wroughtoni Thomas, 1913. J. Bombay Nat. Hist. Soc., 22:87.
TYPE LOCALITY: India, Mysore, Kanara, near Talewadi.
DISTRIBUTION: S. India.
ISIS NUMBER: 5301405016007005001.

Promops Gervais, 1855. Mammiferes *in* Castelnau, Exped. Partes Cent. Amer. Sud., Part 7, p. 58.
  ISIS NUMBER: 5301405016009000000.

Promops centralis Thomas, 1915. Ann. Mag. Nat. Hist., ser. 8, 16:62.
  REVIEWED BY: B. Villa-Ramirez (BVR).
  TYPE LOCALITY: Mexico, Yucatan.
  DISTRIBUTION: Jalisco (Mexico) to W. Peru, N. Argentina, Venezuela and Surinam; Trinidad.
  COMMENT: Includes *occultus* and *davisoni;* see Ojasti and Linares, 1971, Acta Biol. Venez., 7:433–434.
  ISIS NUMBER: 5301405016009001001.

Promops nasutus (Spix, 1823). Sim. Vespert. Brasil., p. 58.
  TYPE LOCALITY: Brazil, Bahia, Sao Francisco River.
  DISTRIBUTION: Venezuela; Trinidad; Surinam; Brazil; Paraguay; N. Argentina; Peru.
  COMMENT: Includes *pamana;* see Goodwin, 1962, Am. Mus. Novit., 2080:11–18.
  ISIS NUMBER: 5301405016009002001 as *P. nasutus.*
            5301405016009003001 as *P. pamana.*

Tadarida Rafinesque, 1814. Precis Som., p. 55.
  REVIEWED BY: F. R. Allison (FRA)(Australia); D. L. Harrison (DLH); J. I. Menzies (JIM)(New Guinea); B. Villa-Ramirez (BVR); A. C. Ziegler (ACZ)(New Guinea).
  COMMENT: Formerly included *Chaerephon, Mops, Mormopterus, and Nyctinomops;* see Freeman, 1981:133.
  ISIS NUMBER: 5301405016011000000.

Tadarida aegyptiaca (E. Geoffroy, 1818). Descrip. de L'Egypte, 2:128.
  TYPE LOCALITY: Egypt, Giza.
  DISTRIBUTION: South Africa to Nigeria, Algeria, and Egypt to Yemen and Oman, east to India and Sri Lanka.
  COMMENT: Includes *tragata;* see Corbet, 1978:63; Freeman, 1981:165. Records from Saudi Arabia and Oman by Jennings, 1979, J. Saudi Ar. Nat. Hist. Soc., 24.
  ISIS NUMBER: 5301405016011002001 as *T. aegyptiaca.*
            5301405016011043001 as *T. tragata.*

Tadarida africana (Dobson, 1876). Ann. Mag. Nat. Hist., ser. 4, 17:348.
  TYPE LOCALITY: South Africa, Transvaal.
  DISTRIBUTION: Ethiopia; S. Sudan; E. Zaire; Kenya; Malawi; Mozambique; South Africa. Tanzania (DLH).
  COMMENT: *T. ventralis* may be the oldest name for this species; see Kock, 1975, Stuttgarter Beitrage Naturk., ser. A (Biol.), 272:1–7.
  ISIS NUMBER: 5301405016011003001.

Tadarida australis (Gray, 1839). Mag. Zool. Bot., 2:501.
  TYPE LOCALITY: Australia, New South Wales.
  DISTRIBUTION: S. and C. Australia.
  COMMENT: This species does not occur in New Guinea or eastward; see McKean and Calaby, 1968, Mammalia, 32:377. May include *kuboriensis* (KFK).
  ISIS NUMBER: 5301405016011007001.

Tadarida brasiliensis (I. Geoffroy, 1824). Ann. Sci. Nat. Zool., 1:343.
  TYPE LOCALITY: Brazil, Parana, Curitiba.
  DISTRIBUTION: S. Brazil, C. Argentina, and C. Chile to S. Oregon, S. Nebraska and South Carolina (U.S.A.); Greater and Lesser Antilles.
  ISIS NUMBER: 5301405016011011001 as *T. brasillensis (sic).*

Tadarida espiritosantensis (Ruschi, 1951). Bol. Mus. Biol. Prof. Mello-Leitao Zool., Santa Teresa, Espirito Santo, 7:19.
  TYPE LOCALITY: Brazil, Espirito Santo, Santa Teresa, Treis Barras.
  DISTRIBUTION: Brazil.
  COMMENT: Mentioned in Pine and Ruschi, 1976, An. Inst. Biol. Univ. Nac. Auton. Mex., 47:183–196; listed as a species by Freeman, 1981:166. Specific status extremely dubious (KFK).

Tadarida fulminans (Thomas, 1903). Ann. Mag. Nat. Hist., ser. 7, 12:501.
  TYPE LOCALITY: Madagascar, Betsilio, Fianarantsoa.
  DISTRIBUTION: E. Zaire; Rwanda; Kenya; Tanzania; Zambia; Zimbabwe; Madagascar.
  ISIS NUMBER: 5301405016011017001.

Tadarida kuboriensis McKean and Calaby, 1968. Mammalia, 32:375.
  TYPE LOCALITY: Papua New Guinea, Western Highlands Prov., Kubor Range, Minj-
    Nona Divide (6° 2' S. and 144° 45' E.), ca. 2750 m.
  DISTRIBUTION: New Guinea.
  COMMENT: Probably a subspecies of *australis*, in which it was formerly included (KFK).

Tadarida lobata (Thomas, 1891). Ann. Mag. Nat. Hist., ser. 6, 7:303.
  TYPE LOCALITY: Kenya, West Pokot, Turkwell Gorge.
  DISTRIBUTION: Kenya; Zimbabwe.
  ISIS NUMBER: 5301405016011024001.

Tadarida teniotis (Rafinesque, 1814). Precis Som., p. 12.
  REVIEWED BY: E. Vernier (EV).
  TYPE LOCALITY: Italy, Sicily.
  DISTRIBUTION: Portugal and Morocco to Japan, S. China, and Taiwan; Madeira and
    Canary Isls.
  ISIS NUMBER: 5301405016011041001.

ORDER PRIMATES

PROTECTED STATUS: CITES - Appendix II as Order Primates (all species not listed in Appendix I).

ISIS NUMBER: 5301406000000000000.

Family Cheirogaleidae

REVIEWED BY: C. P. Groves (CPG); C. A. Hill (CAH).

COMMENT: Formerly included in Lemuridae. For status of this taxon, see Rumpler, 1975:25–40.

PROTECTED STATUS: U.S. ESA - Endangered as Cheirogaleidae.

Allocebus Petter-Rousseaux and Petter, 1967. Mammalia, 31:574.

COMMENT: Previously included in *Cheirogaleus*, but very distinct (CPG).

PROTECTED STATUS: CITES - Appendix I as *Allocebus* spp. and U.S. ESA - Endangered as Cheirogaleidae and *Allocebus* spp.

Allocebus trichotis (Gunther, 1875). Proc. Zool. Soc. Lond., 1875:78.

TYPE LOCALITY: Madagascar, between Tamatave and Morondava.

DISTRIBUTION: E. Madagascar, vicinity of Morondava Bay.

PROTECTED STATUS: CITES - Appendix I as *Allocebus* spp. and U.S. ESA - Endangered as Cheirogaleidae and *Allocebus* spp.

ISIS NUMBER: 5301406001004004001 as *Cheirogaleus trichotis*.

Cheirogaleus E. Geoffroy, 1812. Ann. Mus. Hist. Nat. Paris, 19:172.

COMMENT: Revised by Petter *et al.*, 1977, Faune de Madagascar, 44:27–79.

PROTECTED STATUS: CITES - Appendix I as *Cheirogaleus* spp. and U.S. ESA - Endangered as Cheirogaleidae and *Cheirogaleus* spp.

ISIS NUMBER: 5301406001004000000.

Cheirogaleus major E. Geoffroy, 1812. Ann. Mus. Hist. Nat. Paris, 19:172.

TYPE LOCALITY: Madagascar, Fort Dauphin.

DISTRIBUTION: E. Madagascar.

PROTECTED STATUS: CITES - Appendix I as *Cheirogaleus* spp. and U.S. ESA - Endangered as Cheirogaleidae and *Cheirogaleus* spp.

ISIS NUMBER: 5301406001004002001.

Cheirogaleus medius E. Geoffroy, 1812. Ann. Mus. Hist. Nat. Paris, 19:172.

TYPE LOCALITY: Madagascar, Fort Dauphin.

DISTRIBUTION: W. and S. Madagascar.

PROTECTED STATUS: CITES - Appendix I as *Cheirogaleus* spp. and U.S. ESA - Endangered as Cheirogaleidae and *Cheirogaleus* spp.

ISIS NUMBER: 5301406001004003001.

Microcebus E. Geoffroy, 1834. Cours Hist. Nat. Mamm., lecon 11, 1828:24.

COMMENT: Revised by Petter *et al.*, 1977, Faune de Madagascar, 44:27–79.

PROTECTED STATUS: CITES - Appendix I as *Microcebus* spp. and U.S. ESA - Endangered as Cheirogaleidae and *Microcebus* spp.

ISIS NUMBER: 5301406001005000000.

Microcebus coquereli A. Grandidier, 1867. Rev. Mag. Zool. Paris, ser. 2, 19:85.

TYPE LOCALITY: Madagascar, Morondava.

DISTRIBUTION: W. Madagascar.

PROTECTED STATUS: CITES - Appendix I as *Microcebus* spp. and U.S. ESA - Endangered as Cheirogaleidae and *Microcebus* spp.

ISIS NUMBER: 5301406001005001001.

Microcebus murinus (J. F. Miller, 1777). Cimelia Physica, p. 25.

TYPE LOCALITY: Madagascar.

DISTRIBUTION: W. and S. Madagascar.

PROTECTED STATUS: CITES - Appendix I as *Microcebus* spp. and U.S. ESA - Endangered as Cheirogaleidae and *Microcebus* spp.

ISIS NUMBER: 5301406001005002001.

Microcebus rufus E. Geoffroy, 1834. Cours Hist. Nat. Mamm., lecon 11, 1828:24.
TYPE LOCALITY: Madagascar.
DISTRIBUTION: E. and N. Madagascar.
COMMENT: Separated from *murinus* by Petter, *et al.*, 1977, Faune de Madagascar, 44:30.
PROTECTED STATUS: CITES - Appendix I as *Microcebus* spp. and U.S. ESA - Endangered as Cheirogaleidae and *Microcebus* spp.

Phaner Gray, 1870. Cat. Monkeys, Lemurs, and Fruit-eating Bats Br. Mus., p. 135.
PROTECTED STATUS: CITES - Appendix I as *Phaner* spp. and U.S. ESA - Endangered as Cheirogaleidae and *Phaner* spp.

Phaner furcifer (Blainville, 1841). Osteogr. Mamm., Primates, p. 35.
TYPE LOCALITY: Madagascar, Morondava.
DISTRIBUTION: W., N.E., and extreme S.E. Madagascar.
PROTECTED STATUS: CITES - Appendix I as *Phaner* spp. and U.S. ESA - Endangered as Cheirogaleidae and *Phaner* spp.
ISIS NUMBER: 5301406001004001001 as *Cheirogaleus furcifer*.

Family Lemuridae
REVIEWED BY: C. P. Groves (CPG); C. A. Hill (CAH).
COMMENT: Reviewed by Petter *et al.*, 1977, Faune de Madagascar, 44:1–543.
PROTECTED STATUS: U.S. ESA - Endangered as Lemuridae.
ISIS NUMBER: 5301406001000000000.

Hapalemur I. Geoffroy, 1851. L'Inst. Paris, 19(929):341.
PROTECTED STATUS: CITES - Appendix I as *Hapalemur* spp. and U.S. ESA - Endangered as Lemuridae and *Hapalemur* spp.
ISIS NUMBER: 5301406001001000000.

Hapalemur griseus (Link, 1795). Beytr. Naturg., 1:65.
TYPE LOCALITY: Madagascar.
DISTRIBUTION: E., N.W., and Antsalova Dist. of W. Madagascar.
PROTECTED STATUS: CITES - Appendix I as *Hapalemur* spp. and U.S. ESA - Endangered as Lemuridae and *Hapalemur* spp.
ISIS NUMBER: 5301406001001001001.

Hapalemur simus Gray, 1870. Cat. Monkeys, Lemurs, and Fruit-eating Bats Br. Mus., p. 133.
TYPE LOCALITY: Madagascar.
DISTRIBUTION: E. Madagascar, inland from Mananjary.
PROTECTED STATUS: CITES - Appendix I as *Hapalemur* spp. and U.S. ESA - Endangered as Lemuridae and *Hapalemur* spp.
ISIS NUMBER: 5301406001001002001.

Lemur Linnaeus, 1758. Syst. Nat., 10th ed., 1:24.
COMMENT: Revised by Petter *et al.*, 1977, Faune de Madagascar, p. 128–213.
PROTECTED STATUS: CITES - Appendix I as *Lemur* spp. and U.S. ESA - Endangered as Lemuridae and *Lemur* spp.
ISIS NUMBER: 5301406001002000000.

Lemur catta Linnaeus, 1758. Syst. Nat., 10th ed., 1:30.
TYPE LOCALITY: Madagascar.
DISTRIBUTION: S. Madagascar.
PROTECTED STATUS: CITES - Appendix I as *Lemur* spp. and U.S. ESA - Endangered as Lemuridae and *Lemur* spp.
ISIS NUMBER: 5301406001002001001.

Lemur coronatus Gray, 1842. Ann. Mag. Nat. Hist., 10:257.
TYPE LOCALITY: Madagascar.
DISTRIBUTION: Mt. Ambre (N. Madagascar).
COMMENT: Separated from *mongoz* by Petter, *et al.*, 1977, Faune de Madagascar, p. 151.
PROTECTED STATUS: CITES - Appendix I as *Lemur* spp. and U.S. ESA - Endangered as Lemuridae and *Lemur* spp.

Lemur fulvus E. Geoffroy, 1796. Mag. Encyclop., 1:47.
    TYPE LOCALITY: Madagascar, Tamatave.
    DISTRIBUTION: Coastal Madagascar, except extreme south; Mayotte (Comoro Isls.).
    COMMENT: Sympatric with *macaco;* see Tattersall, 1976, Anthropol. Pap. Am. Mus. Nat.
        Hist., 53:255–262. Includes *rufus, albifrons, collaris, sanfordi;* see Petter and Petter,
        1977, Part 3.1:6; some of these may, however, prove to be distinct species (CPG).
    PROTECTED STATUS: CITES - Appendix I as *Lemur* spp. and U.S. ESA - Endangered as
        Lemuridae and *Lemur* spp.

Lemur macaco Linnaeus, 1766. Syst. Nat., 12th ed., 1:34.
    TYPE LOCALITY: Madagascar.
    DISTRIBUTION: Nosi Be and N.W. Madagascar.
    PROTECTED STATUS: CITES - Appendix I as *Lemur* spp. and U.S. ESA - Endangered as
        Lemuridae and *Lemur* spp.
    ISIS NUMBER: 5301406001002002001.

Lemur mongoz Linnaeus, 1766. Syst. Nat., 12th ed., 1:44.
    TYPE LOCALITY: Comoros, Anjouan Isl.
    DISTRIBUTION: N.W. Madagascar, between Majunga and Betsiboka; Anjouan, Moheli
        (Comoro Isls.).
    PROTECTED STATUS: CITES - Appendix I as *Lemur* spp. and U.S. ESA - Endangered as
        Lemuridae and *Lemur* spp.
    ISIS NUMBER: 5301406001002003001.

Lemur rubriventer I. Geoffroy, 1850. C. R. Acad. Sci. Paris, 31:876.
    TYPE LOCALITY: Madagascar, Tamatave.
    DISTRIBUTION: E. Madagascar.
    PROTECTED STATUS: CITES - Appendix I as *Lemur* spp. and U.S. ESA - Endangered as
        Lemuridae and *Lemur* spp.
    ISIS NUMBER: 5301406001002004001.

Lepilemur I. Geoffroy, 1851. Cat. Method. Mamm. Mus. Paris, p. 75.
    COMMENT: Revised by Petter *et al.,* 1977, Faune de Madagascar, 44:274–318. Six of the
        seven species are known to be karyotypically distinct (CPG). CAH follows
        Rumpler, 1975:25–40, and Corbet and Hill, 1980:83, in placing this genus in a
        subfamily of Lemuridae; Petter and Petter, 1977, Part 3.1:6, placed it in its own
        family Lepilemuridae.
    PROTECTED STATUS: CITES - Appendix I as *Lepilemur* spp. and U.S. ESA - Endangered as
        Lemuridae and *Lepilemur* spp.
    ISIS NUMBER: 5301406001003000000.

Lepilemur dorsalis Gray, 1870. Cat. Monkeys, Lemurs, and Fruit-eating Bats Br. Mus., p.
    135.
    TYPE LOCALITY: N.W. Madagascar.
    DISTRIBUTION: Nosi Be and Ambanja Region (N.W. Madagascar).
    PROTECTED STATUS: CITES - Appendix I as *Lepilemur* spp. and U.S. ESA - Endangered as
        Lemuridae and *Lepilemur* spp.

Lepilemur edwardsi Forbes, 1894. Handbook of Primates, 1:87.
    TYPE LOCALITY: Madagascar, N. coast of Bombetoka Bay.
    DISTRIBUTION: E. Madagascar, between Ankarafantsika and Antsalova.
    PROTECTED STATUS: CITES - Appendix I as *Lepilemur* spp. and U.S. ESA - Endangered as
        Lemuridae and *Lepilemur* spp.

Lepilemur leucopus Major, 1894. Ann. Mag. Nat. Hist., ser. 6, 13:211.
    TYPE LOCALITY: Madagascar, Fort Dauphin.
    DISTRIBUTION: Arid zone of S. Madagascar.
    PROTECTED STATUS: CITES - Appendix I as *Lepilemur* spp. and U.S. ESA - Endangered as
        Lemuridae and *Lepilemur* spp.

Lepilemur microdon Forbes, 1894. Handbook of Primates, 1:88.
TYPE LOCALITY: Madagascar, East of Betsileo.
DISTRIBUTION: E. Madagascar, between Prinet and Ft. Dauphin.
PROTECTED STATUS: CITES - Appendix I as *Lepilemur* spp. and U.S. ESA - Endangered as Lemuridae and *Lepilemur* spp.

Lepilemur mustelinus I. Geoffroy, 1851. Cat. Meth. Coll. Mamm., Primates, p. 76.
TYPE LOCALITY: Madagascar, North of Tamatave.
DISTRIBUTION: E. Madagascar, between Tamatave and Antalaha.
PROTECTED STATUS: CITES - Appendix I as *Lepilemur* spp. and U.S. ESA - Endangered as Lemuridae and *Lepilemur* spp.
ISIS NUMBER: 5301406001003001001.

Lepilemur ruficaudatus A. Grandidier, 1867. Rev. Mag. Zool. Paris, ser. 2, 19:256.
TYPE LOCALITY: Madagascar, Morondava.
DISTRIBUTION: S.W. Madagascar, between Morondava and Sakaraha.
COMMENT: Includes *rufescens*, which Petter and Petter, 1977, Part 3.1:7, considered a distinct species.
PROTECTED STATUS: CITES - Appendix I as *Lepilemur* spp. and U.S. ESA - Endangered as Lemuridae and *Lepilemur* spp.

Lepilemur septentrionalis Rumpler and Albignac, 1975. Am. J. Phys. Anthropol., 42:425.
TYPE LOCALITY: Madagascar, Sahafary Forest.
DISTRIBUTION: Extreme N. Madagascar.
PROTECTED STATUS: CITES - Appendix I as *Lepilemur* spp. and U.S. ESA - Endangered as Lemuridae and *Lepilemur* spp.

Varecia Gray, 1863. Proc. Zool. Soc. Lond., 1863:135.
COMMENT: Separated from *Lemur* by Petter, 1962, Mem. Mus. Hist. Nat. Paris, N.S., A. Zool., 27:1–146.
PROTECTED STATUS: CITES - Appendix I as *Lemur* spp. and U.S. ESA - Endangered as Lemuridae and *Varecia* spp.

Varecia variegata (Kerr, 1792). Anim. Kingdom, p. 85.
TYPE LOCALITY: Madagascar.
DISTRIBUTION: East coast of Madagascar, to 19° S. latitude.
PROTECTED STATUS: CITES - Appendix I as *Lemur* spp. and U.S. ESA - Endangered as Lemuridae and *Varecia* spp.
ISIS NUMBER: 5301406001002005001 as *Lemur variegatus*.

Family Indriidae
REVIEWED BY: C. P. Groves (CPG); C. A. Hill (CAH).
COMMENT: Reviewed by Petter *et al.*, 1977, Faune de Madagascar, 44:1–543.
ISIS NUMBER: 5301406002000000000.

Indri E. Geoffroy and G. Cuvier, 1796. Mag. Encyclop., 1:46.
PROTECTED STATUS: CITES - Appendix I and U.S. ESA - Endangered as *Indri* spp.
ISIS NUMBER: 5301406002002000000.

Indri indri (Gmelin, 1788). Syst. Nat., 13th ed., 1:42.
TYPE LOCALITY: Madagascar.
DISTRIBUTION: N.E. to E.C. Madagascar.
PROTECTED STATUS: CITES - Appendix I and U.S. ESA - Endangered as *Indri* spp.
ISIS NUMBER: 5301406002002001001.

Lichanotus Illiger, 1811. Prodr. Syst. Mamm. et Avium., p. 11.
COMMENT: *Avahi* Jourdan, 1834, is a junior synonym; see Anderson, 1967, *in* Anderson and Jones, p. 158.
PROTECTED STATUS: CITES - Appendix I and U.S. ESA - Endangered as *Avahi* spp.
ISIS NUMBER: 5301406002001000000 as *Avahi*.

Lichanotus laniger (Gmelin, 1788). Syst. Nat., 13th ed., 1:44.
TYPE LOCALITY: Madagascar.
DISTRIBUTION: E. coast and Ankarafantsika Dist. in N.W. Madagascar.
PROTECTED STATUS: CITES - Appendix I and U.S. ESA - Endangered as *Avahi* spp.
ISIS NUMBER: 5301406002001001001 as *Avahi laniger.*

Propithecus Bennett, 1832. Proc. Zool. Soc. Lond., 1832:20.
PROTECTED STATUS: CITES - Appendix I and U.S. ESA - Endangered as *Propithecus* spp.
ISIS NUMBER: 5301406002003000000.

Propithecus diadema Bennett, 1832. Proc. Zool. Soc. Lond., 1832:20.
TYPE LOCALITY: Madagascar
DISTRIBUTION: N. and E. Madagascar.
PROTECTED STATUS: CITES - Appendix I and U.S. ESA - Endangered as *Propithecus* spp.
ISIS NUMBER: 5301406002003001001.

Propithecus verreauxi A. Grandidier, 1867. Rev. Mag. Zool. Paris, ser. 2, 19:84.
TYPE LOCALITY: Madagascar, Tsifanihy (N. of Cape Ste.-Marie).
DISTRIBUTION: N.C. to S.W. Madagascar.
PROTECTED STATUS: CITES - Appendix I and U.S. ESA - Endangered as *Propithecus* spp.
ISIS NUMBER: 5301406002003002001.

Family Daubentoniidae
REVIEWED BY: C. P. Groves (CPG).
ISIS NUMBER: 5301406003000000000.

Daubentonia E. Geoffroy, 1795. Decad. Philos. Litt., 28:195.
PROTECTED STATUS: CITES - Appendix I and U.S. ESA - Endangered as *Daubentonia* spp.
ISIS NUMBER: 5301406003001000000.

Daubentonia madagascariensis (Gmelin, 1788). Syst. Nat., 13th ed., 1:152.
TYPE LOCALITY: N.W. Madagascar.
DISTRIBUTION: N.E. and N.W. Madagascar (discontinuous).
PROTECTED STATUS: CITES - Appendix I as *D. madagascariensis* and U.S. ESA -
Endangered as *Daubentonia* spp.
ISIS NUMBER: 5301406003001001001.

Family Lorisidae
REVIEWED BY: C. P. Groves (CPG); S. Wang (SW)(China).
ISIS NUMBER: 5301406004000000000.

Arctocebus Gray, 1863. Proc. Zool. Soc. Lond., 1863:150.
ISIS NUMBER: 5301406004001000000.

Arctocebus calabarensis (J. A. Smith, 1860). Proc. Roy. Phys. Soc. Edinburgh, 2:177.
TYPE LOCALITY: Nigeria, Old Calabar.
DISTRIBUTION: C. Africa, between Niger and Congo Rivers.
PROTECTED STATUS: CITES - Appendix II as Order Primates.
ISIS NUMBER: 5301406004001001001.

Loris E. Geoffroy, 1796. Mag. Encyclop., 1:48.
ISIS NUMBER: 5301406004002000000.

Loris tardigradus (Linnaeus, 1758). Syst. Nat., 10th ed., 1:29.
TYPE LOCALITY: Sri Lanka.
DISTRIBUTION: Sri Lanka; S. India.
PROTECTED STATUS: CITES - Appendix II as Order Primates.
ISIS NUMBER: 5301406004002001001.

Nycticebus E. Geoffroy, 1812. Ann. Mus. Hist. Nat. Paris, 19:163.
ISIS NUMBER: 5301406004003000000.

Nycticebus coucang (Boddaert, 1785). Elench. Anim., p. 67.
TYPE LOCALITY: Malaysia, Malacca.
DISTRIBUTION: S. Philippines; Assam (India) to Vietnam and Malay Peninsula; W. Indonesia. Yunnan, perhaps Kwangai (China)(SW).
PROTECTED STATUS: CITES - Appendix II as Order Primates..
ISIS NUMBER: 5301406004003001001.

Nycticebus pygmaeus Bonhote, 1907. Abstr. Proc. Zool. Soc. Lond., 38:2.
TYPE LOCALITY: Vietnam, Nhatrang.
DISTRIBUTION: Laos; Kampuchea; Vietnam, east of Mekong River.
COMMENT: Includes *intermedius* see Lekagul and McNeely, 1977:270.
PROTECTED STATUS: CITES - Appendix II as Order Primates and U.S. ESA - Threatened as *N. pygmaeus*.
ISIS NUMBER: 5301406004003002001.

Perodicticus Bennett, 1831. Proc. Zool. Soc. Lond., 1830:109.
ISIS NUMBER: 5301406004004000000.

Perodicticus potto (Muller, 1766). Linn. Natursyst. Suppl., p. 12.
TYPE LOCALITY: Ghana, Elmina.
DISTRIBUTION: Cameroun to Guinea; Congo Republic; Gabon; Zaire to W. Kenya.
PROTECTED STATUS: CITES - Appendix II as Order Primates.
ISIS NUMBER: 5301406004004001001.

Family Galagidae
REVIEWED BY: C. P. Groves (CPG).
COMMENT: Formerly considered a subfamily of Lorisidae; see Hill and Meester, 1977, Part 3.2:2.

Galago E. Geoffroy, 1796. Mag. Encyclop., 1:49.
REVIEWED BY: C. A. Hill (CAH).
COMMENT: Includes *Galagoides*; see Groves, 1974:463; and *Euoticus*; see Petter and Petter-Rousseaux, 1979, *in* Doyle and Martin, eds., The Study of Prosimian Behavior.
ISIS NUMBER: 5301406004006000000 as *Galago*.
ISIS NUMBER: 5301406004005000000 as *Euoticus*.

Galago alleni Waterhouse, 1838. Proc. Zool. Soc. Lond., 1837:87.
TYPE LOCALITY: Equatorial Guinea, Bioko.
DISTRIBUTION: Bioko; Gabon; Cameroun; Congo Republic.
PROTECTED STATUS: CITES - Appendix II as Order Primates.
ISIS NUMBER: 5301406004006001001.

Galago demidovii (Fischer, 1806). Mem. Soc. Nat. Moscow, 1:24.
TYPE LOCALITY: Senegal.
DISTRIBUTION: Senegal to Uganda; Bioko; isolated forests of Kenya and Tanzania south to Malawi.
COMMENT: Formerly included in *Galagoides*; see Groves, 1974:463.
PROTECTED STATUS: CITES - Appendix II as Order Primates.
ISIS NUMBER: 5301406004006003001.

Galago elegantulus Le Conte, 1857. Proc. Acad. Nat. Sci. Phila., p. 10.
TYPE LOCALITY: West Africa.
DISTRIBUTION: Cameroun; Gabon; Congo Republic; Bioko.
COMMENT: Placed in the subgenus *Euoticus* by Petter and Petter-Rousseaux, 1979, *in* Doyle and Martin, eds., The Study of Prosimian Behavior.
PROTECTED STATUS: CITES - Appendix II as Order Primates.
ISIS NUMBER: 5301406004005001001 as *Euoticus elegantulus*.

Galago granti Thomas and Wroughton, 1907. Proc. Zool. Soc. Lond., 1907:286.
TYPE LOCALITY: Mozambique, Inhambane, Coguno.
DISTRIBUTION: Mozambique; Malawi; E. Zimbabwe.

COMMENT: Formerly included in *senegalensis*; see Smithers and Wilson, 1979, Mus. Mem. Nat. Mus. Rhodesia, p. 1–193.
PROTECTED STATUS: CITES - Appendix II as Order Primates.

Galago inustus Schwartz, 1930. Rev. Zool. Bot. Afr., 19:391.
TYPE LOCALITY: Zaire, Itun, Djugu S. of Mahagi.
DISTRIBUTION: E. Zaire; perhaps Uganda.
COMMENT: Placed in the subgenus *Euoticus* by Petter and Petter-Rousseaux, 1979, *in* Doyle and Martin, eds., The Study of Prosimian Behavior. Relationship uncertain (CPG).
PROTECTED STATUS: CITES - Appendix II as Order Primates.
ISIS NUMBER: 5301406004005002001 as *Euoticus inustus*.

Galago senegalensis E. Geoffroy, 1796. Mag. Encyclop., 1:38.
TYPE LOCALITY: Senegal.
DISTRIBUTION: Senegal to Ethiopia, south to South Africa and Angola. Perhaps Namibia.
COMMENT: Hill and Meester, 1977, Part 3.2:2, included *granti* in this species; but see Smithers and Wilson, 1979, Mus. Mem. Nat. Mus. Rhodesia, p. 1–193. Probably contains more than one species (CPG).
PROTECTED STATUS: CITES - Appendix II as Order Primates.
ISIS NUMBER: 5301406004006004001.

Galago zanzibaricus Matschie, 1893. Sitzb. Ges. Naturf. Fr. Berlin, p. 111.
TYPE LOCALITY: Tanzania, Zanzibar, Yambiani.
DISTRIBUTION: E. African coast from Tana River, south to S. Mozambique; Zanzibar.
COMMENT: Separated from *senegalensis* by Kingdon, 1971:309; Groves, 1974:463.
PROTECTED STATUS: CITES - Appendix II as Order Primates.

Otolemur Coquerel, 1859. Rev. Zool. Paris, 11:458.
COMMENT: Removed from *Galago* by Groves, 1974:461–463.

Otolemur crassicaudatus (E. Geoffroy, 1812). Ann. Mus. Hist. Nat. Paris, 19:166.
TYPE LOCALITY: Mozambique, Quelimane.
DISTRIBUTION: S. Somalia and Kenya to Natal (South Africa) and Angola; Zanzibar; Pemba; Mafia.
COMMENT: Probably contains more than one species; see Groves, 1974:463. Includes *garnetti*; but see Olson, 1979, Unpubl. Ph.D. Dissertation, Univ. London, who treated *garnetti* as a distinct species. See Thomas, 1917, Ann. Mag. Nat. Hist., 20:48 for discussion of type locality.
PROTECTED STATUS: CITES - Appendix II as Order Primates.
ISIS NUMBER: 5301406004006002001 as *Galago crassicaudatus*.

Family Tarsiidae
REVIEWED BY: C. P. Groves (CPG).
ISIS NUMBER: 5301406005000000000.

Tarsius Storr, 1780. Prodr. Meth. Mamm., p. 33.
ISIS NUMBER: 5301406005001000000.

Tarsius bancanus Horsfield, 1821. Zool. Res. Java, 2, pl.
TYPE LOCALITY: Indonesia, S.E. Sumatra, Bangka Isl.
DISTRIBUTION: Bangka Isl.; Sumatra; Borneo; Karimata Isl.; Billiton Isl.; Sirhassen Isl. (South Natuna Isls.).
PROTECTED STATUS: CITES - Appendix II as Order Primates.
ISIS NUMBER: 5301406005001001001.

Tarsius spectrum (Pallas, 1779). Nova Spec. Quadruped, p. 275.
TYPE LOCALITY: Indonesia, Sulawesi Selatan, Ujung Pandang (=Makasar).
DISTRIBUTION: Sulawesi; Peleng Isl.; Sanghir Isls.; Savu Isl.; Saleyer Isl.
PROTECTED STATUS: CITES - Appendix II as Order Primates.
ISIS NUMBER: 5301406005001002001.

Tarsius syrichta (Linnaeus, 1758). Syst. Nat., 10th ed., 1:29.
TYPE LOCALITY: Philippine Isls., Samar Isl.
DISTRIBUTION: Mindanao, Bohol Isl., Samar Isl., Leyte Isl. (Philippines).
PROTECTED STATUS: CITES - Appendix II as Order Primates and U.S. ESA - Threatened
as *T. syrichta.*
ISIS NUMBER: 5301406005001003001.

Family Callithricidae
REVIEWED BY: D. K. Candland (DKC); R. Fontaine (RF); C. P. Groves (CPG); C. A. Hill
(CAH).
ISIS NUMBER: 5301406007000000000.

Callithrix Erxleben, 1777. Syst. Regn. Anim., p. 55.
COMMENT: Includes *Mico;* see Cabrera, 1958:185.
ISIS NUMBER: 5301406007002000000.

Callithrix argentata (Linnaeus, 1766). Syst. Nat., 12th ed., 1:40.
TYPE LOCALITY: Brazil, Para, Cameta, on banks of Rio Tocantins.
DISTRIBUTION: N. and C. Brazil; E. Bolivia.
COMMENT: Includes *melanura* and *lucippe;* see Hershkovitz, 1977:436.
PROTECTED STATUS: CITES - Appendix II as Order Primates.
ISIS NUMBER: 5301406007002001001.

Callithrix humeralifer (E. Geoffroy, 1812). Ann. Mus. Hist. Nat. Paris, 19:120.
TYPE LOCALITY: Brazil, left bank of Rio Tapajos, Paricatuba.
DISTRIBUTION: Brazil, between Madeira and Tapajos rivers, south of the Amazon.
COMMENT: Includes *chrysoleuca;* see Hershkovitz, 1977:597.
PROTECTED STATUS: CITES - Appendix II as Order Primates.
ISIS NUMBER: 5301406007002006001 as *C. humeralifer.*
5301406007002003001 as *C. chrysoleuca.*

Callithrix jacchus (Linnaeus, 1758). Syst. Nat., 10th ed., 1:27.
TYPE LOCALITY: Brazil, Pernambuco.
DISTRIBUTION: Brazil.
COMMENT: Includes *aurita, flaviceps, geoffroyi,* and *penicillata;* see Hershkovitz,
1977:489–527.
PROTECTED STATUS: CITES - Appendix II as Order Primates.
CITES - Appendix I as *C. aurita.*
CITES - Appendix I as *C. flaviceps.*
ISIS NUMBER: 5301406007002007001 as *C. jacchus.*
5301406007002002001 as *C. aurita.*
5301406007002004001 as *C. flaviceps.*
5301406007002005001 as *C. geoffroyi.*
5301406007002008001 as *C. penicillata.*

Cebuella Gray, 1866. Proc. Zool. Soc. Lond., 1865:734.

Cebuella pygmaea (Spix, 1823). Sim. Vespert. Brasil., p. 32.
TYPE LOCALITY: Brazil, Amazonas, Solimoes River, Tabatinga.
DISTRIBUTION: N. and W. Brazil; N. Peru; Ecuador.
COMMENT: Revised by Hershkovitz, 1977:462–464.
PROTECTED STATUS: CITES - Appendix II as Order Primates.
ISIS NUMBER: 5301406007002009001.

Leontopithecus Lesson, 1840. Spec. Mamm. Bim. et Quadrum., 1844:184, 200.
COMMENT: Includes *Leontideus* and *Leontocebus;* see Hershkovitz, 1977:807–808.
PROTECTED STATUS: CITES - Appendix I as *Leontopithecus* (=*Leontideus*) spp. and U.S.
ESA - Endangered as *Leontideus* spp.
ISIS NUMBER: 5301406007003000000 as *Leontideus.*

Leontopithecus rosalia (Linnaeus, 1766). Syst. Nat., 12th ed., 1:41.
  TYPE LOCALITY: Brazil, Sao Paulo, Ipanema, east coast between 22° and 23° S.
  DISTRIBUTION: S.E. Brazil.
  COMMENT: Includes *chrysomelas* and *chrysopygus*; see Hershkovitz, 1977:825–846.
    Reviewed by Kleiman, 1981, Mamm. Species, 148:1–7.
  PROTECTED STATUS: CITES - Appendix I as *Leontopithecus* (=*Leontideus*) spp. and U.S.
    ESA - Endangered as *Leontideus* spp.
  ISIS NUMBER: 5301406007003001001 as *Leontideus rosalia*.

Saguinus Hoffmannsegg, 1807. Mag. Ges. Naturf. Fr., 1:101.
  COMMENT: Includes *Tamarin, Tamarinus, Marikina,* and *Oedipomidas*; see Hershkovitz,
    1977:601–603.
  ISIS NUMBER: 5301406007004000000.

Saguinus bicolor (Spix, 1823). Sim. Vespert. Brasil., p. 30.
  TYPE LOCALITY: Brazil, Manaus, Barra de Rio Negro.
  DISTRIBUTION: N. Brazil; perhaps N.E. Peru.
  COMMENT: Includes *martinsi*; see Hershkovitz, 1977:744.
  PROTECTED STATUS: CITES - Appendix I and U.S. ESA - Endangered.
  ISIS NUMBER: 5301406007004001001 as *S. bicolor*.
          5301406007004013001 as *S. martinsi*.

Saguinus fuscicollis (Spix, 1823). Sim. Vespert. Brasil., p. 27.
  TYPE LOCALITY: Brazil, between Solimoes River and Ica River, Sao Paulo de Olivenca.
  DISTRIBUTION: N. and W. Brazil; N. Bolivia; E. Peru; E. Ecuador; S.W. Colombia.
  COMMENT: Includes *devillei, fuscus, lagonotus, illigeri, melanoleucus,* and *weddelli;* see
    Hershkovitz, 1977:640–642.
  PROTECTED STATUS: CITES - Appendix II as Order Primates.
  ISIS NUMBER: 5301406007004003001 as *S. fusciollis (sic).*
          5301406007004002001 as *S. devillei.*
          5301406007004004001 as *S. fuscus.*
          5301406007004007001 as *S. illigeri.*
          5301406007004011001 as *S. lagonotus.*
          5301406007004014001 as *S. melanoleucus.*
          5301406007004022001 as *S. weddelli.*

Saguinus imperator (Goeldi, 1907). Proc. Zool. Soc. Lond., 1907:93.
  TYPE LOCALITY: Brazil, Rio Acre.
  DISTRIBUTION: W. Brazil; E. Peru.
  PROTECTED STATUS: CITES - Appendix II as Order Primates.
  ISIS NUMBER: 5301406007004008001.

Saguinus inustus Schwartz, 1951. Am. Mus. Novit., 1508:1.
  TYPE LOCALITY: Brazil, Amazonas, Tabocal.
  DISTRIBUTION: N.W. Brazil; S.W. Colombia.
  PROTECTED STATUS: CITES - Appendix II as Order Primates.
  ISIS NUMBER: 5301406007004009001.

Saguinus labiatus (E. Geoffroy, 1812). Rec. Observ. Zool., 1:361.
  TYPE LOCALITY: Brazil, Amazonas, Lake Joanacan.
  DISTRIBUTION: W. Brazil; E. Peru.
  PROTECTED STATUS: CITES - Appendix II as Order Primates.
  ISIS NUMBER: 5301406007004010000.

Saguinus leucopus (Gunther, 1877). Proc. Zool. Soc. Lond., 1876:746.
  TYPE LOCALITY: Colombia, Antioquia, Medellin.
  DISTRIBUTION: N. Colombia.
  PROTECTED STATUS: CITES - Appendix I and U.S. ESA - Threatened.
  ISIS NUMBER: 5301406007004012001.

Saguinus midas (Linnaeus, 1758). Syst. Nat., 10th ed., 1:28.
  TYPE LOCALITY: French Guiana.
  DISTRIBUTION: N. Brazil; Guyana; French Guiana; Surinam.

COMMENT: Includes *tamarin;* see Hershkovitz, 1977:711.
PROTECTED STATUS: CITES - Appendix II as Order Primates.
ISIS NUMBER: 5301406007004015001 as *S. midas.*
5301406007004021001 as *S. tamarin.*

Saguinus mystax (Spix, 1823). Sim. Vespert. Brasil., p. 29.
TYPE LOCALITY: Brazil, Amazonas, between Solimoes River and Ica River.
DISTRIBUTION: W. Brazil; Peru.
COMMENT: Includes *pluto* and *pileatus;* see Hershkovitz, 1977:700.
PROTECTED STATUS: CITES - Appendix II as Order Primates.
ISIS NUMBER: 5301406007004016001 as *S. mystax.*
5301406007004019001 as *S. pileatus.*
5301406007004020001 as *S. pluto.*

Saguinus nigricollis (Spix, 1823). Sim. Vespert. Brasil., p. 28.
TYPE LOCALITY: Brazil, Amazonas, Sao Paulo de Olivenca.
DISTRIBUTION: W. Brazil; E. Peru; E. Ecuador.
COMMENT: Includes *graellsi;* see Hershkovitz, 1977:628.
PROTECTED STATUS: CITES - Appendix II as Order Primates.
ISIS NUMBER: 5301406007004017001 as *S. nigricollis.*
5301406007004006001 as *S. graellsi.*

Saguinus oedipus (Linnaeus, 1758). Syst. Nat., 10th ed., 1:28.
TYPE LOCALITY: Colombia, Bolivar, lower Rio Sinu.
DISTRIBUTION: N. Colombia; Panama.
COMMENT: Includes *geoffroyi;* see Hershkovitz, 1977:757.
PROTECTED STATUS: CITES - Appendix I as *S. oedipus (geoffroyi)* and U.S. ESA -
Endangered as *S. oedipus* including *S. geoffroyi.*
ISIS NUMBER: 5301406007004018001 as *S. oedipus.*
5301406007004005001 as *S. geoffroyi.*

Family Callimiconidae
REVIEWED BY: D. K. Candland (DKC); R. Fontaine (RF); C. P. Groves (CPG); C. A. Hill
(CAH).
COMMENT: This family was recognized by Hershkovitz, 1977, but was placed in the
Callithricidae by Pocock, 1920, Proc. Zool. Soc. Lond., 1920:113, and Napier, 1976.

Callimico Miranda-Ribeiro, 1911. Brasil. Rundsch., p. 21.
ISIS NUMBER: 5301406007001000000.

Callimico goeldii (Thomas, 1904). Ann. Mag. Nat. Hist., ser. 7, 14:189.
TYPE LOCALITY: Brazil, Acre, Xapuruy River.
DISTRIBUTION: W. Brazil; N. Bolivia; E. Peru; Colombia.
PROTECTED STATUS: CITES - Appendix I and U.S. ESA - Endangered.
ISIS NUMBER: 5301406007001001001.

Family Cebidae
REVIEWED BY: C. P. Groves (CPG); J. Ramirez-Pulido (JRP) (Mexico).
ISIS NUMBER: 5301406006000000000.

Alouatta Lacepede, 1799. Tabl. Div. Subd. Orders Genres Mamm., p. 4.
REVIEWED BY: C. A. Hill (CAH).
ISIS NUMBER: 5301406006006000000.

Alouatta belzebul (Linnaeus, 1766). Syst. Nat., 12th ed., 1:37.
TYPE LOCALITY: Brazil, Para, Rio Capim.
DISTRIBUTION: N. Brazil; Mexiana Isl. (Brazil).
PROTECTED STATUS: CITES - Appendix II as Order Primates.
ISIS NUMBER: 5301406006006001001.

Alouatta caraya (Humboldt, 1812). Rec. Observ. Zool., 1:355.
TYPE LOCALITY: Paraguay.
DISTRIBUTION: N. Argentina to Mato Grosso (Brazil).
PROTECTED STATUS: CITES - Appendix II as Order Primates.
ISIS NUMBER: 5301406006006002001.

Alouatta fusca (E. Geoffroy, 1812). Ann. Mus. Hist. Nat. Paris, 19:108.
TYPE LOCALITY: Brazil.
DISTRIBUTION: N. Bolivia; S.E. and E.C. Brazil.
COMMENT: *A. guariba* (Humboldt, 1812) may be the correct name for this species (K. F. Koopman).
PROTECTED STATUS: CITES - Appendix II as Order Primates.
ISIS NUMBER: 5301406006006003001.

Alouatta palliata (Gray, 1849). Proc. Zool. Soc. Lond., 1848:138.
TYPE LOCALITY: Nicaragua, Lake Nicaragua.
DISTRIBUTION: W. Ecuador to Veracruz and Oaxaca (Mexico).
PROTECTED STATUS: CITES - Appendix I as *A. palliata* (=*villosa*).

Alouatta pigra Lawrence, 1933. Bull. Mus. Comp. Zool., 75:333.
TYPE LOCALITY: Guatemala, Peten, Uaxactun.
DISTRIBUTION: Yucatan and Chiapas (Mexico) to Belize and Guatemala.
COMMENT: The name *villosa* has been applied to this species (see Napier, 1976, Cat. Primates Br. Mus. (Nat. Hist.), Part 1, p. 76) but Lawrence, 1933, regarded it as a *nomen dubium*; see Smith, 1970, J. Mammal., 51:366, and Hall, 1981:260, 263.
PROTECTED STATUS: CITES - Appendix II as Order Primates and U.S. ESA - Endangered as *A. pigra.*
ISIS NUMBER: 5301406006006005001 as *A. villosa.*

Alouatta seniculus (Linnaeus, 1766). Syst. Nat., 12th ed., 1:37.
TYPE LOCALITY: Colombia, Bolivar, Rio Magdalena, Cartagena.
DISTRIBUTION: Trinidad; Bolivia to Ecuador; Colombia to Guyana, French Guiana, Surinam, and N.C. Brazil.
PROTECTED STATUS: CITES - Appendix II as Order Primates.
ISIS NUMBER: 5301406006006004001.

Aotus Illiger, 1811. Prodr. Syst. Mamm. et Avium., p. 71.
ISIS NUMBER: 5301406006001000000.

Aotus trivirgatus (Humboldt, 1811). Rec. Observ. Zool., 1:306.
TYPE LOCALITY: Venezuela, Duida Range, Rio Casiquiare.
DISTRIBUTION: Panama to Guyana, French Guiana, Surinam, Amazonian Brazil and N. Argentina; Ecuador to Peru.
COMMENT: May be divisible into 2–3 species; see Brumback, 1974, J. Hered., 65:321–323. Hall, 1981:259, listed *bipunctatus* as a distinct species but stated it is almost surely no more than a subspecies of *trivirgatus.*
PROTECTED STATUS: CITES - Appendix II as Order Primates.
ISIS NUMBER: 5301406006001001001.

Ateles E. Geoffroy, 1806. Ann. Mus. Hist. Nat. Paris, 7:262.
REVIEWED BY: C. A. Hill (CAH).
COMMENT: Hershkovitz, 1972, Int. Zoo Yearb., 12:1–12, united all forms in this genus in *paniscus.* However, see Hall, 1981:266, and Handley, 1976, Brigham Young Univ. Sci. Bull., 20(5):43.
ISIS NUMBER: 5301406006009000000.

Ateles belzebuth E. Geoffroy, 1806. Ann. Mus. Hist. Nat. Paris, 7:27.
TYPE LOCALITY: Venezuela, Orinoco, Esmeralda.
DISTRIBUTION: N.C. Brazil; Venezuela; Colombia; Ecuador; N. Peru.
COMMENT: Probably conspecific with *paniscus* (CPG).
PROTECTED STATUS: CITES - Appendix II as Order Primates.
ISIS NUMBER: 5301406006009001001.

Ateles fusciceps Gray, 1866. Proc. Zool. Soc. Lond., 1865:733.
TYPE LOCALITY: Ecuador, Imbabura, Penaherrera, Chinipamba, 1500 m.
DISTRIBUTION: Panama; W. Colombia; W. Ecuador.
PROTECTED STATUS: CITES - Appendix II as Order Primates.
ISIS NUMBER: 5301406006009002001.

Ateles geoffroyi Kuhl, 1820. Beitr. Zool. Vergl. Anat., Abt., 1:26.
TYPE LOCALITY: Nicaragua, San Juan del Norte.
DISTRIBUTION: N.E. Colombia to Tamaulipas and Oaxaca; possibly to Jalisco (Mexico).
PROTECTED STATUS: CITES - Appendix II as Order Primates.
    CITES - Appendix I and U.S. ESA - Endangered as *A. g. frontatus* subspecies only.
    CITES - Appendix I and U.S. ESA - Endangered as *A. g. panamensis* subspecies
        only.
ISIS NUMBER: 5301406006009003001.

Ateles paniscus (Linnaeus, 1758). Syst. Nat., 10th ed., 1:1–26.
TYPE LOCALITY: French Guiana.
DISTRIBUTION: N. and W. Brazil; Guyana; French Guiana; Surinam; E. Peru; N. and C.
    Bolivia.
PROTECTED STATUS: CITES - Appendix II as Order Primates.
ISIS NUMBER: 5301406006009004001.

Brachyteles Spix and Martins, 1823. Sim. Vespert. Brasil., p. 36.
REVIEWED BY: C. A. Hill (CAH).
ISIS NUMBER: 5301406006010000000.

Brachyteles arachnoides (E. Geoffroy, 1806). Ann. Mus. Hist. Nat. Paris, 7:271.
TYPE LOCALITY: Brazil, Rio de Janeiro.
DISTRIBUTION: E. Brazil.
PROTECTED STATUS: CITES - Appendix I and U.S. ESA - Endangered.
ISIS NUMBER: 5301406006010001001.

Cacajao Lesson, 1840. Spec. Mamm. Bim. et Quadrum., p. 181.
REVIEWED BY: C. A. Hill (CAH).
PROTECTED STATUS: CITES - Appendix I and U.S. ESA - Endangered as *Cacajao* spp.
ISIS NUMBER: 5301406006003000000.

Cacajao calvus (I. Geoffroy, 1847). C. R. Acad. Sci. Paris, 24:576.
TYPE LOCALITY: Brazil, Amazonas, Fonte Boa.
DISTRIBUTION: N.W. Brazil; E. Peru.
COMMENT: Includes *rubicundus*; see Hershkovitz, 1972, Int. Zoo Yearb., 12:1–12; but
    also see Szalay and Delson, 1979:290, who listed it as a distinct species.
PROTECTED STATUS: CITES - Appendix I and U.S. ESA - Endangered as *Cacajao* spp.
ISIS NUMBER: 5301406006003001001 as *C. calvus.*
    5301406006003003001 as *C. rubicundus.*

Cacajao melanocephalus (Humboldt, 1812). Rec. Observ. Zool., 1:317.
TYPE LOCALITY: Venezuela, Casiquiare Forests, Mision de San Francisco Solano.
DISTRIBUTION: S.W. Venezuela; N.W. Brazil.
PROTECTED STATUS: CITES - Appendix I and U.S. ESA - Endangered as *Cacajao* spp.
ISIS NUMBER: 5301406006003002001.

Callicebus Thomas, 1903. Ann. Mag. Nat. Hist., ser. 7, 12:456.
COMMENT: A key to the genus was published by Jones and Anderson, 1978, Mamm.
    Species, 112:1–5.
ISIS NUMBER: 5301406006002000000.

Callicebus moloch (Hoffmannsegg, 1807). Mag. Ges. Naturf. Fr., 9:97.
TYPE LOCALITY: Brazil, Grand Para.
DISTRIBUTION: W. Brazil; Paraguay; Bolivia; Peru; E. Ecuador; E. Colombia.
COMMENT: Includes *bruneus, cinerascens, cupreus,* and *ornatus;* see review by Jones and
    Anderson, 1978, Mamm. Species, 112:1–5; but see also Cabrera, 1958:137–142.
PROTECTED STATUS: CITES - Appendix II as Order Primates.
ISIS NUMBER: 5301406006002001001.

Callicebus personatus (E. Geoffroy, 1812). Rec. Observ. Zool. 1:357.
TYPE LOCALITY: Brazil, Espirito Santo.
DISTRIBUTION: S.E. and E.C. Brazil.
COMMENT: Includes *melanocher* and *nigrifrons;* but also see Cabrera, 1958:139–143.
PROTECTED STATUS: CITES - Appendix II as Order Primates.
ISIS NUMBER: 5301406006002002001.

Callicebus torquatus (Hoffmannsegg, 1807). Mag. Ges. Naturf. Fr., 10:86.
TYPE LOCALITY: Brazil, E. bank of R. Tocantins.
DISTRIBUTION: Peru; Colombia; N. Brazil; S. Venezuela.
PROTECTED STATUS: CITES - Appendix II as Order Primates.
ISIS NUMBER: 5301406006002003001.

Cebus Erxleben, 1777. Syst. Regn. Anim., p. 44.
ISIS NUMBER: 5301406006007000000.

Cebus albifrons (Humboldt, 1812). Rec. Observ. Zool., 1:324.
TYPE LOCALITY: Venezuela, Orinoco River.
DISTRIBUTION: Venezuela; Colombia; Ecuador; N. Peru; N.W. Brazil; Trinidad.
PROTECTED STATUS: CITES - Appendix II as Order Primates.
ISIS NUMBER: 5301406006007001001.

Cebus apella (Linnaeus, 1758). Syst. Nat., 10th ed., 1:28.
TYPE LOCALITY: French Guiana.
DISTRIBUTION: N. and C. South America.
PROTECTED STATUS: CITES - Appendix II as Order Primates.
ISIS NUMBER: 5301406006007002001.

Cebus capucinus (Linnaeus, 1758). Syst. Nat., 10th ed., 1:29.
TYPE LOCALITY: N. Colombia.
DISTRIBUTION: W. Ecuador to Honduras.
PROTECTED STATUS: CITES - Appendix II as *C. capucinus.*
ISIS NUMBER: 5301406006007003001.

Cebus olivaceus Schomburgk, 1848. Reise Br. Guiana, 2:247.
TYPE LOCALITY: Venezuela, Bolivar, southern base of Mt. Roraima, 930 m.
DISTRIBUTION: Guyana; French Guiana; Surinam; N. Brazil; Venezuela; N. Colombia.
COMMENT: Replaces *nigrivittatus;* see Husson, 1978, Mammals of Suriname, p. 223.
PROTECTED STATUS: CITES - Appendix II as Order Primates.
ISIS NUMBER: 5301406006007004001 as *Cebus nigrivittatus.*

Chiropotes Lesson, 1840. Spec. Mamm. Bim. et Quadrum., p. 178.
ISIS NUMBER: 5301406006004000000.

Chiropotes albinasus (I. Geoffroy and Deville, 1848). C. R. Acad. Sci. Paris, 27:498.
TYPE LOCALITY: Brazil, Para, Santorem.
DISTRIBUTION: N.C. Brazil.
PROTECTED STATUS: CITES - Appendix I and U.S. ESA - Endangered.
ISIS NUMBER: 5301406006004001001.

Chiropotes satanas (Hoffmannsegg, 1807). Mag. Ges. Naturf. Fr., 10:93.
TYPE LOCALITY: Brazil, Para, lower Tocantins River, Cameta.
DISTRIBUTION: N. Brazil; Guyana; French Guiana; Surinam; S. Venezuela.
PROTECTED STATUS: CITES - Appendix II as Order Primates.
ISIS NUMBER: 5301406006004002001.

Lagothrix E. Geoffroy, 1812. Rec. Observ. Zool. 1:354.
ISIS NUMBER: 5301406006011000000.

Lagothrix flavicauda (Humboldt, 1812). Rec. Observ. Zool., 1:343.
TYPE LOCALITY: Peru, San Martin, Puca Tambo (50 mi. (80 km) E. of Chachapoyas).
DISTRIBUTION: N. Peru.
PROTECTED STATUS: CITES - Appendix II as Order Primates and U.S. ESA - Endangered
    as *L. flavicauda.*
ISIS NUMBER: 5301406006011001001.

Lagothrix lagothricha (Humboldt, 1812). Rec. Observ. Zool., 1:322.
TYPE LOCALITY: Colombia, Vaupes, Guaviare River near confluence with Amanaveni River.
DISTRIBUTION: N. and W. Brazil; Colombia; E. Ecuador; N.E. Peru.
PROTECTED STATUS: CITES - Appendix II as Order Primates.
ISIS NUMBER: 5301406006011002001.

Pithecia Desmarest, 1804. Nouv. Dict. Hist. Nat. Paris, p. 24, Tabl. Meth., p. 8.
REVIEWED BY: D. K. Candland (DKC); R. Fontaine (RF); C. A. Hill (CAH).
ISIS NUMBER: 5301406006005000000.

Pithecia albicans Gray, 1860. Proc. Zool. Soc. Lond., 1860:231.
TYPE LOCALITY: Brazil, Amazonas, Tefe (south bank of Solimoes River).
DISTRIBUTION: South bank of Amazon, between lower Jurua and lower Purus rivers.
COMMENT: Separated from *monachus* by Hershkovitz, 1979, Folia Primatol., 31:1–22.
PROTECTED STATUS: CITES - Appendix II as Order Primates.

Pithecia hirsuta Spix, 1823. Sim. Vespert. Brasil., p. 14.
TYPE LOCALITY: Brazil, Amazonas, Rio Solimoes below Tabatinga.
DISTRIBUTION: From Andean slopes in Colombia, Ecuador and Peru, to west bank of Rio Tapajoz and Rio Negro (Brazil).
COMMENT: Separated from *monachus* by Hershkovitz, 1979, Folia Primatol., 31:1–22.
PROTECTED STATUS: CITES - Appendix II as Order Primates.

Pithecia monachus (E. Geoffroy, 1812). Rec. Observ. Zool. 1:359.
TYPE LOCALITY: Brazil, Rio Madeira.
DISTRIBUTION: West of Rio Jurua and Rio Japura-Caqueta (in Brazil), Colombia, Ecuador and Peru.
COMMENT: Does not include *albicans* or *hirsuta*, see Hershkovitz, 1979, Folia Primatol., 31:1–22, and also for discussion of type locality.
PROTECTED STATUS: CITES - Appendix II as Order Primates.
ISIS NUMBER: 5301406006005001001.

Pithecia pithecia (Linnaeus, 1766). Syst. Nat., 12th ed., 1:40.
TYPE LOCALITY: French Guiana, Cayenne.
DISTRIBUTION: Guyana; French Guiana; Surinam; N. Amazon, east of Rio Negro and Rio Orinoco (N. Brazil, S. Venezuela).
PROTECTED STATUS: CITES - Appendix II as Order Primates.
ISIS NUMBER: 5301406006005002001.

Saimiri Voigt, 1831. Cuvier's Thierreich, 1:95.
REVIEWED BY: D. K. Candland (DKC); R. Fontaine (RF); C. A. Hill (CAH).
ISIS NUMBER: 5301406006008000000.

Saimiri oerstedii Reinhardt, 1872. Vidensk. Medd. Nat. Hist. Kjobenhaven, p. 157.
TYPE LOCALITY: Panama, Chiriqui, David.
DISTRIBUTION: Panama; Costa Rica.
COMMENT: Hershkovitz, 1972, Int. Zoo Yearb., 12:1–12, considered *oerstedii* a subspecies of *sciureus*; but see Hill, 1960, Comparative Anatomy and Taxonomy: Primates, IV. Cebidae (Part A), p. 300–302; Hall, 1981:265.
PROTECTED STATUS: CITES - Appendix I and U.S. ESA - Endangered.
ISIS NUMBER: 5301406006008001001.

Saimiri sciureus (Linnaeus, 1758). Syst. Nat., 10th ed., 1:29.
TYPE LOCALITY: Guyana, Kartabo.
DISTRIBUTION: N. Brazil; Marajo Isl.(Brazil); Guyana; French Guiana; Surinam; Venezuela; Colombia; E. Ecuador; E. Peru; E. Bolivia.
PROTECTED STATUS: CITES - Appendix II as Order Primates.
ISIS NUMBER: 5301406006008002001.

Family Cercopithecidae
REVIEWED BY: E. Delson (ED); C. P. Groves (CPG); P. Grubb (PG); C. A. Hill (CAH); S. Wang (SW)(China).
COMMENT: CAH would divide this family into the Colobidae and Cercopithecidae.
ISIS NUMBER: 5301406008000000000.

Allenopithecus Lang, 1923. Am. Mus. Novit., 87:1.
COMMENT: Separated from *Cercopithecus* by Thorington and Groves, 1970:638, and Szalay and Delson, 1979.

Allenopithecus nigroviridis (Pocock, 1907). Proc. Zool. Soc. Lond., 1907:739.
TYPE LOCALITY: Zaire, upper Congo River.
DISTRIBUTION: N.W. Zaire; N.E. Angola.
PROTECTED STATUS: CITES - Appendix II as Order Primates.
ISIS NUMBER: 5301406008002015001 as *Cercopithecus nigroviridis*.

Cercocebus E. Geoffroy, 1812. Ann. Mus. Hist. Nat. Paris, 19:97.
COMMENT: *Lophocebus* was considered a distinct genus by Groves, 1978, 19:1–34, but see Szalay and Delson, 1979, who considered *Lophocebus* a subgenus of *Cercocebus*. Van Gelder, 1977, Am. Mus. Novit., 2635:8, included *Cercocebus* in *Cercopithecus*.
ISIS NUMBER: 5301406008001000000.

Cercocebus agilis Milne-Edwards, 1886. Rev. Scient., 12:15.
TYPE LOCALITY: Zaire, Republic Poste des Ouaddas (junction of Oubangui and Congo Rivers).
DISTRIBUTION: Equatorial Guinea; Cameroun; N.E. Gabon; Central African Republic; N. Congo Republic; Zaire.
COMMENT: Includes *chrysogaster*; separated from *galeritus* by Groves, 1978, Primates, 19:1–34.
PROTECTED STATUS: CITES - Appendix II as Order Primates.

Cercocebus albigena (Gray, 1850). Proc. Zool. Soc. Lond., 1850:77.
TYPE LOCALITY: Zaire, Mayombe.
DISTRIBUTION: S.E. Nigeria, Cameroun, Congo Republic, Gabon, Equatorial Guinea, N.E. Angola, Central African Republic, Zaire, W. Uganda, Burundi; W. Kenya; W. Tanzania.
COMMENT: Type of subgenus *Lophocebus*; includes *aterrimus* and *opdenboschi*; see Groves, 1978, Primates, 19:1–34, but see also Szalay and Delson, 1979.
PROTECTED STATUS: CITES - Appendix II as Order Primates.
ISIS NUMBER: 5301406008001001001 as *C. albigena*.
             5301406008001002001 as *C. aterrimus*.

Cercocebus galeritus Peters, 1879. Monatsb. Preuss. Akad. Wiss. Berlin, p. 830.
TYPE LOCALITY: Kenya, Tana River, Mitole (2.10° S. 40.10° E.).
DISTRIBUTION: Lower Tana River (Kenya).
COMMENT: Formerly included *agilis*; but see Groves, 1978, Primates, 19:1–34. A mangabey, probably of this species, was recently discovered in Uzungwa Mtns., E. Tanzania (CPG).
PROTECTED STATUS: CITES - Appendix II as Order Primates.
         CITES - Appendix I as *C. g. galeritus* (subspecies only) and U.S. ESA - Endangered as *C. galeritus*.
ISIS NUMBER: 5301406008001003001.

Cercocebus torquatus (Kerr, 1792). Anim. Kingdom, p. 67.
TYPE LOCALITY: West Africa.
DISTRIBUTION: Guinea to Gabon.
COMMENT: Includes *atys*; see Dandelot, 1974, Part 3:12; and Groves, 1978, Primates, 19:1–34.
PROTECTED STATUS: CITES - Appendix II as Order Primates and U.S. ESA - Endangered as *C. torquatus*.
ISIS NUMBER: 5301406008001004001.

Cercopithecus Linnaeus, 1758. Syst. Nat., 10th ed., 1:26.
>   COMMENT: Dandelot, 1974, Part 3:14, included *Allenopithecus, Erythrocebus*, and
>   *Miopithecus* in this genus; but see Groves, 1978, Primates, 19:1–34, who considered
>   them to be distinct genera; Szalay and Delson, 1979, considered *Miopithecus* a
>   subgenus of *Cercopithecus*, and are followed here. Van Gelder, 1977, Am. Mus.
>   Novit., 2635:8, included *Cercocebus, Papio,* and *Theropithecus* in this genus, but see
>   Groves, 1978, Primates, 19:1–34; Ansell, 1978:33; and Cronin and Meikle, 1979,
>   Syst. Zool., 28:259. Designated as a subgroup of *Simia* by Linnaeus; type species, *S.
>   diana,* designated by Stiles and Orleman, 1926, J. Mammal., 7:52. *Simia* was
>   suppressed by Opinion 114 of the ICZN.
>   ISIS NUMBER: 5301406008002000000.

Cercopithecus aethiops (Linnaeus, 1758). Syst. Nat., 10th ed., 1:28.
>   TYPE LOCALITY: Sudan, Sennaar.
>   DISTRIBUTION: Senegal to Ethiopia, south to South Africa; Zanzibar; Pemba; Mafia.
>   Introduced into Lesser Antilles (Caribbean).
>   COMMENT: Includes *pygerythrus, sabaeus, and tantalus;* see Struhsaker, 1970:376.
>   PROTECTED STATUS: CITES - Appendix II as Order Primates.
>   ISIS NUMBER: 5301406008002001001.

Cercopithecus ascanius (Audebert, 1799). Hist. Nat. Singes Makis, 4(2):13.
>   TYPE LOCALITY: Angola (Northwest, by lower Congo River).
>   DISTRIBUTION: Uganda; Zaire; Zambia; Angola; marginally in Central African
>   Republic; W. Kenya.
>   COMMENT: See Machado, 1969, Publ. Cult. Comp. Diamantes Angola, 46:123 for
>   discussion of type locality.
>   PROTECTED STATUS: CITES - Appendix II as Order Primates.
>   ISIS NUMBER: 5301406008002002001.

Cercopithecus campbelli Waterhouse, 1838. Proc. Zool. Soc. Lond., 1838:61.
>   TYPE LOCALITY: Sierra Leone.
>   DISTRIBUTION: Gambia to Ghana.
>   PROTECTED STATUS: CITES - Appendix II as Order Primates.
>   ISIS NUMBER: 5301406008002003001.

Cercopithecus cephus (Linnaeus, 1758). Syst. Nat., 10th ed., 1:27.
>   TYPE LOCALITY: Africa.
>   DISTRIBUTION: Gabon; Congo Republic; S. Cameroun; Equatorial Guinea; S.W. Central
>   African Republic; N.W. Angola.
>   COMMENT: May include *erythrotis;* see Struhsaker, 1970:374–376; but also see Dandelot,
>   1974, Part 3:23.
>   PROTECTED STATUS: CITES - Appendix II as Order Primates.
>   ISIS NUMBER: 5301406008002004001.

Cercopithecus diana (Linnaeus, 1758). Syst. Nat., 10th ed., 1:26.
>   TYPE LOCALITY: Liberia.
>   DISTRIBUTION: Sierra Leone to Ghana.
>   COMMENT: Type species; see comment under *Cercopithecus.*
>   PROTECTED STATUS: CITES - Appendix I as *C. diana (roloway)* and U.S. ESA -
>   Endangered.
>   ISIS NUMBER: 5301406008002006001.

Cercopithecus dryas Schwartz, 1932. Rev. Zool. Bot. Afr., 21:251.
>   TYPE LOCALITY: Zaire, Ikela zone, Yapatsi.
>   DISTRIBUTION: Known only from the type locality.
>   COMMENT: Not a subspecies of *diana;* related to *aethiops;* see Thys van den
>   Audenaerde, 1977, Rev. Zool. Afr., 91:1007, who also discussed the type locality
>   (p. 1006).
>   PROTECTED STATUS: CITES - Appendix II as Order Primates.
>   ISIS NUMBER: 5301406008002007001.

Cercopithecus erythrogaster Gray, 1866. Proc. Zool. Soc. Lond., 1866:169.
TYPE LOCALITY: Nigeria, Lagos.
DISTRIBUTION: S. Nigeria.
PROTECTED STATUS: CITES - Appendix II as Order Primates and U.S. ESA - Endangered
as *C. erythrogaster*.
ISIS NUMBER: 5301406008002008001.

Cercopithecus erythrotis Waterhouse, 1838. Proc. Zool. Soc. Lond., 1838:59.
TYPE LOCALITY: Equatorial Guinea, Bioko.
DISTRIBUTION: S. and E. Nigeria; Cameroun; Bioko.
COMMENT: Considered a subspecies of *cephus* by Struhsaker, 1970:374–376; but also see
Dandelot, 1974, Part 3:23.
PROTECTED STATUS: CITES - Appendix II as Order Primates and U.S. ESA - Endangered
as *C. erythrotis*.

Cercopithecus hamlyni Pocock, 1907. Ann. Mag. Nat. Hist., ser. 7, 20:521.
TYPE LOCALITY: Zaire, Ituri Forest.
DISTRIBUTION: E. Zaire; Rwanda.
PROTECTED STATUS: CITES - Appendix II as Order Primates.
ISIS NUMBER: 5301406008002009001.

Cercopithecus lhoesti Sclater, 1899. Proc. Zool. Soc. Lond., 1898:586.
TYPE LOCALITY: Zaire, Tschepo River, near Stanleyville.
DISTRIBUTION: E. Zaire; W. Uganda; Rwanda; Burundi; Bioko; Mt. Cameroun area.
COMMENT: Includes *preussi*; see Dandelot, 1974, Part 3:26.
PROTECTED STATUS: CITES - Appendix II as Order Primates and U.S. ESA - Endangered
as *C. lhoesti*.
ISIS NUMBER: 5301406008002010000.

Cercopithecus mitis Wolf, 1822. Abbild. Beschreib. Merkw. Naturg. Gegenstands, 2:145.
TYPE LOCALITY: Angola.
DISTRIBUTION: Ethiopia to South Africa; S. and E. Zaire; N.W. Angola.
COMMENT: Includes *albogularis*; see Booth, 1968, Nat. Geogr. Soc. Research Reps. for
1963, p. 37–51; but also see Dandelot, 1974, Part 3:19.
PROTECTED STATUS: CITES - Appendix II as Order Primates.
ISIS NUMBER: 5301406008002011001.

Cercopithecus mona (Schreber, 1774). Saugethiere, 1:103.
TYPE LOCALITY: "Guinea."
DISTRIBUTION: Ghana to Cameroun; introduced into Lesser Antilles (Caribbean).
PROTECTED STATUS: CITES - Appendix II as Order Primates.
ISIS NUMBER: 5301406008002012001.

Cercopithecus neglectus Schlegel, 1876. Mus. Hist. Nat. Pays-Bas. Simiae, p. 70.
TYPE LOCALITY: Sudan, "White Nile."
DISTRIBUTION: S.E. Cameroun to Uganda and N. Angola, W. Kenya, S.W. Ethiopia, and
S. Sudan.
PROTECTED STATUS: CITES - Appendix II as Order Primates.
ISIS NUMBER: 5301406008002013001.

Cercopithecus nictitans (Linnaeus, 1766). Syst. Nat., 12th ed., 1:40.
TYPE LOCALITY: Equatorial Guinea, Benito River.
DISTRIBUTION: Liberia; Ivory Coast; Nigeria to N.W. Zaire, Central African Republic;
Rio Muni and Bioko (Equatorial Guinea).
PROTECTED STATUS: CITES - Appendix II as Order Primates.
ISIS NUMBER: 5301406008002014001.

Cercopithecus petaurista (Schreber, 1774). Saugethiere, 1:97, 185.
TYPE LOCALITY: "Guinea".
DISTRIBUTION: Gambia to Benin.
PROTECTED STATUS: CITES - Appendix II as Order Primates.
ISIS NUMBER: 5301406008002017001.

Cercopithecus pogonias Bennett, 1833. Proc. Zool. Soc. Lond., 1833:67.
TYPE LOCALITY: Equatorial Guinea, Bioko.
DISTRIBUTION: S.E. Nigeria; Cameroun; Bioko; Rio Muni (Equatorial Guinea); N. and
W. Gabon; W. Zaire; Congo Republic.
PROTECTED STATUS: CITES - Appendix II as Order Primates.
ISIS NUMBER: 5301406008002018001.

Cercopithecus salongo Thys van den Audenaerde, 1977. Rev. Zool. Afr., 91:1001.
TYPE LOCALITY: Zaire, zone de Djolu, Wamba (22.31-.33° E., and 0.01° N. - 0.01° S.).
DISTRIBUTION: Known only from the type locality.
COMMENT: Relationships unknown, but evidently a valid species (CPG). Validity as a
species is uncertain (ED).
PROTECTED STATUS: CITES - Appendix II as Order Primates.

Cercopithecus talapoin Schreber, 1774. Saugethiere, 1:101, 186, pl. 17.
TYPE LOCALITY: Angola.
DISTRIBUTION: Angola; S.W. Zaire.
COMMENT: Subgenus *Miopithecus;* see Szalay and Delson, 1979, but also see Groves,
1978, Primates, 19:1-34. A second species of *Miopithecus,* described by Machado,
1969, Publ. Cult. Comp. Diamantes Angola, 46:146-154, has not yet been named;
its distribution is Gabon, Equatorial Guinea, and Cameroun (CPG).
PROTECTED STATUS: CITES - Appendix II as Order Primates.
ISIS NUMBER: 5301406008002019001.

Cercopithecus wolfi Meyer, 1891. Notes Leyden Mus., 13:63.
TYPE LOCALITY: "Central West Africa."
DISTRIBUTION: Zaire; N.E. Angola; W. Uganda; Central African Republic.
COMMENT: Includes *denti;* see Dandelot, 1974, Part 3:25.
PROTECTED STATUS: CITES - Appendix II as Order Primates.
ISIS NUMBER: 5301406008002020000 as *C. wolfi.*
5301406008002005001 as *C. denti.*

Colobus Illiger, 1811. Prodr. Syst. Mamm. et Avium., p. 69.
COMMENT: Includes *Procolobus* and *Piliocolobus;* see Dandelot, 1974, Part 3:37; Szalay
and Delson, 1979.
ISIS NUMBER: 5301406008006000000 as *Colobus.*
5301406008009000000 as *Procolobus.*

Colobus angolensis Sclater, 1860. Proc. Zool. Soc. Lond., 1860:245.
TYPE LOCALITY: Angola, 300 mi. (483 km) inland from Bembe.
DISTRIBUTION: N.E. Angola; S. and E. Zaire; Rwanda; Burundi; N.E. Zambia; S.E.
Kenya; E. Tanzania.
COMMENT: Dandelot, 1974, Part 3:37; Thorington and Groves, 1970:629-647; and
Corbet and Hill, 1980:89, listed *angolensis* as a distinct species. Considered
synonymous with *polykomos* (ED).
PROTECTED STATUS: CITES - Appendix II as Order Primates.

Colobus badius (Kerr, 1792). Anim. Kingdom, p. 74.
TYPE LOCALITY: Sierra Leone.
DISTRIBUTION: Senegal to Ghana.
COMMENT: Subgenus *Piliocolobus;* includes *waldroni* and *temmincki;* see Dandelot, 1974,
Part 3:33; but also see Rahm, 1970:589-626.
PROTECTED STATUS: CITES - Appendix II as Order Primates.
ISIS NUMBER: 5301406008006001001.

Colobus guereza Ruppell, 1835. Neue Wirbelt. Fauna Abyssin. Gehorig. Saugeth., p. 1.
TYPE LOCALITY: Ethiopia, Gojjam and Kulla.
DISTRIBUTION: Nigeria to Ethiopia; Kenya; Uganda; Tanzania.
COMMENT: Subgenus *Colobus;* see Dandelot, 1974, Part 3:37.
PROTECTED STATUS: CITES - Appendix II as Order Primates.
ISIS NUMBER: 5301406008006002001.

Colobus pennantii Waterhouse, 1838. Proc. Zool. Soc. Lond., 1838:57.
TYPE LOCALITY: Equatorial Guinea, Bioko.
DISTRIBUTION: Congo Republic; Bioko; Zaire; W. Uganda; W. and S.E. Tanzania; Zanzibar.
COMMENT: Subgenus *Poliocolobus*; includes *bouvieri, foai, oustaleti, tephrosceles, gordonorum, kirki, ellioti* and *tholloni*; but also see Rahm, 1970:589–626, who considered this species a subspecies of *badius,* and Dandelot, 1974, Part 3:35, who considered *kirki* a distinct species.
PROTECTED STATUS: CITES - Appendix II as Order Primates.
CITES - Appendix II as *C. badius gordonorum.*
CITES - Appendix I as *C. badius kirkii (sic)* and U.S. ESA - Endangered as *C. kirkii (sic).*
ISIS NUMBER: 5301406008006003001 as *C. kirkii (sic).*

Colobus polykomos (Zimmermann, 1780). Geogr. Gesch. Mensch. Vierf. Thiere, 2:202.
TYPE LOCALITY: Sierra Leone.
DISTRIBUTION: Gambia to Benin. Nigeria (PG).
COMMENT: Subgenus *Colobus*; see Dandelot, 1974, Part 3:37.
PROTECTED STATUS: CITES - Appendix II as Order Primates.
ISIS NUMBER: 5301406008006004001.

Colobus preussi (Matschie, 1900). Sitzb. Ges. Naturf. Fr. Berlin, 1900:183.
TYPE LOCALITY: Cameroun, Barombi (on Elephant Lake).
DISTRIBUTION: Yabassi Dist. (Cameroun).
COMMENT: Considered by Rahm, 1970:589–626, to be a subspecies of *badius,* but see Dandelot, 1974, Part 3:37.
PROTECTED STATUS: CITES - Appendix II as Order Primates.

Colobus rufomitratus Peters, 1879. Monatsb. Preuss. Akad. Wiss. Berlin, p. 829.
TYPE LOCALITY: Kenya, Tana River, Muniuni.
DISTRIBUTION: Lower Tana River (Kenya).
COMMENT: Subgenus *Piliocolobus*. Considered by Rahm, 1970:589–626, to be a subspecies of *badius;* but see Dandelot, 1974, Part 3:35.
PROTECTED STATUS: CITES - Appendix I and U.S. ESA - Endangered as *C. badius rufomitratus.*

Colobus satanas Waterhouse, 1838. Proc. Zool. Soc. Lond., 1837:87.
TYPE LOCALITY: Equatorial Guinea, Bioko.
DISTRIBUTION: S.W. Gabon; Rio Muni (Equatorial Guinea); S.W. Cameroun; Bioko.
COMMENT: Subgenus *Colobus*; see Dandelot, 1974, Part 3:33.
PROTECTED STATUS: CITES - Appendix II as Order Primates and U.S. ESA - Endangered as *C. satanas.*

Colobus verus Van Beneden, 1838. Bull. Acad. Sci. Belles-Letters Bruxelles, 5:347.
REVIEWED BY: D. K. Candland (DKC); R. Fontaine (RF).
TYPE LOCALITY: Africa.
DISTRIBUTION: Sierra Leone to Togo; Idah Dist. (E. Nigeria).
COMMENT: See Menzies, 1970, J. West Afr. Sci. Assoc., 15:83–84, for comments on range in Nigeria. Subgenus *Procolobus*; see Dandelot, 1974, Part 3:37.
PROTECTED STATUS: CITES - Appendix II as Order Primates.
ISIS NUMBER: 5301406008009001001 as *Procolobus verus.*

Erythrocebus Trouessart, 1897. Cat. Mamm. Viv. Foss., 1:19.
COMMENT: Recognized as a distinct genus by Thorington and Groves, 1970:638–639; and Szalay and Delson, 1979.

Erythrocebus patas (Schreber, 1775). Saugethiere, 1:98.
TYPE LOCALITY: Senegal.
DISTRIBUTION: Savannahs, from W. Africa to Ethiopia, Kenya, and Tanzania.
PROTECTED STATUS: CITES - Appendix II as Order Primates.
ISIS NUMBER: 5301406008002016001 as *Cercopithecus patas.*

Macaca Lacepede, 1799. Tabl. Mamm., p. 4.
REVIEWED BY: D. K. Candland (DKC); R. Fontaine (RF); J. Fooden (JF).
COMMENT: Includes *Gymnopyga* (*Cynomacaca* is a junior synonym) and *Cynopithecus*; see
Hill, 1974:191; Walker *et al.*, 1968, Mammals of the World, 2nd ed., p. 450;
Fooden, 1969, Bibl. Primatol, No. 10; Albrecht, 1978, Contrib. Primatol., 13;
Fooden, 1980:1–9.
ISIS NUMBER: 5301406008003000000.

Macaca arctoides (I. Geoffroy, 1831). Zool. Voy. de Belanger Indes Orient., p. 61.
TYPE LOCALITY: "Cochin-China" (Indochina).
DISTRIBUTION: Assam (India) to S. China and Malay Peninsula.
COMMENT: *M. speciosa* Blyth, 1875 is a junior synonym; see Fooden, 1969, Bibl.
Primatol, No. 10, and Fooden, 1976, Folia Primatol., 25:225–236.
PROTECTED STATUS: CITES - Appendix II as Order Primates and U.S. ESA - Threatened
as *M. arctoides*.
ISIS NUMBER: 5301406008003001001.

Macaca assamensis (M'Clelland, 1840). Proc. Zool. Soc. Lond., 1839:148.
TYPE LOCALITY: India, Assam.
DISTRIBUTION: Nepal to N. Vietnam. S. China (SW).
PROTECTED STATUS: CITES - Appendix II as Order Primates.
ISIS NUMBER: 5301406008003002001.

Macaca cyclopis (Swinhoe, 1863). Proc. Zool. Soc. Lond., 1862:350.
TYPE LOCALITY: Taiwan, Jusan, Takao Pref.
DISTRIBUTION: Taiwan.
COMMENT: Probably conspecific with *mulatta* (CPG); but see Fooden, 1980:1–9.
PROTECTED STATUS: CITES - Appendix II as Order Primates and U.S. ESA - Threatened
as *M. cyclopis*.
ISIS NUMBER: 5301406008003003001.

Macaca fascicularis (Raffles, 1821). Trans. Linn. Soc. Lond., 13:246.
TYPE LOCALITY: Indonesia, Sumatra, Bengkulen.
DISTRIBUTION: Indochina and Burma to Borneo and Timor; Philippine Isls.; Nicobar
Isls.
COMMENT: Includes *irus*; see Medway, 1977:70–71. Includes *cynomolgos*; see Hill, 1974,
7:476–477. Intergrades, and possibly conspecific, with *mulatta*; see Fooden, 1964,
Science, 143:363–365; Fooden, 1971, Fieldiana Zool., 59:24–32; but also see
Fooden, 1980:1–9.
PROTECTED STATUS: CITES - Appendix II as Order Primates.
ISIS NUMBER: 5301406008003004001.

Macaca fuscata (Blyth, 1875). J. Asiat. Soc. Bengal, 44:6.
TYPE LOCALITY: Japan.
DISTRIBUTION: Honshu, Shikoku, Kyushu, and adjacent small isls. (Japan); Yaku isl.
(Ryukyu Isls.).
COMMENT: Includes *speciosa* I. Geoffroy, 1826 (not *speciosa* Blyth, 1875) which was
suppressed by opinion 920 of ICZN; see Fooden, 1976, Folia Primatol.,
25:225–236.
PROTECTED STATUS: CITES - Appendix II as Order Primates and U.S. ESA - Threatened
as *M. fuscata*.
ISIS NUMBER: 5301406008003005001.

Macaca maura F. Schinz, 1825. Das Thierreich, p. 257.
TYPE LOCALITY: Indonesia, Sulawesi Selatan.
DISTRIBUTION: S. Sulawesi, south of Tempe Depression.
COMMENT: Type species of *Gymnopyga*; see Fooden, 1969, Bibl. Primatol, 10:79.
PROTECTED STATUS: CITES - Appendix II as Order Primates.

Macaca mulatta (Zimmermann, 1780). Geogr. Gesch. Mensch. Vierf. Thiere, 2:195.
TYPE LOCALITY: India, Nepal Terai.
DISTRIBUTION: Afghanistan and India to N. Thailand, China, and Hainan Isl. (China).
COMMENT: See comments under *fascicularis*.
PROTECTED STATUS: CITES - Appendix II as Order Primates.
ISIS NUMBER: 5301406008003006001.

Macaca nemestrina (Linnaeus, 1766). Syst. Nat., 12th ed., 1:35.
TYPE LOCALITY: Indonesia, Sumatra.
DISTRIBUTION: Malay Peninsula; Sumatra; Borneo; Bangka Isl.; Burma; Thailand; Yunnan (China)(SW); Laos; Junk Seylon; Mergui Archipelago.
COMMENT: Includes *pagensis*; see Fooden, 1975, Fieldiana Zool., p. 67; Fooden, 1980:7; Szalay and Delson, 1979. Wilson and Wilson, 1977, Yearb. Phys. Anthro., 20:216, considered *pagensis* a distinct species.
PROTECTED STATUS: CITES - Appendix II as Order Primates.
ISIS NUMBER: 5301406008003007001.

Macaca nigra (Desmarest, 1822). Encyclop. Method. Mamm., Suppl., 2:534.
TYPE LOCALITY: Indonesia, Sulawesi, Maluku, Bacan Isl.
DISTRIBUTION: Sulawesi, northeast of Gorontolo; Bacan; adjacent islands.
COMMENT: Type species of *Cynopithecus*; see Fooden, 1969, Bibl. Primatol, No. 10. Includes *nigrescens*; see Groves, 1980; but see also Fooden, 1969, Bibl. Primatol., No. 10, p. 1–9.
PROTECTED STATUS: CITES - Appendix II as Order Primates.
ISIS NUMBER: 5301406008003008001.

Macaca ochreata (Ogilby, 1841). Proc. Zool. Soc. Lond., 1841:56.
TYPE LOCALITY: Unknown.
DISTRIBUTION: S.E. Sulawesi; Muna; Butung.
COMMENT: Perhaps conspecific with *maura*. Fooden, 1969, Bibl. Primatol., No. 10, recognized *brunnescens*; but Groves, 1980, *in* Lindburg, ed., The Macaques, p. 1–9, included it in *ochreata*.
PROTECTED STATUS: CITES - Appendix II as Order Primates.

Macaca radiata (E. Geoffroy, 1812). Ann. Mus. Hist. Nat. Paris, 19:98.
TYPE LOCALITY: India.
DISTRIBUTION: S. India.
COMMENT: Probably conspecific with *sinica* (CPG); but see Fooden, 1980:1–9. Revised by Fooden, 1981, Fieldiana Zool., N.S., 9. See Hill, 1974, for discussion of type locality.
PROTECTED STATUS: CITES - Appendix II as Order Primates.
ISIS NUMBER: 5301406008003009001.

Macaca silenus (Linnaeus, 1758). Syst. Nat., 10th ed., 1:26.
TYPE LOCALITY: "Ceylon" India, W. Ghats.
DISTRIBUTION: S.W. India; W. Ghats.
COMMENT: See Hill, 1974, p. 652, for discussion of type locality.
PROTECTED STATUS: CITES - Appendix I and U.S. ESA - Endangered.
ISIS NUMBER: 5301406008003010001.

Macaca sinica (Linnaeus, 1771). Mant. Plant., p. 521.
TYPE LOCALITY: Probably Sri Lanka.
DISTRIBUTION: Sri Lanka.
COMMENT: See Fooden, 1979, Primates, 20(2), for review and type locality information.
PROTECTED STATUS: CITES - Appendix II as Order Primates and U.S. ESA - Threatened as *M. sinica*.
ISIS NUMBER: 5301406008003011001.

Macaca sylvanus (Linnaeus, 1758). Syst. Nat., 10th ed., 1:25.
TYPE LOCALITY: North Africa, "Barbary coast."
DISTRIBUTION: Morocco; Algeria; Gibraltar (introduced).
COMMENT: See Fooden, 1976, Folia Primatol., 25:226, for the use of this name.
PROTECTED STATUS: CITES - Appendix II as Order Primates.
ISIS NUMBER: 5301406008003012001.

Macaca thibetana (Milne-Edwards, 1870). C. R. Acad. Sci. Paris, 70:341.
TYPE LOCALITY: China, Szechwan, Moupin.
DISTRIBUTION: E. Tibet; Szechwan to Kwangtung (China).
COMMENT: Includes *esau*; see Fooden, 1966, Folia Primatol., 5:160.
PROTECTED STATUS: CITES - Appendix II as Order Primates.
ISIS NUMBER: 5301406008003013001.

Macaca tonkeana (Meyer, 1899). Abh. Zool. Anthrop.-Ethnology. Mus. Dresden, 7(7):3.
TYPE LOCALITY: Indonesia, Sulawesi Tengah, Tonkean.
DISTRIBUTION: C. Sulawesi, south to Latimojong, northeast to Gorontolo; Togean Isls.
COMMENT: Includes *hecki;* see Groves, 1980; but see also Fooden, 1969, Bibl. Primatol.,
No. 10, p. 1–9. Formerly included in *Cynopithecus;* see Fooden, 1969, Bibl.
Primatol, 10:106–115.
PROTECTED STATUS: CITES - Appendix II as Order Primates.

Nasalis E. Geoffroy, 1812. Ann. Mus. Hist. Nat. Paris, 19:89.
REVIEWED BY: D. K. Candland (DKC); R. Fontaine (RF).
COMMENT: Includes *Simias;* see Groves, 1970:639. Szalay and Delson, 1979; and Delson,
1975, Contrib. Primatol., 5:217, considered *Simias* a subgenus; but also see
Krumbiegel, 1978, Saugetierk. Mitt., 26:59–75.
ISIS NUMBER: 5301406008007000000 as *Nasalis.*
5301406008012000000 as *Simias.*

Nasalis concolor Miller, 1903. Smithson. Misc. Coll., 45:67.
TYPE LOCALITY: Indonesia, W. Sumatra, S. Pagai Isl.
DISTRIBUTION: Mentawi Isls.
COMMENT: Type species of *Simias* (Miller, 1903) which was considered a subgenus by
Delson, 1975, Contrib. Primatol., 5:217.
PROTECTED STATUS: CITES - Appendix I as *Simias concolor* and U.S. ESA - Endangered
as *Nasalis (=Simias) concolor.*
ISIS NUMBER: 5301406008012001001 as *Simias concolor.*

Nasalis larvatus (Wurmb, 1787). Verh. Batav. Genootsch., 3:353.
TYPE LOCALITY: Indonesia, W. Kalimantan, Pontianak.
DISTRIBUTION: Borneo.
COMMENT: Subgenus *Nasalis;* see Delson, 1975, Contrib. Primatol., 5:217.
PROTECTED STATUS: CITES - Appendix I and U.S. ESA - Endangered.
ISIS NUMBER: 5301406008007001001.

Papio Muller, 1773. Des Ritters. . . Linne. . .Vollstandiges Natursyst.
REVIEWED BY: D. K. Candland (DKC); R. Fontaine (RF).
COMMENT: Includes *Mandrillus;* see Dandelot, 1974, Part 3:9. CAH and PG consider
Mandrillus a distinct genus. In view of intergradation data, it might be preferable
to combine most of the species in this genus (except *leucophaeus* and *sphinx*); see
Thorington and Groves, 1970:639 (CPG). Includes *Chaeropithecus* as a subgenus;
see Szalay and Delson, 1979:403. Van Gelder, 1977, Am. Mus. Novit., 2635:8
included *Papio* in *Cercopithecus.* For authorship of generic name, see Delson and
Napier, 1976, Bull. Zool. Nomencl., 33:46.
ISIS NUMBER: 5301406008004000000.

Papio hamadryas (Linnaeus, 1758). Syst. Nat., 10th ed., 1:27.
TYPE LOCALITY: Egypt.
DISTRIBUTION: Senegal to Somalia and S. Arabia, south to South Africa.
COMMENT: Includes *anubis, cynocephalus, papio,* and *ursinus;* see Szalay and Delson,
1979:336, and Jolly and Brett, 1973, J. Med. Primatol., 2:85. CAH and PG follow
Dandelot, 1974, Part 3:9, and others in recognizing these as distinct species.
Subgenus *Chaeropithecus;* see Szalay and Delson, 1979. Nagel, 1973, Folia
Primatol., 19:104–165; and Maples and McKern, 1967, *in* Vagtborg, ed., The
Baboon in Medical Research, p. 2, discuss intergradation in these forms.
PROTECTED STATUS: CITES - Appendix II as Order Primates.
ISIS NUMBER: 5301406008004002001 as *P. hamadryas.*
5301406008004001001 as *P. cynocephalus.*

Papio leucophaeus (F. Cuvier, 1807). Ann. Mus. Hist. Nat. Paris, 9:477.
TYPE LOCALITY: Africa.
DISTRIBUTION: S.E. Nigeria; Cameroun, north of the Sanaga River and just south of it;
Bioko.

COMMENT: Delson and Napier, 1976, Bull. Zool. Nomencl., 33:46, considered this
species in the subgenus *Papio*. Placed in subgenus *Mandrillus* by Dandelot, 1974,
Part 3:9. See Grubb, 1973, Folia Primatol., 20:161–177, for details of distribution.
PROTECTED STATUS: CITES - Appendix I as *P. (=Mandrillus) leucophaeus* and U.S. ESA -
Endangered as *P. leucophaeus*.
ISIS NUMBER: 5301406008004003001.

Papio sphinx (Linnaeus, 1758). Syst. Nat., 10th ed., 1:25.
TYPE LOCALITY: Cameroun, Ja River, Bitye.
DISTRIBUTION: Cameroun, south of the Sanaga River; Rio Muni (Equatorial Guinea);
Gabon; Congo Republic.
COMMENT: Delson and Napier, 1976, Bull. Zool. Nomencl., 33:46, considered *sphinx*
the type of subgenus *Papio*. Placed in subgenus *Mandrillus* by Dandelot, 1974, Part
3:9. See Grubb, 1973, Folia Primatol., 20:161–177, for details of distribution.
PROTECTED STATUS: CITES - Appendix I as *P. (=Mandrillus) sphinx* and U.S. ESA -
Endangered as *P. sphinx*.
ISIS NUMBER: 5301406008004004001.

Presbytis Eschscholtz, 1821. Reise (Kotzebue), 3:196.
REVIEWED BY: A. C. Ziegler (ACZ).
COMMENT: *Semnopithecus* and *Trachypithecus* were recognized as valid genera by
Hooijer, 1962, Zool. Meded. Rijks. Mus. Nat. Hist. Leiden, No. 55:20–24; but see
Thorington and Groves, 1970:629–647. Szalay and Delson, 1979:402 included
these and *Kasi* as subgenera.
ISIS NUMBER: 5301406008008000000.

Presbytis aygula (Linnaeus, 1758). Syst. Nat., 10th ed., 1:27.
TYPE LOCALITY: Indonesia, W. Java.
DISTRIBUTION: W. and C. Java.
COMMENT: Type species of *Presbytis*; see Szalay and Delson, 1979:402.
PROTECTED STATUS: CITES - Appendix II as Order Primates.
ISIS NUMBER: 5301406008008001001.

Presbytis cristata (Raffles, 1821). Trans. Linn. Soc. Lond., 13:244.
TYPE LOCALITY: Indonesia, Sumatra, Bengkulen.
DISTRIBUTION: Burma and Indochina to Borneo and Lombok.
COMMENT: Type species of *Trachypithecus*; see Szalay and Delson, 1979:402.
PROTECTED STATUS: CITES - Appendix II as Order Primates.
ISIS NUMBER: 5301406008008002001 as *P. cristatus (sic)*.

Presbytis entellus (Dufresne, 1797). Bull. Sci. Soc. Philom. Paris, ser. 1, 7:49.
TYPE LOCALITY: India, Bengal.
DISTRIBUTION: India; Nepal; S. Tibet; Sri Lanka; and perhaps Kashmir.
COMMENT: Type species of *Semnopithecus*; perhaps a valid genus (CPG); but also see
Szalay and Delson, 1979, who considered it a subgenus.
PROTECTED STATUS: CITES - Appendix I and U.S. ESA - Endangered.
ISIS NUMBER: 5301406008008003001.

Presbytis francoisi (Pousargues, 1898). Bull. Mus. Hist. Nat. Paris, 4:319.
TYPE LOCALITY: China, Kwangsi, Lungchow.
DISTRIBUTION: N. Vietnam; C. Laos; Kwangsi (China)(SW).
COMMENT: Formerly included in *Trachypithecus*; considered a subgenus by Szalay and
Delson, 1979:402. Includes *leucocephalus* (CPG).
PROTECTED STATUS: CITES - Appendix II as Order Primates and U.S. ESA - Endangered
as *P. francoisi*.
ISIS NUMBER: 5301406008008004001.

Presbytis frontata (Muller, 1838). Tijdschr. Nat. Gesch. Physiol., 5:136.
TYPE LOCALITY: Indonesia, S. Kalimantan.
DISTRIBUTION: N.W. and E. Borneo, except southwest.
COMMENT: Subgenus *Presbytis*; see Szalay and Delson, 1979:402.
PROTECTED STATUS: CITES - Appendix II as Order Primates.
ISIS NUMBER: 5301406008008005001 as *P. frontatus (sic)*.

Presbytis geei Khajuria, 1956. Ann. Mag. Nat. Hist., ser. 12, 9:86.
TYPE LOCALITY: India, Assam, Goalpara Dist., Jamduar Forest Rest House, E. bank of Sankosh River.
DISTRIBUTION: Between Sankosh and Manas Rivers, Indo-Bhutan border.
COMMENT: Formerly included in *Trachypithecus;* considered a subgenus by Szalay and Delson, 1979:402. For authorship of the name, see Biswas, 1967, J. Bombay Nat. Hist. Soc., 63:429–431. May prove to be a subspecies of *pileata* (CAH).
PROTECTED STATUS: CITES - Appendix I and U.S. ESA - Endangered.
ISIS NUMBER: 5301406008008006001.

Presbytis hosei (Thomas, 1889). Proc. Zool. Soc. Lond., 1889:159.
TYPE LOCALITY: Malaysia, Sarawak, Niah.
DISTRIBUTION: N. and E. Borneo.
COMMENT: Subgenus *Presbytis;* see Szalay and Delson, 1979:402. Separated from *aygula* by Medway, 1970:544.
PROTECTED STATUS: CITES - Appendix II as Order Primates.
ISIS NUMBER: 5301406008008007001.

Presbytis johnii (Fischer, 1829). Synops. Mamm., p. 25.
TYPE LOCALITY: India, Tellicherry.
DISTRIBUTION: S. India.
COMMENT: Subgenus *Kasi;* see Szalay and Delson, 1979:402. Probably conspecific with *senex* (CPG).
PROTECTED STATUS: CITES - Appendix II as Order Primates.
ISIS NUMBER: 5301406008008008001.

Presbytis melalophos (Raffles, 1821). Trans. Linn. Soc. Lond., 13:245.
TYPE LOCALITY: Indonesia, Sumatra, Bengkulen.
DISTRIBUTION: Malay Peninsula; Borneo; Sumatra; Junk Seylon; Kundur Isl, Bintang Isl. (Rhio Arch.); Batu Isls. (W. Sumatra); Bunguran Isl. (N. Natuna Isls.).
COMMENT: Subgenus *Presbytis;* see Szalay and Delson, 1979:402. Includes *femoralis* which was regarded as a separate species by Wilson and Wilson, 1977, Yearb. Phys. Anthro., 20:217–222.
PROTECTED STATUS: CITES - Appendix II as Order Primates.
ISIS NUMBER: 5301406008008009001.

Presbytis obscura (Reid, 1837). Proc. Zool. Soc. Lond., 1837:14.
TYPE LOCALITY: Malaysia, Malacca.
DISTRIBUTION: S. Thailand and Malay Peninsula, and small adjacent islands.
COMMENT: Formerly included in *Trachypithecus;* considered a subgenus by Szalay and Delson, 1979:402.
PROTECTED STATUS: CITES - Appendix II as Order Primates.
ISIS NUMBER: 5301406008008010000 as *P. obscurus (sic).*

Presbytis phayrei Blyth, 1847. J. Asiat. Soc. Bengal, 16:733.
TYPE LOCALITY: Burma, Arakan.
DISTRIBUTION: Laos; Burma; C. Vietnam; C. and N. Thailand. Yunnan (China) (SW).
COMMENT: Formerly included in *Trachypithecus;* considered a subgenus by Szalay and Delson, 1979:402.
PROTECTED STATUS: CITES - Appendix II as Order Primates.
ISIS NUMBER: 5301406008008011001.

Presbytis pileata (Blyth, 1843). J. Asiat. Soc. Bengal, 12:174.
TYPE LOCALITY: India, Assam.
DISTRIBUTION: Assam (India); N.W. Burma.
COMMENT: Formerly included in *Trachypithecus;* considered a subgenus by Szalay and Delson, 1979:402.
PROTECTED STATUS: CITES - Appendix I and U.S. ESA - Endangered as *P. pileatus (sic).*
ISIS NUMBER: 5301406008008012001 as *P. pileatus (sic).*

Presbytis potenziani (Bonaparte, 1856). C. R. Acad. Sci. Paris, 43:412.
TYPE LOCALITY: Indonesia, W. Sumatra, Sipora Isl.
DISTRIBUTION: Mentawi Isls.

COMMENT: Perhaps subgenus *Presbytis*; see Szalay and Delson, 1979:402.
PROTECTED STATUS: CITES - Appendix I and U.S. ESA - Threatened.
ISIS NUMBER: 5301406008008013001.

Presbytis rubicunda (Muller, 1838). Tijdschr. Nat. Gesch. Physiol., 5:137.
TYPE LOCALITY: Indonesia, S. Kalimantan, Mt. Sekumbang (S.E. of Banjermassin).
DISTRIBUTION: Borneo; Karimata Isl. (S.W. Borneo).
COMMENT: Subgenus *Presbytis*; see Szalay and Delson, 1979:402.
PROTECTED STATUS: CITES - Appendix II as Order Primates.
ISIS NUMBER: 5301406008008014001 as *P. rubicundus (sic)*.

Presbytis senex (Erxleben, 1777). Regn. Anim., p. 24.
TYPE LOCALITY: Sri Lanka, hill country of South.
DISTRIBUTION: Sri Lanka.
COMMENT: Type of subgenus *Kasi*; see Szalay and Delson, 1979:402. Probably includes
    johnii (ED).
PROTECTED STATUS: CITES - Appendix II as Order Primates and U.S. ESA - Threatened
    as *P. senex*.
ISIS NUMBER: 5301406008008015001.

Presbytis thomasi (Collett, 1893). Proc. Zool. Soc. Lond., 1892:613.
TYPE LOCALITY: Indonesia, Sumatra, Aceh, Langkat.
DISTRIBUTION: Sumatra.
COMMENT: Subgenus *Presbytis*; see Szalay and Delson, 1979:402. Separated from *aygula*
    by Medway, 1970:544.
PROTECTED STATUS: CITES - Appendix II as Order Primates.
ISIS NUMBER: 5301406008008016001.

Pygathrix E. Geoffroy, 1812. Ann. Mus. Hist. Nat. Paris, 19:90.
COMMENT: Includes *Rhinopithecus*; see Groves, 1970:555–587; and Szalay and Delson,
    1979:404. CAH considers *Rhinopithecus* a distinct genus.
ISIS NUMBER: 5301406008010000000 as *Pygathrix*.
    5301406008011000000 as *Rhinopithecus*.

Pygathrix avunculus (Dollman, 1912). Abstr. Proc. Zool. Soc. Lond., 106:18.
TYPE LOCALITY: Vietnam, Songkoi River, Yen Bay.
DISTRIBUTION: N. Vietnam.
COMMENT: Subgenus *Rhinopithecus*; see Szalay and Delson, 1979:404.
PROTECTED STATUS: CITES - Appendix II as Order Primates and U.S. ESA - Threatened
    as *Rhinopithecus avunculus*.
ISIS NUMBER: 5301406008011001001 as *Pygathrix (=Rhinopithecus) avunculus*.

Pygathrix brelichi (Thomas, 1903). Proc. Zool. Soc. Lond., 1903:224.
TYPE LOCALITY: China, N. Kweichow, Van Gin Shan Range.
DISTRIBUTION: Van Gin Shan Range (Kweichow, China).
COMMENT: Subgenus *Rhinopithecus*; considered a valid species, by Groves, 1970:569.
    Probably a subspecies of *roxellana* (SW).
PROTECTED STATUS: CITES - Appendix II as Order Primates.
ISIS NUMBER: 5301406008011002001 as *Rhinopithecus brelichi*.

Pygathrix nemaeus (Linnaeus, 1771). Mant. Plant., p. 521.
TYPE LOCALITY: Cochin-China (Indo-China).
DISTRIBUTION: S. and C. Vietnam; E. Laos; E. Kampuchea; Hainan Isl. (China).
COMMENT: Type species of subgenus *Pygathrix*. Includes *nigripes* (ED).
PROTECTED STATUS: CITES - Appendix I and U.S. ESA - Endangered.
ISIS NUMBER: 5301406008010001001.

Pygathrix roxellana (Milne-Edwards, 1870). C. R. Acad. Sci. Paris, 70:341.
TYPE LOCALITY: China, Szechwan, Moupin.
DISTRIBUTION: N.W. Yunnan, Szechwan, S. Kansu (China).
COMMENT: Type species of subgenus *Rhinopithecus*; see Groves, 1970:569; Szalay and
    Delson, 1979:404. Includes *bieti*; see Napier and Napier, 1967:353. Probably
    includes *brelichi* (SW).
PROTECTED STATUS: CITES - Appendix II as Order Primates.
ISIS NUMBER: 5301406008011003001 as *Rhinopithecus roxellanae (sic)*.

Theropithecus I. Geoffroy, 1843. Arch. Mus. Hist. Nat. Paris, 1841, 2:576.
  COMMENT: Considered a distinct genus by Cronin and Meikle, 1979, Syst. Zool.,
    28:259. Van Gelder, 1977, Am. Mus. Novit., 2635:8, included this genus in
    Cercopithecus.
  ISIS NUMBER: 5301406008005000000.

  Theropithecus gelada (Ruppell, 1835). Neue Wirbelt. Fauna Abyssin. Gehorig. Saugeth., p.
    5.
    TYPE LOCALITY: Ethiopia, Semyen (Simien).
    DISTRIBUTION: N. and E. Ethiopia.
    PROTECTED STATUS: CITES - Appendix II as Order Primates and U.S. ESA - Threatened
      as *T. gelada.*
    ISIS NUMBER: 5301406008005001001.

Family Hylobatidae
    REVIEWED BY: C. P. Groves (CPG); C. A. Hill (CAH); J. T. Marshall (JTM).
    COMMENT: Vaughan, 1978:39-40, included this family in Pongidae; but see Delson
      and Andrews, 1975:441. Szalay and Delson, 1979:461, included Hylobatidae in
      Hominidae.

  Hylobates Illiger, 1811. Prodr. Syst. Mamm. et Avium., p. 67.
    COMMENT: Includes *Symphalangus;* see Anderson, 1967, *in* Anderson and Jones, p. 175;
      and *Nomascus;* see Corbet and Hill, 1980:91. Revised by Groves, 1972, 1:1-89.
      Reviewed by Marshall and Marshall, 1976, Science, 193:235-237.
    PROTECTED STATUS: CITES - Appendix I as *Hylobates* spp. and U.S. ESA - Endangered as
      Hylobates spp. (including *Nomascus* spp.).
    ISIS NUMBER: 5301406009001000000.

  Hylobates agilis F. Cuvier, 1821. Hist. Nat. Mamm. (Geoffroy and F. Cuvier), p. 3.
    TYPE LOCALITY: Indonesia, West Sumatra.
    DISTRIBUTION: Malay Peninsula from the Mudah and Thepha rivers on the N. to the
      Perak and Kelanton rivers on the S.; Sumatra, S.E. of Lake Toba and the Singkil
      River; Kalimantan (Indonesian Borneo) between the Kapuas and Barito rivers.
    COMMENT: Possibly conspecific with *lar* (CPG).
    PROTECTED STATUS: CITES - Appendix I as *Hylobates* spp. and U.S. ESA - Endangered as
      Hylobates spp. (including *Nomascus* spp.).
    ISIS NUMBER: 5301406009001001001.

  Hylobates concolor (Harlan, 1826). J. Acad. Nat. Sci. Phila., ser. 5, 4:231.
    TYPE LOCALITY: Vietnam, Tonkin.
    DISTRIBUTION: Hainan Isl. (China); E. of the Mekong River in S. Yunnan (China), Laos,
      Vietnam and Kampuchea, N. to Red River in Vietnam.
    COMMENT: Type species of *Nomascus* which was recognized as a separate genus by
      Lekagul and McNeely, 1977:308; but see Corbet and Hill, 1980:91, and Szalay and
      Delson, 1979:461 .
    PROTECTED STATUS: CITES - Appendix I as *Hylobates* spp. and U.S. ESA - Endangered as
      Hylobates spp. (including *Nomascus* spp.).
    ISIS NUMBER: 5301406009001002001.

  Hylobates hoolock (Harlan, 1834). Trans. Am. Philos. Soc., 4:52.
    TYPE LOCALITY: India, Assam, Garo Hills.
    DISTRIBUTION: Between the Brahmaputra and Salween rivers in Assam (India); Burma;
      and Yunnan (China).
    PROTECTED STATUS: CITES - Appendix I as *Hylobates* spp. and U.S. ESA - Endangered as
      Hylobates spp. (including *Nomascus* spp.).
    ISIS NUMBER: 5301406009001003001.

  Hylobates klossii (Miller, 1903). Smithson. Misc. Coll., 45:70.
    TYPE LOCALITY: Indonesia, West Sumatra, S. Pagai Isl.
    DISTRIBUTION: Mentawi Isls. (off W. Sumatra).
    PROTECTED STATUS: CITES - Appendix I as *Hylobates* spp. and U.S. ESA - Endangered as
      Hylobates spp. (including *Nomascus* spp.).
    ISIS NUMBER: 5301406009001004001.

Hylobates lar (Linnaeus, 1771). Mant. Plant., p. 521.
  TYPE LOCALITY: Malaysia, Malacca.
  DISTRIBUTION: Between the Salween and Mekong rivers from S. Yunnan (China) south
    to the Mun River (Thailand) and the Mudah and Thepha rivers on the Malay
    Peninsula; S. Malay Peninsula S. of the Perak and Kelantan rivers; Sumatra N.W.
    of Lake Toba and the Singkil River; E. and S. Burma.
  COMMENT: Type locality restricted by Kloss, 1929, Proc. Zool. Soc. Lond., 1929:117.
  PROTECTED STATUS: CITES - Appendix I as *Hylobates* spp. and U.S. ESA - Endangered as
    Hylobates spp. (including *Nomascus* spp.).
  ISIS NUMBER: 5301406009001005001.

Hylobates moloch (Audebert, 1798). Hist. Nat. Singes Makis, 1st fasc., sect. 2, pl. 2.
  TYPE LOCALITY: Indonesia, W. Java, Mt. Salak.
  DISTRIBUTION: Java.
  COMMENT: Possibly conspecific with *lar* (CPG). Type locality restricted by Sody, 1949,
    Treubia, 20:121–190.
  PROTECTED STATUS: CITES - Appendix I as *Hylobates* spp. and U.S. ESA - Endangered as
    Hylobates spp. (including *Nomascus* spp.).
  ISIS NUMBER: 5301406009001006001.

Hylobates muelleri Martin, 1841. Nat. Hist. Mamm. Anim., p. 444.
  TYPE LOCALITY: Indonesia, Kalimantan, "Southeast Borneo."
  DISTRIBUTION: Borneo from the N. bank of the Kapuas River clockwise around the
    island to the E. bank of the Barito River.
  COMMENT: Probably conspecific with *lar* (CPG). Type locality restricted by Lyon, 1911,
    Proc. U.S. Nat. Mus., 40:142.
  PROTECTED STATUS: CITES - Appendix I as *Hylobates* spp. and U.S. ESA - Endangered as
    Hylobates spp. (including *Nomascus* spp.).

Hylobates pileatus (Gray, 1861). Proc. Zool. Soc. Lond., 1861:136.
  TYPE LOCALITY: Kampuchea ("Cambodia").
  DISTRIBUTION: S.E. Thailand and Kampuchea S. of the Mun and Takhrong rivers and
    W. of the Mekong River.
  PROTECTED STATUS: CITES - Appendix I as *Hylobates* spp. and U.S. ESA - Endangered as
    Hylobates spp. (including *Nomascus* spp.).
  ISIS NUMBER: 5301406009001007001.

Hylobates syndactylus (Raffles, 1821). Trans. Linn. Soc. Lond., 13:241.
  TYPE LOCALITY: Indonesia, W. Sumatra, Bengkeulen.
  DISTRIBUTION: Barisan Mountains of Sumatra; mountains of Malay Peninsula S. of
    Perak River.
  COMMENT: Type species of *Symphalangus*; see Groves, 1972, *in* Rumbaugh, ed., Gibbon
    and Siamang, 1:1–89. CAH considers *Symphalangus* a distinct genus.
  PROTECTED STATUS: CITES - Appendix I as *Hylobates* spp. and U.S. ESA - Endangered as
    Hylobates spp. (including *Nomascus* spp.).
    CITES - Appendix I and U.S. ESA - Endangered as *Symphalangus syndactylus*.
  ISIS NUMBER: 5301406009001008001.

Family Pongidae
  REVIEWED BY: C. P. Groves (CPG).
  COMMENT: May be artificial; some or all of the genera should probably be placed in
    Hominidae; see Delson and Andrews, 1975:441, and Szalay and Delson, 1979:466.
  PROTECTED STATUS: CITES - Appendix I as Pongidae.
  ISIS NUMBER: 5301406009000000000.

Gorilla I. Geoffroy, 1852. C. R. Acad. Sci. Paris, 36:933.
  COMMENT: Perhaps a subgenus of *Pan*; see Tuttle, 1967, Am. J. Phys. Anthropol.,
    26:171–206.
  ISIS NUMBER: 5301406009002000000.

Gorilla gorilla (Savage and Wyman, 1847). Boston J. Nat. Hist., 5:417.
  TYPE LOCALITY: Gabon, Gabon Estuary, Mpongwe country.
  DISTRIBUTION: S.E. Nigeria; Cameroun; Rio Muni; Congo Republic; Gabon; N. and E.
    Zaire; S.W. Uganda; N. Rwanda.
  PROTECTED STATUS: CITES - Appendix I as Pongidae and U.S. ESA - Endangered as *G.
    gorilla*.
  ISIS NUMBER: 5301406009002001001.

Pan Oken, 1816. Lehrb. Naturgesch., ser. 3, 2:xi.
  REVIEWED BY: C. A. Hill (CAH).
  COMMENT: In accordance with the provisions of Article 80 of the ICZN Code, *Pan* is
    used instead of *Chimpansee.*
  ISIS NUMBER: 5301406009003000000.

Pan paniscus Schwartz, 1929. Rev. Zool. Bot. Afr., 16:4.
  REVIEWED BY: H. J. Coolidge (HJC).
  TYPE LOCALITY: Zaire, S. of the upper Maringa River, 30 km S. of Befale.
  DISTRIBUTION: Congo Basin of Zaire, on S. side of Congo River.
  PROTECTED STATUS: CITES - Appendix I as Pongidae and U.S. ESA - Endangered as *P.
    paniscus*.
  ISIS NUMBER: 5301406009003001001.

Pan troglodytes (Gmelin, 1788). Syst. Nat., 13th ed., 1:26.
  TYPE LOCALITY: Gabon, Mayoumba.
  DISTRIBUTION: S. Cameroun; Gabon; S. Congo Republic; Uganda; W. Tanzania; E. and
    N. Zaire; W. Central African Republic; Guinea to W. Nigeria, south to Congo
    River in W. Africa.
  COMMENT: For authorship of this name, see Simonetta, 1957, Cat. Sin. Annot.
    Ominoidi Fas., p. 78.
  PROTECTED STATUS: CITES - Appendix I as Pongidae and U.S. ESA - Endangered as *P.
    troglodytes*.
  ISIS NUMBER: 5301406009003002001.

Pongo Lacepede, 1799. Tabl. Div. Subd. Orders Genres Mamm., p. 4.
  ISIS NUMBER: 5301406009004000000.

Pongo pygmaeus (Linnaeus, 1760). Amoenit. Acad., 6:68.
  TYPE LOCALITY: "Borneo."
  DISTRIBUTION: Sumatra, N.W. of Lake Toba; discontinuous in Borneo.
  COMMENT: Reviewed by Groves, 1971, Mamm. Species, 4:1-6.
  PROTECTED STATUS: CITES - Appendix I as Pongidae and U.S. ESA - Endangered as *P.
    pygmaeus*.
  ISIS NUMBER: 5301406009004001001.

Family Hominidae
  REVIEWED BY: C. P. Groves (CPG); C. A. Hill (CAH).
  COMMENT: Should probably include *Pan* and *Gorilla,* and possibly also *Pongo;* see
    Delson and Andrews, 1975:441.
  ISIS NUMBER: 5301406010000000000.

Homo Linnaeus, 1758. Syst. Nat., 10th ed., 1:20.
  ISIS NUMBER: 5301406010001000000.

Homo sapiens Linnaeus, 1758. Syst. Nat., 10th ed., 1:20.
  TYPE LOCALITY: Sweden, Uppsala.
  DISTRIBUTION: Cosmopolitan.
  PROTECTED STATUS: CITES - Appendix II as Order Primates.
  ISIS NUMBER: 5301406010001001001.

ORDER CARNIVORA

COMMENT: Includes Pinnipedia (Otariidae, Odobenidae, Phocidae); see Tedford, 1977; but also see Hall, 1981, and Corbet and Hill, 1980.
ISIS NUMBER: 5301412000000000000.

Family Canidae

REVIEWED BY: A. Langguth (AL); J. Ramirez-Pulido (JRP)(Mexico); O. L. Rossolimo (OLR)(U.S.S.R.); H. J. Stains (HJS); R. G. Van Gelder (RGVG).

COMMENT: Several alternative classifications have been published in recent years; see Langguth, 1969, Z. Wiss. Zool., 179:1–188; Langguth, 1975, in Fox, The Wild Canids, pp. 192–206; Clutton-Brock et al., 1976, 29(3):177–199; Van Gelder, 1978:1–10. Three groups of closely related Recent genera or subgenera may be distinguished on the basis of anatomical as well as chromosomal evidence: 1) Canis, Lycaon, Cuon; 2) Vulpes, Fennecus, Alopex, Otocyon and Urocyon; 3) all other recent South American taxa (AL).
ISIS NUMBER: 5301412001000000000.

Alopex Kaup, 1829. Skizz. Europ. Theirwelt, 1:85.
COMMENT: Included in Vulpes as a subgenus by Bobrinskii et al., 1965, [Key to the mammals of the U.S.S.R.], "Proveshchenie," Moscow, p. 127.
ISIS NUMBER: 5301412001001000000.

Alopex lagopus (Linnaeus, 1758). Syst. Nat., 10th ed., 1:40.
TYPE LOCALITY: N. Scandinavia, Lapland.
DISTRIBUTION: Circumpolar, entire tundra zone of the Holarctic, including most of the Arctic islands.
COMMENT: Placed in genus Canis, subgenus Alopex by Van Gelder, 1978:8.
ISIS NUMBER: 5301412001001001001.

Canis Linnaeus, 1758. Syst. Nat., 10th ed., 1:38.
REVIEWED BY: A. C. Ziegler (ACZ).
COMMENT: Van Gelder, 1978:1–10 included Alopex, Atelocynus, Cerdocyon, Pseudalopex, Lycalopex, Dusicyon, and Vulpes as subgenera. However, this arrangement is not currently employed by most mammalogists; see Corbet, 1978:161; Corbet and Hill, 1980; Hall, 1981:923; Gromov and Baranova, 1981:239.
ISIS NUMBER: 5301412001003000000.

Canis adustus Sundevall, 1846. Ofv. Kongl. Svenska Vet.-Akad. Forhandl. Stockholm, 3:121.
TYPE LOCALITY: South Africa, Transvaal, Magaliesberg (W. Pretoria), "Caffraria Interiore."
DISTRIBUTION: Open woodland and semi-arid grassland from Senegal to Ethiopia, south to N. Namibia, N. Botswana, Zimbabwe, Mozambique and N. South Africa.
ISIS NUMBER: 5301412001003001001.

Canis aureus Linnaeus, 1758. Syst. Nat., 10th ed., 1:40.
TYPE LOCALITY: Iran, Lar Prov.
DISTRIBUTION: N. and E. Africa, south to Senegal, Nigeria, and Tanzania; S. W. Asia; S. E. Europe; Transcaucasia; C. Asia; Iran; Afghanistan; S. Asia to Thailand, including Sri Lanka.
ISIS NUMBER: 5301412001003002001.

Canis latrans Say, 1823. In Long, Account of an Exped. from Pittsburgh to the Rocky Mtns., 1:168.
REVIEWED BY: M. Bekoff (MB).
TYPE LOCALITY: U.S.A., Nebraska, Washington Co., Engineer Cantonment, about 12 mi. (19.2 km) S. E. Blair.
DISTRIBUTION: Originally may have occurred W. of the Mississippi R., N. to about 55° N. lat., and south to Mexico City. Has extended its range north to N. Alaska (U.S.A.), Northwest Territories and Hudson Bay (Canada), south to Costa Rica, and east to the Atlantic coast.
COMMENT: Reviewed by Bekoff, 1977, Mamm. Species, 79:1–9. See Young, 1951, The clever coyote, Wildl. Mng. Inst., Washington, D.C., p. 29, for a discussion of distribution.
ISIS NUMBER: 5301412001003004001.

Canis lupus Linnaeus, 1758. Syst. Nat., 10th ed., 1:39.
TYPE LOCALITY: Sweden, Uppsala.
DISTRIBUTION: North America south to 20° N., in Oaxaca (Mexico); Europe; Asia, including the Arabian Peninsula and Japan; not in the Indochinese peninsula or southern India.
COMMENT: Probably ancestor of and conspecific with the domestic dog, *familiaris*, (RGVG); *familiaris* has page priority over *lupus* in Linnaeus, 1758, but both were published simultaneously, and *lupus* has been universally used for this species. Extinct in much of its former range (AL). Reviewed by Mech, 1974, Mamm. Species, 37:1–6.
PROTECTED STATUS: CITES - Appendix I as *C. lupus* (Indian, Pakistan, Bhutan, and Nepal populations only).
CITES - Appendix II (all populations not listed on Appendix I) U.S. ESA - Endangered, 48 conterminous states (other than Minnesota) and Mexico. U.S. ESA - Threatened, Minnesota populations only.
ISIS NUMBER: 5301412001003005001 as *C. lupus*.
5301412001003003001 as *C. familiaris*.

Canis mesomelas Schreber, 1778. Saugethiere, pl. 95; text 3:370, 568.
TYPE LOCALITY: South Africa, Cape Prov., Cape of Good Hope.
DISTRIBUTION: Africa, south of the tropical rain-forest in the west and as far north as Ethiopia and Sudan in the east.
ISIS NUMBER: 5301412001003006001.

Canis rufus Audubon and Bachman, 1851. The Viviparous Quadrupeds of North America, 2:240.
TYPE LOCALITY: U.S.A., Texas, 15 mi. (24 km) W. Austin.
DISTRIBUTION: S. E. and S. C. U.S.A., from Florida to C. Texas and north to S. Indiana and Missouri.
COMMENT: The widely used name *niger* has been rejected by opinion 447 of the ICZN. The validity of *rufus* as a species has been questioned by Clutton-Brock *et al.*, 1976, 29(3):143, due to the existence of natural hybrids with *lupus* and *latrans*. Natural hybridization may be a consequence of habitat disruption by man; see Paradiso and Nowak, 1972, U.S. Fish and Wild. Serv., Spec. Sci. Rep. Wildl., 145; Nowak, 1979, Univ. Kans. Mus. Nat. Hist. Monogr., 6:1–154, who provided evidence for specific distinctness. Reviewed by Paradiso and Nowak, 1972, Mamm. Species, 22:1–4. Type locality restricted by Goldman, 1937, J. Mammal., 18:38.
PROTECTED STATUS: U.S. ESA - Endangered.
ISIS NUMBER: 5301412001003007001.

Canis simensis Ruppell, 1835. Neue Wirbelt. Fauna Abyssin. Gehorig. Saugeth., 1:39, pl. 14.
TYPE LOCALITY: Ethiopia, mountains of Simen.
DISTRIBUTION: C. Ethiopia.
PROTECTED STATUS: U.S. ESA - Endangered as *C. (=Simia) simensis*.
ISIS NUMBER: 5301412001003008001.

Chrysocyon H. Smith, 1839. Jardine's Natur. Libr., 9:241.
REVIEWED BY: K. L. Anderson (KLA).
ISIS NUMBER: 5301412001005000000.

Chrysocyon brachyurus (Illiger, 1815). Abh. Preuss. Akad. Wiss., 1804–1811:121.
TYPE LOCALITY: Paraguay, "swamps of Paraguay."
DISTRIBUTION: N. E. Argentina; Paraguay; lowlands of Bolivia; Brazil from Rio Grande do Sul to Minas Gerais, Goias and Mato Grosso.
PROTECTED STATUS: CITES - Appendix II and U.S. ESA - Endangered.
ISIS NUMBER: 5301412001005001001.

Cuon Hodgson, 1838. Ann. Mag. Nat. Hist., 1:152.
ISIS NUMBER: 5301412001006000000.

Cuon alpinus (Pallas, 1811). Zoogr. Rosso-Asiat., 1:34.
  TYPE LOCALITY: U.S.S.R., Amurskaya Obl., Udskii-Ostrog.
  DISTRIBUTION: Java; Sumatra; Malaysia; montane forest areas of the Indian peninsula
     and N. Pakistan through Tibet; Indochina and China to Korea and Ussuri region
     (U.S.S.R.), and mountains of S. Siberia and N. Mongolia.
  COMMENT: Reviewed by Cohen, 1978, Mamm. Species, 100:1–3.
  PROTECTED STATUS: CITES - Appendix II and U.S. ESA - Endangered.
  ISIS NUMBER: 5301412001006001001.

Dusicyon H. Smith, 1839. Jardine's Natur. Libr., 9:243.
  REVIEWED BY: K. L. Anderson (KLA).
  COMMENT: Includes *Atelocynus*, *Cerdocyon*, *Lycalopex*, and *Pseudalopex*; see
     comments under *culpaeus*, *microtis*, *thous*, and *vetulus*.
  ISIS NUMBER: 5301412001007000000 as *Dusicyon*.
                5301412001002000000 as *Atelocynus*.
                5301412001004000000 as *Cerdocyon*.
                5301412001009000000 as *Lycalopex*.

Dusicyon australis (Kerr, 1792). Anim. Kingdom, p. 144.
  TYPE LOCALITY: Falkland Islands (U.K.).
  DISTRIBUTION: Falkland Islands.
  COMMENT: Subgenus *Dusicyon*; see Cabrera, 1958:229, and Van Gelder, 1978:7.
     Exterminated by the end of the 19th century (AL).

Dusicyon culpaeus (Molina, 1782). Sagg. Stor. Nat. Chile, p. 293.
  TYPE LOCALITY: Chile, Santiago Prov.
  DISTRIBUTION: From Tierra del Fuego along the Andes of Chile and Argentina, the
     highlands of Bolivia, Peru, Ecuador, and Colombia.
  COMMENT: Subgenus *Dusicyon* or *Pseudalopex*; see Cabrera, 1958:230; Van Gelder,
     1978:7. *Pseudalopex* has been considered a synonym of *Dusicyon* by Cabrera,
     1958:229, and Clutton-Brock *et al.*, 1976, 29(3):166; a subgenus of *Dusicyon* by
     Langguth, 1969, Z. Wiss. Zool., 179:176; a subgenus of *Canis* by Langguth,
     1975:193, and Van Gelder, 1978:7. Includes *culpaeolus* (part) and *inca* (part),
     mismatched skin and skull; see Langguth, 1967, Neotropica, 13:21–28. AL
     considers *Pseudalopex* a distinct genus. Type locality restricted by Cabrera, 1931, J.
     Mammal., 12:62.
  PROTECTED STATUS: CITES - Appendix II.
  ISIS NUMBER: 5301412001007001001.

Dusicyon griseus (Gray, 1837). Mag. Nat. Hist., 1:578.
  TYPE LOCALITY: Chile, Coast of Magellan Straits.
  DISTRIBUTION: Atacama (Chile) and Santiago del Estero (Argentina) south to Tierra del
     Fuego.
  COMMENT: Subgenus *Dusicyon* or *Pseudalopex*; see Cabrera, 1958:233; Van Gelder,
     1978:7. *Pseudalopex* has been considered a synonym of *Dusicyon* by Cabrera,
     1958:229, and Clutton-Brock *et al.*, 1976, 29(3):166; a subgenus of *Dusicyon* by
     Langguth, 1969, Z. Wiss. Zool., 179:176; a subgenus of *Canis* by Langguth,
     1975:193, and Van Gelder, 1978:7. Includes *fulvipes*; see Corbet and Hill, 1980:93;
     Langguth, 1969, Z. Wiss. Zool., p. 179.
  PROTECTED STATUS: CITES - Appendix II as *D. griseus*.
     CITES - Appendix II as *D. fulvipes*.
  ISIS NUMBER: 5301412001007002001.

Dusicyon gymnocercus (G. Fischer, 1814). Zoognosia, 3:178.
  TYPE LOCALITY: Paraguay, vicinity of Asuncion.
  DISTRIBUTION: Argentina, north of Rio Negro; Paraguay; Uruguay; S. Brazil; E. Bolivia.
  COMMENT: Subgenus *Dusicyon* or *Pseudalopex*; see Cabrera, 1958:234; Van Gelder,
     1978:7. *Pseudalopex* has been considered a synonym of *Dusicyon* by Cabrera,
     1958:229; and Clutton-Brock *et al.*, 1976, 29(3):166; a subgenus of *Dusicyon* by
     Langguth, 1969, Z. Wiss. Zool., 179:176; a subgenus of *Canis* by Langguth,
     1975:193; and Van Gelder, 1978:7. Includes *culpaeolus (part)* and *inca* (part),

mismatched skin and skull; see Langguth, 1967, Neotropica, 13:21–38. AL considers *inca* a synonym of *culpaeus*. Type locality restricted by Cabrera, 1958:235.
ISIS NUMBER: 5301412001007003001.

Dusicyon microtis (Sclater, 1883). Proc. Zool. Soc. Lond., 1882:631.
REVIEWED BY: H. E. Evans (HEE).
TYPE LOCALITY: Brazil, Para, S. bank Rio Amazonas.
DISTRIBUTION: Amazonian region of Brazil, Peru, Ecuador, and Colombia; perhaps Bolivia.
COMMENT: Subgenus *Atelocynus*; see Cabrera, 1958:236; Van Gelder, 1978:8, but also see Hershkovitz, 1961, Fieldiana Zool., 39:505–523. Corbet and Hill, 1980:93, listed this species in *Dusicyon*. HEE, KLA, and AL consider *Atelocynus* a distinct genus.
ISIS NUMBER: 5301412001002001001 as *Atelocynus microtis*.

Dusicyon sechurae Thomas, 1900. Ann. Mag. Nat. Hist., ser. 7, 5:148.
TYPE LOCALITY: Peru, Piura Dept., Sechura Desert, Sullana.
DISTRIBUTION: N. W. Peru and S. W. Ecuador.
COMMENT: Subgenus *Dusicyon* or *Pseudalopex*; see Cabrera, 1958:235; Van Gelder, 1978:7. *Pseudalopex* has been considered a synonym of *Dusicyon* by Cabrera, 1958:229; and Clutton-Brock *et al.*, 1976, 29(3):166; a subgenus of *Dusicyon* by Langguth, 1969, Z. Wiss. Zool., 179:176; a subgenus of *Canis* by Langguth, 1975:193; and Van Gelder, 1978:7.
ISIS NUMBER: 5301412001007004001.

Dusicyon thous (Linnaeus, 1766). Syst. Nat., 12th ed., 1:60.
REVIEWED BY: H. E. Evans (HEE).
TYPE LOCALITY: Surinam.
DISTRIBUTION: Uruguay; N. Argentina; Paraguay; lowlands of Bolivia, Brazil except Amazonia; Guianas; Venezuela; Colombia.
COMMENT: Subgenus *Cerdocyon*; see Cabrera, 1958:237; Van Gelder, 1978:7. KLA and AL consider *Cerdocyon* a distinct genus.
ISIS NUMBER: 5301412001004001001 as *Cerdocyon thous*.

Dusicyon vetulus Lund, 1842. Kongl. Dansk. Vid. Selsk. Naturv. Math. Afhandl., 9:4.
TYPE LOCALITY: Brazil, Minas Gerais, Lagoa Santa.
DISTRIBUTION: C. Brazilian Highlands in the States of Mato Grosso, Goias, Minas Gerais, Bahia, and Sao Paulo.
COMMENT: Subgenus *Dusicyon* or *Lycalopex*; see Cabrera, 1958:236; Van Gelder, 1978:8. *Lycalopex* has also been considered a synonym of *Dusicyon* by Clutton-Brock *et al.*, 1976, 29(3):186. AL considers *Lycalopex* a distinct genus.
ISIS NUMBER: 5301412001009001001 as *Lycalopex vetulus*.

Lycaon Brookes, 1827. *In* Griffith's Cuvier Anim. Kingd., 5:151.
ISIS NUMBER: 5301412001010000000.

Lycaon pictus (Temminck, 1820). Ann. Gen. Sci. Phys., 3:54, pl. 35.
TYPE LOCALITY: Mozambique, on the coast.
DISTRIBUTION: Subsaharan Africa, except in the desert area of the north and the tropical rain-forest of the west.
ISIS NUMBER: 5301412001010001001.

Nyctereutes Temminck, 1839. *In* van der Hoeven, Tijdschr. Nat. Ges. Phys., 5:285.
ISIS NUMBER: 5301412001011000000.

Nyctereutes procyonoides (Gray, 1834). Illustr. Indian Zool., 2:pl. 1.
TYPE LOCALITY: China, near Canton.
DISTRIBUTION: Ussuri region (U.S.S.R.); China; Japan; N. Indochina.
COMMENT: Introduced and feral in eastern and middle Europe (AL).
ISIS NUMBER: 5301412001011001001.

Otocyon Muller, 1836. Arch. Anat. Physiol., Jahresber. Fortschr. Wiss., 1835:1.
ISIS NUMBER: 5301412001012000000.

Otocyon megalotis (Desmarest, 1821). Encyclop. Method. Mamm., Suppl., p. 538.
TYPE LOCALITY: South Africa, Cape Prov., Cape of Good Hope.
DISTRIBUTION: S. and E. Africa to Ethiopia.
ISIS NUMBER: 5301412001012001001.

Speothos Lund, 1839. Ann. Sci. Nat. Zool., ser. 2, 11:224.
REVIEWED BY: K. L. Anderson (KLA).
COMMENT: Includes *Icticyon;* see Berta and Marshall, 1978, Foss. Cata., 125:10.
ISIS NUMBER: 5301412001013000000.

Speothos venaticus (Lund, 1842). Kongl. Dansk. Vid. Selsk. Naturv. Math. Afhandl., 9:67.
TYPE LOCALITY: Brazil, Minas Gerais, Lagoa Santa.
DISTRIBUTION: Forested areas of Bolivia, Paraguay and Brazil (except the semiarid N.
    E.); E. Peru; Ecuador; Colombia; Venezuela; Guyana; French Guiana; Surinam;
    Panama.
PROTECTED STATUS: CITES - Appendix I.
ISIS NUMBER: 5301412001013001001.

Urocyon Baird, 1858. Mammals, *in* Repts. Expl. Surv...., 8(1):121.
COMMENT: Included in *Vulpes* as a subgenus by Clutton-Brock *et al.,* 1976,
    29(3):117–199 but see Hall, 1981:941.
ISIS NUMBER: 5301412001014000000.

Urocyon cinereoargenteus (Schreber, 1775). Saugethiere, 2(13):pl. 92.
TYPE LOCALITY: E. North America.
DISTRIBUTION: North America from Oregon, Nevada, Utah and Colorado in the West
    and the U.S.A.-Canadian border in the East through Central America to N.
    Colombia and Venezuela.
COMMENT: Placed in genus *Canis,* subgenus *Vulpes* which includes *Urocyon* by Van
    Gelder, 1978:6. Clutton-Brock *et al.,* 1976, 29(3):117–199, retained *Vulpes,*
    including *Urocyon.*
ISIS NUMBER: 5301412001014001001.

Urocyon littoralis (Baird, 1858). Mammals, *in* Repts. Expl. Surv., 8(1):143.
TYPE LOCALITY: U.S.A., California, San Miguel Isl.
DISTRIBUTION: Islands off the Pacific coast of S. California (U.S.A.).
COMMENT: Placed in genus *Canis,* subgenus *Vulpes* which includes *Urocyon;* may be a
    subspecies of *cinereoargenteus;* see Van Gelder, 1978:8. Clutton-Brock *et al.,* 1976,
    29(3):117–199, retained *Vulpes,* including *Urocyon.*
ISIS NUMBER: 5301412001014002001.

Vulpes Bowdich, 1821. An analysis of the natural classification of Mammalia, p. 40.
COMMENT: Included in *Canis* as a subgenus by Van Gelder, 1978:1–10. However, this
    arrangement is not currently employed by most mammalogists; see Corbet,
    1978:162–163; Corbet and Hill, 1980; Hall, 1981:936; Gromov and Baranova,
    1981:242–243. Includes *Fennecus;* see comments under *zerda* and *Urocyon.*
ISIS NUMBER: 5301412001015000000 as *Vulpes.*
    5301412001008000000 as *Fennecus.*

Vulpes bengalensis (Shaw, 1800). Gen. Zool. Syst. Nat. Hist., 1(2):330.
TYPE LOCALITY: Pakistan, Bangladesh, "Bengal."
DISTRIBUTION: India; Pakistan; S. Nepal.
COMMENT: Placed in genus *Canis,* subgenus *Vulpes* by Van Gelder, 1978:8.
ISIS NUMBER: 5301412001015001001.

Vulpes cana Blanford, 1877. J. Asiat. Soc. Bengal, 2:321.
TYPE LOCALITY: Pakistan, Baluchistan, Gwadar.
DISTRIBUTION: Turkmenia (S.E. U.S.S.R.); Afghanistan; N. E. Iran; Pakistan.
COMMENT: Placed in genus *Canis,* subgenus *Vulpes;* by Van Gelder, 1978:8.
PROTECTED STATUS: CITES - Appendix II.
ISIS NUMBER: 5301412001015002001.

Vulpes chama (A. Smith, 1833). S. Afr. J., 2:89.
    TYPE LOCALITY: Namibia, Namaqualand and country on both sides of the Orange River.
    DISTRIBUTION: Africa, S. of Angola and Zambia.
    COMMENT: Placed in genus *Canis*, subgenus *Vulpes* by Van Gelder, 1978:8.
    ISIS NUMBER: 5301412001015003001.

Vulpes corsac (Linnaeus, 1768). Syst. Nat., 12th ed., 3: appendix 223.
    TYPE LOCALITY: U.S.S.R., N. Kazakhstan, steppes between Ural and Irtysh rivers, near Petropavlovsk.
    DISTRIBUTION: Steppes of S.E. U.S.S.R., Kazakhstan, C. Asia, Mongolia, Transbaikalia, N.E. China; N. Afghanistan.
    COMMENT: Placed in genus *Canis*, subgenus *Vulpes* by Van Gelder, 1978:8.
    ISIS NUMBER: 5301412001015004001.

Vulpes ferrilata Hodgson, 1842. J. Asiat. Soc. Bengal, 11:278.
    TYPE LOCALITY: China, Tibet, near Lhasa.
    DISTRIBUTION: Tibet, Tsinghai, Kansu, and Yunnan (China)(SW); Nepal.
    COMMENT: Placed in genus *Canis*, subgenus *Vulpes* by Van Gelder, 1978:8.
    ISIS NUMBER: 5301412001015005001.

Vulpes macrotis Merriam, 1888. Proc. Biol. Soc. Wash., 4:136.
    REVIEWED BY: H. J. Egoscue (HJE).
    TYPE LOCALITY: U.S.A., California, Riverside Co., vicinity of Box Springs, within 10 mi. (16 km) S. E of Riverside.
    DISTRIBUTION: Western U.S.A. from S. Oregon to Zacatecas (Mexico) and east to W. Texas and E. New Mexico.
    COMMENT: Partly revised by Waithman and Roest, 1977, J. Mammal., 58:157–164. Distinct from *velox*; see Thornton and Creel, 1975, Texas J. Sci., 26:127–136; and Rohwer and Kilgore, 1973, Syst. Zool., 22:157–165, who found hybrids between *velox* and *macrotis* but concluded they were of reduced viability, and retained both as species; but also see Hall, 1981:940. Clutton-Brock, *et al.*, 1976, 29(3):117–199, combined *velox* and *macrotis* for analysis only; see also Van Gelder, 1978:8, who placed this species in genus *Canis*, subgenus *Vulpes*. Reviewed by McGrew, 1979, Mamm. Species, 123:1–6. See Miller and Kellogg, 1955, Bull. U.S. Nat. Mus., 205:685. for discussion of type locality.
    PROTECTED STATUS: U.S. ESA - Endangered as *V. m. mutica* subspecies only.
    ISIS NUMBER: 5301412001015006001.

Vulpes pallida (Cretzschmar, 1826). *In* Ruppell, Atlas Reise Nordl. Afr., Saugeth., p. 33, pl. 11.
    TYPE LOCALITY: Sudan, Kordofan.
    DISTRIBUTION: Semiarid sahelian region of Africa from Senegal through Nigeria, Cameroun and Sudan to Somalia.
    COMMENT: Placed in genus *Canis*, subgenus *Vulpes* by Van Gelder, 1978:8.
    ISIS NUMBER: 5301412001015007001.

Vulpes rueppelli (Schinz, 1825). Cuvier's Thierreich, 4:508.
    TYPE LOCALITY: Sudan, Dongola.
    DISTRIBUTION: Arid areas of N. Africa from Morocco to Somalia; Egypt; Sinai; Arabia; Iran; parts of Pakistan and Afghanistan.
    COMMENT: Placed in genus *Canis*, subgenus *Vulpes* by Van Gelder, 1978:8.
    ISIS NUMBER: 5301412001015008001.

Vulpes velox (Say, 1823). *In* Long, Account of an Exped. from Pittsburgh to the Rocky Mtns., 1:487.
    REVIEWED BY: H. J. Egoscue (HJE).
    TYPE LOCALITY: U.S.A., Colorado, South Platte River.
    DISTRIBUTION: Central North America from C. Alberta (Canada) to N. W. Texas and E. New Mexico (U.S.A.).
    COMMENT: Placed in subgenus *Canis*, subgenus *Vulpes* by Van Gelder, 1978:1–10. Distinct from *macrotis*; see Thornton and Creel, 1975, Texas J. Sci., 26:127–136; Rohwer and Kilgore, 1973, Syst. Zool., 22:157; but also see Hall, 1981:940.

Clutton-Brock, et al., 1976, 29(3):117–199, combined *velox* and *macrotis* for analysis only. Reviewed by Egoscue, 1979, Mamm. Species, 122:1–5.
PROTECTED STATUS: CITES - Appendix I and U.S. ESA - Endangered as *V. v. hebes* subspecies only.
ISIS NUMBER: 5301412001015009001.

Vulpes vulpes (Linnaeus, 1758). Syst. Nat., 10th ed., 1:40.
TYPE LOCALITY: Sweden, Uppsala.
DISTRIBUTION: Europe and continental Asia except the tundra; N. India and peninsular Indochina; Japan; Palearctic Africa; N. America as far south as Texas and New Mexico (U.S.A.), but absent in part of the Central Plains and the Arctic.
COMMENT: Includes *fulva*; see Hall, 1981:936. Placed in genus *Canis*, subgenus *Vulpes* by Van Gelder, 1978:8.
ISIS NUMBER: 5301412001015010000.

Vulpes zerda (Zimmermann, 1780). Geogr. Gesch. Mensch. Vierf. Thiere, 2:247.
REVIEWED BY: H. E. Evans (HEE).
TYPE LOCALITY: Sahara.
DISTRIBUTION: Morocco to Arabia.
COMMENT: : Placed in genus *Canis*, subgenus *Vulpes* by Van Gelder, 1978:8. *Vulpes* includes *Fennecus* which is often treated as a separate genus; see Clutton-Brock *et al.*, 1976, 29(3):117–199; Clutton-Brock and Corbet, 1979, Bull. Zool. Nomencl., 36.
PROTECTED STATUS: CITES - Appendix III as *Fennecus zerda* (Tunisia).
ISIS NUMBER: 5301412001008001001 as *Fennecus zerda*.

Family Ursidae
REVIEWED BY: B. McGillivray (BM); P. Roben (PR); O. L. Rossolimo (OLR)(U.S.S.R.); S. Wang (SW)(China).
COMMENT: Ailuropodidae *(Ailurus* and *Ailuropoda)* was considered a separate family by Corbet and Hill, 1980:95; but see Chorn and Hoffmann, 1978, Mamm. Species, 110:1–6, who included *Ailuropoda* in the Ursidae. Thenius, 1979, Z. Saugetierk., 44:286–305, placed *Ailuropoda* in the monotypic family Ailuropodidae, and *Ailurus* in the subfamily Ailurinae of family Procyonidae. Reviewed by Stains, 1967, *in* Anderson and Jones, p. 329–331; revised by Hendey, 1980, Ann. S. Afr. Mus., 81:1–109.
ISIS NUMBER: 5301412002000000000.

Ailuropoda Milne-Edwards, 1870. Ann. Sci. Nat. Zool., ser. 5, 13(10):1.
COMMENT: Chorn and Hoffmann, 1978, Mamm. Species, 110:1–6; and Hendey, 1980, Ann. S. Afr. Mus., 81:1–109, discussed reasons for placing *Ailuropoda* in Ursidae.
ISIS NUMBER: 5301412002001000000.

Ailuropoda melanoleuca (David, 1869). Nouv. Arch. Mus. Hist. Nat. Paris, Bull., 5:12–13.
REVIEWED BY: S. Wang (SW).
TYPE LOCALITY: China, Szechwan, Moupin.
DISTRIBUTION: Szechwan, Shensi, Kansu; perhaps Tsinghai, on E. edge of Tibetan plateau (China).
COMMENT: Regarded by Hendey, 1980, as the only surviving species in the subfamily Agriotheriinae. Placed in family Ailuropodidae by Thenius, 1979, Z. Saugetierk., 44:286–305. Reviewed by Chorn and Hoffmann, 1978, Mamm. Species, 110:1–6.
ISIS NUMBER: 5301412002001001001.

Helarctos Horsfield, 1825. J. Zool. Lond., 2:221.
REVIEWED BY: S. Wang (SW).
COMMENT: Van Gelder, 1977:13, included *Helarctos* as a synonym of *Melursus. Helarctos* was considered a subgenus of *Ursus* by Hall, 1981:947, and a distinct genus by Corbet and Hill, 1980:94.
ISIS NUMBER: 5301412002002000000.

Helarctos malayanus (Raffles, 1821). Trans. Linn. Soc. Lond., 13:254.
TYPE LOCALITY: Indonesia, Sumatra, Bencoolen.
DISTRIBUTION: Burma; Thailand; Vietnam; Sumatra; Borneo; N.E. India; Malaysia; Yunnan and Szechwan (China) (SW).
PROTECTED STATUS: CITES - Appendix I.
ISIS NUMBER: 5301412002002001001.

Melursus Meyer, 1793. Zool. Entdeck., p. 155.
    COMMENT: See comment under *Helarctos*. *Melursus* was considered a subgenus of *Ursus*
    by Hall, 1981:947, and a distinct genus by Corbet and Hill, 1980:94.
    ISIS NUMBER: 5301412002003000000.

Melursus ursinus (Shaw, 1791). Nat. Misc., 2 (unpaged) pl. 58.
    TYPE LOCALITY: India, Bihar, Patna.
    DISTRIBUTION: Sri Lanka; India, north to the Indian desert and to the foothills of the
    Himalayas.
    ISIS NUMBER: 5301412002003001001.

Tremarctos Gervais, 1855. Hist. Nat. Mamm., 2:20.
    REVIEWED BY: J. P. Jorgenson (JPJ).
    ISIS NUMBER: 5301412002005000000.

Tremarctos ornatus (F. Cuvier, 1825). Hist. Nat. Mamm., 3(50):2, pl.
    TYPE LOCALITY: "Chile, Cordillera de Chile."
    DISTRIBUTION: Mountainous regions of W. Venezuela, Colombia, Ecuador, Peru, and
    W. Bolivia; perhaps Panama.
    COMMENT: *T. ornatus* is not known from Chile; type locality probably east of Trujillo,
    Libertad, Peru; see Cabrera, 1958:242.
    PROTECTED STATUS: CITES - Appendix I.
    ISIS NUMBER: 5301412002005001001.

Ursus Linnaeus, 1758. Syst. Nat., 10th ed., 1:47.
    COMMENT: Includes *Euarctos, Melanarctos, Mylarctos, Selenarctos, Thalarctos, Thalassarctus,*
    Thalassiarchus, Myrmarctos, Ursarctos, and *Vetularctos;* see Ellerman and Morrison-
    Scott, 1951:235; Stains, 1967, *in* Anderson and Jones, p. 331; Hall, 1981:947;
    Kurten and Anderson, 1980, Pleistocene mammals of North America, Columbia
    Univ. Press, New York, p. 183; but also see Corbet, 1978:165–166, and Hendey,
    1980:98.
    ISIS NUMBER: 5301412002006000000.

Ursus americanus Pallas, 1780. Spicil. Zool., 14:5.
    TYPE LOCALITY: Eastern North America.
    DISTRIBUTION: N. C. Alaska to Labrador and Newfoundland (Canada), south to C.
    California, N. Nevada (U.S.A.), N. Nayarit and S. Tamaulipas (Mexico), and
    Florida (U.S.A.).
    COMMENT: Placed in subgenus *Euarctos* by Hall, 1981:947; and Thenius, 1979:293.
    Hendy, 1980:98, regarded *Euarctos* as a distinct genus.
    ISIS NUMBER: 5301412002006002001.

Ursus arctos Linnaeus, 1758. Syst. Nat., 10th ed., 1:47.
    TYPE LOCALITY: Sweden.
    DISTRIBUTION: Formerly, N. W. Africa, all of Palearctic from W. Europe, Near and
    Middle East through N. Himalayas to W. and N. China and Chukotka (U.S.S.R.);
    Hokkaido (Japan). Western North America, north from N. Mexico.
    COMMENT: Hall, 1981:951–958, listed 87 taxa (including 76 other species names) in the
    "*Ursus arctos* -group", which are here included in *arctos;* see Corbet and Hill,
    1980:94; see also Couturier, 1954, L'Ours brun, *Ursus arctos* L., Grenoble.
    PROTECTED STATUS: CITES - Appendix II (all North American subspecies).
        CITES - Appendix I (Italian populations only).
        CITES - Appendix I as *U. a. isabellinus* subspecies only.
        CITES - Appendix I and U.S. ESA - Endangered as *U. a. nelsoni* subspecies only.
        CITES - Appendix I and U.S. ESA - Endangered as *U. a. pruinosus* subspecies only.
            U.S. ESA - Endangered as *U. a. arctos* (Italian populations only). U.S. ESA -
            Threatened as *U. a. horribilis* subspecies only.
    ISIS NUMBER: 5301412002006001001.

Ursus maritimus Phipps, 1774. A Voyage Towards North Pole, p. 185.
    TYPE LOCALITY: Norway, Spitzbergen Isls.
    DISTRIBUTION: Circumpolar in the Arctic.
    COMMENT: Placed in genus *Ursus,* subgenus *Thalarctos* by Gromov and Baranova,

1981:253, and in genus *Thalarctos* by Corbet, 1978:166. Reviewed by DeMaster and Stirling, 1981, Mamm. Species, 145:1–7.
PROTECTED STATUS: CITES - Appendix II as *U. (=Thalarctos) maritimus.*
ISIS NUMBER: 5301412002006003001 as *Ursus maritimus.*

Ursus thibetanus G. Cuvier, 1823. Rech. Oss. Foss., 4:325.
TYPE LOCALITY: India, Assam, Sylhet.
DISTRIBUTION: Afghanistan; Pakistan; N. India; China; S.E. Primorsky Krai (U.S.S.R.); Korea; Japan; Indochina.
COMMENT: Placed in genus *Ursus,* subgenus *Selenarctos* by Gromov and Baranova, 1981:249; and in subgenus *Euarctos* by Thenius, 1979:293. Stains, 1967, *in* Anderson and Jones, p. 331; Corbet, 1978:165; and Hendey, 1980:98, considered it a distinct genus, *Selenarctos.*
PROTECTED STATUS: CITES - Appendix I as *Selenarctos thibetanus.*
ISIS NUMBER: 5301412002004001001.

Family Procyonidae
REVIEWED BY: S. Kortlucke (SK); J. Ramirez-Pulido (JRP) (Mexico).
ISIS NUMBER: 5301412003000000000.

Ailurus F. Cuvier, 1825. Hist. Nat. Mamm., 5(50):3.
COMMENT: Immunological evidence suggests that *Ailurus* is more closely related to ursids than to procyonids; see Todd and Pressmann, 1968, Carniv. Genet. Newsl., 5:105–108, and Sarich, 1976, Trans. Zool. Soc. Lond., 33:165–171. Corbet and Hill, 1980:95, placed this monotypic genus in Ailuropodidae, but see comment under *fulgens,* and under Ursidae.
ISIS NUMBER: 5301412003001000000.

Ailurus fulgens F. Cuvier, 1825. Hist. Nat. Mamm., 5(50):3.
TYPE LOCALITY: India, S. E. Himalayas.
DISTRIBUTION: Yunnan and Szechwan (China); N. Burma; Sikkim; Nepal.
COMMENT: Although this species is usually included in the Procyonidae because of its ringed tail, facial markings, and superficial resemblance of teeth and skull to *Procyon,* this species has no shared derived morphological characters that would place it here (SK). Thenius, 1979, Z. Saugetierk., 44:386–305, and Gromov and Baranova, 1981:255, placed it in the monotypic subfamily Ailurinae.
PROTECTED STATUS: CITES - Appendix II.
ISIS NUMBER: 5301412003001001001.

Bassaricyon J. A. Allen, 1876. Proc. Acad. Nat. Sci. Phila., 28:20, pl. 1; text 1877, 267–268, pl. 2.
COMMENT: The several named forms of *Bassaricyon* are probably conspecific; see Stains, 1967, *in* Anderson and Jones, p. 335. This genus is being revised by SK.
ISIS NUMBER: 5301412003002000000.

Bassaricyon alleni Thomas, 1880. Proc. Zool. Soc. Lond., 1880:397.
TYPE LOCALITY: Ecuador, Napo-Pastaza Prov., Sarayacu, near the Rio Bobonazo.
DISTRIBUTION: Ecuador east of the Andes, and Peru to Cuzco Prov.; Bolivia; possibly into Venezuela.
COMMENT: See comment under *Bassaricyon.*
ISIS NUMBER: 5301412003002001001.

Bassaricyon beddardi Pocock, 1921. Ann. Mag. Nat. Hist., ser. 9, 7:229.
TYPE LOCALITY: Guyana, Lower Essequibo, Bartica.
DISTRIBUTION: Guyana, and possibly adjacent Venezuela and Brasil.
COMMENT: A renaming of *alleni* Sclater, 1895, Proc. Zool. Soc. Lond., 1895:521 (not Thomas, 1880). Cabrera, 1958:253, erroneously listed this species under the genus *Bassaricyon* as *Bassariscus beddardi* (SK).
ISIS NUMBER: 5301412003002002001.

Bassaricyon gabbii J. A. Allen, 1876. Proc. Acad. Nat. Sci. Phila., 28:21.
TYPE LOCALITY: Costa Rica, Limon, Talamanca.
DISTRIBUTION: C. Nicaragua; Costa Rica; Panama; W. Colombia; W. Ecuador.

COMMENT: Includes *medius, richardsoni,* and *siccatus;* see Cabrera, 1958:253–254, and Hall, 1981:979–980.
PROTECTED STATUS: CITES - Appendix III (Costa Rica).
ISIS NUMBER: 5301412003002003001.

Bassaricyon lasius Harris, 1932. Occas. Pap. Mus. Zool. Univ. Michigan, 248:3.
TYPE LOCALITY: Costa Rica, Cartago, Estrella de Cartago, near source of Rio Estrella, 6 or 8 mi. (10 to 13 km) S. Cartago, about 4500 ft. (1372 m).
DISTRIBUTION: Known only from the type locality.

Bassaricyon pauli Enders, 1936. Proc. Acad. Nat. Sci. Phila., 88:365.
TYPE LOCALITY: Panama, Chiriqui, Cerro Pando, between Rio Chiriqui Viejo and Rio Colorado (about 10 mi. (16 km) from El Volcan.)
DISTRIBUTION: Known only from the type locality.

Bassariscus Coues, 1887. Science, 9:516.
COMMENT: Includes *Jentinkia;* see Corbet and Hill, 1980:94; Hall, 1981:961. This genus is being revised by SK.
ISIS NUMBER: 5301412003003000000.

Bassariscus astutus (Lichtenstein, 1830). Abh. Konigl. Akad. Wiss., Berlin, for 1827, p. 119.
TYPE LOCALITY: Mexico, near city of Mexico.
DISTRIBUTION: S. W. Oregon, N. Nevada, Utah, S. W. Wyoming and W. Colorado, south through California, Arizona, New Mexico and Texas (U.S.A.), to the Isthmus of Tehuantepec (Mexico); Tiburon Isl. and several other islands in the Gulf of California.
COMMENT: Includes *albipes, raptor* and *saxicola;* see Hall, 1981:962–965. Records from E. Colorado, Kansas, Ohio, Louisiana, and South Carolina are probably not indicative of the present range (SK).
ISIS NUMBER: 5301412003003001001.

Bassariscus sumichrasti (Saussure, 1860). Rev. Mag. Zool. Paris, ser. 2, 12:7.
TYPE LOCALITY: Mexico, Veracruz, Mirador.
DISTRIBUTION: Guerrero and S. Veracruz (Mexico) to W. Panama.
COMMENT: Includes *monticola* and *variabilis;* see Hall, 1981:965.
PROTECTED STATUS: CITES - Appendix III (Costa Rica).
ISIS NUMBER: 5301412003003002001.

Nasua Storr, 1780. Prodr. Meth. Mamm., p. 35, tabl. A.
ISIS NUMBER: 5301412003004000000.

Nasua nasua (Linnaeus, 1766). Syst. Nat., 12th ed., 1:64.
TYPE LOCALITY: Mexico, Veracruz, Isthmus of Tehuantepec, Achotal.
DISTRIBUTION: S. Arizona and S. W. New Mexico, south through Mexico (except Baja California), Central America, and South America (except Chile) to Argentina.
COMMENT: Includes *narica;* see Handley, 1966, *in* Wenzel and Tipton, eds., Ectoparasites of Panama, p. 789. Includes *aricana, boliviensis, candace, cinerascens, dorsalis, manium, montana, quichua, solitaria,* and *vittata;* see Cabrera, 1958:244–249, and Hall, 1981:975–976.
PROTECTED STATUS: CITES - Appendix III as *N. n. solitaria* subspecies only (Uruguay).
ISIS NUMBER: 5301412003004002001 as *N. nasua.*
5301412003004001001 as *N. narica.*

Nasua nelsoni Merriam, 1901. Proc. Biol. Soc. Wash., 14:100.
TYPE LOCALITY: Mexico, Quintana Roo, Cozumel Isl.
DISTRIBUTION: Known only from the type locality.
COMMENT: Includes *thersites;* see Hall, 1981:976.
ISIS NUMBER: 5301412003004003001.

Nasuella Hollister, 1915. Proc. U.S. Nat. Mus., 49:118.
ISIS NUMBER: 5301412003005000000.

Nasuella olivacea (Gray, 1865). Proc. Zool. Soc. Lond., 1864:703.
TYPE LOCALITY: Colombia, Bogota. Interpreted to be the mountains near the capital.
DISTRIBUTION: Andes of Colombia, W. Venezuela, and Ecuador.

COMMENT: *Nasua olivacea*, Gray, 1843, is a *nomen nudum*. Includes *olivacea, meridensis,* and *quitensis;* see Cabrera, 1958:249, and also for discussion of type locality.
ISIS NUMBER: 5301412003005001001.

Potos E. Geoffroy and G. Cuvier, 1795. Mag. Encyclop., 2:187.
COMMENT: Hernandez-Camacho, 1977, Caldasia, 11:148, resurrected Bonaparte's 1838, family Cercoleptidae for this taxon, as a provisional arrangement, but Hall, 1981:977, and SK consider it a procyonid.
ISIS NUMBER: 5301412003006000000.

Potos flavus (Schreber, 1774). Saugethiere, 1:42.
TYPE LOCALITY: Guyana.
DISTRIBUTION: S. Tamaulipas and Guerrero (possibly Michoacan) (Mexico) to the Mato Grosso of Brasil.
COMMENT: Includes *chapadensis, megalotus, meridensis, modestus, nocturnus* and *prehensilis;* see Cabrera, 1958:249–252, and Hall, 1981:977–978. A revision of South American forms is in preparation by R. M. Wetzel. Central American forms were reviewed by Kortlucke, 1972, Occas. Pap. Mus. Nat. Hist. Univ. Kansas, 17:1–36; also see Cabrera, 1958:250, and Husson, 1978:288–289.
ISIS NUMBER: 5301412003006001001.

Procyon Storr, 1780. Prodr. Meth. Mamm., p. 35.
COMMENT: Revised by Goldman, 1950, N. Am. Fauna, 60:1–153.
ISIS NUMBER: 5301412003007000000.

Procyon cancrivorus (F. Cuvier, 1798). Tabl. Elem. Hist. Nat. Anim., p. 113.
TYPE LOCALITY: French Guiana, Cayenne.
DISTRIBUTION: E. Costa Rica, Panama, and South America to N. E. Argentina.
COMMENT: Includes *aequatorialis, nigripes,* and *proteus;* see Cabrera, 1958:243–244, and Hall, 1981:974.
ISIS NUMBER: 5301412003007001001.

Procyon gloveralleni Nelson and Goldman, 1930. J. Mammal., 11:453.
TYPE LOCALITY: Barbados (Lesser Antilles).
DISTRIBUTION: Known only from the type locality.
COMMENT: See Lotze and Anderson, 1979, Mamm. Species, 119:1–8, who stated that *gloveralleni, maynardi,* and *minor,* may be conspecific with *lotor.* May be extinct; see Hall, 1981:973.

Procyon insularis Merriam, 1898. Proc. Biol. Soc. Wash., 12:17.
TYPE LOCALITY: Mexico, Tres Marias Isls., Maria Madre Isl.
DISTRIBUTION: Maria Madre Isl. *(P. i. insularis)* and Maria Magdalena Isl. *(P. i. vicinus).*
ISIS NUMBER: 5301412003007002001.

Procyon lotor (Linnaeus, 1758). Syst. Nat., 10th ed., 1:48.
TYPE LOCALITY: U.S.A., Pennsylvania.
DISTRIBUTION: S. Canada throughout U.S.A. (except parts of Rocky Mtns.), Mexico, and Central America to C. Panama.
COMMENT: Includes *hernandezii, hudsonicus, nivea, pallidus, psora,* and *pumilus;* see Hall, 1981:967–972. Reviewed by Lotze and Anderson, 1979, Mamm. Species, 119:1–8.
ISIS NUMBER: 5301412003007003001.

Procyon maynardi Bangs, 1898. Proc. Biol. Soc. Wash., 12:92.
TYPE LOCALITY: Bahamas, New Providence Isl.
DISTRIBUTION: Known only from the type locality.
COMMENT: May be conspecific with *lotor;* see Lotze and Anderson, 1979, Mamm. Species, 119:1–8; but also see Hall, 1981:973.
ISIS NUMBER: 5301412003007004001.

Procyon minor Miller, 1911. Proc. Biol. Soc. Wash., 24:4.
TYPE LOCALITY: Guadeloupe (Lesser Antilles), Pointe-a-Pitre (France).
DISTRIBUTION: Guadeloupe Isl. (Lesser Antilles).

COMMENT: May be conspecific with *lotor*; see Lotze and Anderson, 1979, Mamm.
Species, 119:1–8; but also see Hall, 1981:973.
ISIS NUMBER: 5301412003007005001.

Procyon pygmaeus Merriam, 1901. Proc. Biol. Soc. Wash., 14:101.
TYPE LOCALITY: Mexico, Quintana Roo, Cozumel Isl.
DISTRIBUTION: Known only from the type locality.
ISIS NUMBER: 5301412003007006001.

Family Mustelidae

REVIEWED BY: T. Holmes (TH); J. Ramirez-Pulido (JRP)(Mexico); P. Roben (PR); O. L.
Rossolimo (OLR)(U.S.S.R.); H. J. Stains (HJS).
ISIS NUMBER: 5301412004000000000.

Aonyx Lesson, 1827. Manuel Mamm., p. 157.
REVIEWED BY: Nicole Duplaix (ND).
COMMENT: Subfamily Lutrinae. Includes *Amblonyx* and *Paraonyx*; see Lekagul and
McNeely, 1977:559, Ellerman and Morrison-Scott, 1951:278, and Coetzee, 1977,
Part 8:9. Also see Medway, 1977:133, who considered *Amblonyx* a distinct genus.
PROTECTED STATUS: CITES - Appendix II as Lutrinae.
ISIS NUMBER: 5301412004021000000.

Aonyx capensis (Schinz, 1821). Das Thierreich, 1:211–214.
TYPE LOCALITY: South Africa, probably Port Elizabeth.
DISTRIBUTION: Africa from Cape of Good Hope, north to Ethiopia; west to Senegal.
PROTECTED STATUS: CITES - Appendix II as Lutrinae.
ISIS NUMBER: 5301412004021001001.

Aonyx cinerea (Illiger, 1815). Abh. Preuss. Akad. Wiss., 1815:99.
TYPE LOCALITY: Indonesia, W. Java, Batavia.
DISTRIBUTION: India; Burma; S. China and Hainan Isl.; Taiwan; S. E. Asia to Sumatra;
Java; Borneo; Palawan Isl. (Philippines).
COMMENT: Subgenus *Amblonyx*; Ellerman and Morrison-Scott, 1951:278, treated
Amblonyx as a subgenus of *Aonyx*, as did Corbet and Hill, 1980:99, and Lekagul
and McNeely, 1977:559; but see Medway, 1977:133, who considered *Amblonyx* a
distinct genus; see also Davis, 1978, *in* Otters, IUCN Publ., N. S., 1 .
PROTECTED STATUS: CITES - Appendix II as Lutrinae.
ISIS NUMBER: 5301412004021002001.

Aonyx congica Lonnberg, 1910. Ark. Zool., 7(9):1–8.
TYPE LOCALITY: Zaire, "Lower Congo."
DISTRIBUTION: Zaire, Congo Basin to Uganda and Niger.
COMMENT: Includes *philippsi* and *microdon*; subgenus *Paraonyx*; see Coetzee, 1977, Part
8:9.
PROTECTED STATUS: CITES - Appendix I as *A. microdon*. U.S. ESA - Endangered as *A.*
(=Paraonyx) congica (=microdon).
ISIS NUMBER: 5301412004021003001 as *A. congica*.
5301412004021004001 as *A. microdon*.
5301412004021005001 as *A. philippsi*.

Arctonyx F. Cuvier, 1825. Hist. Nat. Mamm., 3(51), pl. and 2 pp. text.
COMMENT: Subfamily Melinae.
ISIS NUMBER: 5301412004012000000.

Arctonyx collaris F. Cuvier, 1825. Hist. Nat. Mamm., 3(51), pl. and 2 pp. text.
TYPE LOCALITY: India, Bhutan Duars, E. Himalayas.
DISTRIBUTION: Widespread in China; ranges from Assam (India) and Burma to
Indochina, Thailand, Sumatra, and probably Perak in Malaya.
COMMENT: Reviewed by Long, 1978, Rep. Fauna Flora Wisc., 14:1–6.
ISIS NUMBER: 5301412004012001001.

Conepatus Gray, 1837. Charlesworth's Mag. Nat. Hist., 1:581.
    COMMENT: Subfamily Mephitinae.
    ISIS NUMBER: 5301412004018000000.

Conepatus chinga (Molina, 1782). Sagg. Stor. Nat. Chile, p. 288.
    TYPE LOCALITY: Chile, near Valparaiso.
    DISTRIBUTION: Chile; Peru; N. Argentina; Bolivia; Uruguay; S. Brazil.
    COMMENT: Includes *rex;* see Kipp, 1965, Z. Saugetierk., 30:193–232; but also see Corbet
        and Hill, 1980:98, who listed *rex* as a distinct species.
    ISIS NUMBER: 5301412004018001001.

Conepatus humboldtii Gray, 1837. Mag. Nat. Hist., 1:581.
    TYPE LOCALITY: Chile, Straits of Magellan.
    DISTRIBUTION: N. E. Argentina and Paraguay south to the Straits of Magellan.
    COMMENT: Includes *castaneus;* see Kipp, 1965, Z. Saugetierk., 30:193–232; but also see
        Corbet and Hill, 1980:98, who list *castaneus* as a distinct species.
    PROTECTED STATUS: CITES - Appendix II as *C. humboldti (sic).*
    ISIS NUMBER: 5301412004018002001.

Conepatus leuconotus (Lichtenstein, 1832). Darst. Saugeth., pl. 44. fig. 1.
    TYPE LOCALITY: Mexico, Veracruz, Rio Alvarado.
    DISTRIBUTION: S. Gulf coast of Texas (U.S.A.), south along coast to Veracruz (Mexico).
    COMMENT: May be only subspecifically distinct from *mesoleucus;* see Hall, 1981:1025.
    ISIS NUMBER: 5301412004018003001.

Conepatus mesoleucus (Lichtenstein, 1832). Darst. Saugeth., pl. 44, fig. 2.
    TYPE LOCALITY: Mexico, Hidalgo, near El Chico.
    DISTRIBUTION: Arizona, Colorado, Texas (U.S.A.), south to Nicaragua.
    COMMENT: May be conspecific with *leuconotus;* see Hall, 1981:1025.
    ISIS NUMBER: 5301412004018004001.

Conepatus semistriatus (Boddaert, 1784). Elench. Anim., p. 84.
    TYPE LOCALITY: Colombia.
    DISTRIBUTION: Veracruz, Tabasco, and Yucatan (Mexico) to Peru and E. Brazil.
    ISIS NUMBER: 5301412004018005001.

Eira H. Smith, 1842. *In* The Nat. Libr., Jardine, ed., 35:201.
    REVIEWED BY: A. I. Roest (AIR).
    COMMENT: *Tayra* Oken, 1816, is invalid; *Galera* Gray, 1843, is preoccupied by *Galera*
        Browne, 1789; see Stains, 1967, *in* Anderson and Jones, p. 337. Subfamily
        Mustelinae.
    ISIS NUMBER: 5301412004001000000.

Eira barbara (Linnaeus, 1758). Syst. Nat., 10th ed., 1:46.
    TYPE LOCALITY: Panama, Veraguas, Calovevora.
    DISTRIBUTION: Sinaloa and Tamaulipas (Mexico), south to Argentina.
    ISIS NUMBER: 5301412004001001001.

Enhydra Fleming, 1822. Philos. Zool., 2:187.
    REVIEWED BY: Nicole Duplaix (ND); T. R. Loughlin (TRL); A. I. Roest (AIR).
    COMMENT: Subfamily Lutrinae.
    PROTECTED STATUS: CITES - Appendix II as Lutrinae.
    ISIS NUMBER: 5301412004022000000.

Enhydra lutris (Linnaeus, 1758). Syst. Nat., 10th ed., 1:45.
    TYPE LOCALITY: U.S.S.R., Kamchatka, Commander Isls.
    DISTRIBUTION: Formerly, coastal Hokkaido (Japan); Sakhalin Isl., Kurile Isls.,
        Commander Isls., Kamchatka (U.S.S.R.), Aleutian Isls., and S. Alaska (U.S.A.) to
        Baja California (Mexico).
    COMMENT: Revised by Roest, 1973, Los. Ang. Cty. Mus. Contrib. Sci., 252:1–17.
        Reviewed by Davis and Lidicker, 1975, Proc. Calif. Acad. Sci., 4:429–437; and
        Estes, 1980, Mamm. Species, 133:1–8.
    PROTECTED STATUS: CITES - Appendix II as Lutrinae.
        CITES - Appendix I and U.S. ESA - Threatened as *E. l. nereis* subspecies only.
    ISIS NUMBER: 5301412004022001001.

Galictis Bell, 1826. Zool. J., 2:552.
  REVIEWED BY: A. I. Roest (AIR).
  COMMENT: *Grison* Oken, 1816, is invalid; includes *Grisonella*; see Stains, 1967, *in*
    Anderson and Jones, p. 337. Subfamily Mustelinae.
  ISIS NUMBER: 5301412004002000000.

Galictis cuja (Molina, 1782). Sagg. Stor. Nat. Chile, p. 291.
  TYPE LOCALITY: Brazil, Minas Gerais, Sao Francisco dos Campos, 1580 m.
  DISTRIBUTION: Peru to E. Brazil, south to S. Chile and Argentina.
  COMMENT: Subgenus *Grisonella*; see Stains, 1967, *in* Anderson and Jones, p. 337.
  ISIS NUMBER: 5301412004002001001.

Galictis vittata (Schreber, 1776). Saugethiere, 3:418.
  TYPE LOCALITY: Surinam.
  DISTRIBUTION: San Luis Potosi and Veracruz (Mexico) to Peru and S. Brazil, at lower
    elevations.
  COMMENT: Includes *allamandi*; see Krumbiegel, 1942, Zool. Anz., 139:105, and Hall,
    1981:1006.
  PROTECTED STATUS: CITES - Appendix III as *G. allamandi* only (Costa Rica).
  ISIS NUMBER: 5301412004002002001.

Gulo Pallas, 1780. Spicil. Zool., 14:25.
  REVIEWED BY: E. Anderson (EA); A. I. Roest (AIR).
  COMMENT: Corbet, 1978:174, attributed *Gulo* to Storr, 1780, Prodr. Meth. Mamm., p. 34.
    Ellerman and Morrison-Scott, 1951:250, dated *Gulo* from Frisch, 1775, which has
    been rejected by the ICZN, Opinion 258, 1954. Subfamily Mustelinae.
  ISIS NUMBER: 5301412004003000000.

Gulo gulo (Linnaeus, 1758). Syst. Nat., 10th ed., 1:45.
  TYPE LOCALITY: Sweden, Lapland.
  DISTRIBUTION: Holarctic taiga and S. tundra, south to 50° N. (Eurasia) and 37° N.
    (North America).
  COMMENT: Includes *luscus*; see Kurten and Rausch, 1959, Acta Arct., 11:1-44; but also
    see Hall, 1981:1007.
  ISIS NUMBER: 5301412004003001001.

Ictonyx Kaup, 1835. Das Thierreich, 1:352.
  REVIEWED BY: A. I. Roest (AIR).
  COMMENT: *Zorilla* Oken, 1816, is invalid; see Stains, 1967, *in* Anderson and Jones, p.
    337. Subfamly Mustelinae.
  ISIS NUMBER: 5301412004004000000.

Ictonyx striatus (Perry, 1810). Arcana Mus. Nat. Hist., Signature YT, Fig. 41.
  TYPE LOCALITY: South Africa, Cape Prov., Cape of Good Hope.
  DISTRIBUTION: Senegal to S. E. Egypt and Ethiopia, south to South Africa.
  ISIS NUMBER: 5301412004004001001.

Lutra Brunnich, 1771. Zool. Fundamenta, p. 34, 42.
  REVIEWED BY: N. Duplaix (ND); A. I. Roest (AIR).
  COMMENT: Subfamily Lutrinae. Corbet, 1978:176, dated *Lutra* from Brisson, 1762, Regn.
    Anim., 2nd ed., p. 201, which was ruled an unavailable work; see Hemming,
    1955, Bull. ICZN, 11(6):197. Includes *Lutrogale* (see Corbet, 1978:176; Hall,
    1981:1029), and *Hydrictis* (see Ansell, 1978:36). Van Zyll de Jong, 1972, R. Ont.
    Mus. Life Sci. Contrib., 80:1-104, referred New World otters to *Lontra*; see Gray,
    1843, Ann. Mag. Nat. Hist., 11:118; but also see revision by Sokolov, 1973, Byull.
    Mosk. Ova. Ispyt. Prir. Otd. Biol., 78:45-52, who regarded *Lontra* as a subgenus of
    Lutra, as did Hall, 1981:1028.
  PROTECTED STATUS: CITES - Appendix II as Lutrinae.
  ISIS NUMBER: 5301412004024000000 as *Lutra*.
             5301412004023000000 as *Lontra*.

Lutra canadensis (Schreber, 1776). Saugethiere, pl. 126b.
TYPE LOCALITY: Eastern Canada.
DISTRIBUTION: North America south to Arizona, Texas, and Florida (U.S.A.).
COMMENT: Includes *mira;* see Van Zyll de Jong, 1972, Life Sci. Contrib. R. Ont. Mus., 80:1–104.
PROTECTED STATUS: CITES - Appendix II as Lutrinae.
ISIS NUMBER: 5301412004023001001 as *Lontra canadensis.*

Lutra felina (Molina, 1782). Sagg. Stor. Nat. Chile, p. 284.
TYPE LOCALITY: Chile.
DISTRIBUTION: West coast of South America from N. Peru to Straits of Magellan.
PROTECTED STATUS: CITES - Appendix I and U.S. ESA - Endangered.
ISIS NUMBER: 5301412004023002001 as *Lontra felina.*

Lutra longicaudis (Olfers, 1818). *In* Eschwege, Neue Bibl. Reisenb., 15(2):233.
TYPE LOCALITY: Brazil.
DISTRIBUTION: Mexico, Central America, W. South America south to Peru, E. South America south to Uruguay.
COMMENT: Includes *annectens, enudris, incarum,* and *platensis;* see Van Zyll de Jong, 1972, Life Sci. Contrib. R. Ont. Mus., 80:1–104; Genoways and Jones, 1975, Occas. Pap. Mus. Texas Tech Univ., 26:1–22.
PROTECTED STATUS: CITES - Appendix I and U.S. ESA - Endangered as *L. longicaudis.* U.S. ESA - Endangered as *L. platensis.*
ISIS NUMBER: 5301412004023003001 as *Lontra longicaudis.*

Lutra lutra (Linnaeus, 1758). Syst. Nat., 10th ed., 1:45.
TYPE LOCALITY: Sweden, Uppsala.
DISTRIBUTION: Eurasia (excl. tundra and desert); British and Japanese Isls.; N. W. Africa; S. India; Sri Lanka; Taiwan; Vietnam; Sumatra; Java.
PROTECTED STATUS: CITES - Appendix I.
ISIS NUMBER: 5301412004024001001.

Lutra maculicollis Lichtenstein, 1835. Arch. Naturgesch., 1:89.
TYPE LOCALITY: Namibia, Orange River, Bambusbergen.
DISTRIBUTION: Liberia to Ethiopia south to South Africa, except east coast and S. W. African deserts.
COMMENT: Davis, 1978, *in* Otters, IUCN Publ., N.S., 1:14–33, placed this species in genus *Hydrictis,* as does ND; but see Ansell, 1978:36, who considered *Hydrictis* a subgenus.
PROTECTED STATUS: CITES - Appendix II as Lutrinae.
ISIS NUMBER: 5301412004024002001.

Lutra perspicillata I. Geoffroy, 1826. Dict. Class. Hist. Nat., 9:519.
TYPE LOCALITY: Indonesia, Sumatra.
DISTRIBUTION: Iraq to Indochina, Malaya, and Sumatra. Possibly Borneo.
COMMENT: Davis, 1978, *in* Otters, IUCN Publ., N.S., 1:14–33, placed this species in genus *Lutrogale,* as does ND; but see Corbet, 1978:176, and Hall, 1981:1029.
PROTECTED STATUS: CITES - Appendix II
ISIS NUMBER: 5301412004024003001.

Lutra provocax Thomas, 1908. Ann. Mag. Nat. Hist., ser. 1:391.
TYPE LOCALITY: Argentina, S. of Lake Nahuel Huapi.
DISTRIBUTION: Patagonia (C. and S. Chile; W. Argentina).
PROTECTED STATUS: CITES - Appendix I and U.S. ESA - Endangered.
ISIS NUMBER: 5301412004023004001 as *Lontra provocax.*

Lutra sumatrana (Gray, 1865). Proc. Zool. Soc. Lond., 1865:123.
TYPE LOCALITY: Indonesia, Sumatra.
DISTRIBUTION: Thailand; Malaya; S. Kampuchea; S. Vietnam; Sumatra; Borneo.
PROTECTED STATUS: CITES - Appendix II as Lutrinae.
ISIS NUMBER: 5301412004024004001.

Lyncodon Gervais, 1845. *In* d'Orbigny, Dict. Univ. Hist. Nat., 4:685.
REVIEWED BY: A. I. Roest (AIR).
COMMENT: Subfamily Mustelinae.
ISIS NUMBER: 5301412004005000000.

Lyncodon patagonicus (Blainville, 1842). Osteogr. Mamm., ser. 2, 10:1.
TYPE LOCALITY: Argentina, Patagonia, Rio Negro.
DISTRIBUTION: Argentina and S. Chile.
ISIS NUMBER: 5301412004005001001.

Martes Pinel, 1792. Actes Soc. Hist. Nat. Paris, 1:55.
REVIEWED BY: E. Anderson (EA); R. A. Powell (RAP); A. I. Roest (AIR).
COMMENT: Subfamily Mustelinae. Includes *Charronia* and *Lamprogale*; see Ellerman and
Morrison-Scott, 1951:244.
ISIS NUMBER: 5301412004006000000.

Martes americana (Turton, 1806). A Gen. Syst. of Nat., 1:60.
TYPE LOCALITY: E. North America.
DISTRIBUTION: Alaska and Canada south to N. California, south in the Sierra Nevada
and Rocky Mtns. to 35° N.
COMMENT: May be conspecific with *martes, melampus,* and *zibellina;* see Anderson, 1970,
Acta Zool. Fenn., 130:132; Hagmeier, 1961, Can. Field Nat., 75:122–138; and
Corbet, 1978:172; but see also Hall, 1981:981; Gromov and Baranova,
1981:256–258.
ISIS NUMBER: 5301412004006001001.

Martes flavigula (Boddaert, 1785). Elench. Anim., 1:88.
TYPE LOCALITY: Nepal.
DISTRIBUTION: Primorsky Krai (U.S.S.R.); Korea; China, west along Himalayan foothills
to N. W. Pakistan; isolates in S. India, S. E. Asia, Taiwan, Sumatra, Java, and
Borneo.
COMMENT: Includes *gwatkinsi;* see Corbet, 1978:174; but also see Anderson, 1970, Acta
Zool. Fenn., 130:108, who considered *gwatkinsi* a distinct species.
PROTECTED STATUS: U.S. ESA - Endangered as *M. f. chrysospila* subspecies only.
ISIS NUMBER: 5301412004006002001 as *M. flavigula.*
5301412004006004001 as *M. gwatkinsi.*

Martes foina (Erxleben, 1777). Regn. Anim., 1:458.
TYPE LOCALITY: Germany.
DISTRIBUTION: S. and C. Europe, through Caucasus, to the Altai (U.S.S.R., Mongolia)
and Himalayas; adjacent China; islands of Corfu, Crete and Rhodes.
ISIS NUMBER: 5301412004006003001.

Martes martes (Linnaeus, 1758). Syst. Nat., 10th ed., 1:46.
TYPE LOCALITY: Sweden, Uppsala.
DISTRIBUTION: Britain; N. and W. Europe to W. Siberia, south to Sicily, Sardinia,
Corsica, the Elburz, and the Caucasus Mts.
COMMENT: May be conspecific with *americana, melampus,* and *zibellina;* see Anderson,
1970, Acta Zool. Fenn., 130:132; Hagmeier, 1961, Can. Field Nat., 75:122–138; and
Corbet, 1978:172; but see also Hall, 1981:981; Gromov and Baranova,
1981:256–258.
ISIS NUMBER: 5301412004006005001.

Martes melampus (Wagner, 1841). Schreber's Saugethiere., Suppl., 2:229.
TYPE LOCALITY: Japan.
DISTRIBUTION: Honshu, Kyushu, Shikoku, Tsushima, intro. on Sado Isl. (Japan); Korea;
perhaps parts of China.
COMMENT: May be conspecific with *americana, martes,* and *zibellina;* see Anderson, 1970,
Acta Zool. Fenn., 130:132; Hagmeier, 1961, Can. Field Nat., 75:122–138; and
Corbet, 1978:172; but see also Hall, 1981:981; Gromov and Baranova,
1981:256–258. Heptner and Naumov, eds., 1967, [Mammals of the Soviet Union],
2(1) p. 507, included Japanese and Korean *melampus* in *zibellina;* but also see
Corbet, 1978:173, 174.
ISIS NUMBER: 5301412004006006001.

Martes pennanti (Erxleben, 1777). Regn. Anim., 1:470.
TYPE LOCALITY: E. Canada, Quebec.
DISTRIBUTION: Yukon to E. Quebec (Canada), N. W. U.S.A., Sierra Nevadas, N. Rocky
   Mtns., Great Lakes Region, New England, to N. Carolina (U.S.A.).
COMMENT: Reviewed by Powell, 1981, Mamm. Species, 156:1–6.
ISIS NUMBER: 5301412004006007001.

Martes zibellina (Linnaeus, 1758). Syst. Nat., 10th ed., 1:46.
TYPE LOCALITY: U.S.S.R., Siberia, Tomsk Dist., N. of Tobolsk.
DISTRIBUTION: Ural Mtns. to Siberia, Kamchatka, Sakhalin (U.S.S.R.); Mongolia;
   Sinkiang and N. E. China; N. Korea; Hokkaido (Japan); originally west to N.
   Scandanavia and W. Poland.
COMMENT: May be conspecific with *americana, martes,* and *melampus;* see Anderson,
   1970, Acta Zool. Fenn., 130:132; Hagmeier, 1961, Can. Field Nat., 75:122–138; and
   Corbet, 1978:172; but see also Hall, 1981:981; Gromov and Baranova,
   1981:256–258. Revised by Pavlinin, 1966, Der Zobel *Martes zibellina* L.,
   Wittenberg. See also comments under *melampus.*
ISIS NUMBER: 5301412004006008001.

Meles Boddaert, 1785. Elench. Anim., 1:45.
REVIEWED BY: A. I. Roest (AIR).
COMMENT: Subfamily Melinae.
ISIS NUMBER: 5301412004013000000.

Meles meles (Linnaeus, 1758). Syst. Nat., 10th ed., 1:48.
TYPE LOCALITY: Sweden, Uppsala.
DISTRIBUTION: Scandinavia to S. Siberia, south to Israel; China; Korea; and Japan, and
   on Ireland, Britain, Balearic Isls., Crete, and Rhodes.
COMMENT: Includes *amurensis;* see Corbet, 1978:175.
ISIS NUMBER: 5301412004013001001.

Mellivora Storr, 1780. Prodr. Meth. Mamm., p. 34, tabl. A.
REVIEWED BY: A. I. Roest (AIR).
COMMENT: Subfamily Mellivorinae.
ISIS NUMBER: 5301412004011000000.

Mellivora capensis (Schreber, 1776). Saugethiere, pl. 125, text 1777, 3:450.
TYPE LOCALITY: South Africa, Cape Prov., Cape of Good Hope.
DISTRIBUTION: Savanna and steppe from Nepal, E. India, and Turkmenia (U.S.S.R.)
   west to the Mediterranean, south to South Africa.
PROTECTED STATUS: CITES - Appendix III (Ghana and Botswana).
ISIS NUMBER: 5301412004011001001.

Melogale I. Geoffroy, 1831. Zool. Voy. de Belanger Indes Orient., 129, Mamm., pl. 5, 1834.
COMMENT: Subfamily Melinae. Includes *Helictis;* see Long, 1978, Rep. Fauna Flora
   Wisc., 14:1–6.
ISIS NUMBER: 5301412004014000000.

Melogale everetti (Thomas, 1895). Ann. Mag. Nat. Hist., ser. 6, 15:331–332.
TYPE LOCALITY: Malaysia, Sabah (N. Borneo), Kinabalu, 4000 ft. (1219 m).
DISTRIBUTION: Borneo.
COMMENT: Although resembling *M. personata orientalis,* this species is regarded as
   distinct by Long, 1978, Rep. Fauna Flora Wisc., 14:1–6, but as a subspecies of
   orientalis by Medway, 1977:132.

Melogale moschata (Gray, 1831). Proc. Zool. Soc. Lond., 1831:94.
TYPE LOCALITY: China, Canton.
DISTRIBUTION: Naga Hills near Manipur, Assam (India), to Taiwan; Hainan Isl., C., and
   S.E. China (SW), N. Laos, and N. Vietnam.
COMMENT: Includes *sorella,* which may be a distinct species; see Long, 1978, Rep.
   Fauna Flora Wisc., 14:1–6.
ISIS NUMBER: 5301412004014001001.

Melogale personata I. Geoffroy, 1831. Zool. Voy. de Belanger Indes Orient., p. 137.
TYPE LOCALITY: S. Burma, near Rangoon.
DISTRIBUTION: Rangoon (S. Burma) to Nepal and Assam (India), through Thailand and Vietnam; Java.
COMMENT: Includes *orientalis*, which may be a distinct species; see Long, 1978, Rep. Fauna Flora Wisc., 14:1–6.
ISIS NUMBER: 5301412004014003001 as *M. personata*.
5301412004014002001 as *M. orientalis*.

Mephitis E. Geoffroy and G. Cuvier, 1795. Mag. Encyclop., 2:187.
COMMENT: Subfamily Mephitinae.
ISIS NUMBER: 5301412004019000000.

Mephitis macroura Lichtenstein, 1832. Darst. Saugeth., pl. 46.
TYPE LOCALITY: Mexico, mountains N. W. of Mexico City.
DISTRIBUTION: S. Arizona, S. New Mexico, and W. Texas (U.S.A.), through Mexico to N. Nicaragua.
ISIS NUMBER: 5301412004019002001.

Mephitis mephitis (Schreber, 1776). Saugethiere, 3(17), pl. 121.
TYPE LOCALITY: Eastern Canada (=Quebec).
DISTRIBUTION: S. W. Northwest Territories to Hudson Bay and S. Quebec (Canada), south to Florida (U.S.A.), N. Tamaulipas, N. Durango, and N. Baja California (Mexico).
ISIS NUMBER: 5301412004019001001.

Mustela Linnaeus, 1758. Syst. Nat., 10th ed., 1:45.
COMMENT: Subfamily Mustelinae. Includes *Grammogale*; see Hall, 1951, Univ. Kans. Publ. Mus. Nat. Hist., 4:1–466; but see also comment under *africana*. Includes *Putorius* and *Lutreola*; see Corbet, 1978:168.
ISIS NUMBER: 5301412004007000000.

Mustela africana Desmarest, 1818. Nouv. Dict. Hist. Nat. Paris, 9:376.
TYPE LOCALITY: Brazil, probably Para.
DISTRIBUTION: Amazon Basin in Brazil, Ecuador, and Peru.
COMMENT: Izor and de la Torre, 1978, J. Mammal., 59:92–102, hypothesized a relationship between *africana* and *felipei*, and reopened the question of whether *Grammogale* is a distinct genus.
ISIS NUMBER: 5301412004007001001.

Mustela altaica Pallas, 1811. Zoogr. Rosso-Asiat., 3:98.
TYPE LOCALITY: U.S.S.R., S. W. Siberia, Altai Mtns.
DISTRIBUTION: E. Kazakhstan, S. and S. E. Siberia, Primorsky Krai (U.S.S.R.); Mongolia; Tibet, W. and N. China; Korea.
ISIS NUMBER: 5301412004007002001.

Mustela erminea Linnaeus, 1758. Syst. Nat., 10th ed., 1:46.
TYPE LOCALITY: Sweden.
DISTRIBUTION: Tundra and forested regions of Palearctic, south to the Pyrenees, Alps, Caucasus, W. Himalayas, Sinkiang (China); N. Mongolia, N. China; and C. Honshu (Japan), and in the Nearctic south to C. California, N. New Mexico, N. Iowa, and Maryland (U.S.A.).
ISIS NUMBER: 5301412004007003001.

Mustela eversmanni Lesson, 1827. Manuel Mamm., p. 144.
TYPE LOCALITY: U.S.S.R., Orenburgsk. Obl., S. of Orenburg, mouth of Khobda River, a tributary of Ilek River.
DISTRIBUTION: Steppes and subdeserts of E. Europe, U.S.S.R.; Mongolia; W., C. and N. E. China.
COMMENT: Possible ancestor of the domestic ferret, *furo*; see Corbet, 1978:171; Corbet and Hill, 1980:96. See Stroganov, 1962, [Mamm. Siberia], 2, [Carnivores], p. 335, for discussion of type locality.

Mustela felipei Izor and de la Torre, 1978. J. Mammal., 59:92.
TYPE LOCALITY: Colombia, Huila, Santa Marta near San Augustin.
DISTRIBUTION: Santa Marta (Huila, Colombia); Cauca (Colombia); (two localities
separated by 70 km, on opposite sides of Cordillera Central).

Mustela frenata Lichtenstein, 1831. Darst. Saugeth., p. 42.
TYPE LOCALITY: Mexico, Dist. Federal, Cuidad de Mexico.
DISTRIBUTION: S. Canada to Venezuela and Bolivia, excluding the S. W. deserts of the
U.S.A.
ISIS NUMBER: 5301412004007004001.

Mustela kathiah Hodgson, 1835. J. Asiat. Soc. Bengal, 4:702.
TYPE LOCALITY: Nepal, Kachar region.
DISTRIBUTION: Himalayas from northern Pakistan through Nepal to Burma; S. and E.
China; Indochinese Peninsula.
ISIS NUMBER: 5301412004007005001.

Mustela lutreola (Linnaeus, 1761). Fauna Suec., p. 5.
TYPE LOCALITY: Finland.
DISTRIBUTION: France; E. Europe to (originally) Ural Mtns. in W. Siberia (U.S.S.R.),
south to Hungary and the Caucasus. In the last century, has spread east to the
middle courses of the Irtysh and Ob rivers.
ISIS NUMBER: 5301412004007006001.

Mustela lutreolina Robinson and Thomas, 1917. Ann. Mag. Nat. Hist., ser. 8, 20:261–262.
TYPE LOCALITY: Indonesia, Java.
DISTRIBUTION: Java.
COMMENT: Ellerman and Morrison-Scott, 1951:260, and Lekagul and McNeely,
1977:536, listed Java as part of the distribution of *sibirica*, inferring that *lutreolina*
is included in that species; and were followed by Corbet, 1978:170; but see Ewer,
1973, The Carnivores, p. 394; Heptner and Naumov, eds., 1967, [Mammals of the
Soviet Union], 2(1), pp. 700–712, who considered *lutreolina* a distinct species.
ISIS NUMBER: 5301412004007007001.

Mustela nigripes (Audubon and Bachman, 1851). The Viviparous Quadrupeds of North
America, 2:297.
TYPE LOCALITY: U.S.A., Wyoming, Goshen Co., Fort Laramie.
DISTRIBUTION: Formerly, S. Alberta and Saskatchewan (Canada) south to Arizona,
Oklahoma, and N.W. Texas (U.S.A.).
COMMENT: Reviewed by Hillman and Clark, 1980, Mamm. Species, 126:1–3.
PROTECTED STATUS: CITES - Appendix I and U.S. ESA - Endangered.
ISIS NUMBER: 5301412004007008001.

Mustela nivalis Linnaeus, 1766. Syst. Nat., 12th ed., 1:69.
TYPE LOCALITY: Sweden, Vesterbotten.
DISTRIBUTION: The Palearctic (excluding Ireland, the Arabian Pen., and Arctic Isls.);
Japan; the Nearctic in Alaska (U.S.A.), Canada, and U.S.A., south to Wyoming
and North Carolina.
COMMENT: Includes *rixosa*; see Corbet, 1978:169; but see also Kurten and Anderson,
1980:150.
ISIS NUMBER: 5301412004007009001 as *M. nivalis*.
5301412004007012001 as *M. rixosa*.

Mustela nudipes Desmarest, 1822. Mammalogie, Part 2, Suppl., p. 537.
TYPE LOCALITY: Indonesia, Java.
DISTRIBUTION: Thailand; Malaya; Sumatra; Java; Borneo.
COMMENT: Includes *leucocephalus*; see Hill, 1960, Bull. Raffles Mus., 29:36. Lekagul and
McNeely, 1977:540, credited the species name to F. Cuvier, 1821.
ISIS NUMBER: 5301412004007010001.

Mustela putorius Linnaeus, 1758. Syst. Nat., 10th ed., 1:46.
TYPE LOCALITY: Sweden, Uppsala.
DISTRIBUTION: Europe (excluding Ireland and most of Scandinavia), east to the Ural
Mtns. (U.S.S.R.).

COMMENT: Probable ancestor of domestic ferret, *furo.;* see Corbet, 1978:171; Corbet and Hill, 1980:96.
ISIS NUMBER: 5301412004007011001.

Mustela sibirica Pallas, 1773. Reise Prov. Russ. Reichs., 2:701.
TYPE LOCALITY: U.S.S.R., E. Kazakhstan, vic. of Ust-Kamenogorsk, Tigeretskoie.
DISTRIBUTION: Tataria, Urals, Siberia, Far East (U.S.S.R.); Pakistan east to N. Burma; N. Thailand; Taiwan; China; Korea; Japan. Sakhalin (U.S.S.R.)(introduced).
COMMENT: Includes *itatsi;* see Corbet, 1978:170. May include *lutreolina;* see Ellerman and Morrison-Scott, 1951:260–262, and Lekagul and McNeely, 1977:536, who listed Java in the distribution for this species; but also see comments under *lutreolina.*
ISIS NUMBER: 5301412004007013001.

Mustela strigidorsa Gray, 1853. Proc. Zool. Soc. Lond., 1853:191.
TYPE LOCALITY: India, Sikkim.
DISTRIBUTION: Nepal east through Burma, Yunnan (China)(SW) and Thailand to Laos.
ISIS NUMBER: 5301412004007014001.

Mustela vison Schreber, 1777. Saugethiere, Suppl., 2:239.
TYPE LOCALITY: Eastern Canada (=Quebec).
DISTRIBUTION: North America from Alaska and Canada through all of U.S.A. except southwestern deserts.
COMMENT: Includes *macrodon;* see Manville, 1966, Proc. U.S. Nat. Mus., 122:10; but also see Hall, 1981:1004, who considered *macrodon* a distinct species. Introduced in Iceland, N. C. Europe, and Siberia; see Corbet, 1978:170.
ISIS NUMBER: 5301412004007015001.

Mydaus F. Cuvier, 1821. Hist. Nat. Mamm., 3(27), pl. and 2 pp. text.
COMMENT: Subfamily Melinae. Includes *Suillotaxus;* see Long, 1978, Rep. Fauna Flora Wisc., 14:1–6.
ISIS NUMBER: 5301412004015000000 as *Mydaus.*
5301412004016000000 as *Suillotaxus.*

Mydaus javanensis (Desmarest, 1820). Encyclop. Method. Mamm., 210.
TYPE LOCALITY: Indonesia, Java.
DISTRIBUTION: Java, Borneo, Sumatra and the Natuna Isls.
COMMENT: Includes *ollula* and *lucifer;* see Long, 1978, Rep. Fauna Flora Wisc., 14:1–6.
ISIS NUMBER: 5301412004015001001.

Mydaus marchei (Huet, 1887). Le Naturaliste, II, 9 annee, 13:149–151.
TYPE LOCALITY: Philippine Isls., Palawan.
DISTRIBUTION: Palawan and Calamian Isls. (Philippine Isls.).
COMMENT: Referred to the genus *Suillotaxus* by Lawrence, 1939, *in* Barbour *et al.,* Mustelidae in Mammal Collections from the Philippine Islands, 86(2):63–65. Considered a subgenus of *Mydaus* by Long, 1978, Rep. Fauna Flora Wisc., 14:1–6.
ISIS NUMBER: 5301412004016001001 as *Suillotaxus marchei.*

Poecilictis Thomas and Hinton, 1920. Ann. Mag. Nat. Hist., 5:367.
COMMENT: Subfamily Mustelinae.
ISIS NUMBER: 5301412004008000000.

Poecilictis libyca (Hemprich and Ehrenberg, 1833). Symb. Phys. Mamm., 2, sig. K.
TYPE LOCALITY: Libya.
DISTRIBUTION: Fringes of the Sahara Desert from Morocco and Egypt on the north, and from Mauretania and N. Nigeria to Sudan on the south.
ISIS NUMBER: 5301412004008001001.

Poecilogale Thomas, 1883. Ann. Mag. Nat. Hist., ser. 5, 11:370.
    COMMENT: Subfamily Mustelinae.
    ISIS NUMBER: 5301412004009000000.

Poecilogale albinucha (Gray, 1864). Proc. Zool. Soc. Lond., 1864:69.
    TYPE LOCALITY: South Africa, presumed to be Cape Colony (uncertain).
    DISTRIBUTION: Zaire, Uganda and Tanzania south to South Africa.
    ISIS NUMBER: 5301412004009001001.

Pteronura Gray, 1837. Charlesworth's Mag. Nat. Hist., 1:580.
    REVIEWED BY: N. Duplaix (ND).
    COMMENT: Subfamily Lutrinae.
    ISIS NUMBER: 5301412004025000000.

Pteronura brasiliensis (Gmelin, 1788). Linn. Syst. Nat., 1:93.
    TYPE LOCALITY: Guyana (=British Guiana), Demerara.
    DISTRIBUTION: Major river systems of South America east of the Andes as far south as
        N. Argentina.
    COMMENT: Reviewed by Harris, 1968, Otters, Weidenfeld and Nicolson, London.
    PROTECTED STATUS: CITES - Appendix I and U.S. ESA - Endangered.
    ISIS NUMBER: 5301412004025001001.

Spilogale Gray, 1865. Proc. Zool. Soc. Lond., 1865:150.
    COMMENT: Subfamily Mephitinae. Revised by Van Gelder, 1959, Bull. Am. Mus. Nat.
        Hist., 117:229–392.
    ISIS NUMBER: 5301412004020000000.

Spilogale putorius (Linnaeus, 1758). Syst. Nat., 10th ed., 1:44.
    TYPE LOCALITY: U.S.A., South Carolina.
    DISTRIBUTION: S. W. Canada east to Minnesota and S. Pennsylvania (U.S.A.); south to
        Costa Rica.
    COMMENT: Includes *gracilis* and *angustifrons*; see Van Gelder, 1959, Bull. Am. Mus. Nat.
        Hist., 117:279; but also see Mead, 1968, J. Mammal., 49:373–390.
    ISIS NUMBER: 5301412004020002001 as *S. putorius*.
                 5301412004020001001 as *S. gracilis*.

Spilogale pygmaea Thomas, 1898. Proc. Zool. Soc. Lond., 1897:898.
    TYPE LOCALITY: Mexico, Sinaloa, Rosario.
    DISTRIBUTION: Sinaloa to Oaxaca (Mexico).
    COMMENT: Includes *australis*; see Genoways and Jones, 1968, An. Inst. Biol. Univ. Nac.
        Auton. Mex., 39:123–132.
    ISIS NUMBER: 5301412004020003001.

Taxidea Waterhouse, 1838. Proc. Zool. Soc. Lond., 1838:154.
    COMMENT: Subfamily Melinae.
    ISIS NUMBER: 5301412004017000000.

Taxidea taxus (Schreber, 1778). Saugethiere, 3:520.
    TYPE LOCALITY: Canada, "Labrador or Hudson Bay."
    DISTRIBUTION: S. W. Canada to Baja California and Puebla (Mexico), east to Lake
        Ontario.
    COMMENT: Reviewed by Long, 1972, J. Mammal., 59:725–759, and Long, 1973, Mamm.
        Species, 26:1–4. Type locality restricted by Long, 1972, J. Mammal., 59:738.
    ISIS NUMBER: 5301412004017001001.

Vormela Blasius, 1884. Ber. Naturforsch Ges. Bemberg, 13, p. 9.
    COMMENT: Subfamliy Mustelinae.
    ISIS NUMBER: 5301412004010000000.

Vormela peregusna (Guldenstaedt, 1770). Nova Comm. Acad. Sci. Petrop., 14(1):441.
    TYPE LOCALITY: U.S.S.R., Rostov Obl., steppes at lower Don River.
    DISTRIBUTION: Steppes and deserts of S. E. Europe, Caucasus, Kazakhstan, Middle Asia;
        S. W. Asia (excl. Arabia); N. China and S. Mongolia.
    ISIS NUMBER: 5301412004010001001.

Family Viverridae

REVIEWED BY: P. Grubb (PG); H. J. Stains (HJS); S. Wang (SW); W. C. Wozencraft (WCW).

COMMENT: Does not include subfamilies Herpestinae Gill, 1872 (=Mungotidae Pocock, 1916) or Galidiinae Gill, 1872, (=Galidiina Gray, 1865); see Pocock, 1916, Proc. Zool. Soc. Lond., 1916:349; Gregory and Hellman, 1939, Proc. Am. Philos. Soc., 81(3):309–392; Wurster and Benirschke, 1968, Chromosoma, 24:336–382; Thenius, 1972, Grund. Verbreit. Saugetierk., pp. 223–224; Radinsky, 1975, J. Mammal., 56:130–150; Wemmer, 1977, Smithson. Contrib. Zool., 239:2. The Viverridae differ from the Herpestidae in the shape and structure of the basicranial region, presence of perineal glands, civetone secretions, ear bursa, and marker chromosomes (WCW).

ISIS NUMBER: 5301412005000000000.

Arctictis Temminck, 1824. Monogr. Mamm., 1:21.
COMMENT: Subfamily Paradoxurinae; see Gill, 1872, 11(1):4.
ISIS NUMBER: 5301412005007000000.

Arctictis binturong (Raffles, 1821). Trans. Linn. Soc. Lond., 13:253.
TYPE LOCALITY: Malaysia, Malacca.
DISTRIBUTION: S. E. Asia; W. Indonesia to Borneo and Java; Palawan (Philippine Isls.); Yunnan (China)(SW).
ISIS NUMBER: 5301412005007001001.

Arctogalidia Merriam, 1897. Science, 2:302.
COMMENT: Subfamily Paradoxurinae; see Gill, 1872, 11(1):4.
ISIS NUMBER: 5301412005008000000.

Arctogalidia trivirgata (Gray, 1832). Proc. Zool. Soc. Lond., 1832:68.
TYPE LOCALITY: Indonesia, Java, Buitenzorg (=Bogor).
DISTRIBUTION: Assam (India) through S. E. Asia and W. Indonesia to Borneo and Java; Yunnan (China)(SW).
COMMENT: Reviewed by Van Bemmel, 1952, Beaufortia, 16:23–41.
ISIS NUMBER: 5301412005008001001.

Chrotogale Thomas, 1912. Abstr. Proc. Zool. Soc. Lond. 106:17.
COMMENT: Subfamily Hemigalinae; see Gill, 1872, 11(1):4.
ISIS NUMBER: 5301412005013000000.

Chrotogale owstoni Thomas, 1912. Abstr. Proc. Zool. Soc. Lond., 106:17.
TYPE LOCALITY: Vietnam, Tonkin, Songkoi River, Yen Bai.
DISTRIBUTION: Laos; N. Vietnam; Yunnan, possibly Kwangsi (China)(SW).
ISIS NUMBER: 5301412005013001001.

Civettictis Pocock, 1915. Proc. Zool. Soc. Lond., 1915:134.
COMMENT: Subfamily Viverrinae; see Gill, 1872, 11(1):4. Formerly included in *Viverra* by Coetzee, 1977, Part 8:17; but see Ansell, 1978:38; Rosevear, 1974:167–168; Kingdon, 1977:159.

Civettictis civetta (Schreber, 1777). Saugethiere, 3:418, 587.
TYPE LOCALITY: Guinea.
DISTRIBUTION: Zululand and Transvaal (South Africa) to Tanzania, N. Namibia, and Angola; C. and E. Africa to S. Somalia through Ethiopia to Kordofan (Sudan), and to Senegal in W. Africa.
COMMENT: Some authors have placed this species in *Viverra*; see Coetzee, 1977, Part 8:17; but also see Ansell, 1978:38, who placed this species in *Civettictis*. Rosevear, 1974:167, noted differences in scent glands; Petter, 1969, Mammalia, 33:607–635, discussed dental differences.
PROTECTED STATUS: CITES - Appendix III as *Viverra civetta* (Botswana).
ISIS NUMBER: 5301412005005001001 as *Viverra civetta*.

Cryptoprocta Bennett, 1833. Proc. Zool. Soc. Lond., 1833:46.
    COMMENT: Subfamily Cryptoproctinae; see Gill, 1872, 11(1):4.
    ISIS NUMBER: 5301412005034000000.

  Cryptoprocta ferox Bennett, 1833. Proc. Zool. Soc. Lond., 1833:46.
    TYPE LOCALITY: Madagascar.
    DISTRIBUTION: Madagascar.
    COMMENT: Beaumont, 1964, Ecolgae Geol. Helv., 57:837–845, placed *ferox* with the
        Felidae because of similarities in dental characters. Coetzee, 1977, Part 8:35;
        Thenius, 1972, Grund. Verbreit. Saugetierk., pp. 205–215, Radinsky, 1975, J.
        Mammal., 56:140–142, and Albignac, 1970, La Terre et la Vie, 24:395–402, placed
        ferox in the Viverridae. Hemmer, 1978, Carnivore, 1:72, suggested an intermediate
        position.
    PROTECTED STATUS: CITES - Appendix II.
    ISIS NUMBER: 5301412005034001001.

Cynogale Gray, 1837. Proc. Zool. Soc. Lond., 1836:88.
    COMMENT: Subfamily Hemigalinae; see Gill, 1872, 11(1):4.
    ISIS NUMBER: 5301412005014000000.

  Cynogale bennettii Gray, 1837. Proc. Zool. Soc. Lond., 1836:88.
    TYPE LOCALITY: Indonesia, Sumatra.
    DISTRIBUTION: Malaya; Indochina; Sumatra; Borneo.
    COMMENT: Includes *lowei;* see Pocock, 1933, Proc. Zool. Soc. Lond., 1933:1035. WCW
        believes *lowei* may be a distinct species.
    PROTECTED STATUS: CITES - Appendix II.
    ISIS NUMBER: 5301412005014001001.

Eupleres Doyere, 1835. Bull. Soc. Sci. Nat., 3:45.
    COMMENT: Subfamily Euplerinae; see Gill, 1872, 11(1):4. Gregory and Hellman, 1939,
        Proc. Am. Philos. Soc., 81(3):809–892, suggested placing *Eupleres* in a separate
        family; but see Coetzee, 1977, Part 8:33.
    ISIS NUMBER: 5301412005015000000.

  Eupleres goudotii Doyere, 1835. Bull. Soc. Sci. Nat., 3:45.
    TYPE LOCALITY: Madagascar, Tamatave.
    DISTRIBUTION: Madagascar.
    COMMENT: Includes *major;* see Coetzee, 1977, Part 8:33; and Albignac, 1974, La Terre et
        la Vie, 28:321–251.
    PROTECTED STATUS: CITES - Appendix II as *E. goudotii* and
        CITES - Appendix II as *E. major.*
    ISIS NUMBER: 5301412005015001001 as *E. goudotii.*
              5301412005015002001 as *E. major.*

Fossa Gray, 1865. Proc. Zool. Soc. Lond., 1864:518.
    COMMENT: Subfamily Fossinae; see Albignac, 1970, La Terre et la Vie, 24:383; Petter,
        1974, Mammalia, 38:605–636.
    ISIS NUMBER: 5301412005016000000.

  Fossa fossa (Schreber, 1777). Saugethiere, 3:424, 587.
    TYPE LOCALITY: Madagascar.
    DISTRIBUTION: Madagascar.
    COMMENT: Correct name may be *F. fossana;* see Coetzee, 1977, Part. 8:33.
    PROTECTED STATUS: CITES - Appendix II.
    ISIS NUMBER: 5301412005016001001.

Genetta G. Cuvier, 1816. Regn. Anim., 1:156.
    REVIEWED BY: L. Schlawe (LS).
    COMMENT: Subfamily Viverrinae; see Gill, 1872, 11(1):4. For reviews, see Schlawe,
        1980, Faun. Abh. Staat. Mus. J. Tierk., Dresden, 7(15):147–161; Schlawe, 1981,
        Zool. Abh. Staat. Mus. Tierk., Dresden, 37(4):85–182, Coetzee, 1977, Part 8:20;
        Rosevear, 1974:177–220, Cabral, 1966, Zool. Afr., 2:25,26, Cabral, 1969, Bol. Inst.

Invest. Cient. Angola., 6(1):3–33, Cabral, 1970, Bol. Inst. Invest. Cient. Angola., 7(2):3–23, and Cabral, 1973, As genetas da Guine Portuguesa e de Mocambique, Lisboa. Under revision by Cabral and Schlawe (LS).
ISIS NUMBER: 5301412005001000000.

Genetta abyssinica (Ruppell, 1836). Neue Wirbelt. Fauna Abyssin. Gehorig. Saugeth., 1:33.
TYPE LOCALITY: Ethiopia, Gondar.
DISTRIBUTION: Ethiopia; N. Somalia
ISIS NUMBER: 5301412005001001001.

Genetta angolensis Bocage, 1882. J. Sci. Math. Phys. Nat. Lisboa, ser. 1, 9:29.
TYPE LOCALITY: Angola, Caconda.
DISTRIBUTION: N. Angola; N. W. Zambia; S. Zaire; S. Tanzania; Mozambique; Malawi.
COMMENT: Includes *mossambica* and *hintoni*; see Schlawe, 1980, Faun. Abh. Staat. Mus. J. Tierk., Dresden, 7(15):147–161; Cabral, 1969, Bol. Inst. Invest. Cient. Angola, 6(1):3–33; Ansell, 1978:39.
ISIS NUMBER: 5301412005001002001.

Genetta felina (Thunberg, 1811). Kongl. Svenska Vet.-Akad. Nya Handl., 32:165.
TYPE LOCALITY: South Africa, Cape of Good Hope.
DISTRIBUTION: Africa, south of the Sahara; S. Arabian Peninsula.
COMMENT: Includes *guardafuensis, senegalensis, dongolana, granti, neumanni, bella, pulchra,* and *ludia*; see Schlawe, 1980, Faun. Abh. Staat. Mus. J. Tierk., Dresden, 7(15):147–161. Rosevear, 1974:193, recognized *senegalensis* as a distinct species. See also comments under *genetta*.

Genetta genetta (Linnaeus, 1758). Syst. Nat., 10th ed., 1:45.
TYPE LOCALITY: Spain, El Pardo, near Madrid.
DISTRIBUTION: Africa, N. W. of the Sahara to Libya; Egypt; Palestine; Iberia; Balearic Isls.; and S. France to the Rhone River.
COMMENT: Rosevear, 1974:193, separated *senegalensis* from *genetta*; but see Coetzee, 1977, Part 8:20, Kingdon, 1977:136–143, and Cabral, 1969, Bol. Inst. Invest. Cient. Angola, 6(1):3–33. Schlawe, 1980, considered *senegalensis* a subspecies of *G. felina*. Includes *afra, bonapartei, barbara, terraesanctae, rhodanica,* and *isabelae*; see Schlawe, 1980, Faun. Abh. Staat. Mus. J. Tierk., Dresden, 7(15):147–161, who also provisionally separated *felina* from *genetta*; but also see Coetzee, 1977, Part 8:20; Ansell, 1978:39; and Cabral, 1969, Bol. Inst. Invest. Cient. Angola, 6(1):3–33.
ISIS NUMBER: 5301412005001003001.

Genetta johnstoni Pocock, 1908. Proc. Zool. Soc. Lond., 1907:1041.
TYPE LOCALITY: Liberia, W. of Putu Mtns., W. of Cavally River.
DISTRIBUTION: Liberia; Ghana.
COMMENT: Includes *lehmanni*; see Coetzee, 1977, Part 8:20; Schlawe, 1980, Faun. Abh. Staat. Mus. J. Tierk., Dresden, 7(15):147–161.
ISIS NUMBER: 5301412005001004001 as *G. lehmanni*.

Genetta maculata (Gray, 1830). Spicil. Zool., 2:9.
TYPE LOCALITY: C. Senegal.
DISTRIBUTION: Africa, south of the Sahara to Bioko; Angola, Botswana and Natal (South Africa).
COMMENT: May be conspecific with *tigrina*; see Coetzee, 1977, Part 8:22; Schlawe, 1980, Faun. Abh. Staat. Mus. J. Tierk., Dresden, 7(15):147–161; Pringle, 1977, Ann. Natal Mus., 23(1):93–115; but also see Ansell, 1978:39. Includes *pardina, amer (sensu* Schwartz, 1930, Rev. Zool. Bot. Afr., 19:275–286), *poensis, deorum, genettoides, rubiginosa, suahelica, fieldiana, aequatorialis, stuhlmanni,* and *bini* (not a synonym of *servalina*); see Schlawe, 1980, Faun. Abh. Staat. Mus. J. Tierk., Dresden, 7(15):147–161. See also Cabral, 1969, Bol. Inst. Invest. Cient. Angola, 6(1):3–33, who considered *maculata* to consist of two species, *poensis,* and *pardina* (including *rubiginosa*).
ISIS NUMBER: 5301412005001005001 as *G. pardina*.

Genetta servalina Pucheran, 1855. Rev. Mag. Zool. Paris, 7(2):154.
TYPE LOCALITY: Gabon.
DISTRIBUTION: Equatorial Africa, E. of the mouth of the Niger River, possibly to Mt.
Elgon (Kenya), south to the Congo River and 7° S.
COMMENT: Includes *cristata*, and *aubryana*, but not *bini (sensu* Rosevear, 1974:200–201),
which may actually be a junior synonym of *maculata*; see Wenzel and Haltenorth,
1972, Saugetierk. Mitt., 20 :110–127; Coetzee, 1977, Part 8:21–22; Schlawe, 1980,
Faun. Abh. Staat. Mus. J. Tierk., Dresden, 7(15):147–161; Cabral, 1970, Bol. Inst.
Invest. Cient. Angola, 7(2):323.
ISIS NUMBER: 5301412005001006001.

Genetta thierryi Matschie, 1902. Verh. V. Internat. Zool. Congr., 1901:1142.
TYPE LOCALITY: Togo, Borugu, 10° 46′ N., 0° 34′ E.
DISTRIBUTION: S. Mauritania to Cameroun.
COMMENT: : Includes *villiersi* Dekeyser 1949; see Kuhn, 1960, Saugetierk. Mitt.,
8:154–160; Schlawe, 1980, Faun. Abh. Staat. Mus. J. Tierk., Dresden, 7(15):147–161;
Cabral, 1969, Bol. Inst. Invest. Cient. Angola, 6(1):3–33; Rosevear, 1974:214–217.
ISIS NUMBER: 5301412005001009001 as *G. villiersi.*

Genetta tigrina (Schreber, 1776). Saugethiere, text 1777, 3:425, pl. 115.
TYPE LOCALITY: South Africa, Cape Prov., Cape of Good Hope.
DISTRIBUTION: From the Cape of Good Hope to S. Natal (South Africa) and Lesotho.
COMMENT: May be conspecific with *maculata* or with *angolensis*; see Schlawe, 1980,
Faun. Abh. Staat. Mus. J. Tierk., Dresden, 7(15):147–161; Ansell, 1978:40; Pringle,
1977, Ann. Natal Mus., 23(1):100–101; Cabral, 1966, Zool. Afr., 2:25–26. See also
comments under *maculata.*
ISIS NUMBER: 5301412005001007001.

Genetta victoriae Thomas, 1901. Proc. Zool. Soc. Lond., 1901:87.
TYPE LOCALITY: Zaire, Semliki River, 17 km W. of Beni.
DISTRIBUTION: E. Zaire, mountainous regions.
COMMENT: Thomas listed the type locality as "Entebbe, Uganda" from a specimen
collected by Sir Harry Johnston. Allen, 1924, Bull. Am. Mus. Nat. Hist., 47:132,
noted that according to Johnston, 1902, The Uganda Protectorate, 1:205, it was
collected on the Semlike River in Zaire, not at Entebbe.
ISIS NUMBER: 5301412005001008001.

Hemigalus Jourdan, 1837. C. R. Acad. Sci. Paris, 5:442.
COMMENT: Subfamily Hemigalinae; see Gill, 1872, 11(1):4. Includes *Diplogale*; see Ewer,
1973, The Carnivores, p. 402, and Medway, 1977:137; but also see Bourliere, 1955,
in Grasse, Traite de Zool., 17:260; Walker *et al.*, 1968, Mammals of the World, p.
124; and Wenzel and Haltenorth, 1972, Saugetierk. Mitt., 20:118, who separated
D. hosei from *Hemigalus.*
ISIS NUMBER: 5301412005017000000.

Hemigalus derbyanus (Gray, 1837). Charlesworth's Mag. Nat. Hist., 1:579.
TYPE LOCALITY: Malaysia.
DISTRIBUTION: Sipora Isl., South Pagi Isl. (W. Sumatra); Borneo; Sumatra; Malaya;
Thailand; peninsular Burma.
PROTECTED STATUS: CITES - Appendix II.
ISIS NUMBER: 5301412005017001001.

Hemigalus hosei (Thomas, 1892). Proc. Zool. Soc. Lond., 1892:222.
TYPE LOCALITY: Malaysia, Sarawak, Mt. Dulit, 4000 ft. (1219 m).
DISTRIBUTION: N. Borneo.
COMMENT: Formerly included in *Diplogale*; see Stains, 1967, *in* Anderson and Jones, p.
341; Ewer, 1973, The Carnivores, p. 402; Medway, 1977:137; but also see Bouliere,
1955, *in* Grasse, Traite de Zool., 17:260; Ducker, 1965, Hand. Zool. Berlin,
10(20a):1–48, Walker *et al.*, 1968, Mammals of the World, p. 1241, and Wenzel and
Haltenorth, 1972, Saugetierk. Mitt., 20:118.
ISIS NUMBER: 5301412005017002001.

Macrogalidia Schwartz, 1910. Ann. Mag. Nat. Hist., ser. 5, 8:423.
COMMENT: Subfamily Paradoxurinae; see Gill, 1872, 11(1):4.
ISIS NUMBER: 5301412005009000000.

Macrogalidia musschenbroekii (Schlegel, 1879). Notes Leyden Mus., 1:43.
TYPE LOCALITY: Indonesia, N. Sulawesi ("N. Celebes").
DISTRIBUTION: Sulawesi.
ISIS NUMBER: 5301412005009001001.

Nandinia Gray, 1843. Spec. Mamm. Br. Mus., p. 54.
COMMENT: Subfamily Paradoxurinae; see Gill, 1872, 11(1):4.
ISIS NUMBER: 5301412005010000000.

Nandinia binotata (Reinhardt, 1830). *In* Gray, Spicil. Zool., 2:9.
TYPE LOCALITY: Equatorial Guinea, Bioko ("Fernando Po").
DISTRIBUTION: Guinea-Bissau to Kenya, Tanzania and Mozambique; Bioko; Angola; S. Sudan.
ISIS NUMBER: 5301412005010001001.

Osbornictis J. A. Allen, 1919. J. Mammal., 1:25.
COMMENT: Subfamily Viverrinae; see Gill, 1872, 11(1):4. Probably congeneric with *Genetta* (WCW).
ISIS NUMBER: 5301412005002000000.

Osbornictis piscivora J. A. Allen, 1919. J. Mammal., 1:25.
TYPE LOCALITY: Zaire ("Congo Belge"), Niapu.
DISTRIBUTION: Zaire.
ISIS NUMBER: 5301412005002001001.

Paguma Gray, 1831. Proc. Zool. Soc. Lond., 1831:95.
COMMENT: Subfamily Paradoxurinae; see Gill, 1872, 11(1):4.
ISIS NUMBER: 5301412005011000000.

Paguma larvata (H. Smith, 1827). *In* Griffith, Cuvier's Anim. Kingd., 2:281.
TYPE LOCALITY: China, Canton.
DISTRIBUTION: India and Nepal, China north to Hopei and Shansi; vic. Peking; Hainan Isl. (China); Taiwan; Thailand; Malaya; Vietnam; Sumatra; N. Borneo; Viper Isl. (S. Andaman Isls.).
COMMENT: Hamilton-Smith type locality unknown. Gray first assigned the name to two specimens collected by J. R. Reeve from Canton, China; see Pocock, 1934, Proc. Zool. Soc. Lond., 1934:666(WCW). Probably includes *lanigera* (SW); see Ellerman and Morrison-Scott, 1951:290.
ISIS NUMBER: 5301412005011001001.

Paradoxurus F. Cuvier, 1821. *In* F. Cuvier and E. Geoffroy, Hist. Nat. Mamm., 24:5.
COMMENT: Subfamily Paradoxurinae; see Gill, 1872, 11(1):4.
ISIS NUMBER: 5301412005012000000.

Paradoxurus hermaphroditus (Pallas, 1777). *In* Schreber, Die Saugethiere, 3:426.
TYPE LOCALITY: India.
DISTRIBUTION: Sri Lanka and India to S. China, Hainan Isl., and Malaysia; Philippine Isls.; W. Indonesia (incl. Lesser Sunda Isls.).
COMMENT: Scattered records in Sulawesi, Moluccas, and Aru Isls. probably result from introduction (WCW).
ISIS NUMBER: 5301412005012001001.

Paradoxurus jerdoni Blanford, 1885. Proc. Zool. Soc. Lond., 1885:613.
TYPE LOCALITY: India, Madras, Palni Hills, Kodaikanal.
DISTRIBUTION: S. India.
ISIS NUMBER: 5301412005012002001.

Paradoxurus zeylonensis (Pallas, 1778). *In* Schreber, Die Saugethiere, 3:451.
    TYPE LOCALITY: Sri Lanka ("Ceylon").
    DISTRIBUTION: Sri Lanka.
    ISIS NUMBER: 5301412005012003001.

Poiana Gray, 1865. Proc. Zool. Soc. Lond., 1864:520.
    COMMENT: Subfamily Viverrinae; see Gill, 1872, 11(1):4. May be congeneric with
        Prionodon (PG, WCW).
    ISIS NUMBER: 5301412005003000000.

Poiana richardsoni (Thomson, 1842). Ann. Mag. Nat. Hist., ser. 10, 1:204.
    TYPE LOCALITY: Equatorial Guinea, Bioko ("Fernando Po").
    DISTRIBUTION: Sierra Leone; Liberia; Ivory Coast; Bioko; Cameroun; Gabon; Congo
        Republic; Zaire.
    COMMENT: Includes *liberiensis* (replaces *leightoni*, a *lapsus)*; see Coetzee, 1977, Part 8:18.
        Rosevear, 1974:227, raised *leightoni* to the species level based on white versus off-
        white ventral coloration and presence of more or less continuous dark spinal line
        despite his comment earlier in the same work, on the unreliability of such
        characters (p. 186–188). He also supported the distinction with a recent record
        and a distributional gap with other known forms. *Poiana* records are few and
        scattered; recent studies of *Poiana* placed *leightoni* as a subspecies; see de Beaufort,
        1965, Mammalia, 29:275–280; Michaelis, 1972, Saugetierk. Mitt., 20:1–110;
        Kingdon, 1977:159. No skulls of *leightoni* are known (WCW).
    ISIS NUMBER: 5301412005003001001.

Prionodon Horsfield, 1822. Zool. Res. Java, Part 5.
    COMMENT: Subfamily Viverrinae; see Gill, 1872, 11(1):4. Probably includes *Poiana* (PG,
        WCW).
    ISIS NUMBER: 5301412005004000000.

Prionodon linsang (Hardwicke, 1821). Trans. Linn. Soc. Lond., 13:256.
    TYPE LOCALITY: Malaysia, Malacca.
    DISTRIBUTION: Bankga Isl.; Java; Borneo; Billiton Isl.; peninsular Burma to Sumatra.
    PROTECTED STATUS: CITES - Appendix II.
    ISIS NUMBER: 5301412005004001001.

Prionodon pardicolor Hodgson, 1842. Calcutta J. Nat. Hist., 2:57.
    TYPE LOCALITY: Nepal.
    DISTRIBUTION: N. E. India to N. Indochina; Nepal; Sikkim.
    PROTECTED STATUS: CITES - Appendix I and U.S. ESA - Endangered.
    ISIS NUMBER: 5301412005004002001.

Viverra Linnaeus, 1758. Syst. Nat., 10th ed., 1:43.
    COMMENT: Subfamily Viverrinae; see Gill, 1872, 11(1):4. Does not include *Civettictis*;
        see Kingdon, 1977:159; Ansell, 1978:38; but also see Coetzee, 1977, Part 8:17.
    ISIS NUMBER: 5301412005005000000.

Viverra megaspila Blyth, 1862. J. Asiat. Soc. Bengal, 31:331.
    TYPE LOCALITY: Burma, Prome.
    DISTRIBUTION: S. India; Burma; Thailand; Indochina; Malaya.
    PROTECTED STATUS: U.S. ESA - Endangered as *V. m. civettina* subspecies only.
    ISIS NUMBER: 5301412005005002001.

Viverra tangalunga Gray, 1832. Proc. Zool. Soc. Lond., 1832:63.
    TYPE LOCALITY: Indonesia, W. Sumatra.
    DISTRIBUTION: Malaya; Sumatra; Rhio-Lingga Arch.; Bangka Isl.; Borneo; Karimata Isl.;
        Sulawesi; Buru; Amboina; Philippine Isls.; Langkawi Isl.
    COMMENT: Introduced to many S. E. Asian Isls (WCW).
    ISIS NUMBER: 5301412005005003001.

Viverra zibetha Linnaeus, 1758. Syst. Nat., 10th ed., 1:43.
TYPE LOCALITY: India.
DISTRIBUTION: N. India; Nepal; Burma; Indochina; Malay Peninsula; north to Anhwei, Shensi, Chekiang and Kiangsu (China)(SW).
ISIS NUMBER: 5301412005005004001.

Viverricula Hodgson, 1838. Ann. Mag. Nat. Hist., ser. 1, 1:152.
COMMENT: Subfamily Viverrinae; see Gill, 1872, 11(1):4.
ISIS NUMBER: 5301412005006000000.

Viverricula indica (Desmarest, 1817). Nouv. Dict. Hist. Nat. Paris, 7:170.
TYPE LOCALITY: India.
DISTRIBUTION: Sri Lanka; India to S. China and Hainan Isl., and Malay Peninsula; Taiwan; Java; Kangean Isl.; Sumbawa; Bali. Introduced to Madagascar.
COMMENT: *V. malaccensis* is invalid; see Pocock, 1933, J. Bombay Nat. Hist. Soc., 36:629. Scattered distribution on many S. E. Asian Isls. due to frequent introductions (WCW).
ISIS NUMBER: 5301412005006001001.

Family Herpestidae
REVIEWED BY: P. Grubb (PG); H. J. Stains (HJS); W. C. Wozencraft (WCW).
COMMENT: Includes Herpestinae Gill, 1872 (=Mungotidae Pocock, 1916) and Galidiinae Gill, 1872 (=Galidiina Gray, 1865); see Pocock, 1916, Proc. Zool. Soc. Lond., 1916:349; Gregory and Hellman, 1939, Proc. Am. Philos. Soc., 81(3):309–392; Wurster and Benirschke, 1968, Chromosoma, 24:336–382; Radinsky, 1975, J. Mammal., 56:130–150; Wemmer, 1977, Smithson. Contrib. Zool., 239:2; Thenius, 1972, Grund. Verbreitierk. Sauget., pp. 223–224.

Atilax F. Cuvier, 1826. *In* E. Geoffroy and F. Cuvier, Hist. Nat. Mamm., 54:1.
ISIS NUMBER: 5301412005022000000.

Atilax paludinosus (G. Cuvier, 1829). Regn. Anim., 2nd ed., 1:158.
TYPE LOCALITY: South Africa, Cape Prov., Cape of Good Hope.
DISTRIBUTION: South Africa to Ethiopia; Nigeria to Gambia.
ISIS NUMBER: 5301412005022001001.

Bdeogale Peters, 1850. Spenersche Z., 25 June, 1850.
COMMENT: Includes *Galeriscus;* see Coetzee, 1977, Part 8:24; Kingdon, 1977:245.
ISIS NUMBER: 5301412005023000000.

Bdeogale crassicauda Peters, 1850. Spenersche Z., 25 June, 1850.
TYPE LOCALITY: Mozambique, Tette, Boror.
DISTRIBUTION: Mozambique; Malawi; Zambia; Kenya; Tanzania; Zanzibar.
COMMENT: See Rosevear, 1974:321, for discussion of type description.
ISIS NUMBER: 5301412005023001001.

Bdeogale jacksoni (Thomas, 1894). Ann. Mag. Nat. Hist., ser. 13, 6:622.
TYPE LOCALITY: Kenya, Mianzini, near Lake Naivasha.
DISTRIBUTION: C. Kenya; S.E. Uganda.
COMMENT: Rosevear, 1974:321, placed *jacksoni* in the genus *Galeriscus;* but also see Kingdon, 1977:245; Sale and Taylor, 1970, J. East Afr. Nat. Hist. Soc., 28(2):10–15; Coetzee, 1977, Part 8:24. Kingdon, 1977, considered *jacksoni* conspecific with *nigripes,* as does PG; but also see Rosevear, 1974:322; Coetzee, 1977, Part 8:24, who noted skull and skin differences.
ISIS NUMBER: 5301412005023002001.

Bdeogale nigripes Pucheran, 1855. Rev. Mag. Zool. Paris, 7(2):111.
TYPE LOCALITY: Gabon.
DISTRIBUTION: Nigeria to N. Angola.
COMMENT: Rosevear, 1974:322, placed *nigripes* in the genus *Galeriscus;* but also see Kingdon, 1977:245; Sale and Taylor, 1970, J. East Afr. Nat. Hist. Soc., 28(2):10-15; Coetzee, 1977, Part 8:24. May include *jacksoni;* see Kingdon, 1977:245.
ISIS NUMBER: 5301412005023003001.

Crossarchus F. Cuvier, 1825. *In* E. Geoffroy and F. Cuvier, Hist. Nat. Mamm., 47:3.
  COMMENT: May include *Liberiictis* (PG).
  ISIS NUMBER: 5301412005024000000.

  Crossarchus alexandri Thomas and Wroughton, 1907. Ann. Mag. Nat. Hist., ser. 19, 7:373.
    TYPE LOCALITY: Zaire, Ubangi, Banzyville.
    DISTRIBUTION: Zaire; Uganda.
    COMMENT: *C. alexandri* and *obscurus* may be conspecific; see Coetzee, 1977, Part 8:29.
      However, *alexandri* is very distinct in cranial proportions, approaching *Liberiictis*
      kuhni (PG).
    ISIS NUMBER: 5301412005024001001.

  Crossarchus ansorgei Thomas, 1910. Ann. Mag. Nat. Hist., ser. 5, 8:195.
    TYPE LOCALITY: Angola, Dalla Tando.
    DISTRIBUTION: N. Angola; S.E. Zaire.
    ISIS NUMBER: 5301412005024002001.

  Crossarchus obscurus F. Cuvier, 1825. *In* E. Geoffroy and F. Cuvier, Hist. Nat. Mamm.,
    47:3.
    TYPE LOCALITY: West Africa.
    DISTRIBUTION: Sierra Leone to Cameroun.
    ISIS NUMBER: 5301412005024003001.

Cynictis Ogilby, 1833. Proc. Zool. Soc. Lond., 1833:48.
  ISIS NUMBER: 5301412005025000000.

  Cynictis penicillata (G. Cuvier, 1829). Regn. Anim., 2nd ed., 1:158.
    TYPE LOCALITY: South Africa, Uitenhage.
    DISTRIBUTION: South Africa; Namibia; S. Angola; Botswana; S.W. Zimbabwe.
    ISIS NUMBER: 5301412005025001001.

Dologale Thomas, 1926. Ann. Mag. Nat. Hist., ser. 17, 9:183.
  COMMENT: Formerly included in *Helogale*; see Hayman, 1936, Ann. Mag. Nat. Hist.,
    ser. 10, 18:625–630.

  Dologale dybowskii (Pousargues, 1893). Bull. Soc. Zool. Fr., 18:51.
    TYPE LOCALITY: Central African Republic, N. of Ubangi on Upper Kemo River. (6° 17'
      N, 17° 12' E.)
    DISTRIBUTION: N.E. Zaire; Central African Republic; S. Sudan; W. Uganda.
    COMMENT: May be congeneric with *Helogale*; see Hayman, 1936, Ann. Mag. Nat. Hist.,
      ser. 10, 18:625–630, who also restricted type locality.
    ISIS NUMBER: 5301412005026001001 as *Helogale dybowskii*.

Galidia I. Geoffroy, 1837. C. R. Acad. Sci. Paris, 5:580.
  ISIS NUMBER: 5301412005018000000.

  Galidia elegans I. Geoffroy, 1837. C. R. Acad. Sci. Paris, 5:580.
    TYPE LOCALITY: Madagascar, Tamatave.
    DISTRIBUTION: Madagascar.
    ISIS NUMBER: 5301412005018001001.

Galidictis I. Geoffroy, 1839. Mag. Zool., Mammal. Art. No. 5, p. 33, footnote, 37.
  ISIS NUMBER: 5301412005019000000.

  Galidictis fasciata (Gmelin, 1788). Syst. Nat., 13th ed., 1:92.
    TYPE LOCALITY: Madagascar.
    DISTRIBUTION: Madagascar.
    COMMENT: Includes *ornata* and *striata*; see Coetzee, 1977, Part 8:34.
    ISIS NUMBER: 5301412005019001001 as *G. fasciata.*
                5301412005019002001 as *G. ornata.*
                5301412005019003001 as *G. striata.*

Helogale Gray, 1862. Proc. Zool. Soc. Lond., 1861:308.
COMMENT: May be congeneric with *Dologale;* see Hayman, 1936, Ann. Mag. Nat. Hist., ser. 10, 18:625–630.
ISIS NUMBER: 5301412005026000000.

Helogale hirtula Thomas, 1904. Ann. Mag. Nat. Hist., ser. 14, 7:97.
TYPE LOCALITY: Somalia, Gabridehari, 60 mi. (96 km) W. Gerlogobi.
DISTRIBUTION: S. Ethiopia; S. and C. Somalia; N. and C. Kenya.
COMMENT: Includes *percivali;* see Coetzee, 1977, Part 8:32.
ISIS NUMBER: 5301412005026002001 as *H. hirtula.*
5301412005026005001 as *H. percivali.*

Helogale parvula (Sundevall, 1846). Ofv. Kongl. Svenska Vet.-Akad. Forhandl. Stockholm, 3(4):121.
TYPE LOCALITY: South Africa, Transvaal, Zoutbansberg.
DISTRIBUTION: Somalia and Ethiopia to South Africa and Namibia.
COMMENT: Includes *macmillani, vetula,* and *victorina;* see Coetzee, 1977, Part 8:31.
ISIS NUMBER: 5301412005026004001 as *H. parvula.*
5301412005026003001 as *H. macmillani.*
5301412005026006001 as *H. vetula.*
5301412005026007001 as *H. victorina.*

Herpestes Illiger, 1811. Prodr. Syst. Mamm. et Avium., p. 135.
COMMENT: Includes *Xenogale;* see Coetzee, 1977, Part 8:26; Hayman, 1940, *in* Sanderson, Trans. Zool. Soc. Lond., 24:623–725; but also see Rosevear, 1974:329; and Ansell, 1978:41. Also includes *Galerella;* see Ewer, 1973, The Carnivores, p. 405; Michaelis, 1972, Saugetierk. Mitt., 20:106; Wenzel and Haltenorth, 1972, Saugetierk. Mitt., 20:122; Coetzee, 1977, Part 8:26; Kingdon, 1977:185; but also see Rosevear, 1974:308; and Ansell, 1978:41. For review of Asiatic forms see Bechthold, 1939, Z. Saugetierk., 14:133–225.
ISIS NUMBER: 5301412005027000000.

Herpestes auropunctatus (Hodgson, 1836). J. Asiat. Soc. Bengal, 5:236.
TYPE LOCALITY: Nepal.
DISTRIBUTION: N. Arabia to S. China, and Hainan Isl. (SW) and Malay Peninsula. Introduced to the West Indies, Hawaiian Isls. (U.S.A.), and many other tropical regions.
COMMENT: Includes *palustris;* see Ewer, 1973, The Carnivores, p. 405; Michaelis, 1972, Saugetierk. Mitt., 20:106; Wenzel and Haltenorth, 1972, Saugetierk. Mitt., 20:121. Probably conspecific with *javanicus;* see Bechthold, 1939, Z. Saugetierk., 14:147; but also see comment under *javanicus.*
ISIS NUMBER: 5301412005027001001.

Herpestes brachyurus Gray, 1837. Proc. Zool. Soc. Lond., 1836:88.
TYPE LOCALITY: Malaysia, Malacca.
DISTRIBUTION: Malaysia; Philippine Isls.; Borneo; Sumatra.
COMMENT: Probably includes *hosei* and *fuscus;* see Bechthold, 1939, Z. Saugetierk., 14:140–141.
ISIS NUMBER: 5301412005027002001.

Herpestes edwardsi (E. Geoffroy, 1818). Descrip. de L'Egypte, 2:139.
TYPE LOCALITY: India, Madras.
DISTRIBUTION: E.C. Arabia to Nepal, India and Sri Lanka.
ISIS NUMBER: 5301412005027004001.

Herpestes fuscus Waterhouse, 1838. Proc. Zool. Soc. Lond., 1838:55.
TYPE LOCALITY: India.
DISTRIBUTION: Sri Lanka; S. India.
ISIS NUMBER: 5301412005027005001.

Herpestes hosei Jentink, 1903. Notes Leyden Mus., 23:226.
TYPE LOCALITY: Malaysia, Sarawak, Baram River.
DISTRIBUTION: Sarawak (Borneo).

COMMENT: Probably conspecific with *brachyurus*; it is only known from the type specimen, which was slightly smaller and more reddish-brown than most *brachyurus* (WCW).

ISIS NUMBER: 5301412005027007001.

Herpestes ichneumon (Linnaeus, 1758). Syst. Nat., 10th ed., 1:43.

TYPE LOCALITY: Egypt, Nile River.

DISTRIBUTION: Africa; Spain; Portugal; north around the E. Mediterranean to extreme S. Turkey.

ISIS NUMBER: 5301412005027008001.

Herpestes javanicus (E. Geoffroy, 1818). Descrip. de L'Egypte, 2:138.

TYPE LOCALITY: Indonesia, West Java.

DISTRIBUTION: Java; Malay Peninsula; Thailand; Kampuchea; C. Vietnam.

COMMENT: May include *auropunctatus*; see Bechthold, 1939, Z. Saugetierk., 14:147; Wenzel and Haltenorth, 1972, Saugetierk. Mitt., 20:121; Michaelis, 1972, Saugetierk. Mitt., 20:106; but also see Ewer, 1973, The Carnivores, p. 405; and Hinton and Dunn, 1967, The Mongooses, who considered *auropunctatus* a distinct species (WCW).

ISIS NUMBER: 5301412005027009001.

Herpestes naso De Winton, 1901. Bull. Liverpool Mus., 3:35.

TYPE LOCALITY: Cameroun, Cameroun River.

DISTRIBUTION: S.E. Nigeria to Gabon and Zaire.

COMMENT: Includes *Xenogale naso*; see Allen, 1919, J. Mammal., 1:26-28; but see also Rosevear, 1974:329. Includes *microdon*; see Hayman, 1940, *in* Sanderson, Trans. Zool. Soc. Lond., 24:623-725, and Coetzee, 1977, Part 8:26.

ISIS NUMBER: 5301412005027010001.

Herpestes pulverulentus (Wagner, 1839) Gelehrte. Anzeign. I. K. Bayer. Akad. Wiss. Munchen., 9:426.

TYPE LOCALITY: South Africa, Cape Prov., Cape of Good Hope.

DISTRIBUTION: S. Angola; Namibia; South Africa.

COMMENT: Included in *Galerella* by Allen, 1924, Bull. Am. Mus. Nat. Hist., 47:176-178; Rosevear, 1974:308; and Ansell, 1978:41. Placed in the subgenus *Galerella* by Coetzee, 1977, Part 8:26.

ISIS NUMBER: 5301412005027012001.

Herpestes sanguineus (Ruppell, 1835). Neue Wirbelt. Fauna Abyssin. Gehorig. Saugeth., 1:27.

TYPE LOCALITY: Sudan, Kordofan.

DISTRIBUTION: Subsaharan Africa.

COMMENT: Included in *Galerella* by Allen, 1924, Bull. Am. Mus. Nat. Hist., 47:176-178; Rosevear, 1974:308; and Ansell, 1978:41; but also see Kingdon, 1977:185. Includes *dentifer, granti,* and *ochraceus*; see Coetzee, 1977, Part 8:27. Reviewed by Taylor, 1975, Mamm. Species, 65:1-5.

ISIS NUMBER: 5301412005027013001 as *H. sanguineus.*
5301412005027003001 as *H. dentifer.*
5301412005027006001 as *H. granti.*
5301412005027011001 as *H. ochracea (sic).*

Herpestes semitorquatus Gray, 1846. Ann. Mag. Nat. Hist., 18:211.

TYPE LOCALITY: Mainland opposite Labuan (=Brunei).

DISTRIBUTION: Borneo; Sumatra.

COMMENT: May be a subspecies of *urva*; see Ewer, 1973, The Carnivores, p. 405.

ISIS NUMBER: 5301412005027014001.

Herpestes smithii Gray, 1837. Charlesworth's Mag. Nat. Hist., 1:578.

TYPE LOCALITY: India, Bombay.

DISTRIBUTION: India; Sri Lanka.

ISIS NUMBER: 5301412005027015001 as *H. smithi (sic).*

Herpestes urva (Hodgson, 1836). J. Asiat. Soc. Bengal, 5:238.
TYPE LOCALITY: Nepal.
DISTRIBUTION: S. China; Taiwan; Nepal to Indochina.
COMMENT: May include *semitorquatus;* see Ewer, 1973, The Carnivores, p. 405.
ISIS NUMBER: 5301412005027016001.

Herpestes vitticollis Bennett, 1835. Proc. Zool. Soc. Lond., 1835:67.
TYPE LOCALITY: India, Travancore.
DISTRIBUTION: S. India; Sri Lanka.
ISIS NUMBER: 5301412005027017001.

Ichneumia I. Geoffroy, 1837. Ann. Sci. Nat. Zool., 8(2):251.
ISIS NUMBER: 5301412005028000000.

Ichneumia albicauda (G. Cuvier, 1829). Regn. Anim., 2nd ed., 21:158.
TYPE LOCALITY: Senegal.
DISTRIBUTION: Subsaharan Africa; Oman.
COMMENT: Reviewed by Taylor, 1972, Mamm. Species, 12:1–4.
ISIS NUMBER: 5301412005028001001.

Liberiictis Hayman, 1958. Ann. Mag. Nat. Hist., ser. 13, 1:449.
COMMENT: Although suggested by Hayman, 1940, *in* Sanderson, Trans. Zool. Soc.
Lond., 24:623–725, to have a close relationship to *Crossarchus,* Coetzee, 1977, Part
8:23, considered *Liberiictis* a distinct genus. Probably should be included in
Crossarchus (PG).
ISIS NUMBER: 5301412005029000000.

Liberiictis kuhni Hayman, 1958. Ann. Mag. Nat. Hist., ser. 13, 1:449.
TYPE LOCALITY: Liberia, Kpeaplay, 6° 36′ N., 8° 30′ W.
DISTRIBUTION: Liberia.
ISIS NUMBER: 5301412005029001001.

Mungos E. Geoffroy and G. Cuvier, 1795. Mag. Encyclop., 2:184, 187.
ISIS NUMBER: 5301412005030000000.

Mungos gambianus (Ogilby, 1835). Proc. Zool. Soc. Lond., 1835:102.
TYPE LOCALITY: Gambia, Cape St. Mary.
DISTRIBUTION: Gambia to Nigeria.
ISIS NUMBER: 5301412005030001001.

Mungos mungo (Gmelin, 1788). Syst. Nat., 13th ed., 1:84.
TYPE LOCALITY: "Asia."
DISTRIBUTION: South Africa, N. to Sudan and W. to Niger and Gambia (PG), excluding
Congo.
COMMENT: Gmelin, 1788, gave the type locality as Asia but Ogilby, 1835, Proc. Zool.
Soc. Lond., 1835:101, restricted it to Gambia. Thomas, 1882, Proc. Zool. Soc.
Lond., 1882:90, believed it to be in the eastern part of South Africa, Cape Prov.,
as did Roberts, 1929, Ann. Transvaal Mus., 13:84, and Allen, 1924, Bull. Am. Mus.
Nat. Hist., 47:155–156 (WCW).
ISIS NUMBER: 5301412005030002001.

Mungotictis Pocock, 1915. Ann. Mag. Nat. Hist., ser. 8, 16:120.
ISIS NUMBER: 5301412005020000000.

Mungotictis decemlineata (A. Grandidier, 1867). Rev. Mag. Zool. Paris, ser. 2, 19:85.
TYPE LOCALITY: Madagascar, W. coast.
DISTRIBUTION: Madagascar.
COMMENT: Includes *lineata* and *substriatus;* see Coetzee, 1977, Part 8:35.
ISIS NUMBER: 5301412005020001001 as *M. decemlineata.*
5301412005020002001 as *M. substriatus.*

Paracynictis Pocock, 1916. Ann. Mag. Nat. Hist., ser. 17, 8:177.
    ISIS NUMBER: 5301412005031000000.

  Paracynictis selousi (De Winton, 1896). Ann. Mag. Nat. Hist., ser. 18, 6:469.
    TYPE LOCALITY: Zimbabwe, near Bulawayo, Essex Vale.
    DISTRIBUTION: Southern Africa.
    ISIS NUMBER: 5301412005031001001.

Rhynchogale Thomas, 1894. Proc. Zool. Soc. Lond., 1894:139.
    ISIS NUMBER: 5301412005032000000.

  Rhynchogale melleri (Gray, 1865). Proc. Zool. Soc. Lond., 1864:575.
    TYPE LOCALITY: Malawi, Zomba.
    DISTRIBUTION: S. Zaire; Tanzania; Malawi; C. and N. Mozambique; Zambia; Zimbabwe,
    N. South Africa.
    ISIS NUMBER: 5301412005032001001.

Salanoia Gray, 1865. Proc. Zool. Soc. Lond., 1864:523.
    ISIS NUMBER: 5301412005021000000.

  Salanoia concolor (I. Geoffroy, 1837). C. R. Acad. Sci. Paris, 5:581.
    TYPE LOCALITY: Madagascar.
    DISTRIBUTION: Madagascar.
    COMMENT: Includes *olivacea* and *unicolor*; Geoffroy, 1839, noted that his 1837 species
    name, *unicolor*, was a typographical error and should have been *concolor*; see
    Coetzee, 1977, Part 8:35.
    ISIS NUMBER: 5301412005021001001 as *S. olivacea.*
                5301412005021002001 as *S. unicolor.*

Suricata Desmarest, 1804. Dict. Class. Hist. Nat., 1st ed., 24:15.
    ISIS NUMBER: 5301412005033000000.

  Suricata suricatta (Erxleben, 1777). Regn. Anim., p. 488.
    TYPE LOCALITY: South Africa, Cape Prov.
    DISTRIBUTION: Angola; Namibia; South Africa; S. Botswana.
    ISIS NUMBER: 5301412005033001001.

Family Protelidae
    COMMENT: Included in the family Hyaenidae by many authors; see Stains, 1967, *in*
    Anderson and Jones, p. 343; Corbet and Hill, 1980:103; but also see Coetzee, 1977,
    Part 8:35; and Ansell, 1978:44.

Proteles I. Geoffroy, 1824. Mem. Mus. Hist. Nat. Paris, 11:355.
    ISIS NUMBER: 5301412006003000000.

  Proteles cristatus (Sparrman, 1783). Bull. Sci. Soc. Philom. Paris, p.139.
    TYPE LOCALITY: South Africa, Cape Prov., Somerset East, near Little Fish River.
    DISTRIBUTION: Sudan; Ethiopia; Somalia; Central African Republic; E. and S. Africa.
    PROTECTED STATUS: CITES - Appendix III (Botswana).
    ISIS NUMBER: 5301412006003001001.

Family Hyaenidae
    REVIEWED BY: I. Rieger (IR); O. L. Rossolimo (OLR)(U.S.S.R.); H. J. Stains (HJS).
    COMMENT: Formerly included Protelidae; see Ansell, 1978:44; Coetzee, 1977, Part 8:35;
    but also see Stains, 1967, *in* Anderson and Jones, p. 343; Corbet and Hill,
    1980:103.
    ISIS NUMBER: 5301412006000000000.

Crocuta Kaup, 1828. Oken's Isis. Encyclop. Zeit, 21(11), column 1145.
    COMMENT: Antedated by *Crocuta* Meigen, 1800 (an insect), but this name has been
    suppressed in Opinion 678 of the ICZN, 1963, Bull. Zool. Nomencl., 20:339–342.
    Reviewed by Ronnefeld, 1969, Saugetierk. Mitt., 17:285–350.
    ISIS NUMBER: 5301412006001000000.

Crocuta crocuta (Erxleben, 1777). Syst. Regn. Anim., p. 578.
 TYPE LOCALITY: Senegambia.
 DISTRIBUTION: S. and E. Africa, west to Senegal.
 COMMENT: Subspecies reviewed by Matthews, 1939, Proc. Zool. Soc. Lond.,
   1939:237–260. Type locality restricted by Cabrera, 1911, Proc. Zool. Soc. Lond.,
   1911:95.
 ISIS NUMBER: 5301412006001001001.

Hyaena Brunnich, 1771. Zool. Fundamenta, p. 34, 42, 43.
 COMMENT: Reviewed by Pocock, 1934, Proc. Zool. Soc. Lond., 1934:825; Ellerman and
   Morrison-Scott, 1951:299. Brisson, 1762, where *Hyaena* was first used, is
   unavailable; see Hemming, 1955, Bull. ICZN, 11(6):197.
 ISIS NUMBER: 5301412006002000000.

Hyaena brunnea Thunberg, 1820. Kongl. Svenska Vet.-Acad. Handl. Stockholm, p. 59.
 TYPE LOCALITY: South Africa, Cape Prov., Cape of Good Hope.
 DISTRIBUTION: Africa south of Zambesi River.
 PROTECTED STATUS: CITES - Appendix I and U.S. ESA - Endangered.
 ISIS NUMBER: 5301412006002001001.

Hyaena hyaena (Linnaeus, 1758). Syst. Nat., 10th ed., 1:40.
 TYPE LOCALITY: S. Iran, Laristan, Benna Mtns.
 DISTRIBUTION: N. and E. Africa south to Tanzania; Asia Minor to Arabia; Iran;
   Transcaucasia; Turkmenia (U.S.S.R.); India; Nepal.
 COMMENT: Type locality restricted by Thomas, 1911, Proc. Zool. Soc. Lond., 1911:134.
   Reviewed by Rieger, 1981, Mamm. Species, 150:105.
 PROTECTED STATUS: U.S. ESA - Endangered as *H. h. barbara* subspecies only.
 ISIS NUMBER: 5301412006002002001.

Family Felidae
 REVIEWED BY: G. E. Glass (GEG); C. P. Groves (CPG); N. A. Neff (NAN); J. Ramirez-
   Pulido (JRP)(Mexico); I. Rieger (IR); O. L. Rossolimo (OLR)(U.S.S.R.); A. H.
   Shoemaker (AHS); H. J. Stains (HJS).
 COMMENT: The generic classification of this family is unstable; see Guggisberg, 1975,
   The Wild Cats of the World, 328 pp. Due to inconsistencies of the dates of
   publication for Schreber's names in the literature, we follow Sherborn, 1892,
   Proc. Zool. Soc. Lond., 1891:587–592 (NAN).
 PROTECTED STATUS: CITES - Appendix II except *Felis catus.*
 ISIS NUMBER: 5301412007000000000.

Acinonyx Brookes, 1828. Cat. Anat. Zool. Mus. J. Brookes, Lond., p. 16, 33.
 COMMENT: Availability of Brookes catalogue names is uncertain (CPG).
 ISIS NUMBER: 5301412007003000000.

Acinonyx jubatus (Schreber, 1776). Saugethiere, 3:pl. 105 (text 3:392, 1777).
 TYPE LOCALITY: South Africa, Cape Prov., Cape of Good Hope.
 DISTRIBUTION: Steppe and savanna zones from Baluchistan through Iran and
   Turkmenia (U.S.S.R.) to N.E. Arabia; widespread in Africa except the Sahara;
   formerly India and Egypt.
 PROTECTED STATUS: CITES - Appendix I and U.S. ESA - Endangered.
 ISIS NUMBER: 5301412007003001001.

Felis Linnaeus, 1758. Syst. Nat., 10th ed., 1:41.
 REVIEWED BY: A. C. Ziegler (ACZ).
 COMMENT: Includes *Herpailurus, Leopardus, Lynchailurus, Oreailurus, Pardofelis,*
   Prionailurus, Profelis, and *Puma,* which are all considered valid genera by CPG; also
   includes *Badiofelis, Chaus, Leptailurus, Mayailurus,* and *Otocolobus;* see Corbet and
   Hill, 1980:104, Corbet, 1978:180, and Hemmer, 1978, Carnivore, 1:71. *Felis* is in
   need of revision. The number and content of genera that should be recognized is
   in dispute; see Hemmer, 1978, Carnivore, 1:71, and Weigel, 1961, Saugetierk.
   Mitt., 9:120. See also comments under *Lynx* and *Panthera.*
 ISIS NUMBER: 5301412007001000000.

Felis aurata Temminck, 1827. Monogr. Mamm., 1:120.
  TYPE LOCALITY: Sierra Leone.
  DISTRIBUTION: Gambia to Kenya, south to Angola, S. Zaire, and Burundi.
  COMMENT: Type locality is uncertain; see Gray, 1838, Ann. Mag. Nat. Hist., ser. 1, 1:27,
    for the Sierra Leone information (IR); and Van Mensch and Van Bree, 1969,
    Biologia Gabon., 4:236–269, for "probably between Cross River and River Congo"
    which was given in a review of the species (CPG).
  PROTECTED STATUS: CITES - Appendix II as Felidae.
  ISIS NUMBER: 5301412007001001001.

Felis badia Gray, 1874. Proc. Zool. Soc. Lond., 1874:322.
  TYPE LOCALITY: Malaysia, Sarawak.
  DISTRIBUTION: Borneo.
  COMMENT: Closely related to and perhaps conspecific with *temmincki* (CPG).
  PROTECTED STATUS: CITES - Appendix II as Felidae.
  ISIS NUMBER: 5301412007001002001.

Felis bengalensis Kerr, 1792. Anim. Kingdom, 1:151.
  TYPE LOCALITY: India, Bengal, Sunderbans.
  DISTRIBUTION: Lower Amur (E. Siberia) through Korea and N.E. China and most of the
    Oriental region west to Baluchistan, and southeast to Taiwan; Philippine Isls.,
    Java, Bali, and Borneo.
  COMMENT: Includes *euptilura* from E. Siberia; see Corbet, 1978:184; but also see
    Gromov and Baranova, 1981:285, who listed *euptilura* as a distinct species.
    Includes *minuta*; see Chasen, 1940:108. See Heptner, 1971, Zool. Zh., 50:1720, for
    discussion of systematics. Subspecies extremely variable (AHS).
  PROTECTED STATUS: CITES - Appendix II as Felidae.
    CITES - Appendix I and U.S. ESA - Endangered as *F. b. bengalensis* subspecies
    only.
  ISIS NUMBER: 5301412007001003001.

Felis bieti Milne-Edwards, 1892. Rev. Gen. Sci. Pures Appl., 3:671.
  TYPE LOCALITY: China, Szechwan, vicinity of Tongolo and Tatsienlu.
  DISTRIBUTION: Dry steppe from the eastern flanks of the Tibetan plateau in C. China
    north to Inner Mongolia; perhaps to Mongolia.
  COMMENT: The distinction between *bieti* and *silvestris* is dubious according to
    Haltenorth, 1953, Die Wildkatzen der alten Welt: eine ubersicht uber die
    Untergattung *Felis*.
  PROTECTED STATUS: CITES - Appendix II as Felidae.
  ISIS NUMBER: 5301412007001004001.

Felis chaus Guldenstaedt, 1776. Nova Comm. Acad. Sci. Petrop., 20:483.
  TYPE LOCALITY: U.S.S.R., Dagestan, Terek River, N. of the Caucasus.
  DISTRIBUTION: Egypt to the Volga delta, east to Sinkiang, Tibet, Szechwan, Yunnan
    (China), and through India to Thailand and Vietnam; Sri Lanka.
  PROTECTED STATUS: CITES - Appendix II as Felidae.
  ISIS NUMBER: 5301412007001007001.

Felis colocolo Molina, 1810. Sagg. Stor. Nat. Chile, p. 295.
  TYPE LOCALITY: Chile, Valparaiso.
  DISTRIBUTION: S.W. Brazil; S. and W. Bolivia; Argentina; Uruguay; Peru; C. Chile;
    Ecuador.
  COMMENT: Includes *pajeros*; see Osgood, 1943, Field Mus. Nat. Hist. Publ. Zool. Ser., p.
    79–84.
  PROTECTED STATUS: CITES - Appendix II as Felidae.
  ISIS NUMBER: 5301412007001008001.

Felis concolor Linnaeus, 1771. Regn. Anim., App. Mant. Plantar., 2:522.
  TYPE LOCALITY: French Guiana, Cayenne.
  DISTRIBUTION: S. Argentina and C. Chile, north to British Columbia, S. Ontario, and
    Nova Scotia (Canada).

COMMENT: *Puma* is considered a valid genus by CPG.
PROTECTED STATUS: CITES - Appendix II as Felidae.
    CITES - Appendix I and U.S. ESA - Endangered as *F. c. coryi* subspecies only.
    CITES - Appendix I and U.S. ESA - Endangered as *F. c. costaricensis* subspecies
      only.
    CITES - Appendix I and U.S. ESA - Endangered as *F. c. cougar* subspecies only.
ISIS NUMBER: 5301412007001009001.

Felis geoffroyi d'Orbigny and Gervais, 1844. Bull. Sci. Soc. Philom. Paris, p. 44.
TYPE LOCALITY: Argentina, Buenos Aires Prov., Rio Negro.
DISTRIBUTION: Uruguay; Paraguay; Argentina; Andean Bolivia; S. Brazil.
COMMENT: Reviewed by Ximenez, 1975, Mamm. Species, 54:1–4. Frequently exhibits melanism (AHS).
PROTECTED STATUS: CITES - Appendix II as Felidae.
ISIS NUMBER: 5301412007001010000.

Felis guigna Molina, 1782. Sagg. Stor. Nat. Chile, p. 295.
TYPE LOCALITY: Chile, Valdivia.
DISTRIBUTION: S. and C. Chile; S. Argentina.
COMMENT: Type locality fixed by Thomas, 1903, Ann. Mag. Nat. Hist., ser. 7, 12:240, based on Philippi, 1870, Arch. Naturg., 36(1):41–43.
PROTECTED STATUS: CITES - Appendix II as Felidae.
ISIS NUMBER: 5301412007001011001.

Felis iriomotensis Imaizumi, 1967. J. Mamm. Soc. Jpn., 3:74.
TYPE LOCALITY: Japan, Ryukyu Isls., Iriomote Isl., Haimida.
DISTRIBUTION: Iriomote Isl. (Ryukyu Isls.)
COMMENT: Possibly conspecific with *bengalensis;* Imaizumi's key characters are polymorphic in *bengalensis* (Glass, 1979, 59th Ann. Meet. Am. Soc. Mammal., p. 5); see Petzsch, 1970, Das Pelzgewerbe, 20:3–7; Guggisberg, 1975, The Wild Cats of the World, p. 89; Glass and Todd, 1977, Z. Saugetierk., 42:36–44.
PROTECTED STATUS: CITES - Appendix II as Felidae and U.S. ESA - Endangered as *F. (=Mayailurus) iriomotensis.*
ISIS NUMBER: 5301412007001012001.

Felis jacobita Cornalia, 1865. Mem. Soc. Ital. Sci. Nat., 1:3.
TYPE LOCALITY: Bolivia, Potosi, near Argentinian border between Huanchaca and Potosi.
DISTRIBUTION: N.E. Chile; S. Peru; S.W. Bolivia; N. Argentina.
COMMENT: Placed in subgenus *Oreailurus* by CPG.
PROTECTED STATUS: CITES - Appendix I and U.S. ESA - Endangered.
ISIS NUMBER: 5301412007001013001.

Felis manul Pallas, 1776. Reise Prov. Russ. Reichs., 3:692.
TYPE LOCALITY: U.S.S.R., S.W. Transbaikalia, Buryat-Mongolsk. A.S.S.R.; S. of Lake Baikal, Kulusutai.
DISTRIBUTION: Steppe and semidesert from Caspian Sea to Kashmir, Transbaikalia, Mongolia, and C. China, north to Inner Mongolia and east to Hopei (SW).
PROTECTED STATUS: CITES - Appendix II as Felidae.
ISIS NUMBER: 5301412007001016001.

Felis margarita Loche, 1858. Rev. Mag. Zool. Paris, ser. 2, 10(2):49.
TYPE LOCALITY: Algeria, N. of Ouargla, near Negonca.
DISTRIBUTION: Deserts from S. Morocco; Senegal; Algeria; Niger; Egypt; probably Sinai; S.E. Arabia; N. Iran; Turkestan; Baluchistan.
COMMENT: Includes *thinobia;* see Corbet, 1978:182. Revised by Hemmer *et al.,* 1976. Z. Saugetierk., 41:286. See also Schauenberg, 1974, Rev. Suisse Zool., 81:949–969 for geographic variation, distribution, systematics and ecology. *F. marginata* is a *lapsus;* see Gray, 1867, Proc. Zool. Soc. Lond., 1867:275.
PROTECTED STATUS: CITES - Appendix II as Felidae.
ISIS NUMBER: 5301412007001017001.

Felis marmorata Martin, 1837. Proc. Zool. Soc. Lond., 1836:108.
  TYPE LOCALITY: Indonesia, Sumatra.
  DISTRIBUTION: N. India and Nepal to Vietnam, Thailand, Malaya, Sumatra, and
    Borneo. Probably S. China.
  COMMENT: *Pardofelis* is considered a valid genus by CPG. Type locality discussed by
    Robinson and Kloss, 1919, J. Fed. Malay St. Mus., 7:261.
  PROTECTED STATUS: CITES - Appendix I and U.S. ESA - Endangered.
  ISIS NUMBER: 5301412007001018001.

Felis nigripes Burchell, 1824. Travels in Interior of Southern Africa, vol. 2:592.
  TYPE LOCALITY: South Africa, Cape Prov., Bechuanaland, near Kuruman.
  DISTRIBUTION: South Africa; S. Namibia; Botswana.
  PROTECTED STATUS: CITES - Appendix I and U.S. ESA - Endangered.
  ISIS NUMBER: 5301412007001019001.

Felis pardalis Linnaeus, 1758. Syst. Nat., 10th ed., 1:42.
  TYPE LOCALITY: Mexico, Veracruz, Hernandez.
  DISTRIBUTION: E. Brazil, Paraguay, N. Argentina, and Peru, north to C. Arizona and
    Texas (U.S.A.). Arkansas (CPG).
  COMMENT: *Leopardus* is considered a valid genus by CPG. Type locality fixed by
    Thomas, 1911, Proc. Zool. Soc. Lond., 1911:136.
  PROTECTED STATUS: CITES - Appendix II as Felidae. U.S. ESA - Endangered as *F.
    pardalis*.
    CITES - Appendix I as *F. p. mearnsi* subspecies only.
    CITES - Appendix I as *F. p. mitis* subspecies only.
  ISIS NUMBER: 5301412007001020000.

Felis planiceps Vigors and Horsfield, 1827. Zool. J., 3:449.
  TYPE LOCALITY: Indonesia, Sumatra.
  DISTRIBUTION: Peninsular Thailand; Malaya; Sumatra; Borneo.
  PROTECTED STATUS: CITES - Appendix I and U.S. ESA - Endangered.
  ISIS NUMBER: 5301412007001022001.

Felis rubiginosa I. Geoffroy, 1831. Zool. Voy. de Belanger Indes Orient. Zool., p. 140.
  TYPE LOCALITY: India, Pondicherry.
  DISTRIBUTION: India; Sri Lanka; Kashmir.
  COMMENT: See Chakraborty, 1978, J. Bombay Nat. Hist. Soc., 75:478–479 for discussion
    of distribution.
  PROTECTED STATUS: CITES - Appendix I (Indian population only).
  ISIS NUMBER: 5301412007001023001.

Felis serval Schreber, 1776. Saugethiere, 3:pl. 108 (text 3:407, 1777).
  TYPE LOCALITY: South Africa, Cape Prov., Cape of Good Hope.
  DISTRIBUTION: S.E. Morocco; Algeria; Subsaharan Africa.
  PROTECTED STATUS: CITES - Appendix II as Felidae and U.S. ESA - Endangered as *F. s.
    constantina* subspecies only.
  ISIS NUMBER: 5301412007001025001.

Felis silvestris Schreber, 1777. Saugethiere, 3:397.
  TYPE LOCALITY: Germany.
  DISTRIBUTION: Deciduous woodland savanna and steppe from W. Europe to N.W.
    China and C. India; Morocco to Egypt, south to Sudan, S. Mauritania and
    Senegal, eastward to Ethiopia, through subsaharan savannas to the Cape (South
    Africa).
  COMMENT: This name was placed on the official list by opinion 465 of the ICZN, 1957.
    Includes *libyca, chutuchta, vellerosa* and *catus* (worldwide), which was domesticated
    from this species; see Corbet, 1978:181. *F. s. vellerosa* is probably a domestic cat
    (CPG).
  PROTECTED STATUS: CITES - Appendix II as Felidae (The domestic cat, *Felis catus*, is
    specifically excluded from protection).
  ISIS NUMBER: 5301412007001026001 as *F. silvestris*.
              5301412007001006001 as *F. catus*.
              5301412007001014001 as *F. libyca*.

Felis temmincki Vigors and Horsfield, 1827. Zool. J., 3:451.
TYPE LOCALITY: Indonesia, Sumatra.
DISTRIBUTION: Nepal, to Shensi and Kiangsi (China)(SW) south to Sumatra.
COMMENT: Includes *tristis*; see Ellerman and Morrison-Scott, 1951:312.
PROTECTED STATUS: CITES - Appendix I and U.S. ESA - Endangered.
ISIS NUMBER: 5301412007001027001.

Felis tigrina Schreber, 1775. Saugethiere, 3:pl. 106 (text 3:396, 1777).
TYPE LOCALITY: French Guiana, Cayenne.
DISTRIBUTION: Costa Rica; Colombia to S. Brazil, Paraguay, and N. Argentina.
PROTECTED STATUS: CITES - Appendix II as Felidae and U.S. ESA - Endangered as *F. tigrina*.
    CITES - Appendix I as *F. t. oncilla* subspecies only.
ISIS NUMBER: 5301412007001028001.

Felis viverrina Bennett, 1833. Proc. Zool. Soc. Lond., 1833:68.
TYPE LOCALITY: India, Kerala, Malabar Coast.
DISTRIBUTION: India, Sri Lanka, Nepal, east to Vietnam and Taiwan (SW); south to Sumatra, Java, and Bali.
COMMENT: Ellerman and Morrison-Scott, 1951, restricted the type locality to "Malabar Coast."
PROTECTED STATUS: CITES - Appendix II as Felidae.
ISIS NUMBER: 5301412007001029001 as *F. viverrinus (sic)*.

Felis wiedii Schinz, 1821. Cuvier's Thierreich., p. 235.
TYPE LOCALITY: Brazil, Bahia, near Rio Mucuri, Morro de Arara.
DISTRIBUTION: S. Brazil, N. Uruguay, and N. Argentina, north to Oaxaca, S. Sinaloa (Mexico) and S. Texas (U.S.A.).
COMMENT: Type locality determined by Allen, 1919, Bull. Am. Mus. Nat. Hist., 41:357.
PROTECTED STATUS: CITES - Appendix II as Felidae and U.S. ESA - Endangered as *F. wiedii*.
    CITES - Appendix I as *F. w. nicaraguae* subspecies only.
    CITES - Appendix I as *F. w. salvinia* subspecies only.
ISIS NUMBER: 5301412007001030000.

Felis yagouaroundi E. Geoffroy, 1803. Cat. Mamm. Mus. Nat. Hist. Nat., p. 124.
TYPE LOCALITY: Argentina, Buenos Aires.
DISTRIBUTION: S. Brazil to Peru, north to S. Arizona and S. Texas (U.S.A.).
COMMENT: *Herpailurus* is considered a valid genus by CPG. First available type locality from Martin, 1833, Proc. Zool. Soc. Lond., 1833:3. Two color phases occur (AHS).
PROTECTED STATUS: CITES - Appendix II as Felidae.
    CITES - Appendix I and U.S. ESA - Endangered as *F. y. cacomitli* subspecies only.
    CITES - Appendix I and U.S. ESA - Endangered as *F. y. fossata* subspecies only.
    CITES - Appendix I and U.S. ESA - Endangered as *F. y. panamensis* subspecies only.
    CITES - Appendix I and U.S. ESA - Endangered as *F. y. tolteca* subspecies only.
ISIS NUMBER: 5301412007001031001.

Lynx Kerr, 1792. Anim. Kingdom, 1:155.
COMMENT: Included in *Felis* by Corbet, 1978:182, but maintained as a separate genus by Matyushkin, 1979, Arch. Zool. Mus. Moscow St. Univ., 18:76–162, and Werdelin, 1981, Ann. Zool. Fenn., 18:37–71, who revised the genus. Includes *Caracal*; Matyushkin, 1979:95, provisionally included the subgenus *Caracal* in this genus, but Werdelin, *(op. cit.)* retained *Caracal* as a distinct genus.

Lynx canadensis Kerr, 1792. Anim. Kingdom, 1:155.
TYPE LOCALITY: Eastern Canada (= Quebec).
DISTRIBUTION: Taiga zone of North America, south to C. Utah and S.W. Colorado, N.E. Nebraska, S. Indiana, and N. Virginia (U.S.A.).
COMMENT: Included in *F. lynx* by Corbet, 1978:183, but given specific status by Matyushkin, 1979, Arch. Zool. Mus. Moscow St. Univ., 18:76–162; and Werdelin, 1981, Ann. Zool. Fenn., 18:37–71.
PROTECTED STATUS: CITES - Appendix II as Felidae.

Lynx caracal (Schreber, 1776). Saugethiere, 3:pl. 106 (text 3:413, 1777).
TYPE LOCALITY: South Africa, Cape Prov., Cape of Good Hope, Table Mountain.
DISTRIBUTION: Steppe and savanna from Turkestan and N.W. India to Egypt, Algeria
and Morocco; Subsaharan Africa in savanna zones.
COMMENT: Provisionally placed by Werdelin, 1981, Ann. Zool. Fenn., 18:37–71, in
genus *Caracal*; see also comment under *Lynx*.
PROTECTED STATUS: CITES - Appendix II as Felidae and
CITES - Appendix I as *Felis caracal*, Asian population only.
ISIS NUMBER: 5301412007001005001 as *Felis caracal*.

Lynx lynx (Linnaeus, 1758). Syst. Nat., 10th ed., 1:43.
TYPE LOCALITY: Sweden, near Uppsala, Wennersborg.
DISTRIBUTION: Taiga forest from Scandinavia to E. Siberia and N.E. China; montane
Europe (formerly widespread, now restricted to Balkans and Carpathians); Asia
Minor through mtns. of C. Asia to Kansu, Tsaidam (Tsinghai), Shansi, and
Szechwan (China); Sakhalin (U.S.S.R.); perhaps Sardinia.
COMMENT: Does not include *canadensis* or *pardinus*; see Matyushkin, 1979, Arch. Zool.
Mus. Moscow St. Univ., 18:76–162, and Werdelin, 1981, Ann. Zool. Fenn., 18:37–
71. Includes *isabellina*; see Corbet, 1978:183.
PROTECTED STATUS: CITES - Appendix II as Felidae.
ISIS NUMBER: 5301412007001015001 as *Felis lynx*.

Lynx pardinus (Temminck, 1824). Monogr. Mamm., 1:116.
TYPE LOCALITY: Portugal, near Lisbon.
DISTRIBUTION: Iberian Peninsula.
COMMENT: Included in *Felis lynx* by Corbet, 1978:182, but given specific status by
Matyushkin, 1979, Arch. Zool. Mus. Moscow St. Univ., 18:76–162, and Werdelin,
1981, Ann. Zool. Fenn., 18:37–71. Probably a race of *lynx* (AHS).
PROTECTED STATUS: CITES - Appendix II as Felidae and U.S. ESA - Endangered as *Felis
lynx pardina* subspecies only.
ISIS NUMBER: 5301412007001021001 as *Felis pardina*.

Lynx rufus (Schreber, 1776). Saugethiere, 3(95):pl. 109b (text 3:412, 1777).
TYPE LOCALITY: U.S.A., New York.
DISTRIBUTION: S. British Columbia to Nova Scotia (Canada), south to Oaxaca (Mexico).
COMMENT: Includes *oaxacensis*; see Goodwin, 1963, Am. Mus. Novit., 2139:1–7.
PROTECTED STATUS: CITES - Appendix II as Felidae.
CITES - Appendix I as *Felis (=Lynx) rufa (sic) escuinapae* subspecies only. U.S. ESA
- Endangered as *Felis rufus escuinapae* subspecies.
ISIS NUMBER: 5301412007001024001 as *Felis rufus*.

Neofelis Gray, 1867. Proc. Zool. Soc. Lond., 1867:265.

Neofelis nebulosa (Griffith, 1821). Descrip. Anim. (Carn.), p. 37, pl.
TYPE LOCALITY: China, Kwangtung, Canton.
DISTRIBUTION: Nepal to Fukien (China) and Indochina; north to Shensi (China); south
to Sumatra and Borneo; Taiwan.
PROTECTED STATUS: CITES - Appendix I and U.S. ESA - Endangered.
ISIS NUMBER: 5301412007002002001 as *Panthera nebulosa*.

Panthera Oken, 1816. Lehrb. Zool., 3(2):1052.
REVIEWED BY: C. A. Hill (CAH).
COMMENT: In accordance with the provisions of Article 80 of the ICZN code, *Panthera*
is used instead of *Leo*. Includes *Uncia*; see Corbet, 1978:184; but also see Hemmer,
1972, Mamm. Species, 20:1–5, and Heptner and Naumov, 1972, [Mammals of the
Soviet Union], Vol. 2(2):211. Van Gelder, 1977:13, included *Panthera* as a synonym
of *Felis*. CAH prefers the use of *Leo* and does not agree with the interpretation of
Article 80 of the ICZN code presented here.
ISIS NUMBER: 5301412007002000000.

Panthera leo (Linnaeus, 1758). Syst. Nat., 10th ed., 1:41.
TYPE LOCALITY: Algeria, Constantine, Barbary.
DISTRIBUTION: Subsaharan Africa (except tropical rainforests); Gir Forest (N.W. India).
COMMENT: Formerly in non-desert N. Africa; S.E. Europe, and S.W. Asia.
PROTECTED STATUS: CITES - Appendix II as Felidae.
        CITES - Appendix I and U.S. ESA - Endangered as *P. l. persica* subspecies only.
ISIS NUMBER: 5301412007002001001.

Panthera onca (Linnaeus, 1758). Syst. Nat., 10th ed., 1:42.
TYPE LOCALITY: Brazil, Pernambuco.
DISTRIBUTION: N. Argentina and S. Brazil to Peru, north to Oaxaca (Mexico) to S. Texas
        and N.C. Arizona, N. Baja California (Mexico), S. California, and S.W. and C.
        New Mexico (U.S.A.).
COMMENT: Type locality subsequently fixed by Thomas, 1911. Proc. Zool. Soc. Lond.,
        1911:135. Frequently exhibits melanism (AHS).
PROTECTED STATUS: CITES - Appendix I and U.S. ESA - Endangered.
ISIS NUMBER: 5301412007002003001.

Panthera pardus (Linnaeus, 1758). Syst. Nat., 10th ed., 1:41.
TYPE LOCALITY: Egypt.
DISTRIBUTION: N. Africa; S.W. Asia, north Caucasus and Kopet Dag, throughout
        Oriental region to Java, north to Korea and Amur region; subsaharan Africa.
COMMENT: Type locality fixed by Thomas, 1911. Proc. Zool. Soc. Lond., 1911:135.
        Frequently exhibits melanism (AHS).
PROTECTED STATUS: CITES - Appendix I and U.S. ESA - Endangered.
ISIS NUMBER: 5301412007002004001.

Panthera tigris (Linnaeus, 1758). Syst. Nat., 10th ed., 1:41.
TYPE LOCALITY: India, Bengal.
DISTRIBUTION: Oriental region (except Borneo and Philippines); north to extreme S.E.
        Siberia (U.S.S.R.); formerly in S. Caucasus, Asia Minor, N. Iran, Afghanistan, Bali,
        and Java.
COMMENT: Type locality fixed by Thomas, 1911, Proc. Zool. Soc. Lond., 1911:135.
        Populations of *sondaica* (including *balica*) and *virgata* probably extinct (AHS).
        Reviewed by Mazak, 1981, Mamm. Species, 152:1–8.
PROTECTED STATUS: CITES - Appendix I (except *P. t. altaica*) and U.S. ESA - Endangered
        as *P. tigris*.
        CITES - Appendix II as *P. t. altaica* subspecies only.
ISIS NUMBER: 5301412007002005001.

Panthera uncia (Schreber, 1775). Saugethiere, 3:pl. 100 (text, 3:386, 1777).
TYPE LOCALITY: Altai Mtns. (U.S.S.R. or Mongolia).
DISTRIBUTION: Mtns. of Central Asia from Altai through the Tien Shan and Pamir
        Ranges to Afghanistan, Kashmir, Nepal, and parts of W. and N.C. China.
COMMENT: Type locality fixed by Pocock, 1930, J. Bombay Nat. Hist. Soc., 34:332.
        Corbet, 1978:185, placed this species in *Panthera*, but see review by Hemmer,
        1972, Mamm. Species, 20:1–5, and Heptner and Naumov, 1972, [Mammals of the
        Soviet Union], Vol. 2(2):211, who retained the genus *Uncia* for this species, as do
        OLR and NAN.
PROTECTED STATUS: CITES - Appendix I and U.S. ESA - Endangered.
ISIS NUMBER: 5301412007002006001.

Family Otariidae
REVIEWED BY: J. E. King (JEK); T. R. Loughlin (TRL); D. W. Rice (DWR); O. L.
        Rossolimo (OLR)(U.S.S.R.); H. J. Stains (HJS).
ISIS NUMBER: 5301413001000000000.

Arctocephalus E. Geoffroy and F. Cuvier, 1826. Dict. Sci. Nat., 39:554.
COMMENT: Includes *Arctophoca* as a subgenus; see Repenning *et al.*, 1971, Antarct. Res.

Ser., 18:1–34, who reviewed *Arctocephalus*. Van Gelder, 1977:13, included *Zalophus* in *Arctocephalus*; but also see Hall, 1981:1059.
PROTECTED STATUS: CITES - Appendix II as *Arctocephalus* spp.
ISIS NUMBER: 5301413001001000000.

Arctocephalus australis (Zimmermann, 1783). Geogr. Gesch. Mensch. Vierf. Thiere, 3:276.
TYPE LOCALITY: Falkland Isls. (U.K.).
DISTRIBUTION: Coasts of South America from Lima (Peru) to Rio de Janeiro (Brazil); Falkland Isls.
PROTECTED STATUS: CITES - Appendix II as *Arctocephalus* spp.
ISIS NUMBER: 5301413001001001001.

Arctocephalus forsteri (Lesson, 1828). Dict. Class. Hist. Nat. Paris, 13:421.
TYPE LOCALITY: New Zealand, Southland, Dusky Sound.
DISTRIBUTION: New Zealand and nearby subantarctic Isls.; S. and W. Australia.
PROTECTED STATUS: CITES - Appendix II as *Arctocephalus* spp.
ISIS NUMBER: 5301413001001002001.

Arctocephalus galapagoensis Heller, 1904. Proc. Calif. Acad. Sci., ser. 3, 3(7):245.
TYPE LOCALITY: Galapagos Isls., Wenman Isl. (Ecuador).
DISTRIBUTION: Galapagos Isls.
COMMENT: Reviewed by Clark, 1975, Mamm. Species, 64:1–2. Considered conspecific with *australis* by Scheffer, 1958, Seals, sealions, and walruses, Stanford Univ. Press, 179 pp.
PROTECTED STATUS: CITES - Appendix II as *Arctocephalus* spp.
ISIS NUMBER: 5301413001001003001.

Arctocephalus gazella (Peters, 1875). Monatsb. Preuss. Akad. Wiss. Berlin, p. 396.
TYPE LOCALITY: Kerguelen Isls. (France).
DISTRIBUTION: Islands S. of Antarctic convergence (Kerguelen, S. Sandwich, S. Orkney, Heard, Bouver, S. Georgia, S. Shetland Isls).
PROTECTED STATUS: CITES - Appendix II as *Arctocephalus* spp.
ISIS NUMBER: 5301413001001004001.

Arctocephalus philippii (Peters, 1866). Monatsb. Preuss. Akad. Wiss. Berlin, p. 396.
TYPE LOCALITY: Juan Fernandez Isl., Robinson Crusoe Isl. (Chile).
DISTRIBUTION: Juan Fernandez Isls.
COMMENT: Formerly included in *Arctophoca*; see Repenning *et al.*, 1971, Antarct. Res. Ser., 18:1–34.
PROTECTED STATUS: CITES - Appendix II as *Arctocephalus* spp.
ISIS NUMBER: 5301413001001005001.

Arctocephalus pusillus (Schreber, 1776). Saugethiere, 3: text to pl. 85.
TYPE LOCALITY: South Africa, Cape Prov., Cape of Good Hope.
DISTRIBUTION: S.W. Africa from Angola to Cape of Good Hope; S.E. Australia; Tasmania.
COMMENT: Includes *doriferus* and *tasmanicus*; see Repenning *et al.*, 1971, Antarct. Res. Ser., 18:1–34.
PROTECTED STATUS: CITES - Appendix II as *Arctocephalus* spp.
ISIS NUMBER: 5301413001001006001.

Arctocephalus townsendi Merriam, 1897. Proc. Biol. Soc. Wash., 11:178.
TYPE LOCALITY: Mexico, Guadalupe Isl., Fondeadero del Oeste (West Anchorage).
DISTRIBUTION: Guadalupe Isl. (Mexico); Channel Isls. (California, U. S. A.).
COMMENT: Formerly included in *Arctophoca*; see Repenning *et al.*, 1971, Antarct. Res. Ser., 18:1–34. Considered conspecific with *philippii* by Scheffer, 1958, Seals, sealions, and walruses, Stanford Univ. Press, 179 pp.
PROTECTED STATUS: CITES - Appendix I.
ISIS NUMBER: 5301413001001007001.

Arctocephalus tropicalis (Gray, 1872). Proc. Zool. Soc. Lond., 1872:659.
TYPE LOCALITY: "Australasian Sea". No exact locality; probably Amsterdam Isl.
DISTRIBUTION: Islands N. of Antarctic Convergence (Tristan, Gough, Marion, Crozet, Amsterdam, Macquarie Isls.).

COMMENT: Considered conspecific with *doriferus* by Scheffer, 1958, Seals, sealions, and walruses, Stanford Univ. Press, 179 pp.
PROTECTED STATUS: CITES - Appendix II as *Arctocephalus* spp.
ISIS NUMBER: 5301413001001008001.

Callorhinus Gray, 1859. Proc. Zool. Soc. Lond., 1859:359.
ISIS NUMBER: 5301413001002000000.

Callorhinus ursinus (Linnaeus, 1758). Syst. Nat., 10th ed., 1:37.
TYPE LOCALITY: U.S.S.R., Commander Isls., Bering Isl.
DISTRIBUTION: North Pacific, in Okhotsk and Bering Seas; Commander and Pribilof Isls., south to Japan, Shantung (China)(SW), and S. California (U.S.A.).
ISIS NUMBER: 5301413001002001001.

Eumetopias Gill, 1866. Proc. Essex Inst. Salem, 5:7.
ISIS NUMBER: 5301413001003000000.

Eumetopias jubatus (Schreber, 1776). Saugethiere, 3:300, pl. 83B.
TYPE LOCALITY: "Northern part of the Pacific Ocean". U.S.S.R., Commander Isls., Bering Isl.
DISTRIBUTION: N. Pacific, from Hokkaido (Japan) to S. California (U.S.A.).
COMMENT: Type locality from Ognev, 1935, [Mamm. U.S.S.R., Adjac. Count.], 3:361; but see Heptner and Naumov, 1976, [Mammals of the Soviet Union], 2(3):718, who gave "E. coast of Kamchatka."
ISIS NUMBER: 5301413001003001001.

Neophoca Gray, 1866. Ann. Mag. Nat. Hist., ser. 3, p. 231.
COMMENT: Does not include *Phocarctos;* see Rice, 1977; King, 1960, Mammalia, 24:445–456.
ISIS NUMBER: 5301413001004000000.

Neophoca cinerea (Peron, 1816). Voy. Terres. Austral., 2:54.
TYPE LOCALITY: Australia, South Australia, Kangaroo Isl.
DISTRIBUTION: Houtmans Abrolhos (Western Australia) to Kangaroo Isl. (South Australia).
ISIS NUMBER: 5301413001004001001.

Otaria Peron, 1816. Voy. Terres. Austral., 2:37.
ISIS NUMBER: 5301413001005000000.

Otaria byronia (Blainville, 1820). J. Phys. Chim. Hist. Nat. Arts Paris, 91:300.
TYPE LOCALITY: Probably Falkland Isls. (U.K.).
DISTRIBUTION: South American coasts from Peru to Uruguay; Falkland Isl. Occasionally north to coast of Brazil.
COMMENT: Includes *flavescens* as a synonym; see King, 1978, J. Mammal., 59:861; but also see Pine *et al.,* 1978, Mammalia, 42:110–111.
ISIS NUMBER: 5301413001005001001 as *O. flavescens.*

Phocarctos Peters, 1866. Monatsb. Preuss. Akad. Wiss. Berlin, p. 269.
COMMENT: Some authors (Stains, 1967, *in* Anderson and Jones, p. 349; Corbet and Hill, 1980) include this genus in *Neophoca;* but see Rice, 1977, and King, 1960, Mammalia, 24:445–456.
ISIS NUMBER: 5301413001006000000.

Phocarctos hookeri (Gray, 1844). Zool. Voy. H.M.S. "Erebus" and "Terror," 1:4.
TYPE LOCALITY: New Zealand, Auckland Isls. ("Falkland Isls. and Cape Horn," obviously an error).
DISTRIBUTION: Subantarctic islands of New Zealand.
ISIS NUMBER: 5301413001006001001.

Zalophus Gill, 1866. Proc. Essex Inst. Salem, 5:7.
    COMMENT: Included in *Arctocephalus* by Van Gelder, 1977:13; but also see Hall,
      1981:1059.
    ISIS NUMBER: 5301413001007000000.

  Zalophus californianus (Lesson, 1828). Dict. Class. Hist. Nat. Paris, 13:420.
    TYPE LOCALITY: U.S.A., California, San Francisco Bay.
    DISTRIBUTION: N. Pacific from Gulf of California (Mexico) to British Columbia
      (Canada); Japan; Galapagos Isls.
    COMMENT: The subspecies *japonicus* may be extinct (TRL).
    ISIS NUMBER: 5301413001007001001.

Family Odobenidae
    REVIEWED BY: J. E. King (JEK); T. R. Loughlin (TRL); D. W. Rice (DWR); O. L.
      Rossolimo (OLR)(U.S.S.R.).
    COMMENT: Hall, 1981:1061, considered Rosmarinae(=Odobenidae) a subfamily of
      Otariidae; but see Corbet, 1978:186, and Corbet and Hill, 1980:106.

Odobenus Brisson, 1762. Regn. Anim., 2nd ed., p. 30.
    COMMENT: Although the names in Brisson, 1762 are invalid, *Odobenus* has been
      retained by Opinion 467 of the ICZN. Hall, 1981:1061, used the generic name
      Rosmarus.
    ISIS NUMBER: 5301413001008000000.

  Odobenus rosmarus (Linnaeus, 1758). Syst. Nat., 10th ed., 1:38.
    TYPE LOCALITY: N. Atlantic.
    DISTRIBUTION: Arctic seas; south as far as New England (U.S.A.), Great Britain,
      Scandinavia, Pribilof Isls., and Honshu (Japan) at least occasionally.
    PROTECTED STATUS: CITES - Appendix III (Canada).
    ISIS NUMBER: 5301413001008001001.

Family Phocidae
    REVIEWED BY: J. E. King (JEK); D. W. Rice (DWR); O. L. Rossolimo (OLR)(U.S.S.R.); H.
      J. Stains (HJS).
    ISIS NUMBER: 5301413002000000000.

Cystophora Nilsson, 1820. Skand. Faun. Dagg. Djur., 1:382.
    ISIS NUMBER: 5301413002001000000.

  Cystophora cristata (Erxleben, 1777). Syst. Regn. Anim., 1:590.
    TYPE LOCALITY: S. Greenland and Newfoundland.
    DISTRIBUTION: N. Atlantic and Arctic oceans, from Newfoundland (Canada) to S.
      Greenland, Svalbard and Novaya Zemlya (U.S.S.R.). Occasionally south to
      Portugal and Florida (U.S.A.).
    ISIS NUMBER: 5301413002001001001.

Erignathus Gill, 1866. Proc. Essex Inst. Salem, 5:5.
    ISIS NUMBER: 5301413002002000000.

  Erignathus barbatus (Erxleben, 1777). Syst. Regn. Anim., 1:590.
    TYPE LOCALITY: North Atlantic, S. Greenland.
    DISTRIBUTION: Circumpolar Arctic seas, south to Hokkaido (Japan) and Newfoundland
      (Canada).
    ISIS NUMBER: 5301413002002001001.

Halichoerus Nilsson, 1820. Skand. Faun. Dagg. Djur., 1:376.
    ISIS NUMBER: 5301413002003000000.

  Halichoerus grypus (Fabricius, 1791). Skr. Nat. Selsk. Copenhagen, 1(2):167.
    TYPE LOCALITY: Greenland (Denmark).
    DISTRIBUTION: Newfoundland area; Iceland; Britain; Norway; Kola Peninsula
      (U.S.S.R.).
    ISIS NUMBER: 5301413002003001001.

Hydrurga Gistel, 1848. Nat. Thier. fur hohere Schulen, p. xi.
    ISIS NUMBER: 5301413002005000000.

Hydrurga leptonyx (Blainville, 1820). J. Phys. Chim. Hist. Nat. Arts Paris, 91:298.
TYPE LOCALITY: Falkland Isls. (U.K.).
DISTRIBUTION: Circumpolar, in southern oceans.
ISIS NUMBER: 5301413002005001001.

Leptonychotes Gill, 1872. Smithson. Misc. Coll., 11(230):70.
ISIS NUMBER: 5301413002006000000.

Leptonychotes weddelli (Lesson, 1826). Bull. Sci. Nat. Geol., 7:437.
TYPE LOCALITY: South Orkney Isl. (Br. Antarct. Trust Terr.).
DISTRIBUTION: Coastal areas of Antarctic continent. Occasionally north to South
America, Australia, and New Zealand.
COMMENT: Reviewed by Stirling, 1971, Mamm. Species, 6:1–5, and Kooyman, 1981,
Weddell seal: Consummate diver. Cambridge Univ. Press, Cambridge, x + 135 pp.
ISIS NUMBER: 5301413002006001001.

Lobodon Gray, 1844. Zool. Voy. H.M.S. "Erebus" and "Terror," 1:2.
ISIS NUMBER: 5301413002007000000.

Lobodon carcinophagus (Hombron and Jacquinot, 1842). Voy. au Pole Sud...sur les
corvettes L'Astrolabe et la Zelee, 1837–1840, Zoologie, pl. 10.
TYPE LOCALITY: Scotia Sea (midway between South Orkney and South Sandwich Isls.)
(Br. Antarct. Trust Terr.).
DISTRIBUTION: Antarctic seas, frequently on pack ice around Antarctic Continent.
ISIS NUMBER: 5301413002007001001.

Mirounga Gray, 1827. *In* Griffith's Cuvier Anim. Kingd., 5:179.
ISIS NUMBER: 5301413002008000000.

Mirounga angustirostris (Gill, 1866). Proc. Essex Inst. Salem, 5:13.
TYPE LOCALITY: Mexico, Baja California, Bahia Tortola.
DISTRIBUTION: S.E. Alaska to California (U.S.A.) and Baja California (Mexico).
PROTECTED STATUS: CITES - Appendix II.
ISIS NUMBER: 5301413002008001001.

Mirounga leonina (Linnaeus, 1758). Syst. Nat., 10th ed., 1:37.
TYPE LOCALITY: Juan Fernandez Isls., Robinson Crusoe Isl. (Chile).
DISTRIBUTION: Subantarctic islands, including Macquarie, Kerguelen, and S. Georgia
Isls. Formerly Juan Fernandez Isls.
PROTECTED STATUS: CITES - Appendix II.
ISIS NUMBER: 5301413002008002001.

Monachus Fleming, 1822. Philos. Zool., 2:187.
PROTECTED STATUS: CITES - Appendix I as *Monachus* spp.
ISIS NUMBER: 5301413002009000000.

Monachus monachus (Hermann, 1779). Beschaft. Berlin Ges. Naturforsch. Fr., 4:501.
TYPE LOCALITY: Yugoslavia, Dalmacija (=Dalmatia), Cherso Isl.
DISTRIBUTION: Mediterranean and Black Seas; N.W. Africa to Cape Blanc.
PROTECTED STATUS: CITES - Appendix I as *Monachus* spp. and U.S. ESA - Endangered
as *Monachus monachus*.
ISIS NUMBER: 5301413002009001001.

Monachus schauinslandi Matschie, 1905. Sitzb. Ges. Naturf. Fr. Berlin, p. 258.
TYPE LOCALITY: U.S.A., Hawaii, Laysan Isl.
DISTRIBUTION: N.W. Hawaiian Isls., from Nihoa to Kure.
PROTECTED STATUS: CITES - Appendix I as *Monachus* spp. and U.S. ESA - Endangered
as *Monachus schauinslandi*.
ISIS NUMBER: 5301413002009002001.

Monachus tropicalis (Gray, 1850). Cat. Spec. Mamm. Coll. Br. Mus., Part 2, p. 28.
TYPE LOCALITY: Jamaica, Pedro Cays, 80 km. south of Jamaica.
DISTRIBUTION: Caribbean Sea and Yucatan.

COMMENT: "Extinct since the early 1950's;" see Kenyon, 1977, J. Mammal., 58:98.
PROTECTED STATUS: CITES - Appendix I as *Monachus* spp. and U.S. ESA - Endangered as *Monachus tropicalis*.
ISIS NUMBER: 5301413002009003001.

Ommatophoca Gray, 1844. Zool. Voy. H.M.S. "Erebus" and "Terror," 1:3.
ISIS NUMBER: 5301413002010000000.

Ommatophoca rossi Gray, 1844. Zool. Voy. H.M.S. "Erebus" and "Terror," 1:3.
TYPE LOCALITY: Antarctica, Ross Sea.
DISTRIBUTION: Circumpolar, Antarctic pack ice, particularly King Haakon VII Sea.
ISIS NUMBER: 5301413002010001001.

Phoca Linnaeus, 1758. Syst. Nat., 10th ed., 1:37.
REVIEWED BY: M. Nishiwaki (MN).
COMMENT: Includes *Pusa*, *Histriophoca*, and *Pagophilus*; see Burns and Fay, 1970, J. Zool. Lond., 161:363–394; Rice, 1977:12–13; Gromov and Baranova, 1981:297–300; but also see Corbet, 1978:187, and Corbet and Hill, 1980:107, who considered *Histriophoca* and *Pagophilus* distinct genera, as did Hall, 1981, who also separated *Pusa*. MN considers these to be distinct genera, but OLR, DWR, and JEK do not.
ISIS NUMBER: 5301413002004000000.

Phoca caspica Gmelin, 1788. *In* Linnaeus, Syst. Nat., 13th ed., 1:64.
TYPE LOCALITY: U.S.S.R., Caspian Sea.
DISTRIBUTION: Caspian Sea (U.S.S.R.).
ISIS NUMBER: 5301413002004001001.

Phoca fasciata Zimmermann, 1783. Geogr. Gesch. Mensch. Vierf. Thiere, 3:277.
TYPE LOCALITY: U.S.S.R., Kurile Isls.
DISTRIBUTION: Okhotsk, W. Bering, Chukchi and Japan Seas.
COMMENT: Formerly in *Histriophoca*; see Burns and Fay, 1970, J. Zool. Lond., 161:363–394; but also see Corbet, 1978:187.
ISIS NUMBER: 5301413002004002001.

Phoca groenlandica Erxleben, 1777. Syst. Regn. Anim., 1:588.
TYPE LOCALITY: Greenland and Newfoundland (Canada).
DISTRIBUTION: N. Atlantic and Arctic oceans from E. Canada to the White Sea (U.S.S.R.).
COMMENT: Formerly included in *Pagophilus*; see Burns and Fay, 1970, J. Zool. Lond., 161:363–394; but also see Corbet, 1978:187.
ISIS NUMBER: 5301413002004003001 as *P. groenlandicus (sic)*.

Phoca hispida Schreber, 1775. Saugethiere, 3:86.
TYPE LOCALITY: Coasts of Greenland and Labrador (Canada).
DISTRIBUTION: Arctic Ocean, Okhotsk, Bering, and Baltic Seas; Lakes Saimaa (Finland) and Ladoga (U.S.S.R.); Nettilling Lake, Baffin Isl. (Canada).
COMMENT: Formerly included in *Pusa*; see Burns and Fay, 1970, J. Zool. Lond., 161:363–394; but also see Corbet, 1978:187.
ISIS NUMBER: 5301413002004004001.

Phoca largha Pallas, 1811. Zoogr. Rosso-Asiat., 1:113.
TYPE LOCALITY: U.S.S.R., E. Kamchatka.
DISTRIBUTION: Associated with pack ice in coastal N. Pacific, Bering and Okhotsk Seas, Aleutian Isls.; south to Kiangsu (China)(SW) and Japan; Aleutian Isls.
COMMENT: Includes *P. insularis* Belkin, 1964; see review by Shaughnessy and Fay, 1977, J. Zool. Lond., 182:385–419; *largha* was considered a subspecies of *vitulina* by Hall, 1981:1181, and many earlier authors.
ISIS NUMBER: 5301413002004006001.

Phoca sibirica Gmelin, 1788. *In* Linnaeus, Syst. Nat., 13th ed., 1:64.
TYPE LOCALITY: U.S.S.R., Lake Baikal; and "Lake Oron."
DISTRIBUTION: Lake Baikal (U.S.S.R.).

COMMENT: Does not occur in Lake Oron; see Scheffer, 1958, Seals, sealions, and walruses, Stanford Univ. Press, p. 101.
ISIS NUMBER: 5301413002004007001.

Phoca vitulina Linnaeus, 1758. Syst. Nat., 10th ed., 1:38.
TYPE LOCALITY: Baltic Sea, Gulf of Bothnia.
DISTRIBUTION: Kurile Isls.; Kamchatka; south to Kiangsu (China)(SW); Alaska to Mexico; Greenland and E. Canada to N.E. U.S.A.; Iceland; Britain and Europe; Seal Lakes, Ungava Pen. (Canada); Iliamna Lake, Alaska (U.S.A.).
COMMENT: Includes *kurilensis* and *richardsi*; see Shaughnessy and Fay, 1977, J. Zool. Lond., 182:385–419.
ISIS NUMBER: 5301413002004008001 as *P. vitulina*.
            5301413002004005001 as *P. kurilensis*.

ORDER CETACEA

COMMENT: Includes Odontoceti and Mysticeti as suborders; see Rice, 1977, and Corbet and Hill, 1980; but also see Vaughan, 1978.

PROTECTED STATUS: CITES - Appendix II as Order Cetacea.

ISIS NUMBER: 5301411000000000000.

Family Platanistidae

REVIEWED BY: R. L. Brownell, Jr. (RLB); J. G. Mead (JGM); D. W. Rice (DWR).

COMMENT: Includes *Lipotes*, Iniidae, and Pontoporiidae (Stenodelphinidae); see Rice, 1977; but also see Pilleri and Gihr, 1977, Invest. Cetacea, 8:11–76, and Pilleri and Gihr, 1980, Invest. Cetacea, 11:33–36. Zhou *et al.*, 1979, Acta Zool. Sin., 25(2):95–100, placed *Lipotes* in a separate family, Lipotidae. RLB, DWR and JGM include the above as subfamilies of Platanistidae.

ISIS NUMBER: 5301411001000000000.

Inia d'Orbigny, 1834. Ann. Mus. Hist. Nat. Paris, 3:31.

ISIS NUMBER: 5301411001001000000.

Inia geoffrensis (de Blainville, 1817). Nouv. Dict. Hist. Nat., 9:151.

TYPE LOCALITY: Brazil, probably upper Amazon River.

DISTRIBUTION: Amazon to the Mamora (Bolivia) and Orinoco river systems.

COMMENT: Includes *boliviensis*; see Casinos and Ocana, 1979, Saugetierk. Mitt., 27:149–206; but also see Pilleri and Gihr, 1977, Invest. Cetacea, 8:11–76, who considered it a distinct species.

PROTECTED STATUS: CITES - Appendix II as Order Cetacea.

ISIS NUMBER: 5301411001001001001.

Lipotes Miller, 1918. Smithson. Misc. Coll., 68(9):2.

ISIS NUMBER: 5301411001002000000.

Lipotes vexillifer Miller, 1918. Smithson. Misc. Coll., 68(9):2.

TYPE LOCALITY: China, Hunan, Dongting-hu (Tungting Lake).

DISTRIBUTION: China, Changjiang (Yangtze) and Quintangjiang River systems.

COMMENT: Revised by Zhou *et al.*, 1978, J. Nanjing Teachers College (Nat. Sci.), pp. 8–13; Zhou *et al.*, 1979, Acta Zool. Sin., 25(2):95–100. Reviewed by Brownell and Herald, 1972, Mamm. Species, 10:1–4.

PROTECTED STATUS: CITES - Appendix I.

ISIS NUMBER: 5301411001002001001.

Platanista Wagler, 1830. Naturliches Syst. Amphibien, p. 35.

COMMENT: *Susu* is a *nomen oblitum*; for remarks on taxonomy, see Pilleri, 1976, Invest. Cetacea, 6:130–137.

PROTECTED STATUS: CITES - Appendix I as *Platanista* spp.

ISIS NUMBER: 5301411001003000000.

Platanista gangetica (Roxburgh, 1801). Asiat. Res. Trans. Soc., 7:171.

TYPE LOCALITY: India, West Bengal, Hooghly River (Ganges River delta) near Calcutta.

DISTRIBUTION: Ganges, Bramaputra, Meghna, Karnaphuli, and Hooghly river systems (India, Nepal and Bangladesh).

COMMENT: Formerly included *minor* (=*indi*, a junior synonym); see Van Bree, 1976, Bull. Zool. Mus. Univ. Amst., 5(17):139–140; Pilleri and Gihr, 1971, Invest. Cetacea, 3:13–21, and Pilleri and Gihr, 1976, 6:105–118.

PROTECTED STATUS: CITES - Appendix I as *Platanista* spp.

ISIS NUMBER: 5301411001003001001.

Platanista minor Owen, 1853. Descrip. Cat. Osteol. R. Mus. Coll. Surgeons, p. 448.

TYPE LOCALITY: Pakistan, Indus River.

DISTRIBUTION: Indus River system of Pakistan.

COMMENT: *indi* Blyth, 1859, is a junior synonym; see Van Bree, 1976, Bull. Zool. Mus. Univ. Amst., 5(17):139–140. Formerly included in *gangetica*; see Pilleri and Gihr, 1971, Invest. Cetacea, 3:13–21, and Pilleri and Gihr, 1976, 6:105–118.

PROTECTED STATUS: CITES - Appendix I as *Platanista* spp.

Pontoporia Gray, 1846. Zool. Voy. H.M.S. "Erebus" and "Terror," 1:46.
  COMMENT: *Stenodelphis* d'Orbigny and Gervais, 1847, is a junior synonym; see Rice, 1977:13.
  ISIS NUMBER: 5301411001004000000.

Pontoporia blainvillei (Gervais and d'Orbigny, 1844). Bull. Sci. Soc. Philom. Paris, p. 39.
  TYPE LOCALITY: Uruguay, mouth of the Rio de La Plata near Montevideo.
  DISTRIBUTION: Coastal waters from Ubatuba (Tropic of Capricorn), Brazil to Peninsula Valdez, Argentina.
  PROTECTED STATUS: CITES - Appendix II as Order Cetacea.
  ISIS NUMBER: 5301411001004001001.

Family Delphinidae
  REVIEWED BY: R. L. Brownell, Jr (RLB); T. Kasuya (TK); J. G. Mead (JGM); M. Nishiwaki (MN); W. F. Perrin (WFP); D. W. Rice (DWR); O. L. Rossolimo (OLR)(U.S.S.R.); S. Wang (SW) (China).
  COMMENT: Includes Stenidae *(Sotalia, Sousa,* and *Steno);* see Kasuya, 1973, Sci. Rep. Whales Res. Inst. (Tokyo), 25:1–103; Mead, 1975, Smithson. Contrib. Zool., 207:1–72; Rice, 1977:7; Barnes, 1978, Bull. Los. Ang. Cty. Mus. Nat. Hist., 28:1–35.
  ISIS NUMBER: 5301411002000000000.

Cephalorhynchus Gray, 1846. Zool. Voy. H.M.S. "Erebus" and "Terror," 1:36.
  COMMENT: Revised by Harmer, 1922. Proc. Zool. Soc. Lond., 1922:627–638.
  PROTECTED STATUS: CITES - Appendix II as Order Cetacea.
  ISIS NUMBER: 5301411002008000000.

Cephalorhynchus commersonii (Lacepede, 1804). Hist. Nat. Cetacees, p. 317.
  TYPE LOCALITY: Chile, Tierra del Fuego, Straits of Magellan.
  DISTRIBUTION: Gulf of San Matias, Argentina, to the Chilean side of the Straits of Magellan, Falkland Isls., S. Georgia Isls., and Kerguelen Isls..
  PROTECTED STATUS: CITES - Appendix II as Order Cetacea.
  ISIS NUMBER: 5301411002008001001.

Cephalorhynchus eutropia Gray, 1846. Zool. Voy. H.M.S. Erebus and Terror, 1:pl. 34.
  TYPE LOCALITY: Pacific Ocean, off the coast of Chile.
  DISTRIBUTION: Coastal waters of Chile (between 37° S. and 55° S.).
  PROTECTED STATUS: CITES - Appendix II as Order Cetacea.
  ISIS NUMBER: 5301411002008002001.

Cephalorhynchus heavisidii (Gray, 1828). Spicil. Zool., 1:2.
  TYPE LOCALITY: South Africa, Cape Prov., Cape of Good Hope.
  DISTRIBUTION: South Atlantic coastal waters from Cape Town (South Africa) to Cape Cross (Namibia).
  PROTECTED STATUS: CITES - Appendix II as Order Cetacea.
  ISIS NUMBER: 5301411002008003001 as *C. heavisidei (sic).*

Cephalorhynchus hectori (Van Beneden, 1881). Bull. R. Acad. Belg., ser. 3, 4:877.
  TYPE LOCALITY: New Zealand, North coast.
  DISTRIBUTION: Coastal waters of New Zealand; possibly S. China Sea (Sarawak).
  PROTECTED STATUS: CITES - Appendix II as Order Cetacea.
  ISIS NUMBER: 5301411002008004001.

Delphinus Linnaeus, 1758. Syst. Nat., 10th ed., 1:77.
  PROTECTED STATUS: CITES - Appendix II as Order Cetacea.
  ISIS NUMBER: 5301411002009000000.

Delphinus delphis Linnaeus, 1758. Syst. Nat., 10th ed., 1:77.
  TYPE LOCALITY: E. North Atlantic ("Oceano Europaeo").
  DISTRIBUTION: Temperate and tropical waters of the world; Black Sea.
  COMMENT: Includes *bairdii;* see Van Bree and Purves, 1972, J. Mammal., 53:372–374;

Rice, 1977:12. Does not include *tropicalis*; see Van Bree and Gallagher, 1978, Beaufortia, 28(342):1–8.
PROTECTED STATUS: CITES - Appendix II as Order Cetacea.
ISIS NUMBER: 5301411002009002001 as *D. delphis.*
5301411002009001001 as *D. bairdii.*

Delphinus tropicalis Van Bree, 1971, Mammalia, 35:345.
TYPE LOCALITY: India, Kerala, Malabar Coast.
DISTRIBUTION: Arabian Sea to South China Sea.
COMMENT: Formerly included in *delphis*; see Van Bree and Gallagher, 1978, Beaufortia, 28(342):1–8.
PROTECTED STATUS: CITES - Appendix II as Order Cetacea.

Feresa Gray, 1870. Proc. Zool. Soc. Lond., 1870:77.
PROTECTED STATUS: CITES - Appendix II as Order Cetacea.
ISIS NUMBER: 5301411002001000000.

Feresa attenuata Gray, 1875. J. Mus. Godeffroy (Hamburg), 8:184.
TYPE LOCALITY: Not designated in original description.
DISTRIBUTION: Tropical and warm temperate waters of the world.
PROTECTED STATUS: CITES - Appendix II as Order Cetacea.
ISIS NUMBER: 5301411002001001001.

Globicephala Lesson, 1828. Compl. Oeuvres Buffon Hist. Nat., 1:441.
COMMENT: Reviewed by Van Bree, 1971, Beaufortia, 19:79–87.
PROTECTED STATUS: CITES - Appendix II as Order Cetacea.
ISIS NUMBER: 5301411002002000000.

Globicephala macrorhynchus Gray, 1846. Zool. Voy. H.M.S. "Erebus" and "Terror," 1:33.
TYPE LOCALITY: "South Seas."
DISTRIBUTION: Tropical and subtropical waters of the world and warm temperate N. Pacific; possibly north to Alaska.
COMMENT: Includes *scammonii* and *sieboldii*; see Van Bree, 1971; Beaufortia, 19(244):21–25.
PROTECTED STATUS: CITES - Appendix II as Order Cetacea.
ISIS NUMBER: 5301411002002001001 as *G. macrorhyncha (sic).*

Globicephala melaena (Traill, 1809). Nicholson's J. Nat. Philos. Chem. Arts, 22:81.
TYPE LOCALITY: U.K., Scotland, Orkney Isls., Pomona, Scapay Bay.
DISTRIBUTION: Cold temperate waters, North Atlantic and southern Oceans.
COMMENT: Includes *leucosagmaphora* and *edwardii*; see Van Bree, 1971, Beaufortia, 19(249):79–87. Kasuya, 1975, Sci. Rep. Whales Res. Inst. (Tokyo), 27:110, described the historic distribution in the N.W. Pacific.
PROTECTED STATUS: CITES - Appendix II as Order Cetacea.
ISIS NUMBER: 5301411002002002001.

Grampus Gray, 1828. Spicil. Zool., 1:2.
PROTECTED STATUS: CITES - Appendix II as Order Cetacea.
ISIS NUMBER: 5301411002010000000.

Grampus griseus (G. Cuvier, 1812). Ann. Mus. Hist. Nat. Paris, 19:13–14.
TYPE LOCALITY: France, Finistere, Brest.
DISTRIBUTION: Temperate and tropical waters of the world.
PROTECTED STATUS: CITES - Appendix II as Order Cetacea.
ISIS NUMBER: 5301411002010001001.

Lagenodelphis Fraser, 1956. Sarawak Mus. J., n.s., 8(7):496.
PROTECTED STATUS: CITES - Appendix II as Order Cetacea.
ISIS NUMBER: 5301411002011000000.

Lagenodelphis hosei Fraser, 1956. Sarawak Mus. J., n.s., 8(7):496.
    TYPE LOCALITY: Indonesia, Sarawak, Baram, mouth of Lutong River.
    DISTRIBUTION: Tropical waters of the world.
    PROTECTED STATUS: CITES - Appendix II as Order Cetacea.
    ISIS NUMBER: 5301411002011001001.

Lagenorhynchus Gray, 1846. Ann. Mag. Nat. Hist., 17:84.
    COMMENT: Reviewed by Fraser, 1966, *in* Norris, ed., Whales, Dolphins, and Porpoises,
        p. 7–31.
    PROTECTED STATUS: CITES - Appendix II as Order Cetacea.
    ISIS NUMBER: 5301411002012000000.

Lagenorhynchus acutus (Gray, 1828). Spicil. Zool., 1:2.
    TYPE LOCALITY: North Sea, Faeroe Isls. (Denmark) (uncertain).
    DISTRIBUTION: Cold temperate to subarctic waters of the North Atlantic, south to
        Massachusetts (U.S.A.) and British Isles.
    COMMENT: Type locality given by OLR as Great Britain, Orkney Isls., *fide* Tomilin,
        1957, [Mamm. U.S.S.R., Adjac. Count.], 9:952. Gromov and Baranova, 1981:221
        stated that the holotype, a skull in the Rijksmuseum, Leiden, is "apparently, from
        the region of the Faeroe Isls." (R. S. Hoffmann).
    PROTECTED STATUS: CITES - Appendix II as Order Cetacea.
    ISIS NUMBER: 5301411002012001001.

Lagenorhynchus albirostris (Gray, 1846). Ann. Mag. Nat. Hist., 17:84.
    TYPE LOCALITY: U.K., England, Norfolk, Great Yarmouth.
    DISTRIBUTION: Cold temperate to subarctic waters of the North Atlantic, south to
        Newfoundland and Portugal.
    PROTECTED STATUS: CITES - Appendix II as Order Cetacea.
    ISIS NUMBER: 5301411002012002001.

Lagenorhynchus australis (Peale, 1848). Mammalia and Ornithology, *in* U.S. Expl. Exped.,
    8:33.
    TYPE LOCALITY: South Atlantic Ocean, off the coast of Patagonia, Argentina, "one day's
        sail north of the Straits of Le Maire between Staten Island and Cape San Diego."
    DISTRIBUTION: Cold temperate waters of Chile, Argentina, and Falkland Isls.
    COMMENT: Included in *cruciger* by Hershkovitz, 1966:67, but considered a distinct
        species by Rice, 1977, Brownell, 1974, *in* Brown, *et al.*, eds., Antarctic Mammals,
        Folio 18, Amer. Geogr. Soc., 19 pp., 11 pls., and IWC, 1975, J. Fish Res. Bd. Can.,
        32:875–983.
    PROTECTED STATUS: CITES - Appendix II as Order Cetacea.
    ISIS NUMBER: 5301411002012003001.

Lagenorhynchus cruciger (Quoy and Gaimard, 1824). Voy. Autour du Monde, sur Uranie et
    la Physicienne, Paris, Zool., p. 87.
    TYPE LOCALITY: Pacific Ocean, 49° S., between Cape Horn and Australia.
    DISTRIBUTION: Antarctic and cold temperate waters of the Atlantic and Pacific coasts of
        South America.
    COMMENT: Formerly included *australis* and *obscurus;* see Rice, 1977.
    PROTECTED STATUS: CITES - Appendix II as Order Cetacea.
    ISIS NUMBER: 5301411002012004001.

Lagenorhynchus obliquidens Gill, 1865. Proc. Acad. Nat. Sci. Phila., 17:177.
    TYPE LOCALITY: U.S.A., California, near San Francisco.
    DISTRIBUTION: North Pacific, from Alaska and Aleutian Isls. to Baja California and
        from Kamchatka to Taiwan.
    COMMENT: *L. thicolea* (Gray, 1946) may be the correct name for this species; see Hall,
        1981:889. May be a Northern Hemisphere form of *obscurus* (RLB and WFP).
    PROTECTED STATUS: CITES - Appendix II as Order Cetacea.
    ISIS NUMBER: 5301411002012005001.

Lagenorhynchus obscurus (Gray, 1828). Spicil. Zool., 1:2.
    TYPE LOCALITY: South Africa, Cape Prov., Cape of Good Hope.
    DISTRIBUTION: Coastal temperate waters of the Southern Hemisphere.

COMMENT: Included in *cruciger* by Hershkovitz, 1966:65, but considered a distinct
  species by Rice, 1977; Brownell, 1974, *in* Brown, *et al.*, eds., Antarctic Mammals,
  Folio 18, Amer. Geogr. Soc., 19 pp. 11 pls.; IWC, 1975, J. Fish Res. Bd. Can.,
  32:875–983. Includes *fitzroyi*; see Rice, 1977.
PROTECTED STATUS: CITES - Appendix II as Order Cetacea.
ISIS NUMBER: 5301411002012006001.

Lissodelphis Gloger, 1841. Gemein. Naturgesch. Thier., 1:169.
  COMMENT: This may be a monotypic genus (WFP and RLB).
  PROTECTED STATUS: CITES - Appendix II as Order Cetacea.
  ISIS NUMBER: 5301411002007000000.

Lissodelphis borealis (Peale, 1848). Mammalia and Ornithology, *in* U.S. Expl. Exped., 8:35.
  TYPE LOCALITY: North Pacific, 10° W. of Astoria, Oregon (46° 6'50" N., 134° 5' W.)
    (U.S.A.).
  DISTRIBUTION: Cold temperate to subarctic waters of the North Pacific, from British
    Columbia to Baja California and Kurile Isls. to Japan.
  COMMENT: May be a Northern Hemisphere form of *peronii* (WFP and RLB); MN
    disagrees.
  PROTECTED STATUS: CITES - Appendix II as Order Cetacea.
  ISIS NUMBER: 5301411002007001001.

Lissodelphis peronii (Lacepede, 1804). Hist. Nat. Cetacees, p. xliii.
  TYPE LOCALITY: Indian Ocean, about 44° S., 141° E. (south of Tasmania).
  DISTRIBUTION: Temperate waters of the Southern Hemisphere.
  PROTECTED STATUS: CITES - Appendix II as Order Cetacea.
  ISIS NUMBER: 5301411002007002001 as *L. peroni (sic)*.

Orcaella Gray, 1866. Cat. Seals and Whales Br. Mus., p. 285.
  PROTECTED STATUS: CITES - Appendix II as Order Cetacea.
  ISIS NUMBER: 5301411002003000000.

Orcaella brevirostris (Gray, 1866). Cat. Seals and Whales Br. Mus., p. 285.
  TYPE LOCALITY: India, Andhra Pradesh (=Madras Presidency), Vishakhapatnam
    (=Vizagapatam) Harbor (in Bay of Bengal).
  DISTRIBUTION: Coastal waters and large rivers of S.E. Asia, Northern Australia and
    Papua New Guinea.
  PROTECTED STATUS: CITES - Appendix II as Order Cetacea.
  ISIS NUMBER: 5301411002003001001.

Orcinus Fitzinger, 1860. Wiss. Naturg. Saugeth., 6:204.
  PROTECTED STATUS: CITES - Appendix II as Order Cetacea.
  ISIS NUMBER: 5301411002004000000.

Orcinus orca (Linnaeus, 1758). Syst. Nat., 10th ed., 1:77.
  TYPE LOCALITY: E. North Atlantic ("Oceano Europaeo").
  DISTRIBUTION: Worldwide, all oceans.
  PROTECTED STATUS: CITES - Appendix II as Order Cetacea.
  ISIS NUMBER: 5301411002004001001.

Peponocephala Nishiwaki and Norris, 1966. Sci. Rep. Whales Res. Inst. (Tokyo), 20:95–99.
  COMMENT: Formerly included in *Lagenorhynchus*; see Rice, 1977. JGM and RLB consider
    Peponocephala a subgenus of *Lagenorhynchus* but DWR, MN, TK, and WFP do not.
  PROTECTED STATUS: CITES - Appendix II as Order Cetacea.
  ISIS NUMBER: 5301411002005000000.

Peponocephala electra (Gray, 1846). Zool. Voy. H.M.S. "Erebus" and "Terror," 1:35.
  TYPE LOCALITY: Unknown.
  DISTRIBUTION: World-wide, in tropical waters.
  COMMENT: Traditionally this species was included in the genus *Lagenorhynchus*, which
    is prefered by JGM and RLB but not DWR, WFP, and TK.
  PROTECTED STATUS: CITES - Appendix II as Order Cetacea.
  ISIS NUMBER: 5301411002005001001.

Pseudorca Reinhardt, 1862. Overs. Danske Vidensk. Selsk. Forh., p. 151.
PROTECTED STATUS: CITES - Appendix II as Order Cetacea.
ISIS NUMBER: 5301411002006000000.

Pseudorca crassidens (Owen, 1846). A History of British Fossil Mammals and Birds, p. 516.
TYPE LOCALITY: U.K., England, near Stanford, Lincolnshire Fens (subfossil).
DISTRIBUTION: Tropical to temperate waters of the world.
PROTECTED STATUS: CITES - Appendix II as Order Cetacea.
ISIS NUMBER: 5301411002006001001.

Sotalia Gray, 1866. Cat. Seals and Whales Br. Mus., p. 401.
COMMENT: Formerly included in Stenidae; see Rice, 1977:7; and comments under Delphinidae.
PROTECTED STATUS: CITES - Appendix I as *Sotalia* spp.
ISIS NUMBER: 5301411002015000000.

Sotalia fluviatilis (Gervais, 1853). *In* Gervais, Bull. Soc. Agric. Herault, p. 148.
TYPE LOCALITY: Peru, Loreto, Rio Maranon above Pebas.
DISTRIBUTION: W. Atlantic from Panama to Santos, Sao Paulo, Brazil; Amazon River system.
COMMENT: Includes *guianensis* and *brasiliensis*; see Rice, 1977:13.
PROTECTED STATUS: CITES - Appendix I as *Sotalia* spp.
ISIS NUMBER: 5301411002015001001.
             5301411002015002001 as *S. guianensis*.

Sousa Gray, 1866. Proc. Zool. Soc. Lond., 1866:213.
COMMENT: Formerly included in Stenidae; see Rice, 1977:7; and comments under Delphinidae.
PROTECTED STATUS: CITES - Appendix I as *Sousa* spp.
ISIS NUMBER: 5301411002016000000.

Sousa chinensis (Osbeck, 1765). Reise nach Ostind. China Rostock, 1:7.
TYPE LOCALITY: China, Guangdong Prov., Zhujiang Kou (mouth of Canton River).
DISTRIBUTION: Coastal waters and rivers from Plettenberg Bay (South Africa) east to S. China and Queensland (Australia).
COMMENT: Includes *borneensis*, *plumbea*, and *lentiginosa*; see Rice, 1977:13. See also Perrin, 1975, Bull. Scripps Inst. Oceanogr. Univ. Calif., 21:1–206, who placed *malayana* in *Stenella attenuata*, as is done here; but also see Rice, 1977:13, who included *malayana* in this species on the advice of P.J. H. van Bree. Zhou *et al.*, 1979, Mar. Mamm. Inf. Oregon St. Univ., Dec., p. 46, stated that *plumbea* (including *lentiginosa*) of the Indian Ocean, may be distinct from *chinensis* (including *borneensis*) of the western Pacific. Pilleri and Gihr, 1972, Invest. Cetacea, 4:107–162, and Pilleri and Gihr, 1973–1974, Invest. Cetacea, 5:95–149, considered *borneensis*, *plumbea*, and *lentiginosa* to be distinct species. JGM, WFP and RLB followed Rice (IWC, 1975 J. Fish. Res. Bd. Can., 32) in including these taxa in *chinensis*.
PROTECTED STATUS: CITES - Appendix I as *Sousa* spp.
ISIS NUMBER: 5301411002016002001 as *S. chinensis*.
             5301411002016001001 as *S. borneensis*.
             5301411002016003001 as *S. lentiginosa*.
             5301411002016004001 as *S. plumbea*.

Sousa teuszii (Kukenthal, 1892). Zool. Jahrb. Syst., 6:442.
TYPE LOCALITY: Cameroun, Cameroun Oriental, Bay of Warships, near Douala.
DISTRIBUTION: Coastal waters in river mouths from Mauritania to N. Angola.
COMMENT: Reviewed by Pilleri and Gihr, 1972, Invest. Cetacea, 4:107–162.
PROTECTED STATUS: CITES - Appendix I as *Sousa* spp.
ISIS NUMBER: 5301411002016005001 as *S. teuszi (sic)*.

Stenella Gray, 1866. Proc. Zool. Soc. Lond., 1866:213.
COMMENT: Reviewed, in part, by Perrin, 1975, Bull. Scripps Inst. Oceanogr. Univ. Calif., 21:1–206.
PROTECTED STATUS: CITES - Appendix II as Order Cetacea.
ISIS NUMBER: 5301411002013000000.

Stenella attenuata (Gray, 1846). Zool. Voy. H.M.S. "Erebus" and "Terror," 1:44.
TYPE LOCALITY: Unknown.
DISTRIBUTION: Tropical waters of Indian, Pacific, and Atlantic Oceans.
COMMENT: Includes *frontalis, graffmani,* and *dubia;* see Rice, 1977:8. Includes *malayana;*
see Perrin, 1975, Bull. Scripps Inst. Oceanogr. Univ. Calif., 21:1–206, but also see
Rice, 1977:13. The taxonomy of the spotted dolphins *(frontalis, dubia, attenuata,* and
plagiodon) remains to be resolved (WFP). The presently recognized nominal forms
(frontalis, attenuata, and *plagiodon)* will probably prove to be two species of spotted
dolphins; *frontalis* is probably conspecific with either *attenuata* (JGM), or *plagiodon*
(WFP); *dubia* (including *asthenops* and *crotaphyscus*), used by Hershkovitz, 1966:31,
may belong to either of the two probable species (WFP and RLB; DWR; JGM).
Hershkovitz, 1966:31, argued that *dubia* is a prior name for *attenuata*. See Hall,
1981:882–883, for additional discussion.
PROTECTED STATUS: CITES - Appendix II as Order Cetacea.
ISIS NUMBER: 5301411002013002001 as *S. dubia.*

Stenella clymene (Gray, 1850). Cat. Mamm. Br. Mus. Cetacea, p. 115.
TYPE LOCALITY: Unknown.
DISTRIBUTION: Tropical and warm temperate waters of the Atlantic Ocean.
COMMENT: Recognized by Hershkovitz, 1966, but not by the IWC, 1975, J. Fish Res.
Bd. Can., 32:875–983 (who included it in *longirostris*). See Perrin *et al.,* 1981, J.
Mammal., 62:583–598, for redescription.
PROTECTED STATUS: CITES - Appendix II as Order Cetacea.

Stenella coeruleoalba (Meyen, 1833). Nova Acta Acad. Caes. Nat. Curios., 16(2):609–610.
TYPE LOCALITY: South Atlantic Ocean near Rio de la Plata (off coast of Argentina and
Uruguay).
DISTRIBUTION: Temperate and tropical waters of the world.
COMMENT: Includes *styx* and *euphrosyne;* see Mitchell, 1970, Can. J. Zool., 48:720.
PROTECTED STATUS: CITES - Appendix II as Order Cetacea.
ISIS NUMBER: 5301411002013001001 as *S. caeruleoalbas (sic).*

Stenella longirostris (Gray, 1828). Spicil. Zool., 1:1.
TYPE LOCALITY: Unknown.
DISTRIBUTION: Tropical to warm temperate waters of the world.
COMMENT: Includes *roseiventris* and *microps;* see Perrin, 1975, Bull. Scripps Inst.
Oceanogr. Univ. Calif., 21:206.
PROTECTED STATUS: CITES - Appendix II as Order Cetacea.
ISIS NUMBER: 5301411002013003001 as *S. longirostris.*
5301411002013004001 as *S. roseiventris.*

Stenella plagiodon (Cope, 1866). Proc. Acad. Nat. Sci. Phila., 18:296.
TYPE LOCALITY: Unknown.
DISTRIBUTION: Tropical to warm temperate waters of the Atlantic Ocean and the Gulf
of Mexico.
COMMENT: Relationships with *frontalis* and *attenuata* are poorly known; see comments
under *attenuata. S. pernettyi* used by Hershkovitz, 1966, was later suppressed by
the International Commission of Zoological Nomenclature (Opinion 1067),
following Van Bree, 1971, Beaufortia, 19(244):21–25, but was employed by Hall,
1981:883. Corbet and Hill, 1980:109, included *Plagiodon* in *attenuata* (as *dubia*).
PROTECTED STATUS: CITES - Appendix II as Order Cetacea.

Steno Gray, 1846. Zool. Voy. H.M.S. "Erebus" and "Terror," 1:30, 43.
COMMENT: Formerly included in Stenidae; see Rice, 1977:7; and comments under
Delphinidae.
PROTECTED STATUS: CITES - Appendix II as Order Cetacea.
ISIS NUMBER: 5301411002017000000.

Steno bredanensis (Lesson, 1828). Compl. Oeuvres Buffon Hist. Nat., 1:206.
TYPE LOCALITY: Coast of France.
DISTRIBUTION: Worldwide, in warm temperate to tropical waters.
PROTECTED STATUS: CITES - Appendix II as Order Cetacea.
ISIS NUMBER: 5301411002017001001.

Tursiops Gervais, 1855. Hist. Nat. Mamm., 2:323.
PROTECTED STATUS: CITES - Appendix II as Order Cetacea.
ISIS NUMBER: 5301411002014000000.

Tursiops truncatus (Montagu, 1821). Mem. Wernerian Nat. Hist. Soc., 3:75.
TYPE LOCALITY: U.K., England, Devonshire, Stoke Gabriel (about 5 mi. (8 km) up the River Dart), Duncannon Pool.
DISTRIBUTION: Tropical and temperate waters of the world, including the Black Sea.
COMMENT: Includes *aduncus, gephyreus, gillii,* and *nuuanu;* see IWC, 1975, J. Fish Res. Bd. Can., 32:968. Ross, 1977, Ann. Cape Prov. Mus. Nat. Hist., 11(9):135–194, stated that *aduncus* (in the tropical Indo-Pacific) may be a distinct species. Hall, 1981:885–887, considered *nesarnack* (synonym of *truncatus*) and *gillii* distinct species.
PROTECTED STATUS: CITES - Appendix II as Order Cetacea.
ISIS NUMBER: 5301411002014002001 as *T. truncatus.*
5301411002014001001 as *T. gilli* (sic).

Family Phocoenidae
REVIEWED BY: J. G. Mead (JGM); D. W. Rice (DWR); O. L. Rossolimo (OLR)(U.S.S.R.); S. Wang (SW)(China).
COMMENT: This family has been considered a subfamily of Delphinidae; see Rice, 1977; Gromov and Baranova, 1981:222.

Neophocaena Palmer, 1899. Proc. Biol. Soc. Wash., 13:23.
ISIS NUMBER: 5301411002018000000.

Neophocaena phocaenoides (G. Cuvier, 1829). Regn. Anim., 1:291.
TYPE LOCALITY: South Africa, Cape Prov., Cape of Good Hope. (Almost certainly erroneous, DWR).
DISTRIBUTION: Persian Gulf to Java, China, and Japan.
COMMENT: Includes *sunameri* and *asiaeorientalis;* see Rice, 1977:10. Reviewed by Pilleri and Gihr, 1975, Mammalia, 39:657:673; Pilleri and Gihr, 1972, Invest. Cetacea, 4:107–162, and Pilleri and Gihr, 1980, Invest. Cetacea, 11:33–36, who considered *sunameri* and *asiaeorientalis* distinct species.
PROTECTED STATUS: CITES - Appendix I.
ISIS NUMBER: 5301411002018001001.

Phocoena G. Cuvier, 1817. Regn. Anim., 1:279.
COMMENT: *Phocaena* and *Phocena* are later spellings.
PROTECTED STATUS: CITES - Appendix II as Order Cetacea.
ISIS NUMBER: 5301411002019000000.

Phocoena dioptrica Lahille, 1912. Ann. Mus. Nat. Hist., Buenos Aires, 23:269.
TYPE LOCALITY: Argentina, Buenos Aires, Punta Colares, near Quilmes.
DISTRIBUTION: Uruguay, Argentina, Falkland Isls., South Georgia, and the Auckland Isls., perhaps Kerguelen and New Zealand.
COMMENT: Perhaps circumpolar; see Baker, 1977, New Zealand J. Mar. Freshw. Res., 11:401–406. Reviewed by Brownell, 1975, Mamm. Species, 66:1–3.
PROTECTED STATUS: CITES - Appendix II as Order Cetacea.
ISIS NUMBER: 5301411002019001001.

Phocoena phocoena (Linnaeus, 1758). Syst. Nat., 10th ed., 1:77.
TYPE LOCALITY: Baltic Sea, "Swedish Seas."
DISTRIBUTION: Temperate to arctic N. Pacific and N. Atlantic; isolated population in Black Sea.
COMMENT: Reviewed by Gaskin *et al.,* 1974, Mamm. Species, 42:1–8.
PROTECTED STATUS: CITES - Appendix II as Order Cetacea.
ISIS NUMBER: 5301411002019002001.

Phocoena sinus Norris and McFarland, 1958. J. Mammal., 39:22.
TYPE LOCALITY: Mexico, Baja California Norte, Punta San Felipe.
DISTRIBUTION: Northern part of the Gulf of California, south to Tres Marias Isls. and N. Jalisco.
PROTECTED STATUS: CITES - Appendix I.
ISIS NUMBER: 5301411002019003001.

Phocoena spinipinnis Burmeister, 1865. Proc. Zool. Soc. Lond., 1865:228.
TYPE LOCALITY: Argentina, Buenos Aires, mouth of the Rio de la Plata.
DISTRIBUTION: Temperate waters of South America, from Rio de la Plata (Uruguay) to Tierra del Fuego to Paita (Peru).
PROTECTED STATUS: CITES - Appendix II as Order Cetacea.
ISIS NUMBER: 5301411002019004001.

Phocoenoides Andrews, 1911. Bull. Am. Mus. Nat. Hist., 30:31.
PROTECTED STATUS: CITES - Appendix II as Order Cetacea.
ISIS NUMBER: 5301411002020000000.

Phocoenoides dalli (True, 1885). Proc. U.S. Nat. Mus., 8:95.
TYPE LOCALITY: U.S.A., Alaska, strait west of Adak, Aleutian Isls.
DISTRIBUTION: North Pacific, temperate to subarctic waters.
COMMENT: Includes *truei;* see Morejohn, 1979, *in* Winn and Olla, eds., Behavior of Marine Animals, 3:45–83.
PROTECTED STATUS: CITES - Appendix II as Order Cetacea.
ISIS NUMBER: 5301411002020001001.

Family Monodontidae
REVIEWED BY: T. Kasuya (TK); J. G. Mead (JGM); D. W. Rice (DWR); O. L. Rossolimo (OLR) (U.S.S.R.).
COMMENT: Some authors include *Orcaella* in this family; see Kasuya, 1973, Sci. Rep. Whales Res. Inst. (Tokyo), 25:1–103; and Barnes, 1976, Syst. Zool., 23:321–343; but see also Rice, 1977. Monodontidae has been considered as a subfamily of Delphinidae; but see also Rice, 1977.
PROTECTED STATUS: CITES - Appendix II as Order Cetacea.
ISIS NUMBER: 5301411003000000000.

Delphinapterus Lacepede, 1804. Hist. Nat. Cetacees, pp. xli, 243.
PROTECTED STATUS: CITES - Appendix II as Order Cetacea.
ISIS NUMBER: 5301411003001000000.

Delphinapterus leucas (Pallas, 1776). Reise Prov. Russ. Reichs., 3(1):85.
TYPE LOCALITY: U.S.S.R., W. Siberia, mouth of Ob River.
DISTRIBUTION: Arctic Ocean; Okhotsk, and Bering Seas; Gulf of Alaska; Gulf of St. Lawrence; occasionally south to Honshu (Japan), France, and Massachusetts (U.S.A.).
PROTECTED STATUS: CITES - Appendix II as Order Cetacea.
ISIS NUMBER: 5301411003001001001.

Monodon Linnaeus, 1758. Syst. Nat., 10th ed., 1:75.
PROTECTED STATUS: CITES - Appendix II as Order Cetacea.
ISIS NUMBER: 5301411003002000000.

Monodon monoceros Linnaeus, 1758. Syst. Nat., 10th ed., 1:75.
TYPE LOCALITY: Arctic Seas.
DISTRIBUTION: Arctic seas, circumpolar; occasionally as far south as the Netherlands.
COMMENT: Reviewed by Reeves and Tracey, 1980, Mamm. Species, 127:1–7.
PROTECTED STATUS: CITES - Appendix II as Order Cetacea.
ISIS NUMBER: 5301411003002001001.

Family Physeteridae
REVIEWED BY: J. G. Mead (JGM); D. W. Rice (DWR); O. L. Rossolimo (OLR)(U.S.S.R.).
PROTECTED STATUS: CITES - Appendix II as Order Cetacea.
ISIS NUMBER: 5301411004000000000.

Kogia Gray, 1846. Zool. Voy. H.M.S. "Erebus" and "Terror," 1:22.
COMMENT: Reviewed by Handley, 1966, *in* Norris, ed., Whales, Dolphins, and

Porpoises, p. 62–69. This genus is sometimes placed in the family Kogiidae; but see Rice, 1977.
PROTECTED STATUS: CITES - Appendix II as Order Cetacea.
ISIS NUMBER: 5301411004001000000.

Kogia breviceps (Blainville, 1838). Ann. Franc. Etr. Anat. Phys., 2:337.
TYPE LOCALITY: South Africa, Cape Prov., Cape of Good Hope.
DISTRIBUTION: Tropical and temperate waters of the world.
COMMENT: Distribution is tentative due to confusion of *breviceps* and *simus* (JGM).
PROTECTED STATUS: CITES - Appendix II as Order Cetacea.
ISIS NUMBER: 5301411004001001001.

Kogia simus (Owen, 1866). Trans. Zool. Soc. Lond., 6(1):30.
TYPE LOCALITY: India, Andhra Pradesh (=Madras Presidency), Waltair.
DISTRIBUTION: Tropical to warm temperate waters of the world.
PROTECTED STATUS: CITES - Appendix II as Order Cetacea.
ISIS NUMBER: 5301411004001002001.

Physeter Linnaeus, 1758. Syst. Nat., 10th ed., 1:76.
ISIS NUMBER: 5301411004002000000.

Physeter macrocephalus Linnaeus, 1758. Syst. Nat., 10th ed., 1:76.
TYPE LOCALITY: Netherlands, Middenpiat.
DISTRIBUTION: Northern Hemisphere, tropical to subarctic waters, Southern Hemisphere, all seas.
COMMENT: Neotype designated by Husson and Holthuis, 1974, Zool. Meded., 48(19):212. Linnaeus used both *catodon* and *macrocephalus* in the 10th edition. *P. catodon* has line priority. In addition, *catodon* is described as "*fistula in rostro*" (blowholes on the nose), which is correct, and is the only diagnostic character in the description that will separate the sperm whale from all other whales. *P. macrocephalus*, however, is described as "*fistula in cervice*" (blowholes on the neck), which is incorrect, and is the same as all other whales (DWR). See Hershkovitz, 1966, for additional discussion; but also see Rice, 1977:11; and Husson and Holthuis, 1974, Zool. Meded., 48(19):205–217.
PROTECTED STATUS: CITES - Appendix I and U.S. ESA - Endangered as *P. catodon* (=*macrocephalus*).
ISIS NUMBER: 5301411004002001001 as *P. catodon*.

Family Ziphiidae

REVIEWED BY: J. G. Mead (JGM); D. W. Rice (DWR); O. L. Rossolimo (OLR)(U.S.S.R.).
COMMENT: Called Hyperoodontidae by some authors, but see Van Bree, 1974, Bijdr. Dierk. D., 44(2):235–238. Family reviewed by Moore, 1968, Fieldiana Zool., 53:209–298.
PROTECTED STATUS: CITES - Appendix II as Order Cetacea.
ISIS NUMBER: 5301411005000000000.

Berardius Duvernoy, 1851. Ann. Sci. Nat. Zool., 15:52, 68.
PROTECTED STATUS: CITES - Appendix II as Order Cetacea.
ISIS NUMBER: 5301411005001000000.

Berardius arnuxii Duvernoy, 1851. Ann. Sci. Nat. Paris Zool., 15:52, 68.
TYPE LOCALITY: New Zealand, Canterbury Prov., Akaroa.
DISTRIBUTION: Southern Hemisphere, circumpolar, temperate to subarctic waters.
PROTECTED STATUS: CITES - Appendix II as Order Cetacea.
ISIS NUMBER: 5301411005001001001 as *B. arnouxi (sic)*.

Berardius bairdii Stejneger, 1883. Proc. U.S. Nat. Mus., 6:75.
TYPE LOCALITY: U.S.S.R., Commander Isls., Bering Isl., Stare Gavan.
DISTRIBUTION: North Pacific, temperate to subarctic waters.
COMMENT: Possibly a subspecies of *arnuxii*; see Davies, 1963, and McLachlan *et al.*, 1966 Ann. Cape Prov. Mus., 5:91–109.
PROTECTED STATUS: CITES - Appendix II as Order Cetacea.
ISIS NUMBER: 5301411005001002001.

Hyperoodon Lacepede, 1804. Hist. Nat. Cetacees, xliv, 319.
  PROTECTED STATUS: CITES - Appendix II as Order Cetacea.
  ISIS NUMBER: 5301411005002000000.

  Hyperoodon ampullatus (Forster, 1770). *In* Kalm, Travels into N. Amer., 1:18.
    TYPE LOCALITY: U.K., England, Essex, Maldon.
    DISTRIBUTION: Atlantic Ocean, cold temperate to subarctic waters; perhaps also N.
      Pacific (Hershkovitz, 1966:146). Possibly Mediterranean Sea (OLR).
    PROTECTED STATUS: CITES - Appendix II as Order Cetacea.
    ISIS NUMBER: 5301411005002001001.

  Hyperoodon planifrons Flower, 1882. Proc. Zool. Soc. Lond., 1882:392.
    TYPE LOCALITY: Australia, Western Australia, Dampier Arch., Lewis Isl.
    DISTRIBUTION: Southern Hemisphere, circumpolar, temperate to antarctic waters;
      occasionally into tropical waters.
    PROTECTED STATUS: CITES - Appendix II as Order Cetacea.
    ISIS NUMBER: 5301411005002002001.

Indopacetus Moore, 1968. Fieldiana Zool., 53(4):111, 254.
  REVIEWED BY: M. Nishiwaki (MN).
  COMMENT: MN includes this genus in *Mesoplodon;* but also see Rice, 1977.
  PROTECTED STATUS: CITES - Appendix II as Order Cetacea.
  ISIS NUMBER: 5301411005003000000.

  Indopacetus pacificus (Longman, 1926). Mem. Queensl. Mus., 8:269.
    TYPE LOCALITY: Australia, Queensland, off Mackay.
    DISTRIBUTION: Tropical Indian Ocean to eastern Australia (Coral Sea).
    COMMENT: Known only from two skulls, the second from Somalia; commonly
      included in *Mesoplodon* (JGM).
    PROTECTED STATUS: CITES - Appendix II as Order Cetacea.
    ISIS NUMBER: 5301411005003001001.

Mesoplodon Gervais, 1850. Ann. Sci. Nat. Paris Zool., ser. 3, 14:16.
  REVIEWED BY: M. Nishiwaki (MN).
  PROTECTED STATUS: CITES - Appendix II as Order Cetacea.
  ISIS NUMBER: 5301411005004000000.

  Mesoplodon bidens (Sowerby, 1804). Trans. Linn. Soc. Lond., 7:310.
    TYPE LOCALITY: U.K., Scotland, Elginshire, Brodie House.
    DISTRIBUTION: North Atlantic and Baltic Sea, temperate to subarctic waters.
    COMMENT: The alleged occurrence of this species in the Mediterranean Sea was
      discussed by P. J. H. van Bree, 1975, Ann. Mus. Civ. Stor. Nat. Genova, 80:226–
      228, who considered the evidence unconvincing.
    PROTECTED STATUS: CITES - Appendix II as Order Cetacea.
    ISIS NUMBER: 5301411005004001001.

  Mesoplodon bowdoini Andrews, 1908. Bull. Am. Mus. Nat. Hist., 24:203.
    TYPE LOCALITY: New Zealand, Canterbury Prov., Brighton Beach.
    DISTRIBUTION: Temperate waters of Australia, New Zealand, and Kerguelen Isls.
    PROTECTED STATUS: CITES - Appendix II as Order Cetacea.
    ISIS NUMBER: 5301411005004002001.

  Mesoplodon carlhubbsi Moore, 1963. Am. Midl. Nat., 70:422.
    TYPE LOCALITY: U.S.A., California, La Jolla.
    DISTRIBUTION: Temperate waters of the North Pacific.
    COMMENT: Very closely related to *bowdoini* (JGM).
    PROTECTED STATUS: CITES - Appendix II as Order Cetacea.
    ISIS NUMBER: 5301411005004003001.

  Mesoplodon densirostris (Blainville, 1817). Nouv. Dict. Hist. Nat., 9:178.
    TYPE LOCALITY: Unknown.
    DISTRIBUTION: Temperate and tropical waters of the world.
    PROTECTED STATUS: CITES - Appendix II as Order Cetacea.
    ISIS NUMBER: 5301411005004004001.

Mesoplodon europaeus (Gervais, 1855). Hist. Nat. Mamm., 2:320.
TYPE LOCALITY: "English Channel."
DISTRIBUTION: Temperate and tropical waters of the North Atlantic.
COMMENT: Aside from the type, and one specimen from Guinea-Bissau, it is known only from the western North Atlantic (JGM, DWR).
PROTECTED STATUS: CITES - Appendix II as Order Cetacea.
ISIS NUMBER: 5301411005004005001.

Mesoplodon ginkgodens Nishiwaki and Kamiya, 1958. Sci. Rep. Whales Res. Inst. (Tokyo), 13:77.
TYPE LOCALITY: Japan, Kanagawa Pref., Oiso Beach.
DISTRIBUTION: Tropical to warm temperate waters of the North Pacific and Indian Oceans off Japan, Taiwan, California (U.S.A.), and Sri Lanka.
COMMENT: Includes *hotaula*; see Moore and Gilmore, 1965, Nature, 205(4977):1239–1240.
PROTECTED STATUS: CITES - Appendix II as Order Cetacea.
ISIS NUMBER: 5301411005004006001.

Mesoplodon grayi Von Haast, 1876. Proc. Zool. Soc. Lond., 1876:9.
TYPE LOCALITY: New Zealand, Chatham Isl., Waitangi Beach.
DISTRIBUTION: Temperate southern oceans, cold temperate waters.
COMMENT: One specimen found in the Netherlands; see Boschma, 1950, Proc. K. Ned. Akad. Wet. Ser. B Palaeontol. Geol. Phys. Chem., 53:779 (JGM, DWR).
PROTECTED STATUS: CITES - Appendix II as Order Cetacea.
ISIS NUMBER: 5301411005004007001.

Mesoplodon hectori (Gray, 1871). Ann. Mag. Nat. Hist., ser. 4, 8:117.
TYPE LOCALITY: New Zealand, Wellington, Titai Bay.
DISTRIBUTION: Southern Hemisphere, temperate waters, and in the northern Pacific; see Mead, 1981, J. Mammal, 62:430–432.
PROTECTED STATUS: CITES - Appendix II as Order Cetacea.
ISIS NUMBER: 5301411005004008001.

Mesoplodon layardii (Gray, 1865). Proc. Zool. Soc. Lond., 1865:358.
TYPE LOCALITY: Probably South Africa.
DISTRIBUTION: Southern Hemisphere, temperate waters.
PROTECTED STATUS: CITES - Appendix II as Order Cetacea.
ISIS NUMBER: 5301411005004009001.

Mesoplodon mirus True, 1913. Smithson. Misc. Coll., 60(25):1.
TYPE LOCALITY: U.S.A., North Carolina, Beaufort Harbor, Bird Island Shoal.
DISTRIBUTION: North Atlantic, temperate waters, South Atlantic coast of South Africa.
PROTECTED STATUS: CITES - Appendix II as Order Cetacea.
ISIS NUMBER: 5301411005004010001.

Mesoplodon stejnegeri True, 1885. Proc. U.S. Nat. Mus., 8:585.
TYPE LOCALITY: U.S.S.R., Commander Isls., Bering Isl.
DISTRIBUTION: North Pacific, cold temperate to subarctic waters.
PROTECTED STATUS: CITES - Appendix II as Order Cetacea.
ISIS NUMBER: 5301411005004011001.

Tasmacetus Oliver, 1937. Proc. Zool. Soc. Lond., 1937:371.
PROTECTED STATUS: CITES - Appendix II as Order Cetacea.
ISIS NUMBER: 5301411005005000000.

Tasmacetus shepherdi Oliver, 1937. Proc. Zool. Soc. Lond., 107:371.
TYPE LOCALITY: New Zealand, North Island, Taranaki, Ohawe.
DISTRIBUTION: Southern Hemisphere, temperate waters, particularly off New Zealand, Chile and Argentina.
PROTECTED STATUS: CITES - Appendix II as Order Cetacea.
ISIS NUMBER: 5301411005005001001.

Ziphius G. Cuvier, 1823. Rech. Oss. Foss., 5:350.
>   PROTECTED STATUS: CITES - Appendix II as Order Cetacea.
>   ISIS NUMBER: 5301411005006000000.

Ziphius cavirostris G. Cuvier, 1823. Rech. Oss. Foss., 5:350.
>   TYPE LOCALITY: France, Bouches-du-Rhone, between Fos and the mouth of the
>   Galegeon River.
>   DISTRIBUTION: Worldwide, tropical, temperate, and subarctic waters.
>   PROTECTED STATUS: CITES - Appendix II as Order Cetacea.
>   ISIS NUMBER: 5301411005006001001.

Family Eschrichtidae
>   REVIEWED BY: C. L. Gray (CLG); J. G. Mead (JGM); D. W. Rice (DWR); R. B. Patten
>   (RBP); O. L. Rossolimo (OLR)(U.S.S.R.).
>   PROTECTED STATUS: CITES - Appendix II as Order Cetacea.
>   ISIS NUMBER: 5301411006000000000.

Eschrichtius Gray, 1864. Ann. Mag. Nat. Hist., ser. 3, 14:350.
>   COMMENT: *Rhachianectes* Cope, 1869, is a junior synonym; see Van Deinse and Junge,
>   1937, Temminckia, 2:161–188.
>   PROTECTED STATUS: CITES - Appendix II as Order Cetacea.
>   ISIS NUMBER: 5301411006001000000.

Eschrichtius robustus (Lilljeborg, 1861). Forh. Skand. Naturf. Ottende Mode, Kopenhagen,
>   1860, 8:602.
>   TYPE LOCALITY: Sweden, Uppland, Graso Isl.
>   DISTRIBUTION: Pacific, temperate to arctic waters.
>   COMMENT: Includes *gibbosus* and *glaucus*; see Rice and Wolman, 1971, Am. Soc. Mamm.
>   Spec. Publ., 3:1–142; Rice, 1977:12. Formerly present in the North Atlantic (JGM).
>   PROTECTED STATUS: CITES - Appendix I as *E. robustus* (=*glaucus*) and U.S. ESA -
>   Endangered as *E. robustus*.
>   ISIS NUMBER: 5301411006001001001 as *E. gibbosus*.

Family Balaenopteridae
>   REVIEWED BY: J. G. Mead (JGM); M. Nishiwaki (MN); D. W. Rice (DWR); O. L.
>   Rossolimo (OLR)(U.S.S.R.).
>   PROTECTED STATUS: CITES - Appendix II as Order Cetacea.
>   ISIS NUMBER: 5301411007000000000.

Balaenoptera Lacepede, 1804. Hist. Nat. Cetacees, p. 36, 114.
>   COMMENT: Includes *Sibbaldus*; see Rice, 1977:13.
>   PROTECTED STATUS: CITES - Appendix II as Order Cetacea.
>   ISIS NUMBER: 5301411007001000000.

Balaenoptera acutorostrata Lacepede, 1804. Hist. Nat. Cetacees, p. 37, 134.
>   TYPE LOCALITY: France, Cherbourg, Mancha.
>   DISTRIBUTION: Worldwide, subtropical to arctic waters.
>   COMMENT: Includes *bonaerensis*; see Omura, 1975, Sci. Rep. Whales Res. Inst. (Tokyo),
>   27:1–36. MN considered *bonaerensis* a distinct species.
>   PROTECTED STATUS: CITES - Appendix II as Order Cetacea.
>   ISIS NUMBER: 5301411007001001001.

Balaenoptera borealis Lesson, 1828. Compl. Oeuvres Buffon Hist. Nat., 1:342.
>   TYPE LOCALITY: Germany(BRD), Schleswig-Holstein, Lubeck Bay, near Gromitz.
>   DISTRIBUTION: Worldwide, subtropical to subarctic waters.
>   PROTECTED STATUS: CITES - Appendix I and U.S. ESA - Endangered.
>   ISIS NUMBER: 5301411007001002001.

Balaenoptera edeni Anderson, 1878. Anat. Zool. Res., p. 551.
TYPE LOCALITY: Burma, Tenasserim, between mouths of Sittang and Bilin Rivers, Thayboo Choung.
DISTRIBUTION: Worldwide, temperate to tropical waters.
COMMENT: Includes *brydei;* see Junge, 1950, Zool. Verh. Rijksmus. Nat. Hist. Leiden, 9:1–26.
PROTECTED STATUS: CITES - Appendix II as Order Cetacea.
ISIS NUMBER: 5301411007001003001.

Balaenoptera musculus (Linnaeus, 1758). Syst. Nat., 10th ed., 1:76.
TYPE LOCALITY: U.K., Scotland, Firth of Forth.
DISTRIBUTION: Worldwide, subtropical to subarctic waters.
PROTECTED STATUS: CITES - Appendix I and U.S. ESA - Endangered.
ISIS NUMBER: 5301411007001004001.

Balaenoptera physalus (Linnaeus, 1758). Syst. Nat., 10th ed., 1:75.
TYPE LOCALITY: Norway, near Svalbard, Spitzbergen Sea.
DISTRIBUTION: Worldwide, tropical to arctic waters.
PROTECTED STATUS: CITES - Appendix I and U.S. ESA - Endangered.
ISIS NUMBER: 5301411007001005001.

Megaptera Gray, 1846. Ann. Mag. Nat. Hist., ser. 1, 17:83.
PROTECTED STATUS: CITES - Appendix II as Order Cetacea.
ISIS NUMBER: 5301411007002000000

Megaptera novaeangliae (Borowski, 1781). Gemein. Naturgesch. Thier., 2(1):21.
TYPE LOCALITY: U.S.A., coast of New England.
DISTRIBUTION: Worldwide, tropical to arctic waters.
COMMENT: Includes *nodosa;* see Kellogg, 1932, Proc. Biol. Soc. Wash., 45:148.
PROTECTED STATUS: CITES - Appendix I and U.S. ESA - Endangered.
ISIS NUMBER: 5301411007002001001.

Family Balaenidae
REVIEWED BY: J. G. Mead (JGM); D. W. Rice (DWR); O. L. Rossolimo (OLR)(U.S.S.R.).
PROTECTED STATUS: CITES - Appendix II as Order Cetacea.
ISIS NUMBER: 5301411008000000000.

Balaena Linnaeus, 1758. Syst. Nat., 10th ed., 1:75.
COMMENT: Includes *Eubalaena;* see Rice, 1977:6–7.
PROTECTED STATUS: CITES - Appendix I as *Eubalaena* spp.
ISIS NUMBER: 5301411008001000000.

Balaena glacialis Muller, 1776. Zool. Danicae Prodr., p. 7.
TYPE LOCALITY: Norway, Finnmark, Nord Kapp (vicinity of North Cape).
DISTRIBUTION: Worldwide in temperate and subarctic seas.
COMMENT: Formerly considered to be three separate species, *glacialis, australis,* and *japonicus;* see Hershkovitz, 1961, Fieldiana Zool., 39(49):547–565. Includes *sieboldi;* see Tomilin, 1957, [Mamm. U.S.S.R., Adjac. Count.], 9:62; Rice, 1977. Gromov and Baranova, 1981:231, placed this species in *Eubalaena.*
PROTECTED STATUS: CITES - Appendix I as *Eubalaena* spp. and U.S. ESA - Endangered as *B. glacialis.*
ISIS NUMBER: 5301411008001001001.

Balaena mysticetus Linnaeus, 1758. Syst. Nat., 10th ed., 1:75.
TYPE LOCALITY: Greenland Sea.
DISTRIBUTION: Northern Hemisphere, arctic seas.
PROTECTED STATUS: CITES - Appendix I as *B. mysticetus* and U.S. ESA - Endangered as *B. mysticetus.*
ISIS NUMBER: 5301411008001002001.

Caperea Gray, 1864. Proc. Zool. Soc. Lond., 1864:202.
    COMMENT: *Neobalaena* Gray, 1870, is a junior synonym; see Van Beneden, 1874, Bull. R.
        Acad. Belg., ser. 2, 37:832.
    PROTECTED STATUS: CITES - Appendix II as Order Cetacea.
    ISIS NUMBER: 5301411008002000000.

Caperea marginata (Gray, 1846). Zool. Voy. H.M.S. "Erebus" and "Terror," 1:48.
    TYPE LOCALITY: Australia, Western Australia.
    DISTRIBUTION: Southern Hemisphere, temperate waters.
    PROTECTED STATUS: CITES - Appendix II as Order Cetacea.
    ISIS NUMBER: 5301411008002001001.

ORDER SIRENIA

ISIS NUMBER: 5301417000000000000.

Family Dugongidae

REVIEWED BY: D. P. Domning (DPD); D. W. Rice (DWR); J. Shoshani (JS).
ISIS NUMBER: 5301417001000000000.

Dugong Lacepede, 1799. Tabl. Mamm., p. 17.
ISIS NUMBER: 5301417001001000000.

Dugong dugon (Muller, 1776). Linne's Vollstand. Natursyst. Suppl., p. 21.
TYPE LOCALITY: Cape of Good Hope to Philippine Isls.
DISTRIBUTION: Red Sea and E. Africa to S. Japanese Archipelago, Caroline Isls.,
Australia, Solomon Islands, New Hebrides, and New Caledonia.
COMMENT: Includes *australis, hemprichi,* and *tabernaculi;* see Corbet, 1978:193. Reviewed
by Husar, 1978, Mamm. Species, 88:1-7.
PROTECTED STATUS: CITES - Appendix I as *D. dugon* (except Australian populations);
CITES - Appendix II as *D. dugon* (Australian populations only). U.S. ESA -
Endangered as *D. dugon.*
ISIS NUMBER: 5301417001001001001.

Hydrodamalis Retzius, 1794. Kongl. Svenska Vet.-Akad. Handl. Stockholm, 15, p. 292.
REVIEWED BY: R. S. Hoffmann (RSH).
COMMENT: Placed in family Dugongidae by Corbet and Hill, 1980:114, and other
authors; but in the monotypic family Hydrodamalidae by some; see Heptner and
Naumov, eds., 1967, [Mammals of the Soviet Union], 2(1):25. Includes *Rytina*
(=*Rhytina,* a later spelling) Illiger, 1811, a junior synonym.

Hydrodamalis gigas (Zimmermann, 1780). Geogr. Gesch. Mensch. Vierf. Thiere, 2, p. 426.
TYPE LOCALITY: U.S.S.R., Kamchatsk. Obl., Commander Isls., Bering Isl.
DISTRIBUTION: Originally coasts of N. Pacific Ocean, but in historic time known only
from Commander Isls. (U.S.S.R.). Extirpated about 1770. Reviewed by Heptner
and Naumov, eds., 1967, [Mammals of the Soviet Union], 2(1):25-46; see also
Domning, 1975, J. Mammal., 56:556-558.

Family Trichechidae

REVIEWED BY: D. P. Domning (DPD); D. W. Rice (DWR); J. Shoshani (JS).
ISIS NUMBER: 5301417002000000000.

Trichechus Linnaeus, 1758. Syst. Nat., 10th ed., 1:34.
COMMENT: *Manatus* Brunwich, 1772, is a junior synonym. Revised by Hatt, 1934, Bull.
Am. Mus. Nat. Hist., 66:533-566.
ISIS NUMBER: 5301417002001000000.

Trichechus inunguis (Natterer, 1883). *In* Pelzeln, Verh. Zool.-Bot. Ges. Wien, 33:89.
TYPE LOCALITY: Brazil, Amazonas, Borba (on lower Rio Madeira).
DISTRIBUTION: Amazon Basin, including Amazon estuaries and Atlantic coast of Ilha
de Marajo.
COMMENT: *T. exunguis* Natterer, 1839, *in* Diesing, Ann. Wiener Mus. Naturgesch.,
2:230, should be regarded as a *nomen oblitum* (DPD). Reports from Orinoco basin
are unsubstantiated (DPD and JS). Reviewed by Husar, 1977, Mamm. Species,
72:1-4.
PROTECTED STATUS: CITES - Appendix I and U.S. ESA - Endangered.
ISIS NUMBER: 5301417002001001001.

Trichechus manatus Linnaeus, 1758. Syst. Nat., 10th ed., 1:34.
TYPE LOCALITY: West Indies.
DISTRIBUTION: S. Atlantic and Caribbean coasts from Virginia (U.S.A.) and West Indies
to Sergipe (formerly to Espirito Santo), (Brazil); Orinoco and Magdalena river
systems.

COMMENT: Includes *latirostris;* see Hatt, 1934, Bull. Am. Mus. Nat. Hist., 66:533–566. Range appears disjunct N. and S. of Amazon estuaries, where *inunguis* occurs; see Domning, 1981, Biol. Conserv., 19:85–97. Reviewed by Husar, 1978, Mamm. Species, 93:1–5. Type locality restricted by Thomas, 1911, Proc. Zool. Soc. Lond., 1911:132.

PROTECTED STATUS: CITES - Appendix I and U.S. ESA - Endangered.

ISIS NUMBER: 5301417002001002001.

Trichechus senegalensis Link, 1795. Beytr. Naturg., 1(2):109.

TYPE LOCALITY: Senegal.

DISTRIBUTION: Senegal to Quanza River (Angola), including Niger-Benue Basin.

COMMENT: Former occurrence in Lake Chad basin uncertain. Not reliably reported above the falls on the lower Congo River; see Hatt, 1934, Bull. Am. Mus. Nat. Hist., 66:554–560, 566. Reviewed by Husar, 1978, Mamm. Species, 89:1–3. Type locality discussed by Hatt, 1934, Bull. Am. Mus. Nat. Hist., 66:535.

PROTECTED STATUS: CITES - Appendix II and U.S. ESA - Threatened.

ISIS NUMBER: 5301417002001003001.

ORDER PROBOSCIDEA

ISIS NUMBER: 5301415000000000000.

Family Elephantidae

REVIEWED BY: D. P. Domning (DPD); J. Shoshani (JS).
ISIS NUMBER: 5301415001000000000.

Elephas Linnaeus, 1758. Syst. Nat., 10th ed., 1:33.
COMMENT: Revised by Maglio, 1973, Trans. Am. Philos. Soc., 63(3):1–149, who gave synonyms for Recent and extinct species.
ISIS NUMBER: 5301415001001000000.

Elephas maximus Linnaeus, 1758. Syst. Nat., 10th ed., 1:33.
TYPE LOCALITY: Sri Lanka (=Ceylon) (domesticated stock).
DISTRIBUTION: Sri Lanka; India to Indochina; Malaysia; Sumatra; N. Borneo.
COMMENT: For discussion of subspecific variation see Deraniyagala, 1955, Ceylon Nat. Mus. Publ., Colombo, and Chasen, 1940. Bornean elephants are believed to be feral descendants of a stock introduced in the 1750's; see de Silva, 1968, Sabah Soc. J., 3(4):169–181, and Olivier, 1978, Unpubl. Ph.D. Dissertation, Cambridge. According to Chasen, 1940, these elephants resemble the continental form rather than the one from Sumatra. See also Ellerman and Morrison-Scott, 1966:336; Corbet, 1978:191–192.
PROTECTED STATUS: CITES - Appendix I and U.S. ESA - Endangered.
ISIS NUMBER: 5301415001001001001.

Loxodonta Anonymous, 1827. Zool. J., 3:140.
COMMENT: The spelling in the original publication (F. Cuvier, 1825, *in* E. Geoffroy Saint-Hilaire and F. Cuvier, Hist. Nat. Mamm., 3(52):2) was "Loxodonte." However, this word has been fully latinized by other authors, at least since 1827, and has been generally accepted by zoologists in that form. The author of the 1827 publication (which is a review of Cuvier's part in the 1825 book) could not be determined. Under Article 11(b) of the ICZN, 1964:9, for a name to be accepted, it must be Latin or latinized. Therefore, neither "Loxodonte" F. Cuvier, 1825, nor "Loxodonta" F. Cuvier, 1827, are acceptable (JS, DPD). Lydekker, 1916, Cat. Ungulate Mammals Br. Mus., 5:85, and others state that *Loxodonta* should be reduced to a subgenus, but *Loxodonta* and *Elephas* are distinct genera based on ongoing anatomical and biochemical/serological studies (JS). Revised by Maglio, 1973, Trans. Am. Philos. Soc., 63:1–149.
ISIS NUMBER: 5301415001002000000.

Loxodonta africana (Blumenbach, 1797). Hand. Hilfsb. Nat., 5th ed., p. 125.
TYPE LOCALITY: South Africa, Orange River.
DISTRIBUTION: S. Mauritania to Ethiopia and South Africa; extinct in N. Africa.
COMMENT: Includes *cyclotis*; see Sikes, 1971, The Natural History of the African Elephant, but also see Haltenorth and Diller, 1977, Saugetiere Africas und Madagaskars, BLV, Munchen, p. 130, who considered *pumilio* (a junior synonym of *cyclotis*) as a distinct species. For alternative discussions of variation in this species, see Ansell, 1971, Part 11:1–5; Ellerman, Morrison-Scott and Hayman, 1953; and Coppens *et al., in* Maglio and Cooke, 1978. Reviewed by Laursen and Bekoff, 1978, Mamm. Species, 92:1–8. Type locality restricted by Pohle, 1926, Z. Saugetierk., 1:63.
PROTECTED STATUS: CITES - Appendix II and U.S. ESA - Threatened.
ISIS NUMBER: 5301415001002001001.

ORDER PERISSODACTYLA
ISIS NUMBER: 5301418000000000000.

Family Equidae
REVIEWED BY: D. K. Bennett (DKB); C. R. Durst (CRD); C. P. Groves (CPG); P. Grubb
(PG); I. U. Kohler (IUK); O. L. Rossolimo (OLR)(U.S.S.R.); A. C. Ziegler (ACZ).
ISIS NUMBER: 5301418001000000000.

Equus Linnaeus, 1758. Syst. Nat., 10th ed., 1:73.
COMMENT: Recent and fossil forms revised by Bennett, 1980, Syst. Zool., 29:272–287;
subgenera reviewed by Groves and Willoughby, 1981, Mammalia, 45:321–354.
ISIS NUMBER: 5301418001001000000.

Equus asinus Linnaeus, 1758. Syst. Nat., 10th ed., 1:73.
TYPE LOCALITY: "Southern Asia" (domesticated stock).
DISTRIBUTION: Formerly N.W. Algeria, adjacent Morocco and Tunisia; Egypt; perhaps
Arabia; N.E. Sudan; survives only in N.E. Ethiopia and N. Somalia; domesticated
worldwide.
COMMENT: Ansell, 1974, Part 14:6, recommended *africanus* replace *asinus*, because
domesticated stock was used for the type material; also see Corbet, 1978:195.
PROTECTED STATUS: U.S. ESA - Endangered as *E. africanus* (=*asinus*).
ISIS NUMBER: 5301418001001001001.

Equus burchelli (Gray, 1824). Zool. J., 1:247.
TYPE LOCALITY: South Africa, Bechuanaland, Little Klibbolikhoni Fontein.
DISTRIBUTION: Blue Nile to Orange River, including S.W. Somalia and S.W. Ethiopia,
to S. Africa to S.E. Zaire and S. and E. Angola.
COMMENT: Does not include the extinct species, *quagga*; see Bennett, 1980, Syst. Zool.,
29:272–287; Groves and Willoughby, 1981, Mammalia, 45:321–354. Many previous
workers regarded *quagga* and *burchelli* as conspecific; see Rau, 1978, Ann. S. Afr.
Mus., 77. At least 21 synonyms of *burchelli* exist, most based on variations in
stripe pattern. Cabrera, 1936, J. Mamm., 17:89–112, synonymized most forms and
demonstrated a cline in stripe pattern. Reviewed by Grubb, 1981, Mamm. Species,
157:1–9.
ISIS NUMBER: 5301418001001002001.

Equus caballus Linnaeus, 1758. Syst. Nat., 10th ed., 1:73.
TYPE LOCALITY: Norway (domesticated stock).
DISTRIBUTION: Domesticated worldwide; wild population survived (at least until
recently) in S.W. Mongolia and adjacent Kansu, Sinkiang and Inner Mongolia
(China)(SW). West to Poland before the 19th Century, in steppe zone.
COMMENT: Groves, 1971, Bull. Zool. Nomencl., 27:269–272., and Corbet, 1978:194, have
proposed that *ferus* replace *caballus*, objecting to use of domesticated stock for
type material. Includes *przewalskii*; despite different chromosome numbers
(Benirschke *et al.*, 1972, Science, 148:382–383), the fundamental chromosome
number of Przewalski and domestic horses is the same and matings produce
fertile offspring; therefore, DKB and IUK consider them conspecific. Gromov and
Baranova, 1981:333–334, recognized the two as separate species and used the
names *gmelini* and *przewalskii*. See Sokolov and Orlov, 1980:249 for discussion of
Mongolian and Chinese wild stocks.
PROTECTED STATUS: CITES - Appendix I and U.S. ESA - Endangered as *Equus przewalskii*
only.
ISIS NUMBER: 5301418001001003001 as *E. caballus*.
5301418001001006001 as *E. przewalskii*.

Equus grevyi Oustalet, 1882. La Nature (Paris), 10(2):12.
TYPE LOCALITY: Ethiopia (=Abyssinian desert), Galla Country.
DISTRIBUTION: Dry desert regions of N. Kenya; S. Somalia; S. and E. Ethiopia.

COMMENT: More different synonymous subgenera (at least five) have been erected to accommodate this monotypic species and its close fossil relatives than any other *Equus* species (DKB); see also Groves and Willoughby, 1981, Mammalia, 45:321–354. Threatened with extinction, except in Kenya (CPG).
PROTECTED STATUS: CITES - Appendix I and U.S. ESA - Threatened.
ISIS NUMBER: 5301418001001004001.

Equus hemionus Pallas, 1775. Nova Comm. Acad. Sci. Petrop., 19:397.
TYPE LOCALITY: U.S.S.R., Transbaikalia, S. Chitinsk. Obl., Tarei-Nor, 50° N, 115° E.
DISTRIBUTION: Formerly much of Mongolia, north to Transbaikalia (U.S.S.R.); east to N.E. Inner Mongolia (China) and possibly W. Manchuria (China); west to Dzhungarian Gate. Survives in S.W. and S.C. Mongolia and adjacent China; see Sokolov and Orlov, 1980:248.
COMMENT: Includes *luteus* (DKB). Groves and Mazak, 1967, Z. Saugetierk., 32:321–355, Corbet, 1978:194, and Gromov and Baranova, 1981:335, included *onager* in this species, but Bennett, 1980, Syst. Zool., 29:272–287, considered it a distinct species. See Sokolov and Orlov, 1980:248 for discussion of Mongolian and Chinese stocks.
PROTECTED STATUS: CITES - Appendix II and U.S. ESA - Endangered as *E. hemionus*.
CITES - Appendix I as *E. h. hemionus* subspecies only.
ISIS NUMBER: 5301418001001005001.

Equus kiang Moorcroft, 1841. Travels in the Himalayan Provinces, 1:312.
TYPE LOCALITY: India, Kashmir, Ladakh.
DISTRIBUTION: Populations exist in dry, intermontane basins of Ladakh (India); Tibet, Tsinghai, Szechwan (China) and adjacent Nepal and Sikkim.
COMMENT: Groves and Mazak, 1967, Z. Saugetierk., 32:321–355; Corbet, 1978:195; and Bennett, 1980, Syst. Zool., 29:272–287, recommended separating *kiang* from *hemionus*.

Equus onager Boddaert, 1785. Elench. Anim., p. 60.
TYPE LOCALITY: N.W. Persia (=Iran), Kasbin, near Caspian.
DISTRIBUTION: Formerly much of Central Asian republics (U.S.S.R.) north to upper Irtysh and Ural Rivers; westward north of the Caucasus and Black Sea at least to Dniestr River; and S.E. of Caspian Sea, Anatolia, N. Iraq, Iran, Afghanistan, and Pakistan to Thar Desert of N.W. India; survives as isolated populations in Rann of Kutch (India), Badkhys Preserve, Turkmenia (U.S.S.R.); and central Iran; also reestablished on Barsa-khelmes Isl. (Aral Sea) (U.S.S.R.).
COMMENT: Includes *hemippus*, *khur*, and *kulan* (DKB); most previous workers regarded *onager* as a synonym of *hemionus*, but Bennett, 1980, Syst. Zool., 29:272–287, considered it a distinct species; see comments under *hemionus*.
PROTECTED STATUS: CITES - Appendix I as *E. hemionus khur* subspecies only.

Equus quagga Gmelin, 1788. *In* Linnaeus, Syst. Nat., 13th ed., 1:213.
TYPE LOCALITY: South Africa.
DISTRIBUTION: South Africa, south of the Vaal River.
COMMENT: Extinct; last specimen, a captive, died in 1872. See also comments under *burchelli*.

Equus zebra Linnaeus, 1758. Syst. Nat., 10th ed., 1:74.
TYPE LOCALITY: South Africa, S.W. Cape Prov., Paardeburg, near Malmesbury.
DISTRIBUTION: Formerly in S. Angola; Namibia; S.W. and S.C. Cape Prov. (South Africa).
COMMENT: *E. z. zebra* is nearly extinct in the wild; recent counts indicate fewer than 70 alive in their habitat within remote mountain basins of the Cape Province, South Africa.
PROTECTED STATUS: CITES - Appendix I and U.S. ESA - Endangered as *E. z. zebra* subspecies only.
CITES - Appendix II as *E. z. hartmannae* subspecies only.
ISIS NUMBER: 5301418001001007001.

Family Tapiridae
REVIEWED BY: C. R. Durst (CRD); C. P. Groves (CPG); J. Ramirez-Pulido (JRP).
ISIS NUMBER: 5301418002000000000.

Tapirus Brunnich, 1772. Zool. Fundamenta, pp. 44, 45.
ISIS NUMBER: 5301418002001000000.

Tapirus bairdii (Gill, 1865). Proc. Acad. Nat. Sci. Phila., 17:183.
TYPE LOCALITY: Panama, Isthmus of Panama.
DISTRIBUTION: S. Veracruz and S. Oaxaca (Mexico) east of Isthmus of Tehuantepec to
    Colombia west of the Rio Cauca and Equador west of the Andes to the Gulf of
    Guayaquil.
COMMENT: Revised by Hershkovitz, 1954, Proc. U.S. Nat. Mus., 103:465–496. Type
    locality restricted by Hershkovitz, 1954, Proc. U.S. Nat. Mus., 103:465–496, to
    Canal Zone.
PROTECTED STATUS: CITES - Appendix I and U.S. ESA - Endangered.
ISIS NUMBER: 5301418002001001001 as *T. bairdi (sic)*.

Tapirus indicus Desmarest, 1819. Nouv. Dict. Hist. Nat., 32:458.
TYPE LOCALITY: Malaysia, Malay Peninsula.
DISTRIBUTION: Burma and Thailand south of 18° N., south through Malay Peninsula;
    Sumatra. Formerly, S. China, and probably parts of Kampuchea, Laos, and
    Vietnam.
PROTECTED STATUS: CITES - Appendix I and U.S. ESA - Endangered.
ISIS NUMBER: 5301418002001002001.

Tapirus pinchaque (Roulin, 1829). Ann. Sci. Nat. Zool., 18:46.
TYPE LOCALITY: Colombia, Cundinamarca, Paramo de Sumapaz.
DISTRIBUTION: Andes of Colombia; Ecuador; perhaps W. Venezuela and N. Peru.
COMMENT: Revised by Hershkovitz, 1954, Proc. U.S. Nat. Mus., 103:465–496.
PROTECTED STATUS: CITES - Appendix I and U.S. ESA - Endangered.
ISIS NUMBER: 5301418002001003001.

Tapirus terrestris (Linnaeus, 1758). Syst. Nat., 10th ed., 1:74.
TYPE LOCALITY: Brazil, Pernambuco.
DISTRIBUTION: Venezuela and Colombia south to S. Brazil, N. Argentina and Paraguay,
    mostly east of the Andes.
COMMENT: Revised by Hershkovitz, 1954, Proc. U.S. Nat. Mus., 103:465–496.
PROTECTED STATUS: CITES - Appendix II and U.S. ESA - Endangered.
ISIS NUMBER: 5301418002001004001.

Family Rhinocerotidae
REVIEWED BY: C. R. Durst (CRD); C. P. Groves (CPG); B. R. Stein (BRS).
PROTECTED STATUS: CITES - Appendix I as Rhinocerotidae.
ISIS NUMBER: 5301418003000000000.

Ceratotherium Gray, 1868. Proc. Zool. Soc. Lond., 1867:1027.
PROTECTED STATUS: CITES - Appendix I as Rhinocerotidae.
ISIS NUMBER: 5301418003001000000.

Ceratotherium simum (Burchell, 1817). Bull. Sci. Soc. Philom. Paris, p. 97.
TYPE LOCALITY: Botswana, Makuba Range, Chue Spring (about 26° 15′ S, 23° 10′ E).
DISTRIBUTION: Formerly much of N.W. Africa; Nile Valley; C. Africa between Lake
    Chad and White Nile River; perhaps portions of E. Africa; S. Africa south of the
    Zambezi River (except in W. Zambia, where it perhaps occurred between the
    Mashi and Zambezia rivers), south to Orange River, and west to E. Namibia.
    Survives mostly as scattered populations in reserves in South Africa, Zimbabwe,
    Mozambique, Uganda, Zaire, and S. Sudan. Extinct in N.W. Africa and lower Nile
    Valley.
COMMENT: Reviewed by Groves, 1972, Mamm. Species, 8:1–6. Revised by Groves,
    1975, Saugetierk. Mitt., 23:200–212.

PROTECTED STATUS: CITES - Appendix I as Rhinocerotidae.
CITES - Appendix I and U.S. ESA - Endangered as *C. s. cottoni* subspecies only.
ISIS NUMBER: 5301418003001001001.

Dicerorhinus Gloger, 1841. Hand. Hilfsb. Nat., p. 125.
COMMENT: *Didermocerus* Brookes, 1828, has been rejected, and *Dicerorhinus* validated;
Bull. Zool. Nomencl., 34:21–24.
PROTECTED STATUS: CITES - Appendix I as Rhinocerotidae.
ISIS NUMBER: 5301418003003000000 as *Didermocerus.*

Dicerorhinus sumatrensis (Fischer, 1814). Zoognosia, 3:301.
TYPE LOCALITY: Indonesia, Sumatra, Bencoolen (=Bintuhan) Dist., Fort Marlborough.
DISTRIBUTION: Formerly Assam (India), Chittagong Hills (Bangladesh), Burma,
Thailand, and Vietnam south through Malay Peninsula to Sumatra; probably also
S. China, Laos, and Cambodia (Kampuchea); Borneo; Mergui Isl. Survives in
Tenasserim Range (Thailand-Burma), Petchabun Range (Thailand); and other
scattered localities in Burma, Malay Peninsula, Sumatra, and Borneo.
COMMENT: Reviewed by Groves and Kurt, 1972, Mamm. Species, 21:1–6. Revised by
Groves, 1967, Saugetierk. Mitt., 15:221–237.
PROTECTED STATUS: CITES - Appendix I as Rhinocerotidae and U.S. ESA - Endangered
as *Dicerorhinus (=Didermocerus) sumatrensis.*
ISIS NUMBER: 5301418003003001001 as *Didermocerus sumatrensis.*

Diceros Gray, 1821. Lond. Med. Repos., 15:306.
PROTECTED STATUS: CITES - Appendix I as Rhinocerotidae.
ISIS NUMBER: 5301418003002000000.

Diceros bicornis (Linnaeus, 1758). Syst. Nat., 10th ed., 1:56.
TYPE LOCALITY: South Africa, Cape Prov. (="India").
DISTRIBUTION: Formerly in suitable open habitats in Africa south of about 10° N. from
Chad, S. Sudan and N. Somalia, and from Angola, south to Cape Province (South
Africa). Survives in reserves and relatively undisturbed areas in much of the
northern three-quarters of its historic range, and in places, to South Africa.
PROTECTED STATUS: CITES - Appendix I as Rhinocerotidae and U.S. ESA - Endangered.
ISIS NUMBER: 5301418003002001001.

Rhinoceros Linnaeus, 1758. Syst. Nat., 10th ed., 1:56.
PROTECTED STATUS: CITES - Appendix I as Rhinocerotidae.
ISIS NUMBER: 5301418003004000000.

Rhinoceros sondaicus Desmarest, 1822. Mammalogie, 2:399.
TYPE LOCALITY: Probably Java (Indonesia).
DISTRIBUTION: Formerly Bangladesh, Burma, Thailand, Laos, Kampuchea, Vietnam,
and probably S. China through Malay Peninsula to Sumatra and Java. Survives in
Ujung Kulon (W. Java), and perhaps in small areas of Burma, Thailand, Laos, and
Kampuchea.
COMMENT: Revised by Groves, 1967, Saugetierk. Mitt., 15:221–237.
PROTECTED STATUS: CITES - Appendix I as Rhinocerotidae and U.S. ESA - Endangered.
ISIS NUMBER: 5301418003004001001.

Rhinoceros unicornis Linnaeus, 1758. Syst. Nat., 10th ed., 1:56.
TYPE LOCALITY: India, Assam, Terai.
DISTRIBUTION: Formerly Indus Valley, Gangetic Plain, Brahmaputra Valley, and
Himalayan foothills from Pakistan to Assam (India). Survives in reserves in
Nepal and Assam; perhaps Bangladesh.
PROTECTED STATUS: CITES - Appendix I as Rhinocerotidae and U.S. ESA - Endangered.
ISIS NUMBER: 5301418003004002001.

ORDER HYRACOIDEA

ISIS NUMBER: 5301416000000000000.

Family Procaviidae

REVIEWED BY: D. P. Domning (DPD); H. N. Hoeck (HNH); J. Shoshani (JS).

COMMENT: Hyracidae Gray, 1821, is a group name based on *Hyrax* Hermann, 1783. The number of valid species is uncertain; see Bothma, 1971, Part 12:1–8; Corbet, 1979, Bull. Br. Mus. (Nat. Hist.) Zool., 36:251–259. Revised by Hahn, 1934, Z. Saugetierk., 9:207. Also see Allen, 1939, A Checklist of African Mammals; Roberts, 1951, The Mammals of South Africa; Meyer, 1978, *in* Maglio and Cooke. The generic definitions are also controversial; Roche, 1972, Mammalia, 36:22–49 retained only *Procavia* and *Dendrohyrax* but Hoeck, 1978, Carnegie Mus. Nat. Hist., 6:146–151 retained *Procavia*, *Heterohyrax*, and *Dendrohyrax* as separate genera, and is followed here.

ISIS NUMBER: 5301416001000000000.

Dendrohyrax Gray, 1868. Ann. Mag. Nat. Hist., ser. 4, 1:48.
COMMENT: A key to the genus was published by Jones, 1978, Mamm. Species, 113:1–4.
ISIS NUMBER: 5301416001001000000.

Dendrohyrax arboreus (A. Smith, 1827). Trans. Linn. Soc. Lond., 15:468.
TYPE LOCALITY: South Africa, Cape Prov., forests of Cape of Good Hope.
DISTRIBUTION: Cape Prov., Natal (South Africa); Mozambique; Zambia; Malawi; Zaire; Tanzania to Kenya and the Sudan.
ISIS NUMBER: 5301416001001001001.

Dendrohyrax dorsalis (Fraser, 1855). Proc. Zool. Soc. Lond., 1854:99.
TYPE LOCALITY: Equatorial Guinea, Bioko.
DISTRIBUTION: Gambia to N. Angola; Bioko; C. and N.E. Zaire; N. Uganda.
COMMENT: Reviewed by Jones, 1978, Mamm. Species, 113:1–4.
ISIS NUMBER: 5301416001001002001.

Dendrohyrax validus True, 1890. Proc. U.S. Nat. Mus., 13:228.
TYPE LOCALITY: Tanzania (=Tanganyika), Mt. Kilimanjaro.
DISTRIBUTION: E. Tanzania; S. Kenya; Pemba Isl.; Tumbatu Isl.; Zanzibar.
COMMENT: Differences between this species and *arboreus* need further study (JS).
ISIS NUMBER: 5301416001001003001.

Heterohyrax Gray, 1868. Ann. Mag. Nat. Hist., ser. 4, 1:50.
COMMENT: Included as a subgenus of *Dendrohyrax* by Roche, 1972, Mammalia, 36:22–49.
ISIS NUMBER: 5301416001002000000.

Heterohyrax antineae (Heim de Balsac and Begouen, 1932). Bull. Mus. Hist. Nat. Paris, ser. 2, 4:479.
TYPE LOCALITY: Algeria, C. Sahara, Ahaggar.
DISTRIBUTION: Ahaggar Mtns. (S. Algeria).
COMMENT: Schwarz, 1933, Ann. Mag. Nat. Hist., (10)12:625–626; Hatt, 1936, Bull. Am. Mus. Nat. Hist., 72:117–141; and Bothma, 1971, Part 12:1–8, stated that this species may be conspecific with *brucei*, and it was so placed by Roche, 1972, Mammalia, 36:41.

Heterohyrax brucei (Gray, 1868). Ann. Mag. Nat. Hist., ser. 4, 1:44.
TYPE LOCALITY: Ethiopia (=Abyssinia).
DISTRIBUTION: Egypt to W. Somalia to N. South Africa to E.C. Angola.
COMMENT: Allen, 1939 considered *Hyrax syriacus* a prior name for this species; Bothma, 1971, Part 12:1–8, referred *syriacus* to *Procavia*; see comment under *P. capensis*.
ISIS NUMBER: 5301416001002001001.

Heterohyrax chapini (Hatt, 1933). Am. Mus. Novit., 594:1.

TYPE LOCALITY: Zaire, Bas Congo Dist., Summit of Loadi Hill, 5 km S.W. of Matadi.

DISTRIBUTION: Known only from the vicinity of the type locality.

COMMENT: Hatt, 1936, Bull. Am. Mus. Nat. Hist., 72:132, and Bothma, 1971, Part 12:1–8, stated that this species may be conspecific with *brucei*, and it was so placed by Roche, 1972, Mammalia, 36:41.

Procavia Storr, 1780. Prodr. Meth. Mamm., p. 40.

COMMENT: Includes *Euhyrax* and *Hyrax*; see Bothma, 1971, Part 12:1–8.

ISIS NUMBER: 5301416001003000000.

Procavia capensis (Pallas, 1766). Misc. Zool., p. 30.

TYPE LOCALITY: South Africa, Cape Prov., Cape of Good Hope.

DISTRIBUTION: Syria; Lebanon; Israel; Arabia; N.E. Africa; Senegal to Somalia to N. Tanzania; S. Malawi to S. Angola, Namibia and South Africa; isolated mountains in Algeria and Libya.

COMMENT: Includes *habessinica, johnstoni, ruficeps, syriaca,* and *welwitchii*; see Roche, 1972, Mammalia, 36:22–49, and Corbet, 1979, Bull. Br. Mus. (Nat. Hist.) Zool., 36:251–259; but see also Bothma, 1971, Part 12:1–8.

ISIS NUMBER: 5301416001003001001.

ORDER TUBULIDENTATA

ISIS NUMBER: 5301414000000000000.

Family Orycteropodidae

REVIEWED BY: D. P. Domning (DPD); J. Shoshani (JS).
ISIS NUMBER: 5301414001000000000.

Orycteropus E. Geoffroy, 1796. Bull. Sci. Soc. Philom. Paris, 1:102.
> COMMENT: Sherborn, 1902, Index Animalium, p. 701, gave "*Orycteropus* Geoffroy,
> Decad. Phil. et Litt. XXVIII. 1795", which neither he nor JS were able to trace.
ISIS NUMBER: 5301414001001000000.

Orycteropus afer (Pallas, 1766). Misc. Zool., p. 64.
> TYPE LOCALITY: South Africa, Cape Prov., Cape of Good Hope.
> DISTRIBUTION: Senegal; Togo; Cameroun; Sudan; Ethiopia; Somalia; N.W. and perhaps
> S.W. Uganda to Tanzania; Rwanda; N., E., and C. Zaire; W. Angola; Namibia;
> South Africa; N.E. Botswana; Zimbabwe; N.E. Zambia; Mozambique.
> COMMENT: Reviewed by Pocock, 1924, Proc. Zool. Soc. Lond., 1924:697–706; Melton,
> 1976, Mammal Rev., 6(2):75–88.
> PROTECTED STATUS: CITES - Appendix II.
ISIS NUMBER: 5301414001001001001.

ORDER ARTIODACTYLA
ISIS NUMBER: 5301419000000000000.

Family Suidae
REVIEWED BY: J. J. Mayer (JJM); O. L. Rossolimo (OLR)(U.S.S.R.); R. G. Van Gelder
(RGVG); S. Wang (SW)(China).
ISIS NUMBER: 5301419001000000000.

Babyrousa Perry, 1811. Arcana, sig. C, Recto.
ISIS NUMBER: 5301419001001000000.

Babyrousa babyrussa (Linnaeus, 1758). Syst. Nat., 10th ed., 1:50.
TYPE LOCALITY: "Borneo" (=Buru Isl., Indonesia).
DISTRIBUTION: N. and C. Sulawesi; Buru (N. Molucca Isls.); Sula Isls.; Malengi Isl.
(Togean Isls.).
PROTECTED STATUS: CITES - Appendix I and U.S. ESA - Endangered.
ISIS NUMBER: 5301419001001001001.

Hylochoerus Thomas, 1904. Nature, 70:577.
ISIS NUMBER: 5301419001002000000.

Hylochoerus meinertzhageni Thomas, 1904. Nature, 70:577.
TYPE LOCALITY: Kenya, Kakumegas (Kakamega) Forest, near Kaimosi, 7,000 ft.
DISTRIBUTION: Liberia through N. Zaire to W.C. Kenya, N. Tanzania and S.W.
Ethiopia.
COMMENT: See Thomas, 1904, Proc. Zool. Soc. Lond., 1903(2):193–199, for designation
of the type specimen.
ISIS NUMBER: 5301419001002001001.

Phacochoerus F. Cuvier, 1826. Dict. Sci. Nat., 39:383.
ISIS NUMBER: 5301419001003000000.

Phacochoerus aethiopicus (Pallas, 1767). Spicil. Zool., 2:2.
TYPE LOCALITY: South Africa, Cape of Good Hope.
DISTRIBUTION: Subsaharan Africa, except equatorial and coastal W. Africa, and S. Cape
Prov. (South Africa).
ISIS NUMBER: 5301419001003001001.

Potamochoerus Gray, 1854. Proc. Zool. Soc. Lond., 1852:129.
ISIS NUMBER: 5301419001004000000.

Potamochoerus porcus (Linnaeus, 1758). Syst. Nat., 10th ed., 1:50.
TYPE LOCALITY: West Africa, Guinea.
DISTRIBUTION: Subsaharan Africa, excluding most of Namibia; Madagascar; Comoro
Isls. (probably introduced).
ISIS NUMBER: 5301419001004001001.

Sus Linnaeus, 1758. Syst. Nat., 10th ed., 1:49.
ISIS NUMBER: 5301419001005000000.

Sus barbatus Muller, 1838. Tijdschr. Nat. Gesch. Physiol., 5:149.
TYPE LOCALITY: Indonesia, Kalimantan, Banjarmasin.
DISTRIBUTION: Sumatra; Banka Isl.; Rhio Arch.; Malay Peninsula; Borneo; Palawan,
Balabac Isl., Calamian Isls. (Philippines).
COMMENT: Heptner and Naumov, eds., 1961, [Mammals of the Soviet Union], Vyssha
Shkola, p. 27, included *verrucosus* in this species, but Corbet and Hill, 1980:117,
listed it as a distinct species without comment.
ISIS NUMBER: 5301419001005001001.

Sus salvanius (Hodgson, 1847). J. Asiat. Soc. Bengal, 16:423.
TYPE LOCALITY: India, Sikkim Terai.
DISTRIBUTION: Sikkim; Bhutan; S. Nepal; N. India.

COMMENT: Subgenus *Porcula*; see Ellerman and Morrison-Scott, 1951:348.
PROTECTED STATUS: CITES - Appendix I and U.S. ESA - Endangered.
ISIS NUMBER: 5301419001005002001.

Sus scrofa Linnaeus, 1758. Syst. Nat., 10th ed., 1:49.
REVIEWED BY: A. C. Ziegler (ACZ).
TYPE LOCALITY: Germany.
DISTRIBUTION: Steppe and broadleaved forest regions of the Palearctic through S.E.
    Asia to Java, Flores, and Solomon Isls.; populations east of Bali are probably all
    introduced feral stock (JJM). Extinct in the British Isles, Scandinavia and Egypt;
    range generally more fragmented than formerly; introduced worldwide.
COMMENT: Includes *cristatus, vittatus,* and domestic pig, *"domestica"*; see Heptner and
    Naumov, eds., 1961, [Mammals of the Soviet Union], Vyssha Shkola, p. 27, and
    Corbet, 1978:196.
ISIS NUMBER: 5301419001005003001.

Sus verrucosus Muller and Schlegel, 1842. Verh. Zoogd. Ind. Arch., 1:107.
TYPE LOCALITY: Indonesia, Java.
DISTRIBUTION: Java; Flores; Sulawesi; Saleyer Isl. (Sulawesi); Halmahera and nearby
    islands; Seram; Philippines; Madoera Isl. (Java).
COMMENT: Includes *celebensis*; see Laurie and Hill, 1954:87. According to Lekagul and
    McNeely, 1977, Mammals of Thailand, pg. 661, known only from Java (*i.e.,*
    excluding *celebensis*), where it occurs with *scrofa* without interbreeding. Regarded
    as conspecific with *barbatus* by Heptner and Naumov, eds., 1961, [Mammals of the
    Soviet Union], Vyssha Shkola, p. 27.
ISIS NUMBER: 5301419001005004001.

Family Tayassuidae
REVIEWED BY: J. J. Mayer (JJM); R. G. Van Gelder (RGVG); R. M. Wetzel (RMW).
ISIS NUMBER: 5301419002000000000.

Catagonus Ameghino, 1904. An. Mus. Soc. Cient. Argent., 58:188.

Catagonus wagneri (Rusconi, 1930). An. Mus. Nac. Hist. Nat. "Bernardino Rivadavia,"
    36:231.
TYPE LOCALITY: Argentina, Santiago del Estero, Llajta Manca.
DISTRIBUTION: Gran Chaco of Paraguay, Argentina, and Bolivia.
COMMENT: Originally described from pre-Hispanic and subfossil remains;
    subsequently discovered alive (Wetzel *et al.,* 1975, Science, 189:379–381; see also
    Wetzel, 1977, Bull. Carnegie Mus. Nat. Hist., 3:1–36); Wetzel, 1981, Carnegie
    Mag., 55:24–32.

Tayassu G. Fischer, 1814. Zoognosia, 3:284.
REVIEWED BY: P. Grubb (PG); J. Ramirez-Pulido (JRP).
COMMENT: Includes *Dicotyles*; see Wetzel, 1977, Bull. Carnegie Mus. Nat. Hist., 3:7;
    Van Gelder, 1977, Am. Mus. Novit., 2635:13–14; Hershkovitz, 1963, Proc. Biol.
    Soc. Wash., 76:85–87.
ISIS NUMBER: 5301419002001000000.

Tayassu pecari (Link, 1795). Beitr. Naturgesch., 2:104.
TYPE LOCALITY: French Guiana, Cayenne.
DISTRIBUTION: Oaxaca and Veracruz (Mexico) to W. Ecuador, Brazil and N.E.
    Argentina.
COMMENT: Includes *albirostris*; see Husson, 1978, The Mammals of Suriname, p. 353.
ISIS NUMBER: 5301419002001001001 as *T. albirostris.*

Tayassu tajacu (Linnaeus, 1758). Syst. Nat., 10th ed., 1:50.
TYPE LOCALITY: Brazil, Pernambuco.
DISTRIBUTION: N. Argentina and N.W. Peru to N.C. Texas, S.W. New Mexico and
    Arizona (U.S.A.).

COMMENT: See Cabrera, 1961:319; see also Thomas, 1911, Proc. Zool. Soc. Lond., 1911:140, for discussion of type locality. JJM and RMW follow Thomas in designating Mexico as the type locality of this species.
PROTECTED STATUS: CITES - Appendix III (Guatemala).
ISIS NUMBER: 5301419002001002001.

Family Hippopotamidae
REVIEWED BY: R. G. Van Gelder (RGVG).
ISIS NUMBER: 5301419003000000000.

Choeropsis Leidy, 1853. Proc. Acad. Nat. Sci. Phila., 6:52.
REVIEWED BY: P. Grubb (PG).
COMMENT: Coryndon, 1977, Proc. K. Ned. Akad. Wet. Ser. B Palaeontol. Geol. Phys. Chem., 80:61–71, argued that *Hexaprotodon* is the oldest valid name for this genus.
ISIS NUMBER: 5301419003001000000.

Choeropsis liberiensis (Morton, 1844). Proc. Acad. Nat. Sci. Phila., 2:14.
TYPE LOCALITY: Liberia, St. Paul's River.
DISTRIBUTION: Discontinuous, in forests of West Africa, from Sierra Leone to Nigeria.
PROTECTED STATUS: CITES - Appendix II.
ISIS NUMBER: 5301419003001001001.

Hippopotamus Linnaeus, 1758. Syst. Nat., 10th ed., 1:74.
ISIS NUMBER: 5301419003002000000.

Hippopotamus amphibius Linnaeus, 1758. Syst. Nat., 10th ed., 1:74.
TYPE LOCALITY: Egypt, Nile River.
DISTRIBUTION: River systems of subsaharan Africa; formerly to lower Nile River.
COMMENT: Extirpated in some localities, especially South Africa (except N.E.).
PROTECTED STATUS: CITES - Appendix III (Ghana).
ISIS NUMBER: 5301419003002001001.

Family Camelidae
REVIEWED BY: I. U. Kohler (IUK); R. G. Van Gelder (RGVG); B. R. Stein (BRS).
ISIS NUMBER: 5301419004000000000.

Camelus Linnaeus, 1758. Syst. Nat., 10th ed., 1:65.
ISIS NUMBER: 5301419004001000000.

Camelus bactrianus Linnaeus, 1758. Syst. Nat., 10th ed., 1:65.
REVIEWED BY: P. Grubb (PG).
TYPE LOCALITY: "Bactria" (=U.S.S.R., Uzbekistan, Bokhara) (domesticated stock).
DISTRIBUTION: Exists in the wild only in S.W. Mongolia, Kansu, Tsinghai, and Sinkiang (China)(SW); domesticated in Iran, Afghanistan, Pakistan, U.S.S.R., Mongolia, and China.
COMMENT: Includes *ferus* Przewalski, 1883, based on wild specimen; *bactrianus* Linnaeus, 1758, has priority (IUK). Produces fertile hybrids with *dromedarius*; though their distributions merge, breeding is regulated (as all individuals are domesticated in the zone of contact) (IUK). Corbet, 1978:197, citing Gray, 1954, stated that male hybrids are sterile.
PROTECTED STATUS: U.S. ESA - Endangered as *C. ferus* (=*bactrianus*).
ISIS NUMBER: 5301419004001001001.

Camelus dromedarius Linnaeus, 1758. Syst. Nat., 10th ed., 1:65.
TYPE LOCALITY: "Africa," deserts of Libya and Arabia. (domesticated stock).
DISTRIBUTION: Extinct in the wild and unknown as fossil; domesticated in North Africa, Arabia, Mediterranean, Balkans, and Middle East; introduced into Australia.
COMMENT: Produces fertile hybrids with *bactrianus*; though their distributions merge, breeding is regulated (as all individuals are domesticated in the zone of contact)

(IUK). Corbet, 1978:197, citing Gray, 1954, stated that male hybrids are sterile. Bohldeen, 1961, Z. Tierzucht. Zuchtungsbiol., 76:107–113, considered *dromedarius* a synonym of *bactrianus.*
ISIS NUMBER: 5301419004001002001.

Lama G. Cuvier, 1800. Lecon's Anat. Comp., I, tab. 1.
REVIEWED BY: P. Grubb (PG).
COMMENT: Should probably include *Vicugna* (PG).
ISIS NUMBER: 5301419004002000000.

Lama glama (Linnaeus, 1758). Syst. Nat., 10th ed., 1:65.
TYPE LOCALITY: Peru, Andes (domesticated stock).
DISTRIBUTION: Domesticated in S. Peru; W. Bolivia; N.W. Argentina.
COMMENT: Should probably include *guanicoe* (PG).
ISIS NUMBER: 5301419004002001001.

Lama guanicoe (Muller, 1776). Linne's Vollstand. Natursyst. Suppl., p. 50.
TYPE LOCALITY: Chile, Andes.
DISTRIBUTION: Cordilleras of the Andes, in S. Peru, Bolivia, Argentina, and Chile; Patagonia, Tierra del Fuego; Navarino Isl.
PROTECTED STATUS: CITES - Appendix II.

Lama pacos (Linnaeus, 1758). Syst. Nat., 10th ed., 1:65.
TYPE LOCALITY: Peru (domesticated stock).
DISTRIBUTION: Domesticated in S. Peru; W. Bolivia.
COMMENT: Often regarded as a synonym for *glama;* see Corbet and Hill, 1980:118.

Vicugna Lesson, 1842. Nouv. Tabl. Regn. Anim. Mammal., p. 167.

Vicugna vicugna (Molina, 1782). Sagg. Stor. Nat. Chile, p. 313.
TYPE LOCALITY: Peru, to S. Ecuador and C. Bolivia. (= Chilean Andes, from Coquimo to Copiapo.)
DISTRIBUTION: S. Peru; W. Bolivia; N.W. Argentina; N. Chile.
COMMENT: Should probably be included in *Lama* (PG).
PROTECTED STATUS: CITES - Appendix I and U.S. ESA - Endangered.
ISIS NUMBER: 5301419004002002001 as *Lama vicugna.*

Family Tragulidae
REVIEWED BY: B. R. Stein (BRS); S. Wang (SW)(China).
ISIS NUMBER: 5301419005000000000.

Hyemoschus Gray, 1845. Ann. Mag. Nat. Hist., ser. 1, 16:350.
REVIEWED BY: R. G. Van Gelder (RGVG).
ISIS NUMBER: 5301419005001000000.

Hyemoschus aquaticus (Ogilby, 1841). Proc. Zool. Soc. Lond., 1840:35.
TYPE LOCALITY: Sierra Leone, Bulham Creek.
DISTRIBUTION: Sierra Leone to Gabon to Zaire and Uganda.
PROTECTED STATUS: CITES - Appendix III (Ghana).
ISIS NUMBER: 5301419005001001001.

Tragulus Pallas, 1779. Spicil. Zool., 13:27.
COMMENT: Original citation, Brisson, 1762, Regn. Anim., 2nd ed., 12:65–68, is not available.
ISIS NUMBER: 5301419005002000000.

Tragulus javanicus (Osbeck, 1765). Reise nach Ostind. China Rostock, p. 357.
TYPE LOCALITY: Indonesia, W. Java, Udjon Kulon Peninsula.
DISTRIBUTION: Indochina; Thailand; Yunnan (China)(SW); Malaysia; Sumatra; Borneo; Java; many adjacent islands.
ISIS NUMBER: 5301419005002001001.

Tragulus meminna (Erxleben, 1777). Syst. Regn. Anim., 1:322.
TYPE LOCALITY: Sri Lanka.
DISTRIBUTION: Sri Lanka; peninsular India; Nepal.
ISIS NUMBER: 5301419005002002001.

Tragulus napu (F. Cuvier, 1822). *In* E. Geoffroy and F. Cuvier, Hist. Nat. Mamm., 37:2.
TYPE LOCALITY: Indonesia, S. Sumatra.
DISTRIBUTION: Indochina; Thailand; Malaysia; Borneo; Balabac Isl. (Philippines);
    Sumatra; many adjacent isls.
ISIS NUMBER: 5301419005002003001.

Family Cervidae

REVIEWED BY: C. R. Durst (CRD); O. L. Rossolimo (OLR) (U.S.S.R.); B. R. Stein (BRS); R.
    G. Van Gelder (RGVG); S. Wang (SW)(China).
COMMENT: Includes Moschidae; see Groves, 1975, J. Bombay Nat. Hist. Soc.,
    72:662–682. Reviewed by Whitehead, 1972, Deer of the World, 194 pp.
ISIS NUMBER: 5301419006000000000.

Alces Gray, 1821. Lond. Med. Repos., 15:307.
ISIS NUMBER: 5301419006006000000.

Alces alces (Linnaeus, 1758). Syst. Nat., 10th ed., 1:66.
TYPE LOCALITY: Sweden.
DISTRIBUTION: N. Eurasia from Scandinavia and Baltic region east to Anadyr region, E.
    Siberia (U.S.S.R); south to Ukraine, S. Siberia, Mongolia, Manchuria;
    Heilungkiang (China); Alaska (U.S.A.); Canada; N. U.S.A.
ISIS NUMBER: 5301419006006001001.

Blastocerus Wagner, 1844. Schreber's Saugthiere, Suppl., 4:366.
COMMENT: Included in *Odocoileus* by Haltenorth, 1963:44–45; but see Koopman, 1967,
    in Anderson and Jones, p. 397. Hershkovitz, 1958, Proc. Biol. Soc. Wash.,
    71:13–16, argued that the first valid use of the name was Gray, 1850.
ISIS NUMBER: 5301419006008000000.

Blastocerus dichotomus (Illiger, 1815). Abh. Preuss. Akad. Wiss., p. 117.
TYPE LOCALITY: Paraguay.
DISTRIBUTION: C. Brazil to Paraguay and N. Argentina.
PROTECTED STATUS: CITES - Appendix I and U.S. ESA - Endangered.
ISIS NUMBER: 5301419006008001001.

Capreolus Gray, 1821. Lond. Med. Repos., 15:307.
ISIS NUMBER: 5301419006009000000.

Capreolus capreolus (Linnaeus, 1758). Syst. Nat., 10th ed., 1:68.
TYPE LOCALITY: Sweden.
DISTRIBUTION: W. Europe; W. and S. U.S.S.R; Turkey; Iraq; N. Iran, east through S.
    Siberia to N. and C. China; Korea.
COMMENT: Includes *pygargus*; Stubbe and Bruholz, 1979, Zool. Zh., 58:1402, provided
    evidence that *pygargus* (including *tianschanicus* and *bedfordi*) is a distinct species.
ISIS NUMBER: 5301419006009001001.

Cervus Linnaeus, 1758. Syst. Nat., 10th ed., 1:66.
REVIEWED BY: P. Grubb (PG).
COMMENT: Includes *Axis*, *Dama* and *Hyelaphus*; see Corbet, 1978:199. Van Gelder, 1977,
    Am. Mus. Novit., 2635:1–25, also included *Elaphurus*.
ISIS NUMBER: 5301419006004000000.

Cervus albirostris Przewalski, 1883. Third Journey in Central Asia, p. 124.
TYPE LOCALITY: China, Kansu, 3 km above mouth of Kokusu River, Humboldt Mtns.,
    Nan Shan.
DISTRIBUTION: Tibet, Tsinghai, Kansu, Szechwan (China) (SW).
ISIS NUMBER: 5301419006004001001.

Cervus axis Erxleben, 1777. Syst. Regn. Anim., 1:312.
    REVIEWED BY: A. C. Ziegler (ACZ).
    TYPE LOCALITY: India, Bihar, "banks of the Ganges."
    DISTRIBUTION: India; Sri Lanka; Nepal; Sikkim; introduced in N.E. New Guinea;
        Hawaii, and New Zealand.
    COMMENT: Formerly placed in *Axis*.
    ISIS NUMBER: 5301419006004002001.

Cervus dama Linnaeus, 1758. Syst. Nat., 10th ed., 1:67.
    REVIEWED BY: A. C. Ziegler (ACZ).
    TYPE LOCALITY: Sweden (introduced).
    DISTRIBUTION: Europe; Iraq; W. Iran; formerly N. Africa; introduced in Australia, New
        Zealand, N. E. New Guinea (now possibly extirpated), and New World.
    COMMENT: Includes *mesopotamicus*; see Corbet, 1978:199. Formerly placed in *Dama*
    PROTECTED STATUS: CITES - Appendix I as *Dama mesopotamica* only. U.S. ESA -
        Endangered as *Dama dama mesopotamica* subspecies only.
    ISIS NUMBER: 5301419006004003001 as *C. dama*.
           5301419006004008001 as *C. mesopotamicus* (*sic*).

Cervus duvauceli G. Cuvier, 1823. Oss. Foss., 2nd ed., 4:505.
    TYPE LOCALITY: N. India.
    DISTRIBUTION: N. and C. India; S.W. Nepal.
    PROTECTED STATUS: CITES - Appendix I and U.S. ESA - Endangered.
    ISIS NUMBER: 5301419006004004001.

Cervus elaphus Linnaeus, 1758. Syst. Nat., 10th ed., 1:67.
    REVIEWED BY: D. E. Babb (DEB).
    TYPE LOCALITY: S. Sweden.
    DISTRIBUTION: Forest, moorland and alpine habitats of Europe, Caucasus and C. Asia to
        W. and N. China and Ussuri region (U.S.S.R.); N.W. Tunisia; N.E. Algeria;
        Corsica; Sardinia; North America from British Columbia (Canada) south to
        California, New Mexico, and Louisiana, east to New York and S. Carolina (U.S.A.)
        (now restricted to western areas and reserves in N. America). Introduced to South
        America and New Zealand.
    COMMENT: Includes *affinis, bactrianus, hanglu, macneilli, wallichi, xanthopygus,*
        yarkandensis; see Corbet, 1978:201. Includes *canadensis, merriami,* and *nannodes*; see
        McCullough, 1969, Univ. Calif. Publ. Zool., 88:1–209; Hall, 1981:1084–1087.
    PROTECTED STATUS: CITES - Appendix I and U.S. ESA - Endangered as *C. e. hanglu*
        subspecies only.
        CITES - Appendix II and U.S. ESA - Endangered as *C. e. bactrianus* subspecies
           only.
        CITES - Appendix III (Tunisia) and U.S. ESA -Endangered as *C. e. barbarus*
           subspecies only. U.S. ESA - Endangered as *C. e. corsicanus* subspecies only.
           U.S. ESA - Endangered as *C. e. macneilli* subspecies only. U.S. ESA -
           Endangered as *C. e. wallichi* subspecies only. U.S. ESA - Endangered as *C. e.*
           yarkandensis subspecies only.
    ISIS NUMBER: 5301419006004005001.

Cervus eldi M'Clelland, 1842. Calcutta J. Nat. Hist., 2:417.
    TYPE LOCALITY: India, Assam, Manipur.
    DISTRIBUTION: India to Vietnam; Hainan Isl. (China).
    PROTECTED STATUS: CITES - Appendix I and U.S. ESA - Endangered.
    ISIS NUMBER: 5301419006004006001.

Cervus nippon Temminck, 1838. Coup d'oeil sur la faune des iles de la Sonde et de
    L'empire du Japon, p. 22.
    TYPE LOCALITY: Japan.
    DISTRIBUTION: Taiwan; E. China; Manchuria; Korea; adjacent E. Siberia; Japan;
        Quelpart Isls; Ryukyu Isls., Tsushima Isls. (Japan); Vietnam.
    COMMENT: Includes *hortulorum, taiouanus* and *pulchellus*, which were considered species

by Imaizumi, 1970, Bull. Sci. Mus. Tokyo, 13:185–194. Reviewed by Feldhamer, 1980, Mamm. Species, 128:1–7.
PROTECTED STATUS: U.S. ESA - Endangered as *C. n. grassianus* subspecies only. U.S. ESA - Endangered as *C. n. keramae* subspecies only. U.S. ESA - Endangered as *C. n. kopschi* subspecies only. U.S. ESA - Endangered as *C. n. mandarinus* subspecies only. U.S. ESA - Endangered as *C. n. taiouanus* subspecies only.
ISIS NUMBER: 5301419006004009001.

Cervus porcinus (Zimmermann, 1780). Spec. Zool. Geogr., p. 532.
TYPE LOCALITY: India, Bengal.
DISTRIBUTION: India to Vietnam; Yunnan (China) (SW); Sri Lanka; Bawean Isl. (Indonesia); Calamian Isls. (Philippines).
COMMENT: Includes *calamianensis* and *kuhli*; see Haltenorth, 1963:54. Formerly placed in *Axis* and *Hyelaphus*; see comment under genus.
PROTECTED STATUS: CITES - Appendix I as *Axis (=Hyelaphus) calamianensis* only.
CITES - Appendix I as *Axis (=Hyelaphus) kuhli* only.
CITES - Appendix I and U.S. ESA - Endangered as *Axis (=Cervus) porcinus annamiticus* subspecies only. U.S. ESA - Endangered as *Axis calamianensis* only.
ISIS NUMBER: 5301419006004010000.

Cervus schomburgki Blyth, 1863. Proc. Zool. Soc. Lond., 1863:155.
TYPE LOCALITY: Thailand.
DISTRIBUTION: Thailand.
COMMENT: Included in *duvauceli* by Haltenorth, 1963:58; but see Corbet and Hill, 1980:120. Probably extinct; last specimen taken in 1932; see Harper, 1945, Extinct and vanishing mammals of the Old World, Spec. Publ. No 12, Amer. Comm. Intern. Wildl. Prot., New York, p. 436.

Cervus timorensis Blainville, 1822. J. Phys. Chim. Hist. Nat. Arts Paris, 94:267.
REVIEWED BY: A. C. Ziegler (ACZ).
TYPE LOCALITY: Indonesia, Lesser Sunda Isls., Timor Isl.
DISTRIBUTION: Sulawesi; Timor; Flores; Java; Bali; Buru; Seram; Molucca Isls.; various adjacent islands. Introduced into N. Australia, New Zealand, New Britain Isl.; S.C. New Guinea, and islands off N.E. coast of New Guinea.
ISIS NUMBER: 5301419006004011001.

Cervus unicolor (Kerr, 1792). Anim. Kingdom, p. 300.
TYPE LOCALITY: Sri Lanka.
DISTRIBUTION: Philippines; Bonin Isls.; Borneo; Sumatra to India and S. and C. China; Hainan Isls. (China); Taiwan; Guam (Marianna Isls.); Siberut Isl.; Sipora Isl.; Pagi Isls.; Nias Isl.; Sri Lanka.
COMMENT: Includes *alfredi* and *mariannus*; see Whitehead, 1972, Deer of the World, p. 160. PG believes *mariannus* and *alfredi* are valid species.
ISIS NUMBER: 5301419006004012001 as *C. unicolor*.
5301419006004007001 as *C. mariannus*.

Elaphodus Milne-Edwards, 1871. Nouv. Arch. Mus. Hist. Nat. Paris, 7(Bull.):93.
ISIS NUMBER: 5301419006002000000.

Elaphodus cephalophus Milne-Edwards, 1871. Nouv. Arch. Mus. Hist. Nat. Paris, 7(Bull.):93.
TYPE LOCALITY: China, Szechwan, Moupin.
DISTRIBUTION: S. and C. China; N. Burma.
ISIS NUMBER: 5301419006002001001.

Elaphurus Milne-Edwards, 1866. Ann. Sci. Nat. Zool., 5:382.
    ISIS NUMBER: 5301419006005000000.

  Elaphurus davidianus (Milne-Edwards, 1866). Ann. Sci. Nat. Zool., 5:382.
    TYPE LOCALITY: China, Chihli, Pekin, Imperial Hunting Park.
    DISTRIBUTION: Formerly N. China; now present only in parks and zoos
    COMMENT: Included in *Elaphurus* by Corbet, 1978:201; but see Van Gelder, 1977, Am.
      Mus. Novit., 2635:1–25.
    ISIS NUMBER: 5301419006005001001 as *E. davidanus (sic)*.

Hippocamelus Leuckart, 1816. Diss. Inaug. de Equo bisulco Molinae, p. 24.
    COMMENT: Included in *Odocoileus* by Haltenorth, 1963:44, 46.
    ISIS NUMBER: 5301419006010000000.

  Hippocamelus antisensis (d'Orbigny, 1834). Ann. Mus. Hist. Nat. Paris, 3:91.
    TYPE LOCALITY: Bolivian Andes, near La Paz.
    DISTRIBUTION: Andes from Ecuador to N.W. Argentina.
    PROTECTED STATUS: CITES - Appendix I and U.S. ESA - Endangered.
    ISIS NUMBER: 5301419006010001001 as *H. antisiensis (sic)*.

  Hippocamelus bisulcus (Molina, 1882). Sagg. Stor. Nat. Chile, p. 320.
    TYPE LOCALITY: Chilean Andes, Colchagua Prov.
    DISTRIBUTION: Andes of S. Chile and S. Argentina.
    PROTECTED STATUS: CITES - Appendix I and U.S. ESA - Endangered.
    ISIS NUMBER: 5301419006010002001.

Hydropotes Swinhoe, 1870. Proc. Zool. Soc. Lond., 1870:90.
    ISIS NUMBER: 5301419006011000000.

  Hydropotes inermis Swinhoe, 1870. Proc. Zool. Soc. Lond., 1870:89.
    TYPE LOCALITY: China, Kiangsu, Chingkiang, Yangtze River, Deer Isl.
    DISTRIBUTION: Lower Yangtze Basin, west to Hupei (China); Korea; introduced in
      England and France.
    ISIS NUMBER: 5301419006011001001.

Mazama Rafinesque, 1817. Am. Mon. Mag., 1(5):363.
    REVIEWED BY: P. Grubb (PG); K. R. Kranz (KRK).
    ISIS NUMBER: 5301419006012000000.

  Mazama americana (Erxleben, 1777). Syst. Regn. Anim., 1:324.
    REVIEWED BY: J. Ramirez-Pulido (JRP).
    TYPE LOCALITY: Brazil, Amazonia.
    DISTRIBUTION: S. Tamaulipas and Yucatan (Mexico) to S. Brazil, N. Argentina, S.
      Bolivia, and Paraguay; Trinidad; Tobago.
    PROTECTED STATUS: CITES - Appendix III (Guatemala) as *M. a. cerasina* subspecies only.
    ISIS NUMBER: 5301419006012001001.

  Mazama chunyi (Hershkovitz, 1959). Proc. Biol. Soc. Wash., 72:45.
    TYPE LOCALITY: Bolivia, La Paz, Cocopunco, 3200 m.
    DISTRIBUTION: Bolivian Andes; S. Peru.
    COMMENT: Prior to 1959 this species was known as *Pudu mephistophiles* (De Winton,
      1896); see comment under that species.

  Mazama gouazoubira (G. Fischer, 1814). Zoognosia, 3:465.
    TYPE LOCALITY: Brazil, Amazonia.
    DISTRIBUTION: Colombia; Venezuela; Guianas; Brazil; Bolivia; Paraguay; Uruguay; N.
      Argentina; Peru; Ecuador; Mexiana Isl. (Brazil); San Jose Isl. (Panama).
    ISIS NUMBER: 5301419006012003001.

  Mazama rufina (Bourcier and Pucheran, 1852). Rev. Zool., p. 561.
    TYPE LOCALITY: Ecuador, Pichincha, Pichincha Mts., Lloa valley.
    DISTRIBUTION: W. Venezuela; perhaps adjacent Colombia; Ecuador; Peru; N. Bolivia;
      S.E. Brazil; N.E. Argentina; E. Paraguay.

COMMENT: Includes *bricenii* and *nana* (Hensel, 1872) (not *nana* Lesson, 1842, or *nana*
Lund, 1839, which Cabrera, 1961:341–342, regarded as *nomina nuda);* see Cabrera,
1961:341; Haltenorth, 1963:48.

ISIS NUMBER: 5301419006012002001 as *M. bricenii.*
5301419006012004001 as *M. nana.*

Moschus Linnaeus, 1758. Syst. Nat., 10th ed., 1:66.
REVIEWED BY: P. Grubb (PG).
COMMENT: Included in family Moschidae by Flerov, 1952, [Musk deer and deer],
Akad. Nauk, Moscow, 225 pp.; tentatively followed by Corbet, 1978:198. Here
regarded as a subfamily of Cervidae, as revised by Groves, 1975, J. Bombay Nat.
Hist. Soc., 72:662–682.
PROTECTED STATUS: CITES - Appendix II as *Moschus* spp.
ISIS NUMBER: 5301419006001000000.

Moschus berezovskii Flerov, 1929. C. R. Acad. Sci. U.S.S.R., 1928A:519.
TYPE LOCALITY: China, Szechwan, near Lungan, Ho-tsi-how Pass.
DISTRIBUTION: S.W. China, N. Vietnam.
COMMENT: A valid species according to Cai and Feng, 1981, Acta Zootax. Sin.,
6:106–110.
PROTECTED STATUS: CITES - Appendix II as *Moschus* spp.

Moschus chrysogaster (Hodgson, 1839). J. Asiat. Soc. Bengal, 8:203.
TYPE LOCALITY: "Nepal" (Probably Tibetan Plateau).
DISTRIBUTION: Wooded slopes of Himalayas in India, Nepal, Bhutan, Sikkim, S. Tibet,
C. and S.E. China (SW).
COMMENT: Formerly included *berezovskii;* see Groves, 1975, J. Bombay Nat. Hist. Soc.,
72:662–682. Regarded as separate species by Cai and Feng, 1981, Acta Zootax. Sin.,
6:106–110. According to PG, *chrysogaster* is the prior name for *sifanicus* (see
below), and the next available name for this species is *leucogaster* Hodgson, 1839.
PROTECTED STATUS: CITES - Appendix II as *Moschus* spp.

Moschus moschiferus Linnaeus, 1758. Syst. Nat., 10th ed., 1:66.
TYPE LOCALITY: U.S.S.R., S.W. Siberia, Altai Mtns. ("Tataria versus Chineum").
DISTRIBUTION: Forests of E. Siberia; N. Mongolia; N. China west to Kansu; Anwei
(SW); Korea; Sakhalin Isl.
COMMENT: Includes *sibiricus;* see Corbet, 1978:198. Type locality restricted by Heptner
and Naumov, eds., 1961, [Mammals of the Soviet Union], p. 83.
PROTECTED STATUS: CITES - Appendix II as *Moschus* spp.
CITES - Appendix I as *M. moschiferus* (Himalayan population only) and U.S. ESA -
Endangered as *M. m. moschiferus* subspecies only.
ISIS NUMBER: 5301419006001001001.

Moschus sifanicus (Buchner, 1891). Melanges Biol. Soc. St. Petersb., 13:162.
TYPE LOCALITY: China, S. Kansu, Hsifan.
DISTRIBUTION: Alpine zone of Tibetan plateau region from E. Afghanistan and N.
Pakistan to W. China and N. Burma; also Honan (China) (SW).
COMMENT: According to PG, correct name probably *chrysogaster* Hodgson, 1839; see
comment under *chrysogaster.*
PROTECTED STATUS: CITES - Appendix II as *Moschus* spp.

Muntiacus Rafinesque, 1815. Analyse de la Nature, p. 56.
REVIEWED BY: P. Grubb (PG); A. C. Ziegler (ACZ).
COMMENT: *Muntiacus* Rafinesque is a *nomen nudum,* but was conserved by Opinion 450
of the International Commission on Zoological Nomenclature.
ISIS NUMBER: 5301419006003000000.

Muntiacus crinifrons (Sclater, 1885). Proc. Zool. Soc. Lond., 1885:1, pl. 1.
TYPE LOCALITY: China, Chekiang, near Ningpo.
DISTRIBUTION: Yunnan, Chekiang, Anwei, Kwangtung (China) (SW).
COMMENT: Included in *muntjak* by Haltenorth, 1963:42; but see Shou, 1964:454.
ISIS NUMBER: 5301419006003001001.

Muntiacus feai (Thomas and Doria, 1889). Ann. Mus. Civ. Stor. Nat. Genova, 7:92.
   TYPE LOCALITY: Burma, Tenasserim, Thagata Juva, southeast of Mt. Mulaiyit.
   DISTRIBUTION: Thailand; peninsular Burma.
   COMMENT: Included in *muntjak* by Haltenorth, 1963:42. Should probably include
      rooseveltorum (PG). For spelling of the species name, see Grubb, 1977, Ann. Mus.
      Civ. Stor. Nat. Genova, 81:202–207.
   PROTECTED STATUS: U.S. ESA - Endangered.
   ISIS NUMBER: 5301419006003002001 as *M. feae (sic)*.

Muntiacus muntjak (Zimmermann, 1780). Geogr. Gesch. Mensch. Vierf. Thiere, 2:131.
   TYPE LOCALITY: Indonesia, Java.
   DISTRIBUTION: Sri Lanka; India to S. China and Hainan Isl.; Indochina; Borneo and
      Lombok.
   COMMENT: Includes *pleiharicus*, listed as a distinct species by Chasen, 1940:203; but see
      Medway, 1977:149. May include *reevesi, feai, rooseveltorum* and *crinifrons*; see
      Haltenorth, 1963:40. PG believes this species is a composite, including two
      sympatric species.
   ISIS NUMBER: 5301419006003003001.

Muntiacus reevesi (Ogilby, 1839). Proc. Zool. Soc. Lond., 1838:105.
   TYPE LOCALITY: China, Kwangtung, near Canton.
   DISTRIBUTION: Shensi, Kansu (SW) and S. China; Taiwan; introduced to England and
      France.
   COMMENT: Included in *muntjak* by Haltenorth, 1963:42; but see Corbet, 1978:199.
   ISIS NUMBER: 5301419006003004001.

Muntiacus rooseveltorum (Osgood, 1932). Field Mus. Nat. Hist. Publ. Zool. Ser., 18:332.
   TYPE LOCALITY: Laos, Muong Yo.
   DISTRIBUTION: Known only from type locality.
   COMMENT: Haltenorth, 1963:40, included this form in *muntjak*. PG believes this species
      should be included in *feai*. Corbet, 1978:199, implied that it was distinct, and it
      was so listed by Corbet and Hill, 1980:119.
   ISIS NUMBER: 5301419006003005001.

Odocoileus Rafinesque, 1832. Atl. J., 1:109.
   REVIEWED BY: J. Ramirez-Pulido (JRP).
   COMMENT: Hall, 1981:1087, employed *Dama* Zimmerman, 1780, an invalid name, for
      this genus; see Bull. Zool. Nomencl., 1960:267–275.
   ISIS NUMBER: 5301419006013000000.

Odocoileus hemionus (Rafinesque, 1817). Am. Mon. Mag., 1:436.
   TYPE LOCALITY: U.S.A., South Dakota, mouth of Big Sioux River.
   DISTRIBUTION: Baja California and Sonora to N. Tamaulipas (Mexico); W. U.S.A. (to
      Minnesota); W. Canada; Alaskan Panhandle (U.S.A.).
   PROTECTED STATUS: U.S. ESA - Endangered as *O. h. cedrosensis* subspecies only.
   ISIS NUMBER: 5301419006013001001.

Odocoileus virginianus (Zimmermann, 1780). Geogr. Gesch. Mensch. Vierf. Thiere, 2:24,
   129.
   TYPE LOCALITY: U.S.A., Virginia.
   DISTRIBUTION: W. and S. Canada; N.W., S.W., C. and E. U.S.A. to Bolivia, Guianas and
      N. Brazil.
   PROTECTED STATUS: CITES - Appendix III (Guatemala) as *O. v. mayensis* subspecies only.
      U.S. ESA - Endangered as *O. v. clavium* subspecies only. U.S. ESA - Endangered as
      O. v. leucurus subspecies only.
   ISIS NUMBER: 5301419006013002001.

Ozotoceros Ameghino, 1891. Rev. Argent. Hist. Nat., 1:243.
   COMMENT: *Ozotoceros* is the name to be used for *Blastoceros*, Fitzinger, 1860, which has
      been used with reference to both *Ozotoceros* and the genus *Blastocerus* (Cabrera,
      1961:329), if *Blastoceros* is regarded as an invalid emendation of *Blastocerus*; see
      Hershkovitz, 1958, Proc. Biol. Soc. Wash., 71:13–16. Included in *Odocoileus* by
      Haltenorth, 1963:46; but see Corbet and Hill, 1980:120.
   ISIS NUMBER: 5301419006007000000 as *Blastoceros*.

Ozotoceros bezoarticus (Linnaeus, 1758). Syst. Nat., 10th ed., 1:67.
TYPE LOCALITY: Brazil, Pernambuco.
DISTRIBUTION: Brazil; N. Argentina; Paraguay; Uruguay; S. Bolivia.
COMMENT: Includes *campestris;* see Cabrera, 1961:330. Type locality restricted by
Thomas, 1911, Proc. Zool. Soc. Lond., 1911:151.
PROTECTED STATUS: CITES - Appendix I and U.S. ESA - Endangered.
ISIS NUMBER: 5301419006007001001 as *Blastoceros campestris.*

Pudu Gray, 1852. Proc. Zool. Soc. Lond., 1850:242.
REVIEWED BY: K. R. Kranz (KRK).
COMMENT: Included in *Mazama* by Haltenorth, 1963:48; but see Pine *et al.,* 1979,
Mammalia, 43:372, and Corbet and Hill, 1980:121.
ISIS NUMBER: 5301419006014000000.

Pudu mephistophiles (De Winton, 1896). Proc. Zool. Soc. Lond., 1896:508.
TYPE LOCALITY: Ecuador, Napo-Pastaza Prov., Paramo de Papallacta.
DISTRIBUTION: Andes of Colombia and Ecuador.
COMMENT: Formerly included *mephistophiles* of Matschie and Sanborn (not DeWinton),
which was described by Hershkovitz, 1959, Proc. Biol. Soc. Wash., 72:45–54, as
Mazama chunyi.
PROTECTED STATUS: CITES - Appendix II.
ISIS NUMBER: 5301419006014001001.

Pudu pudu (Molina, 1782). Sagg. Stor. Nat. Chile, p. 310.
TYPE LOCALITY: Chile, Chiloe Prov.
DISTRIBUTION: S. Chile; S.W. Argentina.
PROTECTED STATUS: CITES - Appendix I and U.S. ESA - Endangered.
ISIS NUMBER: 5301419006014002001.

Rangifer H. Smith, 1827. Griffith's Cuvier Anim. Kingd., Mamm. Syn., p. 304.
ISIS NUMBER: 5301419006015000000.

Rangifer tarandus (Linnaeus, 1758). Syst. Nat., 10th ed., 1:67.
TYPE LOCALITY: Sweden, Alpine Lapland (domesticated stock).
DISTRIBUTION: Circumboreal, south to Altai Mtns. (U.S.S.R.) , N. Mongolia,
Heilungkiang (China) (SW) and Sakhalin Isl.; N. Idaho and Great Lakes region
(U.S.A.); most arctic islands. Domesticated stock introduced into Alaska, Scotland
and South Georgia Isl. (S. Atlantic Ocean).
ISIS NUMBER: 5301419006015001001.

Family Giraffidae
REVIEWED BY: C. R. Durst (CRD); R. G. Van Gelder (RGVG).
COMMENT: Affinities with *Antilocapra* on molecular grounds noted by Beintema *et al.,*
1979, J. Mol. Evol., 13:305–316.
ISIS NUMBER: 5301419007000000000.

Giraffa Brunnich, 1772. Zool. Fundamenta, p. 36. ed., p. 12, 37.
ISIS NUMBER: 5301419007001000000.

Giraffa camelopardalis (Linnaeus, 1758). Syst. Nat., 10th ed., 1:66.
TYPE LOCALITY: Ethiopia, Sennar.
DISTRIBUTION: Senegal to Somalia to South Africa to S. Angola.
COMMENT: Includes *reticulata;* reviewed by Dagg, 1971, Mamm. Species, 5:1–8.
ISIS NUMBER: 5301419007001001001.

Okapia Lankester, 1901. Nature, 64:24.
ISIS NUMBER: 5301419007002000000.

Okapia johnstoni (Sclater, 1901). Proc. Zool. Soc. Lond., 1901(1):50.
TYPE LOCALITY: Zaire, Semliki Forest, Mundala.
DISTRIBUTION: E.C. Zaire; perhaps adjacent areas.
ISIS NUMBER: 5301419007002001001.

Family Bovidae

REVIEWED BY: C. R. Durst (CRD); D. C. Gordon (DCG); P. Grubb (PG); O. L. Rossolimo (OLR)(U.S.S.R.); R. G. Van Gelder (RGVG); S. Wang (SW)(China).

COMMENT: Includes Antilocapridae; see O'Gara and Matson, 1975, J. Mammal., 58:829–846; but see also Corbet and Hill, 1980:121, and Hall, 1981:1106.

ISIS NUMBER: 5301419009000000000.

Addax Rafinesque, 1815. Analyse de la Nature, p. 56.

ISIS NUMBER: 5301419009011000000.

Addax nasomaculatus (Blainville, 1816). Bull. Sci. Soc. Philom. Paris, p. 75.

TYPE LOCALITY: Probably Senegambia.

DISTRIBUTION: Mauritania to Sudan; formerly Egypt to Tunisia.

COMMENT: Nearly extinct in wild (RGVG).

PROTECTED STATUS: CITES - Appendix II.

ISIS NUMBER: 5301419009011001001.

Aepyceros Sundevall, 1847. Kongl. Svenska Vet.-Akad. Handl. Stockholm, for 1845, p. 271.

ISIS NUMBER: 5301419009020000000.

Aepyceros melampus (Lichtenstein, 1812). Reise Prov. Russ. Reichs., 2:544.

TYPE LOCALITY: South Africa, Cape Prov., Little Namaqualand, Klipfontein.

DISTRIBUTION: South Africa to Zaire, Rwanda, Uganda, and N.E. Kenya.

COMMENT: Includes *petersi;* see Ansell, 1972, Part 15:57.

PROTECTED STATUS: U.S. ESA - Endangered as *Aepyceros melampus petersi* subspecies only.

ISIS NUMBER: 5301419009020001001.

Alcelaphus Blainville, 1816. Bull. Sci. Soc. Philom. Paris, p. 75.

COMMENT: Haltenorth, 1963:99, and Van Gelder, 1977, included *Beatragus, Sigmoceros* and *Damaliscus* in this genus, but have not been followed by recent authors; see Swanepoel *et al.,* 1980, Ann. Transvaal Mus., 32(7):187. Formerly included *Sigmoceros lichtensteini;* see comment under *Sigmoceros.*

ISIS NUMBER: 5301419009012000000.

Alcelaphus buselaphus (Pallas, 1766). Misc. Zool., p. 7.

TYPE LOCALITY: Probably Morocco.

DISTRIBUTION: Senegal to W. Somalia to N. South Africa to S. Angola; formerly N. Africa.

COMMENT: Includes *tora;* which according to PG is probably a separate species. Includes *caama;* see Ansell, 1972, Part. 15:53–54.

PROTECTED STATUS: U.S. ESA - Endangered as *A. b. swaynei* subspecies only. U.S. ESA - Endangered as *A. b. tora* subspecies only.

ISIS NUMBER: 5301419009012001001.

Ammodorcas Thomas, 1891. Proc. Zool. Soc. Lond., 1891:207, pl. 21, 22.

REVIEWED BY: D. Ernst (DE).

ISIS NUMBER: 5301419009021000000.

Ammodorcas clarkei (Thomas, 1891). Ann. Mag. Nat. Hist., ser. 6, 7:304.

TYPE LOCALITY: Somalia, Habergerhagi's Country, Buroa Wells.

DISTRIBUTION: E. Ethiopia; N. Somalia.

PROTECTED STATUS: U.S. ESA - Endangered as *A. clarki (sic).*

ISIS NUMBER: 5301419009021001001.

Ammotragus Blyth, 1840. Proc. Zool. Soc. Lond., 1840:13.

COMMENT: Ansell, 1972, Part 15:70, included *Ammotragus* in *Capra;* but see comment under *Capra.*

ISIS NUMBER: 5301419009033000000.

Ammotragus lervia (Pallas, 1777). Spicil. Zool., 12:12.

TYPE LOCALITY: Algeria, Oran.

DISTRIBUTION: N. Egypt to Morocco; Niger to Sudan; perhaps S. Israel. Introduced to New Mexico and Texas (U.S.A.).

COMMENT: Should be included in *Hemitragus* (PG). Reviewed by Gray and Simpson, 1980, Mamm. Species, 144:1–7.
PROTECTED STATUS: CITES - Appendix III (Tunisia).
ISIS NUMBER: 5301419009033001001.

Antidorcas Sundevall, 1847. Kongl. Svenska Vet.-Akad. Handl. Stockholm, for 1845, p. 271.
ISIS NUMBER: 5301419009022000000.

Antidorcas marsupialis (Zimmermann, 1780). Geogr. Gesch. Mensch. Vierf. Thiere, 2:427.
TYPE LOCALITY: South Africa, Cape of Good Hope.
DISTRIBUTION: Former range included N. Namibia; S. Angola; South Africa; Botswana.
COMMENT: Range now much diminished (RGVG).
ISIS NUMBER: 5301419009022001001.

Antilocapra Ord, 1818. J. Phys. Chim. Hist. Nat. Arts Paris, 87:149.
REVIEWED BY: J. Ramirez-Pulido (JRP).
COMMENT: Included in a separate family, Antilocapridae, by Corbet and Hill, 1980:121, but see comment under Bovidae.
ISIS NUMBER: 5301419008001000000.

Antilocapra americana (Ord, 1815). *In* Guthrie, A New Geogr., Hist. and Comml. Grammar..., Philadelphia, 2nd ed., 2:292.
TYPE LOCALITY: U.S.A., Plains and Highlands of the Missouri River.
DISTRIBUTION: S. Alberta and S. Saskatchewan (Canada) south through W. U.S.A. to Hidalgo, Baja California, W. Sonora (Mexico).
COMMENT: Reviewed by O'Gara, 1978, Mamm. Species, 90:1–7.
PROTECTED STATUS: CITES - Appendix II as *A. a. mexicana* subspecies only.
CITES - Appendix I and U.S. ESA - Endangered as *A. a. peninsularis* subspecies only.
CITES - Appendix I and U.S. ESA - Endangered as *A. a. sonoriensis* subspecies only.
ISIS NUMBER: 5301419008001001001.

Antilope Pallas, 1766. Misc. Zool., p. 1.
ISIS NUMBER: 5301419009023000000.

Antilope cervicapra (Linnaeus, 1758). Syst. Nat., 10th ed., 1:69.
TYPE LOCALITY: India, Travancore, inland of Trivandrum.
DISTRIBUTION: Pakistan and India, from Punjab to Nepal and Assam south to Cape Comorin.
COMMENT: Native range now much diminished; introduced into Texas (U.S.A.).
PROTECTED STATUS: CITES - Appendix III (Nepal).
ISIS NUMBER: 5301419009023001001.

Bison H. Smith, 1827. *In* Griffith's Cuvier Anim. Kingd., 5:373.
COMMENT: Revised by McDonald, 1981, North American Bison, their classification and evolution, Univ. California Press, Berkeley, 316 pp.
ISIS NUMBER: 5301419009001000000.

Bison bison (Linnaeus, 1758). Syst. Nat., 10th ed., 1:71.
REVIEWED BY: D. Van Vuren (DVV); M. S. Rich (MSR); J. Ramirez-Pulido (JRP).
TYPE LOCALITY: "Mexico" (= C. Kansas,"Quivera"); redesignated as Canadian River valley, E. New Mexico (U.S.A.).
DISTRIBUTION: Formerly N.W. and C. Canada, south through U.S.A., to Chihuahua, Coahuila (Mexico).
COMMENT: Exterminated in wild except in Yellowstone Park, Wyoming (U.S.A.) and Wood Buffalo Park, Northwest Terr. (Canada). Reintroduced widely within native range; also C. Alaska. See Hershkovitz, 1957, Proc. Biol. Soc. Wash., 79:31–32 for discussion of "Mexico" type locality. See MacDonald, 1981, North American

Bison, their classification and evolution, Univ. California Press, Berkeley, p. 62 for redesignation of type locality.

PROTECTED STATUS: CITES - Appendix I and U.S. ESA - Endangered as *B. b. athabascae* subspecies only.

ISIS NUMBER: 5301419009001001001.

Bison bonasus (Linnaeus, 1758). Syst. Nat., 10th ed., 1:71.

TYPE LOCALITY: "Africa, Asia."

DISTRIBUTION: Europe, originally from S. Sweden south to the Pyrennees Mtns., Balkans, and Caucasus Mts.; now reintroduced to E. Poland, W. Soviet Union, and Caucasus Mtns.

COMMENT: Considered conspecific with *bison* by some authors; see Corbet, 1978:206; but see also Corbet and Hill, 1980:122. According to Heptner *et al.*, 1961, [Mammals of the Soviet Union], Vyssha Shkola, p. 388, type locality probably Belovezhskaya Pushcha (= Forest), Poland.

ISIS NUMBER: 5301419009001002001.

Bos Linnaeus, 1758. Syst. Nat., 10th ed., 1:71.

COMMENT: Includes *Bibos* and *Novibos*; see Ellerman and Morrison-Scott, 1951:380.

ISIS NUMBER: 5301419009002000000.

Bos frontalis Lambert, 1804. Trans. Linn. Soc. Lond., 7:57.

TYPE LOCALITY: Pakistan, N.E. Chittagong (domesticated stock).

DISTRIBUTION: Pakistan; India; Nepal; Burma; Thailand; S. Tibet, Yunnan (China) (SW); S. Vietnam; Kampuchea; Malay Peninsula.

COMMENT: Includes *gaurus*; see Ellerman and Morrison-Scott, 1951:380; but see also Corbet and Hill, 1980:122.

PROTECTED STATUS: CITES - Appendix I and U.S. ESA - Endangered as *Bos gaurus* only.

ISIS NUMBER: 5301419009002001001 as *B. gaurus*.

Bos grunniens Linnaeus, 1766. Syst. Nat., 12th ed., 1:99.

TYPE LOCALITY: Boreal Asia (domesticated stock).

DISTRIBUTION: Sinkiang, Tibet, Tsinghai, Szechwan (China) (SW); N. India; Nepal; domesticated in Central Asia.

COMMENT: Includes *mutus*; but see Corbet, 1978:206.

PROTECTED STATUS: CITES - Appendix I as *B. (grunniens) mutus* and U.S. ESA - Endangered as *B. g. mutus* subspecies only.

ISIS NUMBER: 5301419009002003001 as *B. mutus*.

Bos javanicus D'Alton, 1823. Die Skelete der Wiederkauer, abgebildt und verglichen, p. 7.

TYPE LOCALITY: Indonesia, Java (domesticated stock).

DISTRIBUTION: Burma; Thailand; Indochina; Malaysia; Java; Borneo.

COMMENT: For use of *javanicus* instead of *banteng*, see Hooijer, 1956, Zool. Meded., 34:223–226; Medway, 1977:150.

PROTECTED STATUS: U.S. ESA - Endangered as *B. javanicus* (=*banteng*).

ISIS NUMBER: 5301419009002002001.

Bos sauveli Urbain, 1937. Bull. Soc. Zool. Fr., 62:307.

TYPE LOCALITY: Kampuchea (Cambodia), near Tchep Village.

DISTRIBUTION: Kampuchea.

COMMENT: Included in *Novibos* by Coolidge, 1940, Mem. Mus Comp. Zool. Harv. Univ., 54(6):421–531, 11 pl; but see Ellerman and Morrison-Scott, 1951:380. Perhaps extinct.

PROTECTED STATUS: CITES - Appendix I as *Novibos* (=*Bos*) *sauveli*. U.S. ESA - Endangered as *B. sauveli*.

ISIS NUMBER: 5301419009002005001.

Bos taurus Linnaeus, 1766. Syst. Nat., 10th ed., 1:71.

TYPE LOCALITY: Poland (domesticated stock).

DISTRIBUTION: Originally, from Scotland, S. Sweden, and Baltic south to Iberia, N. Africa, and Near East. Perhaps east to W. Siberia and Kazakhstan (U.S.S.R.). Extinct in the wild; under domestication world wide.

COMMENT: Includes *primigenius* (extinct wild ancestor) and *indicus*; but see Corbet,

1978:206. Last wild individual, a captive, died in 1627; see Harper, 1945, Extinct and vanishing mammals of the Old World, Spec. Publ. No. 12, Amer. Comm. Intern. Wildl. Prot., New York, p. 511.
ISIS NUMBER: 5301419009002006001 as *B. taurus.*
5301419009002004001 as *B. primigenius.*

Boselaphus Blainville, 1816. Bull. Sci. Soc. Philom. Paris, p. 75.
REVIEWED BY: D. Ernst (DE).
ISIS NUMBER: 5301419009003000000.

Boselaphus tragocamelus (Pallas, 1766). Misc. Zool., p. 5.
TYPE LOCALITY: India, no exact locality ("Plains of Peninsular India").
DISTRIBUTION: From base of Himalayas to Bombay and Mysore Prov. (India).
ISIS NUMBER: 5301419009003001001.

Bubalus H. Smith, 1827. *In* Griffith's Cuvier Anim. Kingd., 5:371.
COMMENT: Includes *Anoa;* see Groves, 1969, Beaufortia, 17(223):1–12.
ISIS NUMBER: 5301419009004000000.

Bubalus bubalis (Linnaeus, 1758). Syst. Nat., 10th ed., 1:72.
TYPE LOCALITY: Italy, Rome (domesticated stock).
DISTRIBUTION: India to Indochina; perhaps Sri Lanka and Borneo; widespread as domesticated or feral animal in S.E. Asia, S. Europe, N. Africa, N. Australia, and E. South America.
COMMENT: Includes *arnee;* see Ellerman and Morrison-Scott, 1951:383; but see also Corbet and Hill, 1980:122.
PROTECTED STATUS: CITES - Appendix III (Nepal).
ISIS NUMBER: 5301419009004001001 as *B. arnee.*

Bubalus depressicornis (H. Smith, 1827). *In* Griffith's Cuvier Anim. Kingd., 4:331, 334.
TYPE LOCALITY: Indonesia, Sulawesi (=Celebes).
DISTRIBUTION: Sulawesi.
COMMENT: Formerly included in *Anoa* but placed in *Bubalus,* (subgenus *Anoa*) by Groves, 1969, Beaufortia, 223:3.
PROTECTED STATUS: CITES - Appendix I as *B. (=Anoa) depressicornis.* U.S. ESA - Endangered as *B. anoa depressicornis.*
ISIS NUMBER: 5301419009004002001.

Bubalus mindorensis (Heude, 1888). Mem. Hist. Nat. Emp. Chin., 2:4.
REVIEWED BY: D. W. Kuehn (DWK).
TYPE LOCALITY: Philippines, Mindoro.
DISTRIBUTION: Mindoro (Philippines).
COMMENT: Subgenus *Bubalus;* see Groves, 1969, Beaufortia, 223:10.
PROTECTED STATUS: CITES - Appendix I as *B. (=Anoa) mindorensis.* U.S. ESA - Endangered as *B. mindorensis.*
ISIS NUMBER: 5301419009004003001.

Bubalus quarlesi (Ouwens, 1910). Bull. Dept. Agric. Indes Neerl., 38:7.
TYPE LOCALITY: Indonesia, Sulawesi, mountains of C. Toradja Dist.
DISTRIBUTION: Mountains of Sulawesi.
COMMENT: Subgenus *Anoa;* see Groves, 1969, Beaufortia, 223:1–12. Formerly included in *A. depressicornis;* see Haltenorth, 1963:131.
PROTECTED STATUS: CITES - Appendix I as *B. (=Anoa) quarlesi.* U.S. ESA - Endangered as *B. anoa quarlesi.*
ISIS NUMBER: 5301419009004004001.

Budorcas Hodgson, 1850. J. Asiat. Soc. Bengal, 19:65.
ISIS NUMBER: 5301419009034000000.

Budorcas taxicolor Hodgson, 1850. J. Asiat. Soc. Bengal, 19:65.
TYPE LOCALITY: India, Assam, Mishmi Hills.
DISTRIBUTION: Mishmi Hills (India); Bhutan; Burma; Shensi, Szechwan, Yunnan, Kansu, Tsinghai, S. Tibet (China) (SW).
ISIS NUMBER: 5301419009034001001.

Capra Linnaeus, 1758. Syst. Nat., 10th ed., 1:68.
REVIEWED BY: R. S. Hoffmann (RSH).
COMMENT: Various authors have included *Ammotragus* and *Ovis*; see Ansell, 1972; Part 15:70; Van Gelder, 1977, Am. Mus. Novit., 2635:1–25. Probably should also include *Pseudois* (RGVG). However, most authors have not followed this arrangement; see Corbet and Hill, 1980; Hall, 1981; Gromov and Baranova, 1981; Gray and Simpson, 1980. There is no consensus concerning the number of species to be recognized in this genus; some would recognize only two *(hircus, falconeri)*; see Haltenorth, 1963, while others would recognize up to nine; Corbet, 1978:213–217, is followed here (RSH).
ISIS NUMBER: 5301419009035000000.

Capra caucasica Guldenstaedt and Pallas, 1783. Acta Acad. Sci. Petrop., for 1779, 2:273.
TYPE LOCALITY: U.S.S.R., Caucasus Mtns., between Malka and Baksan Rivers, east of Mt. Elbrus.
DISTRIBUTION: W. Caucasus Mtns. (U.S.S.R.).
COMMENT: Probably should be included in *ibex* (RGVG). Possibly *caucasica* is the prior name for *cylindricornis*, in which case this species should be termed *severtzovi*; see Ellerman and Morrison-Scott, 1951:407. See also comment under *cylindricornis*.

Capra cylindricornis (Blyth, 1841). Proc. Zool. Soc. Lond., 1840:68.
TYPE LOCALITY: U.S.S.R., E. Caucasus Mtns., Mt. Kasber.
DISTRIBUTION: E. Caucasus Mtns. (U.S.S.R.).
COMMENT: Possibly a junior synonym of *caucasica*; see Gromov and Baranova, 1981:404. See also comment under *caucasica*.

Capra falconeri (Wagner, 1839). Munch. Gelehrt. Anz., 9:430.
TYPE LOCALITY: India, Kashmir, Astor.
DISTRIBUTION: Afghanistan; N. Pakistan; N. India; Kashmir; S. Uzbekistan; Tadzhikistan (U.S.S.R.).
PROTECTED STATUS: CITES - Appendix II as *C. falconeri*.
CITES - Appendix I and U.S. ESA - Endangered as *C. f. chialtanensis* subspecies only.
CITES - Appendix I and U.S. ESA - Endangered as *C. f. jerdoni* subspecies only.
CITES - Appendix I and U.S. ESA - Endangered as *C. f. megaceros* subspecies only.
ISIS NUMBER: 5301419009035002001.

Capra hircus Linnaeus, 1758. Syst. Nat., 10th ed., 1:68.
TYPE LOCALITY: Sweden (domesticated stock).
DISTRIBUTION: Greek islands; Turkey; Iran; S.W. Afghanistan; Oman; Caucasus; Turkmenia (U.S.S.R.); Pakistan; adjacent India; domesticated worldwide.
COMMENT: Includes *aegagrus*, but see Corbet, 1978:214.
ISIS NUMBER: 5301419009035006001 as *C. hircus*.
5301419009035001001 as *C. aegagrus*.

Capra ibex Linnaeus, 1758. Syst. Nat., 10th ed., 1:68.
TYPE LOCALITY: Switzerland, Valais.
DISTRIBUTION: C. Europe; Afghanistan and Kashmir to Mongolia and C. China; N. Ethiopia to Syria and Arabia.
COMMENT: Includes *nubiana, sibirica*, and *walie*; see Corbet, 1978:215. PG believes *sibirica* and *nubiana* are valid species, as do most Russian authors; see Gromov and Baranova, 1981:403. Ansell, 1972, Part 15:69–70, listed *walie* as a species of uncertain status.
PROTECTED STATUS: U.S. ESA - Endangered as *C. walie* only.
ISIS NUMBER: 5301419009035003001 as *C. ibex*.
5301419009035005001 as *C. walie*.

Capra pyrenaica Schinz, 1838. N. Denkschr. Schneiz. Ges. Natur. Wiss., 2:9.
TYPE LOCALITY: Spain, Pyrenees Mtns., Huesca, near Maldetta Pass.
DISTRIBUTION: Iberian Peninsula.
PROTECTED STATUS: U.S. ESA - Endangered as *C. p. pyrenaica* subspecies only.
ISIS NUMBER: 5301419009035004001.

Capricornis Ogilby, 1837. Proc. Zool. Soc. Lond., 1836:139.
ISIS NUMBER: 5301419009036000000.

Capricornis crispus (Temminck, 1845). Fauna Japonica, 1(Mamm.), p. 55, pl. 18,19.
TYPE LOCALITY: Japan, Honshu.
DISTRIBUTION: Honshu, Shikoku and Kyushu (Japan); Taiwan.
COMMENT: Includes *swinhoei*; PG considers it a valid species. Included in *sumatraensis*
     by Haltenorth, 1963:119; but see Corbet, 1978:212.
ISIS NUMBER: 5301419009036001001.

Capricornis sumatraensis (Bechstein, 1799). Ueber Vierfuss. Thiere, 1:98.
TYPE LOCALITY: Indonesia, Sumatra.
DISTRIBUTION: Indochina; Burma; Thailand; China, north to Kansu and Anwei;
     Kashmir; N. India; Nepal; Sikkim; Malay Peninsula; Sumatra.
PROTECTED STATUS: CITES - Appendix I and U.S. ESA - Endangered.
ISIS NUMBER: 5301419009036002001.

Cephalophus H. Smith, 1827. *In* Griffith's Cuvier Anim. Kingd., 5:344.
COMMENT: Haltenorth, 1963:71, and Van Gelder, 1977, Am. Mus. Novit., 2635:1–25;
     included *Sylvicapra*. However, recent authors have not followed this arrangement;
     see Swanepoel *et al.*, 1980:188.
ISIS NUMBER: 5301419009009000000.

Cephalophus adersi Thomas, 1918. Ann. Mag. Nat. Hist., ser. 9, 2:151.
TYPE LOCALITY: Tanzania, Zanzibar.
DISTRIBUTION: Zanzibar; adjacent coast of Kenya and probably Tanzania.
COMMENT: Possibly conspecific with *natalensis* and/or *callipygus*; see Ansell, 1972, Part
     15:33.
ISIS NUMBER: 5301419009009001001.

Cephalophus callipygus Peters, 1876. Monatsb. Preuss. Akad. Wiss. Berlin, p. 483, pl. 3,4.
TYPE LOCALITY: Gabon.
DISTRIBUTION: West of Congo and Ubangi Rivers in Congo Republic; Central African
     Republic; Zaire; Gabon and Cameroun.
COMMENT: Possibly conspecific with *natalensis* and/or *adersi*; see Ansell, 1972, Part
     15:33. According to Groves and Grubb, 1974, Rev. Zool. Afr., 88:189–196, not
     related to *natalensis*.
ISIS NUMBER: 5301419009009002001.

Cephalophus dorsalis Gray, 1846. Ann. Mag. Nat. Hist., ser. 1, 18:165.
TYPE LOCALITY: Sierra Leone.
DISTRIBUTION: Guinea-Bissau to Central African Republic and N.E. Zaire; Gabon to N.
     Angola.
ISIS NUMBER: 5301419009009003001.

Cephalophus jentinki Thomas, 1892. Proc. Zool. Soc. Lond., 1892:417.
TYPE LOCALITY: Liberia.
DISTRIBUTION: Liberia; W. Ivory Coast; possibly Sierra Leone.
PROTECTED STATUS: U.S. ESA - Endangered.
ISIS NUMBER: 5301419009009004001.

Cephalophus leucogaster Gray, 1873. Ann. Mag. Nat. Hist., ser. 4, 12:43.
TYPE LOCALITY: Gabon.
DISTRIBUTION: S. Cameroun south to mouth of Congo River, east to E. Zaire.
ISIS NUMBER: 5301419009009005001.

Cephalophus maxwelli (H. Smith, 1827). *In* Griffith's Cuvier Anim. Kingd., 4:267.
TYPE LOCALITY: Sierra Leone.
DISTRIBUTION: Senegal and Gambia to Nigeria.
COMMENT: Included in *monticola* by Haltenorth and Diller, 1977:43. Reviewed by
     Ralls, 1973, Mamm. Species, 31:1–4.
ISIS NUMBER: 5301419009009006001.

Cephalophus monticola (Thunberg, 1789). Resa uti Europa African, Asia..., 2:66.
TYPE LOCALITY: South Africa, Cape Region, Lange Kloof (32° 10' S, 20° 10' E).
DISTRIBUTION: Nigeria to Gabon to Kenya to South Africa to Angola; Zanzibar; Bioko; Pemba Isl.
COMMENT: May include *maxwelli;* see Haltenorth and Diller, 1977:43.
PROTECTED STATUS: CITES - Appendix II.
ISIS NUMBER: 5301419009009007001.

Cephalophus natalensis A. Smith, 1834. S. Afr. Quart. J., 2:217.
TYPE LOCALITY: South Africa, Natal, Port Natal.
DISTRIBUTION: S. Somalia; Kenya; Tanzania; Malawi; Mozambique; Natal; E. Zimbabwe; E. Zambia.
COMMENT: Includes *harveyi;* possibly also includes *adersi* and/or *callipygus;* see Ansell, 1972, Part 15:34, who also included *weynsi.*
ISIS NUMBER: 5301419009009008001.

Cephalophus niger Gray, 1846. Ann. Mag. Nat. Hist., ser. 1, 18:165.
TYPE LOCALITY: "Guinea Coast" =Ghana.
DISTRIBUTION: Guinea to Nigeria, west of lower Niger River.
ISIS NUMBER: 5301419009009009001.

Cephalophus nigrifrons Gray, 1871. Proc. Zool. Soc. Lond., 1871:598, pl. 46.
TYPE LOCALITY: Gabon.
DISTRIBUTION: Cameroun; Gabon; N.E. Angola; Congo; Zaire; W. Uganda; Mt. Elgon, Mt. Kenya (Kenya).
ISIS NUMBER: 5301419009009010000.

Cephalophus ogilbyi (Waterhouse, 1838). Proc. Zool. Soc. Lond., 1838:60.
TYPE LOCALITY: Equatorial Guinea, Bioko (=Fernando Po).
DISTRIBUTION: Sierra Leone; Liberia; W. Ivory Coast; W. Ghana; Togo; S.E. Nigeria; S. Cameroun; Bioko; Gabon.
ISIS NUMBER: 5301419009009011001.

Cephalophus rufilatus Gray, 1846. Ann. Mag. Nat. Hist., ser. 1, 18:166.
TYPE LOCALITY: Sierra Leone, Waterloo Village.
DISTRIBUTION: Senegal to S.W. Sudan and N.E. Uganda south to Cameroun.
ISIS NUMBER: 5301419009009012001.

Cephalophus spadix True, 1890. Proc. U.S. Nat. Mus., 13:227.
TYPE LOCALITY: Tanzania, Mt. Kilimanjaro.
DISTRIBUTION: Tanzania, at higher elevations.
COMMENT: Possibly a subspecies of *sylvicultor* (RGVG).
ISIS NUMBER: 5301419009009013001.

Cephalophus sylvicultor (Afzelius, 1815). Nova Acta Reg. Soc. Sci. Upsala, 7:265, pl. 8, fig. l.
TYPE LOCALITY: "Sierra Leone and region of Pongas and Quia Rivers (Guinea)."
DISTRIBUTION: Guinea-Bissau and Gambia to Gabon; Angola and Zambia to S.W. Sudan and W. Uganda; W. Kenya.
ISIS NUMBER: 5301419009009014001.

Cephalophus weynsi Thomas, 1901. Ann. Mus. Congo Zool., 2(1):15.
TYPE LOCALITY: Zaire, near Stanley Falls.
DISTRIBUTION: Zaire, Uganda, Rwanda, and W. Kenya.
COMMENT: Formerly included in *callipygus,* but separate according to Groves and Grubb, 1974, Rev. Zool. Afr., 88:189–196.

Cephalophus zebra Gray, 1838. Ann. Mag. Nat. Hist., ser. 1, 1:27.
TYPE LOCALITY: Sierra Leone.
DISTRIBUTION: W. Sierra Leone; Liberia; C. Ivory Coast.
ISIS NUMBER: 5301419009009015001.

Connochaetes Lichtenstein, 1814. Mag. Ges. Naturf. Fr., 6:152.
  ISIS NUMBER: 5301419009013000000.

Connochaetes gnou (Zimmermann, 1780). Spec. Zool. Geogr., for 1777, p. 382.
  TYPE LOCALITY: South Africa, Cape of Good Hope.
  DISTRIBUTION: Originally from Transvaal and Natal south to Cape Prov. (South
    Africa); now only in captivity.
  COMMENT: Reviewed by Von Richter, 1974, Mamm. Species, 50:1-6.
  ISIS NUMBER: 5301419009013001001.

Connochaetes taurinus (Burchell, 1824). Travels in Interior of Southern Africa,
    2(footnote):278.
  TYPE LOCALITY: Botswana, (=Bechuanaland).
  DISTRIBUTION: S. Angola; Namibia; N. South Africa; Botswana; C. Mozambique and E.
    Zambia to N.E. Tanzania and S.E. and S.C. Kenya; formerly S. and C. Malawi.
  COMMENT: Formerly in genus *Gorgon*, see Ellerman *et al.*, 1953; but also see Ansell,
    1972, Part 15:50.
  ISIS NUMBER: 5301419009013002001.

Damaliscus Sclater and Thomas, 1894. Book of Antelopes, 1(part 1):3, 51.
  COMMENT: Includes *Beatragus*; see Ansell, 1972, Part 15:54. Placed in *Alcelaphus* by Van
    Gelder, 1977:18; but see also Swanepoel *et al.*, 1980:164.
  ISIS NUMBER: 5301419009014000000.

Damaliscus dorcas (Pallas, 1766). Misc. Zool., p. 6.
  TYPE LOCALITY: South Africa, Cape Prov., Caffer Kayls River, between Mussel Bay and
    Swellendam.
  DISTRIBUTION: Formerly from S.W. Cape Prov. to E. Transvaal (South Africa); now
    only in captivity.
  COMMENT: Includes *phillipsi*; see Ansell, 1972, Part 15:55.
  PROTECTED STATUS: CITES - Appendix II and U.S. ESA - Endangered as *D. d. dorcas*
    subspecies only.
  ISIS NUMBER: 5301419009014001001.

Damaliscus hunteri (Sclater, 1889). Proc. Zool. Soc. Lond., 1889:58.
  TYPE LOCALITY: Kenya, E. bank of Tana River.
  DISTRIBUTION: Somalia to N. Kenya.
  COMMENT: Included in *lunatus* by Haltenorth, 1963:100. Formerly in *Beatragus*; see
    Ansell, 1972, Part 15:54.
  ISIS NUMBER: 5301419009014002001.

Damaliscus lunatus (Burchell, 1823). Travels in Interior of Southern Africa, 2:334.
  TYPE LOCALITY: South Africa, Mathlowing River, N.E. Kuruman.
  DISTRIBUTION: Senegal to N. Nigeria to Sudan and Central African Republic to S.
    Somalia to N. South Africa.
  COMMENT: Includes *korrigum*; see Ansell, 1972, Part 15:56; but PG believes it is a valid
    species.
  PROTECTED STATUS: CITES - Appendix III (Ghana).
  ISIS NUMBER: 5301419009014003001.

Dorcatragus Noack, 1894. Zool. Anz., 17:202.
  REVIEWED BY: D. Ernst (DE); K. R. Kranz (KRK).
  ISIS NUMBER: 5301419009024000000.

Dorcatragus megalotis (Menges, 1894). Zool. Anz., 17:130.
  TYPE LOCALITY: Somalia, Hekebo Plateau, 35 mi. (56 km) S.W. of Berbera.
  DISTRIBUTION: N.E. and N.W. Somalia to E. Ethiopia and Djibouti.
  ISIS NUMBER: 5301419009024001001.

Gazella Blainville, 1816. Bull. Sci. Soc. Philom. Paris, p. 75.
COMMENT: Revised in part by Groves, 1969, Z. Saugetierk., 34:38–60.
ISIS NUMBER: 5301419009025000000.

Gazella cuvieri (Ogilby, 1841). Proc. Zool. Soc. Lond., 1840:35.
TYPE LOCALITY: Morocco, Mogador.
DISTRIBUTION: Morocco; N. Algeria; C. Tunisia.
COMMENT: Nearly extinct (RGVG). According to Groves, 1969, Z. Saugetierk.,
34:38–60, includes *rufifrons* and *thomsoni*; but also see comments under those
species.
PROTECTED STATUS: CITES - Appendix III (Tunisia) as *G. gazella cuvieri* subspecies only.
U.S. ESA - Endangered as *G. cuvieri*.
ISIS NUMBER: 5301419009025001001.

Gazella dama (Pallas, 1766). Misc. Zool., p. 5.
TYPE LOCALITY: Africa, near Lake Chad.
DISTRIBUTION: Senegal to Sudan; S.W. Morocco to Egypt. Present range much reduced
(RGVG).
PROTECTED STATUS: U.S. ESA - Endangered as *G. d. lozanoi* subspecies only. U.S. ESA -
Endangered as *G. d. mhorr* subspecies only.
ISIS NUMBER: 5301419009025002001.

Gazella dorcas (Linnaeus, 1758). Syst. Nat., 10th ed., 1:69.
TYPE LOCALITY: Lower Egypt.
DISTRIBUTION: Morocco to Egypt and N. Somalia; Sinai; Arabia; Israel; possibly Syria
and W. Iraq; E. Iran to C. India.
COMMENT: Includes *bennetti*; see Corbet, 1978:209. Includes *pelzelni*; see Gentry, 1972,
Part 15.1:89; but also see Haltenorth, 1963:112. Numbers and range much reduced
(RGVG).
PROTECTED STATUS: CITES - Appendix III (Tunisia) as *G. dorcas*. U.S. ESA - Endangered
as *G. d. massaesyla* subspecies only. U.S. ESA - Endangered as *G. d. pelzelni*
subspecies only. U.S. ESA - Endangered as *G. d. saudiya* subspecies only.
ISIS NUMBER: 5301419009025003001.

Gazella gazella (Pallas, 1766). Misc. Zool., p. 7.
TYPE LOCALITY: Syria.
DISTRIBUTION: Syria; Sinai to Arabia.
PROTECTED STATUS: U.S. ESA - Endangered.
ISIS NUMBER: 5301419009025004001.

Gazella granti Brooke, 1872. Proc. Zool. Soc. Lond., 1872:602.
TYPE LOCALITY: Tanzania, W. Kinyenye, Ugogo.
DISTRIBUTION: S. Sudan and N.E. Uganda to Juba River, Somalia to N. Tanzania.
ISIS NUMBER: 5301419009025005001.

Gazella leptoceros (F. Cuvier, 1842). *In* E. Geoffroy and F. Cuvier, Hist. Nat. Mamm., 4, pt.
72, pl. 424, 425.
TYPE LOCALITY: Sudan, Sennar, probably between Giza and Wadi Natron.
DISTRIBUTION: Algeria, Tunisia, Libya and W. Egypt.
PROTECTED STATUS: CITES - Appendix III (Tunisia) and U.S. ESA - Endangered.
ISIS NUMBER: 5301419009025006001.

Gazella rufifrons Gray, 1846. Ann. Mag. Nat. Hist., ser. 1, 18:214.
TYPE LOCALITY: Senegal.
DISTRIBUTION: Senegal to N.E. Ethiopia, south to Ghana, Nigeria, and N. Cameroun.
COMMENT: Includes *tilonura*; see Gentry, 1972, Part 15.1:90; but also see Haltenorth,
1963:112. Groves, 1969, Z. Saugetierk., 34:38–60, included *rufifrons* in *cuvieri*, but
subsequently (Groves, 1975, Z. Saugetierk., 40:308–319) separated them.
ISIS NUMBER: 5301419009025007001.

Gazella rufina Thomas, 1894. Proc. Zool. Soc. Lond., 1894:467.
TYPE LOCALITY: Interior of Algeria.
DISTRIBUTION: Algeria.
COMMENT: Thought to be extinct; see Corbet, 1978:210.

Gazella soemmerringi (Cretzschmar, 1826). Ruppell Atlas Reise Nordl. Afr., p. 49, pl. 19.
TYPE LOCALITY: E. Ethiopia.
DISTRIBUTION: N. Somalia; Ethiopia; E.C. Sudan.
ISIS NUMBER: 5301419009025008001.

Gazella spekei Blyth, 1863. Cat. Mamm. Mus. Asiat. Soc. Calcutta, p. 172.
TYPE LOCALITY: Somalia.
DISTRIBUTION: Somalia; E. Ethiopia.
ISIS NUMBER: 5301419009025009001.

Gazella subgutturosa (Guldenstaedt, 1780). Acta Acad. Sci. Petrop., for 1778, 1:251.
TYPE LOCALITY: U.S.S.R., Azerbaidzhan, Steppes of E. Transcaucasia.
DISTRIBUTION: Israel; C. Arabia and E. Caucasus through Iran; Afghanistan; Pakistan;
    Soviet Central Asia; Mongolia; W. China.
PROTECTED STATUS: U.S. ESA - Endangered as *G. s. marica* subspecies only.
ISIS NUMBER: 5301419009025010000.

Gazella thomsoni Gunther, 1884. Ann. Mag. Nat. Hist., ser. 5, 14:427.
TYPE LOCALITY: Kenya, Kilimanjaro Dist.
DISTRIBUTION: S. Sudan to N. Tanzania.
COMMENT: Groves, 1969, Z. Saugetierk., 34:38–60, included *thomsoni* in *cuvieri*; but see
    also Gentry, 1972, Part 15.1:88, 90–91.
ISIS NUMBER: 5301419009025011001.

Hemitragus Hodgson, 1841. Calcutta J. Nat. Hist., 2:218.
COMMENT: Should include *lervia* according to PG; see comment under *Ammotragus*.
ISIS NUMBER: 5301419009037000000.

Hemitragus hylocrius (Ogilby, 1838). Proc. Zool. Soc. Lond., 1837:81.
TYPE LOCALITY: India, Nilgiri Hills.
DISTRIBUTION: S.W. and S. India.
COMMENT: Included in *jemlahicus* by Haltenorth, 1963:125 but see Harrison, 1968:324,
    and Corbet and Hill, 1980:126.
ISIS NUMBER: 5301419009037001001.

Hemitragus jayakari Thomas, 1894. Ann. Mag. Nat. Hist., 13:365.
TYPE LOCALITY: Oman, Jebel Akhdar Range, Jebel Taw.
DISTRIBUTION: Oman; S.E. Arabia.
COMMENT: Included in *jemlahicus* by Haltenorth, 1963:125 but see Harrison, 1968:324,
    and Corbet, 1978:213.
PROTECTED STATUS: U.S. ESA - Endangered.
ISIS NUMBER: 5301419009037002001.

Hemitragus jemlahicus (H. Smith, 1826). *In* Griffith's Cuvier Anim. Kingd., 4:308.
TYPE LOCALITY: Nepal, Jemla Hills.
DISTRIBUTION: Himalayas, from Kashmir through N. India to Nepal and Sikkim; Tibet
    (China). Introduced in New Zealand.
ISIS NUMBER: 5301419009037003001.

Hippotragus Sundevall, 1846. Kongl. Svenska Vet.-Akad. Handl. Stockholm, for 1844, p.
    196.
ISIS NUMBER: 5301419009015000000.

Hippotragus equinus (Desmarest, 1804). Dict. Class. Hist. Nat. Paris, 1st ed., p. 24.
TYPE LOCALITY: Unknown; perhaps N. South Africa, Lataku.
DISTRIBUTION: Senegal to Ethiopia; N. Angola to Kenya and W. Mozambique, south to
    N. South Africa, Botswana and Namibia.
PROTECTED STATUS: CITES - Appendix II.
ISIS NUMBER: 5301419009015001001.

Hippotragus leucophaeus (Pallas, 1766). Misc. Zool., p. 4.
TYPE LOCALITY: South Africa, Cape of Good Hope, Swellendam Dist.
DISTRIBUTION: S. Cape Province (South Africa).
COMMENT: Extirpated about 1799.

Hippotragus niger (Harris, 1838). Proc. Zool. Soc. Lond., p. 2.
TYPE LOCALITY: South Africa, Transvaal, Cashan Range, near Pretoria.
DISTRIBUTION: S.E. Kenya to N. South Africa, west to N. Botswana, also Angola,
    between Cuanza and Loando Rivers.

COMMENT: Includes *variani;* see Ansell 1972, Part 15:47.
PROTECTED STATUS: CITES - Appendix I and U.S. ESA - Endangered as *H. n. variani*
subspecies only.
ISIS NUMBER: 5301419009015002001.

Kobus A. Smith, 1840. Illustr. Zool. S. Afr. Mamm., Part 12, pl. 28.
ISIS NUMBER: 5301419009016000000.

Kobus ellipsiprymnus (Ogilby, 1833). Proc. Zool. Soc. Lond., 1833:47.
TYPE LOCALITY: South Africa, between Lataku (near Kuruman) and W. coast of Africa,
N. of Orange River.
DISTRIBUTION: Senegal to Somalia to N. South Africa to Angola.
COMMENT: Includes *defassa;* see Ansell, 1972, Part 15:42.
ISIS NUMBER: 5301419009016001001.

Kobus kob (Erxleben, 1777). Syst. Regn. Anim., 1:293.
TYPE LOCALITY: Upper Guinea, towards Senegal.
DISTRIBUTION: Senegal to W. Ethiopia and Sudan; N. Zaire to W. Kenya; N.W.
Tanzania.
ISIS NUMBER: 5301419009016002001.

Kobus leche Gray, 1850. Gleanings from Menagerie at Knowsley Hall, 2:23.
TYPE LOCALITY: Botswana (= Bechuanaland), Zoaga River, near Lake Ngami.
DISTRIBUTION: N. Botswana, N.E. Namibia; S.E. Angola, S.E. Zaire and Zambia.
PROTECTED STATUS: CITES - Appendix II and U.S. ESA - Endangered.
ISIS NUMBER: 5301419009016003001.

Kobus megaceros (Fitzinger, 1855). Sitzb. Akad. Wiss. Wien, 17:247.
TYPE LOCALITY: Sudan, Bahr-el-Ghazal, Sobat.
DISTRIBUTION: S. Sudan; W. Ethiopia.
ISIS NUMBER: 5301419009016004001.

Kobus vardoni (Livingstone, 1857). Missionary Travels and Researches in South Africa, p.
256.
TYPE LOCALITY: Zambia, Barotseland, Chobe Valley, near Libonta, 40° 30' S, 23° 15' E.
DISTRIBUTION: S. Zaire and Zambia to Zambesi and Chobe rivers, north to Tanzania
and Malawi; Zaire and E. Angola.
COMMENT: Included in *kob* by Haltenorth, 1963:92 but see Ansell, 1972, Part 15:44.
ISIS NUMBER: 5301419009016005001.

Litocranius Kohl, 1886. Ann. K. K. Natur. Hist. Hofmus. Wien, 1:79.
REVIEWED BY: D. Ernst (DE).
ISIS NUMBER: 5301419009026000000.

Litocranius walleri (Brooke, 1879). Proc. Zool. Soc. Lond., 1878:929, pl. 56.
TYPE LOCALITY: Kenya, Tana River Valley, 3° S., 38° E.
DISTRIBUTION: Somalia to Tana and Galana Rivers in Kenya; N.E. Tanzania and E.
Ethiopia.
ISIS NUMBER: 5301419009026001001.

Madoqua Ogilby, 1837. Proc. Zool. Soc. Lond., 1836:137.
REVIEWED BY: K. R. Kranz (KRK).
COMMENT: Includes *Rhynchotragus;* see Ansell, 1972, Part 15:61.
ISIS NUMBER: 5301419009027000000.

Madoqua guentheri Thomas, 1894. Proc. Zool. Soc. Lond., 1894:324.
TYPE LOCALITY: Ethiopia, Ogaden, 6° 30' N., 42° 30' E., 3000 ft. (914 m).
DISTRIBUTION: S. and C. Somalia; S. Ethiopia; S.E. Sudan; N.E. Uganda; N. Kenya.
ISIS NUMBER: 5301419009027001001.

Madoqua kirki (Gunther, 1880). Proc. Zool. Soc. Lond., 1880:17.
TYPE LOCALITY: Somalia, Brava.
DISTRIBUTION: Kenya; N.W. and E. Tanzania; S. Somalia; S.W. Angola; Namibia.
COMMENT: Includes *damarensis;* see Ansell, 1972, Part 15:64.
ISIS NUMBER: 5301419009027002001.

Madoqua piacentinii Drake-Brockman, 1911. Proc. Zool. Soc. Lond., 1911:56.
TYPE LOCALITY: Somalia, Gharabwein, near Obbia, 5° 25' N., 48° 25' E.
DISTRIBUTION: E. Somalia.
COMMENT: Included in *swaynei* byAnsell, 1972, Part 15:62; but see Yalden, 1979,
Monitore Zool. Ital., n.s., suppl. 11:262. See also comment under *saltiana.*

Madoqua saltiana (Desmarest, 1816). Nouv. Dict. Hist. Nat., 2nd ed., 2:192.
TYPE LOCALITY: Ethiopia (=Abyssinia).
DISTRIBUTION: N.E. Sudan; N. Ethiopia; Somalia; perhaps Djbouti.
COMMENT: Includes *cordeauxi;* see Ansell, 1972, Part 15:63. Includes *elangeri, phillipsi*
and *swaynei;* see Yalden, 1979, Monitore Zool. Ital., n.s., suppl. 11:262; but see also
Ansell, 1972, Part 15:62.
ISIS NUMBER: 5301419009027004001 as *M. saltiana.*
5301419009027003001 as *M. phillipsi.*
5301419009027005001 as *M. swaynei.*

Nemorhaedus H. Smith, 1827. *In* Griffith's Cuvier Anim. Kingd., 5:352.
COMMENT: *Naemorhaedus, Nemorhedus, Nemorrhedus,* and *Naemorhedus* are later
spellings.
ISIS NUMBER: 5301419009038000000.

Nemorhaedus goral (Hardwicke, 1825). Trans. Linn. Soc. Lond., 14:518.
TYPE LOCALITY: Nepal, in the Himalayas.
DISTRIBUTION: N. India and Burma to S.E. Siberia; south to Thailand.
COMMENT: The taxa *baileyi, caudatus* and perhaps *cranbrooki* are probably valid species
(PG). See also Lekagul and McNeely, 1977:699; Corbet and Hill, 1980:126.
PROTECTED STATUS: CITES - Appendix I and U.S. ESA - Endangered.
ISIS NUMBER: 5301419009038001001.

Neotragus H. Smith, 1827. *In* Griffith's Cuvier Anim. Kingd., 5:349.
REVIEWED BY: K. R. Kranz (KRK).
COMMENT: Includes *Nesotragus;* see Ansell, 1972, Part 15:68.
ISIS NUMBER: 5301419009028000000.

Neotragus batesi De Winton, 1903. Proc. Zool. Soc. Lond., 1903(1):192, pl. 19.
TYPE LOCALITY: Cameroun, Bulu Country, Efulen.
DISTRIBUTION: S. Nigeria to Gabon, N.E. Zaire and W. Uganda.
ISIS NUMBER: 5301419009028001001.

Neotragus moschatus (Von Dueben, 1846). *In* Sundevall, Ofvers. Kongl. Svenska Vet.-
Akad. Forhandl., Stockholm, 3(7):221.
TYPE LOCALITY: Tanzania, Chapani Isl., 2 mi. (3 km) from Zanzibar.
DISTRIBUTION: N. South Africa to Kenya; Zanzibar; Mafia Isl.
COMMENT: Includes *livingstonianus;* see Ansell, 1972, Part 15:69.
PROTECTED STATUS: U.S. ESA - Endangered as *Nesotragus m. moschatus* subspecies only.
ISIS NUMBER: 5301419009028002001.

Neotragus pygmaeus (Linnaeus, 1758). Syst. Nat., 10th ed., 1:69.
TYPE LOCALITY: "Guinea, India" (=west coast of Africa).
DISTRIBUTION: Sierra Leone to Ghana.
ISIS NUMBER: 5301419009028003001.

Oreamnos Rafinesque, 1817. Am. Mon. Mag., 2:44.
ISIS NUMBER: 5301419009039000000.

Oreamnos americanus (Blainville, 1816). Bull. Sci. Soc. Philom. Paris, p. 80.
TYPE LOCALITY: U.S.A., Washington, Mt. Adams.
DISTRIBUTION: S.E. Alaska (U.S.A.), S. Yukon and S.W. Mackenzie (Canada) to N.C.
Oregon, C. Idaho, and Montana (U.S.A.). Introduced to Kodiak, Chichagof, and
Baranof Isls. (Alaska), Olympic Peninsula, Washington; C. Montana; Black Hills,
South Dakota; Colorado (U.S.A.).
COMMENT: Reviewed by Rideout and Hoffmann, 1975, Mamm. Species, 63:1–6.
ISIS NUMBER: 5301419009039001001.

Oreotragus A. Smith, 1834. S. Afr. Quart. J., 2:212.
REVIEWED BY: K. R. Kranz (KRK).
ISIS NUMBER: 5301419009029000000.

Oreotragus oreotragus (Zimmermann, 1783). Geogr. Gesch. Mensch. Vierf. Thiere, 3:269.
TYPE LOCALITY: South Africa, Cape of Good Hope.
DISTRIBUTION: South Africa to Nigeria, Sudan, Ethiopia and Somalia.
ISIS NUMBER: 5301419009029001001.

Oryx Blainville, 1816. Bull. Sci. Soc. Philom. Paris, p. 75.
ISIS NUMBER: 5301419009017000000.

Oryx dammah (Cretzschmar, 1826). Ruppell Atlas Reise Nordl. Afr., p. 22.
TYPE LOCALITY: Sudan, Haraza, "probably Kordofan."
DISTRIBUTION: Mauritania to Ethiopia; formerly Egypt to Tunisia.
COMMENT: Includes *tao;* see Ansell, 1972, Part 15:48. The name *algazel* Oken, 1816, has been declared invalid in Opinion 417 of the International Commission on Zoological Nomenclature.
PROTECTED STATUS: CITES - Appendix II as *O. (tao) dammah.*
ISIS NUMBER: 5301419009017001001.

Oryx gazella (Linnaeus, 1758). Syst. Nat., 10th ed., 1:69.
TYPE LOCALITY: "India" (=South Africa).
DISTRIBUTION: Namibia and Angola to Transvaal (South Africa); N.E. Ethiopia to N. Tanzania, west to S.E. Sudan and N.E. Uganda.
COMMENT: Includes *beisa;* see Ansell, 1972, Part 15:49. PG believes *beisa* is a valid species.
ISIS NUMBER: 5301419009017002001.

Oryx leucoryx (Pallas, 1777). Spicil. Zool., 12:17.
TYPE LOCALITY: Arabia.
DISTRIBUTION: S.E. Arabian Peninsula; formerly Iraq.
COMMENT: Included in *O. gazella* by Haltenorth, 1963:88; but see Harrison, 1968:344, and Ansel, 1972, Part 15:48. Probably became extinct in wild, but maintained in captivity, and recently reintroduced into Arabia.
PROTECTED STATUS: CITES - Appendix I and U.S. ESA - Endangered.
ISIS NUMBER: 5301419009017003001.

Ourebia Laurillard, 1842. d'Orbigny's Dict. Univ. D'Hist. Nat., 1:622.
REVIEWED BY: K. R. Kranz (KRK).
ISIS NUMBER: 5301419009030000000.

Ourebia ourebi (Zimmermann, 1783). Geogr. Gesch. Mensch. Vierf. Thiere, 3:268.
TYPE LOCALITY: South Africa, Cape of Good Hope.
DISTRIBUTION: Senegal to W. and C. Ethiopia to N. Mozambique, west to S. Uganda; Angola to N. Namibia, N. Botswana, W. Zambia; S. Mozambique and N.E. South Africa.
COMMENT: Includes *haggardi* and *kenyae;* see Ansell, 1972, Part 15:66.
ISIS NUMBER: 5301419009030001001.

Ovibos Blainville, 1816. Bull. Sci. Soc. Philom. Paris, p. 76.
REVIEWED BY: M. S. Rich (MSR).
ISIS NUMBER: 5301419009040000000.

Ovibos moschatus (Zimmermann, 1780). Geogr. Gesch. Mensch. Vierf. Thiere, 2:86.
TYPE LOCALITY: Canada, Manitoba, between Seal and Churchill rivers.
DISTRIBUTION: Formerly Point Barrow, Alaska (U.S.A.) east to N.E. Greenland, south to N.E. Manitoba (Canada). Range now much reduced. Introduced to Seward Peninsula and Nunivak Isl., Alaska (U.S.A.); Taimyr Peninsula and Wrangel Isl. (U.S.S.R.); Norway.
ISIS NUMBER: 5301419009040001001.

Ovis Linnaeus, 1758. Syst. Nat., 10th ed., 1:70.
REVIEWED BY: R. S. Hoffmann (RSH); J. Ramirez-Pulido (JRP).
COMMENT: Placed in *Capra* by Van Gelder, 1977, Am. Mus. Novit., 2635:1–25; see comments under *Capra*. There is no consensus concerning the number of species to be recognized in this genus; some would recognize only one *(ammon;* see Haltenorth, 1963:126–128); others two *(ammon, canadensis;* see Corbet, 1978:218), while others recognize up to seven, as do the most recent reviews (Nadler *et al.,* 1973, Z. Saugetierk., 30:109–125; Korobitsyna *et al.,* 1974, Quat. Res., 4:235–245).
ISIS NUMBER: 5301419009041000000.

Ovis ammon (Linnaeus, 1758). Syst. Nat., 10th ed., 1:70.
TYPE LOCALITY: U.S.S.R., Kazakh S.S.R., Vostochno-Kazakhstansk. Obl., Altai Mtns., Bukhtarma; alternatively, Ust-Kamenogorsk.
DISTRIBUTION: Turkestan to Mongolia; south to Afghanistan, N. India; Ladak; Nepal; Sikkim; W. and N. China.
COMMENT: Corbet, 1978:218, and Haltenorth, 1963:121, included *orientalis (=aries), musimon,* and *vignei;* but see also review by Nadler *et al.,* 1973, Z. Saugetierk., 38:109–125; Corbet and Hill, 1980:127.
PROTECTED STATUS: CITES - Appendix II as *O. ammon.*
CITES - Appendix I and U.S. ESA - Endangered as *O. a. hodgsoni* subspecies only.
ISIS NUMBER: 5301419009041001001.

Ovis aries Linnaeus, 1758. Syst. Nat., 10th ed., 1:70.
TYPE LOCALITY: Sweden (domesticated stock).
DISTRIBUTION: S. Turkey; E. Turkey to C. and S.E. Iran; possibly Cyprus.
COMMENT: Includes *orientalis;* see Nadler *et al.,* 1973, Z. Saugetierk., 38:109–125. Probably includes *musimon* (RSH); see Corbet and Hill, 1980:127. Hybridizes with *vignei* in C. Iran; see Valdez *et al.,* 1978, Evolution, 32:56–72 and Nadler *et al.,* 1971, Cytogenetics, 10:137–152. Introduced worldwide as domesticate; feral in many places. See also comment under *musimon.*
PROTECTED STATUS: CITES - Appendix I and U.S. ESA - Endangered as *O. orientalis ophion* subspecies only.
ISIS NUMBER: 5301419009041007001 as *O. aries.*
　　　　　　　　5301419009041006001 as *O. orientalis.*

Ovis canadensis Shaw, 1804. Nat. Misc., 51, text to pl. 610.
TYPE LOCALITY: Canada, Alberta, Mountains on Bow River, near Exshaw.
DISTRIBUTION: S. British Columbia and S.W. Alberta (Canada) to Coahuila, Chihuahua, Sonora and Baja California (Mexico).
COMMENT: Corbet, 1978:218, included *nivicola;* but see also Korobitsyna *et al.,* 1974, Quat. Res., 4:235–245; Corbet and Hill, 1980:127.
PROTECTED STATUS: CITES - Appendix II.
ISIS NUMBER: 5301419009041002001.

Ovis dalli Nelson, 1884. Proc. U.S. Nat. Mus., 7:13.
TYPE LOCALITY: Alaska, West bank of Yukon River.
DISTRIBUTION: Alaska to N. British Columbia and W. Mackenzie (Canada).
ISIS NUMBER: 5301419009041003001.

Ovis musimon Pallas, 1811. Zoogr. Rosso-Asiat., 1:230.
TYPE LOCALITY: Italy, Sardinia.
DISTRIBUTION: Originally restricted to islands of Sardinia and Corsica; probably Cyprus. Widely introduced into Europe.
COMMENT: The status of the Cyprian wild sheep, *ophion,* is controversial. It was allocated to *O. orientalis* (= *aries*) by Ellerman and Morrison-Scott, 1951:418, and Gromov and Baranova, 1981:406, but provisionally assigned to *musimon* by Nadler *et al.,* 1973, Z. Saugetierk., 38:119. Payne, 1968, Proc. Prehist. Soc., 34:368–384, considered *musimon* and *ophion* to be primitive feral stocks of domesticated *O. aries.* See also comment under *aries.*
ISIS NUMBER: 5301419009041004001.

Ovis nivicola Eschscholtz, 1829. Zool. Atlas, Part 1, p. 1, pl. 1.
  TYPE LOCALITY: U.S.S.R., E. Kamchatka (U.S.S.R.).
  DISTRIBUTION: Putorana Mts., N.C. Siberia; N.E. Siberia from Lena River east to
    Chukotka and Kamchatka (U.S.S.R.)
  COMMENT: Corbet, 1978:218, and others included *nivicola* in *canadensis*; but see review
    by Korobitsyna *et al.*, 1974, Quat. Res., 4:235–245; Corbet and Hill, 1980:127; and
    Gromov and Baranova, 1981:407.
  ISIS NUMBER: 5301419009041005001.

Ovis vignei Blyth, 1841. Proc. Zool. Soc. Lond., 1840:70.
  TYPE LOCALITY: India, Kashmir, Astor
  DISTRIBUTION: Uzbekistan, Tadzhikistan (U.S.S.R.) and N.E. Iran to Afghanistan,
    Pakistan, and N.W. India; Oman.
  COMMENT: Hybridizes with *aries* (=*orientalis*) in C. Iran; see Valdez *et al.*, 1978,
    Evol., 32:56–72, and Nadler *et al.*, 1971, Cytogenetics, 10:137–152.
  PROTECTED STATUS: CITES - Appendix I and U.S. ESA - Endangered.

Pantholops Hodgson, 1834. Proc. Zool. Soc. Lond., 1834:81.
  ISIS NUMBER: 5301419009042000000.

Pantholops hodgsoni (Abel, 1826). Calcutta Gov't Gazette., 68:234.
  TYPE LOCALITY: China, Tibet, Kooti Pass in Arrun Valley, Tingri Maiden.
  DISTRIBUTION: Tibet, Tsinghai, Szechwan (China) (SW); Ladak (N. India).
  PROTECTED STATUS: CITES - Appendix I.
  ISIS NUMBER: 5301419009042001001.

Pelea Gray, 1851. Proc. Zool. Soc. Lond., 1850:126.
  REVIEWED BY: D. Ernst (DE).
  ISIS NUMBER: 5301419009018000000.

Pelea capreolus (Forster, 1790). *In* Levaillant, Erste Reise Afrika, p. 71.
  TYPE LOCALITY: South Africa, Cape of Good Hope.
  DISTRIBUTION: Cape Prov. to Natal and Transvaal (South Africa).
  ISIS NUMBER: 5301419009018001001.

Procapra Hodgson, 1846. J. Asiat. Soc. Bengal, 15:334.
  COMMENT: Revised by Groves, 1967, Z. Saugetierk., 32:144–149. Gromov and Baranova,
    1981:393, considered *Procapra* a subgenus of *Gazella*; but also see Corbet and Hill,
    1980:126.
  ISIS NUMBER: 5301419009031000000.

Procapra gutturosa (Pallas, 1777). Spicil. Zool., 12:46.
  TYPE LOCALITY: U.S.S.R., S.E. Transbaikalia, Chitinsk. Obl., upper Onon River.
  DISTRIBUTION: Formerly, Mongolia except mts. and S.W. desert, N. China, Zaisan, and
    Transbaikalia (U.S.S.R.); now restricted to E. Mongolia and Inner Mongolia
    (China).
  COMMENT: See Sokolov and Orlov, 1980:280 for discussion of restricted range.
  ISIS NUMBER: 5301419009031001001.

Procapra picticaudata Hodgson, 1846. J. Asiat. Soc. Bengal, 15:334, pl. 2.
  TYPE LOCALITY: China, Tibet, Hundes Dist.
  DISTRIBUTION: Szechwan, Tsinghai, Tibet (China), adjacent Indian Himalayas; possibly
    Sinkiang (China) and Mongolia.
  COMMENT: According to Groves, 1967, Z. Saugetierk., 32:144–149, type locality is more
    likely the district N. of Sikkim.
  ISIS NUMBER: 5301419009031002001.

Procapra przewalskii (Buchner, 1891). Melanges Biol. Soc. St.Petersb., 13:161.
  TYPE LOCALITY: Mongolia, Chagrin-Gol (=Steppe).
  DISTRIBUTION: S. Mongolia; N.W. China.
  COMMENT: Considered a subspecies of *picticaudata* by Ellerman and Morrison-Scott,

1951:388. China, S. Ordos Desert was erroneously given as type locality by Ellerman and Morrison-Scott, 1951, and followed by Gromov and Baranova, 1981.

Pseudois Hodgson, 1846. J. Asiat. Soc. Bengal, 15:343.
COMMENT: Probably should be included in *Capra* on basis of hybridization data (RGVG).
ISIS NUMBER: 5301419009043000000.

Pseudois nayaur (Hodgson, 1833). Asiat. Res., 18(2):135.
TYPE LOCALITY: Nepal, Tibetan frontier.
DISTRIBUTION: Tibet, Himalayas from Ladak eastward to Yunnan, Kansu, Szechwan, and Shensi (China).
ISIS NUMBER: 5301419009043001001.

Pseudois schaeferi Haltenorth, 1963, Handb. Zool., 32:126.
TYPE LOCALITY: China, upper Yangtze Gorge, Drupalong, south of Batang.
DISTRIBUTION: Upper Yangtze Gorge (W. China).
COMMENT: Revised by Groves, 1978, Saugetierk. Mitt., 26:183.

Raphicerus H. Smith, 1827. *In* Griffith's Cuvier Anim. Kingd., 5:342.
REVIEWED BY: K. R. Kranz (KRK).
ISIS NUMBER: 5301419009032000000.

Raphicerus campestris (Thunberg, 1811). Mem. Acad. Imp. Sci. St. Petersb., 3:311.
TYPE LOCALITY: South Africa, Cape of Good Hope.
DISTRIBUTION: Angola to S. Africa to S. Kenya.
ISIS NUMBER: 5301419009032001001.

Raphicerus melanotis (Thunberg, 1811). Mem. Acad. Imp. Sci. St. Petersb., 3:312.
TYPE LOCALITY: South Africa, Cape of Good Hope.
DISTRIBUTION: S. Cape Prov. (South Africa).
ISIS NUMBER: 5301419009032002001.

Raphicerus sharpei Thomas, 1897. Proc. Zool. Soc. Lond., 1896:796, pl. 34.
TYPE LOCALITY: Malawi, Nyasaland, S. Angoniland.
DISTRIBUTION: South Africa to Tanzania and S.E. Zaire.
COMMENT: Included in *melanotis* by Haltenorth, 1963:78 but see Ansell, 1972, Part 15:67.
ISIS NUMBER: 5301419009032003001.

Redunca H. Smith, 1827. *In* Griffith's Cuvier Anim. Kingd., 5:337.
REVIEWED BY: D. Ernst (DE).
ISIS NUMBER: 5301419009019000000.

Redunca arundinum (Boddaert, 1785). Elench. Anim., p. 141.
TYPE LOCALITY: South Africa, Cape of Good Hope.
DISTRIBUTION: South Africa to Gabon and Zaire; Tanzania; Zambia; Mozambique; perhaps Malawi.
ISIS NUMBER: 5301419009019001001.

Redunca fulvorufula (Afzelius, 1815). Nova Acta Reg. Soc. Sci. Upsala, 7:250.
TYPE LOCALITY: South Africa, E. Cape of Good Hope.
DISTRIBUTION: Discontinuous in N. Cameroun; S.W. Ethiopia; S.E. Sudan; N.E. Uganda; W. Kenya; N.E. Tanzania; S. Mozambique; S.E. Botswana; South Africa.
ISIS NUMBER: 5301419009019002001.

Redunca redunca (Pallas, 1767). Spicil. Zool., Part 1, p. 8.
TYPE LOCALITY: Senegal, Goree (=Gori) Isl.
DISTRIBUTION: Senegal to C. Ethiopia; south to E. Zaire and S.W. Tanzania.
ISIS NUMBER: 5301419009019003001.

Rupicapra Blainville, 1816. Bull. Sci. Soc. Philom. Paris, p. 75.
    ISIS NUMBER: 5301419009044000000.

Rupicapra rupicapra (Linnaeus, 1758). Syst. Nat., 10th ed., 1:68.
    TYPE LOCALITY: Switzerland.
    DISTRIBUTION: N.W. Spain, Pyrenees to Caucasus, Carpathian, and Tatra Mtns.; Taurus
        Mts.; N.E. Turkey; "N. Asia Minor" (Kurdistan?).
    PROTECTED STATUS: CITES - Appendix I and U.S. ESA - Endangered as *R. r. ornata*
        subspecies only.
    ISIS NUMBER: 5301419009044001001.

Saiga Gray, 1843. List Mamm. Br. Mus., p. 26.
    ISIS NUMBER: 5301419009045000000.

Saiga tatarica (Linnaeus, 1766). Syst. Nat., 12th ed., 1:97.
    TYPE LOCALITY: U.S.S.R., W. Kazakhstan, "Ural Steppes."
    DISTRIBUTION: N. Caucasus (Kalmyk Steppe), Kazakhstan (U.S.S.R.); S.W. Mongolia;
        Sinkiang (China). Formerly west to Poland.
    COMMENT: Reviewed by Sokolov, 1974, Mamm. Species, 28:1-4.
    PROTECTED STATUS: U.S. ESA - Endangered as *S. t. mongolica* subspecies only.
    ISIS NUMBER: 5301419009045001001.

Sigmoceros Heller, 1912. Smithson. Misc. Coll., 60(8):4.
    COMMENT: Formerly included in *Alcelaphus*; see Vrba, 1979, Biol. J. Linn. Soc., 11:273.

Sigmoceros lichtensteini (Peters, 1852). Reise nach Mossambique, Saugeth., p. 190, pl. 43,
    44.
    TYPE LOCALITY: Mozambique, Tette.
    DISTRIBUTION: C. Tanzania, S.W. Zaire and N.E. Angola to N.E. Zimbabwe and C.
        Mozambique.
    COMMENT: Included in *A. buselaphus* by Haltenorth, 1963:102, but placed in separate
        genus, *Sigmoceros* by Vrba, 1979, J. Linn. Soc., 11:207-228.
    ISIS NUMBER: 5301419009012002001 as *Alcelaphus lichtensteini.*

Sylvicapra Ogilby, 1837. Proc. Zool. Soc. Lond., 1836:138.
    ISIS NUMBER: 5301419009010000000.

Sylvicapra grimmia Linnaeus, 1758. Syst. Nat., 10th ed., 1:70.
    TYPE LOCALITY: South Africa, Cape Prov., Capetown.
    DISTRIBUTION: Subsaharan Africa, except tropical lowlands of Zaire basin.
    COMMENT: Included in *Alcelaphus buselaphus* by Haltenorth, 1963:102, but placed in
        separate genus, *Sigmoceros* by Vrba, 1979, J. Linn. Soc., 11:207-228.
    ISIS NUMBER: 5301419009010001001.

Syncerus Hodgson, 1847. J. Asiat. Soc. Bengal, ser. 2, 16:709.
    COMMENT: Considered a subgenus of *Bubalus* by Haltenorth, 1963:133.
    ISIS NUMBER: 5301419009005000000.

Syncerus caffer (Sparrman, 1779). Kongl. Svenska Vet.-Akad. Handl. Stockholm, 40:79.
    TYPE LOCALITY: South Africa, Cape of Good Hope, Sunday River, Algoa Bay.
    DISTRIBUTION: Senegal to S. Ethiopia to S. Africa.
    COMMENT: Includes *nanus*; see Ansell, 1972, Part 15:18.
    ISIS NUMBER: 5301419009005001001.

Tetracerus Leach, 1825. Trans. Linn. Soc. Lond., 14:524.
    ISIS NUMBER: 5301419009007000000.

Tetracerus quadricornis (Blainville, 1816). Bull. Sci. Soc. Philom. Paris, p. 75.
    TYPE LOCALITY: India, no exact locality ("Plains of Peninsular India").
    DISTRIBUTION: Peninsular India, north to Nepal.
    PROTECTED STATUS: CITES - Appendix III (Nepal).
    ISIS NUMBER: 5301419009007001001.

Tragelaphus Blainville, 1816. Bull. Sci. Soc. Philom. Paris, p. 75.
    REVIEWED BY: D. Ernst (DE).

COMMENT: Includes *Boocercus*; see Ansell, 1972, Part 15:20. Also includes *Taurotragus*; see Van Gelder, 1977, Lammergeyer, 23:1-6; Van Gelder, 1977, Am. Mus. Novit., 2635:17-18, and Ansell, 1978, The Mammals of Zambia, p. 53; but see also Swanepoel *et al.*, 1980:164.
ISIS NUMBER: 5301419009008000000 as *Tragelaphus.*
5301419009006000000 as *Taurotragus.*

Tragelaphus angasi Gray, 1849. Proc. Zool. Soc. Lond., 1848:89.
TYPE LOCALITY: South Africa, Natal, Zululand, St. Lucia Bay.
DISTRIBUTION: N.E. South Africa; Mozambique; S. Malawi; S.E. Zimbabwe.
ISIS NUMBER: 5301419009008001001.

Tragelaphus buxtoni (Lydekker, 1910). Nature (Lond.), 84:397.
TYPE LOCALITY: Ethiopia (=Abyssinia), Arussi Prov., Gallaland, Sahatu Mtns., southeast of Lake Zwai, 9000 ft. (2743 m).
DISTRIBUTION: Ethiopia, east of Rift Valley.
ISIS NUMBER: 5301419009008002001.

Tragelaphus eurycerus (Ogilby, 1837). Proc. Zool. Soc. Lond., 1836:120.
TYPE LOCALITY: West Africa.
DISTRIBUTION: Sierra Leone to Togo; Cameroun, Central African Republic, Congo Republic; Zaire; Kenya; perhaps Gabon, N. Tanzania, and S.W. Ethiopia.
COMMENT: *T. euryceros* is a later spelling; see Haltenorth, 1963:86. Formerly placed in *Boocercus*; reviewed by Ralls, 1978, Mamm. Species, 111:1-4.
PROTECTED STATUS: CITES - Appendix III (Ghana) as *Boocercus (Taurotragus) euryceros* (*sic*).
ISIS NUMBER: 5301419009008003001.

Tragelaphus imberbis (Blyth, 1869). Proc. Zool. Soc. Lond., 1869:55.
TYPE LOCALITY: Ethiopia (=Abyssinia).
DISTRIBUTION: Kenya; E. Tanzania; N.E. Uganda; Somalia; S.E. Ethiopia; S.E. Sudan; Yemen.
ISIS NUMBER: 5301419009008004001.

Tragelaphus oryx (Pallas, 1766). Misc. Zool., p. 9.
TYPE LOCALITY: South Africa, Cape of Good Hope.
DISTRIBUTION: Gambia, Senegal, Guinea-Bissau, Guinea, and Nigeria to Sudan; Angola to S. Africa; thence north through Botswana, S. Zaire, Malawi, Zambia, and Tanzania to S.E. Sudan and S.W. Ethiopia; west to Senegal.
COMMENT: Includes *derbianus*; see Van Gelder, 1977, Lammergeyer, 23:1-6. PG believes *derbianus* is a valid species. Formerly included in *Taurotragus*; see Ansell, 1978:53; but also see Swanepoel *et al.*, 1980:187.
PROTECTED STATUS: U.S. ESA - Endangered as *Taurotragus derbianus derbianus* subspecies only.
ISIS NUMBER: 5301419009006002001 as *Taurotragus oryx.*
5301419009006001001 as *Taurotragus derbianus.*

Tragelaphus scriptus (Pallas, 1766). Misc. Zool., p. 8.
TYPE LOCALITY: Senegal.
DISTRIBUTION: S. Mauritania to Ethiopia and S. Somalia, to South Africa to N. Namibia.
ISIS NUMBER: 5301419009008005001.

Tragelaphus spekei Speke, 1863, *in* Journal of the Discovery of the Source of the Nile, p. 223.
TYPE LOCALITY: Tanzania, Karaque, west of Lake Victoria.
DISTRIBUTION: Gambia to S.W. Ethiopia, south to Angola, Namibia, N.W. Botswana, and South Africa.
COMMENT: The specific name is usually attributed to Sclater, 1864.
PROTECTED STATUS: CITES - Appendix III (Ghana).
ISIS NUMBER: 5301419009008006001.

Tragelaphus strepsiceros (Pallas, 1766). Misc. Zool., p. 9.
TYPE LOCALITY: South Africa, Cape of Good Hope.
DISTRIBUTION: Somalia and Ethiopia to N. South Africa, west to Namibia and S.E. Zaire, N.E. Uganda, S.W. Sudan, Central African Republic and Chad.
ISIS NUMBER: 5301419009008007001.

ORDER PHOLIDOTA
ISIS NUMBER: 5301408000000000000.

Family Manidae
REVIEWED BY: J. G. Hallett (JGH); S. Wang (SW)(China).
ISIS NUMBER: 5301408001000000000.

Manis Linnaeus, 1758. Syst. Nat., 10th ed., 1:36.
  COMMENT: Morphological evidence suggests subdivision of the genus; see Patterson,
    1978, *in* Maglio and Cooke, eds. Evolution of African Mammals, Harvard Univ.
    Press.
  ISIS NUMBER: 5301408001001000000.

Manis crassicaudata Gray, 1827. *In* Griffith's Cuvier Anim. Kingd., 5:282.
  TYPE LOCALITY: India.
  DISTRIBUTION: Pakistan, east to W. Bengal (India) and Yunnan (China)(SW), south to
    Sri Lanka.
  COMMENT: Formerly erroneously called *pentadactyla;* see Emry, 1970, Bull. Am. Mus.
    Nat. Hist., 142:460; Ellerman and Morrison-Scott, 1951.
  PROTECTED STATUS: CITES - Appendix II.
  ISIS NUMBER: 5301408001001001001.

Manis gigantea Illiger, 1815. Abh. Konigl. Akad. Wiss. Berlin, p. 84.
  TYPE LOCALITY: Not found.
  DISTRIBUTION: Senegal to W. Kenya, south to Rwanda, C. Zaire and S.W. Angola.
  PROTECTED STATUS: CITES - Appendix III (Ghana).
  ISIS NUMBER: 5301408001001002001.

Manis javanica Desmarest, 1822. Encyclop. Method. Mamm., 2:377.
  TYPE LOCALITY: Indonesia, Java.
  DISTRIBUTION: Burma; Thailand; Indochina; Sumatra; Java; Borneo; S.W. Philippines;
    adjacent small islands.
  COMMENT: Reviewed by Ellerman and Morrison-Scott, 1951; includes *culionensis,*
    considered a separate species by Sanborn, 1952, Fieldiana Zool., 33:114.
  PROTECTED STATUS: CITES - Appendix II.
  ISIS NUMBER: 5301408001001003001.

Manis pentadactyla Linnaeus, 1758. Syst. Nat., 10th ed., 1:36.
  TYPE LOCALITY: Taiwan (=Formosa).
  DISTRIBUTION: Nepal to S. China, Hainan Isl. (China), and N. Indochina; Taiwan.
  COMMENT: Includes *aurita;* see Emry, 1970, Bull. Am. Mus. Nat. Hist., 142:460;
    Ellerman and Morrison-Scott, 1951.
  PROTECTED STATUS: CITES - Appendix II.
  ISIS NUMBER: 5301408001001005001.

Manis temminckii Smuts, 1832. Enumer. Mamm. Cap., p. 54.
  TYPE LOCALITY: South Africa, N. Cape Prov., Latukou (Lataku?) (=Litakun), near
    Kuruman.
  DISTRIBUTION: N. South Africa; N. and E. Namibia; Zimbabwe; Mozambique;
    Botswana; Angola; Kenya; S. Zaire; S. Sudan; Chad.
  COMMENT: Reviewed by Stuart, 1980, Saugetierk. Mitt., 28:123–129.
  PROTECTED STATUS: CITES - Appendix I and U.S. ESA - Endangered.
  ISIS NUMBER: 5301408001001006001.

Manis tetradactyla Linnaeus, 1766. Syst. Nat., 12th ed., 1:53.
  TYPE LOCALITY: West Africa.
  DISTRIBUTION: Senegal and Gambia to W. Uganda, south to S.W. Angola.
  COMMENT: *M. longicaudata* is a synonym; see Meester, 1972, Part 4:2.
  PROTECTED STATUS: CITES - Appendix III (Ghana) as *M. longicaudata* only.
  ISIS NUMBER: 5301408001001004001 as *M. longicaudata.*

Manis tricuspis Rafinesque, 1821. Ann. Sci. Phys. Brux., 7:215.
  TYPE LOCALITY: West Africa, "Guinee."
  DISTRIBUTION: Senegal to W. Kenya, south to N.E. Zambia and S.W. Angola; Bioko.
  PROTECTED STATUS: CITES - Appendix III (Ghana).
  ISIS NUMBER: 5301408001001007001.

ORDER RODENTIA
ISIS NUMBER: 5301410000000000000.

Family Aplodontidae
REVIEWED BY: H. Levenson (HL).
ISIS NUMBER: 5301410001000000000.

Aplodontia Richardson, 1829. Zool. J., 4:334.
COMMENT: Revised by Taylor, 1918, Univ. Calif. Publ. Zool., 17:455.
ISIS NUMBER: 5301410001001000000.

Aplodontia rufa (Rafinesque, 1817). Am. Mon. Mag., 2:45.
TYPE LOCALITY: U.S.A., Oregon, neighborhood Columbia River.
DISTRIBUTION: S.W. British Columbia (Canada) to C. California (U.S.A.).
COMMENT: Specimens from U.S.A., Oregon, Clackamas Co., Marmot, regarded as
typical; see Taylor, 1918, Univ. Calif. Publ. Zool., 17:455.
ISIS NUMBER: 5301410001001001001.

Family Sciuridae
REVIEWED BY: L. R. Heaney (LRH)(Asia); R. S. Hoffmann (RSH); S. L. Lindsay (SLL); G.
McGrath (GM); T. J. McIntyre (TJM)(Asia); T. Pearson (TP); J. Ramirez-Pulido
(JRP)(Mexico); O. L. Rossolimo (OLR)(U.S.S.R.); S. Wang (SW)(China).
COMMENT: Reviewed by Moore, 1959, Bull. Am. Mus. Nat. Hist., 118:153–206, except
flying squirrels.
ISIS NUMBER: 5301410002000000000.

Aeretes G. M. Allen, 1940. Nat. Hist. Cent. Asia, II, 2:745.
ISIS NUMBER: 5301410002039000000.

Aeretes melanopterus (Milne-Edwards, 1867). Ann. Sci. Nat. Zool., 8:375.
TYPE LOCALITY: China, Hopei.
DISTRIBUTION: Hopei and Szechwan (China).
ISIS NUMBER: 5301410002039001001.

Aeromys Robinson and Kloss, 1915. J. Fed. Malay St. Mus., 6:23.
ISIS NUMBER: 5301410002040000000.

Aeromys tephromelas (Gunther, 1873). Proc. Zool. Soc. Lond., 1873:413.
TYPE LOCALITY: Malaysia, Penang Isl., Wellesley.
DISTRIBUTION: Malaysian subregion, except Java and S.W. Philippines.
COMMENT: Includes *phaeomelas*; see Medway, 1977:101.
ISIS NUMBER: 5301410002040001001.

Aeromys thomasi (Hose, 1900). Ann. Mag. Nat. Hist., 5:215.
TYPE LOCALITY: Malaysia, Sarawak, Baram, Silat River.
DISTRIBUTION: Borneo, except S.E.

Ammospermophilus Merriam, 1892. Proc. Biol. Soc. Wash., 7:27.
REVIEWED BY: D. J. Hafner (DJH).
COMMENT: Formerly included in *Spermophilus* (=*Citellus*) by Hershkovitz, 1949, J.
Mammal., 30:296. Bryant, 1945, Am. Midl. Nat., 33:374–375, considered
Ammospermophilus a distinct genus. Recent biochemical and chromosomal data
support Bryant's recognition of generic status; currently being revised by DJH.
ISIS NUMBER: 5301410002001000000.

Ammospermophilus harrisii (Audubon and Bachman, 1854). The Viviparous Quadrupeds of
North America, 3:267.
TYPE LOCALITY: U.S.A., Arizona, Santa Cruz Co., Santa Cruz Valley at the Mexican
boundary.
DISTRIBUTION: Arizona to S.W. New Mexico (U.S.A.) and adjoining Sonora (Mexico).
ISIS NUMBER: 5301410002001001001.

Ammospermophilus insularis Nelson and Goldman, 1909. Proc. Biol. Soc. Wash., 22:24.
TYPE LOCALITY: Mexico, Baja California Sur, Isla Espiritu Santo.
DISTRIBUTION: Known only from the type locality.

COMMENT: Considered a distinct species by Hall, 1981:381. Specific validity unclear, due to conflicting chromosomal, allozymic, morphological and zoogeographic data (DJH).
ISIS NUMBER: 5301410002001002001.

Ammospermophilus interpres (Merriam, 1890). N. Am. Fauna, 4:21.
TYPE LOCALITY: U.S.A., Texas, El Paso Co., El Paso.
DISTRIBUTION: New Mexico and W. Texas (U.S.A.) to Coahuila, Chihuahua, and Durango (Mexico).
ISIS NUMBER: 5301410002001003001.

Ammospermophilus leucurus (Merriam, 1889). N. Am. Fauna, 2:20.
TYPE LOCALITY: U.S.A., California, Riverside Co., San Gorgonio Pass.
DISTRIBUTION: E. California and S.E. Oregon to Colorado and New Mexico (U.S.A.), south to Baja California Sur (Mexico).
COMMENT: Ongoing investigation indicates that populations south of approx. 29° 30′ N. lat. on the Baja California peninsula may constitute a separate species (oldest available name: *A. canfieldae* Huey) (DJH).
ISIS NUMBER: 5301410002001004001.

Ammospermophilus nelsoni (Merriam, 1893). Proc. Biol. Soc. Wash., 8:129.
TYPE LOCALITY: U.S.A., California, Tulare Co., Tipton.
DISTRIBUTION: San Joaquin Valley (S. California, U.S.A.).
COMMENT: May now be restricted to southern half of its former range (DJH).
ISIS NUMBER: 5301410002001005001.

Atlantoxerus Forsyth Major, 1893. Proc. Zool. Soc. Lond., 1893:189.
COMMENT: Closely related to *Xerus*; see Corbet, 1978:79.
ISIS NUMBER: 5301410002002000000.

Atlantoxerus getulus (Linnaeus, 1758). Syst. Nat., 10th ed., 1:64.
TYPE LOCALITY: Morocco, Agadir.
DISTRIBUTION: Morocco, Algeria.
COMMENT: Includes *trivittatus*; see Corbet, 1978:79.
ISIS NUMBER: 5301410002002001001.

Belomys Thomas, 1908. Ann. Mag. Nat. Hist., ser. 8, 1:2.
ISIS NUMBER: 5301410002041000000.

Belomys pearsoni (Gray, 1842). Ann. Mag. Nat. Hist., 10:263.
TYPE LOCALITY: India, Assam, Darjeeling.
DISTRIBUTION: Sikkim and Assam (India) to S. China, Hainan, Taiwan, and Indochina.
ISIS NUMBER: 5301410002041001001.

Callosciurus Gray, 1867. Ann. Mag. Nat. Hist., 20:277.
COMMENT: Formerly included *Sundasciurus* and *Prosciurillus* (in part; see Moore, 1958, Am. Mus. Novit., 1890:1–5) and *Tamiops* (see Moore and Tate, 1965, Fieldiana Zool., 48:1–351).
ISIS NUMBER: 5301410002003000000.

Callosciurus adamsi (Kloss, 1921). J. Str. Br. Roy. Asiat. Soc., 83:151.
TYPE LOCALITY: Malaysia, Sarawak, Baram, Long Mujan, 150 mi. (241 km) up Baram River, 700–900 ft. (213–274 m).
DISTRIBUTION: N. Borneo.
COMMENT: Distinct from *albescens*; see Medway, 1977:93.

Callosciurus albescens (Bonhote, 1901). Ann. Mag. Nat. Hist., ser. 7, 7:446.
TYPE LOCALITY: Indonesia, Sumatra, Acheen (=Atjeh).
DISTRIBUTION: Sumatra.
COMMENT: Considered a distinct species by Medway, 1977:93.

Callosciurus baluensis (Bonhote, 1901). Ann. Mag. Nat. Hist., ser. 7, 7:174.
TYPE LOCALITY: Malaysia, Sabah, Mt. Kinabalu.
DISTRIBUTION: Higher elevations on Mt. Kinabalu (Sabah, Malaysia).
COMMENT: Often considered a subspecies of *prevosti*; see Medway, 1977:86.

Callosciurus caniceps (Gray, 1842). Ann. Mag. Nat. Hist., ser. 1, 10:263.
TYPE LOCALITY: Burma, Tenasserim.
DISTRIBUTION: Thailand, peninsular Burma, peninsular Malaysia, and adjacent islands.
COMMENT: Revised by Moore and Tate, 1965, Fieldiana Zool., 48:1–351. Closely related
    to *erythraeus, inornatus,* and *phayrei* (LRH).
ISIS NUMBER: 5301410002003001001.

Callosciurus erythraeus (Pallas, 1778). Nova Spec. Quad. Glir. Ord., p. 377.
TYPE LOCALITY: India, Assam, Garo Hills.
DISTRIBUTION: West of Irrawaddy River in India; Burma; S.E. China.
COMMENT: Includes *sladeni;* see Moore and Tate, 1965, Fieldiana Zool., 48:113. May
    include *flavimanus* (LRH).
ISIS NUMBER: 5301410002003002001 as *C. erythraeus.*
    5301410002003013001 as *C. sladeni.*

Callosciurus ferrugineus (F. Cuvier, 1829). Tabl. Meth. Mamm., 3:238.
TYPE LOCALITY: "India."
DISTRIBUTION: S.C. Burma.
COMMENT: Often considered conspecific with *finlaysoni,* but see Moore and Tate, 1965,
    Fieldiana Zool., 48:158, who considered *ferrugineus* a distinct species.
ISIS NUMBER: 5301410002003003001.

Callosciurus finlaysoni (Horsfield, 1824). Zool. Res. Java, 7:7.
TYPE LOCALITY: Thailand, Si Chang Isl. (N. Gulf of Thailand).
DISTRIBUTION: Thailand; Kampuchea; Laos; Vietnam.
COMMENT: Revised by Moore and Tate, 1965, Fieldiana Zool., 48:161.
ISIS NUMBER: 5301410002003004001.

Callosciurus flavimanus (I. Geoffroy, 1831). *In* Zool. Voy. de Belanger Indes Orient., 1:148.
TYPE LOCALITY: Vietnam, Quang Nam Prov., Da Nang (=Tourane).
DISTRIBUTION: East of Irrawaddy River in Burma, Thailand, Malaya, Indochina, S.
    China, and Taiwan.
COMMENT: May be conspecific with *erythraeus* (LRH).
ISIS NUMBER: 5301410002003005001.

Callosciurus inornatus (Gray, 1867). Ann. Mag. Nat. Hist., ser. 3, 20:282.
TYPE LOCALITY: "Mountains in Laos."
DISTRIBUTION: Laos; N. Vietnam.
COMMENT: Revised by Moore and Tate, 1965, Fieldiana Zool., 48:209.
ISIS NUMBER: 5301410002003006001.

Callosciurus melanogaster (Thomas, 1896). Ann. Mus. Civ. Stor. Nat. Genova, 14:668.
TYPE LOCALITY: Indonesia, Mentawi Isls., Sipora Isl.
DISTRIBUTION: Mentawi Isls. (Indonesia).
COMMENT: Member of the *notatus* species group (LRH).
ISIS NUMBER: 5301410002003007001.

Callosciurus nigrovittatus (Horsfield, 1823). Zool. Res. Java, 7:149.
TYPE LOCALITY: Indonesia, West Java.
DISTRIBUTION: Malay Peninsula; Sumatra; Borneo; Java; adjacent small islands.
COMMENT: Member of the *notatus* species group (LRH).
ISIS NUMBER: 5301410002003008001.

Callosciurus notatus (Boddaert, 1785). Elench. Anim., p. 119.
TYPE LOCALITY: Indonesia, West Java.
DISTRIBUTION: Widespread on islands from Malaya to Java, Bali, and Borneo; Salayer
    Isl. (south of Sulawesi).
ISIS NUMBER: 5301410002003009001.

Callosciurus phayrei (Blyth, 1855). J. Asiat. Soc. Bengal, 24:472.
TYPE LOCALITY: Burma, Martaban.
DISTRIBUTION: S. Burma.
COMMENT: Considered a distinct species by Moore and Tate, 1965, Fieldiana Zool.,
    48:201. Closely related to *caniceps* (LRH).

Callosciurus prevosti (Desmarest, 1822). Encyclop. Method. Mamm., p. 335.
TYPE LOCALITY: Malaysia, Malacca Prov., Malacca.
DISTRIBUTION: Malaysian subregion except Java; Sulawesi.
COMMENT: May include *baluensis*; see Medway, 1977:86. Reviewed by Heaney, 1978, Evolution, 32:29–44.
ISIS NUMBER: 5301410002003010001.

Callosciurus pygerythrus (I. Geoffroy, 1832). Mag. Zool. Paris, p. 5, pl. 4–6.
TYPE LOCALITY: Burma, Syriam, near Pegu.
DISTRIBUTION: Nepal to Burma and Yunnan (China).
COMMENT: Revised by Moore and Tate, 1965, Fieldiana Zool., 48:209.
ISIS NUMBER: 5301410002003011001.

Callosciurus quinquestriatus (Anderson, 1871). Proc. Zool. Soc. Lond., 1871:142.
TYPE LOCALITY: Burma, Kakhven Hills, Ponsee (East of Bhamo).
DISTRIBUTION: N.E. Burma; Yunnan (China).
COMMENT: Included in *flavimanus* by Moore and Tate, 1965, Fieldiana Zool., 48:209, but see Corbet and Hill, 1980:132. LRH considers *quinquestriatus* a distinct species.
ISIS NUMBER: 5301410002003012001.

Cynomys Rafinesque, 1817. Am. Mon. Mag., 2:43.
REVIEWED BY: S. D. Shalaway (SDS).
COMMENT: Revised by Pizzimenti, 1975, Occas. Pap. Mus. Nat. Hist. Univ. Kans., 39:1–73. Clark *et al.*, 1971, Mamm. Species, 7:1–4, published a key to the genus.
ISIS NUMBER: 5301410002004000000.

Cynomys gunnisoni (Baird, 1855). Proc. Acad. Nat. Sci. Phila., 7:334.
TYPE LOCALITY: U.S.A., Colorado, Saguache Co., Cochetopa Pass.
DISTRIBUTION: S.E. Utah, S.W. Colorado, N.E. Arizona, and N.W. New Mexico (U.S.A.).
COMMENT: Subgenus *Leucocrossuromys*; see Hall, 1981. Reviewed by Pizzimenti and Hoffmann, 1973, Mamm. Species, 25:1–4.
ISIS NUMBER: 5301410002004001001.

Cynomys leucurus Merriam, 1890. N. Am. Fauna, 3:59.
TYPE LOCALITY: U.S.A., Wyoming, Uinta Co., Fort Bridger.
DISTRIBUTION: S.C. Montana, W. and C. Wyoming, N.E. Utah, and N.W. Colorado (U.S.A.)
COMMENT: Subgenus *Leucocrossuromys*; see Hall, 1981. Reviewed by Clark *et al.*, 1971, Mamm. Species, 7:1–4.
ISIS NUMBER: 5301410002004002001.

Cynomys ludovicianus (Ord, 1815). *In* Guthrie, A New Geogr., Hist. Coml. Grammar, Phila., 2nd ed., 2:292.
TYPE LOCALITY: U.S.A., "Upper Missouri River."
DISTRIBUTION: Saskatchewan (Canada); Montana to E. Nebraska, W. Texas, New Mexico, and S.E. Arizona (U.S.A.); N.E. Sonora, and N. Chihuahua (Mexico).
COMMENT: Subgenus *Cynomys*; see Hall, 1981:411.
ISIS NUMBER: 5301410002004003001.

Cynomys mexicanus Merriam, 1892. Proc. Biol. Soc. Wash., 7:157.
TYPE LOCALITY: Mexico, Coahuila, La Ventura.
DISTRIBUTION: Coahuila, and San Luis Potosi; perhaps Nuevo Leon, and Zacatecas (N.C. Mexico).
COMMENT: Subgenus *Cynomys*; see Hall, 1981:412.
PROTECTED STATUS: CITES - Appendix I and U.S. ESA - Endangered.
ISIS NUMBER: 5301410002004004001.

Cynomys parvidens J. A. Allen, 1905. Mus. Brooklyn Inst. Arts and Sci., Sci. Bull., 1:119.
TYPE LOCALITY: U.S.A., Utah, Iron Co., Buckskin Valley.
DISTRIBUTION: S.C. Utah (U.S.A.).
COMMENT: Subgenus *Leucocrossuromys*; see Hall, 1981. Reviewed by Pizzimenti and Collier, 1975, Mamm. Species, 52:1–3.

PROTECTED STATUS: U.S. ESA - Endangered.
ISIS NUMBER: 5301410002004005001.

Dremomys Heude, 1898. Mem. Hist. Nat. Emp. Chin., 4(2):54.
ISIS NUMBER: 5301410002005000000.

Dremomys everetti (Thomas, 1890). Ann. Mag. Nat. Hist., ser. 6, 6:171.
TYPE LOCALITY: Malaysia, Sarawak, Mt. Penrisen.
DISTRIBUTION: Mountains of Borneo.
ISIS NUMBER: 5301410002005001001.

Dremomys lokriah (Hodgson, 1836). J. Asiat. Soc. Bengal, 5:232.
TYPE LOCALITY: Nepal.
DISTRIBUTION: C. Nepal east to Salween River; Tibet (China); N. Burma; mountains in
E. India.
COMMENT: Revised by Moore and Tate, 1965, Fieldiana Zool., 48:263.
ISIS NUMBER: 5301410002005002001.

Dremomys pernyi (Milne-Edwards, 1867). Rev. Mag. Zool. Paris, ser. 2, 19:19.
TYPE LOCALITY: China, Szechwan, mountains near Moupin.
DISTRIBUTION: Burma and Tibet to C. and S.E. China; Taiwan.
ISIS NUMBER: 5301410002005003001.

Dremomys pyrrhomerus (Thomas, 1895). Ann. Mag. Nat. Hist., ser. 6, 16:242.
TYPE LOCALITY: China, Hupei, Ichang.
DISTRIBUTION: C. and S. China; N. Indochina.
COMMENT: May be conspecific with *rufigenis;* see Moore and Tate, 1965, Fieldiana
Zool., 48:284, who treated them as separate species.
ISIS NUMBER: 5301410002005004001.

Dremomys rufigenis (Blanford, 1878). J. Asiat. Soc. Bengal, 47(2):156.
TYPE LOCALITY: Burma, Tenasserim, Mt. Moolevit.
DISTRIBUTION: N. Burma to Vietnam and peninsular Malaysia.
COMMENT: May include *pyrrhomerus;* see Moore and Tate, 1965, Fieldiana Zool., 48:284,
who treated them as separate species.
ISIS NUMBER: 5301410002005005001.

Epixerus Thomas, 1909. Ann. Mag. Nat. Hist., 8(3):472.
ISIS NUMBER: 5301410002006000000.

Epixerus ebii (Temminck, 1853). Esquisses Zool. sur la Cote de Guine, p. 129.
TYPE LOCALITY: Ghana.
DISTRIBUTION: Sierra Leone to Ghana.
COMMENT: Includes *jonesi;* formerly included *wilsoni;* see Robbins and Schlitter, 1977,
Saugetierk. Mitt., 25:22–23.
PROTECTED STATUS: CITES - Appendix III (Ghana).
ISIS NUMBER: 5301410002006001001.

Epixerus wilsoni du Chaillu, 1860. Proc. Boston Soc. Nat. Hist., 7:364.
TYPE LOCALITY: Gabon.
DISTRIBUTION: Cameroun to Gabon.
COMMENT: Amtmann, 1975, Part 6.1:6–7, included *wilsoni* with *ebii,* but Robbins and
Schlitter, 1977, Saugetierk. Mitt., 25:22–23, recognized *wilsoni* as a distinct species.
ISIS NUMBER: 5301410002006002001.

Eupetaurus Thomas, 1888. J. Asiat. Soc. Bengal, 57:256.
ISIS NUMBER: 5301410002042000000.

Eupetaurus cinereus Thomas, 1888. J. Asiat. Soc. Bengal, 57:258.
TYPE LOCALITY: Pakistan, Gilgit Valley.
DISTRIBUTION: High elevations in N. Pakistan and Kashmir.
COMMENT: Reviewed by McKenna, 1962, Am. Mus. Novit., 2104:1–38.
ISIS NUMBER: 5301410002042001001.

Exilisciurus Moore, 1958. Am. Mus. Novit., 1914:4.
COMMENT: The species included in *Exilisciurus* were formerly included in *Nannosciurus;* see Moore, 1958, Am. Mus. Novit., 1914:1-10, and Moore, 1959, Bull. Am. Mus. Nat. Hist., 118(4):203.
ISIS NUMBER: 5301410002008000000.

Exilisciurus concinnus (Thomas, 1888). Ann. Mag. Nat. Hist., ser. 6, 2:407.
TYPE LOCALITY: Philippines, Zamboanga Prov., Basilan Isl.
DISTRIBUTION: Basilan Isl. (Philippines).
COMMENT: Possibly conspecific with *luncefordi, samaricus,* and *surrutilus* (LRH).
ISIS NUMBER: 5301410002008001001.

Exilisciurus exilis (Muller, 1838). Tijdschr. Nat. Gesch. Physiol., 5:138.
TYPE LOCALITY: Indonesia, Kalimantan, Kapuas River Basin.
DISTRIBUTION: Borneo.
ISIS NUMBER: 5301410002008002001.

Exilisciurus luncefordi (Taylor, 1934). Monogr. Bur. Sci. Manila, 30:373.
TYPE LOCALITY: Philippines, Mindanao, Cotabato Dist., Saub.
DISTRIBUTION: S. Mindanao (Philippines).
COMMENT: See comment under *concinnus.*
ISIS NUMBER: 5301410002020001001 as *Nannosciurus luncefordi.*

Exilisciurus samaricus (Thomas, 1897). Trans. Zool. Soc. Lond., 14(6):389.
TYPE LOCALITY: Philippines, Samar Isl.
DISTRIBUTION: Samar, Leyte, and adjacent islands (Philippines).
COMMENT: See comment under *concinnus.*
ISIS NUMBER: 5301410002020003001 as *Nannosciurus samaricus.*

Exilisciurus surrutilus (Hollister, 1913). Proc. U.S. Nat. Mus., 46:313.
TYPE LOCALITY: Philippines, Zamboanga Prov., Mindanao Isl., Mt. Bliss.
DISTRIBUTION: Mindanao, except southern portion.
COMMENT: See comment under *concinnus.*
ISIS NUMBER: 5301410002020004001 as *Nannosciurus surrutilus.*

Exilisciurus whiteheadi (Thomas, 1887). Ann. Mag. Nat. Hist., ser. 5, 20:127.
TYPE LOCALITY: Malaysia, Sabah, Mt. Kinabalu.
DISTRIBUTION: Mountains of Sabah and Sarawak (Malaysia), above 1000 m.
ISIS NUMBER: 5301410002008003001.

Funambulus Lesson, 1835. Illustr. de Zool., pl. 43.
COMMENT: Reviewed by Moore, 1960, Syst. Zool., 9:1-17.
ISIS NUMBER: 5301410002009000000.

Funambulus layardi (Blyth, 1849). J. Asiat. Soc. Bengal, 18:602.
TYPE LOCALITY: Sri Lanka (=Ceylon), "uplands."
DISTRIBUTION: S. and C. Sri Lanka, and mountains of S. India.
ISIS NUMBER: 5301410002009001001.

Funambulus palmarum (Linnaeus, 1766). Syst. Nat., 12th ed., 1:86.
TYPE LOCALITY: India, Madras.
DISTRIBUTION: C. and S. India; Sri Lanka.
ISIS NUMBER: 5301410002009002001.

Funambulus pennanti Wroughton, 1905. J. Bombay Nat. Hist. Soc., 16:411.
TYPE LOCALITY: India, Surat Dist., Bundha, Mandvi Taluka.
DISTRIBUTION: E. Iran to Nepal and C. India.
ISIS NUMBER: 5301410002009003001.

Funambulus sublineatus (Waterhouse, 1838). Proc. Zool. Soc. Lond., 1838:19.
TYPE LOCALITY: India, Madras, Nilgiri Hills.
DISTRIBUTION: S. India; C. Sri Lanka.
ISIS NUMBER: 5301410002009004001.

Funambulus tristriatus (Waterhouse, 1837). Mag. Nat. Hist., n.s., 1:499.
TYPE LOCALITY: India, Western Ghats, south of 12° N.
DISTRIBUTION: West coast of India, from 20° N. to S. tip.
ISIS NUMBER: 5301410002009005001.

Funisciurus Trouessart, 1880. Le Naturaliste, 2(37):293.
ISIS NUMBER: 5301410002010000000.

Funisciurus anerythrus (Thomas, 1890). Proc. Zool. Soc. Lond., 1890:47.
TYPE LOCALITY: Uganda, Buquera, S. of Lake Albert.
DISTRIBUTION: S.W. Nigeria to Central African Republic, N.E. Zaire and Uganda, south
to S.W. Zaire and N. Shaba Prov. (Zaire).
COMMENT: Includes *mystax* and *raptorum;* see Amtmann, 1975, Part 6.1:9.
ISIS NUMBER: 5301410002010001001 as *F. anerythrus.*
5301410002010010001 as *F. mystax.*
5301410002010012001 as *F. raptorum.*

Funisciurus bayonii (Bocage, 1890). J. Sci. Math. Phys. Nat. Lisboa, 2:3.
TYPE LOCALITY: N. Angola, Duque de Braganca.
DISTRIBUTION: N.E. Angola; S.W. Zaire.
ISIS NUMBER: 5301410002010003001.

Funisciurus carruthersi Thomas, 1906. Ann. Mag. Nat. Hist., ser. 7, 18:140.
TYPE LOCALITY: Uganda, E. Ruwenzori.
DISTRIBUTION: Ruwenzori (S. Uganda) to Burundi.
ISIS NUMBER: 5301410002010004001.

Funisciurus congicus (Kuhl, 1820). Beitr. Zool. Vergl. Anat. Abt., 2:66.
TYPE LOCALITY: Congo.
DISTRIBUTION: Zaire to Namibia.
ISIS NUMBER: 5301410002010005001.

Funisciurus isabella (Gray, 1862). Proc. Zool. Soc. Lond., 1862:180.
TYPE LOCALITY: Cameroun, Cameroun Mtn.
DISTRIBUTION: Cameroun to Congo Republic and Central African Republic.
ISIS NUMBER: 5301410002010006001.

Funisciurus lemniscatus (Le Conte, 1857). Proc. Acad. Nat. Sci. Phila., p. 11.
TYPE LOCALITY: Equatorial Guinea, Rio Muni.
DISTRIBUTION: S. of Sanaga River (Cameroun) to Zaire and Central African Republic.
ISIS NUMBER: 5301410002010007001.

Funisciurus leucogenys (Waterhouse, 1842). Ann. Mag. Nat. Hist., ser. 1, 10:202.
TYPE LOCALITY: Equatorial Guinea, Bioko (=Fernando Po).
DISTRIBUTION: Ghana to Central African Republic to Rio Muni; Bioko.
COMMENT: Includes *auriculatus;* see Amtmann, 1975, Part 6.1:8.
ISIS NUMBER: 5301410002010002001 as *F. auriculatus.*

Funisciurus pyrrhopus (F. Cuvier, 1833). *In* E. Geoffroy and F. Cuvier, Hist. Nat. Mamm.,
4:Tab. Gen. 4.
TYPE LOCALITY: Gabon.
DISTRIBUTION: Gambia to Uganda to Angola.
COMMENT: Includes *leucostigma* and *mandingo;* see Amtmann, 1975, Part 6.1:8.
ISIS NUMBER: 5301410002010011001 as *F. pyrrhopus.*
5301410002010008001 as *F. leugostigma (sic).*
5301410002010009001 as *F. mandingo.*

Funisciurus substriatus De Winton, 1899. Ann. Mag. Nat. Hist., ser. 7, 4:357.
TYPE LOCALITY: Ghana, Kintampo.
DISTRIBUTION: Ivory Coast to Nigeria.
ISIS NUMBER: 5301410002010013001.

Glaucomys Thomas, 1908. Ann. Mag. Nat. Hist., ser. 8, 1:5.
REVIEWED BY: D. W. Linzey (DWL).
COMMENT: Revised by Howell, 1918, N. Am. Fauna, 44:1–60.
ISIS NUMBER: 5301410002043000000.

Glaucomys sabrinus (Shaw, 1801). Gen. Zool., 2:157.
TYPE LOCALITY: Canada, Ontario, mouth of Severn River.
DISTRIBUTION: Alaska and Canada, N.W. U.S.A. to S. California and W. South Dakota
(Black Hills), N.E. U.S.A. to S. Appalachian Mtns. (U.S.A.).
ISIS NUMBER: 5301410002043001001.

Glaucomys volans (Linnaeus, 1758). Syst. Nat., 10th ed., 1:63.
TYPE LOCALITY: U.S.A., Virginia.
DISTRIBUTION: Texas, Kansas, and Minnesota (U.S.A.) to Nova Scotia (Canada) and E.
    U.S.A.; montane populations scattered from N.W. Mexico to Honduras.
COMMENT: Reviewed by Dolan and Carter, 1973, Mamm. Species, 78:1–6.
ISIS NUMBER: 5301410002043002001.

Glyphotes Thomas, 1898. Ann. Mag. Nat. Hist., ser. 6, 2:251.
ISIS NUMBER: 5301410002011000000.

Glyphotes canalvus Moore, 1959. Am. Mus. Novit., 1944:5.
TYPE LOCALITY: Malaysia, Sabah, Mt. Dulit.
DISTRIBUTION: Known only from the type locality.
ISIS NUMBER: 5301410002011001001.

Glyphotes simus Thomas, 1898. Ann. Mag. Nat. Hist., ser. 6, 2:251.
TYPE LOCALITY: Malaysia, Sabah, Mt. Kinabalu.
DISTRIBUTION: Mountains of Sabah and Sarawak (Malaysia), above 1000 m.
ISIS NUMBER: 5301410002011002001.

Heliosciurus Trouessart, 1880. Le Naturaliste, 2nd year, 1:292.
COMMENT: Includes *Aethosciurus ruwenzorii*; see Moore, 1959, Bull. Am. Mus. Nat. Hist.,
    118:153–206.
ISIS NUMBER: 5301410002013000000.

Heliosciurus gambianus (Ogilby, 1835). Proc. Zool. Soc. Lond., 1835:103.
TYPE LOCALITY: Gambia, probably near Fort St. Mary.
DISTRIBUTION: Senegal to Ethiopia and Zambia.
COMMENT: Includes *punctatus* and *rhodesiae*; see Amtmann, 1975, Part 6.1:4.
ISIS NUMBER: 5301410002013001001 as *H. gambianus*.
    5301410002013004001 as *H. punctatus*.
    5301410002013005001 as *H. rhodesiae*.

Heliosciurus rufobrachium (Waterhouse, 1842). Ann. Mag. Nat. Hist., ser. 1, 10:202.
TYPE LOCALITY: Equatorial Guinea, Bioko (Fernando Po).
DISTRIBUTION: Senegal to S.E. Sudan and Kenya, south to Mozambique and S.E.
    Zimbabwe.
COMMENT: Includes *mutabilis* and *undulatus*; see Amtmann, 1975, Part 6.1:5.
ISIS NUMBER: 5301410002013006001 as *H. rufobrachium*.
    5301410002013003001 as *H. mutabilis*.
    5301410002013007001 as *H. undulatus*.

Heliosciurus ruwenzorii (Schwann, 1904). Ann. Mag. Nat. Hist., ser. 7, 13:71.
TYPE LOCALITY: Uganda, Ruwenzori Mtns.
DISTRIBUTION: E. Zaire; Rwanda; Burundi; W. Uganda.
COMMENT: Formerly included in *Aethosciurus*, which is here included in *Paraxerus*; see
    Moore, 1959, Bull. Am. Mus. Nat. Hist., 118:153–206.

Hylopetes Thomas, 1908. Ann. Mag. Nat. Hist., 1:6.
COMMENT: Includes *Eoglaucomys*; see McLaughlin, 1967, *in* Anderson and Jones, p. 215,
    but also see McKenna, 1962, Am. Mus. Novit., 2104:1–38.
ISIS NUMBER: 5301410002044000000.

Hylopetes alboniger (Hodgson, 1836). J. Asiat. Soc. Bengal, 5:231.
TYPE LOCALITY: Nepal.
DISTRIBUTION: Nepal to Szechwan, Yunnan, and Hainan (China) and Indochina.
ISIS NUMBER: 5301410002044001001.

Hylopetes electilis (G. M. Allen, 1925). Am. Mus. Novit., 163:16.
TYPE LOCALITY: China, Hainan Isl., Namfong.
DISTRIBUTION: Hainan Isl. and Fukien (China).
COMMENT: Included in *Petinomys* by most authors, but see McKenna, 1962, Am. Mus.
    Novit., 2104:35.
ISIS NUMBER: 5301410002048002001.

Hylopetes fimbriatus (Gray, 1837). Ann. Mag. Nat. Hist., 1:584.
TYPE LOCALITY: India, Punjab, Simla.
DISTRIBUTION: Afghanistan to Kashmir and Punjab (India).
COMMENT: Placed in monotypic genus *Eoglaucomys* by McKenna, 1962, Am. Mus.
Novit., 2104:1–38, and others, but included in *Hylopetes* by McLaughlin, 1967, *in*
Anderson and Jones, p. 215.
ISIS NUMBER: 5301410002044002001.

Hylopetes lepidus (Horsfield, 1822). Zool. Res. Java, 5.
TYPE LOCALITY: Indonesia, Java.
DISTRIBUTION: Thailand to Java; Borneo.
COMMENT: New name for the species formerly called *sagitta;* see Medway, 1977:104.
Includes *platyurus;* see Hill, 1961, Ann. Mag. Nat. Hist., ser. 13, 4:730, and
Medway, 1977:104. Corbet and Hill, 1980:131, listed *platyurus* as a distinct species,
without comment.
ISIS NUMBER: 5301410002044004001 as *H. lepidus.*
5301410002044007001 as *H. sagitta.*

Hylopetes mindanensis (Rabor, 1939). Philipp. J. Sci., 69:389.
TYPE LOCALITY: Philippines, Mindanao Prov., Misamis Oriental, Gingoog, Badiangan.
DISTRIBUTION: Known only from the type locality.
COMMENT: Holotype destroyed during World War II; status uncertain, possibly allied
to *Petinomys crinitus* (LRH).

Hylopetes nigripes Thomas, 1893. Ann. Mag. Nat. Hist., ser. 6, 12:30.
TYPE LOCALITY: Philippines, Palawan, Puerta Princesa.
DISTRIBUTION: Palawan (Philippines).
ISIS NUMBER: 5301410002044005001.

Hylopetes phayrei (Blyth, 1859). J. Asiat. Soc. Bengal, 28:278.
TYPE LOCALITY: Burma, Rangoon.
DISTRIBUTION: Burma; Thailand; Laos.
ISIS NUMBER: 5301410002044006001.

Hylopetes sipora Chasen, 1940. Bull. Raffles Mus., 15:117.
TYPE LOCALITY: Indonesia, Sumatra, Mentawi Isls., Sipora Isl.
DISTRIBUTION: Known only from the type locality.
COMMENT: Formerly included in *spadiceus;* see Hill, 1961, Ann. Mag. Nat. Hist., ser.
13, 4:731–733, who noted that an adult specimen is needed to clarify the status of
this form.

Hylopetes spadiceus (Blyth, 1847). J. Asiat. Soc. Bengal, 16:867.
TYPE LOCALITY: Burma, Arakan.
DISTRIBUTION: Burma, Thailand, Indochina and Malaysia.
COMMENT: Includes *harrisoni;* see Medway, 1977:105. Formerly included *sipora;* see
Hill, 1961, Ann. Mag. Nat. Hist., ser. 13, 4:731–733. Corbet and Hill, 1980:137,
listed *spadiceus* in *lepidus,* without comment.
ISIS NUMBER: 5301410002044008001 as *H. spadiceus.*
5301410002044003001 as *H. harrisoni.*

Hyosciurus Tate and Archbold, 1935. Am. Mus. Novit., 801:2.
ISIS NUMBER: 5301410002014000000.

Hyosciurus heinrichi Tate and Archbold, 1935. Am. Mus. Novit., 801:2.
TYPE LOCALITY: Indonesia, S. Sulawesi, Latimodjong Mtns.
DISTRIBUTION: Mountains of Sulawesi.
ISIS NUMBER: 5301410002014001001.

Iomys Thomas, 1908. Ann. Mag. Nat. Hist., ser. 8, 1:1.
ISIS NUMBER: 5301410002045000000.

Iomys horsfieldii (Waterhouse, 1838). Proc. Zool. Soc. Lond., 1837:87.
TYPE LOCALITY: Indonesia, Java.
DISTRIBUTION: Malay Peninsula to Java; Mentawi Isls.; Borneo.
ISIS NUMBER: 5301410002045001001.

Lariscus Thomas and Wroughton, 1909. Proc. Zool. Soc. Lond., 1909:389.
ISIS NUMBER: 5301410002015000000.

Lariscus hosei (Thomas, 1892). Ann. Mag. Nat. Hist., ser. 6, 10:215.
TYPE LOCALITY: Malaysia, Sarawak, Baram Dist., Mt. Batu Song.
DISTRIBUTION: Mountains of Sarawak and Sabah (Malaysia).
COMMENT: Sometimes put in the genus *Paralariscus;* see Ellerman, 1947, J. Mammal., 28:249–278; but also see Medway, 1977:97.
PROTECTED STATUS: CITES - Appendix II.
ISIS NUMBER: 5301410002015001001.

Lariscus insignis (F. Cuvier, 1821). *In* E. Geoffroy St. Hilaire and F. Cuvier, Hist. Nat. Mamm., 2:2.
TYPE LOCALITY: Indonesia, Sumatra, "lowlands".
DISTRIBUTION: Malay Peninsula; Sumatra; Java; Borneo; adjacent Isls.
COMMENT: Includes *niobe* and *obscurus* from the Mentawi Isls.; see Chasen, 1940:145–146. Corbet and Hill, 1980:133, listed *niobe* as a distinct species, without comment; LRH considers *niobe* a distinct species.
ISIS NUMBER: 5301410002015002001 as *L. insignis.*
5301410002015003001 as *L. niobe.*
5301410002015004001 as *L. obscurus.*

Marmota Blumenbach, 1779. Hand. Hilfsb. Nat., 1:79.
COMMENT: North American species reviewed by Howell, 1915, N. Am. Fauna, 37:1–80; Eurasian species revised by Gromov *et al.,* 1965; amphiberingian species reviewed by Hoffmann *et al.,* 1979, Occas. Pap. Mus. Nat. Hist. Univ. Kans., 83:1–56. Frase and Hoffmann, 1980, Mamm. Species, 135:1–8, provided a key to N. American species.
ISIS NUMBER: 5301410002016000000.

Marmota baibacina Kastschenko, 1899. Res. Altaisk. Exp., p. 63.
TYPE LOCALITY: U.S.S.R., R.S.F.S.R., Altaisk. Krai, Gorno-Altaisk. A.O., near Cherga, or, alternatively, Multa River, near Nizhne-Uimon.
DISTRIBUTION: Altai Mtns., S.E. Kazakhstan, Kirgizia, S.W. Siberia (U.S.S.R.); Mongolia; Sinkiang (China). Introduced into Caucasus Mtns. (U.S.S.R.).
COMMENT: Placed by Ellerman and Morrison-Scott, 1951:514, in *marmota,* and by Corbet, 1978:81, in *bobak;* Kapitonov, 1966, Trud. Inst. Zool. Kazakh. Akad. Nauk, 26:94–134, analyzed purported hybridization between *baibacina* and *bobak.* Most Soviet authors retain both as distinct species; see Gromov *et al.,* 1965:337–387, and Zimina, ed., 1978, [Marmots. Distribution and Ecology], "Nauka," Moscow, who included *centralis* in this species. Kapitonov, 1966, Trud. Inst. Zool. Kazakh. Akad. Nauk, 26:94–134, indicated that the population called *aphanasievi* is included in this species; but also see Corbet, 1978:81. Includes *lewisi,* a *nomen oblitum;* see Hoffmann, 1977, Proc. Biol. Soc. Wash., 90:291–301. See Bobrinskii *et al.,* 1965, [Key to the mammals of the U.S.S.R.], "Proveshchenie," Moscow, p. 259, and Ognev, 1947, [Mamm. U.S.S.R., Adjac. Count.], Acad. Sci., Moscow, p. 292, for discussion of alternative type localities.

Marmota bobak (Muller, 1776). Natursyst. Suppl., p. 40.
TYPE LOCALITY: U.S.S.R., Ukrainsk. S.S.R., right (W) bank of Dnepr River, (originally "Poland").
DISTRIBUTION: Steppes of E. Europe, east to N. and C. Kazakhstan (U.S.S.R.).
COMMENT: See comments under *baibacina, himalayana,* and *sibirica.* Includes *kozlovi;* see Fokanov, 1966, Zool. Zh., 45:1864; and *tschaganensis;* see Gromov *et al.,* 1965:337–387.
ISIS NUMBER: 5301410002016001001.

Marmota broweri Hall and Gilmore, 1934. Can. Field Nat., 48:57.
TYPE LOCALITY: U.S.A., Alaska, head of Kukpowruk River, "Point Lay."
DISTRIBUTION: Brooks Range of N. Alaska (U.S.A.) from near coast of Chukchi Sea to Alaska-Yukon border; perhaps also N. Yukon (Canada).
COMMENT: Regarded by Hall and Kelson, 1959, and Hall, 1981, as a synonym of

caligata, but Rausch and Rausch, 1971, Mammalia, 35:85–101, and Hoffmann *et al.*, 1979, Occas. Pap. Mus. Nat. Hist. Univ. Kans., 83:1–56, considered *broweri* a distinct species. Type locality restricted by Rausch, 1953, Arctic, 6:117.

Marmota caligata (Eschscholtz, 1829). Zool. Atlas, Part 2, p. 1, pl. 6.
TYPE LOCALITY: U.S.A., Alaska, near Bristol Bay.
DISTRIBUTION: C. Alaska (U.S.A.), Yukon and Northwest Territories (Canada) south to W. and N.E. Washington, C. Idaho, and W. Montana (U.S.A.).
ISIS NUMBER: 5301410002016002001.

Marmota camtschatica (Pallas, 1811). Zoogr. Rosso-Asiat., p. 156.
TYPE LOCALITY: U.S.S.R., R.S.F.S.R., Kamchatsk. Obl.
DISTRIBUTION: E. Siberia from Transbaikalia to Chukotka and Kamchatka (U.S.S.R.).
COMMENT: Hoffmann *et al.*, 1979, Occas. Pap. Mus. Nat. Hist. Univ. Kans., 83:1–56, reviewed this species. Regarded by Ellerman and Morrison-Scott, 1951, and Rausch, 1953, Arctic, 6:91–148, as a synonym of *marmota*.

Marmota caudata (Geoffroy, 1842–1843). *In* Jacquemont, Voy. dans l'Inde, 4, Zool., p. 66.
TYPE LOCALITY: India, Kashmir, upper Indus River, Ghombur.
DISTRIBUTION: W. Tien Shan through the Pamirs to Hindu Kush and Kashmir.
COMMENT: Includes *dichrous*; see Corbet, 1978:82; but also see Gromov *et al.*, 1965:440.
ISIS NUMBER: 5301410002016003001.

Marmota flaviventris (Audubon and Bachman, 1841). Proc. Acad. Nat. Sci. Phila., 1:99.
TYPE LOCALITY: U.S.A., Oregon, Mt. Hood, "mountains between Texas and California."
DISTRIBUTION: S.C. British Columbia and S. Alberta (Canada) south to N. New Mexico, S. Utah, Nevada, and California (U.S.A.).
COMMENT: Reviewed by Frase and Hoffmann, 1980, Mamm. Species, 135:1–8.
ISIS NUMBER: 5301410002016004001.

Marmota himalayana (Hodgson, 1841). J. Asiat. Soc. Bengal, 10:777.
TYPE LOCALITY: Nepal.
DISTRIBUTION: Montane regions of W. China, Nepal, and N. India to Ladak.
COMMENT: Placed by Ellerman and Morrison-Scott, 1951:515, and Corbet, 1978:81, in *bobak*, but geographically separated from that species; evidence for specific status in Gromov *et al.*, 1965:332–356; see comment under *baibacina*.

Marmota marmota (Linnaeus, 1758). Syst. Nat., 10th ed., 1:60.
TYPE LOCALITY: "Alps".
DISTRIBUTION: Swiss, Italian, and French Alps; Austria, S. Germany; Carpathian and Tatra Mtns.; introduced into Pyrennes.
ISIS NUMBER: 5301410002016005001.

Marmota menzbieri (Kashkarov, 1925). Trans. Turk. Sci. Soc., 2:47.
TYPE LOCALITY: U.S.S.R., Kazakh. S.S.R., Yuzhno-Kazakhstansk. Obl., Talassk. Alatau Mtns., Chigyr-Tash, on upper Ugama River.
DISTRIBUTION: W. Tien Shan Mtns., in S. Kazakhstan and N.W. Kirgizia (U.S.S.R.).
COMMENT: Regarded by Ellerman and Morrison-Scott, 1951:514, as a synonym of *marmota*, but see Corbet, 1978:81.

Marmota monax (Linnaeus, 1758). Syst. Nat., 10th ed., 1:60.
TYPE LOCALITY: U.S.A., Maryland.
DISTRIBUTION: Alaska (U.S.A.) through S. Canada to S. Labrador to N.E. and S.C. U.S.A.; south in Rocky Mtns., possibly to N. Idaho.
ISIS NUMBER: 5301410002016006001.

Marmota olympus (Merriam, 1898). Proc. Acad. Nat. Sci. Phila., 50:352.
TYPE LOCALITY: U.S.A., Washington, timberline at head of Soleduc River, Olympic Mtns.
DISTRIBUTION: Olympic Mtns. of W. Washington (U.S.A.).
COMMENT: Reviewed by Hoffmann *et al.*, 1979, Occas. Pap. Mus. Nat. Hist. Univ. Kans., 83:1–56.
ISIS NUMBER: 5301410002016007001.

Marmota sibirica (Radde, 1862). Reise Sud. Ost. Sibir., p. 159.
TYPE LOCALITY: U.S.S.R., R.S.F.S.R., Chitinsk Obl., near Tarei-Nor (Lake).
DISTRIBUTION: S.W. Siberia, Tuva, Transbaikalia (U.S.S.R.); N. and W. Mongolia;
Heilungkiang and Inner Mongolia; formerly Manchuria (China).
COMMENT: Placed (with *baibacina* and *himalayana*) by Ellerman and Morrison-Scott,
1951:515, and Corbet, 1978:81, in *bobak*. Gromov *et al.*, 1965:337–415, and Zimina,
ed., 1978, [Marmots. Distribution and Ecology], "Nauka," Moscow, provided
evidence of specific distinctness and included *caliginosus* in this species; see
comment under *baibacina*. Nikol'skii, 1974, Zool. Zh., 53:436–444, analyzed contact
between *baibacina* and *sibirica* in Tuva; Sokolov and Orlov, 1980:329, also indicated
sympatry in N.W. Mongolia.

Marmota vancouverensis Swarth, 1911. Univ. Calif. Publ. Zool., 7:201.
TYPE LOCALITY: Canada, British Columbia, Vancouver Isl., Mt. Douglas.
DISTRIBUTION: Mountains of Vancouver Isl. (British Columbia, Canada).
COMMENT: Reviewed by Hoffmann *et al.*, 1979, Occas. Pap. Mus. Nat. Hist. Univ.
Kans., 83:1–56.
ISIS NUMBER: 5301410002016008001.

Menetes Thomas, 1908. J. Bombay Nat. Hist. Soc., 18:244.
ISIS NUMBER: 5301410002017000000.

Menetes berdmorei (Blyth, 1849). J. Asiat. Soc. Bengal, 18(30):600–603.
TYPE LOCALITY: Burma, Tenasserim, Thougyeen Dist.
DISTRIBUTION: Indochina; Thailand; Burma; Yunnan (China).
ISIS NUMBER: 5301410002017001001.

Microsciurus J. A. Allen, 1895. Bull. Am. Mus. Nat. Hist., 7:332.
ISIS NUMBER: 5301410002018000000.

Microsciurus alfari (J. A. Allen, 1895). Bull. Am. Mus. Nat. Hist., 7:333.
TYPE LOCALITY: Costa Rica, Jimenez.
DISTRIBUTION: S. Central America; perhaps in N. South America.
COMMENT: Apparently included in *flaviventer* by Cabrera, 1961:355; but see Hall,
1981:439–440.
ISIS NUMBER: 5301410002018001001.

Microsciurus flaviventer (Gray, 1867). Ann. Mag. Nat. Hist., ser. 20, 3:432.
TYPE LOCALITY: Brazil.
DISTRIBUTION: Upper Amazon River Basin, from the Rio Negro to Peru, the Andes of
Peru, Ecuador and Colombia.
COMMENT: Formerly included *alfari* and *mimulus;* see comments under those species.
ISIS NUMBER: 5301410002018003001.

Microsciurus mimulus (Thomas, 1898). Ann. Mag. Nat. Hist., ser. 7, 2:266.
TYPE LOCALITY: Ecuador, Esmeraldas, Cachavi, 665 ft. (203 m).
DISTRIBUTION: N.W. Ecuador, W. Colombia, and Panama.
COMMENT: Includes *boquetensis* and *isthmius;* see Handley, 1966, *in* Wenzel and Tipton,
Ectoparasites of Panama, p. 777. Cabrera, 1961:356, included *mimulus* in *flaviventer;*
but see Hall, 1981:440.
ISIS NUMBER: 5301410002018002001 as *M. boquetensis*.
            5301410002018004001 as *M. isthmius*.

Microsciurus santanderensis (Hernandez-Camacho, 1957). An. Soc. Biol. Bogota, 7:219.
TYPE LOCALITY: Colombia, Santander Dept., Meseta de los Caballeros, N.E. of La
Albania.
DISTRIBUTION: Colombia, between the Magdalena River and the Cordillera Oriental.
COMMENT: Originally described as a subspecies of *Sciurus pucheranii;* see Hernandez-
Camacho, 1960, Caldasia, 8:359–368.

Myosciurus Thomas, 1909. Ann. Mag. Nat. Hist., ser. 8, 3:474.
ISIS NUMBER: 5301410002019000000.

Myosciurus pumilio (Le Conte, 1857). Proc. Acad. Nat. Sci. Phila., 9:11.
TYPE LOCALITY: Gabon.
DISTRIBUTION: S.E. Nigeria; Cameroun; Gabon; Bioko.
ISIS NUMBER: 5301410002019001001.

Nannosciurus Trouessart, 1880. Le Naturaliste, p. 292.
COMMENT: Formerly included the species now placed in *Exilisciurus*; see Moore, 1958, Am. Mus. Novit., 1914:1–5.
ISIS NUMBER: 5301410002020000000.

Nannosciurus melanotis (Muller, 1844). *In* Temminck, Verhandl. Nat. Gesch. Nederland Overz. Bezitt., Zool., pp. 87, 98.
TYPE LOCALITY: Indonesia, Java.
DISTRIBUTION: Sumatra; Java; Borneo; adjacent small islands.
ISIS NUMBER: 5301410002020002001.

Paraxerus Forsyth Major, 1893. Proc. Zool. Soc. Lond., 1893:189.
COMMENT: Includes *Aethosciurus* (with the exception of *ruwenzorii*, here included in *Heliosciurus*) and *Tamiscus*; see Moore, 1959, Bull. Am. Mus. Nat. Hist., 118:153–206.
ISIS NUMBER: 5301410002021000000.

Paraxerus alexandri (Thomas and Wroughton, 1907). Ann. Mag. Nat. Hist., ser. 7, 19:376.
TYPE LOCALITY: Zaire, Gudima, River Iri, Upper Welle.
DISTRIBUTION: N.E. Zaire, Uganda.
ISIS NUMBER: 5301410002021001001.

Paraxerus boehmi (Reichenow, 1886). Zool. Anz., 9:315.
TYPE LOCALITY: S. Zaire, Marungu.
DISTRIBUTION: E. and C. Africa.
COMMENT: Includes *antoniae*, *emini*, and *vulcanorum*; see Amtmann, 1975, Part 6.1:10.
ISIS NUMBER: 5301410002021003001 as *P. boehmi*.
5301410002021002001 as *P. antoniae*.
5301410002021006001 as *P. emini*.
5301410002021012001 as *P. vulcanorum*.

Paraxerus cepapi (A. Smith, 1836). Rept. Exped. Expl. C. Afr...., p. 43.
TYPE LOCALITY: South Africa, W. Transvaal, Rustenberg Dist., Marico River.
DISTRIBUTION: S. Angola to S. Tanzania to Transvaal (South Africa).
ISIS NUMBER: 5301410002021005001.

Paraxerus cooperi Hayman, 1950. Ann. Mag. Nat. Hist., ser. 12, 3:262.
TYPE LOCALITY: Cameroun, Kumba Div., Rumpi Hills, 5° N., 9° 15′ E.
DISTRIBUTION: Cameroun.
COMMENT: Eisentraut, 1976, Bonn. Zool. Monogr., 8, put *cooperi* in a separate genus, *Montisciurus*; but see Corbet and Hill, 1980:131.

Paraxerus flavivittis (Peters, 1852). Monatsb. Preuss. Akad. Wiss. Berlin, p. 274.
TYPE LOCALITY: N.E. Mozambique, Mocamboa (=Mossimboa), 11° S., on the coast.
DISTRIBUTION: N. Mozambique to S. Kenya.
ISIS NUMBER: 5301410002021007001.

Paraxerus lucifer (Thomas, 1897). Proc. Zool. Soc. Lond., 1897:430.
TYPE LOCALITY: Malawi, Kombe Forest, Misuku Mtns., 9° 43′ S., 33° 31′E.
DISTRIBUTION: N. Malawi; S.W. Tanzania; E. Zambia.
ISIS NUMBER: 5301410002013002001 as *Heliosciurus lucifer*.

Paraxerus ochraceus (Huet, 1880). Nouv. Arch. Mus. Hist. Nat. Paris, p. 154.
TYPE LOCALITY: Tanzania, Bagamoyo, near Dar-es-Salaam.
DISTRIBUTION: Tanzania to S. Sudan and Ethiopia.
ISIS NUMBER: 5301410002021008001.

Paraxerus palliatus (Peters, 1852). Monatsb. Preuss. Akad. Wiss. Berlin, p. 273.
TYPE LOCALITY: Mozambique, "Quintangonha," mainland near Mocambique Isl.
DISTRIBUTION: Natal (South Africa) to Somalia.

COMMENT: Includes *sponsus;* see Amtmann, 1975, Part 6.1:10.
ISIS NUMBER: 5301410002021009001 as *P. palliatus.*
                            5301410002021010001 as *P. sponsus.*

Paraxerus poensis (A. Smith, 1830). S. Afr. Quart. J., 2:128.
TYPE LOCALITY: Equatorial Guinea, Bioko (=Fernando Po).
DISTRIBUTION: Sierra Leone to Zaire.

Paraxerus vexillarius (Kershaw, 1923). Ann. Mag. Nat. Hist., ser. 9, 11:591.
TYPE LOCALITY: Tanzania, Usambara.
DISTRIBUTION: C. and E. Tanzania.
COMMENT: Includes *byatti;* see Amtmann, 1975, Part 6.1:10.
ISIS NUMBER: 5301410002021004001 as *P. byatti.*

Paraxerus vincenti Hayman, 1950. Ann. Mag. Nat. Hist., 3:263.
TYPE LOCALITY: Mozambique, Namuli Mtn., north of the Zambezi, 15° 21′ S., 37° 4′ E.
DISTRIBUTION: N. Mozambique.
COMMENT: Perhaps a race of *vexillarius;* see Corbet and Hill, 1980:131.
ISIS NUMBER: 5301410002021011001 as *P. vencenti (sic).*

Petaurillus Thomas, 1908. Ann. Mag. Nat. Hist., ser. 8, 1:3.
ISIS NUMBER: 5301410002046000000.

Petaurillus emiliae Thomas, 1908. Ann. Mag. Nat. Hist., ser. 8, 1:8.
TYPE LOCALITY: Malaysia, Sarawak, Baram.
DISTRIBUTION: Sarawak (Malaysia).
ISIS NUMBER: 5301410002046001001.

Petaurillus hosei (Thomas, 1900). Ann. Mag. Nat. Hist., ser. 7, 5:275.
TYPE LOCALITY: Malaysia, Sarawak, Baram, Toyul River.
DISTRIBUTION: Sarawak (Malaysia).
ISIS NUMBER: 5301410002046002001.

Petaurillus kinlochii (Robinson and Kloss, 1911). J. Fed. Malay St. Mus., 4:171.
TYPE LOCALITY: Malaysia, Selangor, Kapar.
DISTRIBUTION: Selangor (Malay Peninsula).
ISIS NUMBER: 5301410002046003001.

Petaurista Link, 1795. Zool. Beytr., 1(2):52, 78.
ISIS NUMBER: 5301410002047000000.

Petaurista alborufus (Milne-Edwards, 1870). C. R. Acad. Sci. Paris, 70:342.
TYPE LOCALITY: China, Szechwan, Moupin.
DISTRIBUTION: Taiwan; S. and C. China to Thailand and N.E. India.
COMMENT: Includes *lena* which was treated as a separate species by Kuntz and Ming,
     1970, Quart. J. Taiwan Mus., 23; see Jones, 1975, Quart. J. Taiwan Mus., 28:192,
     207. Provisionally includes *pectoralis;* see Ellerman and Morrison-Scott, 1966:465.
ISIS NUMBER: 5301410002047001001.

Petaurista elegans (Muller, 1839). *In* Temminck, Verhandl. Nat. Gesch. Nederland. Overz.
     Bezitt., Zool., pp. 107, 112.
TYPE LOCALITY: Indonesia, Nusa Kumbangan Isl., off S. Java.
DISTRIBUTION: Nepal to Szechwan and Yunnan (China) and Java; Borneo.
COMMENT: Includes *clarkei* and *marica;* see Ellerman and Morrison-Scott, 1966:460–461.
ISIS NUMBER: 5301410002047002001.

Petaurista leucogenys (Temminck, 1827). Monogr. Mamm., 1:27.
TYPE LOCALITY: Japan, Kyushu, Higo.
DISTRIBUTION: Japan, except Hokkaido; Kansu to Yunnan (China).
COMMENT: Includes *xanthotis,* which was considered a separate species by McKenna,
     1962, Am. Mus. Novit., 2104:1–38; see Corbet, 1978:86.
ISIS NUMBER: 5301410002047003001.

Petaurista magnificus (Hodgson, 1836). J. Asiat. Soc. Bengal, 5:231.
TYPE LOCALITY: Nepal.
DISTRIBUTION: Tibet (China); Nepal; Sikkim (India).
COMMENT: Includes *nobilis*, see Ellerman and Morrison-Scott, 1966:464; but also see
Ghose and Saha, 1981, J. Bombay Nat. Hist. Soc., 78:95.
ISIS NUMBER: 5301410002047004001.

Petaurista petaurista (Pallas, 1766). Misc. Zool., p. 54.
TYPE LOCALITY: Indonesia, W. Java.
DISTRIBUTION: Kashmir (India) to Sri Lanka; C. and S.E. China; Taiwan and Java.
COMMENT: Includes *hainana* and *yunanensis*; see Ellerman and Morrison-Scott, 1966:462.
Includes *grandis* which was treated as a separate species by Kuntz and Ming,
1970, Quart. J. Taiwan Mus., 23; see Jones, 1975, Quart. J. Taiwan Mus., 28:193,
207.
ISIS NUMBER: 5301410002047005001.

Petinomys Thomas, 1908. Ann. Mag. Nat. Hist., 1:6.
COMMENT: Formerly included *electilis*, here included in *Hylopetes*; see McKenna, 1962,
Am. Mus. Novit., 2104:35.
ISIS NUMBER: 5301410002048000000.

Petinomys bartelsi Chasen, 1939. Treubia, 17:185.
TYPE LOCALITY: Indonesia, Java, Mt. Pangrango.
DISTRIBUTION: Java.

Petinomys crinitus Hollister, 1911. Proc. Biol. Soc. Wash., 24:185.
TYPE LOCALITY: Philippines, Basilan Isl.
DISTRIBUTION: Basilan and Mindanao Isls. (Philippines).
ISIS NUMBER: 5301410002048001001.

Petinomys fuscocapillus (Jerdon, 1847). J. Asiat. Soc. Bengal, 16:867.
TYPE LOCALITY: S. India, Travancore.
DISTRIBUTION: S. India; Sri Lanka.
ISIS NUMBER: 5301410002048003001.

Petinomys genibarbis (Horsfield, 1824). Zool. Res. Java, (descrip. and plate).
TYPE LOCALITY: Indonesia, E. Java.
DISTRIBUTION: Malaya to Sumatra, Java, and Borneo.
COMMENT: Probably conspecific with *sagitta*; see Medway, 1977:102.
ISIS NUMBER: 5301410002048004001.

Petinomys hageni (Jentink, 1888). Notes Leyden Mus., 11:26.
TYPE LOCALITY: Indonesia, Sumatra, Deli.
DISTRIBUTION: Borneo; Sumatra; Mentawi Isls.
ISIS NUMBER: 5301410002048005001.

Petinomys sagitta (Linnaeus, 1766). Syst. Nat., 12th ed., 1:88.
TYPE LOCALITY: Indonesia, Java.
DISTRIBUTION: Java.
COMMENT: Formerly included in *Hylopetes*; see Medway, 1977:102. May be conspecific
with *genibarbis*; see Ellerman and Morrison-Scott, 1955, Suppl. to Chasen, (1940),
Handlist of Malaysian Mamm., p. 31. Medway, 1977:102, considered it
inadvisable to synonymize *sagitta* and *genibarbis* until the relationship has been
certainly established.

Petinomys setosus (Temminck, 1845). Fauna Japonica, 1(Mamm.), p. 49.
TYPE LOCALITY: Indonesia, W. Sumatra, Padang.
DISTRIBUTION: Burma; Malaya; Sumatra; Borneo.
COMMENT: Includes *morrisi*; see Muul and Thonglongya, 1971, J. Mammal., 52:362–369,
and Corbet and Hill, 1980:137. McKenna, 1962, Am. Mus. Novit., 2104:1–38,
considered *morrisi* assignable to a distinct genus, *Olisthomys*.
ISIS NUMBER: 5301410002048006001.

Petinomys vordermanni (Jentink, 1890). Notes Leyden Mus., 12:150.
  TYPE LOCALITY: Indonesia, Belitung (=Billiton) Isl.
  DISTRIBUTION: S. Burma; Malaya; Borneo.
  COMMENT: McKenna, 1962, Am. Mus. Novit., 2104:1–38, considered *vordermanni*
    representative of an undescribed genus, but Muul and Thonglongya, 1971, J.
    Mammal., 52:362, and Hill, 1961, Ann. Mag. Nat. Hist., ser. 13, 4:733–735,
    retained it in *Petinomys*.
  ISIS NUMBER: 5301410002048007001.

Prosciurillus Ellerman, 1947. Proc. Zool. Soc. Lond., 1947–1948:259.
  COMMENT: Reviewed by Moore, 1958, Am. Mus. Novit., 1890:1–5, who transferred
    leucomus from *Callosciurus* to this genus.
  ISIS NUMBER: 5301410002022000000.

Prosciurillus abstrusus Moore, 1958. Am. Mus. Novit., 1890:3.
  TYPE LOCALITY: Indonesia, S.E. Sulawesi, Gunong Tanke Salokko, 3° 40′ S., 121° 13′ E.
  DISTRIBUTION: Known only from the type locality.
  COMMENT: Includes *obscurus*; see Moore, 1959, Bull. Am. Mus. Nat. Hist., 118:203. *P.*
    obscurus is probably a *nomen nudum* (RSH).
  ISIS NUMBER: 5301410002022003001 as *P. obscurus*.

Prosciurillus leucomus (Forsten, 1839). *In* Temminck, Verhandl. Nat. Gesch. Nederland
    Overz. Bezitt., Zool., 1:87.
  TYPE LOCALITY: Indonesia, N.E. Sulawesi, Minahassa.
  DISTRIBUTION: Sulawesi; Sangihe (=Sanghir) Isls.
  COMMENT: Includes *mowewensis* and *serassinorum*, which were formerly included in
    Callosciurus; see Moore, 1958, Am. Mus. Novit., 1890:2. Includes *elbertae* (LRH);
    Moore, 1958, Am. Mus. Novit., 1890:1–5, did not specifically mention *elbertae*, but
    probably intended to include it in *leucomus*.
  ISIS NUMBER: 5301410002022001001.

Prosciurillus murinus (Forsten, 1844). *In* Temminck, Verhandl. Nat. Gesch. Nederland
    Overz. Bezitt., Zool., 1:87.
  TYPE LOCALITY: Indonesia, N.E. Sulawesi.
  DISTRIBUTION: N.E. and C. Sulawesi.
  ISIS NUMBER: 5301410002022002001.

Protoxerus Forsyth Major, 1893. Proc. Zool. Soc. Lond., 1893:189.
  ISIS NUMBER: 5301410002023000000.

Protoxerus aubinnii (Gray, 1873). Ann. Mag. Nat. Hist., ser. 4, 12:65.
  TYPE LOCALITY: Ghana, Ashanti, Fantree.
  DISTRIBUTION: Liberia to Ghana.
  COMMENT: Subgenus *Allosciurus*; see Amtmann, 1975, Part 6.1:6.

Protoxerus stangeri (Waterhouse, 1843). Proc. Zool. Soc. Lond., 1842:127.
  TYPE LOCALITY: Equatorial Guinea, Bioko (=Fernando Po).
  DISTRIBUTION: Sierra Leone to Kenya and Angola; Bioko.
  COMMENT: Subgenus *Protoxerus*; see Amtmann, 1975, Part 6.1:6.
  ISIS NUMBER: 5301410002023001001.

Pteromys G. Cuvier, 1800. Lecon's Anat. Comp., I, tab. 1.
  ISIS NUMBER: 5301410002049000000.

Pteromys momonga Temminck, 1845. Fauna Japonica, 1(Mamm.), p. 47.
  TYPE LOCALITY: Japan, Kyushu.
  DISTRIBUTION: Kyushu and Honshu (Japan).
  ISIS NUMBER: 5301410002049001001.

Pteromys volans (Linnaeus, 1758). Syst. Nat., 10th ed., 1:64.
  TYPE LOCALITY: Sweden, middle part.
  DISTRIBUTION: Palearctic taiga, from Scandinavia east to Chukotka; south to E. Baltic,
    S. Ural Mtns., Altai Mtns., Mongolia, N. China, Korea, Sakhalin Isl. (U.S.S.R.),
    and Hokkaido (Japan); perhaps mountains of W. China.
  ISIS NUMBER: 5301410002049002001.

Pteromyscus Thomas, 1908. Ann. Mag. Nat. Hist., ser. 8, 1:3.
ISIS NUMBER: 5301410002050000000.

Pteromyscus pulverulentus (Gunther, 1873). Proc. Zool. Soc. Lond., 1873:413.
TYPE LOCALITY: Malaysia, Penang.
DISTRIBUTION: S. Thailand to Sumatra; Borneo.
ISIS NUMBER: 5301410002050001001.

Ratufa Gray, 1867. Ann. Mag. Nat. Hist., ser. 3, 20:273.
ISIS NUMBER: 5301410002024000000.
PROTECTED STATUS: CITES - Appendix II.

Ratufa affinis (Raffles, 1821). Trans. Linn. Soc. Lond., 13:259.
TYPE LOCALITY: Malaysia, Singapore.
DISTRIBUTION: Malaysian subregion (except S.W. Philippines).
PROTECTED STATUS: CITES - Appendix II as *Ratufa* spp.
ISIS NUMBER: 5301410002024001001.

Ratufa bicolor (Sparrman, 1778). Samhelle Hand. (Wet. Afd.), 1:70.
TYPE LOCALITY: Indonesia, W. Java, Anjer.
DISTRIBUTION: E. Nepal and Tibet to Hainan (China), south through the Malay
Peninsula to Java and Bali.
PROTECTED STATUS: CITES - Appendix II as *Ratufa* spp.
ISIS NUMBER: 5301410002024002001.

Ratufa indica (Erxleben, 1777). Syst. Regn. Anim., 1:420.
TYPE LOCALITY: India, near Bombay.
DISTRIBUTION: S. and C. India, excluding the central lowlands.
PROTECTED STATUS: CITES - Appendix II as *Ratufa* spp.
ISIS NUMBER: 5301410002024003001.

Ratufa macroura (Pennant, 1769). Indian Zool., 1, pl. 1.
TYPE LOCALITY: Sri Lanka, highlands of Central and Uva Provs.
DISTRIBUTION: Sri Lanka and S. India.
PROTECTED STATUS: CITES - Appendix II as *Ratufa* spp.
ISIS NUMBER: 5301410002024004001.

Rheithrosciurus Gray, 1867. Ann. Mag. Nat. Hist., ser. 3, 20:273.
ISIS NUMBER: 5301410002025000000.

Rheithrosciurus macrotis (Gray, 1857). Proc. Zool. Soc. Lond., 1856:341.
TYPE LOCALITY: Malaysia, Sarawak.
DISTRIBUTION: Borneo.
ISIS NUMBER: 5301410002025001001.

Rhinosciurus Blyth, 1855. J. Asiat. Soc. Bengal, 14:477.
ISIS NUMBER: 5301410002026000000.

Rhinosciurus laticaudatus (Muller and Schlegel, 1844). *In* Temminck, Verhandl. Nat. Gesch.
Nederland Overz. Bezitt., Zool., 1:100.
TYPE LOCALITY: Indonesia, W. Kalimantan, Pontianak.
DISTRIBUTION: Malaysian subregion except Java and S.W. Philippines.
ISIS NUMBER: 5301410002026001001.

Rubrisciurus Ellerman, 1954. *In* Laurie and Hill, 1954, Land Mammals of New Guinea,
Celebes, and Adjacent Islands, p. 94.
COMMENT: Sometimes considered a subgenus of *Callosciurus;* see Laurie and Hill,
1954:93; but see also Moore, 1959, Bull. Am. Mus. Nat. Hist., 118(4):176, Moore,
1961, Bull. Am. Mus. Nat. Hist., 122(1):8, and McLaughlin, 1967, *in* Anderson and
Jones, p. 214, who considered it a distinct genus.
ISIS NUMBER: 5301410002027000000.

Rubrisciurus rubriventer (Forsten, 1839). *In* Temminck, Verhandl. Nat. Gesch. Nederland
Overz. Bezitt., Zool., 1:86.
TYPE LOCALITY: Indonesia, N.E. Sulawesi, Minahassa.
DISTRIBUTION: N. and C. Sulawesi.
ISIS NUMBER: 5301410002027001001.

Sciurillus Thomas, 1914. Proc. Zool. Soc. Lond., 1914:416.
ISIS NUMBER: 5301410002028000000.

Sciurillus pusillus (Desmarest, 1817). Nouv. Dict. Hist. Nat., 10:109.
TYPE LOCALITY: French Guiana, Cayenne.
DISTRIBUTION: South America, north of the Amazon, and south of the Amazon in the Tapajoz river basin.
ISIS NUMBER: 5301410002028001001.

Sciurotamias Miller, 1901. Proc. Biol. Soc. Wash., 14:23.
REVIEWED BY: J. R. Callahan (JRCL).
COMMENT: Includes *Rupestes;* reviewed by Moore and Tate, 1965, Fieldiana Zool., 48:1–351. *S. davidianus* has a penile duct and Cowper's glands, and its glans and baculum are similar to those of *Ratufa;* therefore, Callahan and Davis, 1982, J. Mammal., 63:42–47, removed this taxon from the Tamiasciurini where it had been placed by Moore, 1959, Bull. Am. Mus. Nat. Hist., 118:182, and tentatively referred it to the Ratufini (JRCL). Gromov *et al.*, 1965, placed this genus in the tribe Tamiini with the genus *Tamias.*
ISIS NUMBER: 5301410002029000000.

Sciurotamias davidianus (Milne-Edwards, 1867). Rev. Mag. Zool. Paris, ser. 2, 19:196.
TYPE LOCALITY: China, Hopei, "Mountains of Pekin."
DISTRIBUTION: Hopei to Szechwan and Hupei; Kweichow (China).
ISIS NUMBER: 5301410002029001001.

Sciurotamias forresti (Thomas, 1922). Ann. Mag. Nat. Hist., ser. 9, 10:399.
TYPE LOCALITY: China, Yunnan Prov., Mekong-Yangtze Divide, 27° 20′ N.
DISTRIBUTION: Yunnan and Szechwan Provs. (China).
COMMENT: Subgenus *Rupestes;* see Moore and Tate, 1965, Fieldiana Zool., 48:309.
ISIS NUMBER: 5301410002029002001.

Sciurus Linnaeus, 1758. Syst. Nat., 10th ed., 1:63.
COMMENT: Includes *Guerlinguetus, Hesperosciurus, Otosciurus, Sciurus* (see Hall, 1981:417–436), *Tenes* (see Corbet, 1978:76), and, *Hadrosciurus* (see Cabrera, 1961:374) as subgenera. Moore, 1959, Bull. Am. Mus. Nat. Hist., 118:153–206, considered *Guerlinguetus* a distinct genus, which included *Hadrosciurus* and *Urosciurus* as subgenera.
ISIS NUMBER: 5301410002030000000 as *Sciurus.*
5301410002012000000 as *Guerlinguetus.*

Sciurus aberti Woodhouse, 1853. Proc. Acad. Nat. Sci. Phila., (1852), 6:110, 220.
TYPE LOCALITY: U.S.A., Arizona, Coconino Co., San Francisco Mtn.
DISTRIBUTION: S.E. Utah, W. Colorado, extreme S.E. Wyoming, W. and C. New Mexico, and Arizona (U.S.A.); Chihuahua, Durango, and Sonora (N.W. Mexico).
COMMENT: Subgenus *Otosciurus;* see Hall, 1981:434. Moore, 1959, Bull. Am. Mus. Nat. Hist., 118:153–206, included *Otosciurus* in subgenus *Sciurus.* Includes *kaibabensis;* see Hoffmeister and Diersing, 1978, J. Mammal., 59:402. Reviewed by Nash and Seaman, 1977, Mamm. Species, 80:1–5.
ISIS NUMBER: 5301410002030001001 as *S. aberti.*
5301410002030013001 as *S. kaibabensis.*

Sciurus aestuans Linnaeus, 1766. Syst. Nat., 12th ed., 1:88.
TYPE LOCALITY: Surinam.
DISTRIBUTION: Venezuela to N. Argentina.
COMMENT: Subgenus *Guerlinguetus;* see Cabrera, 1961:359. Formerly included *gilvigularis;* see Avila Pires, 1964, Bol. Mus. Para. Emilio Goeldi, n. s., Zool., 42:1–23.
ISIS NUMBER: 5301410002012001001 as *Guerlinguetus aestuans.*

Sciurus alleni Nelson, 1898. Proc. Biol. Soc. Wash., 12:147.
TYPE LOCALITY: Mexico, Nuevo Leon, Monterrey.
DISTRIBUTION: S.E. Coahuila through C. Nuevo Leon, south through W. Tamaulipas to extreme N. San Luis Potosi (Mexico).
COMMENT: Subgenus *Sciurus*; see Hall, 1981:430.
ISIS NUMBER: 5301410002030002001.

Sciurus anomalus Gmelin, 1778. Syst. Nat., 13th ed., 1:148.
TYPE LOCALITY: U.S.S.R., Georgia, 25 km S.W. of Kutaisi, Sabeka.
DISTRIBUTION: Turkey; Soviet Transcaucasia; N. and W. Iran; Syria; Israel.
COMMENT: Subgenus *Tenes*; see Corbet, 1978:76.
ISIS NUMBER: 5301410002030003001.

Sciurus arizonensis Coues, 1867. Am. Nat., 1:357.
TYPE LOCALITY: U.S.A., Arizona, Yavapai Co., Fort Whipple.
DISTRIBUTION: C. and S.E. Arizona and W.C. New Mexico (U.S.A.); N.E. Sonora (Mexico).
COMMENT: Subgenus *Sciurus*; see Hall, 1981:432.
ISIS NUMBER: 5301410002030005001.

Sciurus aureogaster F. Cuvier, 1829. *In* E. Geoffroy and F. Cuvier, Hist. Nat. Mamm., 6, livr. 59.
TYPE LOCALITY: Mexico, Tamaulipas, Altamira.
DISTRIBUTION: S.W. and C. Guatemala to Guanajuato to Nayarit and Nuevo Leon (Mexico).
COMMENT: Subgenus *Sciurus*; see Hall, 1981:418. Includes *griseoflavus, nelsoni, poliopus,* and *socialis;* see Musser, 1968, Misc. Publ. Mus. Zool. Univ. Mich., 137:1–112. Introduced to Elliot Key, Dade County, Florida (U.S.A.); see Brown and McGuire, 1975, J. Mammal., 56:405–419.
ISIS NUMBER: 5301410002030006001 as *S. aureogaster.*
5301410002030011001 as *S. griseoflavus.*
5301410002030015001 as *S. nelsoni.*
5301410002030018001 as *S. poliopus.*
5301410002030021001 as *S. socialis.*

Sciurus carolinensis Gmelin, 1788. Syst. Nat., 13th ed., 1:148.
TYPE LOCALITY: U.S.A., "Carolina."
DISTRIBUTION: E. Texas (U.S.A.) to Saskatchewan (Canada) and east to Atlantic Coast. Introduced into Britain and various localities in W. North America.
COMMENT: Subgenus *Sciurus*; see Hall, 1981:417.
ISIS NUMBER: 5301410002030007001.

Sciurus colliaei Richardson, 1839. *In* Beechey, The Zool. of Capt. Beechey's Voy., p. 8.
TYPE LOCALITY: Mexico, Nayarit, San Blas.
DISTRIBUTION: Mexico, W.C. coast including Sonora, Chihuahua, Sinaloa, Durango, Nayarit, Jalisco, and Colima.
COMMENT: Subgenus *Sciurus*; see Hall, 1981:421. Includes *sinaloensis* and *truei;* see Anderson, 1962, Am. Mus. Novit., 2093:9–12.
ISIS NUMBER: 5301410002030009001 as *S. colliaei.*
5301410002030020001 as *S. sinaloensis.*
5301410002030022001 as *S. truei.*

Sciurus deppei Peters, 1863. Monatsb. Preuss. Akad. Wiss. Berlin, p. 654.
TYPE LOCALITY: Mexico, Veracruz, Papantla.
DISTRIBUTION: Tamaulipas (Mexico) to Costa Rica.
COMMENT: Subgenus *Sciurus*; see Hall, 1981:426.
PROTECTED STATUS: CITES - Appendix III (Costa Rica).
ISIS NUMBER: 5301410002030010001.

Sciurus flammifer Thomas, 1904. Ann. Mag. Nat. Hist., ser. 7, 14:33.
  TYPE LOCALITY: Venezuela, Bolivar, Caura Valley, La Union.
  DISTRIBUTION: Venezuela along the Orinoco River from the Colombian border to
    Cuidad Bolivar; perhaps adjacent Colombia.
  COMMENT: Subgenus *Hadrosciurus;* see Cabrera, 1961:374. Moore, 1959, Bull. Am. Mus.
    Nat. Hist., 118:153-206, considered *flammifer* to be in the subgenus *Hadrosciurus* of
    genus *Guerlinguetus.*
  ISIS NUMBER: 5301410002012003001 as *Guerlinguetus flammifer.*

Sciurus gilvigularis Wagner, 1842. Arch. Naturg., 2:43.
  TYPE LOCALITY: Brazil, Borba, Rio Madeira.
  DISTRIBUTION: N. Brazil.
  COMMENT: Subgenus *Guerlinguetus;* includes *paraensis;* see Avila Pires, 1964, Bol. Mus.
    Para. Emilio Goeldi, n. s., Zool., 42:1-23. Cabrera, 1961:359, included *gilvigularis* in
    aestuans.

Sciurus granatensis Humboldt, 1811. Rec. Observ. Zool., 1(1805):8.
  TYPE LOCALITY: Colombia, Dept. Bolivar, Cartagena.
  DISTRIBUTION: Costa Rica to Colombia, Ecuador, and N. and C. Venezuela; Margarita
    Isl.; Trinidad; Tobago.
  COMMENT: Subgenus *Guerlinguetus;* see Hall, 1981:436. Includes *saltuensis;* see Cabrera,
    1961:368.
  ISIS NUMBER: 5301410002018007001 as *Microsciurus saleutenisi (sic).*

Sciurus griseus Ord, 1818. J. Phys. Chim. Hist. Nat. Arts Paris, 87:152.
  TYPE LOCALITY: U.S.A., Oregon, Wasco Co., The Dalles, Columbia River.
  DISTRIBUTION: C. Washington, W. Oregon, and California (U.S.A.) to Baja California
    Norte (Mexico).
  COMMENT: Subgenus *Hesperosciurus;* see Hall, 1981:433. Moore, 1959, Bull. Am. Mus.
    Nat. Hist., 118:153-206, included *Hesperosciurus* in subgenus *Sciurus.*
  ISIS NUMBER: 5301410002030012001.

Sciurus ignitus (Gray, 1867). Ann. Mag. Nat. Hist., ser. 3, 20:429.
  TYPE LOCALITY: Bolivia, near Yungas, upper Rio Beni.
  DISTRIBUTION: Andes of Bolivia and S.E. Peru; Ucayali Valley (E. Peru); N.E. Mato
    Grosso (Brazil); extreme N.W. Argentina.
  COMMENT: Subgenus *Guerlinguetus;* includes *cuscinus;* see Cabrera, 1961:370-371.
  ISIS NUMBER: 5301410002012004001 as *Guerlinguetus ignitus.*
    5301410002012002001 as *Guerlinguetus cuscinus.*

Sciurus igniventris Wagner, 1842. Wiegmann's Arch. Naturg., 1:360.
  TYPE LOCALITY: Brazil, Amazonas, north of the Rio Negro, Marabitanos.
  DISTRIBUTION: N.W. Brazil; S.E. Colombia; S. Venezuela.
  COMMENT: Subgenus *Hadrosciurus;* see Cabrera, 1961:375. Moore, 1959, Bull. Am. Mus.
    Nat. Hist., 118:153-206, placed *igniventris* in the subgenus *Hadrosciurus* of genus
    Guerlinguetus. Formerly included *tricolor;* see Hershkovitz, 1959, J. Mammal.,
    40:346; Cabrera, 1961:376, and footnote, listed *tricolor* in *igniventris* but
    Hershkovitz, 1959, was not incorporated. See comment under *spadiceus.*
  ISIS NUMBER: 5301410002012005001 as *Guerlinguetus igniventris.*

Sciurus lis Temminck, 1844. Fauna Japonica, 1(Mamm.), p. 45.
  TYPE LOCALITY: Japan, Honshu.
  DISTRIBUTION: Honshu, Shikoku, and Kyushu (Japan).
  COMMENT: Subgenus *Sciurus;* see Corbet, 1978:77.

Sciurus nayaritensis J. A. Allen, 1890. Bull. Am. Mus. Nat. Hist., 2:7, footnote.
  TYPE LOCALITY: Mexico, Zacatecas, Sierra Valparaiso.
  DISTRIBUTION: Jalisco (Mexico) north to S.E. Arizona (U.S.A.).
  COMMENT: Subgenus *Sciurus;* see Hall, 1981:431. Includes *chiricahuae* and *apache;* see
    Lee and Hoffmeister, 1963, Proc. Biol. Soc. Wash., 76:188-189, and Hall, 1981:431.
  ISIS NUMBER: 5301410002030014001 as *S. nayaritensis.*
    5301410002030004001 as *S. apache.*
    5301410002030008001 as *S. chiricahuae.*

Sciurus niger Linnaeus, 1758. Syst. Nat., 10th ed., 1:64.
TYPE LOCALITY: U.S.A., S. South Carolina (probably).
DISTRIBUTION: Texas (U.S.A.) and adjacent Mexico, north to Manitoba (Canada) east to the Atlantic Coast.
COMMENT: Subgenus *Sciurus;* see Hall, 1981:427.
PROTECTED STATUS: U.S. ESA - Endangered as *S. n. cinereus* subspecies only.
ISIS NUMBER: 5301410002030016001.

Sciurus oculatus Peters, 1863. Monatsb. Preuss. Akad. Wiss. Berlin, p. 653.
TYPE LOCALITY: Mexico, Veracruz, probably near Las Vigas.
DISTRIBUTION: San Luis Potosi, Hidalgo, Veracruz, Puebla, Mexico, Queretaro and Guanajuato (Mexico).
COMMENT: Subgenus *Sciurus;* see Hall, 1981:430.
ISIS NUMBER: 5301410002030017001.

Sciurus pucheranii (Fitzinger, 1867). Sitzb. Math. Naturw. Cl., 55, Abth., 1:487.
TYPE LOCALITY: Colombia, near Bogota.
DISTRIBUTION: Colombia in the W., C., and E. ranges of the Andes.
COMMENT: Subgenus *Guerlinguetus;* see Cabrera, 1961:372. Moore, 1959, Bull. Am. Mus. Nat. Hist., 118:203, placed *pucheranii* in the genus *Microsciurus*. Includes *medellinensis;* see Cabrera, 1961:372.
ISIS NUMBER: 5301410002018006001 as *Microsciurus pucherani* (sic).
5301410002018005001 as *Microsciurus medellineusis* (sic).

Sciurus pyrrhinus Thomas, 1898. Ann. Mag. Nat. Hist., ser. 7, 2:265.
TYPE LOCALITY: Peru, Junin Dept., Vitoc, Garita del Sol.
DISTRIBUTION: Eastern slopes of the Andes of Peru.
COMMENT: Subgenus *Hadrosciurus;* see Cabrera, 1961:378. Allen, 1915, Bull. Am. Mus. Nat. Hist., 34:215, placed *pyrrhinus* in the genus *Mesosciurus,* some members of which (not including *pyrrhinus*) Moore, 1959, Bull. Am. Mus. Nat. Hist., 118:204, placed as a subgenus in the genus *Syntheosciurus.*

Sciurus richmondi Nelson, 1898. Proc. Biol. Soc. Wash., 12:146.
TYPE LOCALITY: Nicaragua, Escondido River, 50 mi. (80 km) above Bluefields.
DISTRIBUTION: Nicaragua.
COMMENT: Subgenus *Guerlinguetus;* see Hall, 1981:436. Reviewed by Jones and Genoways, 1975, Mamm. Species, 53:1–2.
ISIS NUMBER: 5301410002030019001.

Sciurus sanborni Osgood, 1944. Field Mus. Nat. Hist. Publ. Zool. Ser., 29:191.
TYPE LOCALITY: Peru, Madre de Dios Dept., La Pampa, between the Rio Inambari and Rio Tambopata, 33 km N. of Santo Domingo, 570 m.
DISTRIBUTION: Madre de Dios Dept. (Peru).
COMMENT: Subgenus *Guerlinguetus;* see Cabrera, 1961:373.

Sciurus spadiceus Olfers, 1818. *In* Eschwege, J. Bras., 15(2):208.
TYPE LOCALITY: Brazil, Mato Grosso, Cuyaba.
DISTRIBUTION: S.E. Colombia; Ecuador; N.E. Peru; Brazil; Bolivia.
COMMENT: Subgenus *Hadrosciurus;* see Cabrera, 1961:374–379. Includes *langsdorffi, pyrrhonotus,* and *tricolor;* see Hershkovitz, 1959, J. Mammal., 40:346–347. Cabrera, 1961:376, and footnote, listed *langsdorffi* and *pyrrhonotus* as distinct species and included *tricolor* in *igniventris,* but Hershkovitz, 1959, was not incorporated. Moore, 1959, Bull. Am. Mus. Nat. Hist., 118:153–206, placed *langsdorffi* in *Guerlinguetus (Hadrosciurus) urucumus* (here included in *spadiceus*). Allen, 1915, Bull. Am. Mus. Nat. Hist., 34:269, included *pyrrhonotus* in *Urosciurus,* which Moore, 1959, Bull. Am. Mus. Nat. Hist., 118:153–206, placed as a subgenus in the genus *Guerlinguetus.*

Sciurus stramineus Eydoux and Souleyet, 1841. Voy. "La Bonite" Zool., 1:73.
TYPE LOCALITY: Peru, Piura Dept., Omatoe (=Anotape).
DISTRIBUTION: Extreme N.E. Peru and S.E. Ecuador in the area surrounding the Gulf of Guayaquil.
COMMENT: Subgenus *Guerlinguetus;* see Cabrera, 1961:373. Moore, 1959, Bull. Am. Mus. Nat. Hist., 118:179, placed *stramineus* in the subgenus *Simosciurus* of the genus *Microsciurus.*

Sciurus variegatoides Ogilby, 1839. Proc. Zool. Soc. Lond., 1839:117.
   TYPE LOCALITY: El Salvador.
   DISTRIBUTION: S. Chiapas (Mexico), through Central America to Panama.
   COMMENT: Includes *goldmani;* subgenus *Sciurus;* see Hall, 1981:424. Revised by Harris,
      1937, Misc. Publ. Mus. Zool. Univ. Mich., 38:7–39.
   ISIS NUMBER: 5301410002030023001.

Sciurus vulgaris Linnaeus, 1758. Syst. Nat., 10th ed., 1:63.
   TYPE LOCALITY: Sweden, Uppsala.
   DISTRIBUTION: Forested regions of Palearctic, from Iberia and Great Britain east to
      Kamchatka Peninsula and Sakhalin Isl. (U.S.S.R.); south to Mediterranean and
      Black Seas, N. Mongolia and N.E. China.
   COMMENT: Subgenus *Sciurus;* see Hall, 1981:417.
   ISIS NUMBER: 5301410002030024001.

Sciurus yucatanensis J. A. Allen, 1877. *In* Coues and Allen, Mongr. N. Amer. Rodentia, p.
      705.
   TYPE LOCALITY: Mexico, Yucatan, Merida.
   DISTRIBUTION: Yucatan Peninsula (Mexico); N. and S.W. Belize; N. Guatemala.
   COMMENT: Subgenus *Sciurus;* see Hall, 1981:422.
   ISIS NUMBER: 5301410002030025001.

Spermophilopsis Blasius, 1884. Tageblatt. Versamml. Deutsch. Naturf. Magdeburg, 57:325.
   ISIS NUMBER: 5301410002031000000.

Spermophilopsis leptodactylus (Lichtenstein, 1823). Naturh. Abh. Eversmann's Reise, p. 119.
   TYPE LOCALITY: U.S.S.R., Uzbekistan, 140 km N.W. of Bukhara, near Kara-Ata.
   DISTRIBUTION: Turkmenistan and Uzbekistan (U.S.S.R.); N.E. Iran; N.W. Afghanistan.
   COMMENT: Related to African xerine squirrels *(Atlantoxerus* and *Xerus);* see Nadler and
      Hoffmann, 1974, Experientia, 30:889–890.
   ISIS NUMBER: 5301410002031001001.

Spermophilus F. Cuvier, 1825. Des Dentes des Mammiferes, p. 255.
   REVIEWED BY: C. L. Elliott (CLE); D. J. Hafner (DJH); D. A. Sutton (DAS)(N. America,
      in part).
   COMMENT: *(=Citellus);* see Corbet, 1978:82, and Hershkovitz, 1949, J. Mammal., 30:82.
      Includes *Otospermophilus, Xerospermophilus, Poliocitellus, Ictidomys,* and
      Callospermophilus as subgenera; see Hall, 1981:382. North American species revised
      by Howell, 1938, N. Am. Fauna, 56:1–256. Eurasian species revised by Gromov *et
      al.,* 1965, who also recognized *Colobotis* and *Urocitellus* as subgenera; but see Hall,
      1981:382.
   ISIS NUMBER: 5301410002032000000.

Spermophilus adocetus (Merriam, 1903). Proc. Biol. Soc. Wash., 16:79.
   TYPE LOCALITY: Mexico, Michoacan, La Salada, 40 mi. (64 km) S. Uruapan.
   DISTRIBUTION: E. Jalisco, Michoacan, and N. Guerrero (W.C. Mexico).
   COMMENT: Subgenus *Otospermophilus;* see Hall, 1981:399. Gromov *et al.,* 1965:145,
      considered *Otospermophilus* a distinct genus. Includes *infernatus;* see Alvarez and
      Ramirez-Pulido, 1968, Rev. Soc. Mex. Hist. Nat., 29:181–190.
   ISIS NUMBER: 5301410002032001001.

Spermophilus alashanicus Buchner, 1888. Wiss. Res. Przewalski Cent. Asien Zool. Th.
      I: Saugeth., p. 11.
   TYPE LOCALITY: N. China, S. Alashan Desert.
   DISTRIBUTION: S.C. Mongolia; Alashan and E. Nan Shan (N. China).
   COMMENT: Subgenus *Spermophilus;* see Gromov *et al.,* 1965:208. Placed by Corbet, 1978,
      in *dauricus;* Orlov and Davaa, 1975, pp. 8–9, *in* [Systematics and cytogenetics of
      mammals], Orlov, ed., "Nauka," Moscow, provided evidence of specific
      distinctness.

Spermophilus annulatus Audubon and Bachman, 1842. J. Acad. Nat. Sci. Phila., 8:319.
TYPE LOCALITY: Mexico, Colima, Manzanillo.
DISTRIBUTION: Nayarit to N. Guerrero (Mexico).
COMMENT: Subgenus *Otospermophilus;* see Hall, 1981:403. Gromov *et al.,* 1965:145, considered *Otospermophilus* a distinct genus. Type locality designated by Howell, 1938, N. Am. Fauna, 56:163.
ISIS NUMBER: 5301410002032002001.

Spermophilus armatus Kennicott, 1863. Proc. Acad. Nat. Sci. Phila., 15:158.
TYPE LOCALITY: U.S.A., Wyoming, Uinta Co., foothills of Uinta Mtns., near Fort Bridger.
DISTRIBUTION: S.C. Utah to S. Montana, S.E. Idaho to W. Wyoming (U.S.A.).
COMMENT: Subgenus *Spermophilus;* see Hall, 1981:386.
ISIS NUMBER: 5301410002032003001.

Spermophilus atricapillus W. E. Bryant, 1889. Proc. Calif. Acad. Sci., ser. 2, 2:26.
TYPE LOCALITY: Mexico, Baja California Sur, Comondu.
DISTRIBUTION: Baja California (Mexico).
COMMENT: Subgenus *Otospermophilus;* see Hall, 1981:402. Gromov *et al.,* 1965:145, considered *Otospermophilus* a distinct genus.
ISIS NUMBER: 5301410002032004001.

Spermophilus beecheyi (Richardson, 1829). Fauna Boreali-Americana, 1:170.
TYPE LOCALITY: U.S.A., California, Monterey Co., Monterey.
DISTRIBUTION: W. Washington (U.S.A.) to Baja California Norte (Mexico).
COMMENT: Subgenus *Otospermophilus;* see Hall, 1981:401. Gromov *et al.,* 1965:145, considered *Otospermophilus* a distinct genus. Type locality restricted by Grinnell, 1933, Univ. Calif. Publ. Zool., 40:120.
ISIS NUMBER: 5301410002032005001.

Spermophilus beldingi Merriam, 1888. Ann. N.Y. Acad. Sci., 4:317.
TYPE LOCALITY: U.S.A., California, Placer Co., Donner.
DISTRIBUTION: E. Oregon, S.W. Idaho, N.E. California, N. Nevada, and N.W. Utah (U.S.A.).
COMMENT: Subgenus *Spermophilus;* see Hall, 1981:387.
ISIS NUMBER: 5301410002032006001.

Spermophilus brunneus (A. H. Howell, 1928). Proc. Biol. Soc. Wash., 41:211.
TYPE LOCALITY: U.S.A., Idaho, Adams Co., New Meadows.
DISTRIBUTION: W.C. Idaho (U.S.A.).
COMMENT: Subgenus *Spermophilus;* see Hall, 1981:385.
ISIS NUMBER: 5301410002032007001.

Spermophilus citellus (Linnaeus, 1766). Syst. Nat., 12th ed., 1:80.
TYPE LOCALITY: Austria.
DISTRIBUTION: S.E. Germany and S.W. Poland to European Turkey, Rumania, and Ukraine (U.S.S.R.).
COMMENT: Subgenus *Spermophilus;* see Gromov *et al.,* 1965:232. Formerly included *dauricus* and *xanthoprymnus* as in Ellerman and Morrison-Scott, 1951:506; but see Orlov and Davaa, 1975, pp. 8–9, *in* [Systematics and cytogenetics of mammals], Orlov, ed., "Nauka," Moscow, Gromov *et al.,* 1965:208, 237, and Vorontsov and Lyapunova, 1970, Byull. Mosk. Ova. Ispyt. Prir., Otd. Biol., 75:122–136.
ISIS NUMBER: 5301410002032008001.

Spermophilus columbianus (Ord, 1815). *In* Guthrie, A New Geogr., Hist., Coml. Grammar, Philadelphia, 2nd ed., 2:292.
TYPE LOCALITY: U.S.A., Idaho, Idaho Co., between the forks of the Clearwater and Kooskooskie Rivers.
DISTRIBUTION: S.E. British Columbia and W. Alberta (Canada) to N.E. Oregon, C. Idaho, and C. Montana (U.S.A.).
COMMENT: Subgenus *Urocitellus;* see Gromov *et al.,* 1965:196. Hall, 1981:381, included *Urocitellus* in subgenus *Spermophilus.*
ISIS NUMBER: 5301410002032009001.

Spermophilus dauricus Brandt, 1843. Bull. Phys. Math. Acad. Sci. St. Petersb., 2:379.
TYPE LOCALITY: U.S.S.R., R.S.F.S.R., Chitinsk. Obl., Torei-Nor (Lake).
DISTRIBUTION: Transbaikalia (U.S.S.R.); Mongolia; N. China.
COMMENT: Subgenus *Spermophilus*; see Gromov *et al.*, 1965:244. Corbet, 1978:83, tentatively included *alashanicus* in this species, but see Orlov and Davaa, 1975, pp. 8–9, *in* [Systematics and cytogenetics of mammals], Orlov, ed., "Nauka," Moscow, who provided evidence of specific distinctness. See comment under *alashanicus*. Ellerman and Morrison-Scott, 1951:506, included *dauricus* in *citellus*; but see Gromov *et al.*, 1965:244, who considered *dauricus* a distinct species.

Spermophilus elegans Kennicott, 1863. Acad. Nat. Sci. Philad., p. 158.
TYPE LOCALITY: U.S.A., Wyoming, Uinta Co., Fort Bridger.
DISTRIBUTION: N.E. Nevada, S.E. Oregon, S. Idaho, and S.W. Montana to C. Colorado and W. Nebraska (U.S.A.).
COMMENT: Subgenus *Spermophilus*; see Hall, 1981:385. Regarded by Howell, 1938, N. Am. Fauna, 56:1–256, and Hall, 1981:385, as a subspecies of *richardsonii*; but Nadler *et al.*, 1971, Syst. Zool., 20:298–305, Robinson and Hoffmann, 1975, Syst. Zool., 28:79–88, and Koeppl *et al.*, 1978, J. Mammal., 59:677–696, provided evidence of specific distinctness and included *aureus* and *nevadensis* in *elegans*.

Spermophilus erythrogenys Brandt, 1841. Bull. Acad. Sci. St. Petersb., p. 43.
TYPE LOCALITY: U.S.S.R., Tomsk. Obl., foothills of Altai Mtns., near Barnaul.
DISTRIBUTION: E. Kazakhstan and S.W. Siberia (U.S.S.R.); Sinkiang (China). Isolated population *(pallidicauda)* in Mongolia and Inner Mongolia (China).
COMMENT: Includes *brevicauda* (=*intermedius*), *carruthersi*, and *pallidicauda*; subgenus *Colobotis*; see Gromov *et al.*, 1965:315. Hall, 1981:381, included *Colobotis* in subgenus *Spermophilus*. Ellerman and Morrison-Scott, 1951:508, 511, regarded *brevicauda* and *carruthersi* as synonyms of *pygmaeus*, and *pallidicauda* as a full species. Sludskii *et al.*, 1969, [Mammals of Kazakhstan], vol. 1, "Nauka", Alma-Ata, p. 177, considered *intermedius* (=*brevicauda*) a full species. Provisionally included in *major* by Corbet, 1978:84; but see Gromov *et al.*, 1965:315, Vorontsov and Lyapunova, 1970, Byull. Mosk. Ova. Ispyt. Prir. Otd. Biol., 75:122–136, and 1976, pp. 337–353, *in* Beringia in Cenozoic, V. Kontrimavichus, ed., Acad. Sci., Vladivostok, for evidence of specific distinctness.
ISIS NUMBER: 5301410002032017001 as *S. pallidicauda*.

Spermophilus franklinii (Sabine, 1822). Trans. Linn. Soc. Lond., 13:587.
TYPE LOCALITY: Canada, Saskatchewan, Carlton House.
DISTRIBUTION: N. Great Plains; Alberta, Saskatchewan, and Manitoba (Canada), south to Kansas, Illinois, and Indiana (U.S.A.).
COMMENT: Subgenus *Poliocitellus*; see Hall, 1981:397. Gromov *et al.*, 1965:155, considered *Poliocitellus* a subgenus of the genus *Ictidomys*.
ISIS NUMBER: 5301410002032010001.

Spermophilus fulvus (Lichtenstein, 1823). Naturh. Abh. Eversmann's Reise, p. 119.
TYPE LOCALITY: U.S.S.R., Kazakh. S.S.R., E. of Mugodzhary Mtns., near Kuvandzhur River.
DISTRIBUTION: Kazakhstan (U.S.S.R.), from the Caspian Sea and the Volga River to Lake Balkash south to N.E. Iran, and N. Afghanistan; Sinkiang (China).
COMMENT: Subgenus *Colobotis*; see Gromov *et al.*, 1965:276. Hall, 1981:381, included *Colobotis* in subgenus *Spermophilus*.
ISIS NUMBER: 5301410002032011001.

Spermophilus lateralis (Say, 1823). *In* Long, Account of an Exped. from Pittsburgh to the Rocky Mtns., 2:46.
TYPE LOCALITY: U.S.A., Colorado, Fremont Co., Arkansas River, about 26 mi. (42 km) below Canyon City.
DISTRIBUTION: Montane W. North America, from C. British Columbia to S. New Mexico in the Rocky Mtns., and the Columbia River south to S. California and Nevada.
COMMENT: Subgenus *Callospermophilus*; see Hall, 1981:406. Gromov *et al.*, 1965:150, considered *Callospermophilus* a subgenus of the genus *Otospermophilus*.
ISIS NUMBER: 5301410002032012001.

Spermophilus madrensis (Merriam, 1901). Proc. Wash. Acad. Sci., 3:563.
TYPE LOCALITY: Mexico, Chihuahua, Sierra Madre, near Guadalupe y Calvo, 7000 ft. (2134 m).
DISTRIBUTION: S.W. Chihuahua (Mexico).
COMMENT: Subgenus *Callospermophilus*; see Hall, 1981:410. Gromov *et al.*, 1965:150, considered *Callospermophilus* a subgenus of the genus *Otospermophilus*.
ISIS NUMBER: 5301410002032013001.

Spermophilus major (Pallas, 1778). Nova Spec. Quad. Glir. Ord., p. 125.
TYPE LOCALITY: U.S.S.R., R.S.F.S.R., Kuibyshevsk. Obl., near Kuibyshev (=Samara).
DISTRIBUTION: Steppe between Volga and Irtysh Rivers (U.S.S.R.).
COMMENT: Subgenus *Colobotis*; see Gromov *et al.*, 1965:290. Hall, 1981:381, included *Colobotis* in subgenus *Spermophilus*. Occasionally hybridizes with *erythrogenys* and *fulvus*; see Denisov, 1963, Zool. Zh., 42:1887–1899, and Ognev, 1947, [Mamm. U.S.S.R., Adjac. Count.], Acad. Sci. Moscow, p. 24. Corbet, 1978:84, provisionally included *erythrogenys* and *brevicauda* in this species, but Gromov *et al.*, 1965:290, Vorontsov and Lyapunova, 1970, Byull. Mosk. Ova. Ispyt. Prir. Otd. Biol., 75:122–136, and 1976, pp. 337–353, *in* Beringia in Cenozoic, V. Kontrimavichus, ed., Acad. Sci., Vladivostok, considered *erythrogenys* a distinct species, and Gromov *et al.*, 1965:315, included *brevicauda* in *erythrogenys*; see comment under *erythrogenys*.
ISIS NUMBER: 5301410002032014001.

Spermophilus mexicanus (Erxleben, 1777). Syst. Regn. Anim., 1:428.
TYPE LOCALITY: Mexico, Toluca.
DISTRIBUTION: S. New Mexico and W. Texas (U.S.A.) to Jalisco and S. Puebla (C. Mexico).
COMMENT: Subgenus *Ictidomys*; see Hall, 1981:394. Gromov *et al.*, 1965:153, considered *Ictidomys* a distinct genus. Known to hybridize at several localities with *tridecemlineatus*; see Cothran *et al.*, 1977, J. Mammal., 58:610–622.
ISIS NUMBER: 5301410002032015001.

Spermophilus mohavensis Merriam, 1889. N. Am. Fauna, 2:15.
TYPE LOCALITY: U.S.A., California, San Bernardino Co., near Rabbit Springs, about 15 mi. (24 km) E. Hesperia.
DISTRIBUTION: N.W. Mohave Desert and Owens Valley (S. California, U.S.A.).
COMMENT: Subgenus *Xerospermophilus*; see Hall, 1981:405. Gromov *et al.*, 1965:151, considered *Xerospermophilus* a distinct genus. Type locality designated by Grinnell and Dixon, 1918, Mon. Bull. Calif. Comm. Hort., 7:667.
ISIS NUMBER: 5301410002032016001.

Spermophilus musicus Menetries, 1832. Cat. Raisonne Mus. Peale Phila., p. 21.
TYPE LOCALITY: U.S.S.R., Gruzinsk. S.S.R., Uchkulan ("Caucasus").
DISTRIBUTION: N. Caucasus Mtns. (U.S.S.R.).
COMMENT: Subgenus *Spermophilus*; see Gromov *et al.*, 1965:249. Regarded by Ellerman and Morrison-Scott, 1951:508, and Corbet, 1978:83, as a subspecies of *pygmaeus*. Gromov *et al.*, 1965:249, and Vorontsov and Lyapunova, 1970, Byull. Mosk. Ova. Ispyt. Prir. Otd. Biol., 75:122–136, and 1976, pp. 337–353, *in* Beringia in Cenozoic, V. Kontrimavichus, ed., Acad. Sci., Vladivostok, provided evidence of specific distinctness.

Spermophilus parryii (Richardson, 1825). *In* Parry, 1825 (1827), appendix to Parry's J. of Second Voy., p. 316.
TYPE LOCALITY: Canada, Hudson Bay, Melville Peninsula, Lyon Inlet, Five Hawser Bay.
DISTRIBUTION: N.W. Canada; Alaska (U.S.A.); N.E. Yakutia, Anadyrsk. Krai, and Chukotka (U.S.S.R.).
COMMENT: Subgenus *Urocitellus*; see Gromov *et al.*, 1965:184. Hall, 1981:381, included *Urocitellus* in subgenus *Spermophilus*. Regarded by Ellerman and Morrison-Scott, 1951:511, and Hall and Kelson, 1959:343, as a synonym of *undulatus*. Gromov *et al.*, 1965:184, and Nadler *et al.*, 1974, Comp. Biochem. Physiol., 47A:663–681, provided evidence of specific distinctness. Reviewed by Chernyavskii, 1972, Trud. Mosk. Soc. Nat., 48:199–214, Serdyuk, 1979, Zool. Zh., 58:1692–1702,

(Palearctic), and Pearson, 1981, Geographic and intraspecific cranial variation in North American arctic ground squirrels, M.A. Thesis, Univ. Kansas, Lawrence, 96 pp. (Nearctic).
ISIS NUMBER: 5301410002032018001.

Spermophilus perotensis Merriam, 1893. Proc. Biol. Soc. Wash., 8:131.
   TYPE LOCALITY: Mexico, Veracruz, Perote.
   DISTRIBUTION: Veracruz and Puebla (E.C. Mexico).
   COMMENT: Subgenus *Ictidomys;* see Hall, 1981:397. Gromov *et al.,* 1965:153, considered *Ictidomys* a distinct genus.
   ISIS NUMBER: 5301410002032019001.

Spermophilus pygmaeus (Pallas, 1778). Nova Spec. Quad. Glir. Ord., p. 122.
   TYPE LOCALITY: U.S.S.R., Kazakh. S.S.R., between Ural and Emba Rivers.
   DISTRIBUTION: Kazakhstan and S. Ural Mtns. to Crimea (U.S.S.R.).
   COMMENT: Subgenus *Spermophilus;* see Gromov *et al.,* 1965:257. See comment under *musicus.* Hybridizes rarely with *erythrogenys, fulvus,* and *major;* see Bazhanov, 1944, Dokl. Akad. Nauk S.S.S.R., 42(7):321–322, Bazhanov, 1945, Vestn. Kazakh. Fil. Akad. Nauk S.S.S.R., 5(8):37–39, and Denisov, 1964, Nauchin. Dokl. Vyssh. Shk. Biol. Nauki, 2:49–54. Also hybridizes with *suslicus;* see Denisov and Smirnova, 1976, Acta Theriol., 21:267–278.
   ISIS NUMBER: 5301410002032020001.

Spermophilus relictus (Kashkarov, 1923). Trans. Turk. Sci. Soc., 1:185.
   TYPE LOCALITY: U.S.S.R., Kirgiz. S.S.R., Talassk. Obl., Talassk. Alatau, Kara-Bura, and Kumysh-Tagh Valleys.
   DISTRIBUTION: Tien Shan Mtns. in Kirgizia and S.E. Kazakhstan (U.S.S.R.).
   COMMENT: Subgenus *Spermophilus;* see Gromov *et al.,* 1965:198.

Spermophilus richardsonii (Sabine, 1822). Trans. Linn. Soc. Lond., 13:589.
   TYPE LOCALITY: Canada, Saskatchewan, Carleton House.
   DISTRIBUTION: N. Great Plains in S. Alberta, S. Saskatchewan, S. Manitoba (Canada), Montana, North Dakota, N.E. South Dakota, W. Minnesota, and N.W. Iowa (U.S.A.).
   COMMENT: Subgenus *Spermophilus;* see Hall, 1981:385. Formerly included *elegans;* see comment under that species.
   ISIS NUMBER: 5301410002032021001.

Spermophilus saturatus (Rhoads, 1895). Proc. Acad. Nat. Sci. Phila., 47:43.
   TYPE LOCALITY: U.S.A., Washington, Kittitas Co., Lake Kichelos (=Keechelus), 8000 ft. (2438 m).
   DISTRIBUTION: Cascade Mtns. of W. Washington (U.S.A.) and S.W. British Columbia (Canada).
   COMMENT: Subgenus *Callospermophilus;* see Hall, 1981:409. Gromov *et al.,* 1965:150, considered *Callospermophilus* a subgenus of the genus *Otospermophilus.*
   ISIS NUMBER: 5301410002032022001.

Spermophilus spilosoma Bennett, 1833. Proc. Zool. Soc. Lond., 1833:40.
   TYPE LOCALITY: Mexico, Durango, Durango (City).
   DISTRIBUTION: C. Mexico to S. Texas, S.W. South Dakota, and N.W. Arizona (U.S.A.).
   COMMENT: Subgenus *Ictidomys;* see Hall, 1981:395. Gromov *et al.,* 1965:153, considered *Ictidomys* a distinct genus. Reviewed by Streubel and Fitzgerald, 1978, Mamm. Species, 101:1–4. Howell, 1938, N. Am. Fauna, 56:122, fixed the type locality.
   ISIS NUMBER: 5301410002032023001.

Spermophilus suslicus (Guldenstaedt, 1770). Nova Comm. Acad. Sci. Petrop., 14:389.
   TYPE LOCALITY: U.S.S.R., Voronezhsk. Obl., near Voronezh (City).
   DISTRIBUTION: Steppes of C. and S. Europe, including Poland, E. Rumania, Ukraine north to Oka River and east to the Volga River (U.S.S.R.).
   COMMENT: Subgenus *Spermophilus;* see Gromov *et al.,* 1965:212. See also comment under *pygmaeus.*
   ISIS NUMBER: 5301410002032024001.

Spermophilus tereticaudus Baird, 1858. Mammals *in* Repts. Expl. Surv...., 8(1):315.
TYPE LOCALITY: U.S.A., California, Imperial Co., Old Fort Yuma.
DISTRIBUTION: Deserts of S.E. California, S. Nevada, W. Arizona (U.S.A.), N.E. Baja California and Sonora (Mexico).
COMMENT: Subgenus *Xerospermophilus*; see Hall, 1981:405. Gromov *et al.*, 1965:151, considered *Xerospermophilus* a distinct genus.
ISIS NUMBER: 5301410002032025001.

Spermophilus townsendii Bachman, 1839. J. Acad. Nat. Sci. Phila., 8:61.
TYPE LOCALITY: U.S.A., Washington, Walla Walla Co., near Wallula.
DISTRIBUTION: Great Basin region of E. Washington, Oregon, S. Idaho, E. California, Nevada, and W. Utah (U.S.A.).
COMMENT: Subgenus *Spermophilus*; see Hall, 1981:382. Includes several chromosomally different groups *(mollis, vigilis)* that may be distinct species; see Vorontsov and Lyapunova, 1970, Byull. Mosk. Ova. Ispyt. Prir. Otd. Biol., 75:122–136.
ISIS NUMBER: 5301410002032026001.

Spermophilus tridecemlineatus (Mitchill, 1821). Med. Repos. (N.Y.), (n.s.), 6(21):248.
TYPE LOCALITY: U.S.A., C. Minnesota.
DISTRIBUTION: Great Plains, from C. Texas to E. Utah, Ohio (U.S.A.) and S.C. Canada.
COMMENT: Subgenus *Ictidomys*; see Hall, 1981:391. Gromov *et al.*, 1965:153, considered *Ictidomys* a distinct genus. Reviewed by Streubel and Fitzgerald, 1978, Mamm. Species, 103:1–5. See also comment under *mexicanus*.
ISIS NUMBER: 5301410002032027001.

Spermophilus undulatus (Pallas, 1778). Nova Spec. Quad. Glir. Ord., p. 122.
TYPE LOCALITY: U.S.S.R., Buryat A.S.S.R., Selenga River.
DISTRIBUTION: E. Kazakhstan, S. Siberia, Transbaikalia (U.S.S.R.), N. Mongolia, Heilungkiang and Sinkiang (China).
COMMENT: Subgenus *Urocitellus*; see Gromov *et al.*, 1965:162. Hall, 1981:381, included *Urocitellus* in subgenus *Spermophilus*. Formerly included *parryii*; see Hall and Kelson, 1959:343; but also see Nadler *et al.*, 1974, Comp. Biochem. Physiol., 47A:663–681.
ISIS NUMBER: 5301410002032028001.

Spermophilus variegatus (Erxleben, 1777). Syst. Regn. Anim., 1:421.
TYPE LOCALITY: Mexico, Distrito Federal, Valley of Mexico, near Mexico City.
DISTRIBUTION: S. Nevada to S.W. Texas and Utah (U.S.A.) to Puebla (C. Mexico).
COMMENT: Subgenus *Otospermophilus*; see Hall, 1981:399. Gromov *et al.*, 1965:145, considered *Otospermophilus* a distinct genus. Nelson, 1898, Science, n.s., 8:897, fixed the type locality.
ISIS NUMBER: 5301410002032029001.

Spermophilus washingtoni (A. H. Howell, 1938). N. Am. Fauna, 56:69.
TYPE LOCALITY: U.S.A., Washington, Walla Walla Co., Touchet.
DISTRIBUTION: S.E. Washington, N.E. Oregon (U.S.A.).
COMMENT: Subgenus *Spermophilus*; see Hall, 1981:384.
ISIS NUMBER: 5301410002032030001.

Spermophilus xanthoprymnus (Bennett, 1835). Proc. Zool. Soc. Lond., 1835:90.
TYPE LOCALITY: N.E. Turkey, Armenia, Erzerum (=Erzurum).
DISTRIBUTION: Soviet Transcaucasia, Turkey, Syria, and Palestine.
COMMENT: Subgenus *Spermophilus*; see Gromov *et al.*, 1965:237. Regarded by Ellerman and Morrison-Scott, 1951:506, and Corbet, 1978:83, as a synonym of *citellus*; but see Gromov *et al.*, 1965:237, and Vorontsov and Lyapunova, 1970, Byull. Mosk. Ova. Ispyt. Prir. Otd. Biol., 75:122–136, who provided evidence of specific distinctness.

Sundasciurus Moore, 1958. Am. Mus. Novit., 1914:2.
COMMENT: Reviewed, in part, by Heaney, 1979, Proc. Biol. Soc. Wash., 92:280–286.
ISIS NUMBER: 5301410002033000000.

Sundasciurus brookei (Thomas, 1892). Ann. Mag. Nat. Hist., ser. 6, 9:253.
TYPE LOCALITY: Malaysia, Sarawak, Mt. Dulit.
DISTRIBUTION: Mountains of Borneo.
ISIS NUMBER: 5301410002033001001.

Sundasciurus davensis (Sanborn, 1952). Fieldiana Zool., 33:117.
TYPE LOCALITY: Philippines, Mindanao, Davas Prov., Taguon Mumic, Madaum.
DISTRIBUTION: Known only from the type locality.
COMMENT: May be conspecific with *mindanensis*, *philippinensis*, and *samarensis* (LRH).

Sundasciurus hippurus (I. Geoffroy, 1831). *In* Zool. Voy. de Belanger Indes Orient., 2:149.
TYPE LOCALITY: Malaysia, Malacca.
DISTRIBUTION: Malaysian subregion except Java and S.W. Philippines.
ISIS NUMBER: 5301410002033002001.

Sundasciurus hoogstraali (Sanborn, 1952). Fieldiana Zool., 33:115.
TYPE LOCALITY: Philippines, Calamian Isls., Busuanga Isl., Dimaniang.
DISTRIBUTION: Busuanga Isl. (Philippines).
COMMENT: Member of the *steerii* group; see comment under *steerii* (LRH).

Sundasciurus jentinki (Thomas, 1887). Ann. Mag. Nat. Hist., ser. 5, 20:128.
TYPE LOCALITY: Malaysia, Sabah, Mt. Kinabalu.
DISTRIBUTION: Mountains of N. Borneo.

Sundasciurus juvencus (Thomas, 1908). Ann. Mag. Nat. Hist., ser. 8, 2:498.
TYPE LOCALITY: Philippines, Palawan Isl.
DISTRIBUTION: N. Palawan Isl. (Philippines).
COMMENT: Member of the *steerii* group; see comment under *steerii* (LRH).

Sundasciurus lowii (Thomas, 1892). Ann. Mag. Nat. Hist., ser. 6, 2:253.
TYPE LOCALITY: Malaysia, Sarawak, Lumbidon.
DISTRIBUTION: Malaysian subregion, except Java and S.W. Philippines.
COMMENT: Includes *fraterculus*, from the Mentawi Isls.; see Chasen, 1940, Bull. Raffles Mus., 15:144.
ISIS NUMBER: 5301410002033003001.

Sundasciurus mindanensis (Steere, 1890). List of the Birds and Mammals collected by the Steere Expedition to the Philippines, p. 29.
TYPE LOCALITY: Philippines, Mindanao.
DISTRIBUTION: Mindanao and adjacent small islands (Philippines).
COMMENT: May be conspecific with *davensis*, *philippinensis*, and *samarensis* (LRH).
ISIS NUMBER: 5301410002033004001.

Sundasciurus mollendorffi Matschie, 1898. Sitzb. Ges. Naturf. Fr. Berlin, 5:41.
TYPE LOCALITY: Philippines, Calamian Isls., Culion Isl.
DISTRIBUTION: Known only from the type locality.
COMMENT: Includes *albicauda*; see Matschie, 1898; *albicauda* was listed as a distinct species by Corbet and Hill, 1980, without comment. Member of the *steerii* group (LRH).
ISIS NUMBER: 5301410002033005001.

Sundasciurus philippinensis (Waterhouse, 1839). Proc. Zool. Soc. Lond., 1839:117.
TYPE LOCALITY: Philippines, Mindanao Isl.
DISTRIBUTION: S. and W. Mindanao, and Basilan (Philippines).
COMMENT: May be conspecific with *davensis*, *mindanensis*, and *samarensis* (LRH).
ISIS NUMBER: 5301410002033006001.

Sundasciurus rabori Heaney, 1979. Proc. Biol. Soc. Wash., 92:280–286.
TYPE LOCALITY: Philippines, Palawan Isl., Mt. Mantalingajan, Magtaguimbong.
DISTRIBUTION: Known only from type locality, but probably widespread in mountains on Palawan (Philippines).

Sundasciurus samarensis (Steere, 1890). List of the Birds and Mammals Collected by the Steere Expedition to the Philippines, p. 30.
TYPE LOCALITY: Philippines, Samar.
DISTRIBUTION: Samar and Leyte Isls. (Philippines).
COMMENT: May be conspecific with *davensis*, *mindanensis*, and *philippinensis* (LRH).

Sundasciurus steerii (Gunther, 1877). Proc. Zool. Soc. Lond., 1876:735.
  TYPE LOCALITY: Philippines, Balabac Isl.
  DISTRIBUTION: Balabac and S. Palawan Isl. in lowlands (Philippines).
  COMMENT: Member of the *steerii* group, which contains 1–3 species; see Heaney, 1979,
    Proc. Biol. Soc. Wash., 92:280–286.
  ISIS NUMBER: 5301410002033007001.

Sundasciurus tenuis (Horsfield, 1824). Zool. Res. Java, 1824:153.
  TYPE LOCALITY: Singapore.
  DISTRIBUTION: Malaysian subregion, except Java and S.W. Philippines.
  ISIS NUMBER: 5301410002033008001.

Syntheosciurus Bangs, 1902. Bull. Mus. Comp. Zool. Harv. Univ., 39:25.
  REVIEWED BY: D. A. Sutton (DAS).
  COMMENT: Goodwin, 1946, Bull. Am. Mus. Nat. Hist., 87:275–473, suggested that
    Syntheosciurus might be a subgenus of *Sciurus;* but see Heaney and Hoffmann,
    1978, J. Mammal., 59:854–855, Enders, 1980, J. Mammal., 61:725–727, and Hall,
    1981:438.
  ISIS NUMBER: 5301410002034000000.

Syntheosciurus brochus Bangs, 1902. Bull. Mus. Comp. Zool. Harv. Univ., 39:25.
  TYPE LOCALITY: Panama, Chiriqui, 8 mi. (13 km) N. of Boquete, 7000 ft. (2134 m).
  DISTRIBUTION: Costa Rica to N. Panama.
  COMMENT: Reviewed by Enders, 1953, J. Mammal., 34:509. Includes *poasensis;* see
    Heaney and Hoffmann, 1978, J. Mammal., 59:854–855, Enders, 1980, J. Mammal.,
    61:726, and Hall, 1981:438.
  ISIS NUMBER: 5301410002034001001.

Tamias Illiger, 1811. Prodr. Syst. Mamm. et Avium., p. 83.
  REVIEWED BY: F. J. Brenner (FJB); H. E. Broadbooks (HEB); J. R. Callahan (JRCL); D. P.
    Snyder (DPS); D. A. Sutton (DAS).
  COMMENT: N. American forms revised by Howell, 1929, N. Am. Fauna, 52:11–23.
    Includes *Eutamias (T. sibiricus), Tamias (T. striatus),* and *Neotamias* as subgenera; see
    Ellerman, 1940:428, Corbet, 1978:85, and Nadler *et al.,* 1977, Am. Midl. Nat.,
    98:343–353. Disagreement exists regarding the status of *Eutamias* and *Neotamias;*
    see Ellis and Maxson, 1979, J. Mammal., 60:331–334, White, 1953, Univ. Kans.
    Publ. Mus. Nat. Hist., 5(32):543–561, and Hall, 1981:337. DPS considered this
    arrangement tentative until additional biochemical and cytological data can be
    evaluated.
  ISIS NUMBER: 5301410002035000000 as *Tamias.*
            5301410002007000000 as *Eutamias.*

Tamias alpinus Merriam, 1893. Proc. Biol. Soc. Wash., 8:137.
  TYPE LOCALITY: U.S.A., California, Tulare Co., Big Cottonwood Meadows, just south of
    Mt. Whitney, 3050 m.
  DISTRIBUTION: Alpine zone in Sierra Nevada, from Tuolumne to Tulare Counties (E.C.
    California, U.S.A.).
  ISIS NUMBER: 5301410002007001001 as *Eutamias alpinus.*

Tamias amoenus J. A. Allen, 1890. Bull. Am. Mus. Nat. Hist., 3:90.
  TYPE LOCALITY: U.S.A., Oregon, Klamath Co., Fort Klamath.
  DISTRIBUTION: C. British Columbia (Canada) south to C. California east to C. Montana
    and W. Wyoming (U.S.A.).
  ISIS NUMBER: 5301410002007002001 as *Eutamias amoenus.*

Tamias bulleri J. A. Allen, 1889. Bull. Am. Mus. Nat. Hist., 2:173.
  TYPE LOCALITY: Mexico, Zacatecas, Sierra de Valparaiso.
  DISTRIBUTION: Sierra Madre, in S. Durango, W. Zacatecas, and N. Jalisco (Mexico).
  COMMENT: Monotypic; formerly included *durangae* and *solivagus,* which were
    considered *incertae sedis* by Callahan, 1980, Southwest. Nat., 25:1–8; see *durangae.*
  ISIS NUMBER: 5301410002007003001 as *Eutamias bulleri.*

Tamias canipes (V. Bailey, 1902). Proc. Biol. Soc. Wash., 15:117.
    TYPE LOCALITY: U.S.A., Texas, Culberson Co., Guadalupe Mtns., head of Dog Canyon,
        2130 m.
    DISTRIBUTION: Mountains of S.E. New Mexico and W. Texas (U.S.A.).
    COMMENT: Elevated from subspecies of *cinereicollis* by Fleharty, 1960, J. Mammal.,
        41:235–242. Reviewed by Findley *et al.*, 1975, Mammals of New Mexico, Univ. of
        New Mexico Press, Albuquerque, p. 103–112.
    ISIS NUMBER: 5301410002007004001 as *Eutamias canipes.*

Tamias cinereicollis J. A. Allen, 1890. Bull. Am. Mus. Nat. Hist., 3:94.
    TYPE LOCALITY: U.S.A., Arizona, Coconino Co., San Francisco Mtns.
    DISTRIBUTION: Mountains of C. and E. Arizona and C. and S.W. New Mexico (U.S.A.).
    ISIS NUMBER: 5301410002007005001 as *Eutamias cinereicollis.*

Tamias dorsalis Baird, 1855. Proc. Acad. Nat. Sci. Phila., 7:332.
    TYPE LOCALITY: U.S.A., New Mexico, Grant Co., Fort Webster, copper mines of the
        Mimbres (=near Santa Rita).
    DISTRIBUTION: E. Nevada, S. Idaho, Utah, S.W. Wyoming, and N.W. Colorado south
        through Arizona and W. New Mexico (U.S.A.) to N.W. Durango, W. Coahuila,
        and coastal Sonora (Mexico).
    COMMENT: Reviewed by Callahan and Davis, 1977, Southwest. Nat., 22:71, who
        included *sonoriensis* in this species.
    ISIS NUMBER: 5301410002007006001 as *Eutamias dorsalis.*

Tamias durangae (J. A. Allen, 1903). Bull. Am. Mus. Nat. Hist., 19:594.
    TYPE LOCALITY: Mexico, Durango, Sierra de Candella, Arroyo de Bucy, 7000 ft. (2134
        m).
    DISTRIBUTION: S.W. Chihuahua to W.C. Durango; S.E. Coahuila (Mexico).
    COMMENT: Formerly included in *bulleri;* includes *solivagus;* probably conspecific with
        canipes; see Callahan, 1980, Southwest. Nat., 25:1–8, who considered both *durangae*
        and *solivagus, incertae sedis.*

Tamias merriami J. A. Allen, 1889. Bull. Am. Mus. Nat. Hist., 2:176.
    TYPE LOCALITY: U.S.A., California, San Bernardino Co., San Bernardino Mtns., N. of
        San Bernardino, 4500 ft. (1372 m).
    DISTRIBUTION: From San Francisco Bay south in the Coast Range, and south of
        Columbia (California, U.S.A.) in the Sierra Nevada, to extreme N. Baja California
        (Mexico).
    COMMENT: Formerly included *meridionalis* and *obscurus;* see Callahan, 1977, J.
        Mammal., 58:188–201. See also comment under *obscurus.*
    ISIS NUMBER: 5301410002007007001 as *Eutamias merriami.*

Tamias minimus Bachman, 1839. J. Acad. Nat. Sci. Phila., 8:71.
    TYPE LOCALITY: U.S.A., Wyoming, Sweetwater Co., Green River, near mouth of Big
        Sandy Creek.
    DISTRIBUTION: C. Yukon (Canada) south through Sierra Nevada and S. New Mexico,
        east to Michigan (U.S.A.) and W. Quebec (Canada).
    ISIS NUMBER: 5301410002007008001 as *Eutamias minimus.*

Tamias obscurus J. A. Allen, 1890. Bull. Am. Mus. Nat. Hist., 3:70.
    TYPE LOCALITY: Mexico, Baja California, Sierra San Pedro Martir, near Vallecitos.
    DISTRIBUTION: S. California (San Bernardino Co., U.S.A.) to C. Baja California (Mexico).
    COMMENT: Regarded by Howell, 1929, as a synonym of *merriami;* but see Callahan,
        1977, J. Mammal., 58:188–201, who provided evidence of specific distinctness and
        included *meridionalis* and *davisi* in this species.

Tamias ochrogenys (Merriam, 1897). Proc. Biol. Soc. Wash., 11:195.
    TYPE LOCALITY: U.S.A., California, Mendocino Co., Mendocino.
    DISTRIBUTION: Coast of N. California from Van Duzen River south to S. Sonoma Co.
        (U.S.A.).
    COMMENT: Elevated from subspecies of *townsendii* by Sutton and Nadler, 1974,
        Southwest. Nat., 19:199–212.

Tamias palmeri (Merriam, 1897). Proc. Biol. Soc. Wash., 11:208.
TYPE LOCALITY: U.S.A., Nevada, Clark Co., Charleston Peak, 8000 ft. (2438 m).
DISTRIBUTION: Charleston Mtns. (S. Nevada, U.S.A.).
ISIS NUMBER: 5301410002007009001 as *Eutamias palmeri*.

Tamias panamintinus Merriam, 1893. Proc. Biol. Soc. Wash., 8:134.
TYPE LOCALITY: U.S.A., California, Inyo Co., Panamint Mtns., Johnson Canyon, vicinity of Hungry Bill's Ranch, about 5000 ft. (1524 m).
DISTRIBUTION: Mountains of S.E. California and S.W. Nevada (U.S.A.).
ISIS NUMBER: 5301410002007010001 as *Eutamias panamintinus*.

Tamias quadrimaculatus Gray, 1867. Ann. Mag. Nat. Hist., ser. 3, 20:435.
TYPE LOCALITY: U.S.A., California, Placer Co., Michigan Bluff.
DISTRIBUTION: Sierra Nevada of E.C. California (Plumas to Mariposa Cos.) and adjacent W.C. Nevada (U.S.A.).
ISIS NUMBER: 5301410002007011001 as *Eutamias quadrimaculatus*.

Tamias quadrivittatus (Say, 1823). *In* James, Account of an Exped., to Rocky Mtns..., 2:45.
TYPE LOCALITY: U.S.A., Colorado, Fremont Co., Arkansas River, about 26 mi. (42 km) below Canon City.
DISTRIBUTION: Mountains of Colorado and E. Utah south to N.E. Arizona and S. New Mexico (U.S.A.).
COMMENT: Includes *australis* from the Organ Mtns., New Mexico; see Patterson, 1980, J. Mammal., 61:455–464.
ISIS NUMBER: 5301410002007012001 as *Eutamias quadrivittatus*.

Tamias ruficaudus (A. H. Howell, 1920). Proc. Biol. Soc. Wash., 33:91.
TYPE LOCALITY: U.S.A., Montana, Glacier Co., Upper St. Mary's Lake.
DISTRIBUTION: N.E. Washington to W. Montana (U.S.A.), and S.E. British Columbia (Canada).
ISIS NUMBER: 5301410002007013001 as *Eutamias ruficaudus*.

Tamias senex J. A. Allen, 1890. Bull. Am. Mus. Nat. Hist., 3:83.
TYPE LOCALITY: U.S.A., California, Placer Co., summit of Donner Pass.
DISTRIBUTION: Sierra Nevada of E.C. California and W.C. Nevada to N. coast of California, and N.C. Oregon (U.S.A.).
COMMENT: Elevated from subspecies of *townsendii* by Sutton and Nadler, 1974, Southwest. Nat., 19:199–212.

Tamias sibiricus (Laxmann, 1769). Sibirische Briefe, Gottingen, p. 69.
TYPE LOCALITY: U.S.S.R., R.S.F.S.R., Altaisk Krai, near Barnaul.
DISTRIBUTION: N. European U.S.S.R. and Siberia, to China and Korea; Sakhalin; S. Kurile Isls. (U.S.S.R.); Hokkaido (Japan).
COMMENT: Subgenus *Eutamias*; see Gromov *et al.*, 1965:125. See Chaworth-Musters, 1937, Ann. Mag. Nat. Hist., 19:158, for clarification of the type locality.
ISIS NUMBER: 5301410002007014001 as *Eutamias sibericus* (sic).

Tamias siskiyou (A. H. Howell, 1922). J. Mammal., 3:180.
TYPE LOCALITY: U.S.A., California, Siskiyou Co., Siskiyou Mtns., summit of White Mtn., 6000 ft. (1829 m).
DISTRIBUTION: Siskiyou Mtns. and coast of N. California to C. Oregon (U.S.A.).
COMMENT: Elevated from subspecies of *townsendii* by Sutton and Nadler, 1974, Southwest. Nat., 19:199–212.

Tamias sonomae (Grinnell, 1915). Univ. Calif. Publ. Zool., 12:321.
TYPE LOCALITY: U.S.A., California, Sonoma Co., 1 mi. (1.6 km) W. of Guerneville.
DISTRIBUTION: N.W. California, from San Francisco Bay north to Siskiyou Co. (U.S.A.).
ISIS NUMBER: 5301410002007015001 as *Eutamias sonomae*.

Tamias speciosus Merriam, 1890. *In* J. A. Allen, Bull. Am. Mus. Nat. Hist., 3:86.
TYPE LOCALITY: U.S.A., California, San Bernardino Co., San Bernardino Mtns., Whitewater Creek, 7500 ft. (2255 m).
DISTRIBUTION: Sierra Nevada from Mt. Lassen to San Bernardino Mtns. (California); W. Nevada (U.S.A.).
ISIS NUMBER: 5301410002007016001 as *Eutamias speciosus*.

Tamias striatus (Linnaeus, 1758). Syst. Nat., 10th ed., 1:64.
   TYPE LOCALITY: U.S.A., South Carolina, upper Savannah River.
   DISTRIBUTION: S. Manitoba and Nova Scotia (Canada) to Louisiana, Alabama, and
      Georgia, east to Atlantic Coast (U.S.A.).
   COMMENT: Subgenus *Tamias* see Gromov *et al.*, 1965:134. Type locality restricted by
      Howell, 1929, N. Am. Fauna, 52:14.
   ISIS NUMBER: 5301410002035001001.

Tamias townsendii Bachman, 1839. J. Acad. Nat. Sci. Phila., 8(1):68.
   TYPE LOCALITY: U.S.A., Oregon, Multnomah Co., lower Columbia River, near lower
      mouth of Willamette River.
   DISTRIBUTION: S.W. British Columbia (Canada), W. Washington and Oregon to the
      Rogue River (U.S.A.).
   COMMENT: Formerly included *ochrogenys*, *siskiyou* and *senex*; see comments under those
      species.
   ISIS NUMBER: 5301410002007017001 as *Eutamias townsendii*.

Tamias umbrinus J. A. Allen, 1890. Bull. Am. Mus. Nat. Hist., 3:96.
   TYPE LOCALITY: U.S.A., Utah, Summit Co., Uinta Mtns., Blacks Fork, about 8000 ft.
      (2438 m).
   DISTRIBUTION: E. California and N. Arizona to N. Colorado, S.E. and N.W. Wyoming,
      and extreme S.W. Montana (U.S.A.).
   ISIS NUMBER: 5301410002007018001 as *Eutamias umbrinus*.

Tamiasciurus Trouessart, 1880. Le Naturaliste, 2(37):292.
   REVIEWED BY: F. J. Brenner (FJB); D. A. Sutton (DAS).
   ISIS NUMBER: 5301410002036000000.

Tamiasciurus douglasii (Bachman, 1839). Proc. Zool. Soc. Lond., 1838:99.
   TYPE LOCALITY: U.S.A., Oregon, Clatsop Co., mouth of Columbia River (="shores of
      Columbia River").
   DISTRIBUTION: Coast and Cascade Ranges and Sierra Nevada of S.W. British Columbia
      (Canada) to S. California (U.S.A.).
   COMMENT: Formerly included *mearnsi*; see Lindsay, 1981, J. Mammal., 62:673–682.
   ISIS NUMBER: 5301410002036001001.

Tamiasciurus hudsonicus (Erxleben, 1777). Syst. Regn. Anim., 1:416.
   TYPE LOCALITY: Canada, Ontario, mouth of the Severn River.
   DISTRIBUTION: Alaska, throughout Canada (south of the tundra), W. U.S.A. in
      mountain states; N.E. U.S.A.
   COMMENT: Includes *fremonti*; see Hardy, 1950, Proc. Biol. Soc. Wash., 63:13–14.
   ISIS NUMBER: 5301410002036002001.

Tamiasciurus mearnsi (Townsend, 1897). Proc. Biol. Soc. Wash., 11:146.
   TYPE LOCALITY: Mexico, Baja California Norte, San Pedro Martir Mountains.
   DISTRIBUTION: Sierra San Pedro Martir Mountains (Baja California Norte, Mexico).
   COMMENT: Formerly included in *douglasii*; see Lindsay, 1981, J. Mammal., 62:673–682.

Tamiops J. A. Allen, 1906. Bull. Am. Mus. Nat. Hist., 22:475.
   COMMENT: Formerly included in *Callosciurus*; see Moore and Tate, 1965, Fieldiana
      Zool., 48:1–351, who revised *Tamiops*.
   ISIS NUMBER: 5301410002037000000.

Tamiops maritimus (Bonhote, 1900). Ann. Mag. Nat. Hist., ser. 7, 5:51.
   TYPE LOCALITY: China, Fukien, Foochow.
   DISTRIBUTION: Hupei and Chekiang (China), through S. Vietnam and Laos; Hainan
      (China); Taiwan.
   ISIS NUMBER: 5301410002037002001.

Tamiops macclellandi (Horsfield, 1840). Proc. Zool. Soc. Lond., 1839:152.
   TYPE LOCALITY: India, Assam.
   DISTRIBUTION: E. Nepal to N. Vietnam and Yunnan (China), south through Indochina
      to the S. Malay Peninsula.
   ISIS NUMBER: 5301410002037001001.

Tamiops rodolphei (Milne-Edwards, 1867). Rev. Mag. Zool. Paris, ser. 2, 19:227.
TYPE LOCALITY: Vietnam, Saigon.
DISTRIBUTION: E. Thailand; Kampuchea; S. Vietnam.
ISIS NUMBER: 5301410002037003001.

Tamiops swinhoei (Milne-Edwards, 1874). Recherches...des Mammif., 1:308.
TYPE LOCALITY: China, Szechwan, Moupin (=Muping).
DISTRIBUTION: E. China to N. Burma and N. Vietnam.
ISIS NUMBER: 5301410002037004001.

Trogopterus Heude, 1898. Mem. Hist. Nat. Emp. Chin., 4(1):46–47.
ISIS NUMBER: 5301410002051000000.

Trogopterus xanthipes (Milne-Edwards, 1867). Ann. Sci. Nat. Zool., 8:376.
TYPE LOCALITY: China, N.E. of Hopei.
DISTRIBUTION: Montane forests, from Yunnan to C. and W. China.
COMMENT: Includes *edithae* (Yunnan), *himalaicus* (S. Tibet), *minax* (Szechwan), and
mordax (Hupei); closely related to *Belomys* from S.E. Asia; see Corbet, 1978:87.
ISIS NUMBER: 5301410002051001001.

Xerus Hemprich and Ehrenberg, 1833. Symb. Phys. Mamm., 1.
COMMENT: Includes *Euxerus* and *Geosciurus*; see Amtmann, 1975, Part 6.1:2. Moore,
1959, Bull. Am. Mus. Nat. Hist., 118:153–206, considered *Euxerus* and *Geosciurus*
distinct genera. *Euxerus* was considered distinct by Yalden *et al.*, 1976, Monitore
Zool. Ital., n.s., suppl. 8(1):1–118, who were followed by Rupp, 1980, Saugetierk.
Mitt., 28:87.
ISIS NUMBER: 5301410002038000000.

Xerus erythropus (E. Geoffroy, 1803). Cat. Mamm. Mus. Nat. Hist. Nat. Paris, p. 178.
TYPE LOCALITY: Senegal.
DISTRIBUTION: Morocco and Senegal to Ethiopia and Kenya.
COMMENT: Subgenus *Euxerus*; see Corbet, 1978:79, and Amtmann, 1975, Part 6.1:2; but
also see comment under genus.
ISIS NUMBER: 5301410002038001001.

Xerus inauris (Zimmermann, 1780). Geogr. Gesch. Mensch. Vierf. Thiere, 2:344.
TYPE LOCALITY: South Africa, Kaffirland, 100 mi. (161 km) N. of Cape of Good Hope.
DISTRIBUTION: South Africa; S. Namibia; Botswana; W. Zimbabwe.
COMMENT: Subgenus *Geosciurus*; see Ellerman, 1940:422, and Amtmann, 1975, Part
6.1:2. Moore, 1959, Bull. Am. Mus. Nat. Hist., 118:153–206, considered *Geosciurus* a
distinct genus.
ISIS NUMBER: 5301410002038002001.

Xerus princeps (Thomas, 1929). Proc. Zool. Soc. Lond., 1929:106.
TYPE LOCALITY: N. Namibia, C. Koakoveld, Otjitundua.
DISTRIBUTION: Namibia; S.W. desert region of Angola.
COMMENT: Subgenus *Geosciurus*; see Ellerman, 1940:422, and Amtmann, 1975, Part
6.1:2. Moore, 1959, Bull. Am. Mus. Nat. Hist., 118:153–206, considered *Geosciurus* a
distinct genus.
ISIS NUMBER: 5301410002038003001.

Xerus rutilus (Cretzschmar, 1826). Atlas Ruppell's Reise Nordl. Africa Saugeth., p. 59.
TYPE LOCALITY: Ethiopia, "Eastern slope of Abyssinia."
DISTRIBUTION: N.E. Sudan; Ethiopia; Somalia; Kenya; N.E. Uganda; N.E. Tanzania.
COMMENT: Subgenus *Xerus*; see Ellerman, 1940:420, and Amtmann, 1975, Part 6.1:2.
ISIS NUMBER: 5301410002038004001.

Family Geomyidae
REVIEWED BY: C. A. McLaughlin (CAM).
COMMENT: Genera revised by Russell, 1968, Univ. Kans. Publ. Mus. Nat. Hist.,
16:473–579.
ISIS NUMBER: 5301410003000000000.

Geomys Rafinesque, 1817. Am. Mon. Mag., 2(1):45.
REVIEWED BY: L. R. Heaney (LRH); K. T. Wilkins (KTW).
COMMENT: A key to the species of *Geomys* was published by Baker and Williams, 1974, Mamm. Species, 35:1–4.
ISIS NUMBER: 5301410003002000000.

Geomys arenarius Merriam, 1895. N. Am. Fauna, 8:139.
REVIEWED BY: J. Ramirez-Pulido (JRP).
TYPE LOCALITY: U.S.A., Texas, El Paso Co., El Paso.
DISTRIBUTION: Extreme W. Texas, S.W. and S.C. New Mexico (U.S.A.); N. Chihuahua (Mexico).
COMMENT: Reviewed by Williams and Baker, 1974, Mamm. Species, 36:1–3, and Williams and Genoways, 1978, Ann. Carnegie Mus., 47:541–570.
ISIS NUMBER: 5301410003002001001.

Geomys bursarius (Shaw, 1800). Trans. Linn. Soc. Lond., 5:227.
REVIEWED BY: E. B. Hart (EBH).
TYPE LOCALITY: U.S.A., Minnesota, Sherburne Co., Elk River.
DISTRIBUTION: S.C. Manitoba (Canada) to N.W. Indiana, S.W. Louisiana, S.C. Texas, E.C. New Mexico (U.S.A.).
COMMENT: Includes *breviceps* and *lutescens;* see Hall, 1981:499–500. Revised by Merriam, 1895, N. Am. Fauna, 8:120. A superspecies group composed of 3 to 5 species; under revision by Zimmerman, Heaney, and others (LRH). EBH believes *bursarius* probably contains three distinct species: (1) a species consisting of *ammophilus* and *attwateri,* (2) a species consisting of *brazensis, sagittalis, terricolus, pratincola,* and *ludemani,* (3) a species consisting of the remaining subspecies. LRH disagrees with 2 and 3 above. Tucker and Schmidly, 1981 also considered *attwateri* distinct.
ISIS NUMBER: 5301410003002002001.

Geomys personatus True, 1889. Proc. U.S. Nat. Mus., 11:159.
REVIEWED BY: J. Ramirez-Pulido (JRP).
TYPE LOCALITY: U.S.A., Texas, Cameron Co., Padre Isl.
DISTRIBUTION: S. Texas, south of San Antonio and Del Rio, including Padre and Mustang Isl. (U.S.A.); barrier beaches of extreme N.E. Tamaulipas (Mexico).
COMMENT: Reviewed by Davis, 1940, Bull. Texas Agric. Exper. Sta., 590:1–38, and Williams and Genoways, 1981, Ann. Carnegie Mus., 50(19):435–473.
ISIS NUMBER: 5301410003002006001.

Geomys pinetis Rafinesque, 1806. Am. Mon. Mag., 2:45.
TYPE LOCALITY: U.S.A., Georgia, Screven Co.
DISTRIBUTION: N. and C. Florida; S. Georgia; S. Alabama (U.S.A.).
COMMENT: *G. tuza* is a junior synonym; see Harper, 1952, Proc. Biol. Soc. Wash., 65:36. Includes *colonus, cumberlandius,* and *fontanelus;* see Williams and Genoways, 1980, Ann. Carnegie Mus., 49:405–453. Reviewed by Pembleton and Williams, 1978, Mamm. Species, 86:1–3.
ISIS NUMBER: 5301410003002007001 as *G. pinetis.*
5301410003002003001 as *G. colonus.*
5301410003002004001 as *G. cumberlandius.*
5301410003002005001 as *G. fontanelus.*

Geomys tropicalis Goldman, 1915. Proc. Biol. Soc. Wash., 28:134.
TYPE LOCALITY: Mexico, Tamaulipas, Altamira.
DISTRIBUTION: Vicinity of Altamira and Tampico in S.E. Tamaulipas (Mexico).
COMMENT: Elevated to species status by Alvarez, 1963, Univ. Kans. Publ. Mus. Nat. Hist., 14:426–427. Reviewed by Baker and Williams, 1974, Mamm. Species, 35:1–4, and Williams and Genoways, 1977, Ann. Carnegie Mus., 46:245–264.

Orthogeomys Merriam, 1895. N. Am. Fauna, 8:172.
REVIEWED BY: S. H. Jenkins (SHJ); R. D. Orr (RDO); J. Ramirez-Pulido (JRP); K. T. Wilkins (KTW).

COMMENT: Includes *Heterogeomys* and *Macrogeomys*; see Russell, 1968, Univ. Kans. Publ. Mus. Nat. Hist., 16(6):528, who revised this genus.
ISIS NUMBER: 5301410003005000000 as *Orthogeomys*.
5301410003003000000 as *Heterogeomys*.
5301410003004000000 as *Macrogeomys*.

Orthogeomys cavator (Bangs, 1902). Bull. Mus. Comp. Zool. Harv. Univ., 39:42.
TYPE LOCALITY: Panama, Chiriqui Prov., Boquete, 4800 ft. (1463 m).
DISTRIBUTION: N.W. Panama to C. Costa Rica.
COMMENT: Reviewed by Goodwin, 1946, Bull. Am. Mus. Nat. Hist., 87:378–379.
ISIS NUMBER: 5301410003004001001 as *Macrogeomys cavator*.

Orthogeomys cherriei (J. A. Allen, 1893). Bull. Am. Mus. Nat. Hist., 5:337.
TYPE LOCALITY: Costa Rica, Limon Prov., Santa Clara.
DISTRIBUTION: E. lowlands of Costa Rica.
COMMENT: Revised by Goodwin, 1946, Bull. Am. Mus. Nat. Hist., 87:380–381.
ISIS NUMBER: 5301410003004002001 as *Macrogeomys cherriei*.

Orthogeomys cuniculus Elliot, 1905. Proc. Biol. Soc. Wash., 18:234.
TYPE LOCALITY: Mexico, Oaxaca, Zanatepec.
DISTRIBUTION: Known only from the type locality.
COMMENT: Reviewed by Nelson and Goldman, 1930, J. Mammal., 11:317.
ISIS NUMBER: 5301410003005001001.

Orthogeomys dariensis (Goldman, 1912). Smithson. Misc. Coll., 60(2):8.
TYPE LOCALITY: Panama, Darien Prov., Cana, upper Rio Tuyra, 2000 ft. (610 m).
DISTRIBUTION: E. Panama.
ISIS NUMBER: 5301410003004003001 as *Macrogeomys dariensis*.

Orthogeomys grandis (Thomas, 1893). Ann. Mag. Nat. Hist., ser. 6, 12:270.
TYPE LOCALITY: Guatemala, Sacatepequez Prov., Duenas.
DISTRIBUTION: Honduras to Jalisco (Mexico).
COMMENT: Burt and Stirton, 1961, Misc. Publ. Mus. Zool. Univ. Mich., 117:52, included *pygacanthus* in *grandis*; Russell, 1968, Univ. Kans. Publ. Mus. Nat. Hist., 16(6):477–579, considered *pygacanthus* a distinct species and is followed here.
ISIS NUMBER: 5301410003005002001.

Orthogeomys heterodus (Peters, 1865). Monatsb. Preuss. Akad. Wiss. Berlin, p. 177.
TYPE LOCALITY: Costa Rica, San Jose Prov., Escazu Hts.
DISTRIBUTION: C. Costa Rica.
COMMENT: Reviewed by Goodwin, 1946, Bull. Am. Mus. Nat. Hist., 87:376–378.
ISIS NUMBER: 5301410003004004001 as *Macrogeomys heterodus*.

Orthogeomys hispidus (Le Conte, 1852). Proc. Acad. Nat. Sci. Phila., 6:158.
TYPE LOCALITY: Mexico, Veracruz, near Jalapa.
DISTRIBUTION: Yucatan Peninsula, Belize, Guatemala, and N.W. Honduras, to S. Tamaulipas (Mexico).
COMMENT: Includes *hondurensis*; may include *lanius*; see Hall, 1981:511, 512. Revised by Nelson and Goldman, 1929, Proc. Biol. Soc. Wash., 42:147–152.
ISIS NUMBER: 5301410003003001001 as *Heterogeomys hispidus*.

Orthogeomys lanius (Elliot, 1905). Proc. Biol. Soc. Wash., 18:235.
TYPE LOCALITY: Mexico, Veracruz, Xuchil, S.E. side of Mt. Orizaba.
DISTRIBUTION: Known only from the type locality.
COMMENT: May be conspecific with *hispidus*; see Hall, 1981:512.
ISIS NUMBER: 5301410003003002001 as *Heterogeomys lanius*.

Orthogeomys matagalpae (J. A. Allen, 1910). Bull. Am. Mus. Nat. Hist., 28:97.
TYPE LOCALITY: Nicaragua, Matagalpa, Pena Blanca.
DISTRIBUTION: N.C. Nicaragua to S.C. Honduras.
ISIS NUMBER: 5301410003004005001 as *Macrogeomys matagalpae*.

Orthogeomys pygacanthus Dickey, 1928. Proc. Biol. Soc. Wash., 41:9.
TYPE LOCALITY: El Salvador, San Miguel Dept., Mt. Cacaguatique, 3500 ft. (1067 m).
DISTRIBUTION: El Salvador.

COMMENT: This species was included in *grandis* by Burt and Stirton, 1961, Misc. Publ. Mus. Zool. Univ. Mich., 117:52, but was considered a distinct species by Russell, 1968, Univ. Kans. Publ. Mus. Nat. Hist., 16(6):477–579, and Corbet and Hill, 1980:139. Hall, 1981:509, followed Burt and Stirton, 1961, in including this species in *grandis*; Russell, 1968, was not mentioned.
ISIS NUMBER: 5301410003005003001.

Orthogeomys underwoodi (Osgood, 1931). Field Mus. Nat. Hist. Publ. Zool. Ser., 295, 18:143.
TYPE LOCALITY: Costa Rica, San Jose Prov., Alto de Jabillo Pirris, between San Geronimo and Pozo Azul.
DISTRIBUTION: W. Costa Rica.
ISIS NUMBER: 5301410003004006001 as *Macrogeomys underwoodi*.

Pappogeomys Merriam, 1895. N. Am. Fauna, 8:145.
REVIEWED BY: S. H. Jenkins (SHJ); J. Ramirez-Pulido (JRP); K. T. Wilkins (KTW).
COMMENT: Includes *Cratogeomys* and *Platygeomys*; see Hooper, 1946, J. Mammal., 27:397; Russell, 1968, Univ. Kans. Publ. Mus. Nat. Hist., 16(7):592, revised the genus.
ISIS NUMBER: 5301410003006000000 as *Pappogeomys*.
5301410003001000000 as *Cratogeomys*.

Pappogeomys alcorni Russell, 1957. Univ. Kans. Publ. Mus. Nat. Hist., 9:359.
TYPE LOCALITY: Mexico, Jalisco, 4 mi. (6 km) W. of Mazamitla, 6600 ft. (2012 m).
DISTRIBUTION: S.E. Jalisco (Mexico), in Sierra del Tigre Mtns.
ISIS NUMBER: 5301410003006001001.

Pappogeomys bulleri (Thomas, 1892). Ann. Mag. Nat. Hist., ser. 6, 10:196.
TYPE LOCALITY: Mexico, Jalisco, Talpa, W. slope Sierra de Mascota, 8500 ft. (2591 m).
DISTRIBUTION: Nayarit and Jalisco to Colima (Mexico).
COMMENT: Reviewed by Genoways and Jones, 1969, J. Mammal., 50:748–755.
ISIS NUMBER: 5301410003006002001.

Pappogeomys castanops (Baird, 1852). Rept. Stansbury's Expl. Surv. Grt. Salt Lake Utah, App. C, p. 313.
TYPE LOCALITY: U.S.A., Colorado, Bent Co., along prairie road to Bent's Fort, near present town of Las Animas.
DISTRIBUTION: S.E. Colorado and S.W. Kansas (U.S.A.) to N. Tamaulipas and C. San Luis Potosi (Mexico). Colima (Mexico) (JRP).
COMMENT: May represent two species; see Berry and Baker, 1972, J. Mammal., 53:303–309.
ISIS NUMBER: 5301410003006003001.

Pappogeomys fumosus (Merriam, 1892). Proc. Biol. Soc. Wash., 7:165.
TYPE LOCALITY: Mexico, Colima, 3 mi. (5 km) W. of Colima, 1700 ft. (518 m).
DISTRIBUTION: Plain of Colima (Mexico).
ISIS NUMBER: 5301410003001001001 as *Cratogeomys fumosus*.

Pappogeomys gymnurus (Merriam, 1892). Proc. Biol. Soc. Wash., 7:166.
TYPE LOCALITY: Mexico, Jalisco, Zapotlan (=Ciudad Guzman), 4000 ft. (1219 m).
DISTRIBUTION: S. and C. Jalisco, N.E. Michoacan (Mexico).
ISIS NUMBER: 5301410003001002001 as *Cratogeomys gymnurus*.

Pappogeomys merriami (Thomas, 1893). Ann. Mag. Nat. Hist., ser. 6, 12:271.
TYPE LOCALITY: Mexico, "Southern Mexico," probably Valley of Mexico.
DISTRIBUTION: W.C. Veracruz to Distrito Federal, Morelos, and surrounding areas, including S.E. central plateau and S. Sierra Madre Oriental (Mexico).
COMMENT: Includes *perotensis*, *oreocetes*, and *peregrinus*; see Russell, 1968, Univ. Kans. Publ. Mus. Nat. Hist., 16(7):706–712.
ISIS NUMBER: 5301410003001003001 as *Cratogeomys merriami*.
5301410003001005001 as *Cratogeomys oreocetes*.
5301410003001006001 as *Cratogeomys peregrinus*.
5301410003001007001 as *Cratogeomys perotensis*.

Pappogeomys neglectus (Merriam, 1902). Proc. Biol. Soc. Wash., 15:68.
TYPE LOCALITY: Mexico, Queretaro, Cerro de la Calentura, about 8 mi. (13 km) N.W. of Pinal de Amoles, 9500 ft. (2896 m).
DISTRIBUTION: Known only from the type locality.
ISIS NUMBER: 5301410003001004001 as *Cratogeomys neglectus*.

Pappogeomys tylorhinus (Merriam, 1895). N. Am. Fauna, 8:167.
TYPE LOCALITY: Mexico, Hidalgo, Tula, 6800 ft. (2073 m).
DISTRIBUTION: Distrito Federal, and Hidalgo to C. Jalisco (Mexico).
COMMENT: Includes *planiceps;* see Russell, 1968, Univ. Kans. Publ. Mus. Nat. Hist., 16(7):735.
ISIS NUMBER: 5301410003001009001 as *Cratogeomys tylorhinus*.
5301410003001008001 as *Cratogeomys planiceps*.

Pappogeomys zinseri (Goldman, 1939). J. Mammal., 20:91.
TYPE LOCALITY: Mexico, Jalisco, Lagos, 6150 ft. (1875 m).
DISTRIBUTION: N.E. Jalisco (Mexico).
ISIS NUMBER: 5301410003001010001 as *Cratogeomys zinseri*.

Thomomys Wied, 1839. Nova Acta Phys.-Med. Acad. Caes. Leop.-Carol., 19(1):377.
REVIEWED BY: R. D. Orr (RDO).
ISIS NUMBER: 5301410003007000000.

Thomomys bottae (Eydoux and Gervais, 1836). Mag. Zool. Paris, 6:23.
REVIEWED BY: J. Ramirez-Pulido (JRP).
TYPE LOCALITY: U.S.A., California, Monterey Co., Monterey, along coast.
DISTRIBUTION: S.W. U.S.A., north to Oregon and Colorado, south to Sinaloa and Nuevo Leon (Mexico).
COMMENT: Includes *baileyi, fulvus, suboles, harquahalae, pectoralis,* and *lachuguilla;* see Anderson, 1966, Syst. Zool., 15:189–198, and Hoffmeister, 1969, Misc. Publ. Univ. Kans. Mus. Nat. Hist., 51:75–91, who revised this species. Hoffmeister, 1969, found evidence of hybridization between *bottae* and *umbrinus* in Sycamore Canyon, Arizona; he provisionally considered *bottae* and *umbrinus* distinct species. Hall, 1981:469, reported evidence of intergradation in Nuevo Leon (Mexico), reviewed Hoffmeister, 1969, and included *bottae* in *umbrinus*.
ISIS NUMBER: 5301410003007002001 as *T. bottae*.
5301410003007001001 as *T. baileyi*.

Thomomys bulbivorus (Richardson, 1829). Fauna Boreali-Americana, 1:206.
TYPE LOCALITY: U.S.A., Oregon, banks of Columbia River, probably near Portland.
DISTRIBUTION: Willamette Valley (N.W. Oregon, U.S.A.).
COMMENT: Revised by Bailey, 1915, N. Am. Fauna, 39:1–136.
ISIS NUMBER: 5301410003007003001 as *T. bulbivorous (sic)*.

Thomomys clusius Coues, 1875. Proc. Acad. Nat. Sci. Phila., 27:138.
TYPE LOCALITY: U.S.A., Wyoming, Carbon Co., Bridger Pass, 18 mi. (29 km) S.W. Rawlins.
DISTRIBUTION: Carbon and Sweetwater Cos. (C. Wyoming, U.S.A.).
COMMENT: Formerly included in *talpoides;* revised by Thaeler and Hinesley, 1979, J. Mammal., 60:480–488.

Thomomys idahoensis Merriam, 1901. Proc. Biol. Soc. Wash., 14:114.
TYPE LOCALITY: U.S.A., Idaho, Clark Co., Birch Creek.
DISTRIBUTION: E.C. Idaho, adjacent Montana and W. Wyoming; N. Utah (U.S.A.).
COMMENT: Includes *pygmaeus* and *confinus;* see Thaeler, 1972, J. Mammal., 53:417–428, Murrelet, 1977, and J. Mammal., 58:49–51, who revised this species. Formerly included in *talpoides* by Hall and Kelson, 1959:441, but see Hall, 1981:1179.
ISIS NUMBER: 5301410003007004001.

Thomomys mazama Merriam, 1897. Proc. Biol. Soc. Wash., 11:214.
TYPE LOCALITY: U.S.A., Oregon, Klamath Co., Anna Creek near Crater Lake, Mt. Mazama, 6000 ft. (1829 m).
DISTRIBUTION: N.W. Washington through C. Oregon to N. California (U.S.A.).
COMMENT: Revised by Thaeler, 1968, Univ. Calif. Publ. Zool., 86:1–46.
ISIS NUMBER: 5301410003007005001.

Thomomys monticola J. A. Allen, 1893. Bull. Am. Mus. Nat. Hist., 5:48.
  TYPE LOCALITY: U.S.A., California, El Dorado Co., Mt. Tallac, 7500 ft. (2286 m).
  DISTRIBUTION: Sierra Nevada Mtns. of N. and E.C. California; W.C. Nevada (U.S.A.).
  COMMENT: Revised by Thaeler, 1968, Univ. Calif. Publ. Zool., 86:1–46.
  ISIS NUMBER: 5301410003007006001.

Thomomys talpoides (Richardson, 1828). Zool. J., 3:518.
  TYPE LOCALITY: Canada, Saskatchewan, North Saskatchewan River, Carlton House near
    Fort Carlton.
  DISTRIBUTION: S. British Columbia and C. Alberta to S.W. Manitoba (Canada), south to
    C. and W. South Dakota, N. New Mexico, N. Arizona, C. and W. Nevada, and
    E.C. California (U.S.A.).
  COMMENT: Formerly included *idahoensis* and *clusius;* see Thaeler, 1972, J. Mammal.,
    53:417–428, and Thaeler and Hinesley, 1979, J. Mammal., 60:480–488. Revised by
    Bailey, 1915, N. Am. Fauna, 39:1–136, Thaeler, 1968, Chromosoma, 25:172–183,
    and Thaeler, 1974, J. Mammal., 55:885–889.
  ISIS NUMBER: 5301410003007007001.

Thomomys townsendii (Bachman, 1839). J. Acad. Nat. Sci. Phila., 8:105.
  TYPE LOCALITY: U.S.A., Idaho, Canyon Co., near Nampa.
  DISTRIBUTION: S.E. Idaho; S.W. Idaho; S.E. Oregon, N. Nevada, and N.E. California
    (U.S.A.).
  COMMENT: Revised by Davis, 1937, J. Mammal., 18:145–158. Considered a distinct
    species by Thaeler, 1968, Evolution, 22:543–555, who provided evidence of a zone
    of hybridization between *bottae* and *townsendii.* Hall, 1981:469, 495, reviewed this
    evidence and included *townsendii* in *umbrinus (sensu* Hall). Includes *similis* which
    may be a distinct species (RDO).
  ISIS NUMBER: 5301410003007008001.

Thomomys umbrinus (Richardson, 1829). Fauna Boreali-Americana, 1:202.
  REVIEWED BY: J. Ramirez-Pulido (JRP).
  TYPE LOCALITY: Mexico, Veracruz (probably vicinity of Boca del Norte).
  DISTRIBUTION: S.W. U.S.A. to Puebla and Veracruz (Mexico).
  COMMENT: Revised by Hoffmeister, 1969, Misc. Publ. Univ. Kans. Mus. Nat. Hist.,
    51:75–91. Hall, 1981:469, included *baileyi, bottae,* and *townsendii* in this species
    based on a reevaluation of data in Hoffmeister, 1969, Thaeler, 1968, Evolution,
    22:543–555, and reported intergradation (between *umbrinus* and *bottae)* in Nuevo
    Leon (Mexico). Hoffmeister, 1969, considered *bottae* a distinct species; Thaeler,
    1968, considered *townsendii* a distinct species. Further studies of these taxa are
    needed.
  ISIS NUMBER: 5301410003007009001.

Zygogeomys Merriam, 1895. N. Am. Fauna, 8:195.
  REVIEWED BY: R. D. Orr (RDO); J. Ramirez-Pulido (JRP).
  ISIS NUMBER: 5301410003008000000.

Zygogeomys trichopus Merriam, 1895. N. Am. Fauna, 8:196.
  TYPE LOCALITY: Mexico, Michoacan, Nahuatzen, 8000–8500 ft. (2438–2591 m).
  DISTRIBUTION: Known from the type locality, and 6 mi. (10 km) S.E. of Patzcuaro (C.
    Michoacan, Mexico).
  ISIS NUMBER: 5301410003008001001.

Family Heteromyidae
  REVIEWED BY: D. F. Williams (DFW).
  ISIS NUMBER: 5301410004000000000.

Dipodomys Gray, 1841. Ann. Mag. Nat. Hist., ser. 1, 7:521.
  REVIEWED BY: T. L. Best (TLB); J. Ramirez-Pulido (JRP) (Mexico).
  COMMENT: Interspecific relationships discussed by Schnell *et al.,* 1978, Syst. Zool.,
    27:34–48, and Johnson and Selander, 1971, Syst. Zool., 20:377–405.
  ISIS NUMBER: 5301410004001000000.

Dipodomys agilis Gambel, 1848. Proc. Acad. Nat. Sci. Phila., 4:77.
TYPE LOCALITY: U.S.A., California, Los Angeles Co., Los Angeles.
DISTRIBUTION: S.W. and S.C. California (U.S.A.); Baja California (Mexico).
COMMENT: Includes *antiquarius, australis, eremoecus, paralius, pedionomus,* and *peninsularis;* see Best, 1978, J. Mammal., 59:174, Lackey, 1967, Trans. San Diego Soc. Nat. Hist., 14:313–344, and Hall, 1981:1179.
ISIS NUMBER: 5301410004001001001 as *D. agilis.*
　　　　　　　5301410004001016001 as *D. paralius.*
　　　　　　　5301410004001017001 as *D. peninsularis.*

Dipodomys californicus Merriam, 1890. N. Am. Fauna, 4:49.
TYPE LOCALITY: U.S.A., California, Mendocino Co., Ukiah.
DISTRIBUTION: N. California; S.C. Oregon.
COMMENT: Considered distinct from *heermanni;* includes *eximius* and *saxatilis;* see Patton *et al.,* 1976, J. Mammal., 57:159. Hall, 1981:578, listed *californicus* as a subspecies of *heermanni,* without discussion of Patton *et al.,* 1976.

Dipodomys compactus True, 1889. Proc. U.S. Nat. Mus., 11:160.
TYPE LOCALITY: U.S.A., Texas, Cameron Co., Padre Isl.
DISTRIBUTION: S. Texas (U.S.A.); N. Tamaulipas (Mexico).
COMMENT: Considered distinct from *ordii;* includes *largus, parvabullatus, and sennetti;* see Schmidly and Hendricks, 1976, Bull. South. Calif. Acad. Sci., 75:225, and Johnson and Selander, 1971, Syst. Zool., 20:377–405. Hall, 1981:565, provisionally listed this species as a subspecies of *ordii.*

Dipodomys deserti Stephens, 1887. Am. Nat., 21:42.
TYPE LOCALITY: U.S.A., California, San Bernardino Co., Mohave River, 3 to 4 mi. (5–7 km) from, and opposite, Hesperia.
DISTRIBUTION: E. California, to S. and W. Nevada, S.W. Utah, W. to S.C. Arizona (U.S.A.); N.W. Sonora and N.E. Baja California Norte (Mexico).
COMMENT: Revised by Nader, 1978, Ill. Biol. Monogr., 49:1–116.
ISIS NUMBER: 5301410004001002001.

Dipodomys elator Merriam, 1894. Proc. Biol. Soc. Wash., 9:109.
TYPE LOCALITY: U.S.A., Texas, Clay Co., Henrietta.
DISTRIBUTION: N.C. Texas and S.W. Oklahoma (U.S.A.).
COMMENT: Probably no longer occurs in Oklahoma (TLB).
ISIS NUMBER: 5301410004001003001.

Dipodomys elephantinus (Grinnell, 1919). Univ. Calif. Publ. Zool., 21:43.
TYPE LOCALITY: U.S.A., California, San Benito Co., Bear Valley, 1 mi. (1.6 km) N. Cook P.O., 1300 ft. (396 m).
DISTRIBUTION: W.C. California (U.S.A.).
COMMENT: May be conspecific with *venustus;* see Stock, 1974, J. Mammal., 55:505, and Schnell *et al.,* 1978, Syst. Zool., 27:34.
ISIS NUMBER: 5301410004001004001.

Dipodomys gravipes Huey, 1925. Proc. Biol. Soc. Wash., 38:83.
TYPE LOCALITY: Mexico, Baja California Norte, 2 mi. (3 km) W. Santo Domingo Mission, 30° 45′ N., 115° 58′ W.
DISTRIBUTION: W. Baja California Norte (Mexico).
COMMENT: Considered distinct from *agilis* by Best, 1978, J. Mammal., 59:174.
ISIS NUMBER: 5301410004001005001.

Dipodomys heermanni Le Conte, 1853. Proc. Acad. Nat. Sci. Phila., 6:224.
TYPE LOCALITY: U.S.A., California, Calaveras Co., Sierra Nevada, probably in the Upper Sonoran Life Zone on Calaveras River.
DISTRIBUTION: C. California (U.S.A.).
COMMENT: Revised by Grinnell, 1922, Univ. Calif. Publ. Zool., 24:1–124. Does not include *californicus;* see Patton *et al.,* 1976, J. Mammal., 57:159; see comment under *californicus.*
PROTECTED STATUS: U.S. ESA - Endangered.
ISIS NUMBER: 5301410004001006001.

Dipodomys ingens (Merriam, 1904). Proc. Biol. Soc. Wash., 17:141.
  TYPE LOCALITY: U.S.A., California, San Luis Obispo Co., Carrizo Plain, Painted Rock, 20 mi. (32 km) S.E. of Simmler.
  DISTRIBUTION: S.W. San Joaquin Valley and adjacent areas (W.C. California, U.S.A.).
  ISIS NUMBER: 5301410004001007001.

Dipodomys insularis Merriam, 1907. Proc. Biol. Soc. Wash., 20:77.
  TYPE LOCALITY: Mexico, Baja California Sur, Gulf of California, San Jose Isl.
  DISTRIBUTION: San Jose Isl. (Baja California Sur, Mexico).
  COMMENT: Probably a subspecies of *merriami;* see Lidicker, 1960, Univ. Calif. Publ. Zool., 67:204 (TLB).
  ISIS NUMBER: 5301410004001008001.

Dipodomys margaritae Merriam, 1907. Proc. Biol. Soc. Wash., 20:76.
  TYPE LOCALITY: Mexico, Baja California Sur, Santa Margarita Isl.
  DISTRIBUTION: Known only from the type locality.
  COMMENT: Lidicker, 1960, Univ. Calif. Publ. Zool., 67(2):204, included *margaritae* in *merriami,* but Huey, 1964, Trans. San Diego Soc. Nat. Hist., 13:123, and Hall, 1981:587, considered *margaritae* as a distinct species.
  ISIS NUMBER: 5301410004001009001.

Dipodomys merriami Mearns, 1890. Bull. Am. Mus. Nat. Hist., 2:290.
  TYPE LOCALITY: U.S.A., Arizona, Maricopa Co., New River, between Phoenix and Prescott.
  DISTRIBUTION: N.W. Nevada and N.E. California to Texas (U.S.A.), S. Baja California Sur, and San Luis Potosi (Mexico).
  COMMENT: Formerly included *margaritae;* see Huey, 1964, Trans. San Diego Soc. Nat. Hist., 13:123.
  ISIS NUMBER: 5301410004001010001.

Dipodomys microps (Merriam, 1904). Proc. Biol. Soc. Wash., 17:145.
  TYPE LOCALITY: U.S.A., California, Inyo Co., Owens Valley, Lone Pine.
  DISTRIBUTION: N.W. Arizona, W. Utah, Nevada, E.C. and N.E. California, S.E. Oregon, and S.W. Idaho (U.S.A.).
  COMMENT: Revised by Csuti, 1979, Univ. Calif. Publ. Zool., 111:1–69.
  ISIS NUMBER: 5301410004001011001.

Dipodomys nelsoni Merriam, 1907. Proc. Biol. Soc. Wash., 20:75.
  TYPE LOCALITY: Mexico, Coahuila, La Ventura.
  DISTRIBUTION: Mexican Plateau from N. Coahuila and S.E. Chihuahua to N. San Luis Potosi and S. Nuevo Leon (Mexico).
  COMMENT: Further study of the relationship of *nelsoni* to *spectabilis* is needed; see Hall, 1981:581; also see Nader, 1978, Ill. Biol. Monogr., 49:1–116, who included *nelsoni* in *spectabilis* and Matson, 1980, J. Mammal., 61:563–566, who presented evidence of specific distinctness.
  ISIS NUMBER: 5301410004001012001.

Dipodomys nitratoides Merriam, 1894. Proc. Biol. Soc. Wash., 9:112.
  TYPE LOCALITY: U.S.A., California, Tulare Co., San Joaquin Valley, Tipton.
  DISTRIBUTION: W.C. California (U.S.A.).
  COMMENT: Except for bacular morphology, very similar to *merriami;* see Best and Schnell, 1974, Am. Midl. Nat., 91:257. Revised by Grinnell, 1922, Univ. Calif. Publ. Zool., 24:1–124.
  ISIS NUMBER: 5301410004001013001.

Dipodomys ordii Woodhouse, 1853. Proc. Acad. Nat. Sci. Phila., 6:224.
  TYPE LOCALITY: U.S.A., Texas, El Paso Co., El Paso.
  DISTRIBUTION: Hidalgo (Mexico) to S.C. Washington (U.S.A.), S.E. Alberta, and S.W. Saskatchewan (Canada).
  COMMENT: Includes *pullus;* see Anderson, 1972, Bull. Am. Mus. Nat. Hist., 148(2):317. Does not include *compactus;* see Schmidly and Hendricks, 1976, Bull. South. Calif. Acad. Sci., 75:225; also see comment under *compactus.*
  ISIS NUMBER: 5301410004001014001.

Dipodomys panamintinus (Merriam, 1894). Proc. Biol. Soc. Wash., 9:114.
TYPE LOCALITY: U.S.A., California, Inyo Co., Panamint Mtns., head of Willow Creek.
DISTRIBUTION: W.C. Nevada to S.C. California; isolated population in S. Nevada and adjacent California (U.S.A.).
ISIS NUMBER: 5301410004001015001.

Dipodomys phillipsii Gray, 1841. Ann. Mag. Nat. Hist., ser. 1, 7:522.
TYPE LOCALITY: Mexico, Mexico, Valley of Mexico.
DISTRIBUTION: N.W. Oaxaca to Durango (Mexico).
COMMENT: Includes *ornatus;* see Genoways and Jones, 1971, J. Mammal., 52:265. Reviewed by Jones and Genoways, 1975, Mamm. Species, 51:1–3.
PROTECTED STATUS: CITES - Appendix II as *D. p. phillipsii* subspecies only.
ISIS NUMBER: 5301410004001018001.

Dipodomys spectabilis Merriam, 1890. N. Am. Fauna, 4:46.
TYPE LOCALITY: U.S.A., Arizona, Cochise Co., Dos Cabezos.
DISTRIBUTION: San Luis Potosi (Mexico) to S.C. Arizona, N.W. New Mexico, N.E. Arizona, and W.C. Texas (U.S.A.).
COMMENT: Revised by Nader, 1978, Ill. Biol. Monogr., 49:1–116, who included *nelsoni;* but also see Hall, 1981:581, and Matson, 1980, J. Mammal., 61:563–566, who presented evidence of specific distinctness.
ISIS NUMBER: 5301410004001019001.

Dipodomys stephensi (Merriam, 1907). Proc. Biol. Soc. Wash., 20:78.
TYPE LOCALITY: U.S.A., California, Riverside Co., San Jacinto Valley, W. of Winchester.
DISTRIBUTION: S.W. California (U.S.A.).
COMMENT: Includes *cascus;* relationships to other species of the *heermanni* group studied by Lackey, 1967, Trans. San Diego Soc. Nat. Hist., 14:313. Reviewed by Bleich, 1977, Mamm. Species, 73:1–3.
ISIS NUMBER: 5301410004001020001.

Dipodomys venustus (Merriam, 1904). Proc. Biol. Soc. Wash., 17:142.
TYPE LOCALITY: U.S.A., California, Santa Cruz Co., Santa Cruz.
DISTRIBUTION: From San Francisco Bay to Estero Bay (W.C. California, U.S.A.).
COMMENT: Revised by Grinnell, 1922, Univ. Calif. Publ. Zool., 29:1–124. May be conspecific with *elephantinus* (TLB); see comment under *elephantinus;* also see Hall, 1981:576. May be conspecific with *agilis* (DFW).
ISIS NUMBER: 5301410004001021001.

Heteromys Desmarest, 1817. Nouv. Dict. Hist. Nat., 2nd ed., 14:181.
REVIEWED BY: S. Anderson (SA); J. Ramirez-Pulido (JRP)(Mexico).
COMMENT: Revised by Goldman, 1911, N. Am. Fauna, 34:14–32. Hall and Kelson, 1959:543, and Hall, 1981:596, commented on the need for revision of this genus.
ISIS NUMBER: 5301410004002000000.

Heteromys anomalus (Thompson, 1815). Trans. Linn. Soc. Lond., 11:161.
TYPE LOCALITY: Trinidad and Tobago, Trinidad.
DISTRIBUTION: W. and N. Colombia; N. Venezuela; Trinidad; Tobago; Margarita Isl.

Heteromys australis Thomas, 1901. Ann. Mag. Nat. Hist., ser. 7, 7:174.
TYPE LOCALITY: Ecuador, Esmeraldas Prov., Cachabi River, San Javier, below Cachabi.
DISTRIBUTION: E. Panama; S.W. Colombia; N.W. Ecuador.
ISIS NUMBER: 5301410004002001001.

Heteromys desmarestianus Gray, 1868. Proc. Zool. Soc. Lond., 1868:204.
TYPE LOCALITY: Guatemala, Coban.
DISTRIBUTION: N.W. Colombia to E. Oaxaca (Mexico).
ISIS NUMBER: 5301410004002002001.

Heteromys gaumeri J. A. Allen and Chapman, 1897. Bull. Am. Mus. Nat. Hist., 9:9.
TYPE LOCALITY: Mexico, Yucatan, Chichen-Itza.
DISTRIBUTION: N. Guatemala and Yucatan Peninsula (Mexico).
ISIS NUMBER: 5301410004002003001.

Heteromys goldmani Merriam, 1902. Proc. Biol. Soc. Wash., 15:41.
TYPE LOCALITY: Mexico, Chiapas, Chicharras.
DISTRIBUTION: N.W. Guatemala; S.E. Chiapas (Mexico).
ISIS NUMBER: 5301410004002004001.

Heteromys lepturus Merriam, 1902. Proc. Biol. Soc. Wash., 15:42.
TYPE LOCALITY: Mexico, Oaxaca, Mountains near Santo Domingo, W. of Guichicovi.
DISTRIBUTION: S. Veracruz and E. Oaxaca (Mexico).
COMMENT: Includes *nigricaudatus;* see Goodwin, 1969, Bull. Am. Mus. Nat. Hist.,
    141:1–269. Hall, 1981:600, listed *nigricaudatus* as a distinct species but did not
    mention Goodwin, 1969.
ISIS NUMBER: 5301410004002005001 as *H. lepturus.*
    5301410004002008001 as *H. nigricaudatus.*

Heteromys longicaudatus Gray, 1868. Proc. Zool. Soc. Lond., 1868:204.
TYPE LOCALITY: Mexico, Tabasco, Montecristo.
DISTRIBUTION: S.E. Tabasco (Mexico).
ISIS NUMBER: 5301410004002006001.

Heteromys nelsoni Merriam, 1902. Proc. Biol. Soc. Wash., 15:43.
TYPE LOCALITY: Mexico, Chiapas, Pinabete, 8,200 ft. (2499 m).
DISTRIBUTION: S. Chiapas (Mexico).
ISIS NUMBER: 5301410004002007001.

Heteromys oresterus Harris, 1932. Occas. Pap. Mus. Zool. Univ. Mich., 248:4.
TYPE LOCALITY: Costa Rica, Cordillera de Talamanca, El Copey de Dota, 6000 ft. (1829
    m).
DISTRIBUTION: Talamanca Range of Costa Rica.
ISIS NUMBER: 5301410004002009001.

Heteromys temporalis Goldman, 1911. N. Am. Fauna, 34:26.
TYPE LOCALITY: Mexico, Veracruz, Motzorongo.
DISTRIBUTION: C. Veracruz (Mexico).
ISIS NUMBER: 5301410004002010001.

Liomys Merriam, 1902. Proc. Biol. Soc. Wash., 15:44.
REVIEWED BY: J. Ramirez-Pulido (JRP).
COMMENT: Revised by Genoways, 1973, Spec. Publ. Mus. Texas Tech Univ., 5:1–368. A
    key to the species of *Liomys* was published by Dowler and Genoways, 1978,
    Mamm. Species, 82:1–6.
ISIS NUMBER: 5301410004003000000.

Liomys adspersus (Peters, 1874). Monatsb. Preuss. Akad. Wiss. Berlin, p. 357.
TYPE LOCALITY: Panama, City of Panama.
DISTRIBUTION: C. Panama.
ISIS NUMBER: 5301410004003001001.

Liomys irroratus (Gray, 1868). Proc. Zool. Soc. Lond., 1868:205.
TYPE LOCALITY: Mexico, Oaxaca, Oaxaca.
DISTRIBUTION: Oaxaca to S.C. Chihuahua (Mexico) and S. Texas (U.S.A.).
COMMENT: Includes *bulleri* and *guerrerensis;* see Genoways, 1973, Spec. Publ. Mus.
    Texas Tech Univ., 5:106–108. Reviewed by Dowler and Genoways, 1978, Mamm.
    Species, 82:1–6.
ISIS NUMBER: 5301410004003008001 as *L. irroratus.*
    5301410004003004001 as *L. bulleri.*
    5301410004003006001 as *L. guerrerensis.*

Liomys pictus (Thomas, 1893). Ann. Mag. Nat. Hist., ser. 6, 12:233.
TYPE LOCALITY: Mexico, Jalisco, San Sebastian, 4300 ft. (1311 m).
DISTRIBUTION: W. coast of Mexico from Sonora to Chiapas, and E. coast in Veracruz;
    extreme N.W. Guatemala.
COMMENT: Includes *annectens* and *pinetorum;* see Genoways, 1973, Spec. Publ. Mus.

Texas Tech Univ., 5:175, 185. Reviewed by McGhee and Genoways, 1978, Mamm. Species, 83:1–5.

ISIS NUMBER: 5301410004003009001 as *L. pictus*.
5301410004003002001 as *L. annectens*.
5301410004003010001 as *L. pinetorum*.

Liomys salvini (Thomas, 1893). Ann. Mag. Nat. Hist., ser. 6, 11:331.

TYPE LOCALITY: Guatemala, Sacatepequez, Duenas.

DISTRIBUTION: C. Costa Rica to E. Oaxaca (Mexico).

COMMENT: Includes *anthonyi, crispus,* and *heterothrix;* see Genoways, 1973, Spec. Publ. Mus. Texas Tech Univ., 5:235–236. Reviewed by Carter and Genoways, 1978, Mamm. Species, 84:1–5.

ISIS NUMBER: 5301410004003011001 as *L. salvini*.
5301410004003003001 as *L. anthonyi*.
5301410004003005001 as *L. crispus*.
5301410004003007001 as *L. heterothrix*.

Liomys spectabilis Genoways, 1971. Occas. Pap. Mus. Nat. Hist. Univ. Kans., 5:1.

TYPE LOCALITY: Mexico, Jalisco, 2.2 mi. (3.5 km) N.E. Contla, 3850 ft. (1173 m).

DISTRIBUTION: S.E. Jalisco (Mexico).

ISIS NUMBER: 5301410004003012001.

Microdipodops Merriam, 1891. N. Am. Fauna, 5:115.

COMMENT: Revised by Hall, 1941, Field Mus. Nat. Hist. Publ. Zool. Ser., 27:233–277; also see Hafner *et al.,* 1979, J. Mammal., 60:1–10.

ISIS NUMBER: 5301410004004000000.

Microdipodops megacephalus Merriam, 1891. N. Am. Fauna, 5:116.

TYPE LOCALITY: U.S.A., Nevada, Elko Co., Halleck.

DISTRIBUTION: N.E. and E.C. California, S.E. Oregon, N. and C. Nevada, and W.C. Utah (U.S.A.).

COMMENT: Reviewed by O'Farrell and Blaustein, 1974, Mamm. Species, 46:1–3.

ISIS NUMBER: 5301410004004001001.

Microdipodops pallidus Merriam, 1901. Proc. Biol. Soc. Wash., 14:127.

TYPE LOCALITY: U.S.A., Nevada, Churchill Co., Mountain Well.

DISTRIBUTION: E.C. California, W. and S.C. Nevada (U.S.A.).

COMMENT: Reviewed by O'Farrell and Blaustein, 1974, Mamm. Species, 47:1–2.

ISIS NUMBER: 5301410004004002001.

Perognathus Wied, 1839. Nova Acta Phys.-Med. Acad. Caes. Leop.-Carol., 19(1):368.

REVIEWED BY: J. Ramirez-Pulido (JRP)(Mexico).

COMMENT: Revised by Osgood, 1900, N. Am. Fauna, 18:1–73.

ISIS NUMBER: 5301410004005000000.

Perognathus alticola Rhoads, 1894. Proc. Acad. Nat. Sci. Phila., 45:412.

TYPE LOCALITY: U.S.A., California, San Bernardino Co., San Bernardino Mtns., Squirrel Inn, Little Bear Valley, 5,500 ft. (1576 m).

DISTRIBUTION: S.C. California (U.S.A.).

COMMENT: Probably a subspecies of *parvus* (DFW).

ISIS NUMBER: 5301410004005001001.

Perognathus amplus Osgood, 1900. N. Am. Fauna, 18:32.

TYPE LOCALITY: U.S.A., Arizona, Yavapai Co., Fort Verde.

DISTRIBUTION: W. and C. Arizona (U.S.A.) to N.W. Sonora (Mexico).

ISIS NUMBER: 5301410004005002001.

Perognathus anthonyi Osgood, 1900. N. Am. Fauna, 18:56.

TYPE LOCALITY: Mexico, Baja California Norte, South Bay, Cedros (=Cerros) Isl.

DISTRIBUTION: Cedros Isl. (Baja California Norte, Mexico).

COMMENT: Probably a subspecies of *fallax* (DFW).

ISIS NUMBER: 5301410004005003001.

Perognathus arenarius Merriam, 1894. Proc. Calif. Acad. Sci., ser. 2, 4:461.
TYPE LOCALITY: Mexico, Baja California Sur, San Jorge, near Comondu.
DISTRIBUTION: Baja California (Mexico).
COMMENT: Reviewed by Huey, 1964, Trans. San Diego Soc. Nat. Hist., 13(7):85–168.
ISIS NUMBER: 5301410004005005001.

Perognathus artus Osgood, 1900. N. Am. Fauna, 18:55.
TYPE LOCALITY: Mexico, Chihuahua, Batopilas.
DISTRIBUTION: Coastal W. Mexico from Sonora to Sinaloa; S.W. Chihuahua; W.
    Durango.
COMMENT: Revised by Anderson, 1964, Am. Mus. Novit., 2184:1–27.
ISIS NUMBER: 5301410004005006001.

Perognathus baileyi Merriam, 1894. Proc. Acad. Nat. Sci. Phila., 46:262.
TYPE LOCALITY: Mexico, Sonora, Magdalena.
DISTRIBUTION: S. California, S. Arizona, and S.W. New Mexico (U.S.A.) to N. Sinaloa
    and Baja California Sur (Mexico).
ISIS NUMBER: 5301410004005007001.

Perognathus californicus Merriam, 1889. N. Am. Fauna, 1:26.
TYPE LOCALITY: U.S.A., California, Alameda Co., Berkeley.
DISTRIBUTION: S. and C. California (U.S.A.); N.C. Baja California Norte (Mexico).
ISIS NUMBER: 5301410004005008001.

Perognathus dalquesti Roth, 1976. J. Mammal., 57:562.
TYPE LOCALITY: Mexico, Baja California Sur, 4 mi. (6 km) S.E. Migrino (32 km S.S.E.
    Todos Santos).
DISTRIBUTION: S.W. tip of Baja California Sur (Mexico).
COMMENT: Similar to *arenarius*; see Roth, 1976.

Perognathus fallax Merriam, 1889. N. Am. Fauna, 1:19.
TYPE LOCALITY: U.S.A., California, San Bernardino Co., Reche Canyon (3 mi. (5 km)
    S.E. Colton), 1,250 ft. (381 m).
DISTRIBUTION: S.W. California (U.S.A.) to W.C. Baja California (Mexico).
COMMENT: Reviewed by Huey, 1964, Trans. San Diego Soc. Nat. Hist., 13(7):85–168.
ISIS NUMBER: 5301410004005009001.

Perognathus fasciatus Wied, 1839. Nova Acta Phys.-Med. Acad. Caes. Leop.-Carol.,
    19(1):369.
TYPE LOCALITY: U.S.A., North Dakota, Williams Co., Buford.
DISTRIBUTION: Great Plains from S.E. Alberta, Saskatchewan, and S.W. Manitoba
    (Canada) to N.E. Utah, S. Colorado and E. South Dakota (U.S.A.).
COMMENT: Includes *infraluteus, litus,* and *olivaceogriseus;* revised by Williams and
    Genoways, 1979, Ann. Carnegie Mus., 48:73–102.
ISIS NUMBER: 5301410004005010001.

Perognathus flavescens Merriam, 1889. N. Am. Fauna, 1:11.
TYPE LOCALITY: U.S.A., Nebraska, Cherry Co., Kennedy.
DISTRIBUTION: Great Plains and intermountain basins from Minnesota and N. Utah
    (U.S.A.) to N. Chihuahua (Mexico).
COMMENT: Includes *apache* (see Williams, 1978, Bull. Carnegie Mus. Nat. Hist.,
    10:1–57) and *melanotis* (see Hall, 1981:531).
ISIS NUMBER: 5301410004005011001 as *P. flavescens.*
    5301410004005004001 as *P. apache.*

Perognathus flavus Baird, 1855. Proc. Acad. Nat. Sci. Phila., 7:332.
TYPE LOCALITY: U.S.A., Texas, El Paso Co., El Paso.
DISTRIBUTION: S.W. Great Plains and intermountain plateaus from S. Dakota, E.
    Wyoming, and S. Utah (U.S.A.) to Sonora and Puebla (Mexico).
COMMENT: Includes *merriami;* see Wilson, 1973, Proc. Biol. Soc. Wash., 86:191. Revised
    by Baker, 1954, Univ. Kans. Publ. Mus. Nat. Hist., 7:341–347.
ISIS NUMBER: 5301410004005012001 as *P. flavus.*
    5301410004005020001 as *P. merriami.*

Perognathus formosus Merriam, 1889. N. Am. Fauna, 1:17.
TYPE LOCALITY: U.S.A., Utah, Washington Co., St. George.
DISTRIBUTION: W. Utah, Nevada, N.E. California (U.S.A.), and E. coast of Baja
California to Concepcion Bay (Baja California Sur, Mexico).
COMMENT: Reviewed by Huey, 1964, Trans. San Diego Soc. Nat. Hist., 13(7):85–168.
ISIS NUMBER: 5301410004005013001.

Perognathus goldmani Osgood, 1900. N. Am. Fauna, 18:54.
TYPE LOCALITY: Mexico, Sinaloa, Sinaloa.
DISTRIBUTION: Lowlands of N.W. Mexico from N.E. Sonora to N. Sinaloa.
COMMENT: Revised by Anderson, 1964, Am. Mus. Novit., 2184:1–27.
ISIS NUMBER: 5301410004005014001.

Perognathus hispidus Baird, 1858. Mammals *in* Repts. Expl. Surv...., 8(1):421.
TYPE LOCALITY: Mexico, Tamaulipas, Charco Escondido.
DISTRIBUTION: Great Plains and plateaus from S. North Dakota to S.E. Arizona, W.
Louisiana (U.S.A.), Tamaulipas and Hidalgo (Mexico).
COMMENT: Revised by Glass, 1947, J. Mammal., 28:174–179.
ISIS NUMBER: 5301410004005015001.

Perognathus inornatus Merriam, 1889. N. Am. Fauna, 1:15.
TYPE LOCALITY: U.S.A., California, Fresno Co., Fresno.
DISTRIBUTION: Central and Salinas Valleys of California (U.S.A.).
COMMENT: Revised by Osgood, 1918, Proc. Biol. Soc. Wash., 31:93–100.
ISIS NUMBER: 5301410004005016001.

Perognathus intermedius Merriam, 1889. N. Am. Fauna, 1:18.
TYPE LOCALITY: U.S.A., Arizona, Mohave Co., Mud Spring.
DISTRIBUTION: S.C. Utah and Arizona to W. Texas (U.S.A.), south to N.W. Sonora and
S. Chihuahua (Mexico).
COMMENT: Includes *minimus,* which was transferred from *penicillatus* by Hoffmeister
and Lee, 1967, J. Mammal., 48:361. Reviewed by Hoffmeister, 1974, Southwest.
Nat., 19(2):213–214.
ISIS NUMBER: 5301410004005017001.

Perognathus lineatus Dalquest, 1951. J. Wash. Acad. Sci., 41:362.
TYPE LOCALITY: Mexico, San Luis Potosi, 1 km S. of Arriaga.
DISTRIBUTION: San Luis Potosi (Mexico).
COMMENT: Probably conspecific with *nelsoni* (DFW).
ISIS NUMBER: 5301410004005018001.

Perognathus longimembris (Coues, 1875). Proc. Acad. Nat. Sci. Phila., 27:305.
TYPE LOCALITY: U.S.A., California, Kern Co., Tehachapi Mtns., Old Fort Tejon.
DISTRIBUTION: S.E. Oregon and W. Utah (U.S.A.) to N. Sonora and Baja California
Norte (Mexico).
COMMENT: Revised by Osgood, 1918, Proc. Biol. Soc. Wash., 31:93–100.
ISIS NUMBER: 5301410004005019001.

Perognathus nelsoni Merriam, 1894. Proc. Acad. Nat. Sci. Phila., 46:266.
TYPE LOCALITY: Mexico, San Luis Potosi, Hacienda La Parada, about 25 mi. (40 km)
N.W. Ciudad San Luis Potosi.
DISTRIBUTION: Chihuahuan desert plateau from S.E. New Mexico (U.S.A.) to Jalisco
and San Luis Potosi (Mexico).
COMMENT: May include *lineatus* (DFW).
ISIS NUMBER: 5301410004005021001.

Perognathus parvus (Peale, 1848). Mammalia and Ornithology *in* U.S. Expl. Exped...., 8:53.
TYPE LOCALITY: U.S.A., Oregon, Wasco Co., probably near the Dalles.
DISTRIBUTION: Great Basin desert and grasslands from S. British Columbia (Canada),
south to E. California and east to S.E. Wyoming and N. Arizona (U.S.A.).
ISIS NUMBER: 5301410004005022001.

Perognathus penicillatus Woodhouse, 1852. Proc. Acad. Nat. Sci. Phila., 6:200.
TYPE LOCALITY: U.S.A., Arizona, Yuma Co., 1 mi. (1.6 km) S.W. Parker.
DISTRIBUTION: S.E. California and S. Nevada (U.S.A.) to San Luis Potosi (Mexico).
COMMENT: Revised by Hoffmeister and Lee, 1967, J. Mammal., 48:361–380.
ISIS NUMBER: 5301410004005023001.

Perognathus pernix J. A. Allen, 1898. Bull. Am. Mus. Nat. Hist., 10:149.
TYPE LOCALITY: Mexico, Sinaloa, Rosario.
DISTRIBUTION: Coastal lowlands from Sonora to Nayarit (Mexico).
ISIS NUMBER: 5301410004005024001.

Perognathus spinatus Merriam, 1889. N. Am. Fauna, 1:21.
TYPE LOCALITY: U.S.A., California, San Bernardino Co., 25 mi. (40 km) below the
    Needles, Colorado River.
DISTRIBUTION: S.E. California (U.S.A.); Baja California (Mexico).
ISIS NUMBER: 5301410004005025001.

Perognathus xanthonotus Grinnell, 1912. Proc. Biol. Soc. Wash., 25:128.
TYPE LOCALITY: U.S.A., California, Kern Co., E. slope Walker Pass, Freeman Canyon,
    4,900 ft. (1494 m).
DISTRIBUTION: Vicinity of type locality in N.E. Tehachapi Mtns. (California, U.S.A.).
COMMENT: Probably a subspecies of *parvus* (DFW).
ISIS NUMBER: 5301410004005026001.

Family Castoridae
REVIEWED BY: M. S. Boyce (MSB); F. J. Brenner (FJB); R. Guenzel (RG); S. H. Jenkins
    (SHJ).
ISIS NUMBER: 5301410005000000000.

Castor Linnaeus, 1758. Syst. Nat., 10th ed., 1:58.
ISIS NUMBER: 5301410005001000000.

Castor canadensis Kuhl, 1820. Beitr. Zool. Vergl. Anat. Abt., 1:64.
REVIEWED BY: J. Ramirez-Pulido (JRP).
TYPE LOCALITY: Canada, Hudson Bay.
DISTRIBUTION: Brooks Range in Alaska (U.S.A.) to Labrador (Canada), Tamaulipas
    (Mexico), and N. Florida (U.S.A.). Introduced in Europe and Asia.
COMMENT: Lavrov and Orlov, 1973, Zool. Zh., 52:734–742, considered this species
    distinct from *fiber*. Reviewed by Jenkins and Busher, 1979, Mamm. Species,
    120:1–8.
ISIS NUMBER: 5301410005001001001.

Castor fiber Linnaeus, 1758. Syst. Nat., 10th ed., 1:58.
REVIEWED BY: O. L. Rossolimo (OLR); S. Wang (SW).
TYPE LOCALITY: Sweden.
DISTRIBUTION: N.W. and N.C. Eurasia, east to Lake Baikal, and south to France and
    Mongolia.
COMMENT: Lavrov and Orlov, 1973, Zool. Zh., 52:734–742, clarified the distinction
    between this species and *canadensis*. Lavrov, 1979, Zool. Zh., 58(1):88–96,
    considered *albicus* a distinct species, but also see Corbet and Hill, 1980:142. OLR
    includes *albicus* in *fiber*.
PROTECTED STATUS: U.S. ESA - Endangered as *C. f. birulai* subspecies only.
ISIS NUMBER: 5301410005001002001.

Family Anomaluridae
REVIEWED BY: C. A. McLaughlin (CAM).
ISIS NUMBER: 5301410006000000000.

Anomalurus Waterhouse, 1843. Proc. Zool. Soc. Lond., 1842:124.
COMMENT: Includes *Anomalurops*; see Misonne, 1974, Part 6:4.
PROTECTED STATUS: CITES - Appendix III as *Anomalurus* spp. native to Ghana only.
ISIS NUMBER: 5301410006002000000 as *Anomalurus*.
            5301410006001000000 as *Anomalurops*.

Anomalurus beecrofti Fraser, 1853. Proc. Zool. Soc. Lond., 1852:17.
TYPE LOCALITY: Equatorial Guinea, Bioko (=Fernando Po).
DISTRIBUTION: Senegal to Uganda and Kasai (Zaire); Bioko.
COMMENT: Includes *fulgens;* formerly included in *Anomalurops;* see Misonne, 1974, Part 6:4.
PROTECTED STATUS: CITES - Appendix III (Ghana).
ISIS NUMBER: 5301410006001001001 as *Anomalurops beecrofti.*

Anomalurus derbianus (Gray, 1842). Ann. Mag. Nat. Hist., ser. 1, 10:262.
TYPE LOCALITY: Sierra Leone.
DISTRIBUTION: Sierra Leone to Angola, east to Kenya, south to Mozambique and Zambia.
COMMENT: Includes *cinereus, erythronotus, fraseri, imperator, jacksonii, neavei,* and *orientalis;* see Misonne, 1974, Part 6:5.
PROTECTED STATUS: CITES - Appendix III (Ghana).
ISIS NUMBER: 5301410006002002001 as *A. derbianus.*
5301410006002001001 as *A. cinereus.*
5301410006002003001 as *A. jacksoni* (sic).
5301410006002004001 as *A. orientalis.*

Anomalurus peli (Temminck, 1845). Verhandl. Nat. Gesch. Nederland Bezitt., Zool., 1(2):109.
TYPE LOCALITY: Ghana (=Gold Coast), Dabocrom.
DISTRIBUTION: Sierra Leone to Ghana.
COMMENT: Includes *auzembergeri;* see Misonne, 1974, Part 6:4.
PROTECTED STATUS: CITES - Appendix III (Ghana).
ISIS NUMBER: 5301410006002005001.

Anomalurus pusillus Thomas, 1887. Ann. Mag. Nat. Hist., ser. 5, 20:440.
TYPE LOCALITY: Zaire, Monbuttu, Bellima and Tingasi.
DISTRIBUTION: S. Cameroun; Gabon; Zaire.
COMMENT: Includes *batesi;* see Misonne, 1974, Part 6:5.
ISIS NUMBER: 5301410006002006001.

Idiurus Matschie, 1894. Sitzb. Ges. Naturf. Fr. Berlin, p. 194.
PROTECTED STATUS: CITES - Appendix III as *Idiurus* spp. native to Ghana only.
ISIS NUMBER: 5301410006003000000.

Idiurus macrotis Miller, 1898. Proc. Biol. Soc. Wash., 12:73.
TYPE LOCALITY: Cameroun, Efulen.
DISTRIBUTION: Sierra Leone to E. Zaire.
COMMENT: Includes *langi* and *panga;* see Misonne, 1974, Part 6:4.
PROTECTED STATUS: CITES - Appendix III (Ghana).
ISIS NUMBER: 5301410006003002001 as *I. macrotis.*
5301410006003001001 as *I. langi.*
5301410006003003001 as *I. panga.*

Idiurus zenkeri Matschie, 1894. Sitzb. Ges. Naturf. Fr. Berlin, p. 197.
TYPE LOCALITY: Cameroun, Yaunde.
DISTRIBUTION: S. Cameroun to Uganda.
ISIS NUMBER: 5301410006003004001.

Zenkerella Matschie, 1898. Sitzb. Ges. Naturf. Fr. Berlin, 4:23.
COMMENT: Includes *Aethurus;* see Walker, 1968:754.
ISIS NUMBER: 5301410006004000000.

Zenkerella insignis Matschie, 1898. Sitzb. Ges. Naturf. Fr. Berlin, 4:24.
TYPE LOCALITY: Cameroun, Yaunde.
DISTRIBUTION: S.W. Cameroun; Rio Muni (Equatorial Guinea); Gabon; Central African Republic.
ISIS NUMBER: 5301410006004001001.

Family Pedetidae
REVIEWED BY: C. A. McLaughlin (CAM).
ISIS NUMBER: 5301410007000000000.

Pedetes Illiger, 1811. Prodr. Syst. Mamm. et Avium., p. 81.
ISIS NUMBER: 5301410007001000000.

Pedetes capensis (Forster, 1778). Kongl. Svenska Vet.-Akad. Handl. Stockholm, 39:109.
TYPE LOCALITY: South Africa, Cape Prov., Cape of Good Hope.
DISTRIBUTION: South Africa; Namibia; Angola; Zimbabwe; Mozambique; Zambia; S.
    Zaire; Tanzania; Kenya.
COMMENT: Includes *surdaster*; see Misonne, 1974, Part 6:8.
ISIS NUMBER: 5301410007001001001 as *P. capensis*.
    5301410007001002001 as *P. surdaster*.

Family Cricetidae
REVIEWED BY: J. H. Honacki (JH).
COMMENT: Assignment of this family, separate from the Muridae and Arvicolidae,
    follows Chaline and Mein, 1979, and Reig, 1980, J. Zool. Lond., 192:257–281. For
    review of the taxonomic treatments of this family see Swanepoel *et al.*, 1980,
    Ann. Transvaal Mus., 32(7):155–196, Arata, 1967, *in* Anderson and Jones, pp.
    226–232, and Carleton, 1980:1–146. Inclusion of the Dendromurinae,
    Cricetomyinae, Gerbillinae, Nesomyinae, Petromyscinae, and Otomyinae in
    Cricetidae follows Misonne, 1974, Part 6:9; Chaline and Mein, 1979, considered
    these taxa as distinct families except Otomyinae, which they placed in
    Nesomyidae, and Petromyscinae, which they placed in Dendromuridae; they also
    included Myospalacinae (as is done here) and Spalacinae (here considered a
    separate family) in Cricetidae. Arata, 1967, *in* Anderson and Jones, pp. 241–242,
    included Dendromurinae and Otomyinae in Muridae, subfamily Murinae. Reig,
    1980, J. Zool. Lond., 192:258–260, recognized, as families, Gerbillidae,
    Lophiomyidae, and Myospalacidae (here included in Cricetidae, as subfamilies, as
    in Corbet, 1978:93), and Nesomyinae, Dendromurinae, Cricetomyinae,
    Petromyscinae, Cricetinae, Neotominae, Sigmodontinae, and Platacanthomyinae,
    as subfamilies of Cricetidae. Carleton, 1980, considered recognition of two
    separate subfamilies (Sigmodontinae and Neotominae or Peromyscinae) of New
    World cricetines premature; thus, the subfamily Hesperomyinae is used here (see
    Reig, 1980, for a discussion of the validity of the name). Chaline *et al.*, 1977,
    Mammalia, 41:245–252, and Chaline and Mein, 1979, incorporated recent fossil
    evidence in their classification; their recognition of the major phyletic units of
    Muroid rodents as families (see above) "may anticipate a future trend" (Carleton,
    1980:2–7). The treatment adopted here is not completely consistent with either
    Chaline and Mein, 1979, or Reig, 1980 (R.S. Hoffmann, JH).
ISIS NUMBER: 5301410008000000000 as Cricetidae.
    5301410013000000000 as Platacanthomyidae.

Abrawayaomys Cunha and Cruz, 1979. Bol. Mus. Biol. Prof. Mello-Leitao Zool., 96:2.
REVIEWED BY: A. Langguth (AL); M. A. Mares (MAM); R. S. Voss (RSV).
COMMENT: Authors note external and cranial similarities with *Neacomys*, and dental
    similarities with *Oryzomys* and *Akodon*. Subfamily Hesperomyinae; see comment
    under Cricetidae.

Abrawayaomys ruschii Cunha and Cruz, 1979. Bol. Mus. Biol. Prof. Mello-Leitao Zool.,
    96:2.
TYPE LOCALITY: Brazil, Espirito Santo, Forno Grande, Castelo.
DISTRIBUTION: Known only from the type locality.

Aepeomys Thomas, 1898. Ann. Mag. Nat. Hist., ser. 7, 1:452.
REVIEWED BY: M. D. Carleton (MDC); A. Langguth (AL); M. A. Mares (MAM); O. A.
    Reig (OAR); R. S. Voss (RSV).
COMMENT: Included in *Thomasomys* as a synonym by Osgood, 1933, J. Mammal.,

14:161, Ellerman, 1941:366, Cabrera, 1961:425, Walker *et al.*, 1975:766, Handley, 1976, Brigham Young Univ. Sci. Bull. Biol. Ser., 20(5):51, and Corbet and Hill, 1980:146. Considered a distinct genus by Gardner and Patton, 1976, Occas. Pap. Mus. Zool. La. St. Univ., 49:26, 32, 34. Subfamily Hesperomyinae; see comment under Cricetidae.

Aepeomys fuscatus J. A. Allen, 1912. Bull. Am. Mus. Nat. Hist., 31:89.
  TYPE LOCALITY: Colombia, Valle de Cauca, San Antonio, 2040 m.
  DISTRIBUTION: W. and C. Andes of Colombia.
  COMMENT: Formerly included in *lugens* as a subspecies by Cabrera, 1961:431; see Gardner and Patton, 1976, Occas. Pap. Mus. Zool. La. St. Univ., 49:26, 32.

Aepeomys lugens (Thomas, 1896). Ann. Mag. Nat. Hist., ser. 6, 18:306.
  TYPE LOCALITY: Venezuela, Merida, Loma del Morro, 900 m.
  DISTRIBUTION: W. Venezuela to Andean Ecuador.
  COMMENT: Includes *vulcani* as a subspecies; see Cabrera, 1961:432. Formerly included *fuscatus*; see Gardner and Patton, 1976, Occas. Pap. Mus. Zool. La. St. Univ., 49:26, 32. Formerly included in *Thomasomys*; see comment under *Aepeomys*. *A. vulcani* may be a separate species (OAR).
  ISIS NUMBER: 5301410008055015001 as *Thomasomys lugens*.

Akodon Meyen, 1833. Verhandl. Kais. Leop.-Carol. Akad. Wiss., 16(2):599.
  REVIEWED BY: M. D. Carleton (MDC); A. Langguth (AL); M. A. Mares (MAM); R. A. Ojeda (RAO); O. A. Reig (OAR).
  COMMENT: Ellerman, 1941, included in *Akodon*, as subgenera, the following genera named by Thomas: *Thalpomys, Thaptomys, Bolomys, Chroeomys, Deltamys, Hypsimys,* and also *Abrothrix* Waterhouse. Cabrera, 1961:457–458, followed Ellerman but also included *Microxus* as a subgenus and *Chalcomys* as a synonym of *Akodon*. Reig, 1978, Publ. Mus. Munic. Cienc. Nat. Mar del Plata "Lorenzo Scaglia," 2(8):164–190, included *Abrothrix, Chroeomys, Deltamys, Hypsimys,* and *Thalpomys* as subgenera and *Chalcomys* and *Thaptomys* as synonyms but considered *Bolomys* and *Microxus* distinct genera; this was a reevaluation of his views in Bianchi *et al.*, 1971, Evolution, 25:724–736, in which *Abrothrix, Bolomys, Chroeomys, Hypsimys, Microxus, Thalpomys,* and *Thaptomys* were considered distinct genera; also see Reig, 1980, J. Zool. Lond., 192:257–281. Corbet and Hill, 1980:150–151, followed Bianchi *et al.*, 1971, Evolution, 25:724–736. Massoia, 1964, Physis, ser. 24, 68:299–305, included *Deltamys* as a subgenus of *Akodon*; but also see Massoia, 1980, Hist. Nat., 1(25):179, who listed *Deltamys* as a distinct genus, but did not mention Reig, 1978:164–190, (followed here) who included it in *Akodon* as a subgenus. *Thalpomys*, based on *T. lasiotis* (which is a synonym of *Bolomys lasiurus*), is a subjective synonym of *Bolomys*; see Langguth, 1975, Papeis Avulsos Zool. Sao Paulo, 29(8):45–54. Gardner and Patton, 1976, Occas. Pap. Mus. Zool. La. St. Univ., 49:1–48, listed *Abrothrix* as a distinct genus, following Bianchi *et al.*, 1971, Evolution, 25:724–736, but included it in the oxymycterine group following Hershkovitz, 1966, Z. Saugetierk., 31:127. Arata, 1967, *in* Anderson and Jones, p. 229, included *Microxus* in *Akodon*; Hershkovitz, 1966, Z. Saugetierk., 31:86, placed *Microxus* in *Abrothrix* as a synonym. Reig, 1978:164–190, and Reig, 1980, J. Zool. Lond., 192:257–281, are followed here (except *Thalpomys* is included in *Bolomys*). Formerly included *arviculoides, tapirapoanus* (which were included in species of *Bolomys* by Reig, 1978:167), and *chacoensis* (which was transferred to *Bibimys* by Massoia, 1980, Ameghiniana, 17:280–287). This genus is in need of revision. Subfamily Hesperomyinae; see comment under Cricetidae.
  ISIS NUMBER: 5301410008001000000.

Akodon aerosus Thomas, 1913. Ann. Mag. Nat. Hist., ser. 8, 11:406.
  TYPE LOCALITY: Equador, upper Rio Pastaza, Mirador.
  DISTRIBUTION: Equador; Peru; Bolivia.
  COMMENT: Formerly included in *urichi* by Cabrera, 1961:448, but see Gardner and Patton, 1976, Occas. Pap. Mus. Zool. La. St. Univ., 49:27.

Akodon affinis (J. A. Allen, 1912). Bull. Am. Mus. Nat. Hist., 31:89.
TYPE LOCALITY: Colombia, Valle del Cauca Dept., San Antonio, near Cali, 2400 m.
DISTRIBUTION: Mountains of W. Colombia.
ISIS NUMBER: 5301410008001001001.

Akodon albiventer Thomas, 1897. Ann. Mag. Nat. Hist., ser. 6, 20:217.
REVIEWED BY: R. H. Pine (RHP).
TYPE LOCALITY: Argentina, Salta Prov., Bajo Rio Cachi.
DISTRIBUTION: S. and W. Bolivia; N.W. Argentina; N. Chile; S.E. Peru.
COMMENT: Formerly included in *Bolomys* by Bianchi *et al.*, 1971, Evolution, 25:724–736, and Gardner and Patton, 1976, Occas. Pap. Mus. Zool. La. St. Univ., 49:28. Includes *berlepschii*; placed in *Akodon* by Pine *et al.*, 1979, Mammalia, 43:347–348; also see Reig, 1978, Publ. Mus. Munic. Cienc. Nat. Mar del Plata "Lorenzo Scaglia," 2(8):167, who discussed the scope of *Bolomys*.
ISIS NUMBER: 5301410008001002001 as *A. albiventer.*
5301410008001007001 as *A. beriepschii (sic).*

Akodon andinus (Philippi, 1858). Arch. Naturgesch., 23(1):77.
TYPE LOCALITY: Chile, Santiago Prov., Altos Andes.
DISTRIBUTION: Peru; Bolivia; N. Chile; N. Argentina.
ISIS NUMBER: 5301410008001004001.

Akodon azarae (Fischer, 1829). Synopsis Mammal., p. 325.
TYPE LOCALITY: Argentina, Entre Rios Prov., 30° 30′ S. latitude, between the Uruguay and Parana Rivers.
DISTRIBUTION: C. and N.E. Argentina; Uruguay; Bolivia; Paraguay; S. Brazil.
ISIS NUMBER: 5301410008001006001.

Akodon boliviensis Meyen, 1833. Verhandl. Kais. Leop.-Carol. Akad. Wiss., 16(2):600, pl. 43, fig.1.
TYPE LOCALITY: Peru, Chucuito Prov., Puno Dept., Pichu-Pichun.
DISTRIBUTION: S. Peru; N.W. Argentina; Bolivia.
COMMENT: Karyology and morphometrics reviewed by Barquez *et al.*, 1980, Ann. Carnegie Mus., 49(22):379–403.
ISIS NUMBER: 5301410008001009001.

Akodon budini (Thomas, 1918). Ann. Mag. Nat. Hist., ser. 9, 1:191.
TYPE LOCALITY: Argentina, Jujuy Prov., Leon, about 1500 m.
DISTRIBUTION: Mountains of N.W. Argentina.
COMMENT: Includes *deceptor*; subgenus *Hypsimys*; see Reig, 1978, Publ. Mus. Munic. Cienc. Nat. Mar del Plata "Lorenzo Scaglia," 2(8):176, and Cabrera, 1961:451.
ISIS NUMBER: 5301410008001010001.

Akodon caenosus Thomas, 1918. Ann. Mag. Nat. Hist., ser. 9, 1:189.
TYPE LOCALITY: Argentina, Jujuy Prov., Leon, about 1500 m.
DISTRIBUTION: N.W. Argentina; S. Bolivia.
COMMENT: Reviewed by Barquez *et al.*, 1980, Ann. Carnegie Mus., 49(22):379–403.
ISIS NUMBER: 5301410008001011001.

Akodon cursor (Winge, 1887). E. Mus. Lundii, 1(3):25.
TYPE LOCALITY: Brazil, Minas Gerais, Lagoa Santa, Rio das Velhas.
DISTRIBUTION: S.E. and C. Brasil; Uruguay; Paraguay; N. Argentina (AL).
COMMENT: Formerly included in *arviculoides* (which was included in *Bolomys lasiurus* by Reig, 1978, Publ. Mus. Munic. Cienc. Nat. Mar del Plata "Lorenzo Scaglia," 2(8):167); includes *montensis*; see Ximenez and Langguth, 1970, Communic. Zool. Mus. Hist. Nat. Montevideo, 10(128):1–7, and Cabrera, 1961:439; also see Massoia, 1979, Physis, 38(95):1–7, and Massoia and Fornes, 1962, Physis, 23(65):185–194; subgenus *Akodon*. Karyology reviewed by Yonenaga *et al.*, 1975, Cytogenet. Cell Genet., 15:388–399. The status of this species is provisional (AL).

Akodon dolores Thomas, 1916. Ann. Mag. Nat. Hist., ser. 8, 18:324.
TYPE LOCALITY: Argentina, Cordoba Prov., Yacanto, 900 m.
DISTRIBUTION: Sierra de Cordoba (C. Argentina).
COMMENT: Karyology reviewed by Bianchi *et al.*, 1979, Genetica, 50:99–104.
ISIS NUMBER: 5301410008001013001.

Akodon illuteus (Thomas, 1925). Ann. Mag. Nat. Hist., ser. 9, 15:582.
TYPE LOCALITY: Argentina, Tucuman Prov., Sierra de Aconquija, between 3000–4000 m.
DISTRIBUTION: N.W. Argentina.
COMMENT: Placed in *Akodon (Abrothrix)* by Cabrera, 1961:455; Osgood, 1943, Publ.
Field Mus. Nat. Hist. Zool., ser. 30, 542:1–268, considered the inclusion of this
species in *Abrothrix* provisional. Bianchi *et al.*, 1971, Evolution, 25:724–736,
considered *Abrothrix* a distinct genus. Gardner and Patton, 1976, Occas. Pap. Mus.
Zool. La. St. Univ., 49:30, followed Bianchi *et al.*, 1971, Evolution, 25:724–736, but
listed this species in *Akodon*. Reig, 1978, Publ. Mus. Munic. Cienc. Nat. Mar del
Plata "Lorenzo Scaglia," 2(8):173, included *illuteus* in subgenus *Abrothrix*.
ISIS NUMBER: 5301410008001015001.

Akodon iniscatus Thomas, 1919. Ann. Mag. Nat. Hist., ser. 9, 3:205.
TYPE LOCALITY: Argentina, Chubut Prov., Valle del Lago Blanco.
DISTRIBUTION: S. and C. Argentina.
ISIS NUMBER: 5301410008001016001.

Akodon jelskii (Thomas, 1894). Ann. Mag. Nat. Hist., ser. 6, 16:360.
TYPE LOCALITY: Peru, Junin Dept., Junin.
DISTRIBUTION: S. Peru to Bolivia and Argentina.
COMMENT: Subgenus *Chroeomys*; see Cabrera, 1961:458, and Reig, 1978, Publ. Mus.
Munic. Cienc. Nat. Mar del Plata "Lorenzo Scaglia," 2(8):176. Bianchi *et al.*, 1971,
Evolution, 25:724–736, considered *Chroeomys* a distinct genus. Gardner and
Patton, 1976, Occas. Pap. Mus. Zool. La. St. Univ., 49:30, listed this species in
Akodon.
ISIS NUMBER: 5301410008001017001.

Akodon kempi (Thomas, 1917). Ann. Mag. Nat. Hist., ser. 8, 20:99.
TYPE LOCALITY: Argentina, Buenos Aires Prov., Isla Ella, Rio Parana.
DISTRIBUTION: Islands of the Parana Estuary (E. Argentina); S. Uruguay (AL).
COMMENT: Subgenus *Deltamys*; see Cabrera, 1961:451, Massoia, 1964, Physis, ser. 24,
68:299–305, and Reig, 1978, Publ. Mus. Munic. Cienc. Nat. Mar del Plata
"Lorenzo Scaglia," 2(8):176; also see Massoia, 1980, Hist. Nat., 1(25):179, who
considered *Deltamys* a distinct genus.
ISIS NUMBER: 5301410008001018001.

Akodon lanosus (Thomas, 1897). Ann. Mag. Nat. Hist., ser. 6, 20:218.
TYPE LOCALITY: Argentina, Tierra del Fuego Prov., Bahia Monteith, Strait of Magellan.
DISTRIBUTION: S. Argentina; S. Chile.
COMMENT: Included as a distinct species in *Abrothrix* by Pine *et al.*, 1978, Mammalia,
42:105–114, and Bianchi *et al.*, 1971, Evolution, 25:724–736. Reig, 1978, Publ. Mus.
Munic. Cienc. Nat. Mar del Plata "Lorenzo Scaglia," 2(8):173, considered *lanosus* a
distinct species and *Abrothrix* a subgenus of *Akodon*. Corbet and Hill, 1980:151,
listed *lanosus* as a subspecies of *longipilis*, without comment.
ISIS NUMBER: 5301410008001020001.

Akodon llanoi Pine, 1976. Mammalia, 40(1):63.
REVIEWED BY: R. H. Pine (RHP).
TYPE LOCALITY: Argentina, Tierra del Fuego Prov., Isla de los Estados, Bahia Capitan
Canepa.
DISTRIBUTION: Isla de los Estados (Argentina).
COMMENT: Closely related to *xanthorhinus*; see Pine, 1976, Mammalia, 40:63–68.

Akodon longipilis (Waterhouse, 1837). Proc. Zool. Soc. Lond., 1837:16.
TYPE LOCALITY: Chile, Coquimbo Prov., Coquimbo.
DISTRIBUTION: Chile; W. and C. Argentina.
COMMENT: Included in subgenus *Abrothrix* of *Akodon* by Reig, 1978, Publ. Mus. Munic.
Cienc. Nat. Mar del Plata "Lorenzo Scaglia," 2(8):173, and Pine *et al.*, 1979,
Mammalia, 43:349; see comment under genus for alternative treatments of
Abrothrix. Corbet and Hill, 1980:151, listed *lanosus* and *sanborni* as conspecific with
longipilis. Pine *et al.*, 1979, Mammalia, 43:351, and Osgood, 1943, Field Mus. Nat.
Hist. Publ. Zool. Ser., 30:194, presented evidence that *longipilis* and *sanborni* may

hybridize; Pine *et al.*, 1978, Mammalia, 42:105–114, considered *lanosus* a distinct species.
ISIS NUMBER: 5301410008001023001.

Akodon mansoensis De Santis and Justo, 1980. Neotropica, 26(75):121.
TYPE LOCALITY: Argentina, Rio Negro Prov., Bariloche Dept., Estacion Aforo, Rio Manso Superior.
DISTRIBUTION: Andes of Rio Negro Prov. (Argentina).
COMMENT: Subgenus *Abrothrix*; see De Santis and Justo, 1980, Neotropica, 26(75):121–127.

Akodon markhami Pine, 1973. An. Inst. Patagonia, 4(1–3):423–426.
TYPE LOCALITY: Chile, Magallanes, Isla Wellington, 1.2 km W.N.W. or 1.2 km N.N.W., Puerto Eden.
DISTRIBUTION: Isla Wellington (S. Chile).

Akodon molinae Contreras, 1968. Zool. Platense, 1(2):9–12.
TYPE LOCALITY: Argentina, Buenos Aires Prov., Partido de Villarino, Laguna Chasico.
DISTRIBUTION: E.C. Argentina.
COMMENT: Karyology reviewed by Bianchi *et al.*, 1973, Can. J. Genet. Cytol., 15:855–861.

Akodon mollis Thomas, 1894. Ann. Mag. Nat. Hist., ser. 6, 14:363.
TYPE LOCALITY: Peru, Piura Dept., Tumbes.
DISTRIBUTION: Ecuador; Peru; Bolivia.
ISIS NUMBER: 5301410008001025001.

Akodon nigrita (Lichtenstein, 1829). Darst. Saugeth., 7:pl. 35, fig. 1.
TYPE LOCALITY: Brazil, vicinity of Rio de Janeiro.
DISTRIBUTION: E. Brazil; Paraguay; N.E. Argentina.
COMMENT: Included in subgenus *Thaptomys* of *Akodon* by Massoia, 1963, Physis, 24:73–80, and Myers and Wetzel, 1979, J. Mammal., 60:640; Reig, 1978, Publ. Mus. Munic. Cienc. Nat. Mar del Plata "Lorenzo Scaglia," 2(8):176, considered *Thaptomys* a synonym of *Akodon*; also see comment under genus. Includes *subterraneus*; see Massoia, 1963, Physis, 24:73–80, and Myers and Wetzel, 1979:640.
ISIS NUMBER: 5301410008001026001.

Akodon olivaceus (Waterhouse, 1837). Proc. Zool. Soc. Lond., 1837:16.
TYPE LOCALITY: Chile, Aconcagua Prov., Valparaiso.
DISTRIBUTION: Chile; S. and W. Argentina.
ISIS NUMBER: 5301410008001028001.

Akodon orophilus Osgood, 1913. Field Mus. Nat. Hist. Publ. Zool. Ser., 10:98.
TYPE LOCALITY: Peru, Amazonas Dept., Leimabamba, Alto Utcubamba, 2400 m.
DISTRIBUTION: Mountains of S. and N. Peru.
ISIS NUMBER: 5301410008001029001.

Akodon pacificus Thomas, 1902. Ann. Mag. Nat. Hist., ser. 7, 9:135.
TYPE LOCALITY: Bolivia, La Paz Dept., La Paz, 4000 m.
DISTRIBUTION: Andes of W. Bolivia.
ISIS NUMBER: 5301410008001030001.

Akodon puer Thomas, 1902. Ann. Mag. Nat. Hist., ser. 7, 9:136.
TYPE LOCALITY: Bolivia, Choquecamate, Rio Secure, 4000 m.
DISTRIBUTION: W. Bolivia; S. and C. Peru.
ISIS NUMBER: 5301410008001031001.

Akodon reinhardti Langguth, 1975. Papeis Avulsos Zool. Sao Paulo, 29(8):45–54.
REVIEWED BY: R. H. Pine (RHP).
TYPE LOCALITY: Brazil, Minas Gerais, Lagoa Santa.
DISTRIBUTION: E. Brazil.
COMMENT: *A. reinhardti* is the new name for *lasiotis* of several authors, but not Lund; the holotype of *lasiotis* Lund, proved to be a *Bolomys lasiurus*; see Langguth, 1975, Papeis Avulsos Zool. Sao Paulo, 29(8):45–54; *Thalpomys*, based on *lasiotis* Lund, is a subjective synonym of *Bolomys*.
ISIS NUMBER: 5301410008001021001 as *A. lasiotis*.

Akodon sanborni Osgood, 1943. Field Mus. Nat. Hist. Publ. Zool. Ser., 30:194.
 TYPE LOCALITY: Chile, Isla Chiloe, mouth of Rio Inio.
 DISTRIBUTION: S. Chile; Argentina.
 COMMENT: Included in subgenus *Abrothrix* of *Akodon* by Cabrera, 1961:457, Reig, 1978,
  Publ. Mus. Munic. Cienc. Nat. Mar del Plata "Lorenzo Scaglia," 2(8):173, and
  Pine *et al.*, 1979, Mammalia, 43:351; see comment under *Akodon* for alternative
  treatments of *Abrothrix*.
 ISIS NUMBER: 5301410008001032001.

Akodon serrensis Thomas, 1902. Ann. Mag. Nat. Hist., ser. 7, 9:61.
 TYPE LOCALITY: Brazil, Parana, Serra do Mar, Roca Nova, 1000 m.
 DISTRIBUTION: S.E. Brazil; N. Argentina.
 ISIS NUMBER: 5301410008001033001.

Akodon surdus Thomas, 1917. Smithson. Misc. Coll., 68(4):2.
 TYPE LOCALITY: Peru, Cuzco Dept., Huadquina, 1500 m.
 DISTRIBUTION: Andes of S.E. Peru.
 ISIS NUMBER: 5301410008001034001.

Akodon urichi J. A. Allen and Chapman, 1897. Bull. Am. Mus. Nat. Hist., 9:19.
 TYPE LOCALITY: Trinidad and Tobago, Trinidad, Caparo.
 DISTRIBUTION: Trinidad; Tobago; Venezuela; Colombia.
 COMMENT: Gardner and Patton, 1976, Occas. Pap. Mus. Zool. La. St. Univ., 49:1–48,
  considered *urichi* distinct from *aerosus*.
 ISIS NUMBER: 5301410008001036001.

Akodon varius Thomas, 1902. Ann. Mag. Nat. Hist., ser. 7, 9:134.
 TYPE LOCALITY: Bolivia, Cochabamba Dept., Cochabamba, 2400 m.
 DISTRIBUTION: W. Argentina; Bolivia; Paraguay.
 COMMENT: Karyology and morphometrics reviewed by Barquez *et al.*, 1980, Ann.
  Carnegie Mus., 49(22):379–403.
 ISIS NUMBER: 5301410008001037001.

Akodon xanthorhinus (Waterhouse, 1837). Proc. Zool. Soc. Lond., 1837:17.
 TYPE LOCALITY: Argentina, Tierra del Fuego Prov., Hoste Isl., Hardy Peninsula.
 DISTRIBUTION: Patagonia and Tierra del Fuego (S. Argentina; S. Chile).
 COMMENT: This species and *longipilis* have very similar karyotypes; Bianchi *et al.*, 1971,
  Evolution, 25:728, considered it probable that it is an *Abrothrix*; Gardner and
  Patton, 1976, Occas. Pap. Mus. Zool. La. St. Univ., 49:30, placed it in *Abrothrix*
  which they considered an oxymycterine genus, following Hershkovitz, 1966, Z.
  Saugetierk., 31:127. Reig, 1978, Publ. Mus. Munic. Cienc. Nat. Mar del Plata
  "Lorenzo Scaglia," 2(8):173, included *Abrothrix* in *Akodon* as a subgenus but did
  not mention *xanthorhinus*; Pine *et al.*, 1978, Mammalia, 42:106, and Pine *et al.*,
  1979, Mammalia, 43:351, listed *xanthorhinus* as a distinct species of *Akodon*.
 ISIS NUMBER: 5301410008001038001.

Ammodillus Thomas, 1904. Ann. Mag. Nat. Hist., ser. 7, 14:102.
 REVIEWED BY: C. B. Robbins (CBR); C. J. Terry (CJT).
 COMMENT: Reviewed by Pavlinov, 1981, Zool. Zh., 60(3):472–474. Subfamily
  Gerbillinae; see comment under Cricetidae.
 ISIS NUMBER: 5301410008087000000.

Ammodillus imbellis (De Winton, 1898). Ann. Mag. Nat. Hist., ser. 7, 1:249.
 TYPE LOCALITY: Somalia, Goodar.
 DISTRIBUTION: Somalia; S. Ethiopia.
 ISIS NUMBER: 5301410008087001001.

Andalgalomys Williams and Mares, 1978. Ann. Carnegie Mus., 47(9):197.
 REVIEWED BY: D. F. Williams (DFW).
 COMMENT: Subfamily Hesperomyinae; see comment under Cricetidae.

Andalgalomys olrogi Williams and Mares, 1978. Ann. Carnegie Mus., 47(9):203.
 TYPE LOCALITY: Argentina, Catamarca Prov., W. Bank Rio Amanao, about 15 km (by
  road) W. of Andalgala.
 DISTRIBUTION: C. Catamarca Prov. (Argentina).

Andalgalomys pearsoni (Myers, 1977). Occas. Pap. Mus. Zool. Univ. Mich., 676:1.
TYPE LOCALITY: Paraguay, Boqueron Dept., 410 km (by road) N.W. of Villa Hayes.
DISTRIBUTION: Nueva Asuncion Dept. and Boqueron Dept. (Paraguay).
COMMENT: Formerly included in *Graomys;* see Williams and Mares, 1978, Ann.
Carnegie Mus., 47:193–221. Male accessory glands and taxonomy reviewed by
Voss and Linzey, 1981, Misc. Publ. Mus. Zool. Univ. Mich., 159:1–41.

Andinomys Thomas, 1902. Proc. Zool. Soc. Lond., 1902(1):116.
REVIEWED BY: M. D. Carleton (MDC); O. P. Pearson (OPP); R. H. Pine (RHP); C. J.
Terry (CJT).
COMMENT: Karyology reviewed by Pearson and Patton, 1976, J. Mammal., 57:339–350;
Simonetti and Spotorno, 1980, An. Mus. Hist. Nat. Valparaiso, 13:285–297,
compared karyotypes and morphometrics of *Andinomys, Auliscomys* and *Phyllotis.*
Revised by Hershkovitz, 1962, Fieldiana Zool., 46:472–483. Subfamily
Hesperomyinae; see comment under Cricetidae.
ISIS NUMBER: 5301410008002000000.

Andinomys edax Thomas, 1902. Proc. Zool. Soc. Lond., 1902 (1):116.
TYPE LOCALITY: Bolivia, Potosi Dept., between Potosi and Sucre, El Cabrado, 3700 m.
DISTRIBUTION: Peru; Bolivia; N.W. Argentina; N. Chile.
COMMENT: Spotorno, 1976, An. Mus. Hist. Nat. Valparaiso, 9:141–161, and Pine *et al.,*
1979, Mammalia, 43:359, reported the species from N. Chile.
ISIS NUMBER: 5301410008002001001.

Anotomys Thomas, 1906. Ann. Mag. Nat. Hist., ser. 17, 7:86.
REVIEWED BY: C. J. Terry (CJT); R. S. Voss (RSV).
COMMENT: Male accessory reproductive glands and taxonomy reviewed by Voss and
Linzey, 1981, Misc. Publ. Mus. Zool. Univ. Mich., 159:1–41. Subfamily
Hesperomyinae; see comment under Cricetidae.
ISIS NUMBER: 5301410008003000000.

Anotomys leander Thomas, 1906. Ann. Mag. Nat. Hist., ser. 7, 17:87.
TYPE LOCALITY: Ecuador, Pinchincha Prov., Mt. Pichincha; 11,500 ft. (3450 m).
DISTRIBUTION: N. Ecuador, montane streams.
COMMENT: Karyology reviewed by Gardner, 1971, Experientia, 27:1088–1089, and
Gardner and Patton, 1976, Occas. Pap. Mus. Zool. La. St. Univ., 49:1–48. The
specimen reported by Gardner, 1971, and Gardner and Patton, 1976, was
misidentified; it is not *leander* and probably not an *Anotomys* (RSV).
ISIS NUMBER: 5301410008003001001.

Anotomys trichotis (Thomas, 1897). Ann. Mag. Nat. Hist., ser. 6, 20:220.
TYPE LOCALITY: Colombia, Cundinamarca Dept., near Rio Magdalena.
DISTRIBUTION: Colombia; W. Venezuela.
COMMENT: Handley, 1976, Brigham Young Univ. Sci. Bull. Biol. Ser., 20(5):53,
included this species in *Anotomys,* but also see Corbet and Hill, 1980:156, Hooper,
1968, J. Mammal., 49:550–553, and Cabrera, 1961:510, who included *trichotis* in
Rheomys.
ISIS NUMBER: 5301410008049005001 as *Rheomys trichotis.*

Auliscomys Osgood, 1915. Field Mus. Nat. Hist. Publ. Zool. Ser., 10:190.
REVIEWED BY: M. D. Carleton (MDC); A. Langguth (AL); O. P. Pearson (OPP); J. J.
Pizzimenti (JJP); C. J. Terry (CJT); D. F. Williams (DFW).
COMMENT: Revised by Hershkovitz, 1962, Fieldiana Zool., 46:408–524, who included
Auliscomys in *Phyllotis* as a subgenus; Pearson and Patton, 1976, J. Mammal.,
57:341, 346, Spotorno, 1976, An. Mus. Hist. Nat. Valparaiso, 9:141–161, and Reig,
1978, Publ. Mus. Munic. Cienc. Nat. Mar del Plata "Lorenzo Scaglia,"
2(8):176–180, considered *Auliscomys* a distinct genus as did Simonetti and
Spotorno, 1980, An. Mus. Hist. Nat. Valparaiso, 13:285–297, who compared
karyotypes and morphometrics of *Andinomys, Auliscomys,* and *Phyllotis.* JJP
considers *Auliscomys* a subgenus of *Phyllotis;* MDC and OPP consider *Auliscomys* a
distinct genus. Subfamily Hesperomyinae; see comment under Cricetidae.

Auliscomys boliviensis (Waterhouse, 1846). Proc. Zool. Soc. Lond., 1846:9.
TYPE LOCALITY: Bolivia, Potosi Dept., near Potosi.
DISTRIBUTION: W. Bolivia, N. Chile, and S. Peru, at high elevations.
COMMENT: Formerly included in *Phyllotis*; see Pearson and Patton, 1976, J. Mammal., 57:341. Pine *et al.*, 1979, Mammalia, 43:353, listed this species in *Phyllotis*, without comment.
ISIS NUMBER: 5301410008042003001 as *Phyllotis boliviensis*.

Auliscomys micropus (Waterhouse, 1837). Proc. Zool. Soc. Lond., 1837:17.
TYPE LOCALITY: Argentina, Santa Cruz Prov., in the interior plains of Patagonia, 50° N., near the banks of the Rio Santa Cruz.
DISTRIBUTION: S. Argentina; S. Chile.
COMMENT: Pearson, 1958, Univ. Calif. Publ. Zool., 56(4):452, Pearson and Patton, 1976, J. Mammal., 57:339, and Spotorno and Walker, 1979, Arch. Biol. Med. Exp., 12:83–90, included this species in subgenus *Loxodontomys* of *Phyllotis*; Hershkovitz, 1962, Fieldiana Zool., 46:399, did not consider *micropus* in a separate subgenus; Pine *et al.*, 1979, Mammalia, 43:356–357, considered *Loxodontomys* a *nomen nudum*, and reviewed the species' distribution. Transferred to *Auliscomys* by Simonetti and Spotorno, 1980, An. Mus. Hist. Nat. Valparaiso, 13:285–297.
ISIS NUMBER: 5301410008042008001 as *Phyllotis micropus*.

Auliscomys pictus (Thomas, 1884). Proc. Zool. Soc. Lond., 1884:457.
TYPE LOCALITY: Peru, Junin Dept., Junin, 13,700 ft. (4176 m).
DISTRIBUTION: High Andes from C. Peru to La Paz Dept. (Bolivia).
COMMENT: Formerly included in *Phyllotis*; see Pearson and Patton, 1976, J. Mammal., 57:341.
ISIS NUMBER: 5301410008042010001 as *Phyllotis pictus*.

Auliscomys sublimis Thomas, 1900. Ann. Mag. Nat. Hist., ser. 7, 6:467.
TYPE LOCALITY: Peru, Arequipa Dept., Rinconado Malo Pass, between Caylloma and Calalla, 18,000 ft. (5486 m).
DISTRIBUTION: N.W. Argentina; W. Bolivia; S. Peru; N.E. Chile.
COMMENT: Formerly included in *Phyllotis*; see Pearson and Patton, 1976, J. Mammal., 57:341.
ISIS NUMBER: 5301410008042011001 as *Phyllotis sublimis*.

Baiomys True, 1894. Proc. U.S. Nat. Mus., (1893), 16:758.
REVIEWED BY: M. D. Carleton (MDC); J. Ramirez-Pulido (JRP); C. J. Terry (CJT); G. Urbano- V. (GUV).
COMMENT: Revised by Packard, 1960, Univ. Kans. Mus. Nat. Hist. Misc. Publ., 9(23):579–670; reviewed by Carleton, 1980. Formerly included *hummelincki* which was transferred to *Calomys* by Hershkovitz, 1962, Fieldiana Zool., 46:152; also see Handley, 1976, Brigham Young Univ. Sci. Bull. Biol. Ser., 20(5):53. Subfamily Hesperomyinae; see comment under Cricetidae.
ISIS NUMBER: 5301410008004000000.

Baiomys musculus (Merriam, 1892). Proc. Biol. Soc. Wash., 7:170.
TYPE LOCALITY: Mexico, Colima, Colima.
DISTRIBUTION: N.W. Nicaragua to S. Nayarit and C. Veracruz (Mexico).
COMMENT: Reviewed by Packard and Montgomery, 1978, Mamm. Species, 102:1–3.
ISIS NUMBER: 5301410008004001001.

Baiomys taylori (Thomas, 1887). Ann. Mag. Nat. Hist., ser. 5, 19:66.
TYPE LOCALITY: U.S.A., Texas, Duval Co., San Diego.
DISTRIBUTION: S.E. Arizona, S.W. New Mexico, S.E. and N.C. Texas (U.S.A.), to Michoacan and C. Veracruz (Mexico).
COMMENT: Includes *allex* and *analogous*; see Packard, 1960, Univ. Kans. Mus. Nat. Hist. Misc. Publ., 9(23):579–670.
ISIS NUMBER: 5301410008004002001.

Beamys Thomas, 1909. Ann. Mag. Nat. Hist., ser. 8, 4:107.
  REVIEWED BY: M. D. Carleton (MDC); C. J. Terry (CJT).
  COMMENT: Reviewed by Ansell, 1978:77–78; Ansell and Ansell, 1973, Puku, 7:21–69.
    Subfamily Cricetomyinae; see comment under Cricetidae.
  ISIS NUMBER: 5301410008100000000.

Beamys hindei Thomas, 1909. Ann. Mag. Nat. Hist., ser. 8, 4:108.
  TYPE LOCALITY: Kenya, Taveta.
  DISTRIBUTION: S. Kenya coast; Tanzania; Malawi; N.E. Zambia.
  COMMENT: Includes *major* which was listed as a distinct species by Misonne, 1974,
    Part 6:11; see Ansell, 1978:76. After examining specimens of both, MDC would
    retain *major* as a separate species pending demonstration of conspecificity with
    hindei.   •
  ISIS NUMBER: 5301410008100001001 as *B. hindei.*
            5301410008100002001 as *B. major.*

Bibimys Massoia, 1979. Physis, sec. C., 38(95):2.
  REVIEWED BY: M. D. Carleton (MDC); K. F. Koopman (KFK); A. Langguth (AL); M. A.
    Mares (MAM); O. A. Reig (OAR); R. S. Voss (RSV).
  COMMENT: Scapteromyine forms with dental similarities to *Kunsia* and *Scapteromys.*
    Subfamily Hesperomyinae; see comment under Cricetidae.

Bibimys chacoensis (Shamel, 1931). J. Wash. Acad. Sci., 21:247.
  TYPE LOCALITY: Argentina, Chaco Prov., Las Palmas.
  DISTRIBUTION: N. and C. Argentina.
  COMMENT: Included in *Akodon* by Cabrera, 1961:442, but Massoia, 1980, Ameghiniana,
    17:280–287, transferred *chacoensis* to *Bibimys.*

Bibimys labiosus (Winge, 1887). E. Mus. Lundii, 1(3):25.
  TYPE LOCALITY: Brazil, Minas Gerais, Lagoa Santa.
  DISTRIBUTION: Minas Gerais (Brazil).
  COMMENT: Tentatively included in *Scapteromys* by Cabrera, 1961:475, and in *Akodon* by
    Hershkovitz, 1966, Z. Saugetierk., 31:96; but see Massoia, 1980, Ameghiniana,
    17:280–287, who included *labiosus* in *Bibimys.*

Bibimys torresi Massoia, 1979. Physis, sec. C., 38(95):3.
  TYPE LOCALITY: Argentina, Buenos Aires, Campana, Parana River delta, at the
    confluence of Arroyo Las Piedras with Arroyo Cucarachas.
  DISTRIBUTION: Known only from the Parana River delta (N.E. Argentina).

Blarinomys Thomas, 1896. Ann. Mag. Nat. Hist., ser. 6, 18:310.
  REVIEWED BY: R. H. Pine (RHP); C. J. Terry (CJT).
  COMMENT: Subfamily Hesperomyinae; see comment under Cricetidae.
  ISIS NUMBER: 5301410008005000000.

Blarinomys breviceps (Winge, 1887). E. Mus. Lundii, 1(3):34.
  TYPE LOCALITY: Brazil, S.W. Minas Gerais Prov., Rio das Velhas, Lagoa Santa.
  DISTRIBUTION: S.E. Brazil.
  COMMENT: Reviewed by Matson and Abravaya, 1977, Mamm. Species, 74:1–3.
  ISIS NUMBER: 5301410008005001001.

Bolomys Thomas, 1916. Ann. Mag. Nat. Hist., ser. 8, 18:339.
  REVIEWED BY: M. D. Carleton (MDC); A. Langguth (AL); M. A. Mares (MAM); R. A.
    Ojeda (RAO); O. A. Reig (OAR).
  COMMENT: Included as a subgenus in *Akodon* by Ellerman, 1941:406, and Cabrera,
    1961:453. Bianchi *et al.*, 1971, Evolution, 25:724–736, and Reig, 1978, Publ. Mus.
    Munic. Cienc. Nat. Mar del Plata "Lorenzo Scaglia," 2(8):164–190, considered
    Bolomys a distinct genus. Includes *Cabreramys*; see Reig, 1978:164–190; Massoia,
    1980, Hist. Nat., 1(25):179, used the name *Cabreramys* but did not mention Reig,
    1978:164–190. Includes *Thalpomys* as a subjective synonym; see Langguth, 1975,
    Papeis Avulsos Zool. Sao Paulo, 29(8):45–54, and comments under *Akodon.*

Gardner and Patton, 1976, Occas. Pap. Mus. Zool. La. St. Univ., 25:27–30, Cabrera, 1961:453, Reig, 1978:164–190, and Pine *et al.*, 1979, Mammalia, 43:348, differed regarding the contents of *Bolomys*. Formerly included *Akodon albiventer* which includes *berlepschii;* see Pine *et al.*, 1979, Mammalia, 43:347–348; also see Corbet and Hill, 1980:151. There is no consensus concerning the content of this genus. Subfamily Hesperomyinae; see comment under Cricetidae.

Bolomys amoenus (Thomas, 1900). Ann. Mag. Nat. Hist., ser. 7, 6:468.
TYPE LOCALITY: Peru, Arequipa Dept., Calalla, Rio Colca, 3500 m.
DISTRIBUTION: Andes of S.E. Peru.
COMMENT: Included in *Bolomys* (type species) by Reig, 1978, Publ. Mus. Munic. Cienc. Nat. Mar del Plata "Lorenzo Scaglia," 2(8):164–190, Bianchi *et al.*, 1971, Evolution, 25:724–736, Spotorno, *in* Pine *et al.*, 1979, Mammalia, 43:348, and Corbet and Hill, 1980:151. Listed in *Akodon* by Gardner and Patton, 1976, Occas. Pap. Mus. Zool. La. St. Univ., 49:28, 30.
ISIS NUMBER: 5301410008001003001 as *Akodon amoenus.*

Bolomys lactens (Thomas, 1918). Ann. Mag. Nat. Hist., ser. 9, 1:188.
TYPE LOCALITY: Argentina, Jujuy Prov., Leon about 1500 m.
DISTRIBUTION: N.W. Argentina.
COMMENT: Includes *orbus, negrito,* and *leucolimnaeus;* see Reig, 1978, Publ. Mus. Munic. Cienc. Nat. Mar del Plata "Lorenzo Scaglia," 2(8):164–190, who considered *Bolomys* a distinct genus.
ISIS NUMBER: 5301410008001019001 as *Akodon lactens.*

Bolomys lasiurus (Lund, 1841). Kongl. Dansk. Vid. Selsk. Naturv. Math. Afhandl., 8:50.
TYPE LOCALITY: Brazil, Minas Gerais, Rio das Velhas, Lagoa Santa.
DISTRIBUTION: E. Brazil; Paraguay.
COMMENT: Includes *lasiotis* Lund as a synonym; see Langguth, 1975, Papeis Avulsos Zool. Sao Paulo, 29(8):45–54; see also comment under *Akodon reinhardti.* Gardner and Patton, 1976, Occas. Pap. Mus. Zool. La. St. Univ., 49:25, transferred *lasiurus* from *Zygodontomys* to *Akodon,* and were followed by Voss and Linzey, 1981, Misc. Publ. Mus. Zool. Univ. Mich., 159:29. Reig, 1978, Publ. Mus. Munic. Cienc. Nat. Mar del Plata "Lorenzo Scaglia," 2(8):164–190, and Maia and Langguth, 1981, Z. Saugetierk., 46:241–249, included *lasiurus* in *Bolomys.* Includes *arviculoides, brachyurus, fuscinus,* and *pixuna;* see Reig, 1978:164–190. Massoia, 1962, Physis, 23(65):185–194, reviewed *arviculoides* and considered it a distinct species, and Massoia, 1980, Hist. Nat., 1(25):179, listed *arviculoides* and *pixuna* as distinct species (without comment) in *Cabreramys,* which is included in *Bolomys;* see comment under genus. Ximenez and Langguth, 1970, Communic. Zool. Mus. Hist. Nat. Montevideo, 10(128):1–5, considered *cursor* a distinct species of *Akodon* and separated it from *arviculoides.* Karyology reviewed by Maia and Langguth, 1981, Z. Saugetierk., 46:241–249.
ISIS NUMBER: 5301410008059003001 as *Zygodontomys lasiurus.*
       5301410008001005001 as *Akodon arviculoides.*

Bolomys lenguarum (Thomas, 1898). Ann. Mag. Nat. Hist., ser. 7, 2:271.
TYPE LOCALITY: Paraguay, Chaco Boreal, Waikthlatingwaialwa.
DISTRIBUTION: Bolivia; S.W. Brazil; Paraguay; Argentina.
COMMENT: Includes *tapirapoanus* which was transferred from *Akodon;* see Reig, 1978, Publ. Mus. Munic. Cienc. Nat. Mar del Plata "Lorenzo Scaglia," 2(8):164–190. Hershkovitz, 1966, Z. Saugetierk., 31:127, included *tapirapoanus* in *Zygodontomys;* Cabrera, 1961:439, 447, included it in *Akodon. B. lenguarum* was formerly included in *Cabreramys* where it had been transferred from *Akodon* by Massoia and Fornes, 1967, Acta Zool. Lilloana, 23:407–430; see Reig, 1978:167.
ISIS NUMBER: 5301410008001035001 as *Akodon tapirapoanus.*

Bolomys obscurus (Waterhouse, 1837). Proc. Zool. Soc. Lond., 1837:16.
TYPE LOCALITY: Uruguay, Maldonado.
DISTRIBUTION: S. Uruguay; N.E. and E.C. Argentina.
COMMENT: Formerly included in *Akodon* by Cabrera, 1961:444, and Gardner and

Patton, 1976, Occas. Pap. Mus. Zool. La. St. Univ., 49:28, and in *Cabreramys* by
Massoia and Fornes, 1967, Acta Zool. Lilloana, 23:407–430; includes *benefactus;* see
Reig, 1978, Publ. Mus. Munic. Cienc. Nat. Mar del Plata "Lorenzo Scaglia,"
2(8):167.
ISIS NUMBER: 5301410008001027001 as *Akodon obscurus.*

Bolomys temchuki Massoia, 1980. Hist. Nat., 1(25):179.
TYPE LOCALITY: Argentina, Misiones, Depto. Capital, en terrenos del INTA, Arroyo
Zaiman.
DISTRIBUTION: Misiones (Argentina).
COMMENT: Description based on external characters compared with *obscurus;* described
in *Cabreramys* (a synonym of *Bolomys);* see Reig, 1978, Publ. Mus. Munic. Cienc.
Nat. Mar del Plata "Lorenzo Scaglia," 2(8):164–190.

Brachiones Thomas, 1925. Ann. Mag. Nat. Hist., ser. 9, 16:548.
REVIEWED BY: C. B. Robbins (CBR); C. J. Terry (CJT); S. Wang (SW).
COMMENT: Subfamily Gerbillinae; see comment under Cricetidae.
ISIS NUMBER: 5301410008088000000.

Brachiones przewalskii (Buchner, 1889). Wiss. Res. Przewalski Cent. Asian Zool., Th. 1,
Saugeth., p. 51.
TYPE LOCALITY: China, Sinkiang, Lob Nor.
DISTRIBUTION: Sinkiang to Kansu (China).
COMMENT: Not yet found in Mongolia; see Corbet, 1978:128. Sokolov and Orlov,
1980:156, included this species in *Meriones,* without comment.
ISIS NUMBER: 5301410008088001001.

Brachytarsomys Gunther, 1875. Proc. Zool. Soc. Lond., 1875:79.
REVIEWED BY: C. J. Terry (CJT).
COMMENT: Subfamily Nesomyinae; see comment under Cricetidae.
ISIS NUMBER: 5301410008060000000.

Brachytarsomys albicauda Gunther, 1875. Proc. Zool. Soc. Lond., 1875:80.
TYPE LOCALITY: Madagascar, between Tamatave and Morondava.
DISTRIBUTION: E. Madagascar.
ISIS NUMBER: 5301410008060001001.

Brachyuromys Forsyth Major, 1896. Ann. Mag. Nat. Hist., ser. 6, 18:322.
REVIEWED BY: C. J. Terry (CJT).
COMMENT: Subfamily Nesomyinae; see comment under Cricetidae.
ISIS NUMBER: 5301410008061000000.

Brachyuromys betsileoensis (Bartlett, 1880). Proc. Zool. Soc. Lond., 1879:770.
TYPE LOCALITY: Madagascar, S.E. Betsileo, Ampitambe Forest.
DISTRIBUTION: S.E. Betsileo Country, Andringitra (Madagascar).
ISIS NUMBER: 5301410008061001001.

Brachyuromys ramirohitra Forsyth Major, 1896. Ann. Mag. Nat. Hist., ser. 6, 18:323.
TYPE LOCALITY: Madagascar, border of N.E. Betsileo, Ampitambe Forest.
DISTRIBUTION: Betsileo Country, Andringitra (E. Madagascar), possibly in most of the
eastern forest.
ISIS NUMBER: 5301410008061002001.

Calomys Waterhouse, 1837. Proc. Zool. Soc. Lond., 1837:21.
REVIEWED BY: A. Langguth (AL); M. A. Mares (MAM); R. A. Ojeda (RAO); O. P.
Pearson (OPP); C. J. Terry (CJT); D. F. Williams (DFW).
COMMENT: Includes *Hesperomys* as a synonym; revised by Hershkovitz, 1962, Fieldiana
Zool., 46:122–174. Reviewed by Pearson and Patton, 1976, J. Mammal.,
57:339–350. Includes *Baiomys hummelincki;* see Hershkovitz, 1962:152, and
Handley, 1976, Brigham Young Univ. Sci. Bull. Biol. Ser., 20(5):53. Subfamily
Hesperomyinae; see comment under Cricetidae.
ISIS NUMBER: 5301410008006000000.

Calomys callosus (Rengger, 1830). Naturg. Saugeth. Paraguay, p. 231.
TYPE LOCALITY: Paraguay, Villa Pilar Dept., Rio Paraguay, opposite mouth of Rio Bermejo.
DISTRIBUTION: E. and S.W. Brazil; Bolivia; Paraguay; N. Argentina.
COMMENT: Includes *venustus* and *expulsus*; see Hershkovitz, 1962, Fieldiana Zool., 46:171–174. Formerly included *fecundus* (see Pearson and Patton, 1976, J. Mammal., 57:343) and *muriculus* (see Williams and Mares, 1978, Ann. Carnegie Mus., 47(9):197, 200, 214). AL included *muriculus* in *callosus*.
ISIS NUMBER: 5301410008006001001 as *C. callosus*.
5301410008006003001 as *C. expulsus*.
5301410008006010001 as *C. venustus*.

Calomys fecundus (Thomas, 1926). Ann. Mag. Nat. Hist., ser. 9, 17:321.
TYPE LOCALITY: S. Bolivia, Tarija Dist., Tablada.
DISTRIBUTION: Bolivia.
COMMENT: Included in *callosus* by Hershkovitz, 1962, Fieldiana Zool., 46:172; considered a distinct species by Pearson and Patton, 1976, J. Mammal., 57:343. Probably synonymous with *muriculus* (DFW); see Williams and Mares, 1978, Ann. Carnegie Mus., 47:197.

Calomys hummelincki (Husson, 1960). Stud. Faun. Curacao Carib. Isl., 43:34.
REVIEWED BY: R. H. Pine (RHP).
TYPE LOCALITY: N.W. Curacao, Klein Santa Martha (Netherlands).
DISTRIBUTION: Curacao; Aruba; Venezuela.
COMMENT: Originally named in *Baiomys*, included in *C. laucha* by Hershkovitz, 1962, Fieldiana Zool., 46;152; considered a distinct species by Handley, 1976, Brigham Young Univ. Sci. Bull. Biol. Ser., 20(5):53.

Calomys laucha (Olfers, 1818). *In* Eschwege, Neue Bibl. Reisenb., 15(2):209.
TYPE LOCALITY: Paraguay, Vicinity of Asuncion.
DISTRIBUTION: S.E. Brazil; Paraguay; Uruguay; C. Argentina; S. Bolivia.
COMMENT: Includes *dubius*, *gracilipes*, and *tener*; see Hershkovitz, 1962, Fieldiana Zool., 46:150–153. Formerly included *hummelincki* (See Handley, 1976, Brigham Young Univ. Sci. Bull. Biol. Ser., 20(5):53) and *musculinus* (see Massoia *et al.*, 1968, Rev. Invest. Agro. INTA, ser. 1, Biol. Prod. Anim., 5:63–92). AL attributed *laucha* to Fischer, 1814; see comment under *Reithrodon physodes*.
ISIS NUMBER: 5301410008006006001 as *C. laucha*.
5301410008006002001 as *C. dubius*.
5301410008006005001 as *C. gracilipes*.
5301410008006009001 as *C. tener*.

Calomys lepidus (Thomas, 1884). Proc. Zool. Soc. Lond., 1884:454.
TYPE LOCALITY: Peru, Junin Dept., Junin.
DISTRIBUTION: S. and C. Peru; N.E. Chile; W. Bolivia; Jujuy (Argentina).
COMMENT: Includes *ducillus*; see Hershkovitz, 1962, Fieldiana Zool., 46:163.
ISIS NUMBER: 5301410008006007001.

Calomys muriculus (Thomas, 1921). Ann. Mag. Nat. Hist., ser. 9, 8:623.
TYPE LOCALITY: Bolivia, Santa Cruz Dept., San Antonio, Parapiti.
DISTRIBUTION: Lowlands and E. slope of Andes in Bolivia.
COMMENT: Included in *callosus* by Hershkovitz, 1962, Fieldiana Zool., 46:172; see Williams and Mares, 1978, Ann. Carnegie Mus., 47:197, 200, 214. Probably a synonym of *fecundus* of Pearson and Patton, 1976, J. Mammal., 57:343 (DFW). AL includes *muriculus* in *callosus*.
ISIS NUMBER: 5301410008006008001.

Calomys musculinus (Thomas, 1913). Ann. Mag. Nat. Hist., ser. 8, 11:138.
TYPE LOCALITY: Argentina, Jujuy Prov., Maimara, 2230 m.
DISTRIBUTION: Jujuy, Salta, Tucuman, Catamarca, Mendoza, and Buenos Aires, and probably Cordoba and San Luis Provs. (Argentina).
COMMENT: Included in *laucha* by Hershkovitz, 1962, Fieldiana Zool., 46:152; but see Massoia *et al.*, 1968, Rev. Invest. Agro. INTA, ser. 1, Biol. Prod. Anim., 5:63–92,

and Pearson and Patton, 1976, J. Mammal., 57:348, who considered *musculinus* a distinct species. Distribution in Buenos Aires Prov. reviewed by Massoia and Fornes, 1967, Acta Zool. Lilloana, 23:407–430.

Calomys sorellus (Thomas, 1900). Ann. Mag. Nat. Hist., ser. 7, 6:297.
  TYPE LOCALITY: Peru, Libertad Dept., 13 km south of Huamachuco, 3500 m.
  DISTRIBUTION: Peru.
  COMMENT: Includes *frida;* see Hershkovitz, 1962, Fieldiana Zool., 46:137.
  ISIS NUMBER: 5301410008006004001 as *C. frida.*

Calomyscus Thomas, 1905. Abstr. Proc. Zool. Soc. Lond., 24:23.
  REVIEWED BY: O. L. Rossolimo (OLR); C. J. Terry (CJT).
  COMMENT: Reviewed by Vorontsov and Potapova, 1979, Zool. Zh., 58(9):1391–1397. More material from Iran and Afghanistan is needed to determine the number of species in this genus with certainty (OLR). Subfamily Cricetinae; see comment under Cricetidae.
  ISIS NUMBER: 5301410008007000000.

Calomyscus bailwardi Thomas, 1905. Abstr. Proc. Zool. Soc. Lond., 24:23.
  TYPE LOCALITY: Iran, Husistan, 120 km. southeast of Ahwaz, Mala-i-Mir (=Jzeh).
  DISTRIBUTION: Iran; Transcaucasia (U.S.S.R.); S. Turkmenia; Afghanistan; N. Pakistan.
  COMMENT: Vorontsov *et al.,* 1979, Zool. Zh., 58(8):1213–1224, considered *baluchi, hotsoni, mystax,* and *urartensis* distinct species and included *mustersi* in *baluchi.* Corbet, 1978:89, included all of the above forms in *bailwardi.*
  ISIS NUMBER: 5301410008007001001.

Calomyscus baluchi Thomas, 1920. J. Bombay Nat. Hist. Soc., 26:938.
  TYPE LOCALITY: Pakistan, Baluchistan, Kelat Dist.
  DISTRIBUTION: Baluchistan (Pakistan); E. Afghanistan.
  COMMENT: Includes *mustersi;* revised by Vorontsov *et al.,* 1979, Zool. Zh., 58(8):1213–1224.

Calomyscus hotsoni Thomas, 1920. J. Bombay Nat. Hist. Soc., 26:938.
  TYPE LOCALITY: Pakistan, Baluchistan, Gwambuk Kaul, 50 km S.W. of Panjgur, 26° 30' N., 63° 50' E.
  DISTRIBUTION: Known only from the vicinity of the type locality.

Calomyscus mystax Kashkarov, 1925. Trans. Turkestansk. Nauch. ob-va pri Sredniaziatsk. Univ. (Tashkent), 2:43.
  TYPE LOCALITY: U.S.S.R., Turkmenistan, Great Balkhan Mtns., Bashi-Mugur.
  DISTRIBUTION: Great and Little Balkhan and Kopet Dag Mtns. and Badkhiz desert, S. Turkmenia (U.S.S.R.), probably adjacent Iran and Afghanistan.

Calomyscus urartensis Vorontsov and Kartavseva, 1979. Zool. Zh., 58:1218.
  TYPE LOCALITY: U.S.S.R., Azerbaidzhan S.S.R., Nakhichevansk. A.S.S.R., Alindzhachai R., 7 km N. Dzhul'ta.
  DISTRIBUTION: Extreme S. Transcaucasus (Azerbaidzhan, U.S.S.R.); N.W. Iran; probably Armenia (U.S.S.R.).
  COMMENT: May be conspecific with *hotsoni;* see Gromov and Baranova, 1981:161.

Chilomys Thomas, 1897. Ann. Mag. Nat. Hist., ser. 6, 19:500.
  REVIEWED BY: A. Langguth (AL); R. H. Pine (RHP); C. J. Terry (CJT).
  COMMENT: Subfamily Hesperomyinae; see comment under Cricetidae.
  ISIS NUMBER: 5301410008008000000.

Chilomys instans (Thomas, 1895). Ann. Mag. Nat. Hist., ser. 6, 16:368.
  TYPE LOCALITY: Colombia, Bogota region, Cundinamarca Dept., Hacienda de La Selva, 1380 m.
  DISTRIBUTION: N. and C. Colombia; Ecuador; Andes of Venezuela.
  COMMENT: Reviewed by Handley, 1976, Brigham Young Univ. Sci. Bull. Biol. Ser., 20(5):51; Voss and Linzey, 1981, Misc. Publ. Mus. Zool. Univ. Mich., 159:10, 13, reviewed male accessory reproductive glands.
  ISIS NUMBER: 5301410008008001001.

Chinchillula Thomas, 1898. Ann. Mag. Nat. Hist., ser. 1, 7:280.
REVIEWED BY: M. A. Mares (MAM); R. H. Pine (RHP); C. J. Terry (CJT).
COMMENT: Karyology reviewed by Pearson and Patton, 1976, J. Mammal., 57:339–350. Revised by Hershkovitz, 1962, Fieldiana Zool., 46:484–491. Subfamily Hesperomyinae; see comment under Cricetidae.
ISIS NUMBER: 5301410008009000000.

Chinchillula sahamae Thomas, 1898. Ann. Mag. Nat. Hist., ser. 1, 7:280.
TYPE LOCALITY: Bolivia, S.W. La Paz Esperanza, 50 km north of Mt. Sajama, Pacajes, 4200 m.
DISTRIBUTION: S. Peru; N. Chile; W. Bolivia; N.W. Argentina.
ISIS NUMBER: 5301410008009001001.

Cricetomys Waterhouse, 1840. Proc. Zool. Soc. Lond., 1840:2.
REVIEWED BY: C. J. Terry (CJT).
COMMENT: Subfamily Cricetomyinae; see comment under Cricetidae.
ISIS NUMBER: 5301410008101000000.

Cricetomys emini Wroughton, 1910. Ann. Mag. Nat. Hist., ser. 8, 5:106.
TYPE LOCALITY: Zaire(=N.E. Congo), Monbuttu, Gadda.
DISTRIBUTION: Sierra Leone to Gabon, Uganda, and Zaire; Bioko.
COMMENT: Separated from *gambianus* by Genest-Villard, 1967, Mammalia, 31:390–455, and Rosevear, 1969, The Rodents of West Africa, Br. Mus. Lond., 677:221.

Cricetomys gambianus Waterhouse, 1840. Proc. Zool. Soc. Lond., 1840:2.
TYPE LOCALITY: Gambia, River Gambia.
DISTRIBUTION: Senegal to Sudan to Angola and South Africa.
COMMENT: Formerly included *emini*; see Genest-Villard, 1967, Mammalia, 31:390–455; Rosevear, 1969, The Rodents of West Africa, Br. Mus. Lond., 677:221. Reviewed by Misonne, 1974, Part 6:12; Ansell, 1978:75.
ISIS NUMBER: 5301410008101001001.

Cricetulus Milne-Edwards, 1867. Ann. Sci. Nat. Paris, 7:376.
REVIEWED BY: R. S. Hoffmann (RSH); O. L. Rossolimo (OLR); C. J. Terry (CJT); S. Wang (SW)(China).
COMMENT: Includes *Allocricetulus* and *Tscherskia*; see Corbet, 1978:90. Reviewed by Flint, 1966, Die Zwerghamster der Palaearktischen Fauna, Stuttgart. Gromov and Baranova, 1981:153–154, retained *Allocricetulus* and *Tscherskia* as genera, without comment. Subfamily Cricetinae; see comment under Cricetidae.
ISIS NUMBER: 5301410008010000000.

Cricetulus alticola Thomas, 1917. Ann. Mag. Nat. Hist., 19:455.
TYPE LOCALITY: India, Ladak, Shushul, 13,500 ft. (4115 m).
DISTRIBUTION: Kashmir and Ladak (N. India).
COMMENT: Formerly included *tibetanus* which was transferred to *kamensis*; see Wang and Cheng, 1973, Acta Zool. Sin., 19:61–68, and Corbet, 1978:91.
ISIS NUMBER: 5301410008010001001.

Cricetulus barabensis (Pallas, 1773). Reise Prov. Russ. Reichs., 2:704.
TYPE LOCALITY: U.S.S.R., Altaisk. Krai, near Barnaul, Pavlovsk, Kasmalinsky-Bor.
DISTRIBUTION: Steppes of S. Siberia from Irtysh River to Ussuri region (U.S.S.R.), south to N.W. Mongolia, N.E. China, and Korea.
COMMENT: Includes *fumatus*; see Orlov and Iskhakova, 1975, Zool. Zh., 54(4):597–604. Corbet, 1978:91, included *griseus* and *pseudogriseus* in *barabensis*; but also see Flint, 1966, Die Zwerghamster der Palaearktischen Fauna, Stuttgart, who considered *griseus* a distinct allopatric species, and Orlov and Iskhakova, 1975, Zool. Zh., 54(4):597–604, Matthey, 1960, Caryologia, 13:199–223, Orlov et al., 1978, *in*, [Geography and Dynamics of Plants and Animals in Mongolian P.R.], pp. 149–164, and Sokolov and Orlov, 1980, [Guide to the Mammals of Mongolian P.R.], p. 123, who considered *griseus* and *pseudogriseus* as distinct species. Formerly included *obscurus*; see Sokolov and Orlov, 1980, [Guide to the Mammals of Mongolian P.R.], p. 123.
ISIS NUMBER: 5301410008010002001.

Cricetulus curtatus Allen, 1925. Am. Mus. Novit., 179:3.
TYPE LOCALITY: China, W. Inner Mongolia, Iren Dabasu (=Ehrlien).
DISTRIBUTION: Steppes of S. Mongolia north of the Altai Mtns. to Inner Mongolia, Honan, and Ningsia (China).
COMMENT: Ellerman and Morrison-Scott, 1966:626, considered this a subspecies of *eversmanni*, but see Orlov *et al.*, 1978, *in* [Geography and Dynamics of Plants and Animals in Mongolian P.R.], pp. 149–164. Formerly included in *Allocricetulus*; see Corbet, 1978:90; but also see Gromov and Baranova, 1981:154.

Cricetulus eversmanni Brandt, 1859. Melanges. Biol. Acad. St. Petersb., p. 210.
TYPE LOCALITY: U.S.S.R., Orenburgsk. Obl., near Orenburg.
DISTRIBUTION: Kazakstan, steppes from Volga River to Tarbagatai (U.S.S.R.).
COMMENT: Formerly included in *Allocricetulus*; formerly included *curtatus*; see Corbet, 1978:90, 92, and Gromov and Baranova, 1981:154.
ISIS NUMBER: 5301410008010003001.

Cricetulus griseus Milne-Edwards, 1867. Ann. Sci. Nat. Paris, 7:376.
TYPE LOCALITY: China, "Inner Mongolia," Suenhoafu, near Kalgan (in Hopei).
DISTRIBUTION: N.E. China (Anhwei north to Liaoning, west to Shansi); perhaps Inner Mongolia.
COMMENT: Orlov and Iskhakova, 1975, Zool. Zh., 54:597, and Corbet, 1978:91, included *obscurus* in *griseus*; but also see Sokolov and Orlov, 1980, [Guide to the Mammals of Mongolian P.R.], p. 123, who considered *obscurus* to be a distinct species; see also comment under *barabensis*.

Cricetulus kamensis (Satunin, 1903). Ann. Zool. Mus. St. Petersb., 7:574.
TYPE LOCALITY: China, N.E. Tibet, Mekong Dist., River Moktschjun.
DISTRIBUTION: Tibet, Tsinghai, and Kansu (China).
COMMENT: Includes *kozlovi*, *lama*, and *tibetanus*; see Wang and Cheng, 1973, Acta Zool. Sin., 19:61–68, and Corbet, 1978:91.
ISIS NUMBER: 5301410008010004001 as *C. lama*.

Cricetulus longicaudatus (Milne-Edwards, 1867). Rech. Mamm., p. 136.
TYPE LOCALITY: China, N. Shansi, near Saratsi.
DISTRIBUTION: Altai and Tuva (U.S.S.R.); W. and S. Mongolia; Sinkiang to Hopei and Szechwan (China).
COMMENT: Flint, 1966, Die Zwerghamster der Palaearktischen Fauna, Stuttgart, suggested that *kamensis* might be conspecific, but Wang and Cheng, 1973, Acta Zool. Sin., 19:61–68, considered it a distinct species.
ISIS NUMBER: 5301410008010005001.

Cricetulus migratorius (Pallas, 1773). Reise Prov. Russ. Reichs., 2:703.
REVIEWED BY: M. Andera (MA).
TYPE LOCALITY: U.S.S.R., W. Kazakhstan, lower Ural River.
DISTRIBUTION: S.E. Europe through Asia Minor, Transcaucasia, and Kazakhstan to Mongolia and Ningsia (China), south to Israel, Iraq, Iran, and Pakistan.
ISIS NUMBER: 5301410008010006001.

Cricetulus obscurus Milne-Edwards, 1867. Rech. Mamm., p. 136.
TYPE LOCALITY: China, N. Shansi, Saratsi.
DISTRIBUTION: Kansu to Shansi (China) and S.C. and S.E. Mongolia.
COMMENT: Included in *barabensis* by Ellerman and Morrison-Scott, 1951:624; but see Sokolov and Orlov, 1980, [Guide to the Mammals of Mongolian P.R.], p. 123, who considered *obscurus* a karyotypically and morphologically distinct species.

Cricetulus pseudogriseus Orlov and Iskhakova, 1975. Zool. Zh., 54(4):599.
TYPE LOCALITY: U.S.S.R., S. Buryatskaya A.S.S.R., Kyakhta Dist., Naushki.
DISTRIBUTION: Transbaikalia (U.S.S.R.); N.C. and N.E. Mongolia; N. Inner Mongolia (China).
COMMENT: Considered a distinct species by Orlov *et al.*, 1978, *in* [Geography and Dynamics of Plants and Animals in Mongolian P.R.], p. 149–164, and Sokolov and Orlov, 1980, [Guide to the Mammals of Mongolian P.R.], p. 123; Corbet, 1978:91, included this form in *barabensis*.

Cricetulus triton De Winton, 1899. Proc. Zool. Soc. Lond., 1899:575.
TYPE LOCALITY: China, N. Shantung.
DISTRIBUTION: Kansu and Kiangsu to N.E. China, Korea, and upper Ussuri region
(U.S.S.R.).
COMMENT: Subgenus *Tscherskia*; see Corbet, 1978:90, and Ellerman and Morrison-Scott,
1951:626. Includes *albipes*; see Kartavseva *et al.*, 1980, Zool. Zh., 59:899–904.
ISIS NUMBER: 5301410008010007001.

Cricetus Leske, 1779. Anfansgr. Naturg., 1:168.
REVIEWED BY: M. Andera (MA); O. L. Rossolimo (OLR); S. Wang (SW).
COMMENT: Kuznetsov, 1965, *in* Bobrinskii *et al.*, 1965, Key to the Mammals of the
U.S.S.R., Moscow, included *Mesocricetus* in this genus, but Corbet, 1978:90, 92,
considered *Mesocricetus* a distinct genus; he stated that these taxa are in need of
revision. Subfamily Cricetinae; see comment under Cricetidae.
ISIS NUMBER: 5301410008011000000.

Cricetus cricetus (Linnaeus, 1758). Syst. Nat., 10th ed., 1:60.
TYPE LOCALITY: Germany.
DISTRIBUTION: Belgium, C. Europe, W. Siberia and N. Kazakhstan to the upper Yenesei
and the Altai (U.S.S.R.) and Sinkiang (China).
ISIS NUMBER: 5301410008011001001.

Daptomys Anthony, 1929. Am. Mus. Novit., 383:1.
REVIEWED BY: A. Langguth (AL); C. J. Terry (CJT); R. S. Voss (RSV).
COMMENT: Subfamily Hesperomyinae; see comment under Cricetidae.
ISIS NUMBER: 5301410008012000000.

Daptomys oyapocki Dubost and Petter, 1978. Mammalia, 42:436.
TYPE LOCALITY: French Guiana, Trois Sauts, not far from the banks of Oyapock River,
2° 10' N., 53° 11' W.
DISTRIBUTION: Known only from the type locality.
COMMENT: Known only by the holotype (RSV).

Daptomys peruviensis Musser and Gardner, 1974. Am. Mus. Novit., 2537:7.
TYPE LOCALITY: Peru, Loreto Dept., Balta, 10° 08' S., 17° 13' W., 300 m.
DISTRIBUTION: Known only from the type locality.
COMMENT: Known only by the holotype.

Daptomys venezuelae Anthony, 1929. Am. Mus. Novit., 383:2.
TYPE LOCALITY: Venezuela, Sucre, Rio Neveri, about 15 mi. W. of Cumanacoa, 2400 ft.
(720 m).
DISTRIBUTION: Venezuela.
COMMENT: Reviewed by Musser and Gardner, 1974, Am. Mus. Novit., 2537:1–23.
ISIS NUMBER: 5301410008012001001.

Delanymys Hayman, 1962. Rev. Zool. Bot. Afr., 65:1–2.
REVIEWED BY: C. J. Terry (CJT).
COMMENT: Subfamily Dendromurinae; see comment under Cricetidae.
ISIS NUMBER: 5301410008103000000.

Delanymys brooksi Hayman, 1962. Rev. Zool. Bot. Afr., 65:1–2.
TYPE LOCALITY: S.W. Uganda, Kigezi, near Kanaba, Echuya (or Muchuya) Swamp.
DISTRIBUTION: S.W. Uganda; Kivu (Zaire).
ISIS NUMBER: 5301410008103001001.

Dendromus Smith, 1829. Zool. J. Lond., 4:438.
REVIEWED BY: M. D. Carleton (MDC); C. J. Terry (CJT).
COMMENT: Includes *Poemys*; see Ansell, 1978:76. Reviewed, in part, by Dieterlen, 1971,
Saugetierk. Mitt., 19:97–132. This genus is in need of revision; see Misonne, 1974,
Part 6:13. Subfamily Dendromurinae; see comment under Cricetidae.
ISIS NUMBER: 5301410008104000000.

Dendromus kahuziensis Dieterlen, 1969. Z. Saugetierk., 34:348–353.
  TYPE LOCALITY: Zaire, Kivu, Mt. Kahuzi, 2100 m.
  DISTRIBUTION: Kivu region (Zaire).
  ISIS NUMBER: 5301410008104002001.

Dendromus lovati De Winton, 1900. Proc. Zool. Soc. Lond., 1899:986.
  TYPE LOCALITY: Ethiopia, Managasha, near Addis Ababa.
  DISTRIBUTION: Ethiopia.
  ISIS NUMBER: 5301410008104003001.

Dendromus melanotis Smith, 1834. S. Afr. J., 2:158.
  TYPE LOCALITY: South Africa, near Port Natal (Durban).
  DISTRIBUTION: Ethiopia and Uganda to South Africa to N. Nigeria; Guinea.
  COMMENT: Includes *exoneratus*; see Misonne, 1974, Part 6:13.
  ISIS NUMBER: 5301410008104004001 as *D. melanotis*.
              5301410008104001001 as *D. exoneratus*.

Dendromus mesomelas Brants, 1827. Het. Gesl. Muiz., p. 122.
  TYPE LOCALITY: South Africa, E. Cape Prov., Sunday's River, east of Port Elizabeth.
  DISTRIBUTION: Ethiopia to South Africa to Cameroun.
  COMMENT: Includes *oreas*; see Misonne, 1974, Part 6:13.
  ISIS NUMBER: 5301410008104005001 as *D. mesomelas*.
              5301410008104009001 as *D. oreas*.

Dendromus messorius Thomas, 1903. Ann. Mag. Nat. Hist., ser. 7, 12:340.
  TYPE LOCALITY: Cameroun, Efulen.
  DISTRIBUTION: S. Nigeria to Zaire.
  COMMENT: Included in *mystacalis* by Misonne, 1974, Part 6:13, but considered distinct
    by Dieterlen, 1971, Saugetierk. Mitt., 19:97–132 (MDC).
  ISIS NUMBER: 5301410008104006001.

Dendromus mystacalis Heuglin, 1863. Nova Acta Acad. Caes. Leop.-Carol., Halle, 30:2,
    suppl.:5.
  TYPE LOCALITY: Ethiopia, Baschlo region (? Bashilo River).
  DISTRIBUTION: Ethiopia to Zaire and South Africa.
  COMMENT: Includes *pumilio*; see Misonne, 1974, Part 6:13, who also included *messorius*
    in this species; but see Dieterlen, 1971, Saugetierk. Mitt., 19:97–132, who
    considered *messorius* a distinct species.
  ISIS NUMBER: 5301410008104007001 as *D. mystacalis*.
              5301410008104010001 as *D. pumilo (sic)*.

Dendromus nyikae Wroughton, 1909. Ann. Mag. Nat. Hist., ser. 8, 3:248.
  TYPE LOCALITY: Malawi, Nyika Plateau.
  DISTRIBUTION: C. Angola; E. Transvaal (South Africa); S.E. Zimbabwe; E. Zambia; N.
    and C. Malawi; S. W. Tanzania.
  COMMENT: Includes *angolensis, bernardi,* and *longicaudatus*; see Misonne, 1974, Part 6:13.
  ISIS NUMBER: 5301410008104008001.

Dendroprionomys Petter, 1966. Mammalia, 30:131.
  REVIEWED BY: C. J. Terry (CJT).
  COMMENT: Subfamily Dendromurinae; see comment under Cricetidae.
  ISIS NUMBER: 5301410008105000000.

Dendroprionomys rousseloti Petter, 1966. Mammalia, 30:131.
  TYPE LOCALITY: Central African Republic, La Maboke (Mbaiki) Forest (pres de
    M'Baiki).
  DISTRIBUTION: Vicinity of Brazzaville (Congo).
  COMMENT: Known only from four specimens; see Misonne, 1974, Part 6:13.
  ISIS NUMBER: 5301410008105001001.

Deomys Thomas, 1888. Proc. Zool. Soc. Lond., 1888:130.
  REVIEWED BY: C. J. Terry (CJT).
  COMMENT: Subfamily Dendromurinae; see comment under Cricetidae.
  ISIS NUMBER: 5301410008106000000.

Deomys ferrugineus Thomas, 1888. Proc. Zool. Soc. Lond., 1888:130.
TYPE LOCALITY: Lower Congo.
DISTRIBUTION: Uganda; Zaire; S.W. Central African Republic; S. Cameroun; Gabon;
Congo; Rio Muni; Bioko.
ISIS NUMBER: 5301410008106001001.

Desmodilliscus Wettstein, 1916. Anz. Akad. Wiss. Wien, 53(14):153.
REVIEWED BY: C. B. Robbins (CBR); C. J. Terry (CJT).
COMMENT: Subfamily Gerbillinae; see comment under Cricetidae.
ISIS NUMBER: 5301410008089000000.

Desmodilliscus braueri Wettstein, 1916. Anz. Akad. Wiss. Wien, 53(14):153.
TYPE LOCALITY: Sudan, south of El Obeid.
DISTRIBUTION: Sudan; N. Nigeria; Senegal; S. Mauritania; Upper Volta; S. Niger.
COMMENT: Reviewed by Setzer, 1969, Misc. Publ. Univ. Kans. Mus. Nat. Hist.,
51:238–288.
ISIS NUMBER: 5301410008089001001.

Desmodillus Thomas and Schwann, 1904. Abstr. Proc. Zool. Soc. Lond., 2:6.
REVIEWED BY: C. B. Robbins (CBR); C. J. Terry (CJT).
COMMENT: Subfamily Gerbillinae; see comment under Cricetidae.
ISIS NUMBER: 5301410008090000000.

Desmodillus auricularis (A. Smith, 1834). S. Afr. Quart. J., 2:160.
TYPE LOCALITY: South Africa, Mtns. of Little Namaqualand.
DISTRIBUTION: South Africa; Namibia; Botswana.
COMMENT: Includes *caffer*; see Ellerman *et al.*, 1953, South African Mammals..., p. 315.
ISIS NUMBER: 5301410008090001001 as *D. auricularis*.
5301410008098005001 as *Tatera caffer*.

Dipodillus Lataste, 1881. Le Naturaliste, 3(64):506.
REVIEWED BY: D. M. Lay (DML); C. B. Robbins (CBR); C. J. Terry (CJT).
COMMENT: Reviewed by Cockrum *et al.*, 1976, Mammalia, 40:313–326, and Osborn and
Helmy, 1980, Fieldiana Zool., n.s., 5:1–579, who both considered *Dipodillus* a
distinct genus but differed regarding its content; the treatment here follows
Cockrum *et al.*, 1976, Mammalia, 40:313–326; See comment under *Gerbillus*.
Considered a subgenus of *Gerbillus* by Schlitter, 1976, Diss. Abstr. Int., 37(6).
There is no consensus regarding the content of *Dipodillus*; see also Ellerman, 1941,
Allen, 1939, Bull. Mus. Comp. Zool. Harv. Univ., 83:319, Petter, 1959, Mammalia,
23:308, and Harrison, 1972, The Mammals of Arabia, 3:382–670. This genus is in
need of comprehensive revision (JH, DML). Subfamily Gerbillinae; see comment
under Cricetidae.

Dipodillus maghrebi Schlitter and Setzer, 1972. Proc. Biol. Soc. Wash., 84(45):387.
TYPE LOCALITY: Morocco, Fes Prov., 15 km W.S.W. Taounate, 34° 29' N., 4° 48' W.
DISTRIBUTION: Known only from the type locality.

Dipodillus simoni (Lataste, 1881). Le Naturaliste, 1:499.
TYPE LOCALITY: Algeria, N. of Hodna, Oued-Magra.
DISTRIBUTION: N.E. Algeria; Tunisia; coastal plains of Egypt and Libya; perhaps
Morocco.
COMMENT: Includes *kaiseri*; see Petter, 1975, Part 6.3:2, Cockrum *et al.*, 1976,
Mammalia, 40:313–326, and Osborn and Helmy, 1980, Fieldiana Zool., n.s., 5:161.
Also see Corbet, 1978:123, who listed *kaiseri* as a distinct species; *simoni* and *kaiseri*
have disjunct distributions. May include *zakariai*; see Cockrum *et al.*, 1976,
Mammalia, 40:313–326 (DML).
ISIS NUMBER: 5301410008091034001 as *Gerbillus simoni*.

Dipodillus zakariai Cockrum, Vaughan, and Vaughan, 1976. Mammalia, 40:320.
TYPE LOCALITY: Tunisia, Kerkennah Isls., 500 m. east of Kellabine, about 42 km east of
Sfax.
DISTRIBUTION: Kerkennah Isl. (Tunisia).
COMMENT: May be conspecific with *simoni* (DML).

Eligmodontia F. Cuvier, 1837. Ann. Sci. Nat. Paris, ser. 2, 7:169.
REVIEWED BY: A. Langguth (AL); M. A. Mares (MAM); R. A. Ojeda (RAO); C. J. Terry (CJT); D. F. Williams (DFW).
COMMENT: Revised by Hershkovitz, 1962, Fieldiana Zool., 46:1–524. Karyology reviewed by Pearson and Patton, 1976, J. Mammal., 57:339–350. Subfamily Hesperomyinae; see comment under Cricetidae.
ISIS NUMBER: 5301410008013000000.

Eligmodontia typus F. Cuvier, 1837. Ann. Sci. Nat. Paris, ser. 2, 7:169.
TYPE LOCALITY: Argentina, Corrientes Prov.
DISTRIBUTION: Argentina; N. Chile; W. Bolivia; S. Peru.
COMMENT: Includes *elegans* and *puerulus;* see Hershkovitz, 1962, Fieldiana Zool., 46:186–187, and Pine *et al.,* 1979, Mammalia, 43:352; but also see Reise, 1973, Guyana Inst. Biol. Univ. Concepcion, 27:1–20, and Corbet and Hill, 1980:153, who listed *puerulus* as a distinct species; *puerulus* is probably a separate species (DFW). Includes *hypogaeus* which may be a composite; see Massoia, 1976–1977, Rev. Invest. Agro. INTA, ser. 5, Patalogia Vegetal, 13:15–20, and Williams and Mares, 1978, Ann. Carnegie Mus., 47(9):201, 218; also see Pearson and Patton, 1976, J. Mammal., 57:339–350, who listed *hypogaeus* as a distinct species of *Graomys,* Cabrera, 1961:495, who included *hypogaeus* in *Phyllotis (Graomys) griseoflavus,* and Hershkovitz, 1962, Fieldiana Zool., 46:462–463, who regarded it as a distinct species of *Phyllotis.* Type locality is uncertain; see Hershkovitz, 1962, Fieldiana Zool., 46:185.
ISIS NUMBER: 5301410008013001001.

Eliurus Milne-Edwards, 1885. Ann. Sci. Nat. Paris, 63, Ser. Zool., 20(1):1.
REVIEWED BY: C. J. Terry (CJT).
COMMENT: Subfamily Nesomyinae; see comment under Cricetidae.
ISIS NUMBER: 5301410008062000000.

Eliurus minor Forsyth Major, 1896. Ann. Mag. Nat. Hist., ser. 6, 18:462.
TYPE LOCALITY: Madagascar, N.E. Betsileo, Ampitambe Forest.
DISTRIBUTION: N.E. Betsileo Country (E. Madagascar).
ISIS NUMBER: 5301410008062002001.

Eliurus myoxinus Milne-Edwards, 1885. Ann. Sci. Nat. Paris, 63, Ser. Zool., 20(1):1.
TYPE LOCALITY: Madagascar, West coast.
DISTRIBUTION: Madagascar.
COMMENT: Includes *majori* and *tanala;* see Petter, 1975, Part 6.2:3.
ISIS NUMBER: 5301410008062003001 as *E. myoxinus.*
5301410008062001001 as *E. majori.*
5301410008062004001 as *E. tanala.*

Euneomys Coues, 1874. Proc. Acad. Nat. Sci. Phila., 26:185.
REVIEWED BY: C. J. Terry (CJT).
COMMENT: Reviewed by Hershkovitz, 1962, Fieldiana Zool., 46:492–502, who included *Chelemyscus* in this genus. Subfamily Hesperomyinae; see comment under Cricetidae.
ISIS NUMBER: 5301410008014000000.

Euneomys chinchilloides (Waterhouse, 1839). Zool. Voy. H.M.S. "Beagle," Mammalia, p. 72.
TYPE LOCALITY: Chile, Tierra del Fuego, Straits of Magellan, near E. entrance.
DISTRIBUTION: S. Chile; S. Argentina.
COMMENT: Includes *petersoni;* see Hershkovitz, 1962, Fieldiana Zool., 46:499.
ISIS NUMBER: 5301410008014001001 as *E. chinchilloides.*
5301410008014004001 as *E. petersoni.*

Euneomys fossor (Thomas, 1899). Ann. Mag. Nat. Hist., ser. 7, 4:280.
TYPE LOCALITY: Argentina, Salta Prov.
DISTRIBUTION: N.W. Argentina.
COMMENT: Known from a single specimen which may be a composite; see Hershkovitz, 1962, Fieldiana Zool., 46:498, 500.
ISIS NUMBER: 5301410008014002001.

Euneomys mordax Thomas, 1912. Ann. Mag. Nat. Hist., ser. 8, 10:410.
TYPE LOCALITY: Argentina, Mendoza, San Rafael.
DISTRIBUTION: W.C. Argentina; adjacent Chile.
COMMENT: Hershkovitz, 1962, Fieldiana Zool., 46:499–502, suggested that *noei, mordax,* and *chinchilloides* may be conspecific.
ISIS NUMBER: 5301410008014003001.

Euneomys noei Mann, 1944. Biologica, 1:95.
TYPE LOCALITY: Chile, Santiago, Valle de la Junta, Canyon of the Rio Volcan, 2400 m.
DISTRIBUTION: Chile.
COMMENT: Hershkovitz, 1962, Fieldiana Zool., 46:499–500, listed this form as a distinct species but doubted that it was separable from *mordax*. Reviewed by Pine *et al.,* 1979, Mammalia, 43:358.

Galenomys Thomas, 1916. Ann. Mag. Nat. Hist., ser. 8, 17:143.
REVIEWED BY: O. P. Pearson (OPP); J. J. Pizzimenti (JJP); D. F. Williams (DFW).
COMMENT: Revised by Hershkovitz, 1962, Fieldiana Zool., 46:464, who considered *Galenomys* a distinct genus, as did Pearson, 1958, Univ. Calif. Publ. Zool., 56(4):393, 395. Pearson and Patton, 1976, J. Mammal., 57:339, listed it as a subgenus of *Phyllotis*, without discussion. Subfamily Hesperomyinae; see comment under Cricetidae.
ISIS NUMBER: 5301410008015000000.

Galenomys garleppi (Thomas, 1898). Ann. Mag. Nat. Hist., ser. 7, 1:279.
TYPE LOCALITY: Bolivia, La Paz Dept., Esperanza, northeast of Mt. Sajama, 4140 m.
DISTRIBUTION: Extreme N. Chile and adjacent Bolivia and Peru.
COMMENT: Listed in subgenus *Galenomys* of *Phyllotis* by Pearson and Patton, 1976, J. Mammal., 57:339, without comment; see Hershkovitz, 1962, Fieldiana Zool., 46:468.
ISIS NUMBER: 5301410008015001001.

Gerbillurus Shortridge, 1942. Ann. S. Afr. Mus., 36(1):27–100.
REVIEWED BY: C. B. Robbins (CBR); C. J. Terry (CJT).
COMMENT: *Gerbillurus* was considered a subgenus of *Gerbillus* by Shortridge, 1942, Ann. S. Afr. Mus., 36(1):27–100, and Ellerman *et al.,* 1953, Southern Afr. Mammals ... A Reclassification, Trustees Brit. Mus. Nat. Hist., 363 pp. However, Roberts, 1951, The Mammals of South Africa, 700 pp., Lundholm, 1955, Ann. Transvaal Mus., 22:279–303, Schlitter, 1976, Diss. Abstr. Int., 37(6), Davis, 1975, Part 6.4:5, and Swanepoel *et al.,* 1980, Ann. Transvaal Mus., 32(7):175, considered it a distinct genus. Davis, 1975, Part 6.4:5, also suggested that *paeba* and *vallinus* may not be congeneric. Subfamily Gerbillinae; see comment under Cricetidae.

Gerbillurus paeba (A. Smith, 1836). Rep. Expl. Int. S. Afr., app., p. 43.
TYPE LOCALITY: South Africa, Vryberg.
DISTRIBUTION: South Africa to S.W. Angola and Mozambique.
COMMENT: Includes *calidus* and *swalius;* see Davis, 1975, Part 6.4:5. Type locality restricted by Roberts, 1951, The Mammals of South Africa, p. 26.
ISIS NUMBER: 5301410008091029001 as *Gerbillus paeba.*
5301410008091006001 as *Gerbillus calidus.*
5301410008091036001 as *Gerbillus swalius.*

Gerbillurus setzeri (Schlitter, 1973). Bull. South. Calif. Acad. Sci., 72(1):13–18.
TYPE LOCALITY: Namibia, Gobabeb, 1 mi. (1.6 km) east of Namib Desert Research Station.
DISTRIBUTION: Namibia.

Gerbillurus tytonis (Bauer and Niethammer, 1960). Bonn. Zool. Beitr., (1959), 10:236–260.
TYPE LOCALITY: Namibia, Sossus Vlei.
DISTRIBUTION: Namibia.
COMMENT: Formerly included in *vallinus;* see Schlitter, 1973, Bull. South. Calif. Acad. Sci., 72(1):13–18, and Davis, 1975, Part 6.4:5, who considered it a distinct species.

Gerbillurus vallinus (Thomas, 1918). Ann. Mag. Nat. Hist., ser. 9, 2:148.
    TYPE LOCALITY: South Africa, Bushmanland, Kenhart, Tuin.
    DISTRIBUTION: S.W. Angola; Namibia; Cape Prov. (South Africa).
    COMMENT: Formerly included *tytonis*; see Schlitter, 1973, Bull. South. Calif. Acad. Sci.,
      72(1):17 and Davis, 1975, Part 6.4:5.
    ISIS NUMBER: 5301410008091037001 as *Gerbillus vallinus.*

Gerbillus Desmarest, 1804. Nouv. Dict. Hist. Nat. Paris, 1:24, Tabl. Meth., p. 22.
    REVIEWED BY: E. L. Cockrum (ELC); D. L. Harrison (DLH); D. M. Lay (DML); C. B.
      Robbins (CBR).
    COMMENT: Includes *Monodia*; formerly included *Gerbillurus* and *Microdillus*; see Petter,
      1975, Part 6.3:11, 12, Davis, 1975, Part 6.4:5, Schlitter, 1976, Unpubl. Ph.D.
      Dissertation Univ. Maryland, 558 pp., and Schlitter, 1976, Diss. Abstr. Int., 37(6).
      Schlitter, in both 1976 treatments, included *Dipodillus* as a subgenus of *Gerbillus.*
      Petter, 1959, Mammalia, 23:308, Petter, 1975, Part 6.3:2, and Cockrum *et al.*, 1976,
      Mammalia, 40:313–326, considered *Dipodillus* a distinct genus. Wassif, 1956, Ain
      Shams Sci. Bull., No. 1:173–194 and Osborn and Helmy, 1980, Fieldiana Zool.,
      n.s., 5:1–579, regarded *Dipodillus* as a distinct genus, but in a different sense from
      Petter, 1959, and Cockrum *et al.*, 1976, who are followed here; revision and clear
      definition of *Hendecapleura*, *Gerbillus*, and *Dipodillus* (*sensu* Osborn and Helmy,
      1980) is needed; see also Ellerman, 1941:500–504. If *Dipodillus* is considered a
      distinct genus then the other naked-footed forms of *Gerbillus* should also be
      considered a distinct genus; the name *Hendecapleura* is available (ELC). This
      genus and related forms are in need of revision. Subfamily Gerbillinae; see
      comment under Cricetidae.
    ISIS NUMBER: 5301410008091000000.

Gerbillus agag Thomas, 1903. Proc. Zool. Soc. Lond., 1903(1):296.
    TYPE LOCALITY: Sudan, W. Kordofan, Agageh Wells.
    DISTRIBUTION: N. Nigeria, Mali, and Niger to Chad, Sudan, and Kenya.
    COMMENT: Includes *cosensi*, *nigeriae*, and *sudanensis*; see Kock, 1978, Senckenberg. Biol.,
      58:127, who also included *rosalinda* in this species; but see Petter, 1975, Part 6.3:9,
      Schlitter, 1976, Diss. Abstr. Int., 37(6), and Corbet and Hill, 1980:164, who
      considered *rosalinda* a distinct species. Petter, 1975, Part 6.3:9, considered *cosensi* as
      possibly distinct and included *dalloni* in *agag*, but Kock, 1978, Senckenberg. Biol.,
      58:126, included *dalloni* in *gerbillus*. DLH and DML followed Schlitter, 1976, who
      considered *nigeriae* a distinct species, as did Tranier, 1975, Mammalia, 39:703–704;
      DML also considers *cosensi* distinct pending revision.
    ISIS NUMBER: 5301410008091009001 as *G. cosensi.*
              5301410008091027001 as *G. nigeriae.*

Gerbillus amoenus (De Winton, 1902). Ann. Mag. Nat. Hist., ser. 7, 9:46.
    TYPE LOCALITY: Egypt, Giza Prov.
    DISTRIBUTION: Egypt and Libya, possibly across Tunisia and Algeria to Mauritania.
    COMMENT: Wassif, 1956, Ain Shams Sci. Bull., No. 1:185, and Ranck, 1968, Bull. U.S.
      Nat. Mus., 275:129, regarded *amoenus* as distinct. Reviewed by Osborn and
      Helmy, 1980, Fieldiana Zool., n.s., 5:167, and regarded in genus *Dipodillus* and as
      distinct from *nanus* and *dasyurus*; formerly included in *campestris* by Petter, 1975,
      Part 6.3:11, and in *nanus* by Corbet, 1978:120; see comment under *Gerbillus.*
    ISIS NUMBER: 5301410008091002001.

Gerbillus andersoni De Winton, 1902. Ann. Mag. Nat. Hist., ser. 7, 9:45.
    TYPE LOCALITY: Egypt, E. Alexandria, Mandara.
    DISTRIBUTION: Jordan; Israel; Sinai; N. Egypt; Libya; Tunisia.
    COMMENT: Includes *allenbyi*, *eatoni*, *inflatus*, and *versicolor*; see Cockrum *et al.*, 1976,
      Mammalia, 40:470. Includes *bonhotei*; see Osborn and Helmy, 1980, Fieldiana
      Zool., n.s., 5:129. Harrison, 1972, The Mammals of Arabia, 3:536, considered
      allenbyi a distinct species.
    ISIS NUMBER: 5301410008091003001 as *G. andersoni.*
              5301410008091013001 as *G. eatoni.*

Gerbillus aquilus Schlitter and Setzer, 1973. Proc. Biol. Soc. Wash., 86:167.
TYPE LOCALITY: Iran, 60 km west of Kerman.
DISTRIBUTION: S.E. Iran; W. Pakistan; S. Afghanistan.
COMMENT: Includes *subsolanus* as a junior synonym; formerly this species was included in *cheesmani* by Schlitter and Setzer, 1973; but see Lay and Nadler, 1975, Mammalia, 39:437, 440; also see Corbet, 1978:123.

Gerbillus bottai Lataste, 1882. Le Naturaliste, 4(5):36.
TYPE LOCALITY: Sudan, Sennar.
DISTRIBUTION: Sudan; Kenya.
COMMENT: Includes *harwoodi* and *luteolus*; see Kock, 1978, Bull. Carnegie Mus. Nat. Hist., 6:31–37. Petter, 1975, Part 6.3:10–11, considered *harwoodi* a distinct species and included *luteolus* in *campestris*. This species is a composite; the holotype of *harwoodi* has no accessory tympanum, whereas that of *luteolus* has an accessory tympanum, thus *harwoodi* and *luteolus* are regarded as distinct pending revision (DML). Roche, 1975, Monitore Zool. Ital., suppl. 6:263–268, considered *harwoodi* as probably distinct from *pusillus*. DML includes this species in genus *Dipodillus*; but see comment under *Gerbillus*.
ISIS NUMBER: 5301410008091005001 as *G. bottai*.
5301410008091018001 as *G. harwoodi*.

Gerbillus campestris Le Vaillant, 1857. Atlas Expl. Sci. Alg. Mamm., pl. V, fig. 2.
TYPE LOCALITY: Algeria, Constantine Prov., Philipeville.
DISTRIBUTION: Morocco to Libya; Sudan; N. Somalia.
COMMENT: Included in genus *Dipodillus* by Osborn and Helmy, 1980, Fieldiana Zool., n.s., 5:141–154; but see comment under *Gerbillus*. Includes *somalicus*, *quadrimaculatus*, *lowei*, and *stigmonyx*; see Petter, 1975, Part 6.3:11, who also included *amoenus* in this species but noted that *amoenus* might be a distinct species. Yalden *et al.*, 1976, Monitore Zool. Ital., suppl. 8:27, discussed *somalicus*. Formerly included *luteolus* (here included in *bottai*; see Kock, 1978, Bull. Carnegie Mus. Nat. Hist., 6:31–37, who also considered *stigmonyx* a *nomen dubium* and *somalicus* a subspecies of *campestris*) and *amoenus* (which was considered a distinct species by Osborn and Helmy, 1980, Fieldiana Zool., n.s., 5:167). Corbet, 1978:120, tentatively included *amoenus* in *nanus* and listed *jamesi* (without comment) in *campestris*; but see Petter, 1975, Part 6.3:11, who listed *jamesi* as a distinct species. Includes *dodsoni, haymani, patrizii, wassifi,* and *venustus*; see Osborn and Helmy, 1980, Fieldiana Zool., n.s., 5:141–155. DML regards *lowei* and *somalicus* as separate species pending revision.
ISIS NUMBER: 5301410008091007001 as *G. campestris*.
5301410008091035001 as *G. stigmonyx*.

Gerbillus cheesmani Thomas, 1919. J. Bombay Nat. Hist. Soc., 26:748.
TYPE LOCALITY: Iraq, Lower Euphrates, near Basra.
DISTRIBUTION: S.W. Iran; Iraq; Arabia; North Yemen; South Yemen; Oman.
COMMENT: Formerly included *aquilus*; see Lay and Nadler, 1975, Mammalia, 39:423–445, and Corbet, 1978:123. Includes *arduus*; see Ellerman and Morrison-Scott, 1951:636, Harrison, 1972, The Mammals of Arabia, 3:548, and Schlitter, 1976, Unpubl. Ph.D. Dissertation, Univ. Maryland, 558 pp. Petter, 1975, Part 6.3:9, included *cheesmani* in *gerbillus* but see Corbet, 1978:123, Schlitter, 1976, Diss. Abstr. Int., 37(6), and Harrison, 1972, The Mammals of Arabia, 3:546, who considered *cheesmani* a distinct species.
ISIS NUMBER: 5301410008091008001.

Gerbillus dasyurus (Wagner, 1842). Arch. Naturgesch., 8(1):20.
TYPE LOCALITY: Egypt, Sinai.
DISTRIBUTION: W. Iraq to Israel and Sinai; Arabian Peninsula.
COMMENT: Formerly included *mesopotamiae*; see Harrison, 1972, The Mammals of Arabia, 3:532, Lay and Nadler, 1975, Mammalia, 39:423–445, and Osborn and Helmy, 1980, Fieldiana Zool., n.s., 5:155, who included *dasyurus* in genus *Dipodillus*; but see comment under *Gerbillus*.
ISIS NUMBER: 5301410008091011001.

Gerbillus dunni Thomas, 1904. Ann. Mag. Nat. Hist., ser. 7, 14:101.
  TYPE LOCALITY: Somalia (Somaliland), Gerlogobi.
  DISTRIBUTION: Ethiopia; Somalia; Djibouti.
  COMMENT: May be conspecific with *latastei;* see Cockrum, 1977, Mammalia, 41:78, and
    Schlitter, 1976, Diss. Abstr. Int., 37(6). Yalden, 1976, Monitore Zool. Ital., suppl.
    8:25, included *dunni* in *pyramidum.*
  ISIS NUMBER: 5301410008091012001.

Gerbillus famulus Yerbury and Thomas, 1895. Proc. Zool. Soc. Lond., 1895:551.
  TYPE LOCALITY: South Yemen, Aden, Lahej.
  DISTRIBUTION: South Yemen, North Yemen.
  COMMENT: Formerly included in *nanus* by Petter, 1975, Part 6.3:10, but considered a
    distinct species by Harrison, 1972, The Mammals of Arabia, 3:513, and Corbet,
    1978:120. DML includes this species in genus *Dipodillus;* but see comment under
    Gerbillus.
  ISIS NUMBER: 5301410008091014001.

Gerbillus gerbillus (Olivier, 1801). Bull. Sci. Soc. Philom. Paris, 2:121.
  TYPE LOCALITY: Egypt, Giza Prov.
  DISTRIBUTION: Israel to Morocco and Mauritania; Chad.
  COMMENT: Includes *foleyi* and *longicaudus* (see Petter, 1975, Part 6.3:9), *dalloni* (see
    Kock, 1978, Senckenberg. Biol., 58:126), *hirtipes* which was formerly included in
    pyramidum by Petter, 1975, Part 6.3:8 (see Cockrum, 1976, Mammalia, 40:523–524,
    and Corbet, 1978:121). Lay *et al.*, 1975, Z. Saugetierk., 40:148, suggested that
    hirtipes may be a distinct species. Schlitter, 1976, Unpubl. Ph.D. Dissertation,
    Univ. Maryland, 558 pp., included *hirtipes* in *pyramidum.* Formerly included
    latastei; see Cockrum, 1977, Mammalia, 41:78. Formerly included *cheesmani,*
    gleadowi, and *sudanensis;* see Corbet, 1978:121–122, Schlitter, 1976, Diss. Abstr. Int.,
    37(6), Harrison, 1972, The Mammals of Arabia, 3:546, and Kock, 1978,
    Senckenberg. Biol., 58(3):113. DML considers *dalloni* and *hirtipes* distinct species
    pending revision.
  ISIS NUMBER: 5301410008091016001 as *G. gerbillus.*
               5301410008091010001 as *G. dallonii (sic).*
               5301410008091015001 as *G. foleyi.*
               5301410008091021001 as *G. longicaudus.*

Gerbillus gleadowi Murray, 1886. Ann. Mag. Nat. Hist., 17:246.
  TYPE LOCALITY: Pakistan (Upper Sin), Rohri Dist., Mirpur-Drahrki Taluka, 15 mi. (24
    km) southwest of Rehti, Beruto.
  DISTRIBUTION: N.W. India, Pakistan.
  COMMENT: Formerly included in *gerbillus* by Petter, 1975, Part 6.3:9; Corbet, 1978:122,
    Schlitter, 1976, Diss. Abstr. Int., 37(6), and Lay and Nadler, 1975, Mammalia,
    39:423–443, considered *gleadowi* a distinct species.
  ISIS NUMBER: 5301410008091017001.

Gerbillus henleyi (De Winton, 1903). Novit. Zool., 10:284.
  TYPE LOCALITY: Lower Egypt, Wadi Natron, Zaghig.
  DISTRIBUTION: Algeria to Israel and Jordan; North Yemen.
  COMMENT: Includes *mariae* and *jordani;* see Ellerman and Morrison-Scott, 1951:634,
    Kock, 1978, Bull. Carnegie Mus. Nat. Hist., 6:31–37, and Osborn and Helmy, 1980,
    Fieldiana Zool., n.s., 5:179–181, who included *henleyi* in genus *Dipodillus;* but see
    comment under *Gerbillus.* See Bahmangar and Lay, 1975, Mammalia, 39:323, for
    discussion of distribution.
  ISIS NUMBER: 5301410008091019001.

Gerbillus hesperinus Cabrera, 1936. Bol. Real. Soc. Esp. Hist. Nat., p. 365.
  TYPE LOCALITY: Morocco, Mogador (=Essouira).
  DISTRIBUTION: Coastal Morocco north of Middle Atlas Mtns.
  COMMENT: Corbet, 1978:122, included *hesperinus* in *pyramidum,* and Schlitter, 1976,
    Unpubl. Ph.D. Dissertation, Univ. Maryland, 558 pp., included it in *riggenbachi;*
    but see Lay, 1975, Fieldiana Zool., 65:95–98, and Benazzou and Genest-Villard,
    1980, Mammalia, 44:410, who considered *hesperinus* a distinct species.

Gerbillus hoogstraali Lay, 1975. Fieldiana Zool., 65:90.
TYPE LOCALITY: Morocco, 7 km S. Taroudannt.
DISTRIBUTION: Known only from the type locality.

Gerbillus jamesi Harrison, 1967. Mammalia, 31(3):383.
TYPE LOCALITY: Tunisia, between Bou Ficha and Enfidaville.
DISTRIBUTION: Tunisia.
COMMENT: Corbet, 1978:120, included *jamesi* in *campestris*, without comment; Petter, 1975, Part 6.3:11, considered *jamesi* a distinct species. DML includes *jamesi* in genus *Dipodillus*; but see comment under *Gerbillus*.

Gerbillus latastei Thomas and Trouessart, 1903. Bull. Soc. Zool. Fr., 28:172.
TYPE LOCALITY: Tunisia, Kebili.
DISTRIBUTION: Tunisia to Sinai and Somalia.
COMMENT: Includes *aureus* (sensu Ranck, 1968, The Rodents of Libya, Bull. U.S. Nat. Mus., 275:1–264), *favillus*, and *nalutensis*; may include *bonhotei*, *dunni*, *perpallidus*, *riggenbachi*, and *rosalinda*; see Cockrum, 1977, Mammalia, 41:78. Included as a synonym of *pyramidum* by Petter, 1975, Part 6.3:8, who considered *aureus* a subspecies of *pyramidum*. Corbet, 1978:121, 122, included *latastei* in *gerbillus* and considered *aureus* a distinct species but did not cite Cockrum, 1977. Schlitter, 1976, Diss. Abstr. Int., 37(6), listed *aureus* as a distinct species; Schlitter, 1976, Unpubl. Ph.D. Dissertation, Univ. Maryland, 558 pp., included *latastei* in *pyramidum*. DLH regards *latastei* as a species but considers the status of *aureus* and *latastei* uncertain.
ISIS NUMBER: 5301410008091020001.

Gerbillus mackilligini Thomas, 1904. Ann. Mag. Nat. Hist., ser. 7, 14:158.
TYPE LOCALITY: Egypt, Wadi Allaqi, E. desert of Nubia, ca. 22° N., 35° E.
DISTRIBUTION: S. Egypt, in E. desert; probably adjacent Sudan.
COMMENT: Placed in *nanus* by Petter, 1975, Part 6.3:10, who considered it possibly distinct; considered a distinct species in genus *Dipodillus* by Osborn and Helmy, 1980, Fieldiana Zool., n.s., 5:159; but see comments under *Gerbillus* and *G. nanus*.
ISIS NUMBER: 5301410008091022001.

Gerbillus mauritaniae (Heim de Balsac, 1943). Bull. Mus. Hist. Nat. Paris, 15:287.
TYPE LOCALITY: Mauritania, Aouker Region, S. of Archane Titarek.
DISTRIBUTION: Known only from the type locality.
COMMENT: Formerly included in *Monodia*; see Petter, 1975, Part 6.3:11; this species is known only by the holotype. DML includes this species in genus *Dipodillus*; but see comment under *Gerbillus*.
ISIS NUMBER: 5301410008093001001 as *Monodia mauritaniae*.

Gerbillus mesopotamiae Harrison, 1956. J. Mammal., 37:417.
TYPE LOCALITY: Iraq, southwest of Faluja, W. bank of Euphrates River, near Amiriya.
DISTRIBUTION: Valleys of the Tigris, Euphrates, and Karun Rivers (Iraq; S.W. Iran).
COMMENT: Formerly included in *dasyurus*; see Harrison, 1972, The Mammals of Arabia, 3:532. Reviewed by Lay and Nadler, 1975, Mammalia, 39:423–445. DML includes *mesopotamiae* in genus *Dipodillus*; but see comment under *Gerbillus*.

Gerbillus muriculus Thomas and Hinton, 1923. Proc. Zool. Soc. Lond., 1923:263.
TYPE LOCALITY: Sudan, Darfur, 80 mi. (129 km) northeast of El Fasher, Madu.
DISTRIBUTION: Darfur (Sudan).
COMMENT: Petter, 1975, Part 6.3:10, included this species in *nanus*, but on page 12 listed it as a distinct species. Corbet and Hill, 1980:164, listed *muriculus* as a distinct species, as do DLH and DML. DML includes *muriculus* in genus *Dipodillus*; but see comment under *Gerbillus*.
ISIS NUMBER: 5301410008091024001.

Gerbillus nancillus Thomas and Hinton, 1923. Proc. Zool. Soc. Lond., 1923:260.
TYPE LOCALITY: Sudan, Darfur, 45 mi. (72 km) north of El Fasher.
DISTRIBUTION: Sudan.
ISIS NUMBER: 5301410008091025001.

Gerbillus nanus Blanford, 1875. Ann. Mag. Nat. Hist., 16:312.
   TYPE LOCALITY: Pakistan, Baluchistan, west of Gwadar, Gedrosia.
   DISTRIBUTION: Baluchistan to the Arabian Peninsula and Israel to Algeria; Somalia.
   COMMENT: Included in genus *Dipodillus* by Osborn and Helmy, 1980, Fieldiana Zool.,
      n.s., 5:167; but see comment under *Gerbillus*. Includes *brockmani, garamantis,*
      principulus, and perhaps *muriculus* (treated here as distinct); see Petter, 1975, Part
      6.3:10, 12, and Kock, 1978, Bull. Carnegie Mus. Nat. Hist., 6:34, who also included
      mackilligini in *nanus.* Corbet, 1978:120, included *amoenus* and *quadrimaculatus* in
      nanus; but see Petter, 1975, Part 6.3:11, who included *amoenus* and *quadrimaculatus*
      in *campestris,* and Osborn and Helmy, 1980, Fieldiana Zool., n.s., 5:167 (followed
      here), who considered *amoenus* and *mackilligini* distinct species. Includes *arabium,*
      indus, and *lixa;* see Ellerman and Morrison-Scott, 1951:633, Lay *et al.,* 1975, Z.
      Saugetierk., 40:141–150, and Harrison, 1972, The Mammals of Arabia, 3:522.
      Yalden *et al.,* 1976, Monitore Zool. Ital., suppl. 8:27, discussed *brockmani.* May
      include *grobbeni, hilda,* and *quadrimaculatus* (DML), who regards *brockmani* and
      garamantis as distinct species pending revision.
   ISIS NUMBER: 5301410008091026001.

Gerbillus occiduus Lay, 1975. Fieldiana Zool., 65:94.
   TYPE LOCALITY: Morocco, Aoreora, 80 km W.S.W. Goulimine.
   DISTRIBUTION: Known only from the type locality.

Gerbillus perpallidus Setzer, 1958. J. Egypt Public Health. Assoc., 33:221.
   TYPE LOCALITY: Egypt, Bir Victoria.
   DISTRIBUTION: N. Egypt, west of the Nile.
   COMMENT: Petter, 1975, Part 6.3:8, included *perpallidus* in *pyramidum,* but Corbet,
      1978:122, and Schlitter, 1976, Diss. Abstr. Int., 37(6), considered *perpallidus* a
      distinct species. May be conspecific with *latastei;* see Cockrum, 1977, Mammalia,
      41:78. Karyotypically distinct from *latastei* in the sense of Cockrum, 1977 (DML);
      see Jordan, 1974, Mammalia, 38:62, and Lay *et al.,* 1975, Z. Saugetierk., 40:141–150.

Gerbillus poecilops Yerbury and Thomas, 1895. Proc. Zool. Soc. Lond., 1895:549.
   TYPE LOCALITY: South Yemen, Aden, Lahej.
   DISTRIBUTION: South Yemen; North Yemen; Saudi Arabia.
   COMMENT: DML includes this species in genus *Dipodillus;* but see comment under
      Gerbillus.
   ISIS NUMBER: 5301410008091030001.

Gerbillus pulvinatus Rhoads, 1896. Proc. Acad. Nat. Sci. Phila., p. 537.
   TYPE LOCALITY: Ethiopia, Lake Rudolf, Rusia.
   DISTRIBUTION: Ethiopia.
   COMMENT: Includes *bilensis;* see Petter, 1975, Part 6.3:8; also see Yalden *et al.,* 1976,
      Monitore Zool. Ital., suppl. 8:25, who included *bilensis* and *pulvinatus* in
      pyramidum; but see Benazzou and Genest-Villard, 1980, Mammalia, 44:412. DML
      regards *bilensis* as a distinct species pending revision.
   ISIS NUMBER: 5301410008091031001 as *G. pulvinatus.*
              5301410008091004001 as *G. bilensis.*

Gerbillus pusillus Peters, 1878. Monatsb. Preuss. Akad. Wiss. Berlin, p. 201.
   TYPE LOCALITY: Kenya, Ndi and Kitui.
   DISTRIBUTION: Kenya.
   COMMENT: Includes *diminutus* and *percivali;* see Petter, 1975, Part 6.3:10. DML regards
      diminutus and *percivali* as distinct species pending revision. Roche, 1975, Monitore
      Zool. Ital., suppl. 6:263–268, included *ruberrimus* in this species; but see Yalden *et*
      al., 1976, Monitore Zool. Ital., suppl. 8:26, and Rupp, 1980, Saugetierk. Mitt.,
      28:87, who considered *ruberrimus* a distinct species, and are followed here. DML
      included *pusillus* in genus *Dipodillus;* but see comment under *Gerbillus.*

Gerbillus pyramidum I. Geoffroy, 1825. Dict. Class. Hist. Nat., 7:321.
   TYPE LOCALITY: Egypt, Giza Prov.
   DISTRIBUTION: Jordan and Israel to Algeria and N. Somalia; Chad; Senegal.
   COMMENT: Includes *floweri;* see Osborn and Helmy, 1980, Fieldiana Zool., n.s., 5:113.

Includes *acticola* and *dongolanus*; see Petter, 1975, Part 6.3:8, who also listed *hirtipes* as a synonym; *acticola* and *dongolanus* are considered distinct by DML pending revision. Lay *et al.*, 1975, Z. Saugetierk., 40:148, suggested that *hirtipes* may be a distinct species; Corbet, 1978:121, included *hirtipes* in *gerbillus* and is followed here; see also comment under *gerbillus*. Formerly included *Taterillus pygargus* (see Petter, 1975, Part 6.3:5 and Petter, 1952, Mammalia, 16:37), *latastei, aureus* (see Cockrum, 1977, Mammalia, 41:78), *perpallidus, riggenbachi* (see Lay *et al.*, 1975, Z. Saugetierk., 40:148, Schlitter, 1976, Diss. Abstr. Int., 37(6), and Corbet and Hill, 1980:164), and *hesperinus* (see Lay, 1975, Fieldiana Zool., 65:95–98, and Benazzou and Genest-Villard, 1980, Mammalia, 44:410). Corbet, 1978:122, included *riggenbachi* and *hesperinus* in *pyramidum*, but considered *perpallidus* a distinct species. See Hubert, 1978, Bull. Carnegie Mus. Nat. Hist., 6:38–40, for discussion of distribution.

ISIS NUMBER: 5301410008091032001 as *G. pyramidum*.
5301410008091001001 as *G. acticola*.

Gerbillus riggenbachi Thomas, 1903. Novit. Zool., 10:301.
TYPE LOCALITY: Western Sahara (Rio de Oro).
DISTRIBUTION: Western Sahara.
COMMENT: Considered a distinct species by Schlitter, 1976, Diss. Abstr. Int., 37(6), Lay *et al.*, 1975, Z. Saugetierk., 40:148, and Corbet and Hill, 1980:164; Corbet, 1978:122, and Petter, 1975, Part 6.3:8, included *riggenbachi* in *pyramidum*, without comment. May be conspecific with *latastei*; see Cockrum, 1977, Mammalia, 41:78.

Gerbillus rosalinda St. Leger, 1929. Ann. Mag. Nat. Hist., ser. 10, 4:295.
TYPE LOCALITY: Sudan, Kordofan, 145 km southwest of El Obeid Abu Zabad.
DISTRIBUTION: Sudan.
COMMENT: Considered a distinct species by Schlitter, 1976, Diss. Abstr. Int., 37(6), Petter, 1975, Part 6.3:9, and Corbet and Hill, 1980:164; but also see Kock, 1978, Senckenberg. Biol., 58:127, who included *rosalinda* in *agag*. May be conspecific with *latastei*; see Cockrum, 1977, Mammalia, 41:78. DML considers *rosalinda* distinct pending revision.

ISIS NUMBER: 5301410008091033001.

Gerbillus ruberrimus Rhoads, 1896. Proc. Acad. Nat. Sci. Phila., p. 538.
TYPE LOCALITY: Ethiopia, Finik, near Webi Shebeli.
DISTRIBUTION: Somalia and Kenya.
COMMENT: Listed as a distinct species by Yalden *et al.*, 1976, Monitore Zool. Ital., suppl. 8:26, and Rupp, 1980, Saugetierk. Mitt., 28:87; but also see Roche, 1975, Monitore Zool. Ital., suppl. 6:263–268, who included *ruberrimus* in *pusillus*. DML includes *ruberrimus* in genus *Dipodillus*; but see comment under *Gerbillus*.

Gerbillus syrticus Misonne, 1974. Bull. Inst. R. Sci. Nat. Belg., 50(6):1–6.
TYPE LOCALITY: Coast of Libya, 12 km N. of Nofilia.
DISTRIBUTION: Libya.
COMMENT: DML includes this species in genus *Dipodillus*; but see comment under *Gerbillus*.

Gerbillus watersi De Winton, 1901. Novit. Zool., 8:399.
TYPE LOCALITY: Sudan, Upper Nile, Shendi.
DISTRIBUTION: Somalia; Djibouti; Sudan.
COMMENT: Petter, 1975, Part 6.3:11, listed *watersi* both as a distinct species and as a subspecies of *nanus*; includes *Monodia juliani*; see Roche, 1975, Monitore Zool. Ital., suppl. 6(13):263–268. Kock, 1978, Bull. Carnegie Mus. Nat. Hist., 6:31–37, and Corbet and Hill, 1980:164, considered *watersi* a distinct species. DML considers *juliani* distinct from *watersi* pending revision; he places both forms in genus *Dipodillus*; but see comment under *Gerbillus*.

ISIS NUMBER: 5301410008091038001.

Graomys Thomas, 1916. Ann. Mag. Nat. Hist., ser. 8, 17:141.
REVIEWED BY: M. A. Mares (MAM); D. F. Williams (DFW).
COMMENT: Formerly included in *Phyllotis* by Hershkovitz, 1962, Fieldiana Zool., 46:217; but considered a distinct genus by Pearson and Patton, 1976, J. Mammal., 57:341, Reig, 1978, Publ. Mus. Munic. Cienc. Nat. Mar del Plata "Lorenzo Scaglia," 2(8):180–194, Spotorno, 1976, An. Mus. Hist. Nat. Valparaiso, 9:141–161, and Williams and Mares, 1978, Ann. Carnegie Mus., 47:193–221; the latter transferred *pearsoni* from *Graomys* to *Andalgalomys*; they also included *hypogaeus* in *Eligmodontia typus*, as did Massoia, 1976–1977, Rev. Invest. Agro. INTA, ser. 5, Patalogia Vegetal, 13:15–20. Subfamily Hesperomyinae; see comment under Cricetidae.

Graomys domorum (Thomas, 1902). Ann. Mag. Nat. Hist., ser. 7, 9:132.
TYPE LOCALITY: Bolivia, Cochabamba Dept., Tapacari, 3000 m.
DISTRIBUTION: E. Andes of Bolivia and N.W. Argentina.
COMMENT: Included in *griseoflavus* by Hershkovitz, 1962, Fieldiana Zool., 46:458, but considered a distinct species by Pearson and Patton, 1976, J. Mammal., 57:341. Includes *taterona*; see Reig, 1978, Publ. Mus. Munic. Cienc. Nat. Mar del Plata "Lorenzo Scaglia," 2(8):180, and Cabrera, 1961:494. Hershkovitz, 1962, Fieldiana Zool., 46:458, assigned *taterona* to *griseoflavus*.

Graomys edithae Thomas, 1919. Ann. Mag. Nat. Hist., ser. 9, 3:495.
TYPE LOCALITY: Argentina, La Rioja Prov., Otro Cerro, about 45 km W. Chumbicha.
DISTRIBUTION: Known only from the type locality.
COMMENT: Cabrera, 1961:495, included *edithae* in *griseoflavus*, but Hershkovitz, 1962, Fieldiana Zool., 46:461, considered *edithae* a distinct species of *Phyllotis*; Williams and Mares, 1978, Ann. Carnegie Mus., 47:201, listed *edithae* as a species of *Graomys*.

Graomys griseoflavus (Waterhouse, 1837). Proc. Zool. Soc. Lond., 1837:28.
TYPE LOCALITY: Argentina, Rio Negro Prov., mouth of Rio Negro.
DISTRIBUTION: Argentina; Bolivia; Paraguay; perhaps S.W. Brazil.
COMMENT: Includes *cachinus* and *centralis*; see Hershkovitz, 1962, Fieldiana Zool., 46:452, who also included *taterona* in *griseoflavus*, but see Reig, 1978, Publ. Mus. Munic. Cienc. Nat. Mar del Plata "Lorenzo Scaglia," 2(8):180, who included *taterona* in *domorum*. Formerly included *domorum*; see Pearson and Patton, 1976, J. Mammal., 57:341. Formerly included *hypogaeus*, which was included in *Eligmodontia typus* by Massoia, 1976–1977, Rev. Invest. Agro. INTA, ser. 5, Patalogia Vegetal, 13:15–20, and Williams and Mares, 1978, Ann. Carnegie Mus., 47(9):201, 218; also see Pearson and Patton, 1976, J. Mammal., 57:339–350, who listed *hypogaeus* as a distinct species of *Graomys*. Includes *chacoensis* J. A. Allen, 1901 (see Cabrera, 1961:495, and Hershkovitz, 1962, Fieldiana Zool., 46:452) which may be a separate species (DFW).
ISIS NUMBER: 5301410008042006001 as *Phyllotis griseoflavus*.

Gymnuromys Forsyth Major, 1896. Ann. Mag. Nat. Hist., ser. 6, 18:324.
REVIEWED BY: R. D. Owen (RDON).
COMMENT: Subfamily Nesomyinae; see comment under Cricetidae.
ISIS NUMBER: 5301410008063000000.

Gymnuromys roberti Major, 1896. Ann. Mag. Nat. Hist., ser. 6, 18:324.
TYPE LOCALITY: Madagascar, N.E. Betsileo, Ampitambe Forest.
DISTRIBUTION: E. Madagascar.
ISIS NUMBER: 5301410008063001001.

Habromys Hooper and Musser, 1964. Occas. Pap. Mus. Zool. Univ. Mich., 635:12.
REVIEWED BY: S. Anderson (SA); W. Caire (WC); R. P. Canham (RPC); L. N. Carraway (LNC); D. G. Huckaby (DGH); C. W. Kilpatrick (CWK); T. E. Lawlor (TL); R. H. Pine (RHP); J. Ramirez-Pulido (JRP); G. Urbano-V. (GUV).
COMMENT: Included in *Peromyscus* as a subgenus by Hall, 1981:718, Hooper, 1968, *in* King, ed., The Biology of *Peromyscus* (Rodentia), p. 38, and Musser, 1969, Am. Mus. Novit., 2357:1–23. Linzey and Layne, 1974, Am. Mus. Novit., 2532:1–20, studied the morphology of the spermatozoa. Reviewed by Carleton, 1980:118,

125, who considered *Habromys* a distinct genus. Pine *et al.*, 1979, Mammalia, 43:357, considered *Habromys* Hooper and Musser, a *nomen nudum*; Carleton, 1980, may have been first to make *Habromys* available (RHP). Subfamily Hesperomyinae; see comment under Cricetidae.

Habromys chinanteco (Robertson and Musser, 1976). Occas. Pap. Mus. Nat. Hist. Univ. Kans., 47:1.
TYPE LOCALITY: Mexico, Oaxaca, Cerro Pelon, 31.6 km S. Vista Hermosa, 2650 m.
DISTRIBUTION: Vicinity of the type locality.

Habromys lepturus (Merriam, 1898). Proc. Biol. Soc. Wash., 12:118.
TYPE LOCALITY: Mexico, Oaxaca, Cerro Zempoaltepec, 8200 ft. (2499 m).
DISTRIBUTION: Cerro Zempoaltepec and Sierra de Juarez (Oaxaca, Mexico).
COMMENT: Includes *ixtlani*; see Musser, 1969, Am. Mus. Novit., 2357:17.
ISIS NUMBER: 5301410008039029001 as *Peromyscus lepturus.*
5301410008039027001 as *Peromyscus ixtlani.*

Habromys lophurus (Osgood, 1904). Proc. Biol. Soc. Wash., 17:72.
TYPE LOCALITY: Guatemala, Huehuetenango, Todos Santos, 10,000 ft. (3048 m).
DISTRIBUTION: Highlands of Chiapas (Mexico), S.W. Guatemala, and El Salvador.
COMMENT: Reviewed by Musser, 1969, Am. Mus. Novit., 2357:1-23.
ISIS NUMBER: 5301410008039031001 as *Peromyscus lophurus.*

Habromys simulatus (Osgood, 1904). Proc. Biol. Soc. Wash., 17:72.
TYPE LOCALITY: Mexico, Veracruz, near Jico (Xico), 6000 ft. (1829 m).
DISTRIBUTION: Type locality and near Zacualpan (Veracruz, Mexico).
COMMENT: Reviewed by Musser, 1969, Am. Mus. Novit., 2357:1-23, and Robertson and Musser, 1976, Occas. Pap. Mus. Nat. Hist. Univ. Kans., 47:4.
ISIS NUMBER: 5301410008039051001 as *Peromyscus simulatus.*

Hodomys Merriam, 1894. Proc. Acad. Nat. Sci. Phila., 46:232.
REVIEWED BY: D. G. Huckaby (DGH); J. Ramirez-Pulido (JRP); G. Urbano-V. (GUV).
COMMENT: Included in *Neotoma* as a subgenus by Hall, 1981:771, and Genoways and Birney, 1974, Mamm. Species, 41:1, 2. Carleton, 1980:121, considered *Hodomys* a distinct genus. Subfamily Hesperomyinae; see comment under Cricetidae.

Hodomys alleni (Merriam, 1892). Proc. Biol. Soc. Wash., 7:168.
TYPE LOCALITY: Mexico, Colima, Manzanillo.
DISTRIBUTION: S. Sinaloa to Oaxaca; Balsas Basin of C. Puebla (Mexico).
COMMENT: Reviewed by Genoways and Birney, 1974, Mamm. Species, 41:1-4, Hall, 1981:771, and Burt and Barkalow, 1942, J. Mammal., 23:287-297, as *Neotoma* (*Hodomys*); but see Carleton, 1980:121, who considered *Hodomys* a distinct genus.
ISIS NUMBER: 5301410008027002001 as *Neotoma alleni.*

Holochilus Brandt, 1835. Mem. Acad. Imp. Sci. St. Petersb., ser. 6, 3(2):428.
REVIEWED BY: A. Langguth (AL); M. A. Mares (MAM); R. A. Ojeda (RAO); R. H. Pine (RHP); O. A. Reig (OAR).
COMMENT: Revised by Hershkovitz, 1955, Fieldiana Zool., 37:639-673. Reviewed by Gardner and Patton, 1976, Occas. Pap. Mus. Zool. La. St. Univ., 49:1-48, and Massoia, 1980, Ameghiniana, 17:280-287, who also included the Pleistocene form *molitor* in *Holochilus*. Subfamily Hesperomyinae; see comment under Cricetidae.
ISIS NUMBER: 5301410008016000000.

Holochilus brasiliensis (Desmarest, 1819). Nouv. Dict. Hist. Nat. Paris, 2nd ed., 29:62.
TYPE LOCALITY: Brazil, Minas Gerais, Lagoa Santa.
DISTRIBUTION: S. and E. Brazil; Uruguay; N.E. and C. Argentina.
COMMENT: Reviewed by Hershkovitz, 1955, Fieldiana Zool., 37:639-673, and Massoia, 1971, Rev. Invest. Agropec. INTA, Buenos Aires, ser. 1, Biol. Prod. Anim., 8(1):13-40. Gardner and Patton, 1976, Occas. Pap. Mus. Zool. La. St. Univ., 49:31, suggested that *brasiliensis* was a composite, as defined by Hershkovitz, 1955, Fieldiana Zool., 37:639-673, who included *chacarius, sciureus, guianae, berbicensis, incarum, venezuelae, amazonicus*, and *nanus* in this species; also see Massoia, 1980,

Ameghiniana, 17:280–287, who considered *chacarius* and *sciureus* (which includes the remaining forms above) distinct species, and included *leucogaster* in *brasiliensis*. OAR retains *nanus* in *brasiliensis* and considers *amazonicus* and *venezuelae* distinct species.
ISIS NUMBER: 5301410008016001001.

Holochilus chacarius Thomas, 1906, Ann. Mag. Nat. Hist., ser. 7, 18:446.
TYPE LOCALITY: Paraguay, Chaco, one league N.W. of Concepcion.
DISTRIBUTION: Paraguay; N.E. Argentina.
COMMENT: Formerly included in *brasiliensis* by Hershkovitz, 1955, Fieldiana Zool., 37:665; includes *balnearum*; see Massoia, 1976, *in* Vidal *et al.*, Physis, 35(90):76, and Massoia, 1980, Ameghiniana, 17:280–287. Karyology reviewed by Riva *et al.*, 1977, Physis, 36(92):215–218. OAR considers *balnearum* a distinct species.

Holochilus magnus Hershkovitz, 1955. Fieldiana Zool., 37:657.
TYPE LOCALITY: Uruguay, about 40 km south Treinta y Tres, Rio Cebollati, Paso de Averias.
DISTRIBUTION: Uruguay; S.E. Brazil.
COMMENT: Does not occur in Argentina (AL); also see Hershkovitz, 1955, Fieldiana Zool., 37:657, and Cabrera, 1961:507.
ISIS NUMBER: 5301410008016002001.

Holochilus sciureus Wagner, 1842. Arch. Naturgesch., ser. 8, 1:16.
TYPE LOCALITY: Brazil, Minas Gerais, Rio Sao Francisco.
DISTRIBUTION: Colombia; Venezuela; Guianas; Amazonian Brazil; E. Peru; N. Bolivia; Minas Gerais (Brazil).
COMMENT: Hershkovitz, 1955, Fieldiana Zool., 37:639–673, included this species in *brasiliensis*; includes *guianae* Thomas, 1901, *berbicensis*, *incarum*, *venezuelae*, *amazonicus*, and *nanus*; see Massoia, 1980, Ameghiniana, 17:280–287. Husson, 1978:419, included *nanus* in *brasiliensis*. Reviewed by Alencar, 1969, Rev. Brasil. Biol., 29(4):567–570, and Twigg, 1965, Proc. Zool. Soc. Lond., 145(2):263–283. OAR includes *nanus* in *brasiliensis* and considers *amazonicus* and *venezuelae* distinct species.

Hypogeomys A. Grandidier, 1869. Rev. Mag. Zool. Paris, 21:338.
REVIEWED BY: R. D. Owen (RDON).
COMMENT: Subfamily Nesomyinae; see comment under Cricetidae.
ISIS NUMBER: 5301410008064000000.

Hypogeomys antimena A. Grandidier, 1869. Rev. Mag. Zool. Paris, 21:339.
TYPE LOCALITY: Madagascar, Menabe, banks of the Tsidsibon and Andranoumene.
DISTRIBUTION: W. Madagascar.
ISIS NUMBER: 5301410008064001001.

Ichthyomys Thomas, 1893. Proc. Zool. Soc. Lond., 1893:337.
REVIEWED BY: R. S. Voss (RSV).
COMMENT: Subfamily Hesperomyinae; see comment under Cricetidae.
ISIS NUMBER: 5301410008017000000.

Ichthyomys hydrobates (Winge, 1891). Vidensk. Medd. Nat. Foren., ser. 5, 3:20.
TYPE LOCALITY: Venezuela, Merida, Sierra de Merida.
DISTRIBUTION: Ecuador; Andean Colombia; W. Venezuela.
ISIS NUMBER: 5301410008017001001.

Ichthyomys pittieri Handley and Mondolfi, 1963. Acta Biol. Venez., 3:417.
REVIEWED BY: R. H. Pine (RHP).
TYPE LOCALITY: Venezuela, Aragua, Rancho Grande Nat. Park, near the head of the Rio Limon.
DISTRIBUTION: Known only from the type locality.

Ichthyomys stolzmanni Thomas, 1893. Proc. Zool. Soc. Lond., 1893:339.
TYPE LOCALITY: Peru, Junin Dept., Chanchamayo, about 3000 ft. (900 m).
DISTRIBUTION: E. Ecuador; Andean Peru.
ISIS NUMBER: 5301410008017002001.

Irenomys Thomas, 1919. Ann. Mag. Nat. Hist., ser. 9, 3:201.
  REVIEWED BY: R. D. Owen (RDON).
  COMMENT: Subfamily Hesperomyinae; see comment under Cricetidae.
  ISIS NUMBER: 5301410008018000000.

  Irenomys tarsalis (Philippi, 1900). An. Mus. Nac. Chile Zool., 14:10.
    TYPE LOCALITY: Chile, Fundo San Juan, Valdivia Prov., near la Union.
    DISTRIBUTION: Argentina; Chile; Chiloe Isl. and Guaitecas Isl. (Chile).
    COMMENT: Includes *longicaudatus;* see Cabrera, 1961:498. Reise and Venegas, 1974, Bol.
      Soc. Biol. Concepcion, 47:71–85, extended the known distribution in Chile.
    ISIS NUMBER: 5301410008018001001.

Isthmomys Hooper and Musser, 1964. Occas. Pap. Mus. Zool. Univ. Mich., 635:12.
  REVIEWED BY: S. Anderson (SA); W. Caire (WC); R. P. Canham (RPC); M. D. Carleton
    (MDC); L. N. Carraway (LNC); D. G. Huckaby (DGH); C. W. Kilpatrick (CWK); T.
    E. Lawlor (TL); R. H. Pine (RHP).
  COMMENT: Included as a subgenus of *Peromyscus* by Hall, 1981:717, and Hooper, 1968,
    in King, ed., The Biology of *Peromyscus* (Rodentia), p. 38. Linzey and Layne, 1974,
    Am. Mus. Novit., 2532:1–20, studied the morphology of the spermatozoa.
    Reviewed by Carleton, 1980:118, 124, who considered *Isthmomys* a distinct genus.
    Pine *et al.,* 1979, Mammalia, 43:357, considered *Isthmomys* Hooper and Musser, a
    nomen nudum; Carleton, 1980, may have been first to make *Isthmomys* available
    (RHP). Subfamily Hesperomyinae; see comment under Cricetidae.

  Isthmomys flavidus (Bangs, 1902). Bull. Mus. Comp. Zool. Harv. Univ., 39:27.
    TYPE LOCALITY: Panama, Volcan de Chiriqui, Boquete, 3000–5000 ft. (914–1524 m).
    DISTRIBUTION: Known from the type locality and the upper Rio Changena (W.
      Panama).
    COMMENT: Handley, 1966, *in* Wenzel and Tipton, eds., Ectoparasites of Panama, pp.
      753–795, reported the additional locality for this species, and stated that
      specimens from the Azuero Peninsula may belong in *pirrensis;* but also see
      Carleton, 1980:24.
    ISIS NUMBER: 5301410008039016001 as *Peromyscus flavidus.*

  Isthmomys pirrensis (Goldman, 1912). Smithson. Misc. Coll., 60(2):5.
    TYPE LOCALITY: Panama, Darien, Mt. Pirri, head Rio Limon, 4500 ft. (1372 m).
    DISTRIBUTION: E. Panama; perhaps adjacent Colombia.
    COMMENT: Tentatively included in *flavidus* by Corbet and Hill, 1980:148. Considered a
      distinct species by Handley, 1966, *in* Wenzel and Tipton, eds., Ectoparasites of
      Panama, pp. 753–795, Carleton, 1980:1–146, and Hall, 1981:718. Walker *et al.,*
      1975, stated that this species occurs in N. Colombia.
    ISIS NUMBER: 5301410008039046001 as *Peromyscus pirrensis.*

Juscelinomys Moojen, 1965. Rev. Brasil. Biol., 25:281–285.
  REVIEWED BY: A. Langguth (AL).
  COMMENT: Subfamily Hesperomyinae; see comment under Cricetidae.

Juscelinomys candango Moojen, 1965. Rev. Brasil. Biol., 25:281–285.
  TYPE LOCALITY: Brazil, Brasilia.
  DISTRIBUTION: C. Brazil.

Kunsia Hershkovitz, 1966. Z. Saugetierk., 31(2):112.
  REVIEWED BY: A. Langguth (AL).
  COMMENT: Formerly included in *Scapteromys;* see revision by Hershkovitz, 1966, Z.
    Saugetierk., 31:1–149. Subfamily Hesperomyinae; see comment under Cricetidae.

  Kunsia fronto (Winge, 1887). E. Mus. Lundii, 1(3):44.
    TYPE LOCALITY: Brazil, Minas Gerais, Rio das Velhas, Lagoa Santa.
    DISTRIBUTION: Chaco (Argentina); Minais Gerais, Distrito Federale (Brazil); perhaps
      Paraguay.
    COMMENT: Includes *chacoensis* (Gyldenstolpe, 1932); see Hershkovitz, 1966, Z.

Saugetierk., 31:116; the type is from Pleistocene cave deposits. Reviewed by Avila-Pires, 1972, Rev. Brasil. Biol., 32(3):419–422.
ISIS NUMBER: 5301410008051002001 as *Scapteromys chacoensis.*

Kunsia tomentosus (Lichtenstein, 1830). Darst. Saugeth., 7(15):33.
TYPE LOCALITY: Brazil, Rio Uruguay.
DISTRIBUTION: E.C. Brazil; Bolivia.
COMMENT: Includes *gnambiquarae* and *principalis;* see Hershkovitz, 1966, Z. Saugetierk., 31:119; Gardner and Patton, 1976, Occas. Pap. Mus. Zool. La. St. Univ., 49:26, listed *tomentosus* in *Scapteromys,* following Brum, 1965, An. Congr. Lat. Am. Zool., 2:315–320. See also Massoia and Fornes, 1965, Neotropica, 11(34):1–7.
ISIS NUMBER: 5301410008051004001 as *Scapteromys tomentosus.*
5301410008051003001 as *Scapteromys gnambiquarae.*

Leimacomys Matschie, 1893. Sitzb. Ges. Naturf. Fr. Berlin, 4:107–109.
REVIEWED BY: B. R. Stein (BRS).
COMMENT: Subfamily Dendromurinae; see comment under Cricetidae.
ISIS NUMBER: 5301410011027000000.

Leimacomys buettneri Matschie, 1893. Sitzb. Ges. Naturf. Fr. Berlin, 4:107–109.
TYPE LOCALITY: Togo, Bismarckburg.
DISTRIBUTION: Known only from the type locality.
COMMENT: Reviewed by Misonne, 1966, Ann. Mus. R. Afr. Cent., 144:42, and Misonne, 1974, Part 6:14.
ISIS NUMBER: 5301410011027001001.

Lenoxus Thomas, 1909. Ann. Mag. Nat. Hist., ser. 8, 4:236.
REVIEWED BY: R. D. Owen (RDON).
COMMENT: Subfamily Hesperomyinae; see comment under Cricetidae.
ISIS NUMBER: 5301410008019000000.

Lenoxus apicalis (J. A. Allen, 1900). Bull. Am. Mus. Nat. Hist., 13:224.
TYPE LOCALITY: Peru, Puno Dept., Rio Inambari, Inca Mines.
DISTRIBUTION: S.E. Peru; W. Bolivia.
ISIS NUMBER: 5301410008019001001.

Lophiomys Milne-Edwards, 1867. L'Institut, 35:46.
REVIEWED BY: R. D. Owen (RDON).
COMMENT: Subfamily Lophiomyinae; see comment under Cricetidae.
ISIS NUMBER: 5301410008067000000.

Lophiomys imhausi Milne-Edwards, 1867. L'Institut, 35:46.
TYPE LOCALITY: Somalia (=Somaliland).
DISTRIBUTION: E. Sudan; Ethiopia; Somalia; Kenya.
ISIS NUMBER: 5301410008067001001.

Macrotarsomys Milne-Edwards and G. Grandidier, 1898. Bull. Mus. Hist. Nat. Paris, 4:179.
REVIEWED BY: R. D. Owen (RDON).
COMMENT: Subfamily Nesomyinae; see comment under Cricetidae.
ISIS NUMBER: 5301410008065000000.

Macrotarsomys bastardi Milne-Edwards and G. Grandidier, 1898. Bull. Mus. Hist. Nat. Paris, 4:179.
TYPE LOCALITY: Madagascar, south of Mangoky.
DISTRIBUTION: W. Madagascar.
ISIS NUMBER: 5301410008065001001.

Macrotarsomys ingens Petter, 1959. Mammalia, 23:140.
TYPE LOCALITY: Madagascar, between Tananarive and Majunga 200 m. from d'Ampijoroa.
DISTRIBUTION: N.W. Madagascar.
ISIS NUMBER: 5301410008065002001.

Malacothrix Wagner, 1843. Schreber's Saugethiere, Suppl., 3:496.
REVIEWED BY: R. D. Owen (RDON).
COMMENT: Subfamily Dendromurinae; see comment under Cricetidae.
ISIS NUMBER: 5301410008107000000.

Malacothrix typica (A. Smith, 1834). S. Afr. Quart. J., 2 (in 8 installments).
TYPE LOCALITY: South Africa, Cape Prov., Graaff Reinet Dist.
DISTRIBUTION: South Africa; Namibia; S. Angola; Botswana.
ISIS NUMBER: 5301410008107001001 as *M. typicus (sic)*.

Megadendromus Dieterlen and Rupp, 1978. Z. Saugetierk., 43(3):129.
COMMENT: Subfamily Dendromurinae; see comment under Cricetidae.

Megadendromus nikolausi Deiterlen and Rupp, 1978. Z. Saugetierk., 43(3):131.
TYPE LOCALITY: Ethiopia, S. Goba, Bale Mtns.
DISTRIBUTION: E. Ethiopia.

Megadontomys Merriam, 1898. Proc. Biol. Soc. Wash., 12:115.
REVIEWED BY: S. Anderson (SA); W. Caire (WC); R. P. Canham (RPC); M. D. Carleton
(MDC); L. N. Carraway (LNC); D. G. Huckaby (DGH); C. W. Kilpatrick (CWK); T.
E. Lawlor (TL); J. Ramirez-Pulido (JRP); G. Urbano-V. (GUV).
COMMENT: Included as a subgenus of *Peromyscus* by Hall, 1981:716, Musser, 1964,
Occas. Pap. Mus. Zool. Univ. Mich., 636:13–19, and Hooper, 1968, *in* King, ed.,
The Biology of *Peromyscus* (Rodentia), p. 38. Linzey and Layne, 1974, Am. Mus.
Novit., 2532:1–20, studied the morphology of the spermatozoa. Reviewed by
Carleton, 1980:118, 124, who considered *Megadontomys* a distinct genus.
Subfamily Hesperomyinae; see comment under Cricetidae.

Megadontomys thomasi (Merriam, 1898). Proc. Biol. Soc. Wash., 12:116.
TYPE LOCALITY: Mexico, Guerrero, mountains near Chilpancingo, 9700 ft. (2957 m).
DISTRIBUTION: C. Guerrero, C. Veracruz, and C. Oaxaca (Mexico).
COMMENT: Includes *cryophilus* and *nelsoni*; see Musser, 1964, Occas. Pap. Mus. Zool.
Univ. Mich., 636:13–19, and Hall, 1981:717.
ISIS NUMBER: 5301410008039056001 as *Peromyscus thomasi*.

Megalomys Trouessart, 1881. Le Naturaliste, 1:357.
REVIEWED BY: R. H. Pine (RHP).
COMMENT: Included in *Oryzomys* by Forsyth Major, 1901, Ann. Mag. Nat. Hist., ser. 7,
7:204–206. *M. audreyae*, known only as subfossil from Barbuda, Lesser Antilles,
may be Holocene; see Hall, 1981:624–625. Subfamily Hesperomyinae; see
comment under Cricetidae.
ISIS NUMBER: 5301410008020000000 as *Megalomys*.
5301410008020001001 for *M. audreyae*.

Megalomys desmarestii (Fischer, 1829). Synopsis Mammal., p. 316.
TYPE LOCALITY: Martinique (Lesser Antilles) (France).
DISTRIBUTION: Known only from the type locality.
COMMENT: Apparently extinct.

Megalomys luciae (Forsyth Major, 1901). Ann. Mag. Nat. Hist., ser. 7, 7:206.
TYPE LOCALITY: Santa Lucia (Lesser Antilles) (U.K.).
DISTRIBUTION: Known only from the type locality.
COMMENT: Apparently extinct.
ISIS NUMBER: 5301410008020002001.

Meriones Illiger, 1811. Prodr. Syst. Mamm. et Avium., p. 82.
REVIEWED BY: D. M. Lay (DML); C. B. Robbins (CBR); O. L. Rossolimo (OLR); S. Wang
(SW).
COMMENT: Formerly included *Sekeetamys*; see Petter, 1956, Mammalia, 20:419–426,
Petter, 1975, Part 6.3:4, and Corbet, 1978:124. Reviewed by Chaworth-Musters and
Ellerman, 1947, Proc. Zool. Soc. Lond., 1947–1948:478–504; Arabian species

reviewed by Harrison, 1972, The Mammals of Arabia, 3:559–593; African species are in need of revision (see Corbet, 1978:125). Sokolov and Orlov, 1980:156, included *Brachiones przewalskii* in *Meriones*, without comment; but see Corbet, 1978:128. Subfamily Gerbillinae; see comment under Cricetidae.
ISIS NUMBER: 5301410008092000000.

Meriones chengi Wang, 1964. Acta Zootax. Sin., 1:9.
   TYPE LOCALITY: China, Sinkiang, Turfan, 42° 55' N., 89° 06' E.
   DISTRIBUTION: Known only from the type locality.
   COMMENT: Possibly a form of *meridianus*; but should be regarded as distinct until comparisons are possible (DML).

Meriones crassus Sundevall, 1842. Svenska Vet. Akad., ser. 3, p. 233.
   TYPE LOCALITY: Egypt, Sinai, Fount of Moses (Ain Musa), 29° 53' N., 32° 39' E.
   DISTRIBUTION: N. Africa south to Niger and Sudan; Israel; Jordan; Saudi Arabia; Syria; Iraq; Iran; S. Afghanistan; W. Pakistan.
   COMMENT: Includes *longifrons*; see Chaworth-Musters and Ellerman, 1947, Proc. Zool. Soc. Lond., 1947–1948:482, and Corbet, 1978:127. Formerly included *sacramenti* and *zarudnyi*; see Harrison, 1972, The Mammals of Arabia, 3:590, and Lay, 1967, Fieldiana Zool., 54:1–282. Reviewed by Koffler, 1972, Mamm. Species, 9:1–4. *M. longifrons* may be a valid species (DML).
   ISIS NUMBER: 5301410008092003001.

Meriones hurrianae Jerdon, 1867. Mamm. India, p. 186.
   TYPE LOCALITY: India, Hurriana Dist. (=Hariana).
   DISTRIBUTION: Extreme S.E. Iran, Pakistan, and N.W. India (principally in the Thar Desert); perhaps Afghanistan.
   COMMENT: Occurrence in Afghanistan is dubious (DML).
   ISIS NUMBER: 5301410008092004001.

Meriones libycus Lichtenstein, 1823. Verz. Doublet. Zool. Mus. Univ. Berlin, p. 5.
   TYPE LOCALITY: Egypt, near Alexandria.
   DISTRIBUTION: Western Sahara to Egypt, through N. Arabia, Iraq, Iran, and Afghanistan, to Sinkiang (China).
   COMMENT: Includes *caudatus*; see Petter, 1975, Part 6.3:3, Lay and Nadler, 1969, Cytogenetics, 8:45, and Corbet and Hill, 1980:166; but also see Ranck, 1968, Bull. U.S. Nat. Mus., 275:164, who considered *caudatus* a distinct species. Includes *arimalius* (see Corbet, 1978:127, who also discussed the status of *caudatus*), *iranensis* (see Lay, 1967, Fieldiana Zool., 54:1–282), and *erythrourus* (which Gromov and Baranova, 1981:165, listed as possibly distinct). Type locality restricted by Chaworth-Musters and Ellerman, 1947, Proc. Zool. Soc. Lond., 1947–1948:485.
   ISIS NUMBER: 5301410008092005001 as *M. libycus*.
                5301410008092001001 as *M. arimalius*.

Meriones meridianus (Pallas, 1773). Reise Prov. Russ. Reichs., p. 702.
   TYPE LOCALITY: U.S.S.R., Kazakhstan S.S.R., Novo-Bogatinsk, 47° 33' N., 51° 11' E.
   DISTRIBUTION: Lower Don R., and N. of the Caucasus (U.S.S.R.) to Mongolia and Hopei (China), south to E. Iran, N. Afghanistan, Tsinghai and Shansi (China); isolated population in Armenia (U.S.S.R.).
   COMMENT: Type locality restricted by Chaworth-Musters and Ellerman, 1947, Proc. Zool. Soc. Lond., 1947–1948:483.
   ISIS NUMBER: 5301410008092006001.

Meriones persicus (Blanford, 1875). Ann. Mag. Nat. Hist., 16:132.
   TYPE LOCALITY: Iran, Kohrud, 72 mi. (116 km) north of Isfahan, 33° 40' N., 51° 25' E.
   DISTRIBUTION: Iran; adjacent Transcaucasian U.S.S.R., Turkey, and Iraq; Turkmenia (U.S.S.R.); Afghanistan; Pakistan, west of the Indus River.
   ISIS NUMBER: 5301410008092007001.

Meriones rex Yerbury and Thomas, 1895. Proc. Zool. Soc. Lond., 1895:552.
   TYPE LOCALITY: South Yemen, Lahej, 13° 01' N., 44° 54' E.
   DISTRIBUTION: S.W. Saudi Arabia; Yemen.
   ISIS NUMBER: 5301410008092008001.

Meriones sacramenti Thomas, 1922. Ann. Mag. Nat. Hist., ser. 9, 10:552.
TYPE LOCALITY: Israel, 10 mi. (16 km) S. Beersheba, 31° 15' N., 34° 47' E.
DISTRIBUTION: Jaffa south to Beersheba (Israel).
COMMENT: Formerly included in *crassus* by Ellerman and Morrison-Scott, 1966:647, but
considered a distinct species by Harrison, 1972, The Mammals of Arabia, 3:590,
Zavahi and Wahrman, 1957, Mammalia, 21:341–380, and Petter, 1957, Mammalia,
21:241–257.

Meriones shawi (Duvernoy, 1842). Mem. Soc. Sci. Nancy, 3:22.
TYPE LOCALITY: Algeria, Oran, 35° 42' N., 0° 38' W.
DISTRIBUTION: Morocco to N. Algeria, Tunisia, and Egypt.
COMMENT: Considered a distinct species by Lay and Nadler, 1969, Cytogenetics,
8:35–50, Harrison, 1972, The Mammals of Arabia, 3:593, and Petter, 1975, Part
6.3:4. Includes *grandis* and *isis;* see Petter, 1975, Part 6.3:4, and Corbet, 1978:127.
Lay, 1981, *in litt.,* considered *grandis* a distinct species. Setzer, 1961, J. Egypt
Public Health. Assoc., 36:81–90, and Ranck, 1968, Bull. U.S. Nat. Mus., 275:1–264,
referred to specimens of *shawi* as *libycus.* DML considers *shawi* a valid species,
readily distinguishable from *libycus.* Formerly included *tristrami;* see Corbet,
1978:126.
ISIS NUMBER: 5301410008092009001.

Meriones tamariscinus (Pallas, 1773). Reise Prov. Russ. Reichs., 2:702.
TYPE LOCALITY: U.S.S.R., W. Kazakhstan S.S.R., Saraitschikowski (=Saraichik), 47° 30'
N., 51° 47' E.
DISTRIBUTION: N. Caucasus and Kazakhstan to the Altai Mtns., and through N.
Sinkiang to W. Kansu (China).
COMMENT: Type locality restricted by Chaworth-Musters and Ellerman, 1947, Proc.
Zool. Soc. Lond., 1947–1948:482.
ISIS NUMBER: 5301410008092010001.

Meriones tristrami Thomas, 1892. Ann. Mag. Nat. Hist., ser. 6, 9:148.
TYPE LOCALITY: Israel, Dead Sea.
DISTRIBUTION: Israel and W. Jordan to Turkey, N.W. Syria, E. Iraq, N.W. Iran, and
Transcaucasian U.S.S.R.
COMMENT: Includes *blackleri; tristrami* was formerly included in *shawi;* see Petter, 1961,
Mammalia, suppl., p. 48, Harrison, 1972, The Mammals of Arabia, 3:572, and
Corbet, 1978:126. Gromov and Baranova, 1981:164, listed *blackleri* as a distinct
species, without comment.
ISIS NUMBER: 5301410008092011001 as *M. tristrami.*
5301410008092002001 as *M. blackleri.*

Meriones unguiculatus (Milne-Edwards, 1867). Ann. Sci. Nat. Zool., ser. 5, 7:377.
TYPE LOCALITY: China, N. Shansi, 10 mi. (16 km) N.E. of Tschang-Kur, Eul-che san hao
(=Ershi san hao), 45° 04' N., 126° 03' E.
DISTRIBUTION: N. China west to Kansu and Sinkiang; Mongolia; Tuva and Buryat.-
Mongolsk. A.S.S.R. (U.S.S.R.).
COMMENT: Reviewed by Gulotta, 1971, Mamm. Species, 3:1–5. Type locality restricted
by Chaworth-Musters and Ellerman, 1947, Proc. Zool. Soc. Lond., 1947–1948:483.
The pet and laboratory strains of Mongolian gerbil were derived from this
species; see Corbet, 1978:126.
ISIS NUMBER: 5301410008092012001.

Meriones vinogradovi Heptner, 1931. Zool. Anz., 94:122.
TYPE LOCALITY: Iran, Persian Azarbaidjan.
DISTRIBUTION: Armenia and Azerbaidzhan (U.S.S.R.); adjacent Asia Minor and N.W.
Iran; N. Syria; probably N. Iraq and S.E. Turkey.
ISIS NUMBER: 5301410008092013001.

Meriones zarudnyi Heptner, 1937. Byull. Mosk. Ova. Ispyt. Prir. Otd. Biol., 46:19.
TYPE LOCALITY: U.S.S.R., Turkmen. S.S.R., Kushka, 35° 16' N., 62° 20' E.
DISTRIBUTION: N. Afghanistan; S.E. Turkmenia (U.S.S.R.); probably E. Iran.
COMMENT: Formerly included in *crassus* by Ellerman and Morrison-Scott, 1966:647; but

considered specifically distinct by Lay, 1967, Fieldiana Zool., 54:1–282, and Heptner *et al.*, 1958, Trudy Inst. Zool. Parasit. Acad. Sci. Turkmenia S.S.R., 3:141–147.

ISIS NUMBER: 5301410008092014001.

Mesocricetus Nehring, 1898. Zool. Anz., 21:494.

REVIEWED BY: M. Andera (MA); O. L. Rossolimo (OLR); F. Spitzenberger (FS).

COMMENT: Kuznetsov, 1965, *in* Bobrinskii *et al.*, 1965, Key to the Mammals of the U.S.S.R., Moscow, included *Mesocricetus* in *Cricetus*; Corbet, 1978:90, 92, considered both distinct genera, in need of revision. Reviewed by Vorontsov, 1960, C. R. Acad. Sci. U.S.S.R., 132(6):1448–1451. Subfamily Cricetinae; see comment under Cricetidae.

ISIS NUMBER: 5301410008021000000.

Mesocricetus auratus (Waterhouse, 1839). Proc. Zool. Soc. Lond., 1839:57.

TYPE LOCALITY: Syria, Aleppo.

DISTRIBUTION: Asia Minor (Turkey) to Syria; perhaps Lebanon and Israel.

COMMENT: Formerly included *brandti;* see Zilfian *et al.*, 1975, [Abstr. Symp. Syst. and Cytogenet. Mamm.], Moscow, p. 18–19. Corbet, 1978:92, and FS included *brandti* in this species, but stated that it may be distinct. Records from Lebanon and Israel were doubted by Atallah, 1977, Saugetierk. Mitt., 25:320. A detailed bibliography was presented in Kittel, 1969, Z. Verzuchstierk., 11:1–115. The pet and laboratory strains of golden hamster were derived from this species; see Corbet, 1978:92.

ISIS NUMBER: 5301410008021001001.

Mesocricetus brandti (Nehring, 1898). Zool. Anz., 21:331.

TYPE LOCALITY: U.S.S.R., Georgia, near Tbilisi.

DISTRIBUTION: N. Transcaucasia (U.S.S.R.); Kurdistan; possibly Lebanon, and Israel; see comment under *auratus*.

COMMENT: Zilfian *et al.*, 1975, [Abstr. Symp. Syst. and Cytogenet. Mamm.], Moscow, p. 18–19, considered *brandti* distinct from *auratus*; this had been considered possible by Corbet, 1978:92, Todd *et al.*, 1972, J. Hered., 63:73–77, and others.

Mesocricetus newtoni (Nehring, 1898). Zool. Anz., 21:329.

TYPE LOCALITY: Bulgaria, Kolarovgrad (=Schumla or Shumen).

DISTRIBUTION: E. Bulgaria; E. Rumania.

COMMENT: Considered a distinct species by Corbet, 1978:93; breeding experiments by Raicu and Bratosin, 1968, Genet. Res. Camb., 11:113–114, produced sterile offspring between *newtoni* and *auratus*.

Mesocricetus raddei (Nehring, 1894). Zool. Anz., 18:148.

TYPE LOCALITY: U.S.S.R., N. Caucasus, Daghestan, Samur River.

DISTRIBUTION: N. Caucasus to Don River and Sea of Azov (U.S.S.R.).

Microdillus Thomas, 1910. Ann. Mag. Nat. Hist., ser. 8, 5:197.

REVIEWED BY: D. L. Harrison (DLH); C. B. Robbins (CBR).

COMMENT: Formerly included in *Gerbillus* by Walker *et al.*, 1975:850, but see Petter, 1975, Part 6.3:12, and Corbet and Hill, 1980:165, who considered *Microdillus* a distinct genus. Subfamily Gerbillinae; see comment under Cricetidae.

Microdillus peeli (De Winton, 1898). Ann. Mag. Nat. Hist., ser. 7, 1:250.

TYPE LOCALITY: Somalia, Eyk.

DISTRIBUTION: Somalia; E. Ethiopia.

COMMENT: Formerly included in *Gerbillus;* see Petter, 1975, Part 6.3:12.

Microxus Thomas, 1909. Ann. Mag. Nat. Hist., ser. 8, 4:237.

REVIEWED BY: A. Langguth (AL); R. A. Ojeda (RAO).

COMMENT: Included in *Akodon* by Cabrera, 1961:458, and Arata, 1967, *in* Anderson and Jones, p. 229. Reig, 1978, Publ. Mus. Munic. Cienc. Nat. Mar del Plata "Lorenzo Scaglia," 2(8):176, and Reig, 1980, J. Zool. Lond., 192:257–281, considered *Microxus* a distinct genus. Bianchi *et al.*, 1971, Evolution, 25:724–736, and Ellerman, 1941:419, considered *Microxus* a distinct genus and were followed by

Corbet and Hill, 1980:151. Gardner and Patton, 1976, Occas. Pap. Mus. Zool. La. St. Univ., 49:30, followed Hershkovitz, 1966, Z. Saugetierk., 31:86, who considered *Microxus* a synonym of *Abrothrix* which he placed in the oxymycterine group. Formerly included *Oxymycterus iheringi;* see Massoia, 1963, Physis, 24:129–136, and Hershkovitz, 1966, Z. Saugetierk., 31:86, 127. This taxon is in need of revision and its relationship with the oxymycterines and akodonts needs clarification. Subfamily Hesperomyinae; see comment under Cricetidae.

Microxus bogotensis (Thomas, 1895). Ann. Mag. Nat. Hist., ser. 6, 16:369.
    TYPE LOCALITY: Colombia, Bogota Region, 2620 m.
    DISTRIBUTION: Colombia; Venezuela.
    COMMENT: Formerly included in *Akodon,* but see comment under genus.
    ISIS NUMBER: 5301410008001008001 as *Akodon bogotensis.*

Microxus latebricola Anthony, 1924. Am. Mus. Novit., 139:3.
    TYPE LOCALITY: Ecuador, Ambato Prov., Hacienda San Francisco, E. of Ambato, Rio Cusutagua, about 2400 m.
    DISTRIBUTION: Andean Ecuador.
    COMMENT: Formerly included in *Akodon,* but see comment under genus.
    ISIS NUMBER: 5301410008001022001 as *Akodon latebricola.*

Microxus mimus (Thomas, 1901). Ann. Mag. Nat. Hist., ser. 7, 7:183.
    TYPE LOCALITY: Peru, Puno Dept., Limbane, 2600 m.
    DISTRIBUTION: S.E. Peruvian Andes.
    COMMENT: Formerly included in *Akodon;* Hershkovitz, 1966, Z. Saugetierk., 31:86, assigned this species to *Abrothrix* without comment; but see comment under genus.
    ISIS NUMBER: 5301410008001024001 as *Akodon mimus.*

Myospalax Laxmann, 1769. Sibirische Briefe, Gottingen, p. 75.
    REVIEWED BY: R. D. Owen (RDON); O. L. Rossolimo (OLR).
    COMMENT: Reviewed by Martynova, 1976, Zool. Zh., 55(8):1265–1275, and Corbet, 1978:93. Subfamily Myospalacinae; see comment under Cricetidae.
    ISIS NUMBER: 5301410008022000000.

Myospalax fontanieri (Milne-Edwards, 1867). Ann. Sci. Nat. Paris, 7:376.
    TYPE LOCALITY: China, Kansu.
    DISTRIBUTION: Dry grasslands from Hopei to Kansu, E. Tsinghai, Szechwan, and Anhwei (China).
    ISIS NUMBER: 5301410008022001001.

Myospalax myospalax (Laxmann, 1773). Kongl. Svenska. Vet.-Akad. Hand. Stockholm, 34:134.
    TYPE LOCALITY: U.S.S.R., Altaisky Krai, 100 km S.E. of Barnaul, Sommaren, near Paniusheva on Alei River.
    DISTRIBUTION: Altai and Tarbagatai Mtns. and adjacent valleys of the Ob and Irtysh rivers (U.S.S.R.); Sinkiang (China).
    COMMENT: Formerly included *psilurus* and *aspalax;* see Martynova, 1976, Zool. Zh., 55(8):1267.
    ISIS NUMBER: 5301410008022002001.

Myospalax psilurus (Milne-Edwards, 1874). Rech. Mamm., p. 126.
    TYPE LOCALITY: China, Chihli, south of Peking.
    DISTRIBUTION: Transbaikalia and Ussuri region (U.S.S.R.) to E. Mongolia; N.E. and C. China.
    COMMENT: Includes *aspalax* which may be a distinct species; see Martynova, 1976, Zool. Zh., 55(8):1267. Gromov and Baranova, 1981:170, considered *psilurus* a distinct species, but Corbet, 1978:94, included *psilurus* in *myospalax.*
    ISIS NUMBER: 5301410008022003001.

Myospalax rothschildi Thomas, 1911. Ann. Mag. Nat. Hist., ser. 8, 8:722.
    TYPE LOCALITY: China, Kansu, 40 mi. (64 km) S.E. Tao-chou.
    DISTRIBUTION: Kansu and Hupeh (China).

Myospalax smithi Thomas, 1911. Ann. Mag. Nat. Hist., ser. 8, 8:720.
  TYPE LOCALITY: China, Kansu, 30 mi. (48 km) S.E. Tao-chou.
  DISTRIBUTION: Kansu (China).

Mystromys Wagner, 1841. Arch. Naturgesch., p. 132.
  REVIEWED BY: R. D. Owen (RDON).
  COMMENT: Reviewed by Vorontsov, 1966, Zool. Zh., 45:436–446, and Voss and Linzey,
    1981, Misc. Publ. Mus. Zool. Univ. Mich., 159:19. Arata, 1967, *in* Anderson and
    Jones, p. 230, included this genus in Cricetini and Swanepoel *et al.*, Ann.
    Transvaal Mus., 32(7):174, included it in Cricetinae; Corbet and Hill, 1980:158,
    included it in Nesomyinae.
  ISIS NUMBER: 5301410008023000000.

Mystromys albicaudatus (A. Smith, 1834). S. Afr. Quart. J., 2:148.
  TYPE LOCALITY: South Africa, Cape Prov., Albany district.
  DISTRIBUTION: Cape Prov. to Transvaal (South Africa).
  COMMENT: Formerly included *longicaudatus* Noack, 1887, which was transferred to
    Praomys natalensis; see Misonne, 1974, Part 6:15, 25.
  ISIS NUMBER: 5301410008023001001.

Neacomys Thomas, 1900. Ann. Mag. Nat. Hist., ser. 7, 5:153.
  REVIEWED BY: R. D. Owen (RDON).
  COMMENT: Subfamily Hesperomyinae; see comment under Cricetidae.
  ISIS NUMBER: 5301410008024000000.

Neacomys guianae Thomas, 1905. Ann. Mag. Nat. Hist., ser. 7, 16:310.
  TYPE LOCALITY: Guyana, Demerara River, 120 ft. (37 m).
  DISTRIBUTION: Surinam; Guyana; S. Venezuela; N. Brazil.
  COMMENT: Distribution in Brazil reported by Peterson *et al.*, 1981, Bol. Of. Sanit.
    Panam., 91(4):324–339.
  ISIS NUMBER: 5301410008024001001.

Neacomys spinosus (Thomas, 1882). Proc. Zool. Soc. Lond., 1882:105.
  TYPE LOCALITY: N. Peru, Amazonas Dept., Huambo, 1100 m.
  DISTRIBUTION: S.W. Brazil to Colombia, E. Ecuador, and Peru; perhaps Bolivia.
  COMMENT: Systematics and karyology reviewed by Gardner and Patton, 1976, Occas.
    Pap. Mus. Zool. La. St. Univ., 49:1–48.
  ISIS NUMBER: 5301410008024004001.

Neacomys tenuipes Thomas, 1900. Ann. Mag. Nat. Hist., ser. 7, 5:153.
  REVIEWED BY: R. H. Pine (RHP).
  TYPE LOCALITY: Colombia, Cundinamarca, Bogota region, Guaquimay.
  DISTRIBUTION: E. Panama; Colombia; E. Ecuador; Venezuela.
  COMMENT: Includes *pictus* and *pusillus*; see Cabrera, 1961:411–412. Venezuelan records
    discussed by Handley, 1976, Brigham Young Univ. Sci. Bull. Biol. Ser., 20(5):49.
  ISIS NUMBER: 5301410008024002001 as *N. pictus*.
               5301410008024003001 as *N. pusillus*.

Nectomys Peters, 1861. Abh. Preuss. Akad. Wiss., (1860), p. 151.
  REVIEWED BY: A. Langguth (AL); R. D. Owen (RDON); R. H. Pine (RHP); O. A. Reig
    (OAR).
  COMMENT: Formerly included *N. (Sigmodontomys) alfari* which was included in
    Oryzomys by Gardner and Patton, 1976, Occas. Pap. Mus. Zool. La. St. Univ.,
    49:16–17. Revised by Hershkovitz, 1944, Misc. Publ. Mus. Zool. Univ. Mich.,
    58:1–101. Subfamily Hesperomyinae; see comment under Cricetidae.
  ISIS NUMBER: 5301410008025000000.

Nectomys parvipes Petter, 1979. Mammalia, 43:507.
  TYPE LOCALITY: French Guiana, Comte River, Cacao, 4° 35' N., 52° 28' W.
  DISTRIBUTION: Known only from the type locality.

Nectomys squamipes (Brants, 1827). Het. Gesl. Muiz., p. 138.
TYPE LOCALITY: Brazil, Sao Paulo Prov., Sao Sebastiao.
DISTRIBUTION: Guianas to Colombia to Peru; Brazil; Paraguay; N.E. Argentina.
COMMENT: May contain more than one species; karyology reviewed by Gardner and
    Patton, 1976, Occas. Pap. Mus. Zool. La. St. Univ., 49:1–48. Revised by
    Hershkovitz, 1944, Misc. Publ. Mus. Zool. Univ. Mich., 58:1–101. This species is a
    composite according to OAR, who retains *aquaticus, olivaceous, pollens,* and
    mattensis in *squamipes,* and considers *melanius* (including *fulvinus, montanus,* and
    tarrensis), apicalis (including *napensis, grandis, magdalenae,* and *saturatus), palmipes*
    (including *tatei),* and *garleppii* (including *vallensis*) distinct species on
    cytotaxonomic grounds; all of the above forms were included in *squamipes* by
    Cabrera, 1961:412–417.
ISIS NUMBER: 5301410008025002001.

Nelsonia Merriam, 1897. Proc. Biol. Soc. Wash., 11:277.
REVIEWED BY: J. Ramirez-Pulido (JRP); G. Urbano-V. (GUV).
COMMENT: Revised by Hooper, 1954, Occas. Pap. Mus. Zool. Univ. Mich., 558:1–12;
    reviewed by Carleton, 1980. Subfamily Hesperomyinae; see comment under
    Cricetidae.
ISIS NUMBER: 5301410008026000000.

Nelsonia neotomodon Merriam, 1897. Proc. Biol. Soc. Wash., 11:278.
TYPE LOCALITY: Mexico, Zacatecas, mountains near Plateado, 8200 ft. (2499 m).
DISTRIBUTION: S. Durango to S.C. Jalisco and N.C. Michoacan (Mexico).
COMMENT: Reviewed by Genoways and Jones, 1968, Proc. Biol. Soc. Wash., 81:97–100.
ISIS NUMBER: 5301410008026001001.

Neotoma Say and Ord, 1825. J. Acad. Nat. Sci. Phila., 4:345.
REVIEWED BY: D. G. Huckaby (DGH); J. Ramirez-Pulido (JRP)(Mexico); G. Urbano-V.
    (GUV).
COMMENT: Revised by Goldman, 1910, N. Am. Fauna, 31:1–124, and Goldman, 1932, J.
    Mammal., 13:59–67. Burt and Barkalow, 1942, J. Mammal., 23:287–297, Genoways
    and Birney, 1974, Mamm. Species, 41:3, and Hall, 1981:771, included *N. (Hodomys)*
    alleni in this genus, but Carleton, 1980:121, considered *Hodomys* a distinct genus.
    Karyotypic variation reviewed by Mascarello and Hsu, 1976, Evolution,
    30:152–169. Subfamily Hesperomyinae; see comment under Cricetidae.
ISIS NUMBER: 5301410008027000000.

Neotoma albigula Hartley, 1894. Proc. Calif. Acad. Sci., ser. 2, 4:157.
TYPE LOCALITY: U.S.A., Arizona, Pima Co., vicinity of Fort Lowell, near Tucson.
DISTRIBUTION: S.E. California to S.E. Utah, W. and S.E. Colorado, and C. Texas (U.S.A.),
    south to N.E. Michoacan and Hidalgo (Mexico).
COMMENT: Includes *latifrons* and *montezumae;* see Hall and Genoways, 1970, J.
    Mammal., 51:504–516, and Hall, 1981:753. Hybridizes with *micropus* in S.E.
    Colorado and possibly E. Coahuila; see Finley, 1958, Univ. Kans. Mus. Nat. Hist.
    Misc. Publ., 10:213–552, Anderson, 1969, Univ. Kans. Mus. Nat. Hist. Misc. Publ.,
    51:25–50, and Hall, 1981:750–752.
ISIS NUMBER: 5301410008027001001.

Neotoma angustapalata Baker, 1951. Univ. Kans. Mus. Nat. Hist. Misc. Publ., 5:217.
TYPE LOCALITY: Mexico, Tamaulipas, 70 km (by highway) S. Ciudad Victoria, 6 km W.
    Panamerican Highway at El Carrizo.
DISTRIBUTION: S. Tamaulipas and adjacent San Luis Potosi (Mexico).
COMMENT: Revised by Birney, 1973, Univ. Kans. Mus. Nat. Hist. Misc. Publ., 58:1–173;
    also see Hall, 1981:765.
ISIS NUMBER: 5301410008027003001.

Neotoma anthonyi J. A. Allen, 1898. Bull. Am. Mus. Nat. Hist., 10:151.
TYPE LOCALITY: Mexico, Baja California Norte, Todos Santos Isl.
DISTRIBUTION: Known only from the type locality.
ISIS NUMBER: 5301410008027004001.

Neotoma bryanti Merriam, 1887. Am. Nat., 21:191.
  REVIEWED BY: R. D. Owen (RDON).
  TYPE LOCALITY: Mexico, Baja California Norte, Cedros (=Cerros) Isl.
  DISTRIBUTION: Known only from the type locality.
  ISIS NUMBER: 5301410008027005001.

Neotoma bunkeri Burt, 1932. Trans. San Diego Soc. Nat. Hist., 7:181.
  TYPE LOCALITY: Mexico, Baja California Sur, Coronados Isl., 26° 06' N., 111° 18' W.
  DISTRIBUTION: Known only from the type locality.
  ISIS NUMBER: 5301410008027006001.

Neotoma chrysomelas J. A. Allen, 1908. Bull. Am. Mus. Nat. Hist., 24:653.
  TYPE LOCALITY: Nicaragua, Matagalpa Prov., Matagalpa.
  DISTRIBUTION: N.W. Nicaragua; Honduras.
  COMMENT: May be conspecific with *mexicana*; see Hall, 1981:765.
  ISIS NUMBER: 5301410008027007001.

Neotoma cinerea (Ord, 1815). *In* Guthrie, A new geogr. hist. comml. grammar...., 2nd
    Amer. ed., Philadelphia, 2:292.
  TYPE LOCALITY: U.S.A., Montana, Cascade Co., Great Falls.
  DISTRIBUTION: S.E. Yukon, extreme S.W. Northwest Terr., W. Alberta, and extreme
    S.W. Saskatchewan (Canada), south to N. New Mexico, N. Arizona, and C.
    California (U.S.A.).
  COMMENT: Sole member of subgenus *Teonoma*; see Hall, 1981:767.
  ISIS NUMBER: 5301410008027008001.

Neotoma floridana (Ord, 1818). Bull. Sci. Soc. Philom. Paris, p. 181.
  TYPE LOCALITY: U.S.A., Florida, Duval Co., St. Johns River, near Jacksonville.
  DISTRIBUTION: C. Florida to Connecticut, S. Illinois, S.W. South Dakota, E.C. Colorado,
    and E.C. Texas (U.S.A.).
  COMMENT: Includes *magister*; see Birney, 1976, J. Mammal., 57:108-109, who stated that
    further study may show that *magister* is a distinct species. Reviewed by Wiley,
    1980, Mamm. Species, 139:1-7. Birney, 1973, Misc. Publ., Univ. Kans. Mus. Nat.
    Hist., 58:1-173, showed that this species hybridizes with *micropus* at one locality
    but considered *floridana* and *micropus* distinct species.
  ISIS NUMBER: 5301410008027009001.

Neotoma fuscipes Baird, 1858. Mammals, *in* Repts. Expl. Surv...., 8(1):495.
  TYPE LOCALITY: U.S.A., California, Sonoma Co., Petaluma.
  DISTRIBUTION: W. Oregon through California (U.S.A.) to N. Baja California (Mexico).
  ISIS NUMBER: 5301410008027010001.

Neotoma goldmani Merriam, 1903. Proc. Biol. Soc. Wash., 16:48.
  TYPE LOCALITY: Mexico, Coahuila, Saltillo, 5000 ft. (1524 m).
  DISTRIBUTION: S.E. Chihuahua to C. San Luis Potosi (Mexico).
  ISIS NUMBER: 5301410008027011001.

Neotoma lepida Thomas, 1893. Ann. Mag. Nat. Hist., ser. 6, 12:235.
  TYPE LOCALITY: U.S.A., "Simpson's Route" between Camp Floyd (=Fairfield), Utah and
    Carson City, Nevada.
  DISTRIBUTION: Baja California and N.W. Sonora (Mexico) to C. California, S.E. Oregon,
    S.W. Idaho, W. Arizona, W. Utah, and extreme W.C. Colorado (U.S.A.).
  COMMENT: May include two or more distinct species; see Mascarello, 1978, J.
    Mammal., 59:477-495. Includes *devia*; see Hall, 1981:756.
  ISIS NUMBER: 5301410008027012001.

Neotoma martinensis Goldman, 1905. Proc. Biol. Soc. Wash., 18:28.
  TYPE LOCALITY: Mexico, Baja California Norte, San Martin Isl.
  DISTRIBUTION: Known only from the type locality.
  ISIS NUMBER: 5301410008027013001.

Neotoma mexicana Baird, 1855. Proc. Acad. Nat. Sci. Phila., 7:333.
  TYPE LOCALITY: Mexico, Chihuahua, mountains near Chihuahua.
  DISTRIBUTION: El Salvador and E.C. Honduras to S. and W. Coahuila and E. Sonora
    (Mexico), E. Arizona, W. and C. New Mexico, S.E. Utah, and S. and N.C. Colorado
    (U.S.A.).
  COMMENT: Revised by Hall, 1955, J. Wash. Acad. Sci., 45:328–332. May include
    chrysomelas; see Hall, 1981:765.
  ISIS NUMBER: 5301410008027014001.

Neotoma micropus Baird, 1855. Proc. Acad. Nat. Sci. Phila., 7:333.
  TYPE LOCALITY: Mexico, Tamaulipas, Charco Escondido.
  DISTRIBUTION: S.W. Kansas to S.W. New Mexico (U.S.A.), N. Veracruz, and S.E. San
    Luis Potosi (Mexico).
  COMMENT: Includes *canescens* and *planiceps;* see Birney, 1973, Misc. Publ. Univ. Kans.
    Mus. Nat. Hist., 58:173. This species produces hybrids with *floridana* at one
    known locality (see Birney, 1973, who considered *micropus* and *floridana* distinct
    species), and may hybridize with *albigula* in S.E. Colorado (according to Finley,
    1958, Misc. Publ., Univ. Kans. Mus. Nat. Hist., 10:213–552) and possibly in E.
    Coahuila, Mexico (see Anderson, 1969, Univ. Kans. Mus. Nat. Hist. Misc. Publ.,
    51:25–50); see Hall, 1981:748, 750–752.
  ISIS NUMBER: 5301410008027015001.

Neotoma nelsoni Goldman, 1905. Proc. Biol. Soc. Wash., 18:29.
  TYPE LOCALITY: Mexico, Veracruz, Perote, 7800 ft. (2377 m).
  DISTRIBUTION: Known only from the type locality.
  ISIS NUMBER: 5301410008027016001.

Neotoma palatina Goldman, 1905. Proc. Biol. Soc. Wash., 18:27.
  TYPE LOCALITY: Mexico, Jalisco, Bolanos, 2800 ft. (853 m).
  DISTRIBUTION: N.C. Jalisco (Mexico).
  COMMENT: Revised by Hall and Genoways, 1970, J. Mammal., 51:504–516.
  ISIS NUMBER: 5301410008027017001.

Neotoma phenax (Merriam, 1903). Proc. Biol. Soc. Wash., 16:81.
  TYPE LOCALITY: Mexico, Sonora, Rio Mayo, Camoa.
  DISTRIBUTION: S.W. Sonora and N.W. Sinaloa (Mexico).
  COMMENT: Sole member of subgenus *Teanopus;* reviewed by Jones and Genoways,
    1978, Mamm. Species, 108:1–3; Burt and Barkalow, 1942, J. Mammal., 23:296.
  ISIS NUMBER: 5301410008027018001.

Neotoma stephensi Goldman, 1905. Proc. Biol. Soc. Wash., 18:32.
  TYPE LOCALITY: U.S.A., Arizona, Mohave Co., Hualapai Mtns., 6300 ft. (1920 m).
  DISTRIBUTION: N.W. and W.C. New Mexico; N.E. and C. Arizona; S.C. Utah (U.S.A.).
  COMMENT: Revised by Hoffmeister and de la Torre, 1960, J. Mammal., 41:476–491.
  ISIS NUMBER: 5301410008027019001.

Neotoma varia Burt, 1932. Trans. San Diego Soc. Nat. Hist., 7:178.
  TYPE LOCALITY: Mexico, Sonora, Turner Isl., 28° 43′ N., 112° 19′ W.
  DISTRIBUTION: Known only from the type locality.
  ISIS NUMBER: 5301410008027020001.

Neotomodon Merriam, 1898. Proc. Biol. Soc. Wash., 12:127.
  REVIEWED BY: R. D. Owen (RDON); J. Ramirez-Pulido (JRP); G. Urbano-V. (GUV).
  COMMENT: Included in *Peromyscus* as a subgenus by Yates *et al.,* 1979, Syst. Zool.,
    28:40–48, and Patton *et al.,* 1981:288–308, *in* Smith and Joule, eds., Mammalian
    Population Genetics, Univ. Georgia Press, Athens; but see Carleton, 1980:118, 126,
    and Hall, 1981:745, who considered *Neotomodon* a distinct genus. Subfamily
    Hesperomyinae; see comment under Cricetidae.
  ISIS NUMBER: 5301410008028000000.

Neotomodon alstoni Merriam, 1898. Proc. Biol. Soc. Wash., 12:128.
  TYPE LOCALITY: Mexico, Michoacan, Nahuatzin, 8500 ft. (2591 m).
  DISTRIBUTION: C. Michoacan to C. Puebla and C. Veracruz (Mexico).

COMMENT: Includes *orizabae* and *perotensis*; see Yates *et al.*, 1979, Syst. Zool., 28:40–48, and Hall, 1981:745–746.
ISIS NUMBER: 5301410008028001001.

Neotomys Thomas, 1894. Ann. Mag. Nat. Hist., ser. 6, 14:346.
REVIEWED BY: R. D. Owen (RDON); R. H. Pine (RHP).
COMMENT: Reviewed by Pearson, 1951, Bull. Mus. Comp. Zool. Harv. Univ., 106(3):117–174, and Sanborn, 1947, Fieldiana Zool., 31(7):51–57. Also see Gardner and Patton, 1976, Occas. Pap. Mus. Zool. La. St. Univ., 49:31, 37, and Pearson and Patton, 1976, J. Mammal., 57:339–350. Subfamily Hesperomyinae; see comment under Cricetidae.
ISIS NUMBER: 5301410008029000000.

Neotomys ebriosus Thomas, 1894. Ann. Mag. Nat. Hist., ser. 6, 14:348.
TYPE LOCALITY: C. Peru, Junin Dept., Vitoc Valley.
DISTRIBUTION: Peru; Bolivia; N. Chile; N.W. Argentina.
COMMENT: Karyology reviewed by Pearson and Patton, 1976, J. Mammal., 57:339–350. Pine *et al.*, 1979, Mammalia, 43:357, recorded this species in N. Chile.
ISIS NUMBER: 5301410008029001001.

Nesomys Peters, 1870. Sitzb. Ges. Naturf. Fr. Berlin, p. 54.
REVIEWED BY: R. D. Owen (RDON).
COMMENT: Subfamily Nesomyinae; see comment under Cricetidae.
ISIS NUMBER: 5301410008066000000.

Nesomys rufus Peters, 1870. Sitzb. Ges. Naturf. Fr. Berlin, p. 55.
TYPE LOCALITY: Madagascar, Vohima.
DISTRIBUTION: Madagascar.
COMMENT: Includes *audeberti* and *lambertoni*; see Petter, 1975, Part 6.2:3.
ISIS NUMBER: 5301410008066003001 as *N. rufus.*
               5301410008066001001 as *N. audeberti.*
               5301410008066002001 as *N. lambertoni.*

Nesoryzomys Heller, 1904. Proc. Calif. Acad. Sci., 3:241.
REVIEWED BY: M. S. Boyce (MSB); R. Guenzel (RG); A. Langguth (AL); R. A. Ojeda (RAO); J. Ramirez-Pulido (JRP).
COMMENT: Considered a distinct genus by Gardner and Patton, 1976, Occas. Pap. Mus. Zool. La. St. Univ., 49:20, and Voss and Linzey, 1981, Misc. Publ. Mus. Zool. Univ. Mich., 159:24. Ellerman, 1941:406, and Corbet and Hill, 1980:142, considered *Nesoryzomys* a subgenus of *Oryzomys*. Subfamily Hesperomyinae; see comment under Cricetidae.

Nesoryzomys darwini Osgood, 1929. Field Mus. Nat. Hist. Publ. Zool. Ser., 17:23.
TYPE LOCALITY: Galapagos Isls., Santa Cruz Isl., Academia Bay (Ecuador).
DISTRIBUTION: Santa Cruz Isl. (Galapagos Isls.).
COMMENT: Formerly included in *Oryzomys*; see comment under genus. Probably extinct; see Corbet and Hill, 1980:143.
ISIS NUMBER: 5301410008035021001 as *Oryzomys darwini.*

Nesoryzomys fernandinae Hutterer and Hirsch, 1979. Bonn. Zool. Beitr., 30:276.
TYPE LOCALITY: Galapagos Isls., Fernandina Isl. (Ecuador).
DISTRIBUTION: Known only from the type locality.
COMMENT: Formerly included in *Oryzomys*; see comment under genus.

Nesoryzomys indefessus (Thomas, 1899). Ann. Mag. Nat. Hist., ser. 7, 4:280.
TYPE LOCALITY: Galapagos Isls., Santa Cruz Isl., Academia Bay (Ecuador).
DISTRIBUTION: Santa Cruz Isl. (Galapagos Isls.).
COMMENT: Formerly included in *Oryzomys*; see comment under genus. Probably extinct; see Corbet and Hill, 1980:144.
ISIS NUMBER: 5301410008035032001 as *Oryzomys indefessus.*

Nesoryzomys narboroughi Heller, 1904. Proc. Calif. Acad. Sci., 3(3):242.
TYPE LOCALITY: Galapagos Isls., Fernandina Isl., Punta Mangle (Ecuador).
DISTRIBUTION: Fernandina Isl. (Colon Arch., Galapagos Isls.).
COMMENT: Formerly included in *Oryzomys*; see comment under genus. Cabrera, 1961:410, considered *narboroughi* possibly conspecific with *indefessus*; but see Gardner and Patton, 1976, Occas. Pap. Mus. Zool. La. St. Univ., 49:20, and Voss and Linzey, 1981, Misc. Publ. Mus. Zool. Univ. Mich., 159:24, who considered *narboroughi* a distinct species.
ISIS NUMBER: 5301410008035046001 as *Oryzomys narboroughi*.

Nesoryzomys swarthi Orr, 1938. Proc. Calif. Acad. Sci., 23(21):304.
TYPE LOCALITY: Galapagos Isls., James Isl., Sullivan Bay (Ecuador).
DISTRIBUTION: Known only from James Isl. (Galapagos Isls.).
COMMENT: Reviewed by Peterson, 1966, Mammalia, 30:441-445.

Neusticomys Anthony, 1921. Am. Mus. Novit., 20:2.
REVIEWED BY: A. Langguth (AL); R. D. Owen (RDON); R. S. Voss (RSV).
COMMENT: Male reproductive anatomy reviewed by Voss and Linzey, 1981, Misc. Publ. Mus. Zool. Univ. Mich., 159:1-41. Subfamily Hesperomyinae; see comment under Cricetidae.
ISIS NUMBER: 5301410008030000000.

Neusticomys monticolus Anthony, 1921. Am. Mus. Novit., 20:2.
TYPE LOCALITY: Ecuador, Pichincha Prov., Nono, Hacienda San Francisco, 10,500 ft. (3150 m).
DISTRIBUTION: N. Ecuador; Andean Colombia.
ISIS NUMBER: 5301410008030001001.

Notiomys Thomas, 1890. *In* Milne-Edwards, Mission Sci. Cap. Horn, 1882-3, 6, Mamm., p. 23.
REVIEWED BY: R. D. Owen (RDON).
COMMENT: Revised by Osgood, 1943, Field Mus. Nat. Hist. Publ. Zool. Ser., 30:151-166; reviewed by Cabrera, 1961:470-474. Subfamily Hesperomyinae; see comment under Cricetidae.
ISIS NUMBER: 5301410008031000000.

Notiomys angustus (Thomas, 1927). Ann. Mag. Nat. Hist., ser. 9, 19:654.
TYPE LOCALITY: Argentina ("N.W. Patagonia"), Rio Negro Prov., Lake Nahuel Huapi, Bariloche.
DISTRIBUTION: W. Argentina.
ISIS NUMBER: 5301410008031001001.

Notiomys delfini (Cabrera, 1905). Rev. Chil. Hist. Nat., 9:15.
TYPE LOCALITY: Chile, Straits of Magellan, Punta Arenas.
DISTRIBUTION: S. Chile; S. Argentina.
COMMENT: Cabrera, 1961:470, considered *delfini* a distinct species; Corbet and Hill, 1980:152, listed it as a subspecies of *megalonyx*.
ISIS NUMBER: 5301410008031002001.

Notiomys edwardsii (Thomas, 1890). *In* Milne-Edwards, Mission Sci. Cap. Horn, 1882-3, 6, Mamm., p. 24.
TYPE LOCALITY: Argentina ("southern Patagonia"), Santa Cruz Prov., south of Santa Cruz.
DISTRIBUTION: S. Argentina.
ISIS NUMBER: 5301410008031003001.

Notiomys macronyx (Thomas, 1894). Ann. Mag. Nat. Hist., ser. 6, 14:362.
TYPE LOCALITY: Argentina, Mendoza Prov., Fort San Rafael.
DISTRIBUTION: S. and W. Argentina; E. and S. Chile.
COMMENT: Reviewed by Pine *et al.*, 1979, Mammalia, 43:351.
ISIS NUMBER: 5301410008031004001.

Notiomys megalonyx (Waterhouse, 1845). Proc. Zool. Soc. Lond., 1844:154.
TYPE LOCALITY: Chile, Valparaiso Prov., Lake Quintero.
DISTRIBUTION: C. Chile.
COMMENT: Reviewed by Pine *et al.*, 1979, Mammalia, 43:352. May include *delfini;* see
   Corbet and Hill, 1980:152.
ISIS NUMBER: 5301410008031005001.

Notiomys valdivianus (Philippi, 1858). Arch. Naturgesch., 24(1):303.
TYPE LOCALITY: Chile, Valdivia Prov.
DISTRIBUTION: S. and C. Chile; Mocha and Chiloe Isls. (Chile); S. and W. Argentina.
ISIS NUMBER: 5301410008031006001.

Nyctomys Saussure, 1860. Rev. Mag. Zool. Paris, ser. 2, 12:106.
REVIEWED BY: R. D. Owen (RDON); J. Ramirez-Pulido (JRP).
COMMENT: Reviewed by Carleton, 1980:139. Subfamily Hesperomyinae; see comment
   under Cricetidae.
ISIS NUMBER: 5301410008032000000.

Nyctomys sumichrasti (Saussure, 1860). Rev. Mag. Zool. Paris, ser. 2, 12:107.
TYPE LOCALITY: Mexico, Veracruz, Uvero, 20 km N.W. of Santiago Tuxtla.
DISTRIBUTION: W. and C. Panama to Jalisco and S. Veracruz (Mexico).
COMMENT: Type locality restricted by Alvarez, 1963, J. Mammal., 44:583. Reviewed by
   Voss and Linzey, 1981, Misc. Publ. Mus. Zool. Univ. Mich., 159:23–24.
ISIS NUMBER: 5301410008032001001.

Ochrotomys Osgood, 1909. N. Am. Fauna, 28:222.
REVIEWED BY: R. E. Barry, Jr. (REB); M. D. Carleton (MDC); L. N. Carraway (LNC); D.
   W. Linzey (DWL); R. D. Owen (RDON).
COMMENT: Included in *Peromyscus* by Hooper and Musser, 1964, Occas. Pap. Mus.
   Zool. Univ. Mich., 635:8–9. Considered distinct from *Peromyscus* by Blair, 1942, J.
   Mammal., 23:196–204, Manville, 1961, J. Mammal., 42:103–104, Hooper, 1968, *in*
   King, ed., The Biology of *Peromyscus* (Rodentia), p. 27, and Carleton, 1980:118,
   122. Subfamily Hesperomyinae; see comment under Cricetidae.
ISIS NUMBER: 5301410008033000000.

Ochrotomys nuttalli (Harlan, 1832). Mon. Am. J. Geol. Nat. Sci. Phila., p. 446.
TYPE LOCALITY: U.S.A., Virginia, Norfolk Co., Norfolk.
DISTRIBUTION: E. Texas to S. Illinois, east to S. Virginia and C. Florida (U.S.A.).
COMMENT: Reviewed by Linzey and Packard, 1977, Mamm. Species, 75:1–6; revised by
   Packard, 1969, Misc. Publ. Univ. Kans. Mus. Nat. Hist., 51:373–406, and Patton
   and Hsu, 1967, J. Mammal., 48:637.
ISIS NUMBER: 5301410008033001001.

Onychomys Baird, 1858. Mammals, *in* Repts. Expl. Surv...., 8(1):458.
REVIEWED BY: M. S. Boyce (MSB); M. D. Carleton (MDC); R. Guenzel (RG); J. Ramirez-
   Pulido (JRP); G. Urbano-V. (GUV).
COMMENT: Revised by Hollister, 1914, Proc. U.S. Nat. Mus., 47(2057):427–489. Recent
   and fossil forms reviewed by Carleton and Eshelman, 1979, Univ. Mich. Pap.
   Paleontol., 7(21):1–60, and Carleton, 1980:118, 122. Karylogy reviewed by Baker *et
   al.*, 1979, J. Mammal., 60:297–306. Subfamily Hesperomyinae; see comment under
   Cricetidae.
ISIS NUMBER: 5301410008034000000.

Onychomys arenicola Mearns, 1896. Preliminary diagnosis of new mammals from the
   Mexican border of the United States, p. 3 (preprint of Proc. U.S. Natl. Mus., 19:137–
   140).
TYPE LOCALITY: U.S.A., Texas, El Paso Co., 6 mi. (9.7 km) above El Paso.
DISTRIBUTION: C. and S.W. New Mexico and extreme W. Texas (U.S.A.); N.W.
   Chihuahua and N.E. Sonora (Mexico); perhaps S.E. New Mexico, W. Texas, and C.
   Mexico.

COMMENT: Formerly included in *torridus;* considered a distinct species by Hinesley, 1979, J. Mammal., 60:117–128, and Baker *et al.,* 1979, J. Mammal., 60:297–306; reviewed by Hall, 1981:1180.

Onychomys leucogaster (Wied-Neuwied, 1841). Reise Nord-America, 2:99.
TYPE LOCALITY: U.S.A., North Dakota, Oliver Co., Mandan Indian Village, near Fort Clark.
DISTRIBUTION: N. Tamaulipas (Mexico) to E. California, S.E. Washington, W. Minnesota (U.S.A.), S. Alberta, S. Saskatchewan, and S.W. Manitoba (Canada).
COMMENT: Reviewed by McCarty, 1978, Mamm. Species, 87:1–6, and Engstrom and Choate, 1979, J. Mammal., 60:723–739.
ISIS NUMBER: 5301410008034001001.

Onychomys torridus (Coues, 1874). Proc. Acad. Nat. Sci. Phila., 26:183.
TYPE LOCALITY: U.S.A., Arizona, Graham Co., Camp Grant.
DISTRIBUTION: C. California, S. Nevada, and S.W. Utah to W. Texas (U.S.A.), N. Baja California, and San Luis Potosi (Mexico).
COMMENT: Partially reviewed by Matson and Friesen, 1978, Bull. South. Calif. Acad. Sci., 77(3):116–123. Reviewed by McCarty, 1975, Mamm. Species, 59:1–5. Formerly included *arenicola;* see Hinesley, 1979, J. Mammal., 60:117–128; range will require adjustment if *arenicola* is found to occur in S.E. New Mexico, W. Texas (U.S.A.) and C. Mexico; see also comment under *arenicola.*
ISIS NUMBER: 5301410008034002001.

Oryzomys Baird, 1858. Mammals, *in* Repts. Expl. Surv...., 8(1):458.
REVIEWED BY: M. S. Boyce (MSB); M. D. Carleton (MDC); A. L. Gardner (ALG); R. Guenzel (RG); A. Langguth (AL); R. A. Ojeda (RAO); J. Ramirez-Pulido (JRP)(Mexico); O. A. Reig (OAR).
COMMENT: Includes *Macruroryzomys, Melanomys, Micronectomys, Microryzomys, Oligoryzomys,* and *Sigmodontomys;* see Gardner and Patton, 1976, Occas. Pap. Mus. Zool. La. St. Univ., 49:1–48, who considered *Oecomys* a distinct genus, as does RAO; but also see Hershkovitz, 1960, Proc. U.S. Nat. Mus., 110(3420):515–532, Husson, 1978:402, Hall, 1981:618, and Voss and Linzey, 1981, Misc. Publ. Mus. Zool. Univ. Mich., 159:13, who considered *Oecomys* a subgenus of *Oryzomys.* Ellerman, 1941:340, included *Oligoryzomys* in *Oryzomys* as a subgenus; Cabrera, 1961:380, included *Oligoryzomys* in subgenus *Oryzomys* as a synonym. Gardner and Patton, 1976, Occas. Pap. Mus. Zool. La. St. Univ., 49:2, included *Macruroryzomys* and *Micronectomys* in *Oryzomys;* these names were proposed as subgenera by Hershkovitz, 1948, Proc. U.S. Nat. Mus., 98(3221):49–56, but Hershkovitz, 1970, J. Mammal., 51:791, and Pine and Wetzel, 1975, Mammalia, 39:653, considered *Macruroryzomys* and *Micronectomys, nomina nuda.* Walker *et al.,* 1975:759, considered *Melanomys* a distinct genus. Formerly included *Nesoryzomys* (see Gardner and Patton, 1976:20, and Voss and Linzey, 1981, Misc. Publ. Mus. Zool. Univ. Mich., 159:24), *borreroi* (which Gardner and Patton, 1976:41, transferred to *Zygodontomys), Wiedomys pyrrhorhinos* (see Hershkovitz, 1959, Proc. Biol. Soc. Wash., 72:5), *Pseudoryzomys wavrini* (see Hershkovitz, 1962, Fieldiana Zool., 46:208), *Rhipidomys maculipes* (see Hershkovitz, 1960, Proc. U.S. Nat. Mus., 110:519), and *simplex* (which Massoia, 1980, Ameghiniana, 17:280–287, transferred to *Pseudoryzomys).* North American species revised by Goldman, 1918, N. Am. Fauna, 43:1–100. *Oryzomys* is in need of revision. Subfamily Hesperomyinae; see comment under Cricetidae.
ISIS NUMBER: 5301410008035000000.

Oryzomys albigularis (Tomes, 1860). Proc. Zool. Soc. Lond., 1860:264.
TYPE LOCALITY: Ecuador, Chimborazo, Pallatanga, 1485 m.
DISTRIBUTION: Costa Rica and W. Panama; E. Panama through the Andes to Peru; N.W. Bolivia; W. Venezuela.
COMMENT: Includes *devius* (see Handley, 1966, *in* Wenzel and Tipton, eds., Ectoparasites of Panama, p. 779), *maculiventer* (see Gardner and Patton, 1976, Occas. Pap. Mus. Zool. La. St. Univ., 49:19, 38), and *pirrensis* (see Cabrera,

1961:383, and Hershkovitz, 1966, Z. Saugetierk., 31:137). Formerly included
auriventer and *boliviae* (which is included in *nitidus*); see Gardner and Patton,
1976:24, 38, who considered *albigularis* a composite; subgenus *Oryzomys*.
ISIS NUMBER: 5301410008035001001 as *O. albigularis*.
              5301410008035024001 as *O. devius*.

Oryzomys alfari (J. A. Allen, 1897). Bull. Am. Mus. Nat. Hist., 9:39.
    TYPE LOCALITY: Costa Rica, Limon, Jimenez, 700 ft. (213 m).
    DISTRIBUTION: E. Honduras to Panama to N. Ecuador and W. Venezuela.
    COMMENT: Formerly included in *Nectomys*; Gardner and Patton, 1976, Occas. Pap.
        Mus. Zool. La. St. Univ., 49:16–17, placed *alfari* in *Oryzomys (Sigmodontomys)*.
        Corbet and Hill, 1980:144, included *alfari* in *russulus*, without comment. Hall,
        1981:623, considered *alfari* a distinct species. Includes *russulus*; see Cabrera,
        1961:418, and Hershkovitz, 1970, J. Mammal., 51:791.
    ISIS NUMBER: 5301410008025001001 as *Nectomys alfari*.

Oryzomys alfaroi (J. A. Allen, 1891). Bull. Am. Mus. Nat. Hist., 3:214.
    REVIEWED BY: G. Urbano-V. (GUV).
    TYPE LOCALITY: Costa Rica, San Carlos.
    DISTRIBUTION: S. Tamaulipas and Guerrero (Mexico) to W. Panama; E. Panama;
        Colombia; Ecuador.
    COMMENT: Subgenus *Oryzomys*; see Hall, 1981:608, 615. See also review by
        Hershkovitz, 1966, Z. Saugetierk., 31:138.
    ISIS NUMBER: 5301410008035002001.

Oryzomys altissimus Osgood, 1933. Field Mus. Nat. Hist. Publ. Zool. Ser., 20:5.
    TYPE LOCALITY: Peru, Mt. Pasco, La Quinua, 3480 m.
    DISTRIBUTION: Andean Ecuador and Peru.
    COMMENT: Subgenus *Microryzomys*; see Voss and Linzey, 1981, Misc. Publ. Mus. Zool.
        Univ. Mich., 159:13. Kiblisky, 1969, Experientia, 25:1338, and Gardner and Patton,
        1976, Occas. Pap. Mus. Zool. La. St. Univ., 49:11, 24, considered *altissimus* a
        subspecies of *minutus*.
    ISIS NUMBER: 5301410008035003001.

Oryzomys andinus Osgood, 1914. Field Mus. Nat. Hist. Publ. Zool. Ser., 10:156.
    TYPE LOCALITY: Peru, Libertad, Chicama, Hacienda Llagueda, 1800 m.
    DISTRIBUTION: N. Peru, west of the Andes.
    COMMENT: Subgenus *Oligoryzomys*; see Gardner and Patton, 1976, Occas. Pap. Mus.
        Zool. La. St. Univ., 49:19, 24.
    ISIS NUMBER: 5301410008035004001.

Oryzomys aphrastus Harris, 1932. Occas. Pap. Mus. Zool. Univ. Mich., 248:5.
    TYPE LOCALITY: Costa Rica, San Jose, San Joaquin de Dota, 4000 ft. (1219 m).
    DISTRIBUTION: Known only from the type locality.
    COMMENT: Status uncertain; subgenus *Oryzomys*; see Hall, 1981:618.
    ISIS NUMBER: 5301410008035006001.

Oryzomys arenalis Thomas, 1913. Ann. Mag. Nat. Hist., ser. 8, 12:571.
    TYPE LOCALITY: Peru, Lambayeque Dept., Eten.
    DISTRIBUTION: N.E. Peru.
    COMMENT: Gyldenstolpe, 1932, Kongl. Svenska Vet.-Akad. Handl. Stockholm, 11(3):27,
        placed *arenalis* in *Oligoryzomys* which was considered a subgenus by Gardner and
        Patton, 1976, Occas. Pap. Mus. Zool. La. St. Univ., 49:19; Cabrera, 1961:384, placed
        both taxa in subgenus *Oryzomys*.
    ISIS NUMBER: 5301410008035007001.

Oryzomys argentatus Spitzer and Lazell, 1978. J. Mammal., 59:787.
    TYPE LOCALITY: U.S.A., Florida, Monroe Co., Cudjoe Key.
    DISTRIBUTION: Known only from the type locality.
    COMMENT: Subgenus *Oryzomys*; see Spitzer and Lazell, 1978, J. Mammal., 59:787–792.

Oryzomys auriventer Thomas, 1890. Ann. Mag. Nat. Hist., ser. 4, 7:379.
    TYPE LOCALITY: Ecuador, Napo-Pastaza Prov., Mirador, 1500 m.
    DISTRIBUTION: Peru; Ecuador.

COMMENT: Considered distinct from *albigularis* by Gardner and Patton, 1976, Occas. Pap. Mus. Zool. La. St. Univ., 49:19, 24, 38, who included this species in subgenus *Oryzomys*.

Oryzomys balneator Thomas, 1900. Ann. Mag. Nat. Hist., ser. 7, 5:273.
TYPE LOCALITY: Ecuador, Napo-Pastaza Prov., Alto Pastaza, Mirador, 1500 m.
DISTRIBUTION: E. and S. Ecuador.
COMMENT: Subgenus *Oryzomys*; see Cabrera, 1961:384.
ISIS NUMBER: 5301410008035009001.

Oryzomys bauri J. A. Allen, 1892. Bull. Am. Mus. Nat. Hist., 4:48.
TYPE LOCALITY: Galapagos Isls., Barrington Isl. (Ecuador).
DISTRIBUTION: Santa Fe Isl. (Galapagos Isls.).
COMMENT: Subgenus *Oryzomys*; see Gardner and Patton, 1976, Occas. Pap. Mus. Zool. La. St. Univ., 49:16. Considered a synonym of *galapagoensis* by Cabrera, 1961:389.

Oryzomys bicolor (Tomes, 1860). Proc. Zool. Soc. Lond., 1860:217.
TYPE LOCALITY: Ecuador, Santiago-Zamora Prov., Rio Gualaquiza, Gualaquiza, 885 m.
DISTRIBUTION: Panama to the Guianas, S.C. Brazil, N.W. Bolivia, and S.E. Peru.
COMMENT: Includes *endersi*, *phaeotis*, and *trabeatus*; subgenus *Oecomys*; see Hershkovitz, 1960, Proc. U.S. Nat. Mus., 110:533–544, Husson, 1978:406, and Hall, 1981:619; also see comment under genus. Cabrera, 1961:407, and Corbet and Hill, 1980:144, listed *phaeotis* as a distinct species, without comment.
ISIS NUMBER: 5301410008035010001 as *O. bicolor*.
                      5301410008035025001 as *O. endersi*.
                      5301410008035052001 as *O. phaeotis*.
                      5301410008035062001 as *O. trabeatus*.

Oryzomys bombycinus Goldman, 1912. Smithson. Misc. Coll., 56(36):6.
REVIEWED BY: R. H. Pine (RHP).
TYPE LOCALITY: Panama, Cerro Azul, near headwaters of Chagres River, 770 m.
DISTRIBUTION: E. Nicaragua to N. Ecuador.
COMMENT: Revised by Pine, 1971, J. Mammal., 52:590–596. May be conspecific with *rivularis*; see Gardner and Patton, 1976, Occas. Pap. Mus. Zool. La. St. Univ., 49:40–41, and Hall, 1981:617, who listed *bombycinus* in subgenus *Oryzomys*. *O. bombycinus* is conspecific with *rivularis* (ALG).
ISIS NUMBER: 5301410008035011001.

Oryzomys buccinatus (Olfers, 1818). *In* Eschwege, Neue Bibl. Reisenb., 15:209.
TYPE LOCALITY: Paraguay, Caraguatay, 45 km east of Asuncion, Atyra (=Atira).
DISTRIBUTION: Paraguay; N.E. Argentina.
COMMENT: Hershkovitz, 1959, J. Mammal., 40:347, included *ratticeps* in *buccinatus*; but see Avila-Pires, 1960, Acta Trab. Prim. Cong. Sudam. Zool., sec. 5, Vertebrados, 4:3–7, Corbet and Hill, 1980:144, Voss and Linzey, 1981, Misc. Publ. Mus. Zool. Univ. Mich., 159:13, and Myers and Carleton, 1981, Misc. Publ. Mus. Zool. Univ. Mich., 161:1–41, who all listed *ratticeps* as a distinct species. Subgenus *Oryzomys*; see Cabrera, 1961:385, 395. AL employs the name *angouya* (Fischer, 1814), for this species; see comment under *Reithrodon physodes*.
ISIS NUMBER: 5301410008035013001.

Oryzomys caliginosus (Tomes, 1860). Proc. Zool. Soc. Lond., 1860:263.
TYPE LOCALITY: Ecuador, Esmeraldas Prov., Esmeraldas.
DISTRIBUTION: E. Honduras to S. Ecuador.
COMMENT: Subgenus *Melanomys*; see Gardner and Patton, 1976, Occas. Pap. Mus. Zool. La. St. Univ., 49:11, 16, and Voss and Linzey, 1981, Misc. Publ. Mus. Zool. Univ. Mich., 159:13.
ISIS NUMBER: 5301410008035014001.

Oryzomys capito (Olfers, 1818). *In* Eschwege, Neue Bibl. Reisenb., 15:209.
TYPE LOCALITY: Paraguay, San Ignacio Guazu.
DISTRIBUTION: E. Costa Rica to the Guianas, S. Brazil, Bolivia, E. Peru, and N.W. Argentina; Trinidad.
COMMENT: Formerly included *yunganus*; see Gardner and Patton, 1976, Occas. Pap.

Mus. Zool. La. State Univ., 49:24, 40. Includes *carrikeri* and *talamancae*; see Handley, 1966, *in* Wenzel and Tipton, eds., Ectoparasites of Panama, p. 780, and Gardner and Patton, 1976:38–40, who also tentatively included *goeldii, perenensis, castaneus, magdalenae, medius, modestus, mollipilosus, oniscus* and *velutinus* in this species, as is done here. Includes *laticeps*; subgenus *Oryzomys*; see Hershkovitz, 1966, Z. Saugetierk., 31:137, and Husson, 1978:406. Cabrera, 1961:338, included *bolivaris, intermedius*, and *legatus* in this species, but see Gardner and Patton, 1976:40, who indicated that they may be synonymous with *nitidus* which they considered a distinct species; the relationships between *capito, nitidus, alfaroi*, and *laticeps* require additional resolution. *O. bolivaris* is a synonym of *rivularis; intermedius* has no status being a renaming of *laticeps*, which is a subjective synonym of *capito* (ALG). AL employs the name *megacephalus* (Fischer, 1814), for this species; see comment under *Reithrodon physodes*.
ISIS NUMBER: 5301410008035015001 as *O. capito*.
                   5301410008035060001 as *O. talamaccae (sic)*.

Oryzomys caudatus Merriam, 1901. Proc. Wash. Acad. Sci., 3:289.
TYPE LOCALITY: Mexico, Oaxaca, Comaltepec, 3500 ft. (1067 m).
DISTRIBUTION: N.C. Oaxaca (Mexico).
COMMENT: Reviewed by Haiduk *et al.*, 1980, J. Mammal., 60:610–614; Goodwin, 1969, Bull. Am. Mus. Nat. Hist., 141:157; subgenus *Oryzomys*.
ISIS NUMBER: 5301410008035017001.

Oryzomys chacoensis Myers and Carleton, 1981. Misc. Publ. Mus. Zool. Univ. Michigan, 161:19.
TYPE LOCALITY: Paraguay, Boqueron Dept., 419 km, by road, N.W. of Villa Hayes (along the trans-Chaco Highway).
DISTRIBUTION: Chaco of Paraguay, Bolivia, Argentina, and S.W. Mato Grosso (Brazil).
COMMENT: Subgenus *Oligoryzomys*; see Myers and Carleton, 1981:20.

Oryzomys chaparensis Osgood, 1916. Field Mus. Nat. Hist. Publ. Zool. Ser., 10:205.
TYPE LOCALITY: Bolivia, Cochabamba, Todos Santos, above the Chapare River, 360 m.
DISTRIBUTION: E. and C. Bolivia.
COMMENT: Subgenus *Oryzomys*; see Cabrera, 1961:388.
ISIS NUMBER: 5301410008035018001.

Oryzomys concolor (Wagner, 1845). Arch. Naturgesch., 11(1):147.
TYPE LOCALITY: Brazil, Amazonas, Rio Curicuriari, below Sao Gabriel.
DISTRIBUTION: S.C. Costa Rica to Guianas, Brazil, N. Bolivia, and Peru; N. Argentina (AL); Trinidad.
COMMENT: Includes *helvolus, mamorae, marmosurus, melleus, mincae, osgoodi, roberti, tectus, trinitatis*, and *vicencianus*; subgenus *Oecomys*; see Hershkovitz, 1960, Proc. U.S. Nat. Mus., 110:545–559, Husson, 1978:406, Hall, 1981:619, and Voss and Linzey, 1981, Misc. Publ. Mus. Zool. Univ. Mich., 159:13. Gardner and Patton, 1976, Occas. Pap. Mus. Zool. La. St. Univ., 49:20, 41, considered *Oecomys* a distinct genus. Reviewed by Massoia and Fornes, 1965, Physis, 25(70):319–324.
ISIS NUMBER: 5301410008035037001 as *O. mamorae*.
                   5301410008035038001 as *O. marmosurus*.
                   5301410008035041001 as *O. melleus*.
                   5301410008035043001 as *O. mincae*.
                   5301410008035049001 as *O. osgoodi*.
                   5301410008035055001 as *O. roberti*.
                   5301410008035061001 as *O. tectus*.
                   5301410008035063001 as *O. trinitatis*.

Oryzomys couesi (Alston, 1877). Proc. Zool. Soc. Lond., 1876:756.
TYPE LOCALITY: Guatemala, Coban.
DISTRIBUTION: C. Panama to Hidalgo to S. Sonora (Mexico) and to S. Texas (U.S.A.); Cozumel Isl. (Mexico); Jamaica.
COMMENT: Formerly included in *palustris*; see Benson and Gehlbach, 1979, J. Mammal., 60:225–228, who considered *O. couesi aquaticus* distinct from *palustris*; also see Haiduk *et al.*, 1980, J. Mammal., 60:610–614, and Hall, 1981:1179. Forms not

investigated by Benson and Gehlbach, but which are included in this species, were placed in *couesi* by Hall and Kelson, 1959:556-560. Hall, 1960, Southwest. Nat., 5:173, considered *O. palustris texensis* and *O. couesi aquaticus* conspecific and included *couesi* in *palustris*. Subsequently *azuerensis, gatunensis* (see Handley, 1966, in Wenzel and Tipton, eds., Ectoparasites of Panama, p. 781), *cozumelae* (see Jones and Lawlor, 1965, Univ. Kans. Mus. Nat. Hist. Misc. Publ., 16:413), and *antillarum* (see Hershkovitz, 1966, in Wenzel and Tipton, eds., Ectoparasites of Panama, p. 736) were included in *palustris* but are here included in *couesi* on the basis of its recognition by Benson and Gehlbach, 1979. Does not include *peninsulae*. Also see Hall, 1981:611, 1179; subgenus *Oryzomys*.

ISIS NUMBER: 5301410008035019001 as *O. couesi.*
                5301410008035005001 as *O. antillarum.*
                5301410008035008001 as *O. azuerensis.*
                5301410008035020001 as *O. cozumelae.*
                5301410008035030001 as *O. gatunensis.*

Oryzomys delicatus J. A. Allen and Chapman, 1897. Bull. Am. Mus. Nat. Hist., 9:19.
REVIEWED BY: R. H. Pine (RHP).
TYPE LOCALITY: Trinidad and Tobago, Trinidad, Caparo.
DISTRIBUTION: N.C. Brazil; Surinam; Guyana; Trinidad; Venezuela; N.E. Colombia.
COMMENT: Includes *microtis* (see Pine, 1973, Acta Amazonica, 3(2):63), *navus* and *messorius*; subgenus *Oligoryzomys* (see Husson, 1978:398). Cabrera, 1961:388-389, included *tenuipes* in this species and included *delicatus* in subgenus *Oryzomys*. The relationship of *delicatus* and *tenuipes* to *fornesi* and *flavescens* needs further investigation; see Myers and Carleton, 1981, Misc. Publ. Mus. Zool. Univ. Mich., 161:36.
ISIS NUMBER: 5301410008035022001 as *O. delicatus.*
                5301410008035042001 as *O. microtis.*

Oryzomys delticola Thomas, 1917. Ann. Mag. Nat. Hist., ser. 8, 20:96.
TYPE LOCALITY: Argentina, Buenos Aires Prov., Ella Isl.
DISTRIBUTION: Uruguay and N.E. Argentina.
COMMENT: Subgenus *Oligoryzomys*; see Langguth, 1963, Communic. Zool. Mus. Hist. Nat. Montevideo, 7(99):1-19, Massoia, 1973, Rev. Invest. Agro. INTA, ser. 1, 10(1):21-37, and Gardner and Patton, 1976, Occas. Pap. Mus. Zool. La. St. Univ., 49:19. Possibly conspecific with *longicaudatus*; see Cabrera, 1961:389, who placed this species in subgenus *Oryzomys*.
ISIS NUMBER: 5301410008035023001.

Oryzomys dimidiatus (Thomas, 1905). Ann. Mag. Nat. Hist., ser. 7, 15:586.
REVIEWED BY: R. D. Owen (RDON).
TYPE LOCALITY: Nicaragua, Rio Escondido, 7 mi. (11 km) below Rama.
DISTRIBUTION: Nicaragua.
COMMENT: Subgenus *Oryzomys*; see Hall, 1981:608, 612. Formerly included in subgenus *Micronectomys* which Hershkovitz, 1970, J. Mammal., 51:791, and Pine and Wetzel, 1975, Mammalia, 39:353, considered a *nomen nudum*. Known only from two specimens; see Genoways and Jones, 1971, J. Mammal., 52:833-834, who considered *dimidiatus* in subgenus *Micronectomys*.

Oryzomys flavescens (Waterhouse, 1837). Proc. Zool. Soc. Lond., 1837:19.
TYPE LOCALITY: Uruguay, Maldonado.
DISTRIBUTION: S. Brazil; Uruguay; N. and C. Argentina.
COMMENT: Formerly included in genus *Oligoryzomys* which was considered a subgenus of *Oryzomys* by Langguth, 1963, Communic. Zool. Mus. Hist. Nat. Montevideo, 7:3, and Massoia, 1973, Rev. Invest. Agro. INTA, ser. 1, 10(1):21-37, and by Gardner and Patton, 1976, Occas. Pap. Mus. Zool. La. St. Univ., 49:19, 24, and Myers and Carleton, 1981, Misc. Publ. Mus. Zool. Univ. Mich., 161:1-41, who considered *flavescens* a distinct species. Hershkovitz, 1966, Z. Saugetierk., 31:137, considered *flavescens* a subspecies of *nigripes*. Cabrera, 1961:380, 389, included *flavescens* in subgenus *Oryzomys* (within which he included *Oligoryzomys* as a synonym).
ISIS NUMBER: 5301410008035026001.

Oryzomys fornesi Massoia, 1973. Rev. Invest. Agro. INTA, ser. 1, 10(1):21–37.
  TYPE LOCALITY: Argentina, Formosa Prov., Naineck.
  DISTRIBUTION: N. Argentina; E. Paraguay; S. Brazil; Beni (Bolivia).
  COMMENT: Subgenus *Oligoryzomys*; see Massoia, 1973, Rev. Invest. Agro. INTA, ser. 1,
    10(1):21–37, and Myers and Carleton, 1981, Misc. Publ. Mus. Zool. Univ. Mich.,
    161:25–28, who also recommended that *longitarsus* (which may represent the
    earliest name for this species) be considered a *nomen dubium*.

Oryzomys fulgens Thomas, 1893. Ann. Mag. Nat. Hist., ser. 6, 11:403.
  REVIEWED BY: G. Urbano-V. (GUV).
  TYPE LOCALITY: "Mexico," probably in or near the Valley of Mexico.
  DISTRIBUTION: Known only from the type locality.
  COMMENT: Status doubtful, possibly conspecific with *couesi*; subgenus *Oryzomys*; see
    Hall, 1981:608, 612, and comment under *couesi*.
  ISIS NUMBER: 5301410008035027001.

Oryzomys fulvescens (Saussure, 1860). Rev. Mag. Zool. Paris, ser. 2, 12:102.
  REVIEWED BY: G. Urbano-V. (GUV).
  TYPE LOCALITY: Mexico, Veracruz, Orizaba.
  DISTRIBUTION: Venezuela to Panama to Tamaulipas, Nuevo Leon, and Nayarit
    (Mexico).
  COMMENT: Subgenus *Oligoryzomys*; see Gardner and Patton, 1976, Occas. Pap. Mus.
    Zool. La. St. Univ., 49:19, 24, and Hall, 1981:620, who considered *fulvescens* a
    distinct species. Cabrera, 1961:380, included *Oligoryzomys* in subgenus *Oryzomys*.
    Possibly conspecific with *nigripes*; see Hershkovitz, 1966, Z. Saugetierk., 31:137.
    Handley, 1976, Brigham Young Univ. Sci. Bull. Biol. Ser., 20(5):48, gave the
    distribution in Venezuela. Haiduk *et al.*, 1980, J. Mammal., 60:610–614, reviewed
    the karyology of this species.
  ISIS NUMBER: 5301410008035028001.

Oryzomys galapagoensis (Waterhouse, 1839). Zool. Voy. H.M.S. "Beagle," Mammalia, p. 66.
  TYPE LOCALITY: Galapagos Isls., Chatham Isl. (= San Cristobal) (Ecuador).
  DISTRIBUTION: San Cristobal Isl. (Galapagos Isls.).
  COMMENT: Closely related to *bauri*; subgenus *Oryzomys*; see Gardner and Patton, 1976,
    Occas. Pap. Mus. Zool. La. St. Univ., 49:16. *O. galapagoensis* is a synonym of *bauri*
    (ALG). Probably extinct; see Corbet and Hill, 1980:143.
  ISIS NUMBER: 5301410008035029001.

Oryzomys gorgasi Hershkovitz, 1971. J. Mammal., 52:700–709.
  TYPE LOCALITY: Colombia, Antioquia Dept., Loma Teguerre, between Rio Atrato and
    mouth of a channel of the east bank cienaga, just below and opposite Sautata
    (Choco), 7° 54′ N., 77° W.
  DISTRIBUTION: N.W. Colombia.
  COMMENT: Subgenus *Oryzomys*; see Hershkovitz, 1971.

Oryzomys hammondi (Thomas, 1913). Ann. Mag. Nat. Hist., ser. 8, 12:570.
  TYPE LOCALITY: Ecuador, Pichincha Prov., Mindo, 1284 m.
  DISTRIBUTION: Andean Ecuador.
  COMMENT: Formerly included in *Macruroryzomys*; see Cabrera, 1961:410, and Walker *et*
    al., 1975:764. Hershkovitz, 1970, J. Mammal., 51:789–794, considered
    Macruroryzomys a *nomen nudum*; *hammondi* is a possible precursor to the extinct
    Antillean genus *Megalomys*. Also see Pine and Wetzel, 1975, Mammalia, 39:353.
  ISIS NUMBER: 5301410008035031001.

Oryzomys intectus Thomas, 1921. Ann. Mag. Nat. Hist., ser. 9, 8:356.
  TYPE LOCALITY: Colombia, Antioquia, Santa Elena, near Medellin, 2700 m.
  DISTRIBUTION: C. Colombia.
  COMMENT: Status uncertain; provisionally placed in subgenus *Oryzomys*; see Cabrera,
    1961:390.
  ISIS NUMBER: 5301410008035033001.

Oryzomys kelloggi Avila-Pires, 1959. Atas Soc. Biol. Rio de J., 3(4):2.
REVIEWED BY: R. H. Pine (RHP).
TYPE LOCALITY: Brazil, Minas Gerais, Alem Paraiba.
DISTRIBUTION: S.E. Brazil.
COMMENT: Subgenus uncertain; probably sugenus *Oryzomys* (RHP); probably
subgenus *Oecomys* (AL).

Oryzomys lamia Thomas, 1901. Ann. Mag. Nat. Hist., ser. 7, 8:528.
TYPE LOCALITY: Brazil, Minas Gerais, Rio Jordao.
DISTRIBUTION: S.E. Brazil.
COMMENT: According to Thomas, 1901, Ann. Mag. Nat. Hist., ser. 7, 8:528, nearest to
"*laticeps* var. *intermedia*," here included in *capito*; also see Hershkovitz, 1966, Z.
Saugetierk., 31:138.

Oryzomys longicaudatus (Bennett, 1832). Proc. Zool. Soc. Lond., 1832:2.
TYPE LOCALITY: Chile, Valparaiso.
DISTRIBUTION: Chile; adjacent Argentina; Peru.
COMMENT: Subgenus *Oligoryzomys*; see Massoia, 1973, Rev. Invest. Agro. INTA, ser. 1,
10(1):21–37, Gardner and Patton, 1976, Occas. Pap. Mus. Zool. La. St. Univ., 49:19,
and Myers and Carleton, 1981, Misc. Publ. Mus. Zool. Univ. Mich., 161:1–41.
Hershkovitz, 1966, Z. Saugetierk., 31:137, included *longicaudatus* in *nigripes*.
Gardner and Patton, 1976:19–20, and Myers and Carleton, 1981, Misc. Publ. Mus.
Zool. Univ. Mich., 161:35, considered *longicaudatus* and *flavescens* distinct species;
further investigation of the relationship of *longicaudatus* with *nigripes, chacoensis*,
and *delticola* is needed. Cabrera, 1961:380, 390, included *Oligoryzomys* in subgenus
Oryzomys, and included *destructor* in *longicaudatus*. OAR considers *destructor* a
separate species.
ISIS NUMBER: 5301410008035034001.

Oryzomys macconnelli Thomas, 1910. Ann. Mag. Nat. Hist., ser. 8, 6:186.
REVIEWED BY: R. H. Pine (RHP).
TYPE LOCALITY: Guyana, Rio Supinaam, a tributary of the Lower Essequibo.
DISTRIBUTION: Surinam; Guyana; adjacent S. Venezuela; S. Colombia; E. Ecuador; E.
Peru; N. Brazil.
COMMENT: Included in *capito* by Hershkovitz, 1960, Proc. U.S. Nat. Mus., 110:513–568,
but regarded as a distinct species by Pine, 1973, Acta Amazonica, 3(2):63, Gardner
and Patton, 1976, Occas. Pap. Mus. Zool. La. St. Univ., 49:16, and Husson,
1978:392, 406, who included *macconnelli* in subgenus *Oryzomys*.
ISIS NUMBER: 5301410008035035001.

Oryzomys melanostoma (Tschudi, 1844). Fauna Peruana, p. 182.
TYPE LOCALITY: E. Peru, "Oriental Region," probably Huanuco Dept.
DISTRIBUTION: E. Peru.
COMMENT: Subgenus *Oryzomys*; see Cabrera, 1961:393. Subgenus *Oligoryzomys* (OAR).
ISIS NUMBER: 5301410008035039001.

Oryzomys melanotis Thomas, 1893. Ann. Mag. Nat. Hist., ser. 6, 11:404.
REVIEWED BY: G. Urbano-V. (GUV).
TYPE LOCALITY: Mexico, Jalisco, Mineral San Sebastian.
DISTRIBUTION: El Salvador; N. Belize and N. Honduras to Yucatan and Oaxaca to C.
Tamaulipas and S. Sinaloa (Mexico).
COMMENT: Includes *rostratus*; subgenus *Oryzomys*; see Hooper, 1953, Occas. Pap. Mus.
Zool. Univ. Mich., 544:8, and Hall, 1981:613.
ISIS NUMBER: 5301410008035040001.

Oryzomys minutus (Tomes, 1860). Proc. Zool. Soc. Lond., 1860:215.
REVIEWED BY: R. D. Owen (RDON); R. H. Pine (RHP).
TYPE LOCALITY: Ecuador, Chimborazo Prov., Pallatanga.
DISTRIBUTION: N. Venezuela, W. and C. Colombia to Ecuador and Peru.
COMMENT: Subgenus *Microryzomys*; see Gardner and Patton, 1976, Occas. Pap. Mus.

Zool. La. St. Univ., 49:11, 24, who included *altissimus* in this species; but also see Voss and Linzey, 1981, Misc. Publ. Mus. Zool. Univ. Mich., 159:13, who considered *altissimus* a distinct species. Subgenus *Microryzomys* reviewed by Myers and Carleton, 1981, Misc. Publ. Mus. Zool. Univ. Mich., 161:11–12.
ISIS NUMBER: 5301410008035044001.

Oryzomys munchiquensis J. A. Allen, 1912. Bull. Am. Mus. Nat. Hist., 31:85.
TYPE LOCALITY: Colombia, Cauca, Florida, 2300 m.
DISTRIBUTION: W. Colombia, in the basins of Cauca and Patia.
COMMENT: Formerly included in genus *Oligoryzomys*, which was considered a subgenus of *Oryzomys* by Gardner and Patton, 1976, Occas. Pap. Mus. Zool. La. St. Univ., 49:19, and Myers and Carleton, 1981, Misc. Publ. Mus. Zool. Univ. Mich., 161:36. Cabrera, 1961:380, included *Oligoryzomys* in subgenus *Oryzomys* as a synonym. The relationship of *munchiquensis* to *fornesi* and *flavescens* needs further investigation; see Myers and Carleton, 1981, Misc. Publ. Mus. Zool. Univ. Mich., 161:36.
ISIS NUMBER: 5301410008035045001.

Oryzomys nelsoni Merriam, 1898. Proc. Biol. Soc. Wash., 12:15.
REVIEWED BY: G. Urbano-V. (GUV).
TYPE LOCALITY: Mexico, Nayarit, Tres Marias Isls., Maria Madre Isl.
DISTRIBUTION: Known only from the type locality.
COMMENT: Subgenus *Oryzomys*; see Hall, 1981:608, 612.
ISIS NUMBER: 5301410008035047001.

Oryzomys nigripes (Olfers, 1818). *In* Eschwege, Neue Bibl. Reisenb., 15:209.
TYPE LOCALITY: Paraguay, Paraguari Dept., Ybycui Nat. Park, 85 km S.S.E. Atyra.
DISTRIBUTION: Paraguay; E. Brazil; Argentina.
COMMENT: Includes *eliurus*; see Cabrera, 1961:394, Massoia and Fornes, 1967, Acta Zool. Lilloana, 23:407–430, and Myers and Carleton, 1981, Misc. Publ. Mus. Zool. Univ. Mich., 161:35; but see Massoia, 1973, Rev. Invest. Agro. INTA, ser. 1, 10(1):21–37. *O. eliurus* may be a distinct species, based on karyotypic differences (OAR). Subgenus *Oligoryzomys*; see Hershkovitz, 1966, Z. Saugetierk., 31:137, who included *flavescens*, *fulvescens*, and *longicaudatus* in this species, but Gardner and Patton, 1976, Occas. Pap. Mus. Zool. La. St. Univ., 49:20, 24, considered them distinct species and indicated that the relationship between *nigripes* and *longicaudatus* requires elucidation. Myers and Carleton, 1981, Misc. Publ. Mus. Zool. Univ. Mich., 161:1–41, considered *longicaudatus* and *flavescens* distinct species, and also considered *tarso nigro* Fischer, 1814, invalid (see comments under *Reithrodon physodes*); AL employs the name *tarsonigro* for this species. Cabrera, 1961:380, placed *Oligoryzomys* in subgenus *Oryzomys* as a synonym.
ISIS NUMBER: 5301410008035048001.

Oryzomys nitidus (Thomas, 1884). Proc. Zool. Soc. Lond., 1884:452.
TYPE LOCALITY: Peru, Junin Dept. (Tulumayo Valley), Amable Maria (about 10 km south of San Ramon, 11° 10' S., 75° 19' W.).
DISTRIBUTION: Ecuador; Peru; Bolivia; N.W. Argentina; S. Brazil.
COMMENT: Includes *boliviae* and *legatus*; *nitidus* was formerly included in *capito*; subgenus *Oryzomys*; see Gardner and Patton, 1976, Occas. Pap. Mus. Zool. La. St. Univ., 49:38–40; but also see Massoia, 1974, Rev. Invest. Agro. INTA, ser. 5, Patalogia Vegetal., 11(1). Hershkovitz, 1966, Z. Saugetierk., 31:138, included *nitidus* in *alfaroi*. Gardner and Patton, 1976, Occas. Pap. Mus. Zool. La. St. Univ., 49:38–40, stated that *bolivaris* and *intermedius* probably belonged in *nitidus*. ALG states that *bolivaris* is a synonym of *rivularis* and that *intermedius* has no status, being a renaming of *laticeps*, which is a subjective synonym of *capito*.

Oryzomys palustris (Harlan, 1837). Am. J. Sci., 31:385.
REVIEWED BY: G. Urbano-V. (GUV).
TYPE LOCALITY: U.S.A., New Jersey, Salem Co., "Fastland," near Salem.
DISTRIBUTION: E. Texas to S.E. Kansas, east to New Jersey and Florida (U.S.A.).
COMMENT: Formerly included *couesi*, *azuerensis*, *gatunensis*, *cozumelae*, and *antillarum*; see

Benson and Gehlbach, 1979, J. Mammal., 60:225–228, and comments under *couesi*. Subgenus *Oryzomys;* see Hall, 1981:608.
ISIS NUMBER: 5301410008035050001.

Oryzomys peninsulae Thomas, 1897. Ann. Mag. Nat. Hist., ser. 6, 20:548.
REVIEWED BY: G. Urbano-V. (GUV).
TYPE LOCALITY: Mexico, Baja California Sur, Santa Anita.
DISTRIBUTION: S. Baja California (Mexico).
COMMENT: May be conspecific with *couesi;* considered a subspecies of *palustris* by Hall, 1981:610; subgenus *Oryzomys;* see also comments under *couesi.*
ISIS NUMBER: 5301410008035051001.

Oryzomys polius Osgood, 1913. Field Mus. Nat. Hist. Publ. Zool. Ser., 10:97.
TYPE LOCALITY: Peru, Amazonas, Tambo Carrizal, mountain slopes east of Balsas, 1500 m.
DISTRIBUTION: Alto Maranon (N. Peru).
COMMENT: Subgenus *Oryzomys;* see Cabrera, 1961:394.
ISIS NUMBER: 5301410008035053001.

Oryzomys ratticeps (Hensel, 1873). Abh. Preuss. Akad. Wiss., (1872), p. 36.
TYPE LOCALITY: Brazil, Rio Grande do Sul.
DISTRIBUTION: S. Brazil; Paraguay.
COMMENT: Included in *buccinatus* by Hershkovitz, 1959, J. Mammal., 40:347, but considered a distinct species in subgenus *Oryzomys* by Avila-Pires, 1960, Acta Trab. Prim. Cong. Sudam. Zool., sec. 5, Vertebrados, 4:3–8, Avila-Pires, 1959, Atas Soc. Biol. Rio de J., 3(4):3, Cabrera, 1961:395, Corbet and Hill, 1980:144, Voss and Linzey, 1981, Misc. Publ. Mus. Zool. Univ. Mich., 159:13, and Myers and Carleton, 1981, Misc. Publ. Mus. Zool. Univ. Mich., 161:19.
ISIS NUMBER: 5301410008035054001.

Oryzomys rivularis J. A. Allen, 1901. Bull. Am. Mus. Nat. Hist., 14:406.
TYPE LOCALITY: Ecuador, Esmeraldas Prov., Rio Verde.
DISTRIBUTION: N.W. Ecuador.
COMMENT: May include *bombycinus;* subgenus *Oryzomys;* see Gardner and Patton, 1976, Occas. Pap. Mus. Zool. La. St. Univ., 39:40–41. Hershkovitz, 1960, Proc. U.S. Nat. Mus., 110:544, considered *rivularis* a synonym of *laticeps* (=*capito*). *O. rivularis* includes *bombycinus* and *bolivaris* (ALG).

Oryzomys robustulus (Thomas, 1914). Ann. Mag. Nat. Hist., ser. 8, 14:243.
REVIEWED BY: R. D. Owen (RDON).
TYPE LOCALITY: Ecuador, Gualaquiza.
DISTRIBUTION: E. Ecuador.
COMMENT: Subgenus *Melanomys;* see Cabrera, 1961:402.
ISIS NUMBER: 5301410008035056001.

Oryzomys spodiurus Hershkovitz, 1940. J. Mammal., 21:79.
TYPE LOCALITY: Ecuador, Imbabura Prov., Hacienda Chinipamba, Intag, near Penaherrera, 1500 m.
DISTRIBUTION: W. Ecuador.
COMMENT: Subgenus *Oligoryzomys;* see Hershkovitz, 1940, J. Mammal., 21:79. Probably a subspecies of *longicaudatus;* see Cabrera, 1961:380, 395, who included *Oligoryzomys* in subgenus *Oryzomys* as a synonym.
ISIS NUMBER: 5301410008035058001.

Oryzomys subflavus (Wagner, 1842). Arch. Naturgesch., 8(1):362.
TYPE LOCALITY: Brazil, Minas Gerais, probably Lagoa Santa.
DISTRIBUTION: E. Brazil; Guianas.
COMMENT: Includes *catherinae* and *regalis* (a renaming of *rex*); see Hershkovitz, 1960, Proc. U.S. Nat. Mus., 110:519, 543. Subgenus *Oryzomys;* see Cabrera, 1961:404, and Corbet and Hill, 1980:143, who listed *catherinae* as a distinct species, without comment.
ISIS NUMBER: 5301410008035059001 as *O. subflavus.*
5301410008035016001 as *O. catherinae.*

Oryzomys utiaritensis J. A. Allen, 1916. Bull. Am. Mus. Nat. Hist., 35:527.
TYPE LOCALITY: Brazil, Mato Grosso, Utiariti, along Rio Papagaio.
DISTRIBUTION: C. Brazil.
COMMENT: Subgenus *Oligoryzomys;* see Gardner and Patton, 1976, Occas. Pap. Mus.
    Zool. La. St. Univ., 49:19, and Myers and Carleton, 1981, Misc. Publ. Mus. Zool.
    Univ. Mich., 161:36–37. Cabrera, 1961:380, included *Oligoryzomys* in subgenus
    Oryzomys, as a synonym. The type may be a composite based on *nigripes* skin and
    fornesi skull; see Myers and Carleton, 1981, Misc. Publ. Mus. Zool. Univ. Mich.,
    161:36–37.
ISIS NUMBER: 5301410008035064001.

Oryzomys victus Thomas, 1898. Ann. Mag. Nat. Hist., ser. 7, 1:178.
TYPE LOCALITY: St. Vincent (Lesser Antilles) (U.K.).
DISTRIBUTION: St. Vincent (Lesser Antilles).
COMMENT: Considered extinct by Corbet and Hill, 1980:144. Subgenus *Oligoryzomys;*
    see Hall, 1981:620.
ISIS NUMBER: 5301410008035065001.

Oryzomys villosus J. A. Allen, 1899. Bull. Am. Mus. Nat. Hist., 12:210.
TYPE LOCALITY: Colombia, Magdalena Dept., Sierra de Santa Marta, Valparaiso, 1350 m.
DISTRIBUTION: N. Colombia.
COMMENT: Possibly conspecific with *albigularis;* see Hershkovitz, 1966, Z. Saugetierk.,
    31:137; subgenus *Oryzomys.* Also see Gardner and Patton, 1976, Occas. Pap. Mus.
    Zool. La. St. Univ., 49:38, who believed that *villosus* is closely related to *capito.*
ISIS NUMBER: 5301410008035066001.

Oryzomys xantheolus Thomas, 1894. Ann. Mag. Nat. Hist., ser. 6, 14:354.
TYPE LOCALITY: Peru, Piura Dept., Tumbez.
DISTRIBUTION: W. Peru; perhaps S.W. Ecuador.
COMMENT: Subgenus *Oryzomys;* see Gardner and Patton, 1976, Occas. Pap. Mus. Zool.
    La. St. Univ., 49:24.
ISIS NUMBER: 5301410008035067001.

Oryzomys yunganus Thomas, 1902. Ann. Mag. Nat. Hist., ser. 7., 9:130.
TYPE LOCALITY: Bolivia, Beni Dept., Churuplaga, Rio Secure, 1300 m.
DISTRIBUTION: Bolivia; Peru.
COMMENT: Considered distinct from *capito* by Gardner and Patton, 1976, Occas. Pap.
    Mus. Zool. La. St. Univ., 49:24, 40; subgenus *Oryzomys.*

Oryzomys zunigae Sanborn, 1949. Publ. Mus. Hist. Nat. Javier Prado Zool., 1(3):2.
TYPE LOCALITY: Peru, Lima Dept., Lomas de Atocongo.
DISTRIBUTION: W.C. Peru.
COMMENT: Subgenus *Melanomys;* see Cabrera, 1961:402.
ISIS NUMBER: 5301410008035068001.

Osgoodomys Hooper and Musser, 1964. Occas. Pap. Mus. Zool. Univ. Mich., 635:12.
REVIEWED BY: S. Anderson (SA); W. Caire (WC); R. P. Canham (RPC); M. D. Carleton
    (MDC); L. N. Carraway (LNC); D. G. Huckaby (DGH); C. W. Kilpatrick (CWK); T.
    E. Lawlor (TL); R. H. Pine (RHP); J. Ramirez-Pulido (JRP); G. Urbano-V. (GUV).
COMMENT: Included in *Peromyscus* as a subgenus by Hall, 1981:720, and Hooper, 1968,
    in King, ed., The Biology of *Peromyscus* (Rodentia), p. 38. Linzey and Layne, 1974,
    Am. Mus. Novit., 2532:1–20, studied the morphology of the spermatozoa.
    Reviewed by Carleton, 1980:118, 123, who considered *Osgoodomys* a distinct
    genus. Pine *et al.,* 1979, Mammalia, 43:357, considered *Osgoodomys* Hooper and
    Musser, a *nomen nudum;* Carleton, 1980, may have been first to make the name
    available (RHP). Subfamily Hesperomyinae; see comment under Cricetidae.

Osgoodomys banderanus (J. A. Allen, 1897). Bull. Am. Mus. Nat. Hist., 9:51.
TYPE LOCALITY: Mexico, Nayarit, Valle de Banderas.
DISTRIBUTION: Nayarit to Guerrero (Mexico).
COMMENT: Formerly included *angelensis, coatlanensis,* and *sloeops,* which are included
    in *Peromyscus mexicanus;* see Musser, 1969, Am. Mus. Novit., 2357:2–7. Karyotype
    reported by by Lee and Elder, 1977, J. Mammal., 58:479–487.
ISIS NUMBER: 5301410008039004001 as *Peromyscus banderanus.*

Otomys F. Cuvier, 1824. Des Dentes des Mammiferes, 255, pl. 60.
REVIEWED BY: R. D. Owen (RDON).
COMMENT: Includes *Myotomys*; see Misonne, 1974, Part 6:34, and Swanepoel *et al.*,
1980, Ann. Transvaal Mus., 32(7):177; Walker *et al.*, 1975:931, considered
Myotomys a distinct genus. Subfamily Otomyinae; see comment under Cricetidae.
ISIS NUMBER: 5301410008111000000.

Otomys anchietae Bocage, 1882. J. Sci. Math. Phys. Nat. Lisboa, 9:26.
TYPE LOCALITY: Angola, Caconda.
DISTRIBUTION: Angola; Mt. Elgon (Uganda); S. Tanzania.
ISIS NUMBER: 5301410008111001001.

Otomys angoniensis Wroughton, 1906. Ann. Mag. Nat. Hist., ser. 7, 18:274.
TYPE LOCALITY: Malawi (=Nyasaland), Angoniland, M'Kombhuie, 8,000 ft. (2438 m).
DISTRIBUTION: Cape Prov. (South Africa) to Kenya.
COMMENT: Includes *tugelensis*; see Misonne, 1974, Part 6:33, who also included
maximus in this species; but Swanepoel *et al.*, 1980, Ann. Transvaal Mus.,
32(7):161, considered *maximus* a distinct species.
ISIS NUMBER: 5301410008111007001 as *O. tugelensis*.

Otomys denti Thomas, 1906. Ann. Mag. Nat. Hist., ser. 7, 18:142.
TYPE LOCALITY: Uganda, E. Mt. Ruwenzori, Mubuku Valley.
DISTRIBUTION: Zaire; Rwanda; Uganda; Tanzania; Malawi; Zambia.
COMMENT: Includes *kempi*; see Misonne, 1974, Part 6:34, and Ansell, 1978:78.
ISIS NUMBER: 5301410008111002001.

Otomys irroratus Brants, 1827. Het. Gesl. Muiz., p. 94.
TYPE LOCALITY: South Africa, Cape Town Dist., Constantia.
DISTRIBUTION: South Africa to Kenya and Cameroun.
COMMENT: Includes *tropicalis*; see Delany, 1975, The Rodents of Uganda, Br. Mus. Nat.
Hist., 764:1–165, who also included *typus* in this species. Misonne, 1974, Part 6:33,
and Ansell, 1978:78, considered *typus* a distinct species.
ISIS NUMBER: 5301410008111003001.

Otomys laminatus Thomas and Schwann, 1905. Abstr. Proc. Zool. Soc. Lond., 18:23.
TYPE LOCALITY: South Africa, Natal, Zululand, Sibundeni.
DISTRIBUTION: South Africa.
ISIS NUMBER: 5301410008111005001.

Otomys maximus Roberts, 1924. Ann. Transvaal Mus., 10:70.
TYPE LOCALITY: Zambia, Machili River.
DISTRIBUTION: N.W. Zimbabwe; Zambia; Angola.
COMMENT: Formerly included in *angoniensis* by Misonne, 1974, Part 6:33, and Ansell,
1978:78, but considered a distinct species by Smithers and Wilson, 1979, Mus.
Mem. Nat. Mus. Rhodesia, 9:1–193, and Swanepoel *et al.*, 1980, Ann. Transvaal
Mus., 32(7):161.

Otomys saundersiae Roberts, 1929. Ann. Transvaal Mus., 13:115.
TYPE LOCALITY: South Africa, Cape Prov., Grahamstown.
DISTRIBUTION: South Africa.
COMMENT: Includes *karoensis*; see Misonne, 1974, Part 6:34.
ISIS NUMBER: 5301410008111004001 as *O. karoensis*.

Otomys sloggetti Thomas, 1902. Ann. Mag. Nat. Hist., ser. 7, 10:311.
TYPE LOCALITY: South Africa, Cape Prov., Deelfontein.
DISTRIBUTION: South Africa.
ISIS NUMBER: 5301410008111006001.

Otomys typus (Heuglin, 1877). Reise Nordost Afrika, 2:77.
TYPE LOCALITY: Ethiopia, highlands of Simien.
DISTRIBUTION: Ethiopia to S. Sudan and Zambia, in montane areas.
COMMENT: Formerly included in *irroratus* by Dieterlen, 1968, Z. Saugetierk.,
33:321–352, and Delany, 1975, The Rodents of Uganda, Br. Mus. Nat. Hist.,
764:1–165. Considered a distinct species by Misonne, 1974, Part 6:33, and Ansell,
1978:78.
ISIS NUMBER: 5301410008111008001.

Otomys unisulcatus Cuvier, 1829. *In* Geoffroy and Cuvier, Hist. Nat., Mamm., 3: liv. 60 (pl. and text on *"Otomys cafre"*).
TYPE LOCALITY: South Africa, Cape Prov., S.W. Karroo, Matjesfontein, southwest of Laingsburg.
DISTRIBUTION: South Africa.
COMMENT: Formerly included in *Myotomys;* see Misonne, 1974, Part 6:34, and Swanepoel *et al.,* 1980, Ann. Transvaal Mus., 32(7):177.
ISIS NUMBER: 5301410008111009001.

Otonyctomys Anthony, 1932. Am. Mus. Novit., 586:1.
REVIEWED BY: R. D. Owen (RDON); J. Ramirez-Pulido (JRP); G. Urbano-V. (GUV).
COMMENT: Subfamily Hesperomyinae; see comment under Cricetidae.
ISIS NUMBER: 5301410008036000000.

Otonyctomys hatti Anthony, 1932. Am. Mus. Novit., 586:1.
TYPE LOCALITY: Mexico, Yucatan, Chichen-Itza.
DISTRIBUTION: Yucatan Peninsula (Mexico); Belize; N.W. Guatemala.
ISIS NUMBER: 5301410008036001001.

Ototylomys Merriam, 1901. Proc. Wash. Acad. Sci., 3:561.
REVIEWED BY: T. E. Lawlor (TL); R. D. Owen (RDON).
COMMENT: Revised by Lawlor, 1969, J. Mammal., 50:28–42; reviewed by Carleton, 1980:118–119. Subfamily Hesperomyinae; see comment under Cricetidae.
ISIS NUMBER: 5301410008037000000.

Ototylomys phyllotis Merriam, 1901. Proc. Wash. Acad. Sci., 3:562.
REVIEWED BY: J. Ramirez-Pulido (JRP); G. Urbano-V. (GUV).
TYPE LOCALITY: Mexico, Yucatan, Tunkas.
DISTRIBUTION: Costa Rica to S.C. Tabasco and Yucatan; N.C. Guerrero (Mexico).
COMMENT: Includes *brevirostris* and *connectans;* see Lawlor, 1969, J. Mammal., 50:36. Ramirez-Pulido and Sanchez-H., 1974, Rev. Soc. Mex. Hist. Nat., 33:110–111, reported the possible occurrence of this species in Guerrero, Mexico. Male accessory glands reviewed by Voss and Linzey, 1981, Misc. Publ. Mus. Zool. Univ. Mich., 159:1–41.
ISIS NUMBER: 5301410008037001001.

Oxymycterus Waterhouse, 1837. Proc. Zool. Soc. Lond., 1837:20.
REVIEWED BY: J. K. Braun (JKB); M. D. Carleton (MDC); A. Langguth (AL); R. A. Ojeda (RAO).
COMMENT: Massoia, 1963, Physis, 24:132, included *Akodon (Abrothrix) sanborni* in this genus, but see Pine *et al.,* 1979, Mammalia, 43:351. This genus needs revision; its relationships with *Akodon (Abrothrix)* and *Microxus* need clarification; see Hershkovitz, 1966, Z. Saugetierk., 31:127, Bianchi *et al.,* 1971, Evolution, 25:724, Gardner and Patton, 1976, Occas. Pap. Mus. Zool. La. St. Univ., 49:30, Massoia, 1963, Physis, 24:132, Reig, 1978, Publ. Mus. Munic. Cienc. Nat. Mar del Plata "Lorenzo Scaglia," 2(8):164–190, Corbet and Hill, 1980:151, 152, and Voss and Linzey, 1981, Misc. Publ. Mus. Zool. Univ. Mich., 159:1–41. Subfamily Hesperomyinae; see comment under Cricetidae.
ISIS NUMBER: 5301410008038000000.

Oxymycterus akodontius Thomas, 1921. Ann. Mag. Nat. Hist., ser. 9, 8:615.
TYPE LOCALITY: Argentina, Valle Grande Dept., Jujuy Prov., Higuerilla, 20 km east of Tilcara, 2000 m.
DISTRIBUTION: N.W. Argentina.
COMMENT: Possibly conspecific with *paramensis;* see Cabrera, 1961:466.
ISIS NUMBER: 5301410008038001001.

Oxymycterus angularis Thomas, 1909. Ann. Mag. Nat. Hist., ser. 8, 4:237.
TYPE LOCALITY: Brazil, Pernambuco, Sao Lourenco.
DISTRIBUTION: E. Brazil.
ISIS NUMBER: 5301410008038002001.

Oxymycterus delator Thomas, 1903. Ann. Mag. Nat. Hist., ser. 7, 11:489.
TYPE LOCALITY: Paraguay, Sapucai.
DISTRIBUTION: Paraguay.
COMMENT: Male accessory glands reviewed by Voss and Linzey, 1981, Misc. Publ.
Mus. Zool. Univ. Mich., 159:4, 14.
ISIS NUMBER: 5301410008038003001.

Oxymycterus hispidus Pictet, 1843. Mem. Soc. Phys. Hist. Nat. Geneve, 10:212.
TYPE LOCALITY: E. Brazil, Bahia.
DISTRIBUTION: N.E. Argentina to E. Brazil.
ISIS NUMBER: 5301410008038004001.

Oxymycterus iheringi Thomas, 1896. Ann. Mag. Nat. Hist., ser. 6, 18:308.
TYPE LOCALITY: Brasil, Rio Grande do Sul, Taquara do Mundo Novo, Rio dos Sinos.
DISTRIBUTION: Brasil; N.E. Argentina.
COMMENT: Formerly included in *Microxus* and *Akodon*; included in *Oxymycterus* by
Massoia, 1963, Physis, ser. 24, 67:73–80. Hershkovitz, 1966, Z. Saugetierk., 31:86,
considered this species "a local form of *Oxymycterus rutilans*."
ISIS NUMBER: 5301410008001014001 as *Akodon iheringi*.

Oxymycterus inca Thomas, 1900. Ann. Mag. Nat. Hist., ser. 7, 6:298.
TYPE LOCALITY: C. Peru, Junin Prov., Rio Perene, 800 m.
DISTRIBUTION: W. Bolivia; C. and S.E. Peru.
ISIS NUMBER: 5301410008038005001.

Oxymycterus paramensis Thomas, 1902. Ann. Mag. Nat. Hist., ser. 7, 9:139.
REVIEWED BY: O. A. Reig (OAR).
TYPE LOCALITY: Bolivia, Rio Secure, Choquecamate.
DISTRIBUTION: Bolivia; adjacent Argentina; E. Peru.
COMMENT: Included in *rutilans* by Hershkovitz, 1966, Z. Saugetierk., 31:127, without
comment; but see Cabrera, 1961:467–469, Hooper and Musser, 1964, Misc. Publ.
Mus. Zool. Univ. Mich., 123:45, and Corbet and Hill, 1980:152, who considered
paramensis a distinct species, as do MDC and OAR. May be conspecific with
akodontius; see Cabrera, 1961:466.
ISIS NUMBER: 5301410008038006001.

Oxymycterus roberti Thomas, 1901. Ann. Mag. Nat. Hist., ser. 7, 8:530.
TYPE LOCALITY: E. Brazil, S.W. Minas Gerais, Rio Jordao, Paranaiba.
DISTRIBUTION: E. Brazil.
ISIS NUMBER: 5301410008038007001.

Oxymycterus rutilans (Olfers, 1818). *In* Eschwege, Neue Bibl. Reisenb., 15:209.
REVIEWED BY: O. A. Reig (OAR).
TYPE LOCALITY: Paraguay.
DISTRIBUTION: Argentina; Uruguay; Paraguay; E. and S. Brazil.
COMMENT: Includes *rufus*; see Hershkovitz, 1959, J. Mammal., 40:339, Hershkovitz,
1966, Z. Saugetierk., 31:127, Gardner and Patton, 1976, Occas. Pap. Mus. Zool. La.
St. Univ., 49:28, and Voss and Linzey, 1981, Misc. Publ. Mus. Zool. Univ. Mich.,
159:14. May include *iheringi*; see Hershkovitz, 1966, Z. Saugetierk., 31:86, 127,
who also included *paramensis* in *rutilans*, without comment. Cabrera, 1961:467–469,
Hooper and Musser, 1964, Misc. Publ. Mus. Zool. Univ. Mich., 123:45, and Corbet
and Hill, 1980:152, considered *paramensis* and *rufus* (a synonym of *rutilans*) distinct
species. AL employs the name *rufus* (Fischer, 1814), for this species; see comment
under *Reithrodon physodes*.
ISIS NUMBER: 5301410008038008001 as *O. rufus*.

Pachyuromys Lataste, 1880. Le Naturaliste, 1:313.
REVIEWED BY: J. K. Braun (JKB); C. B. Robbins (CBR).
COMMENT: Subfamily Gerbillinae; see comment under Cricetidae.
ISIS NUMBER: 5301410008094000000.

Pachyuromys duprasi Lataste, 1880. Le Naturaliste, 1:314.
TYPE LOCALITY: Algerian Sahara, Laghouat.
DISTRIBUTION: W. Morocco and Mauritania to Egypt in the Sahara.
ISIS NUMBER: 5301410008094001001.

Parotomys Thomas, 1918. Ann. Mag. Nat. Hist., ser. 7, 2:204.
  REVIEWED BY: R. D. Owen (RDON).
  COMMENT: Subfamily Otomyinae; see comment under Cricetidae.
  ISIS NUMBER: 5301410008112000000.

Parotomys brantsi (Smith, 1834). S. Afr. J., 2:150.
  TYPE LOCALITY: South Africa, Cape Prov., Little Namaqualand, Port Nolloth.
  DISTRIBUTION: Cape Prov. (South Africa); S. Namibia; Botswana.
  ISIS NUMBER: 5301410008112001001.

Parotomys littledalei Thomas, 1918. Ann. Mag. Nat. Hist., ser. 7, 2:205.
  TYPE LOCALITY: South Africa, Cape Prov., Bushmanland, Kenhardt, Tuin.
  DISTRIBUTION: Namibia; Cape Prov. (South Africa).
  ISIS NUMBER: 5301410008112002001.

Peromyscus Gloger, 1841. Gem. Hand. Hilfs. Nat., 1:95.
  REVIEWED BY: S. Anderson (SA); W. Caire (WC); R. P. Canham (RPC); M. D. Carleton
    (MDC); L. N. Carraway (LNC); D. G. Huckaby (DGH); C. W. Kilpatrick (CWK); T.
    E. Lawlor (TL); R. H. Pine (RHP); J. Ramirez-Pulido (JRP); G. Urbano-V. (GUV).
  COMMENT: Revised by Osgood, 1909, N. Am. Fauna, 28:1–285; reviewed by Hooper,
    1968, *in* King, ed., The Biology of *Peromyscus* (Rodentia), p. 27, and Carleton,
    1980. Includes *Peromyscus* and *Haplomylomys* as subgenera; formerly included
    Habromys, Isthmomys, Megadontomys, Ochrotomys, Osgoodomys, and *Podomys* which
    were considered distinct genera by Carleton, 1980:118, 122–127. Morphology of
    the spermatozoa studied by Linzey and Layne, 1974, Amer. Mus. Novit.,
    2532:1–20. Yates *et al.,* 1979, Syst. Zool., 28:40–48, and Patton *et al.,* 1981:288–308,
    in Smith and Joule, eds., Mammalian Population Genetics, Univ. Georgia Press,
    Athens, included *Neotomodon* in *Peromyscus,* but Carleton, 1980:118, 126, and Hall,
    1981:745, considered *Neotomodon* a distinct genus. The *mexicanus* group of
    subgenus *Peromyscus* was revised by Huckaby, 1980, Los Ang. Cty. Mus. Contrib.
    Sci., 326:1–24. The *boylii* group was revised by Carleton, 1977:1–47, and Carleton,
    1979, J. Mammal., 60:280–296. The *eremicus* group was revised by Lawlor, 1971,
    Occas. Pap. Mus. Zool. Univ. Mich., 661:1–22. Subfamily Hesperomyinae; see
    comment under Cricetidae.
  ISIS NUMBER: 5301410008039000000.

Peromyscus attwateri J. A. Allen, 1895. Bull. Am. Mus. Nat. Hist., 7:330.
  REVIEWED BY: R. E. Barry, Jr. (REB).
  TYPE LOCALITY: U.S.A., Texas, Kerr Co., Turtle Creek.
  DISTRIBUTION: S.E. Kansas, S.W. Missouri, and N. Arkansas to Texas (U.S.A.).
  COMMENT: Formerly included in *boylii;* see Schmidly, 1973, J. Mammal., 54:125.
    Reviewed by Schmidly, 1974, Mamm. Species, 48:1–3.
  ISIS NUMBER: 5301410008039002001.

Peromyscus aztecus (Saussure, 1860). Rev. Mag. Zool. Paris, ser. 2, 12:105.
  TYPE LOCALITY: Mexico, Veracruz, 10 km east of Mirador, about 3800 ft. (1158 m).
  DISTRIBUTION: Jalisco and Veracruz (Mexico) to Honduras and El Salvador.
  COMMENT: Formerly included in *boylii;* see Alvarez, 1961, Univ. Kans. Mus. Nat. Hist.
    Misc. Publ., 14:113. Includes *cordillerae, evides, hondurensis, hylocetes,* and *oaxacensis;*
    see Carleton, 1979, J. Mammal., 60:293–295, Carleton, 1977:1–47, Musser, 1969,
    Am. Mus. Novit., 2357:8, and Hall, 1981:1179–1180.
  ISIS NUMBER: 5301410008039003001 as *P. aztecus.*
               5301410008039015001 as *P. evides.*
               5301410008039024001 as *P. hondurensis.*
               5301410008039025001 as *P. hylocetes.*
               5301410008039041001 as *P. oaxacensis.*

Peromyscus boylii (Baird, 1855). Proc. Acad. Nat. Sci. Phila., 7:335.
  REVIEWED BY: R. E. Barry, Jr. (REB).
  TYPE LOCALITY: U.S.A., California, Eldorado Co., Middle Fork of American River, near
    Auburn.
  DISTRIBUTION: California to W. Oklahoma (U.S.A.), south to Honduras.

COMMENT: Formerly included *attwateri* (see Schmidly, 1973, J. Mammal., 54:125), *aztecus, cordillerae* (see Carleton, 1979, J. Mammal., 60:293–295), *madrensis, simulus,* and *spicilegus* (see Carleton, 1977:1–47, but also see Hall, 1981:696). Also see Alvarez, 1961, Univ. Kans. Mus. Nat. Hist. Misc. Publ., 14(7):111–120, and Diersing, 1976, Proc. Biol. Soc. Wash., 89(39):451–466.
ISIS NUMBER: 5301410008039005001 as *P. boyli (sic).*

Peromyscus bullatus Osgood, 1904. Proc. Biol. Soc. Wash., 17:63.
TYPE LOCALITY: Mexico, Veracruz, Perote.
DISTRIBUTION: Vicinity of type locality and 3 km W. Limon (Mexico).
COMMENT: Considered a distinct species by Osgood, 1909, N. Am. Fauna, 28:183–184, and Hoffmeister, 1951, Ill. Biol. Monogr., 21:25. Thought to be a subspecies of *truei* by Hooper, 1968, *in* King, ed., The Biology of *Peromyscus* (Rodentia), p. 55. Hall, 1981:706, and Carleton, 1980:16, listed *bullatus* as a distinct species.
ISIS NUMBER: 5301410008039006001.

Peromyscus californicus (Gambel, 1848). Proc. Acad. Nat. Sci. Phila., 4:78.
REVIEWED BY: R. E. Barry, Jr. (REB).
TYPE LOCALITY: U.S.A., California, Monterey Co., Monterey.
DISTRIBUTION: C. and S. California (U.S.A.) to N. Baja California (Mexico).
COMMENT: Reviewed by Avise *et al.*, 1974, Syst. Zool., 23:226–238, Merritt, 1978, Mamm. Species, 85:1–6, and Smith, 1979, J. Mammal., 60:705–722.
ISIS NUMBER: 5301410008039007001.

Peromyscus caniceps Burt, 1932. Trans. San Diego Soc. Nat. Hist., 7:174.
TYPE LOCALITY: Mexico, Baja California Sur, Monserrate Isl., 25° 38′ N., 111° 02′ W.
DISTRIBUTION: Known only from the type locality.
COMMENT: Reviewed by Lawlor, 1971, Trans. San Diego Soc. Nat. Hist., 16:91–124. Probably a subspecies of *eva*; see Lawlor, 1971, Occas. Pap. Mus. Zool. Univ. Mich., 661:17.
ISIS NUMBER: 5301410008039008001.

Peromyscus crinitus (Merriam, 1891). N. Am. Fauna, 5:53.
REVIEWED BY: R. E. Barry, Jr. (REB).
TYPE LOCALITY: U.S.A., Idaho, Jerome Co., Shoshone Falls, north side of Snake River.
DISTRIBUTION: C. Oregon to W. Colorado (U.S.A.), south to Baja California Norte and N.W. Sonora (Mexico).
COMMENT: Revised by Hall and Hoffmeister, 1942, J. Mammal., 23:51–56.
ISIS NUMBER: 5301410008039011001.

Peromyscus dickeyi Burt, 1932. Trans. San Diego Soc. Nat. Hist., 7:176.
TYPE LOCALITY: Mexico, Baja California Sur, Tortuga Isl., 27° 21′ N., 111° 54′ W.
DISTRIBUTION: Known only from the type locality.
COMMENT: Reviewed by Avise *et al.*, 1974, Syst. Zool., 23:226–238.
ISIS NUMBER: 5301410008039012001.

Peromyscus difficilis (J. A. Allen, 1891). Bull. Am. Mus. Nat. Hist., 3:298.
TYPE LOCALITY: Mexico, Zacatecas, Sierra de Valparaiso.
DISTRIBUTION: N. Colorado (U.S.A.) to Oaxaca (Mexico).
COMMENT: Includes *nasutus*; see Hoffmeister and de la Torre, 1961, J. Mammal., 42:1–13, Hooper, 1968, *in* King, ed., The Biology of *Peromyscus* (Rodentia), p. 27, and Hall, 1981:706. Reviewed by Diersing, 1976, Proc. Biol. Soc. Wash., 89:451–466. May comprise two distinct species (WC).
ISIS NUMBER: 5301410008039013001.

Peromyscus eremicus (Baird, 1858). Mammals, *in* Repts. Expl. Surv...., 8(1):479.
REVIEWED BY: R. E. Barry, Jr. (REB).
TYPE LOCALITY: U.S.A., California, Imperial Co., Colorado River, Old Fort Yuma, opposite Yuma, Arizona.
DISTRIBUTION: S. Nevada to Baja California, east to W. Texas and south to San Luis Potosi (Mexico).
COMMENT: Includes *collatus*; see Lawlor, 1971, Trans. San Diego Soc. Nat. Hist., 16:121, but also see Hall, 1981:662, who listed *collatus* as a distinct species. Formerly

included *eva;* see Lawlor, 1971, Occas. Pap. Mus. Zool. Univ. Mich., 661:15–17. Reviewed by Veal and Caire, 1979, Mamm. Species, 118:1–6, and Avise *et al.*, 1974, Syst. Zool., 23:226–238.
ISIS NUMBER: 5301410008039014001 as *P. eremicus.*
5301410008039009001 as *P. collatus.*

Peromyscus eva Thomas, 1898. Ann. Mag. Nat. Hist., ser. 7, 1:44.
TYPE LOCALITY: Mexico, S. Baja California, San Jose del Cabo.
DISTRIBUTION: Baja California Sur and Carmen Isl. (Mexico).
COMMENT: Formerly included in *eremicus;* see Lawlor, 1971, Occas. Pap. Mus. Zool. Univ. Mich., 661:15–17.

Peromyscus furvus J. A. Allen and Chapman, 1897. Bull. Am. Mus. Nat. Hist., 9:201.
TYPE LOCALITY: Mexico, Veracruz, 1.5 mi. (2 km) east of Jalapa, 4400 ft. (1341 m).
DISTRIBUTION: C. Veracruz and S. San Luis Potosi to N.W. Oaxaca (Mexico).
COMMENT: Includes *latirostris* and *angustirostris;* see Hall, 1971, An. Inst. Biol. Univ. Nac. Auton. Mex., 39:154, and Musser, 1964, Occas. Pap. Mus. Zool. Univ. Mich., 636:12. Reviewed by Huckaby, 1980, Los Ang. Cty. Mus. Contrib. Sci., 326:1–24.
ISIS NUMBER: 5301410008039018001 as *P. furvus.*
5301410008039028001 as *P. latirostris.*

Peromyscus gossypinus (Le Conte, 1853). Proc. Acad. Nat. Sci. Phila., 6:411.
REVIEWED BY: R. E. Barry, Jr. (REB).
TYPE LOCALITY: U.S.A., Georgia, Liberty Co., near Riceboro, probably Le Conte Plantation.
DISTRIBUTION: S. Illinois and S.E. Virginia to Gulf of Mexico and S. Florida (U.S.A.).
COMMENT: Hybridizes with *leucopus;* reviewed by Wolfe and Linzey, 1977, Mamm. Species, 70:1–5.
ISIS NUMBER: 5301410008039019001.

Peromyscus grandis Goodwin, 1932. Am. Mus. Novit., 560:4.
TYPE LOCALITY: Guatemala, Alta Verapaz, Finca Concepcion, 3 mi. (5 km) S. San Miguel Tucuru, 3750 ft. (1143 m).
DISTRIBUTION: Vicinity of the type locality.
COMMENT: Reviewed by Huckaby, 1980, Los Ang. Cty. Mus. Contrib. Sci., 326:1–24.
ISIS NUMBER: 5301410008039020001.

Peromyscus guardia Townsend, 1912. Bull. Am. Mus. Nat. Hist., 31:126.
TYPE LOCALITY: Mexico, Baja California Norte, Angel de la Guarda Isl., 29° 33′ N., 113° 35′ W.
DISTRIBUTION: Angel de la Guarda Isl., Granito and Mejia Isls. (Gulf of California, Mexico).
COMMENT: Reviewed by Lawlor, 1971, Trans. San Diego Soc. Nat. Hist., 16:91–124. Formerly included *interparietalis;* see Banks, 1967, J. Mammal., 48:210–218.
ISIS NUMBER: 5301410008039021001.

Peromyscus guatemalensis Merriam, 1898. Proc. Biol. Soc. Wash., 12:118.
TYPE LOCALITY: Guatemala, Huehuetenango, Todos Santos, 10,000 ft. (3048 m).
DISTRIBUTION: S.W. Guatemala and adjacent Chiapas (Mexico).
COMMENT: Includes *altilaneus;* see Huckaby, 1980, Los Ang. Cty. Mus. Contrib. Sci., 326:1–24; Hall, 1981:714, listed *altilaneus* as a distinct species. Formerly included *tropicalis* (here included in *mexicanus*); see Musser, 1969, Am. Mus. Novit., 2357:9.
ISIS NUMBER: 5301410008039022001 as *P. guatemalensis.*
5301410008039001001 as *P. altilaneus.*

Peromyscus gymnotis Thomas, 1894. Ann. Mag. Nat. Hist., ser. 6, 14:365.
TYPE LOCALITY: "Guatemala."
DISTRIBUTION: Pacific coast of S.W. Guatemala and adjacent Chiapas (Mexico).
COMMENT: Formerly included in *mexicanus;* includes *allophylus;* see Musser, 1971, Am. Mus. Novit., 2453:1–10. Reviewed by Huckaby, 1980, Los Ang. Cty. Mus. Contrib. Sci., 326:1–24.
ISIS NUMBER: 5301410008039023001.

Peromyscus hooperi Lee and Schmidly, 1977. J. Mammal., 58:263.
  TYPE LOCALITY: Mexico, Coahuila, 2.5 mi. (4 km) W., 21 mi. (34 km) S. Ocampo, 3500 ft. (1067 m).
  DISTRIBUTION: C. and S.E. Coahuila, from the Sierra del Pino south-southeastward to El Gorrion (Mexico).

Peromyscus interparietalis Burt, 1932. Trans. San Diego Soc. Nat. Hist., 7:175.
  TYPE LOCALITY: Mexico, Baja California Norte, San Lorenzo Sur Isl., 28° 36' N., 112° 51' W.
  DISTRIBUTION: North and South San Lorenzo Isls., and Salsipuedes Isl. (Gulf of California, Mexico).
  COMMENT: Revised by Banks, 1967, J. Mammal., 48:210–218. Reviewed by Lawlor, 1971, Trans. San Diego Soc. Nat. Hist., 16:90–124; includes *lorenzi* and *ryckmani*.
  ISIS NUMBER: 5301410008039026001.

Peromyscus leucopus (Rafinesque, 1818). Am. Monthly Mag., 3:446.
  REVIEWED BY: R. E. Barry, Jr. (REB).
  TYPE LOCALITY: U.S.A., W. Kentucky, "pine barrens," near mouth of Ohio River.
  DISTRIBUTION: Nova Scotia; S. Maine (U.S.A.) to S.E. Alberta (Canada); and (excluding Florida) to Arizona and Yucatan Peninsula (Mexico).
  COMMENT: Hybridizes with *gossypinus*; see Wolfe and Linzey, 1977, Mamm. Species, 70:1–5. Reviewed by Osgood, 1909, N. Am. Fauna, 28:113–117, and Hall, 1981:685–689.
  ISIS NUMBER: 5301410008039030001.

Peromyscus madrensis Merriam, 1898. Proc. Biol. Soc. Wash., 12:16.
  TYPE LOCALITY: Mexico, Nayarit, Tres Marias Isls., Maria Madre Isl.
  DISTRIBUTION: Tres Marias Isls. (Mexico).
  COMMENT: Considered a distinct species by Carleton, 1977:37, and Carleton, 1980:16. The relationship of *madrensis* to *spicilegus* and *boylii* requires further study; see Carleton, 1977:37. Hall, 1981:696, listed *madrensis* in *boylii*.

Peromyscus maniculatus (Wagner, 1845). Wiegmann's Archive Naturgesch., 11, 1:148.
  TYPE LOCALITY: Canada, Labrador, Moravian settlements.
  DISTRIBUTION: S.E. Alaska and N. Canada to C. Oaxaca (Mexico), except S.E. U.S.A.
  COMMENT: Includes *oreas*; see Hooper, 1968, *in* King, ed., The Biology of *Peromyscus* (Rodentia), p. 45. Sheppe, 1961, Proc. Am. Philos. Soc., 105:421–446, considered *oreas* a distinct species, reproductively isolated from *maniculatus* in most places. Relationship with *sitkensis* reviewed by Thomas, 1973, Cytologia, 38:485–495, and with *melanotis* reviewed by Bowers, 1974, J. Mammal., 55:720–737. LNC considers *oreas* a distinct species.
  ISIS NUMBER: 5301410008039032001.

Peromyscus mayensis Carleton and Huckaby, 1975. J. Mammal., 56:444.
  TYPE LOCALITY: Guatemala, Huehuetenango, about 7 km N.W. Santa Eulalia, Yaiquich, 2950 m.
  DISTRIBUTION: Known only from the type locality.

Peromyscus megalops Merriam, 1898. Proc. Biol. Soc. Wash., 12:119.
  TYPE LOCALITY: Mexico, Oaxaca, La Cieneguilla ranch, near Santa Maria Ozolotepec, 10,000 ft. (3048 m).
  DISTRIBUTION: Sierra Madre del Sur of Guerrero and Oaxaca (Mexico).
  COMMENT: Hall, 1981:715, listed *melanurus* as a subspecies of *megalops*; but see Huckaby, 1980, Los Ang. Cty. Mus. Contrib. Sci., 326:1–24, who considered *melanurus* a distinct species. Reviewed by Musser, 1964, Occas. Pap. Mus. Zool. Univ. Mich., 636:1–22.
  ISIS NUMBER: 5301410008039033001.

Peromyscus mekisturus Merriam, 1898. Proc. Biol. Soc. Wash., 12:124.
  TYPE LOCALITY: Mexico, Puebla, Chalchicomula, 8400 ft. (2560 m).
  DISTRIBUTION: S.E. Puebla (Mexico).
  COMMENT: Reviewed by Osgood, 1909, N. Am. Fauna, 28:189.
  ISIS NUMBER: 5301410008039034001.

Peromyscus melanocarpus Osgood, 1904. Proc. Biol. Soc. Wash., 17:73.
TYPE LOCALITY: Mexico, Oaxaca, Cerro Zempoaltepec, above Yacochi.
DISTRIBUTION: N.E. slopes of Sistema Montanoso Poblano Oaxaqueno (N.C. Oaxaca, Mexico).
COMMENT: Reviewed by Huckaby, 1980, Los Ang. Cty. Mus. Contrib. Sci., 326:12.
ISIS NUMBER: 5301410008039035001.

Peromyscus melanophrys (Coues, 1874). Proc. Acad. Nat. Sci. Phila., 26:181.
TYPE LOCALITY: Mexico, Oaxaca, Santa Efigenia.
DISTRIBUTION: S. Durango and S. Coahuila, south through interior Mexico to Chiapas.
COMMENT: Includes *xenurus;* see Baker, 1952, Univ. Kans. Mus. Nat. Hist. Misc. Publ., 5:251–258.
ISIS NUMBER: 5301410008039036001.

Peromyscus melanotis J. A. Allen and Chapman, 1897. Bull. Am. Mus. Nat. Hist., 9:203.
TYPE LOCALITY: Mexico, Veracruz, Las Vigas, 8000 ft. (2438 m).
DISTRIBUTION: S.E. Arizona (U.S.A.) and W. Chihuahua to Morelos and Veracruz (Mexico).
COMMENT: Reviewed by Bowers *et al.,* 1973, Evolution, 27:378–386; includes populations previously included in *maniculatus rufinus.* Also see Bowers, 1974, J. Mammal., 55:720–737, and Hall, 1981:684.
ISIS NUMBER: 5301410008039037001.

Peromyscus melanurus Osgood, 1909. N. Am. Fauna, 28:215.
TYPE LOCALITY: Mexico, Oaxaca, below Pluma Hidalgo, 3000 ft. (914 m).
DISTRIBUTION: Pacific slopes of the Sierra Madre del Sur of Oaxaca (Mexico).
COMMENT: Included in *megalops* by Hall, 1981:715, but Huckaby, 1980, Los Ang. Cty. Mus. Contrib. Sci., 326:1–24, considered it a distinct species.

Peromyscus merriami Mearns, 1896. Preliminary diagnoses of new mammals from the Mexican border of the United States, p. 2; preprint of Proc. U.S. Nat. Mus., 19:138.
TYPE LOCALITY: Mexico, Sonora, Sonoyta, on Sonoyta River (Mexico).
DISTRIBUTION: S. Arizona (U.S.A.) through Sonora to N. Sinaloa (Mexico).
COMMENT: Includes *goldmani;* see Hoffmeister and Diersing, 1973, Southwest. Nat., 18:354–357. Also see Hoffmeister and Lee, 1963, J. Mammal., 44:201–213, Hall and Kelson, 1952, Univ. Kans. Mus. Nat. Hist. Misc. Publ., 5:368, and Lawlor, 1971, Occas. Pap. Mus. Zool. Univ. Mich., 661:1–22.
ISIS NUMBER: 5301410008039038001.

Peromyscus mexicanus (Saussure, 1860). Rev. Mag. Zool. Paris, ser. 2, 12:103.
TYPE LOCALITY: Mexico, Veracruz, 10 km east of Mirador.
DISTRIBUTION: S. San Luis Potosi and Veracruz through E. and S. Mexico to W. Panama.
COMMENT: Includes *nudipes* (see Huckaby, 1980, Los Ang. Cty. Mus. Contrib. Sci., 326:1–24), *angelensis, coatlanensis, sloeops,* and *tropicalis* (see Musser, 1969, Am. Mus. Novit., 2357:1–23). Hall, 1981:713, listed *nudipes* as a distinct species, and *tropicalis* as a subspecies of *guatemalensis.* Formerly included *gymnotis* and *allophylus;* see Musser, 1971, Am. Mus. Novit., 2453:1. Type locality reviewed by Dalquest, 1950, Occas. Pap. Mus. Zool. La. St. Univ., 23:1–15.
ISIS NUMBER: 5301410008039039001 as *P. mexicanus.*
                 5301410008039040001 as *P. nudipes.*

Peromyscus ochraventer Baker, 1951. Univ. Kans. Mus. Nat. Hist. Misc. Publ., 5:213.
TYPE LOCALITY: Mexico, Tamaulipas, 70 km (by highway) S. Ciudad Victoria, 6 km W. Panamerican Highway, El Carrizo, 2800 ft. (853 m).
DISTRIBUTION: S. Tamaulipas and adjacent San Luis Potosi (Mexico).
COMMENT: Revised by Huckaby, 1980, Los Ang. Cty. Mus. Contrib. Sci., 326:7–9. Reviewed by Carleton, 1977:1, 37, and Carleton, 1980, Misc. Publ. Mus. Zool. Univ. Mich., 157:1–146.
ISIS NUMBER: 5301410008039042001.

Peromyscus pectoralis Osgood, 1904. Proc. Biol. Soc. Wash., 17:59.
TYPE LOCALITY: Mexico, Queretaro, Jalpan.
DISTRIBUTION: S.E. New Mexico and C. Texas (U.S.A.) to Jalisco and Hidalgo (Mexico).
COMMENT: Includes *collinus* and *laceianus*; see Schmidly, 1972, Southwest. Nat.,
17:113–138, Kilpatrick and Zimmerman, 1975, Syst. Zool., 24:143–162, and
Kilpatrick and Zimmerman, 1976, J. Mammal., 57:506–522. Possibly a composite
of two or more species; see Avise *et al.*, 1974, J. Mammal., 55:761. Reviewed by
Schmidly, 1974, Mamm. Species, 49:1–3, and Carleton, 1977:1–47.
ISIS NUMBER: 5301410008039043001.

Peromyscus pembertoni Burt, 1932. Trans. San Diego Soc. Nat. Hist., 7:176.
TYPE LOCALITY: Mexico, Sonora, San Pedro Nolasco Isl., 27° 58′ N., 111° 24′ W.
DISTRIBUTION: Known only from the type locality.
COMMENT: Reviewed by Lawlor, 1971, Occas. Pap. Mus. Zool. Univ. Mich., 661:1–22.
ISIS NUMBER: 5301410008039044001.

Peromyscus perfulvus Osgood, 1945. J. Mammal., 26:299.
TYPE LOCALITY: Mexico, Michoacan, 10 km W. Apatzingan, 1040 ft. (317 m).
DISTRIBUTION: S. Jalisco through Michoacan to W. Guerrero (Mexico).
COMMENT: Reviewed by Hooper, 1968, *in* King, ed., The Biology of *Peromyscus*
(Rodentia), pp. 27–74.
ISIS NUMBER: 5301410008039045001.

Peromyscus polionotus (Wagner, 1843). Wiegmann's Archive Naturgesch., 9, 2:52.
REVIEWED BY: R. E. Barry, Jr. (REB).
TYPE LOCALITY: U.S.A., Georgia.
DISTRIBUTION: N.E. Mississippi and N. Alabama through Georgia to W. South
Carolina, N. and E. (coast) Florida (U.S.A.).
COMMENT: Reviewed by Selander *et al.*, 1971, Studies in Genetics, VI, Univ. Texas
Publ., 7103:49–90.
ISIS NUMBER: 5301410008039047001.

Peromyscus polius Osgood, 1904. Proc. Biol. Soc. Wash., 17:61.
TYPE LOCALITY: Mexico, Chihuahua, Colonia Garcia.
DISTRIBUTION: W.C. Chihuahua (Mexico).
COMMENT: Reviewed by Carleton, 1977, Occas. Pap. Mus. Zool. Univ. Mich., 675:1, 37,
Kilpatrick and Zimmerman, 1975, Syst. Zool., 24:143–162, and Anderson, 1972,
Bull. Am. Mus. Nat. Hist., 148:348–349.
ISIS NUMBER: 5301410008039048001.

Peromyscus pseudocrinitus Burt, 1932. Trans. San Diego Soc. Nat. Hist., 7:173.
TYPE LOCALITY: Mexico, Baja California Sur, Coronados Isl., 26° 06′ N., 111° 18′ W.
DISTRIBUTION: Known only from the type locality.
COMMENT: Reviewed by Lawlor, 1971, Occas. Pap. Mus. Zool. Univ. Mich., 661:1–22.
ISIS NUMBER: 5301410008039049001.

Peromyscus sejugis Burt, 1932. Trans. San Diego Soc. Nat. Hist., 7:171.
TYPE LOCALITY: Mexico, Baja California Sur, Santa Cruz Isl., 25° 17′ N., 110° 43′ W.
DISTRIBUTION: Santa Cruz Isl. and San Diego Isl. (Gulf of California, Mexico).
COMMENT: Reviewed by Avise *et al.*, 1974, Syst. Zool., 23:226–238.
ISIS NUMBER: 5301410008039050001.

Peromyscus simulus Osgood, 1904. Proc. Biol. Soc. Wash., 17:64.
TYPE LOCALITY: Mexico, Nayarit, San Blas.
DISTRIBUTION: Nayarit and S. Sinaloa (Mexico).
COMMENT: Formerly included in *boylii*; see Carleton, 1977:1–47; Hall, 1981:696, listed
simulus in *boylii*.

Peromyscus sitkensis Merriam, 1897. Proc. Biol. Soc. Wash., 11:223.
TYPE LOCALITY: U.S.A., Alaska, Baranof Isl., Sitka.
DISTRIBUTION: Alexander Arch. of S. Alaska (U.S.A.) and Queen Charlotte Isls. of
British Columbia (Canada).
COMMENT: Reviewed by Thomas, 1973, Cytologia, 38:485–495.
ISIS NUMBER: 5301410008039052001.

Peromyscus slevini Mailliard, 1924. Proc. Calif. Acad. Sci., ser. 4, 12:1221.
   TYPE LOCALITY: Mexico, Baja California Sur, Santa Catalina Isl., 17 mi. (27 km) N.E.
      Punta San Marcial, 25° 43′ 50″ N.
   DISTRIBUTION: Known only from the type locality.
   ISIS NUMBER: 5301410008039053001.

Peromyscus spicilegus J. A. Allen, 1897. Bull. Am. Mus. Nat. Hist., 9:50.
   TYPE LOCALITY: Mexico, Jalisco, Mascota, Mineral San Sebastian.
   DISTRIBUTION: S. Sonora and Chihuahua to Michoacan (W. Mexico).
   COMMENT: Formerly included in *boylii*; see Carleton, 1977:1–47. Hall, 1981:696, listed
      spicilegus in *boylii*.

Peromyscus stephani Townsend, 1912. Bull. Am. Mus. Nat. Hist., 31:126.
   TYPE LOCALITY: Mexico, Sonora, San Esteban Isl., 28° 34′ N., 113° 21′ W.
   DISTRIBUTION: Known only from the type locality.
   COMMENT: Reviewed by Lawlor, 1971, Trans. San Diego Soc. Nat. Hist., 16:91–124,
      Hooper and Musser, 1964, Occas. Pap. Mus. Zool. Univ. Mich., 635:12, and
      Carleton, 1977:1, 37.
   ISIS NUMBER: 5301410008039054001.

Peromyscus stirtoni Dickey, 1928. Proc. Biol. Soc. Wash., 41:5.
   TYPE LOCALITY: El Salvador, La Union, Rio Goascoran, 13° 30′ N., 100 ft. (30 m).
   DISTRIBUTION: S.E. Guatemala; El Salvador; S. Honduras.
   COMMENT: Reviewed by Huckaby, 1980, Los Ang. Cty. Mus. Contrib. Sci., 326:1–24.
   ISIS NUMBER: 5301410008039055001.

Peromyscus truei (Shufeldt, 1885). Proc. U.S. Nat. Mus., 8:407.
   REVIEWED BY: R. E. Barry, Jr. (REB).
   TYPE LOCALITY: U.S.A., New Mexico, McKinley Co., Fort Wingate.
   DISTRIBUTION: N.C. Texas; C. and S.W. Oregon to Colorado (U.S.A.) south to Baja
      California and to Oaxaca (Mexico).
   COMMENT: Includes *comanche*; see Schmidly, 1973, Southwest. Nat., 18:276, Hall,
      1981:701, and Hoffmeister, 1981, Mamm. Species, 161:1–5; but also see Johnson
      and Packard, 1974, Occas. Pap. Mus. Texas Tech Univ., 24:1–16, who considered
      comanche a distinct species. Revised by Hoffmeister, 1951, Ill. Biol. Monogr.,
      21:1–104; reviewed by Hoffmeister, 1981, Mamm. Species, 161:1–5. Zimmerman *et
      al.*, 1978, Evolution, 32:565–579, considered *gentilis* a sibling species on the basis
      of a revision that has not been published. May include more than one species
      (CWK).
   ISIS NUMBER: 5301410008039057001 as *P. truei*.
                       5301410008039010001 as *P. comanche*.

Peromyscus winkelmanni Carleton, 1977. Occas. Pap. Mus. Zool. Univ. Mich., 675:2.
   TYPE LOCALITY: Mexico, Michoacan, 6.3 mi. (10 km) (by road) W.S.W. Dos Aguas, 8000
      ft. (2438 m).
   DISTRIBUTION: Known only from the vicinity of the type locality.

Peromyscus yucatanicus J. A. Allen and Chapman, 1897. Bull. Am. Mus. Nat. Hist., 9:8.
   TYPE LOCALITY: Mexico, Yucatan, Chichen-Itza.
   DISTRIBUTION: Yucatan Peninsula (Mexico).
   COMMENT: Reviewed by Lawlor, 1965, Univ. Kans. Mus. Nat. Hist. Misc. Publ.,
      16:421–438, and Huckaby, 1980, Los Ang. Cty. Mus. Contrib. Sci., 326:1–24.
   ISIS NUMBER: 5301410008039058001.

Peromyscus zarhynchus Merriam, 1898. Proc. Biol. Soc. Wash., 12:117.
   TYPE LOCALITY: Mexico, Chiapas, mountains above Tumbala, 5500 ft. (1676 m).
   DISTRIBUTION: Mountains of N.C. Chiapas (Mexico).
   COMMENT: Revised by Hooper and Musser, 1964, Occas. Pap. Mus. Zool. Univ. Mich.,
      635:1–13, and Huckaby, 1980, Los Ang. Cty. Mus. Contrib. Sci., 326:1–24.
   ISIS NUMBER: 5301410008039059001.

Petromyscus Thomas, 1926. Ann. Mag. Nat. Hist., ser. 9, 17:179.
REVIEWED BY: R. D. Owen (RDON).
COMMENT: Taxonomy discussed by Swanepoel *et al.*, 1980, Ann. Transvaal Mus., 32(7):159. Subfamily Dendromurinae; see comment under Cricetidae.
ISIS NUMBER: 5301410008108000000.

Petromyscus collinus (Thomas and Hinton, 1925). Proc. Zool. Soc. Lond., 1925:237.
TYPE LOCALITY: Namibia, Damaraland, Karibib.
DISTRIBUTION: S.W. Cape Prov. (South Africa); Namibia; S. Angola.
ISIS NUMBER: 5301410008108001001 as *P. collinus*.
5301410011056004001 as *Praomys collinus*.

Petromyscus monticularis (Thomas and Hinton, 1925). Proc. Zool. Soc. Lond., 1925:238.
TYPE LOCALITY: Namibia, Great Namaqualand, Great Brukkaros Mtn., near Berseba.
DISTRIBUTION: Namibia.
ISIS NUMBER: 5301410008108002001 as *P. monticularis*.
5301410011056011001 as *Praomys monticularis*.

Phaenomys Thomas, 1917. Ann. Mag. Nat. Hist., ser. 8, 20:196.
REVIEWED BY: J. K. Braun (JKB).
COMMENT: Subfamily Hesperomyinae; see comment under Cricetidae.
ISIS NUMBER: 5301410008040000000.

Phaenomys ferrugineus (Thomas, 1894). Ann. Mag. Nat. Hist., ser. 6, 14:352.
TYPE LOCALITY: E. Brazil, Rio de Janeiro.
DISTRIBUTION: E. Brazil.
ISIS NUMBER: 5301410008040001001.

Phodopus Miller, 1910. Smithson. Misc. Coll., 52:498.
REVIEWED BY: O. L. Rossolimo (OLR); S. Wang (SW)(China).
COMMENT: Subfamily Cricetinae; see comment under Cricetidae.
ISIS NUMBER: 5301410008041000000.

Phodopus campbelli (Thomas, 1905). Ann. Mag. Nat. Hist., 15:322.
TYPE LOCALITY: Mongolia, Shaborte, 42° 40' N.
DISTRIBUTION: Transbaikalia (U.S.S.R.); Mongolia; Heilungkiang, Inner Mongolia, Hopei, and Sinkiang (China).
COMMENT: Formerly included in *sungorus* by Ellerman and Morrison-Scott, 1951:627, but see Vorontsov *et al.*, 1967, C. R. Acad. Sci. U.S.S.R., 172(3):703–705, and Galkina *et al.*, 1977, *in* [Fauna and Systematics of Siberian Vertebrates], Nauka, Novosibirsk, pp. 60–80. Corbet, 1978:89, listed *campbelli* as a subspecies of *sungorus*, without comment; Yudin *et al.*, 1979, [Mammals of Altai-Sayan Montane Region], Novosibirsk, p. 156, believed *campbelli* to be a full species. Gromov and Baranova, 1981:158, included *campbelli* in *sungorus*, without comment.

Phodopus roborovskii (Satunin, 1903). Ann. Zool. Mus. St. Petersb., 7:571.
TYPE LOCALITY: China, Nan Shan Mtns., upper part of Shargol Dzhin River.
DISTRIBUTION: Tuva and E. Kazakhstan (U.S.S.R.); Mongolia; N.E. and N.C. China.
COMMENT: *P. przhewalskii* may be a separate species; see Sludskii, 1977, [Mammals of Kazakhstan], 1(2):467.
ISIS NUMBER: 5301410008041001001.

Phodopus sungorus (Pallas, 1773). Reise Prov. Russ. Reichs., 2:703.
TYPE LOCALITY: U.S.S.R., E. Kazakhstan, 100 km west of Semipalatinsk, near Grachevsk.
DISTRIBUTION: E. Kazakhstan and S.W. Siberia (U.S.S.R.).
COMMENT: Formerly included *campbelli*; see Galkina *et al.*, 1977, *in* [Fauna and Systematics of Siberian Vertebrates], Nauka, Novosibirsk, pp. 60–80. Gromov and Baranova, 1981, included *campbelli* in *sungorus*, without comment. See also comment under *campbelli*.
ISIS NUMBER: 5301410008041002001.

Phyllotis Waterhouse, 1837. Proc. Zool. Soc. Lond., 1837:27.

REVIEWED BY: M. D. Carleton (MDC); A. Langguth (AL); M. A. Mares (MAM); O. P. Pearson (OPP); R. H. Pine (RHP)(Chile); J. J. Pizzimenti (JJP); D. F. Williams (DFW).

COMMENT: Revised by Pearson, 1958, Univ. Calif. Publ. Zool., 56(4):391–496, and Hershkovitz, 1962, Fieldiana Zool., 49:217–463. Includes *Paralomys*; see Hershkovitz, 1962:217, and Pearson and Patton, 1976, J. Mammal., 57:339–350. *Galenomys* was listed as a subgenus of *Phyllotis*, without comment, by Pearson and Patton, 1976:339, but see Hershkovitz, 1962:468, and Pearson, 1958, Univ. Calif. Publ. Zool., 56(4):393, 395, who considered *Galenomys* a distinct genus. Hershkovitz, 1962:217, included *Auliscomys* and *Graomys* in this genus but see Pearson and Patton, 1976:341, and Gardner and Patton, 1976, Occas. Pap. Mus. Zool. La. St. Univ., 49:1–48, who considered them distinct genera. Formerly included *micropus*, which was transferred to *Auliscomys* by Simonetti and Spotorno, 1980, An. Mus. Hist. Nat. Valparaiso, 13:285–297, who also compared karyotypes and morphometrics of *Phyllotis*, *Auliscomys*, and *Andinomys*. *Loxodontomys* was considered a synonym of *Phyllotis* (represented by *micropus*) by Hershkovitz, 1962:217; but *Loxodontomys* was considered a *nomen nudum* by Pine *et al.*, 1979, Mammalia, 43:356–357. Hershkovitz, 1962:462–463, included *hypogaeus* as a distinct species of *Phyllotis*; but see Massoia, 1976–1977, Rev. Invest. Agro. INTA, ser. 5, Patalogia Vegetal, 13:15–20, and Williams and Mares, 1978, Ann. Carnegie Mus., 47(9):201, 218, who included *hypogaeus* in *Eligmodontia typus* as a synonym. Subfamily Hesperomyinae; see comment under Cricetidae.

ISIS NUMBER: 5301410008042000000.

Phyllotis amicus Thomas, 1900. Ann. Mag. Nat. Hist., ser. 7, 5:355.

TYPE LOCALITY: Peru, Cajamarca Dept., Tolon, 100 m.

DISTRIBUTION: N. and W. Peru.

ISIS NUMBER: 5301410008042001001.

Phyllotis andium Thomas, 1912. Ann. Mag. Nat. Hist., ser. 8, 10:409.

TYPE LOCALITY: Ecuador, Canar Prov., Canar, 8500 ft. (2591 m).

DISTRIBUTION: Andean S.C. Ecuador and N. Peru.

ISIS NUMBER: 5301410008042002001.

Phyllotis bonaeriensis Crespo, 1964. Neotropica, 10:99.

TYPE LOCALITY: Argentina, Buenos Aires, Sierra de la Ventana.

DISTRIBUTION: Buenos Aires Prov. (Argentina).

COMMENT: Formerly included in *darwini*; see Reig, 1978, Publ. Mus. Munic. Cienc. Nat. Mar del Plata "Lorenzo Scaglia," 2(8):180.

Phyllotis caprinus Pearson, 1958. Univ. Calif. Publ. Zool., 56:435.

TYPE LOCALITY: Argentina, Jujuy Prov., Tilcara, 8000 ft. (2438 m).

DISTRIBUTION: N. Argentina and S. Bolivia on E. slope of Andes.

COMMENT: Reviewed by Pearson and Patton, 1976, J. Mammal., 57:339, who considered *caprinus* a distinct species. Hershkovitz, 1962, Fieldiana Zool., 46:330, included it as a subspecies of *darwini*.

Phyllotis darwini (Waterhouse, 1837). Proc. Zool. Soc. Lond., 1837:28.

TYPE LOCALITY: Chile, Coquimbo Prov., Coquimbo.

DISTRIBUTION: C. and S. Peru; N. and C. Chile; W. Argentina; S. and C. Bolivia.

COMMENT: Formerly included *caprinus*, *definitus*, *magister*, *wolffsohni* (see Pearson, 1972, J. Mammal., 53:677–686, Pearson and Patton, 1976, J. Mammal., 57:339, and Pearson and Ralph, 1978, Mem. Mus. Hist. Nat. "Javier Prado," 18:1–97), and *bonaeriensis* (see Reig, 1978, Publ. Mus. Munic. Cienc. Nat. Mar del Plata "Lorenzo Scaglia," 2(8):180). Pearson, 1958, Univ. Calif. Publ. Zool., 56(4):415, Hershkovitz, 1962, Fieldiana Zool., 46:304, and Pine *et al.*, 1979, Mammalia, 43:353–354, included *osgoodi* in *darwini*; but also see Spotorno, 1976, An. Mus. Hist. Nat. Valparaiso, 9:141–161, who considered *osgoodi* a distinct species.

ISIS NUMBER: 5301410008042004001.

Phyllotis definitus Osgood, 1915. Field Mus. Nat. Hist. Publ. Zool. Ser., 10:189.
   TYPE LOCALITY: Peru, Ancash Dept., Macate, 9000 ft. (2743 m).
   DISTRIBUTION: Andes of Ancash Dept. (Peru).
   COMMENT: Formerly included in *magister* (see Pearson, 1958, Univ. Calif. Publ. Zool.,
      56(4):431) and in *darwini* (see Hershkovitz, 1962, Fieldiana Zool., 46:296), but see
      Pearson, 1972, J. Mammal., 53:680–682, and Pearson and Patton, 1976, J.
      Mammal., 57:339–350.

Phyllotis gerbillus Thomas, 1900. Ann. Mag. Nat. Hist., ser. 7, 5:151.
   TYPE LOCALITY: Peru, Piura Dept., Piura.
   DISTRIBUTION: N.W. Peru.
   COMMENT: Formerly included in *Paralomys;* see Hershkovitz, 1962, Fieldiana Zool.,
      46:217, 399, and Pearson and Patton, 1976, J. Mammal., 57:346, who considered
      gerbillus close to *darwini,* and not in a separate genus or subgenus.
   ISIS NUMBER: 5301410008042005001.

Phyllotis haggardi Thomas, 1908. Ann. Mag. Nat. Hist., ser. 7, 2:270.
   TYPE LOCALITY: Ecuador, Pichincha Prov., Monte Pichincha, above Quito, 4000 m.
   DISTRIBUTION: Andean Ecuador.
   COMMENT: Reviewed by Pearson, 1972, J. Mammal., 53:683, and Pearson and Patton,
      1976, J. Mammal., 57:339–350.
   ISIS NUMBER: 5301410008042007001.

Phyllotis magister Thomas, 1912. Ann. Mag. Nat. Hist., ser. 8, 10:406.
   TYPE LOCALITY: Peru, Arequipa Dept., Arequipa, 2300 m.
   DISTRIBUTION: Andes of S. Peru and N. Chile.
   COMMENT: Formerly included *definitus;* see Pearson, 1972, J. Mammal., 53:680–682.
      Hershkovitz, 1962, Fieldiana Zool., 46:288, included *magister* in *darwini,* but see
      Pearson, 1972, J. Mammal., 53:682, and Pearson and Patton, 1976, J. Mammal.,
      57:339, 346, who considered *magister* a distinct species. Reported from N. Chile by
      Spotorno, 1976, An. Mus. Hist. Nat. Valparaiso, 9:141–161, and Pine *et al.,* 1979,
      Mammalia, 43:355.

Phyllotis osgoodi Mann, 1945. Biologica, 2:81.
   TYPE LOCALITY: Chile, Puna of Tarapaca, Parinacota.
   DISTRIBUTION: Altiplano of Arica Prov. (Chile).
   COMMENT: Included in *darwini* by Pearson, 1958, Univ. Calif. Publ. Zool., 56(4):415,
      Hershkovitz, 1962, Fieldiana Zool., 46:304, and Pine *et al.,* 1979, Mammalia,
      43:353–354; but see Spotorno, 1976, An. Mus. Hist. Nat. Valparaiso, 9:141–161,
      Spotorno and Walker, 1979, Arch. Biol. Med. Exp., 12:83–90, and Walker *et al.,*
      1979, Cytogenet. and Cell Genet., 24:209–216, who considered *osgoodi* a distinct
      species.

Phyllotis osilae J. A. Allen, 1901. Bull. Am. Mus. Nat. Hist., 14:44.
   TYPE LOCALITY: Peru, Puno, Osila (=Asillo), (17 mi. (27 km) E.N.E. Ayaviri), 13,000 ft.
      (3962 m).
   DISTRIBUTION: S.E. Peru; W. Bolivia; N. Argentina.
   ISIS NUMBER: 5301410008042009001.

Phyllotis wolffsohni Thomas, 1902. Ann. Mag. Nat. Hist., ser. 7, 9:131.
   TYPE LOCALITY: Bolivia, Cochabamba Dept., Tapacari, 9900 ft. (3018 m).
   DISTRIBUTION: E. slopes of Andes in C. Bolivia.
   COMMENT: Hershkovitz, 1962, Fieldiana Zool., 46:339, included *wolffsohni* in *darwini.*
      Pearson and Patton, 1976, J. Mammal., 57:341, considered *wolffsohni* a distinct
      species.

Platacanthomys Blyth, 1859. J. Asiat. Soc. Bengal, 28:288.
   REVIEWED BY: J. E. Bucher (JEB).
   COMMENT: Formerly considered in a separate family and subfamily of Gliridae; see
      Arata, 1967, *in* Anderson and Jones, pp. 244–245. Transferred to subfamily

Platacanthomyinae of Cricetidae by Chaline and Mein, 1979, Chaline *et al.*, 1977, Mammalia, 41:245–252, Mein and Freudenthal, 1971, Scr. Geol., 2:137, and Reig, 1980, J. Zool. Lond., 192:260.
ISIS NUMBER: 5301410013001000000.

Platacanthomys lasiurus Blyth, 1859. J. Asiat. Soc. Bengal, 28:289.
TYPE LOCALITY: India, Malabar, Alipi.
DISTRIBUTION: S. India.
ISIS NUMBER: 5301410013001001001.

Podomys Osgood, 1909. N. Am. Fauna, 28:226.
REVIEWED BY: S. Anderson (SA); R. E. Barry, Jr. (REB); W. Caire (WC); R. P. Canham (RPC); M. D. Carleton (MDC); L. N. Carraway (LNC); D. G. Huckaby (DGH); C. W. Kilpatrick (CWK); T. E. Lawlor (TL).
COMMENT: Included as a subgenus of *Peromyscus* by Hall, 1981:720; Hooper, 1968, *in* King, ed., The Biology of *Peromyscus* (Rodentia), p. 38, 66. Linzey and Layne, 1974, Am. Mus. Novit., 2532:1–20, studied the morphology of the spermatozoa. Reviewed by Carleton, 1980:118, 125, who considered *Podomys* a distinct genus. Subfamily Hesperomyinae; see comment under Cricetidae.

Podomys floridanus (Chapman, 1889). Bull. Am. Mus. Nat. Hist., 2:117.
TYPE LOCALITY: U.S.A., Florida, Alachua Co., Gainesville.
DISTRIBUTION: Peninsular Florida (U.S.A.).
COMMENT: Reviewed by Smith *et al.*, 1973, J. Mammal., 54:1–13.
ISIS NUMBER: 5301410008039017001 as *Peromyscus floridanus.*

Podoxymys Anthony, 1929. Am. Mus. Novit., 383:4.
REVIEWED BY: J. K. Braun (JKB).
COMMENT: Subfamily Hesperomyinae; see comment under Cricetidae.
ISIS NUMBER: 5301410008043000000.

Podoxymys roraimae Anthony, 1929. Am. Mus. Novit., 383:4.
TYPE LOCALITY: Guyana, Mt. Roraima, 2580 m.
DISTRIBUTION: Guyana; Venezuela; adjacent Brazil.
ISIS NUMBER: 5301410008043001001.

Prionomys Dollman, 1910. Ann. Mag. Nat. Hist., ser. 8, 6:226.
REVIEWED BY: J. K. Braun (JKB).
COMMENT: Subfamily Dendromurinae; see comment under Cricetidae.
ISIS NUMBER: 5301410008109000000.

Prionomys batesi Dollman, 1910. Ann. Mag. Nat. Hist., ser. 8, 6:228.
TYPE LOCALITY: Cameroun, Ja River, Bitye.
DISTRIBUTION: S. Cameroun; Central African Republic.
ISIS NUMBER: 5301410008109001001.

Psammomys Cretzschmar, 1828. Ruppell Atlas, Reise Nordl. Afr., p. 56.
REVIEWED BY: J. K. Braun (JKB); C. B. Robbins (CBR).
COMMENT: Subfamily Gerbillinae; see comment under Cricetidae.
ISIS NUMBER: 5301410008095000000.

Psammomys obesus Cretzschmar, 1826. Ruppell Atlas, Reise Nordl. Afr., p. 58, pl. 22.
TYPE LOCALITY: Egypt, Alexandria.
DISTRIBUTION: Algeria to Arabia, Jordan and Israel; south to Mauritania and Sudan.
COMMENT: Formerly included *vexillaris*; see Cockrum *et al.*, 1977, Mammalia, 41:321–326.
ISIS NUMBER: 5301410008095001001.

Psammomys vexillaris Thomas, 1925. Ann. Mag. Nat. Hist., ser. 9, 16:198.
TYPE LOCALITY: Libya, Tripolitania Prov., Bu Ngem (Bondjem).
DISTRIBUTION: Algeria; Tunisia; Libya.
COMMENT: Included in *obesus* by Petter, 1975, Part 6.3:3, and Corbet, 1978:128; but see Cockrum *et al.*, 1977, Mammalia, 41:321–326.

Pseudoryzomys Hershkovitz, 1962. Fieldiana Zool., 46:208.
REVIEWED BY: J. K. Braun (JKB); A. Langguth (AL); M. A. Mares (MAM); R. H. Pine (RHP).
COMMENT: Subfamily Hesperomyinae; see comment under Cricetidae.
ISIS NUMBER: 5301410008044000000.

Pseudoryzomys simplex (Winge, 1887). E. Mus. Lundii, 1(3):11.
TYPE LOCALITY: Brazil, Minas Gerais, Lagoa Santa.
DISTRIBUTION: E. Brazil.
COMMENT: Formerly included *incertae sedis* in *Oryzomys* by Cabrera, 1961:395; but see Massoia, 1980, Ameghiniana, 17:280–287.
ISIS NUMBER: 5301410008035057001 as *Oryzomys simplex*.

Pseudoryzomys wavrini (Thomas, 1921). Ann. Mag. Nat. Hist., ser. 9, 7:177.
TYPE LOCALITY: Paraguay, N. Chaco, Jesematathla, west of Concepcion, 100 m.
DISTRIBUTION: Bolivia; Paraguay; N. Argentina.
COMMENT: Formerly included in *Oryzomys*; see Hershkovitz, 1962, Fieldiana Zool., 46:208. Pine and Wetzel, 1975, Mammalia, 39:649–655, reviewed this species.
ISIS NUMBER: 5301410008044001001.

Punomys Osgood, 1943. J. Mammal., 24:369.
REVIEWED BY: O. P. Pearson (OPP).
COMMENT: Subfamily Hesperomyinae; see comment under Cricetidae.
ISIS NUMBER: 5301410008045000000.

Punomys lemminus Osgood, 1943. J. Mammal., 24:369.
TYPE LOCALITY: Peru, Puno Dept., San Antonio de Esquilache, 4500 m.
DISTRIBUTION: S. Peru.
ISIS NUMBER: 5301410008045001001.

Reithrodon Waterhouse, 1837. Proc. Zool. Soc. Lond., 1837:29.
REVIEWED BY: J. K. Braun (JKB); A. L. Gardner (ALG); A. Langguth (AL); M. A. Mares (MAM).
COMMENT: Karyology and systematics reviewed by Pearson and Patton, 1976, J. Mammal., 57:339–350, and Gardner and Patton, 1976, Occas. Pap. Mus. Zool. La. St. Univ., 49:27, 30, 36. Reviewed by Reig, 1978, Publ. Mus. Munic. Cienc. Nat. Mar del Plata "Lorenzo Scaglia," 2(8):184–185. Subfamily Hesperomyinae; see comment under Cricetidae.
ISIS NUMBER: 5301410008046000000.

Reithrodon physodes (Olfers, 1818). *In* Eschwege, Neue Bibl. Reisenb., 15:209.
TYPE LOCALITY: Argentina, pampas south of Buenos Aires, south bank of the Rio de la Plata.
DISTRIBUTION: Argentina; adjacent Chile; Uruguay.
COMMENT: Includes *auritus, caurinus, cuniculoides, evae, pachycephalus,* and *typicus*; see Hershkovitz, 1959, J. Mammal., 40:348, and Cabrera, 1961:501. The name *auritus* was given to this species by Fischer, 1814, Zoognosia, 3:71; see Langguth, 1966, Bull. Zool. Nomencl., 23(6):285–287. Reig, 1978, Publ. Mus. Munic. Cienc. Nat. Mar del Plata "Lorenzo Scaglia," 2(8):185, referred to this species using Fischer's name, as did Dalby and Mares, 1974, Am. Midl. Nat., 92:205–206. The use of *physodes* here follows Myers and Carleton, 1981, Misc. Publ. Mus. Zool. Univ. Mich., 161:1–41, who considered some names used by Fischer, 1814, inconsistently binomial (which, if true, would make the work unavailable); ALG also considered Fischer, 1814, an invalid work. However, Langguth, 1981, *in litt.* responded to Myers and Carleton, 1981, and to Sabrosky, 1967, Case ZN(S)1774, Bull. Zool. Nomencl., 24:141, and concluded that Fischer, 1814, is an available work, and that *auritus* is the name to be used for this species. Other names from Fischer, 1814, are employed in this volume pending resolution of this problem (the editors).
ISIS NUMBER: 5301410008046001001.

Reithrodontomys Giglioli, 1874. Bull. Soc. Geogr. Ital., Roma, 11:326.
 REVIEWED BY: M. D. Carleton (MDC); A. V. Linzey (AVL); J. Ramirez-Pulido
  (JRP)(Mexico); G. Urbano-V. (GUV).
 COMMENT: Includes *Aporodon* as a subgenus; see Hooper, 1952, Misc. Publ. Mus. Zool.
  Univ. Mich., 77:1–255, Hershkovitz, 1966, *in* Wenzel and Tipton, eds.,
  Ectoparasites of Panama, pp. 725–752, and Carleton, 1980:125–126. Karyology
  reviewed by Carleton and Myers, 1979, J. Mammal., 60:307–313. Revised by
  Howell, 1914, N. Am. Fauna, 36:1–97; Hooper, 1952, Misc. Publ. Mus. Zool. Univ.
  Mich., 77:1–255, revised the Central American species. Subfamily Hesperomyinae;
  see comment under Cricetidae.
 ISIS NUMBER: 5301410008047000000.

Reithrodontomys brevirostris Goodwin, 1943. Am. Mus. Novit., 1231:1.
 TYPE LOCALITY: Costa Rica, Alajuela, canyons above Villa Quesada, 5000 ft. (1524 m).
 DISTRIBUTION: N.C. Nicaragua; C. Costa Rica.
 COMMENT: Subgenus *Aporodon*; see Hooper, 1952, Misc. Publ. Mus. Zool. Univ. Mich.,
  77:1–255. Includes *nicaraguae*; see Jones and Genoways, 1970, Occas. Pap. W.
  Found. Vert. Zool., 2:10.
 ISIS NUMBER: 5301410008047001001.

Reithrodontomys burti Benson, 1939. Proc. Biol. Soc. Wash., 52:147.
 TYPE LOCALITY: Mexico, Sonora, Rio Sonora, Rancho de Costa Rica.
 DISTRIBUTION: C. Sinaloa to W.C. Sonora (Mexico).
 COMMENT: Subgenus *Reithrodontomys*; see Hooper, 1952, Misc. Publ. Mus. Zool. Univ.
  Mich., 77:1–255.
 ISIS NUMBER: 5301410008047002001.

Reithrodontomys chrysopsis Merriam, 1900. Proc. Biol. Soc. Wash., 13:152.
 TYPE LOCALITY: Mexico, Volcan Popocatepetl, 11,500 ft. (3505 m).
 DISTRIBUTION: S.E. Jalisco to W.C. Veracruz (Mexico).
 COMMENT: Subgenus *Reithrodontomys*; see Hooper, 1952, Misc. Publ. Mus. Zool. Univ.
  Mich., 77:1–255.
 ISIS NUMBER: 5301410008047003001.

Reithrodontomys creper Bangs, 1902. Bull. Mus. Comp. Zool. Harv. Univ., 39:39.
 TYPE LOCALITY: Panama, Chiriqui, Volcan de Chiriqui, 11,000 ft. (3353 m).
 DISTRIBUTION: C. Costa Rica; W. Panama.
 COMMENT: Subgenus *Aporodon*; see Hooper, 1952, Misc. Publ. Mus. Zool. Univ. Mich.,
  77:1–255.
 ISIS NUMBER: 5301410008047004001.

Reithrodontomys darienensis Pearson, 1939. Not. Naturae Acad. Nat. Sci. Phila., 6:1.
 TYPE LOCALITY: Panama, Darien, Santa Cruz de Cana, upper Rio Tuyra, 2000 ft. (610
  m).
 DISTRIBUTION: Panama; perhaps N.W. Colombia.
 COMMENT: Subgenus *Aporodon*; see Hooper, 1952, Misc. Publ. Mus. Zool. Univ. Mich.,
  77:1–255.
 ISIS NUMBER: 5301410008047005001.

Reithrodontomys fulvescens J. A. Allen, 1894. Bull. Am. Mus. Nat. Hist., 6:319.
 TYPE LOCALITY: Mexico, Sonora, Oposura, 2000 ft. (610 m).
 DISTRIBUTION: W. Nicaragua to S.C. Arizona, S.E. Kansas, S.W. Missouri, and
  Mississippi (U.S.A.).
 COMMENT: Subgenus *Reithrodontomys*; see Hooper, 1952, Misc. Publ. Mus. Zool. Univ.
  Mich., 77:1–255.
 ISIS NUMBER: 5301410008047006001.

Reithrodontomys gracilis J. A. Allen and Chapman, 1897. Bull. Am. Mus. Nat. Hist., 9:9.
 TYPE LOCALITY: Mexico, Yucatan, Chichen-Itza.
 DISTRIBUTION: N.W. Costa Rica to Chiapas and Yucatan (Mexico).
 COMMENT: Includes *insularis*; see Jones, 1964, Proc. Biol. Soc. Wash., 77:123–124.
  Subgenus *Aporodon*; see Hooper, 1952, Misc. Publ. Mus. Zool. Univ. Mich.,
  77:1–255.
 ISIS NUMBER: 5301410008047007001.

Reithrodontomys hirsutus Merriam, 1901. Proc. Wash. Acad. Sci., 3:553.
TYPE LOCALITY: Mexico, Jalisco, Ameca, 4000 ft. (1219 m).
DISTRIBUTION: S. Nayarit and N.W. Jalisco (Mexico).
COMMENT: Subgenus *Reithrodontomys*; see Hooper, 1952, Misc. Publ. Mus. Zool. Univ. Mich., 77:1–255.
ISIS NUMBER: 5301410008047008001.

Reithrodontomys humulis (Audubon and Bachman, 1841). Proc. Acad. Nat. Sci. Phila., 1:97.
TYPE LOCALITY: U.S.A., South Carolina, Charleston Co., Charleston.
DISTRIBUTION: E. Texas and S.E. Oklahoma to S. Ohio, east to the Atlantic coast (U.S.A.).
COMMENT: Subgenus *Reithrodontomys*; see Carleton, 1980:15, and Hall, 1981:637.
ISIS NUMBER: 5301410008047009001.

Reithrodontomys megalotis (Baird, 1858). Mammals, *in* Repts. Expl. Surv...., 8(1):451.
TYPE LOCALITY: Mexico or U.S.A., type from between Janos, Chihuahua, and San Luis Springs, Grant Co., New Mexico.
DISTRIBUTION: C. Oaxaca to N.C. and N.W. Mexico; W. and N.C. U.S.A.; S.C. British Columbia and S.E. Alberta (Canada).
COMMENT: Subgenus *Reithrodontomys*; see Hooper, 1952, Misc. Publ. Mus. Zool. Univ. Mich., 77:1–255.
ISIS NUMBER: 5301410008047010001.

Reithrodontomys mexicanus (Saussure, 1860). Rev. Mag. Zool. Paris, ser. 2, 12:109.
TYPE LOCALITY: Mexico, Veracruz, Mirador.
DISTRIBUTION: Andes of W. Colombia and N. Ecuador; W. Panama to Oaxaca to Michoacan and S. Tamaulipas (Mexico).
COMMENT: Subgenus *Aporodon*; see Hooper, 1952, Misc. Publ. Mus. Zool. Univ. Mich., 77:1–255.
ISIS NUMBER: 5301410008047011001.

Reithrodontomys microdon Merriam, 1901. Proc. Wash. Acad. Sci., 3:548.
TYPE LOCALITY: Guatemala, Huehuetenango, Todos Santos, 10,000 ft. (3048 m).
DISTRIBUTION: S. Guatemala to C. Chiapas; N.C. Oaxaca; N.E. Michoacan to Distrito Federal (Mexico).
COMMENT: Subgenus *Aporodon*; see Hooper, 1952, Misc. Publ. Mus. Zool. Univ. Mich., 77:1–255.
ISIS NUMBER: 5301410008047012001.

Reithrodontomys montanus (Baird, 1855). Proc. Acad. Nat. Sci. Phila., 7:335.
TYPE LOCALITY: U.S.A., Colorado, either Saguache Co., upper end of San Luis Valley or Alamosa Co., Medano Creek.
DISTRIBUTION: N. Durango and N.E. Sonora (Mexico) to N.W. South Dakota, E.C. Texas, and S.W. Missouri (U.S.A.).
COMMENT: Subgenus *Reithrodontomys*; see Hooper, 1952, Misc. Publ. Mus. Zool. Univ. Mich., 77:1–255. For type locality information see Hall, 1981:636.
ISIS NUMBER: 5301410008047013001.

Reithrodontomys paradoxus Jones and Genoways, 1970. Occas. Pap. W. Found. Vert. Zool., 2:12.
TYPE LOCALITY: Nicaragua, Carazo, 3 mi. (5 km) N.N.W. Diriamba, about 660 m.
DISTRIBUTION: W.C. Nicaragua; C. Costa Rica.
COMMENT: Subgenus *Aporodon*; see Carleton, 1980:15.

Reithrodontomys raviventris Dixon, 1908. Proc. Biol. Soc. Wash., 21:197.
TYPE LOCALITY: U.S.A., California, San Mateo Co., Redwood City.
DISTRIBUTION: Vicinity of San Francisco Bay (California, U.S.A.).
COMMENT: Subgenus *Reithrodontomys*; see Carleton, 1980:15. Subspecies *raviventris* and *halicoetes* thought to be in terminal stages of speciation; see Fisler, 1965, Univ. Calif. Publ. Zool., 77:1–108, and Shellhammer, 1965, J. Mammal., 48:549–556.
PROTECTED STATUS: U.S. ESA - Endangered.
ISIS NUMBER: 5301410008047014001.

Reithrodontomys rodriguezi Goodwin, 1943. Am. Mus. Novit., 1231:1.
TYPE LOCALITY: Costa Rica, Cartago, Volcan de Irazu, 9400 ft. (2865 m).
DISTRIBUTION: Known only from the type locality.
COMMENT: Subgenus *Aporodon;* see Hooper, 1952, Misc. Publ. Mus. Zool. Univ. Mich., 77:1–255.
ISIS NUMBER: 5301410008047015001.

Reithrodontomys spectabilis Jones and Lawlor, 1965. Univ. Kans. Mus. Nat. Hist. Misc. Publ., 16(3):413.
TYPE LOCALITY: Mexico, Quintana Roo, Cozumel Isl., 2.5 km N. San Miguel.
DISTRIBUTION: Cozumel Isl. (Quintana Roo, Mexico).
COMMENT: Subgenus *Aporodon;* see Carleton, 1980:15.

Reithrodontomys sumichrasti (Saussure, 1861). Rev. Mag. Zool. Paris, ser. 2, 13:3.
TYPE LOCALITY: Mexico, Veracruz, Mirador.
DISTRIBUTION: W. Panama to N. Costa Rica; N.W. Nicaragua to S. Tabasco (Mexico); C. Guerrero to S. Oaxaca to C. Veracruz to S. San Luis Potosi and S.W. Jalisco (Mexico).
COMMENT: Subgenus *Reithrodontomys;* see Hooper, 1952, Misc. Publ. Mus. Zool. Univ. Mich., 77:1–255.
ISIS NUMBER: 5301410008047016001.

Reithrodontomys tenuirostris Merriam, 1901. Proc. Wash. Acad. Sci., 3:547.
TYPE LOCALITY: Guatemala, Todos Santos, 10,000 ft. (3048 m).
DISTRIBUTION: S. Guatemala.
COMMENT: Subgenus *Aporodon;* see Hooper, 1952, Misc. Publ. Mus. Zool. Univ. Mich., 77:1–255.
ISIS NUMBER: 5301410008047017001.

Rhagomys Thomas, 1917. Ann. Mag. Nat. Hist., ser. 8, 20:192.
REVIEWED BY: J. K. Braun (JKB).
COMMENT: Subfamily Hesperomyinae; see comment under Cricetidae.
ISIS NUMBER: 5301410008048000000.

Rhagomys rufescens (Thomas, 1886). Ann. Mag. Nat. Hist., ser. 5, 17:250.
TYPE LOCALITY: E. Brazil, Rio de Janeiro.
DISTRIBUTION: Rio de Janeiro (E. Brazil).
ISIS NUMBER: 5301410008048001001.

Rheomys Thomas, 1906. Ann. Mag. Nat. Hist., ser. 7, 17:421.
REVIEWED BY: R. S. Voss (RSV).
COMMENT: Reviewed by Hooper, 1968, J. Mammal., 49:550–553, Carleton, 1973, Misc. Publ. Mus. Zool. Univ. Mich., 146:1–43, Voss and Linzey, 1981, Misc. Publ. Mus. Zool. Univ. Mich., 159:19–21, and Goodwin, 1946, Bull. Am. Mus. Nat. Hist., 87:271–473. Hall, 1981:775, stated that *Rheomys* should possibly be treated as a subgenus of *Ichthyomys.* Formerly included *trichotis,* which was included in *Anotomys* by Handley, 1976, Brigham Young Univ. Sci. Bull. Biol. Ser., 20(5):53. Subfamily Hesperomyinae; see comment under Cricetidae.
ISIS NUMBER: 5301410008049000000.

Rheomys hartmanni Enders, 1939. Proc. Acad. Nat. Sci. Phila., 90:295.
TYPE LOCALITY: Panama, Chiriqui, Hot Springs on Rio Cotito, 4900 ft. (1494 m).
DISTRIBUTION: Costa Rica; W. Panama.
ISIS NUMBER: 5301410008049001001.

Rheomys mexicanus Goodwin, 1959. Am. Mus. Novit., 1967:4.
TYPE LOCALITY: Mexico, Oaxaca, Miahuatlan Dist., San Jose Lachiguiri.
DISTRIBUTION: S.E. Oaxaca (Mexico).
ISIS NUMBER: 5301410008049002001 as *R. mexicana (sic).*

Rheomys raptor Goldman, 1912. Smithson. Misc. Coll., 60(2):7.
TYPE LOCALITY: Panama, near head Rio Limon, Mt. Pirri, 4500 ft. (1372 m).
DISTRIBUTION: Known only from the type locality.

COMMENT: Treated as a subspecies of *Rheomys trichotis* (here included in *Anotomys*) by Cabrera, 1961:511, but see Hall, 1981:776, who considered *raptor* a distinct species of *Rheomys*.
ISIS NUMBER: 5301410008049003001.

Rheomys thomasi Dickey, 1928. Proc. Biol. Soc. Wash., 41:11.
REVIEWED BY: J. Ramirez-Pulido (JRP); G. Urbano-V. (GUV).
TYPE LOCALITY: El Salvador, San Miguel, Finca San Felipe, Mt. Cacaquatique, 3500 ft. (1077 m).
DISTRIBUTION: El Salvador; W. Guatemala; S.E. Chiapas (Mexico).
COMMENT: Reviewed by Hooper, 1968, J. Mammal., 49:550–553.
ISIS NUMBER: 5301410008049004001.

Rheomys underwoodi Thomas, 1906. Ann. Mag. Nat. Hist., ser. 7, 17:422.
TYPE LOCALITY: Costa Rica, Cartago, Tres Rios.
DISTRIBUTION: Costa Rica; W. Panama.
COMMENT: Reviewed by Hooper, 1968, J. Mammal., 49:552, and Goodwin, 1946, Bull. Am. Mus. Nat. Hist., 87:403.
ISIS NUMBER: 5301410008049006001.

Rhipidomys Tschudi, 1844. Wiegmann's Archive Naturgesch., 10, 1:252.
REVIEWED BY: J. K. Braun (JKB).
COMMENT: Reviewed in part by Hooper and Musser, 1964, Misc. Publ. Mus. Zool. Univ. Mich., 123:20–21, Carleton, 1973, Misc. Publ. Mus. Zool. Univ. Mich., 146:1–43, and Handley, 1976, Brigham Young Univ. Sci. Bull. Biol. Ser., 20(5):50–51. This genus is in need of revision at the species level and its relationship to other thomasomyines needs clarification; see Hall, 1981:626, Pine, 1973, Acta Amazonica, 3(2):64, Hershkovitz, 1966, Z. Saugetierk., 31:125, and Cabrera, 1961:419. Subfamily Hesperomyinae; see comment under Cricetidae.
ISIS NUMBER: 5301410008050000000.

Rhipidomys latimanus (Tomes, 1860). Proc. Zool. Soc. Lond., 1860:213.
TYPE LOCALITY: C. Ecuador, Chimborazo Prov., Pallatanga, 1485 m.
DISTRIBUTION: Venezuela; Colombia; Ecuador.
COMMENT: Includes *fulviventer* and *venustus*; see Cabrera, 1961:420; Handley, 1976, Brigham Young Univ. Sci. Bull. Biol. Ser., 20(5):50, listed *fulviventer* and *venustus* as distinct species, without comment. Karology reviewed by Gardner and Patton, 1976, Occas. Pap. Mus. Zool. La. St. Univ., 49:26, 32–33.
ISIS NUMBER: 5301410008050001001.

Rhipidomys leucodactylus (Tschudi, 1845). Fauna Peruana, 1:183.
TYPE LOCALITY: E. Peru.
DISTRIBUTION: S. Venezuela; N.W. and E. Ecuador; N.W. and E. Peru; N.W. Argentina; S. Bolivia.
COMMENT: Includes *rex*; see Cabrera, 1961:421; Walker *et al.*, 1975:765, listed *rex* as a distinct species, without comment.
ISIS NUMBER: 5301410008050002001.

Rhipidomys macconnelli De Winton, 1900. Trans. Linn. Soc. Lond., 8:52.
TYPE LOCALITY: Venezuela, Bolivar, Mt. Roraima.
DISTRIBUTION: S.E. Venezuela; Amazonas (Brazil); perhaps adjacent Guyana.
COMMENT: Formerly included in *Thomasomys*; see Hershkovitz, 1959, Proc. Biol. Soc. Wash., 72:9, and Cabrera, 1961:422.
ISIS NUMBER: 5301410008050003001.

Rhipidomys maculipes (Pictet and Pictet, 1844). Notice Anim. Nouv. Mus. Geneve, p. 67.
TYPE LOCALITY: Brazil, Bahia.
DISTRIBUTION: Bahia (Brazil).
COMMENT: Formerly included in *Oryzomys*; see Hershkovitz, 1960, Proc. U.S. Nat. Mus., 110:519. Cabrera, 1961:405, listed this species in *Oryzomys*.
ISIS NUMBER: 5301410008035036001 as *Oryzomys maculipes*.

Rhipidomys mastacalis (Lund, 1840). Kongl. Dansk. Vid. Selsk. Naturv. Math. Afhandl., p. 24.
TYPE LOCALITY: Brazil, S.W. Minas Gerais, Rio das Velhas, Lagoa Santa.
DISTRIBUTION: N.E. and E.C. Brazil; Guianas; Venezuela; Margarita and Tobago Isls.
COMMENT: Includes *emiliae* and *venezuelae*; see Cabrera, 1961:423, and Pine, 1973, Acta Amazonica, 3(2):64. Carvalho and Toccheton, 1969, Rev. Biol. Trop. San Jose, 15(2):215–226, listed *emiliae* as a distinct species without comment; Handley, 1976, Brigham Young Univ. Sci. Bull. Biol. Ser., 20(5):50, listed *venezuelae* as a distinct species, without comment. Reviewed by Mares *et al.*, 1981, Ann. Carnegie Mus., 50(4):117, and Hooper and Musser, 1964, Misc. Publ. Mus. Zool. Univ. Mich., 123:20.
ISIS NUMBER: 5301410008050004001.

Rhipidomys scandens Goldman, 1913. Smithson. Misc. Coll., 60(22):8.
TYPE LOCALITY: E. Panama, near head of Rio Limon, Mt. Pirri, 5000 ft. (1524 m).
DISTRIBUTION: Known only from the type locality.
COMMENT: Known only by the holotype.
ISIS NUMBER: 5301410008050005001.

Rhipidomys sclateri (Thomas, 1887). Proc. Zool. Soc. Lond., 1887:152.
TYPE LOCALITY: Guyana, Demerara Dist., Maccassima.
DISTRIBUTION: Guyana; adjacent Brazil and Venezuela; Trinidad.
COMMENT: Husson, 1978:414, stated that *sclateri* may occur in Surinam. Includes *couesi*; see Cabrera, 1961:424; but also see Handley, 1976, Brigham Young Univ. Sci. Bull. Biol. Ser., 20(5):50, who listed *couesi* as a distinct species, without comment.
ISIS NUMBER: 5301410008050006001.

Rhombomys Wagner, 1841. Arch. Naturgesch., 7, 1:129.
REVIEWED BY: C. B. Robbins (CBR); O. L. Rossolimo (OLR); S. Wang (SW).
COMMENT: Subfamily Gerbillinae; see comment under Cricetidae.
ISIS NUMBER: 5301410008096000000.

Rhombomys opimus (Lichtenstein, 1823). Naturh. Abh. Eversmann's Reise, p. 122.
TYPE LOCALITY: U.S.S.R., Kazakhstan, ("between Orenburg and Bokhara") "pre-Aral" Kara-Kum Desert.
DISTRIBUTION: S. Mongolia to Ningsia, Kansu, and Sinkiang (China), Kazakhstan (U.S.S.R.), Iran, Afghanistan, and N. Pakistan.
COMMENT: Reviewed by Kuznetsov, 1944, *in* Bobrinskii *et al.*, 1965, [Key to the Mammals of the U.S.S.R.], Moscow, p. 325–327, and Gromov and Baranova, 1981:167; *giganteus* may be a distinct species. Corbet, 1978:129, included *giganteus* in *opimus*, and also gave the type locality as near Bokhara, Uzbekistan.
ISIS NUMBER: 5301410008096001001.

Saccostomus Peters, 1846. Monatsb. Preuss. Akad. Wiss. Berlin, p. 258.
REVIEWED BY: R. D. Owen (RDON).
COMMENT: Revised by Hubert, 1978, *in* Schlitter, ed., Bull. Carnegie Mus. Nat. Hist., 6:48–52. Subfamily Cricetomyinae; see comment under Cricetidae.
ISIS NUMBER: 5301410008102000000.

Saccostomus campestris Peters, 1846. Monatsb. Preuss. Akad. Wiss. Berlin, p. 258.
TYPE LOCALITY: Mozambique, Tete.
DISTRIBUTION: Transvaal (South Africa); Namibia; S. Angola; Botswana; E. Zimbabwe; Mozambique; Malawi; S. Zaire.
COMMENT: Formerly included *mearnsi*; see Hubert, 1978, *in* Schlitter, ed., Bull. Carnegie Mus. Nat. Hist., 6:48–52.
ISIS NUMBER: 5301410008102001001.

Saccostomus mearnsi Heller, 1910. Smithson. Misc. Coll., 54 (1924):3.
TYPE LOCALITY: Kenya, Changamwe.
DISTRIBUTION: Tanzania and Uganda to Ethiopia and Somalia.
COMMENT: Formerly included in *campestris* by Misonne, 1974, Part 6:14. Includes *cricetulus*, *isiolae*, and *umbriventer*; see Hubert, 1978, *in* Schlitter, ed., Bull. Carnegie Mus. Nat. Hist., 6:48–52.

Scapteromys Waterhouse, 1837. Proc. Zool. Soc. Lond., 1837:20.
 REVIEWED BY: J. K. Braun (JKB); A. Langguth (AL).
 COMMENT: Formerly included *tomentosus* and *fronto* (which were placed in *Kunsia* by
  Hershkovitz, 1966, Z. Saugetierk., 31:81–149, who revised *Scapteromys*) and *labiosus*
  (which was placed in *Bibimys* by Massoia, 1980, Ameghiniana, 17:280–287); also
  see comments under *Kunsia*. Subfamily Hesperomyinae; see comment under
  Cricetidae.
 ISIS NUMBER: 5301410008051000000.

Scapteromys tumidus (Waterhouse, 1837). Proc. Zool. Soc. Lond., 1837:15.
 TYPE LOCALITY: Uruguay, Maldonado.
 DISTRIBUTION: Uruguay; adjacent Brazil; E. Argentina; S. Paraguay.
 COMMENT: Includes *aquaticus*; see Massoia and Fornes, 1964, Physis, 24(68):279–297,
  Hershkovitz, 1966, Z. Saugetierk., 31:102, and Voss and Linzey, 1981, Misc. Publ.
  Mus. Zool. Univ. Mich., 159:10, 15; but also see Gardner and Patton, 1976, Occas.
  Pap. Mus. Zool. La. St. Univ., 49:26, 30, who listed *aquaticus* as a distinct species.
 ISIS NUMBER: 5301410008051005001 as *S. tumidus*.
  5301410008051001001 as *S. aquaticus*.

Scolomys Anthony, 1924. Am. Mus. Novit., 139:1.
 REVIEWED BY: J. K. Braun (JKB).
 COMMENT: Subfamily Hesperomyinae; see comment under Cricetidae.
 ISIS NUMBER: 5301410008052000000.

Scolomys melanops Anthony, 1924. Am. Mus. Novit., 139:2.
 TYPE LOCALITY: E. Ecuador, Tungurahua, Mera.
 DISTRIBUTION: Known only from the vicinity of the type locality.
 COMMENT: Known only by six specimens collected in 1924; see Walker *et al.*, 1975:763.
 ISIS NUMBER: 5301410008052001001.

Scotinomys Thomas, 1913. Ann. Mag. Nat. Hist., ser. 8, 11:408.
 REVIEWED BY: J. Ramirez-Pulido (JRP).
 COMMENT: Revised by Hooper, 1972, Occas. Pap. Mus. Zool. Univ. Mich., 655:1–32;
  also see Hooper and Carleton, 1976, Misc. Publ. Mus. Zool. Univ. Mich., 151:1–52,
  Carleton, 1980:118, 123, and Hall, 1981:733–735. Subfamily Hesperomyinae; see
  comment under Cricetidae.
 ISIS NUMBER: 5301410008053000000.

Scotinomys teguina (Alston, 1877). Proc. Zool. Soc. Lond., 1876:755.
 TYPE LOCALITY: Guatemala, Alta Verapaz, Coban.
 DISTRIBUTION: W. Panama to E. Oaxaca (Mexico).
 ISIS NUMBER: 5301410008053003001.

Scotinomys xerampelinus (Bangs, 1902). Bull. Mus. Comp. Zool. Harv. Univ., 39:41.
 TYPE LOCALITY: Panama, Chiriqui, Volcan de Chiriqui, 10,300 ft. (3139 m).
 DISTRIBUTION: W. Panama; C. Costa Rica.
 COMMENT: Includes *harrisi* and *longipilosus*; see Hooper, 1972, Occas. Pap. Mus. Zool.
  Univ. Mich., 665:24.
 ISIS NUMBER: 5301410008053004001 as *S. xerampelinus*.
  5301410008053001001 as *S. harrisi*.
  5301410008053002001 as *S. longipilosus*.

Sekeetamys Ellerman, 1947. Proc. Zool. Soc. Lond., 1947–1948:271.
 REVIEWED BY: J. K. Braun (JKB); C. B. Robbins (CBR).
 COMMENT: Formerly included in *Meriones* by Ellerman and Morrison-Scott, 1966:638.
  Considered a distinct genus by Petter, 1956, Mammalia, 20:419–426, Petter, 1975,
  Part 6.3:4, and Corbet, 1978:124. Subfamily Gerbillinae; see comment under
  Cricetidae.
 ISIS NUMBER: 5301410008097000000.

Sekeetamys calurus (Thomas, 1892). Ann. Mag. Nat. Hist., 9:76.
 TYPE LOCALITY: Egypt, Sinai, near Tor.
 DISTRIBUTION: E. Egypt through the Sinai to Israel, Jordan, and C. Saudi Arabia.

COMMENT: Includes *makrami;* see Petter, 1975, Part 6.3:4, and Harrison, 1972, The
Mammals of Arabia, 3:596. Nader, 1974, Mammalia, 38:347–348, reported this
species from Saudi Arabia.
ISIS NUMBER: 5301410008097001001.

Sigmodon Say and Ord, 1825. J. Acad. Nat. Sci. Phila., 4(2):352.
REVIEWED BY: B. R. Blood (BRB); J. Ramirez-Pulido (JRP); G. Urbano-V. (GUV).
COMMENT: Revised by Bailey, 1902, Proc. Biol. Soc. Wash., 15:101–116, Zimmerman,
1970, Publ. Mus. Mich. St. Univ. Biol. Ser., 4(9):385–454, and Baker, 1969, Misc.
Publ. Univ. Kans. Mus. Nat. Hist., 51:177–232; karyology reviewed by Gardner
and Patton, 1976, Occas. Pap. Mus. Zool. La. St. Univ., 49:28, 30. Includes
Sigmomys as a subgenus; see Husson, 1978:427–429, and Hershkovitz, 1966, Z.
Saugetierk., 31:129; Hershkovitz, 1955, Fieldiana Zool., 37:639–673, and Cabrera,
1961:507–508, included *Sigmomys* in *Sigmodon* as a synonym. Handley, 1976,
Brigham Young Univ. Sci. Bull. Biol. Ser., 20(5):53, listed *Sigmomys* as a distinct
genus. A key to the species of *Sigmodon* was published by Baker and Shump,
1978, Mamm. Species, 94:1–4. Subfamily Hesperomyinae; see comment under
Cricetidae.
ISIS NUMBER: 5301410008054000000.

Sigmodon alleni Bailey, 1902. Proc. Biol. Soc. Wash., 15:112.
TYPE LOCALITY: Mexico, Jalisco, Mascota, Mineral San Sebastian.
DISTRIBUTION: S. Sinaloa to S. Oaxaca (Mexico).
COMMENT: Includes *guerrerensis, macdougalli, macrodon, planifrons, setzeri,* and *vulcani;*
see review by Shump and Baker, 1978, Mamm. Species, 95:1, Baker, 1969, Misc.
Publ. Univ. Kans. Mus. Nat. Hist., 51:186–196, and Hall, 1981:743–744.
ISIS NUMBER: 5301410008054001001 as *S. alleni.*
5301410008054005001 as *S. guerrerensis.*
5301410008054008001 as *S. macdougalli.*
5301410008054009001 as *S. macrodon.*
5301410008054013001 as *S. planifrons.*
5301410008054014001 as *S. vulcani.*

Sigmodon alstoni (Thomas, 1881). Proc. Zool. Soc. Lond., 1880:691.
TYPE LOCALITY: Venezuela, Sucre, Cumana Dist.
DISTRIBUTION: N. and E. Venezuela; S. Guyana; Surinam; N. Brazil.
COMMENT: Included in *hispidus* by Hershkovitz, 1955, Fieldiana Zool., 37:639–673; but
considered a distinct species by Hershkovitz, 1966, Z. Saugetierk., 31:129,
Handley, 1976, Brigham Young Univ. Sci. Bull. Biol. Ser., 20(5):53, and Husson,
1978:427–429. Included in genus *Sigmomys* by Handley, 1976, Brigham Young
Univ. Sci. Bull. Biol. Ser., 20(5):53; but see Husson, 1978:427–429, and
Hershkovitz, 1966, Z. Saugetierk., 31:129, who considered *Sigmomys* a subgenus.

Sigmodon arizonae Mearns, 1890. Bull. Am. Mus. Nat. Hist., 2:287.
TYPE LOCALITY: U.S.A., Arizona, Yavapai Co., Fort Verde.
DISTRIBUTION: Nayarit (Mexico) to S. and C. Arizona, S. Nevada, and S.E. California
(U.S.A.).
COMMENT: Formerly included in *hispidus;* see Zimmerman, 1970, Publ. Mus. Mich. St.
Univ. Biol. Ser., 4(9):385–454, and Severinghaus and Hoffmeister, 1978, J.
Mammal., 59:868–870.
ISIS NUMBER: 5301410008054003001.

Sigmodon fulviventer J. A. Allen, 1889. Bull. Am. Mus. Nat. Hist., 2:180.
TYPE LOCALITY: Mexico, Zacatecas, Zacatecas.
DISTRIBUTION: Guanajuato and Michoacan (Mexico) to S.W. and C. New Mexico and
S.E. Arizona (U.S.A.).
COMMENT: Includes *melanotis* and *minimus;* see review by Baker and Shump, 1978,
Mamm. Species, 94:1–4, Hall, 1981:742, and Baker, 1969, Misc. Publ. Univ. Kans.
Mus. Nat. Hist., 51:196–213.
ISIS NUMBER: 5301410008054004001 as *S. fulviventer.*
5301410008054010001 as *S. melanotis.*
5301410008054011001 as *S. minimus.*

Sigmodon hispidus Say and Ord, 1825. J. Acad. Nat. Sci. Phila., 42:354.
TYPE LOCALITY: U.S.A., Florida, St. Johns River.
DISTRIBUTION: Peru and N. Venezuela to S.E. California, S. Nebraska, S. Virginia, and Florida (U.S.A.).
COMMENT: Formerly included *arizonae* and *mascotensis*; see Zimmerman, 1970, Publ. Mus. Mich. St. Univ. Biol. Ser., 4(9):389, and Severinghaus and Hoffmeister, 1978, J. Mammal., 59(4):868–870. Formerly included *alstoni*; see Husson, 1978:429. The status of the forms, assigned to *hispidus*, from C. and S. America is unclear; see Zimmerman, 1970:445, Dalby and Lillevik, 1969, Publ. Mus. Mich. St. Univ. Biol. Ser., 4:65–104, and Baker and Shump, 1978, Mamm. Species, 94:1–4. Reviewed by Cameron and Spencer, 1981, Mamm. Species, 158:1–9.
ISIS NUMBER: 5301410008054006001.

Sigmodon leucotis Bailey, 1902. Proc. Biol. Soc. Wash., 15:115.
TYPE LOCALITY: Mexico, Zacatecas, Sierra de Valparaiso, 2653 m.
DISTRIBUTION: S. Nuevo Leon and S.W. Chihuahua to C. Oaxaca (Mexico).
COMMENT: Includes *alticola*; see review by Shump and Baker, 1978, Mamm. Species, 96:1, and Baker, 1969, Misc. Publ. Univ. Kans. Mus. Nat. Hist., 51:213–223.
ISIS NUMBER: 5301410008054007001 as *S. leucotis*.
5301410008054002001 as *S. alticola*.

Sigmodon mascotensis J. A. Allen, 1897. Bull. Am. Mus. Nat. Hist., 9:54.
TYPE LOCALITY: Mexico, Jalisco, Mascota, Mineral San Sebastian.
DISTRIBUTION: Oaxaca to Jalisco and probably S. Nayarit (Mexico).
COMMENT: Formerly included in *hispidus*; see Zimmerman, 1970, Publ. Mus. Mich. St. Univ. Biol. Ser., 4(9):389. Severinghaus and Hoffmeister, 1978, J. Mammal., 59:870, reported that specimens from S. Nayarit are probably *mascotensis*.

Sigmodon ochrognathus Bailey, 1902. Proc. Biol. Soc. Wash., 15:115.
TYPE LOCALITY: U.S.A., Texas, Brewster Co., Chisos Mtns., 8000 ft. (2438 m).
DISTRIBUTION: S.E. Arizona, S.W. New Mexico, and W. Texas (U.S.A.) to W. Coahuila and C. Durango (Mexico).
COMMENT: Reviewed by Baker and Shump, 1978, Mamm. Species, 97:1–2, Baker, 1969, Misc. Publ. Univ. Kans. Mus. Nat. Hist., 51:223–230, and Hall, 1981:744–745.
ISIS NUMBER: 5301410008054012001.

Steatomys Peters, 1846. Monatsb. Preuss. Akad. Wiss. Berlin, p. 258.
REVIEWED BY: R. D. Owen (RDON).
COMMENT: Reviewed by Swanepoel and Schlitter, 1978, *in* Schlitter, ed., Bull. Carnegie Mus. Nat. Hist., 6:53–76, and Coetzee, 1977, Part 6.8:1–4. This genus is in need of further revision; see Swanepoel and Schlitter, 1978:53–54, and Schlitter, 1978, Bull. Carnegie Mus. Nat. Hist., 6:212. Subfamily Dendromurinae; see comment under Cricetidae.
ISIS NUMBER: 5301410008110000000.

Steatomys caurinus Thomas, 1912. Ann. Mag. Nat. Hist., ser. 9, 9:271.
TYPE LOCALITY: Nigeria, Panyam, 4000 ft. (1219 m).
DISTRIBUTION: C. Nigeria; S. Niger; N. Benin; Togo; W. Ghana; upper Volta; Ivory Coast; W. Senegal.
COMMENT: Included in *pratensis* by Coetzee, 1977, Part 6.8:1; considered a distinct species by Swanepoel and Schlitter, 1978, *in* Schlitter, ed., Bull. Carnegie Mus. Nat. Hist., 6:53–76.
ISIS NUMBER: 5301410008110003001.

Steatomys cuppedius Thomas and Hinton, 1920. Novit. Zool., 27:318.
TYPE LOCALITY: Nigeria, Farniso (=Panisau), near Kano, 1700 ft. (518 m).
DISTRIBUTION: N. Nigeria; S.C. Niger; Senegal.
COMMENT: Included in *parvus* by Coetzee, 1977, Part 6.8:2; considered a distinct species by Swanepoel and Schlitter, 1978, *in* Schlitter, ed., Bull. Carnegie Mus. Nat. Hist., 6:53–76.
ISIS NUMBER: 5301410008110004001 as *S. cupredius (sic)*.

Steatomys jacksoni Hayman, 1936. Proc. Zool. Soc. Lond., 1935:930.
   TYPE LOCALITY: Ghana, Ashanti, Wenchi.
   DISTRIBUTION: Known only from the type locality.
   COMMENT: Included in *pratensis* by Coetzee, 1977, Part 6.8:1; considered a distinct
      species by Swanepoel and Schlitter, 1978, *in* Schlitter, ed., Bull. Carnegie Mus.
      Nat. Hist., 6:53-76; the holotype is the only known specimen.

Steatomys krebsii Peters, 1852. Reise nach Mossambique, Saugethiere, p. 165.
   TYPE LOCALITY: South Africa, Kaffraria.
   DISTRIBUTION: S. Angola; W. Zambia; N. Botswana; South Africa.
   COMMENT: Reviewed by Coetzee, 1977, Part 6.8:3, and Ansell, 1978:77.
   ISIS NUMBER: 5301410008110006001.

Steatomys minutus Thomas and Wroughton, 1905. Ann. Mag. Nat. Hist., ser. 7, 16:174.
   TYPE LOCALITY: Angola, Benguela Dist., Quillenges.
   DISTRIBUTION: S. Angola; N. Namibia; Zambia.
   COMMENT: Included in *parvus* by Coetzee, 1977, Part 6.8:2; considered a distinct
      species by Ansell, 1978:77, who included *swalius* as a subspecies. Swanepoel *et al.*,
      1980, Ann. Transvaal Mus., 32(7):160, 175, followed Coetzee, 1977, Part 6.8.
   ISIS NUMBER: 5301410008110007001 as *S. minutus*.
               5301410008110010001 as *S. swalius*.

Steatomys parvus Rhoads, 1896. Proc. Acad. Nat. Sci. Phila., p. 529.
   TYPE LOCALITY: Ethiopia (=Abyssinia), Lake Rudolf, Rusia.
   DISTRIBUTION: S. Sudan; N.E. Uganda; S. Ethiopia and Somalia to Natal (South Africa);
      N. Botswana.
   COMMENT: Includes *aquilo* and *thomasi*; see Coetzee, 1977, Part 6.8:2-3. Formerly
      included *cuppedius* (see Swanepoel and Schlitter, 1978, *in* Schlitter, ed., Bull.
      Carnegie Mus. Nat. Hist., 6:53-76) and *minutus* (see Ansell, 1978:77). Yalden *et al.*,
      1976, Monitore Zool. Ital., n.s., 8(1):1-118, included *parvus* in *pratensis*, but
      Coetzee, 1977, Part 6.8:2, and Swanepoel *et al.*, 1980, Ann. Transvaal Mus.,
      32(7):160, did not.
   ISIS NUMBER: 5301410008110008001 as *S. parvus*.
               5301410008110001001 as *S. aquilo*.
               5301410008110011001 as *S. thomasi*.

Steatomys pratensis Peters, 1846. Monatsb. Preuss. Akad. Wiss. Berlin, p. 258.
   TYPE LOCALITY: Mozambique, on the Zambezi River, Tete.
   DISTRIBUTION: Cameroun to S. Sudan, south to Transvaal, Natal, and perhaps E.
      Orange Free State (South Africa).
   COMMENT: Includes *bocagei*, *gazellae*, and *opimus*; see Coetzee, 1977, Part 6.8:1-2, who
      also included *caurinus* and *jacksoni* in this species, but Swanepoel and Schlitter,
      1978, *in* Schlitter, ed., Bull. Carnegie Mus. Nat. Hist., 6:53-76, considered *caurinus*
      and *jacksoni* distinct species.
   ISIS NUMBER: 5301410008110009001 as *S. pratensis*.
               5301410008110002001 as *S. bocagei*.
               5301410008110005001 as *S. gazellae*.

Tatera Lataste, 1882. Le Naturaliste, 2:126.
   REVIEWED BY: R. DeBry (RDB); C. B. Robbins (CBR).
   COMMENT: Revised, in part, by Davis, 1966, Ann. Mus. R. Afr. Cent., 144:49-65;
      reviewed by Davis, 1975, Part 6.4:1-5. Karyology reviewed by Matthey and
      Petter, 1970, Mammalia, 34:585-597. Subfamily Gerbillinae; see comment under
      Cricetidae.
   ISIS NUMBER: 5301410008098000000.

Tatera afra (Gray, 1830). Spicil. Zool., 2:10.
   TYPE LOCALITY: South Africa, Cape of Good Hope.
   DISTRIBUTION: S.W. Cape Prov. (South Africa).
   COMMENT: Includes *gilli*; see Davis, 1975, Part 6.4:3.
   ISIS NUMBER: 5301410008098001001.

Tatera boehmi (Noack, 1888). Zool. Jahrb. Syst., 2:241.
TYPE LOCALITY: S. Zaire, Marungu, Qua Mpala.
DISTRIBUTION: Mozambique; N. Zimbabwe and Angola to Kenya and Uganda.
COMMENT: Includes *fraterculus* and *fallax*; see Davis, 1975, Part 6.4:2.
ISIS NUMBER: 5301410008098003001 as *T. boehmi.*
5301410008098008001 as *T. fallax.*
5301410008098010001 as *T. fraterculus.*

Tatera brantsii (A. Smith, 1834). Rept. Exped. Expl. C. Afr...., p. 43.
TYPE LOCALITY: Lesotho border, Orange River Colony, 29° 20′ S., 27° 10′ E.
DISTRIBUTION: Cape Prov. to N. Zululand (South Africa), S.W. Zimbabwe, W. Zambia,
and S. Angola.
COMMENT: Includes *joanae*; see Davis, 1975, Part 6.4:4. Type locality restricted by
Ansell, 1978:80.
ISIS NUMBER: 5301410008098004001 as *T. brantsi (sic).*
5301410008098015001 as *T. joanie (sic).*

Tatera inclusa Thomas and Wroughton, 1908. Proc. Zool. Soc. Lond., 1908:169.
TYPE LOCALITY: Mozambique, Gorongoza Dist., Tambarara.
DISTRIBUTION: Tanzania to E. Zimbabwe and Mozambique.
COMMENT: Includes *cosensi* and *pringlei*; see Davis, 1975, Part 6.4:4.
ISIS NUMBER: 5301410008098006001 as *T. cosensi.*
5301410008098023001 as *T. pringlei.*

Tatera indica (Hardwicke, 1807). Trans. Linn. Soc. Lond., 8:279.
TYPE LOCALITY: N. India, United Provs., between Benares and Hardwar.
DISTRIBUTION: Sri Lanka; India; Pakistan; Afghanistan; Nepal; Iran; Iraq; Kuwait; Syria.
COMMENT: Reviewed by Harrison, 1973, The Mammals of Arabia, 3:554–559, and Lay,
1967, Fieldiana Zool., 54:1–282.
ISIS NUMBER: 5301410008098014001.

Tatera leucogaster (Peters, 1852). Monatsb. Preuss. Akad. Wiss. Berlin, p. 274.
TYPE LOCALITY: Mozambique, Mossuril, on the mainland opposite Mozambique Isl.
DISTRIBUTION: N. South Africa and Namibia to S. Zaire and S.W. Tanzania.
COMMENT: Includes *angolae, nigrotibialis, nyasae, ndolae,* and *schinzi*; see Davis, 1975, Part
6.4:3.
ISIS NUMBER: 5301410008098017001 as *T. leucogaster.*
5301410008091028001 as *Gerbillus nigrotibialis.*

Tatera nigricauda (Peters, 1878). Monatsb. Preuss. Akad. Wiss. Berlin, p. 200.
TYPE LOCALITY: Kenya, Taita, Ndi.
DISTRIBUTION: S.W. Ethiopia; N.E. Uganda; Kenya; N.E. Tanzania.
COMMENT: Includes *bayeri* and *nyama*; see Davis, 1975, Part 6.4:2–3.
ISIS NUMBER: 5301410008098020001.

Tatera robusta (Cretzschmar, 1826). Ruppell Atlas, Reise Nordl. Afr., 1:75.
TYPE LOCALITY: Sudan, Ambukol.
DISTRIBUTION: Guinea-Bissau to Ethiopia and Somalia; Kenya; N. Tanzania.
COMMENT: Includes *guineae, macropus, minuscula, phillipsi,* and *shoana*; see Davis, 1975,
Part 6.4:3. Hubert *et al.,* 1973, Mammalia, 37:81, listed *guineae* as a distinct species.
ISIS NUMBER: 5301410008098024001 as *T. robusta.*
5301410008098012001 as *T. guineae.*
5301410008098018001 as *T. macropus.*
5301410008098019001 as *T. minuscula.*
5301410008098022001 as *T. phillipsi.*
5301410008098025001 as *T. shoana.*
5301410008091023001 as *Gerbillus macropus.*

Tatera valida (Bocage, 1890). J. Sci. Math. Phys. Nat. Lisboa, 2:6.
TYPE LOCALITY: W. Angola, Rio Cuando.
DISTRIBUTION: Senegal to Cameroun, Sudan, Ethiopia, and Kenya to S.W. Tanzania,
Zambia, and Angola.

COMMENT: Includes *benvenuta, dichrura, flavipes, gambiana, hopkinsoni, kempi, nigrita, taborae,* and *wellmani;* see Davis, 1975, Part 6.4:3–4. Hubert *et al.,* 1973, Mammalia, 37:76–87, suggested that *gambiana* and *kempi* may be conspecific, but listed the junior name *gambiana* as a distinct species. Rosevear, 1969, The Rodents of West Africa, Br. Mus. Lond., included *gambiana* in *kempi,* and also listed *hopkinsoni* and *wellmani* as distinct species. Matthey and Petter, 1970, listed *valida* in *Taterillus.*

ISIS NUMBER: 5301410008098027001 as *T. valida.*
5301410008098002001 as *T. benvenuta.*
5301410008098007001 as *T. dichrura.*
5301410008098009001 as *T. flavipes.*
5301410008098011001 as *T. gambiana.*
5301410008098013001 as *T. hopkinsoni.*
5301410008098016001 as *T. kempii (sic).*
5301410008098021001 as *T. nigrita.*
5301410008098026001 as *T. taborae.*
5301410008098028001 as *T. welmanni (sic).*

Taterillus Thomas, 1910. Ann. Mag. Nat. Hist., ser. 8, 6:222.
REVIEWED BY: R. DeBry (RDB); C. B. Robbins (CBR).
COMMENT: Revised by Robbins, 1977, *in* Sokolov, ed., [Advances in Modern Theriology], pp. 178–194. Karyology reviewed by Matthey and Petter, 1970, Mammalia, 34:585–597. Subfamily Gerbillinae; see comment under Cricetidae.
ISIS NUMBER: 5301410008099000000.

Taterillus arenarius Robbins, 1974. Proc. Biol. Soc. Wash., 87(35):399.
TYPE LOCALITY: Mauritania, Trarza Region, Tiguent.
DISTRIBUTION: Mauritania; Niger; Mali.
COMMENT: Sympatric in S. Mauritania with *pygargus;* see Robbins, 1975, Unpubl. Ph.D. Dissertation, Univ. Arizona, p. 33.

Taterillus congicus Thomas, 1915. Ann. Mag. Nat. Hist., ser. 8, 16:147.
TYPE LOCALITY: Zaire, Upper Welle, Poko.
DISTRIBUTION: Cameroun; Chad; Central African Republic; Zaire; Sudan; Uganda.
COMMENT: Includes *clivosus;* see Robbins, 1977, *in* Sokolov, ed., [Advances in Modern Theriology], p. 189.
ISIS NUMBER: 5301410008099001001.

Taterillus emini (Thomas, 1892). Ann. Mag. Nat. Hist., ser. 6, 9:78.
TYPE LOCALITY: N. Uganda, Wadelai, 2° 42' N., 31° 22' E.
DISTRIBUTION: Sudan; W. Ethiopia; Uganda; N.W. Kenya; N.E. Zaire.
COMMENT: Includes *butleri, gyas,* and *anthonyi;* see Robbins, 1977, *in* Sokolov, ed., [Advances in Modern Theriology], p. 189. Robbins, 1973, Mammalia, 37:642–645, provisionally included *harringtoni, nubilus, osgoodi, tenebricus, lowei, melanops, illustris,* and *zammarani* in *emini,* and Petter, 1975, Part 6.3:6, included *osgoodi* in *emini* and *harringtoni* in *lowei;* but see Robbins, 1977, *in* Sokolov, ed., [Advances in Modern Theriology], pp.189–190, who recognized *harringtoni* as distinct from *emini* and transferred, to *harringtoni,* the above forms from Robbins, 1973:644.
ISIS NUMBER: 5301410008099002001.

Taterillus gracilis (Thomas, 1892). Ann. Mag. Nat. Hist., ser. 6, 9:77.
TYPE LOCALITY: Gambia.
DISTRIBUTION: N. Nigeria and Niger to Gambia and Senegal.
COMMENT: *T. gracilis* and *pygargus* are sympatric in W. Senegal where their karyotypes were studied by Matthey and Jotterand, 1972, Mammalia, 36:193–209. Includes *nigeriae;* see Robbins, 1977, *in* Sokolov, ed., [Advances in Modern Theriology], p. 190, and Robbins, 1975, Unpubl. Ph.D. Dissertation, Univ. Arizona, p. 52, who also included *angelus* in *gracilis.*
ISIS NUMBER: 5301410008099003001 as *T. gracilis.*
5301410008099008001 as *T. nigeriae.*

Taterillus harringtoni (Thomas, 1906). Ann. Mag. Nat. Hist., ser. 7, 18:303.
 TYPE LOCALITY: Ethiopia, East of Lake Rudolf, Mutti Galeb, near 5° 40′ N., 36° 20′ E.
 DISTRIBUTION: Central African Republic; Sudan; Ethiopia; Somalia; E. Uganda; Kenya;
  Tanzania.
 COMMENT: Includes *nubilus*, *osgoodi*, *tenebricus*, *lowei*, *melanops*, *illustris*, *zammarani*,
  kadugliensis, *lorenzi*, *meneghetti*, *perluteus*, and *rufus*; see Robbins, 1977, *in* Sokolov,
  ed., [Advances in Modern Theriology], p. 190, and Robbins, 1975, Unpubl. Ph.D.
  Dissertation, Univ. Arizona, p. 75; see also comment under *emini*.
 ISIS NUMBER: 5301410008099004001 as *T. harringtoni.*
              5301410008099006001 as *T. lowei.*
              5301410008099007001 as *T. melanops.*
              5301410008099009001 as *T. nubilus.*
              5301410008099010001 as *T. osgoodi.*
              5301410008099012001 as *T. tenebricus.*

Taterillus lacustris (Thomas and Wroughton, 1907). Ann. Mag. Nat. Hist., ser. 7, 19:37.
 TYPE LOCALITY: Nigeria, Lake Chad (=Kaddai).
 DISTRIBUTION: N.E. Nigeria; Cameroun; possibly Chad and Niger.
 COMMENT: Reviewed by Robbins, 1975, Unpubl. Ph.D. Dissertation, Univ. Arizona, p.
  33.
 ISIS NUMBER: 5301410008099005001.

Taterillus pygargus (F. Cuvier, 1838). Trans. Zool. Soc. Lond., 2:142.
 TYPE LOCALITY: Senegal, probably St. Louis.
 DISTRIBUTION: Gambia; Senegal; S. Mauritania; W. Mali.
 COMMENT: Formerly included in *Gerbillus pyramidum*; see Petter, 1952, Mammalia,
  16:37, and Petter, 1975, Part 6.3:5.
 ISIS NUMBER: 5301410008099011001.

Thomasomys Coues, 1884. Am. Nat., 18:1275.
 REVIEWED BY: R. DeBry (RDB); A. Langguth (AL); R. H. Pine (RHP).
 COMMENT: Includes *Wilfredomys*; see Hershkovitz, 1962, Fieldiana Zool., 46:21, Arata,
  1967, *in* Anderson and Jones, p. 230, and Pine, 1980, Mammalia, 44:195–202.
  Barlow, 1969, Life Sci. Contrib. R. Ontario Mus., 75:24–26, considered *Wilfredomys*
  a distinct genus. Includes *Delomys*; see Ellerman, 1941:366, and Cabrera, 1961:425;
  but also see Avila Pires, 1960, Bol. Mus. Nac. Rio de Janeiro, 220:1–6. Formerly
  included *Rhipidomys macconnelli*, *Wiedomys pyrrhorhinos* (see Hershkovitz, 1959,
  Proc. Biol. Soc. Wash., 72:5, 9), *Aepeomys lugens*, and *A. fuscatus* (see Gardner and
  Patton, 1976, Occas. Pap. Mus. Zool. Louisiana State Univ., 49:26, 32). Cabrera,
  1961:431, and Walker *et al.*, 1975:766, included *Aepeomys* in *Thomasomys*. This
  genus is in need of revision; its relationship to other thomasomyines needs
  clarification; see Hooper and Musser, 1964, Misc. Publ. Mus. Zool. Univ.
  Michigan, 123:21–22, Hershkovitz, 1966, Z. Saugetierk., 31:125, and Cabrera,
  1961:425. Reviewed, in part, by Carleton, 1973, Misc. Publ. Mus. Zool. Univ.
  Michigan, 146:1–43, Gardner and Patton, 1976:26, 32, and Voss and Linzey, 1981,
  Misc. Publ. Mus. Zool. Univ. Michigan, 159:1–41. Subfamily Hesperomyinae; see
  comment under Cricetidae.
 ISIS NUMBER: 5301410008055000000.

Thomasomys aureus (Tomes, 1860). Proc. Zool. Soc. Lond., 1860:219.
 TYPE LOCALITY: Ecuador, Chimborazo Prov., Pallatanga, 1350 m.
 DISTRIBUTION: C. and W. Colombia; N.C. Venezuela; Andean Ecuador; N.W. and E.
  Peru.
 COMMENT: Hooper and Musser, 1964, Misc. Publ. Mus. Zool. Univ. Mich., 123:21,
  considered *aureus* well differentiated from other representatives of this genus;
  also see Carleton, 1973, Misc. Publ. Mus. Zool. Univ. Mich., 146:1–43. Karyology
  reviewed by Gardner and Patton, 1976, Occas. Pap. Mus. Zool. La. St. Univ.,
  49:26, 32; Voss and Linzey, 1981, Misc. Publ. Mus. Zool. Univ. Mich., 159:1–41,
  reviewed the accessory male reproductive glands.
 ISIS NUMBER: 5301410008055001001.

Thomasomys baeops (Thomas, 1899). Ann. Mag. Nat. Hist., ser. 7, 3:152.
    TYPE LOCALITY: Ecuador, El Oro Prov., Chilla Valley, Rio Pita, 3500 m.
    DISTRIBUTION: W. Ecuador.
    COMMENT: Stomach morphology reviewed by Carleton, 1973, Misc. Publ. Mus. Zool.
        Univ. Mich., 146:1–43; accessory male reproductive glands reviewed by Voss and
        Linzey, 1981, Misc. Publ. Mus. Zool. Univ. Mich., 159:1–41. Sometimes spelled
        boeops, but see Thomas, 1899, in which the ligature "ae" was used.
    ISIS NUMBER: 5301410008055002001 as *T. boeops (sic)*.

Thomasomys bombycinus Anthony, 1925. Am. Mus. Novit., 178:1.
    TYPE LOCALITY: Colombia, Antioquia Dept., Paramillo, 3750 m.
    DISTRIBUTION: W. Colombia.
    ISIS NUMBER: 5301410008055003001 as *T. bombyvinus (sic)*.

Thomasomys cinereiventer J. A. Allen, 1912. Bull. Am. Mus. Nat. Hist., 31:80.
    TYPE LOCALITY: Colombia, Cauca Dept., 64 km west of Popayan, 3070 m.
    DISTRIBUTION: Colombia; Ecuador.
    ISIS NUMBER: 5301410008055004001.

Thomasomys cinereus (Thomas, 1882). Proc. Zool. Soc. Lond., 1882:108.
    TYPE LOCALITY: Peru, Cajamarca Dept., Cutervo, 2760 m.
    DISTRIBUTION: S.W. Ecuador; N.W. Peru.
    COMMENT: Reviewed by Carleton, 1973, Misc. Publ. Mus. Zool. Univ. Mich., 146:1–43,
        and Hooper and Musser, 1964, Misc. Publ. Mus. Zool. Univ. Mich., 123:21.
    ISIS NUMBER: 5301410008055005001.

Thomasomys daphne Thomas, 1917. Smithson. Misc. Coll., 68(4):2.
    TYPE LOCALITY: Peru, Cuzco Dept., Ocabamba Valley, 2730 m.
    DISTRIBUTION: S.W. Peru; C. Bolivia.
    ISIS NUMBER: 5301410008055006001.

Thomasomys dorsalis (Hensel, 1872). Abh. Konigl. Preuss. Akad. Wiss. Berlin, p. 42.
    TYPE LOCALITY: Brazil, Rio Grande do Sul Prov.
    DISTRIBUTION: S. and E. Brazil; N.E. Argentina (AL).
    COMMENT: Includes *collinus* and *sublineatus*; see Cabrera, 1961:428; Walker *et al.*,
        1975:766, listed *collinus* and *sublineatus* as distinct species, without comment.
        Reviewed by Massoia, 1962, Physis, 23(64):27–34. *Delomys* was considered a
        distinct genus by Avila-Pires, 1960, Arq. Mus. Nac. Rio de J., 50:32–33, and Avila-
        Pires, 1960, Bol. Mus. Nac. Rio de J., 220:1–6; Cabrera, 1961:425, 428–429,
        included *Delomys* in *Thomasomys* as a synonym. RHP considers *Delomys* a
        subgenus which probably merits generic status.
    ISIS NUMBER: 5301410008055007001.

Thomasomys gracilis Thomas, 1917. Smithson. Misc. Coll., 68(4):2.
    TYPE LOCALITY: Peru, Cuzco Dept., Machu Picchu, 3600 m.
    DISTRIBUTION: Ecuador; S.E. Peru.
    ISIS NUMBER: 5301410008055008001.

Thomasomys hylophilus Osgood, 1912. Field Mus. Nat. Hist. Publ. Zool. Ser., 10(5):50.
    TYPE LOCALITY: Colombia, Santander Dept., Upper Rio Tachira, Paramo de Tama.
    DISTRIBUTION: N. Colombia; W. Venezuela.
    COMMENT: Stomach morphology reviewed by Carleton, 1973, Misc. Publ. Mus. Zool.
        Univ. Mich., 146:1–43. Reviewed by Handley, 1976, Brigham Young Univ. Sci.
        Bull. Biol. Ser., 20(5):51.
    ISIS NUMBER: 5301410008055009001.

Thomasomys incanus (Thomas, 1894). Ann. Mag. Nat. Hist., ser. 6, 14:350.
    TYPE LOCALITY: Peru, Junin, Vitoc Valley.
    DISTRIBUTION: Andean Peru.
    ISIS NUMBER: 5301410008055010001.

Thomasomys ischyurus Osgood, 1914. Field Mus. Nat. Hist. Publ. Zool. Ser., 10(12):162.
    TYPE LOCALITY: Peru, Amazonas, 65 km east of Chachapoyas, Tambo Almirante, about
        1500 m.
    DISTRIBUTION: N.W. Peru; W. Ecuador.

COMMENT: Stomach morphology reviewed by Carleton, 1973, Misc. Publ. Mus. Zool.
Univ. Mich., 146:1–43.
ISIS NUMBER: 5301410008055011001.

Thomasomys kalinowskii (Thomas, 1894). Ann. Mag. Nat. Hist., ser. 6, 14:349.
TYPE LOCALITY: Peru, Junin, Vitoc Valley.
DISTRIBUTION: C. Peru.
COMMENT: Karyology reviewed by Gardner and Patton, 1976, Occas. Pap. Mus. Zool.
La. St. Univ., 49:26, 32.
ISIS NUMBER: 5301410008055012001.

Thomasomys ladewi Anthony, 1926. Am. Mus. Novit., 239:1.
TYPE LOCALITY: Bolivia, La Paz Prov., Rio Aceramarca, 3240 m.
DISTRIBUTION: N.W. Bolivia.
ISIS NUMBER: 5301410008055013001.

Thomasomys laniger (Thomas, 1895). Ann. Mag. Nat. Hist., ser. 6, 16:59.
TYPE LOCALITY: Colombia, Cundinamarca, Bogota Region, 2600 m.
DISTRIBUTION: W. Venezuela; C. Colombia.
COMMENT: Formerly included *monochromos;* see Gardner and Patton, 1976, Occas. Pap.
Mus. Zool. La. St. Univ., 49:26, 32. Reviewed by Handley, 1976, Brigham Young
Univ. Sci. Bull. Biol. Ser., 20(5):51, Hooper and Musser, 1964, Misc. Publ. Mus.
Zool. Univ. Mich., 123:21, and Voss and Linzey, 1981, Misc. Publ. Mus. Zool.
Univ. Mich., 159:1–41.
ISIS NUMBER: 5301410008055014001.

Thomasomys monochromos Bangs, 1900. Proc. N. Engl. Zool. Club, 1:97.
TYPE LOCALITY: Colombia, Magdalena, Sierra Nevada de Santa Marta, Macotama, 3300
m.
DISTRIBUTION: N.E. Colombia.
COMMENT: Formerly included in *laniger* by Cabrera, 1961:431; but also see Gardner
and Patton, 1976, Occas. Pap. Mus. Zool. La. St. Univ., 49:26, 32, who considered
monochromos a distinct species.

Thomasomys notatus Thomas, 1917. Smithson. Misc. Coll., 68(4):2.
TYPE LOCALITY: Peru, Cuzco, Torontoy, 2850 m.
DISTRIBUTION: S.E. Peru.
COMMENT: Karyology reviewed by Gardner and Patton, 1976, Occas. Pap. Mus. Zool.
La. St. Univ., 49:26, 32.
ISIS NUMBER: 5301410008055016001.

Thomasomys oenax Thomas, 1928. Ann. Mag. Nat. Hist., ser. 10, 1:154.
REVIEWED BY: R. H. Pine (RHP).
TYPE LOCALITY: Brazil, Rio Grande do Sul, San Lorenzo.
DISTRIBUTION: S. Brazil; Uruguay.
COMMENT: Formerly included in *Wilfredomys;* reviewed by Pine, 1980, Mammalia,
44:196–198; but also see Barlow, 1969, Life Sci. Contrib. R. Ont. Mus., 75:1–59.
ISIS NUMBER: 5301410008055017001.

Thomasomys oreas Anthony, 1926. Am. Mus. Novit., 239:2.
TYPE LOCALITY: Bolivia, La Paz Prov., Cocopunco.
DISTRIBUTION: Andean Bolivia.
ISIS NUMBER: 5301410008055018001.

Thomasomys paramorum Thomas, 1898. Ann. Mag. Nat. Hist., ser. 7, 1:453.
TYPE LOCALITY: Ecuador, Chimborazo, south of Mt. Chimborazo, Paramo.
DISTRIBUTION: Andean Ecuador.
COMMENT: Stomach morphology reviewed by Carleton, 1973, Misc. Publ. Mus. Zool.
Univ. Mich., 146:1–43.
ISIS NUMBER: 5301410008055019001.

Thomasomys pictipes Osgood, 1933. Field Mus. Nat. Hist. Publ. Zool. Ser., 20(2):11.
REVIEWED BY: R. H. Pine (RHP).
TYPE LOCALITY: Argentina, Misiones, Caraguatay, Rio Parana, 100 mi. (161 km) S. Rio Iguazu, 26° 37′ S., 54° 46′ W.
DISTRIBUTION: N.E. Argentina; S.E. Brazil.
COMMENT: Reviewed by Pine, 1980, Mammalia, 44:199–200.
ISIS NUMBER: 5301410008055020001.

Thomasomys pyrrhonotus Thomas, 1886. Ann. Mag. Nat. Hist., ser. 5, 18:421.
REVIEWED BY: R. H. Pine (RHP).
TYPE LOCALITY: Peru, Cajamarca Dist., Rio Malleta, Tambillo, 5800 ft. (1768 m).
DISTRIBUTION: S. Ecuador; N.W. Peru.
COMMENT: Includes *auricularis;* reviewed by Pine, 1980, Mammalia, 44:200.
ISIS NUMBER: 5301410008055021001.

Thomasomys rhoadsi Stone, 1914. Proc. Acad. Nat. Sci. Phila., 66:12.
TYPE LOCALITY: Ecuador, Pichincha Prov., Mt. Pichincha, Hacienda Garzon, 3100 m.
DISTRIBUTION: Ecuador.
COMMENT: Stomach morphology reviewed by Carleton, 1973, Misc. Publ. Mus. Zool. Univ. Mich., 146:1–43; male accessory glands reviewed by Voss and Linzey, 1981, Misc. Publ. Mus. Zool. Univ. Mich., 159:1–41.
ISIS NUMBER: 5301410008055022001.

Thomasomys rosalinda Thomas and St. Leger, 1926. Ann. Mag. Nat. Hist., ser. 9, 18:345.
TYPE LOCALITY: Peru, Amazonas, Goncha, 2550 m.
DISTRIBUTION: N.W. Peru.
ISIS NUMBER: 5301410008055023001.

Thomasomys taczanowskii (Thomas, 1882). Proc. Zool. Soc. Lond., 1882:109.
TYPE LOCALITY: Peru, Cajamarca Dist., Rio Malleta, Tambillo.
DISTRIBUTION: N.W. Peru.
COMMENT: Karyology reviewed by Gardner and Patton, 1976, Occas. Pap. Mus. Zool. La. St. Univ., 49:26, 32.
ISIS NUMBER: 5301410008055024001.

Thomasomys vestitus (Thomas, 1898). Ann. Mag. Nat. Hist., ser. 7, 1:454.
TYPE LOCALITY: Venezuela, Merida, Rio Milla, 1630 m.
DISTRIBUTION: W. Venezuela.
COMMENT: Reviewed by Handley, 1976, Brigham Young Univ. Sci. Bull. Biol. Ser., 20(5):51.
ISIS NUMBER: 5301410008055025001.

Tylomys Peters, 1866. Monatsb. Preuss. Akad. Wiss. Berlin, p. 404.
REVIEWED BY: R. DeBry (RDB); J. Ramirez-Pulido (JRP)(Mexico).
COMMENT: Reviewed by Carleton, 1980:1–146, Lawlor, 1969, J. Mammal., 50:28–42, and Hall, 1981:626–628. Subfamily Hesperomyinae; see comment under Cricetidae.
ISIS NUMBER: 5301410008056000000.

Tylomys bullaris Merriam, 1901. Proc. Wash. Acad. Sci., 3:561.
TYPE LOCALITY: Mexico, Chiapas, Tuxtla (= Tuxtla Gutierrez).
DISTRIBUTION: Known only from the type locality.
ISIS NUMBER: 5301410008056001001 as *T. bularis (sic).*

Tylomys fulviventer Anthony, 1916. Bull. Am. Mus. Nat. Hist., 35:366.
TYPE LOCALITY: Panama, Darien, Tacarcuna, 4200′.
DISTRIBUTION: E. Panama.
COMMENT: Possibly a subspecies of *mirae;* see Cabrera, 1961:435. Reviewed by Handley, 1966, *in* Wenzel and Tipton, eds., Ectoparasites of Panama, Field Mus. Nat. Hist., p. 782.
ISIS NUMBER: 5301410008056002001.

Tylomys mirae Thomas, 1899. Ann. Mag. Nat. Hist., ser. 7, 4:278.
TYPE LOCALITY: Ecuador, Rio Mira, Paramba.
DISTRIBUTION: N. Ecuador; S.W. and C. Colombia.
COMMENT: May be conspecific with *fulviventer*; see Cabrera, 1961:435; but also see Hall, 1981:627.
ISIS NUMBER: 5301410008056003001.

Tylomys nudicaudus (Peters, 1866). Monatsb. Preuss. Akad. Wiss. Berlin, p. 404.
REVIEWED BY: G. Urbano-V. (GUV).
TYPE LOCALITY: Guatemala, Alta Verapaz, La Primavera, 10 mi. (16 km) S.W. Coban, about 3200 ft. (975 m).
DISTRIBUTION: Nicaragua to C. Guerrero and Veracruz (Mexico).
COMMENT: Includes *gymnurus*; see Goodwin, 1969, Bull. Am. Mus. Nat. Hist., 141:160, Schaldach, 1966, Saugetierk. Mitt., 4:295, and Hall, 1981:627.
ISIS NUMBER: 5301410008056004001.

Tylomys panamensis (Gray, 1873). Ann. Mag. Nat. Hist., ser. 4, 12:417.
TYPE LOCALITY: Panama.
DISTRIBUTION: E. Panama.
COMMENT: Reviewed by Handley, 1966, *in* Wenzel and Tipton, eds., Ectoparasites of Panama, Field Mus. Nat. Hist., p. 782.
ISIS NUMBER: 5301410008056005001.

Tylomys tumbalensis Merriam, 1901. Proc. Wash. Acad. Sci., 3:560.
TYPE LOCALITY: Mexico, Chiapas, Tumbala.
DISTRIBUTION: Known only from the type locality.
ISIS NUMBER: 5301410008056006001.

Tylomys watsoni Thomas, 1899. Ann. Mag. Nat. Hist., ser. 7, 4:278.
TYPE LOCALITY: Panama, Chiriqui, Vulcan de Chiriqui, Bogava (=Bugaba), 800 ft. (244 m).
DISTRIBUTION: Costa Rica; Panama.
COMMENT: Reviewed by Handley, 1966, *in* Wenzel and Tipton, eds., Ectoparasites of Panama, Field Mus. Nat. Hist., p. 782.
ISIS NUMBER: 5301410008056007001.

Typhlomys Milne-Edwards, 1877. Bull. Sci. Soc. Philom. Paris, (1876), ser. 6, 12(2):9.
REVIEWED BY: J. E. Bucher (JEB); S. Wang (SW).
COMMENT: Formerly considered in a separate family and in a subfamily of Gliridae; see Arata, 1967, *in* Anderson and Jones, pp. 244–245. Transferred to subfamily Platacanthomyinae of Cricetidae by Chaline and Mein, 1979, Chaline *et al.*, 1977, Mammalia, 41:245–252, Mein and Freudenthal, 1971, Scr. Geol., 2:137, and Reig, 1980, J. Zool. Lond., 192:260.
ISIS NUMBER: 5301410013002000000.

Typhlomys cinereus Milne-Edwards, 1877. Bull. Sci. Soc. Philom. Paris, (1876), ser. 6, 12(2):9.
TYPE LOCALITY: China, W. Fukien.
DISTRIBUTION: Yunnan, Fukien, and Kwangsi (China); N. Vietnam.
ISIS NUMBER: 5301410013002001001.

Wiedomys Hershkovitz, 1959. Proc. Biol. Soc. Wash., 72:5.
REVIEWED BY: R. DeBry (RDB); R. H. Pine (RHP).
COMMENT: Reviewed by Reig, 1980, J. Zool. Lond., 192:257–281. Subfamily Hesperomyinae; see comment under Cricetidae.
ISIS NUMBER: 5301410008057000000.

Wiedomys pyrrhorhinos (Wied-Neuwied, 1821). Reise nach Brasilien, 2:177.
TYPE LOCALITY: Brazil, Bahia Prov., Rio Ressaro.
DISTRIBUTION: E. Brazil; perhaps Paraguay.
COMMENT: This species has been included in *Oryzomys* and in *Thomasomys*; see Hershkovitz, 1959, Proc. Biol. Soc. Wash., 72:5–10; also see Pine, 1980, Mammalia, 44:195.
ISIS NUMBER: 5301410008057001001.

Xenomys Merriam, 1892. Proc. Biol. Soc. Wash., 7:160.
REVIEWED BY: R. DeBry (RDB); M. D. Carleton (MDC); J. Ramirez-Pulido (JRP); G. Urbano-V. (GUV).
COMMENT: Reviewed by Carleton, 1980:118, 121. Subfamily Hesperomyinae; see comment under Cricetidae.
ISIS NUMBER: 5301410008058000000.

Xenomys nelsoni Merriam, 1892. Proc. Biol. Soc. Wash., 7:161.
TYPE LOCALITY: Mexico, Colima, Hacienda Magdalena, between cities of Colima and Manzanillo.
DISTRIBUTION: Colima, W. Jalisco, perhaps W. Nayarit (Mexico).
ISIS NUMBER: 5301410008058001001.

Zygodontomys J. A. Allen, 1897. Bull. Am. Mus. Nat. Hist., 9:38.
REVIEWED BY: M. D. Carleton (MDC); A. Langguth (AL); O. A. Reig (OAR); D. F. Williams (DFW).
COMMENT: Revised by Hershkovitz, 1962, Fieldiana Zool., 46:196–207, who included *Zygodontomys* in the phyllotine group; but see Pearson and Patton, 1976, J. Mammal., 57:349; also see Hooper and Musser, 1964, Misc. Publ. Mus. Zool. Univ. Mich., 123:57. Formerly included *lasiurus* which was transferred to *Bolomys* by Reig, 1978, Publ. Mus. Munic. Cienc. Nat. Mar del Plata "Lorenzo Scaglia," 2(8):167. Reviewed by Maia and Langguth, 1981, Z. Saugetierk., 46:241–249. This genus is in need of revision. Subfamily Hesperomyinae; see comment under Cricetidae.
ISIS NUMBER: 5301410008059000000.

Zygodontomys borreroi (Hernandez-Camacho), 1957. An. Soc. Biol. Bogota, 7:223.
TYPE LOCALITY: Colombia, Santander, Botulia, Hacienda Montebello, Mt. San Pablo, between 350 and 500 m.
DISTRIBUTION: Known only from the type locality.
COMMENT: Placed in *Zygodontomys* by Gardner and Patton, 1976, Occas. Pap. Mus. Zool. La. St. Univ., 49:41; but placed in *Oryzomys* by Hershkovitz, 1970, J. Mammal., 51:792. Known only by the holotype.
ISIS NUMBER: 5301410008035012001 as *Oryzomys borreroi*.

Zygodontomys brevicauda (J. A. Allen and Chapman, 1893). Bull. Am. Mus. Nat. Hist., 5:215.
TYPE LOCALITY: Trinidad and Tobago, Trinidad, Princestown.
DISTRIBUTION: S.E. Costa Rica to Ecuador, Colombia, Venezuela, Guianas, and adjacent Brazil; Trinidad and Tobago.
COMMENT: Includes *cherriei, microtinus, seorsus,* and *punctulatus*; see Hershkovitz, 1962, Fieldiana Zool., 49:203–204, Hall, 1981:733, and Husson, 1978:415–419. Handley, 1966, *in* Wenzel and Tipton, eds., Ectoparasites of Panama, Field Mus. Nat. Hist., p. 783, listed *microtinus* and *seorsus* as distinct species, without comment. Kiblisky et al., 1970, Acta Cient. Venez., 21(1):35, Gardner and Patton, 1976, Occas. Pap. Mus. Zool. La. St. Univ., 49:28, and Pearson and Patton, 1976, J. Mammal., 57:349, listed *microtinus* as a distinct species. *Z. brevicauda* is probably a composite according to OAR who retains *cherriei* and *tobagi* in *brevicauda* and considers *brunneus* (including *sanctaemartae*), *microtinus* (including *thomasi* and *stellae*), and *punctulatus* (including *griseus* and *fraterculus*) distinct species based on comparisons of most holotypes and distributions; all of the forms considered above by OAR were included in *brevicauda* by Hershkovitz, 1962, Fieldiana Zool., 46:196–205.
ISIS NUMBER: 5301410008059001001 as *Z. brevicauda*.
          5301410008059002001 as *Z. cherriei*.
          5301410008059004001 as *Z. microtinus*.
          5301410008059005001 as *Z. punctulatus*.

Zygodontomys reigi Tranier, 1976. C. R. Acad. Sci. Paris, ser. D., 283:1201.
TYPE LOCALITY: Guyana, vicinity of Cayenne.
DISTRIBUTION: Guyana.
COMMENT: Probably conspecific with *brevicauda* (AL). May be a *nomen nudum* (OAR).

Family Spalacidae

REVIEWED BY: O. L. Rossolimo (OLR).

COMMENT: Reviewed by Corbet, 1978:129–130, and Topachevskii, 1969, *in* [Fauna S.S.S.R., Mammals], n.s., 3(3):99; also see Reig, 1980, J. Zool. Lond., 192:257–281, and comment under Cricetidae.

ISIS NUMBER: 5301410009000000000.

Spalax Guldenstaedt, 1770. Nova Comm. Acad. Sci. Petrop., ser. 14, 1:410.

COMMENT: Includes *Microspalax*; see Corbet, 1978:129. Also see Topachevskii, 1969, *in* [Fauna S.S.S.R., Mammals], n.s., 3(3):114, who considered *Microspalax* a distinct genus. Gromov and Baranova, 1981:128, employed the name *Nannospalax* in place of *Microspalax*. For a discussion of the content of this genus, see Martynova *et al.*, 1975, Abstr. Symp. Syst. and Cytogenet. Mamm., Moscow, p. 12–13, and Vorontsov *et al.*, 1977, Zool. Zh., 56:1207–1214.

ISIS NUMBER: 5301410009001000000.

Spalax giganteus Nehring, 1898. Sitzb. Ges. Naturf. Fr. Berlin, p. 169.

TYPE LOCALITY: U.S.S.R., Daghestan, W. shore of Caspian Sea, Petrovsky-Port near Makhach-Kala.

DISTRIBUTION: Steppe N.W. of Caspian Sea and in Kazakhstan east of Ural River (U.S.S.R.).

COMMENT: Includes *uralensis*; see Corbet, 1978:130.

Spalax leucodon Nordmann, 1840. Demidoff Voy., 3:34.

TYPE LOCALITY: U.S.S.R., S. Ukraine, near Odessa.

DISTRIBUTION: Danube basin to Greece and S. Ukraine; Caucasus through Asia Minor, N. Iraq, and Syria to Israel; coastal Egypt and Libya, west to Benghazi.

COMMENT: Includes *ehrenbergi* and *nehringi*; see Corbet, 1978:130, Harrison, 1972, The Mammals of Arabia, 3:438–440, and Orlov, 1969, *in* Vorontsov, [The Mammals:Evolution, Karyology, Taxonomy, Fauna], Novosibirsk. Wahrman *et al.*, 1969, Comp. Mamm. Cytogen., pp. 30–48, Topachevskii, 1969, *in* [Fauna S.S.S.R., Mammals], n.s., 3(3):99, and others presented evidence that *ehrenbergi* and *nehringi* are specifically distinct and that this superspecies may contain 2–5 species. OLR follows Topachevskii, 1969, in regarding these forms as distinct species. See also Lyapunova *et al.*, 1974, Symp. Ther. II, Prague, pp. 203–215, and Soldatovic and Savic, 1978, Saugetierk. Mitt., 26:252–256.

ISIS NUMBER: 5301410009001002001 as *S. leucodon (sic)*.
5301410009001001001 as *S. ehrenbergi*.

Spalax microphthalmus Guldenstaedt, 1770. Nova Comm. Acad. Sci. Petrop., 14:1.

TYPE LOCALITY: U.S.S.R., Voronezhsk. Obl., Novokhoper Steppe.

DISTRIBUTION: Ukraine and S. Russia, southwest to Bulgaria and Greece, east to Volga and N.C. Caucasus.

COMMENT: Includes *arenarius*, *polonicus* (=*zemni*), and *graecus*; see Corbet, 1978:130. OLR follows Topachevskii, 1969, *in* [Fauna S.S.S.R., Mammals], n.s., 3(3):99, in regarding these forms as distinct species, as did Gromov and Baranova, 1981:131–132.

ISIS NUMBER: 5301410009001003001.

Family Rhizomyidae

REVIEWED BY: J. E. Bucher (JEB).

COMMENT: Family status reviewed by Arata, 1967 *in* Anderson and Jones, p. 236, and Reig, 1980, J. Zool. Lond., 192:257–281.

ISIS NUMBER: 5301410010000000000.

Cannomys Thomas, 1915. Ann. Mag. Nat. Hist., ser. 8, 16:57.

ISIS NUMBER: 5301410010001000000.

Cannomys badius (Hodgson, 1841). Calcutta J. Nat. Hist., 2:60.

TYPE LOCALITY: Nepal.

DISTRIBUTION: Nepal; Assam (India); Burma; E. and S. Thailand.

ISIS NUMBER: 5301410010001001001.

Rhizomys Gray, 1831. Proc. Zool. Soc. Lond., 1831:95.
    REVIEWED BY: S. Wang (SW).
    ISIS NUMBER: 5301410010002000000.

Rhizomys pruinosus Blyth, 1851. J. Asiat. Soc. Bengal, 20:519.
    TYPE LOCALITY: India, Assam, Khasi Hills, Cherrapunji.
    DISTRIBUTION: S. China and Assam (India) to Malaya.
    ISIS NUMBER: 5301410010002001001.

Rhizomys sinensis Gray, 1831. Proc. Zool. Soc. Lond., 1831:95.
    TYPE LOCALITY: China, Kwangtung, near Canton.
    DISTRIBUTION: S. and C. China to N. Burma.
    ISIS NUMBER: 5301410010002002001.

Rhizomys sumatrensis (Raffles, 1821). Trans. Linn. Soc. Lond., 13:258.
    TYPE LOCALITY: Malaysia, Malacca.
    DISTRIBUTION: Yunnan (China); Indochina; Thailand; Burma; Malaya; Sumatra.
    ISIS NUMBER: 5301410010002003001.

Tachyoryctes Ruppell, 1835. Neue Wirbelt. Fauna Abyssin. Gehorig. Saugeth., p. 35.
    ISIS NUMBER: 5301410010003000000.

Tachyoryctes macrocephalus Ruppell, 1842. Mus. Senckenbergianum Abh., 3:97.
    TYPE LOCALITY: Ethiopia (=Abyssinia), Shoa.
    DISTRIBUTION: Ethiopia.
    ISIS NUMBER: 5301410010003006001.

Tachyoryctes splendens Ruppell, 1835. Neue Wirbelt. Fauna Abyssin. Gehorig. Saugeth., p. 36.
    TYPE LOCALITY: Ethiopia (=Abyssinia), Gondar.
    DISTRIBUTION: Ethiopia; Somalia; Kenya; Uganda; E. Zaire; Rwanda; Burundi; Mt. Kilimanjaro (Tanzania).
    COMMENT: Includes *ankoliae, annectens, audax, cheesmani, daemon, naivashae, pontifex, rex, ruandae, ruddi, spalacinus,* and *storeyi;* see Misonne, 1974, Part 6:7.
    ISIS NUMBER: 5301410010003013001 as *T. splendens.*
                 5301410010003001001 as *T. ankoliae.*
                 5301410010003002001 as *T. annectens.*
                 5301410010003003001 as *T. audax.*
                 5301410010003004001 as *T. cheesmani.*
                 5301410010003005001 as *T. daemon.*
                 5301410010003007001 as *T. naivashae.*
                 5301410010003008001 as *T. pontifex.*
                 5301410010003009001 as *T. rex.*
                 5301410010003010001 as *T. ruandae.*
                 5301410010003011001 as *T. ruddi.*
                 5301410010003012001 as *T. spalacinus.*
                 5301410010003014001 as *T. storeyi.*

Family Arvicolidae
    REVIEWED BY: J. Chaline (JC); R. S. Hoffmann (RSH); L. D. Martin (LDM); C. A. Repenning (CAR); O. L. Rossolimo (OLR)(U.S.S.R.); W. von Koenigswald (WVK); S. Wang (SW) (China).
    COMMENT: For assignment of name and family rank, see Kretzoi, 1962, Vert. Hungar. (Budapest), 4:171–175, Kretzoi, 1969, Vert. Hungar. (Budapest), 11:155–193, Van der Meulen, 1978, Ann. Carnegie Mus., 47:101–145, Chaline and Mein, 1979, and Reig, 1980, J. Zool. Lond., 192:260. For a review of alternative classifications of the rodents, see Swanepoel *et al.*, 1980, Ann. Transvaal Mus., 32(7):155–196, and Anderson and Jones, 1967. CAR retained this group in Cricetidae.

Alticola Blanford, 1881. J. Asiat. Soc. Bengal, 50:96.
    COMMENT: Revised, in part, by Heptner and Rossolimo, 1968, Sbor. Tr. Zool. Mus. Moscow State Univ., 10:53–93. Closely related to *Dinaromys, Eothenomys,* and

Hyperacrius (JC), or possibly to *Lagurus* (CAR). Placed in tribe Clethrionomyini by Gromov and Polyakov, 1977:126.
ISIS NUMBER: 5301410008068000000.

Alticola macrotis Radde, 1862. Reise in den Suden von Ost-Sibierien, 1:196, Die Saugethierfauna im Suden von Ost-Sibirien. Besobrasoff, St. Petersburg.
TYPE LOCALITY: U.S.S.R., R.S.F.S.R., S. Krasnoyarsk. Krai, E. Sayan Mtns.
DISTRIBUTION: Sayan and Altai Mtns.; mountains S.E. of Lake Baikal and head of Shilka River (U.S.S.R.); Khangai Mtns. (Mongolia).
COMMENT: See comment under *Eothenomys lemminus.*
ISIS NUMBER: 5301410008068002001.

Alticola roylei Gray, 1842. Ann. Mag. Nat. Hist., ser. 1, 10:265.
TYPE LOCALITY: India, "Cashmere."
DISTRIBUTION: N.W. Afghanistan, N. Pakistan, probably adjacent Tibet and N. India; Pamir, Tien Shan, Altai, and other ranges, north to Lake Baikal, and east through Mongolia to Sinkiang (China).
COMMENT: Includes *argentatus* and *barakschin;* see Heptner and Rossolimo, 1968, Sbor. Tr. Zool. Mus. Moscow State Univ., 10:53–93; but *argentatus* may be a distinct species and *barakschin* may be a subspecies of *stoliczkanus;* see Gromov and Polyakov, 1977, [Voles (Microtinae), Fauna U.S.S.R.], Nauka, Moscow and Leningrad, 3(8):129. Sokolov and Orlov, 1980:133, and Gromov and Baranova, 1981:176, both treated *argentatus,* as a distinct species and the latter excluded *barakschin* from *roylei.*
ISIS NUMBER: 5301410008068003001.

Alticola stoliczkanus Blanford, 1875. J. Asiat. Soc. Bengal, 44:107.
TYPE LOCALITY: India, Ladak, Kunlun Mtns.
DISTRIBUTION: Mustang Dist. (Nepal); N. Ladak (Tibet). N.W. Kansu (China) (SW).
COMMENT: Includes *nanshanicus; lama,* provisionally included following Gromov and Polyakov, 1977:130, and Corbet, 1978:103, was placed in *roylei* by Heptner and Rossolimo, 1968, Sbor. Tr. Zool. Mus. Moscow State Univ., 10:53–93.
ISIS NUMBER: 5301410008068004001.

Alticola stracheyi (Thomas, 1880). Ann. Mag. Nat. Hist., ser. 5, 6:332.
TYPE LOCALITY: India, Ladak; originally cited as Kumaon.
DISTRIBUTION: Kashmir; Ladak; N. Nepal; Tibet (China) (SW).
COMMENT: Considered a subspecies of *stoliczkanus* by Gromov and Polyakov, 1977:130, and Corbet, 1978, and placed with *roylei* by Heptner and Rossolimo, 1968, Sbor. Tr. Zool. Mus. Moscow State Univ., 10:53–93; but Mitchell, 1977, Accounts of Nepalese Mammals, considered it a distinct species and included *bhatnagari* as a synonym; he also discussed the type locality. Gromov and Baranova, 1981:177, allied *barakschin* with *stoliczkanus* and *stracheyi.*
ISIS NUMBER: 5301410008068001001 as *A. bhatnagari.*

Alticola strelzowi (Kastchenko, 1899). Izv. Imp. Tomsk. Univ., 16:50.
TYPE LOCALITY: U.S.S.R., R.S.F.S.R., Altaisk. Krai, Altai Mtns., near Lake Tenga.
DISTRIBUTION: Altai Mtns. of N.W. Mongolia, west through Gorno-Altaisk. A.O. and Kazakh S.S.R. north of Lake Balkhash to beyond Karaganda (U.S.S.R.).
ISIS NUMBER: 5301410008068005001.

Arborimus Taylor, 1915. Proc. Calif. Acad. Sci., ser. 4, 5(5):119.
REVIEWED BY: M. L. Johnson (MLJ).
COMMENT: Formerly included in *Phenacomys;* see Johnson, 1973, J. Mammal., 54:239–244, and Johnson and Maser, in press, Northwest. Sci. Hall, 1981:788, considered *Arborimus* a subgenus of *Phenacomys.*

Arborimus albipes Merriam, 1901. Proc. Biol. Soc. Wash., 14:125.
TYPE LOCALITY: U.S.A., California, Humboldt Co., Arcata, in redwoods.
DISTRIBUTION: W. Oregon, N.W. California, south to Humbolt Co. (U.S.A.).
COMMENT: Close relationship of this species with *longicaudus* documented by Johnson and Maser, in press, Northwest, Sci., who separated *Arborimus* from *Phenacomys;* Hall, 1981:787, placed this species in the subgenus *Phenacomys.*
ISIS NUMBER: 5301410008084001001 as *Phenacomys albipes.*

Arborimus longicaudus True, 1890. Proc. U.S. Nat. Mus., 13:303.
TYPE LOCALITY: U.S.A., Oregon, Coos Co., Marshfield.
DISTRIBUTION: W. Oregon, N.W. California (U.S.A.), south to Freestone (Sonoma Co.).
COMMENT: Includes *silvicola*; see Johnson, 1968, Syst. Zool., 17:27. Formerly included
in *Phenacomys*; see comment under *Arborimus*.
ISIS NUMBER: 5301410008084003001 as *Phenacomys longicaudus*.
5301410008084004001 as *Phenacomys silvicola*.

Arvicola Lacepede, 1799. Tabl. Mamm., p. 10.
REVIEWED BY: J.-P. Airoldi (JPA); M. Andera (MA); M. L. Johnson (MLJ).
COMMENT: Considered by Ellerman and Morrison-Scott, 1951, Gromov and Polyakov,
1977:235, Hibbard *et al.*, 1978, Contrib. Mus. Paleontol. Univ. Mich., 25(2):11–44,
Chaline and Mein, 1979, Martin, R. A., 1979, Evol. Monogr., 2:30, and Repenning,
1979, Abstr. Pap. Geol. Soc. Am., to be Palearctic only by excluding Nearctic
Microtus richardsoni. However, Hooper and Hart, 1962, Misc. Publ. Mus. Zool.
Univ. Mich., 120, 68 pp., Jannett, 1974, Am. Midl. Nat., 92:230–234, Corbet and
Hill, 1980:161, and others included *richardsoni* as well (RSH). Placed in tribe
Microtini (= Arvicolini) by Gromov and Polyakov, 1977:203.
ISIS NUMBER: 5301410008069000000.

Arvicola sapidus Miller, 1908. Ann. Mag. Nat. Hist., ser. 8, 1:195.
TYPE LOCALITY: Spain, Burgos Prov., Santo Domingo de Silos.
DISTRIBUTION: Iberian Peninsula; W. France.
COMMENT: Formerly included in *terrestris*, but separated by Matthey, 1956, Mammalia,
20:93–123, and Reichstein, 1963, Z. Zool. Syst. Evolutionforsch., 1:155–204.

Arvicola terrestris (Linnaeus, 1758). Syst. Nat., 10th ed., 1:61.
TYPE LOCALITY: Sweden, Uppsala.
DISTRIBUTION: Europe (except Iberia, W. France, and S. Italy), south to Israel and Iran,
north to Arctic Ocean, east through most of Siberia, south to Lake Baikal and N.
Tien Shan Mtns. (U.S.S.R. and N.W. China).
COMMENT: Formerly included *sapidus*; see comment under *sapidus*. The populations of
small, fossorial forms of central Europe are sometimes separated as *scherman*; see
Meylan, 1977, EPPO Bull., 7:209–221.
ISIS NUMBER: 5301410008069002001.

Clethrionomys Tilesius, 1850. Isis, 2:28.
REVIEWED BY: R. P. Canham (RPC); M. L. Johnson (MLJ); S. R. Leffler (SRL).
COMMENT: Includes *Caryomys*, *Craseomys*, *Evotomys*, and *Neoaschizomys*; see Corbet,
1978:97, and Ellerman and Morrison-Scott, 1951:659. Corbet, 1978:97–98, 100,
included *Eothenomys andersoni* in this genus; but see Imaizumi, 1960, Coloured
Illustrations of the Mammals of Japan, 196 pp., and Aimi, 1980, Mem. Fac. Sci.
Kyoto Univ. Ser. Biol., 8:35–84.
ISIS NUMBER: 5301410008072000000.

Clethrionomys californicus (Merriam, 1890). N. Am. Fauna, 4:26.
TYPE LOCALITY: U.S.A., California, Humboldt Co., Eureka.
DISTRIBUTION: Pacific coast coniferous forest from the Columbia River south through
W. Oregon and N. California (U.S.A.).
COMMENT: The name *occidentalis* was formerly applied, but populations north of the
Columbia River, including *occidentalis* and *caurinus* are now assigned to *gapperi*;
see Johnson and Ostenson, 1959, J. Mammal., 40:574–577, and Cowan and Guiget,
1965, The Mammals of British Columbia, Brit. Columbia Prov. Mus. Handbook,
11:1–413.

Clethrionomys centralis Miller, 1906. Ann. Mag. Nat. Hist., ser. 7, 17:373.
TYPE LOCALITY: U.S.S.R., Kazakh. S.S.R., Taldy-Kurgansk. Obl., Koksu valley,
Dzhungarsk. Alatau Mtns.
DISTRIBUTION: Restricted to Tien Shan and Dzhungarsk. Alatau Mtns. (U.S.S.R. and
Sinkiang, China).

COMMENT: Earlier literature refers to these voles as *frater* (Thomas, 1908). Corbet, 1978:99, determined that *frater* is a junior synonym of *centralis*, which had been incorrectly referred to *rutilus* by Ellerman and Morrison-Scott, 1951:661. Corbet, 1978:99, placed *centralis* as a synonym of *glareolus*, but Orlov, 1968, [Probl. Evolution], 1:184–194, Orlov and Kryukova, 1975, [Systematics and cytogenetics of mammals], Moscow, p. 26, and Gromov and Polyakov, 1977, [Voles (Microtinae), Fauna U.S.S.R.], Nauka, Moscow and Leningrad, 3(8):1–502, provided evidence of its specific distinctness. Still referred to in Russian literature as *frater* (RSH, OLR).

Clethrionomys gapperi (Vigors, 1830). Zool. J., 5:204.
TYPE LOCALITY: Canada, Ontario, between York (=Toronto) and Lake Simcoe.
DISTRIBUTION: From Labrador to N. British Columbia (Canada), south in the Appalachians to N. Georgia; Great Lakes, Northern Plains, and Rocky Mtns. (New Mexico and Arizona) to Columbia River in Washington (U.S.A.).
COMMENT: Includes *occidentalis* and *caurinus*; see Johnson and Ostenson, 1959, J. Mammal., 40:571–577, Cowan and Guiguet, 1965, The Mammals of British Columbia, Br. Columbia Prov. Mus. Handbook, 11:1–413, and Hall, 1981:780, 782–783. Closely related to Eurasian *glareolus*; captive mating produces fertile hybrids, but of reduced fertility; see Grant, 1974, J. Zool. Lond., 174:245–254. Youngman, 1975, Mammals of the Yukon Terr., Nat. Mus. Can. Publ. Zool., 10:85, considered *gapperi* and *rutilus* conspecific; but see Nadler *et al.*, 1978, Can. J. Zool., 56(7):1564–1575, and Hall, 1981:778–784; also see comment under *rutilus*. Reviewed by Merritt, 1981, Mamm. Species, 146:1–9.
ISIS NUMBER: 5301410008072001001 as *C. gapperi*.
ISIS NUMBER: 5301410008072003001 as *C. occidentalis*.

Clethrionomys glareolus (Schreber, 1780). Die Saugethiere, 4:680.
REVIEWED BY: M. Andera (MA).
TYPE LOCALITY: Denmark, Lollard Isl.
DISTRIBUTION: W. Eurasia, from Britain and Scandinavia south to Pyrennes, Italy, the Balkans, Transcaucasus, and east to C. Siberia (Altai and Sayan Mtns., U.S.S.R.).
COMMENT: See comments under *gapperi* and *centralis*.
ISIS NUMBER: 5301410008072002001.

Clethrionomys rufocanus (Sundevall, 1846). Ofv. Kongl. Svenska Vet.-Akad. Forhandl. Stockholm, 3:122.
REVIEWED BY: M. Andera (MA).
TYPE LOCALITY: Sweden, Lappmark.
DISTRIBUTION: N. Palearctic from Scandinavia to Chukotka and to Mongolia, Transbaikalia, N.E. China; Korea; Kamchatka, Sakhalin (U.S.S.R.), Hokkaido and Rishiri Isl. (Japan).
COMMENT: Includes *bedfordiae*; see Gromov and Polyakov, 1977:147, and Corbet, 1978:99, but also see Imaizumi, 1972, Mem. Nat. Sci. Mus. Tokyo, 5:131–139. Includes *montanus* and *rex*; see Aimi, 1980, Mem. Fac. Sci. Kyoto Univ. Ser. Biol., 8:35–84. Corbet, 1978:99, tentatively included *montanus* in *rex*, which he listed as a species; Imaizumi, 1972, Mem. Nat. Sci. Mus. Tokyo, 5:131–139, regarded both as distinct species; he believed them to be close to *rufocanus*, which also occurs on Hokkaido (as *C. r. bedfordiae*) and possibly Rishiri (as *sikotanensis*); also see Imaizumi, 1971, J. Mamm. Soc. Jpn., 5:99–103. Revision of this species is needed for the Japanese, Korean, and Chinese forms; more than one species may be present (MLJ).
ISIS NUMBER: 5301410008072004001.

Clethrionomys rutilus (Pallas, 1778). Nova Spec. Quad. Glir. Ord., p. 246.
REVIEWED BY: M. Andera (MA).
TYPE LOCALITY: U.S.S.R., R.S.F.S.R., center of Ob River delta.
DISTRIBUTION: N. Scandinavia east to Chukotka, south to N. Kazakhstan (U.S.S.R.), Mongolia, Transbaikalia, N.E. China, Sakhalin and Hokkaido; St. Lawrence Isl. (Bering Sea); Alaska east to Hudson Bay, south to N. British Columbia and Manitoba (Canada).

COMMENT: Revised by Hinton, 1926, Monograph of the voles and lemmings (Microtinae), Br. Mus. Nat. Hist., pp. 1–488, Manning, 1956, Bull. Nat. Mus. Can., 144:1–67, and Orr, 1945, J. Mammal., 26:67–74. Youngman, 1975, Mammals of the Yukon Terr., Nat. Mus. Can. Publ. Zool., 10:85, considered *rutilus* and *gapperi* conspecific; but also see Nadler *et al.*, 1978, Can. J. Zool., 56(7):1564–1575, and Hall, 1981:778–784, and comments under *gapperi*. Rossolimo, 1971, Byull. Mosk. Ova. Ispyt. Prir. Otd. Biol., 76:63–68, considered *rjabovi* to be a color mutant of *rutilus*, but Corbet, 1978:98, listed it as a subspecies.

ISIS NUMBER: 5301410008072005001.

Clethrionomys sikotanensis (Tokuda, 1935). Mem. Coll. Sci. Kyoto, 106:241.

TYPE LOCALITY: U.S.S.R., R.S.F.S.R., Sakhalin Obl., Sikotan Isl.

DISTRIBUTION: Poronaisk Raion and S.W. Sakhalin; Sikotan and Kunashir (Kurile Isls.); possibly Rishiri Isl.

COMMENT: Allocated by Corbet, 1978:99, and Aimi, 1980, Mem. Fac. Sci. Kyoto Univ., Ser. Biol., 8:35–84, to *rufocanus*; Gromov and Polyakov, 1977, [Voles (Microtinae), Fauna U.S.S.R.], Nauka, Moscow and Leningrad, 3(8):1–502, and Imaizumi, 1972, Mem. Nat. Sci. Mus. Tokyo, 5:131–139, provided evidence of specific status. See Imaizumi, 1971, J. Mamm. Soc. Jpn., 5:99–103, Gromov and Polyakov, 1977:149, and Gromov and Baranova, 1981:179, for discussion of distribution.

Dicrostonyx Gloger, 1841. Hand. Hilfsb. Nat., 1:97.

REVIEWED BY: M. L. Johnson (MLJ).

COMMENT: This genus is more complex than previously believed; see Rausch and Rausch, 1972, Z. Saugetierk., 37:372–384; Rausch, 1977, *in* Sokolov, ed., [Adv. Mod. Theriol.], Acad. Sci. U.S.S.R., Nauka, Moscow, pp. 162–177; compare with Corbet, 1978:96, and Hall, 1981:835–837. Many previously nominate subspecies are here considered full species. According to experimental breeding and karyotypic evidence the *torquatus* group represents a superspecies comprising at least nine allospecies (MLJ). The geographically isolated taxon *nunatakensis* was described as a subspecies of *torquatus* by Youngman, 1967, Proc. Biol. Soc. Wash., 80:31–34, and listed as a subspecies of *groenlandicus* by Hall, 1981:836. It is morphologically distinct from geographically adjacent *kilangmiutak* and *rubricatus*; see Youngman, 1975, Mammals of the Yukon Terr., Natl. Mus. Can. Publ. Zool., 10:116, and is here left unassigned; known only from the Ogilvie Mtns., C. Yukon (Canada) (RSH). Two subgenera *(Dicrostonyx* and *Misothermus)* have been recognized; *D. hudsonius* represents the latter. Placed in tribe Dicrostonyxini by Gromov and Polyakov, 1977:177.

ISIS NUMBER: 5301410008073000000.

Dicrostonyx exsul G. M. Allen, 1919. Bull. Mus. Comp. Zool. Harv. Univ., 62:532.

TYPE LOCALITY: U.S.A., Alaska, Bering Sea, St. Lawrence Isl.

DISTRIBUTION: Known only from the type locality.

COMMENT: Female diploid number is 34; fundamental number is 54. Formerly included in *torquatus*; see Rausch, 1977, *in* Sokolov, ed., [Adv. Mod. Theriol.], Acad. Sci. U.S.S.R., Nauka, Moscow, pp. 162–177.

Dicrostonyx groenlandicus Traill, 1823. *In* Scoresby, Jour. Voy. to Northern Whale-Fishery, p. 416.

TYPE LOCALITY: Greenland, Jameson's Land (Denmark).

DISTRIBUTION: N. Greenland, west to Baffin, Southampton, Axel Heiberg, Melville and Prince Patrick Isls. (Canada).

COMMENT: Includes *clarus* and, tentatively, *lentus*; see Youngman, 1975, Mammals of the Yukon Terr., Nat. Mus. Can. Publ. Zool., 10:115 (neither taxon has been kayotyped (RSH)). Limits of distribution uncertain; diploid number is 46 for male and female, and the fundamental number is 52. Formerly included in *torquatus*; see Rausch, 1977, *in* Sokolov, ed., [Adv. Mod. Theriol.], Acad. Sci. U.S.S.R., Nauka, Moscow, pp. 162–177. Hall, 1981:835–837, included all N. American taxa except *hudsonius* under this name. See also comments under genus *Dicrostonyx*.

Dicrostonyx hudsonius (Pallas, 1778). Nova Spec. Quad. Glir. Ord., p. 208.
TYPE LOCALITY: Canada, Labrador.
DISTRIBUTION: Labrador and N. Quebec (Canada).
COMMENT: A distinctive species placed in its own monotypic subgenus, *Misothermus;* see Guilday, 1968, Univ. Colo. Stud. Ser. Earth Sci., No. 6, and Rausch, 1977, *in* Sokolov, ed., [Adv. Mod. Theriol.], Acad. Sci. U.S.S.R., Nauka, Moscow, pp. 162–177.
ISIS NUMBER: 5301410008073001001.

Dicrostonyx kilangmiutak Anderson and Rand, 1945. J. Mammal., 26:305.
TYPE LOCALITY: Canada, W. side of Victoria Strait, S.E. point of Victoria Isl., DeHaven Point.
DISTRIBUTION: Victoria and Banks Isl. and the adjacent mainland (Canada).
COMMENT: Distribution poorly known. Diploid number is 47 for males and females and the fundamental number is 56; formerly included in *torquatus;* see Rausch, 1977, *in* Sokolov, ed., [Adv. Mod. Theriol.], Acad. Sci. U.S.S.R., Nauka, Moscow, pp. 162–177; or *groenlandicus;* see Hall, 1981:836.

Dicrostonyx nelsoni Merriam, 1900. Proc. Wash. Acad. Sci., 2:25.
TYPE LOCALITY: U.S.A., Alaska, Norton Sound, St. Michael.
DISTRIBUTION: W. coastal Alaska and Alaska Peninsula.
COMMENT: Tentatively includes *peninsulae;* see Youngman, 1975, Mammals of the Yukon Terr., Nat. Mus. Can. Publ. Zool., 10:115 (has not been karyotyped (RSH)). Range poorly known; diploid number is 30 for males and females and the fundamental number is 54; formerly included in *torquatus;* see Rausch, 1977, *in* Sokolov, ed., [Adv. Mod. Theriol.], Acad. Sci. U.S.S.R., Nauka, Moscow, pp. 162–177; or *groenlandicus;* see Hall, 1981:836.

Dicrostonyx richardsoni Merriam, 1900. Proc. Wash. Acad. Sci., 2:26.
TYPE LOCALITY: Canada, Manitoba, Fort Churchill.
DISTRIBUTION: Canada west of Hudson Bay.
COMMENT: The extent of westward distribution is unknown. The diploid number is 44 for males and 42 for females and the fundamental number is 50; formerly included in *torquatus;* see Rausch, 1977, *in* Sokolov, ed., [Adv. Mod. Theriol.], Acad. Sci. U.S.S.R., Nauka, Moscow, pp. 162–177; or *groenlandicus;* see Hall, 1981:836.

Dicrostonyx rubricatus (Richardson, 1889). The Zool. of Captain Beechey's Voy., p. 7.
TYPE LOCALITY: U.S.A., Alaska, Bering Strait.
DISTRIBUTION: N. Alaska (U.S.A.).
COMMENT: The diploid numbers are 33 and 42 for males, 32 and 34 for females; the fundamental number is 55; formerly included in *torquatus;* see Rausch, 1977, *in* Sokolov, ed., [Adv. Mod. Theriol.], Acad. Sci. U.S.S.R., Nauka, Moscow, pp. 162–177; or *groenlandicus;* see Hall, 1981:836.

Dicrostonyx torquatus (Pallas, 1778). Nova Spec. Quad. Glir. Ord., p. 206.
REVIEWED BY: M. Andera (MA).
TYPE LOCALITY: U.S.S.R., R.S.F.S.R., mouth of Ob River.
DISTRIBUTION: Palearctic tundra from White Sea to Chukotka (U.S.S.R.), Novaya Zemlya, and New Siberian Isls.
COMMENT: Formerly included *groenlandicus* and *vinogradovi;* see Corbet, 1978:96, but see Rausch, 1977, *in* Sokolov, ed., [Adv. Mod. Theriol.], Acad. Sci. U.S.S.R., Nauka, Moscow, pp. 162–177, who considered them distinct species. Includes *chionopaes* which may be a distinct species; see Gileva, 1975, Dokl. Akad. Nauk, 224:697–700, Gileva, 1980:99–103, *in* Vorontsov and Van Brink, eds., Animal genetics and evolution, Junk, The Hague, and Kozlovskii, 1974, Dokl. Akad. Nauk S.S.S.R., 219:981–984. See also comments under genus *Dicrostonyx.*
ISIS NUMBER: 5301410008073002001.

Dicrostonyx unalascensis Merriam, 1900. Proc. Wash. Acad. Sci., 2:25.
TYPE LOCALITY: U.S.A., Alaska, Umnak Island.
DISTRIBUTION: Umnak and Unalaska Isls. (U.S.A.).

COMMENT: Includes *stevensoni* Nelson, 1929; the diploid number is 34 for males and females, and the fundamental number is 54; see Rausch and Rausch, 1972, Z. Saugetierk., 37:372–384. Formerly included in *torquatus*; see Rausch, 1977, *in* Sokolov, ed., [Adv. Mod. Theriol.], Acad. Sci. U.S.S.R., Nauka, Moscow, pp. 162–177; or *groenlandicus*; see Hall, 1981:837.

Dicrostonyx vinogradovi Ognev, 1948. Mammals of Eastern Europe and Northern Asia, 6:509.
TYPE LOCALITY: U.S.S.R., Wrangel Isl.
DISTRIBUTION: Known only from the type locality.
COMMENT: Formerly included in *torquatus*; karyological and morphological analyses, and experimental breeding indicated species distinction; see Chernyavskii and Kozlovskii, 1980, Zool. Zh., 59(2):266–273. The diploid number is 28 and the fundamental number is 54. See also Rausch, 1977, *in* Sokolov, ed., [Adv. Mod. Theriol.], Acad. Sci. U.S.S.R., Nauka, Moscow, pp. 162–177.

Dinaromys Kretzoi, 1955. Acta Geol. Acad. Sci. Hung., 3:347–353.
REVIEWED BY: M. Andera (MA).
COMMENT: For use of this name rather than *Dolomys*, see Corbet, 1978:104. Closely related to *Hyperacrius*, *Alticola*, and *Eothenomys* (JC). Placed in tribe Pliomyini of subfamily Ondatrinae by CAR. Placed in tribe Clethrionomyini by Gromov and Polyakov, 1977:114. Koenigswald, 1980, Abh. Senckenb. Naturforsch. Ges., 539, believed that no close relationship with any other living arvicolid is probable on the basis of the internal structure of the molar enamel; there is a probable relationship with the fossil *Propliomys hungaricus*.

Dinaromys bogdanovi (Martino, 1922). Ann. Mag. Nat. Hist., ser. 9, 9:413.
TYPE LOCALITY: Yugoslavia, Rijeka Prov., Montenegro, Cetinje.
DISTRIBUTION: Mountains of Yugoslavia; perhaps N. Albania.
COMMENT: Previously referred to the extinct genus *Dolomys*, and sometimes to *D. milleri*; see Corbet, 1978:105. CAR referred *Dolomys* to tribe Ondatrinae, as did Kretzoi, 1955, Acta Geol. Acad. Sci. Hung., 3:347–353.
ISIS NUMBER: 5301410008074001001 as *Dolomys bogdanovi*.

Ellobius Fischer, 1814. Zoognosia, 3:72.
COMMENT: Placed with *Prometheomys* in subfamily Prometheomyinae by CAR. Gromov, 1972, Sbor. Tr. Zool. Mus. M.G.U., 13:8–32, excluded *Ellobius* from Arvicolidae as a microtodontine cricetid. OLR follows Corbet, 1978:117, in recognizing only two species *fuscocapillus* and *talpinus*; but see Lyapunova and Vorontsov, 1978, Genetika, 14:2012–2024, and Gromov and Baranova, 1981:151. See comment under *Reithrodon physodes* (Cricetidae), for discussion of Fischer, 1814, names.
ISIS NUMBER: 5301410008075000000.

Ellobius alaicus Vorontsov et al., 1969. *In* [The Mammals: Evolution, karyology, taxonomy, fauna], Sib. Otd., Akad. Nauk., Novosibirsk, p. 127.
TYPE LOCALITY: U.S.S.R., Kirgiz S.S.R., Oshsk. Obl., Alai Valley between Sary-Tash and Bardabo.
DISTRIBUTION: Alai Valley and Alai Mtns. (Kirgizia, U.S.S.R.).
COMMENT: Included in *talpinus* by Corbet, 1978:117, but Lyapunova and Vorontsov, 1978, Genetika, 14:2012–2024, provided evidence of specific distinctness, although they agreed it is closely related.

Ellobius fuscocapillus Blyth, 1843. J. Asiat. Soc. Bengal, 11:887.
TYPE LOCALITY: Pakistan, Baluchistan Reg., Quetta Div., Quetta.
DISTRIBUTION: Baluchistan and Afghanistan through Iran and S. Turkmenistan to Kurdistan.
COMMENT: Formerly included *lutescens*; see Corbet, 1978:117, but also see Lyapunova and Vorontsov, 1978, Genetika, 14:2012–2024, and Lyapunova *et al.*, 1980:239–247, *in* Vorontsov and Van Brink, eds., Animal genetics and evolution, Junk, The Hague, 383 pp.
ISIS NUMBER: 5301410008075001001.

Ellobius lutescens Thomas, 1897. Ann. Mag. Nat. Hist., ser. 6, 20:308.
TYPE LOCALITY: Turkey, Kurdistan, Van.
DISTRIBUTION: S. Caucasus Mtns. (U.S.S.R.), south through E. Turkey and W. Iran;
perhaps Iraq.
COMMENT: Corbet, 1978:117, and others considered *lutescens* a synonym of *fuscocapillus;*
but Lyapunova and Vorontsov, 1978, Genetika, 14:2012–2024, and Lyapunova *et
al.*, 1980:293–247, *in* Vorontsov and Van Brink, eds., Animal genetics and
evolution, Junk, The Hague, 383 pp., provided evidence of specific distinctness.
ISIS NUMBER: 5301410008075002001.

Ellobius talpinus (Pallas, 1770). Nova Comm. Acad. Sci. Petrop., 14(2):568.
REVIEWED BY: M. Andera (MA).
TYPE LOCALITY: U.S.S.R., R.S.F.S.R., Kuibyshevsk. Obl., W. Bank of Volga River,
between Kuibyshev (=Samara) and Kostychi.
DISTRIBUTION: Steppes from S. Ukraine and Crimea, east through Kazakhstan
(U.S.S.R.) to Sinkiang and Inner Mongolia (China) and Mongolia, north to
Sverdlovsk and south to N. Afghanistan, N.W. Iran and N. Pakistan.
COMMENT: Includes *tancrei,* which was considered a distinct species by Gromov and
Baranova, 1981:152. Karyology reviewed by Lyapunova *et al.,* 1980:239–247, *in*
Vorontsov and Van Brink, eds., Animal genetics and evolution, Junk, The Hague.
ISIS NUMBER: 5301410008075003001.

Eolagurus Argyropulo, 1946. Vestn. Akad. Nauk Kazakh. S.S.R., 7–8:44.
REVIEWED BY: M. Andera (MA).
COMMENT: Included in *Lagurus* by Corbet, 1978:116, but considered a distinct genus by
Gromov and Polyakov, 1977, [Voles (Microtinae), Fauna U.S.S.R.], Nauka, Moscow
and Leningrad, 3(8):160, 169, who placed *Lagurus* and *Eolagurus* in the tribe
Lagurini.

Eolagurus luteus (Eversmann, 1840). Bull. Soc. Nat. Moscow, p. 25.
TYPE LOCALITY: U.S.S.R., Kazakh. S.S.R., Aktyubinsk. Obl., N.W. of Aral Sea.
DISTRIBUTION: Kazakhstan to N. Sinkiang (China) and W. Mongolia.
COMMENT: Now extinct in Kazakhstan; see Corbet, 1978:117. Range changes in recent
times reviewed by Kalabukhov, 1970, Ekologiya, 1:69–76. Formerly included in
Lagurus; Gromov and Polyakov, 1977, [Voles (Microtinae), Fauna U.S.S.R.], Nauka,
Moscow and Leningrad, 3(8):169, elevated *Eolagurus* to full generic rank and
regarded *przewalskii* as a distinct species, *op. cit.,* p. 174; they were followed by
Sokolov and Orlov, 1980:139, and Gromov and Baranova, 1981:184.
ISIS NUMBER: 5301410008078003001 as *Lagurus luteus.*

Eolagurus przewalskii (Buchner, 1888). Wiss. Res. Przewalski Cent. Asien Zool. Th.
I: Saugeth., p. 127.
TYPE LOCALITY: China, Tsinghai (= N. Tsaidam), shore of Iche-zaidemin Nor.
DISTRIBUTION: N. Tibet and Sinkiang (China) to S. and W. Mongolia.
COMMENT: Formerly included in *luteus;* see Gromov and Polyakov, 1977:174, who
were followed by Sokolov and Orlov, 1980:139, and Gromov and Baranova,
1981:184; Corbet, 1978:117, included *przewalskii* in *luteus* without comment.

Eothenomys Miller, 1896. N. Am. Fauna, 12:45.
REVIEWED BY: M. L. Johnson (MLJ).
COMMENT: Includes *Anteliomys* (see Ellerman and Morrison-Scott, 1951:667) and
Phaulomys (see Corbet, 1978:100). Gromov and Polyakov, 1977:137, considered
Anteliomys a distinct genus. Revised by Corbet, 1978, and Aimi, 1980, Mem. Fac.
Sci. Kyoto Univ. Ser. Biol., 8:35–84, who included *Aschizomys* in this genus, and
were followed here (RSH); closely related to *Hyperacrius, Alticola,* and *Dinaromys*
(JC). Placed in tribe Clethrionomyini by Gromov and Polyakov, 1977:158.
ISIS NUMBER: 5301410008076000000 as *Eothenomys.*
5301410008070000000 as *Aschizomys.*

Eothenomys andersoni (Thomas, 1905). Abstr. Proc. Zool. Soc. Lond., 23:18.
TYPE LOCALITY: Japan, N. Honshu, Iwate Ken, Tsunagi, near Morioka.
DISTRIBUTION: Honshu (Japan).

COMMENT: Formerly included in *Aschizomys;* includes *niigatae,* from C. Honshu, and
imaizumii from S. Honshu; see Corbet, 1978:100, and Aimi, 1980, Mem. Fac. Sci.
Kyoto Univ. Ser. Biol., 8:35–84; both were considered distinct species by Ota and
Jameson, 1961, Pacif. Sci., 15:594–604, and Imaizumi, 1960, Coloured Illustrations
of the Mammals of Japan, 196 pp. Ota and Jameson, 1961, Ecology, 42:184–186,
Ota and Jameson, 1961, Pacif. Sci., 15:594–604, Gromov and Polyakov, 1977:147,
and Corbet, 1978:100, placed *andersoni* in *Clethrionomys.*

Eothenomys chinensis (Thomas, 1891). Ann. Mag. Nat. Hist., ser. 6, 8:117.
TYPE LOCALITY: China, Szechwan, Kiatingfu.
DISTRIBUTION: Szechwan and Yunnan (China).
COMMENT: Placed in genus *Anteliomys* by Gromov and Polyakov, 1977:138; but see
Corbet, 1978:101.
ISIS NUMBER: 5301410008076001001.

Eothenomys custos (Thomas, 1912). Ann. Mag. Nat. Hist., ser. 8, 9:517.
TYPE LOCALITY: China, Yunnan, Atunsi.
DISTRIBUTION: Mountains of Szechwan and Yunnan (China).
COMMENT: Placed in genus *Anteliomys* by Gromov and Polyakov, 1977:138, but see
Corbet, 1978:101.
ISIS NUMBER: 5301410008076002001.

Eothenomys eva (Thomas, 1911). Abstr. Proc. Zool. Soc. Lond., 90:4.
TYPE LOCALITY: China, Kansu, S.E. of Taochow.
DISTRIBUTION: Mountains of S. Kansu and adjacent Szechwan, Shensi, and Hupei
(China).
COMMENT: Regarded by Ellerman and Morrison-Scott, 1951:660, and Gromov and
Polyakov, 1977:147, as a synonym of *Clethrionomys rufocanus;* but see Corbet,
1978:102.

Eothenomys inez (Thomas, 1908). Abstr. Proc. Zool. Soc. Lond., 63:45.
TYPE LOCALITY: China, Shansi, 12 mi. (19 km) N.W. Kolanchow.
DISTRIBUTION: Shansi, Shensi and perhaps Hopei (China).
COMMENT: Regarded by Ellerman and Morrison-Scott, 1951:660, and Gromov and
Polyakov, 1977:147, as a synonym of *Clethrionomys rufocanus;* but see Corbet,
1978:102.

Eothenomys lemminus Miller, 1899. Proc. Acad. Nat. Sci. Phila., (1898), p. 369.
TYPE LOCALITY: U.S.S.R., R.S.F.S.R., Khabarovsk. Krai, Chukotsk. Nat. Okr., Zaliv Sv.
Kresta, S.W. of Egvekinom, "Kelsey Station, Plover Bay, Bering Strait".
DISTRIBUTION: Chukotka Peninsula (N.E. Siberia), west to Lena River and south to
Amur River (U.S.S.R.).
COMMENT: Formerly included in *Aschizomys;* see comments under *Eothenomys.* Most
Russian authors have allocated this species to the genus *Alticola,* usually as a
subspecies of *A. macrotis;* see Yudin *et al.,* 1976, [Small mammals of the northern
Far East,] Nauka Sibir. Otd. Novosibirsk, 270 pp., Gromov and Polyakov,
1977:133, and Gromov and Baranova, 1981:177. Corbet, 1978:100, provisionally
placed it in *Eothenomys,* calling it an "enigmatic species"; Aimi, 1980, Mem. Fac.
Sci. Kyoto Univ. Ser. Biol., 8:35–84, provided additional evidence for this
allocation, which is followed here. Bykova *et al.,* 1978, Experientia, 34:1146–1148,
suggested that there may be more than one Siberian species.
ISIS NUMBER: 5301410008070001001 as *Aschizomys lemminus.*

Eothenomys melanogaster (Milne-Edwards, 1871). Nouv. Arch. Mus. Hist. Nat. Paris,
7(Bull.):93.
TYPE LOCALITY: China, Szechwan, Moupin.
DISTRIBUTION: S. China to S. Kansu and Ningsiahui; N. Thailand, and N. Burma;
Taiwan.
COMMENT: Includes *fidelis, miletus,* and *eleusis;* see Corbet, 1978:101.
ISIS NUMBER: 5301410008076004001 as *E. melanogaster.*
          5301410008076003001 as *E. fidelis.*

Eothenomys olitor (Thomas, 1911). Abstr. Proc. Zool. Soc. Lond., 100:50.
TYPE LOCALITY: China, Yunnan, Chaotungfu.
DISTRIBUTION: Yunnan (China).
COMMENT: Placed in genus *Anteliomys* by Gromov and Polyakov, 1977:138; but see Corbet, 1978:101.
ISIS NUMBER: 5301410008076005001.

Eothenomys proditor Hinton, 1923. Ann. Mag. Nat. Hist., ser. 9, 11:152.
TYPE LOCALITY: China, Yunnan, Likiang Range (27° 30' N.), 13,000 ft. (3962 m).
DISTRIBUTION: Mountains of Yunnan and Szechwan (China).
COMMENT: Placed in genus *Anteliomys* by Gromov and Polyakov, 1977:138; but see Corbet, 1978:101.
ISIS NUMBER: 5301410008076006001.

Eothenomys regulus (Thomas, 1907). Proc. Zool. Soc. Lond., 1906:863.
TYPE LOCALITY: Korea, Mingyong, 110 mi. (177 km) S.E. Seoul.
DISTRIBUTION: Hopei (China); Korea.
COMMENT: Regarded by Allen, 1940, The Mammals of China and Mongolia, Am. Mus. Nat. Hist., 2(2):621–1350, Ellerman and Morrison-Scott, 1951:666, and Gromov and Polyakov, 1977:147, as a synonym of *Clethrionomys rufocanus*; but see Corbet, 1978:102.

Eothenomys shanseius (Thomas, 1908). Proc. Zool. Soc. Lond., 1908:643.
TYPE LOCALITY: China, Shansi, 100 mi. (161 km) N.W. Taiyuenfu.
DISTRIBUTION: Shansi and perhaps Hopei (China).
COMMENT: Regarded by Allen, 1940, The Mammals of China and Mongolia, Am. Mus. Nat. Hist., 2(2):621–1350, Ellerman and Morrison-Scott, 1951:666, and Gromov and Polyakov, 1977:147, as a synonym of *Clethrionomys rufocanus*; but see Corbet, 1978:102.

Eothenomys smithi (Thomas, 1905). Ann. Mag. Nat. Hist., ser. 7, 15:493.
TYPE LOCALITY: Japan, Honshu, Kobe.
DISTRIBUTION: Dogo, Honshu, Shikoku, and Kyushu (Japan).
COMMENT: Includes *kageus*; see Ota and Jameson, 1961, Ecology, 42:184–186, and Tanaka, 1971, Jpn. J. Zool., 16:163–176. MLJ by pers. comm. with Y. Imaizumi considers *kageus* a separate species. Placed by Tanaka, 1971, Jpn. J. Zool., 16:163–176, in the monotypic genus *Phaulomys* Thomas 1905; but see Corbet, 1978:102, and Aimi, 1980, Mem. Fac. Sci. Kyoto Univ. Ser. Biol., 8:35–84. Placed by Gromov and Polyakov, 1977:134, in genus *Alticola*, subgenus *Aschizomys*.

Hyperacrius Miller, 1896. N. Am. Fauna, 12:54.
COMMENT: Reviewed by Phillips, 1969, J. Mammal., 50:457–474. Placed in tribe Clethrionomyini by Gromov and Polyakov, 1977:136. Closely related to *Alticola*, *Eothenomys* and *Dinaromys* (JC). Closely related to *Prometheomys* (WVK). Related either to *Prometheomys* or to *Lagurus* (CAR).
ISIS NUMBER: 5301410008077000000.

Hyperacrius fertilis (True, 1894). Proc. U.S. Nat. Mus., 17:10.
TYPE LOCALITY: India, Kashmir, Pir Panjal Mountains.
DISTRIBUTION: Kashmir; N. Pakistan.
ISIS NUMBER: 5301410008077001001.

Hyperacrius wynnei (Blanford, 1881). J. Asiat. Soc. Bengal, 49:244.
TYPE LOCALITY: Pakistan, Punjab, Murree (probably N. of the City).
DISTRIBUTION: Kagan and Swat (N. Pakistan).
COMMENT: Type locality discussed by Phillips, 1969, J. Mammal., 50:462.
ISIS NUMBER: 5301410008077002001.

Lagurus Gloger, 1841. Hand. Hilfsb. Nat., 1:97.
   REVIEWED BY: M. Andera (MA).
   COMMENT: Includes *Lemmiscus*; see Hall, 1981:821. Formerly included *Eolagurus*; see
      Gromov and Polyakov, 1977, [Voles (Microtinae), Fauna U.S.S.R.], Nauka, Moscow
      and Leningrad, 3(8):160, who placed *Lagurus* and *Eolagurus* in the tribe Lagurini.
      Placed in tribe Lagurini with *Alticola* and *Hyperacrius* by CAR. See also Corbet,
      1978:116, who included *Eolagurus* in *Lagurus* without comment.
   ISIS NUMBER: 5301410008078000000.

Lagurus curtatus (Cope, 1868). Abstr. Proc. Acad. Nat. Sci. Philad., p. 2.
   REVIEWED BY: M. D. Carleton (MDC).
   TYPE LOCALITY: U.S.A., Nevada, Esmeralda Co., Pigeon Spring, Mt. Magruder.
   DISTRIBUTION: N.W. and E.C. California to C. Washington, W. North Dakota and N.
      Colorado (U.S.A.); S.E. Alberta and S.W. Saskatchewan (Canada); in sagebrush
      steppe and desert.
   COMMENT: Thomas, 1912, Ann. Mag. Nat. Hist., 9:401, erected the subgenus *Lemmiscus*;
      also see Davis, 1939, The Recent Mammals of Idaho, Caxton Printers: Caldwell,
      Idaho, 400 pp.; Gromov and Polyakov, 1977:176, and MDC considered *Lemmiscus* a
      distinct genus; but see Hall, 1981:821. Reviewed by Carroll and Genoways, 1980,
      Mamm. Species, 124:1–6.
   ISIS NUMBER: 5301410008078001001.

Lagurus lagurus (Pallas, 1773). Reise Prov. Russ. Reichs., 2:704.
   TYPE LOCALITY: U.S.S.R., Kazakh. S.S.R., Gur'evsk. Obl., mouth of Ural River.
   DISTRIBUTION: Steppes from Ukraine to N. Kazakhstan to W. Mongolia and W.
      Sinkiang (China).
   ISIS NUMBER: 5301410008078002001.

Lemmus Link, 1795. Beitr. Naturgesch., 1(2):75.
   REVIEWED BY: K. Curry-Lindahl (KCL); M. L. Johnson (MLJ); G. A. Sidorowicz (GAS).
   COMMENT: Revised by Rausch and Rausch, 1975, Z. Saugetierk., 40:8–34. Includes
      Myopus; see Chaline and Mein, 1979, and Koenigswald and Martin, in press, Bull.
      Carnegie Mus. Nat. Hist. Placed in tribe Lemmini by Gromov and Polyakov,
      1977:189.
   ISIS NUMBER: 5301410008079000000 as *Lemmus*.
                5301410008081000000 as *Myopus*.

Lemmus lemmus (Linnaeus, 1758). Syst. Nat., 10th ed., 1:59.
   REVIEWED BY: M. Andera (MA).
   TYPE LOCALITY: Sweden, Lappmark.
   DISTRIBUTION: Mountains of Scandinavia and tundra from Lapland to White Sea
      (U.S.S.R).
   COMMENT: Hybridization with *sibiricus* produced sterile male offspring; see Rausch,
      1977, *in* Sokolov, ed., [Adv. Mod. Theriol.], Acad. Sci. U.S.S.R., "Nauka," Moscow,
      pp. 162–167. Also see Krivosheev and Rossolimo, 1966, Byull. Mosk. Ova. Ispyt.
      Prir. Otd. Biol., 71:5–17, Rausch and Rausch, 1975, Z. Saugetierk., 40:8–34, and
      Curry-Lindahl, 1980, Der Berglemming *Lemmus lemmus*, 140 pp.; but see
      Sidorowicz, 1960, 1964, Acta Theriol., 4:53–80, 8:217–226. Type locality,
      Lappmark, *(sensu* Linnaeus) means mountains of Swedish Lapland (KCL).
   ISIS NUMBER: 5301410008079001001.

Lemmus novosibiricus Vinogradov, 1924. Nat. Hist., ser. 9, 14:187.
   TYPE LOCALITY: U.S.S.R., New Siberian Arch., Kotelnyi Isl.
   DISTRIBUTION: New Siberian and Lyakhov Isls. (U.S.S.R.).
   COMMENT: Formerly included in *obensis*; see Curry-Lindahl, 1980, Der Berglemming
      Lemmus lemmus, 140 pp.; but also see Sidorowicz, 1964, Acta Theriol., 8:223.

Lemmus schisticolor (Lilljeborg, 1844). Ofv. Kongl. Svenska Vet.-Akad. Forhandl.
      Stockholm, 1:33.
   REVIEWED BY: M. Andera (MA).
   TYPE LOCALITY: Norway, Gulbrandsdal, Mjosen, near Lillehammer.
   DISTRIBUTION: Coniferous forest from Norway and Sweden through Siberia to Kolyma
      River and Kamchatka (U.S.S.R.), south to the Altai, N. Mongolia, Heilungkiang
      (China), and the Sikhote Alin Range; perhaps Sakhalin Isl. (U.S.S.R.).

COMMENT: Transferred to *Lemmus* from *Myopus* by Chaline and Mein, 1979. Koenigswald and Martin, in press, Bull. Carnegie Mus. Nat. Hist., also place this species in *Lemmus* on the basis of the pattern of molar enamel.

ISIS NUMBER: 5301410008081001001 as *Myopus schisticolor.*

Lemmus sibiricus (Kerr, 1792). Anim. Kingdom, p. 241.

TYPE LOCALITY: U.S.S.R., R.S.F.S.R., Yamalo-Nenetsk. Nats. Okr., between Polar Ural Mtns. and lower course of Ob River.

DISTRIBUTION: Siberian tundra from White Sea to Chukotka (N.E. Siberia) and Kamchatka; St. George Isl. (Pribilofs) and Nunivak Isl. (Bering Sea); W. Alaska east to Baffin Isl. and Hudson Bay, south in the Rocky Mtns. to C. British Columbia (Canada).

COMMENT: *L. obensis* is a junior synonym (*L. obensis* is retained as a specific name for *sibiricus* by Gromov *et al.*, 1963, [Mammal Fauna of the U.S.S.R.], 1:1–640, and Flint *et al.*, 1965, [Mammals of the U.S.S.R.], 438 pp.); includes *amurensis, nigripes*, and *trimucronatus*; see Rausch, 1953, Arctic, 6:91–148, Krivosheev, 1971, *in* Tavrovskii *et al.*, [Mammals of Yakutiya], Rausch and Rausch, 1975, Z. Saugetierk., 40:8–34, and Curry-Lindahl, 1980, Der Berglemming *Lemmus lemmus*, 140 pp.; for alternative treatment see Corbet, 1978:97, and Sidorowicz, 1964, Acta Theriol., 8:217–226. *L. amurensis* was retained as a distinct species by Ognev, 1948, [Mammals of the U.S.S.R. and Adj. Count.], vol. 6, Rodents, English Trans., 1963, Gromov *et al.*, 1963, [Mammal Fauna of the U.S.S.R.], 1:1–640, Flint *et al.*, 1965, [Mammals of the U.S.S.R.], 438 pp., and Gromov and Polyakov, 1977:196; while Krivosheev and Rossolimo, 1966, Byull. Mosk. Ova. Ispyt. Prir. Otd. Biol., 71:5–17, and Rubina *et al.*, 1973, *in* Kontrimavichus, ed., [Biol. Probl. North], 2:77–80, expressed uncertainty about its taxonomic status. Khvorostyanskaya, 1980:40, *in* Panteleev, ed., [Rodents. Materials All-Union Conf.], Nauka, Moscow, 471 pp., Chernyavskii, 1980, *op cit.* p. 131, and Pokrovskii and Makaranets, 1980, *op cit.* p. 259, provided chromosomal and breeding data supporting specific distinctness. RSH, MLJ and OLR treat *amurensis* as distinct.

ISIS NUMBER: 5301410008079003001 as *L. sibiricus.*
5301410008079002001 as *L. nigripes.*

Microtus Schrank, 1798. Fauna Boica, 1(1):72.

REVIEWED BY: M. Andera (MA)(Eurasia); S. R. Leffler (SRL).

COMMENT: Includes, as subgenera or as synonyms, *Agricola, Alexandromys, Arvalomys, Aulacomys, Campicola, Chilotus, Chionomys, Euarvicola, Iberomys, Lasiopodomys, Lemmimicrotus, Mynomes, Orthriomys, Stenocranius, Sumeriomys, Suranomys,* and *Sylvicola;* see Chaline, 1974:448, Corbet, 1978:110, Ellerman and Morrison-Scott, 1951, Walker *et al.*, 1975, 2:845, Van den Brink, 1957, Die Saugetiere Europas, and Anderson, 1959, Univ. Kans. Publ. Mus. Nat. Hist., 9:415–511. There is no consensus concerning limits or validity of many of these taxa. May include *Proedromys* (here considered a distinct genus following Wang *et al.*, 1966, Acta Zootax. Sin., 3:85–91; see Chaline, 1974:448, and Corbet, 1978:110), *Pedomys* (here included in *Pitymys;* see Van der Meulen, 1978, Ann. Carnegie Mus., 47:101–145, and comments under *Pitymys ochrogaster),* and *Herpetomys* (here included in *Pitymys* following Martin, R. A., 1974, *in* Webb, ed., Pleistocene Mammals of Florida, p. 60, and Repenning, pers. comm. to RSH; see Anderson, 1960). Opinion also differs as to whether *Arvicola* and *Pitymys* should be retained as distinct genera as is done here (following Chaline and Mein, 1979) or considered subgenera of *Microtus*, as in Hall, 1981 (SRL, WDS). Gromov and Polyakov, 1977:322, 327, considered *Chionomys* and *Lasiopodomys* as distinct genera; but also see Hall, 1981:790.

ISIS NUMBER: 5301410008080000000.

Microtus abbreviatus Miller, 1899. Proc. Biol. Soc. Wash., 13:13.

REVIEWED BY: M. L. Johnson (MLJ).

TYPE LOCALITY: U.S.A., Alaska, Bering Sea, Hall Isl.

DISTRIBUTION: Hall and St. Matthew Isls. (Alaska, U.S.A.).

COMMENT: Subgenus *Stenocranius;* see Rausch and Rausch, 1968, Z. Saugetierk., 36:65–99. Related to *miurus* of the Alaskan mainland; see Fedyk, 1970, Acta

Theriol., 15:143–152, Nelson, 1931, J. Mammal., 12:311, and Rausch, 1977, *in*
Sokolov, ed., [Adv. Mod. Theriol.], Acad. Sci. U.S.S.R., Nauka, Moscow, p. 169.
MLJ considers *miurus* conspecific with *abbreviatus*.

Microtus agrestis (Linnaeus, 1761). Fauna Suec., 11(2):11.
   TYPE LOCALITY: Sweden, Uppsala.
   DISTRIBUTION: Britain, Scandinavia, and France east through Europe and Siberia to
      Lena River; south to Pyrennes, N. Portugal, N. Yugoslavia, Altai Mtns., Sinkiang
      (China) (SW), and Lake Baikal.
   COMMENT: Subgenus *Microtus*; see Anderson, 1959, Univ. Kans. Publ. Mus. Nat. Hist.,
      9:496 *(Arvalomys* of Chaline, 1974:440–450). Sometimes regarded as conspecific
      with *pennsylvanicus*, as by Klimkiewicz, 1970, Mammalia, 34:641–665, but not
      closely related; see Frank, 1959, Z. Saugetierk., 24:91–93, and Vorontsov and
      Lyapunova, 1976, *in* Kontrimavichus, ed., [Beringia in Cenozoic], Acad. Sci.
      Vladivostok.
   ISIS NUMBER: 5301410008080001001.

Microtus arvalis (Pallas, 1778). Nova Spec. Quad. Glir. Ord., p. 78.
   TYPE LOCALITY: Germany, no exact locality; neotype from U.S.S.R., R.S.F.S.R.,
      Leningrad Obl.
   DISTRIBUTION: France and N. Spain east through Europe and Siberia to the upper
      Yenesei River; south to N. Turkey, Caucasus, Lake Balkash, Sinkiang (China)
      (SW), and Altai Mtns.; isolates in Guernsey and Orkney Isls.
   COMMENT: Subgenus *Microtus*; see Anderson, 1960, Univ. Kans. Publ. Mus. Nat. Hist.,
      12:181–216. Includes *ilaeus*; see Meier and Yatsenko, 1980, Zool. Zh.,
      59(2):283–288, who used *kirgisorum* for what was formerly known as *ilaeus*; "*ilaeus*"
      (now *kirgisorum*) was considered distinct by Corbet, 1978:114, on the basis of
      evidence of hybrid sterility, also see Meier, 1976, Trans. I Intl. Theriol. Congr.,
      1:400–401; see also comment under *kirgisorum*. Includes *orcadensis*; see Corbet,
      1961, Nature (Lond.), 191:1037–1040. Includes *igmanensis*; see Zivkovic *et al.*, 1975,
      Arh. Biol. Nauka, 26:123–134, and Corbet, 1978:113. The status of *igmanensis* needs
      further clarification (SRL); see also comment under *epiroticus*. Type locality of
      neotype fixed by Meier *et al.*, 1972, Zool. Zh., 51:157–161.
   ISIS NUMBER: 5301410008080002001 as *M. arvalis*.
                5301410008080033001 as *M. orcadensis*.

Microtus brandti (Radde, 1861). Melanges Biol. Acad. St. Petersb., 3:683.
   TYPE LOCALITY: U.S.S.R., R.S.F.S.R., Chitinsk. Obl., near Tarei-Nor (Lake).
   DISTRIBUTION: Mongolia, Transbaikalia (U.S.S.R.), Inner Mongolia, Heilungkiang,
      Kirin, and Hopei (China) (SW).
   COMMENT: Subgenus *Lasiopodomys*; see Chaline, 1974, and Chaline and Mein, 1979.
      Regarded as full genus by Erbaeva, 1976, Trudy Zool. Inst., Leningr., 66:107–116,
      Gromov and Polyakov, 1977:322, Repenning, 1980, Can. J. Anthro., 1:37–44, and
      Gromov and Baranova, 1981:214; also see Ognev, 1950, [Mamm. U.S.S.R., Adjac.
      Count.], Acad. Sci. Moscow, 7:292–293.
   ISIS NUMBER: 5301410008080004001.

Microtus breweri (Baird, 1858). Mammals, *in* Repts. Expl. Surv...., 8(1):525.
   REVIEWED BY: R. H. Tamarin (RHT).
   TYPE LOCALITY: U.S.A., Massachusetts, Muskeget Isl.
   DISTRIBUTION: Muskeget Isl. (Massachusetts, U.S.A.).
   COMMENT: Subgenus *Microtus*; related to *pennsylvanicus* of the adjacent mainland;
      treated as a full species in a review by Tamarin and Kunz, 1974, Mamm. Species,
      45:1–3, but may be conspecific; see Fivush *et al.*, 1975, J. Mammal., 56:272–273.

Microtus cabrerae Thomas, 1906. Ann. Mag. Nat. Hist., ser. 7, 17:576.
   REVIEWED BY: K. F. Koopman (KFK).
   TYPE LOCALITY: Spain, Madrid Prov., Sierra de Guadarrama, near Rascafria.
   DISTRIBUTION: Spain; Portugal (KFK).
   COMMENT: Subgenus *Iberomys*; regarded as a synonym of *guentheri*, by Van den Brink,
      1957, Die Saugetiere Europas, and as a synonym of *socialis* by Van den Brink,
      1972, Die Saugetiere Europas, 2nd ed., but not by Niethammer *et al.*, 1964, Bonn.

Zool. Beitr., 15:127–148, and Corbet, 1978:114, who also included *dentatus* in this species, as is done here. Chaline, 1974, placed *dentatus* in *Iberomys*, but *cabrarae* in *Arvalomys*. Placed "conf. *Microtus*" by Gromov and Polyakov, 1977:298.
ISIS NUMBER: 5301410008080005001.

Microtus californicus (Peale, 1848). Mammalia and Ornithology *in* U.S. Expl. Exped..., 8:46.
REVIEWED BY: R. E. Barry, Jr. (REB); J. Ramirez-Pulido (JRP); G. Urbano-V. (GUV).
TYPE LOCALITY: U.S.A., California, Santa Clara Co., San Francisco Bay, probably at San Francisquito Creek, near Palo Alto.
DISTRIBUTION: S.W. Oregon, south through California (U.S.A.) to N. Baja California (Mexico).
COMMENT: Subgenus *Microtus*; see Anderson, 1959, Univ. Kans. Publ. Mus. Nat. Hist., 9:496; Gill, 1980:105–117, *in* Vorontsov and Van Brink, eds., Animal Genetics and Evolution, Junk, The Hague, 353 pp., found evidence of sterility in crosses between *M. c. californicus* and *M. c. stephensi*.
ISIS NUMBER: 5301410008080006001.

Microtus canicaudus Miller, 1897. Proc. Biol. Soc. Wash., 11:67.
TYPE LOCALITY: U.S.A., Oregon, Polk Co., McCoy.
DISTRIBUTION: Restricted to Willamette Valley, Oregon, and adjacent Washington (U.S.A.).
COMMENT: Subgenus *Microtus*; closely related to *montanus* and formerly considered conspecific; Hsu and Johnson, 1970, J. Mammal., 51:824–826, provided evidence of specific distinctness. See Maser and Storm, 1970, A Key to the Microtinae of the Pacific Northwest, O.S.U. Book Stores, Corvalis, Oregon, for status and distribution.

Microtus chrotorrhinus (Miller, 1894). Proc. Boston Soc. Nat. Hist., 26:190.
REVIEWED BY: R. E. Barry, Jr. (REB).
TYPE LOCALITY: U.S.A., New Hampshire, Coos Co., Mt. Washington, head of Tuckerman Ravine, 5300 ft. (1615 m).
DISTRIBUTION: S. Labrador southwest through S. Quebec and Ontario (Canada) to N. Minnesota; south in Appalachians to North Carolina (U.S.A.).
COMMENT: Subgenus *Microtus*; see Anderson, 1959, Univ. Kans. Publ. Mus. Nat. Hist., 9:496.
ISIS NUMBER: 5301410008080008001.

Microtus clarkei Hinton, 1923. Ann. Mag. Nat. Hist., ser. 9, 11:158.
TYPE LOCALITY: China, Yunnan, Kui-chiang-Salween divide (28° N.), 11,000 ft. (3353 m).
DISTRIBUTION: Mountains of Tibet (SW) and Yunnan (China) and N. Burma.
COMMENT: Perhaps subgenus *Lasiopodomys*, and related to *M. fortis* (RSH); *Pitymys* according to SW.
ISIS NUMBER: 5301410008080009001.

Microtus coronarius Swarth, 1911. Univ. Calif. Publ. Zool., 7:131.
TYPE LOCALITY: U.S.A., Alaska, Coronation Isl., Egg Harbor.
DISTRIBUTION: S.E. Alaska; Coronation, Forrester, and Warren Isls. (U.S.A.).
COMMENT: Subgenus *Chionomys*; see Anderson, 1960, or *Arvalomys* (=*Microtus*); see Chaline and Mein, 1979. Closely related to *longicaudus*; see Hall, 1981:809; also see comment under *M. longicaudus*.
ISIS NUMBER: 5301410008080010001.

Microtus epiroticus (Ondrias, 1966). Saugetierk. Mitt., 14 (Suppl):59.
TYPE LOCALITY: Greece, Epirus, Perama, near Ionnina.
DISTRIBUTION: C. and E. Europe, from Gulf of Finland south to Yugoslavia, N.W. Greece and Caucasus Mtns., east to Ural Mtns.
COMMENT: Subgenus *Microtus*; see Chaline and Mein, 1979; formerly included in *arvalis*, but differentiated chromosomally (see Malygin and Orlov, 1974, Zool. Zh., 53), and the bacula differ (see Aksenova and Tarasov, 1974, Zool. Zh., 53:609–615). Originally named *subarvalis* Meier, Orlov and Skholl, 1972 (Zool. Zhur., 51:157–161), but this name is preoccupied by *subarvalis* (Heller, 1930) (a

fossil); see Gromov and Polyakov, 1977, [Voles (Microtinae), Fauna U.S.S.R.], Nauka, Moscow and Leningrad, 3(8):308. This is the senior synonym of *Microtus subarvalis*; see Kral *et al.*, 1980, Prirodoved. Pr. Ustavia. Cesk. Acad. Ved. Brno, 14(9):1-29. For evidence of specific distinctness of *epiroticus*, see also Ruzic *et al.*, 1975, Arh. Poljopr. Nauka, 28:153-160.

Microtus evoronensis Kovalskaya and Sokolov, 1980. Zool. Zh., 59(9):1410.
   TYPE LOCALITY: U.S.S.R., Khabarovsk. Krai, Lake Evoron basin, Devyatka River.
   DISTRIBUTION: Known only from the type locality. Probably more widely distributed in Khabarovsk. Krai (U.S.S.R.).
   COMMENT: Provisionally placed in the subgenus *Microtus* (RSH, pers. comm. from V. G. Orlov). Considered a distinct species by Kovalskaya and Sokolov, 1980, Zool. Zh., 59(9):1409-1416, based on chromosomal differences and hybrid sterility in crosses with *maximowiczii*.

Microtus fortis Buchner, 1889. Sci. Res. Przewalski's Exp. Cent. Asia. Zool., ser. 1, 3:99.
   TYPE LOCALITY: China, Inner Mongolia, Ordos Desert, Huang Ho Valley, Sujan.
   DISTRIBUTION: Transbaikalia and Amur region (U.S.S.R.) south to lower Yangtse Valley (China).
   COMMENT: Subgenus *Lasiopodomys*; see Chaline, 1974, and Chaline and Mein, 1979; but also see Ognev, 1950, Mamm. U.S.S.R., Adjac. Count. Engl. transl., 1964:1-626, who placed this species in *Arvicola*; Anderson, 1960, Gromov and Polyakov, 1977:290, Meier, 1978, Trudy Zool. Inst., Leningr., 75:3-62, and Gromov and Baranova, 1981:208, placed it in the subgenus *Microtus*. Includes *calamorum*; reviewed by Meier, 1978, who considered *fortis* close to *maximowiczii* and *sachalinensis*.
   ISIS NUMBER: 5301410008080012001.

Microtus gregalis (Pallas, 1778). Nova Spec. Quad. Glir. Ord., p. 238.
   TYPE LOCALITY: U.S.S.R., R.S.F.S.R., Tomsk. Obl., E. of Chulym River.
   DISTRIBUTION: Palearctic tundra from White Sea, east to Anadyr region (U.S.S.R.); mountains and steppes of N.C. and S. Siberia from S. Urals to upper tributaries of Amur River, and S. to Pamir (U.S.S.R.), Tien Shan (U.S.S.R. and China), and Altai (U.S.S.R. and Mongolia) Mtns., N. Mongolia, N.E. China, and probably North Korea.
   COMMENT: Subgenus *Stenocranius*; see Anderson, 1960:200; see comments under *miurus*.
   ISIS NUMBER: 5301410008080013001.

Microtus gud Satunin, 1909. Izv. Kavkas. Mus., 4:272.
   TYPE LOCALITY: U.S.S.R., Gruzinsk. S.S.R. (Georgia), Gudauri, S. of Krestovyi Pass (Caucasus Range).
   DISTRIBUTION: Caucasus Mtns. (U.S.S.R. and N.E. Turkey).
   COMMENT: Subgenus *Chionomys* of Anderson, 1960:202; related to *nivalis* and *roberti*; see Gromov and Polyakov, 1977, [Voles (Microtinae), Fauna U.S.S.R.], Nauka, Moscow and Leningrad, 3(8):1-502, and Corbet, 1978:112. Chaline, 1974, included *Chionomys* within subgenus *Suranomys*, but Gromov and Polyakov, 1977:327, and Gromov and Baranova, 1981:216, considered *Chionomys* a full genus.
   ISIS NUMBER: 5301410008080015001.

Microtus guentheri (Danford and Alston, 1880). Proc. Zool. Soc. Lond., 1880:62.
   REVIEWED BY: G. Storch (GS).
   TYPE LOCALITY: Turkey, Maras Prov., Taurus Mtns., near Maras (=Marash).
   DISTRIBUTION: S. Bulgaria; S. Yugoslavia; E. Greece; Turkey; and N. Lebanon.
   COMMENT: Subgenus *Iberomys*; see Van den Brink, 1957, Die Saugetiere Europas; Chaline and Mein, 1979:102, or subgenus *Sumeriomys*; see Gromov and Polyakov, 1977:283. Not a synonym of *socialis*, as treated by Corbet, 1978:112; see Morlok, 1978, Senckenberg. Biol., 59;155-162; revised by Felten *et al.*, 1971, Senckenberg. Biol., 52:393-424. Ranck, 1968, The Rodents of Libya, Bull. U.S. Nat. Mus., 275:1-264, included *mustersi* in this species, but Kock *et al.*, 1972, Z. Saugetierk., 37:204-229 included *mustersi* in *irani*.

Microtus irani Thomas, 1921. J. Bombay Nat. Hist. Soc., 27:581.
  REVIEWED BY: G. Storch (GS).
  TYPE LOCALITY: Iran, Fars Prov., Shiraz, Bagh-i-Rezi (=Bagh-e-Razi).
  DISTRIBUTION: E. Turkey; N. Syria; Lebanon; Israel; W. Jordan; Cyrenaica in Libya; N.
    Iraq; W. and N. Iran; Turkmenia (U.S.S.R.).
  COMMENT: Subgenus *Sumeriomys* of Argyropulo, 1933, Z. Saugetierk., 8:180–182 (in
    part *Suranomys* of Chaline, 1974); see Gromov and Polyakov, 1977:283; includes
    mustersi; not a subspecies of *socialis* as treated by Corbet, 1978:112; see Kock *et al.*,
    1972, Z. Saugetierk., 37:204–229; and Morlok, 1978, Senckenberg. Biol.,
    59:155–162; includes *paradoxus* (OLR, GS); see Gromov and Polyakov, 1977:283.

Microtus kikuchii Kuroda, 1920. Zool. Mag. (Tokyo), 32:36.
  TYPE LOCALITY: Taiwan, Mt. Morrison.
  DISTRIBUTION: Taiwan.
  COMMENT: Subgenus *Microtus*; see Ellerman and Morrison-Scott, 1951:702, and
    Gromov and Polyakov, 1977:294. Thought by Zimmermann, 1964, Mitt. Zool.
    Mus. Berlin, 40:87–140, to be related to *maximowiczii* of mainland Asia.
  ISIS NUMBER: 5301410008080019001.

Microtus kirgisorum (Ognev, 1950). [Mamm. U.S.S.R., Adjac. Count.], Acad. Sci. Moscow,
    vol. 7, p. 181.
  TYPE LOCALITY: U.S.S.R., Kazakh. S.S.R., Dzhambulsk. Obl., Kirgizsk.
    (=Aleksandrovsk.) Ridge, Tuyuk Valley.
  DISTRIBUTION: S. Kazakhstan, Kirgiziya, Tadzhikistan, and S.E. Turkmenia (U.S.S.R.);
    probably N. Afghanistan.
  COMMENT: Subgenus *Microtus*; see Ognev, 1950. Considered a synonym of *ilaeus*, (=M.
    arvalis ilaeus) by Corbet, 1978:114, and Malygin and Deulin, 1979, Zool. Zh.,
    58:731–741; but Meier and Yatsenko, 1980, Zool. Zh., 59(2):283–288, provided
    evidence of specific distinctness and validity of the name. See also comment
    under *arvalis*.

Microtus longicaudus (Merriam, 1888). Am. Nat., 22:934.
  REVIEWED BY: R. E. Barry, Jr. (REB).
  TYPE LOCALITY: U.S.A., South Dakota, Custer Co., Custer (Black Hills), 5500 ft. (1676
    m).
  DISTRIBUTION: Rocky Mtns., from N. Yukon and E. Alaska south to S. New Mexico and
    Arizona, and adjacent foothills and plains; east to W. South Dakota; west to N.W.
    Pacific Coast (U.S.A.).
  COMMENT: Subgenus *Chionomys*, as in Anderson, 1960; or *Microtus* (= *Arvalomys*), as in
    Chaline, 1974, and Gromov and Polyakov, 1977:317. This widespread species is in
    need of taxonomic revision (SRL). Revised in part by Long, 1965, Univ. Kans.
    Publ. Mus. Nat. Hist., 14:493–758. Probably includes *coronarius*, an insular
    derivative (RSH, SRL).
  ISIS NUMBER: 5301410008080021001.

Microtus mandarinus (Milne-Edwards, 1871). Rech. Mamm., p. 129.
  TYPE LOCALITY: China, Shansi, probably near Saratsi.
  DISTRIBUTION: Korea; Hopei, Kiangsu, Anwei, Shensi, Shansi, Inner Mongolia (China)
    (SW); N. Mongolia; Transbaikalia (U.S.S.R.).
  COMMENT: Subgenus *Lasiopodomys*; see Chaline, 1974. Includes *vinogradovi*; see Corbet,
    1978:116.
  ISIS NUMBER: 5301410008080023001.

Microtus maximowiczii Schrenk, 1859. Reisen und Forsch. in Amur-Lande St. Petersburg.,
    1:140.
  TYPE LOCALITY: U.S.S.R., R.S.F.S.R., Chitinsk. Obl., upper Amur region, mouth of
    Omutnaya River.
  DISTRIBUTION: Lake Baikal to upper Amur region (U.S.S.R.), E. Mongolia, and N.E.
    China.

COMMENT: Subgenus *Microtus; (=Arvalomys)*, see Chaline, 1974, and Gromov and Polyakov, 1977:291. Includes *ungurensis;* formerly included *fortis;* see Corbet, 1978:114. Reviewed by Meier, 1978, Trudy Zool. Inst., Leningr., 75:3–62, and Orlov *et al.*, 1974, Zool. Zh., 53:1391–1396. See comments under *kikuchii, montebelli,* and *sachalinensis.*

ISIS NUMBER: 5301410008080046001 as *M. ungurensis.*

Microtus mexicanus (Saussure, 1861). Rev. Mag. Zool. Paris, ser. 2, 13:3.

REVIEWED BY: J. Ramirez-Pulido (JRP); G. Urbano V. (GUV).

TYPE LOCALITY: Mexico, Puebla, Mt. Orizaba.

DISTRIBUTION: Extreme S. Utah and S.W. Colorado (U.S.A.), south in mountains to C. Oaxaca (Mexico).

COMMENT: Subgenus *Microtus;* see Anderson, 1959, Univ. Kans. Publ. Mus. Nat. Hist., 9:415–511, and Gromov and Polyakov, 1977:319. Hooper and Hart, 1962, Misc. Publ. Mus. Zool. Univ. Mich., No. 120, 68 pp., placed this species in genus *Microtus,* subgenus *Pitymys,* closely allied with *ochrogaster* and *pinetorum* (RSH), but the dental pattern shows no resemblance to *Pitymys* (CAR). Includes *fulviventer;* see Musser, 1964, Occas. Pap. Mus. Zool. Univ. Mich., 636:122.

ISIS NUMBER: 5301410008080024001.

Microtus middendorffi (Poliakov, 1881). Mem. Acad. Imp. Sci. St. Petersb., 39(2):70.

TYPE LOCALITY: U.S.S.R., R.S.F.S.R., Krasnoyarsk. Krai, Taimyr Peninsula.

DISTRIBUTION: N.C. and N.E. Siberia, from the Polar Ural Mtns. to the Kolyma River, and N. of Magadan.

COMMENT: Subgenus *Microtus;* see Matthey and Zimmerman, 1961, Rev. Suisse Zool., 68:63–72; includes *hyperboreus,* see Gileva, 1972, Dokl. Akad. Nauk S.S.S.R., 203:689–692, and Pokrovskii *et al.*, 1975, [Population fluctuations of animals], Acad. Sci., Sverdlovsk, p. 39–62; see also Corbet, 1978:116; but see Gromov and Polyakov, 1977:312, and Gromov and Baranova, 1981:214; OLR also retains *hyperboreus* as a separate species.

ISIS NUMBER: 5301410008080025001 as *M. middendorffi.*
             5301410008080016001 as *M. hyperboreus.*

Microtus millicens Thomas, 1911. Abstr. Proc. Zool. Soc. Lond., 100:49.

TYPE LOCALITY: China, Szechwan, Weichoe, Si-ho River.

DISTRIBUTION: Mountains of Szechwan and Tibet (China) (SW).

COMMENT: Subgenus *Lasiopodomys,* according to Chaline, 1974, but "conf. *Neodon*" according to Gromov and Polyakov, 1977:257. This is apparently one of the least known species of *Microtus* (SRL).

ISIS NUMBER: 5301410008080026001.

Microtus miurus Osgood, 1901. N. Am. Fauna, 21:64.

TYPE LOCALITY: U.S.A., Alaska, Cook Inlet, Turnagain Arm, Mtns. near Hope City, head of Bear Creek.

DISTRIBUTION: W. Alaska (U.S.A.) east to extreme W. Northwest Territories and south to S.W. Yukon (Canada).

COMMENT: Subgenus *Stenocranius;* see Anderson, 1960:200; placed by Rausch, 1964, Z. Saugetierk., 29:343–358, and Rausch and Rausch, 1968, Z. Saugetierk., 33:65–99, in *gregalis;* reinstated as a species by Fedyk, 1970, Acta Theriol., 15:143–152. Regarded as conspecific with *abbreviatus* by MLJ.

Microtus mongolicus (Radde, 1861). Melanges Biol. Acad. St. Petersb., 3:681.

TYPE LOCALITY: U.S.S.R., R.S.F.S.R., Chitinsk. Obl., Omutnaya River (tributary to Amur River).

DISTRIBUTION: N.E. China (SW), N. Mongolia, and Transbaikalia (U.S.S.R.).

COMMENT: Subgenus *Microtus (=Arvalomys* Chaline, 1974); see Meier, 1978, Trudy Zool. Inst., Leningr., 75:3–62; regarded by Ellerman and Morrison-Scott, 1951:697, as a synonym of *arvalis;* separated by Malygin and Orlov, 1974, Zool. Zh., 53:616–622, and Malygin, 1978, Abstr. Pap., II Congr. Theriol. Internat. Brno, p. 331.

Microtus montanus (Peale, 1848). Mammalia and Ornithology *in* U.S. Expl. Exped..., 8:44.
REVIEWED BY: R. E. Barry, Jr. (REB).
TYPE LOCALITY: U.S.A., California, Siskiyou Co., headwaters of Sacramento River, near Mt. Shasta.
DISTRIBUTION: British Columbia (Canada) and C. Montana south to New Mexico and Arizona, west through the Cascade Mtns. and Sierra Nevada (U.S.A.).
COMMENT: Subgenus *Microtus*; see Anderson, 1960:204; two karyotypically different forms are known, which may represent distinct species; see Judd *et al.*, 1980, J. Mammal., 61:109–113. Revised by Anderson, 1959, Univ. Kans. Publ. Mus. Nat. Hist., 9:415–511.
ISIS NUMBER: 5301410008080027001.

Microtus montebelli (Milne-Edwards, 1872). Rech. Mamm., p. 285.
TYPE LOCALITY: Japan, Honshu, Fusiyama.
DISTRIBUTION: Honshu, Sado, and Kyushu Isls. (Japan); Sikotan Isl., Kuriles (U.S.S.R.).
COMMENT: Subgenus *Microtus* (=*Arvalomys*, Chaline, 1974); see Chaline and Mein, 1979:103; thought by Zimmermann, 1964, Mitt. Zool. Mus. Berlin, 40:87–140, to be related to *maximowiczii* of mainland Asia. Sokolov, 1954, Zool. Zh., 33:947–950, provided distribution information.
ISIS NUMBER: 5301410008080028001.

Microtus mujanensis Orlov and Kovalskaya, 1978. Zool. Zh., 57:1224.
TYPE LOCALITY: U.S.S.R., Buryat-Mongolsk. A.S.S.R., Bauntovsk. Obl., Miyha River at Miyha.
DISTRIBUTION: Known only from the type locality.
COMMENT: Subgenus *Microtus*; (RSH from V. G. Orlov, pers. comm.); Orlov and Kovalskaya, 1978, Zool. Zh., 57:1224–1232, described this species as morphologically close to *maximowiczii*, but the species are karyotypically different and F1 hybrids are sterile (SRL). Gromov and Baranova, 1981:209, indicated that the name is based on an earlier publication (Orlov and Kovalskaya, 1975, [Systematics and cytogenetics of mammals], Moscow, p. 32); which may be a *nomen nudum* (RSH).

Microtus nesophilus V. Bailey, 1898. Science, n.s., 8:783.
REVIEWED BY: R. E. Barry, Jr. (REB).
TYPE LOCALITY: U.S.A., New York, Suffolk Co., Great Gull Isl. off E. tip of Long Island.
DISTRIBUTION: Great and Little Gull islands, off Long Island, New York (U.S.A.).
COMMENT: Subgenus *Microtus*; closely related to *pennsylvanicus*; to which Youngman, 1967, J. Mammal., 48:586, referred it; now extinct; see Hall, 1981:796.

Microtus nivalis (Martins, 1842). Rev. Zool. Paris, p. 331.
TYPE LOCALITY: Switzerland, Bernese Oberlander, Faulhorn.
DISTRIBUTION: Discontinuous, mostly montane, from Spain through the Alps, Appennine, Tatra, Carpathian, and Balkan ranges to Turkey, Palestine, Caucasus, Elburz, Zagros, and Kopet Dag Mtns.
COMMENT: Subgenus *Chionomys*; see Anderson, 1960, or *Suranomys* Chaline, 1974; Gromov and Polyakov, 1977:327, considered *Chionomys* a full genus. Van der Meulen, 1978, Ann. Carnegie Mus., 47:140, considered *Suranomys* a junior synonym of *Chionomys* (LDM). Related to *gud* and *roberti*, but all three occur in the Caucasus. Peshev, 1970, Mammalia, 34:252–268, showed that further study is needed of the populations now included in this species (SRL).
ISIS NUMBER: 5301410008080029001.

Microtus oaxacensis Goodwin, 1966. Am. Mus. Novit., 2243:1–4.
REVIEWED BY: J. Ramirez-Pulido (JRP); G. Urbano V. (GUV).
TYPE LOCALITY: Mexico, Oaxaca, Ixtlan Dist., Tarasbundi Ranch, near Vista Hermosa, 5000 ft. (1524 m).
DISTRIBUTION: Mountains of N.C. Oaxaca (vic. of Vista Hermosa, Mexico).
COMMENT: Subgenus undetermined (RSH). Goodwin, 1966, Am. Mus. Novit., 2243:1–4, stated only that this species is distinct from *fulviventer* and *umbrosus*; his illustrations would place the species in the subgenus *Microtus* (SRL). Martin, R. A., 1974, *in* Webb, ed., Pleistocene Mammals of Florida, p. 60, placed this species in the genus *Pitymys* (LDM).
ISIS NUMBER: 5301410008080030001.

Microtus oeconomus (Pallas, 1776). Reise Prov. Russ. Reichs., 3:693.
  TYPE LOCALITY: U.S.S.R., R.S.F.S.R., Omsk. Obl., Ishim Valley.
  DISTRIBUTION: N. Palearctic, from Scandinavia and the Netherlands east to Chukotka
    (U.S.S.R.); south to S.E. Germany, Hungary, Ukraine, Kazakhstan, Tibet,
    Szechwan, and Heilungkiang (China) (SW), and Ussuri region; Kurile Islands and
    Sakhalin; St. Lawrence Isl. (Bering Sea); Alaska (U.S.A.); Yukon and Northwest
    Territories (Canada).
  COMMENT: Subgenus *Sumeriomys* (=*Suranomys* Chaline, 1974), see Argyropulo, 1933, Z.
    Saugetierk., 8:180–182; or *Microtus*; see Gromov and Polyakov, 1977:295, and
    Gromov and Baranova, 1981:209. The taxon *limnophilus*, an isolate described from
    the Tsaidam, Tsinghai, China, is usually assigned to this species (see Corbet,
    1978:115), but may be a distinct species; see Orlov *et al.*, 1978, [Geogr. and
    Dynamics of Plants and Anim. in the Mongolian People's Republic], p. 149–164;
    also see Sokolov and Orlov, 1980:151, who included Sinkiang, Kansu, N.W.
    China, and W. Mongolia in the distribution of *limnophilus*, which they considered
    a distinct species. Includes *ratticeps*; see Corbet, 1978:115. Ognev, 1950, [Mammals
    of the U.S.S.R. and Adj. Count.], Vol. 7, Rodents, Engl. Trans., 1964:1–626, and
    Ellerman, 1941, differed in the name used for this species; see Hall, 1981:805.
    Ognev's arguments for *ratticeps* are convincing, but would require nomenclatural
    reorganization in two subgenera. This problem might be best submitted to the
    ICZN for resolution (SRL).
  ISIS NUMBER: 5301410008080032001.

Microtus oregoni (Bachman, 1839). J. Acad. Nat. Sci. Phila., 8:60.
  REVIEWED BY: R. E. Barry, Jr. (REB).
  TYPE LOCALITY: U.S.A., Oregon, Clatsop Co., Astoria.
  DISTRIBUTION: Extreme S. British Columbia south to N. California, along Pacific Coast
    of U.S.A.
  COMMENT: Subgenus *Chilotus*; see Anderson, 1959, Univ. Kans. Publ. Mus. Nat. Hist.,
    9:415–511; and Anderson, 1960. Included by Chaline, 1974, in *Suranomys*.
    Reviewed by Maser and Storm, 1970, A key to Microtinae of the Pacific
    Northwest, O.S.U. Book Stores, Corvallis, Oregon.
  ISIS NUMBER: 5301410008080034001.

Microtus pennsylvanicus (Ord, 1815). *In* Guthrie, A new geogr., hist., comml.,
    grammar...Philadelphia, 2nd Amer. ed., 2:292.
  REVIEWED BY: R. E. Barry, Jr. (REB); J. Ramirez-Pulido (JRP).
  TYPE LOCALITY: U.S.A., Pennsylvania, "meadows below Philadelphia."
  DISTRIBUTION: North America from Newfoundland to W. Alaska, north into tundra
    zone, south to Georgia, N. Great Plains, S. Rocky Mtns. (U.S.A.), isolated
    population in N. Chihuahua (Mexico).
  COMMENT: Subgenus *Microtus*; see Anderson, 1960. Includes *chihuahuensis*; see Bradley
    and Cockrum, 1968, Am. Mus. Novit., 2325:3–7. Klimkiewicz, 1970, Mammalia,
    34:640–665, suggested that *pennsylvanicus* is conspecific with *agrestis*. This species
    requires review over its entire range and comparison with related Old and New
    World species (SRL). Closely related insular forms, such as *breweri*, *provectus* (here
    included in *pennsylvanicus*), *nesophilus* and others may best be considered
    subspecies; see Chamberlain, 1954, J. Mammal., 35:587, Fivush *et al.*, 1975, J.
    Mammal., 56:272, and Wheeler, 1956, Evolution, 10:176–186. Hall, 1981,
    considered *breweri* and *nesophilus* distinct species and *provectus* a subspecies of
    pennsylvanicus, whereas Anderson, 1959:496, afforded all full specific status,
    without elaborating; see comments under *breweri* and *nesophilus*. Reviewed by
    Reich, 1981, Mamm. Species, 159:1–8. See Anderson and Hubbard, 1971, Am.
    Mus. Novit., 2460:1–8, for discussion of distribution.
  ISIS NUMBER: 5301410008080035001.

Microtus richardsoni (DeKay, 1842). Zoology of New York, Part I, Mammals, p. 91.
  REVIEWED BY: M. L. Johnson (MLJ).
  TYPE LOCALITY: Canada, Alberta, vic. Jasper House, "Near the foot of the Rocky
    Mountains."
  DISTRIBUTION: N. Rocky Mtns., from S. British Columbia and S.W. Alberta (Canada) to
    C. Wyoming and Utah, and in Cascade Mtns. from S. British Columbia, south
    through Oregon (U.S.A.).
  COMMENT: Allocated by Hooper and Hart, 1962, Misc. Publ. Zool. Univ. Mich. Mus.,
    No. 120, 68 pp., Jannett, 1974, Am. Midl. Nat., 92:230–234, Corbet and Hill,
    1980:161, and others to *Arvicola*; Chaline and Mein, 1979, Repenning, 1979, Abstr.
    Pap. Geol. Soc. Am., and Martin, R. A., 1979, Evol. Monogr., 2:30, considered this
    vole to have arisen in North America and to belong in *Microtus (Aulacomys)* (MLJ,
    CAR). The development of the highly derived molar enamel pattern found in
    this form requires an early immigration and long independent history; a separate
    derivation from *Allophaiomys* seems more likely (LDM).
  ISIS NUMBER: 5301410008069001001 as *Arvicola richardsoni.*

Microtus roberti Thomas, 1906. Ann. Mag. Nat. Hist., ser. 7, 17:418.
  TYPE LOCALITY: Turkey, Pontus Prov., Sumila, 30 mi. (48 km) S. of Trebizond
    (=Trabzon).
  DISTRIBUTION: Caucasus Mtns. (U.S.S.R. and N.E. Turkey).
  COMMENT: Subgenus *Chionomys*; see Anderson, 1960:202; related to *gud* and *nivalis*
    (RSH). Chaline, 1974, included *Chionomys* within the subgenus *Suranomys*, but
    Gromov and Polyakov, 1977:327, regarded *Chionomys* as a full genus.
  ISIS NUMBER: 5301410008080038001.

Microtus sachalinensis Vasin, 1955. Zool. Zh., 34:427–431.
  TYPE LOCALITY: U.S.S.R., R.S.F.S.R., Sakhalin Isl., Poronaisk Dist., Olen River.
  DISTRIBUTION: Sakhalin Isl. (U.S.S.R.).
  COMMENT: Subgenus *Microtus*; see Gromov and Polyakov, 1977:292. Related to
    maximowiczii of mainland Asia; see Meier, 1978, Trudy Zool. Inst., Leningr.,
    75:3–62, *loc. cit.*, 79:85–90, who provided evidence of specific status.

Microtus socialis (Pallas, 1773). Reise Prov. Russ. Reichs., 2:705.
  REVIEWED BY: K. F. Koopman (KFK).
  TYPE LOCALITY: U.S.S.R., Kazakh. S.S.R., probably Gur'evsk Obl., between Volga and
    Ural rivers.
  DISTRIBUTION: Palearctic steppe from Dneper River and Crimea east to Lake Balkhash
    (U.S.S.R.) and N.W. Sinkiang (China) (KFK).
  COMMENT: Subgenus *Sumeriomys* (in part, *Suranomys* of Chaline, 1974); see Gromov
    and Polyakov, 1977:281. Corbet, 1978:112, included *guentheri* and *irani;* but see
    Kock *et al.*, 1972, Z. Saugetierk., 37:204–229, Morlok, 1978, Senckenberg. Biol.,
    59:155–162, and Gromov and Polyakov, 1977:283. Gromov and Baranova,
    1981:210, considered *irani* a subspecies of *socialis.* Ognev, [1950, Mamm. U.S.S.R.,
    Adjac. Count.], Vol. 7, Rodents, Engl. Trans., Ognev, 1964:1–626, proposed a close
    relationship between *socialis* and *Microtus (Chilotus) oregoni.* See also comments
    under *guentheri* and *irani.*
  ISIS NUMBER: 5301410008080041001.

Microtus townsendii (Bachman, 1839). J. Acad. Nat. Sci. Phila., 8:60.
  TYPE LOCALITY: U.S.A., Oregon, Multnomah Co., Wappatoo (Sauvie) Isl. in lower
    Columbia River, near mouth of Willamette River.
  DISTRIBUTION: Vancouver Isl.; extreme S.W. British Columbia (Canada) south to N.
    California, along Pacific coast of U.S.A.
  COMMENT: Subgenus *Microtus*; see Anderson, 1959, Univ. Kans. Publ. Mus. Nat. Hist.,
    9:415–511, and Anderson, 1960, Univ. Kans. Publ. Mus. Nat. Hist., 12:181–216.
    Reviewed by Maser and Storm, 1970, A key to Microtinae of the Pacific
    Northwest, O.S.U. Book Stores, Corvallis, Oregon.
  ISIS NUMBER: 5301410008080043001.

Microtus transcaspicus Satunin, 1905. Izv. Kavkas. Mus., 2:57–60.
    TYPE LOCALITY: U.S.S.R., Turkmen S.S.R., Ashkhabadsk. Obl., Kopet-Dag Mts., Chuli
    Valley, near Ashkhabad.
    DISTRIBUTION: Turkmenia, from Caspian Sea east to N. Afghanistan; south to N. Iran.
    COMMENT: Subgenus *Microtus*; see Aksenova, 1978, Trudy Zool. Inst., Leningr.,
    79:91–101. Meier, 1974, Trans. I Intl. Theriol. Congr., 1:400–401., provided
    evidence of specific distinctness, but see Ognev, 1950, [Mammals of the U.S.S.R.
    and Adj. Count.], Vol. 7, Rodents, Engl. Trans., 1964:1–626, who considered this a
    subspecies of *arvalis*. Provisionally includes *khorkoutensis*; see Corbet, 1978:114;
    reviewed by Malygin, 1978, Zool. Zh., 67(7):1062–1073.
    ISIS NUMBER: 5301410008080044001.

Microtus umbrosus Merriam, 1898. Proc. Biol. Soc. Wash., 12:107.
    REVIEWED BY: J. Ramirez-Pulido (JRP); G. Urbano-V. (GUV).
    TYPE LOCALITY: Mexico, Oaxaca, Mt. Zempoaltepec, 8200 ft. (2499 m).
    DISTRIBUTION: Known only from the vicinity of the type locality.
    COMMENT: This is the only species in the subgenus *Orthriomys*; see Anderson, 1959,
    Univ. Kans. Publ. Mus. Nat. Hist., 9:415–511. Martin, R. A., 1974, *in* Webb, ed.,
    Pleistocene Mammals of Florida, p. 61, suggested that this species "may
    eventually" be included in *Neodon*.
    ISIS NUMBER: 5301410008080045001.

Microtus xanthognathus (Leach, 1815). Zool. Misc., 1:60.
    REVIEWED BY: R. E. Barry, Jr. (REB).
    TYPE LOCALITY: Canada, Manitoba, Ft. Churchill?, "Hudson Bay."
    DISTRIBUTION: C. coast of Hudson Bay in Manitoba (Canada) to C. Alaska (U.S.A.) and
    N.E. British Columbia (Canada).
    COMMENT: Subgenus *Microtus*; see Anderson, 1959, Univ. Kans. Publ. Mus. Nat. Hist.,
    9:415–511. Anderson, 1960, Univ. Kans. Publ. Mus. Nat. Hist., 12:192, suggested
    that this species may be most closely related to *chrotorrhinus*; dental characters
    indicate specific distinctness (SRL), as do size differences (REB); see also Guilday
    et al., 1978, Bull. Carnegie Mus. Nat. Hist., 11:1–67.
    ISIS NUMBER: 5301410008080047001.

Neofiber True, 1884. Science, 4:34.
    COMMENT: Chaline and Mein, 1979, and Kretzoi, 1969, Vert. Hungar. (Budapest),
    11:155–193, placed this genus and *Ondatra*, in the tribe Ondatrini, as does CAR.
    However, WVK places them in separate tribes or subfamilies based on differences
    in the fossil history and the pattern of enamel on the molars; see Koenigswald,
    1980, Abh. Senckenb. Naturforsch. Ges., 539. Martin, L. D., 1975, *in* Smith and
    Friedland, eds., Studies on the Cenozoic Paleontology and Stratigraphy, p. 109,
    and Martin, L. D., 1979, Trans. Nebr. Acad. Sci., 7:91–100, came to a similar
    conclusion based on dental characters, and because the similarity between the
    two genera decreases in the earlier members of the lineages. Placed by Gromov
    and Polyakov, 1977:336, as genus *incertae sedis*.
    ISIS NUMBER: 5301410008082000000.

Neofiber alleni True, 1884. Science, 4:34.
    TYPE LOCALITY: U.S.A., Florida, Brevard Co., Georgiana.
    DISTRIBUTION: S. and E. Florida; extreme S.E. Georgia (U.S.A.).
    COMMENT: Revised by Schwartz, 1953, Occas. Pap. Mus. Zool. Univ. Mich., 547:1–27.
    Reviewed by Birkenholz, 1972, Mamm. Species, 15:1–4.
    ISIS NUMBER: 5301410008082001001.

Ondatra Link, 1795. Beytr. Naturg., 1(2):76.
    REVIEWED BY: M. Andera (MA); M. S. Boyce (MSB); R. Guenzel (RG); J. Ramirez-Pulido
    (JRP); G. Urbano-V. (GUV).
    COMMENT: Revised by Hollister, 1911, N. Am. Fauna, 32. Chaline and Mein, 1979, and
    Kretzoi, 1969, Vert. Hungar. (Budapest), 11:155–193, placed this genus and
    Neofiber in the tribe Ondatrini, as does CAR; however, WVK places them in

separate tribes or subfamilies based on differences in the fossil history and the pattern of enamel on the molars; see Koenigswald, 1980, Abh. Senckenb. Naturforsch. Ges., 539. Martin, L. D., 1975, *in* Smith and Friedland, eds., Studies on the Cenozoic Paleontology and Stratigraphy, p. 109, and Martin, L. D., 1979, Trans. Nebr. Acad. Sci., 7:91–100, also placed *Neofiber* far from *Ondatra*, and considered *Ondatra* an endemic North American arvicolid not closely related to any European form.

ISIS NUMBER: 5301410008083000000.

Ondatra zibethicus (Linnaeus, 1766). Syst. Nat., 12th ed., 1:79.
TYPE LOCALITY: E. Canada, no exact locality, probably E. New Brunswick.
DISTRIBUTION: North America, north to the treeline, south to the Gulf of Mexico, Rio Grande and lower Colorado River valleys. Introduced into Czechoslovakia in 1905; now widely distributed in the Palearctic.
COMMENT: Includes *obscurus*; see Cameron, 1959, Bull. Nat. Mus. Can., 154:85, and Pietsch, 1970, Z. Saugetierk., 35:257–288. Reviewed by Willner *et al.*, 1980, Mamm. Species, 141:1–8. See Hoffmann, 1958, Die Bisamratte, 260 pp., and Corbet, 1978:106 for discussion of distribution.
ISIS NUMBER: 5301410008083001001.

Phenacomys Merriam, 1889. N. Am. Fauna, 2:32.
COMMENT: Revised by Howell, 1926, N. Am. Fauna, 48:66 pp. Formerly included *Arborimus*; see Johnson, 1973, J. Mammal., 54:239–244, and Johnson and Maser, in press, Northwest. Sci. Placed by Gromov and Polyakov, 1977:336, as genus *incertae sedis*.
ISIS NUMBER: 5301410008084000000.

Phenacomys intermedius Merriam, 1889. N. Am. Fauna, 2:32.
TYPE LOCALITY: Canada, British Columbia, 20 mi. (32 km) N.N.W. Kamloops.
DISTRIBUTION: S. Yukon to the south coast of Hudson Bay to Labrador, south to the St. Lawrence River and Great Lakes, New Mexico (in the Rocky Mtns.), and E.C. California (in the Sierra Nevada, U.S.A.).
COMMENT: Includes *ungavae*; see Crowe, 1943, Bull. Am. Mus. Nat. Hist., 80:391–410; but see also Hall, 1981:785, who suggested that *intermedius* "may be a composite of two or three allopatric species."
ISIS NUMBER: 5301410008084002001.

Pitymys McMurtrie, 1831. Cuvier's Anim. Kingdom, Amer. ed., 1:434.
REVIEWED BY: M. Andera (MA); S. R. Leffler (SRL); W. D. Severinghaus (WDS).
COMMENT: Chaline and Mein, 1979, Corbet, 1978:106, and most European authors treat *Pitymys* as a genus; but see Gromov and Polyakov, 1977:259. Includes as subgenera or as synonyms: *Blanfordimys, Herpetomys, Meridiopitymys, Neodon, Parapitymys, Pedomys,* and *Phaiomys*; see Corbet, 1978:106. Palearctic species revised by Kratochvil, 1970, Prirodoved. Pr. Ustavia. Cesk. Acad. Ved. Brno, 4(12):1–63, and Kratochvil and Kral, 1974, Zool. Listy, 23:289–302. WDS, on the basis of examination of North American taxa, treated *Pitymys* as as subgenus of *Microtus*; see Severinghaus, 1981, J. Tenn. Acad. Sci., 56:20–22; SRL and Gromov and Baranova, 1981:201, agree.
ISIS NUMBER: 5301410008071000000 as *Blanfordimys*.

Pitymys afghanus (Thomas, 1912). Ann. Mag. Nat. Hist., ser. 8, 9:349.
TYPE LOCALITY: Afghanistan, Badkhiz, Gulran.
DISTRIBUTION: Mtns. of Afghanistan, S. Turkmenia, and N.E. Iran. Isolated populations in the W. Pamirs and Great Balkhan Mtns. (U.S.S.R.).
COMMENT: Subgenus *Neodon*; see Chaline, 1974; or subgenus *Blanfordimys*; see Gromov and Polyakov, 1977:284. Martin, R. A., 1974, *in* Webb, ed., Pleistocene Mammals of Florida, p. 61, suggested that this species "may eventually" be included in *Neodon*. Includes *bucharicus*; see Neithammer, 1970, Bonn. Zool. Beitr., 21:109–115; but also see Gromov and Baranova, 1981:208, who suggested *bucharicus* may be a distinct species. Formerly placed in the monotypic genus *Blanfordimys*; see Corbet, 1978:106; Gromov and Baranova, 1981:207, retained *Blanfordimys* as a subgenus.
ISIS NUMBER: 5301410008071001001 as *Blanfordimys afghanus*.

Pitymys bavaricus Konig, 1962. Senckenberg. Biol., 43:2.
TYPE LOCALITY: West Germany, Bavaria, Garmisch-Partenkirchen.
DISTRIBUTION: Bavaria (Germany).
COMMENT: Subgenus probably *Pitymys;* see Konig, *op cit.,* pp. 4–10, and Kratochvil, 1970, Prirodoved. Pr. Ustavia. Cesk. Acad. Ved. Brno, 4:1–63.

Pitymys ciscaucasicus (Ognev, 1924). Gryzuny Severnovo Kavkaza, p. 34.
TYPE LOCALITY: U.S.S.R., Gruzinsk. S.S.R. (Georgia), near Vladikavkaz (=Ordzhonikidze).
DISTRIBUTION: N. Caucasus Mtns. (U.S.S.R.).
COMMENT: Subgenus probably *Meridiopitymys;* see Chaline, 1974; regarded by Ellerman and Morrison-Scott, 1951, as a synonym of *subterraneus,* and by Corbet, 1978:109, of *majori;* Kratochvil and Kral, 1974, Zool. Listy, 23:289–302, provided evidence of specific distinctness.

Pitymys daghestanicus (Shidlovskii, 1919). Raboty Zemskoi Opytnoi Stantsi, 2:12.
TYPE LOCALITY: U.S.S.R., R.S.F.S.R., Dagestansk. A.S.S.R., Gunibsk. Okr., Karda.
DISTRIBUTION: Daghestan, E. Caucasus Mts. (U.S.S.R.).
COMMENT: Subgenus probably *Meridiopitymys;* see Chaline, 1974; regarded by Ellerman and Morrison-Scott, 1951, as a synonym of *subterraneus,* and by Corbet, 1978:109, of *majori;* Kratochvil and Kral, 1974, Zool. Listy, 23:289–302, provided evidence of specific distinctness.

Pitymys duodecimcostatus (de Selys-Longchamps, 1839). Rev. Zool. Paris, p. 8.
TYPE LOCALITY: France, Gard, Montpellier.
DISTRIBUTION: S.E. France; E. and S. Spain.
COMMENT: Subgenus *Meridiopitymys;* see Chaline, 1974. Formerly included the Balkan taxa *atticus* and *thomasi,* as in Ellerman and Morrison-Scott, 1951; but see Petrov and Zivkovic, 1972, Saugetierk. Mitt., 20:249–258, and Petrov and Zivkovic, 1979, Biosystematika, 5(1):113–125; also see comments under *thomasi.*
ISIS NUMBER: 5301410008080011001 as *Microtus duodecimcostatus.*

Pitymys felteni Malec and Storch, 1963. Senckenberg. Biol., 44 (3):171.
REVIEWED BY: G. Storch (GS).
TYPE LOCALITY: Yugoslavia, Rep. Makedonija (Macedonia), Pelister Mtns., near Trnovo-Magarevo.
DISTRIBUTION: S. Yugoslavia.
COMMENT: Subgenus *Parapitymys;* see Chaline, 1978, Cahiers de l'Ecole Pratique des Hautes Etudes, Montpellier, 1/4:145–167. Formerly included in *savii* by Corbet, 1978:109, but see the revision by Petrov *et al.,* 1976, Senckenberg. Biol., 57(1/3):1–10; see also comment under *savii.*

Pitymys gerbei (Gerbe, 1879). Le Naturaliste, 1:51.
TYPE LOCALITY: France, Loire-Inferieure, Dreneuf.
DISTRIBUTION: S.W. France; Pyrenees Mtns. of France and Spain.
COMMENT: Subgenus *Parapitymys;* see Chaline, 1978, Cahiers de l'Ecole Pratique des Hautes Etudes, Montpellier, 1/4:145–167. Formerly included in *savii* by Corbet, 1978:109 (as *pyrenaicus*); but see Spitz, 1978, Mammalia, 42:267–304, who employed this name, rather than *pyrenaicus* Longchamp, 1847, which he suggested was a *nomen dubium;* see comments under *savii.*

Pitymys guatemalensis Merriam, 1898. Proc. Biol. Soc. Wash., 12:108.
REVIEWED BY: J. Ramirez-Pulido (JRP).
TYPE LOCALITY: Guatemala, Huehuetenango, Todos Santos, 10,000 ft. (3048 m).
DISTRIBUTION: Sierra Madre from C. Chiapas (Mexico), south through S.W. Guatemala.
COMMENT: Subgenus *Herpetomys,* usually in genus *Microtus;* see Hall, 1981:803; placed by Martin, R. A., 1974, *in* Webb, ed., Pleistocene Mammals of Florida, p. 60., in genus *Pitymys,* and CAR in genus *Pitymys,* subgenus *Herpetomys.* See Anderson, 1960, for a discussion of the subgenus *Herpetomys.*
ISIS NUMBER: 5301410008080014001 as *Microtus guatemalensis.*

Pitymys juldaschi (Severtzov, 1879). Sap. Turk. Otd. Obsh. Lubit. Estestv., 1:63.

TYPE LOCALITY: U.S.S.R., Kirgizsk. S.S.R., Dzhalal-Abadsk. Obl., Kara-Kul'(Lake) basin, near Aksu.

DISTRIBUTION: Tien Shan and Pamir Mtns. (U.S.S.R.), and adjacent China, Pakistan, and Afghanistan.

COMMENT: Subgenus *Neodon,* (CAR); see Ellerman and Morrison-Scott, 1951; but placed in subgenus *Phaiomys* by Chaline, 1974. Includes *carruthersi;* see Corbet, 1978:107. According to OLR, the type locality is in Kirgizia rather than Kara-Kul' in the Pamirs, Tadzhikistan.

ISIS NUMBER: 5301410008080018001 as *Microtus juldaski (sic).*
5301410008080007001 as *Microtus carruthesi (sic).*

Pitymys leucurus (Blyth, 1863). J. Asiat. Soc. Bengal, 32:89.

TYPE LOCALITY: India, Ladak, near Lake Chomoriri (=Tsomoriri).

DISTRIBUTION: Tsinghai, Sinkiang, and Tibet (China) (SW), and the Himalayas west to Kashmir.

COMMENT: Subgenus *Phaiomys;* see Chaline, 1974; or subgenus *Neodon;* see Gromov and Polyakov, 1977:252. Corbet, 1978:107, included *strauchi,* which Van der Meulen, 1978, Ann. Carnegie Mus., 47:101–145, listed as a distinct species. If *Pitymys* were considered a subgenus of *Microtus, leucurus* would be preoccupied by *leucurus* Gerbe, 1852, now included in *M. nivalis; M. leucurus* Blyth would then be called *M. blythi* (Blanford, 1875) (WDS). Martin, R. A., 1974, *in* Webb, ed., Pleistocene Mammals of Florida, p. 60, considered *Phaiomys* a distinct genus. Includes *fuscus,* considered a separate species of *Microtus (Lasiopodomys)* by Zheng and Wang, 1980, Acta Zootax. Sin., 5(1):106–111; but also see Corbet, 1978:107.

ISIS NUMBER: 5301410008080020001 as *Microtus leucurus.*

Pitymys liechtensteini Wettstein, 1927. Anz. Akad. Wiss. Wien, 64:2.

REVIEWED BY: G. Storch (GS).

TYPE LOCALITY: Yugoslavia, Rep. Hrvatska (Croatia), Velebit Mtns., Mt. Mali Rajinac, near Krasno.

DISTRIBUTION: Alps of N.E. Italy and S. Austria to C. Yugoslavia.

COMMENT: Subgenus *Pitymys;* see Chaline, 1974; regarded by Ellerman and Morrison-Scott, 1951, as a synonym of *subterraneus;* Petrov and Zivkovic, 1971, Arh. Biol. Nauka, 23:31–32, provided evidence of specific distinctness. Possibly a subspecies of *multiplex;* see Storch and Winking, 1977, Z. Saugetierk., 42:78–88.

Pitymys lusitanicus (Gerbe, 1879). Rev. Mag. Zool. Paris, ser. 3, 7:44.

TYPE LOCALITY: Portugal, no exact locality.

DISTRIBUTION: N. Iberian Peninsula; S.W. France.

COMMENT: Subgenus *Meridiopitymys;* see Chaline, 1974. Includes *mariae;* see Spitz, 1978, Mammalia, 42:267–304, and Neithammer, 1970, Bonn. Zool. Beitr., 21:109–115; regarded by Ellerman and Morrison-Scott, 1951, as a synonym of *savii;* Almaca, 1973, Revista Fac. Cienc. Univ. Lisboa, 17:383–426, provided evidence of specific distinctness.

Pitymys majori (Thomas, 1906). Ann. Mag. Nat. Hist., ser. 7, 17:419.

REVIEWED BY: G. Storch (GS).

TYPE LOCALITY: Turkey, Trabzon Prov., Sumela (=Meryemana), 30 mi. (48 km) S. of Trebizond (=Trabzon).

DISTRIBUTION: N., E., and W. Turkey; W. Caucasus; S. Yugoslavia (GS).

COMMENT: Subgenus *Meridiopitymys;* see Chaline, 1974; Corbet, 1978:109, included *ciscaucasicus* and *daghestanicus,* for which Kratochvil and Kral, 1974, Zool. Listy, 23:289–302, provided evidence of specific distinctness; see also Felten *et al.,* 1971, Senckenberg. Biol., 52(6):393–424. Includes *nasarovi,* which may be a distinct species; see Tembotov *et al.,* 1976, Nalchik, pp. 3–35. Six different karyotypic forms have been found in the Caucasus ranges (see Hatuhov and Tembotov, 1978:390, *in* II Congr. Theriol. Intern. (Brno), Obrtel, Folk, and Pellantova, eds., Inst. Vert. Zool., Czech. Acad. Sci., Brno, and Khatukhov (= Hatuhov), 1980:382, *in* [Rodents. Materials of the All-Union Conf.], Panteleev, ed., Nauka, Moscow, 471 pp.). Ketenchiev, 1980:21, *in* [Rodents. Materials of the All-Union Conf.], Panteleev, ed., Nauka, Moscow, 471 pp., suggested that all may constitute a single, chromosomally polymorphic, species. See Petrov and Zivkovic, 1979, Biosystematika, 5(1):113–125, for alternative discussion of distribution.

Pitymys multiplex (Fatio, 1905). Arch. Sci. Phys. Nat. Geneve, ser. 4, 19:193.
    REVIEWED BY: G. Storch (GS).
    TYPE LOCALITY: Switzerland, Ticino Canton, near Lugano.
    DISTRIBUTION: S. Alps in Switzerland and France, and N. Apennines and S. Alps east to the Adige River in Italy.
    COMMENT: Subgenus *Pitymys*; see Chaline, 1978, Cahiers de l'Ecole Pratique des Hautes Etudes, Montpellier, 1/4:145–167. Reviewed by Meylan, 1970, Rev. Suisse Zool., 77:562–575, and Storch and Winking, 1977, Z. Saugetierk., 42:78–88. Includes *fatioi* and *druentius*; see Graf and Meylan, 1980, Z. Saugetierk., 45:133–148, and Corbet, 1978:108.

Pitymys ochrogaster (Wagner, 1842). *In* Schreber, Die Saugethiere ..., Suppl. 3:592.
    REVIEWED BY: R. E. Barry, Jr. (REB).
    TYPE LOCALITY: U.S.A., probably Indiana, New Harmony, "America."
    DISTRIBUTION: W. West Virginia, northwest into Alberta, west to S.C. Colorado and N. New Mexico; isolated population in Texas and Louisiana.
    COMMENT: Subgenus *Pedomys*; see Anderson, 1960. See Hooper and Hart, 1962, Misc. Publ. Zool. Univ. Mich. Mus., No. 120, 68 pp., Martin, R. A., 1974, *in* Webb, ed., Pleistocene Mammals of Florida, p. 60, and Chaline and Mein, 1979, for basis of allocation to *Pitymys (Pedomys)*. Van Der Meulen, 1978, Ann. Carnegie Mus., 47:101–145, regarded *Pedomys* as a subgenus of *Microtus*; Hall, 1981:812, placed this species in *Microtus* subgenus *Pitymys*. WDS recommended that, if *Pitymys* is considered a distinct genus, *Pedomys* should also be given generic status. Includes *ludovicianus*; see Raun and Laughlin, 1972, Southwest. Nat., 16:439. Includes *minor*, which Severinghaus, 1977, Proc. Biol. Soc. Wash., 90:49–54, listed as a distinct species without comment; but see Hall, 1981:8–13.
    ISIS NUMBER: 5301410008080031001 as *Microtus ochrogaster*.
                         5301410008080022001 as *Microtus ludovicianus*.

Pitymys pinetorum (Le Conte, 1830). Ann. Lyc. Nat. Hist., 3:133.
    TYPE LOCALITY: U.S.A., Georgia, Liberty Co., Riceboro, probably on the LeConte plantation.
    DISTRIBUTION: C. Texas to N.C. Wisconsin and N. Florida, to Maine (U.S.A.).
    COMMENT: Subgenus *Pitymys*; see Chaline and Mein, 1979, who included *nemoralis* and *parvulus*, in this species, as is done here; Van der Meulen, 1978, Ann. Carnegie Mus., 47:101–145, regarded them as distinct species in subgenus *Pitymys*. Hooper and Hart, 1962, Misc. Publ. Mus. Zool. Univ. Mich., No. 120, 68 pp., regarded *pinetorum* as close to *ochrogaster* and *mexicanus* in subgenus *Pitymys* of genus *Microtus* as did Hall, 1981:816. Reviewed by Smolen, 1981, Mamm. Species, 147:1–7, under the name *Microtus*.
    ISIS NUMBER: 5301410008080036001 as *Microtus pinetorium (sic)*.

Pitymys quasiater (Coues, 1874). Proc. Acad. Nat. Sci. Phila., 26:191.
    REVIEWED BY: J. Ramirez-Pulido (JRP); G. Urbano-V. (GUV).
    TYPE LOCALITY: Mexico, Veracruz, Jalapa.
    DISTRIBUTION: S.E. San Luis Potosi to N.C. Oaxaca (Mexico).
    COMMENT: Placed by most recent authors in *Pitymys*, see Hall, 1981:817, who placed this species in *Microtus* subgenus *Pitymys*.
    ISIS NUMBER: 5301410008080037001 as *Microtus quasiater*.

Pitymys savii (de Selys-Longchamps, 1838). Rev. Zool. Paris, p. 248.
    TYPE LOCALITY: Italy, near Pisa.
    DISTRIBUTION: Italy; Sicily; S. France.
    COMMENT: Subgenus *Parapitymys*; see Chaline, 1978, Cahiers de l'Ecole Pratique des Hautes Etudes, Montpellier, 1/4:145–167. Corbet, 1978:109, and WDS included

pyrenaicus (=*gerbei*) and *felteni*, for which St. Girons, 1973, Les Mammiferes du France et du Benelux, p. 342–344, Winking, 1974, Symp. Ther. II, Prague, p. 267–273, and Petrov *et al.*, 1976, Senckenberg. Biol., 57(1/3):1–10, provided evidence of specific distinctness.
ISIS NUMBER: 5301410008080039001 as *Microtus savii.*

Pitymys schelkovnikovi (Satunin, 1907). Izv. Kavkas. Mus., 3:243.
TYPE LOCALITY: U.S.S.R., Azerbaidzhansk. S.S.R., Talysk Mtns., Lenkoransk. Okr., near Dzhi.
DISTRIBUTION: Talysk and Elburz Mtns.; S. Azerbaidzhan (U.S.S.R.); S. and probably N.W. Iran.
COMMENT: Subgenus uncertain; regarded by Ellerman and Morrison-Scott, 1951, as a synonym of *subterraneus*; evidence for specific distinctness provided by Kratochvil, 1970, Prirodoved. Pr. Ustavia. Cesk. Acad. Ved. Brno, 4:1–63.

Pitymys sikimensis (Hodgson, 1849). Ann. Mag. Nat. Hist., ser. 2, 3:203.
TYPE LOCALITY: India, Sikkim, no exact locality.
DISTRIBUTION: Himalayas from Nepal to N. Burma, Szechwan, and S. Kansu (China).
COMMENT: Subgenus *Neodon*; see Chaline, 1974. Includes *irene*; see Corbet, 1978:107, but Gromov and Polyakov, 1977:256, considered *irene* a distinct species. Martin, R. A., 1974, *in* Webb, ed., Pleistocene Mammals of Florida, p. 60, considered *Neodon* a distinct genus; also see Martin, R. A., 1979, Evol. Monogr., 2.
ISIS NUMBER: 5301410008080040001 as *Microtus sikimensis.*
          5301410008080017001 as *Microtus irene.*

Pitymys subterraneus (de Selys-Longchamps, 1836). Essai Monogr. sur les Campagnols des Env. de Liege., p. 10.
TYPE LOCALITY: Belgium, Liege, Waremme.
DISTRIBUTION: N. and C. France through C. Europe to Ukraine and the Don River, south to Balkans, but true southern limits uncertain. Isolated populations in N.E. Russia.
COMMENT: Subgenus *Pitymys*; see Chaline, 1974. Revised by Niethammer, 1972, Bonn. Zool. Beitr., 23:290–309.
ISIS NUMBER: 5301410008080042001 as *Microtus subterraneus.*

Pitymys tatricus Kratochvil, 1952. Acta Acad. Sci. Nat. Moravo-Siles., 24:155–194.
TYPE LOCALITY: Czechoslovakia, Poprad Dist., Velka Studena Dolina valley, High Tatra Mtns.
DISTRIBUTION: Alpine zone of Tatra Mtns. between Czechoslovakia and Poland; Pilsko Mtn., Beslksid Ziwiecki Mtns. (W. Carpathians).
COMMENT: Subgenus *Pitymys*; see Chaline, 1974; regarded by Ellerman and Morrison-Scott, 1951, as a synonym of *subterraneus*. Revised by Kratochvil, 1970, Prirodoved. Pr. Ustavia. Cesk. Acad. Ved. Brno, 4:1–63.

Pitymys thomasi (Barrett-Hamilton, 1903). Ann. Mag. Nat. Hist., ser. 7, 11:306.
TYPE LOCALITY: Yugoslavia, Montenegro, Vranici.
DISTRIBUTION: S. coastal Yugoslavia; Albania and adjacent Greece.
COMMENT: Subgenus *Meridiopitymys*; see Chaline, 1974. Includes *atticus*; see Petrov and Zivkovic, 1979, Biosystematika, 5(1):113–125; and Corbet, 1978:110. Regarded by Ellerman and Morrison-Scott, 1951, as a synonym of *duodecimcostatus*; evidence of specific distinctness provided by Petrov and Zivkovic, 1972, Saugetierk. Mitt., 20:249–258.

Proedromys Thomas, 1911. Abstr. Proc. Zool. Soc. Lond., 90:4.
REVIEWED BY: S. R. Leffler (SRL).
COMMENT: Monotypic; retained as a genus following Wang *et al.*, 1966, Acta Zootax. Sin., 3:85–91, and Gromov and Polyakov, 1977:243; but also see Corbet, 1978:110, who included *Proedromys* in *Microtus* following Ellerman and Morrison-Scott, 1951, who considered it a subgenus. Martin, R. A., 1974, *in* Webb, ed., Pleistocene Mammals of Florida, p. 61, suggested that this genus "may eventually" be included in *Neodon*.

Proedromys bedfordi Thomas, 1911. Abstr. Proc. Zool. Soc. Lond., 90:4.
   TYPE LOCALITY: China, Kansu, 60 mi. (96 km) S.E. Minchow.
   DISTRIBUTION: Kansu and Szechwan (China) (SW).
   COMMENT: Placed by Chaline, 1974, and Corbet, 1978:116, in genus *Microtus*, but
      retained in monotypic genus by Wang *et al.*, 1966, Acta Zootax. Sin., 3:85–91, and
      by Gromov and Polyakov, 1977:243.
   ISIS NUMBER: 5301410008080003001 as *Microtus bedfordi.*

Prometheomys Satunin, 1901. Zool. Anz., 24:572.
   COMMENT: Placed with *Ellobius* in subfamily Prometheomyinae by CAR; placed in
      monotypic tribe Prometheomyini by Gromov and Polyakov, 1977:102.
   ISIS NUMBER: 5301410008085000000.

Prometheomys schaposchnikowi Satunin, 1901. Zool. Anz., 25:574.
   TYPE LOCALITY: U.S.S.R., Gruzinsk. S.S.R. (Georgia), Gudaur, S. of Krestovyi Pass
      (Caucasus Mtns.).
   DISTRIBUTION: C. and W. Caucasus (U.S.S.R. and extreme N.E. Turkey).
   ISIS NUMBER: 5301410008085001001.

Synaptomys Baird, 1858. Mammals, *in* Repts. Expl. and Surv...., 8(1):558.
   REVIEWED BY: B. R. Blood (BRB); A. V. Linzey (AVL).
   COMMENT: Placed in tribe Lemmini by Gromov and Polyakov, 1977:200.
   ISIS NUMBER: 5301410008086000000.

Synaptomys borealis (Richardson, 1828). Zool. J., 3:517.
   TYPE LOCALITY: Canada, Mackenzie, Great Bear Lake, Ft. Franklin.
   DISTRIBUTION: S.E. and C. Alaska (U.S.A.), to S. shore of Hudson Bay to Labrador
      (Canada), south to N. Minnesota, Montana, and Washington; isolated populations
      from S.E. Quebec to N. New Hampshire (U.S.A.).
   COMMENT: Subgenus *Mictomys*; revised by Howell, 1927, N. Am. Fauna, 50:1–38.
      Koenigswald and Martin, in press, Bull. Carnegie Mus. Nat. Hist., considered
      Mictomys a distinct genus; two extinct subgenera are also referable to *Mictomys.*
   ISIS NUMBER: 5301410008086001001.

Synaptomys cooperi Baird, 1858. Mammals, *in* Repts. Expl. and Surv...., 8(1):558.
   TYPE LOCALITY: U.S.A., New Hampshire, Carroll Co., at Jackson.
   DISTRIBUTION: Midwestern and E. U.S.A., through S.E. Canada and south to North
      Carolina and Arkansas; isolated populations in S.W. Kansas and Nebraska.
   COMMENT: Subgenus *Synaptomys*. Revised by Wetzel, 1955, J. Mammal., 36:120. The
      subspecies *paludis* may be extinct; *helaletes* recently recollected (AVL).
   ISIS NUMBER: 5301410008086002001.

Family Muridae
   REVIEWED BY: J. H. Honacki (JH).
   COMMENT: Assignment of family separate from Cricetidae and Arvicolidae follows
      Chaline and Mein, 1979, and Reig, 1980, J. Zool. Lond., 192:257–281. For a review
      of the taxonomy of this family see Swanepoel *et al.*, 1980, Ann. Transvaal Mus.,
      32(7):155–196, and Arata, 1967, *in* Anderson and Jones.
   ISIS NUMBER: 5301410011000000000.

Acomys I. Geoffroy, 1838. Ann. Sci. Nat. Paris Zool., ser. 2, 10:126.
   REVIEWED BY: M. Andera (MA)(Europe); D. A. Schlitter (DS); E. Van der Straeten
      (EVS)(Africa).
   COMMENT: Reviewed by Setzer, 1975, Part 6.5:1–2.
   ISIS NUMBER: 5301410011001000000.

Acomys cahirinus (Desmarest, 1891). Nouv. Dict. Hist. Nat., 29:70.
   TYPE LOCALITY: Egypt, Cairo.
   DISTRIBUTION: Cyprus; Western Sahara to Egypt; Nigeria; Ethiopia; Somalia; Sudan;
      Kenya; Uganda; Jordan; Lebanon; Syria; Yemen; Oman; S. Iraq; Iran; Pakistan.
   COMMENT: Includes *cineraceus, hawashensis, johannis, kempi,* and *percivali*; see Setzer,

1975, Part 6.5:1. Includes *dimidiatus, albigena, hunteri,* and *ignitus;* see Yalden *et al.,* 1976, Monitore Zool. Ital., n.s., suppl. 8(1):42; Corbet, 1978:142–143, and Osborn and Helmy, 1980, Fieldiana Zool., n.s., 5:297–298; but also see Setzer, 1975, Part 6.5:1, who listed *dimidiatus* as a distinct species which included *hystrella* and *intermedius.* Includes *louisae;* see Yalden *et al.,* 1976, Monitore Zool. Ital., n.s., suppl. 8(1):1–118, but also see Setzer, 1975, Part 6.5:2, who included *louisae* in *subspinosus. A. percivali* is probably a distinct species from Kenya, Uganda and Sudan (DS).

ISIS NUMBER: 5301410011001002001 as *A. cahirinus.*
5301410011001001001 as *A. albigena.*
5301410011001003001 as *A. cinerascens (sic).*
5301410011001004001 as *A. demidiatus (sic).*
5301410011001005001 as *A. hawashensis.*
5301410011001006001 as *A. hunteri.*
5301410011001007001 as *A. hystrella.*
5301410011001008001 as *A. ignitus.*
5301410011001009001 as *A. intermedius.*
5301410011001010001 as *A. johannis.*
5301410011001011001 as *A. kempi.*
5301410011001012001 as *A. louisae.*
5301410011001014001 as *A. percivali.*

Acomys cilicicus Spitzenberger, 1978. Ann. Nat. Hist. Mus. Wien, 81:444.
TYPE LOCALITY: Turkey, Vil Mersin, 17 km E. Silifke.
DISTRIBUTION: Asia Minor.

Acomys minous Bate, 1906. Proc. Zool. Soc. Lond., 1905(2):321.
TYPE LOCALITY: Greece, Crete Isl., Kanea.
DISTRIBUTION: Crete (Greece).
COMMENT: Considered a distinct species by Niethammer and Krapp, 1978, Handbuch der Saugetiere Europas, 1:452; but also see Corbet, 1978:142, who listed *minous* as a subspecies of *cahirinus.*

Acomys russatus (Wagner, 1840). Abh. Akad. Wiss. Munich, 3:195.
TYPE LOCALITY: Egypt, Sinai.
DISTRIBUTION: E. Egypt; Sinai; Jordan; Israel; Arabia.
COMMENT: Includes *lewisi;* see Harrison, 1972, The Mammals of Arabia, 3:489, Corbet, 1978:142, and Osborn and Helmy, 1980, Fieldiana Zool., n.s., 5:286–293.
ISIS NUMBER: 5301410011001015001.

Acomys spinosissimus Peters, 1852. Reise nach Mossambique, Saugeth., p. 160.
TYPE LOCALITY: Mozambique, Tette and Buio.
DISTRIBUTION: S. Tanzania; S.E. Zaire; Zambia; Zimbabwe; E. Botswana; Mozambique; N. South Africa.
COMMENT: Includes *selousi;* see Davis, 1974, Ann. Transvaal Mus., 29(9):159.
ISIS NUMBER: 5301410011001016001.

Acomys subspinosus (Waterhouse, 1838). Proc. Zool. Soc. Lond., 1837:104.
TYPE LOCALITY: South Africa, Cape Prov., Cape of Good Hope.
DISTRIBUTION: South Africa; Sudan.
COMMENT: Setzer, 1975, Part 6.5:2, included *louisae, nubilus,* and *wilsoni* in this species; but see also Yalden *et al.,* 1976, Monitore Zool. Ital., n.s., suppl. 8(1):44–48, and Rupp, 1980, Saugetierk. Mitt., 28(2):87, who included *louisae* in *cahirinus,* and listed *wilsoni* as a distinct species.
ISIS NUMBER: 5301410011001017001 as *A. subspinosus.*
5301410011001013001 as *A. nubilis.*

Acomys wilsoni (Thomas, 1892). Ann. Mag. Nat. Hist., ser. 6, 10:22.
TYPE LOCALITY: Kenya, Mombasa.
DISTRIBUTION: Sudan; Ethiopia; Somalia; Kenya; Uganda.
COMMENT: Formerly included in *subspinosus* by Setzer, 1975, Part 6.5:2; but see Yalden *et al.,* Monitore Zool. Ital., n.s., suppl. 8(1):44–48, and Rupp, 1980, Saugetierk. Mitt., 28:87.
ISIS NUMBER: 5301410011001018001.

Aethomys Thomas, 1915. Ann. Mag. Nat. Hist., ser. 8, 16:477.
REVIEWED BY: E. Van der Straeten (EVS).
COMMENT: Reviewed by Davis, 1975, Part 6.6.
ISIS NUMBER: 5301410011002000000.

Aethomys bocagei (Thomas, 1904). Ann. Mag. Nat. Hist., ser. 7, 13:416.
TYPE LOCALITY: Angola, Pungo Andongo.
DISTRIBUTION: C. and W. Angola; Cabinda; perhaps Zaire and Congo Republic.
ISIS NUMBER: 5301410011002001001.

Aethomys chrysophilus (De Winton, 1897). Proc. Zool. Soc. Lond., 1896:801.
TYPE LOCALITY: S. Zimbabwe, Mashonaland, Mazoe.
DISTRIBUTION: S.E. Kenya and Angola to South Africa.
ISIS NUMBER: 5301410011002002001.

Aethomys granti (Wroughton, 1908). Ann. Mag. Nat. Hist., ser. 8, 1:257.
TYPE LOCALITY: South Africa, Deelfontein.
DISTRIBUTION: C. Cape Prov. (South Africa).
COMMENT: Included in subgenus *Micaelamys* by Ellerman, 1941:213. Distribution
    mapped by Davis, 1974, Ann. Transvaal Mus., 29(9):149.

Aethomys hindei (Thomas, 1902). Ann. Mag. Nat. Hist., ser. 7, 9:278.
TYPE LOCALITY: Kenya, Machakos.
DISTRIBUTION: N. Nigeria to Sudan, Kenya, and N.E. Tanzania.
COMMENT: Includes *stannarius*; see Davis, 1975, Part 6.6:3.
ISIS NUMBER: 5301410011002007001 as *A. stannarius*.

Aethomys kaiseri (Noack, 1887). Zool. Jahrb. Syst., 2:228.
TYPE LOCALITY: Zaire, Marungu.
DISTRIBUTION: Kenya; S.W. Uganda; Rwanda; S. and E. Zaire; N. Angola; C. Malawi;
    Zambia; S.W. Tanzania.
COMMENT: Delany, 1975, The Rodents of Uganda, Trustees Br. Mus. Nat. Hist., 165
    pp., included *nyikae* in this species; but Davis, 1975, Part 6.6:2–4, and Ansell,
    1978:84, considered it a distinct species.
ISIS NUMBER: 5301410011002003001.

Aethomys namaquensis (A. Smith, 1834). S. Afr. Quart. J., 2:160.
TYPE LOCALITY: South Africa, Cape Prov., Namaqualand, Cape of Good Hope.
DISTRIBUTION: South Africa to S. Angola, S. Malawi, and Mozambique.
COMMENT: Included in subgenus *Micaelamys* by Davis, 1975, Part 6.6:4.
ISIS NUMBER: 5301410011002004001.

Aethomys nyikae (Thomas, 1897). Proc. Zool. Soc. Lond., 1897:431.
TYPE LOCALITY: Malawi, Nyika Plateau.
DISTRIBUTION: N. Zambia; Malawi; N.E. Angola; S. Zaire; E. Zimbabwe.
COMMENT: Includes *dollmani*; see Ansell, 1978:84. Delany, 1975, The Rodents of
    Uganda, Trustees Br. Mus. Nat. Hist., 165 pp., considered *nyikae* a synonym of
    kaiseri; but Davis, 1975, Part 6.6:2–4, and Ansell, 1978:84, considered it a distinct
    species.
ISIS NUMBER: 5301410011002005001.

Aethomys silindensis Roberts, 1938. Ann. Transvaal Mus., 19:245.
TYPE LOCALITY: Zimbabwe, Mt. Silinda.
DISTRIBUTION: Known only from the type locality.

Aethomys thomasi (De Winton, 1897). Ann. Mag. Nat. Hist., ser. 6, 20:327.
TYPE LOCALITY: Angola, Galanga.
DISTRIBUTION: W. and C. Angola.
ISIS NUMBER: 5301410011002008001.

Anisomys Thomas, 1904. Proc. Zool. Soc. Lond., 1903(2):199.
REVIEWED BY: J. I. Menzies (JIM); E. Van der Straeten (EVS); A. C. Ziegler (ACZ).
COMMENT: Generic relationships reviewed by Musser, 1981, Bull. Am. Mus. Nat. Hist.,
    169(2):137.
ISIS NUMBER: 5301410011003000000.

Anisomys imitator Thomas, 1904. Proc. Zool. Soc. Lond., 1903(2):200.
TYPE LOCALITY: Papua New Guinea, Central Prov., Aroa River, Avera.
DISTRIBUTION: Interior New Guinea.
ISIS NUMBER: 5301410011003001001.

Anonymomys Musser, 1981. Bull. Am. Mus. Nat. Hist., 168(3):300.

Anonymomys mindorensis Musser, 1981. Bull. Am. Mus. Nat. Hist., 168(3):300.
TYPE LOCALITY: Philippines, Mindoro, Halcon Range, Ilong Peak, 4500 ft. (1372 m).
DISTRIBUTION: Known only from the type locality.
COMMENT: Known only by three specimens; see Musser, 1981, Bull. Am. Mus. Nat. Hist., 168(3):300–301.

Apodemus Kaup, 1829. Skizz. Europ. Thierwelt, 1:154.
REVIEWED BY: M. Andera (MA)(Europe); O. L. Rossolimo (OLR) (U.S.S.R.); E. Van der Straeten (EVS)(Africa); S. Wang (SW)(China).
COMMENT: European species of the genus reviewed in Niethammer and Krapp, eds., 1978, Handbuch der Saugetiere Europas, 1:305–381. Generic relationships reviewed by Musser, 1981, Bull. Am. Mus. Nat. Hist., 168(3):318–329.
ISIS NUMBER: 5301410011004000000.

Apodemus agrarius (Pallas, 1771). Reise Prov. Russ. Reichs., 1:454.
TYPE LOCALITY: U.S.S.R., Ulianovsk. Obl., middle Volga River, Ulianovsk (formerly Simbirsk).
DISTRIBUTION: C. Europe to Lake Baikal, south to Thrace, Caucasus, and Tien Shan Mtns.; Amur River through Korea to Yunnan, Fukien, and Taiwan (China); Quelpart Isl. (Korea).
COMMENT: Includes *chevrieri;* see Corbet, 1978:137. Penis morphology reviewed by Williams *et al.,* 1980, Mammalia, 44:245–258.
ISIS NUMBER: 5301410011004001001 as *A. agarius (sic).*

Apodemus argenteus (Temminck, 1845). Fauna Japonica, 1(Mamm.), p. 51.
TYPE LOCALITY: Japan.
DISTRIBUTION: Japan.

Apodemus draco (Barrett-Hamilton, 1900). Proc. Zool. Soc. Lond., 1900:418.
TYPE LOCALITY: S. China, N.W. Fukien, Kuatun.
DISTRIBUTION: Hopei, Fukien (SW), S. Kansu, and Shensi (China) to Burma and Assam (India).
COMMENT: Formerly included in *sylvaticus;* includes *orestes;* see Corbet, 1978:137.

Apodemus flavicollis (Melchior, 1834). Dansk. Staat. Norg. Pattedyr, p. 99.
TYPE LOCALITY: Denmark, Sieland Isl.
DISTRIBUTION: Europe to the Ural Mtns., south to the Crimea (U.S.S.R.), Israel and the Balkans.
COMMENT: Includes *tauricus;* formerly included *peninsulae;* see Corbet, 1978:134, 136, who discussed the possibility of hybridization between *flavicollis* and *sylvaticus.* Penis morphology reviewed by Williams *et al.,* 1980, Mammalia, 44:245–258.
ISIS NUMBER: 5301410011004002001.

Apodemus gurkha Thomas, 1924. J. Bombay Nat. Hist. Soc., 29:888.
TYPE LOCALITY: Nepal, Gorkha, Laprak.
DISTRIBUTION: Nepal.

Apodemus krkensis Miric, 1968. Z. Saugetierk., 33:375.
REVIEWED BY: K. F. Koopman (KFK).
TYPE LOCALITY: Yugoslavia, Quarner, Krk Isl., Baska.
DISTRIBUTION: Krk Isl. (Yugoslavia).
COMMENT: Possibly conspecific with *mystacinus;* see Neithammer, 1978, *in* Niethammer and Krapp, eds., Handbuch der Saugetiere Europas, 1:311; but also see Williams *et al.,* 1980, Mammalia, 44:245–258, who suggested that *krkensis* is conspecific with *sylvaticus.* KFK agrees with Williams *et al.,* 1980.

Apodemus latronum Thomas, 1911. Abstr. Proc. Zool. Soc. Lond., 100:49.
  TYPE LOCALITY: China, W. Szechwan, Tatsienlu.
  DISTRIBUTION: Szechwan, Yunnan (China); N. Burma.

Apodemus microps Kratochvil and Rosicky, 1952. Zool. Entomol. Listy, Brno, 1:64.
  TYPE LOCALITY: Czechoslovakia, Kosice Dist., Saca.
  DISTRIBUTION: E. Europe; Asia Minor.
  COMMENT: May include *microtis;* see Kratochvil, 1962, Zool. Listy, 11:15–26; but also
    see Corbet, 1978:134–135, who included it in *sylvaticus.*

Apodemus mystacinus (Danford and Alston, 1877). Proc. Zool. Soc. Lond., 1877:279.
  TYPE LOCALITY: Turkey, Adana Prov., Bulgar Dagh Mt., Zebil.
  DISTRIBUTION: S.E. Europe; Israel; Lebanon; Asia Minor; N. Iraq; N.W. Iran; S. Georgia
    (U.S.S.R.); Rhodes; Crete.
  COMMENT: Includes *pohlei;* see Harrison, 1972, The Mammals of Arabia, 3:442; and
    Corbet, 1978:134. See also comment under *krkensis.*
  ISIS NUMBER: 5301410011004003001.

Apodemus peninsulae (Thomas, 1907). Proc. Zool. Soc. Lond., 1906:862.
  TYPE LOCALITY: Korea, 110 km S.E. of Seoul, Mingyoung.
  DISTRIBUTION: Ussuri to Altai Mtns., south to Korea, Kansu, and Shensi (China);
    Sakhalin; Hokkaido; perhaps to Szechwan and Yunnan (China).
  COMMENT: Formerly included in *flavicollis* (by Ellerman and Morrison-Scott, 1966:566),
    and in *speciosus* (by Kusnetzov, 1965, *in* Bobrinskii *et al.*, Key to the Mammals of
    the U.S.S.R., Moscow); includes *praetor;* see Corbet, 1978:136. Reviewed by
    Vorontsov *et al.*, 1977, Zool. Zh., 56(3):437–449.
  ISIS NUMBER: 5301410011004004001.

Apodemus semotus Thomas, 1908. Ann. Mag. Nat. Hist., ser. 8, 1:447.
  REVIEWED BY: K. F. Koopman (KFK).
  TYPE LOCALITY: Taiwan.
  DISTRIBUTION: Taiwan.
  COMMENT: May be conspecific with *draco* (KFK).

Apodemus speciosus (Temminck, 1845). Fauna Japonica, 1(Mamm.), p. 52.
  TYPE LOCALITY: Japan.
  DISTRIBUTION: Japan; Kunashir Isl. (U.S.S.R.).
  ISIS NUMBER: 5301410011004005001.

Apodemus sylvaticus (Linnaeus, 1758). Syst. Nat., 10th ed., 1:62.
  TYPE LOCALITY: Sweden, Uppsala.
  DISTRIBUTION: Europe to Altai Mtns., Pamirs, Himalayas, Afghanistan, and Israel;
    Morocco to Tunisia.
  COMMENT: May hybridize with *flavicollis;* see Corbet, 1978:134, who included *microtis*
    in this species. Also see Kratochvil, 1962, Zool. Listy, 11:15–26, who suggested
    that *microtis* may be conspecific with *microps.* May include *krkensis;* see Williams *et
    al.*, 1980, Mammalia, 44:245–258.
  ISIS NUMBER: 5301410011004006001.

Apomys Mearns, 1905. Proc. U.S. Nat. Mus., 28:455.
  REVIEWED BY: B. R. Stein (BRS).
  COMMENT: This genus was included in *Rattus* by Corbet and Hill, 1980:172; but see
    Musser, 1977, Am. Mus. Novit., 2624:11, and Musser, 1981, Bull. Am. Mus. Nat.
    Hist., 169(2):136, who considered it a distinct genus.

Apomys abrae (Sanborn, 1952). Fieldiana Zool., 33(2):133.
  TYPE LOCALITY: Philippines, Luzon Isl., Abra Prov., Abra, 3500 ft. (1067 m).
  DISTRIBUTION: Luzon (Philippines).

Apomys datae (Meyer, 1899). Abh. Mus. Dresden, ser. 7, 7:25.
  TYPE LOCALITY: Philippines, N. Luzon, Lepanto.
  DISTRIBUTION: Luzon (Philippines).
  COMMENT: Includes *major;* see Sanborn, 1950, Fieldiana Zool., 33:133.

Apomys hylocoetes Mearns, 1905. Proc. U.S. Nat. Mus., 28:456.
TYPE LOCALITY: Philippines, Mindanao, Mt. Apo, 2150 m.
DISTRIBUTION: Mindanao (Philippines).

Apomys insignis Mearns, 1905. Proc. U.S. Nat. Mus., 28:459.
TYPE LOCALITY: Philippines, S. Mindanao, Mt. Apo.
DISTRIBUTION: Mindanao (Philippines).

Apomys littoralis (Sanborn, 1952). Fieldiana Zool., 33(2):134.
TYPE LOCALITY: Philippines, Mindanao, Bugasan, Cotabato, 50 ft. (15 m).
DISTRIBUTION: Mindanao (Philippines).

Apomys microdon Hollister, 1913. Proc. U.S. Nat. Mus., 46:327.
REVIEWED BY: K. F. Koopman (KFK).
TYPE LOCALITY: Philippines, Cataduanes, Biga.
DISTRIBUTION: Cataduanes Isl. (Philippines).
COMMENT: *A. hollisteri* was proposed as a substitute name because of secondary homonymy with *microdon* Peters (now in *Praomys*) (KFK).

Apomys musculus Miller, 1911. Proc. U.S. Nat. Mus., 38:403.
TYPE LOCALITY: Philippines, Luzon, Benguet, Baguio, 1500 m.
DISTRIBUTION: Luzon (Philippines).

Apomys petraeus Mearns, 1905. Proc. U.S. Nat. Mus., 28:458.
TYPE LOCALITY: Philippines, Mindanao, Mt. Apo, 2340 m.
DISTRIBUTION: Mindanao (Philippines).

Arvicanthis Lesson, 1842. Nouv. Tabl. Regn. Anim. Mammal., p. 147.
REVIEWED BY: E. Van der Straeten (EVS).
COMMENT: Reviewed by Misonne, 1974, Part 6:18; but see Yalden *et al.*, 1976, Monitore Zool. Ital., n.s., suppl. 8(1):1–118, Rupp, 1980, Saugetierk. Mitt., 28:81–123, and Corbet and Hill, 1980:170, who listed five species in this genus. Generic relationships reviewed by Musser, 1981, Bull. Am. Mus. Nat. Hist., 168(3):318–329.
ISIS NUMBER: 5301410011005000000.

Arvicanthis niloticus (Desmarest, 1822). Encyclop. Method. Mamm., p. 281.
TYPE LOCALITY: Egypt.
DISTRIBUTION: Egypt to Sudan and Ethiopia to Senegal; E. Zaire; Kenya; Tanzania; Uganda; Zambia.
COMMENT: Includes *abyssinicus, blicki, lacernatus, somalicus,* and *testicularis;* see Misonne, 1974, Part 6:18. Includes *Pelomys dembeensis;* see Dieterlen, 1974, Z. Saugetierk., 39:229–231. Dorst, 1972, Mammalia, 36:182–192, listed *blicki* as a distinct species. Also see Corbet and Hill, 1980:170, who listed *abyssinicus, blicki, somalicus,* and *testicularis* as provisionally distinct species following Ellerman, 1941, The Families and Genera of Living Rodents, 2:124–126, and Yalden *et al.*, 1976, Monitore Zool. Ital., n.s., suppl. 8(1):44–48, who considered *abyssinicus, blicki, dembeensis,* and *somalicus* distinct species, but stated that further work is needed to resolve the status of these taxa; Rupp, 1980, Saugetierk. Mitt., 28:81–123, followed Yalden *et al.*, 1976.
ISIS NUMBER: 5301410011005002001 as *A. niloticus.*
5301410011005001001 as *A. lacernatus.*
5301410011005003001 as *A. somalicus.*
5301410011051002001 as *Pelomys dembeensis.*

Bandicota Gray, 1873. Ann. Mag. Nat. Hist., ser. 4, 12:418.
REVIEWED BY: J. T. Marshall (JTM); B. R. Stein (BRS); S. Wang (SW)(China).
ISIS NUMBER: 5301410011006000000.

Bandicota bengalensis (Gray and Hardwicke, 1833). Illustr. Indian Zool., pl. 21.
TYPE LOCALITY: India, Bengal.
DISTRIBUTION: Sri Lanka; Pakistan and India to Sumatra, Java, and Penang Isl.
ISIS NUMBER: 5301410011006001001.

Bandicota indica (Bechstein, 1800). Ueber Vierfuss. Thiere, 2:713.
TYPE LOCALITY: India, Pondicherry.
DISTRIBUTION: Sri Lanka; India to S. China, Vietnam, Sumatra, and Java.
COMMENT: Introduced into Kedah and Perlis (Malay Peninsula) in 1946 (JTM).
ISIS NUMBER: 5301410011006002001.

Bandicota savilei Thomas, 1916. J. Bombay Nat. Hist. Soc., 24:641.
TYPE LOCALITY: Burma, Mt. Popa.
DISTRIBUTION: C. Burma; Thailand to Vietnam.

Batomys Thomas, 1895. Ann. Mag. Nat. Hist., ser. 6, 16:162.
COMMENT: This genus may contain an additional species; see Musser and Gordon,
1981, J. Mammal., 62:523. Includes *Mindanaomys*; see Misonne, 1969, Ann. Mus. R.
Afr. Cent. Tervuren, ser. 8, Sci. Zool., 172:78, and Musser, 1981, Bull. Am. Mus.
Nat. Hist., 169(2):136; but also see Alcasid, 1969, Checklist of Philippine
Mammals, Manila Nat. Mus., Philippines, p. 28, and Walker *et al.*, 1975, 2:879,
who listed *Mindanaomys* as a distinct genus.
ISIS NUMBER: 5301410011007000000.

Batomys dentatus Miller, 1911. Proc. U.S. Nat. Mus., 38:400.
TYPE LOCALITY: Philippines, Luzon, Benguet.
DISTRIBUTION: N. Luzon (Philippines).
ISIS NUMBER: 5301410011007001001.

Batomys granti Thomas, 1895. Ann. Mag. Nat. Hist., ser. 6, 16:162.
TYPE LOCALITY: Philippines, N. Luzon, Mt. Data.
DISTRIBUTION: N. Luzon (Philippines).
ISIS NUMBER: 5301410011007002001.

Batomys salomonseni (Sanborn, 1953). Vidensk. Medd. Dan. Naturhist. Foren, 115:287.
TYPE LOCALITY: Philippines, Mindanao, Bukidnon Prov., Mt. Katanglad, 1600 m.
DISTRIBUTION: Mindanao (Philippines).
COMMENT: Formerly included in *Mindanaomys*; see Musser, 1981, Bull. Am. Mus. Nat.
Hist., 169(2):137.
ISIS NUMBER: 5301410011042001001 as *Mindanaomys salomonseni*.

Berylmys Ellerman, 1947. Proc. Zool. Soc. Lond., 1947:261.
COMMENT: Considered a subgenus of *Rattus* by Ellerman, 1947, Proc. Zool. Soc. Lond.,
1947:261, Ellerman and Morrison-Scott, 1966:600, and Marshall,1977, *in* Lekagul
and McNeely, p. 405; included in subgenus *Bullimus* of *Rattus* by Misonne, 1969,
Ann. Mus. R. Afr. Cent. Tervuren, ser. 8, Sci. Zool., 172:141; but see Musser and
Boeadi, 1980, J. Mammal., 61:396-397, who considered *Berylmys* distinct, and
Musser, 1981, Bull. Am. Mus. Nat. Hist., 169(2):136, who considered *Bullimus* a
distinct genus from the Philippines; also see Ellerman, 1949, The Families and
Genera of Living Rodents, 3:78, and Musser, 1981, Zool. Verhandelingen, 189:30.

Berylmys berdmorei (Blyth, 1851). J. Asiat. Soc. Bengal, 20:173.
TYPE LOCALITY: Burma, Mergui.
DISTRIBUTION: S. Burma; Thailand except extreme S.; Vietnam.
COMMENT: Formerly included in *Rattus*; see comment under genus.
ISIS NUMBER: 5301410011058016001 as *Rattus berdmorei*.

Berylmys bowersi (Anderson, 1879). Zool. Res. Yunnan, p. 304.
TYPE LOCALITY: China, Yunnan, Hotha.
DISTRIBUTION: N.W. India to Fukien, Chekiang, Anhwei (China) (SW), Indochina, and
peninsular Malaysia.
COMMENT: Misonne, 1969, Ann. Mus. R. Afr. Cent. Tervuren, ser. 8, Sci. Zool.,
172:143, considered this species to be related to species in *Berylmys*. Marshall,
1977, *in* Lekagul and McNeely, p. 405, included it in subgenus *Berylmys* of *Rattus*,
and Musser, 1981, Zool. Verhandelingen, 189:30, included it in genus *Berylmys*;
see comment under genus.
ISIS NUMBER: 5301410011058018001 as *R. bowersi*.

Berylmys mackenziei (Thomas, 1916). J. Bombay Nat. Hist. Soc., 24:40.
TYPE LOCALITY: Burma, Chin Hills, Haingyan, 80 km north of Kindat.
DISTRIBUTION: Assam (India) to Thailand.
COMMENT: Probably not a subspecies of *bowersi;* see Marshall, 1977, *in* Lekagul and
McNeely, pp. 405, 443, who included this species in *Rattus (Berylmys);* see
comment under genus.

Berylmys manipulus (Thomas, 1916). J. Bombay Nat. Hist. Soc., 24, 3:413.
TYPE LOCALITY: W. Burma, Kabaw Valley, Kampat, 20 mi. (32 km) W. of Kindat.
DISTRIBUTION: Assam (India); Burma.
COMMENT: Formerly included in *Rattus;* see comment under genus.
ISIS NUMBER: 5301410011058067001.

Bullimus Mearns, 1905. Proc. U.S. Nat. Mus., 28:450.
REVIEWED BY: L. R. Heaney (LRH).
COMMENT: Formerly included as a subgenus of *Rattus,* with representatives from
Malaya, Borneo, and Sulawesi, as well as the Philippines, by Misonne, 1969, Ann.
Mus. R. Afr. Cent. Tervuren, ser. 8, Sci. Zool., 172:140–142; but see Musser, 1981,
Bull. Am. Mus. Nat. Hist., 169(2):136, who considered *Bullimus* a distinct genus
from the Philippines; also see Sanborn, 1952, Fieldiana Zool., 33(2):129–130.

Bullimus bagobus Mearns, 1905. Proc. U.S. Nat. Mus., 28:450.
TYPE LOCALITY: Philippines, Mindanao, Mt. Apo, Todaya., 1203 m.
DISTRIBUTION: Mindanao (Philippines).
COMMENT: Formerly included in *Rattus;* see comment under genus. The incorrect
spelling *bagopus* is often used; may include *rabori* (LRH).
ISIS NUMBER: 5301410011058012001 as *R. bagopus (sic).*

Bullimus luzonicus (Thomas, 1895). Ann. Mag. Nat. Hist., ser. 6, 16:163.
TYPE LOCALITY: Philippines, Luzon, Benguet, Mt. Data, 2460 m.
DISTRIBUTION: Luzon (Philippines).
COMMENT: Formerly included in *Rattus;* see comment under genus.
ISIS NUMBER: 5301410011058063001.

Bullimus rabori (Sanborn, 1952). Fieldiana Zool., 33(2):130.
TYPE LOCALITY: Philippines, Mindanao, Zamboanga Prov., Katipunam Munic., Sigayan.
DISTRIBUTION: Zamboanga Peninsula, Mindanao (Philippines).
COMMENT: Formerly included in *Rattus;* see comment under genus. Reviewed by
Alcasid, 1969, Checklist of Philippine Mammals, Manila Nat. Mus., Philippines,
p. 33. Closely related to, and probably best considered, a subspecies of *bagobus*
(LRH).

Bunomys Thomas, 1910. Ann. Mag. Nat. Hist., ser. 8, 6:508.
COMMENT: Formerly included in *Rattus* by Ellerman, 1941:148, and in subgenus
Bullimus of *Rattus* by Misonne, 1969, Ann. Mus. R. Afr. Cent. Tervuren, ser. 8, Sci.
Zool., 172:141; but see Musser, 1981, Bull. Am. Mus. Nat. Hist., 169(2):115, 137,
who considered *Bunomys* and *Bullimus* distinct genera.

Bunomys andrewsi (J. A. Allen, 1911). Bull. Am. Mus. Nat. Hist., 30:366.
TYPE LOCALITY: Indonesia, Sulawesi, Sulawesi Tengarra, Butung (=Buton) Isl.
DISTRIBUTION: Sulawesi, except the N.E. peninsula.
COMMENT: Formerly included in *Rattus;* includes *adspersus, inferior,* and *heinrichi;* see
Musser, 1981, Bull. Am. Mus. Nat. Hist., 169(2):117.
ISIS NUMBER: 5301410011058007001 as *Rattus andrewsi.*

Bunomys chrysocomus (Hoffmann, 1887). Abh. Zool. Anthrop.-Ethnology Mus. Dresden,
3:17.
TYPE LOCALITY: Indonesia, N. Sulawesi, Minahassa.
DISTRIBUTION: Sulawesi (Indonesia).
COMMENT: Includes *brevimolaris, coelestis, nigellus, rallus,* and *koka;* see Musser, 1973,
Am. Mus. Novit., 2511:24. Formerly placed in subgenus *Maxomys* of *Rattus* by

Laurie and Hill, 1954:118; but see Musser *et al.*, 1979, J. Mammal., 60:592–606
(who revised *Maxomys*), Musser, 1981, Bull. Am. Mus. Nat. Hist., 168(3):246, and
Musser, 1981, Bull. Am. Mus. Nat. Hist., 169(2):115–117, 137, who placed this
species in *Bunomys*.
ISIS NUMBER: 5301410011058024001 as *R. chrysocomus*.
5301410011058025001 as *R. coelestis*.

Bunomys fratrorum (Thomas, 1896) Ann. Mag. Nat. Hist., 18:246.
TYPE LOCALITY: Indonesia, N. Sulawesi, Rurukan.
DISTRIBUTION: C. and N. Sulawesi.
COMMENT: Formerly included in *Maxomys chrysocomus*; see Musser, 1973, Am. Mus.
Novit., 2511:24, and Musser *et al.*, 1979, J. Mammal., 60:592–605. Formerly
included in *Rattus*; see comment under genus.
ISIS NUMBER: 5301410011058043001.

Bunomys penitus (Miller and Hollister, 1921). Proc. Biol. Soc. Wash., 34:72.
TYPE LOCALITY: Indonesia, Middle Sulawesi, S.W. of Lake Lindoe, Lehio.
DISTRIBUTION: S.W. peninsula and C. Sulawesi.
COMMENT: Formerly included in *Rattus*; see comment under genus. Formerly included
in *adspersus* (which is now in *andrewsi*); includes *sericatus*; see Musser, 1973, Am.
Mus. Novit., 2511:24.

Carpomys Thomas, 1895. Ann. Mag. Nat. Hist., ser. 6, 16:161.
REVIEWED BY: B. R. Stein (BRS).
COMMENT: Reviewed by Musser, 1981, Bull. Am. Mus. Nat. Hist., 169(2):67–176.
ISIS NUMBER: 5301410011008000000.

Carpomys melanurus Thomas, 1895. Ann. Mag. Nat. Hist., ser. 6, 16:162.
TYPE LOCALITY: Philippines, N. Luzon, Mt. Data, 7,000–8,000 ft. (2134–2438 m).
DISTRIBUTION: N. Luzon (Philippines).
ISIS NUMBER: 5301410011008001001.

Carpomys phaeurus Thomas, 1895. Ann. Mag. Nat. Hist., ser. 6, 16:162.
TYPE LOCALITY: Philippines, N. Luzon, Mt. Data.
DISTRIBUTION: N. Luzon (Philippines).
ISIS NUMBER: 5301410011008002001 as *C. phaerus (sic)*.

Celaenomys Thomas, 1898. Trans. Zool. Soc. Lond., 14(6):390.
REVIEWED BY: B. R. Stein (BRS).
COMMENT: Reviewed by Musser, 1981, Bull. Am. Mus. Nat. Hist., 169(2):67–176.
ISIS NUMBER: 5301410011073000000.

Celaenomys silaceus (Thomas, 1895). Ann. Mag. Nat. Hist., ser. 6, 16:161.
TYPE LOCALITY: Philippines, N. Luzon, Mt. Data, Lepanto, 8,000 ft. (2438 m).
DISTRIBUTION: Luzon (Philippines).
ISIS NUMBER: 5301410011073001001.

Chiromyscus Thomas, 1925. Proc. Zool. Soc. Lond., 1925:503.
REVIEWED BY: B. R. Stein (BRS).
COMMENT: Reviewed by Musser, 1981, Bull. Am. Mus. Nat. Hist., 168(3):316–317.
ISIS NUMBER: 5301410011009000000.

Chiromyscus chiropus (Thomas, 1891). Ann. Mus. Civ. Stor. Nat. Genova, ser. 2, 10:884.
TYPE LOCALITY: E. Burma, Karin Hills.
DISTRIBUTION: E. Burma; N. Thailand; Vietnam; C. Laos.
ISIS NUMBER: 5301410011009001001.

Chiropodomys Peters, 1868. Monatsb. Preuss. Akad. Wiss. Berlin, p. 448.
REVIEWED BY: B. R. Stein (BRS); S. Wang (SW)(China).
COMMENT: Revised by Musser, 1979, Bull. Am. Mus. Nat. Hist., 162(6):379–445.
ISIS NUMBER: 5301410011010000000.

Chiropodomys calamianensis (Taylor, 1934). Monogr. Bur. Sci. Manila, 30:470.
TYPE LOCALITY: Philippines, Calamian Isls., Busuanga Isl., Minuit.
DISTRIBUTION: Busuanga, Balabac, and Palawan Isls. (Philippines).
ISIS NUMBER: 5301410011010001001.

Chiropodomys gliroides (Blyth, 1856). J. Asiat. Soc. Bengal, 24:721.
TYPE LOCALITY: India, Assam, Cherrapunjee.
DISTRIBUTION: Kwangsi (China); Assam (India) to Burma, Thailand, Indochina, and
Malaya; Sumatra; Java; Bali; Natuna Isls.; Borneo.
COMMENT: Includes *pusillus, anna,* and *niadis;* see Musser, 1979, Bull. Am. Mus. Nat.
Hist., 162(6):408–409.
ISIS NUMBER: 5301410011010002001 as *C. gliroides.*
5301410011010005001 as *C. niadus (sic).*
5301410011010006001 as *C. pusillus.*

Chiropodomys karlkoopmani Musser, 1979. Bull. Am. Mus. Nat. Hist., 162(6):389.
TYPE LOCALITY: Indonesia, North Pagai Isl.
DISTRIBUTION: Known only from North Pagai Isl.

Chiropodomys major Thomas, 1893. Ann. Mag. Nat. Hist., ser. 6, 11:344.
TYPE LOCALITY: Malaysia, Sarawak, Sadong.
DISTRIBUTION: Borneo.
ISIS NUMBER: 5301410011010003001.

Chiropodomys muroides Medway, 1965. J. Malay. Branch R. Asiat. Soc., 36(3):133.
TYPE LOCALITY: Malaysia, Sabah, Bundu Tuhan, Kinabalu.
DISTRIBUTION: Borneo.
ISIS NUMBER: 5301410011010004001.

Chiruromys Thomas, 1888. Proc. Zool. Soc. Lond., 1888:237.
REVIEWED BY: J. I. Menzies (JIM); A. C. Ziegler (ACZ).
COMMENT: Formerly included in *Pogonomys;* see Dennis and Menzies, 1979, J. Zool.
Lond., 189:315–332.

Chiruromys forbesi Thomas, 1888. Proc. Zool. Soc. Lond., 1888:239.
TYPE LOCALITY: Papua New Guinea, Central Prov., Sogeri.
DISTRIBUTION: E. New Guinea, d'Entrecasteaux Isls., and Louisiade Isls.
COMMENT: Includes *pulcher* and *shawmayeri* Laurie, 1952; see Dennis and Menzies,
1979, J. Zool. Lond., 189:315–332.
ISIS NUMBER: 5301410011055002001 as *Pogonomys forbesi.*
5301410011055007001 as *Pogonomys shawmayeri.*

Chiruromys lamia (Thomas, 1897). Ann. Mus. Civ. Stor. Nat. Genova, 18:615.
TYPE LOCALITY: Papua New Guinea, Central Prov., Upper Kemp Welch River,
Ighibirei.
DISTRIBUTION: S.E. New Guinea.
COMMENT: Includes *kagi;* see Dennis and Menzies, 1979, J. Zool. Lond., 189:315–332.
ISIS NUMBER: 5301410011055004001 as *Pogonomys lamia.*
5301410011055003001 as *Pogonomys kagi.*

Chiruromys vates (Thomas, 1908). Ann. Mag. Nat. Hist., ser. 8, 2:495.
TYPE LOCALITY: Papua New Guinea, Central Prov., Upper St. Joseph's River
(=Angabunga River), Madeu, 610–915 m.
DISTRIBUTION: S.E. New Guinea.
ISIS NUMBER: 5301410011055009001 as *Pogonomys vates.*

Chrotomys Thomas, 1895. Ann. Mag. Nat. Hist., ser. 6, 16:161.
REVIEWED BY: B. R. Stein (BRS).
COMMENT: An additional species is being described; see Musser and Gordon, 1981, J.
Mammal., 62:523.
ISIS NUMBER: 5301410011074000000.

Chrotomys whiteheadi Thomas, 1895. Ann. Mag. Nat. Hist., ser. 6, 16:161.
TYPE LOCALITY: Philippines, N. Luzon, Mt. Data, 8,000 ft. (2438 m).
DISTRIBUTION: Luzon and Mindoro (Philippines).
ISIS NUMBER: 5301410011074001001.

Colomys Thomas and Wroughton, 1907. Ann. Mag. Nat. Hist., ser. 7, 19:379.
REVIEWED BY: E. Van der Straeten (EVS).
COMMENT: Includes *Nilopegamys*; see Misonne, 1974, Part 6:18.
ISIS NUMBER: 5301410011011000000.

Colomys goslingi Thomas and Wroughton, 1907. Ann. Mag. Nat. Hist., ser. 7, 19:380.
TYPE LOCALITY: Zaire, Uele River, Gambi.
DISTRIBUTION: Gabon; Cameroun; Central African Republic; Zaire; Uganda; Kenya; Zambia; Sudan; Ethiopia.
COMMENT: Includes *Nilopegamys plumbeus*; see Hayman, 1966, Ann. Mus. R. Afr. Cent. Zool., 144:29-38.
ISIS NUMBER: 5301410011011001001.

Conilurus Ogilby, 1838. Trans. Linn. Soc. Lond., 18:124.
REVIEWED BY: J. I. Menzies (JIM); C. H. S. Watts (CHSW); A. C. Ziegler (ACZ).
COMMENT: Generic relationships reviewed by Musser, 1981, Bull. Am. Mus. Nat. Hist., 169(2):137.
ISIS NUMBER: 5301410011012000000.

Conilurus albipes (Lichtenstein, 1829). Darst. Saugeth., Part 6, pl. 29.
TYPE LOCALITY: Australia, South Australia.
DISTRIBUTION: E. and S. Australia.
COMMENT: Probably extinct (CHSW).
ISIS NUMBER: 5301410011012001001.

Conilurus penicillatus (Gould, 1842). Proc. Zool. Soc. Lond., 1842:12.
TYPE LOCALITY: Australia, Northern Territory, Port Essington.
DISTRIBUTION: N. Northern Territory and N.E. West Australia (Australia); S.C. New Guinea.
COMMENT: Includes *hemileucurus*; see Tate, 1951, Bull. Am. Mus. Nat. Hist., 97:271, and Ride, 1970:242.
ISIS NUMBER: 5301410011012002001.

Crateromys Thomas, 1895. Ann. Mag. Nat. Hist., ser. 6, 16:163.
REVIEWED BY: B. R. Stein (BRS).
COMMENT: Reviewed by Musser and Gordon, 1981, J. Mammal., 62:513-525.
ISIS NUMBER: 5301410011013000000.

Crateromys paulus Musser and Gordon, 1981. J. Mammal., 62:515.
TYPE LOCALITY: Philippines, Mindoro Occidental Prov., Ilin (=Iling) Isl.
DISTRIBUTION: Known only from the type locality.
COMMENT: Known only by the holotype.

Crateromys schadenbergi (Meyer, 1895). Abh. Mus. Dresden, 6:1.
TYPE LOCALITY: Philippines, N. Luzon, Mt. Data.
DISTRIBUTION: N. Luzon (Philippines).
ISIS NUMBER: 5301410011013001001.

Cremnomys Wroughton, 1912. J. Bombay Nat. Hist. Soc., 21:340.
COMMENT: Formerly included in *Millardia* by Misonne, 1969, Ann. Mus. R. Afr. Cent. Tervuren, ser. 8, Sci. Zool., 172:1-219, 27 pl., and in *Rattus* by Ellerman, 1949, The Families and Genera of Living Rodents, 3:57; see Mishra and Dhanda, 1975, J. Mammal., 56:76.

Cremnomys blanfordi (Thomas, 1881). Ann. Mag. Nat. Hist., 7:24.
TYPE LOCALITY: India, Madras, Kadapa.
DISTRIBUTION: Sri Lanka; S. and C. India.
ISIS NUMBER: 5301410011058017001 as *Rattus blanfordi*.

Cremnomys cutchicus Wroughton, 1912. J. Bombay Nat. Hist. Soc., 21:340.
TYPE LOCALITY: India, Cutch, Dhonsa.
DISTRIBUTION: S. and W. India.
ISIS NUMBER: 5301410011058029001 as *Rattus cutchicus.*

Cremnomys elvira (Ellerman, 1947). Ann. Mag. Nat. Hist., 13:207.
TYPE LOCALITY: India, E. Ghats, Salem Dist., Kurumbapatti.
DISTRIBUTION: S.E. India.
ISIS NUMBER: 5301410011058038001 as *Rattus elvira.*

Crossomys Thomas, 1907. Ann. Mag. Nat. Hist., 20:70.
REVIEWED BY: J. I. Menzies (JIM); A. C. Ziegler (ACZ).
COMMENT: Generic relationships reviewed by Musser, 1981, Bull. Am. Mus. Nat. Hist., 169(2):67–176.
ISIS NUMBER: 5301410011075000000.

Crossomys moncktoni Thomas, 1907. Ann. Mag. Nat. Hist., 20:71.
TYPE LOCALITY: Papua New Guinea, Central Prov., Brown River, Serigina, "not less than" 1373 m.
DISTRIBUTION: E. interior New Guinea.
ISIS NUMBER: 5301410011075001001.

Crunomys Thomas, 1897. Trans. Zool. Soc. Lond., 14(6):393.
REVIEWED BY: B. R. Stein (BRS).
COMMENT: Reviewed by Musser, 1981, Bull. Am. Mus. Nat. Hist., 169(2):67–176; this genus also occurs in Sulawesi, but the species has not yet been published; also see Musser, 1977, Am. Mus. Novit., 2624:7, 12.
ISIS NUMBER: 5301410011014000000.

Crunomys fallax Thomas, 1897. Trans. Zool. Soc. Lond., 14(6):394.
TYPE LOCALITY: Philippines, N.C. Luzon, Isabella Prov., 1000 ft. (308 m).
DISTRIBUTION: N. Luzon (Philippines).
ISIS NUMBER: 5301410011014001001.

Crunomys melanius Thomas, 1907. Proc. Zool. Soc. Lond., 1907:141.
TYPE LOCALITY: Philippines, Mindanao, Davao, Mt. Apo, 924 m.
DISTRIBUTION: Mindanao (Philippines).
ISIS NUMBER: 5301410011014002001.

Dacnomys Thomas, 1916. J. Bombay Nat. Hist. Soc., 24(3):404.
REVIEWED BY: B. R. Stein (BRS).
COMMENT: Reviewed by Musser, 1981, Bull. Am. Mus. Nat. Hist., 168(3):314–316.
ISIS NUMBER: 5301410011015000000.

Dacnomys millardi Thomas, 1916. J. Bombay Nat. Hist. Soc., 24(3):404.
TYPE LOCALITY: India, near Darjeeling, Gopaldhara.
DISTRIBUTION: Nepal; Bengal and Assam (India); N. Laos.
ISIS NUMBER: 5301410011015001001.

Dasymys Peters, 1875. Monatsb. Preuss. Akad. Wiss. Berlin, p. 12.
REVIEWED BY: D. A. Schlitter (DS); E. Van der Straeten (EVS).
COMMENT: Reviewed by Misonne, 1974, Part 6:19.
ISIS NUMBER: 5301410011016000000.

Dasymys incomtus (Sundevall, 1847). Ofv. Kongl. Svenska Vet.-Akad. Forhandl. Stockholm, for 1846, 3(4):120.
TYPE LOCALITY: South Africa, Natal, Durban.
DISTRIBUTION: Senegal to Ethiopia to South Africa.
COMMENT: Includes *nudipes;* see Misonne, 1974, Part 6:19. *D. nudipes* is probably a distinct species (DS).
ISIS NUMBER: 5301410011016001001 as *D. incomtus.*
5301410011016002001 as *D. nudipes.*

Diomys Thomas, 1917. J. Bombay Nat. Hist. Soc., 25(2):203.
    REVIEWED BY: B. R. Stein (BRS).
    ISIS NUMBER: 5301410011017000000.

    Diomys crumpi Thomas, 1917. J. Bombay Nat. Hist. Soc., 25(2):203.
    TYPE LOCALITY: India, Bihar, Hazaribagh, Paresnath.
    DISTRIBUTION: N.E. India.
    ISIS NUMBER: 5301410011017001001.

Diplothrix Thomas, 1916. J. Bombay Nat. Hist. Soc., 24:404.
    COMMENT: Formerly included in *Rattus;* see Musser and Boeadi, 1980, J. Mammal.,
        61:396.

    Diplothrix legatus (Thomas, 1906). Ann. Mag. Nat. Hist., ser. 7, 17:88.
    TYPE LOCALITY: Japan, Ryukyu Isls., Amamioshima Isl.
    DISTRIBUTION: Ryukyu Isls. (Japan).
    ISIS NUMBER: 5301410011058056001 as *Rattus legatus.*

Echiothrix Gray, 1867. Proc. Zool. Soc. Lond., 1867:599.
    REVIEWED BY: B. R. Stein (BRS).
    COMMENT: For use of the name *Echiothrix,* see Thomas, 1898, Trans. Zool. Soc. Lond.,
        16:397. Generic status reviewed by Musser, 1981, Bull. Am. Mus. Nat. Hist.,
        169(2):137.
    ISIS NUMBER: 5301410011018000000.

    Echiothrix leucura Gray, 1867. Proc. Zool. Soc. Lond., 1867:600.
    TYPE LOCALITY: Indonesia, N. Sulawesi, uncertain.
    DISTRIBUTION: N. and C. Sulawesi (Indonesia).
    COMMENT: For type locality information see Thomas, 1896, Ann. Mag. Nat. Hist., ser.
        6, 18:246, and Thomas, 1898, Trans. Zool. Soc. Lond., 16:397.
    ISIS NUMBER: 5301410011018001001.

Eropeplus Miller and Hollister, 1921. Proc. Biol. Soc. Wash., 34:94.
    REVIEWED BY: B. R. Stein (BRS).
    COMMENT: Reviewed by Musser, 1981, Bull. Am. Mus. Nat. Hist., 169(2):137.
    ISIS NUMBER: 5301410011019000000.

    Eropeplus canus Miller and Hollister, 1921. Proc. Biol. Soc. Wash., 34:94.
    TYPE LOCALITY: Indonesia, C. Sulawesi, S.W. of Lake Lindoe, Goenoeng Lehio.
    DISTRIBUTION: C. Sulawesi.
    ISIS NUMBER: 5301410011019001001.

Gatamiya Deraniyagala, 1965. J. R. As. Soc. Ceylon, 9:214.
    REVIEWED BY: K. F. Koopman (KFK).
    COMMENT: The status of this named genus and species is highly dubious (KFK).

    Gatamiya weragami Deraniyagala, 1965. J. R. As. Soc. Ceylon, 9:214.
    TYPE LOCALITY: Sri Lanka, North Central Prov., near Laggala, Dasgiriya.
    DISTRIBUTION: Known only from the type locality.

Golunda Gray, 1837. Charlesworth's Mag. Nat. Hist., 1:586.
    REVIEWED BY: B. R. Stein (BRS).
    ISIS NUMBER: 5301410011020000000.

    Golunda ellioti Gray, 1837. Charlesworth's Mag. Nat. Hist., 1:586.
    TYPE LOCALITY: India, Bombay.
    DISTRIBUTION: Sri Lanka; India; Nepal; Bhutan; Pakistan.
    ISIS NUMBER: 5301410011020001001.

Grammomys Thomas, 1915. Ann. Mag. Nat. Hist., ser. 8, 16:150.
    COMMENT: Formerly included in *Thamnomys* by Misonne, 1974, Part 6:30, but
        considered a distinct genus by Yalden *et al.,* 1976, Monitore Zool. Ital., n.s., suppl.
        8(1):1–118, and Rupp, 1980, Saugetierk. Mitt., 28:87, 92.
    ISIS NUMBER: 5301410011021000000 as *Grammomys.*

Grammomys cometes (Thomas and Wroughton, 1908). Proc. Zool. Soc. Lond., 1908:549.
TYPE LOCALITY: Mozambique, Inhambane.
DISTRIBUTION: Mozambique; Zimbabwe; Malawi; Tanzania; Kenya.
COMMENT: Formerly included in *Thamnomys*; see comment under genus.
ISIS NUMBER: 5301410011021002001 as *Thamnomys cometes*.

Grammomys dolichurus (Smuts, 1832). Enumer. Mamm. Cap., p. 38.
TYPE LOCALITY: South Africa, near Cape Town.
DISTRIBUTION: Guinea and Mali to Cape Prov. and Natal (South Africa).
COMMENT: Includes *buntingi* and *gigas*; see Misonne, 1974, Part 6:31, who included
    dolichurus in *Thamnomys*; but see comment under genus.
ISIS NUMBER: 5301410011021003001 as *Thamnomys dolichurus*.
    5301410011021001001 as *Thamnomys buntingi*.
    5301410011021004001 as *Thamnomys gigas*.

Grammomys rutilans (Peters, 1876). Monatsb. Preuss. Akad. Wiss. Berlin, p. 478.
TYPE LOCALITY: Gabon, Limbareni.
DISTRIBUTION: Uganda to Angola and Guinea; Bioko.
COMMENT: Placed in subgenus *Grammomys* of *Thamnomys* by Misonne, 1974, Part 6:30;
    but see comment under genus.
ISIS NUMBER: 5301410011063001001 as *Thamnomys rutilans*.

Hadromys Thomas, 1911. J. Bombay Nat. Hist. Soc., 20(4):999.
REVIEWED BY: B. R. Stein (BRS).
ISIS NUMBER: 5301410011022000000.

Hadromys humei (Thomas, 1886). Proc. Zool. Soc. Lond., 1886:63.
TYPE LOCALITY: India, Manipur, Moirang.
DISTRIBUTION: Manipur to N.W. Assam (India).
ISIS NUMBER: 5301410011022001001.

Haeromys Thomas, 1911. Ann. Mag. Nat. Hist., ser. 8, 7:207.
REVIEWED BY: B. R. Stein (BRS).
COMMENT: Reviewed by Musser, 1981, Bull. Am. Mus. Nat. Hist., 169(2):137.
ISIS NUMBER: 5301410011023000000.

Haeromys margarettae (Thomas, 1893). Ann. Mag. Nat. Hist., ser. 6, 11:346.
TYPE LOCALITY: Malaysia, Sarawak, Penrisen Hills.
DISTRIBUTION: Borneo.
ISIS NUMBER: 5301410011023001001.

Haeromys minahassae (Thomas, 1896). Ann. Mag. Nat. Hist., ser. 6, 18:247.
TYPE LOCALITY: Indonesia, Sulawesi, Sulawesi Utara, Minahassa, Rurukan.
DISTRIBUTION: Sulawesi Utara (Sulawesi).
ISIS NUMBER: 5301410011023002001.

Haeromys pusillus (Thomas, 1893). Ann. Mag. Nat. Hist., ser. 6, 11:232.
TYPE LOCALITY: Malaysia, Sabah, Mt. Kinabalu.
DISTRIBUTION: Borneo.
ISIS NUMBER: 5301410011023003001.

Hapalomys Blyth, 1859. J. Asiat. Soc. Bengal, 28:296.
REVIEWED BY: B. R. Stein (BRS); S. Wang (SW)(China).
COMMENT: Revised by Musser, 1972, Am. Mus. Novit., 2503:27.
ISIS NUMBER: 5301410011024000000.

Hapalomys delacouri Thomas, 1927. Proc. Zool. Soc. Lond., 1927:55.
TYPE LOCALITY: Vietnam (=Annam), Dak-to.
DISTRIBUTION: Laos; Vietnam; Kwangsi (SW) and Hainan (China).
COMMENT: Formerly included in *longicaudatus*; see Musser, 1972, Am. Mus. Novit.,
    2503:27.

Hapalomys longicaudatus Blyth, 1859. J. Asiat. Soc. Bengal, 28:296.
TYPE LOCALITY: Burma, Tenasserim, Sitang River Valley.
DISTRIBUTION: S.E. Burma; S.W. Thailand; Malaya.

COMMENT: Formerly included *delacouri;* see Ellerman and Morrison-Scott, 1951:559, and Musser, 1972, Am. Mus. Novit., 2503:27.
ISIS NUMBER: 5301410011024001001.

Hybomys Thomas, 1910. Ann. Mag. Nat. Hist., ser. 8, 5:85.
REVIEWED BY: D. A. Schlitter (DS); E. Van der Straeten (EVS).
COMMENT: Reviewed by Misonne, 1974, Part 6.
ISIS NUMBER: 5301410011025000000.

Hybomys trivirgatus (Temminck, 1853). Esquisses Zool. sur la Cote de Guine, p. 159.
TYPE LOCALITY: Ghana (=Gold Coast), Dabocrom.
DISTRIBUTION: Guinea to mid-west Nigeria.
COMMENT: Includes *planifrons,* which DS believes is probably a distinct species; see Misonne, 1974, Part 6:19.
ISIS NUMBER: 5301410011025001001.

Hybomys univittatus (Peters, 1876). Monatsb. Preuss. Akad. Wiss. Berlin, p. 479.
TYPE LOCALITY: Gabon, Dongila.
DISTRIBUTION: Guinea to Ghana; Cameroun and Gabon to Uganda; Bioko.
ISIS NUMBER: 5301410011025002001.

Hydromys E. Geoffroy, 1804. Bull. Sci. Soc. Philom. Paris, 93:353.
REVIEWED BY: J. I. Menzies (JIM)(New Guinea); C. H. S. Watts (CHSW)(Australia); A. C. Ziegler (ACZ)(New Guinea).
COMMENT: Includes *Baiyankamys;* see Mahoney, 1968, Mammalia, 32:64–71. Generic relationships reviewed by Musser, 1981, Bull. Am. Mus. Nat. Hist., 169(2):67–176.
ISIS NUMBER: 5301410011076000000 as *Hydromys.*
5301410011072000000 as *Baiyankamys.*

Hydromys chrysogaster E. Geoffroy, 1804. Bull. Sci. Soc. Philom. Paris, 93:354.
TYPE LOCALITY: Australia, Tasmania, Bruni Isl.
DISTRIBUTION: Tasmania; Australia (non-desert regions); Kei Isls.; Aru Isls.; New Guinea.
COMMENT: Includes *caurinus, fuliginosus, grootensis, lawnensis, longmani, melicertes,* and *moae;* see Ride, 1970:243.
ISIS NUMBER: 5301410011076001001.

Hydromys habbema Tate and Archbold, 1941. Am. Mus. Novit., 1101:3.
TYPE LOCALITY: Indonesia, Irian Jaya, Djajawidjaja Div., 15 km N. Mt. Wilhelmina, Lake Habbema, 3225 m.
DISTRIBUTION: C. New Guinea.
COMMENT: Includes *Baiyankamys shawmayeri* Hinton, 1943; see Mahoney, 1968, Mammalia, 32:64–71.
ISIS NUMBER: 5301410011076002001 as *H. habrema (sic).*
5301410011072001001 as *Baiyankamys shawmayeri.*

Hydromys neobrittanicus Tate and Archbold, 1935. Am. Mus. Novit., 803:8.
TYPE LOCALITY: Papua New Guinea, East New Britain Prov., New Britain Isl., "Wide Bay, Balayang, Bainings," probably = Baining Mtns., vic. Wide Bay, Balayang, about 763 m.
DISTRIBUTION: New Britain Isl. (Papua New Guinea).
COMMENT: Concerning type locality, see similar type locality citations in Mayr, 1934, Am. Mus. Novit., 709:1–15 (ACZ).
ISIS NUMBER: 5301410011076003001.

Hylomyscus Thomas, 1926. Ann. Mag. Nat. Hist., ser. 9, 17:174.
REVIEWED BY: D. A. Schlitter (DS); E. Van der Straeten (EVS).
COMMENT: Included as a subgenus of *Praomys* by Misonne, 1974, Part 6:24, and Ansell, 1978:81–82; but see Robbins *et al.,* 1980, Ann. Carnegie Mus., 49(2):31–48, who also stated *Heimyscus* may be a distinct genus. East African species reviewed by Bishop, 1979, Mammalia, 43:519–530. *Hylomyscus* is in need of revision at the generic and specific levels.

Hylomyscus aeta (Thomas, 1911). Ann. Mag. Nat. Hist., ser. 8, 7:591.
TYPE LOCALITY: Cameroun, Ja River, Bitye.
DISTRIBUTION: Cameroun; Bioko; Gabon; Congo Republic; N. and E. Zaire; W. Uganda.
COMMENT: Reviewed by Robbins *et al.*, 1980, Ann. Carnegie Mus., 49(2):31–48. May prove to be a subspecies of *carillus;* see Misonne, 1974, Part 6:28.
ISIS NUMBER: 5301410011058003001 as *Rattus aeta.*

Hylomyscus alleni (Waterhouse, 1838). Proc. Zool. Soc. Lond., 1837:77.
TYPE LOCALITY: Equatorial Guinea, Bioko (=Fernando Po).
DISTRIBUTION: Guinea to Gabon and the Central African Republic; Bioko.
COMMENT: The relationship of *alleni* to *stella* and to *simus* needs further study; see Misonne, 1974, Part 6:28–29. Considered distinct from *stella* by Robbins *et al.*, 1980, Ann. Carnegie Mus., 49(2):31–48, who included *simus* and *canus* in *alleni* and is followed here.
ISIS NUMBER: 5301410011058005001 as *Rattus alleni.*

Hylomyscus baeri Heim de Balsac and Aellen, 1965. Biol. Gabonica, 1:175–178.
TYPE LOCALITY: Ivory Coast, Adiopodoume (5° 20′ N., 4° 7′ W.).
DISTRIBUTION: Ivory Coast; Ghana.
COMMENT: Reviewed by Robbins and Setzer, 1979, J. Mammal., 60:649–650.

Hylomyscus carillus (Thomas, 1904). Ann. Mag. Nat. Hist., ser. 7, 13:418.
TYPE LOCALITY: Angola, Pungo Andongo.
DISTRIBUTION: Zaire; N. Uganda; N. Angola.
COMMENT: May include *aeta;* see Misonne, 1974, Part 6:28, and Robbins *et al.*, 1980, Ann. Carnegie Mus., 49(2):31–48.

Hylomyscus denniae (Thomas, 1906). Ann. Mag. Nat. Hist., ser. 7, 18:144.
TYPE LOCALITY: Uganda, Mubuku Valley, Ruwenzori East.
DISTRIBUTION: Uganda; Kenya; Zaire; Zambia.
COMMENT: Placed in *Praomys (Hylomyscus)* by Misonne, 1974, Part 6:28, Ansell, 1978:81–82, and Bishop, 1979, Mammalia, 43:527; but see Robbins *et al.*, 1980, Ann. Carnegie Mus., 49(2):31–48.
ISIS NUMBER: 5301410011056006001 as *Praomys denniae.*

Hylomyscus fumosus Brosset, Dubost, and Heim de Balsac, 1965. Biologia Gabon., 1:154–163.
TYPE LOCALITY: Gabon, Makokou.
DISTRIBUTION: Gabon; Cameroun; perhaps Central African Republic.
COMMENT: May represent a distinct genus, *Heimyscus;* see Robbins *et al.*, 1980, Ann. Carnegie Mus., 49(2):33, 45, Misonne, 1974, Part 6:28, and Misonne, 1969, Ann. Kon. Mus. Mid.-Afr. Zool. Wetensch., 172:125.

Hylomyscus parvus Brosset, Dubost and Heim de Balsac, 1965. Biologia Gabon., 1:148–154.
TYPE LOCALITY: Gabon, Belinga.
DISTRIBUTION: Zaire; Congo Republic; Gabon; Cameroun.
COMMENT: Reviewed by Robbins *et al.*, 1980, Ann. Carnegie Mus., 49(2):31–48.

Hylomyscus stella (Thomas, 1911). Ann. Mag. Nat. Hist., ser. 8, 7:590.
TYPE LOCALITY: Zaire, Ituri Forest between Mawambi and Avakubi.
DISTRIBUTION: Nigeria to Kenya and Uganda.
COMMENT: Considered distinct from *alleni;* see Bishop, 1979, Mammalia, 43:527, and Robbins *et al.*, 1980, Ann. Carnegie Mus., 49(2):31–48, but also see Misonne, 1974, Part 6:28. See also comment under *H. alleni.*
ISIS NUMBER: 5301410011056015001 as *Praomys stella.*

Hyomys Thomas, 1904. Proc. Zool. Soc. Lond., 1903(2):198.
REVIEWED BY: J. I. Menzies (JIM); A. C. Ziegler (ACZ).
COMMENT: Generic relationships reviewed by Musser, 1981, Bull. Am. Mus. Nat. Hist., 169(2):67–176.
ISIS NUMBER: 5301410011026000000.

Hyomys goliath (Milne-Edwards, 1900). Bull. Mus. Hist. Nat. Paris, 6:165.
   TYPE LOCALITY: Papua New Guinea, Central Prov., highlands of the Aroa River basin.
   DISTRIBUTION: Interior New Guinea.
   ISIS NUMBER: 5301410011026001001.

Kadarsanomys Musser, 1981. Zool. Verhandelingen, 189:5.

Kadarsanomys sodyi (Bartels, 1937). Treubia, 16:45.
   TYPE LOCALITY: Indonesia, Java, West Java, Mt. Pangrango Gede, 1000 m.
   DISTRIBUTION: Java.
   COMMENT: Formerly included in *Rattus* and *Lenothrix canus*; see Musser, 1981, Zool.
      Verhandelingen, 189:5.

Komodomys Musser and Boeadi, 1980. J. Mammal., 61:397.
   REVIEWED BY: B. R. Stein (BRS).

Komodomys rintjanus (Sody, 1941). Treubia, 18:310.
   TYPE LOCALITY: Indonesia, Rintja Isl., Lohoboeaja.
   DISTRIBUTION: Rintja and Padar Isls. (Lesser Sunda Isls.).
   COMMENT: Formerly included in *Rattus* by Laurie and Hill, 1954:121; see Musser and
      Boeadi, 1980, J. Mammal., 61:395-413.

Leggadina Thomas, 1910. Ann. Mag. Nat. Hist., ser. 8, 6:606.
   REVIEWED BY: C. H. S. Watts (CHSW)(Australia); A. C. Ziegler (ACZ)(New Guinea).
   COMMENT: Included in *Pseudomys* by Ride, 1970:244, but considered a separate genus
      by Watts, 1976, Trans. R. Soc. S. Aust., 100:105-108 (who included *delicatulus* and
      hermannsburgensis in *Leggadina*), Mahoney and Posamentier, 1975, Aust. Mammal.,
      1(4):334, and Fox and Briscoe, 1980, Aust. Mammal., 3(2):109-126, who transferred
      delicatulus and *hermannsburgensis* to *Pseudomys*, and are followed here.

Leggadina forresti (Thomas, 1906). Abstr. Proc. Zool. Soc. Lond., 32:6.
   TYPE LOCALITY: Australia, Northern Territory, Alexandria.
   DISTRIBUTION: Arid parts of Western Australia, Northern Territory, South Australia,
      Queensland and New South Wales (Australia).
   COMMENT: Includes *messorius*, *waitei*, and *berneyi*; see Ride, 1970:246-247.
   ISIS NUMBER: 5301410011057006001 as *Pseudomys forresti*.

Leggadina lakedownensis Watts, 1976. Trans. R. Soc. S. Aust., 100(2):105.
   TYPE LOCALITY: Australia, Queensland, Lakeland Downs.
   DISTRIBUTION: Cooktown region, N. Queensland (Australia).

Lemniscomys Trouessart, 1881. Bull. Soc. d'Etudes Sci. Angers, 10:124.
   REVIEWED BY: E. Van der Straeten (EVS).
   COMMENT: *L. striatus* species complex revised by Van der Straeten, 1975, Unpubl.
      Ph.D. Dissertation, Antwerp, and Van der Straeten and Verheyen, 1980,
      Mammalia, 44:73-82.
   ISIS NUMBER: 5301410011028000000.

Lemniscomys barbarus (Linnaeus, 1766). Syst. Nat., 12th ed., 1:addenda.
   TYPE LOCALITY: Morocco (= "Barbaria").
   DISTRIBUTION: Tunisia; Algeria; Morocco; W. Sahara; Senegal; Gambia; Nigeria;
      Cameroun; Sudan; Ethiopia; Kenya; N. Uganda; Tanzania; E. Zaire.
   ISIS NUMBER: 5301410011028001001.

Lemniscomys bellieri Van der Straeten, 1975. Rev. Zool. Afr., 89:906.
   TYPE LOCALITY: Ivory Coast, Ayeremou (= Lamto).
   DISTRIBUTION: Guinea; Doka Woodland (Ivory Coast).

Lemniscomys griselda (Thomas, 1904). Ann. Mag. Nat. Hist., ser. 7, 13:414.
   TYPE LOCALITY: Angola, Jinga country, Muene Coshi.
   DISTRIBUTION: Angola.
   COMMENT: Formerly included *linulus* and *rosalia*; see Van der Straeten, 1980, Rev. Zool.

Afr., 94:185–201, Van der Straeten and Verheyen, 1980, Mammalia, 44:73, and Van der Straeten, 1980, Ann. Cape Prov. Mus. Nat. Hist., 13(5):55–62.
ISIS NUMBER: 5301410011028002001.

Lemniscomys linulus (Thomas, 1910). Ann. Mag. Nat. Hist., ser. 8, 6:429.
TYPE LOCALITY: Senegal (= French Gambia), Gamon.
DISTRIBUTION: Doka Woodland (Ivory Coast and Senegal).
COMMENT: Formerly included in *griselda* by Misonne, 1974, Part 6:20; but see Van der Straeten, 1980, Rev. Zool. Afr., 94:185–201, and Van der Straeten and Verheyen, 1980, Mammalia, 44:73.

Lemniscomys macculus (Thomas and Wroughton, 1910). Trans. Zool. Soc. Lond., 19:515.
TYPE LOCALITY: Uganda, S.E. Ruwenzori, Mokia.
DISTRIBUTION: Savannahs of S. Sudan, Uganda, Ethiopia, Kenya, Tanzania, Burundi, and N.E. Zaire.
COMMENT: Reviewed by Van der Straeten and Verheyen, 1979, Mammalia, 43:377–389.
ISIS NUMBER: 5301410011028004001.

Lemniscomys mittendorfi Eisentraut, 1968. Bonn. Zool. Beitr., 19:7.
TYPE LOCALITY: Cameroun, Lake Oku.
DISTRIBUTION: Known only from the type locality.
COMMENT: Formerly included in *striatus* by Misonne, 1974, Part 6:20; but see Van der Straeten, 1978, Afr. Small Mamm. Newsl., 1:5–6, and Van der Straeten and Verheyen, 1980, Mammalia, 44:73.

Lemniscomys rosalia (Thomas, 1904). Ann. Mag. Nat. Hist., ser. 7, 13:414.
TYPE LOCALITY: Tanzania (= Tanganyika), Nguru Mtns., Monda.
DISTRIBUTION: Transvaal (South Africa) to N. Namibia; S. Zimbabwe; Malawi; Mozambique; Tanzania; S. Kenya.
COMMENT: Formerly included in *griselda* by Misonne, 1974, Part 6:20; includes *calidior*, *maculosus*, *phaeotis*, *sabiensis*, *sabulatus*, *spinalis*, and *zuluensis*; see Van der Straeten and Verheyen, 1980, Mammalia, 44:73, and Van der Straeten, 1980, Ann. Cape Prov. Mus. Nat. Hist., 13(5):55–62.

Lemniscomys roseveari Van der Straeten, 1980. Ann. Cape Prov. Mus. Nat. Hist., 13(5):55.
TYPE LOCALITY: Zambia, Zambezi (= Balovae), 13° 33' S., 23° 07' E., 1015 m.
DISTRIBUTION: Known only from the type locality.

Lemniscomys striatus (Linnaeus, 1758). Syst. Nat., 10th ed., 1:62.
TYPE LOCALITY: "India" (probably = West Africa).
DISTRIBUTION: Sierra Leone to N. Tanzania; W. Sudan; Ethiopia; Kenya; Uganda; Zaire; N.W. Angola; Zambia; Malawi.
COMMENT: Includes *lynesi*; see Misonne, 1974, Part 6:20. *L. fasciatus* is a synonym; see Van der Straeten, 1975, Unpubl. Ph.D. Dissertation, Antwerp. Formerly included *mittendorfi*; see Van der Straeten and Verheyen, 1980, Mammalia, 44:73. Type locality clarified by Thomas, 1911, Proc. Zool. Soc. Lond., 1911:120–158.
ISIS NUMBER: 5301410011028005001 as *L. striatus*.
5301410011028003001 as *L. lynesi*.

Lenomys Thomas, 1898. Trans. Zool. Soc. Lond., 14(6):409.
REVIEWED BY: B. R. Stein (BRS).
COMMENT: Reviewed by Musser, 1981, Bull. Am. Mus. Nat. Hist., 169(2):137.
ISIS NUMBER: 5301410011029000000.

Lenomys meyeri (Jentink, 1879). Notes Leyden Mus., 1:12.
TYPE LOCALITY: Indonesia, Sulawesi, Sulawesi Utara, Menado.
DISTRIBUTION: Sulawesi.
COMMENT: Includes *longicaudus* and *limpo*; see Musser, 1970, Am. Mus. Novit., 2440:18.
ISIS NUMBER: 5301410011029002001 as *L. meyeri*.
5301410011029001001 as *L. longicaudus*.

Lenothrix Miller, 1903. Proc. U.S. Nat. Mus., 26(1317):466.
    REVIEWED BY: B. R. Stein (BRS).
    COMMENT: Considered a genus distinct from *Rattus* by Misonne, 1969, Ann. Mus. R.
        Afr. Cent. Tervuren, ser. 8, Sci. Zool., 172:122, Musser and Boeadi, 1980, J.
        Mammal., 61:396, and Musser, 1981, Bull. Am. Mus. Nat. Hist., 168(3):310–314.

Lenothrix canus Miller, 1903. Proc. U.S. Nat. Mus., 26(1317):466.
    TYPE LOCALITY: Indonesia, Tuangku Isl. (west of Sumatra).
    DISTRIBUTION: Malaya; Tuangku Isl.; Sarawak.
    COMMENT: Includes *malaisia*; see Medway, 1977:119. Musser, 1981, Zool.
        Verhandelingen, 189:4, 5, excluded the W. Javan *sodyi* from *canus* and placed it in
        Kadarsanomys.
    ISIS NUMBER: 5301410011058022001 as *Rattus canus*.

Leopoldamys Ellerman, 1947. Proc. Zool. Soc. Lond., 1947–1948:267.
    REVIEWED BY: B. McGillivray (BM); J. I. Menzies (JIM).
    COMMENT: Formerly included in *Rattus*; see Musser, 1981, Bull. Am. Mus. Nat. Hist.,
        168(3):256.

Leopoldamys edwardsi (Thomas, 1882). Proc. Zool. Soc. Lond., 1882:587.
    TYPE LOCALITY: China, mountains of W. Fukien (probably Kuatun).
    DISTRIBUTION: W. Bengal, Sikkim, and N. Assam (India) to N. Thailand, S. China, and
        Vietnam; Malaya; Sumatra.
    COMMENT: Includes *ciliatus, garonum, gigas, listeri, melli, milleti,* and *setiger;* may contain
        more than one species; see Musser, 1981, Bull. Am. Mus. Nat. Hist., 168(3):256,
        263.
    ISIS NUMBER: 5301410011058036001 as *Rattus edwardsi*.

Leopoldamys neilli Marshall, 1977. *In* Lekagul and McNeely, Mammals of Thailand, p. 485.
    TYPE LOCALITY: Thailand, Saraburi Prov., Kaengkhoi Dist.
    DISTRIBUTION: C. and W. Thailand.

Leopoldamys sabanus (Thomas, 1887). Ann. Mag. Nat. Hist., 20:269.
    TYPE LOCALITY: Malaysia, Sabah, Mt. Kinabalu.
    DISTRIBUTION: Bangladesh; Thailand; Indochina; Malay Peninsula; Sumatra; Borneo;
        Java; smaller islands on the Sunda Shelf.
    COMMENT: Includes *macrourus* which has priority but is based on a specimen of
        uncertain origin; formerly included *siporanus;* see Musser, 1981, Bull. Am. Mus.
        Nat. Hist., 168(3):256–257, who also stated that *sabanus* may contain more than
        one species.
    ISIS NUMBER: 5301410011058099001 as *Rattus sabanus*.

Leopoldamys siporanus (Thomas, 1895). Ann. Mus. Civ. Stor. Nat. Genova, 34:11.
    TYPE LOCALITY: Indonesia, Mentawi Isls., Sipora.
    DISTRIBUTION: North Pagai, South Pagai, Sipora, and Siberut (Mentawi Isls.).
    COMMENT: Provisionally considered distinct from *sabanus* by Musser, 1981, Bull. Am.
        Mus. Nat. Hist., 168(3):257, 267; includes *soccatus*.

Leporillus Thomas, 1906. Ann. Mag. Nat. Hist., ser. 7, 17:83.
    REVIEWED BY: C. H. S. Watts (CHSW).
    COMMENT: Generic relationships reviewed by Musser, 1981, Bull. Am. Mus. Nat. Hist.,
        169(2):67–176.
    ISIS NUMBER: 5301410011030000000.

Leporillus apicalis (Gould, 1854). Proc. Zool. Soc. Lond., 1853:126.
    TYPE LOCALITY: Australia, South Australia.
    DISTRIBUTION: S. and C. Australia.
    COMMENT: Probably extinct (CHSW).
    ISIS NUMBER: 5301410011030001001.

Leporillus conditor (Sturt, 1848). Narr. Exped. C. Aust., 1:120.
    TYPE LOCALITY: Australia, New South Wales, Laidley's Ponds.
    DISTRIBUTION: W. and S. Australia; New South Wales; Franklin Isl. (South Australia).

COMMENT: Includes *jonesi;* see Ride, 1970:244. Extirpated from the mainland; see Corbet and Hill, 1980:177.
PROTECTED STATUS: CITES - Appendix I and U.S. ESA - Endangered.
ISIS NUMBER: 5301410011030002001.

Leptomys Thomas, 1897. Ann. Mus. Civ. Stor. Nat. Genova, 18:610.
REVIEWED BY: J. I. Menzies (JIM); A. C. Ziegler (ACZ).
COMMENT: Generic relationship reviewed by Musser, 1981, Bull. Am. Mus. Nat. Hist., 169(2):67–176.
ISIS NUMBER: 5301410011077000000.

Leptomys elegans Thomas, 1897. Ann. Mus. Civ. Stor. Nat. Genova, 18:610.
TYPE LOCALITY: Papua New Guinea, (="British New Guinea"), Central Prov., Astrolabe Range.
DISTRIBUTION: New Guinea.
COMMENT: See Tate, 1951, Bull. Am. Mus. Nat. Hist., 97:223, for discussion of type locality.
ISIS NUMBER: 5301410011077001001.

Limnomys Mearns, 1905. Proc. U.S. Nat. Mus., 28:451.
REVIEWED BY: B. R. Stein (BRS).
COMMENT: Revised by Musser, 1977, Am. Mus. Novit., 2636:1–14, who removed it from *Rattus.*

Limnomys sibuanus Mearns, 1905. Proc. U.S. Nat. Mus., 28:452.
TYPE LOCALITY: Philippines, S.E. Mindanao, Mt. Apo, 6600 ft. (2012 m).
DISTRIBUTION: Mindanao (Philippines).
COMMENT: Includes *mearnsi* and *picinus* (in part; a composite); see Musser, 1977, Am. Mus. Novit., 2636:1–14.

Lophuromys Peters, 1874. Monatsb. Preuss. Akad. Wiss. Berlin, p. 234.
REVIEWED BY: E. Van der Straeten (EVS).
COMMENT: Revised by Dieterlen, 1976, Stuttgarten Beitrage Naturk., ser. A. (Biol.), 285:1–96.
ISIS NUMBER: 5301410011031000000.

Lophuromys cinereus Dieterlen and Gelmroth, 1974. Z. Saugetierk., 39:338.
TYPE LOCALITY: Zaire, Marais Mukaba, Nat. Park Kahuzi-Biega.
DISTRIBUTION: Known only from the type locality.
COMMENT: Probably a subspecies of *flavopunctatus* (EVS).

Lophuromys flavopunctatus Thomas, 1888. Proc. Zool. Soc. Lond., 1888:14.
TYPE LOCALITY: Ethiopia (= Abyssinia), Shoa, (= probably Ankober).
DISTRIBUTION: Ethiopia to N.E. Angola and N. Mozambique.
COMMENT: Includes *brevicaudus;* see Misonne, 1974, Part 6:21. Thomas, 1902, Proc. Zool. Soc. Lond., 1902(2):314, discussed the type locality.
ISIS NUMBER: 5301410011031002001 as *L. flavopunctatus.*
5301410011031001001 as *L. brevicaudus.*

Lophuromys luteogaster Hatt, 1934. Am. Mus. Novit., 708:4.
TYPE LOCALITY: Zaire, Ituri, Medje.
DISTRIBUTION: Medje, Irangi, Bafwasende, and Tangula (Zaire).
ISIS NUMBER: 5301410011031003001.

Lophuromys medicaudatus Dieterlen, 1975. Bonn. Zool. Beitr., 26:295.
TYPE LOCALITY: Zaire, Lemera-Nyabutera.
DISTRIBUTION: Montane forests of E. Zaire, above 2000 m.

Lophuromys melanonyx Petter, 1972. Mammalia, 36:177.
TYPE LOCALITY: Ethiopia, Bale Dist., Dinsho (=Gurie).
DISTRIBUTION: C. Ethiopia.
COMMENT: Reviewed by Yalden *et al.,* 1976, Monitore Zool. Ital., n.s., suppl. 8(1):52.

Lophuromys nudicaudus Heller, 1911. Smithson. Misc. Coll., 56(17):11.
TYPE LOCALITY: Cameroun, Bula country, Efulen.
DISTRIBUTION: Cameroun; Bioko; Equatorial Guinea.
COMMENT: *L. naso* is a junior synonym; see Dieterlen, 1978, Bonn. Zool. Beitr.,
29:287–299. Included in *sikapusi* by Misonne, 1974, Part 6:20.
ISIS NUMBER: 5301410011031006001 as *L. nudicaudus.*
5301410011031005001 as *L. naso.*

Lophuromys rahmi Verheyen, 1964. Rev. Zool. Bot. Afr., 69:206.
TYPE LOCALITY: Zaire, Kivu, Bogamanda near Lemera.
DISTRIBUTION: Montane E. Zaire, above 1800 m.
ISIS NUMBER: 5301410011031007001.

Lophuromys sikapusi (Temminck, 1853). Esquisses Zool. sur la Cote de Guine, p. 160.
TYPE LOCALITY: Ghana, Dabacrom.
DISTRIBUTION: Sierra Leone to W. Kenya and N.W. Angola; Bioko.
COMMENT: May include *major*; see Ellerman, 1941:262.
ISIS NUMBER: 5301410011031008001 as *L. sikapusi.*
5301410011031004001 as *L. major.*

Lophuromys woosnami Thomas, 1906. Ann. Mag. Nat. Hist., ser. 7, 18:146.
TYPE LOCALITY: Uganda, Ruwenzori East, Mubuku Valley.
DISTRIBUTION: E. Zaire and W. Uganda above 1700 m.
ISIS NUMBER: 5301410011031009001.

Lorentzimys Jentink, 1911. Nova Guinea, 9:166.
REVIEWED BY: J. I. Menzies (JIM); A. C. Ziegler (ACZ).
COMMENT: Generic relationships reviewed by Musser, 1981, Bull. Am. Mus. Nat. Hist.,
169(2):67–176.
ISIS NUMBER: 5301410011032000000.

Lorentzimys nouhuysi Jentink, 1911. Nova Guinea, 9:166.
TYPE LOCALITY: Indonesia, Irian Jaya, Southern Div. (?), (Noord = Lorentz River),
Bivak 2, about 400 m.
DISTRIBUTION: New Guinea.
COMMENT: Includes *alticola*; see Laurie and Hill, 1954:98, and Lidicker, 1968, J.
Mammal., 49:609–643. For discussion of type locality see Tate, 1951, Bull. Am.
Mus. Nat. Hist., 97:254.
ISIS NUMBER: 5301410011032001001.

Macruromys Stein, 1933. Z. Saugetierk., 8:94.
REVIEWED BY: J. I. Menzies (JIM); A. C. Ziegler (ACZ).
COMMENT: Musser, 1981, Bull. Am. Mus. Nat. Hist., 169(2):67–176, reviewed generic
relationships.
ISIS NUMBER: 5301410011033000000.

Macruromys elegans Stein, 1933. Z. Saugetierk., 8:95.
TYPE LOCALITY: Indonesia, Irian Jaya, Paniai Div., Weyland Mtns., Mt. Kunupi, 1800
m.
DISTRIBUTION: W.C. interior New Guinea.
ISIS NUMBER: 5301410011033001001.

Macruromys major Rummler, 1935. Z. Saugetierk., 10:105.
TYPE LOCALITY: Papua New Guinea; Eastern Highlands Prov. (?), Kratke Mtns.,
Buntibasa Dist., 1220–1525 m.
DISTRIBUTION: N. New Guinea.
ISIS NUMBER: 5301410011033002001.

Malacomys Milne-Edwards, 1877. Bull. Sci. Soc. Philom. Paris, ser. 6, 12:10.
REVIEWED BY: E. Van der Straeten (EVS).
COMMENT: Revised by Rautenbach and Schlitter, 1978, Ann. Carnegie Mus.,
47(17):385–422, and Van der Straeten and Verheyen, 1979, Rev. Zool. Afr.,
93:10–35 (West African species).
ISIS NUMBER: 5301410011034000000.

Malacomys cansdalei Ansell, 1958. Ann. Mag. Nat. Hist., ser. 13, 1:342.
TYPE LOCALITY: Ghana, Oda.
DISTRIBUTION: Forest zone of Ghana, Ivory Coast, and Liberia.
COMMENT: Includes *giganteus;* see Van der Straeten and Verheyen, 1979, Rev. Zool.
Afr., 93:10–35. Rautenbach and Schlitter, 1978, Ann. Carnegie Mus.,
47(17):385–422, considered this species a subspecies of *longipes.*

Malacomys edwardsi Rochebrune, 1885. Bull. Sci. Soc. Philom. Paris, ser. 7, 9:87.
TYPE LOCALITY: Liberia (=Rep. of Guinea), Mellacoree River.
DISTRIBUTION: Guinea to Nigeria.
ISIS NUMBER: 5301410011034001001.

Malacomys longipes Milne-Edwards, 1877. Bull. Sci. Soc. Philom. Paris, ser. 6, 12:10.
TYPE LOCALITY: Gabon, Gaboon River.
DISTRIBUTION: Liberia to Sudan south to Zimbabwe and Angola; Bioko.
ISIS NUMBER: 5301410011034002001.

Malacomys verschureni Verheyen and Van der Straeten, 1977. Rev. Zool. Afr., 91:739.
TYPE LOCALITY: Zaire, Mamiki.
DISTRIBUTION: Known only from the type locality.

Mallomys Thomas, 1898. Novit. Zool., 5:1.
REVIEWED BY: J. I. Menzies (JIM); A. C. Ziegler (ACZ).
COMMENT: Includes *Dendrosminthus;* generic relationships reviewed by Musser, 1981,
Bull. Am. Mus. Nat. Hist., 169(2):67–176.
ISIS NUMBER: 5301410011035000000.

Mallomys rothschildi Thomas, 1898. Novit. Zool., 5:2.
TYPE LOCALITY: Papua New Guinea, Central Prov., Owen Stanley Range, between Mt.
Musgrave and Mt. Scratchley, 1525–1830 m.
DISTRIBUTION: Interior New Guinea.
COMMENT: Type locality discussed by Tate, 1951, Bull. Am. Mus. Nat. Hist., 97:275.
ISIS NUMBER: 5301410011035001001.

Margaretamys Musser, 1981. Bull. Am. Mus. Nat. Hist., 168(3):275.
REVIEWED BY: B. McGillivray (BM); J. I. Menzies (JIM).

Margaretamys beccarii (Jentink, 1880). Notes Leyden Mus., 2:11.
TYPE LOCALITY: Indonesia, Sulawesi, Sulawesi Utara, Menado.
DISTRIBUTION: N.E. and C. Sulawesi.
COMMENT: Includes *thysanurus;* see Musser, 1971, Zool. Meded., 45:147–157. Formerly
placed in subgenus *Maxomys* of *Rattus* by Laurie and Hill, 1954:118, and in *Rattus;*
but see Musser *et al.,* 1979, J. Mammal., 60:592–606, who revised *Maxomys,* and
Musser, 1981, Bull. Am. Mus. Nat. Hist., 168(3):274–286.
ISIS NUMBER: 5301410011058015001 as *Rattus beccarii.*

Margaretamys elegans Musser, 1981. Bull. Am. Mus. Nat. Hist., 168(3):286.
TYPE LOCALITY: Indonesia, Sulawesi, Gunung Nokilalaki, 6500 ft. (1981 m).
DISTRIBUTION: C. Sulawesi.

Margaretamys parvus Musser, 1981. Bull. Am. Mus. Nat. Hist., 168(3):294.
TYPE LOCALITY: Indonesia, Sulawesi, Gunung Nokilalaki, 7400 ft. (2256 m).
DISTRIBUTION: Known only from the vicinity of the type locality.

Mastacomys Thomas, 1882. Ann. Mag. Nat. Hist., ser. 5, 4:413.
REVIEWED BY: C. H. S. Watts (CHSW).
COMMENT: Generic relationships reviewed by Musser, 1981, Bull. Am. Mus. Nat. Hist.,
169(2):67–176.
ISIS NUMBER: 5301410011036000000.

Mastacomys fuscus Thomas, 1882. Ann. Mag. Nat. Hist., ser. 5, 4:413.
TYPE LOCALITY: Australia, Tasmania.
DISTRIBUTION: E. New South Wales, S. Victoria, and Tasmania (Australia).
ISIS NUMBER: 5301410011036001001.

Maxomys Sody, 1936. Natuurk. Tidjschr. Ned.-Ind., 96:55.
  REVIEWED BY: J. I. Menzies (JIM)(Sulawesi).
  COMMENT: Revised by Musser *et al.*, 1979, J. Mammal., 60:592–606, who included some
    species formerly included in *Rattus* and transferred some species formerly
    included in *Maxomys* to *Rattus*. Reviewed by Musser, 1981, Bull. Am. Mus. Nat.
    Hist., 168(3):245–330, and Musser, 1981, Bull. Am. Mus. Nat. Hist., 169(2):67–176.

Maxomys alticola (Thomas, 1888). Ann. Mag. Nat. Hist., ser. 6, 2:408.
  TYPE LOCALITY: Malaysia, Sabah, Kinabalu.
  DISTRIBUTION: N. Borneo.
  ISIS NUMBER: 5301410011058006001 as *Rattus alticola*.

Maxomys baeodon (Thomas, 1894). Ann. Mag. Nat. Hist., ser. 6, 14:452.
  TYPE LOCALITY: Malaysia, Sabah, Kinabalu.
  DISTRIBUTION: N. Borneo.
  ISIS NUMBER: 5301410011058011001 as *Rattus baeodon*.

Maxomys bartelsii (Jentink, 1910). Notes Leyden Mus., 33:69.
  TYPE LOCALITY: Indonesia, W. Java, Gunung Pangerango, 1800 m.
  DISTRIBUTION: W. and C. Java.

Maxomys dollmani (Ellerman, 1941). The Families and Genera of Living Rodents, p. 218.
  TYPE LOCALITY: Indonesia, Sulawesi, Sulawesi Selatan, Quarles Mtns., Rantekaroa.
  DISTRIBUTION: Sulawesi Selatan (Sulawesi).
  ISIS NUMBER: 5301410011058034001 as *Rattus dollmani*.

Maxomys hellwaldi (Jentink, 1878). Notes Leyden Mus., 1:11.
  TYPE LOCALITY: Indonesia, Sulawesi, Sulawesi Utara, Menado.
  DISTRIBUTION: Sulawesi.
  ISIS NUMBER: 5301410011058048001 as *Rattus hellwaldi*.

Maxomys hylomyoides (Robinson and Kloss, 1916). J. Str. Br. Roy. Asiat. Soc., 73:273.
  TYPE LOCALITY: Indonesia, W. Sumatra, Korinchi Peak.
  DISTRIBUTION: Sumatra.

Maxomys inas (Bonhote, 1906). Proc. Zool. Soc. Lond., 1906:9.
  TYPE LOCALITY: Malaysia, Perak, Gunong Inas.
  DISTRIBUTION: Malaya, in montane forests.
  ISIS NUMBER: 5301410011058052001 as *Rattus inas*.

Maxomys inflatus (Robinson and Kloss, 1916). J. Str. Br. Roy. Asiat. Soc., 73:273.
  TYPE LOCALITY: Indonesia, W. Sumatra, Korinchi Peak, Sungei Kumbang.
  DISTRIBUTION: Sumatra.

Maxomys moi (Robinson and Kloss, 1922). Ann. Mag. Nat. Hist., ser. 9, 9:95.
  TYPE LOCALITY: Vietnam, Lang Bian Mtns.
  DISTRIBUTION: Vietnam; Laos.
  ISIS NUMBER: 5301410011058073001 as *Rattus moi*.

Maxomys musschenbroeki (Jentink, 1878). Notes Leyden Mus., 1:10.
  TYPE LOCALITY: Indonesia, Sulawesi, Sulawesi Utara, Menado.
  DISTRIBUTION: Sulawesi.
  ISIS NUMBER: 5301410011058076001 as *Rattus musschenbroekii*.

Maxomys ochraceiventer (Thomas, 1894). Ann. Mag. Nat. Hist., ser. 6, 14:451.
  TYPE LOCALITY: Malaysia, Sabah, Kinabalu.
  DISTRIBUTION: N. and E. Borneo.
  ISIS NUMBER: 5301410011058083001 as *Rattus ochraceiventer*.

Maxomys pagensis (Miller, 1903). Smithson. Misc. Coll., 45:39.
  TYPE LOCALITY: Indonesia, W. Sumatra, S. Pagai Isl.
  DISTRIBUTION: N. Pagai, S. Pagai, Sipora, and Siberut (Mentawai Isls.).

Maxomys panglima (Robinson, 1921). Ann. Mag. Nat. Hist., ser. 9, 7:235.
  TYPE LOCALITY: Philippines, Palawan Isl.
  DISTRIBUTION: Balabac, Palawan, Busuanga, and Culion Isls. (Philippines).
  COMMENT: Includes *palawanensis*; see Musser *et al.*, 1979, J. Mammal., 60:604.

Maxomys rajah (Thomas, 1894). Ann. Mag. Nat. Hist., ser. 6, 14:451.
TYPE LOCALITY: Malaysia, Sarawak, Mt. Batu Song.
DISTRIBUTION: Peninsular Thailand; Malay Peninsula; Riau Arch.; Sumatra; Borneo.
COMMENT: Includes *hidongis, lingensis, pellax,* and *similis;* this species was formerly
   included in *Rattus;* see Musser *et al.,* 1979, J. Mammal., 60:593.
ISIS NUMBER: 5301410011058092001 as *Rattus rajah.*

Maxomys surifer (Miller, 1900). Proc. Biol. Soc. Wash., 13:148.
TYPE LOCALITY: Thailand, Trang.
DISTRIBUTION: S. Burma, Thailand, and Indochina to Sumatra, Java, Borneo, and
   adjacent islands; North Natuna Isls.
ISIS NUMBER: 5301410011058104001 as *Rattus surifer.*

Maxomys whiteheadi (Thomas, 1894). Ann. Mag. Nat. Hist., ser. 6, 14:452.
TYPE LOCALITY: Malaysia, Sabah, Kinabalu.
DISTRIBUTION: Peninsular Thailand, Malay Peninsula, Sumatra, Borneo, and various
   adjacent isls.; Sirhassen (South Natuna Isls.).
ISIS NUMBER: 5301410011058114001 as *Rattus whiteheadi.*

Mayermys Laurie and Hill, 1954. List Land Mammals New Guinea,..., p. 133.
REVIEWED BY: J. I. Menzies (JIM); A. C. Ziegler (ACZ).
COMMENT: Generic relationships reviewed by Musser, 1981, Bull. Am. Mus. Nat. Hist.,
   169(2):67–176.
ISIS NUMBER: 5301410011078000000.

Mayermys ellermani Laurie and Hill, 1954. List Land Mammals New Guinea,..., p. 134.
TYPE LOCALITY: Papua New Guinea, Chimbu Prov., Bismarck Range, N. slopes of Mt.
   Wilhelm, 2440 m.
DISTRIBUTION: E. New Guinea.
ISIS NUMBER: 5301410011078001001.

Melasmothrix Miller and Hollister, 1921. Proc. Biol. Soc. Wash., 34:93.
REVIEWED BY: B. R. Stein (BRS).
COMMENT: Reviewed by Musser, 1981, Bull. Am. Mus. Nat. Hist., 169(2):137.
ISIS NUMBER: 5301410011037000000.

Melasmothrix naso Miller and Hollister, 1921. Proc. Biol. Soc. Wash., 34:93.
TYPE LOCALITY: Indonesia, C. Sulawesi, Rano Rano.
DISTRIBUTION: Sulawesi.
ISIS NUMBER: 5301410011037001001.

Melomys Thomas, 1922. Ann. Mag. Nat. Hist., ser. 9, 9:261.
REVIEWED BY: J. I. Menzies (JIM); C. H. S. Watts (CHSW) (Australia); A. C. Ziegler
   (ACZ)(New Guinea).
COMMENT: Includes *Paramelomys;* see Ellerman, 1941:226, and Musser, 1981, Bull. Am.
   Mus. Nat. Hist., 169(2):137. Currently being revised by J. I. Menzies.
ISIS NUMBER: 5301410011038000000.

Melomys aerosus (Thomas, 1920). Ann. Mag. Nat. Hist., ser. 9, 6:428.
TYPE LOCALITY: Indonesia, Seram, Mt. Manusela, 1800 m.
DISTRIBUTION: Seram Isl. (Indonesia).
ISIS NUMBER: 5301410011038001001.

Melomys albidens Tate, 1951. Bull. Am. Mus. Nat. Hist., 97:286.
TYPE LOCALITY: Indonesia, Irian Jaya, Djajawidjaja Div., 24 km N. Mt. Wilhelmina,
   Lake Habbema, 3225 m.
DISTRIBUTION: W.C. interior New Guinea.
ISIS NUMBER: 5301410011038002001.

Melomys arcium (Thomas, 1913). Ann. Mag. Nat. Hist., 12:214.
TYPE LOCALITY: Papua New Guinea, Milne Bay Prov., Rossel Isl.
DISTRIBUTION: Rossel Isl. (Louisiade Arch.).

Melomys burtoni (Ramsay, 1887). Proc. Linn. Soc. N.S.W., ser. 2, 2:533.
TYPE LOCALITY: Australia, Western Australia, near Derby.
DISTRIBUTION: Australian coast from N. New South Wales through Queensland to N.E. Western Australia.
COMMENT: Includes *littoralis*; see Knox, 1978, J. Zool. Lond., 185:276–277. Probably includes *australius, callopes, M. cervinipes albiventer, melicus, mixtus,* and *murinus; M. lutillus* from S. New Guinea and Cape York may prove to be a subspecies of *burtoni* (CHSW).
ISIS NUMBER: 5301410011038009001 as *M. littoralis.*

Melomys capensis Tate, 1951. Bull. Am. Mus. Nat. Hist., 97:295.
TYPE LOCALITY: Australia, Queensland, Nesbit River, Rocky Scrub (E. of Coen), 1500 ft.
DISTRIBUTION: McIlwraith Ranges of N.E. Queensland (Australia).
COMMENT: Formerly included in *cervinipes* by Tate, 1951, Bull. Am. Mus. Nat. Hist., 97:295; considered a distinct species by Baverstock *et al.,* 1980, Aust. J. Zool., 28:569, based on blood protein differences. Probably includes *banfieldi* and *limicauda* (CHSW).

Melomys cervinipes (Gould, 1852). Mamm. Aust., Part 4, 3:pl. 14.
TYPE LOCALITY: Australia, Queensland, Stradbroke Isl.
DISTRIBUTION: C. Queensland and N. New South Wales (Australia).
COMMENT: Populations north of Cooktown (Queensland) considered in a different species *(capensis)* by Baverstock *et al.,* 1980, Aust. J. Zool., 28:553–574.
ISIS NUMBER: 5301410011038003001.

Melomys fellowsi Hinton, 1943. Ann. Mag. Nat. Hist., 10:554.
TYPE LOCALITY: Papua New Guinea, Madang Prov. (?), Purari-Ramu Divide, Baiyanka, 2440 m.
DISTRIBUTION: E. interior New Guinea.
ISIS NUMBER: 5301410011038004001.

Melomys fraterculus (Thomas, 1920). Ann. Mag. Nat. Hist., ser. 9, 6:428.
TYPE LOCALITY: Indonesia, Seram, Mt. Manusela, 1800 m.
DISTRIBUTION: Seram Isl. (Indonesia).
ISIS NUMBER: 5301410011038005001.

Melomys fulgens (Thomas, 1920). Ann. Mag. Nat. Hist., ser. 9, 6:426.
TYPE LOCALITY: Indonesia, Seram, Teloeti Bay.
DISTRIBUTION: Talaud Isls.; Seram Isl. (Indonesia).
ISIS NUMBER: 5301410011038006001.

Melomys leucogaster (Jentink, 1908). Nova Guinea, 9:3.
TYPE LOCALITY: Indonesia, Irian Jaya, Southern Div., (Lorentz River), Alkmaar, 300 m.
DISTRIBUTION: S. and N.C. New Guinea.
COMMENT: Type locality discussed by Tate, 1951, Bull. Am. Mus. Nat. Hist., 97:307.
ISIS NUMBER: 5301410011038007001.

Melomys levipes (Thomas, 1897). Ann. Mus. Civ. Stor. Nat. Genova, 18:617.
TYPE LOCALITY: Papua New Guinea, Central Prov., Haveri, 9° 25' S., 147° 35' E., 700 m.
DISTRIBUTION: New Guinea; Thornton Peak (N. Queensland, Australia).
COMMENT: Does not include *lorentzi*; see comment under *lorentzi*. Baverstock *et al.,* 1980, Aust. J. Zool., 28:553–574, considered the Thornton Peak, Queensland, population a distinct, undescribed, species as does CHSW.
ISIS NUMBER: 5301410011038008001.

Melomys lorentzi (Jentink, 1908). Nova Guinea, 9:3.
TYPE LOCALITY: Indonesia, Irian Jaya, Southern Div., (Lorentz River), Resi Camp, 900 m.
DISTRIBUTION: New Guinea.
COMMENT: Formerly included in *levipes*; see Ziegler, 1971, Occas. Pap. Bernice P. Bishop Mus., 24:149. Type locality discussed by Tate, 1951, Bull. Am. Mus. Nat. Hist., 97:289.

Melomys lutillus (Thomas, 1913). Ann. Mag. Nat. Hist., 12:216.
  TYPE LOCALITY: Papua New Guinea, Central Prov., Angabunga River, Owgarra.
  DISTRIBUTION: N. and C. Queensland and Hinchinbrook Isl. (Australia); New Guinea; Murray Isl. (Torres Strait).
  COMMENT: Baverstock *et al.*, 1980, Aust. J. Zool., 28:553, 568, presented evidence that *lutillus* in Australia is probably conspecific with *burtoni*. May prove to be a subspecies of *burtoni* (CHSW).
  ISIS NUMBER: 5301410011038010001.

Melomys moncktoni (Thomas, 1904). Ann. Mag. Nat. Hist., 14:399.
  TYPE LOCALITY: Papua New Guinea, Northern Prov., about 8° 30' S., 148° E. (=8° 20' S., 148° 20' E., near Ioma and Buna; alternatively 8° 30' S., 148° 20' E., Kumusi River).
  DISTRIBUTION: S. and E. New Guinea; D'Entrecasteaux Isls.
  COMMENT: Possibly two specifically distinct forms are presently included under this name (ACZ). Type locality discussed by Tate, 1951, Bull. Am. Mus. Nat. Hist., 97:296, and Laurie and Hill, 1954:123.
  ISIS NUMBER: 5301410011038011001.

Melomys obiensis (Thomas, 1911). Ann. Mag. Nat. Hist., ser. 8, 7:208.
  TYPE LOCALITY: Indonesia, Moluccas, Obi Isl.
  DISTRIBUTION: Obi Isl. (Halmahera Isls., Indonesia).
  ISIS NUMBER: 5301410011038012001.

Melomys platyops (Thomas, 1906). Ann. Mag. Nat. Hist., 17:327.
  TYPE LOCALITY: Papua New Guinea, Central Prov., head of Aroa River.
  DISTRIBUTION: New Guinea.
  COMMENT: Probably at least two specifically distinct forms are presently included under this name (ACZ).
  ISIS NUMBER: 5301410011038013001.

Melomys porculus (Thomas, 1904). Ann. Mag. Nat. Hist., 14:400.
  TYPE LOCALITY: Solomon Isls., Guadalcanal Isl., Aola.
  DISTRIBUTION: Guadalcanal Isl. (Solomon Isls.).
  COMMENT: Tate, 1951, Bull. Am. Mus. Nat. Hist., 97:314, placed this species in *Uromys*, but Laurie and Hill, 1954:126, and subsequent authors included it in *Melomys*.
  ISIS NUMBER: 5301410011038014001.

Melomys rubex Thomas, 1922. Ann. Mag. Nat. Hist., 9:263.
  TYPE LOCALITY: Indonesia, Irian Jaya, Djajawidjaja Div., Doormanpad-bivak (3° 30' S., 138° 30' E., 1410 m).
  DISTRIBUTION: Interior New Guinea.
  COMMENT: For information supplementing the original description, see Thomas, 1922, Nova Guinea, 13:730 (JIM). Type locality discussed by Laurie and Hill, 1954:124.
  ISIS NUMBER: 5301410011038015001.

Melomys rubicola Thomas, 1924. Ann. Mag. Nat. Hist., 13:298.
  TYPE LOCALITY: Australia, Torres Strait, Bramble Cay, about 9° S., 144° E.
  DISTRIBUTION: Bramble Cay (Torres Strait).
  COMMENT: Possibly represents either *leucogaster* or an Australian species (ACZ).
  ISIS NUMBER: 5301410011038016001.

Melomys rufescens (Alston, 1877). Proc. Zool. Soc. Lond., 1877:124.
  TYPE LOCALITY: Papua New Guinea, New Ireland or East New Britain Prov., "Duke of York Isl. or adjacent parts of New Britain or New Ireland" Islands.
  DISTRIBUTION: New Guinea; Bismarck Arch.; Bougainville Isl.
  ISIS NUMBER: 5301410011038017001.

Mesembriomys Palmer, 1906. Proc. Biol. Soc. Wash., 19:97.
  REVIEWED BY: C. H. S. Watts (CHSW).
  COMMENT: Includes *Ammomys*; see Musser, 1981, Bull. Am. Mus. Nat. Hist., 169(2):67–176, who reviewed generic relationships.
  ISIS NUMBER: 5301410011039000000.

Mesembriomys gouldii (Gray, 1843). List Mamm. Br. Mus., p. 116.
   TYPE LOCALITY: Australia, Northern Territory, Port Essington.
   DISTRIBUTION: N. Western Australia, N. Northern Territory, N. Queensland, Melville
      Isl., Bathurst Isl. (Australia).
   ISIS NUMBER: 5301410011039001001.

Mesembriomys macrurus (Peters, 1876). Monatsb. Preuss. Akad. Wiss. Berlin, p. 355.
   TYPE LOCALITY: Australia, Western Australia, near Roebourne.
   DISTRIBUTION: N. Western Australia; N. Northern Territory.
   ISIS NUMBER: 5301410011039002001.

Microhydromys Tate and Archbold, 1941. Am. Mus. Novit., 1101:2.
   REVIEWED BY: J. I. Menzies (JIM); A. C. Ziegler (ACZ).
   COMMENT: Generic relationships reviewed by Musser, 1981, Bull. Am. Mus. Nat. Hist.,
      169(2):67–176.
   ISIS NUMBER: 5301410011079000000.

Microhydromys richardsoni Tate and Archbold, 1941. Am. Mus. Novit., 1101:2.
   TYPE LOCALITY: Indonesia, Irian Jaya, Djajawidjaja Div., Idenburg River, 4 km S.W.
      Bernhard Camp, 850 m.
   DISTRIBUTION: C. and E. New Guinea.
   COMMENT: Known from three specimens (ACZ).
   ISIS NUMBER: 5301410011079001001.

Micromys Dehne, 1841. *Micromys agilis*, Ein Neues Saugethier der Fauna von Dresden, p. 1.
   REVIEWED BY: M. Andera (MA); O. L. Rossolimo (OLR)(U.S.S.R.); S. Wang (SW)(China).
   COMMENT: This genus may contain more than one species; see Musser, 1979, Bull.
      Am. Mus. Nat. Hist., 162(6):436.
   ISIS NUMBER: 5301410011040000000.

Micromys minutus (Pallas, 1771). Reise Prov. Russ. Reichs., 1:454.
   TYPE LOCALITY: U.S.S.R., Ulyanovsk. Obl. middle Volga River, Simbirsk (now
      Ulyanovsk).
   DISTRIBUTION: N. Spain and British Isles to Finland and Caucasus to Yakutia (U.S.S.R.),
      N.E. China, and Korea; S. and C. Japan; Taiwan; Shensi to Chekiang, Kwangtung,
      and Yunnan (China) to Assam (India).
   COMMENT: Includes *danubialis;* described by Simionescu, 1971, Studii si Comunicari
      Bacau, p. 389; see Corbet, 1978:132.
   ISIS NUMBER: 5301410011040001001.

Millardia Thomas, 1911. J. Bombay Nat. Hist. Soc., 20(4):998.
   REVIEWED BY: B. R. Stein (BRS).
   COMMENT: Formerly included *Cremnomys;* see Mishra and Dhanda, 1975, J. Mammal.,
      56:76–80; but also see Misonne, 1969, Ann. Mus. R. Afr. Cent. Tervuren, ser. 8,
      Sci. Zool., 172:1–219, who considered *Cremnomys* a subgenus of *Millardia.*
   ISIS NUMBER: 5301410011041000000.

Millardia gleadowi (Murray, 1886). Proc. Zool. Soc. Lond., 1885:809.
   TYPE LOCALITY: Pakistan, Sind, Karachi.
   DISTRIBUTION: Pakistan; adjacent India and Afghanistan.
   ISIS NUMBER: 5301410011041001001.

Millardia kathleenae Thomas, 1914. J. Bombay Nat. Hist. Soc., 23:29.
   TYPE LOCALITY: Burma, Pagan.
   DISTRIBUTION: C. Burma.
   ISIS NUMBER: 5301410011041002001.

Millardia kondana Mishra and Dhanda, 1975. J. Mammal., 56:76.
   TYPE LOCALITY: India, Maharashtra State, Poona Dist., Sinhgarh, 18° 23′ N., 73° 42′ E.
   DISTRIBUTION: Maharashtra region (India).

Millardia meltada (Gray, 1837). Ann. Mag. Nat. Hist., 1:586.
   TYPE LOCALITY: India, S. Mahratta, Dharwar.
   DISTRIBUTION: Sri Lanka; India; Pakistan; Nepal.
   ISIS NUMBER: 5301410011041003001.

Muriculus Thomas, 1903. Proc. Zool. Soc. Lond., 1902(2):314.
REVIEWED BY: E. Van der Straeten (EVS).
ISIS NUMBER: 5301410011043000000.

Muriculus imberbis (Ruppell, 1845). Mus. Senckenbergianum Abh., 3:110.
TYPE LOCALITY: Ethiopia, Simien (Simen).
DISTRIBUTION: Ethiopia.
COMMENT: Closely related to *Mus* and will probably be included in it; see Misonne, 1974, Part 6:21.
ISIS NUMBER: 5301410011043001001 as *M. imberbis*.
5301410011044013001 as *Mus imberbis*.

Mus Linnaeus, 1766. Syst. Nat., 12th ed., 1:138.
REVIEWED BY: J. T. Marshall (JTM); E. Van der Straeten (EVS)(Africa); S. Wang (SW)(China).
COMMENT: Includes *Nannomys* (see Petter and Matthey, 1975, Part 6.7, and Ellerman, 1941:240), *Coelomys*, *Mycteromys* (see Misonne, 1969, Ann. Mus. R. Afr. Cent. Tervuren, ser. 8, Sci. Zool., 172:1–219), *Pyromys* (see Bishop, 1974, Mammalia, 38:139–141), *Leggada* and *Leggadilla* (see Marshall, 1977, Bull. Am. Mus. Nat. Hist., 158(3):211) who also (pp. 173–220) provided a synopsis of Asian species.
ISIS NUMBER: 5301410011044000000.

Mus abbotti Waterhouse, 1838. Proc. Zool. Soc. Lond., 1837:77.
TYPE LOCALITY: Turkey, Trebizond.
DISTRIBUTION: Macedonia (Yugoslavia, Greece); Turkey; N. Iran.
COMMENT: Considered distinct from *musculus* by Marshall and Sage, 1981, Symp. Zool. Soc. Lond., 47:15–25. Probably includes *spretus*, with intermediate specimens from three localities in Egypt (JTM).

Mus baoulei Vermeiren and Verheyen, 1980. Rev. Zool. Afr., 94:573.
TYPE LOCALITY: Ivory Coast, Lamto (= Ayeremou).
DISTRIBUTION: Known only from the type locality.

Mus booduga (Gray, 1837). Charlesworth's Mag. Nat. Hist., 1:586.
TYPE LOCALITY: India, S. Mahratta.
DISTRIBUTION: India; Sri Lanka; C. Burma.
COMMENT: Includes *fulvidiventris* and *lepidoides*; see Marshall, 1977, Bull. Am. Mus. Nat. Hist., 158(3):204.
ISIS NUMBER: 5301410011044003001.

Mus bufo (Thomas, 1906). Ann. Mag. Nat. Hist., ser. 7, 18:145.
TYPE LOCALITY: Uganda, Ruwenzori East.
DISTRIBUTION: E. Zaire; adjacent Uganda.
ISIS NUMBER: 5301410011044004001.

Mus callewaerti (Thomas, 1925). Ann. Mag. Nat. Hist., ser. 9, 15:668.
TYPE LOCALITY: Zaire, Luluabourg.
DISTRIBUTION: S. Zaire; Angola.
ISIS NUMBER: 5301410011044005001.

Mus caroli Bonhote, 1902. Novit. Zool., 9:627.
TYPE LOCALITY: Japan, Ryukyu (=Liukiu) Isls.
DISTRIBUTION: Okinawa Isl. (Ryukyu Isls.); Taiwan; Hainan, Fukien, and Yunnan (China); Indochina to Sumatra and Java; Flores (Lesser Sunda Isls.).
COMMENT: *M. kakhyensis* is unidentifiable, but was applied erroneously to specimens of *caroli*; see Marshall, 1977, *in* Lekagul and McNeely, p. 434.
ISIS NUMBER: 5301410011044014001 as *M. kakhyensis*.

Mus castaneus Waterhouse, 1843. Ann. Mag. Nat. Hist., 12:134.
TYPE LOCALITY: Philippines.
DISTRIBUTION: Cities of S. and E. Asia.
COMMENT: Formerly included in *musculus*; includes *tytleri* as a subspecies; see Marshall and Sage, 1981, Symp. Zool. Soc. Lond., 47:20–21.

Mus cervicolor Hodgson, 1845. Ann. Mag. Nat. Hist., 15:268.
  TYPE LOCALITY: Nepal.
  DISTRIBUTION: Nepal and Assam (India), to Indochina; Sumatra; Java.
  ISIS NUMBER: 5301410011044006001.

Mus cookii Ryley, 1914. J. Bombay Nat. Hist. Soc., 22:664.
  TYPE LOCALITY: Burma, N. Shan States.
  DISTRIBUTION: India; Nepal; Burma; S.C. China; N. Thailand; Laos; Vietnam.
  COMMENT: Includes *nagarum* and *palnica*; see Marshall, 1977, Bull. Am. Mus. Nat. Hist.,
    158(3):212; also see Marshall and Sage, 1981, Symp. Zool. Soc. Lond., 47:17, 23,
    who listed *nagarum* as a distinct species, without comment.

Mus crociduroides (Robinson and Kloss, 1916). J. Str. Br. Roy. Asiat. Soc., 73:271.
  TYPE LOCALITY: Indonesia, W. Sumatra, Korinchi Peak.
  DISTRIBUTION: W. Sumatra.
  ISIS NUMBER: 5301410011044007001.

Mus domesticus Rutty, 1772. Essay Nat. Hist. Co. Dublin, 1:281.
  TYPE LOCALITY: Ireland, Dublin.
  DISTRIBUTION: Europe, W. and S. of Elbe River; N. Africa; east through deserts of Iran
    and Pakistan to Himalayas. Introduced to the British Isles, Americas, and
    Australia.
  COMMENT: Includes *bactrianus, brevirostris, homourus,* and *praetextus*; see Marshall and
    Sage, 1981, Symp. Zool. Soc. Lond., 47:19–20, who considered *domesticus* distinct
    from *musculus*.

Mus dunni Wroughton, 1912. J. Bombay Nat. Hist. Soc., 21:339.
  TYPE LOCALITY: India, Punjab, Ambala, 900 ft. (274 m).
  DISTRIBUTION: India.
  COMMENT: Reviewed by Marshall, 1977, Bull. Am. Mus. Nat. Hist., 158(3):204.

Mus famulus Bonhote, 1898. J. Bombay Nat. Hist. Soc., 12:99.
  TYPE LOCALITY: S. India, Coonoor.
  DISTRIBUTION: S. India.
  ISIS NUMBER: 5301410011044009001.

Mus fernandoni (Phillips, 1932). Spolia Zeylan., 16:325.
  TYPE LOCALITY: Sri Lanka, Central Prov.
  DISTRIBUTION: Sri Lanka.
  ISIS NUMBER: 5301410011044010001.

Mus goundae Petter and Genest, 1970. Mammalia, 34:455.
  TYPE LOCALITY: Central African Republic, Gounda River.
  DISTRIBUTION: N. Central African Republic.

Mus gratus (Thomas and Wroughton, 1910). Trans. Zool. Soc. Lond., 19:507.
  TYPE LOCALITY: Uganda, Mubuku Valley.
  DISTRIBUTION: W. Uganda.
  COMMENT: May be conspecifc with *minutoides*; see Delany, 1975, The Rodents of
    Uganda, Trustees Br. Mus. Nat. Hist., p. 95.
  ISIS NUMBER: 5301410011044011001.

Mus haussa (Thomas and Hinton, 1920). Novit. Zool., 27:319.
  TYPE LOCALITY: Nigeria, Farniso.
  DISTRIBUTION: Senegal; Mali; Niger; Nigeria.
  ISIS NUMBER: 5301410011044012001.

Mus hortulanus Nordmann, 1840. *In* Demidoff Voy. Russie, 3:45.
  TYPE LOCALITY: U.S.S.R., N. Caucasus.
  DISTRIBUTION: Austria; Yugoslavia; trans-Caucasian U.S.S.R.
  COMMENT: Considered a species distinct from *musculus* by Marshall and Sage, 1981,
    Symp. Zool. Soc. Lond., 47:18.

Mus indutus (Thomas, 1910). Ann. Mag. Nat. Hist., ser. 8, 5:89.
TYPE LOCALITY: Botswana, Molopo.
DISTRIBUTION: Botswana; Transvaal and Orange Free State (South Africa).
COMMENT: Includes *deserti*; see Petter and Matthey, 1975, Part 6.7:3.
ISIS NUMBER: 5301410011044008001 as *M. deserti*.

Mus mahomet Rhoads, 1896. Proc. Acad. Nat. Sci. Phila., p. 532.
TYPE LOCALITY: Somalia, Sheikh Mahomet.
DISTRIBUTION: Ethiopia; Somalia.
COMMENT: Reviewed by Yalden *et al.*, Monitore Zool. Ital., n.s., suppl. 8(1):1–118, and
    Rupp, 1980, Saugetierk. Mitt., 28:87, 89–90.
ISIS NUMBER: 5301410011044015001.

Mus mattheyi Petter, 1969. Mammalia, 33:118.
TYPE LOCALITY: Ghana, Accra.
DISTRIBUTION: Senegal; Ivory Coast; Ghana.

Mus mayori (Thomas, 1915). J. Bombay Nat. Hist. Soc., 23:415.
TYPE LOCALITY: Sri Lanka, Pattipola.
DISTRIBUTION: S. and C. Sri Lanka.
COMMENT: Musser would place this species in a monotypic genus, *Coelomys*, which
    was considered a subgenus of *Mus* by Marshall, 1977, Bull. Am. Mus. Nat. Hist.,
    158(3):209 (JTM).
ISIS NUMBER: 5301410011044016001.

Mus minutoides Smith, 1834. S. Afr. Quart. J., 2:157.
TYPE LOCALITY: South Africa, near Cape Town.
DISTRIBUTION: Africa, south of the Sahara.
COMMENT: Includes *musculoides* and *bellus*; see Petter and Matthey, 1975, Part 6.7:3.
ISIS NUMBER: 5301410011044017001 as *M. minutoides*.
               5301410011044001001 as *M. bellus*.
               5301410011044018001 as *M. musculoides*.

Mus musculus Linnaeus, 1766. Syst. Nat., 12th ed., 1:138.
REVIEWED BY: M. Andera (MA); O. L. Rossolimo (OLR)(U.S.S.R.); G. Urbano-V. (GUV);
    A. C. Ziegler (ACZ)(New Guinea).
TYPE LOCALITY: Sweden, Uppsala County, Uppsala.
DISTRIBUTION: Eurasia from Sweden, N. tip of Jutland, and E. bank of Elbe River to
    Japan; nearly worldwide commensally with man, often feral where introduced.
COMMENT: Revised by Schwartz and Schwartz, 1943, J. Mammal., 24:59–72, and
    Marshall and Sage, 1981, Symp. Zool. Soc. Lond., 47:15–25. Musser, 1977, Am.
    Mus. Novit., 2624:12, included *castaneus* and *commissarius* from the Philippines in
    this species. Does not include *poschiavinus* which Capanna *et al.*, 1977, Boll. Zool.,
    44:213–246, considered a distinct species. Corbet, 1978:141, included *spretus* in
    this species but Pelz and Niethammer, 1978, Z. Saugetierk., 43:302–304,
    considered *spretus* a distinct species. Includes *molossinus*; see Marshall, 1977, Bull.
    Am. Mus. Nat. Hist., 158(3):211; listed as provisionally distinct by Marshall and
    Sage, 1981, Symp. Zool. Soc. Lond., 47:17, 24, but recent biochemical evidence
    suggests that it is conspecific with *musculus* (JTM). Includes *wagneri*; see Marshall
    and Sage, 1981, Symp. Zool. Soc. Lond., 47:15–25, who listed *abbotti, castaneus,*
    domesticus, hortulanus, poschiavinus, and *spretus* as distinct species; also see
    comments under *poschiavinus* and *spretus*.
ISIS NUMBER: 5301410011044019001.

Mus oubanguii Petter and Genest, 1970. Mammalia, 34:454.
TYPE LOCALITY: Central African Republic, Ippy La Maboke, Bangassow.
DISTRIBUTION: Central African Republic.

Mus pahari Thomas, 1916. J. Bombay Nat. Hist. Soc., 24:414.
TYPE LOCALITY: India, Sikkim.
DISTRIBUTION: Sikkim and Assam (India) to Yunnan (China) and Indochina.
ISIS NUMBER: 5301410011044021001.

Mus phillipsi Wroughton, 1912. J. Bombay Nat. Hist. Soc., 21:772.
TYPE LOCALITY: India, Nimur Dist.
DISTRIBUTION: India.

Mus platythrix Bennett, 1832. Proc. Zool. Soc. Lond., 1832:121.
TYPE LOCALITY: India, Dukhun.
DISTRIBUTION: India.
ISIS NUMBER: 5301410011044023001.

Mus poschiavinus Fatio, 1869. Vert. Suisse, 1:207.
REVIEWED BY: M. Andera (MA).
TYPE LOCALITY: Switzerland, Grisons, Poschiavo.
DISTRIBUTION: Poschiavo Valley (Swiss Alps).
COMMENT: Considered a distinct species by Capanna *et al.*, 1977, Boll. Zool.,
    44:213–246; also see Marshall, 1977, Bull. Am. Mus. Nat. Hist., 158(3):216, who
    discussed the relationship of *poschiavinus* to *domesticus*. Corbet, 1978:141, listed it
    provisionally as a subspecies of *musculus*. Gropp *et al.*, 1972, Chromosoma,
    39:269–288, considered *poschiavinus* chromosomally distinct. Marshall and Sage,
    1981, Symp. Zool. Soc. Lond., 47:19, 24, listed *poschiavinus* as biologically and
    chromosomally distinct from *domesticus* and *musculus*, but cranially and
    biochemically inseparable from *domesticus*.

Mus procondon Rhoads, 1896. Proc. Acad. Nat. Sci. Phila., p. 531.
TYPE LOCALITY: Somalia, Sheikh Husein.
DISTRIBUTION: Ethiopia; Somalia; N.E. Zaire.
COMMENT: Reviewed by Yalden *et al.*, Monitore Zool. Ital., n.s., suppl. 8(1):1–118, and
    Rupp, 1980, Saugetierk. Mitt., 28(2):87, 89–90; includes *pasha*; possibly conspecific
    with *minutoides*.
ISIS NUMBER: 5301410011044024001.

Mus saxicola Elliot, 1839. Madras J. Litt. Sci., 10:215.
TYPE LOCALITY: India, Madras.
DISTRIBUTION: India; Pakistan; Nepal.

Mus setulosus Peters, 1876. Monatsb. Preuss. Akad. Wiss. Berlin, p. 480.
TYPE LOCALITY: Cameroun, Victoria.
DISTRIBUTION: Guinea to Gabon and Central African Republic; Ethiopia.

Mus setzeri Petter, 1978. Mammalia, 42:377.
TYPE LOCALITY: Botswana, 82 km W. of Mohembo.
DISTRIBUTION: Namibia; Botswana; Zambia.

Mus shortridgei (Thomas, 1914). J. Bombay Nat. Hist. Soc., 23:30.
TYPE LOCALITY: Burma, Mt. Popa.
DISTRIBUTION: Burma; Thailand; Indochina.

Mus sorella (Thomas, 1909). Ann. Mag. Nat. Hist., ser. 8, 4:548.
TYPE LOCALITY: Kenya, Mt. Elgon, Kirui.
DISTRIBUTION: Uganda; Kenya; Malawi; Tanzania; Zambia.
COMMENT: Includes *neavei*; see Ansell, 1978:81, and Swanepoel *et al.*, 1980, Ann.
    Transvaal Mus., 32(7):160. Petter and Matthey, 1975, Part 6.7:1–4, considered
    neavei a distinct species.
ISIS NUMBER: 5301410011044025001 as *M. sorellus* (sic).
    5301410011044020001 as *M. neavei*.

Mus spretus Lataste, 1883. Acta Linn. Soc. Bordeaux, ser. 7, 4:27.
TYPE LOCALITY: Algeria, Oued Magra, N. of Hodna.
DISTRIBUTION: S. France; Iberian Peninsula; Balearic Isls. (Spain); Morocco; Algeria;
    Tunisia; Libya.
COMMENT: Considered distinct from *musculus* by Britton *et al.*, 1976, C. R. Hebd. Seanc.
    Acad. Sci. Paris, 283:515–518, and Pelz and Niethammer, 1978, Z. Saugetierk.,
    43:302–304. Marshall and Sage, 1981, Symp. Zool. Soc. Lond., 47:17–24, listed
    spretus as distinct; but probably conspecific with *abbotti* because of intermediate
    specimens from three places in Egypt (JTM); see also comment under *abbotti*.

Mus tenellus (Thomas, 1903). Proc. Zool. Soc. Lond., 1:298.
TYPE LOCALITY: Sudan, Roseires.
DISTRIBUTION: Sudan; N. Tanzania; Somalia; Ethiopia; perhaps Kenya.
COMMENT: May include *wamae*; see Ellerman, 1941:252–253.
ISIS NUMBER: 5301410011044026001 as *M. tenellus*.
5301410011044028001 as *M. wamae*.

Mus terricolor Blyth, 1851. J. Asiat. Soc. Bengal, 20:172.
TYPE LOCALITY: S. India, Bengal, neighborhood of Calcutta.
DISTRIBUTION: Nepal to Pakistan and adjacent India.
COMMENT: Reviewed by Marshall, 1977, Bull. Am. Mus. Nat. Hist., 158(3):214.
Probably represents very small specimens of *dunni* and *booduga* (JTM).

Mus triton (Thomas, 1909). Ann. Mag. Nat. Hist., ser. 8, 4:548.
TYPE LOCALITY: Kenya, Mt. Elgon, Kirui.
DISTRIBUTION: Uganda; Kenya; Zaire; Malawi; Mozambique.
COMMENT: Petter and Matthey, 1975, Part 6.7:2, suggested that two species may be
involved based on differing karyotypes. Possibly conspecific with *mahomet*; see
Yalden *et al.*, 1976, Monitore Zool. Ital., n.s., suppl. 8(1):30.
ISIS NUMBER: 5301410011044027001.

Mus vulcani Robinson and Kloss, 1919. Ann. Mag. Nat. Hist., ser. 9, 4:378.
TYPE LOCALITY: Indonesia, W. Java, Gede, Kandang Badak.
DISTRIBUTION: Java.

Mylomys Thomas, 1906. Ann. Mag. Nat. Hist., ser. 7, 18:224.
REVIEWED BY: E. Van der Straeten (EVS).
ISIS NUMBER: 5301410011045000000.

Mylomys dybowskii (Pousargues, 1893). Bull. Soc. Zool. Fr., 18:163.
TYPE LOCALITY: Central African Republic (=French Congo), Kemo River.
DISTRIBUTION: Ivory Coast to Kenya and Zaire.
COMMENT: Includes *cuninghamei* and *lowei*; see Misonne, 1974, Part 6:22.
ISIS NUMBER: 5301410011051003001 as *Pelomys dybowskii*.
5301410011045001001 as *M. cuninghamei*.
5301410011045002001 as *M. lowei*.

Neohydromys Laurie, 1952. Bull. Br. Mus. (Nat. Hist.) Zool., 1:311.
REVIEWED BY: J. I. Menzies (JIM); A. C. Ziegler (ACZ).
COMMENT: Generic relationships reviewed by Musser, 1981, Bull. Am. Mus. Nat. Hist.,
169(2):67–176.
ISIS NUMBER: 5301410011080000000.

Neohydromys fuscus Laurie, 1952. Bull. Br. Mus. (Nat. Hist.) Zool., 1:311.
TYPE LOCALITY: Papua New Guinea, Chimbu Prov., high N. slopes of Mt. Wilhelm,
2745–3050 m.
DISTRIBUTION: E. New Guinea.
ISIS NUMBER: 5301410011080001001.

Nesokia Gray, 1842. Ann. Mag. Nat. Hist., ser. 7, 10:264.
REVIEWED BY: O. L. Rossolimo (OLR); E. Van der Straeten (EVS); S. Wang (SW).
ISIS NUMBER: 5301410011046000000.

Nesokia indica (Gray and Hardwicke, 1830). Illustr. Indian Zool., 1:pl. 11.
TYPE LOCALITY: India (uncertain).
DISTRIBUTION: Asia Minor; Egypt to Pakistan, N. India, and Sinkiang (China);
Turkmenia, Uzbekistan, Tadzhikistan (U.S.S.R.).
ISIS NUMBER: 5301410011046001001.

Niviventer Marshall, 1976. Family Muridae: rats and mice. Government Printing Office,
Bangkok, p. 402.
REVIEWED BY: B. McGillivray (BM); S. Wang (SW)(China).

COMMENT: Formerly included in *Rattus*; see Musser, 1981, Bull. Am. Mus. Nat. Hist., 168(3):236, who revised the genus.

Niviventer andersoni (Thomas, 1911). Abstr. Proc. Zool. Soc. Lond., 90:4.
TYPE LOCALITY: China, Szechwan, Omi San.
DISTRIBUTION: S.E. Tibet, Yunnan, Szechwan and Shensi (China).
COMMENT: Reviewed by Musser and Chiu, 1979, J. Mammal., 60:581-592.

Niviventer brahma (Thomas, 1914). J. Bombay Nat. Hist. Soc., 23:232.
TYPE LOCALITY: India, N. Assam, Anzong Valley in Mishmi Hills, 6000 ft. (1829 m).
DISTRIBUTION: N. Assam (India), N. Burma.
COMMENT: Revised by Musser, 1970, Am. Mus. Novit., 2406:1-27; also see Musser, 1973, J. Mammal., 54:267-270. Formerly included in *Maxomys*; see Musser *et al.*, 1979, J. Mammal., 60:592-605.
ISIS NUMBER: 5301410011058019001 as *Rattus brahma*.

Niviventer bukit (Bonhote, 1903). Ann. Mag. Nat. Hist., 11:125.
TYPE LOCALITY: Malaysia, Bukit Besar, Jalor.
DISTRIBUTION: S. Burma and S. Thailand to Sumatra, Java and Bali; Con Son Isl. (off Vietnam).
COMMENT: Includes *baturus, besuki, condorensis, jacobsoni, lepturoides, lieftincki, marinus, pan, temmincki,* and *treubii*; Marshall, 1977, *in* Lekagul and McNeely, p. 403, included *huang* in this species, but Musser, 1981, Bull. Am. Mus. Nat. Hist., 168(3):327, included *huang* in *fulvescens* and is followed here; Musser also stated that the relationships between *bukit, fulvescens,* and *rapit* need further clarification. Formerly included in *Maxomys*; see Musser *et al.*, 1979, J. Mammal., 60:592-605.

Niviventer confucianus (Milne-Edwards, 1871). Nouv. Arch. Mus. Hist. Nat. Paris, 7(Bull.):93.
TYPE LOCALITY: China, Szechwan, Moupin.
DISTRIBUTION: N.E. China to N. Thailand, N. Burma and Vietnam; Hainan; Taiwan.
COMMENT: Includes *canorus, champa, chihliensis, culturatus, elegans, littoreus, lotipes, luticolor, mentosus, sacer, sinianus, yaoshanensis,* and *zappeyi*; see Musser and Chiu, 1979, J. Mammal., 60:581-592, and Musser, 1981, Bull. Am. Mus. Nat. Hist., 168(3):237, 251, 255, who reviewed this species and stated that the relationship of *confucianus* to *niviventer* needs clarification. Formerly included in *Maxomys*; see Musser *et al.*, 1979, J. Mammal., 60:592-605.

Niviventer coxingi (Swinhoe, 1864). Proc. Zool. Soc. Lond., 1864:185.
TYPE LOCALITY: Taiwan.
DISTRIBUTION: Taiwan; N. Burma.
COMMENT: Reviewed by Musser and Chiu, 1979, J. Mammal., 60:591. The relationship between *coxingi* and *rapit* needs further study; see Musser, 1981, Bull. Am. Mus. Nat. Hist., 168(3):237, 251. Formerly included in *Maxomys*; see Musser *et al.*, 1979, J. Mammal., 60:592-605.
ISIS NUMBER: 5301410011058027001 as *Rattus coxingi*.

Niviventer cremoriventer (Miller, 1900). Proc. Biol. Soc. Wash., 13:144.
TYPE LOCALITY: Thailand, Trang Prov.
DISTRIBUTION: Peninsular Thailand and Malay Peninsula to Sumatra, Borneo, Java, and Bali.
COMMENT: Includes *barussanus, cretaceiventer, flaviventer, gilbiventer, kina, malawali, mengurus, solus, spatulatus,* and *sumatrae*; formerly included *langbianis*; see Musser, 1973, Am. Mus. Novit., 2525:165, and Musser, 1981, Bull. Am. Mus. Nat. Hist., 168(3):237, 247. Formerly included in *Maxomys*; see Musser *et al.*, 1979, J. Mammal., 60:592-605.
ISIS NUMBER: 5301410011058028001 as *Rattus cremoriventer*.
5301410011058102001 as *Rattus solus*.

Niviventer eha (Wroughton, 1916). J. Bombay Nat. Hist. Soc., 24:428.
TYPE LOCALITY: India, Sikkim, Lachen, 8800 ft. (2682 m).
DISTRIBUTION: Nepal to Tibet (SW), N. Yunnan, Kweichow (SW) (China), and Burma.
COMMENT: Reviewed by Musser, 1970, Am. Mus. Novit., 2406:1-27. Formerly included in *Maxomys*; see Musser *et al.*, 1979, J. Mammal., 60:592. Includes *ninus*; see Musser, 1981, Bull. Am. Mus. Nat. Hist., 168(3):237.

ISIS NUMBER: 5301410011058037001 as *Rattus eha.*

Niviventer excelsior (Thomas, 1911). Abstr. Proc. Zool. Soc. Lond., 1911:4.
TYPE LOCALITY: China, W. Szechwan, Tatsienlu.
DISTRIBUTION: Szechwan (China).
COMMENT: Reviewed by Musser and Chiu, 1979, J. Mammal., 60:581–592.

Niviventer fulvescens (Gray, 1847). Cat. Hodgson Coll. Br. Mus., p. 18.
TYPE LOCALITY: Nepal.
DISTRIBUTION: Himalayas, from N. India and Nepal to Bangladesh, Kansu, and Hainan
(China), and Vietnam.
COMMENT: Formerly included in *Maxomys;* see Musser *et al.,* 1979, J. Mammal., 60:592.
Includes *blythi, caudatior, cinnamomeus, flavipilis, gracilis, huang, jerdoni, lepidus, ling,
mekongis, minor, octomammis, orbus, vulpicolor,* and *wongi;* see Musser, 1981, Bull.
Am. Mus. Nat. Hist., 168(3):237, 251, 255, who stated that the relationships
between *fulvescens* and *niviventer, bukit,* and *rapit* need clarification.
ISIS NUMBER: 5301410011058044001 as *Rattus fulvescens.*
5301410011058051001 as *Rattus huang.*

Niviventer hinpoon Marshall, 1977. *In* Lekagul and McNeely, Mammals of Thailand, p.
459.
TYPE LOCALITY: Thailand, Saraburi Prov., Kaengkhoi Dist.
DISTRIBUTION: Khorat Plateau (Thailand).
COMMENT: Formerly included in *Maxomys;* see Musser *et al.,* 1979, J. Mammal.,
60:592–605.

Niviventer langbianis (Robinson and Kloss, 1922). Ann. Mag. Nat. Hist., ser. 9, 9:96.
TYPE LOCALITY: S. Vietnam, Langbian Peak.
DISTRIBUTION: Assam (India) to Vietnam.
COMMENT: Includes *indosinicus* and *vientianensis;* separated from *cremoriventer* by
Musser, 1973, Am. Mus. Novit., 2525:1–65; also see Musser, 1981, Bull. Am. Mus.
Nat. Hist., 168(3):237.

Niviventer lepturus (Jentink, 1879). Notes Leyden Mus., 2:17.
TYPE LOCALITY: Indonesia, Java.
DISTRIBUTION: W. and C. Java.
COMMENT: Formerly included in *rapit;* includes *fredericae* and *maculipectus;* see Musser,
1981, Bull. Am. Mus. Nat. Hist., 168(3):237.

Niviventer niviventer (Hodgson, 1836). J. Asiat. Soc. Bengal, 5:234.
TYPE LOCALITY: Nepal, Katmandu.
DISTRIBUTION: N.E. Pakistan; Nepal; N. India to Sikkim.
COMMENT: Formerly included in *Maxomys;* see Musser *et al.,* 1979, J. Mammal.,
60:592–606. Musser, 1975, Am. Mus. Novit., 2525:55, discussed the scope of this
species. The limits and relationships of *niviventer* and its allies, in particular
fulvescens and *confucianus,* need clarification; see Marshall, 1977, *in* Lekagul and
McNeely, pp. 402–403, and Musser, 1981, Bull. Am. Mus. Nat. Hist.,
168(3):237–256. Includes *lepcha, monticola,* and *niveiventer;* see Musser, 1981:237.
ISIS NUMBER: 5301410011058081001 as *Rattus niviventer.*

Niviventer rapit (Bonhote, 1903). Ann. Mag. Nat. Hist., ser. 7, 11:123.
TYPE LOCALITY: Malaysia, Sabah, Kinabalu.
DISTRIBUTION: Malay Peninsula; Sumatra; Borneo.
COMMENT: Formerly included in *Maxomys;* see Musser *et al.,* 1979, J. Mammal.,
60:592–605. Includes *atchinensis, cameroni,* and *fraternus;* formerly included *lepturus;*
see Musser, 1981, Bull. Am. Mus. Nat. Hist., 168(3):237, who stated that the
relationships between *rapit, coxingi, bukit,* and *fulvescens* need further study.

Niviventer tenaster (Thomas, 1916). Ann. Mag. Nat. Hist., 17:425.
TYPE LOCALITY: Burma (Tenasserim), Mt. Mulaiyit, 5000–6000 ft. (1524–1829 m).
DISTRIBUTION: Mountains of Assam (India), S. Burma, and Vietnam.
COMMENT: Formerly included in *confucianus* and *cremoriventer;* provisionally
considered a distinct species by Musser, 1981, Bull. Am. Mus. Nat. Hist.,
168(3):237, 253.

Notomys Lesson, 1842. Nouv. Tabl. Regn. Anim. Mammal., p. 129.
  REVIEWED BY: C. H. S. Watts (CHSW).
  COMMENT: Includes *Ascopharynx*, *Podanomalus*, and *Thylacomys*; generic relationships
    reviewed by Musser, 1981, Bull. Am. Mus. Nat. Hist., 169(2):67–176.
  PROTECTED STATUS: CITES - Appendix II as *Notomys* spp.
  ISIS NUMBER: 5301410011048000000.

Notomys alexis Thomas, 1922. Ann. Mag. Nat. Hist., ser. 9, 9:316.
  TYPE LOCALITY: Australia, Northern Territory, Alexandria Downs.
  DISTRIBUTION: Throughout C. Australia in Western Australia, Northern Territory,
    South Australia, and Queensland.
  PROTECTED STATUS: CITES - Appendix II as *Notomys* spp.
  ISIS NUMBER: 5301410011048001001.

Notomys amplus Brazenor, 1936. Mem. Nat. Mus. Melb., 9:7.
  TYPE LOCALITY: Australia, Northern Territory, Charlotte Waters.
  DISTRIBUTION: C. Australia.
  COMMENT: Probably extinct (CHSW).
  PROTECTED STATUS: CITES - Appendix II as *Notomys* spp.
  ISIS NUMBER: 5301410011048002001.

Notomys aquilo Thomas, 1921. Ann. Mag. Nat. Hist., ser. 9, 8:540.
  TYPE LOCALITY: Australia, Queensland, Cape York.
  DISTRIBUTION: N. Queensland and Groot Eylandt off Northern Territory (Australia).
  COMMENT: Includes *carpentarius*; see Ride, 1970:245.
  PROTECTED STATUS: CITES - Appendix II as *Notomys* spp. and U.S. ESA - Endangered.
  ISIS NUMBER: 5301410011048003001.

Notomys cervinus (Gould, 1854). Proc. Zool. Soc. Lond., 1853:127.
  TYPE LOCALITY: Australia, "Interior of South Australia."
  DISTRIBUTION: S.W. Queensland, South Australia, and Northern Territory (Australia).
  PROTECTED STATUS: CITES - Appendix II as *Notomys* spp.
  ISIS NUMBER: 5301410011048004001.

Notomys fuscus (W. Jones, 1925). Rec. S. Aust. Mus., 3:3.
  TYPE LOCALITY: Australia, South Australia, Ooldea.
  DISTRIBUTION: C. Australia in W. South Australia, Northern Territory, and S.W.
    Queensland.
  COMMENT: Includes *filmeri*; see Aitken, 1968, S. Aust. Nat., 43:37–45.
  PROTECTED STATUS: CITES - Appendix II as *Notomys* spp.
  ISIS NUMBER: 5301410011048005001.

Notomys longicaudatus (Gould, 1844). Proc. Zool. Soc. Lond., 1844:104.
  TYPE LOCALITY: Australia, Western Australia, Moore River.
  DISTRIBUTION: S.W. and C. Australia.
  COMMENT: Probably extinct (CHSW).
  PROTECTED STATUS: CITES - Appendix II as *Notomys* spp.
  ISIS NUMBER: 5301410011048006001.

Notomys macrotis Thomas, 1921. Ann. Mag. Nat. Hist., ser. 9, 8:538.
  TYPE LOCALITY: Australia, Western Australia, Moore River.
  DISTRIBUTION: S.W. Australia.
  COMMENT: Includes *megalotis*; see Mahoney, 1975, Aust. Mammal., 1(4):367–374.
    Probably extinct (CHSW).
  PROTECTED STATUS: CITES - Appendix II as *Notomys* spp.
  ISIS NUMBER: 5301410011048007001 as *N. megalotis*.

Notomys mitchellii (Ogilby, 1838). Trans. Linn. Soc. Lond., 18:130.
  TYPE LOCALITY: Australia, New South Wales, junction of Murray and Murrumbidgee
    Rivers.
  DISTRIBUTION: S. Australia.
  COMMENT: Includes *richardsoni*; see Ride, 1970:245.
  PROTECTED STATUS: CITES - Appendix II as *Notomys* spp.
  ISIS NUMBER: 5301410011048008001 as *N. mitchelli (sic)*.

Notomys mordax Thomas, 1922. Ann. Mag. Nat. Hist., ser. 9, 9:317.
TYPE LOCALITY: Australia, Queensland, Darling Downs.
DISTRIBUTION: Known only from the type locality.
COMMENT: Probably extinct (CHSW).
PROTECTED STATUS: CITES - Appendix II as *Notomys* spp.
ISIS NUMBER: 5301410011048009001.

Oenomys Thomas, 1904. Ann. Mag. Nat. Hist., ser. 7, 13:416.
REVIEWED BY: E. Van der Straeten (EVS).
ISIS NUMBER: 5301410011049000000.

Oenomys hypoxanthus (Pucheran, 1855). Rev. Mag. Zool. Paris, ser. 2, 7:206.
TYPE LOCALITY: Gabon.
DISTRIBUTION: Sierra Leone to N. Angola; Zaire to Sudan and Kenya; Ethiopia.
ISIS NUMBER: 5301410011049001001.

Papagomys Sody, 1941. Treubia, 18:322.
ISIS NUMBER: 5301410011050000000.

Papagomys armandvillei (Jentink, 1892). Weber's Zool. Ergebn., 3:79, pl. 5.
TYPE LOCALITY: Indonesia, Flores Isl.
DISTRIBUTION: Flores Isl. (Lesser Sunda Isls.).
COMMENT: Reviewed by Musser, 1981, Bull. Am. Mus. Nat. Hist., 169:69–175.
ISIS NUMBER: 5301410011050001001.

Parahydromys Poche, 1906. Zool. Anz., 30:326.
REVIEWED BY: J. I. Menzies (JIM); A. C. Ziegler (ACZ).
COMMENT: Includes *Drosomys*; generic relationships reviewed by Musser, 1981, Bull.
    Am. Mus. Nat. Hist., 169(2):67–176.
ISIS NUMBER: 5301410011081000000.

Parahydromys asper (Thomas, 1906). Ann. Mag. Nat. Hist., 17:326.
TYPE LOCALITY: Papua New Guinea, Central Prov. (?), (E. part of Owen Stanley Range),
    Richardson Range, Mt. Gayata, 610–1220 m.
DISTRIBUTION: New Guinea.
COMMENT: Type locality discussed by Laurie and Hill, 1954:136.
ISIS NUMBER: 5301410011081001001.

Paraleptomys Tate and Archbold, 1941. Am. Mus. Novit., 1101:1.
REVIEWED BY: J. I. Menzies (JIM); A. C. Ziegler (ACZ).
COMMENT: Generic relationships reviewed by Musser, 1981, Bull. Am. Mus. Nat. Hist.,
    169(2):67–176.
ISIS NUMBER: 5301410011082000000.

Paraleptomys rufilatus Osgood, 1945. Fieldiana Zool., 31:1.
TYPE LOCALITY: Indonesia, Irian Jaya, Jayapura Div., Cyclops Mtns., Mt. Dafonsero,
    1434 m.
DISTRIBUTION: N.C. New Guinea.
ISIS NUMBER: 5301410011082001001.

Paraleptomys wilhelmina Tate and Archbold, 1941. Am. Mus. Novit., 1101:1.
TYPE LOCALITY: Indonesia, Irian Jaya, Djajawidjaja Div., N. of Mt. Wilhelmina, 9 km
    N.E. Lake Habbema, 2800 m.
DISTRIBUTION: C. New Guinea.
ISIS NUMBER: 5301410011082002001.

Paruromys Ellerman, 1954. *In* Laurie and Hill, List of Land Mammals of New Guinea and
    Adj. Isls., 1758–1952, Br. Mus. Nat. Hist., p. 117.
COMMENT: Formerly included as a subgenus of *Rattus* by Ellerman, 1954, *in* Laurie and
    Hill, 1954, p. 117, and in the subgenus *Bullimus* of *Rattus* by Misonne, 1969, Ann.
    Mus. R. Afr. Cent. Tervuren, ser. 8, Sci. Zool., 172:141; but see Musser, 1981, Bull.
    Am. Mus. Nat. Hist., 169(2):137, who considered *Paruromys* and *Bullimus* distinct

genera. Ellerman, 1954, *in* Laurie and Hill, p. 117, also tentatively included
microbullatus (here included in *Rattus callitrichus;* see Musser, 1973, Am. Mus.
Novit., 2511:24) in *Paruromys.*

Paruromys dominator (Thomas, 1921). Ann. Mag. Nat. Hist., 7:244.
TYPE LOCALITY: Indonesia, N. Sulawesi, Minahassa, Mt. Masarang, 1200 m.
DISTRIBUTION: Sulawesi.
COMMENT: Includes *frosti;* see Musser, 1973, Am. Mus. Novit., 2511:24, and Musser,
1971, Am. Mus. Novit., 2454:1-19. Formerly included in *Rattus;* see comment
under genus.
ISIS NUMBER: 5301410011058035001.

Pelomys Peters, 1852. Monatsb. Preuss. Akad. Wiss. Berlin, p. 275.
REVIEWED BY: E. Van der Straeten (EVS).
COMMENT: Divided into 3 subgenera: *Komemys, Desmomys,* and *Pelomys;* see Misonne,
1974, Part 6:22-24. Formerly included *dembeensis* which was transferred to
Arvicanthis by Dieterlen, 1974, Z. Saugetierk., 39:229-231.
ISIS NUMBER: 5301410011051000000.

Pelomys campanae Huet, 1888. Le Naturaliste, ser. 2, 10(31):143.
TYPE LOCALITY: Angola, Cabinda, Landana.
DISTRIBUTION: W. Angola; Zaire.
COMMENT: Subgenus *Pelomys;* see Misonne, 1974, Part 6:24.
ISIS NUMBER: 5301410011051001001.

Pelomys fallax (Peters, 1852). Monatsb. Preuss. Akad. Wiss. Berlin, p. 275.
TYPE LOCALITY: Mozambique, Caya Dist., Zambesi River and Boror, Licuare River.
DISTRIBUTION: Kenya to Mozambique and Botswana; Angola.
COMMENT: Includes *luluae;* subgenus *Pelomys;* see Misonne, 1974, Part 6:24.
ISIS NUMBER: 5301410011051004001 as *P. fallax.*
                         5301410011051008001 as *P. luluae.*

Pelomys harringtoni Thomas, 1903. Proc. Zool. Soc. Lond., 1902(2):313.
TYPE LOCALITY: Ethiopia, Western Shoa, Kutai, Katchisa.
DISTRIBUTION: Ethiopia.
COMMENT: Subgenus *Desmomys;* see Misonne, 1974, Part 6:23. Reviewed by Dieterlen,
1974, Z. Saugetierk., 39:229-231.
ISIS NUMBER: 5301410011051005001.

Pelomys hopkinsi Hayman, 1955. Rev. Zool. Bot. Afr., 52:323.
TYPE LOCALITY: Uganda, Rwamachuchu.
DISTRIBUTION: Rwanda; Uganda.
COMMENT: Subgenus *Komemys;* see Misonne, 1974, Part 6:23.
ISIS NUMBER: 5301410011051006001.

Pelomys isseli (De Beaux, 1924). Ann. Mus. Civ. Stor. Nat. Genova, 51:207.
TYPE LOCALITY: Uganda, Lake Victoria, Kome Isl.
DISTRIBUTION: Lake Victoria islands of Kome, Bugala, and Bunyama (Uganda).
COMMENT: Subgenus *Komemys;* see Misonne, 1974, Part 6:23.
ISIS NUMBER: 5301410011051007001 as *P. isselli (sic).*

Pelomys minor Cabrera and Ruxton, 1926. Ann. Mag. Nat. Hist., ser. 9, 17:601.
TYPE LOCALITY: Zaire, Luluabourg.
DISTRIBUTION: N.E. Angola; N.W. Zambia; S. Zaire.
COMMENT: Subgenus *Pelomys;* see Misonne, 1974, Part 6:23.
ISIS NUMBER: 5301410011051009001.

Pelomys rex (Thomas, 1906). Ann. Mag. Nat. Hist., ser. 7, 18:304.
TYPE LOCALITY: Ethiopia, Kaffa, Charada Forest, 6000 ft. (1829 m).
DISTRIBUTION: Known only from the type locality.
COMMENT: Position is very doubtful; only known from skin of type; see Dieterlen,
1974, Z. Saugetierk., 39:229-231.
ISIS NUMBER: 5301410011051010001.

Phloeomys Waterhouse, 1839. Proc. Zool. Soc. Lond., 1839:107.
REVIEWED BY: B. R. Stein (BRS).
ISIS NUMBER: 5301410011052000000.

Phloeomys cumingi (Waterhouse, 1839). Proc. Zool. Soc. Lond., 1839:108.
TYPE LOCALITY: Philippines, Luzon.
DISTRIBUTION: Luzon (Philippines).
COMMENT: May include *pallidus;* see Musser and Gordon, 1981, J. Mammal., 62:523.
Includes *elegans;* see Ellerman, 1941, The Families and Genera of Living Rodents,
vol. 2, p. 293.
ISIS NUMBER: 5301410011052001001.

Phloeomys pallidus Nehring, 1890. Sitzb. Ges. Naturf. Fr. Berlin, p. 106.
TYPE LOCALITY: Philippines, Luzon.
DISTRIBUTION: Luzon (Philippines).
ISIS NUMBER: 5301410011052002001.

Pithecheir Cuvier, 1838. Hist. Nat. Mamm., 7:livr. 66, two pp. text.
REVIEWED BY: B. R. Stein (BRS).
COMMENT: See Lesson, 1840, Species Mamm., 265, for origin of this name.
ISIS NUMBER: 5301410011053000000.

Pithecheir melanurus Cuvier, 1838. Lesson's Compl. Oeuvres de Buffon, 1:447.
TYPE LOCALITY: Indonesia, W. Sumatra (uncertain).
DISTRIBUTION: Sumatra; Java.
COMMENT: Formerly included *parvus,* which was considered a distinct species by Muul
and Lim Boo Liat, 1971, J. Mammal., 52:436.
ISIS NUMBER: 5301410011053001001.

Pithecheir parvus Kloss, 1916. J. Fed. Malay St. Mus., 6:250.
TYPE LOCALITY: Malaysia, Selangor, Bukit Kutu, near Kuala Kubu, 3400 ft.
DISTRIBUTION: Pahang and Selangor (Malaysia).
COMMENT: Formerly included in *melanurus;* see Muul and Lim Boo Liat, 1971, J.
Mammal., 52:436.

Pogonomelomys Rummler, 1936. Z. Saugetierk., 11:248.
REVIEWED BY: J. I. Menzies (JIM); A. C. Ziegler (ACZ).
COMMENT: Generic relationships reviewed by Musser, 1981, Bull. Am. Mus. Nat. Hist.,
169(2):67–176.
ISIS NUMBER: 5301410011054000000.

Pogonomelomys bruijni (Peters and Doria, 1876). Ann. Mus. Civ. Stor. Nat. Genova, 8:336.
TYPE LOCALITY: Indonesia, Irian Jaya, Sorong Div., Salawati Isl.
DISTRIBUTION: S. and W. New Guinea.
ISIS NUMBER: 5301410011054001001.

Pogonomelomys mayeri (Rothschild and Dollman, 1932). Abstr. Proc. Zool. Soc. Lond.,
353:14.
TYPE LOCALITY: Indonesia, Irian Jaya, Paniai Div., Weyland Range, Gebroeders Mtns.,
1525 m.
DISTRIBUTION: W. New Guinea.
COMMENT: Reviewed by Tate, 1951, Bull. Am. Mus. Nat. Hist., 97:316.
ISIS NUMBER: 5301410011054002001.

Pogonomelomys ruemmleri Tate and Archbold, 1941. Am. Mus. Novit., 1101:6.
TYPE LOCALITY: Indonesia, Irian Jaya, Djajawidjaja Div., N. slope Mt. Wilhelmina, Lake
Habbema, 3225 m.
DISTRIBUTION: Interior New Guinea.
COMMENT: Includes *Rattus shawmayeri* Hinton, 1943; see George, 1978, *in* Tyler, The
Status of Endangered Australasian Wildlife, R. Zool. Soc. S. Aust., p. 95. Probably
merits distinct generic status (ACZ).
ISIS NUMBER: 5301410011054003001.

Pogonomelomys sevia (Tate and Archbold, 1935). Am. Mus. Novit., 803:3.
TYPE LOCALITY: Papua New Guinea, Morobe Prov., Huon Peninsula, Cromwell Range, Sevia, 1400 m.
DISTRIBUTION: E. New Guinea.
ISIS NUMBER: 5301410011054004001.

Pogonomys Milne-Edwards, 1877. C. R. Acad. Sci. Paris, 85:1081.
REVIEWED BY: J. I. Menzies (JIM); A. C. Ziegler (ACZ).
COMMENT: Formerly included *Chiruromys*; see Dennis and Menzies, 1979, J. Zool. Lond., 189:315–332; also see Musser, 1981, Bull. Am. Mus. Nat. Hist., 169(2):67–176, who reviewed generic relationships.
ISIS NUMBER: 5301410011055000000.

Pogonomys loriae Thomas, 1897. Ann. Mus. Civ. Stor. Nat. Genova, 18:613.
TYPE LOCALITY: Papua New Guinea, Central Prov., "among the mountains behind the Astrolabe Range, near Mt. Wori Wori," Haveri, 9° 25′ S., 147° 35′ E., 700 m.
DISTRIBUTION: New Guinea; D'Entrecasteaux Isls.
COMMENT: Includes *fergussoniensis* and *dryas*; *mollipilosus* has been applied to *loriae* but the name is now considered a junior synonym of *macrourus*; see Dennis and Menzies, 1979, J. Zool. Lond., 189:315–332.
ISIS NUMBER: 5301410011055001001 as *P. fergussoniensis*.

Pogonomys macrourus (Milne-Edwards, 1877). C. R. Acad. Sci. Paris, 85:1081.
TYPE LOCALITY: Indonesia, Irian Jaya, Manokwari Div., Amberbaki (or Arfak Mtns. (?)).
DISTRIBUTION: New Guinea; Bismarck Arch.; Cape York (Australia).
COMMENT: Includes *mollipilosus* and *lepidus*; see Dennis and Menzies, 1979, J. Zool. Lond., 189:315–332. Type locality discussed by Tate, 1951, Bull. Am. Mus. Nat. Hist., 97:279.
ISIS NUMBER: 5301410011055005001 as *P. macrourus*.
                5301410011055006001 as *P. mollipilosus*.

Pogonomys sylvestris Thomas, 1920. Ann. Mag. Nat. Hist., ser. 9, 6:534.
TYPE LOCALITY: Papua New Guinea, Morobe Prov., Rawlinson Mtns., 1500 m.
DISTRIBUTION: New Guinea.
ISIS NUMBER: 5301410011055008001.

Praomys Thomas, 1915. Ann. Mag. Nat. Hist., ser. 8, 15:477.
REVIEWED BY: D. A. Schlitter (DS); E. Van der Straeten (EVS).
COMMENT: Includes *Mastomys*, *Myomys*, and *Myomyscus*; see Misonne, 1974, Part 6:24, who also included *Hylomyscus* in this genus; but also see Robbins *et al.*, 1980, Ann. Carnegie Mus., 49(2):31–48, who considered *Hylomyscus* a distinct genus and is followed here. *Mastomys*, *Myomyscus*, and *Praomys* including *Myomys* (here considered subgenera), may be distinct genera; see Van der Straeten, 1979, Afr. Small Mamm. Newsl., 3:27–30; also see Swanepoel *et al.*, 1980, Ann. Transvaal Mus., 32(7):160. This group is in need of revision at the generic and specific levels; see Ansell, 1978:81–82, Misonne, 1974, Part 6, and Van der Straeten and Verheyen, 1978, Z. Saugetierk., 43:31–41.
ISIS NUMBER: 5301410011056000000.

Praomys albipes (Ruppell, 1842). Mus. Senckenbergianum Abh., 3:107.
TYPE LOCALITY: Ethiopia, Massawa.
DISTRIBUTION: Ethiopia; Sudan.
COMMENT: Subgenus *Myomyscus*; see Misonne, 1974, Part 6:26.
ISIS NUMBER: 5301410011056001001.

Praomys angolensis (Bocage, 1890). J. Sci. Math. Phys. Nat. Lisboa, ser. 2, 2(5):12.
TYPE LOCALITY: Angola, Capangombe.
DISTRIBUTION: Angola; S. Zaire.
COMMENT: Subgenus *Mastomys*; see Misonne, 1974, Part 6:25.
ISIS NUMBER: 5301410011056002001.

Praomys butleri (Wroughton, 1907). Ann. Mag. Nat. Hist., ser. 7, 20:503.
TYPE LOCALITY: Sudan, Bahr-El-Ghazal, Chak Chak and Dem Zubeir.
DISTRIBUTION: S.W. Sudan.
COMMENT: Subgenus *Myomyscus;* see Misonne, 1974, Part 6:26. Probably conspecific
with *daltoni* in subgenus *Myomys* (EVS).
ISIS NUMBER: 5301410011056003001.

Praomys daltoni (Thomas, 1892). Ann. Mag. Nat. Hist., ser. 6, 10:181.
TYPE LOCALITY: West Africa (probably Gambia).
DISTRIBUTION: Senegal to Cameroun.
COMMENT: Subgenus *Myomyscus;* see Misonne, 1974, Part 6:27. Includes *Myomys
ingoldbyi;* see Van der Straeten and Verheyen, 1978, Z. Saugetierk., 45:31–41, who
placed *daltoni* in subgenus *Myomys.* May include *butleri* (EVS).
ISIS NUMBER: 5301410011058030001 as *Rattus daltoni.*

Praomys delectorum (Thomas, 1910). Ann. Mag. Nat. Hist., ser. 8, 6:430.
TYPE LOCALITY: Malawi, Mlanji Plateau.
DISTRIBUTION: Malawi; Tanzania; Kenya.
COMMENT: Subgenus *Praomys;* see Misonne, 1974, Part 6:27.
ISIS NUMBER: 5301410011056005001.

Praomys derooi Van der Straeten and Verheyen, 1978. Z. Saugetierk., 43:33.
TYPE LOCALITY: Togo, Borgou.
DISTRIBUTION: Ghana; Togo; Benin; Nigeria.
COMMENT: Subgenus *Myomys;* see Van der Straeten and Verheyen, 1978, Z.
Saugetierk., 43:33.

Praomys erythroleucus (Temminck, 1853). Esquisses Zool. sur la Cote de Guine, p. 160.
TYPE LOCALITY: Guinea.
DISTRIBUTION: Morocco and Senegal to Sudan and Zaire.
COMMENT: Subgenus *Mastomys;* see Misonne, 1974, Part 6:25. Chromosome number
2N=38; formerly included *huberti;* see Petter, 1977, Mammalia, 41:441–444.

Praomys fumatus (Peters, 1878). Monatsb. Preuss. Akad. Wiss. Berlin, p. 200.
TYPE LOCALITY: Kenya, Ukamba.
DISTRIBUTION: E. and N. Tanzania to S. Sudan, Ethiopia, Yemen, and Somalia.
COMMENT: Subgenus *Myomyscus;* see Misonne, 1974, Part 6:26.
ISIS NUMBER: 5301410011056007001.

Praomys hartwigi Eisentraut, 1968. Bonn. Zool. Beitr., 19:8–11.
TYPE LOCALITY: Cameroun, Lake Oku.
DISTRIBUTION: Known only from the type locality.
COMMENT: Subgenus *Praomys;* see Misonne, 1974, Part 6:27.
ISIS NUMBER: 5301410011056008001.

Praomys huberti (Wroughton, 1908). Ann. Mag. Nat. Hist., ser. 8, 1:255.
TYPE LOCALITY: Nigeria, Zungeru.
DISTRIBUTION: Senegal to W. Zaire; Morocco; perhaps also in Zimbabwe and E. Africa.
COMMENT: Subgenus *Mastomys;* see Misonne, 1974, Part 6:25. Chromosome number
2N=32; formerly included in *erythroleucus;* see Petter, 1977, Mammalia,
41:441–444. Probably a synonym of *natalensis;* see Misonne, 1974, Part 6:25.

Praomys jacksoni (De Winton, 1897). Ann. Mag. Nat. Hist., ser. 6, 20:318.
TYPE LOCALITY: Uganda, Entebbe.
DISTRIBUTION: Cameroun to S. Sudan, Kenya, Tanzania, and Zambia.
COMMENT: Subgenus *Praomys;* see Misonne, 1974, Part 6:27. Included in *morio* by
Delany, 1975, Rodents of Uganda; Ansell, 1978:82, considered *jacksoni* a distinct
species, but did not mention Delany, 1975.
ISIS NUMBER: 5301410011056009001.

Praomys morio (Trouessart, 1881). Bull. Soc. d'Etudes Sci. Angers, 10:121.
TYPE LOCALITY: Cameroun, Cameroon Mtns.
DISTRIBUTION: Cameroun; Central African Republic.
COMMENT: Subgenus *Praomys;* see Misonne, 1974, Part 6:27. Delany, 1975, Rodents of
Uganda, included *jacksoni* in this species; but see comment under *jacksoni.*
ISIS NUMBER: 5301410011056012001.

Praomys natalensis (A. Smith, 1834). S. Afr. J., 2:156.
TYPE LOCALITY: South Africa, Port Natal (=Durban).
DISTRIBUTION: W., E., and S. Africa.
COMMENT: Includes *coucha, kulmei,* and *longicaudatus* Noack, 1887; subgenus *Mastomys;* see Misonne, 1974, Part 6:24–25; also see comment under *Mystromys albicaudatus* (Cricetidae). Corbet and Hill, 1980:177, and Swanepoel *et al.,* 1980:160, considered this species an aggregate of at least two sibling species; also see Hallett, 1979, S. Afr. J. Sci., 75:413–415, Gordon, 1978, J. Zool. Lond., 186:397–401, Green *et al.,* 1978, Am. J. Trop. Med. Hyg., 27:627–629, and Lyons *et al.,* 1977, Heredity, 38:197–200.
ISIS NUMBER: 5301410011056013001 as *P. natalensis.*
5301410011056010001 as *P. kulmei.*
5301410011058059001 as *Rattus longicaudatus.*
5301410008023002001 as *Mystromys longicaudatus.*

Praomys pernanus (Kershaw, 1921). Ann. Mag. Nat. Hist., ser. 9, 8:568.
TYPE LOCALITY: Kenya, Amala River.
DISTRIBUTION: Kenya; N. Tanzania; Rwanda.
COMMENT: Subgenus *Mastomys;* see Misonne, 1974, Part 6:25. Probably should be placed in subgenus *Myomys* (DS).

Praomys rostratus (Miller, 1900). Proc. Wash. Acad. Sci., 2:637.
TYPE LOCALITY: Liberia, Mt. Coffee.
DISTRIBUTION: Forest zone of West Africa, west of Dahomey Gap (Senegal to Ghana).
COMMENT: See Van der Straeten and Verheyen, 1979, Afr. Small Mamm. Newsl., 4:13–14, who considered this species distinct from *tullbergi.* Subgenus *Praomys;* see Misonne, 1974, Part 6:27.

Praomys shortridgei (St. Leger, 1933). Proc. Zool. Soc. Lond., p. 411.
TYPE LOCALITY: Namibia, Okavango-Omatako Junction.
DISTRIBUTION: Namibia; Botswana.
COMMENT: Subgenus *Mastomys;* see Misonne, 1974, Part 6:25.
ISIS NUMBER: 5301410011056014001.

Praomys tullbergi (Thomas, 1894). Ann. Mag. Nat. Hist., ser. 6, 13:205.
TYPE LOCALITY: Ghana, Ashanti, Wasa, Ankober River.
DISTRIBUTION: Guinea to Gabon.
COMMENT: Formerly included *rostratus;* see Van der Straeten and Verheyen, 1979, Afr. Small Mamm. Newsl., 4:13–14. Subgenus *Praomys;* see Misonne, 1974, Part 6:27.
ISIS NUMBER: 5301410011056016001.

Praomys verreauxii (Smith, 1834). S. Afr. Quart. J., 2:156.
TYPE LOCALITY: South Africa, near Cape Town.
DISTRIBUTION: S.W. Cape Prov. (South Africa).
COMMENT: Subgenus *Myomyscus;* see Misonne, 1974, Part 6:26. Reviewed by Davis, 1974, Ann. Transvaal Mus., 29(9):160.
ISIS NUMBER: 5301410011056017001 as *P. verreauxi (sic).*

Pseudohydromys Rummler, 1934. Z. Saugetierk., 9:47.
REVIEWED BY: J. I. Menzies (JIM); A. C. Ziegler (ACZ).
COMMENT: Generic relationships reviewed by Musser, 1981, Bull. Am. Mus. Nat. Hist., 169(2):67–176.
ISIS NUMBER: 5301410011083000000.

Pseudohydromys murinus Rummler, 1934. Z. Saugetierk., 9:48.
TYPE LOCALITY: Papua New Guinea, Morobe Prov., Mt. Missim, 2135 m.
DISTRIBUTION: Interior E. New Guinea.
ISIS NUMBER: 5301410011083001001.

Pseudohydromys occidentalis Tate, 1951. Bull. Am. Mus. Nat. Hist., 97:224.
TYPE LOCALITY: Indonesia, Irian Jaya, Djajawidjaja Div., north of Lake Wilhelmina, Lake Habbema, 3225 m.
DISTRIBUTION: Interior W. New Guinea.
ISIS NUMBER: 5301410011083002001.

Pseudomys Gray, 1832. Proc. Zool. Soc. Lond., 1832:39.
REVIEWED BY: C. H. S. Watts (CHSW)(Australia); A. C. Ziegler (ACZ).
COMMENT: Includes *Gyomys* and *Thetomys;* see Ride, 1970:243, 248, Musser, 1981, Bull. Am. Mus. Nat. Hist., 169(2):67–176, and Fox and Briscoe, 1980, Aust. Mammal., 3(2):109–126, who reviewed the genus. Most recent formal revision by Tate, 1951, Bull. Am. Mus. Nat. Hist., 97:187–430. Formerly included *Leggadina;* see Watts, 1976, Trans. R. Soc. S. Aust., 100:105–108, Mahoney and Posamentier, 1975, Aust. Mammal., 1(4):334, and Fox and Briscoe, 1980, Aust. Mammal., 3(2):112.
ISIS NUMBER: 5301410011057000000.

Pseudomys albocinereus (Gould, 1845). Proc. Zool. Soc. Lond., 1845:78.
TYPE LOCALITY: Australia, Western Australia, Moore River.
DISTRIBUTION: S.W. Western Australia; Bernier Isl. (Western Australia).
COMMENT: Ride, 1970:246, included *glaucus* and *apodemoides* in this species but Baverstock *et al.,* 1977, Aust. J. Biol. Sci., 30:471–485, considered *apodemoides* a distinct species, and Fox and Briscoe, 1980, Aust. Mammal., 3(2):112, 121, considered *glaucus* distinct; both are followed here.
ISIS NUMBER: 5301410011057001001.

Pseudomys apodemoides Finlayson, 1932. Trans. Proc. R. Soc. S. Aust., 56:170.
TYPE LOCALITY: Australia, Southern Australia, Coombe.
DISTRIBUTION: S.E. South Australia; W. Victoria; New South Wales.
COMMENT: Ride, 1970, included this species in *albocinereus,* but Baverstock *et al.,* 1977, Aust. J. Biol. Sci., 30:471–485, considered it a distinct species.

Pseudomys australis Gray, 1832. Proc. Zool. Soc. Lond., 1832:39.
TYPE LOCALITY: Australia, New South Wales, Liverpool Plains.
DISTRIBUTION: New South Wales; Queensland; South Australia; Northern Territory.
COMMENT: Includes *auritus* and *minnie;* see Ride, 1970:246. Includes *lineolatus;* see Troughton, 1967, Furred Animals of Australia, 9th ed. Probably includes *rawlinnae;* see comment under *gouldii* (CHSW).
ISIS NUMBER: 5301410011057002001.

Pseudomys chapmani Kitchener, 1980. Rec. W. Aust., 8(3):405.
TYPE LOCALITY: Australia, Western Australia, Pilbara, 31 km 136°, Mt. Meharry (23° 11' S., 118° 48' E.).
DISTRIBUTION: Pilbara Dist. of Western Australia.
COMMENT: Stated to be most closely related to *hermannsburgensis* by Kitchener, 1980, Rec. W. Aust., 8(3):405.

Pseudomys delicatulus (Gould, 1842). Proc. Zool. Soc. Lond., 1842:13.
TYPE LOCALITY: Australia, Northern Territory, Port Essington.
DISTRIBUTION: N. Australia; S.C. New Guinea.
COMMENT: Includes *patria* and *pumilus;* see Ride, 1970:247, and Fox and Briscoe, 1980, Aust. Mammal., 3(2):123. Formerly included in *Leggadina* by Watts, 1976, Trans. R. Soc. S. Aust., 100:105–108; see Baverstock *et al.,* 1976, Trans. R. Soc. S. Aust., 100:109–112, and Fox and Briscoe, 1980, Aust. Mammal., 3(2):109–126. See Waithman, 1979, Aust. Zool., 20:313–326, for discussion of distribution.
ISIS NUMBER: 5301410011057003001 as *Leggadina delicatulus.*

Pseudomys desertor (Troughton, 1932). Rec. Aust. Mus., 18:293.
TYPE LOCALITY: Australia, Northern Territory, Wycliffe Creek.
DISTRIBUTION: W. and C. Australia.
ISIS NUMBER: 5301410011057004001.

Pseudomys fieldi (Waite, 1896). Rept. Horn Sci. Exped. Cent. Aust., Zool., Part 2:403.
TYPE LOCALITY: Australia, Northern Territory, Alice Springs Region.
DISTRIBUTION: Known only from the type locality.
PROTECTED STATUS: U.S. ESA - Endangered.
ISIS NUMBER: 5301410011057005001.

Pseudomys fumeus (Brazenor, 1934). Mem. Nat. Mus. Melb., 8:158.
TYPE LOCALITY: Australia, Victoria Beech Forest.
DISTRIBUTION: C. and E. Victoria (Australia).

COMMENT: Reviewed by Menkhorst and Seebeck, 1981, Aust. Wildl. Res., 8(1):87–96.
PROTECTED STATUS: CITES - Appendix I and U.S. ESA - Endangered.
ISIS NUMBER: 5301410011057007001.

Pseudomys glaucus (Thomas, 1910). Ann. Mag. Nat. Hist., 6:609.
TYPE LOCALITY: Australia, S. Queensland.
DISTRIBUTION: S. Queensland (Australia).
COMMENT: Included in *albocinereus* by Ride, 1970:246, but Fox and Briscoe, 1980, Aust.
Mammal., 3(2):112, 121, considered *glaucus* a distinct species in need of further
study.

Pseudomys gouldii (Waterhouse, 1839). Zool. Voy. H.M.S. "Beagle," Mammalia, 1:67.
TYPE LOCALITY: Australia, New South Wales.
DISTRIBUTION: S. Western Australia, South Australia, and W. New South Wales
(Australia).
COMMENT: Ride, 1970:247, gave *rawlinnae* as a synonym. CHSW considered *rawlinnae* a
synonym of *australis*.
PROTECTED STATUS: U.S. ESA - Endangered.
ISIS NUMBER: 5301410011057008001.

Pseudomys gracilicaudatus (Gould, 1845). Proc. Zool. Soc. Lond., 1845:77.
TYPE LOCALITY: Australia, Queensland, Darling Downs.
DISTRIBUTION: C. and S. Queensland, New South Wales (Australia).
COMMENT: Formerly included *nanus;* see Ride, 1970:246. Baverstock *et al.,* 1977, Aust. J.
Biol. Sci., 30:471–485, considered *nanus* a distinct species.
ISIS NUMBER: 5301410011057009001.

Pseudomys hermannsburgensis (Waite, 1896). Rept. Horn Sci. Exped. Cent. Aust., Zool., Part
2:405.
TYPE LOCALITY: Australia, Northern Territory, Hermannsburg.
DISTRIBUTION: Arid parts of Western Australia, South Australia, S. Northern Territory,
N.W. Victoria, W. New South Wales, and S.W. Queensland (Australia).
COMMENT: Formerly included in *Leggadina* by Watts, 1976, Trans. R. Soc. S. Aust.,
100:105–108; see Baverstock *et al.,* 1976, Trans. R. Soc. S. Aust., 100:109–112, and
Fox and Briscoe, 1980, Aust. Mammal., 3(2):109–126.
ISIS NUMBER: 5301410011057010001 as *Leggadina hermannburgensis (sic).*

Pseudomys higginsi (Trouessart, 1897). Cat. Mamm. Viv. Foss., 1:473.
TYPE LOCALITY: Australia, Tasmania, Kentishbury.
DISTRIBUTION: Tasmania (Australia).
ISIS NUMBER: 5301410011057011001.

Pseudomys nanus (Gould, 1858). Proc. Zool. Soc. Lond., 1858:243.
TYPE LOCALITY: Australia, Western Australia, Moore River.
DISTRIBUTION: Western Australia and Northern Territory (Australia).
COMMENT: Includes *ferculinus;* see Ride, 1970:246. See also comment under
gracilicaudatus.
ISIS NUMBER: 5301410011057012001.

Pseudomys novaehollandiae (Waterhouse, 1844). Proc. Zool. Soc. Lond., 1843:146.
TYPE LOCALITY: Australia, New South Wales, upper Hunter River.
DISTRIBUTION: Coastal region of E. New South Wales, Victoria, and N. Tasmania
(Australia).
PROTECTED STATUS: U.S. ESA - Endangered.
ISIS NUMBER: 5301410011057013001.

Pseudomys occidentalis Tate, 1951. Bull. Am. Mus. Nat. Hist., 97:246.
TYPE LOCALITY: Australia, Western Australia, Tambellup.
DISTRIBUTION: S.W. Western Australia.
PROTECTED STATUS: U.S. ESA - Endangered.
ISIS NUMBER: 5301410011057014001.

Pseudomys oralis Thomas, 1921. Ann. Mag. Nat. Hist., ser. 9, 8:621.
TYPE LOCALITY: Australia, coast of New South Wales.
DISTRIBUTION: N.E. New South Wales and S.E. Queensland (Australia).
ISIS NUMBER: 5301410011057015001.

Pseudomys pilligaensis Fox and Briscoe, 1980. Aust. Mammal., 3(2):112.
TYPE LOCALITY: Australia, New South Wales, Merriwindi State Forest, 3 km W. of Pilliga-Baradine Rd., Cumberdeen Rd. (31° 52' S., 148° 59' E.).
DISTRIBUTION: Pilliga Scrub (New South Wales, Australia).
COMMENT: Further study of the relationships of *pilligaensis, delicatulus, novaehollandiae,* and *hermannsburgensis* is needed; see Fox and Briscoe, 1980, Aust. Mammal., 3(2):124.

Pseudomys praeconis Thomas, 1910. Ann. Mag. Nat. Hist., ser. 8, 6:608.
TYPE LOCALITY: Australia, Western Australia, Peron Peninsula.
DISTRIBUTION: Shark Bay and Bernier Isl. (Western Australia).
PROTECTED STATUS: CITES - Appendix I and U.S. ESA - Endangered.
ISIS NUMBER: 5301410011057016001.

Pseudomys shortridgei (Thomas, 1907). Proc. Zool. Soc. Lond., 1906:765.
TYPE LOCALITY: Australia, Western Australia, East Pingelly.
DISTRIBUTION: W. Victoria; S.W. Western Australia.
PROTECTED STATUS: CITES - Appendix II and U.S. ESA - Endangered.
ISIS NUMBER: 5301410011057017001.

Rattus Fischer, 1803. Natl. Mus. Nat. Paris, 2:128.
REVIEWED BY: K. F. Koopman (KFK); J. T. Marshall (JTM)(S.E. Asia); B. McGillivray (BM); J. I. Menzies (JIM)(New Guinea, Sulawesi); G. G. Musser (GGM); S. Wang (SW) (China); C. H. S. Watts (CHSW)(Australia); A. C. Ziegler (ACZ)(New Guinea).
COMMENT: *Rattus* Frisch, 1775, is unavailable; see Bull. Zool. Nomencl., 1950, 4:549. Australian species revised by Taylor and Horner, 1973, Bull. Am. Mus. Nat. Hist., 150(1):1–130. New Guinea species revised by Dennis and Menzies, 1978, Aust. J. Zool., 26:197–206. Generic limits discussed by Musser and Boeadi, 1980, J. Mammal., 61:396, Musser, 1981, Bull. Am. Mus. Nat. Hist., 168(3):225–334, Musser, 1981, Bull. Am. Mus. Nat. Hist., 169(2):67–176, and Misonne, 1969, Ann. Mus. R. Afr. Cent. Tervuren, ser. 8, Sci. Zool., 172:1–219, 27 pl. Includes *Acanthomys, Geromys, Mollicomys, Nesoromys,* and *Stenomys;* see Musser, 1981, Bull. Am. Mus. Nat. Hist., 169(2):137, 168. Formerly included *Komodomys rintjanus, Diplothrix legatus* (see Musser and Boeadi, 1980, J. Mammal., 61:395–413), *Limnomys* (see Musser, 1977, Am. Mus. Novit., 2636:114), *Maxomys,* in part (see Musser *et al.,* 1979, J. Mammal., 60:592–606), *Cremnomys, Millardia* (see Mishra and Dhanda, 1975, J. Mammal., 56:76), *Tarsomys* (see Musser, 1977, Am. Mus. Novit., 2624:12), *Niviventer, Leopoldamys,* the species *ohiensis,* which was transferred to *Srilankamys,* the species *beccarii,* which was transferred to *Margaretamys* (see Musser, 1981, Bull. Am. Mus. Nat. Hist., 168(3):225–334), *sodyi,* which was transferred to *Kadarsanomys* (see Musser, 1981, Zool. Verhandelingen, 189:5), *Berylmys, Bullimus, Bunomys, Paruromys, Tryphomys,* and *Taeromys* (see Musser, 1981, Bull. Am. Mus. Nat. Hist., 169(2):67–176). Formerly included *shawmayeri,* which was transferred to *Pogonomelomys ruemmleri* by George, 1978, *in* Tyler, The Status of Endangered Australasian Wildlife, Roy. Zool. Soc. South Australia, p. 95.
ISIS NUMBER: 5301410011058000000.

Rattus annandalei (Bonhote, 1903). Fasc. Malayenses Zool., 1:30.
TYPE LOCALITY: Malaysia, S. Perak, Sungkei.
DISTRIBUTION: Malaya; Sumatra.
ISIS NUMBER: 5301410011058008001.

Rattus argentiventer (Robinson and Kloss, 1916). J. Str. Br. Roy. Asiat. Soc., 73:274.
TYPE LOCALITY: Indonesia, "west coast of" Sumatra, Pasir Ganting.
DISTRIBUTION: Thailand and Indochina to Java, Borneo; Mindoro and Mindanao (Philippines); introduced on Lesser Sunda Isls., Sulawesi, and New Guinea.
COMMENT: Includes *pesticulus;* see Musser, 1973, Am. Mus. Novit., 2511:1–30. Reviewed by Musser, 1981, Bull. Am. Mus. Nat. Hist., 169(2):67–176.
ISIS NUMBER: 5301410011058010001.

Rattus atchinus Miller, 1942. Proc. Acad. Nat. Sci. Phila., 94:152.
TYPE LOCALITY: Indonesia, Sumatra, Atjeh, Blangbeke.
DISTRIBUTION: Sumatra.

Rattus baluensis (Thomas, 1894). Ann. Mag. Nat. Hist., ser. 6, 14:454.
TYPE LOCALITY: Malaysia, Sabah, Kinabalu.
DISTRIBUTION: N.E. Borneo; Sumatra.
ISIS NUMBER: 5301410011058013001.

Rattus blangorum Miller, 1942. Proc. Acad. Nat. Sci. Phila., 94:145.
TYPE LOCALITY: Indonesia, N. Sumatra, Atjeh, Blangnanga, 3600 ft. (1097 m).
DISTRIBUTION: N. Sumatra.

Rattus bontanus Thomas, 1921. Ann. Mag. Nat. Hist., 7:246.
TYPE LOCALITY: Indonesia, N. Sulawesi, Minahassa, Mt. Masarang.
DISTRIBUTION: Sulawesi.
COMMENT: Sody, 1941, Treubia, 18(2):260, included this species in *Taeromys;* but see
comment under *Taeromys.*

Rattus burrus (Miller, 1902). Proc. U.S. Nat. Mus., 24:768.
TYPE LOCALITY: India, Nicobar Isls., Trinkut Isl.
DISTRIBUTION: Nicobar Isls.

Rattus callitrichus (Jentink, 1878). Notes Leyden Mus., p. 12.
TYPE LOCALITY: Indonesia, N. Sulawesi, Menado.
DISTRIBUTION: Sulawesi.
COMMENT: Includes *maculipilis, jentinki,* and *microbullatus;* see Musser, 1970, Am. Mus.
Novit., 2440:31. Ellerman, 1954, *in* Laurie and Hill, p. 119, included *microbullatus*
in *Paruromys;* but see Musser, 1970, Am. Mus. Novit., 2440:31.
ISIS NUMBER: 5301410011058021001 as *R. callitrichus.*
5301410011058065001 as *R. maculipilis.*

Rattus ceramicus Thomas, 1920. Ann. Mag. Nat. Hist., ser. 9, 6:425.
TYPE LOCALITY: Indonesia, Seram, Mt. Manusela.
DISTRIBUTION: Seram (Indonesia).
COMMENT: Formerly included in the monotypic genus, *Nesoromys;* see Musser, 1981,
Bull. Am. Mus. Nat. Hist., 169:137, 168, and comment under *Rattus;* closely
related to, or part of, the *R. niobe* group on New Guinea.
ISIS NUMBER: 5301410011047001001.

Rattus culionensis Sanborn, 1952. Fieldiana Zool., 33(2):131.
REVIEWED BY: L. R. Heaney (LRH).
TYPE LOCALITY: Philippines, Culion Isl., Siuk.
DISTRIBUTION: Culion Isl. (Philippines).
COMMENT: Listed as a species by Corbet and Hill, 1980:172, and Alcasid, 1969,
Checklist of Philippine Mammals, Manila Nat. Mus., Philippines, p. 29. Closely
related to, and probably best considered, a subspecies of *muelleri* (LRH).

Rattus dammermani Thomas, 1921. Ann. Mag. Nat. Hist., ser. 9, 7:247.
TYPE LOCALITY: Indonesia, Sulawesi, Wadjo.
DISTRIBUTION: S.W. Sulawesi.
COMMENT: Includes *toxi;* see Musser, 1971, Beaufortia, 18:205–216, for synonymy and
clarification of type locality.
ISIS NUMBER: 5301410011058031001.

Rattus doboensis (Beaufort, 1911). Abh. Senckenb. Naturforsch. Ges., 34:112.
TYPE LOCALITY: Indonesia, Aru Isls., Dobo Isl.
DISTRIBUTION: Aru Isls. (Indonesia).
ISIS NUMBER: 5301410011058033001.

Rattus elephinus Sody, 1941. Treubia, 18:307.
TYPE LOCALITY: Indonesia, Sula Isls., Talubu (= Taliabu, east of Sulawesi).
DISTRIBUTION: Talubu (Sula Isls.).

Rattus enganus (Miller, 1906). Proc. U.S. Nat. Mus., 30:821.
TYPE LOCALITY: Indonesia, Engano Isl. (W. of Sumatra).
DISTRIBUTION: Known only from Engano Isl.

Rattus everetti (Gunther, 1879). Proc. Zool. Soc. Lond., 1879:75.
TYPE LOCALITY: Philippines, Mindanao.
DISTRIBUTION: Luzon, Lubang, and N. Mindanao (Philippines).
COMMENT: Includes *albigularis, gala,* and *tagulayensis;* see Ellerman, 1949, The Families and Genera of Living Rodents, 3:65, but also see Alcasid, 1969, Checklist of Philippine Mammals, Manila Nat. Mus., Philippines, p. 29, who listed *gala* as a distinct species (KFK and GGM). Probably a relative of *Tryphomys adustus* (JH).
ISIS NUMBER: 5301410011058039001 as *R. everetti.*
5301410011058004001 as *R. albigularis.*
5301410011058106001 as *R. tagulayensis.*

Rattus exulans (Peale, 1848). Mammalia and Ornithology, *in* U.S. Expl. Exped., 8:47.
TYPE LOCALITY: Society Isls., Tahiti Isl. (France).
DISTRIBUTION: Bangladesh; Burma; Thailand; Indochina; Malaysia through Philippines, Indonesia, and New Guinea to Micronesia, New Zealand, and Polynesia, including Hawaii and Easter Island. Not on mainland Australia.
COMMENT: Includes *concolor* (see Ellerman and Morrison-Scott, 1951:590), *basilanus, calcis, leucophaetus, luteiventris, mayonicus, negrinus, ornatulus, pantarensis, querceti, vigoratus, vulcani* (see Musser, 1977, Am. Mus. Novit., 2624:12), *micronesiensis* (see Schwarz and Schwarz, 1967, Ann. Esc. Nac. Cienc. Biol. Mex., 14:146), and *bocourti* (see Musser, 1970, Mammalia, 55:489). Human introduction responsible for most of the Pacific insular occurrences (ACZ).
ISIS NUMBER: 5301410011058041001 as *R. exulans.*
5301410011058014001 as *R. basilanus.*
5301410011058020001 as *R. calcis.*
5301410011058057001 as *R. leucophaetus.*
5301410011058061001 as *R. luteiventris.*
5301410011058069001 as *R. mayonicus.*
5301410011058070001 as *R. micronesiensis.*
5301410011058078001 as *R. negrinus.*
5301410011058085001 as *R. ornatulus.*
5301410011058087001 as *R. pantarensis.*
5301410011058091001 as *R. querceti.*
5301410011058112001 as *R. vigoratus.*
5301410011058113001 as *R. vulcani.*

Rattus feliceus (Thomas, 1920). Ann. Mag. Nat. Hist., 6:423.
TYPE LOCALITY: Indonesia, Seram, Mt. Manusela, 6000 ft.
DISTRIBUTION: Seram (Molucca Isls., Indonesia).
COMMENT: Formerly included in *ruber, sensu* Laurie and Hill, 1954:110 (see *praetor);* see Musser, 1981, Bull. Am. Mus. Nat. Hist., 169(2):168, who considered *feliceus* a distinct species.

Rattus foramineus Sody, 1941. Treubia, 18:308.
TYPE LOCALITY: Indonesia, S. Sulawesi, Boeloekoemba.
DISTRIBUTION: Boeloekoemba (S. Sulawesi); Peleng Isl. (E. Sulawesi).

Rattus fuscipes Waterhouse, 1839. Zool. Voy. H.M.S. "Beagle," Mammalia, p. 66.
TYPE LOCALITY: Australia, Western Australia, King George's Sound.
DISTRIBUTION: E. Queensland; E. New South Wales; Victoria; South Australia; S.W. Western Australia.
COMMENT: Includes *assimilis, glauerti, greyi,* and *manicatus;* see Taylor and Horner, 1973, Bull. Am. Mus. Nat. Hist., 150(1):15. Reviewed by Musser, 1981, Bull. Am. Mus. Nat. Hist., 169(2):123.
ISIS NUMBER: 5301410011058045001.

Rattus hamatus Miller and Hollister, 1921. Proc. Biol. Soc. Wash., 34:96.
TYPE LOCALITY: Indonesia, C. Sulawesi, S.W. of Lake Lindoe, Lehio.
DISTRIBUTION: C. Sulawesi.
ISIS NUMBER: 5301410011058047001.

Rattus hoffmanni (Matschie, 1901). Abh. Senckenb. Naturforsch. Ges., 25:284.
TYPE LOCALITY: Indonesia, N. Sulawesi, Minahassa.
DISTRIBUTION: Sulawesi; Malengi (Togian Isls.).
COMMENT: Includes *biformatus, linduensis, mengkoka, mollicomus, mollicomulus,* and *tatei* Ellerman, 1941; see Musser, 1971, Am. Mus. Novit., 2454:1–19.
ISIS NUMBER: 5301410011058049001.

Rattus hoogerwerfi Chasen, 1939. Treubia, 17(3):496.
TYPE LOCALITY: Indonesia, Sumatra, Atjeh, Blang Kedjeren, 2900 ft. (884 m).
DISTRIBUTION: Sumatra.
COMMENT: Listed as a species by Misonne, 1969, Ann. Mus. R. Afr. Cent. Tervuren, ser. 8, Sci. Zool., 172:132, Corbet and Hill, 1980:173, and Musser, 1981, Zool. Verhandelingen, 189:30.
ISIS NUMBER: 5301410011058050001.

Rattus hoxaensis Dao Van Tien, 1960, Zool. Anz., 164:236.
TYPE LOCALITY: Vietnam, Vinh-linh, Hoxa.
DISTRIBUTION: N. Vietnam.

Rattus infraluteus (Thomas, 1888). Ann. Mag. Nat. Hist., ser. 6, 2:409.
TYPE LOCALITY: Malaysia, Sabah, Kinabalu.
DISTRIBUTION: Borneo; Sumatra.
COMMENT: Formerly included *maxi* from Java; see Musser, 1981, Zool. Verhandelingen, 189:29.
ISIS NUMBER: 5301410011058053001.

Rattus latidens Sanborn, 1952. Fieldiana Zool., 33:125.
REVIEWED BY: L. R. Heaney (LRH).
TYPE LOCALITY: Philippines, Luzon, Mountain Prov., Mt. Data, 7500 ft. (2286 m).
DISTRIBUTION: Known only from the type locality.
COMMENT: Musser, 1977, Am. Mus. Novit., 2624:6, and Alcasid, 1969, Checklist of Philippine Mammals, Manila, Nat. Mus., Philippines, p. 31, considered *latidens* a distinct species. A relative of *Tryphomys adustus* (LRH).

Rattus leucopus (Gray, 1867). Proc. Zool. Soc. Lond., 1867:598.
TYPE LOCALITY: Australia, Queensland, Cape York.
DISTRIBUTION: N. Queensland (Australia); New Guinea.
ISIS NUMBER: 5301410011058058001.

Rattus losea (Swinhoe, 1871). Proc. Zool. Soc. Lond., 1870:637.
TYPE LOCALITY: Taiwan.
DISTRIBUTION: Taiwan; Hainan Isl. and Fukien (China) to Vietnam; Thailand; Malaya.
COMMENT: Includes *exiguus* and *sakeratensis;* see Marshall, 1977, *in* Lekagul and McNeely, p. 465.
ISIS NUMBER: 5301410011058060001 as *R. losea.*
5301410011058040001 as *R. exiguus.*

Rattus lutreolus Gray, 1841. J. Two Exped. Aust., 2:409.
TYPE LOCALITY: Australia, South Australia, Torrens River.
DISTRIBUTION: Tasmania and adjacent islands, Victoria, S.E. South Australia, E. New South Wales, and E. Queensland (Australia).
COMMENT: Includes *vellerosus;* see Taylor and Horner, 1973, Bull. Am. Mus. Nat. Hist., 150(1):55. Reviewed by Musser, 1981, Bull. Am. Mus. Nat. Hist., 169(2):123.
ISIS NUMBER: 5301410011058062001.

Rattus macleari (Thomas, 1887). Proc. Zool. Soc. Lond., 1887:533.
TYPE LOCALITY: Christmas Isl. (Australia).
DISTRIBUTION: Christmas Isl. (Indian Ocean).
ISIS NUMBER: 5301410011058064001.

Rattus marmosurus Thomas, 1921. Ann. Mag. Nat. Hist., 7:246.
TYPE LOCALITY: Indonesia, N. Sulawesi, Minahassa, Mt. Masarang.
DISTRIBUTION: N. and Middle Sulawesi.

COMMENT: Includes *facetus* and *tondanus;* see Musser, 1973, Am. Mus. Novit., 2511:24.
Sody, 1941, included *marmosurus* in *Taeromys;* but see comment under *Taeromys.*
ISIS NUMBER: 5301410011058068001 as *R. marmosurus.*
5301410011058042001 as *R. facetus.*
5301410011058108001 as *R. tondanus.*

Rattus maxi Sody, 1932. Natuurh. Maandbl. Maastricht., 21:157.
TYPE LOCALITY: Indonesia, Java, Bandoeng, Tjiboeni.
DISTRIBUTION: Java.
COMMENT: Listed as a distinct species by Musser, 1981, Zool. Verhandelingen, 189:29;
formerly included in *infraluteus* by Ellerman, 1949, The Families and Genera of
Living Rodents, 3:69.

Rattus mindorensis (Thomas, 1898). Trans. Zool. Soc. Lond., 14:402.
TYPE LOCALITY: Philippines, Mindoro, Mt. Dulangan, 1500 m.
DISTRIBUTION: N. Mindoro (Philippines).
COMMENT: Alcasid, 1969, Checklist of Philippine Mammals, Manila Nat. Mus.,
Philippines, p. 32, listed this species as distinct.
ISIS NUMBER: 5301410011058072001.

Rattus montanus Phillips, 1932. Ceylon J. Sci., Sec. B, 16:323.
TYPE LOCALITY: Sri Lanka, West Haputale, Ohiya.
DISTRIBUTION: Sri Lanka.
ISIS NUMBER: 5301410011058074001.

Rattus morotaiensis Kellogg, 1945. Proc. Biol. Soc. Wash., 58:66.
TYPE LOCALITY: Indonesia, Morotai Isl.
DISTRIBUTION: Gilolo group (Molucca Isls., Indonesia).

Rattus muelleri (Jentink, 1879). Notes Leyden Mus., 2:16.
TYPE LOCALITY: Indonesia, Sumatra, Singalur, Batanh.
DISTRIBUTION: S. Burma, Malay Peninsula, Sumatra, Borneo, Palawan, and adjacent
islands; Natuna Isls.; Nicobar Isls.
ISIS NUMBER: 5301410011058075001 as *R. mulleri (sic).*

Rattus nativitatus (Thomas, 1889). Proc. Zool. Soc. Lond., 1888:533.
TYPE LOCALITY: Christmas Isl. (Australia).
DISTRIBUTION: Christmas Isl. (Indian Ocean).
ISIS NUMBER: 5301410011058077001.

Rattus niobe (Thomas, 1906). Ann. Mag. Nat. Hist., 17:327.
TYPE LOCALITY: Papua New Guinea, Central Prov., Angabunga River, Owgarra.
DISTRIBUTION: Interior New Guinea.
ISIS NUMBER: 5301410011058079001.

Rattus nitidus (Hodgson, 1845). Ann. Mag. Nat. Hist., 15:267.
TYPE LOCALITY: Nepal.
DISTRIBUTION: Nepal; Thailand; W. Burma; Kansu and Kiangsu to S. China;
Philippines; introduced in N.W. New Guinea and Sulawesi (Indonesia).
COMMENT: Includes *vanheurni, manuselae* (see Musser, 1973, Am. Mus. Novit.,
2511:1–30, and Musser, 1981, Bull. Am. Mus. Nat. Hist., 169(2):122), *Mus guhai*
(see Marshall, 1977, Bull. Am. Mus. Nat. Hist., 158(3):210), and *subditivus* (see
Musser, 1971, Am. Mus. Novit., 2454:119). Calaby and Taylor, 1980, Zool. Meded.,
55:215–219, considered *ruber* a junior synonym of *nitidus,* and the next available
name for the forms that were formerly included in *ruber* is *praetor.*
ISIS NUMBER: 5301410011058080001 as *R. nitidus.*
5301410011058098001 as *R. ruber.*

Rattus norvegicus (Berkenhout, 1769). Outlines Nat. Hist. Great Britain and Ireland, 1:5
(N.V.).
REVIEWED BY: M. Andera (MA); O. L. Rossolimo (OLR)(U.S.S.R.); G. Urbano-V. (GUV).
TYPE LOCALITY: Great Britain.
DISTRIBUTION: Original distribution assumed to be S.E. Siberia and N. China;
introduced worldwide.

COMMENT: Includes *magnirostris;* see Musser, 1977, Am. Mus. Novit., 2624:12. Includes *insolatus;* see Jones and Johnson, 1965, Univ. Kans. Publ. Mus. Nat. Hist., 16:389.
ISIS NUMBER: 5301410011058082001 as *R. norvegicus.*
5301410011058066001 as *R. magnirostris.*

Rattus omichlodes Misonne, 1979. Bull. Inst. R. Sci. Nat. Belg., 51:1.
TYPE LOCALITY: Indonesia, Irian Jaya, Paniai Div., Ertsberg, (4° 4' S., 137° 7' E.), 3400 m.
DISTRIBUTION: Known only from the type locality.
COMMENT: Misonne, 1979, Bull. Inst. R. Sci. Nat. Belg., 51:1, considered this species to be related to *richardsoni* and *niobe.*

Rattus owiensis Troughton, 1946. Rec. Aust. Mus., 21:374.
TYPE LOCALITY: Papua New Guinea, East Sepik Prov., E. Schouten group, off the mouth of the Sepik River, Owi Isl.
DISTRIBUTION: Owi Isl. (Papua New Guinea).
COMMENT: For discussion of type locality see Tate, 1951, Bull. Am. Mus. Nat. Hist., 97:338.

Rattus palmarum (Zelebor, 1869). Reise der Oesterr. Fregatte Novara. Zool. Th. I, Wirbelth, I, Saugeth., p. 26.
TYPE LOCALITY: India, Nicobar Isls.
DISTRIBUTION: Nicobar Isls.
ISIS NUMBER: 5301410011058086001.

Rattus praetor (Thomas, 1888). Ann. Mag. Nat. Hist., ser. 1, 2:158.
TYPE LOCALITY: Solomon Isls., Guadalcanal Isl., Aola.
DISTRIBUTION: New Guinea; Bismarck Arch.; Solomon Isls.
COMMENT: This name is used in place of *ruber* (now considered a junior synonym of *nitidus);* see Calaby and Taylor, 1980, Zool. Meded., 55:215–219. For information supplementing the original description see Thomas, 1890, Proc. Zool. Soc. Lond., 1889:481. Laurie and Hill, 1954:110, included *feliceus* in *ruber,* but Musser, 1981, Bull. Am. Mus. Nat. Hist., 169(2):168, considered *feliceus* a distinct species from Seram.

Rattus pulliventer (Miller, 1902). Proc. U.S. Nat. Mus., 24:765.
TYPE LOCALITY: India, Nicobar Isls., Great Nicobar Isl.
DISTRIBUTION: Nicobar Isls.
ISIS NUMBER: 5301410011058089001.

Rattus punicans Miller and Hollister, 1921. Proc. Biol. Soc. Wash., 34:98.
TYPE LOCALITY: Indonesia, Middle Sulawesi, Pinedapa.
DISTRIBUTION: C. Sulawesi.
ISIS NUMBER: 5301410011058090001.

Rattus ranjiniae Agrawal and Ghosal, 1969. Proc. Zool. Soc. Calcutta, 22:41.
TYPE LOCALITY: India, Kerala, Trivandrum.
DISTRIBUTION: Known only from the type locality.

Rattus rattus (Linnaeus, 1758). Syst. Nat., 10th ed., 1:61.
REVIEWED BY: M. Andera (MA); G. Urbano-V. (GUV).
TYPE LOCALITY: Sweden, Uppsala County, Uppsala.
DISTRIBUTION: India and S.E. Asia. Introduced worldwide in tropics and warm temperate zones.
COMMENT: Formerly included *simalurensis;* see Musser, 1979, Bull. Am. Mus. Nat. Hist., 162(6):439. Includes *masaretes* (see Musser, 1970, J. Mammal., 51:606–609), *palelae* (see Musser, 1971, Am. Mus. Novit., 2454:1–19), *benguetensis, coloratus, kelleri, lalolis, mindanensis, robiginosus, sapoensis,* and *zamboangae* (see Musser, 1977, Am. Mus. Novit., 2624:12), *sladeni, flavipectus* (see Ellerman and Morrison-Scott, 1951:583), and *mansorius* (see Johnson, 1962, Bull. Bernice P. Bishop Mus., p. 225). Revised by Schwarz and Schwarz, 1967, Ann. Esc. Nac. Cienc. Biol. Mex., 14:79–178. Karyology reviewed by Neithammer, 1975, Zool. Anz., 194:405–415; Gomperl, 1980, p. 87–92, *in* Vorontsov and Van Brink, eds., Animal genetics and evolution, Junk, The Hague. The different chromosomal "races" may actually be

separate species; see Yosida, 1980, Cytogenetics of the Black Rat, Univ. Park Press, Baltimore, U.S.A., 256 pp.

ISIS NUMBER: 5301410011058094001 as *R. rattus*.
5301410011058026001 as *R. coloratus*.
5301410011058054001 as *R. kelleri*.
5301410011058055001 as *R. lalolis*.
5301410011058071001 as *R. mindanensis*.
5301410011058096001 as *R. robiginosus*.
5301410011058101001 as *R. sladeni*.
5301410011058116001 as *R. zamboangae*.

Rattus remotus (Robinson and Kloss, 1914). Ann. Mag. Nat. Hist., ser. 8, 18:231.
TYPE LOCALITY: Thailand, Koh Samui.
DISTRIBUTION: Phangan, Tao, and Samui Isls. (Thailand).
COMMENT: Considered a subspecies of *muelleri* by Hill, 1960, Bull. Raffles Mus., 29:1–112, and as a subspecies of *annandalei* by Chasen, 1940; but see Marshall, 1977, *in* Lekagul and McNeely, p. 469. Conspecific with *sikkimensis* (GGM).

Rattus rennelli Troughton, 1946. Rec. Aust. Mus., 21:375.
TYPE LOCALITY: Solomon Isls., Rennell Isl.
DISTRIBUTION: Rennell Isl. (Solomon Isls.).
ISIS NUMBER: 5301410011058095001.

Rattus richardsoni Tate, 1949. Am. Mus. Novit., 1421:1.
TYPE LOCALITY: Indonesia, Irian Jaya, Djajawidjaja Div., N. of Mt. Wilhelmina, near Lake Habbema, 3225 m.
DISTRIBUTION: W.C. interior New Guinea.

Rattus rogersi (Thomas, 1907). Ann. Mag. Nat. Hist., ser. 7, 20:206.
TYPE LOCALITY: India, Andaman Isls. (Bay of Bengal).
DISTRIBUTION: South Andaman Isl.
ISIS NUMBER: 5301410011058097001.

Rattus salocco Tate and Archbold, 1935. Am. Mus. Novit., 802:7.
TYPE LOCALITY: Indonesia, S.E. Sulawesi, Mengkoka Range, Tanka Salocco.
DISTRIBUTION: S.E. Sulawesi.
ISIS NUMBER: 5301410011058100001.

Rattus sikkimensis Hinton, 1919. J. Bombay Nat. Hist. Soc., 26:394.
TYPE LOCALITY: India, Sikkim, Pashok.
DISTRIBUTION: Nepal and N. Burma to Yunnan, Kwangtung, and Hainan (China) and Vietnam.
COMMENT: The synonym *koratensis* was used for this species by Marshall, 1977, *in* Lekagul and McNeely, pp. 471–472; also see Musser *et al.*, 1979, J. Mammal., 60:599. *R. remotus* is considered a conspecific by GGM (JTM).

Rattus simalurensis Miller, 1903. Proc. U.S. Nat. Mus., 26:458.
TYPE LOCALITY: Indonesia, Sumatra, Simalur Isl.
DISTRIBUTION: N. Mentawi Isls. (Indonesia).
COMMENT: Formerly included in *rattus* by Ellerman, 1949, The Families and Genera of Living Rodents, 3:61, but Musser, 1979, Bull. Am. Mus. Nat. Hist., 162(6):439, considered it a distinct species.

Rattus sordidus (Gould, 1858). Proc. Zool. Soc. Lond., 1857:242.
TYPE LOCALITY: Australia, Queensland, "open plains of" Darling Downs.
DISTRIBUTION: S.C. and S.E. New Guinea; Queensland, Northern Territory, New South Wales, South Australia, and Western Australia (Australia).
COMMENT: Includes *contatus*; see Taylor and Horner, 1973, Bull. Am. Mus. Nat. Hist., 150(1):75. Taylor and Horner, 1973, included *villosissimus* and *colletti* in this species, but Baverstock *et al.*, 1977, Chromosoma, 61:227–241, considered them separate species, while Ride, 1970:138, considered *villosissimus* distinct. Dennis and Menzies, 1978, Aust. J. Zool., 26:197–206, considered *bunae* and *gestri* (or *gestroi*) distinct species from New Guinea; but see Musser, 1981, Bull. Am. Mus.

Nat. Hist., 169(2):119, 120, 123–124, who included *bunae, colletti, gestri* (or *gestroi*), and *villosissimus* in *sordidus*. *R. s. sordidus, colletti,* and *villosissimus,* will hybridize but the hybrids exhibit reduced fertility (CHSW).
ISIS NUMBER: 5301410011058103001.

Rattus stoicus (Miller, 1902). Proc. U.S. Nat. Mus., 24:759.
TYPE LOCALITY: India, Andaman Isls., Henry Lawrence Isl.
DISTRIBUTION: Andaman Isls.

Rattus taerae Sody, 1932. Natuurh. Maandbl. Maastricht., 21:158.
TYPE LOCALITY: Indonesia, N. Sulawesi, Lembean, E. of Tondano.
DISTRIBUTION: N. Sulawesi.
COMMENT: Includes *tatei* Sody, 1941; see Musser, 1971, Zool. Meded., 45(11):135.
ISIS NUMBER: 5301410011058105001.

Rattus tiomanicus (Miller, 1900). Proc. Wash. Acad. Sci., 2:212.
TYPE LOCALITY: Malaysia, Pahang, Tioman Isl.
DISTRIBUTION: Malaya to Borneo and Palawan; many small adjacent islands.
COMMENT: Includes *jalorensis*; see Medway, 1977:110.
ISIS NUMBER: 5301410011058107001.

Rattus tunneyi Thomas, 1904. Novit. Zool., 11:223.
TYPE LOCALITY: Australia, Northern Territory, Mary River.
DISTRIBUTION: N. and S.W. Western Australia, Northern Territory, E. Queensland, and N.E. New South Wales (Australia).
COMMENT: Includes *culmorum* and *melvilleus*; see Taylor and Horner, 1973, Bull. Am. Mus. Nat. Hist., 150(1):92. This species will hybridize with *colletti* but the hybrids exhibit reduced fertility (CHSW).
ISIS NUMBER: 5301410011058109001.

Rattus turkestanicus (Satunin, 1903). Ann. Mus. Zool. Acad. Imp. Sci. St. Petersb., 7:588.
REVIEWED BY: O. L. Rossolimo (OLR).
TYPE LOCALITY: U.S.S.R., Russian Turkestan, Ferghana, Assam-bob.
DISTRIBUTION: S. Russian Turkestan, N.E. Iran, and Afghanistan to N. India and S.W. China.
COMMENT: Includes *rattoides*; see Schlitter and Thonglongya, 1971, Proc. Biol. Soc. Wash., 84:171–174, and Musser *et al.*, 1979, J. Mammal., 60:599; also see Corbet, 1978:139. Gromov and Baranova, 1981:143, employed the name *rattoides* for this species, without comment.
ISIS NUMBER: 5301410011058093001 as *R. rattoides.*

Rattus tyrannus (Miller, 1911). Proc. U.S. Nat. Mus., 38:397.
TYPE LOCALITY: Philippines, Ticao.
DISTRIBUTION: Ticao Isl. (Philippines).
ISIS NUMBER: 5301410011058110001.

Rattus verecundus (Thomas, 1904). Novit. Zool., 11:598.
TYPE LOCALITY: Papua New Guinea, Central Prov., Aroa River, Avera.
DISTRIBUTION: Interior New Guinea.
ISIS NUMBER: 5301410011058111001.

Rattus xanthurus (Gray, 1867). Proc. Zool. Soc. Lond., 1867:598.
TYPE LOCALITY: Indonesia, N. Sulawesi, Tondano, 1100 m.
DISTRIBUTION: Sulawesi.
COMMENT: Includes *paraxanthus* and *faberi*; see Musser, 1973, Am. Mus. Novit., 2511:24. Sody, 1941, Treubia, 18(2):260, included *xanthurus* in *Taeromys*; but see comment under *Taeromys*.
ISIS NUMBER: 5301410011058115001 as *Taeromys xanthurus.*
5301410011058088001 as *R. paraxanthurus (sic).*

Rhabdomys Thomas, 1916. Ann. Mag. Nat. Hist., ser. 8, 18:69.
REVIEWED BY: E. Van der Straeten (EVS).
ISIS NUMBER: 5301410011059000000.

Rhabdomys pumilio (Sparrman, 1784). Kongl. Svenska Vet.-Akad. Nya Handl. Stockholm, p. 236.
TYPE LOCALITY: South Africa, Cape of Good Hope, Snake River, Sitzicamma Forest.
DISTRIBUTION: N. Angola; Namibia; South Africa; Mozambique; Malawi; Zambia; Botswana; S. Zimbabwe; Tanzania; Kenya; E.C. Uganda.
ISIS NUMBER: 5301410011059001001.

Rhynchomys Thomas, 1895. Ann. Mag. Nat. Hist., ser. 6, 16:160.
REVIEWED BY: B. R. Stein (BRS).
COMMENT: Revised by Musser and Freeman, 1981, J. Mammal., 62:154.
ISIS NUMBER: 5301410011084000000.

Rhynchomys isarogensis Musser and Freeman, 1981. J. Mammal., 62:154.
TYPE LOCALITY: Philippines, Camarines Sur Prov., S.E. Peninsula of Luzon Isl., Mt. Isarog, 5500 ft. (1576 m).
DISTRIBUTION: Known only from the type locality.
COMMENT: Known only by the holotype.

Rhynchomys soricoides Thomas, 1895. Ann. Mag. Nat. Hist., ser. 6, 16:160.
TYPE LOCALITY: Philippines, N. Luzon, Mt. Data, 8000 ft. (2438 m).
DISTRIBUTION: Luzon (Philippines).
ISIS NUMBER: 5301410011084001001.

Solomys Thomas, 1922. Ann. Mag. Nat. Hist., 9:261.
REVIEWED BY: J. I. Menzies (JIM); A. C. Ziegler (ACZ).
COMMENT: Included in *Uromys* by Tate, 1951, Bull. Am. Mus. Nat. Hist., 97:313, but considered a distinct genus by Laurie and Hill, 1954:128. Generic relationships reviewed by Musser, 1981, Bull. Am. Mus. Nat. Hist., 169(2):67–176.
ISIS NUMBER: 5301410011060000000.

Solomys ponceleti (Troughton, 1935). Rec. Aust. Mus., 19:260.
TYPE LOCALITY: Papua New Guinea, Bougainville Prov., Bougainville Isl., about 16 km inland from Buin.
DISTRIBUTION: Bougainville Isl. (Papua New Guinea) (in Solomon Isls.).
ISIS NUMBER: 5301410011060001001.

Solomys salebrosus Troughton, 1936. Rec. Aust. Mus., 19:346.
TYPE LOCALITY: Papua New Guinea, Bougainville Prov., Bougainville Isl.
DISTRIBUTION: Bougainville Isl. (Papua New Guinea) (in Solomon Isls.).
ISIS NUMBER: 5301410011060002001.

Solomys sapientis (Thomas, 1902). Ann. Mag. Nat. Hist., 9:446.
TYPE LOCALITY: Solomon Isls., Santa Ysabel Isl.
DISTRIBUTION: Santa Ysabel and Choiseul Isls. (Solomon Isls.).
ISIS NUMBER: 5301410011060003001.

Srilankamys Musser, 1981. Bull. Am. Mus. Nat. Hist., 168(3):268.
REVIEWED BY: B. McGillivray (BM).

Srilankamys ohiensis (Phillips, 1929). Ceylon J. Sci., Sec. B, 15:167.
TYPE LOCALITY: Sri Lanka, Ohiya, W. Haputale, 6000 ft. (1829 m).
DISTRIBUTION: Uva Prov. (Sri Lanka).
COMMENT: Formerly included in *Maxomys* (see Musser *et al.*, 1979, J. Mammal., 60:592–605), *Rattus*, and in the *Leopoldamys* and *Niviventer* groups (see Musser, 1981, Bull. Am. Mus. Nat. Hist., 168(3):267–274).
ISIS NUMBER: 5301410011058084001 as *Rattus ohiensis*.

Stenocephalemys Frick, 1914. Ann. Carnegie Mus., 9:7.
REVIEWED BY: D. A. Schlitter (DS); E. Van der Straeten (EVS).
COMMENT: Reviewed by Yalden *et al.*, 1976, Monitore Zool. Ital., n.s., suppl. 8(1):39–40, and Rupp, 1980, Saugetierk. Mitt., 28:81–123.

ISIS NUMBER: 5301410011061000000.

Stenocephalemys albocaudata Frick, 1914. Ann. Carnegie Mus., 9:8.
TYPE LOCALITY: Ethiopia, Chilalo Mtns., Inyala Camp.
DISTRIBUTION: Ethiopia.
ISIS NUMBER: 5301410011061001001.

Stenocephalemys griseicauda Petter, 1972. Mammalia, 36:171.
TYPE LOCALITY: Ethiopia, Bale Mtns., Dinsho.
DISTRIBUTION: Ethiopia.

Stochomys Thomas, 1926. Ann. Mag. Nat. Hist., ser. 9, 17:176.
REVIEWED BY: E. Van der Straeten (EVS).
COMMENT: Includes *Dephomys;* see Misonne, 1974, Part 6:29. Included in *Aethomys* by
Delany, 1975, The Rodents of Uganda; but also see Corbet and Hill, 1980:177.

Stochomys defua (Miller, 1900). Proc. Wash. Acad. Sci., 2:635.
TYPE LOCALITY: Liberia, Mt. Coffee.
DISTRIBUTION: Guinea to Ghana.
COMMENT: Includes *eburnea;* see Misonne, 1974, Part 6:29. Sometimes included in a
separate genus *Dephomys;* see Rosevear, 1969.
ISIS NUMBER: 5301410011058032001 as *Rattus defua.*

Stochomys longicaudatus (Tullberg, 1893). Nova Acta Reg. Soc. Sci. Upsala, ser. 3,
16(12):36.
TYPE LOCALITY: Cameroun.
DISTRIBUTION: True rain forest, from Togo to Gabon, Zaire, and Uganda.
COMMENT: Includes *ituricus;* see Misonne, 1974, Part 6:29.

Taeromys Sody, 1941. Treubia, 18(2):260.
REVIEWED BY: G. G. Musser (GGM).
COMMENT: Formerly included in *Rattus* by Ellerman, 1949, The Families and Genera of
Living Rodents, 3:189, and in subgenus *Bullimus* of genus *Rattus* by Misonne,
1969, Ann. Mus. R. Afr. Cent. Tervuren, ser. 8, Sci. Zool., 172:141; but considered
a distinct genus by Musser, 1981, Bull. Am. Mus. Nat. Hist., 169(2):137, who also
included *Arcuomys* in *Taeromys.* Musser, 1971, Zool. Meded., 45:128, reviewed the
species allocated to *Taeromys* by Sody, 1941, Treubia, 18(2):260, but *bontanus,*
marmosurus, and *xanthurus* are here included in *Rattus* and *dominator* is included in
Paruromys (GGM, JH).

Taeromys arcuatus (Tate and Archbold, 1935). Am. Mus. Novit., 802:9.
TYPE LOCALITY: Indonesia, S.E. Sulawesi, Mengkoka Range, Tanka Salocco, 1500 m.
DISTRIBUTION: S.E. Sulawesi.
COMMENT: Formerly included in *Rattus* and *Arcuomys;* see comment under genus.
ISIS NUMBER: 5301410011058009001 as *Rattus arcuatus.*

Taeromys celebensis (Gray, 1867). Proc. Zool. Soc. Lond., 1867:598.
TYPE LOCALITY: Indonesia, N. Sulawesi, Menado.
DISTRIBUTION: Sulawesi.
COMMENT: Formerly included in *Rattus;* see comment under genus.
ISIS NUMBER: 5301410011058023001 as *Rattus celebensis.*

Tarsomys Mearns, 1905. Proc. U.S. Nat. Mus., 28:453.
COMMENT: Included in *Rattus* by Simpson, 1945, Bull. Am. Mus. Nat. Hist., 85:1–350,
and Misonne, 1969, Ann. Mus. R. Afr. Cent. Tervuren, ser. 8, Sci. Zool., 172:182;
but see Musser, 1977, Am. Mus. Novit., 2624:12, and Musser, 1981, Bull. Am. Mus.
Nat. Hist., 169(2):137, who listed *Tarsomys* as a distinct genus.

Tarsomys apoensis (Mearns, 1905). Proc. U.S. Nat. Mus., 28:453.
TYPE LOCALITY: Philippines, Mindanao, Davao, Mt. Apo, 2075 m.
DISTRIBUTION: Mindanao (Philippines).

Tateomys Musser, 1969. Am. Mus. Novit., 2384:1.
REVIEWED BY: J. I. Menzies (JIM).
COMMENT: Generic status reviewed by Musser, 1981, Bull. Am. Mus. Nat. Hist., 169(2):137.
ISIS NUMBER: 5301410011062000000.

Tateomys rhinogradoides Musser, 1969. Am. Mus. Novit., 2384:3.
TYPE LOCALITY: Indonesia, Sulawesi; Latimodjong, 3° 50' S., 120° 10' E., 2200 m.
DISTRIBUTION: Known only from the type locality.
ISIS NUMBER: 5301410011062001001.

Thallomys Thomas, 1920. Ann. Mag. Nat. Hist., ser. 9, 5:141.
REVIEWED BY: D. A. Schlitter (DS); E. Van der Straeten (EVS).

Thallomys paedulcus (Sundevall, 1846). Ofv. Kongl. Svenska Vet.-Akad. Forhandl. Stockholm, 3(4):120.
TYPE LOCALITY: South Africa, Caffraria Interior, Prope Tropicum.
DISTRIBUTION: S. Africa to C. Angola; S. Zaire and S. Kenya; W. Somalia; S. Ethiopia.
COMMENT: Probably a composite of two species (DS).
ISIS NUMBER: 5301410011002006001 as *Aethomys paedulcus*.

Thamnomys Thomas, 1907. Ann. Mag. Nat. Hist., ser. 7, 19:121.
REVIEWED BY: E. Van der Straeten (EVS).
COMMENT: Misonne, 1974, Part 6:30, included *Grammomys* in *Thamnomys*, but Yalden *et al.*, 1976, Monitore Zool. Ital., n.s., suppl. 8(1):1–118, and Rupp, 1980, Saugetierk. Mitt., 28(2):87, 92, considered *Grammomys* a distinct genus.
ISIS NUMBER: 5301410011063000000 as *Thamnomys*.

Thamnomys venustus Thomas, 1907. Ann. Mag. Nat. Hist., ser. 7, 19:122.
TYPE LOCALITY: Uganda, Ruwenzori East.
DISTRIBUTION: Zaire; Uganda; Rwanda.
ISIS NUMBER: 5301410011063002001.

Tokudaia Kuroda, 1943. Biogeographica, 13(9):61.
REVIEWED BY: B. R. Stein (BRS).
ISIS NUMBER: 5301410011064000000.

Tokudaia osimensis (Abe, 1934). J. Sci. Hiroshima Univ., 3:107.
TYPE LOCALITY: Japan, Ryukyu Isls. (Liukiu Isls.), Amamioshima Isl., Sumiyo-mura.
DISTRIBUTION: N. Okinawa and Amamioshima Isl. (Ryukyu Isls., Japan).
ISIS NUMBER: 5301410011064001001.

Tryphomys Miller, 1910. Proc. U.S. Nat. Mus., 38(1911):399.
COMMENT: Formerly included in *Rattus* by Misonne, 1969, Ann. Mus. R. Afr. Cent. Tervuren, ser. 8, Sci. Zool., 172:140–143; but see Musser, 1981, Bull. Am. Mus. Nat. Hist., 169(2):136, 166. Also see comments under *Rattus everetti* and *R. latidens*.

Tryphomys adustus Miller, 1910. Proc. U.S. Nat. Mus., 38(1911):399.
TYPE LOCALITY: Philippines, Luzon, Benguet, 2460 m.
DISTRIBUTION: Luzon (Philippines).
COMMENT: Listed as a distinct species by Misonne, 1969, Ann. Mus. R. Afr. Cent. Tervuren, ser. 8, Sci. Zool., 172:140, and Musser, 1977, Am. Mus. Novit., 2624:6, who included this species in *Rattus*. *Tryphomys* was considered a distinct genus by Walker *et al.*, 1975, p. 906, and Musser, 1981, Bull. Am. Mus. Nat. Hist., 169(2):137, 166. Does not include *Bunomys penitus*, *B. nigellus*, *B. rallus*, and *B. brevimolaris*; see Musser, 1973, Am. Mus. Novit., 2511:24, and Musser, 1981, Bull. Am. Mus. Nat. Hist., 169(2):115–117, 137.
ISIS NUMBER: 5301410011058002001 as *Rattus adustus*.

Uranomys Dollman, 1909. Ann. Mag. Nat. Hist., ser. 8, 4:552.
REVIEWED BY: E. Van der Straeten (EVS).
ISIS NUMBER: 5301410011065000000.

Uranomys ruddi Dollman, 1909. Ann. Mag. Nat. Hist., ser. 8, 4:552.
TYPE LOCALITY: Kenya, Mt. Elgon, Kirui.
DISTRIBUTION: Senegal to Kenya to Mozambique.
ISIS NUMBER: 5301410011065001001.

Uromys Peters, 1867. Monatsb. Preuss. Akad. Wiss. Berlin, p. 343.
REVIEWED BY: J. I. Menzies (JIM); C. H. S. Watts (CHSW) (Australia); A. C. Ziegler
(ACZ)(New Guinea).
COMMENT: Generic relationships reviewed by Musser, 1981, Bull. Am. Mus. Nat. Hist.,
169(2):67–176; includes *Cyromys* and *Gymnomys*.
ISIS NUMBER: 5301410011066000000.

Uromys anak Thomas, 1907. Ann. Mag. Nat. Hist., 20:72.
TYPE LOCALITY: Papua New Guinea, Central Prov., Brown River, Efogi, "not less than"
1220 m.
DISTRIBUTION: Interior New Guinea.
ISIS NUMBER: 5301410011066001001.

Uromys caudimaculatus (Krefft, 1867). Proc. Zool. Soc. Lond., 1867:316.
TYPE LOCALITY: Australia, Queensland, Cape York.
DISTRIBUTION: New Guinea; Aru Isls.; Kei Isls.; Waigeo Isl.; N. Queensland (Australia).
COMMENT: Includes *sherrini* and *exilis*; see Ride, 1970:248.
ISIS NUMBER: 5301410011066002001.

Uromys imperator (Thomas, 1888). Ann. Mag. Nat. Hist., 1:157.
TYPE LOCALITY: Solomon Isls., Guadalcanal Isl., Aola.
DISTRIBUTION: Guadalcanal Isl. (Solomon Isls.).
ISIS NUMBER: 5301410011066003001.

Uromys neobrittanicus Tate and Archbold, 1935. Am. Mus. Novit., 803:4.
TYPE LOCALITY: Papua New Guinea, New Britain Isl.
DISTRIBUTION: New Britain (Bismarck Arch., Papua New Guinea).
ISIS NUMBER: 5301410011066004001.

Uromys rex (Thomas, 1888). Ann. Mag. Nat. Hist., 1:157.
TYPE LOCALITY: Solomon Isls., Guadalcanal Isl., Aola.
DISTRIBUTION: Guadalcanal Isl. (Solomon Isls.).
ISIS NUMBER: 5301410011066005001.

Uromys salamonis (Ramsay, 1883). Proc. Linn. Soc. N.S.W., 7:43.
TYPE LOCALITY: Solomon Isls., Florida Isl.
DISTRIBUTION: Florida Isl. (Solomon Isls.).
COMMENT: Type locality of "Ugi Island," in error; see Laurie and Hill, 1954:130.
ISIS NUMBER: 5301410011066006001.

Vandeleuria Gray, 1842. Ann. Mag. Nat. Hist., 10:265.
REVIEWED BY: B. R. Stein (BRS); S. Wang (SW)(China).
COMMENT: Revised by Musser, 1979, Bull. Am. Mus. Nat. Hist., 162(6):337–445.
ISIS NUMBER: 5301410011067000000.

Vandeleuria nolthenii Phillips, 1929. Ceylon J. Sci., Sec. B, 15:165.
TYPE LOCALITY: Sri Lanka, Ohiya, West Haputale.
DISTRIBUTION: Highlands of Sri Lanka.
COMMENT: Reviewed by Musser, 1979, Bull. Am. Mus. Nat. Hist., 162(6):437.

Vandeleuria oleracea (Bennett, 1832). Proc. Zool. Soc. Lond., 1832:121.
TYPE LOCALITY: India, Madras.
DISTRIBUTION: India; Sri Lanka; Nepal to Burma; Yunnan (China); Thailand; N.
Indochina.
ISIS NUMBER: 5301410011067001001.

Vernaya Anthony, 1941. Field Mus. Nat. Hist. Publ. Zool. Ser., 27:110.
REVIEWED BY: B. R. Stein (BRS); S. Wang (SW)(China).
ISIS NUMBER: 5301410011068000000.

Vernaya fulva (G. M. Allen, 1927). Am. Mus. Novit., 270:11.
TYPE LOCALITY: China, Yunnan, Yinpankai, Mekong River.
DISTRIBUTION: Yunnan (China); N. Burma.
ISIS NUMBER: 5301410011068001001.

Xenuromys Tate and Archbold, 1941. Am. Mus. Novit., 1101:3.
REVIEWED BY: J. I. Menzies (JIM); A. C. Ziegler (ACZ).
COMMENT: Generic relationships reviewed by Musser, 1981, Bull. Am. Mus. Nat. Hist., 169(2):67–176.
ISIS NUMBER: 5301410011069000000.

Xenuromys barbatus (Milne-Edwards, 1900). Bull. Mus. Hist. Nat. Paris, 6:167.
TYPE LOCALITY: Papua New Guinea, "British New Guinea."
DISTRIBUTION: Interior New Guinea.
COMMENT: Includes *guba;* see Tate, 1951, Bull. Am. Mus. Nat. Hist., 97:284.
ISIS NUMBER: 5301410011069001001.

Xeromys Thomas, 1889. Proc. Zool. Soc. Lond., 1889:248.
REVIEWED BY: C. H. S. Watts (CHSW).
COMMENT: Generic relationships reviewed by Musser, 1981, Bull. Am. Mus. Nat. Hist., 169(2):67–176.
ISIS NUMBER: 5301410011085000000.

Xeromys myoides Thomas, 1889. Proc. Zool. Soc. Lond., 1889:248.
TYPE LOCALITY: Australia, Queensland, Mackay.
DISTRIBUTION: E. Queensland and Northern Territory (Australia).
PROTECTED STATUS: CITES - Appendix I and U.S. ESA - Endangered.
ISIS NUMBER: 5301410011085001001.

Zelotomys Osgood, 1910. Field Mus. Nat. Hist. Publ. Zool. Ser., 10(2):7.
REVIEWED BY: D. A. Schlitter (DS); E. Van der Straeten (EVS).
COMMENT: Reviewed by Misonne, 1974, Part 6:31.
ISIS NUMBER: 5301410011070000000.

Zelotomys hildegardeae (Thomas, 1902). Ann. Mag. Nat. Hist., ser. 7, 9:213.
TYPE LOCALITY: Kenya, Machakos.
DISTRIBUTION: Angola, Zambia, and Malawi, to Sudan, Uganda, and Kenya.
ISIS NUMBER: 5301410011070001001.

Zelotomys woosnami (Schwann, 1906). Proc. Zool. Soc. Lond., 1906:108.
TYPE LOCALITY: Botswana, Molopo River.
DISTRIBUTION: Botswana; Namibia; South Africa.
ISIS NUMBER: 5301410011070002001.

Zyzomys Thomas, 1909. Ann. Mag. Nat. Hist., ser. 8, 3:372.
REVIEWED BY: C. H. S. Watts (CHSW).
COMMENT: Includes *Laomys;* see Ride, 1970:244, and Musser, 1981, Bull. Am. Mus. Nat. Hist., 169(2):67–176.
ISIS NUMBER: 5301410011071000000.

Zyzomys argurus (Thomas, 1889). Ann. Mag. Nat. Hist., ser. 6, 3:433.
TYPE LOCALITY: Australia, "South Australia."
DISTRIBUTION: Pilbara region of Western Australia through Kimberleys to N. coastal Queensland, between Cooktown and Townsville (Australia).
ISIS NUMBER: 5301410011071001001.

Zyzomys pedunculatus (Waite, 1896). Rept. Horn Sci. Exped. Cent. Aust., Zool., Part 2:395.
TYPE LOCALITY: Australia, Northern Territory, Alice Springs.
DISTRIBUTION: MacDonnell and James Ranges of Northern Territory (Australia).

PROTECTED STATUS: CITES - Appendix I and U.S. ESA - Endangered as *Z. pedunculatus*. U.S. ESA - Endangered as *Notomys pedunculatus*.
ISIS NUMBER: 5301410011071002001.

Zyzomys woodwardi (Thomas, 1909). Ann. Mag. Nat. Hist., ser. 8, 3:373.
TYPE LOCALITY: Australia, Western Australia, Wyndham.
DISTRIBUTION: Kimberleys of Western Australia and Arnhemland of Northern Territory (Australia).
ISIS NUMBER: 5301410011071003001.

Family Gliridae
REVIEWED BY: J. E. Bucher (JEB).
COMMENT: Includes Muscardinidae; see Corbet, 1978:143. Formerly included Platacanthomyinae, placed here in Cricetidae; see Chaline and Mein, 1979, Chaline *et al.*, 1977, Mammalia, 41:245–252, and Mein and Freudenthal, 1971, Scr. Geol., 2:137. European species reviewed by Storch, 1978, *in* Niethammer and Krapp, eds., Handb. Saugetiere Europas, 1:201–280.
ISIS NUMBER: 5301410012000000000.

Dryomys Thomas, 1906. Proc. Zool. Soc. Lond., 1905(2):348.
REVIEWED BY: M. Andera (MA); O. L. Rossolimo (OLR)(U.S.S.R.); G. A. Sidorowicz (GAS); F. Spitzenberger (FS); G. Storch (GS); S. Wang (SW)(China).
COMMENT: Includes *Dyromys*; see Ellerman and Morrison-Scott, 1951:544.
ISIS NUMBER: 5301410012001000000.

Dryomys laniger Felten and Storch, 1968. Senckenberg. Biol., 49(6):429.
TYPE LOCALITY: Turkey, Antalya Prov., 20 km S.S.E. Elmali, Bey Mtns., Ciglikara, 2000 m.
DISTRIBUTION: W. and C. Taurus Mtns. (Turkey).
COMMENT: Revised by Spitzenberger, 1976, Z. Saugetierk., 41:237–249.

Dryomys nitedula (Pallas, 1778). Nova Spec. Quad. Glir. Ord., p. 88.
TYPE LOCALITY: U.S.S.R., Tatarsk. A.S.S.R. (=Kazansk. Obl.), Volga River.
DISTRIBUTION: Deciduous woodland from Asia Minor and N. and C. Iran to Tien Shan Mtns. (Sinkiang, China) and through the Caucasus to Moscow (U.S.S.R.), west to Carpathians, E. Alps, and Balkans; isolated populations in Calabria, and Israel.
COMMENT: Reviewed by Rossolimo, 1971, Zool. Zh., 50:247–258; also see Roesler and Witte, 1969, Zool. Anz., 182:27–51.
ISIS NUMBER: 5301410012001001001.

Eliomys Wagner, 1840. Abh. Bayer. Akad. Wiss., 3:176.
REVIEWED BY: M. Andera (MA); O. L. Rossolimo (OLR) (U.S.S.R.); G. A. Sidorowicz (GAS); G. Storch (GS).
ISIS NUMBER: 5301410012002000000.

Eliomys quercinus (Linnaeus, 1766). Syst. Nat., 12th ed., 1:84.
TYPE LOCALITY: Germany.
DISTRIBUTION: Spanish Sahara to N.W. Libya; N.E. Libya; N. Egypt; N. Arabia to S. Asia Minor; Sicily and most of W. Mediterranean islands; Europe from Mediterranean (except S. Balkans) to N. Germany and Finland, east to S. Urals (U.S.S.R.).
COMMENT: Includes *melanurus*; see Corbet, 1978:145; but see also Trainier and Petter, 1978, Mammalia, 42:349–353, and Delibes *et al.*, 1980, Saugetierk. Mitt., 28:289–292.
ISIS NUMBER: 5301410012002002001 as *E. quercinus*.
5301410012002001001 as *E. melanurus*.

Glirulus Thomas, 1906. Proc. Zool. Soc. Lond., 1905(2):347.
REVIEWED BY: J. E. Bucher (JEB).
ISIS NUMBER: 5301410012003000000.

Glirulus japonicus (Schinz, 1845). Syst. Verz. Saug., 2:530.
TYPE LOCALITY: Japan.
DISTRIBUTION: Honshu, Shikoku, and Kyushu (Japan).
ISIS NUMBER: 5301410012003001001.

Graphiurus Smuts, 1832. Enumer. Mamm. Cap., pp. 32–33.
REVIEWED BY: J. E. Bucher (JEB).
COMMENT: Includes *Claviglis;* revised by Genest-Villard, 1978, Mammalia, 42:391–522.
ISIS NUMBER: 5301410012007000000.

Graphiurus crassicaudatus Jentink, 1888. Notes Leyden Mus., p. 41.
TYPE LOCALITY: Liberia, Du Queah River.
DISTRIBUTION: Bioko; Cameroun; E. Nigeria; Ivory Coast; Liberia; Ghana.
ISIS NUMBER: 5301410012007004001.

Graphiurus hueti Roquebrune, 1883. Faune Seneg., p. 109.
TYPE LOCALITY: Senegal, Saint-Louis.
DISTRIBUTION: Senegal to Central African Republic and Angola.
COMMENT: Includes *monardi;* see Genest-Villard, 1978, Mammalia, 42:421.
ISIS NUMBER: 5301410012007005001 as *G. hueti.*
          5301410012007007001 as *G. monardi.*

Graphiurus murinus (Desmarest, 1822). Encyclop. Method. Mamm., Suppl., p. 542.
TYPE LOCALITY: South Africa, Cape of Good Hope.
DISTRIBUTION: Senegal to Sudan to South Africa to Angola.
COMMENT: Includes *ansorgei, johnstoni, orobinus, soleatus, surdus,* and *woosnami;* see
          Misonne, 1974, Part 6:36. Genest-Villard, 1978, Mammalia, 42:404–405, included
          angolensis in *murinus* (see comment under *platyops*) and transferred *brockmani* and
          personatus from this species to *parvus.*
ISIS NUMBER: 5301410012007008001 as *G. murinus.*
          5301410012007001001 as *G. angolensis.*
          5301410012007002001 as *G. ansorgei.*
          5301410012007006001 as *G. johnstoni.*
          5301410012007010001 as *G. orobinus.*
          5301410012007015001 as *G. soleatus.*
          5301410012007016001 as *G. surdas (sic).*
          5301410012007017001 as *G. woosmani (sic).*

Graphiurus ocularis (A. Smith, 1829). Zool. J., 4:439.
TYPE LOCALITY: South Africa, Cape Prov., Plattenberg Bay (east of Knysna).
DISTRIBUTION: South Africa.
ISIS NUMBER: 5301410012007009001.

Graphiurus parvus (True, 1893). Proc. U.S. Nat. Mus., 16(95A):601.
TYPE LOCALITY: Kenya, Tana River.
DISTRIBUTION: Sierra Leone and Mali to Ethiopia and Somalia, south to Zambia and
          Zimbabwe; perhaps Angola.
COMMENT: Includes *brockmani* and *personatus* (formerly included in *murinus* by
          Misonne, 1974, Part 6:36); see Genest-Villard, 1978, Mammalia, 42:399.
ISIS NUMBER: 5301410012007011001 as *G. parvus.*
          5301410012007003001 as *G. brockmani.*
          5301410012007012001 as *G. personatus.*

Graphiurus platyops Thomas, 1897. Ann. Mag. Nat. Hist., 19:388.
TYPE LOCALITY: S. Zimbabwe, Mashonaland, Enkeldorn.
DISTRIBUTION: South Africa; Namibia; Mozambique; Zimbabwe; Zambia; Malawi;
          Angola; S. Zaire.
COMMENT: Includes *rupicola;* see Misonne, 1974, Part 6:36, who also included *angolensis*
          in this species, but Genest-Villard, 1978, Mammalia, 42:404, transferred *angolensis*
          to *murinus.*
ISIS NUMBER: 5301410012007013001 as *G. platyops.*
          5301410012007014001 as *G. rupicola.*

Muscardinus Kaup, 1829. Skizz. Europ. Thierwelt, 1:139.
REVIEWED BY: M. Andera (MA); O. L. Rossolimo (OLR) (U.S.S.R.); G. A. Sidorowicz (GAS); G. Storch (GS).
ISIS NUMBER: 5301410012005000000.

Muscardinus avellanarius (Linnaeus, 1758). Syst. Nat., 10th ed., 1:62.
TYPE LOCALITY: Sweden.
DISTRIBUTION: S. Britain and W. Europe from the Mediterranean (except Iberia) to S. Sweden and the Baltic, east to Kazan region of U.S.S.R.; N. Asia Minor (Turkey); Sicily.
ISIS NUMBER: 5301410012005001001.

Myomimus Ognev, 1924. Priroda Okhota Ukraine [Nat. and Hunting in Ukraine], Kharkov, 1–2:115–116.
REVIEWED BY: M. Andera (MA); O. L. Rossolimo (OLR) (U.S.S.R.); G. Storch (GS).
COMMENT: Includes *Philistomys*; see Kowalski, 1963, Acta Zool. Cracov., 8(14):561.
ISIS NUMBER: 5301410012006000000.

Myomimus personatus Ognev, 1924. Priroda Okhota Ukraine [Nat. and Hunting in Ukraine], Kharkov, 1–2:115.
TYPE LOCALITY: U.S.S.R., Turkmenistan, Kopet Dagh Mtns., Kaine-Kassyr.
DISTRIBUTION: Kopet Dag Mtns. of Turkmenistan (U.S.S.R.); Iran.
COMMENT: Reviewed by Rossolimo, 1976, Zool. Zh., 55, 10:1515–1525.
ISIS NUMBER: 5301410012006001001.

Myomimus roachi (Bate, 1937). Ann. Mag. Nat. Hist., ser. 10, 20:399.
TYPE LOCALITY: Israel, Mt. Carmel, Tabun Cave, upper Pleistocene layers.
DISTRIBUTION: S.E. Bulgaria; W. Turkey.
COMMENT: Includes *bulgaricus*; see Storch, 1978, *in* Niethammer and Krapp, eds., 1:240–241; but also see Corbet, 1978:147, who considered *bulgaricus* conspecific with *personatus*.

Myomimus setzeri Rossolimo, 1976. Vestn. Zool., 4:51.
TYPE LOCALITY: Iran, Kurdistan, 4 km W. of Bane.
DISTRIBUTION: Iran.

Myoxus Zimmermann, 1780. Geogr. Gesch. Mensch. Vierf. Thiere, 2:351.
REVIEWED BY: M. Andera (MA); O. L. Rossolimo (OLR) (U.S.S.R.); G. A. Sidorowicz (GAS); G. Storch (GS).
COMMENT: The Brisson, 1762, name *Glis* is invalid, thus the name *Myoxus* applies.
ISIS NUMBER: 5301410012004000000 as *Glis*.

Myoxus glis (Linnaeus, 1766). Syst. Nat., 12th ed., 1:87.
TYPE LOCALITY: Germany.
DISTRIBUTION: Woodland parts of Europe from the Mediterranean (except S. and C. Iberia) to the Baltic, east to the Volga River, west to S.W. France; Caucasus; N. Asia Minor; N. Iran; Crete; Corfu; Kefallinia and N. Adriatic Isls.; Sicily; Elba; Corsica; Sardinia.
COMMENT: Reviewed by Vietinghoff-Riesch, 1960, Monogr. Wildsaugetiere, 14:1–196, and Corbet, 1978:144.
ISIS NUMBER: 5301410012004001001 as *Glis glis*.

Family Seleviniidae
REVIEWED BY: O. L. Rossolimo (OLR).
ISIS NUMBER: 5301410014000000000.

Selevinia Belosludov and Bashanov, 1938. Uchen. Zap. Kaz. Univ. Alma-Ata, 1, 1:81.
ISIS NUMBER: 5301410014001000000.

Selevinia betpakdalaensis Belosludov and Bashanov, 1938. Uchen. Zap. Kaz. Univ. Alma-Ata, 1, 1:81.
TYPE LOCALITY: U.S.S.R., S. Kazakhstan, N. Betpak-Dala Desert, Kyzyl-Ui.
DISTRIBUTION: C. and S.E. Kazakhstan (U.S.S.R.).
ISIS NUMBER: 5301410014001001001.

Family Zapodidae
REVIEWED BY: G. S. Jones (GSJ); A. E. Muchlinski (AEM); O. L. Rossolimo
(OLR)(U.S.S.R.); S. Wang (SW)(China).
ISIS NMBER: 5301410015000000000.

Eozapus Preble, 1899. N. Am. Fauna, 15:37.
ISIS NUMBER: 5301410015002000000.

Eozapus setchuanus (Pousargues, 1896). Bull. Mus. Hist. Nat. Paris, 2:13.
TYPE LOCALITY: China, Szechwan, Kungding, Tatsienlu.
DISTRIBUTION: W. Szechwan, Kansu, Yunnan, and Tsinghai (China).
COMMENT: Revised by Vinogradov, 1925, Proc. Zool. Soc. Lond., 1925:577.
ISIS NUMBER: 5301410015002001001.

Napaeozapus Preble, 1899. N. Am. Fauna, 15:33.
COMMENT: Revised by Wrigley, 1972, Ill. Biol. Monogr., 47:8–118.
ISIS NUMBER: 5301410015003000000.

Napaeozapus insignis (Miller, 1891). Am. Nat., 25:742.
TYPE LOCALITY: Canada, New Brunswick, Restigouche River.
DISTRIBUTION: S.E. Manitoba to S. Labrador (Canada), south to Rhode Island, N.E.
Georgia, C. Michigan, and C. Wisconsin (U.S.A.).
COMMENT: Revised by Preble, 1899, N. Am. Fauna, 15:33–37. Reviewed by Whitaker,
1972, Mamm. Species, 14:1–7.
ISIS NUMBER: 5301410015003001001.

Sicista Gray, 1827. *In* Griffith's Cuvier Anim. Kingd., 5:228.
REVIEWED BY: B. R. Stein (BRS).
COMMENT: Revised by Ognev, 1948, [Mamm. U.S.S.R., Adjac. Count.], 6:32–45.
ISIS NUMBER: 5301410015001000000.

Sicista betulina Pallas, 1779. Nova Spec. Quad. Glir. Ord., p. 332.
TYPE LOCALITY: U.S.S.R., Novosibirsk. Obl., Baraba Steppe, mouth of Ishim River.
DISTRIBUTION: Boreal and montane forests from Norway and Denmark to Ussuri
region (China and S.E. Siberia, U.S.S.R.), north to Arctic Circle at White Sea,
south to Austria, Carpathian, Caucasus and Sayan Mtns.
COMMENT: Pallas' type specimen is probably not preserved (AEM).
ISIS NUMBER: 5301410015001001001.

Sicista caucasica Vinogradov, 1925. Proc. Zool. Soc. Lond., 1925:584.
TYPE LOCALITY: U.S.S.R., R.S.F.S.R., Krasnodarsk. Krai (=Kuban Prov.), Maikop dist.
DISTRIBUTION: Western Caucasus, Armenia (U.S.S.R.).
COMMENT: Corbet, 1978:149, included *caucasica* in *concolor*, but see Sokolov *et al.*,
1980:38, for evidence of specific distinctness.
ISIS NUMBER: 5301410015001002001.

Sicista caudata Thomas, 1907. Proc. Zool. Soc. Lond., 1907 (2):413.
TYPE LOCALITY: U.S.S.R., Sakhalin Obl., 17 mi. (27 km) N.W. Korsakov.
DISTRIBUTION: Ussuri region and Sakhalin Isl. (U.S.S.R.); N.E. China.
COMMENT: Corbet, 1978:149, included *caudata* in *concolor*, but see Sokolov *et al.*,
1980:38, for evidence of specific distinctness.
ISIS NUMBER: 5301410015001003001.

Sicista concolor (Buchner, 1892). Imp. Sci. St. Petersb., 35(3):107.
TYPE LOCALITY: China, Kansu, N. slope of the mountains of Sining, Guiduisha.
DISTRIBUTION: Kansu, Tsinghai, and Szechwan (W. China); Kashmir.
COMMENT: Corbet, 1978:149, included *caucasica*, *caudata*, and *tianshanica* in this species;
also see Bobrinskii *et al.*, 1965 [Key to the Mammals of the U.S.S.R.], Moscow; but
also see Ellerman and Morrison-Scott, 1951:524 and Ognev, 1948, 6:32–45. OLR,
AEM, and GSJ consider these forms distinct species. See also comments under
caucasica, *caudata*, and *tianshanica*.
ISIS NUMBER: 5301410015001004001.

Sicista kluchorica Sokolov, Kovalskaya, and Baskevich, 1980. Gryzuny Severnovo Kavkaza., p. 38.
TYPE LOCALITY: U.S.S.R., Georgian S.S.R., Tebardin Zapovednik, Upper Severnyi Klukhor River.
DISTRIBUTION: Known only from the type locality.
COMMENT: See Sokolov *et al.*, 1980:38, and Sokolov, *et al.*, 1981, Zool. Zh., 60(9):1391, for evidence of specific distinctness.

Sicista napaea Hollister, 1912. Smithson. Misc. Coll., 60(14):2.
TYPE LOCALITY: U.S.S.R., Altaisk. Krai, Altai Mtns., Seminsk Ridge, about 5 mi. (8 km) S. of Tapuchii.
DISTRIBUTION: N.W. Altai Mtns. (U.S.S.R.).
ISIS NUMBER: 5301410015001005001.

Sicista pseudonapaea Strautman, 1949. Vestn. Akad. Nauk Kazakh. S.S.R., 5:109.
TYPE LOCALITY: U.S.S.R., E. Kazakhstan, Altai Mtns., N. slope of Narym Range, Katon-Karagai.
DISTRIBUTION: Taiga of Altai Mtns. (U.S.S.R.); probably W. China and N.W. Mongolia.
COMMENT: Bobrinskii *et al.*, 1965 [Key to the Mammals of the U.S.S.R.], Moscow, and Sokolov *et al.*, 1980:38, considered this species distinct, but see Gromov *et al.*, 1963, [Mammal Fauna of the U.S.S.R.], Moscow, 2 vols., and Gromov and Baranova, 1981:111, who included *pseudonapaea* in *betulina*.

Sicista subtilis (Pallas, 1773). Reise Prov. Russ. Reichs., 1(2):705.
TYPE LOCALITY: U.S.S.R., Kurgansk. Obl., on road from Zmeinogolovsk to Kurgan, on Tobol River.
DISTRIBUTION: Steppes from E. Austria, Hungary, and Rumania to Altai, Lake Balkhash, and Lake Baikal (U.S.S.R.); possibly N.W. China.
COMMENT: Type specimen has probably not been preserved (AEM).
ISIS NUMBER: 5301410015001006001.

Sicista tianshanica Salensky, 1903. Ezheg. Zool. Muz. Akad. Nauk, 8:17.
TYPE LOCALITY: China, Sinkiang, S. slope Tien Shan Mtns., between Khapchagai-gol and Tsaima Rivers.
DISTRIBUTION: Tien Shan Mtns. (U.S.S.R., China).
COMMENT: Corbet, 1978:149, included *tianshanica* in *concolor*; see Sokolov *et al.*, 1980:38, for evidence of specific distinctness.

Zapus Coues, 1876. Bull. U.S. Geol. Geogr. Surv. Terr., ser. 2, 1:253.
COMMENT: Revised by Krutzsch, 1954, Univ. Kans. Publ. Mus. Nat. Hist., 7:349–472. A key to the genus was published by Whitaker, 1972, Mamm. Species, 11:1–7.
ISIS NUMBER: 5301410015004000000.

Zapus hudsonius (Zimmermann, 1780). Geogr. Gesch. Mensch. Vierf. Thiere, 2:358.
TYPE LOCALITY: Canada, Ontario, Hudson Bay, Fort Severn.
DISTRIBUTION: S. Alaska (U.S.A.) to S. coast Hudson Bay to Labrador (Canada), south to N. South Carolina, N.W. Alabama, N.E. Oklahoma (U.S.A.), and S. British Columbia (Canada). Isolated populations in S. Wyoming, N.C. Colorado, C. New Mexico, and E.C. Arizona (U.S.A.).
COMMENT: The S. Rocky Mtn. subspecies *luteus*, formerly assigned to *princeps*, has been transferred to *hudsonius* by Hafner *et al.*, 1981, J. Mammal., 62:501–512. Reviewed by Whitaker, 1972, Mamm. Species, 11:1–7.
ISIS NUMBER: 5301410015004001001.

Zapus princeps J. A. Allen, 1893. Bull. Am. Mus. Nat. Hist., 5:71.
TYPE LOCALITY: U.S.A., Colorado, La Plata Co., Florida.
DISTRIBUTION: S. Yukon (Canada) to N.E. South Dakota, N. New Mexico, N. Arizona, C. Utah, C. Nevada, and E.C. California (U.S.A.).
COMMENT: Formerly included *luteus*; see comment under *hudsonius*. GSJ considers *trinotatus* a subspecies of *princeps*.
ISIS NUMBER: 5301410015004002001.

Zapus trinotatus Rhoads, 1895. Proc. Acad. Nat. Sci. Phila., (1894), 47:421.
TYPE LOCALITY: Canada, British Columbia, Lulu Isl., mouth of the Frazer River.
DISTRIBUTION: S.W. British Columbia (Canada) along coast to San Francisco Bay (California, U.S.A.).
COMMENT: GSJ considers *trinotatus* to be conspecific with *princeps*.
ISIS NUMBER: 5301410015004003001.

Family Dipodidae
REVIEWED BY: O. L. Rossolimo (OLR)(U.S.S.R.); B. R. Stein (BRS); S. Wang (SW)(China).
ISIS NUMBER: 5301410016000000000.

Alactagulus Nehring, 1897. Sitzb. Ges. Naturf. Fr. Berlin, 9:154.
ISIS NUMBER: 5301410016001000000.

Alactagulus pumilio Kerr, 1792. Anim. Kingdom, p. 275.
TYPE LOCALITY: U.S.S.R., Kazakhstan, between Caspian Sea and Irtysh River.
DISTRIBUTION: S. European Russia from Don River through Kazakhstan to Irtysh River, south to N.E. Iran; S. Mongolia and adjacent Inner Mongolia; Sinkiang and Ningsiahui (China).
COMMENT: Includes *acontion* and *pygmaeus*; Corbet, 1978:156, concluded that *pygmaeus* is invalid and that *acontion* is a junior synonym of *pumilio*.
ISIS NUMBER: 5301410016001001001.

Allactaga (F. Cuvier, 1837). Proc. Zool. Soc. Lond., 1836:141.
ISIS NUMBER: 5301410016002000000.

Allactaga bobrinskii (Kolesnikov, 1937). Bull. Sredne-Az. Gos. Univ., 22(29):255.
TYPE LOCALITY: U.S.S.R., Uzbekistan, 140 km N.W. of Bukhara, Khala-Ata.
DISTRIBUTION: Uzbekistan and Turkmenistan, in the Kizil-Kum and Kara-Kum deserts (U.S.S.R.).
COMMENT: Reviewed by Shenbrot, 1974, Zool. Zh., 53(11):1697–1702.
ISIS NUMBER: 5301410016002001001.

Allactaga bullata G. M. Allen, 1925. Am. Mus. Novit., 161:2.
TYPE LOCALITY: Mongolia, near Tatsyn-Tsagan-Nur.
DISTRIBUTION: Deserts of S. and W. Mongolia; adjacent Inner Mongolia, Kansu, and Sinkiang; Ningsiahui (China).
COMMENT: Reviewed by Bannikov, 1954, [Mamm. Mongol. Peoples Republic], p. 389.
ISIS NUMBER: 5301410016002002001.

Allactaga elater (Lichtenstein, 1825). Abh. Konigl. Akad. Wiss. Berlin, p. 155.
TYPE LOCALITY: U.S.S.R., E. Kazakhstan ("Kirgiz Steppe").
DISTRIBUTION: N. Caucasus, E. Asia Minor and Lower Volga to Turkestan, Iran, Afghanistan, Baluchistan, and Sinkiang (China), in desert and semi-desert zones.
ISIS NUMBER: 5301410016002003001.

Allactaga euphratica Thomas, 1881. Ann. Mag. Nat. Hist., 8:15.
TYPE LOCALITY: Iraq.
DISTRIBUTION: Steppe and semi-desert from Syria, Jordan and N. Saudi Arabia through Iraq to Asia Minor and the Caucasus and east through N. Iran and Afghanistan.
COMMENT: Includes *williamsi*, *laticeps*, and *schmidti*; see Corbet, 1978:155, and Harrison, 1972, The mammals of Arabia, 3:408. Many Russian authors employ the name *williamsi* for this species; see Gromov and Baranova, 1981:118.
ISIS NUMBER: 5301410016002004001 as *A. euphractica (sic)*.
5301410016002010001 as *A. williamsi*.

Allactaga firouzi Womochel, 1978. Fieldiana Zool., 72:65.
TYPE LOCALITY: Iran, Isfahan Prov., 18 mi. (29 km) S. Shah Reza, 2,253 m.
DISTRIBUTION: Known only from the type locality.

Allactaga hotsoni Thomas, 1920. J. Bombay Nat. Hist. Soc., 26(4):936.
TYPE LOCALITY: Iran, Kerman, Kant.
DISTRIBUTION: Known only from the type locality.

COMMENT: Relationship with *euphratica* needs further study; see Corbet, 1978:155, and Lay, 1967, Fieldiana Zool., 54:199.
ISIS NUMBER: 5301410016002005001.

Allactaga major (Kerr, 1792). Anim. Kingdom, p. 274.
TYPE LOCALITY: U.S.S.R., Kazakhstan, steppes between Caspian Sea and Irtysh River.
DISTRIBUTION: Steppes and deserts from Moscow and Kiev to Ob River (W. Siberia), south to the Caucasus, Kazakhstan, and Tien Shan Mtns.
COMMENT: *A. jaculus* Pallas, of Soviet authors, is a junior synonym; see Ognev, 1948:94, and Corbet, 1978:156.
ISIS NUMBER: 5301410016002006001.

Allactaga nataliae Sokolov, 1981. Zool. Zh., 60(5):793.
TYPE LOCALITY: Mongolia, S. Zaaltaisk. Gobi, Dzamyn-Bilgekh-Bulak.
DISTRIBUTION: S. Mongolia, from Altai Sumon east to Bordzon-Gobi.
COMMENT: Closely related to *bullata*; reviewed by Sokolov *et al.*, 1981, Zool. Zh., 60:895–906.

Allactaga severtzovi Vinogradov, 1925. Proc. Zool. Soc. Lond., 1925:583.
TYPE LOCALITY: U.S.S.R., S.E. Kazakhstan, Tamar-Utkul near Taldy-Kurgan (formerly Kopal).
DISTRIBUTION: S. Kazakhstan, N. Uzbekistan, N.E. Turkmenistan, and S.W. Tadzhikistan (U.S.S.R.).
ISIS NUMBER: 5301410016002007001.

Allactaga sibirica (Forster, 1778). Kongl. Svenska Vet.-Akad. Handl. Stockholm, 39:112.
TYPE LOCALITY: U.S.S.R., S.E. Transbaikalia, Chitinsk. Obl., near Tarei-Nur Lake.
DISTRIBUTION: From Ural River and Caspian Sea to N.E. and N.C. China.
COMMENT: Includes *saltator*; see Corbet, 1978:154.
ISIS NUMBER: 5301410016002008001.

Allactaga tetradactyla (Lichtenstein, 1823). Verz. Doublet. Zool. Mus. Univ. Berlin, p. 2.
TYPE LOCALITY: Libyan Desert between Siwa and Alexandria.
DISTRIBUTION: Coastal gravel plains of Egypt and Libya, from near Alexandria to the Gulf of Sidra.
ISIS NUMBER: 5301410016002009001.

Cardiocranius Satunin, 1903. Ann. Mus. Zool. Acad. Imp. Sci. St. Petersb., 7:582.
ISIS NUMBER: 5301410016008000000.

Cardiocranius paradoxus Satunin, 1903. Ann. Mus. Zool. Acad. Imp. Sci. St. Petersb., 7:584.
TYPE LOCALITY: China, N.W. Kansu, Nan Shan, Shargol-Dzhin.
DISTRIBUTION: N.W. and N.C. China; Mongolia and adjacent U.S.S.R.; north of Lake Balkhash (Kazakhstan, U.S.S.R.).
COMMENT: Gromov and Baranova, 1981:114, suggested that the Kazakstan population may be a separate species.
ISIS NUMBER: 5301410016008001001.

Dipus Zimmermann, 1780. Geogr. Gesch. Mensch. Vierf. Thiere, 2:354.
ISIS NUMBER: 5301410016003000000.

Dipus sagitta (Pallas, 1773). Reise Prov. Russ. Reichs., 2:706.
TYPE LOCALITY: U.S.S.R., N. Kazakhstan, Pavlodarsk. Obl., right bank of Irtysh River near Yamyshevskaya at Podpusknoi.
DISTRIBUTION: Desert, steppe and dry woodland from Don River and N.W. coast of Caspian Sea to Shensi and N.E. China.
ISIS NUMBER: 5301410016003001001.

Euchoreutes Sclater, 1891. Proc. Zool. Soc. Lond., 1890:610.
ISIS NUMBER: 5301410016010000000.

Euchoreutes naso Sclater, 1891. Proc. Zool. Soc. Lond., 1890:610.
TYPE LOCALITY: N.W. China, W. Sinkiang, W. of Takla-Makan Desert, near Yarkand.

DISTRIBUTION: Alashan and Gobi deserts in S. Mongolia and Inner Mongolia (China); W. Sinkiang, Tsinghai, Kansu, and Ningsiahui (China).
ISIS NUMBER: 5301410016010001001.

Jaculus Erxleben, 1777. Syst. Regn. Anim., 1:404.
COMMENT: Includes *Scirtopoda* (in part), *Haltomys*, and *Eremodipus*; see Corbet, 1978:151. OLR follows Heptner, 1975, Byull. Mosk. Ova. Ispyt. Prir. Otd. Biol., 80(3):5–15, who considered *Eremodipus* a distinct genus. *Jaculus* was placed on the official List of Generic Names by the International Commission on Zoological Nomenclature (Opinion 730).
ISIS NUMBER: 5301410016004000000.

Jaculus blanfordi (Murray, 1884). Ann. Mag. Nat. Hist., 14:98.
TYPE LOCALITY: Iran, Bushire.
DISTRIBUTION: E. and S. Iran; Baluchistan (Pakistan).
ISIS NUMBER: 5301410016004001001.

Jaculus jaculus (Linnaeus, 1758). Syst. Nat., 10th ed., 1:63.
TYPE LOCALITY: Egypt, Giza Pyramids.
DISTRIBUTION: N. Nigeria; Niger; S.W. Mauritania to Morocco to Somalia; Arabia and Syria to S.W. Iran.
COMMENT: Includes *deserti*; see Corbet, 1978:152, and Harrison, 1972, The Mammals of Arabia, 3:419; but also see Ranck, 1968, Bull. U.S. Nat. Mus., 275:1–264, who considered *deserti* a distinct species.
ISIS NUMBER: 5301410016004002001.

Jaculus lichtensteini (Vinogradov, 1927). Z. Saugetierk., 2:92.
TYPE LOCALITY: U.S.S.R., Turkmenia, vicinity of Merv.
DISTRIBUTION: Uzbekistan, Kazakhstan and S.W. Turkestan from Caspian Sea to Aral Sea; south of Lake Balkhash (U.S.S.R.).
COMMENT: Formerly included in *Eremodipus*; see Prakash and Ghosh, 1975, Rodents in Desert Environments, 28:520. Gromov and Baranova, 1981:127, considered *Eremodipus* a distinct genus.
ISIS NUMBER: 5301410016004003001.

Jaculus orientalis Erxleben, 1777. Syst. Regn. Anim., 1:404.
TYPE LOCALITY: Egypt, in the "mountains separating Egypt from Arabia."
DISTRIBUTION: Morocco to Sinai and S. Israel, except extremely arid habitats.
ISIS NUMBER: 5301410016004004001.

Jaculus turcmenicus Vinogradov and Bondar, 1949. C. R. Acad. Sci. U.S.S.R., 65:559.
TYPE LOCALITY: U.S.S.R., Turkmenia, Nebitagsk. Dist., Tchagil sands, S. coast of Kara-Bogaz Gol.
DISTRIBUTION: S.E. coast of Caspian Sea through Turkmenia to the Kyzyl-Kum Desert, Uzbekistan (U.S.S.R.).
COMMENT: May be a subspecies of *blanfordi*; see Heptner, 1975, Byull. Mosk. Ova. Ispyt. Prir. Otd. Biol., 8(3):5–15. For distribution see Stalmakova, 1957, Zool. Zh., 36:275–279.

Paradipus Vinogradov, 1930. Izv. Acad. Sci. U.S.S.R., p. 333.
ISIS NUMBER: 5301410016005000000.

Paradipus ctenodactylus (Vinogradov, 1929). Izv. Acad. Sci. U.S.S.R., p. 248.
TYPE LOCALITY: U.S.S.R., E. Turkmenia, near Repetek.
DISTRIBUTION: Sand desert of Turkmenistan and Uzbekistan (U.S.S.R.).
ISIS NUMBER: 5301410016005001001.

Pygeretmus Gloger, 1841. Gem. Hand. Hilfs. Nat., 1:106.
COMMENT: Reviewed by Vorontsov, *et al.*, 1969, pp. 74–84, *in* Vorontsov, ed., [The Mammals: Evolution, karyology, taxonomy, fauna], Novosibirsk.
ISIS NUMBER: 5301410016006000000.

Pygeretmus platyurus (Lichtenstein, 1823). Naturh. Abh. Eversmann's Reise, p. 121.
 TYPE LOCALITY: U.S.S.R., Kazakhstan, E. shore of Aral Sea, Kuwan-Darya River.
 DISTRIBUTION: W. and C. Kazakhstan (U.S.S.R.).
 COMMENT: Reviewed by Silvestrov *et al.*, 1969, Byull. Mosk. Ova. Ispyt. Prir. Otd.
   Biol., 74(3):118–133. See also comment under *vinogradovi*.
 ISIS NUMBER: 5301410016006001001.

Pygeretmus shitkovi (Kusnetzov, 1930). C. R. Acad. Sci. U.S.S.R., p. 623.
 TYPE LOCALITY: U.S.S.R., S.E. Kazakhstan, Taldy-Kurgansk. Obl., N.W. of Ala-Kul Lake,
   Rybalnoie.
 DISTRIBUTION: E. Kazakhstan in region of Lake Balkhash (U.S.S.R.).
 COMMENT: The later spelling *zhitkovi* is often used by Russian authors (BRS).
 ISIS NUMBER: 5301410016006002001.

Pygeretmus vinogradovi (Vorontsov, 1958). Zool. Zh., 37:96.
 TYPE LOCALITY: U.S.S.R., Kazakhstan, E. Kazakhstansk. Obl., South of Zaisan Lake,
   near Topolev-Mys.
 DISTRIBUTION: S.E. Kazakhstan, on south side of Lake Zaisan (U.S.S.R).
 COMMENT: Corbet, 1978:157, and Gromov and Baranova, 1981:120, placed this species
   as a synonym of *platyurus* without comment; it was regarded as a distinct species
   by Bobrinskii *et al.*, 1965 [Key to the Mammals of the U.S.S.R], Moscow, and
   Vorontsov, ed., 1969, [The Mammals: Evolution, karyology, taxonomy, fauna],
   Novosibirsk.

Salpingotus Vinogradov, 1922. *In* Kozlov, Mongolia and Amdo, p. 540.
 COMMENT: Formerly included the species *michaelis* which was transferred to
   Salpingotulus by Pavlinov, 1980, Vestn. Zool., 2:47–50.
 ISIS NUMBER: 5301410016009000000.

Salpingotus crassicauda Vinogradov, 1924. Zool. Anz., 61:150.
 TYPE LOCALITY: Mongolia, Gobi Altai Mtns., Shara Sumeh.
 DISTRIBUTION: S. Kazakhstan (U.S.S.R.); Sinkiang (China); S. Mongolia.
 ISIS NUMBER: 5301410016009001001.

Salpingotus heptneri Vorontsov and Smirnov, 1969. *In* Vorontsov, ed., [The Mammals:
   Evolution, karyology, taxonomy, fauna], Novosibirsk, p. 60.
 TYPE LOCALITY: U.S.S.R., Karakalpaksk. A.S.S.R., 80 km N.E. Takahta-Kumpir, N.W.
   Kyzyl-Kum Desert.
 DISTRIBUTION: Known only from the type locality, just south of the Aral Sea in
   Uzbekistan (U.S.S.R.).
 COMMENT: May be a subspecies of *crassicauda*; see Sabilaiev, 1978, *in* [II All-Union
   Theriol. Conf.], Akad. Nauk U.S.S.R., Moscow, p. 37.

Salpingotus kozlovi Vinogradov, 1922. *In* Kozlov, Mongolia and Amdo, p. 542.
 TYPE LOCALITY: N. China, Gobi Desert, Khara-Khoto.
 DISTRIBUTION: Gobi Desert of S. Mongolia and N. China.
 ISIS NUMBER: 5301410016009002001.

Salpingotus thomasi Vinogradov, 1928. Ann. Mag. Nat. Hist., ser. 10, 1:373.
 TYPE LOCALITY: Uncertain. Perhaps Afghanistan or N.E. India.
 DISTRIBUTION: Known only from the type, from Afghanistan, N.E. India, or S. Tibet.
 COMMENT: Type of doubtful origin; see Corbet, 1978:158, and Ognev, 1940, [Mamm.
   U.S.S.R., Adjac. Count.], 6:83.
 ISIS NUMBER: 5301410016009003001.

Salpingotulus Pavlinov, 1980. Vestn. Zool., 2:47–50.

Salpingotulus michaelis (Fitzgibbon, 1966). Mammalia, 30(3):431.
 TYPE LOCALITY: Pakistan, N.W. Baluchistan, Nushki Desert.
 DISTRIBUTION: N.W. Baluchistan.
 COMMENT: Formerly included in *Salpingotus* by Fitzgibbon, 1966, Mammalia, 30:431;
   but see Pavlinov, 1980, Vestn. Zool., 2:47–50.

Stylodipus G. M. Allen, 1925. Am. Mus. Novit., 161:4.
COMMENT: *Scirtopoda* of Soviet authors is a junior synonym; see Corbet, 1978:153.
ISIS NUMBER: 5301410016007000000.

Stylodipus telum (Lichtenstein, 1823). Naturh. Abh. Eversmann's Reise, p. 120.
TYPE LOCALITY: U.S.S.R., Kazakhstan, steppe at N.E. shore of Aral Sea.
DISTRIBUTION: E. Ukraine, N. Caucasus, and Kazakhstan (U.S.S.R.); Sinkiang,
Ningsiahui, and Inner Mongolia (China); S. Mongolia.
COMMENT: Includes *andrewsi,* which was considered a distinct species by Sokolov and
Orlov, 1980:187; see Corbet, 1978:153.
ISIS NUMBER: 5301410016007001001.

Family Hystricidae
REVIEWED BY: J. E. Bucher (JEB).
COMMENT: Reviewed by Mohr, 1965, Altweltliche Stachelschweine, Wittenberg
Lutherstadt: Ziemsen, 164 pp.
ISIS NUMBER: 5301410017000000000.

Atherurus F. Cuvier, 1829. Dict. Sci. Nat., 59:483.
ISIS NUMBER: 5301410017001000000.

Atherurus africanus Gray, 1842. Ann. Mag. Nat. Hist., 10:261.
TYPE LOCALITY: Sierra Leone.
DISTRIBUTION: Gambia; Sierra Leone; Liberia; Ghana; Zaire; Kenya; Uganda; S. Sudan.
COMMENT: Includes *centralis* and *turneri;* see Misonne, 1974, Part 6:8.
ISIS NUMBER: 5301410017001001001 as *A. africanus.*
5301410017001002001 as *A. centralis.*
5301410017001004001 as *A. turneri.*

Atherurus macrourus (Linnaeus, 1758). Syst. Nat., 10th ed., 1:57.
REVIEWED BY: S. Wang (SW).
TYPE LOCALITY: Malaysia, Malacca.
DISTRIBUTION: Assam (India), Szechwan, Yunnan, Hupei, and Hainan (China) to
Malaya, Sumatra, and adjacent islands.
COMMENT: Includes *retardatus* and *angustiramus;* see Van Weers, 1977, Beaufortia,
26(336):205–230.
ISIS NUMBER: 5301410017001003001.

Hystrix Linnaeus, 1758. Syst. Nat., 10th ed., 1:56.
REVIEWED BY: S. Wang (SW)(China).
COMMENT: Includes *Acanthion;* see Starret, 1967, *in* Anderson and Jones, p. 257.
PROTECTED STATUS: CITES - Appendix III as *Hystrix* spp. native to Ghana only.
ISIS NUMBER: 5301410017002000000.

Hystrix africaeaustralis Peters, 1852. Reise nach Mossambique, Saugethiere, p. 170.
TYPE LOCALITY: Mozambique, Querimba coast, about 10° 30′ to 12° S., 40° 30′ E., sea
level.
DISTRIBUTION: Mouth of the Congo River to Rwanda, Uganda, Kenya, W. and S.
Tanzania, Mozambique, and South Africa.
COMMENT: Revised by Corbet and Jones, 1965, Proc. Zool. Soc. Lond., 1965:285–300.
ISIS NUMBER: 5301410017002001001.

Hystrix brachyura Linnaeus, 1758. Syst. Nat., 10th ed., 1:57.
TYPE LOCALITY: Malaysia, Malacca.
DISTRIBUTION: Nepal; Sikkim and Assam (India); C. and S. China; Burma; Thailand;
Indochina; Malaya; Sumatra; Borneo.
COMMENT: Includes *hodgsoni, klossi,* and *subcristata;* see Van Weers, 1979, Beaufortia,
29:215–272; also see Lekagul and McNeely, 1977:492.
ISIS NUMBER: 5301410017002002001 as *H. brachyurus (sic).*

Hystrix cristata Linnaeus, 1758. Syst. Nat., 10th ed., 1:56.
 TYPE LOCALITY: Italy, near Rome.
 DISTRIBUTION: Morocco to Egypt; Senegal to Ethiopia and N. Tanzania; Sicily, Italy,
    Albania, and N. Greece (European populations possibly introduced).
 COMMENT: Includes *galeata;* see Corbet, 1978:159, and Corbet and Jones, 1965, Proc.
    Zool. Soc. Lond., 1965:285–300.
 PROTECTED STATUS: CITES - Appendix III (Ghana).
 ISIS NUMBER: 5301410017002003001 as *H. cristata.*
             5301410017002004001 as *H. galeata.*

Hystrix indica Kerr, 1792. Anim. Kingdom, p. 213.
 TYPE LOCALITY: India.
 DISTRIBUTION: Transcaucasus; Asia Minor; Israel; Arabia to S. Kazakhstan (U.S.S.R.)
    and India; Sri Lanka; Tibet (China).
 COMMENT: Citation based on Smellie's Buffon, 1781, 7: pl. 206. Gromov and Baranova,
    1981:102, employed the name *leucura* for this species without reference to Kerr,
    1792.
 ISIS NUMBER: 5301410017002006001.

Hystrix javanica (F. Cuvier, 1822). Mem. Mus. Hist. Nat. Paris, 9:431.
 TYPE LOCALITY: Indonesia, Java.
 DISTRIBUTION: Java, Bali, Sumbawa, Flores, and S. Sulawesi (Indonesia).
 COMMENT: Formerly included in *brachyura* by Chasen, 1940, but see Van Weers, 1979,
    Beaufortia, 29:215–272.

Thecurus Lyon, 1907. Proc. U.S. Nat. Mus., 32:582.
 ISIS NUMBER: 5301410017003000000.

Thecurus crassispinis (Gunther, 1877). Proc. Zool. Soc. Lond., 1876:736.
 TYPE LOCALITY: Malaysia, Sabah, opposite Labuan Isl.
 DISTRIBUTION: N. Borneo.
 ISIS NUMBER: 5301410017003001001.

Thecurus pumilis (Gunther, 1879). Ann. Mag. Nat. Hist., ser. 5, 4:106.
 TYPE LOCALITY: Philippines, Palawan, Paragua (= Puerto Princesa).
 DISTRIBUTION: Palawan and Busuanga Isls. (Philippines).
 ISIS NUMBER: 5301410017003002001.

Thecurus sumatrae Lyon, 1907. Proc. U.S. Nat. Mus., 32:583.
 TYPE LOCALITY: Indonesia, E. coast of Sumatra, Aru Bay.
 DISTRIBUTION: Sumatra (Indonesia).
 COMMENT: Treated as a subspecies of *crassispinis* by Chasen, 1940, but see Van Weers,
    1978, Beaufortia, 28:17–33.

Trichys Gunther, 1877. Proc. Zool. Soc. Lond., 1876:739.
 COMMENT: Reviewed by Van Weers, 1976, Beaufortia, 25:15–31.
 ISIS NUMBER: 5301410017004000000.

Trichys fasciculata (Shaw, 1801). Gen. Zool., 2:11.
 TYPE LOCALITY: Malaysia, "Runuk Tanjong."
 DISTRIBUTION: Borneo; Sumatra; Malaya.
 COMMENT: Includes *lipura* and *macrotis;* see Van Weers, 1976, Beaufortia, 25:19, who
    designated a neotype. Medway, 1977:122, considered *lipura* a distinct species, but
    did not mention Van Weers, 1976.
 ISIS NUMBER: 5301410017004001001 as *T. lipura.*
             5301410017004002001 as *T. macrotis.*

Family Erethizontidae
 REVIEWED BY: W. E. Dodge (WED); M. A. Mares (MAM); C. A. Woods (CAW).
 COMMENT: Spelled Erithizontidae by Corbet and Hill, 1980:189; see comment under
    Erethizon.
 ISIS NUMBER: 5301410018000000000.

Chaetomys Gray, 1843. List Mamm. Br. Mus., p. 123.
COMMENT: Wood and Patterson, in press, Biol. Comp. Zool., referred *Chaetomys* to the family Echimyidae based on the dental formula.
ISIS NUMBER: 5301410018001000000.

Chaetomys subspinosus (Olfers, 1818). Neue Bibl. Reisenb., p. 211.
TYPE LOCALITY: Brazil, Para, Cameta.
DISTRIBUTION: N.C. and N.E. Brazil.
COMMENT: Avila-Pires, 1967, Rev. Brasil. Biol., 27:178, gave Brazil, Bahia, Ilheus, as the type locality.
PROTECTED STATUS: U.S. ESA - Endangered.
ISIS NUMBER: 5301410018001001001.

Coendou Lacepede, 1799. Tabl. Mamm., p. 11.
COMMENT: Formerly included *Sphiggurus;* see Husson, 1978:484–490. Cabrera, 1961:600, Walker *et al.,* 1975:1012, Corbet and Hill, 1980:189, and Hall, 1981:853, included *Sphiggurus* in *Coendou.*
ISIS NUMBER: 5301410018002000000.

Coendou bicolor (Tschudi, 1844). Fauna Peruana, p. 186.
TYPE LOCALITY: Peru, Junin Dept., between Tulumayo and Chanchamayo Rivers.
DISTRIBUTION: Bolivia; Peru; Andean and W. Ecuador; N. Colombia; perhaps S.W. Colombia.
COMMENT: May include *rothschildi;* see Hall, 1981:854. Corbet and Hill, 1980:189, listed *rothschildi* as a subspecies of *bicolor;* see comment under *rothschildi.*
ISIS NUMBER: 5301410018002001001.

Coendou mexicanus (Kerr, 1792). Anim. Kingdom, 1:214.
REVIEWED BY: K. F. Koopman (KFK); J. Ramirez-Pulido (JRP).
TYPE LOCALITY: Mexico, in the mountains.
DISTRIBUTION: San Luis Potosi and Yucatan (Mexico) to W. Panama.
COMMENT: This species is probably referable to *Sphiggurus* (KFK).
ISIS NUMBER: 5301410018002003001.

Coendou pallidus (Waterhouse, 1848). Nat. Hist. Mammal., 2:434.
TYPE LOCALITY: Unknown, probably West Indies.
DISTRIBUTION: West Indies.
COMMENT: Type believed to be from the West Indies. Known only by the two immature specimens upon which the original description was based; see Hall, 1981:854. CAW includes *pallidus* in *Sphiggurus* based on its pelage.
ISIS NUMBER: 5301410018002004001.

Coendou prehensilis (Linnaeus, 1758). Syst. Nat., 10th ed., 1:57.
TYPE LOCALITY: Brazil, Pernambuco.
DISTRIBUTION: E. Venezuela; Guyanas; C. and E. Brazil; Bolivia; Trinidad.
COMMENT: Trinidad record from Goodwin and Greenhall, 1961, Bull. Am. Mus. Nat. Hist., 122:202.
ISIS NUMBER: 5301410018002005001.

Coendou rothschildi Thomas, 1902. Ann. Mag. Nat. Hist., ser. 7, 10:169.
TYPE LOCALITY: Panama, Chiriqui, Sevilla Isl.
DISTRIBUTION: Panama.
COMMENT: Possibly a subspecies of *bicolor;* see Hall, 1981:854, and Goldman, 1920, Smithson. Misc. Coll., 69:135. Corbet and Hill, 1980:189, listed *rothschildi* as a subspecies of *bicolor.*
ISIS NUMBER: 5301410018002006001.

Echinoprocta Gray, 1865. Proc. Zool. Soc. Lond., 1865:321.

Echinoprocta rufescens (Gray, 1865). Proc. Zool. Soc. Lond., 1865:321.
TYPE LOCALITY: Colombia.
DISTRIBUTION: Colombia.

Erethizon F. Cuvier, 1822. Mem. Mus. Hist. Nat. Paris, 9:432.
COMMENT: *Erethizon* Burnett, 1829, is a later spelling; see Corbet and Hill, 1980:189.

Erethizon dorsatum (Linnaeus, 1758). Syst. Nat., 10th ed., 1:57.
REVIEWED BY: J. Ramirez-Pulido (JRP).
TYPE LOCALITY: E. Canada (=Quebec Prov.).
DISTRIBUTION: C. Alaska (U.S.A.) to S. Hudson Bay and Labrador (Canada), south to E. Tennessee, and C. Iowa, C. Texas (U.S.A.), N. Coahuila, Chihuahua, and Sonora (Mexico), and S. California (U.S.A.).
COMMENT: Reviewed by Miller and Kellogg, 1955, Bull. U.S. Nat. Mus., 205:631, who restricted the type locality; also see Anderson and Rand, 1943, Can. J. Res., 21:292–309. Includes *couesi;* see review by Woods, 1973, Mamm. Species, 29:1–6.

Sphiggurus F. Cuvier, 1825. Des Dentes des Mammiferes, p. 256.
REVIEWED BY: K. F. Koopman (KFK); C. A. Woods (CAW).
COMMENT: Formerly included in *Coendou* by Cabrera, 1961:600, Walker *et al.*, 1975:1012, and Hall, 1981:853. Considered a distinct genus by Husson, 1978:484–490. May include *Coendou pallidus* (CAW); probably includes *C. mexicanus* (KFK).

Sphiggurus insidiosus (Lichtenstein, 1818). Zool. Mus. Univ. Berlin, p. 18–19.
TYPE LOCALITY: Brazil, Bahia, Salvador.
DISTRIBUTION: E. and Amazonian Brazil; Surinam.
COMMENT: Formerly included in *Coendou;* see Husson, 1978:484; also see comment under *Sphiggurus.*
ISIS NUMBER: 5301410018002002001 as *Coendou insidiosus.*

Sphiggurus spinosus (F. Cuvier, 1822). Mem. Mus. Hist. Nat. Paris, 9:433.
TYPE LOCALITY: Paraguay, along the Parana River.
DISTRIBUTION: Paraguay; S. and E. Brazil; N.E. Argentina; Uruguay.
COMMENT: Possibly restricted to the shores of the Parana River (CAW). Formerly included in *Coendou;* see Husson, 1978:484–490. See comment under *Sphiggurus.*
PROTECTED STATUS: CITES - Appendix III as *Coendou spinosus* (Uruguay).
ISIS NUMBER: 5301410018002007001 as *Coendou spinosus.*

Sphiggurus vestitus (Thomas, 1899). Ann. Mag. Nat. Hist., ser. 7, 4:284.
TYPE LOCALITY: Colombia.
DISTRIBUTION: Colombia; W. Venezuela south of Lake Maracaibo.
COMMENT: Includes *pruinosus;* see Cabrera, 1961:284, but also see Handley, 1976, Brigham Young Univ. Sci. Bull. Biol. Ser., 20(5):55, who listed *pruinosus* as a distinct species, without comment. Formerly included in *Coendou;* see comment under *Sphiggurus.*
ISIS NUMBER: 5301410018002008001 as *Coendou vistatus (sic).*

Sphiggurus villosus (F. Cuvier, 1822). Mem. Mus. Hist. Nat. Paris, 9:413–437.
TYPE LOCALITY: Brazil, mountains near Rio de Janeiro, Corcoracto.
DISTRIBUTION: Minas Gerais to Rio Grande do Sul (S.E. Brazil).
COMMENT: Considered a distinct species by Husson, 1978:489. Cabrera, 1961:600–601, included this species in *insidiosus;* see comment under *Sphiggurus;* formerly included in *Coendou.*

Family Caviidae
REVIEWED BY: M. A. Mares (MAM); T. Pearson (TP); C. A. Woods (CAW).
ISIS NUMBER: 5301410019000000000.

Cavia Pallas, 1766. Misc. Zool., p. 30.
COMMENT: Reviewed by Huckinghaus, 1961, Z. Wiss. Zool., 166:34–58, and Cabrera, 1961:575–580.
ISIS NUMBER: 5301410019001000000.

Cavia aperea Erxleben, 1777. Syst. Regn. Anim., 1:348.

TYPE LOCALITY: Brazil, Pernambuco.

DISTRIBUTION: Colombia; Venezuela; Guianas; Brazil; N. Argentina; Uruguay; Paraguay.

COMMENT: Includes *guianae;* see Huckinghaus, 1961, Z. Wiss. Zool., 166:58, and Husson, 1978:449; but also see Cabrera, 1961:578, who placed *guianae* in *porcellus.* Corbet and Hill, 1980:189, listed *guianae* as a distinct species without comment. Includes *pamparum;* see Massoia and Fornes, 1967, Acta Zool. Lilloana, 23:407–430, and Huckinghaus, 1961, Z. Wiss. Zool., 166:57; but see Cabrera, 1961:577, who listed *pamparum* as a distinct species.

ISIS NUMBER: 5301410019001001001 as *C. aperea.*
5301410019001004001 as *C. pamparum.*

Cavia fulgida Wagler, 1831. Isis, 24:512.

TYPE LOCALITY: "Amazonia."

DISTRIBUTION: E. Brazil, between Minas Gerais and Santa Catarina.

COMMENT: Type locality of "Amazonia" probably an error; see Cabrera, 1961:577.

ISIS NUMBER: 5301410019001002001.

Cavia nana Thomas, 1917. Ann. Mag. Nat. Hist., ser. 8, 19:159.

TYPE LOCALITY: Bolivia, La Paz Dept., Chulumani, 2000 m.

DISTRIBUTION: W. Bolivia.

COMMENT: Considered a distinct species by Cabrera, 1961:577; but see Huckinghaus, 1961, Z. Wiss. Zool., 166:58, who included *nana* in *aperea.*

ISIS NUMBER: 5301410019001003001.

Cavia porcellus (Linnaeus, 1758). Syst. Nat., 10th ed., 1:59.

TYPE LOCALITY: Brazil, Pernambuco (questionable).

DISTRIBUTION: Domesticated worldwide; possibly feral in N. South America.

COMMENT: Husson, 1978:451, reserved the use of *porcellus* to denote domesticated guinea pigs, which are probably derived from *tschudii;* see Corbet and Hill, 1980:189, but also see Huckinghaus, 1961, Z. Wiss. Zool., 166:96, who regarded *porcellus* as a synonym of *aperea.* Domesticated animal with no established wild population (CAW). N. South American populations may be feral domestic guinea pigs (K. F. Koopman); but see comment under *aperea.*

ISIS NUMBER: 5301410019001005001.

Cavia tschudii Fitzinger, 1857. Sitzb. Akad. Wiss. Wien, p. 154.

TYPE LOCALITY: Peru, Ica Dept., Ica.

DISTRIBUTION: Peru; S. Bolivia; N.W. Argentina; N. Chile.

COMMENT: Formerly included in *aperea* by Huckinghaus, 1961, Z. Wiss. Zool., 166:57; but see Cabrera, 1961:579, and Pine *et al.,* 1979, Mammalia, 43:361, who considered *tschudii* a distinct species. Includes *stolida;* see Cabrera, 1961:579; but also see Huckinghaus, 1961, Z. Wiss. Zool., 166:58, who considered *stolida* a distinct species.

ISIS NUMBER: 5301410019001006001.

Dolichotis Desmarest, 1820. J. Phys. Chim. Hist. Nat. Arts Paris, 88:205.

COMMENT: Includes *Pediolagus;* see Starrett, 1967, *in* Anderson and Jones, p. 263; also see Cabrera, 1961:580, who considered *Pediolagus* a distinct genus.

ISIS NUMBER: 5301410019005000000.

Dolichotis patagonum (Zimmermann, 1780). Geogr. Gesch. Mensch. Vierf. Thiere, 2:328.

TYPE LOCALITY: Argentina, Santa Cruz Prov., Puerto Deseado.

DISTRIBUTION: Argentina, approx. 28° S. (Bolson de Pipanco, Catamarca Prov.) to 50° S.

ISIS NUMBER: 5301410019005001001 as *D. patagona (sic).*

Dolichotis salinicola (Burmeister, 1876). Proc. Zool. Soc. Lond., 1875:634.

TYPE LOCALITY: Argentina, S.W. Catamarca Prov., between Totoralejos and Recreo.

DISTRIBUTION: Paraguay; N.W. Argentina; extreme S. Bolivia.

COMMENT: Formerly included in *Pediolagus;* see Starrett, 1967, *in* Anderson and Jones, p. 263; also see comment under *Dolichotis.*

ISIS NUMBER: 5301410019005002001.

Galea Meyen, 1832. Nouv. Acta Acad. Caes. Leop.-Carol., 16:597.
COMMENT: Reviewed by Huckinghaus, 1961, Z. Wiss. Zool., 166:58–72, and Cabrera, 1961:573–575.
ISIS NUMBER: 5301410019002000000.

Galea flavidens (Brandt, 1835). Mem. Acad. Imp. Sci. St. Petersb., ser. 6, 3:439.
TYPE LOCALITY: Unknown; possibly Minas Gerais, Brazil.
DISTRIBUTION: Brazil.
COMMENT: Couto, 1950, Mem. Paleontol. Bras., p. 232, considered *flavidens* synonymous with *spixii*, but see Cabrera, 1961:573, who considered both to be distinct species and discussed type locality.
ISIS NUMBER: 5301410019002001001.

Galea musteloides Meyen, 1832. Nouv. Acta Acad. Caes. Leop.-Carol., 16:597.
TYPE LOCALITY: Peru, Paso de Tacna, on road to Lake Titicaca.
DISTRIBUTION: S. Peru; Bolivia; Argentina; N. Chile.
COMMENT: Reviewed by Mann, 1950, Invest. Zool. Chil., 1:2.
ISIS NUMBER: 5301410019002002001.

Galea spixii (Wagler, 1831). Isis, 24:512.
TYPE LOCALITY: Brazil, Minas Gerais, Lagoa Santa.
DISTRIBUTION: Brazil; Bolivia, east of the Andes.
COMMENT: Includes *wellsi*; see Corbet and Hill, 1980:190; also see Cabrera, 1961:575, who fixed the type locality. Huckinghaus, 1961, Z. Wiss. Zool., 166:71, considered *wellsi* a distinct species.
ISIS NUMBER: 5301410019002003001.

Kerodon F. Cuvier, 1825. Des Dentes des Mammiferes, p. 151.
REVIEWED BY: T. E. Lacher, Jr. (TEL); M. R. Willig (MRW).
COMMENT: Reviewed by Huckinghaus, 1961, Z. Wiss. Zool., 166:12–14, 33. J. Moojen and M. Locks are describing a new species in this genus, from Brazil (MRW and TEL).
ISIS NUMBER: 5301410019003000000.

Kerodon rupestris (Wied, 1820). Isis, 6:43.
TYPE LOCALITY: Brazil, Bahia, Rio Belmonte.
DISTRIBUTION: E. Brazil.
COMMENT: MRW and TEL give the type locality as Rio Pardo and Moojen, 1952, Os Roedores do Brasil, as Rio Grande de Belmonte, Rio Pardo, Rio San Francisco; but see Cabrera, 1961:580.
ISIS NUMBER: 5301410019003001001.

Microcavia H. Gervais and Ameghino, 1880. Mamm. Fos. Am. Sud., p. 50.
COMMENT: Reviewed by Huckinghaus, 1961, Z. Wiss. Zool., 166:72–84, and Cabrera, 1961:570–573.
ISIS NUMBER: 5301410019004000000.

Microcavia australis (I. Geoffroy and d'Orbigny, 1833). Mag. Zool. Paris, p. 3.
TYPE LOCALITY: Argentina, Patagonia, vicinity of the lower part of the Rio Negro.
DISTRIBUTION: Argentina between Jujuy and Santa Cruz Provs.; Aisen Prov. (Chile); possibly extreme S. Bolivia.
COMMENT: Reviewed by Thomas, 1921, Ann. Mag. Nat. Hist., 9(7):445.
ISIS NUMBER: 5301410019004001001.

Microcavia niata (Thomas, 1898). Ann. Mag. Nat. Hist., ser. 7, 1:282.
TYPE LOCALITY: Bolivia, Oruro Dept., Monte Sajama, 4000 m.
DISTRIBUTION: S.W. Bolivia in the high Andes.
ISIS NUMBER: 5301410019004002001.

Microcavia shiptoni (Thomas, 1925). Ann. Mag. Nat. Hist., ser. 9, 15:419.
TYPE LOCALITY: Argentina, Catamarca Prov., Laguna Blanca, 3400 m.
DISTRIBUTION: N.W. Argentina.
ISIS NUMBER: 5301410019004003001.

Family Hydrochaeridae
REVIEWED BY: J. P. Jorgenson (JPJ); M. A. Mares (MAM); T. Pearson (TP); C. A. Woods (CAW).
ISIS NUMBER: 5301410020000000000.

Hydrochaeris Brunnich, 1772. Zool. Fundamenta, p. 44.
COMMENT: Husson, 1978:457, discussed the spelling of the generic name.
ISIS NUMBER: 5301410020001000000.

Hydrochaeris hydrochaeris (Linnaeus, 1766). Syst. Nat., 12th ed., 1:103.
TYPE LOCALITY: Surinam.
DISTRIBUTION: Panama, Colombia, Venezuela, the Guyanas and Peru, south through Brazil, Paraguay, N.E. Argentina, and Uruguay.
COMMENT: Includes *isthmius;* see Handley, 1966, *in* Wenzel and Tipton, eds., Ectoparasites of Panama, p. 785. Cabrera, 1961:583, gave Pernambuco, Brazil as the type locality, but see Husson, 1978:451, for restriction.
ISIS NUMBER: 5301410020001001001 as *H. hydrochaeris.*
5301410020001002001 as *H. isthmius.*

Family Heptaxodontidae
COMMENT: Known only from sub-Recent fossils from Greater and N. Lesser Antilles; see Hall, 1981:855–856.

Family Dinomyidae
REVIEWED BY: M. A. Mares (MAM); T. Pearson (TP); C. A. Woods (CAW).
ISIS NUMBER: 5301410021000000000.

Dinomys Peters, 1873. Monatsb. Preuss. Akad. Wiss. Berlin, p. 551.
ISIS NUMBER: 5301410021001000000.

Dinomys branickii Peters, 1873. Monatsb. Preuss. Akad. Wiss. Berlin, p. 551.
TYPE LOCALITY: Peru, Junin Dept., Amable Maria.
DISTRIBUTION: Colombia, Ecuador, Peru, Brazil, and Bolivia.
ISIS NUMBER: 5301410021001001001.

Family Agoutidae
REVIEWED BY: M. A. Mares (MAM); T. Pearson (TP); C. A. Woods (CAW).
COMMENT: There is no general agreement concerning the familial or subfamilial status of this taxon; see Husson, 1978:472, and Cabrera, 1961:593 (Agoutidae), Hall, 1981:858 (Agoutinae of the Dasyproctidae), Starrett, 1967, *in* Anderson and Jones, p. 269 (Cuniculinae of the Dasyproctidae), and Ellerman, 1940:221 (Cuniculidae).

Agouti Lacepede, 1799. Tabl. Mamm., p. 9.
COMMENT: Includes *Cuniculus* and *Stictomys;* see Handley, 1976, Brigham Young Univ. Sci. Bull. Biol. Ser., 20(5):55, Walker, *et al.,* 1975:1026, and Husson, 1978:47. *Cuniculus* Brisson, 1762 is unavailable; see Hopwood, 1947, Proc. Zool. Soc. Lond., 117:533–536, and Hemming, 1955, Bull. ICZN, 11(6):197.
ISIS NUMBER: 5301410022001000000 as *Cuniculus.*
5301410022002000000 as *Stictomys.*

Agouti paca (Linnaeus, 1766). Syst. Nat., 12th ed., 1:81.
REVIEWED BY: J. Ramirez-Pulido (JRP).
TYPE LOCALITY: French Guiana, Cayenne.
DISTRIBUTION: S.E. San Luis Potosi (Mexico) to Paraguay, Guianas, and S. Brazil. Introduced into Cuba (CAW).
ISIS NUMBER: 5301410022001001001 as *Cuniculus paca.*

Agouti taczanowskii (Stolzmann, 1865). Proc. Zool. Soc. Lond., 1865:161.
TYPE LOCALITY: Ecuador, Andes.
DISTRIBUTION: Mountains of Peru, Ecuador, Colombia, and N.W. Venezuela.

COMMENT: Cabrera, 1961:595, placed this species in *Stictomys*, but Handley, 1976,
Brigham Young Univ. Sci. Bull. Biol. Ser., 20(5):55, and Gardner, 1971,
Experientia, 26:1088, included it in *Agouti*.
ISIS NUMBER: 5301410022002001001 as *Stictomys taczanowskii*.

Family Dasyproctidae

REVIEWED BY: M. A. Mares (MAM); T. Pearson (TP); C. A. Woods (CAW).
COMMENT: The Agoutidae has been included in Dasyproctidae by several authors, but
see comments under Agoutidae.
ISIS NUMBER: 5301410022000000000.

Dasyprocta Illiger, 1811. Prodr. Syst. Mamm. et Avium., p. 93.
ISIS NUMBER: 5301410022003000000.

Dasyprocta azarae Lichtenstein, 1823. Verz. Doublet. Zool. Mus. Berlin, p. 3.
TYPE LOCALITY: Brazil, Sao Paulo.
DISTRIBUTION: E.C. and S. Brazil; Paraguay; N.E. Argentina.
ISIS NUMBER: 5301410022003004001.

Dasyprocta coibae Thomas, 1902. Novit. Zool., 9:136.
TYPE LOCALITY: Panama, Coiba Isl.
DISTRIBUTION: Coiba Isl. (Panama).
COMMENT: Reviewed by Hall, 1981:862.
ISIS NUMBER: 5301410022003005001.

Dasyprocta cristata (Desmarest, 1816). Nouv. Dict. Hist. Nat., 2(1):215.
TYPE LOCALITY: Surinam.
DISTRIBUTION: Guianas.
COMMENT: May be synonymous with *leporina*; see Husson, 1978:464–466, and
Hershkovitz, 1972:311–341, *in* Keast *et al.*, eds, Evolution, Mammals, and Southern
Continents.
ISIS NUMBER: 5301410022003006001.

Dasyprocta fuliginosa Wagler, 1832. Isis, 25:1220.
TYPE LOCALITY: Brazil, Amazonas, Borba, on lower Rio Madeira (="Amazon River").
DISTRIBUTION: Colombia; S. Venezuela; Surinam; N. Brazil. Peru (TP).
COMMENT: Reviewed by Allen, 1915, Bull. Am. Mus. Nat. Hist., 34:625.
ISIS NUMBER: 5301410022003007001.

Dasyprocta guamara Ojasti, 1972. Mem. Soc. Cienc. Nat. La Salle, 32:176.
TYPE LOCALITY: Venezuela, Delta Amacuro, Araguabisi, 9° 13′ 27″ N., 61° 0′ 16″ W.
DISTRIBUTION: Orinoco Delta (Venezuela).

Dasyprocta kalinowskii Thomas, 1897. Ann. Mag. Nat. Hist., ser. 6, 20:219.
TYPE LOCALITY: Peru, Cuzco Dept., Santa Ana Valley, Idma.
DISTRIBUTION: S.E. Peru.
ISIS NUMBER: 5301410022003008001.

Dasyprocta leporina (Linnaeus, 1758). Syst. Nat., 10th ed., 1:59.
TYPE LOCALITY: Surinam, Peninka, Peninka Creek and Cennewijne River.
DISTRIBUTION: Lesser Antilles; Venezuela; Guianas; Amazonian and E. Brazil;
introduced into the Virgin Islands.
COMMENT: Recognized as *D. agouti cayana* by Cabrera, 1961:585–586; but see Husson,
1978:457, 466, who included *agouti* in this species; may also include *cristata*.
Includes *albida*, *antillensis*, and *noblei*; see Varona, 1974:75, who included these
forms in *agouti*.

Dasyprocta mexicana Saussure, 1860. Rev. Mag. Zool. Paris, ser. 2, 12:53.
REVIEWED BY: J. Ramirez-Pulido (JRP).
TYPE LOCALITY: Mexico, probably Veracruz (="hot zone of Mexico").
DISTRIBUTION: C. Veracruz and E. Oaxaca (Mexico); introduced into Cuba.
COMMENT: See Hall, 1981:859, for discussion of type locality.
ISIS NUMBER: 5301410022003009001.

Dasyprocta prymnolopha Wagler, 1831. Isis, 24:619.
TYPE LOCALITY: Brazil, Para.
DISTRIBUTION: N.E. Brazil.
COMMENT: Original type locality designation of Guyana was probably an error; see Cabrera, 1961:588. Ojasti, 1972, Mem. Soc. Cienc. Nat. La Salle, 32:164, considered the status of this form uncertain.

Dasyprocta punctata Gray, 1842. Ann. Mag. Nat. Hist., ser. 1, 10:264.
REVIEWED BY: J. Ramirez-Pulido (JRP).
TYPE LOCALITY: Nicaragua, Chinadega, El Realejo.
DISTRIBUTION: Chiapas and Yucatan Peninsula (S. Mexico) to S. Bolivia, N. Argentina, and S.W. Brazil. Introduced into Cuba and the Cayman Isls.
COMMENT: Includes *variegata*; see Goldman, 1913, Smithson. Misc. Coll., 60(22):11; but see Handley, 1976, Brigham Young Univ. Sci. Bull. Biol. Ser., 20(5):56, who listed *variegata* as a distinct species.
ISIS NUMBER: 5301410022003011001.

Dasyprocta ruatanica Thomas, 1901. Ann. Mag. Nat. Hist., ser. 7, 8:272.
TYPE LOCALITY: Honduras, Roatan Isl.
DISTRIBUTION: Roatan Isl. (Honduras).
ISIS NUMBER: 5301410022003012001.

Myoprocta Thomas, 1903. Ann. Mag. Nat. Hist., ser. 7, 12:464.
ISIS NUMBER: 5301410022004000000.

Myoprocta acouchy (Erxleben, 1777). Syst. Regn. Anim., 1:354.
TYPE LOCALITY: French Guiana, Cayenne.
DISTRIBUTION: Guianas; Amazonian Brazil; Ecuador; N. Peru; S. Venezuela; S. Colombia.
COMMENT: Includes *pratti*; see Husson, 1978:471-472. Formerly included *exilis*; see Husson, 1978:468.
ISIS NUMBER: 5301410022004001001.

Myoprocta exilis (Wagner, 1831). Isis, 24:621.
TYPE LOCALITY: Brazil, mouth of Rio Negro.
DISTRIBUTION: Guianas; S. Venezuela and Colombia; E. Ecuador; N. Peru; Amazon Basin (Brazil).
COMMENT: Listed by Cabrera, 1961:591, as *M. acouchy exilis*. Considered a species by Husson, 1978:468-472, and Tate, 1939, Bull. Am. Mus. Nat. Hist., 76:151-229. The type locality was reviewed by Allen, 1916, Bull. Am. Mus. Nat. Hist., 35:559-610.

Family Chinchillidae
REVIEWED BY: M. A. Mares (MAM); T. Pearson (TP); C. A. Woods (CAW).
ISIS NUMBER: 5301410023000000000.

Chinchilla Bennett, 1829. Gard. Menag. Zool. Soc., 1:1.
PROTECTED STATUS: CITES - Appendix I as *Chinchilla* spp., South American populations only.
ISIS NUMBER: 5301410023001000000.

Chinchilla brevicaudata Waterhouse, 1848. Nat. Hist. Mammal., 2:241.
TYPE LOCALITY: Peru.
DISTRIBUTION: Andes of S. Bolivia, S. Peru, N.W. Argentina, and Chile.
COMMENT: Pine *et al.*, 1979, Mammalia, 43:362-363, included *brevicaudata* in *lanigera*, without comment.
PROTECTED STATUS: CITES - Appendix I as *Chinchilla* spp., South American populations only.
CITES - Appendix I and U.S. ESA - Endangered as *C. b. boliviana* subspecies only.
ISIS NUMBER: 5301410023001001001.

Chinchilla lanigera (Molina, 1782). Sagg. Stor. Nat. Chile, p. 301.
    TYPE LOCALITY: Chile, Coquimbo Prov., Coquimbo.
    DISTRIBUTION: N. Chile, in foothills of the Andes and coastal mountains south to
        Coquimbo.
    PROTECTED STATUS: CITES - Appendix I as *Chinchilla* spp., South American populations
        only.
    ISIS NUMBER: 5301410023001002001 as *C. laniger (sic)*.

Lagidium Meyen, 1833. Nouv. Acta Acad. Caes. Leop.-Carol., 16(2):576.
    COMMENT: Revised by Osgood, 1943, Field Mus. Nat. Hist. Publ. Zool. Ser., 30:137.
    ISIS NUMBER: 5301410023002000000.

Lagidium peruanum Meyen, 1833. Nouv. Acta Acad. Caes. Leop.-Carol., 16(2):578.
    TYPE LOCALITY: Peru, Puno Dept., Pisacoma.
    DISTRIBUTION: C. and S. Peru.
    ISIS NUMBER: 5301410023002001001.

Lagidium viscacia (Molina, 1782). Sagg. Stor. Nat. Chile, p. 307.
    TYPE LOCALITY: Chile, Santiago Prov., Cordillera de Santiago.
    DISTRIBUTION: W. Argentina; S. and W. Bolivia; N. Chile; S. Peru.
    ISIS NUMBER: 5301410023002002001 as *L. viscaccia (sic)*.

Lagidium wolffsohni (Thomas, 1907). Ann. Mag. Nat. Hist., ser. 7, 19:440.
    TYPE LOCALITY: Argentina, Santa Cruz, Baguales and Vizcachas Mtns., 50° 50' S., 72° 20'
        W.
    DISTRIBUTION: S.W. Argentina and adjacent Chile.
    ISIS NUMBER: 5301410023002003001.

Lagostomus Brookes, 1828. Trans. Linn. Soc. Lond., 16:96.
    ISIS NUMBER: 5301410023003000000.

Lagostomus maximus (Desmarest, 1817). Nouv. Dict. Hist. Nat., 2(13):117.
    TYPE LOCALITY: Unknown; possibly from pampas of Buenos Aires, Argentina.
    DISTRIBUTION: N., C., and E. Argentina; S. Paraguay.
    COMMENT: Type locality discussed by Cabrera, 1961:559.
    ISIS NUMBER: 5301410023003001001.

Family Capromyidae
    REVIEWED BY: G. C. Clough (GCC); M. A. Mares (MAM); G. S. Morgan (GSM); C. A.
        Woods (CAW).
    COMMENT: Does not include *Myocastor*; see Woods and Howland, 1979, J. Mammal.,
        60:95–116; also see comments under Myocastoridae. Includes the genera
        Hexolobodon and *Hyperplagiodontia* (from Haiti) and *Macrocapromys* (from Cuba),
        which are known only from fossils, possibly Holocene; see Hall, 1981:866–867,
        870.
    ISIS NUMBER: 5301410024000000000.

Capromys Desmarest, 1822. Bull. Sci. Soc. Philom. Paris, p. 185.
    COMMENT: Reviewed by Kratochvil *et al.*, 1978, Acta Sci. Nat. Brno, 12(11):1–60, and
        Varona and Arredondo, 1979, Poeyana, 195:1–51. The status of *Geocapromys*,
        Mysateles, Mesocapromys and *Paracapromys* as genera or subgenera is unresolved;
        see Kratochvil *et al.*, 1978, Woods and Howland, 1979, J. Mammal., 60:112,
        Rodriguez *et al.*, 1979, Folia Zool., 28(2):97–102, and Hall, 1981:863–866. This
        genus is in need of revision and may prove to contain fewer species (CAW).
    ISIS NUMBER: 5301410024001000000.

Capromys angelcabrerai Varona, 1979. Poeyana, 194:6.
    TYPE LOCALITY: Cuba, Camaguey Prov., Cayos de Ana Maria.
    DISTRIBUTION: Cayos de Ana Maria (Cuba).
    COMMENT: Similar to *nanus*; subgenus *Pygmaeocapromys*; see Varona, 1979, Poeyana,
        194:6.

Capromys arboricolus (Kratochvil, Rodriguez, and Barus, 1978). Acta Sci. Nat. Sci. Bohemoslov-Brno, 12(11):48.
TYPE LOCALITY: Cuba, Oriente Prov., 12 km W. of Holguin.
DISTRIBUTION: Known only from the type locality.
COMMENT: Placed in the genus *Mysateles*, subgenus *Leptocapromys* by Kratochvil *et al.*, 1978, Acta Sci. Nat. Sci. Bohemoslov-Brno, 12(11):15. However, retained in *Capromys* by Hall, 1981:863, and Corbet and Hill, 1980:191. GSM, GCC, and CAW include *Mysateles* in *Capromys*.

Capromys auritus Varona, 1970. Poeyana, ser. A, 73:1.
TYPE LOCALITY: Cuba, Las Villas Prov., Archipielago de Sabana, Cayo Fragoso, 79° 27' W., 22° 41' N.
DISTRIBUTION: Known only from the type locality.
COMMENT: Placed in the genus *Mesocapromys* by Kratochvil *et al.*, 1978, Acta Sci. Nat. Sci. Bohemoslov-Brno, 12(11):15. However, retained in *Capromys* by Hall, 1981:865, and Corbet and Hill, 1980:191. GSM, CAW, and GCC include *Mesocapromys* in *Capromys*.

Capromys garridoi Varona, 1970. Poeyana, ser. A, 74:2.
TYPE LOCALITY: Cuba, Archipielago de los Canarreos, Cayo Maja.
DISTRIBUTION: Known only from the type locality.
COMMENT: Placed in the genus *Mysateles*, subgenus *Leptocapromys* by Kratochvil *et al.*, 1978, Acta Sci. Nat. Sci. Bohemoslov-Brno, 12(11):15. However, retained in *Capromys* by Hall, 1981:863, and Corbet and Hill, 1980:191. GSM, GCC, and CAW include *Mysateles* in *Capromys*.

Capromys melanurus Poey, 1865. *In* Peters, 1865, Monatsb. K. Preuss. Akad. Wiss. Berlin, p. 384.
TYPE LOCALITY: Cuba, Oriente Prov., Manzanillo.
DISTRIBUTION: Oriente Prov. (Cuba).
COMMENT: Placed in the genus and subgenus *Mysateles* by Kratochvil *et al.*, 1978, Acta Sci. Nat. Sci. Bohemoslov-Brno, 12(11):15. However, retained in *Capromys* by Hall, 1981:863, and Corbet and Hill, 1980:191. CAW, GSM, and GCC include *Mysateles* in *Capromys*. The author and date of publication for this species are usually given as Peters, 1864, but Varona, 1974:63, established the correct author and date as Poey, 1865.
ISIS NUMBER: 5301410024001001001.

Capromys nanus G. M. Allen, 1917. Proc. N. Engl. Zool. Club, 6:54.
TYPE LOCALITY: Cuba, Matanzas Prov., Sierra de Hato Nuevo.
DISTRIBUTION: Cienaga (swamp) de Zapata (Matanzas Prov., Cuba).
COMMENT: Subgenus *Pygmaeocapromys*; see Varona, 1979, Poeyana, 194:6. Placed in the genus *Mesocapromys*, subgenus *Paracapromys* by Kratochvil *et al.*, 1978, Acta Sci. Nat. Sci. Bohemoslov-Brno, 12(11):15, and Rodriguez *et al.*, 1979, Folia Zool., 28(2):97–102. However, retained in *Capromys* by Hall, 1981:863, Varona, 1979, Poeyana, 194:6, and Corbet and Hill, 1980:191. GSM, GCC, and CAW include *Mesocapromys* in *Capromys*. Originally based on fossil material, but subsequently found living in the Zapata Swamp (GSM).
ISIS NUMBER: 5301410024001002001 as *C. nana (sic)*.

Capromys pilorides (Say, 1822). J. Acad. Nat. Sci. Phila., 2:333.
TYPE LOCALITY: "South America or one of the West Indian islands."
DISTRIBUTION: Cuba and the Isle of Pines.
COMMENT: Subgenus *Capromys*; see Hall, 1981:863.
ISIS NUMBER: 5301410024001003001.

Capromys prehensilis Poeppig, 1824. J. Acad. Nat. Sci. Phila., 4:11.
TYPE LOCALITY: Cuba, wooded south coast.
DISTRIBUTION: Cuba and the Isle of Pines.
COMMENT: Placed in the subgenus *Mysateles* by Varona, 1974, and in the genus and subgenus *Mysateles* by Kratochvil *et al.*, 1978, Acta Sci. Nat. Sci. Bohemoslov-

Brno, 12(11):15. However, retained in *Capromys* by Hall, 1981:863, and Corbet and Hill, 1980:191. GSM, GCC, and CAW include *Mysateles* in *Capromys*.
ISIS NUMBER: 5301410024001004001.

Capromys sanfelipensis Varona and Garrido, 1970. Poeyana, ser. A, 75:3.
TYPE LOCALITY: Cuba, Pinar del Rio Prov., Archipielago de los Canarreos, Cayo Juan Garcia.
DISTRIBUTION: Known only from the type locality.
COMMENT: Placed in subgenus *Mesocapromys* by Varona, 1974, and in genus *Mesocapromys*, subgenus *Paracapromys* by Kratochvil *et al.*, 1978, Acta Sci. Nat. Sci. Bohemoslov-Brno, 12(1):15. However, retained in *Capromys* by Hall, 1981:865, and Corbet and Hill, 1980:191. CAW, GSM, and GCC include *Mesocapromys* in *Capromys*.

Geocapromys Chapman, 1901. Bull. Am. Mus. Nat. Hist., 14:314.
COMMENT: Included in *Capromys* by Mohr, 1939, Mitt. Zool. Mus. Hamburg, 48:75, Varona, 1974:67, and Hall, 1981:865. However, considered a distinct genus by Woods and Howland, 1979, J. Mammal., 60:112, and Miller, 1929, Smithson. Misc. Coll., 82(4):1–3. GSM considers *Geocapromys* a distinct genus. Includes *G. columbianus* and *G. pleistocenicus*, known only from fossils, probably Holocene, from Cuba; see Hall, 1981:866.
ISIS NUMBER: 5301410024002000000 as *Geocapromys*.
          5301410024002002001 for *G. columbianus*.

Geocapromys brownii (Fischer, 1830). Synopis. Mamm., Addenda, p. 389 (=589).
TYPE LOCALITY: Jamaica.
DISTRIBUTION: Jamaica and Little Swan Isl.
COMMENT: Includes *thoracatus* from Little Swan Isl.; see Hall, 1981:866, and Varona, 1974:67. Clough, 1976, Biol. Conserv., 10:43–47, stated that *thoracatus* is extinct. Although most recent workers list *thoracatus* as a subspecies of *brownii*, it is quite different and probably should be regarded as a distinct species (GSM, CAW).
ISIS NUMBER: 5301410024002001001.

Geocapromys ingrahami (J. A. Allen, 1891). Bull. Am. Mus. Nat. Hist., 3:329.
TYPE LOCALITY: Bahamas, East Plana Key.
DISTRIBUTION: Known only from the type locality.
ISIS NUMBER: 5301410024002003001.

Isolobodon Allen, 1916. Ann. N.Y. Acad. Sci., 27:19.
COMMENT: See Hall, 1981:869–870, for recently extinct species, *levir* and *montanus*.

Isolobodon portoricensis Allen, 1916. Ann. N.Y. Acad. Sci., 27:19.
TYPE LOCALITY: Puerto Rico, Jobo Dist., near Utuado.
DISTRIBUTION: Puerto Rico; Hispaniola (Haiti and Dominican Republic) and offshore Isls. Introduced on St. Thomas, St. Croix, and Mona Isls.
COMMENT: Reported as extinct by Hall, 1981:868, but CAW states that this species may still survive in Hispaniola and Puerto Rico.

Plagiodontia F. Cuvier, 1836. Ann. Sci. Nat. Paris Zool., ser. 2, 6:347.
COMMENT: Reviewed by Mohr, 1939, Mitt. Zool. Mus. Hamburg, 48:81–87, Johnson, 1948, Proc. Biol. Soc. Wash., 61:69–76, and Anderson, 1965, Proc. Biol. Soc. Wash., 78:95–98. See Hall, 1981:867–869, for recently extinct species, *ipnaeum*, *spelaeum*, *araeum*, *calatensis*, and *velozi*.
ISIS NUMBER: 5301410024003000000.

Plagiodontia aedium F. Cuvier, 1836. Ann. Sci. Nat. Paris Zool., ser. 2, 6:347.
TYPE LOCALITY: Dominican Republic.
DISTRIBUTION: Hispaniola (Haiti and Dominican Republic).
COMMENT: Includes *hylaeum*; see Varona, 1974:70.
ISIS NUMBER: 5301410024003001001 as *P. aedium*.
          5301410024003002001 as *P. hylaeum*.

Family Myocastoridae
REVIEWED BY: G. C. Clough (GCC); M. A. Mares (MAM); G. S. Morgan (GSM); C. A. Woods (CAW).
COMMENT: Included in Capromyidae by Vaughan, 1978:40, Hall, 1981:871, Corbet and Hill, 1980:192, and others, but see Woods and Howland, 1979, J. Mammal., 60:95–116.
ISIS NUMBER: 5301410025000000000.

Myocastor Kerr, 1792. Anim. Kingdom, p. 225.
COMMENT: *Myocastor* is not a capromyid and was referred to the Family Myocastoridae Ameghino, 1904, by Woods and Howland, 1979, J. Mammal., 60:114. Although Patterson and Pasqual, 1968, Breviora, 301:6, included *Myocastor* and several fossil forms as the subfamily Myocastorinae of the Echimyidae, it is generally agreed that *Myocastor* is not an echimyid (GSM). Includes *Myopotamus;* see Cabrera, 1958:569.
ISIS NUMBER: 5301410025001000000.

Myocastor coypus (Molina, 1782). Sagg. Stor. Nat. Chile, p. 287.
TYPE LOCALITY: Chile, Santiago Prov., Rio Maipo.
DISTRIBUTION: S. Brazil; Paraguay; Uruguay; Bolivia; Argentina; Chile.
COMMENT: Widely introduced into North America, Europe, N. Asia, and E. Africa (GSM, CAW).
ISIS NUMBER: 5301410025001001001.

Family Octodontidae
REVIEWED BY: M. A. Mares (MAM); C. A. Woods (CAW).
ISIS NUMBER: 5301410026000000000.

Aconaemys Ameghino, 1891. Rev. Argent. Hist. Nat., 1:245.
ISIS NUMBER: 5301410026001000000.

Aconaemys fuscus (Waterhouse, 1842). Proc. Zool. Soc. Lond., 1841:91.
TYPE LOCALITY: Argentina, Mendoza Prov., Valle de las Cuevas.
DISTRIBUTION: High Andes of Chile and Argentina (between 33° and 41° S.).
ISIS NUMBER: 5301410026001001001.

Octodon Bennett, 1832. Proc. Zool. Soc. Lond., 1832:46.
ISIS NUMBER: 5301410026002000000.

Octodon bridgesi Waterhouse, 1845. Proc. Zool. Soc. Lond., 1844:155.
TYPE LOCALITY: Chile, Curico Prov., Rio Teno.
DISTRIBUTION: Andes, in Colchagua, Curico, and Concepcion Provs. (Chile).
COMMENT: Massoia, 1979, Neotropica, 25:36, reported *Octodon*, perhaps *bridgesi*, from S. Argentina.
ISIS NUMBER: 5301410026002001001.

Octodon degus (Molina, 1782). Sagg. Stor. Nat. Chile, p. 303.
TYPE LOCALITY: Chile, Santiago Prov., Santiago.
DISTRIBUTION: Chile, west slope of the Andes between Vallenar and Curico, to 1200 m.
COMMENT: Reviewed by Woods and Boraker, 1975, Mamm. Species, 67:1–5.
ISIS NUMBER: 5301410026002002001.

Octodon lunatus Osgood, 1943. Field Mus. Nat. Hist. Publ. Zool. Ser., 30:110.
TYPE LOCALITY: Chile, Valparaiso Prov., Olmue.
DISTRIBUTION: Coastal mountains of Valparaiso, Aconcagua, and Coquimbo Provs. (Chile).
COMMENT: Probably a subspecies of *bridgesi* (CAW).
ISIS NUMBER: 5301410026002003001.

Octodontomys Palmer, 1903. Science, n.s., 17:873.
    ISIS NUMBER: 5301410026003000000.

  Octodontomys gliroides (Gervais and d'Orbigny, 1844). Bull. Sci. Soc. Philom. Paris, p. 22.
    TYPE LOCALITY: Bolivia, La Paz Dept., near La Paz.
    DISTRIBUTION: Andes of N. Chile, S.W. Bolivia, and N.W. Argentina.
    ISIS NUMBER: 5301410026003001001.

Octomys Thomas, 1920. Ann. Mag. Nat. Hist., ser. 9, 6:117.
    COMMENT: Includes *Tympanoctomys*; see Packard, 1967, *in* Anderson and Jones, p. 278.
    MAM believes that *Tympanoctomys* is a distinct genus.
    ISIS NUMBER: 5301410026004000000.

  Octomys barrerae Lawrence, 1941. Proc. N. Engl. Zool. Club, 18:43.
    TYPE LOCALITY: Argentina, Mendoza Prov., La Paz.
    DISTRIBUTION: Arid plains of Mendoza Prov. (Argentina).
    COMMENT: Formerly placed in a separate genus, *Tympanoctomys*, by Yepes, 1940, Rev.
    Inst. Bact., 9:569, and Cabrera, 1961:516, but Packard, 1967, *in* Anderson and
    Jones, p. 278, considered it congeneric with *Octomys*.
    ISIS NUMBER: 5301410026004001001.

  Octomys mimax Thomas, 1920. Ann. Mag. Nat. Hist., ser. 9, 6:118.
    TYPE LOCALITY: Argentina, Catamarca Prov., La Puntilla.
    DISTRIBUTION: Foothills and lower montane slopes of the Andes, and portions of the
    Monte desert of Catamarca, La Rioja, San Juan, and N. Mendoza Provs. (W.
    Argentina).
    ISIS NUMBER: 5301410026004002001.

Spalacopus Wagler, 1832. Isis, 25:1219.
    ISIS NUMBER: 5301410026005000000.

  Spalacopus cyanus (Molina, 1782). Sagg. Stor. Nat. Chile, p. 300.
    TYPE LOCALITY: Chile, Valparaiso Prov.
    DISTRIBUTION: Chile, west of the Andes.
    COMMENT: Includes *tabanus*; see Cabrera, 1961:517.
    ISIS NUMBER: 5301410026005002001 as *S. cyanus*.
                 5301410026005001001 as *S. tabanus*.

Family Ctenomyidae
    REVIEWED BY: M. A. Mares (MAM); C. A. Woods (CAW).
    COMMENT: Considered a distinct family by Packard, 1967, *in* Anderson and Jones, p.
    278–280, and Cabrera, 1961:546; but also see Ellerman, 1940:158, Landry, 1957,
    Univ. Calif. Publ. Zool., 56:1–118, Pasqual *et al.*, 1965, Ameghiniana, 4:19–30, and
    Reig, 1970, J. Mammal., 51:592–601, who placed it as a subfamily of Octodontidae.
    ISIS NUMBER: 5301410027000000000.

Ctenomys Blainville, 1826. Bull. Sci. Soc. Philom. Paris, p. 62.
    COMMENT: Revision of this genus may reduce the number of species recognized; see
    Packard, 1967, *in* Anderson and Jones, p. 280.
    ISIS NUMBER: 5301410027001000000.

  Ctenomys australis Rusconi, 1934. Rev. Chil. Nat. Hist., 38:108.
    TYPE LOCALITY: Argentina, Buenos Aires Prov.
    DISTRIBUTION: E. Argentina.
    COMMENT: Considered distinct from *porteousi* by Contreras and Reig, 1965, Physis,
    25(3):62, and Roig and Reig, 1969, Comp. Biochem. Physiol., 30:670.

  Ctenomys azarae Thomas, 1903. Ann. Mag. Nat. Hist., ser. 7, 11:228.
    TYPE LOCALITY: Argentina, Buenos Aires Prov., 37° 45' S., 65° W.
    DISTRIBUTION: La Pampa Prov. (Argentina).
    COMMENT: Considered distinct from *mendocinus* by Roig and Reig, 1969, Comp.
    Biochem. Physiol., 30:670.

  Ctenomys boliviensis Waterhouse, 1848. Nat. Hist. Mammal., 2:278.
    TYPE LOCALITY: Bolivia, Santa Cruz Dept., Santa Cruz de la Sierra.
    DISTRIBUTION: C. Bolivia at the foot of the Andes.
    ISIS NUMBER: 5301410027001001001.

Ctenomys brasiliensis Blainville, 1826. Bull. Sci. Soc. Philom. Paris, 3:62.
TYPE LOCALITY: Brazil, Minas Gerais.
DISTRIBUTION: E. Brazil.
ISIS NUMBER: 5301410027001002001.

Ctenomys colburni J. A. Allen, 1903. Bull. Am. Mus. Nat. Hist., 19:188.
TYPE LOCALITY: Argentina, Santa Cruz Prov., Arroyo Aiken.
DISTRIBUTION: Extreme W. Santa Cruz Prov. (Argentina).
ISIS NUMBER: 5301410027001003001.

Ctenomys conoveri Osgood, 1946. Fieldiana Zool., 31:47.
TYPE LOCALITY: Paraguay, Boqueron, 16 km W. of Filadelfia.
DISTRIBUTION: Chaco of Paraguay.

Ctenomys dorsalis Thomas, 1900. Ann. Mag. Nat. Hist., ser. 7, 6:385.
TYPE LOCALITY: Paraguay, Boqueron Prov., N. Chaco.
DISTRIBUTION: Paraguay, west of the River Paraguay.
ISIS NUMBER: 5301410027001004001.

Ctenomys emilianus Thomas and St. Leger, 1926. Ann. Mag. Nat. Hist., ser. 9, 18:637.
TYPE LOCALITY: Argentina, Neuquen Prov., Chos Malal.
DISTRIBUTION: Neuquen Prov., at the base of the Andes (Argentina).
ISIS NUMBER: 5301410027001005001.

Ctenomys frater Thomas, 1902. Ann. Mag. Nat. Hist., ser. 9, 7:185.
TYPE LOCALITY: Argentina, Jujuy Prov., Sunchal.
DISTRIBUTION: Mountains south of Jujuy and in Salta Prov. (Argentina); S.W. Bolivia.
ISIS NUMBER: 5301410027001006001.

Ctenomys fulvus Philippi, 1860. Reise Wuste Atacama, p. 157.
TYPE LOCALITY: Chile, Antofagasta Prov., Pingo-Pingo.
DISTRIBUTION: Mountains and Monte desert of N.W. Argentina and N. Chile.
COMMENT: Includes *robustus*, a Rassenkreis subspecies restricted to the oasis of Pica in
  Tarapaca Prov., Chile; see Mann, 1978, Gayana Zool., 40:292. Includes *coludo*,
  famosus, johanis, and *tulduco*; see Cabrera, 1961:548–549; but also see Contreras, *et
  al.*, 1977, Physis, sec. C, 36(92):159–162, who considered these forms provisionally
  distinct; also see comment under *validus*.
ISIS NUMBER: 5301410027001007001 as *C. fulvus*.
  5301410027001021001 as *C. robustus*.

Ctenomys knighti Thomas, 1919. Ann. Mag. Nat. Hist., ser. 9, 3:498.
TYPE LOCALITY: Argentina, Catamarca Prov., Otro Cerro.
DISTRIBUTION: Mountains between Tucuman and La Rioja (W. Argentina), north to
  Salta.
ISIS NUMBER: 5301410027001008001.

Ctenomys latro Thomas, 1918. Ann. Mag. Nat. Hist., ser. 9, 1:38.
TYPE LOCALITY: Argentina, Tucuman Prov., Tapia.
DISTRIBUTION: N.W. Argentina.
COMMENT: Considered distinct from *mendocinus* by Roig and Reig, 1969, Comp.
  Biochem. Physiol., 30:670.

Ctenomys leucodon Waterhouse, 1848. Nat. Hist. Mammal., 2:281.
TYPE LOCALITY: Bolivia, La Paz Dept., San Andres de Machaca.
DISTRIBUTION: W. Bolivia and E. Peru around Lake Titicaca.
ISIS NUMBER: 5301410027001009001.

Ctenomys lewisi Thomas, 1926. Ann. Mag. Nat. Hist., ser. 9, 17:323.
TYPE LOCALITY: Bolivia, Tarija Dept., Sama.
DISTRIBUTION: S. Bolivia.
ISIS NUMBER: 5301410027001010001.

Ctenomys magellanicus Bennett, 1836. Proc. Zool. Soc. Lond., 1835:190.
TYPE LOCALITY: Chile, Magallanes, Bahia de San Gregorio.
DISTRIBUTION: Extreme S. Chile and S. Argentina.
ISIS NUMBER: 5301410027001011001.

Ctenomys maulinus Philippi, 1872. Z. Ges. Naturw., N. F., 6:442.
  TYPE LOCALITY: Chile, Talca Prov., Laguna de Maule.
  DISTRIBUTION: Between Talca and Cautin Provs. (S.C. Chile).
  ISIS NUMBER: 5301410027001012001.

Ctenomys mendocinus Philippi, 1869. Arch. Naturg., 1:38.
  TYPE LOCALITY: Argentina, Mendoza Prov., Mendoza.
  DISTRIBUTION: East of the mountains from Salta Prov. to Chubut (Argentina).
  COMMENT: Formerly included *azarae, latro, occultus,* and *tucumanus;* see Roig and Reig,
    1969, Comp. Biochem. Physiol., 30:670, and Reig *et al.,* 1966, Contr. Cient. Fac.
    Cienc. Exact. Fis. Nat. Univ. Buenos Aires (Zool.), 2:299–352.
  ISIS NUMBER: 5301410027001013001.

Ctenomys minutus Nehring, 1887. Sitzb. Ges. Naturf. Fr. Berlin, p. 47.
  TYPE LOCALITY: Brazil, Rio Grande do Sul, Campos.
  DISTRIBUTION: Rio Grande do Sul and Mato Grosso (S.W. Brazil); Uruguay; N.W.
    Argentina.
  COMMENT: Reviewed by Langguth and Abella, 1970, Communic. Zool. Mus. Hist. Nat.
    Montevideo, 10(129).
  ISIS NUMBER: 5301410027001014001.

Ctenomys nattereri Wagner, 1848. Arch. Naturg., 1:72.
  TYPE LOCALITY: Brazil, Mato Grosso, Caicara.
  DISTRIBUTION: Mato Grosso (Brazil).
  ISIS NUMBER: 5301410027001015001 as *C. natteri (sic).*

Ctenomys occultus Thomas, 1920. Ann. Mag. Nat. Hist., ser. 9, 6:243.
  TYPE LOCALITY: Argentina, Monteagudo, 80 km S.E. Tucuman City.
  DISTRIBUTION: N. Argentina.
  COMMENT: Revised by Reig *et al.,* 1966, Contr. Cient. Fac. Cienc. Exact. Fis. Nat. Univ.
    Buenos Aires (Zool.), 2:299–352; also see Cabrera, 1961:552, who included this
    species in *mendocinus.*

Ctenomys opimus Wagner, 1848. Arch. Naturg., 1:75.
  TYPE LOCALITY: Bolivia, Oruro Dept., Sajama.
  DISTRIBUTION: N.W. Argentina; S.W. Bolivia; S. Peru; N. Chile.
  COMMENT: May be conspecific with *fulvus; opimus* and *fulvus* cannot be separated on
    the basis of karyotypes; see Gallardo, 1979, Arch. Biol. Med. Exp., 12:80.
  ISIS NUMBER: 5301410027001016001.

Ctenomys perrensis Thomas, 1898. Ann. Mag. Nat. Hist., ser. 6, 18:311.
  TYPE LOCALITY: Argentina, Corrientes Prov., Goya.
  DISTRIBUTION: Corrientes and Misiones Provs. (N.E. Argentina).
  ISIS NUMBER: 5301410027001017001.

Ctenomys peruanus Sanborn and Pearson, 1947. Proc. Biol. Soc. Wash., 60:13.
  TYPE LOCALITY: Peru, Puno Dept., Pisacoma.
  DISTRIBUTION: Altiplano of extreme S. Peru.
  ISIS NUMBER: 5301410027001018001.

Ctenomys pontifex Thomas, 1918. Ann. Mag. Nat. Hist., ser. 9, 1:39.
  TYPE LOCALITY: Argentina, Mendoza Prov., San Rafael.
  DISTRIBUTION: W. Argentina, east of the Andes to San Luis.
  ISIS NUMBER: 5301410027001019001.

Ctenomys porteousi Thomas, 1916. Ann. Mag. Nat. Hist., ser. 8, 18:304.
  TYPE LOCALITY: Argentina, Buenos Aires Prov., Bonifacio.
  DISTRIBUTION: Buenos Aires and La Pampa Provs. (E. Argentina).
  COMMENT: Cabrera, 1961:555, included *australis* in this species; but see Contreras and
    Reig, 1965, Physis, 25(3):62, and Roig and Reig, 1969, Comp. Biochem. Physiol.,
    30:670, who considered *australis* a distinct species.
  ISIS NUMBER: 5301410027001020001.

Ctenomys saltarius Thomas, 1912. Ann. Mag. Nat. Hist., ser. 8, 10:639.
TYPE LOCALITY: Argentina, Salta Prov., Salta.
DISTRIBUTION: Salta and Jujuy Provs. (N. Argentina).
ISIS NUMBER: 5301410027001022001.

Ctenomys sericeus J. A. Allen, 1903. Bull. Am. Mus. Nat. Hist., 19:187.
TYPE LOCALITY: Argentina, Santa Cruz Prov., Rio Chico.
DISTRIBUTION: S.W. Argentina.
ISIS NUMBER: 5301410027001023001.

Ctenomys steinbachi Thomas, 1907. Ann. Mag. Nat. Hist., ser. 7, 20:164.
TYPE LOCALITY: Bolivia, Santa Cruz Dept., near Santa Cruz de la Sierra.
DISTRIBUTION: Bolivia, east of the Andes.
ISIS NUMBER: 5301410027001024001.

Ctenomys talarum Thomas, 1898. Ann. Mag. Nat. Hist., ser. 7, 1:285.
TYPE LOCALITY: Argentina, Buenos Aires Prov., Los Talas.
DISTRIBUTION: Along the coast in Buenos Aires Prov. (E. Argentina).
COMMENT: Possibly a subspecies of *mendocinus;* see Cabrera, 1961:556.
ISIS NUMBER: 5301410027001025001.

Ctenomys torquatus Lichtenstein, 1830. Darst. Saugeth., text of pl. 31.
TYPE LOCALITY: Brazil, S. provinces, and banks of Uruguay River.
DISTRIBUTION: Uruguay; N.E. Argentina; extreme S. Brazil.
COMMENT: Considered a distinct species by Ellerman, 1940:167, and Walker *et al.,*
1975:1048; but see Cabrera, 1961:547, who provisionally included it in *brasiliensis.*
See Moojen, 1952, Biblio. Cient. Brasil., A(2):188, for discussion of type locality.
ISIS NUMBER: 5301410027001026001.

Ctenomys tuconax Thomas, 1925. Ann. Mag. Nat. Hist., ser. 9, 15:583.
TYPE LOCALITY: Argentina, Tucuman Prov., Concepcion.
DISTRIBUTION: East of the mountains in Tucuman Prov. (N.W. Argentina).
ISIS NUMBER: 5301410027001027001.

Ctenomys tucumanus Thomas, 1900. Ann. Mag. Nat. Hist., ser. 7, 6:301.
TYPE LOCALITY: Argentina, Tucuman Prov.
DISTRIBUTION: N.W. Argentina.
COMMENT: Considered distinct from *mendocinus* by Roig and Reig, 1969, Comp.
Biochem. Physiol., 30:670.

Ctenomys validus Contreras, Roig, and Suzarte, 1977. Physis, sec. C., 36(92):160.
TYPE LOCALITY: Argentina, Mendoza Prov., Guaymallen Dept., sand banks of El
Borbollon, El Algarrobal (near city of Mendoza).
DISTRIBUTION: Mendoza Prov. (Argentina).
COMMENT: Closely related to *johanis* (here included in *fulvus)* according to Contreras *et
al.,* 1977.

Family Abrocomidae
REVIEWED BY: H. Levenson (HL); M. A. Mares (MAM); C. A. Woods (CAW).
ISIS NUMBER: 5301410028000000000.

Abrocoma Waterhouse, 1837. Proc. Zool. Soc. Lond., 1837:30.
ISIS NUMBER: 5301410028001000000.

Abrocoma bennetti Waterhouse, 1837. Proc. Zool. Soc. Lond., 1837:31.
TYPE LOCALITY: Chile, Aconcagua Prov., vicinity Aconcagua, flanks of Cordillera.
DISTRIBUTION: From Copiapo to the area of Rio Biobio (Chile).
ISIS NUMBER: 5301410028001001001.

Abrocoma cinerea Thomas, 1919. Ann. Mag. Nat. Hist., ser. 9, 4:132.
TYPE LOCALITY: Argentina, Jujuy Prov., Cerro Casabindo.
DISTRIBUTION: S.E. Peru, W. Bolivia, N. Chile, and N.W. Argentina.
ISIS NUMBER: 5301410028001002001.

Family Echimyidae
REVIEWED BY: M. A. Mares (MAM); C. A. Woods (CAW).
COMMENT: Includes *Heteropsomys* from the Greater Antilles; known only from fossils, possibly Holocene; see Hall, 1981:875–876.
ISIS NUMBER: 5301410029000000000.

Carterodon Waterhouse, 1848. Nat. Hist. Mammal., 2:351.
ISIS NUMBER: 5301410029001000000.

Carterodon sulcidens (Lund, 1839). Afh. Kongl. Danske Vid. Selsk., p. 39.
TYPE LOCALITY: Brazil, Minas Gerais, Lagoa Santa.
DISTRIBUTION: E. Brazil.
ISIS NUMBER: 5301410029001001001.

Clyomys Thomas, 1916. Ann. Mag. Nat. Hist., ser. 8, 18:300.
ISIS NUMBER: 5301410029003000000.

Clyomys laticeps (Thomas, 1909). Ann. Mag. Nat. Hist., ser. 8, 4:240.
TYPE LOCALITY: Brazil, Santa Catarina, Joinville.
DISTRIBUTION: Between Minas Gerais and Santa Catarina (E. Brazil).
ISIS NUMBER: 5301410029003001001.

Dactylomys I. Geoffroy, 1838. Ann. Sci. Nat. Paris Zool., ser. 2, 10:126.
COMMENT: Includes *Lachnomys*; see Cabrera, 1961:543.
ISIS NUMBER: 5301410029012000000.

Dactylomys boliviensis Anthony, 1920. J. Mammal., 1:82.
TYPE LOCALITY: Bolivia, Cochabamba Dept., Mision de San Antonio.
DISTRIBUTION: C. Bolivia; S.E. Peru.
ISIS NUMBER: 5301410029012001001 as *D. bolivensis (sic)*.

Dactylomys dactylinus (Desmarest, 1817). Nouv. Dict. Hist. Nat., 2nd ed., 10:57.
TYPE LOCALITY: Upper Amazon area.
DISTRIBUTION: Upper Amazon Basin in N. Brazil, Peru, Ecuador, and perhaps Colombia.
ISIS NUMBER: 5301410029012002001.

Dactylomys peruanus J. A. Allen, 1900. Bull. Am. Mus. Nat. Hist., 13:220.
TYPE LOCALITY: Peru, Puno Dept., Inca Mines.
DISTRIBUTION: S.E. Peru.
COMMENT: Type locality corrected in Allen, 1901, Bull. Am. Mus. Nat. Hist., 14:41–46.

Diplomys Thomas, 1916. Ann. Mag. Nat. Hist., ser. 8, 18:240.
ISIS NUMBER: 5301410029004000000.

Diplomys caniceps (Gunther, 1877). Proc. Zool. Soc. Lond., 1876:745.
TYPE LOCALITY: Colombia, Antioquia Dept., Medellin.
DISTRIBUTION: W. Colombia; N. Ecuador.
ISIS NUMBER: 5301410029004001001.

Diplomys labilis (Bangs, 1901). Am. Nat., 35:638.
TYPE LOCALITY: Panama, San Miguel Isl.
DISTRIBUTION: Panama (including San Miguel Isl.); W. Colombia.
COMMENT: Includes *darlingi*; see Handley, 1966, *in* Wenzel and Tipton, eds., Ectoparasites of Panama, Field Mus. Nat. Hist., p. 787, and Hall, 1981:874.
ISIS NUMBER: 5301410029004002001.

Diplomys rufodorsalis (J. A. Allen, 1899). Bull. Am. Mus. Nat. Hist., 12:197.
TYPE LOCALITY: Colombia, Magdalena Dept., Onaca.
DISTRIBUTION: N.E. Colombia.
ISIS NUMBER: 5301410029004003001.

Echimys G. Cuvier, 1809. Bull. Sci. Soc. Philom. Paris, 24:394.
COMMENT: Formerly included *armatus* which is the type species of the genus *Makalata* Husson, 1978:445.
ISIS NUMBER: 5301410029005000000.

Echimys blainvillei (F. Cuvier, 1837). Ann. Sci. Nat. Paris Zool., ser. 2, 8:371.
TYPE LOCALITY: Brazil, Bahia, Isla de Deos.
DISTRIBUTION: S.E. Brazil, mainland and coastal islands.
ISIS NUMBER: 5301410029005002001.

Echimys braziliensis Waterhouse, 1848. Nat. Hist. Mammal., 2:330.
TYPE LOCALITY: Brazil, Minas Gerais, Lagoa Santa.
DISTRIBUTION: E. Brazil.
ISIS NUMBER: 5301410029005003001.

Echimys chrysurus (Zimmermann, 1780). Geogr. Gesch. Mensch. Vierf. Thiere, 2:352.
TYPE LOCALITY: Surinam.
DISTRIBUTION: Guianas to lower Amazonian N.E. Brazil.
ISIS NUMBER: 5301410029005004001.

Echimys dasythrix (Hensel, 1872). Abh. Konigl. Akad. Wiss. Berlin, p. 49.
TYPE LOCALITY: Brazil, Rio Grande do Sul.
DISTRIBUTION: S.E. and E. Brazil.
ISIS NUMBER: 5301410029005005001.

Echimys grandis (Wagner, 1845). Arch. Naturg., 1:145.
TYPE LOCALITY: Brazil, Amazonas, Manaqueri.
DISTRIBUTION: Amazonian Brazil and N.E. Peru.
ISIS NUMBER: 5301410029005006001.

Echimys macrurus (Wagner, 1842). Arch. Naturg., 1:360.
TYPE LOCALITY: Brazil, Amazonas, Borba.
DISTRIBUTION: Brazil, south of the Amazon River.
ISIS NUMBER: 5301410029005007001.

Echimys nigrispinus (Wagner, 1842). Arch. Naturg., 1:361.
TYPE LOCALITY: Brazil, Sao Paulo, Ipanema.
DISTRIBUTION: E. Brazil.
ISIS NUMBER: 5301410029005008001.

Echimys saturnus Thomas, 1928. Ann. Mag. Nat. Hist., ser. 10, 2:409.
TYPE LOCALITY: Ecuador, Napo-Pastaza Prov., Rio Napo.
DISTRIBUTION: Ecuador, east of the Andes.
ISIS NUMBER: 5301410029005009001.

Echimys semivillosus (I. Geoffroy, 1838). Ann. Sci. Nat. Paris Zool., ser. 2, 10:125.
TYPE LOCALITY: Colombia, Bolivar Dept., Cartagena.
DISTRIBUTION: N. Colombia; Venezuela; Margarita Isl.
ISIS NUMBER: 5301410029005010001.

Echimys unicolor (Wagner, 1842). Arch. Naturg., 1:361.
TYPE LOCALITY: Brazil.
DISTRIBUTION: Brazil.
COMMENT: Exact distribution is unknown (CAW).
ISIS NUMBER: 5301410029005011001.

Euryzygomatomys Goeldi, 1901. Bol. Mus. Para., 3:179.
ISIS NUMBER: 5301410029006000000.

Euryzygomatomys spinosus (G. Fischer, 1814). Zoognosia, 3:105.
TYPE LOCALITY: Paraguay, Cordillera, Atira.
DISTRIBUTION: S. and E. Brazil; N.E. Argentina; Paraguay.
COMMENT: The use of the names derived from Fischer, 1814, is provisional pending clarification of the availability of the work; see comments under *Reithrodon physodes* (Cricetidae).
ISIS NUMBER: 5301410029006001001.

Hoplomys J. A. Allen, 1908. Bull. Am. Mus. Nat. Hist., 24:649.
ISIS NUMBER: 5301410029007000000.

Hoplomys gymnurus (Thomas, 1897). Ann. Mag. Nat. Hist., ser. 6, 20:550.
TYPE LOCALITY: Ecuador, Esmeraldas Prov., Cachavi.
DISTRIBUTION: E.C. Honduras to N.W. Ecuador.
COMMENT: Formerly included *hoplomyoides* which was transferred to *Proechimys* by
Handley, 1976, Brigham Young Univ. Sci. Bull. Biol. Ser., 20(5):57.
ISIS NUMBER: 5301410029007001001.

Isothrix Wagner, 1845. Arch. Naturg., 1:145.
ISIS NUMBER: 5301410029008000000.

Isothrix bistriatus Wagner, 1845. Arch. Naturg., 1:146.
TYPE LOCALITY: Brazil, Rio Guapore.
DISTRIBUTION: S.W. to N.C. Brazil; S. Venezuela; adjacent Colombia.
ISIS NUMBER: 5301410029008001001.

Isothrix pictus (Pictet, 1841). Notice Anim. Nouv. Mus. Geneve, p. 29.
TYPE LOCALITY: Brazil, Bahia.
DISTRIBUTION: E. Brazil.
ISIS NUMBER: 5301410029008002001.

Isothrix villosus (Deville, 1852). Rev. Mag. Zool. Paris, ser. 2, 4:360.
TYPE LOCALITY: Peru, Sarayacu.
DISTRIBUTION: E. Peru.
ISIS NUMBER: 5301410029008003001.

Kannabateomys Jentink, 1891. Notes Leyden Mus., 13:109.
ISIS NUMBER: 5301410029013000000.

Kannabateomys amblyonyx (Wagner, 1845). Arch. Naturg., 1:146.
TYPE LOCALITY: Brazil, Sao Paulo, Ipanema.
DISTRIBUTION: E. Brazil; Paraguay; N.E. Argentina.
ISIS NUMBER: 5301410029013001001.

Lonchothrix Thomas, 1920. Ann. Mag. Nat. Hist., ser. 9, 6:113.
ISIS NUMBER: 5301410029009000000.

Lonchothrix emiliae Thomas, 1920. Ann. Mag. Nat. Hist., ser. 9, 6:114.
TYPE LOCALITY: Brazil, Rio Tapajoz, Villa Braga.
DISTRIBUTION: Brazil, south of the Amazon River.
ISIS NUMBER: 5301410029009001001.

Makalata Husson, 1978. The Mammals of Suriname, p. 445.

Makalata armata (I. Geoffroy, 1830). Rev. Zool., 1:101.
TYPE LOCALITY: French Guiana, Cayenne.
DISTRIBUTION: Andes of N. Ecuador and Colombia; Venezuela; Guianas; Tobago;
Trinidad; perhaps Martinique.
COMMENT: Formerly included in *Echimys*; transferred to *Makalata* by Husson, 1978:445.
Corbet and Hill, 1980:194, listed this species in *Echimys*. Martinique record
probably erroneous; see Hall, 1981:1180.
ISIS NUMBER: 5301410029005001001 as *Echimys armatus*.

Mesomys Wagner, 1845. Arch. Naturg., 1:145.
COMMENT: Revision of this genus is needed; see Husson, 1978:440.
ISIS NUMBER: 5301410029010000000.

Mesomys didelphoides (Desmarest, 1817). Nouv. Dict. Hist. Nat., 2:58.
TYPE LOCALITY: Unknown.
DISTRIBUTION: Brazil.
ISIS NUMBER: 5301410029010001001.

Mesomys hispidus (Desmarest, 1817). Nouv. Dict. Hist. Nat., 2:58.
TYPE LOCALITY: Brazil, Amazonas, Borba.
DISTRIBUTION: N. and E. Peru; E. Ecuador; N. Brazil.
COMMENT: Formerly included *stimulax;* see Husson, 1978:438.
ISIS NUMBER: 5301410029010002001.

Mesomys obscurus (Wagner, 1840). Abh. Akad. Wiss. Munich, 3:196.
TYPE LOCALITY: Unknown.
DISTRIBUTION: Brazil.
COMMENT: Known only from the original description; status uncertain; see Cabrera, 1961:536.
ISIS NUMBER: 5301410029010003001.

Mesomys stimulax Thomas, 1911. Ann. Mag. Nat. Hist., ser. 8, 7:607.
TYPE LOCALITY: N. Brazil, Amazon estuary, "Cameta, Lower Tocantins."
DISTRIBUTION: N. Brazil, Surinam.
COMMENT: Formerly included in *hispidus* by Cabrera, 1961:536; considered a distinct species by Husson, 1978:438.

Proechimys J. A. Allen, 1899. Bull. Am. Mus. Nat. Hist., 12:257.
COMMENT: Reviewed, in part, by Reig *et al.,* 1980:291–312, *in* Voronstov and Van Brink, eds., Animal genetics and evolution, Junk, The Hague.
ISIS NUMBER: 5301410029011000000.

Proechimys albispinus (I. Geoffroy, 1838). Ann. Sci. Nat. Paris Zool., 10:125.
TYPE LOCALITY: Brazil, Bahia, Isla de Deos.
DISTRIBUTION: Bahia and adjacent islands (Brazil).
ISIS NUMBER: 5301410029011001001.

Proechimys amphichoricus Moojen, 1948. Univ. Kans. Publ. Mus. Nat. Hist., 1:344.
TYPE LOCALITY: Venezuela, Amazonas, Cerro Duida, 325 m.
DISTRIBUTION: S. Venezuela; adjacent Brazil.
COMMENT: Formerly included in *semispinosus;* see Reig *et al.,* 1980:291–312, *in* Voronstov and Van Brink, eds., Animal genetics and evolution, Junk, The Hague.

Proechimys brevicauda (Gunther, 1877). Proc. Zool. Soc. Lond., 1876:748.
TYPE LOCALITY: Peru, Loreto, Rio Huallaga, Chamicuros.
DISTRIBUTION: E. Peru; N.W. Brazil.
COMMENT: Formerly included in *longicaudatus;* see Reig *et al.,* 1980:291–312, *in* Voronstov and Van Brink, eds., Animal genetics and evolution, Junk, The Hague, and Patton and Gardner, 1972, Occas. Pap. Mus. Zool. La. St. Univ., 44.

Proechimys canicollis (J. A. Allen, 1899). Bull. Am. Mus. Nat. Hist., 12:200.
TYPE LOCALITY: Colombia, Magdalena Dept., Bonda.
DISTRIBUTION: N. Colombia to S. Guyana; perhaps adjacent Brazil.
ISIS NUMBER: 5301410029011002001.

Proechimys cuvieri Petter, 1978. C. R. Acad. Sci. Paris, ser. D, 287:263.
TYPE LOCALITY: French Guiana, Saul.
DISTRIBUTION: French Guiana; Surinam; Guyana.

Proechimys dimidiatus (Gunther, 1877). Proc. Zool. Soc. Lond., 1876:747.
TYPE LOCALITY: Brazil, Rio de Janeiro.
DISTRIBUTION: E. Brazil.
ISIS NUMBER: 5301410029011003001.

Proechimys goeldii Thomas, 1905. Ann. Mag. Nat. Hist., ser. 7, 15:587.
TYPE LOCALITY: Brazil, Para, Santarem.
DISTRIBUTION: Amazonian Brazil between Jamunda and Tapajoz Rivers; W. Brazil.
ISIS NUMBER: 5301410029011004001.

Proechimys guairae Thomas, 1901. Proc. Biol. Soc. Wash., 14:27.
TYPE LOCALITY: Venezuela, Federal Dist., La Guaira.
DISTRIBUTION: N.C. Venezuela, E. of Lake Maracaibo and the Merida Andes.

COMMENT: Includes *ochraceus; guairae* and *ochraceus* were formerly included in *guyannensis;* see Reig *et al.,* 1980:291–312, *in* Voronstov and Van Brink, eds., Animal genetics and evolution, Junk, The Hague. These authors also mention a closely related, undescribed species ("Barina's") from south of the Merida Andes.

Proechimys guyannensis (E. Geoffroy, 1803). Cat. Mamm. Mus. Nat. Hist. Nat., p. 194.
TYPE LOCALITY: French Guiana, Cayenne.
DISTRIBUTION: Colombia to the Guianas, N.E. Peru, N.W. Bolivia and C. Brazil; Gorgona Isl. (Colombia).
COMMENT: Formerly included *warreni* (see Husson, 1978:436), *guairae, oris, poliopus, trinitatus,* and *urichi* (see Reig *et al.,* 1980:291–312, *in* Voronstov and Van Brink, eds., Animal genetics and evolution, Junk, The Hague). Includes *cherriei;* see Petter, 1978, C.R. Acad. Sci. Paris, ser. D, 287:261–264, and Reig *et al.,* 1980:291–312, *in* Voronstov and Van Brink, eds., Animal genetics and evolution, Junk, The Hague. Patton and Gardner, 1972, Occas. Pap. Mus. Zool. La. St. Univ., 44, reported the karyotype, from E. Peru.
ISIS NUMBER: 5301410029011005001.

Proechimys hendeei Thomas, 1926. Ann. Mag. Nat. Hist., ser. 9, 18:162.
TYPE LOCALITY: Peru, Loreto Dept., Puca Tambo.
DISTRIBUTION: N.E. Peru to S. Colombia.
COMMENT: Patton and Gardner, 1972, Occas. Pap. Mus. Zool. La. St. Univ., 44, reported the karyotype of this species, from E. Peru.
ISIS NUMBER: 5301410029011006001.

Proechimys hoplomyoides (Tate, 1939). Bull. Am. Mus. Nat. Hist., 76:179.
TYPE LOCALITY: Venezuela, Bolivar Prov., Mt. Roraima.
DISTRIBUTION: S.E. Venezuela; adjacent Guyana and Brazil.
COMMENT: Transferred from *Hoplomys* by Handley, 1976, Brigham Young Univ. Sci. Bull. Biol. Ser., 20(5):57.

Proechimys iheringi Thomas, 1911. Ann. Mag. Nat. Hist., ser. 8, 8:252.
TYPE LOCALITY: Brazil, Sao Paulo, Sao Sabastiao Isl.
DISTRIBUTION: E. Brazil.
ISIS NUMBER: 5301410029011007001.

Proechimys longicaudatus (Rengger, 1830). Naturg. Saugeth. Paraguay, p. 236.
TYPE LOCALITY: Paraguay.
DISTRIBUTION: C. and E. Peru; W. Bolivia; Paraguay; Brazil.
COMMENT: Formerly included *brevicauda;* see Reig *et al.,* 1980:291–312, *in* Voronstov and Van Brink, eds., Animal genetics and evolution, Junk, The Hague. Patton and Gardner, 1972, Occas. Pap. Mus. Zool. La. St. Univ., 44, reported the karyotype of this species, from E. Peru.
ISIS NUMBER: 5301410029011008001.

Proechimys myosuros (Lichtenstein, 1820). Abh. Konigl. Akad. Wiss. Berlin, p. 192.
TYPE LOCALITY: Brazil, Bahia.
DISTRIBUTION: Bahia (Brazil).
ISIS NUMBER: 5301410029011009001.

Proechimys oris Thomas, 1904. Ann. Mag. Nat. Hist., ser. 7, 14:105.
TYPE LOCALITY: Brazil, Para, Igarape-assu, near Belem.
DISTRIBUTION: C. Brazil.
COMMENT: Formerly included in *guyannensis;* see Reig *et al.,* 1980:291–312, *in* Voronstov and Van Brink, eds., Animal genetics and evolution, Junk, The Hague.

Proechimys poliopus Osgood, 1914. Field Mus. Nat. Hist. Publ., Zool. Ser., 10:135.
TYPE LOCALITY: Venezuela, Zulia Prov., Rio Aurare, El Panorama.
DISTRIBUTION: N.W. Venezuela, between Lake Maracaibo and the Sierra de Perija; adjacent Colombia.
COMMENT: Formerly included in *guyannensis;* see Reig *et al.,* 1980:291–312, *in* Voronstov and Van Brink, eds., Animal genetics and evolution, Junk, The Hague.

Proechimys quadruplicatus Hershkovitz, 1948. Proc. U.S. Nat. Mus., 97:138.
TYPE LOCALITY: Ecuador, Napo-Pastaza Prov., Rio Napo, Isla Llunchi.
DISTRIBUTION: E. Ecuador and E. Peru.
ISIS NUMBER: 5301410029011010001.

Proechimys semispinosus (Tomes, 1860). Proc. Zool. Soc. Lond., 1860:265.
TYPE LOCALITY: Ecuador, Santiago-Zamora Prov., Gualaquiza.
DISTRIBUTION: S.E. Honduras to N.E. Peru and Amazonian Brazil.
COMMENT: Formerly included *amphichoricus*; see Reig *et al.*, 1980:291–312, *in* Voronstov
and Van Brink, eds., Animal genetics and evolution, Junk, The Hague, who also
employed the name *centralis* for animals assigned to *semispinosus* from N.
Venezuela.
ISIS NUMBER: 5301410029011011001.

Proechimys setosus (Desmarest, 1817). Nouv. Dict. Hist. Nat., ser. 2, 10:59.
TYPE LOCALITY: Brazil, Bahia, Moojen.
DISTRIBUTION: E. Brazil.
ISIS NUMBER: 5301410029011012001.

Proechimys trinitatus (J. A. Allen and Chapman, 1893). Bull. Am. Mus. Nat. Hist., 5:223.
TYPE LOCALITY: Trinidad and Tobago, Trinidad.
DISTRIBUTION: Trinidad.
COMMENT: Formerly included in *guyannensis*; see Reig *et al.*, 1980:291–312, *in*
Voronstov and Van Brink, eds., Animal genetics and evolution, Junk, The Hague.

Proechimys urichi (J. A. Allen, 1899). Bull. Am. Mus. Nat. Hist., 12:199.
TYPE LOCALITY: Venezuela, Sucre, Quebrada Seca.
DISTRIBUTION: N. Venezuela.
COMMENT: Formerly included in *guyannensis*; see Reig *et al.*, 1980:291–312, *in*
Voronstov and Van Brink, eds., Animal genetics and evolution, Junk, The Hague.

Proechimys warreni Thomas, 1905. Ann. Mag. Nat. Hist., ser. 7, 16:312–313.
TYPE LOCALITY: Guyana, 80 mi. (129 km) up the Demerara River, Comackka.
DISTRIBUTION: Guyana; Surinam; exact limits not known.
COMMENT: Considered a distinct species by Husson, 1978:436; considered a subspecies
of *guyannensis* by Cabrera, 1961:520.

Thrichomys Trouessart, 1880. Cat. Mamm. Bull. Soc. Etudes Sci. Angers, 1881:179.
COMMENT: Formerly referred to as *Cercomys* which was based on *Cercomys cunicularius*,
a composite; see Petter, 1973, Mammalia, 37:422–426.
ISIS NUMBER: 5301410029002000000 as *Cercomys*.

Thrichomys apereoides Lund, 1839. Afh. Kongl. Danske Vid. Selsk., p. 38.
TYPE LOCALITY: Brazil, Minas Gerais, Lagoa Santa.
DISTRIBUTION: E. Brazil; Paraguay.
COMMENT: Formerly referred to as *Cercomys cunicularius*, a composite; see Petter, 1973,
Mammalia, 37:422–426, and Mares *et al.*, 1981, Ann. Carnegie Mus., 50(4):120.
ISIS NUMBER: 5301410029002001001 as *Cercomys cunicularis (sic)*.

Thrinacodus Gunther, 1879. Proc. Zool. Soc. Lond., 1879:144.
ISIS NUMBER: 5301410029014000000.

Thrinacodus albicauda Gunther, 1879. Proc. Zool. Soc. Lond., 1879:144.
TYPE LOCALITY: Colombia, Antioquia Dept., Medellin.
DISTRIBUTION: N.W. and C. Colombia, west of the Cordillera Central.
ISIS NUMBER: 5301410029014001001.

Thrinacodus edax Thomas, 1916. Ann. Mag. Nat. Hist., ser. 8, 18:299.
TYPE LOCALITY: Venezuela, Merida, Sierra de Merida.
DISTRIBUTION: W. Venezuela and adjacent N. Colombia.
ISIS NUMBER: 5301410029014003001.

Family Thryonomyidae
REVIEWED BY: Editors.
ISIS NUMBER: 5301410030000000000.

Thryonomys Fitzinger, 1867. Sitzb. Akad. Wiss. Wien, 56(1):141.
ISIS NUMBER: 5301410030001000000.

Thryonomys gregorianus (Thomas, 1894). Ann. Mag. Nat. Hist., ser. 6, 13:202.
TYPE LOCALITY: Kenya, Kiroyo, Luiji Reru River, 00° 35′ S., 37° 05′ E.
DISTRIBUTION: Cameroun; Central African Republic; Zaire; S. Sudan; Ethiopia; Kenya;
Uganda; Tanzania; Malawi; Zambia; Zimbabwe; Mozambique.
COMMENT: Includes *harrisoni, logonensis, rutshuricus,* and *sclateri;* see Misonne, 1974, Part
6:7.
ISIS NUMBER: 5301410030001001001 as *T. gregorianus.*
5301410030001002001 as *T. harrisoni.*
5301410030001003001 as *T. logonensis.*
5301410030001004001 as *T. rutshuricus.*
5301410030001005001 as *T. sclateri.*

Thryonomys swinderianus (Temminck, 1827). Monogr. Mamm., 1:248.
TYPE LOCALITY: Sierra Leone.
DISTRIBUTION: Africa south of the Sahara.
ISIS NUMBER: 5301410030001006001.

Family Petromyidae
REVIEWED BY: H. Levenson (HL).
ISIS NUMBER: 5301410031000000000.

Petromus A. Smith, 1831. S. Afr. Quart. J., 1(5):10.
ISIS NUMBER: 5301410031001000000.

Petromus typicus A. Smith, 1831. S. Afr. Quart. J., 1(5):11.
TYPE LOCALITY: South Africa, mouth of Orange River.
DISTRIBUTION: W. South Africa to S.W. Angola.
ISIS NUMBER: 5301410031001001001.

Family Bathyergidae
REVIEWED BY: H. Levenson (HL).
ISIS NUMBER: 5301410032000000000.

Bathyergus Illiger, 1811. Prodr. Syst. Mamm. et Avium., p. 86.
ISIS NUMBER: 5301410032001000000.

Bathyergus janetta Thomas and Schwann, 1904. Abstr. Proc. Zool. Soc. Lond., 2:6.
TYPE LOCALITY: South Africa, N.W. Cape Prov., coastal Little Namaqualand, Port
Nolloth.
DISTRIBUTION: S.W. South Africa; S. Namibia.
COMMENT: Ellerman *et al.,* 1953, included *janetta* as a subspecies of *suillus,* but de
Graaff, 1975, Part 6.9:2–3, regarded *janetta* and *suillus* as monotypic species.
ISIS NUMBER: 5301410032001001001.

Bathyergus suillus (Schreber, 1782). Saugethiere, 4:715.
TYPE LOCALITY: South Africa, Cape of Good Hope.
DISTRIBUTION: S. South Africa.
COMMENT: Includes *intermedius;* see de Graaff, 1975, Part 6.9:3.
ISIS NUMBER: 5301410032001002001.

Cryptomys Gray, 1864. Proc. Zool. Soc. Lond., 1864:124.
ISIS NUMBER: 5301410032002000000.

Cryptomys hottentotus (Lesson, 1826). Zool., 1:166.
TYPE LOCALITY: South Africa, S.W. Cape Prov., near Paarl (east of Capetown).
DISTRIBUTION: South Africa to Tanzania, S. Zaire, and Namibia.
COMMENT: Includes *bocagei, damarensis, darlingi, holosericeus,* and *natalensis;* see de Graaff, 1975, Part 6.9:3–4.
ISIS NUMBER: 5301410032002006001 as *C. hottentotus.*
5301410032002001001 as *C. bolcagei (sic).*
5301410032002002001 as *C. damarensis.*
5301410032002003001 as *C. darlingi.*
5301410032002005001 as *C. holosericeus.*

Cryptomys mechowi (Peters, 1881). Sitzb. Ges. Naturf. Fr. Berlin, p. 133.
TYPE LOCALITY: Angola, Malange.
DISTRIBUTION: Angola; S. Zaire; Malawi; Zambia; Tanzania.
COMMENT: Includes *ansorgei, blainei,* and *mellandi;* see de Graaff, 1975, Part 6.9:3.
ISIS NUMBER: 5301410032002008001.

Cryptomys ochraceocinereus (Heuglin, 1864). Nouv. Acta Acad. Caes. Leop. Dresden, 31:3.
TYPE LOCALITY: Sudan, Bahr-el-Ghazal.
DISTRIBUTION: Ghana to Sudan and N. Uganda.
COMMENT: Includes *foxi, lechei, kummi,* and *zechi;* see de Graaff, 1975, Part 6.9:3.
ISIS NUMBER: 5301410032002009001 as *C. ochraceocinereus.*
5301410032002004001 as *C. foxi.*
5301410032002007001 as *C. lechei.*
5301410032002010001 as *C. zechi.*

Georychus Illiger, 1811. Prodr. Syst. Mamm. et Avium., p. 87.
ISIS NUMBER: 5301410032003000000.

Georychus capensis (Pallas, 1778). Nova Spec. Quad. Glir. Ord., 76:172.
TYPE LOCALITY: South Africa, Cape of Good Hope.
DISTRIBUTION: South Africa.
COMMENT: Includes *canescens* and *yatesi;* see de Graaff, 1975, Part 6.9:3.
ISIS NUMBER: 5301410032003001001.

Heliophobius Peters, 1846. Monatsb. Preuss. Akad. Wiss. Berlin, p. 259.
ISIS NUMBER: 5301410032004000000.

Heliophobius argenteocinereus Peters, 1846. Monatsb. Preuss. Akad. Wiss. Berlin, p. 259.
TYPE LOCALITY: Mozambique, Tete (on the Zambezi River).
DISTRIBUTION: Zimbabwe, E. Zambia, and N. Mozambique to Zaire and Kenya.
COMMENT: Includes *mottoulei;* see de Graaff, 1975, Part 6.9:2.
ISIS NUMBER: 5301410032004001001 as *H. argenteocinereus.*
5301410032004002001 as *H. mottoulei.*

Heliophobius spalax Thomas, 1910. Ann. Mag. Nat. Hist., ser. 8, 6:315.
TYPE LOCALITY: Kenya, Taveta, near Mt. Kilimanjaro.
DISTRIBUTION: S. Kenya; N. Tanzania.
ISIS NUMBER: 5301410032004003001.

Heterocephalus Ruppell, 1842. Mus. Senckenbergianum Abh., 3(2):99.
ISIS NUMBER: 5301410032005000000.

Heterocephalus glaber Ruppell, 1842. Mus. Senckenbergianum Abh., 3(2):99.
TYPE LOCALITY: Ethiopia, Shoa.
DISTRIBUTION: C. Somalia; C. and E. Ethiopia; C. and S. Kenya.
ISIS NUMBER: 5301410032005001001.

Family Ctenodactylidae
REVIEWED BY: H. Levenson (HL).
COMMENT: Reviewed by George, 1979, Zool. J. Linn. Soc., 65:261–280.
ISIS NUMBER: 5301410033000000000.

Ctenodactylus Gray, 1830. Spicil. Zool., p. 10.
ISIS NUMBER: 5301410033001000000.

Ctenodactylus gundi (Rothmann, 1776). *In* Schloezer, Briefwechsel, p. 339.
TYPE LOCALITY: Libya, Gharian, 80 km S. of Tripoli.
DISTRIBUTION: N. Morocco to N.W. Libya.
COMMENT: Includes *massonii*; see Misonne, 1974, Part 6:5.
ISIS NUMBER: 5301410033001001001.

Ctenodactylus vali Thomas, 1902. Proc. Zool. Soc. Lond., 1902(2):11.
TYPE LOCALITY: Libya, Wadi Bey (northwest of Bonjem, Tripoli).
DISTRIBUTION: S. Morocco; W. Algeria; N.W. Libya.
COMMENT: Includes *joleaudi*; see Misonne, 1974, Part 6:5. Corbet, 1978:160, included
     vali in *gundi*, but Corbet and Hill, 1980:196, listed both as distinct species.
ISIS NUMBER: 5301410033001003001 as *C. vali.*
              5301410033001002001 as *C. joleaudi.*

Felovia Lataste, 1886. Le Naturaliste, 3:287.
ISIS NUMBER: 5301410033002000000.

Felovia vae Lataste, 1886. Le Naturaliste, 3:287.
TYPE LOCALITY: Senegal, upper Senegal River, Medina Dist., Felou.
DISTRIBUTION: Senegal; Mauritania; Mali.
ISIS NUMBER: 5301410033002001001.

Massoutiera Lataste, 1885. Le Naturaliste, 3:21.
ISIS NUMBER: 5301410033003000000.

Massoutiera mzabi (Lataste, 1881). Bull. Soc. Zool. Fr., 6:314.
TYPE LOCALITY: Algeria, Ghardaia.
DISTRIBUTION: S.E. Algeria; S.W. Libya; N. Niger; N. Chad.
COMMENT: Includes *rothschildi* and *harterti*; see Ellerman and Morrison-Scott, 1951:522,
     and Corbet, 1978:160.
ISIS NUMBER: 5301410033003002001 as *M. mzabi.*
              5301410033003001001 as *M. harterti.*
              5301410033003003001 as *M. rothschildi.*

Pectinator Blyth, 1855. J. Asiat. Soc. Bengal, 24:294.
ISIS NUMBER: 5301410033004000000.

Pectinator spekei Blyth, 1855. J. Asiat. Soc. Bengal, 24:294.
TYPE LOCALITY: Somalia, 9° N., 47° E.
DISTRIBUTION: Ethiopia; Somalia; Djibouti.
COMMENT: Includes *legerae* and *meridionalis*; see Misonne, 1974, Part 6:5.
ISIS NUMBER: 5301410033004001001.

ORDER LAGOMORPHA
ISIS NUMBER: 5301409000000000000.

Family Ochotonidae
REVIEWED BY: H. E. Broadbooks (HEB); R. M. Mitchell (RMM); O. L. Rossolimo
(OLR)(U.S.S.R.); A. T. Smith (ATS); M. L. Weston (MLW); S. Wang (SW)(China).
COMMENT: Includes Lagomyidae; see Corbet, 1978:65. Revisions of the family include
Gureev, 1964, and Corbet, 1978. Other useful treatments include Allen, 1938, Am.
Mus. Nat. Hist., 620 pp., Ellerman and Morrison-Scott, 1951, Ognev, 1940,
[Mamm. U.S.S.R., Adjac. Count.], Vol. 4, Rodents, and Hall, 1981.
ISIS NUMBER: 5301409001000000000.

Ochotona Link, 1795. Beytr. Naturg., 2:74.
COMMENT: MLW, RMM, ATS concur with Corbet, 1978:66, that there is little ground
for recognizing subgenera. The subgeneric classifications published (*e.g.*,
Ellerman and Morrison-Scott, 1951; Allen, 1938, Am. Mus. Nat. Hist., 620 pp;
Ognev, 1940, [Mamm. U.S.S.R., Adjac. Count.], Vol. 4, Rodents) differ
dramatically, even when based on the same distinguishing characteristics.
ISIS NUMBER: 5301409001001000000.

Ochotona alpina (Pallas, 1773). Reise Prov. Russ. Reichs., 2:701.
TYPE LOCALITY: U.S.S.R., Kazakh S.S.R., Altai Mtns., Tigeretskoe Range, vic. of Ust-
Kamenogorsk.
DISTRIBUTION: N. Ural, Putorana, Sayan and Altai Mtns. (U.S.S.R. and Mongolia); N.E.
Siberia to Chukotka and Kamchatka; Sakhalin (U.S.S.R.); N. Kansu, N. Ningsia,
and N.E. China; Korea; Hokkaido (Japan).
COMMENT: Includes *hyperborea*; see Corbet, 1978:69; Gromov and Baranova, 1981:74.
Also see Sokolov and Orlov, 1980:79, who considered *hyperborea* a distinct species
with a distribution overlapping that of *alpina*. Does not include *collaris* and
princeps; see Weston, 1981, *in* Myers, Proc. World Lagomorph Conf.
ISIS NUMBER: 5301409001001001000 as *O. alpina.*
5301409001001004001 as *O. hyperborea.*

Ochotona collaris (Nelson, 1893). Proc. Biol. Soc. Wash., 8:117.
TYPE LOCALITY: U.S.A., Alaska, near head of Tanana River.
DISTRIBUTION: W.C. Mackenzie, S. Yukon, N.W. British Columbia (Canada); S.E. Alaska
(U.S.A.).
COMMENT: Broadbooks, 1965, Am. Midl. Nat., 73:299–335, and Youngman, 1975,
Mammals of the Yukon Terr., Nat. Mus. Nat. Sci. (Ottawa), Publ. Zool. 10, 192 pp,
considered *collaris* and *princeps* conspecific. Corbet, 1978, following Gureev, 1964,
included *collaris* in *alpina*. A statistical re-evaluation of craniometric data by
Weston, 1981, *in* Myers, Proc. World Lagomorph Conf., indicated that *collaris*,
princeps, and *alpina* are separate species. Hall, 1981:286, recognized *collaris* as a
distinct species.
ISIS NUMBER: 5301409001001002001.

Ochotona curzoniae (Hodgson, 1858). J. Asiat. Soc. Bengal, 26:207.
TYPE LOCALITY: China, Tibet, Chumbi Valley.
DISTRIBUTION: Tibetan Plateau; adjacent Kansu, Tsinghai, Szechwan (China), Sikkim
and E. Nepal. Perhaps Iran.
COMMENT: Includes *melanostoma* and may include *seiana* from Iran; see Corbet,
1978:69. Treated as a subspecies of *daurica* by Mitchell, 1978, Saugetierk. Mitt.,
26:211, but it is now considered a distinct species (ATS, MLW, RMM).

Ochotona daurica (Pallas, 1776). Reise Prov. Russ. Reichs., 3:692.
TYPE LOCALITY: U.S.S.R., Transbaikalia, Buryat-Mongolsk. A.S.S.R., Onon River,
Kulusutai or Selenga River.
DISTRIBUTION: Steppes from Altai and Transbaikalia through N. China and N.
Mongolia.
COMMENT: Includes *mursaevi*; see Corbet, 1978:68. Formerly included *curzoniae*; see
Corbet, 1978:69; but also see Mitchell, 1978, Saugetierk. Mitt., 26:211. See

Ellerman and Morrison-Scott, 1951:452 and Ognev, 1940, [Mamm. U.S.S.R., Adjac. Count.], p. 62 for discussion of type locality.
ISIS NUMBER: 5301409001001003001.

Ochotona erythrotis (Buchner, 1890). Wiss. Res. Przewalski Cent. Asien Zool. Th. I: Saugeth., p. 165.
TYPE LOCALITY: China, East Tibet, Burchan-Budda.
DISTRIBUTION: Tibet, Yunnan, E. Tsinghai, S. Kansu, and N. Szechwan (China).
COMMENT: Includes *gloveri;* see Corbet, 1978:68.

Ochotona kamensis Argyropulo, 1948. Trudy Zool. Inst. Leningr., 7:124–128.
TYPE LOCALITY: China, W. Szechwan, "Kam".
DISTRIBUTION: Known only from Kam, W. Szechwan and Tibet (China).
COMMENT: The Argyropulo, 1948, citation given by Gureev, 1964:237, and other authors is correct. Argyropulo, 1941, purportedly described *kamensis*, but the paper was never published, although Argyropulo, 1948, Trudy Zool. Inst. Leningr., 7:126, cited the earlier date (HEB, MLW).

Ochotona koslowi (Buchner, 1894). Mamm. Przewalski, 1:187.
TYPE LOCALITY: China, S.E. Sinkiang (N. Tibet), Guldsha Valley.
DISTRIBUTION: N. edge of Tibetan Plateau.
ISIS NUMBER: 5301409001001005001.

Ochotona ladacensis (Gunther, 1875). Ann. Mag. Nat. Hist., 16:231.
TYPE LOCALITY: India, Ladak, Changra Lake, 4300 m.
DISTRIBUTION: S.W. Sinkiang, Tsinghai, E. Tibet (China); Kashmir (India); Pakistan.
ISIS NUMBER: 5301409001001006001.

Ochotona lama Mitchell and Punzo, 1975. Mammalia, 39:422.
TYPE LOCALITY: Nepal, Mustang District, Lupra (28° 48' N., 83° 47' E.), about 3640 m.
DISTRIBUTION: Mustang District (Nepal).
COMMENT: Provisionally included as a subspecies of *roylei* by Corbet, 1978:68, but retained as a species by Mitchell, 1978, Saugetierk. Mitt., 26:212.

Ochotona macrotis (Gunther, 1875). Ann. Mag. Nat. Hist., 16:231.
TYPE LOCALITY: China, S.W. Sinkiang, Kunlun Mtns., Doba (on road from Yarkand to Karakorum Pass).
DISTRIBUTION: Himalayas from Bhutan through Karakorum Range, Kunlun Shan, Pamirs, and W. Tien Shan.
COMMENT: Included in *roylei* by Gureev, 1964, Corbet, 1978:68, and Gromov and Baranova, 1981:72, and by Roberts, 1977, The Mammals of Pakistan, p. 127. Morphological and ecological differences in the area of sympatry strongly suggest that *macrotis* is a distinct species; see Kawamichi, 1971, J. Fac. Hokkaido Univ. Jpn., Ser. VI, Zool. 17:587–609. Treated as a species by Mitchell, 1978, Saugetierk. Mitt., 26:211.

Ochotona pallasi (Gray, 1867). Ann. Mag. Nat. Hist., 20:220.
TYPE LOCALITY: U.S.S.R., probably W. Kazakhstan ("Asiatic Russia, Kirgisien").
DISTRIBUTION: Arid areas (mtns. and high steppes) from Altai (U.S.S.R., Mongolia), to Sinkiang and Inner Mongolia (China); west to north of Lake Balkash (U.S.S.R.) and E. Tien Shan (China).
COMMENT: Commonly referred to as *pricei* in the Soviet literature; see Gureev, 1964:253; and Vorontsov and Ivanitskaya, 1973, Caryologia, 26:213–223. Includes *pricei;* see Corbet, 1978:69.

Ochotona princeps (Richardson, 1828). Zool. J., 3:520.
TYPE LOCALITY: Canada, Alberta, near Athabasca Pass, head of Athabasca River.
DISTRIBUTION: Mountains of W. North America from C. British Columbia (Canada) to N. New Mexico, Utah, C. Nevada, and E.C. California (U.S.A.).

COMMENT: Broadbooks, 1965, Am. Midl. Nat., 299–335, and Youngman, 1975, Mammals of the Yukon Terr., Nat. Mus. Nat. Sci. (Ottawa), Publ. Zool. 10, 192 pp., considered *princeps* and *collaris* conspecific. Corbet, 1978, following Gureev, 1964, included *princeps* in *alpina*. A statistical re-evaluation of craniometric data by Weston, 1981, *in* Myers, Proc. World Lagomorph Conf., indicated that *princeps, collaris,* and *alpina* are separate species. See also comments under *alpina*.
ISIS NUMBER: 5301409001001008001.

Ochotona pusilla (Pallas, 1769). Nova Comm. Imp. Acad. Sci. Petrop., 13:531.
TYPE LOCALITY: U.S.S.R., Orenburgsk. Obl., "Samarsk Steppe," near Buzuluk, left bank of Samara River.
DISTRIBUTION: Steppes from Volga across N. Kazakhstan to upper Irtysh River (U.S.S.R.).
ISIS NUMBER: 5301409001001009001.

Ochotona roylei (Ogilby, 1839). Royle's Illus. Botany Himalaya, 69, pl. 4.
TYPE LOCALITY: India, Punjab, Choor Mountain, 60 mi. (96 km) N. of Saharanpur.
DISTRIBUTION: Mountain arc from the Tien Shan (U.S.S.R., China) through the Pamirs, along the Himalayan Mtns. to Szechwan and Yunnan (China) and N. Burma.
COMMENT: Includes *angdawai, mitchelli, nepalensis,* and *himalayana;* see Corbet, 1978:68; but also see Mitchell, 1978, Saugetierk. Mitt., 26:212. Also includes *nubrica;* see Gureev, 1964:239. *O. angdawai* and *mitchelli* are merely color phases of *roylei* (RMM).
ISIS NUMBER: 5301409001001010001.

Ochotona rufescens (Gray, 1842). Ann. Mag. Nat. Hist., 10:266.
TYPE LOCALITY: Afghanistan, Kabul, Baber's Tomb.
DISTRIBUTION: Mtns. of Afghanistan, Baluchistan, Iran and S.W. Turkmenia (U.S.S.R.).
COMMENT: Includes *shukurovi;* see Corbet, 1978:69.
ISIS NUMBER: 5301409001001011001.

Ochotona rutila (Severtzov, 1873). Mem. Soc. Amis. Sci. Moscow, 8:19.
TYPE LOCALITY: U.S.S.R., S.E. Kazakhstan, Zailiiskii Alatau Mtns., near Alma-Ata (formerly Vernyi), 7000–8000 ft. (2134–2438 m).
DISTRIBUTION: Isolated mountain ranges from the Pamirs to Tien Shan (U.S.S.R. and China); N. Afghanistan.
ISIS NUMBER: 5301409001001012001.

Ochotona thibetana (Milne-Edwards, 1871). Nouv. Arch. Mus. Hist. Nat. Paris, 7(Bull.):93.
TYPE LOCALITY: China, Szechwan, Moupin.
DISTRIBUTION: Mtns. of W. China from S. Shansi and Nan Shan to Szechwan, S.E. Tibet (China); Sikkim. Perhaps N. Burma and Ladak (India).
COMMENT: Includes *sikimaria, osgoodi,* and *cansus;* see Corbet, 1978:67. Also includes *forresti;* see Gureev, 1964:260, and Feng and Kao, 1974, Acta Zool. Sin., 20:76–88. Feng and Kao, *op. cit.,* treated *cansus* as a distinct species.
ISIS NUMBER: 5301409001001013001.

Ochotona thomasi Argyropulo, 1948. Trudy Zool. Inst. Leningr., 7:127.
TYPE LOCALITY: China, Tsinghai, (Lake) Alyk Nor (=Chinghai, Alak Nor, 35° 30′ N., 97° 20′ E.).
DISTRIBUTION: N.E. Tsinghai, Kansu, and Szechwan (China).

Prolagus Pomel, 1853. Cat. Meth. Vert. Foss. Bass. la Loire, p. 43.
REVIEWED BY: R. S. Hoffmann (RSH).

Prolagus sardus Wagner, 1832. Abh. Bayer. Akad. Wiss., 1:763–767.
TYPE LOCALITY: Italy, Sardinia.
DISTRIBUTION: Mediterranean Isls. of Corsica (France) and Sardinia (Italy); adjacent small islands.
COMMENT: Described from fossils, but apparently survived until historic times; see Tobien, 1935, Ber. Freiburger Naturf. Ges., 34:253–344; perhaps as late as 1774; see Kurten, 1968, Pleistocene mammals of Europe, Aldine, p. 226. Reviewed by Dawson, 1969, Paleovertebrata, 2(4):157–190.

Family Leporidae
REVIEWED BY: M. Fitzsimmons (MF); K. Myers (KM); O. L. Rossolimo (OLR) (U.S.S.R.).
ISIS NUMBER: 5301409002000000000.

Bunolagus Thomas, 1929. Proc. Zool. Soc. Lond., 1929:109.
REVIEWED BY: T. J. Robinson (TJR).
COMMENT: Reviewed by Petter, 1972, Part 5:1–7. Karyological evidence supports retention of *Bunolagus* (2n=44) as distinct from *Lepus* (2n=48) (TJR).

Bunolagus monticularis (Thomas, 1903). Ann. Mag. Nat. Hist., ser. 7, 11:78.
TYPE LOCALITY: South Africa, Central Cape Colony, Deelfontein.
DISTRIBUTION: Deelfontein and region east of Calvinia, C. Cape Prov. (South Africa).
ISIS NUMBER: 5301409002002015001 as *Lepus monticularis*.

Caprolagus Blyth, 1845. J. Asiat. Soc. Bengal, 14:247.
REVIEWED BY: R. K. Ghose (RKG).
ISIS NUMBER: 5301409002001000000.

Caprolagus hispidus (Pearson, 1839). *In* M'Clelland, Proc. Zool. Soc. Lond., 1838:152.
TYPE LOCALITY: India, N. Assam, foot of Himalayas.
DISTRIBUTION: N.E. India through S. Himalayas and Nepal and from Gorakhpur to Upper Assam, Tripura (India), and Bangladesh.
COMMENT: Since 1951, there have been few reports of this species from Uttar Pradesh and Assam; see Santapau and Humayun, 1960, J. Bombay Nat. Hist. Soc., 57:400–402; Mallinson, 1971, J. Bombay Nat. Hist. Soc., 68:443–444; and Ghose, 1978, J. Bombay Nat. Hist. Soc., 75:206–209.
PROTECTED STATUS: CITES - Appendix I and U.S. ESA - Endangered.
ISIS NUMBER: 5301409002001001001.

Lepus Linnaeus, 1758. Syst. Nat., 10th ed., 1:57.
REVIEWED BY: J. E. C. Flux (JECF); J. Ramirez-Pulido (JRP) (Mexico).
COMMENT: The taxonomy of this genus is unclear. *L. crawshayi, whytei,* and *peguensis* have been variously treated as separate species or have been included in *nigricollis. L. europaeus, tolai,* and *tibetanus* have been placed in *capensis* or treated as distinct species. Such "lumping" would undoubtedly make *alleni* and *flavigularis* conspecific with *callotis,* and *insularis* with *californicus,* but these have not yet been proposed to my knowledge (JECF).
ISIS NUMBER: 5301409002002000000.

Lepus alleni Mearns, 1890. Bull. Am. Mus. Nat. Hist., 2:294.
TYPE LOCALITY: U.S.A., Arizona, Pima Co., Rillito.
DISTRIBUTION: S.C. Arizona (U.S.A.) to N. Nayarit and Tiburon Isl. (Mexico).
COMMENT: Possibly a subspecies of *callotis* (JECF). Recognized as a distinct species by Hall, 1981:331.
ISIS NUMBER: 5301409002002001001.

Lepus americanus Erxleben, 1777. Syst. Regn. Anim., 1:330.
TYPE LOCALITY: Canada, Ontario, Hudson Bay, Fort Severn.
DISTRIBUTION: S. and C. Alaska (U.S.A.) to S. and C. coasts of Hudson Bay to Newfoundland and Anacosti Isl. (introduced) (Canada), south to the S. Appalachians, S. Michigan, North Dakota, N.C. New Mexico, S.C. Utah, and E.C. California (U.S.A.).
COMMENT: Distinct small species, but subgeneric separation *(Poecilolagus* Lyon, 1904) not supported; see Hall, 1981:314.
ISIS NUMBER: 5301409002002002001.

Lepus brachyurus Temminck, 1845. *In* Siebold's Fauna Japonica, Mamm., p. 44, pl. 11.
TYPE LOCALITY: Japan, Kyushu, Nagasaki.
DISTRIBUTION: Honshu, Shikoku, Kyushu, Oki Isls. and Sado Isl. (Japan).
COMMENT: Reviewed by Imaizumi, 1970, The Handbook of Japanese Land Mammals, p. 310. Gromov and Baranova, 1981:63, placed this species in genus *Caprolagus:* see also comment under *mandshuricus.*
ISIS NUMBER: 5301409002002004001.

Lepus californicus Gray, 1837. Charlesworth's Mag. Nat. Hist., 1:586.
TYPE LOCALITY: U.S.A., California, "St. Antoine" (probably near Mission of San Antonio).
DISTRIBUTION: Hidalgo and S. Queretaro to N. Sonora and Baja California (Mexico), north to S.W. Oregon and C. Washington, S. Idaho, E. Colorado, S. South Dakota, W. Missouri, and N.W. Arkansas (U.S.A.).
COMMENT: Type locality discussed by Hall, 1981:326.
ISIS NUMBER: 5301409002002005001.

Lepus callotis Wagler, 1830. Naturliches Syst. Amphibien, p. 23.
TYPE LOCALITY: "Mexico" (southern end of Mexican Tableland).
DISTRIBUTION: C. Oaxaca (Mexico) to S.W. New Mexico (U.S.A.).
COMMENT: Includes *gaillardi* and *mexicanus;* see Anderson and Gaunt, 1962, Am. Mus. Novit., 2088:5; Hall, 1981:328-330. See comment under *alleni.*
ISIS NUMBER: 5301409002002006001 as *L. callotis.*
5301409002002011001 as *L. gaillardi.*
5301409002002014001 as *L. mexicanus.*

Lepus capensis Linnaeus, 1758. Syst. Nat., 10th ed., 1:58.
TYPE LOCALITY: South Africa, Cape of Good Hope.
DISTRIBUTION: Africa (in non-forested areas); open woodland, steppe and subdesert of the Palearctic from S. Sweden and Finland to Britain (introduced to Ireland), through Europe to the West Siberian Lowlands, Mongolia, China, Iran, and Arabia; also introduced into North and South America, and Australasia.
COMMENT: Includes *arabicus, cyanotis, europaeus, starcki, tibetanus, tolai,* and *atlanticus;* see Corbet, 1978:71. JECF and OLR doubt that *europaeus* is a subspecies of *capensis;* see Angermann, 1972, *in* Grzimek, ed., Anim. Life Encyclop., 12:432. *L. starcki* may also be a full species (JECF). Most Russian authors consider *tolai* (including *tibetanus)* a distinct species; see Gromov and Baranova, 1981:65. Sludskii *et al.,* 1980:58, 85, indicated an area of sympatry between *europaeus* and *tolai* in Kazakhstan. Sokolov and Orlov, 1980:85, considered *tibetanus* a distinct species.
ISIS NUMBER: 5301409002002007001 as *L. capensis.*
5301409002002009001 as *L. europaeus.*

Lepus castroviejoi Palacios, 1977. Donana, Acta Vertebr., 1976, 3(2):205-223.
TYPE LOCALITY: Spain, Leon Prov., San Emiliano, Puerto Ventana.
DISTRIBUTION: Between Sierra de Ancares and Sierra de Pena Labra (N. Spain).
COMMENT: Probably a subspecies of *europaeus* (JECF).

Lepus flavigularis Wagner, 1844. *In* Schreber, Die Saugethiere ..., Suppl. 4:106.
TYPE LOCALITY: Mexico, Oaxaca, probably near Tehuantepec City.
DISTRIBUTION: Coastal plains and bordering foothills on south end of Isthmus of Tehuantepec (Oaxaca, Mexico), along Pacific coast to Chiapas (Mexico).
COMMENT: Closely related to *callotis;* see Anderson and Gaunt, 1962, Am. Mus. Novit., 2088:1-16; also see Hall, 1981:330.
ISIS NUMBER: 5301409002002010001.

Lepus habessinicus Hemprich and Ehrenberg, 1832. Symb. Phys. Mamm., p. 2.
TYPE LOCALITY: Ethiopia, E. coast, near Arkiko.
DISTRIBUTION: Ethiopia; Somalia.
COMMENT: Probably subspecies of *capensis* (JECF). Reviewed by Petter, 1963, Mammalia, 27:238-255.
ISIS NUMBER: 5301409002002012001.

Lepus insularis Bryant, 1891. Proc. Calif. Acad. Sci., ser. 2, 3:92.
TYPE LOCALITY: Mexico, Baja California del Sur, Gulf of California, Espiritu Santo Isl.
DISTRIBUTION: Espiritu Santo Isl., off coast of Baja California (Mexico).
COMMENT: Melanic form, related to *californicus;* see Hall, 1981:328.
ISIS NUMBER: 5301409002002013001.

Lepus mandshuricus Radde, 1861. Melanges Biol. Acad. St. Petersb., 3:684.
TYPE LOCALITY: U.S.S.R., S. Khabarovskii Krai, Bureja Mtns.
DISTRIBUTION: N.E. China; N. Korea; Ussuri region (U.S.S.R.).
COMMENT: Distinct from *brachyurus;* see Angermann, 1966, Mitt. Zool. Mus. Berlin,
42(2):321–335. Placed in *Caprolagus brachyurus* by Gromov and Baranova, 1981:63.

Lepus nigricollis F. Cuvier, 1823. Dict. Sci. Nat., 26:307.
TYPE LOCALITY: India, Madras.
DISTRIBUTION: Pakistan; India; Sri Lanka; introduced into Java and Mauritius.
COMMENT: Includes *ruficaudatus;* see Prater, 1965, Indian Animals, p. 219. *L.
ruficaudatus* is closer to *capensis* according to Petter, 1961, Z. Saugetierk., 26:1–11.
L. ruficaudatus may be a distinct species (JECF). May include *whytei, crawshayi,
peguensis* and *siamensis;* see Petter, 1961, Z. Saugetierk., 26:1–11; but also see
comments under *peguensis* and *saxatilis.*
ISIS NUMBER: 5301409002002016001.

Lepus oiostolus Hodgson, 1840. J. Asiat. Soc. Bengal, 9:1186.
TYPE LOCALITY: Unknown, S. Tibet or Nepal.
DISTRIBUTION: Tibetan Plateau, from Ladak (India) west to W. China (SW).
COMMENT: Reviewed by Angermann, 1967, Mitt. Zool. Mus. Berlin, 43(2):189–203.
ISIS NUMBER: 5301409002002017001.

Lepus peguensis Blyth, 1855. J. Asiat. Soc. Bengal, 24:471.
TYPE LOCALITY: Burma, Upper Pegu.
DISTRIBUTION: Burma to Indochina and Hainan (China).
COMMENT: Includes *siamensis;* see Lekagul and McNeely, 1977:333; but also see
comments under *nigricollis.*
ISIS NUMBER: 5301409002002019001 as *L. peguensis.*
5301409002002021001 as *L. siamensis.*

Lepus saxatilis F. Cuvier, 1823. Dict. Sci. Nat., 26:309.
TYPE LOCALITY: South Africa, Cape of Good Hope.
DISTRIBUTION: Cape Prov. and Zululand (South Africa) north to N. Namibia, Kenya
and S. Sudan; relict populations in N.E. Sahara.
COMMENT: Includes *crawshayi* and *whytei,* see Ansell, 1978:67; Swanepoel *et al.,* 1980,
Ann. Transvaal Mus., 32:159; but see also Petter, 1961, Z. Saugetierk., 26:1–11;
1972, Part 5:4–5.
ISIS NUMBER: 5301409002002020001 as *L. saxatilis.*
5301409002002008001 as *L. crawshyi (sic).*
5301409002002025001 as *L. whytei.*

Lepus sinensis Gray, 1832. Illustr. Indian Zool., 2:20.
TYPE LOCALITY: China, Kwangtung, Canton region.
DISTRIBUTION: S.E. China; Taiwan; disjunct in S. Korea.
COMMENT: Includes *coreanus;* see Corbet, 1978:73.
ISIS NUMBER: 5301409002002022001.

Lepus timidus Linnaeus, 1758. Syst. Nat., 10th ed., 1:57.
TYPE LOCALITY: Sweden, Uppsala.
DISTRIBUTION: Alaska (U.S.A.) to Labrador and Newfoundland (Canada) and
Greenland; Palearctic from Scandinavia to Siberia (U.S.S.R.) south to Hokkaido
(Japan); Sikhote Alin Mtns. (U.S.S.R.); Heilungkiang, N. Sinkiang (China) (SW);
Altai, N. Tien Shan; N. Ukraine, and Lithuania; isolated populations in the Alps,
Scotland, and Ireland.
COMMENT: Includes *othus* and *arcticus;* see Corbet, 1978:73. Corbet and Hill, 1980:197,
listed *arcticus* as a distinct species; Hall, 1981:318, listed both as distinct species.
ISIS NUMBER: 5301409002002023001 as *L. timidus.*
5301409002002003001 as *L. arcticus.*
5301409002002018001 as *L. othus.*

Lepus townsendii Bachman, 1839. J. Acad. Nat. Sci. Phila., 8(1):90.
TYPE LOCALITY: U.S.A., Washington, Walla Walla Co., Fort Walla Walla, near present town of Wallula.
DISTRIBUTION: S. Alberta to S.W. Ontario (Canada), south to S.W. Wisconsin, C. Kansas, N.C. New Mexico, west to C. Nevada, E.C. California (U.S.A.) and S.C. British Columbia (Canada).
ISIS NUMBER: 5301409002002024001.

Lepus yarkandensis Gunther, 1875. Ann. Mag. Nat. Hist., 16:229.
TYPE LOCALITY: China, Sinkiang (Chinese Turkestan), Yarkand.
DISTRIBUTION: Steppes of S. Sinkiang (China).
COMMENT: Reviewed by Angermann, 1967, Mitt. Zool. Mus. Berlin, 43(2):189–203.
ISIS NUMBER: 5301409002002026001.

Nesolagus Major, 1899. Trans. Linn. Soc. Lond., 7:493.
ISIS NUMBER: 5301409002003000000.

Nesolagus netscheri (Schlegel, 1880). Notes Leyden Mus., 2:62.
TYPE LOCALITY: Indonesia, Sumatra, Padang Highlands.
DISTRIBUTION: Sumatra.
PROTECTED STATUS: CITES - Appendix II.
ISIS NUMBER: 5301409002003001001.

Oryctolagus Lilljeborg, 1871. Sverig. Och Norges Ryggradsdjur, 1:417.
ISIS NUMBER: 5301409002004000000.

Oryctolagus cuniculus (Linnaeus, 1758). Syst. Nat., 10th ed., 1:58.
TYPE LOCALITY: Germany.
DISTRIBUTION: W. and S. Europe through the Mediterranean region to Morocco and N. Algeria; original range probably limited to Iberia and N.W. Africa; introduced on all continents except Antarctica and Asia. Worldwide as domesticated forms.
ISIS NUMBER: 5301409002004001001.

Pentalagus Lyon, 1904. Smithson. Misc. Coll., 45:428.
ISIS NUMBER: 5301409002005000000.

Pentalagus furnessi (Stone, 1900). Proc. Acad. Nat. Sci. Phila., p. 460.
TYPE LOCALITY: Japan, Amami Isls., Amami-Oshima.
DISTRIBUTION: Amami Isls. (Amami-Oshima and Tokun-Oshima) (S. Japan).
PROTECTED STATUS: U.S. ESA - Endangered.
ISIS NUMBER: 5301409002005001001.

Poelagus St. Leger, 1932. Proc. Zool. Soc. Lond., 1932:119.
REVIEWED BY: T. J. Robinson (TJR).
COMMENT: Formerly included as a subgenus of *Pronolagus*; see Ellerman and Morrison-Scott, 1951:425; but see also Petter, 1972, Part 5:5.

Poelagus marjorita (St. Leger, 1932). Proc. Zool. Soc. Lond., 1932:119.
TYPE LOCALITY: Africa, Uganda, Bunyuru, near Masindi.
DISTRIBUTION: S. Sudan; N.W. Uganda; N.E. Zaire; Central African Republic; Angola.
ISIS NUMBER: 5301409002006002001 as *Pronolagus marjorita*.

Pronolagus Lyon, 1904. Smithson. Misc. Coll., 45:416.
REVIEWED BY: T. J. Robinson (TJR).
ISIS NUMBER: 5301409002006000000.

Pronolagus crassicaudatus (I. Geoffroy, 1832). Guerin's Mag. Zool., 2:cl. 1, pl. 9 and text.
TYPE LOCALITY: South Africa, Natal, Port Natal (=Durban).
DISTRIBUTION: South Africa.
COMMENT: The relationship of *crassicaudatus* and *randensis* is unclear; see Petter, 1972, Part 5:6.
ISIS NUMBER: 5301409002006001001.

Pronolagus randensis Jameson, 1907. Ann. Mag. Nat. Hist., ser. 7, 20:404.
TYPE LOCALITY: South Africa, Transvaal, Johannesburg, Observatory, 5900 ft. (1798 m).
DISTRIBUTION: South Africa; E. Botswana; Zimbabwe; Namibia.
COMMENT: Included in *crassicaudatus* by Lundholm, 1955, Ann. Transvaal Mus., 22:279–303; but see Petter, 1972, Part 5:6.
ISIS NUMBER: 5301409002006003001.

Pronolagus rupestris (A. Smith, 1834). S. Afr. J., 2:174.
TYPE LOCALITY: South Africa, W. Cape Province (uncertain, probably Van Rhynsdorp).
DISTRIBUTION: South Africa to Kenya.
ISIS NUMBER: 5301409002006004001.

Romerolagus Merriam, 1896. Proc. Biol. Soc. Wash., 10:173.
REVIEWED BY: J. Ramirez-Pulido (JRP).
ISIS NUMBER: 5301409002007000000.

Romerolagus diazi (Diaz, 1893). Catalogo, Comision Geografico-Exploradora de la Republica Mexicana, Exposicion Intern, Columbina de Chicago, p. 42.
TYPE LOCALITY: Mexico, Puebla, E. Slope of Mt. Iztaccihuatl.
DISTRIBUTION: Distrito Federal, Mexico, and W. Puebla (Mexico).
PROTECTED STATUS: CITES - Appendix I and U.S. ESA - Endangered.
ISIS NUMBER: 5301409002007001001.

Sylvilagus Gray, 1867. Ann. Mag. Nat. Hist., ser. 3, 20:221.
REVIEWED BY: F. J. Brenner (FJB); J. A. Chapman (JAC); J. Ramirez-Pulido (JRP)(Mexico).
COMMENT: : Includes *Brachylagus* as a subgenus; see Hall, 1981:294.
ISIS NUMBER: 5301409002008000000.

Sylvilagus aquaticus (Bachman, 1837). J. Acad. Nat. Sci. Phila., 7:319.
TYPE LOCALITY: U.S.A., W. Alabama.
DISTRIBUTION: E. Texas and E. Oklahoma to Alabama and N.W. South Carolina, north to S. Illinois and S.W. Indiana (U.S.A.).
COMMENT: Reviewed by Chapman and Feldhamer, 1981, Mamm. Species, 151:1–4.
ISIS NUMBER: 5301409002008001001.

Sylvilagus audubonii (Baird, 1858). Mammals *in* Repts. Expl. Surv...., 8(8):608.
TYPE LOCALITY: U.S.A., California, San Francisco Co., San Franciso.
DISTRIBUTION: N.E. Puebla and W. Veracruz (Mexico) to C. Montana and S.W. North Dakota, N.C. Utah, C. Nevada, and N.C. California (U.S.A.), south to Baja California and C. Sinaloa (Mexico).
COMMENT: Reviewed by Chapman and Willner, 1978, Mamm. Species, 106:1–4.
ISIS NUMBER: 5301409002008002001.

Sylvilagus bachmani (Waterhouse, 1839). Proc. Zool. Soc. Lond., 1839:103.
TYPE LOCALITY: U.S.A., California, San Luis Obisbo.
DISTRIBUTION: W. Oregon (U.S.A.) S. of the Columbia River to Baja California (Mexico), east to Cascade-Sierra Nevada Range (U.S.A.).
COMMENT: Type locality restricted by Nelson, 1909, N. Am. Fauna, 29:247. Reviewed by Chapman, 1974, Mamm. Species, 34:1–4.
ISIS NUMBER: 5301409002008003001.

Sylvilagus brasiliensis (Linnaeus, 1758). Syst. Nat., 10th ed., 1:58.
TYPE LOCALITY: Brazil, Pernambuco.
DISTRIBUTION: S. Tamaulipas (Mexico) to Peru, Bolivia, N. Argentina, and S. Brazil.
COMMENT: Formerly included *dicei*; revised by Diersing, 1981, J. Mammal., 62:539–556. Type locality restricted by Thomas, 1911, Proc. Zool. Soc. Lond., 1911:146.
ISIS NUMBER: 5301409002008004001.

Sylvilagus cunicularius (Waterhouse, 1848). A Natural History of the Mammalia, 2:132.
TYPE LOCALITY: Mexico, Zacualpan.
DISTRIBUTION: S. Sinaloa to E. Oaxaca and Veracruz (Mexico).
ISIS NUMBER: 5301409002008005001.

Sylvilagus dicei Harris, 1932. Occas. Pap. Mus. Zool. Univ. Mich., 248:1.
TYPE LOCALITY: Costa Rica, El Copey de Dota.
DISTRIBUTION: Cordillera de Talamanca (S.E. Costa Rica, N.W. Panama).
COMMENT: Formerly included in *brasiliensis;* revised by Diersing, 1981, J. Mammal., 62:539–556.

Sylvilagus floridanus (J. A. Allen, 1890). Bull. Am. Mus. Nat. Hist., 3:160.
TYPE LOCALITY: U.S.A., Florida, Brevard Co., Sebastian River.
DISTRIBUTION: Venezuela (including adjacent islands) to (disjunct in parts of C. America) N.W. Arizona, Michigan, Massachusetts, and Florida (U.S.A.) and S. Saskatchewan and S.C. Quebec (Canada).
COMMENT: Widely introduced; see Hall, 1981:301. Range expanding (JAC). Reviewed by Chapman *et al.,* 1980, Mamm. Species, 136:1–8.
ISIS NUMBER: 5301409002008006001.

Sylvilagus graysoni (J.A. Allen, 1877). *In* Coues and Allen, Monog. N. Amer. Rodentia (U.S. Geol. Geograph. Survey Terr., Rep., 11:347).
TYPE LOCALITY: Mexico, Nayarit, Tres Marias Isls., probably Maria Madre Isl.
DISTRIBUTION: Tres Marias Isls., Nayarit (Mexico).
COMMENT: An insular species probably derived from *cunicularius;* see Diersing and Wilson, 1980, Smithson. Contrib. Zool., 297:1–34; Hall, 1981:314. See Nelson, 1899, N. Am. Fauna, 14:16 for discussion of type locality.
ISIS NUMBER: 5301409002008007001.

Sylvilagus idahoensis (Merriam, 1891). N. Am. Fauna, 5:76.
TYPE LOCALITY: U.S.A., Idaho, Custer County, near Goldburg.
DISTRIBUTION: S.W. Oregon to E.C. California, S.W. Utah, north to S.W. Montana (U.S.A.). Isolated population in W.C. Washington (U.S.A.).
COMMENT: The status of the isolated Washington population is uncertain. Placed in the monotypic genus *Brachylagus* by Dawson, 1967, Univ. Kans. Dept. Geol. Spec. Publ. No. 2, p. 303; and, together with *bachmani,* in the genus *Microlagus* by Gureev, 1964:170–173; but also see Hall, 1981:294, who recognized *Brachylagus* as a subgenus. Reviewed by Green and Flinders, 1980, Mamm. Species, 125:1–4, under the name *Brachylagus.*
ISIS NUMBER: 5301409002008008001.

Sylvilagus insonus (Nelson, 1904). Proc. Biol. Soc. Wash., 17:103.
TYPE LOCALITY: Mexico, Guerrero, Omilteme.
DISTRIBUTION: Sierra Madre del Sur, C. Guerrero (Mexico).
ISIS NUMBER: 5301409002008009001.

Sylvilagus mansuetus Nelson, 1907. Proc. Biol. Soc. Wash., 20:83.
TYPE LOCALITY: Mexico, Baja California del Sur, Gulf of California, San Jose Isl.
DISTRIBUTION: Known only from the type locality.
COMMENT: May be a subspecies of *bachmani* (JAC); also see Hall, 1981:299.
ISIS NUMBER: 5301409002008010001.

Sylvilagus nuttallii (Bachman, 1837). J. Acad. Nat. Sci. Phila., 7:345.
TYPE LOCALITY: U.S.A., Oregon, mouth of Malheur River, probably near Vale.
DISTRIBUTION: Intermountain area of N. America from S. British Columbia to S. Saskatchewan (Canada), S. to E. California, Nevada, C. Arizona, and N.W. New Mexico (U.S.A.).
COMMENT: Type locality restricted by Bailey, 1936, N. Am. Fauna, 55:107. *S. floridanus* appears to be displacing *nuttallii* in some areas (JAC); also see Genoways and Jones, Occas. Pap. Mus. Texas Tech Univ., 6:1–36. Reviewed by Chapman, 1975, Mamm. Species, 56:1–3.
ISIS NUMBER: 5301409002008011001.

Sylvilagus palustris (Bachman, 1837). J. Acad. Nat. Sci. Phila., 7:194.
TYPE LOCALITY: U.S.A., E. South Carolina near coast.
DISTRIBUTION: Florida to S.E. Virginia (U.S.A.) on the coastal plain.
COMMENT: Reviewed by Chapman and Willner, 1981, Mamm. Species, 153:1–3.
ISIS NUMBER: 5301409002008012001.

Sylvilagus transitionalis (Bangs, 1895). Proc. Boston Soc. Nat. Hist., 26:405.
TYPE LOCALITY: U.S.A., Connecticut, New Lond. Co., Liberty Hill.
DISTRIBUTION: S. Maine to N. Alabama along the Appalachian Mtns. (U.S.A.).
COMMENT: The distribution of this species has been much reduced (JAC); probably
involves displacement by *S. floridanus* (MF). Reviewed by Chapman, 1975, Mamm.
Species, 55:1–4.
ISIS NUMBER: 5301409002008013001.

ORDER MACROSCELIDEA
COMMENT: Often included in Insectivora, but see McKenna, 1975:41.

Family Macroscelididae
REVIEWED BY: G. B. Corbet (GBC).
COMMENT: Revised by Corbet and Hanks, 1968, Bull. Br. Mus. (Nat. Hist.) Zool., 16:47–111.
ISIS NUMBER: 5301403008000000000.

Elephantulus Thomas and Schwann, 1906. Proc. Zool. Soc. Lond., 1906:577.
COMMENT: Includes *Nasilio;* see Corbet and Hanks, 1968, Bull. Br. Mus. (Nat. Hist.) Zool., 16:47–111.
ISIS NUMBER: 5301403008001000000.

Elephantulus brachyrhynchus (A. Smith, 1836). Rept. Exped. Expl. C. Afr...., p. 42.
TYPE LOCALITY: South Africa, N. Cape Prov. or S. Botswana.
DISTRIBUTION: N. South Africa and Namibia to Kenya and Uganda.
COMMENT: Formerly included *fuscus;* see Corbet, 1974, Part 1.5:5.
ISIS NUMBER: 5301403008001001001.

Elephantulus edwardi (A. Smith, 1839). Illustr. Zool. S. Afr. Mamm., pl. 14.
TYPE LOCALITY: South Africa, Cape Prov., Oliphants River.
DISTRIBUTION: Cape Prov. (South Africa).
ISIS NUMBER: 5301403008001002001.

Elephantulus fuscipes (Thomas, 1894). Ann. Mag. Nat. Hist., ser. 6, 13:68.
TYPE LOCALITY: Zaire, Niam-Niam country, N'doruma.
DISTRIBUTION: Uganda; N.E. Zaire; S. Sudan.
ISIS NUMBER: 5301403008001003001.

Elephantulus fuscus (Peters, 1852). Reise nach Mossambique, Saugeth., p. 87.
TYPE LOCALITY: Mozambique, near Quelimane, Boror.
DISTRIBUTION: Mozambique; S. Malawi; S.E. Zambia.
COMMENT: Formerly included in *brachyrhynchus;* see Corbet, 1974, Part 1.5:5.

Elephantulus intufi (A. Smith, 1836). Rept. Exped. Expl. C. Afr...., p. 42.
TYPE LOCALITY: South Africa, Transvaal, Marico District, flats beyond Kurrichane.
DISTRIBUTION: S.W. Angola; Namibia; Botswana; N.W. Transvaal (South Africa).
ISIS NUMBER: 5301403008001004001.

Elephantulus myurus Thomas and Schwann, 1906. Proc. Zool. Soc. Lond., 1906:586.
TYPE LOCALITY: South Africa, Transvaal, Woodbush.
DISTRIBUTION: Zimbabwe; E. South Africa; Mozambique.
ISIS NUMBER: 5301403008001005001.

Elephantulus revoili (Huet, 1881). Bull. Sci. Soc. Philom. Paris, ser. 7, 5:96.
TYPE LOCALITY: Somalia, Medjourtine.
DISTRIBUTION: N. Somalia.
ISIS NUMBER: 5301403008001006001 as *E. revoilii (sic)*.

Elephantulus rozeti (Duvernoy, 1833). Mem. Soc. Hist. Nat. Strasbourg, 1 (2), art. M:18.
TYPE LOCALITY: Algeria, near Oran.
DISTRIBUTION: Morocco; Algeria; Tunisia; W. Libya.
ISIS NUMBER: 5301403008001007001.

Elephantulus rufescens (Peters, 1878). Monatsb. Preuss. Akad. Wiss. Berlin, p. 198.
TYPE LOCALITY: Kenya, Taita, Ndi.
DISTRIBUTION: S. and E. Ethiopia; N. and S.E. Kenya; Uganda; Sudan; N. Tanzania; N. Somalia.
ISIS NUMBER: 5301403008001008001.

Elephantulus rupestris (A. Smith, 1831). Proc. Zool. Soc. Lond., 1831:11.
TYPE LOCALITY: S. Africa or Namibia, mountains near mouth of Orange River.
DISTRIBUTION: Namibia; Cape Prov. (South Africa).
ISIS NUMBER: 5301403008001009001.

Macroscelides A. Smith, 1829. Zool. J. Lond., 4:435.
ISIS NUMBER: 5301403008002000000.

Macroscelides proboscideus (Shaw, 1800). Gen. Zool., 1 (2), Mammalia, p. 536.
TYPE LOCALITY: South Africa, Cape Prov., Oudtshoorn Div., Roodeval.
DISTRIBUTION: South Africa; Namibia.
ISIS NUMBER: 5301403008002001001.

Petrodromus Peters, 1846. Monatsb. Preuss. Akad. Wiss. Berlin, p. 258.
ISIS NUMBER: 5301403008003000000.

Petrodromus tetradactylus Peters, 1846. Monatsb. Preuss. Akad. Wiss. Berlin, p. 258.
TYPE LOCALITY: Mozambique, Tette.
DISTRIBUTION: Mozambique; Tanzania; S.E. Kenya; Zambia; Malawi; Zimbabwe; Zaire;
    N.E. Angola; Natal and Transvaal (South Africa); Zanzibar.
COMMENT: Includes *sultan, rovumae, and tordayi;* see Corbet, 1974, Part 1.5:2. It is
    possible that *tordayi* is a separate species (GBC).
ISIS NUMBER: 5301403008003001001.

Rhynchocyon Peters, 1847. Monatsb. Preuss. Akad. Wiss. Berlin, p. 36.
ISIS NUMBER: 5301403008004000000.

Rhynchocyon chrysopygus Gunther, 1881. Proc. Zool. Soc. Lond., 1881:164.
TYPE LOCALITY: Kenya, Mombasa.
DISTRIBUTION: E. Kenya.
COMMENT: Reviewed by Rathburn, 1979, Mamm. Species, 117:1–4.
ISIS NUMBER: 5301403008004001001.

Rhynchocyon cirnei Peters, 1847. Monatsb. Preuss. Akad. Wiss. Berlin, p. 37.
TYPE LOCALITY: Mozambique, Bororo Dist., Quelimane.
DISTRIBUTION: Mozambique; Malawi; S. Tanzania; N.E. Zambia; E. Zaire; Uganda.
COMMENT: Includes *stuhlmanni,* which could be a distinct species (GBC); see Corbet,
    1974, Part 1.5:2.
ISIS NUMBER: 5301403008004002001.

Rhynchocyon petersi Bocage, 1880. J. Sci. Math. Phys. Nat. Lisboa, ser. 1, 7:159.
TYPE LOCALITY: Tanzania, mainland opposite Zanzibar.
DISTRIBUTION: Zanzibar; E. Tanzania; S.E. Kenya; Mafia Island.
ISIS NUMBER: 5301403008004003001.

Bibliography and Literature Cited

AFANAS'EV, A. V., V. S. BAZHANOV, M. N. KORELOV, A. A. SLUDSKII, AND E. I. STRAUTMAN. 1953. Zveri Kazakhstana. [Mammals of Kazakhstan.] Acad. Sci. Kazakh SSR, Alma-Ata, 635 pp.

ALBIGNAC, R. 1973. Faune de Madagascar. 36, Mammiferes, Carnivores. Centre O.R.S.T.O.M. de Tananarive, Paris, 206 pp.

ALCASID, G. L. 1969. Checklist of Philippine mammals. Nat. Mus. Philippines, Manila, 51 pp.

ALCOVER, J. A. 1979. Els mamifers de les Balears. Moll, Palmade de Mallorca.

ALLEN, G. M. 1938. The mammals of China and Mongolia (Natural History of Central Asia, vol. 9, Part 1). Am. Mus. Nat. Hist., New York, xxv, 1–620 pp.

ALLEN, G. M. 1939. A checklist of African mammals. Bull. Mus. Comp. Zool., Harvard Coll., 83:1–763.

ALLEN, G. M. 1940. The mammals of China and Mongolia (Natural History of Central Asia, vol. 9, Part 2). Am. Mus. Nat. Hist., New York, xxvi, 621–1350 pp.

ALLEN, G. M. 1942. Extinct and vanishing mammals of the Western Hemisphere. Amer. Comm. Intern. Wildl. Protection, Spec. Publ. No. 11, xv + 620 pp.

ALMACA, C. 1968. La Faune mammalogique du Portugal. Arquivos do Museo Bocage, 11:6–9.

AMTMANN, E. 1975. Part 6.1, Family Sciuridae, in J. Meester and H. W. Setzer, eds., 1971–1977.

ANDERSON, S. 1959. Distribution, variation, and relationships of the montane vole, Microtus montanus. Univ. Kansas Publ. Mus. Nat. Hist., 9:415–511.

ANDERSON, S. 1960. The baculum in microtine rodents. Univ. Kansas Publ. Mus. Nat. Hist., 12:181–216.

ANDERSON, S. 1972. Mammals of Chihuahua. Taxonomy and distribution. Bull. Am. Mus. Nat. Hist., 148:149–410.

ANDERSON, S., AND J. K. JONES, JR., eds. 1967. Recent mammals of the world: a synopsis of families. Ronald Press, New York, 453 pp.

ANGERMANN, R. 1966. Beitrage zur Kenntnis der Gattung Lepus. Mitt. Zool. Mus. Berlin, 42:127–144, 321–335.

ANGERMANN, R. 1967. ———. Mitt. Zool. Mus. Berlin, 43:161–178, 184–203.

ANGERMANN, R. 1972. Pp. 419–462 in Animal life encyclopedia, vol. 12 (Mammals III), B. Grzimek, ed. Van Nostrand and Reinhold Co., New York.

ANONYMOUS. 1967. Provisional list of the fauna of Guyana. J. Guyana Mus. Zool., 42:58–60.

ANSELL, W. F. H. 1972. Part 15, Order Artiodactyla, main text, in J. Meester and H. W. Setzer, eds., 1971–1977.

ANSELL, W. F. H. 1974. Part 14, Order Perissodactyla, in J. Meester and H. W. Setzer, eds., 1971–1977.

ANSELL, W. F. H. 1978. The mammals of Zambia. Nat. Parks and Wildl. Ser., Chilanga, Zambia, ii + 126 pp. + maps.

ARCHER, M. 1981. Results of the Archbold Expeditions. No. 104. Systematic revision of the marsupial dasyurid genus Sminthopsis Thomas. Bull. Am. Mus. Nat. Hist., 168:61–224.

ATALLAH, S. I. 1977. Mammals of the eastern Mediterranean region: their ecology, systematics and geographic relationships. Saugetierk. Mitt., 25:241–320.

ATALLAH, S. I. 1978. ———. Saugetierk. Mitt., 26:1–50.

ATANASSOV, N., AND Z. PESCHEV. 1963. Die Saugetiere Bulgariens. Saugetierk. Mitt., 12:101–112.

BAIRD, S. F. 1858. Mammals: explorations and surveys for a railroad route . . . . Washington, D.C., pp. xlvii + 757, pls. 17–60, figs. 1–35.

BAKER, R. H. 1974. Records of mammals from Ecuador. Publ. Michigan State Univ. (Biol. Ser.), 5:131–146.

BANFIELD, A. W. F. 1974. The mammals of Canada. National Museums of Canada, Univ. Toronto Press, xxv + 438 pp., 46 pls., 113 figs., 176 maps.

BANKS, R. C., AND R. L. BROWNELL. 1969. Taxonomy of the common dolphins of the eastern Pacific Ocean. J. Mammal., 50:262–271.

BANNIKOV, A. G. 1954. Mlekopitayushchie Mongol'skoi Narodnoi Respubliki. [Mammals of the Mongolian People's Republic.] Acad. Sci., Moscow, 669 pp.

BASILIO, A. 1962. La vida animal en la Guinea Espanola. Descripcion y vida de los animales en la selva tropical africana (2nd ed.). Inst. Estudios Africanos, Madrid.

BEE, J. W., AND E. R. HALL. 1956. Mammals of Northern Alaska. Univ. Kansas Publ., Mus. Nat. Hist. Misc. Publ. No. 8, 309 pp.

BERE, R. M. 1962. The wild mammals of Uganda and neighbouring regions of East Africa. Longmans, Green and Co., London, xii + 148 pp.

BLACKWELDER, R. E. 1972. Mammalia, *in* Guide to the taxonomic literature of vertebrates. Iowa St. Univ. Press, Ames, pp. 207–256.

BOBRINSKII, N. A., B. A. KUZNETSOV, AND A. P. KUZYAKIN. 1965. Opredelitl' mlekopitayushchikh SSSR. [Key to the mammals of the USSR.] Prosveshchenie, Moscow, 382 pp. + 40 pls., 111 maps.

BODENHEIMER, F. S. 1958. The present taxonomic status of the terrestrial mammals of Palestine. Bull. Res. Counc. Israel (Zoology), 7B:165–190.

BOTHMA, J. DU P. 1971. Part 12, Order Hyracoidea, *in* J. Meester and H. W. Setzer, eds., 1971–1977.

BROSSET, A. 1963. Mammiferes des iles Galapagos . . . . Mammalia, 27:323–338.

BROSSET, A., AND G. DUBOST. 1967. Chiropteres de la Guyane Francaise. Mammalia, 31:583–594.

BRYANT, M. D. 1945. Phylogeny of Nearctic Sciuridae. Am. Midl. Nat., 33:257–390.

BURT, W. H., AND R. A. STIRTON. 1961. The mammals of El Salvador. Misc. Publ. Mus. Zool. Univ. Michigan, 117:1–69.

CABRERA, A. 1958. Catalogo de los Mamiferos de America del Sur. Rev. Mus. Argent. Cienc. Nat. Bernardino Rivadavia, Inst. Nac. Invest. Cienc. Nat., Cienc. Zool., 4(1):xvi + 1–307.

CABRERA, A. 1961. ———. Rev. Mus. Argent. Cienc. Nat. Bernardino Rivadavia, Inst. Nac. Invest. Cienc. Nat., Cienc. Zool., 4(2):xvii–xxii + 309–732.

CALABY, J. H. 1971. The current status of Australian Macropodidae. Australian Zool., 16:17–29.

CAMPBELL, C. B. G. 1966. Taxonomic status of tree shrews. Science, 153:436.

CARLETON, M. D. 1977. Interrelationships of populations of the *Peromyscus boylii* species group (Rodentia, Muroidea) in western Mexico. Occas. Pap. Mus. Zool. Univ. Michigan, 675:1–47.

CARLETON, M. D. 1980. Phylogenetic relationships in neotomine-peromyscine rodents (Muroidea) and a reappraisal of the dichotomy within New World Cricetinae. Misc. Publ. Mus. Zool. Univ. Michigan, 157:1–146.

CHALINE, J. 1974. Esquisse de l'evolution morphologique, biometrique, et chromosomique du genre *Microtus* (Arvicolidae, Rodentia) dans le Pleistocene de l'hemisphere Nord. Bull. Soc. Geol. France, ser. 7, 16:440–450.

CHALINE, J., AND P. MEIN. 1979. Les rongeurs et l'evolution. Doin, Paris, xi + 235 pp.

CHANG, C., AND TSUNG-YI WANG. 1963. Faunistic studies of mammals of the Chinghai Province. Acta Zool. Sin., 15:125–138 (in Chinese).

CHASEN, F. N. 1940. A handlist of Malaysian mammals. A systematic list of the mammals of the Malay Peninsula, Sumatra, Borneo, and Java, including the adjacent small islands. Bull. Raffles Mus., 15:xx + 1–209.

CHIARELLI, A. B. 1971. Taxonomic atlas of living Primates. Acad. Press, London, vii + 363 pp.

CHOTOLCHU, N., AND M. STUBBE. 1971. Zur Saugetierfauna der Mongolei—II. Erstnachweise von zwei *Sorex*-Arten. Mitt. Zool. Mus. Berlin, 47:43–45.

CLARK, W. E. LE GROS. 1971. The antecedents of man, 3rd ed. Quadrangle Books, Chicago, Ill., vii + 374 pp.

CLEMENS, W. A. 1977. Phylogeny of the marsupials. Pp. 51–68 *in* The biology of the marsupials, B. Stonehouse and D. Gilmore, eds. Macmillan Press, Baltimore.

CLUTTON-BROCK, J., G. B. CORBET, AND M. HILLS. 1976. A review of the family Canidae, with a classification by numerical methods. Bull. Br. Mus. (Nat. Hist.), Zool., 29:117–199.

COCKRUM, E. L., AND H. W. SETZER. 1976. Types and type localities of North African rodents. Mammalia, 40:633–670.

COETZEE, C. G. 1977. Part 6.8, Genus *Steatomys, in* J. Meester and H. W. Setzer, eds., 1971–1977.

COETZEE, C. G. 1977. Part 8, Order Carnivora, main text, *in* J. Meester and H. W. Setzer, eds., 1971–1977.

COLLINS, L. R. 1973. Monotremes and marsupials: a reference for zoological institutions. Smithsonian Institution, Washington, D.C.

COPLEY, H. 1950. Small mammals of Kenya. Longmans Green, London.

CORBET, G. B. 1966. The terrestrial mammals of western Europe. G. T. Foulis, London, xi + 264 pp.

CORBET, G. B. 1974. Part 1.2, Subfamily Potamogalinae, *in* J. Meester and H. W. Setzer, eds., 1971–1977.

CORBET, G. B. 1974. Part 1.4, Family Erinaceidae, *in* J. Meester and H. W. Setzer, eds., 1971–1977.

CORBET, G. B. 1974. Part 1.5, Family Macroscelididae, *in* J. Meester and H. W. Setzer, eds., 1971–1977.

CORBET, G. B. 1978. The mammals of the Palaearctic region: a taxonomic review. Br. Mus. (Nat. Hist.) and Cornell Univ. Press, London and Ithaca, N.Y., 314 pp.

CORBET, G. B., AND J. E. HILL. 1980. A world list of mammalian species. Br. Mus. (Nat. Hist.) and Cornell Univ. Press, London and Ithaca, N.Y., vii + 226 pp.

CORBET, G. B., AND H. N. SOUTHERN. 1977. The handbook of British mammals. Blackwell Sci. Publ., Oxford, xxxii + 520 pp.

CURRY-LINDAHL, K. 1980. Der Berglemming *Lemmus lemmus.* A. Ziemsen, Wittenberg Lutherstadt, 140 pp.

DA CUNHA VIEIRA, C. 1955. Lista remissiva dos mamiferos do Brasil. Arq. Zool. Estado Sao Paulo, 8:341–474.

DAMMERMAN, K. W. 1931. The mammals of Java. I. Rodentia. Treubia, 13:429–470.

DANDELOT, P. 1974. Part 3, Order Primates, main text, *in* J. Meester and H. W. Setzer, eds., 1971–1977.

DAVIES, J. L. 1963. The whales and seals of Tasmania. Tasmanian Mus. and Art Gall., Hobart.

DAVIS, D. D. 1962. Mammals of the lowland rainforest of North Borneo. Bull. Nat. Mus. Singapore, 31:1–129.

DAVIS, D. H. S. 1975. Part 6.4, Genera *Tatera* and *Gerbillurus, in* J. Meester and H. W. Setzer, eds., 1971–1977.

DAVIS, D. H. S. 1975. Part 6.6, Genus *Aethomys, in* J. Meester and H. W. Setzer, eds., 1971–1977.

DAVIS, J. A. 1978. A classification of otters. Pp. 14–33 *in* Otters, N. Duplaix, ed. Intern. Union Cons. Nat. Natur. Res. (IUCN) Publ., n.s., 158 pp.

DEBLASE, A. G. 1980. The bats of Iran. Fieldiana Zool., n.s. 4, 1307:xvii + 1–424.

DE GAMA, M. M. 1957. Mamiferos de Portugal (chaves para a sua determinacao). Memoir. Estudos Mus. Zool. Univ. Coimbra, Portugal, 246:1–246.

DEGERBOL, M. I. 1950. Mammals, *in* List of Danish vertebrates. Dansk Videnskabs Forlag, Copenhagen, pp. 134–150.

DEGERBOL, M., AND P. FREUCHEN. 1935. Mammals, *in* Report of the Fifth Thule Expedition 1921–1924, 2(4–5):1–278 + 1 map. Nordisk Forlag, Copenhagen.

DE GRAAFF, G. 1975. Part 6.9, Family Bathyergidae, *in* J. Meester and H. W. Setzer, eds., 1971–1977.

DELANY, M. J. 1975. The rodents of Uganda. Br. Mus. (Nat. Hist.), No. 764, viii + 165 pp.

DELSON, E., AND P. ANDREWS. 1975. Evolution and interrelationships of the catarrhine primates. Pp. 405–446 *in* Phylogeny of the primates: a multidisciplinary approach, W. P. Luckett and F. S. Szalay, eds. Plenum Publ. Co., New York.

DEVORE, I., ed. 1965. Primate behavior. Field studies of monkeys and apes. Holt, Rinehart and Winston, New York, 654 pp.

DOBSON, G. E. 1878. Catalogue of the Chiroptera in the collection of the British Museum. Taylor and Francis Ltd., London, xlii + 567 pp., 30 pls. (Reprinted 1966).

DORST, J., AND P. DANDELOT. 1970. A field guide to the larger mammals of Africa. Houghton Mifflin Co., Boston, 287 pp.

DOYLE, G. A., AND R. D. MARTIN, eds. 1979. The study of prosimian behavior. Acad. Press, New York, xvii + 696 pp.

DULIC, B., AND D. MIRIC. 1967. Catalogus faunae Jugoslaviae. Mammalia. Consilium Acad. Sci. Reipubl. Socialist. Foederat. Jugoslaviae, 4:1–46.

EISENBERG, J. F., ed. 1979. Vertebrate ecology in the northern Neotropics. Smithson. Inst. Press, Washington, D.C., 271 pp.

EISENBERG, J. F. 1981. The mammalian radiations. An analysis of trends in evolution, adaptation, and behavior. Univ. Chicago Press, xx + 610 pp.

EISENBERG, J. F., AND E. GOULD. 1970. The tenrecs: a study in mammalian behavior and evolution. Smithson. Contr. Zool., 27:vi + 138 pp.

EISENBERG, J. F., AND G. M. McKAY. 1970. An annotated checklist of the Recent mammals of Ceylon with keys to the species. Ceylon J. Sci. (Biol.) 8:69–99.

EISENBERG, J. F., AND K. H. REDFORD. 1979. Biogeographic analysis of the mammalian fauna of Venezuela. Pp. 31–36 in Vertebrate ecology in the northern Neotropics, J. F. Eisenberg, ed. Smithson. Inst. Press, Washington, D.C.

EISENTRAUT, M. 1964. La faune de chiropteres de Fernando-Po. Mammalia, 28:529–552.

ELLERMAN, J. R. 1940. Families and genera of living rodents. Vol. 1, Rodents other than Muridae. Br. Mus. (Nat. Hist.), London, xxvi + 689 pp.

ELLERMAN, J. R. 1941. ———. Vol. 2, Family Muridae, xii + 690 pp.

ELLERMAN, J. R. 1949. ———. Vol. 3, Part 1 [Additions and Corrections], v + 210 pp.

ELLERMAN, J. R. 1961. The fauna of India including Pakistan, Burma and Ceylon, 2nd ed., Vol. 3, Parts 1, 2. Zool. Surv. India, Calcutta, xxx + 482; 483–884 pp.

ELLERMAN, J. R., AND T. C. S. MORRISON-SCOTT. 1951. Checklist of Palaearctic and Indian mammals 1758–1946. Br. Mus. (Nat. Hist.), London, 810 pp.

ELLERMAN, J. R., AND T. C. S. MORRISON-SCOTT. 1955. Supplement to Chasen (1940) Handlist of Malaysian mammals, containing a generic synonymy and a complete index. Br. Mus. (Nat. Hist.), London, 66 pp.

ELLERMAN, J. R., AND T. C. S. MORRISON-SCOTT. 1966. Checklist of Palaearctic and Indian mammals 1758 to 1946. 2nd ed., Br. Mus. (Nat. Hist.), London, 810 pp.

ELLERMAN, J. R., T. C. S. MORRISON-SCOTT, AND R. W. HAYMAN. 1953. Southern African mammals 1758 to 1951: a reclassification. Br. Mus. (Nat. Hist.), London, 363 pp.

ELLIOTT, O. 1971. Bibliography of the tree shrews 1780–1969. Primates, 12:323–414.

EWER, R. F. 1973. The carnivores. Cornell Univ. Press, Ithaca, N.Y., xv + 494 pp.

FELTEN, H., F. SPITZENBERGER, AND G. STORCH. 1971. Zur Kleinsaugetiere West-Anatoliens. Teil I. Senckenberg. Biol., 52(6):393–424.

FELTEN, H. et al. 1973. ———. Teil II, Senckenberg. Biol., 54(4/6):227–290.

FELTEN, H. et al. 1977. ———. Teil IIIa, Senckenberg. Biol., 58(1/2):1–44.

FLEROV, K. K. 1952. Kabargi i oleni [Musk deer and deer.] Fauna SSSR, Mlekopitayushchie [Mammals.] 1(2). Akad. Nauk, Moscow-Leningrad, 255 pp. (Engl. translation, Off. Tech. Serv., Washington, D.C., 1960).

FLINT, W. [=V.] E. 1966. Die Zwerghamster der palaearktischen Fauna. NBB, Stuttgart.

FLINT, V. E., YU. D. CHUGUNOV, AND V. M. SMIRIN. 1965. Mlekopitayushchie SSSR. [Mammals of the USSR.] "Misl", Moscow, 437 pp.

FOKIN, I. M. 1978. Tushkanchiki. [Jerboas.] Leningrad Univ., Leningrad, 183 pp.

FOODEN, J. 1980. Classification and distribution of living macaques (Macaca Lacepede, 1799). Pp. 1–9 in The macaques: studies in ecology, behavior and evolution, D. G. Lindburg, ed. Van Nostrand and Reinhold Co., New York, xii + 384 pp.

FOX, M. W., ed. 1975. The wild canids. Their systematics, behavioral ecology and evolution. Van Nostrand and Reinhold, New York, xvi + 508 pp.

FREEMAN, P. W. 1981. A multivariate study of the family Molossidae (Mammalia, Chiroptera): morphology, ecology, evolution. Fieldiana Zool., n.s. 7, 1316:vii + 1–173.

FRICK, H. 1956. Zur Taxonomie der Tubulidentata. Saugetierk. Mitt., 4:15–17.

FUNAIOLI, U. 1971. Guida breve dei Mammiferi della Somalia. Inst. Agronom. l'Oltremare. Biblioteca Agraria Tropicale, 232 pp.

FUNAIOLI, U., AND A. M. SIMONETTA. 1961. The mammalian fauna of the Somali Republic . . . . Monitore Zool. Ital. (Suppl.), 74:285–295.

GARDNER, A. L., AND J. L. PATTON. 1976. Karyotypic variation in oryzomyine rodents (Cricetinae) with comments on chromosomal evolution in the neotropical cricetine complex. Occas. Pap. Mus. Zool., La. St. Univ., 49:1–48.

GENEST, H., AND F. PETTER. 1971. Part 1.1, Subfamilies Tenrecinae and Oryzorictinae, *in* J. Meester and H. W. Setzer, eds., 1971–1977.

GENTRY, A. W. 1972. Part 15.1, Genus *Gazella, in* J. Meester and H. W. Setzer, eds., 1971–1977.

GILL, T. 1872. Arrangement of the families of mammals with analytical tables. Smithson. Misc. Coll., 11(1):i–vi + 1–98.

GOODWIN, G. G. 1942. Mammals of Honduras. Bull. Am. Mus. Nat. Hist., 79:107–195.

GOODWIN, G. G. 1946. Mammals of Costa Rica. Bull. Am. Mus. Nat. Hist., 87:275–473.

GOODWIN, G. G. 1955. Mammals of Guatemala, with the description of a new little brown bat. Am. Mus. Novit., 1744:1–5.

GOODWIN, G. G., AND A. M. GREENHALL. 1961. A review of the bats of Trinidad and Tobago . . . . Bull. Am. Mus. Nat. Hist., 122:189–301.

GOODWIN, R. 1967. The mammals of Jamaica. Caribbean Biological Centre Newsletter, 2:17–21.

GRAY, A. P. 1954. Mammalian hybrids: a checklist with bibliography. Farnham Royal, Commonwealth Agric. Bur., x + 262 pp.

GRAY, G. G., AND C. D. SIMPSON. 1980. Ammotragus lervia. Mamm. Species, 144:1–7.

GRIFFITHS, M. 1968. Echidnas. Pergamon Press, Oxford, ix + 282 pp.

GRIFFITHS, M. 1978. The biology of the monotremes. Acad. Press, New York, viii + 367 pp.

GROMOV, I. M., AND G. I. BARANOVA, eds. 1981. Katalog mlekopitayushchikh SSSR. [Catalog of mammals of the USSR.] Nauka, Leningrad, 456 pp.

GROMOV, I. M., D. I. BIBIKOV, N. I. KALABUKHOV, AND M. N. MEIER. 1965. Nazemnye belich'i [Ground squirrels] (Marmotinae). Fauna SSSR, Mlekopitayushchie [Mammals]. Nauka, Moscow-Leningrad, 3(2):1–466 + ii pp.

GROMOV, I. M., A. A. GUREEV, G. A. NOVIKOV, I. I. SOKOLOV, P. P. STRELKOV, AND K. K. CHAPSKII. 1963. Mlekopitayushchie fauni SSSR. [Mammal fauna of the USSR.] 2 vols., Acad. Sci., Moscow-Leningrad, 1100 pp.

GROMOV, I. M., AND I. YA. POLYAKOV. 1977. Polevki [Voles] (Microtinae). Fauna SSSR, Mlekopitayushchie [Mammals]. Nauka, Moscow-Leningrad, 3(8):1–504.

GROMOV, I. M., AND A. I. YANUSHEVICH. 1972. Mlekopitayushchie Kirgizii. [Mammals of Kirgiziya.] Ilim, Frunze, 463 pp.

GROVES, C. P. 1970. The forgotten leaf-eaters and the phylogeny of the Colobinae. Pp. 555–586 *in* Old World monkeys: evolution, systematics, and behavior, J. R. Napier and P. H. Napier, eds. Acad. Press, New York.

GROVES, C. P. 1972. Systematics and phylogeny of gibbons. Pp. 1–89 *in* Gibbon and siamang. Vol. 1, D. M. Rumbaugh, ed. Karger, Basel.

GROVES, C. P. 1974. Taxonomy and phylogeny of prosimians. Pp. 449–473 *in* Prosimian biology, R. D. Martin, G. A. Doyle, and A. C. Walker, eds. Duckworth, London.

GROVES, C. P. 1978. Phylogenetic and population systematics of the mangabeys (Primates: Cercopithecoidea). Primates, 19:1–34.

GROVES, C. P. 1980. Speciation in *Macaca*: the view from Sulawesi. Pp. 84–123 *in* The macaques: studies in ecology, behavior and evolution, D. G. Lindburg, ed. Van Nostrand and Reinhold Co., New York, xii + 384 pp.

GUGGISBERG, C. A. W. 1975. Wild cats of the world. Taplinger Publ. Co., New York, 328 pp.

GUREEV, A. A. 1964. Zaitseobraznye (Lagomorpha). Fauna SSSR, Mlekopitayushchie [Mammals]. Nauka, Moscow-Leningrad, 3(10):1–276.

GUREEV, A. A. 1971. Zemleroiki (Soricidae) fauny mira. [Shrew (Soricidae) fauna of the world.] Nauka, Leningrad, 252 pp.

GUREEV, A. A. 1979. Nasekomoyadnie [Insectivores] (Mammalia, Insectivora). Fauna SSSR, Mlekopitayushchie [Mammals]. Nauka, Leningrad, 4(2):1–502 pp.

HALL, E. R. 1981. The mammals of North America, 2nd ed. John Wiley, New York, 2 vols., xv + 1–600 + 90; vi + 601–1181 + 90 pp.

HALL, E. R., AND K. R. KELSON. 1959. The mammals of North America. Ronald Press, New York, 2 vols., xxx + 1–546 + 79; viii + 547–1083 + 79 pp.

HALTENORTH, T. 1953. Die Wildkatzen der alten Welt: eine Ubersicht uber die Untergattung *Felis*. NBB, Leipzig.

HALTENORTH, T. 1958. Klassifikation der Saugetiere. 1. Ordnung Kloakentiere, Monotremata Bonaparte, 1838; 2. Ordnung Beuteltiere, Marsupialia Illiger, 1811. Handbuch der Zoologie, 8(16):1–40.

HALTENORTH, T. 1963. Klassifikation der Saugetiere: Artiodactyla I. Handbuch der Zoologie, 8 (32):1–167.

HALTENORTH, T., AND H. DILLER. 1977. Saugetiere Afrikas und Madagaskars. BLV, Munchen, 403 pp.

HANDLEY, C. O., JR. 1966. Checklist of the mammals of Panama. Pp. 753–795 in Ectoparasites of Panama, R. L. Wenzel and V. J. Tipton, eds. Field Museum, Chicago.

HANDLEY, C. O., JR. 1976. Mammals of the Smithsonian Venezuelan Project. Brigham Young Univ. Sci. Bull. Biol. Ser., 20(5):1–91.

HANDLEY, C. O., JR. 1979. Mammals of the Dismal Swamp: an historical account. Pp. 297–357 in The Great Dismal Swamp, P. W. Kirk, Jr., ed. Univ. Press Virginia, Charlottesville.

HARRINGTON, F. A., JR. 1977. A guide to the mammals of Iran. Dept. Environ. Iran, Tehran, 89 pp., 34 pls., 3 maps.

HARRIS, C. J. 1968. Otters. A study of the Recent Lutrinae. Weidenfeld and Nicolson, London, xiv + 397 pp.

HARRIS, W. P., JR. 1943. A list of mammals from Costa Rica. Univ. Michigan Mus. Zool., Occas. Pap., 476:1–15.

HARRISON, D. L. 1964. The mammals of Arabia. Vol. 1, Insectivora, Chiroptera, Primates. Ernest Benn, London, xx + 1–192 pp.

HARRISON, D. L. 1968. ———. Vol. 2, Carnivora, Hyracoidea, Artiodactyla, xvi + 193–381 pp.

HARRISON, D. L. 1972. ———. Vol. 3, Lagomorpha, Rodentia, xvii + 382–670 pp.

HASSINGER, J. D. 1973. A survey of the mammals of Afghanistan. Fieldiana Zool., 60:1–195.

HATT, R. T. 1959. The mammals of Iraq. Misc. Publ. Mus. Zool. Univ. Michigan, 106:1–113.

HATUHOV [=KHATUKHOV], A. M., AND A. K. TEMBOTOV. 1978. Pg. 390 in II Congr. Theriol. Intern. (Brno), R. Obrtel, C. Folk, and J. Pellantova, eds. Inst. Vert. Zool., Czech. Acad. Sci., Brno.

HAYMAN, R. W. 1955. Mammals of Sierra Leone. Zoo Life (London), 10:2–6.

HAYMAN, R. W., AND J. E. HILL. 1971. Part 2, Order Chiroptera, in J. Meester and H. W. Setzer, eds., 1971–1977.

HEIM DE BALSAC, H. 1972. Insectivores. Pp. 629–660 in Biogeography and ecology in Madagascar, R. Battistina and G. Richard-Vindard, eds. W. Junk, The Hague.

HEIM DE BALSAC, H., AND J. MEESTER. 1977. Part 1, Order Insectivora, main text, in J. Meester and H. W. Setzer, eds., 1971–1977.

HENDEY, Q. B. 1980. Agriotherium (Mammalia, Ursidae) from Langebaanweg, South Africa, and relationships of the genus. Ann. S. Afr. Mus., 81(1):1–109.

HEPTNER, V. G., AND N. P. NAUMOV, eds. 1961. Mlekopitayushchie Sovetskovo Soyuza. [Mammals of the Soviet Union.] Vol. 1, Parnokopytnye i Neparnokopytnye. [Artiodactyla and Perissodactyla.] Vysshaya Shkola, Moscow, 776 pp.

HEPTNER, V. G., AND N. P. NAUMOV, eds. 1967. ———. Vol. 2, Part 1, Morskie korovy i khishchye. [Sea cows and carnivores.] Vysshaya Shkola, Moscow, 1004 pp.

HEPTNER, V. G., AND N. P. NAUMOV, eds. 1972. ———. Vol. 2, Part 2, Khishchye (gieny i koshki). [Carnivores (hyaenas and cats).] Vysshaya Shkola, Moscow, 551 pp.

HEPTNER, V. G., AND N. P. NAUMOV, eds. 1976. ———. Vol. 2, Part 3, Lastonogie i zubatye kity. [Pinnipeds and toothed whales.] Vysshaya Shkola, Moscow, 718 pp.

HERSHKOVITZ, P. 1948. Mammals of northern Colombia, preliminary report no. 3: water rats (genus Nectomys), with supplemental notes on related forms. Proc. U.S. Nat. Mus., 98:49–56.

HERSHKOVITZ, P. 1949. Mammals of northern Colombia, preliminary report no. 4: monkeys (Primates), with taxonomic revisions of some forms. Proc. U.S. Nat. Mus., 98:323–427.

HERSHKOVITZ, P. 1949. Mammals of northern Colombia, preliminary report no. 5: bats (Chiroptera). Proc. U.S. Nat. Mus., 99:429–454.

HERSHKOVITZ, P. 1962. Evolution of Neotropical cricetine rodents (Muridae) with special reference to the phyllotine group. Fieldiana Zool., 46:1–524.

HERSHKOVITZ, P. 1966. Catalog of living whales. Bull. U.S. Nat. Mus., 246:1–259.

HERSHKOVITZ, P. 1972. The Recent mammals of the Neotropical Region: a zoogeographic and ecological review. Pp. 311–431 in Evolution, mammals, and southern continents, A. Keast, F. C. Erk, and B. Glass, eds. State Univ. New York Press, Albany.

HERSHKOVITZ, P. 1972. Notes on New World monkeys. Intern. Zoo Yrb., 12:3–12.

HERSHKOVITZ, P. 1977. Living New World monkeys (Platyrrhini). Volume I. Univ. Chicago Press, 1117 pp.

HICKMAN, G. C. 1981. National mammal guides: a review of references to Recent faunas. Mammal Rev., 11(2):53–85.

HILL, J. E., AND T. D. CARTER. 1941. The mammals of Angola, Africa. Bull. Am. Mus. Nat. Hist., 78:1–211.

HILL, W. C. OSMAN. 1953. Primates: comparative anatomy and taxonomy. Vol. 1, Strepsirhini. Edinburgh Univ. Press, Edinburgh, xxiii + 798 pp.

HILL, W. C. OSMAN. 1955. ———. Vol. 2, Haplorhini: Tarsioidea. Edinburgh Univ. Press, xxii + 347 pp.

HILL, W. C. OSMAN. 1957. ———. Vol. 3, Pithecoidea, Platyrrhini (Families *Hapalidae* and *Callimiconidae*). Edinburgh Univ. Press, xix + 354 pp.

HILL, W. C. OSMAN. 1960. ———. Vol. 4, Cebidae, Part A. Edinburgh Univ. Press, xxii + 523 pp.

HILL, W. C. OSMAN. 1962. ———. Vol. 5, Cebidae, Part B. Edinburgh Univ. Press, xxi + 537 pp.

HILL, W. C. OSMAN. 1966. ———. Vol. 6, Catarrhini, Cercopithecoidea, Cercopithecinae. Edinburgh Univ. Press, xxiii + 757 pp., 19 maps.

HILL, W. C. OSMAN. 1970. ———. Vol. 8, Cynopithecinae. *Papio, Mandrillus, Theropithecus.* John Wiley, New York, xix + 680 pp., 11 maps.

HILL, W. C. OSMAN. 1974. ———. Vol. 7, Cynopithecinae. *Cercocebus, Macaca, Cynopithecus.* John Wiley, New York, xxi + 934 pp., 13 maps.

HILL, W. C. OSMAN, AND J. MEESTER. 1977. Part 3.2, Infraorder Lorisiformes, in J. Meester and H. W. Setzer, eds., 1971–1977.

HINTON, H. E., AND A. M. S. DUNN. 1967. The mongooses: their natural history and behaviour. Oliver and Boyd, Edinburgh, vii + 144 pp.

HOFFMANN, M. 1958. Die Bisamratte. Ihre Lebensgewohnheiten, Verbreitung, Bekampfung und wirtschaftliche bedeutung. Akad. Verlag, Leipzig, viii + 260 pp.

HOOPER, E. T. 1960. The glans penis in *Neotoma* (Rodentia) and allied genera. Univ. Michigan Mus. Zool., Occas. Pap., 618:1–21.

HOOPER, E. T., AND B. S. HART. 1962. A synopsis of Recent North American microtine rodents. Univ. Michigan Mus. Zool., Occas. Pap., 120:1–68.

HUFNAGL, E. 1972. Libyan mammals. Oleander Press, New York.

HUNSAKER, D., II, ed. 1976. The biology of marsupials. Acad. Press, New York, xv + 537 pp.

HUNT, R. M. 1974. The auditory bulla in Carnivora: an anatomical basis for reappraisal of carnivore evolution. J. Morph., 143(1):21–76.

HUSSON, A. M. 1960. De Zoogdieren van de Nederlandse Antillen. Natuurwetens. Werkgroep Nederl. Antillen, 12:1–70.

HUSSON, A. M. 1978. The mammals of Suriname. Brill, Leiden, xxxiv + 569 pp., 161 pls.

IMAIZUMI, Y. 1960. Coloured illustrations of the mammals of Japan. Hoikusha Publ. Co., Osaka, 196 pp.

IMAIZUMI, Y. 1970. The handbook of Japanese land mammals (in Japanese; English summaries). Shin-Shichoa-Sha, Tokyo, xxx + 350 pp.

ISHUNIN, G. I. 1961. Fauna Uzbekskoi SSR. Vol. 3, Mlekopitayushchie (Khishchye i kopytnye). [Mammals (Carnivores and ungulates).] Akad. Nauk Uzbek. SSR, Tashkent, 231 pp.

JENKINS, P. D. 1976. Variation in Eurasian shrews of the genus *Crocidura* (Insectivora: Soricidae). Bull. Br. Mus. (Nat. Hist.), Zool., 30(7):269–309.

JEPSEN, G. L. 1970. Bat origins and evolution. Pp. 1–64 in Biology of bats, Vol. 1, W. A. Wimsatt, ed. Acad. Press, New York.

JONES, G. S., F.-L. HUANG, AND T.-Y. CHANG. 1969. A checklist and the vernacular names of Taiwan mammals (excluding Sirenia, Pinnipedia, and Cetacea): a review of the literature. Chinese J. Microbiol., 2:47–65.

JONES, G. S., AND D. B. JONES. 1976. A bibliography of the land mammals of southeast Asia 1699–1969. Spec. Publ. Dept. Entomol., Bernice P. Bishop Mus., Honolulu, Hawaii, 238 pp.

JONES, J. K., JR. 1966. Bats from Guatemala. Univ. Kansas Publ., Mus. Nat. Hist., 16:439–472.

JONES, J. K., JR., AND D. CARTER. 1976. Pp. 7–38 in Biology of bats of the New World family Phyllostomatidae. Part I, R. J. Baker, J. K. Jones, Jr., and D. Carter, eds. Spec. Publ. Mus. Texas Tech. Univ., 10:1–218.

JONES, J. K., JR., AND D. CARTER. 1979. Pp. 7–11 in Biology of bats of the New World family Phyllostomatidae. Part III, R. J. Baker, J. K. Jones, Jr., and D. Carter, eds. Spec. Publ. Mus. Texas Tech. Univ., 16:1–441.

JONES, J. K., JR., AND D. H. JOHNSON. 1960. Review of the insectivores of Korea. Univ. Kansas Mus. Nat. Hist. Publ., 9:549–578.

JONES, J. K., JR., AND D. H. JOHNSON. 1965. Synopsis of the lagomorphs and rodents of Korea. Univ. Kansas Mus. Nat. Hist. Publ., 16:357–407.

JONES, J. K., JR., AND C. J. PHILLIPS. 1970. Comments on systematics and zoogeography of bats in the Lesser Antilles. Studies on the fauna of Curacao and other Caribbean islands, 32:131–145.

JUNGE, J. A., AND R. S. HOFFMANN. 1981. An annotated key to the long-tailed shrews (genus Sorex) of the United States and Canada, with notes on Middle American Sorex. Occas. Paps. Mus. Nat. Hist., Univ. Kansas, 94:1–48.

KEAST, A. 1972. Australian mammals: zoogeography and evolution. Pp. 195–246 in Evolution, mammals and southern continents, A. Keast, F. C. Erk, and B. Glass, eds. State Univ. New York Press, Albany.

KEAST, A., F. C. ERK, AND B. GLASS, eds. 1972. Evolution, mammals, and southern continents. St. Univ. New York Press, Albany, 543 pp.

KETENCHIEV, KH. A. 1980. Pp. 21–22 in P. A. Panteleev, ed. Nauka, Moscow, 471 pp.

KHATUKHOV, A. M. 1980. Pp. 382–383 in P. A. Panteleev, ed. Nauka, Moscow, 471 pp.

KING, J. A., ed. 1968. Biology of Peromyscus (Rodentia). Spec. Publ. No. 2, Amer. Soc. Mammal., xiii + 593 pp.

KING, J. E. 1954. The otariid seals of the Pacific Coast of America. Bull. Br. Mus. (Nat. Hist.), Zool., 2:309–337.

KINGDON, J. 1971. East African mammals. An atlas of evolution in Africa. Acad. Press, New York, [Vol. 1] x + 446 pp.

KINGDON, J. 1974. ———. Vol. 2, Part A (Insectivores and bats), xi + 1–341 + 1 pp.

KINGDON, J. 1974. ———. Vol. 2, Part B (Hares and Rodents), ix + 343–703 + lvii pp.

KINGDON, J. 1977. ———. Vol. 3, Part A (Carnivores), viii + 476 pp.

KINGDON, J. 1979. ———. Vol. III, Part B (Large Mammals), v + 436 pp.

KIRKPATRICK, R., AND A. CARTWRIGHT. 1975. List of mammals known to occur in Belize. Biotropica, 7:136–140.

KIRSCH, J. A. W. 1977. The classification of marsupials. Pp. 1–50 in The biology of marsupials, D. Hunsaker, II, ed. Acad. Press, New York.

KIRSCH, J. A. W., AND J. H. CALABY. 1977. The species of living marsupials: an annotated list. Pp. 9–26 in The biology of marsupials, B. Stonehouse and D. Gilmore, eds. Macmillan Press, Baltimore.

KONTRIMAVICHUS, V. L., ed. 1976. Beringiya v Kainozoe. [Beringia in Cenozoic.] Acad. Sci., Vladivostok, 594 pp.

KOOPMAN, K. F. 1978. Zoogeography of Peruvian bats with special emphasis on the role of the Andes. Am. Mus. Novit., 2651:1–33.

KOOPMAN, K. F., AND J. K. JONES, JR. 1970. Classification of bats. Pp. 22–28 in About bats: a chiropteran symposium, B. H. Slaughter and D. W. Walton, eds. Southern Methodist Univ. Press, Dallas, Texas, vii + 339 pp.

KOWALSKI, K., ed. 1964. Klucze do oznaczania kregowcow Polski. [Keys to classification of vertebrates of Poland.] Part 5, Skaki-Mammalia. Panstowowe wydawn Naukowe, Warsaw-Krakow, 180 pp.

KUHN, H.-J. 1965. A provisional checklist of the mammals of Liberia. Senckenberg. Biol., 46:321–340.

KURODA, N. 1938. A list of Japanese mammals. Tokyo.

KURODA, N. 1940. A monograph of the Japanese mammals. Sansiedo Co., Ltd., Tokyo and Osaka, 311 pp.

KURTEN, B. 1968. Pleistocene mammals of Europe. Aldine, Chicago, viii + 317 pp.

KURTEN, B., AND E. ANDERSON. 1980. Pleistocene mammals of North America. Columbia Univ. Press, New York, xvii + 442 pp.

LANDRY, S. O. 1957. The interrelationships of the New World and Old World hystricomorph rodents. Univ. California Publ. Zool., 56:1–118.

LANFRANCO, G. G. 1969. Maltese mammals. Progress Press Company, Malta.

LANGGUTH, A. 1975. Ecology and evolution in the South American canids. Pp. 192–206 in The wild canids. Their systematics, behavioral ecology, and evolution, M. W. Fox, ed. Van Nostrand and Reinhold Co., New York.

LARGEN, M. J., D. KOCK, AND D. W. YALDEN. 1974. Catalogue of the mammals of Ethiopia. 1. Chiroptera. Monitore Zool. Ital., n.s. suppl., 5:221–298.

LAURIE, E. M. O., AND J. E. HILL. 1954. List of land mammals of New Guinea, Celebes, and adjacent islands 1758–1952. Br. Mus. (Nat. Hist.), London, 175 pp.

LAVOCAT, R. 1974. What is an hystricomorph? Pp. 7–20 in The biology of hystricomorph rodents, I. W. Rowlands and B. Weir, eds. Symp. Zool. Soc. London. Acad. Press, New York.

LAY, D. M. 1967. A study of the mammals of Iran. Fieldiana Zool., 54:1–282.

LEATHERWOOD, S., D. K. CALDWELL, AND H. E. WINN. 1976. Whales, dolphins, and porpoises of the western North Atlantic. A guide to their identification. Nat. Oc. At. Admin., Nat. Marine Fish. Ser., iv + 176 pp.

LEKAGUL, B., AND J. A. MCNEELY. 1977. Mammals of Thailand. Assoc. Cons. Wildl., Kuruspha Ladprao Press, Bangkok, 758 pp.

LEOPOLD, A. S. 1959. Wildlife of Mexico—The game birds and mammals. Univ. California Press, Berkeley, xvi + 568 pp.

LEWIS, R. E., J. H. LEWIS, AND S. I. ATALLAH. 1967. A review of Lebanese mammals. Lagomorpha and Rodentia. J. Zool., 153:45–70.

LEWIS, R. E., J. H. LEWIS, AND S. I. ATALLAH. 1968. A review of Lebanese mammals. Carnivora, Pinnipedia, Hyracoidea and Artiodactyla. J. Zool., 154:517–531.

LUCKETT, W. P., ed. 1980. Comparative biology and evolutionary relationships of tree shrews. Plenum Publ. Co., New York, xv + 314 pp.

LUCKETT, W. P., AND F. S. SZALAY, eds. 1975. Phylogeny of the primates: a multidisciplinary approach. Plenum Publ. Co., New York, xiv + 483 pp.

MACHADO, A. DE BARROS. 1969. Mamiferos de Angola ainda nao citadoes ou pouco conhecidos. Publ. Cultur. Companhia de Diamante Angola, 46:93–232.

MAGLIO, V. J. 1973. Origin and evolution of the Elephantidae. Trans. Amer. Phil. Soc., 63:1–149.

MAGLIO, V. J., AND H. B. S. COOKE, eds. 1978. Evolution of African mammals. Harvard Univ. Press, Cambridge, xiii + 641 pp.

MANN FISCHER, G. 1978. Los pequinos mamiferos de Chile (marsupiales, quiropteros, edentados y roedores). Guyana (Zool.), 40:342 pp.

MARES, M. A., R. A. OJEDA, AND M. P. KOSCO. 1981. Observations on the distribution and ecology of the mammals of Salta Province, Argentina. Ann. Carnegie Mus., 50(6):151–206.

MARES, M. A., R. R. WILLIG, K. E. STREIKIN, AND T. E. LACHER, JR. 1981. The mammals of northeastern Brazil: a preliminary assessment. Ann. Carnegie Mus., 50(4):81–137.

MARSHALL, P. 1967. Wild mammals of Hong Kong. Oxford Univ. Press, Hong Kong.

MASER, C., AND R. M. STORM. 1970. A key to Microtinae of the Pacific Northwest (Oregon, Washington, Idaho). O.S.U. Book Stores, Corvallis, Oregon, xiii + 163 pp.

MATYUSHKIN, E. 1979. Rysi Golarktiki. [Lynx of the Holarctic.] Pp. 76–162 in Mlekopitayushchie. Issledovaniya po faune Sovetskovo Soyuza [Mammals. Investigations on the fauna of the Soviet Union], O. L. Rossolimo, ed. Sbor. Trud. Zool. Mus., Moscow St. Univ., 28:1–280.

MAXWELL, G. 1967. Seals of the world. Constable, London, xii + 153 pp.

MCKENNA, M. C. 1975. Towards a phylogenetic classification of the Mammalia. Pp. 21–46 in Phylogeny of the primates: a multidisciplinary approach. W. P. Luckett and F. S. Szalay, eds. Plenum Publ. Co., New York.

MEDWAY, LORD. 1969. The wild mammals of Malaya and offshore islands including Singapore. Oxford Univ. Press, London, xix + 127 pp.

MEDWAY, LORD. 1970. The monkeys of Sundaland: ecology and systematics of the cercopithecids of a humid equatorial environment. Pp. 513–553 *in* Old World monkeys: evolution, systematics, and behavior, J. R. Napier and P. H. Napier, eds. Acad. Press, New York.

MEDWAY, LORD. 1977. Mammals of Borneo: field keys and an annotated checklist, 2nd ed. Monogr. Malays. Br. R. As. Soc., Kuala Lumpur, No. 7:xii + 1–172.

MEDWAY, LORD. 1978. The wild mammals of Malaya (Peninsular Malaysia) and Singapore, 2nd ed. Oxford Univ. Press, Kuala Lumpur, 128 pp.

MEESTER, J. 1972. Part 4, Order Pholidota, *in* J. Meester and H. W. Setzer, eds., 1971–1977.

MEESTER, J. 1974. Part 1.3, Family Chrysochloridae, *in* J. Meester and H. W. Setzer, eds., 1971–1977.

MEESTER, J., AND H. W. SETZER, eds. 1971–1977. The mammals of Africa: an identification manual. Parts 1–15. Smithson. Inst. Press, Washington, D.C.

MEYLAN, A. 1966. List des mammiferes de Suisse. Bull. Soc. Vaudoise Sci. Nat., 69:233–245.

MILLER, G. S., JR., AND R. KELLOGG. 1955. List of North American Recent mammals. Bull. U.S. Nat. Mus., 205:xii + 954.

MISONNE, X. 1974. Part 6, Order Rodentia, main text, *in* J. Meester and H. W. Setzer, eds., 1971–1977.

MITCHELL, R. M. 1977. Accounts of Nepalese mammals and analysis of the host-ectoparasite data by computer techniques. Unpubl. Ph.D. Dissertation, Iowa St. Univ., Ames, xii + 558 pp.

MONTGOMERY, G. G., ed. 1982. The evolution and ecology of sloths, anteaters and armadillos (Mammalia, Xenarthra = Edentata). Smithson. Inst. Press, Washington, D.C., *in press.*

MOORE, J. C. 1959. Relationships among living squirrels of the Sciurinae. Bull. Am. Mus. Nat. Hist., 118:157–206.

MOORE, J. C., AND G. H. H. TATE. 1965. A study of the diurnal squirrels, Sciurinae, of the Indian and Indochinese subregions. Fieldiana Zool., 48:1–351.

MOREAU, R. E., G. H. E. HOPKINS, AND R. W. HAYMAN. 1946. The type localities of some African mammals. Proc. Zool. Soc. London, 1945–46:387–447.

NAPIER, J. R., AND P. H. NAPIER. 1967. A handbook of living primates. Acad. Press, London, 456 pp.

NAPIER, J. R., AND P. H. NAPIER, eds. 1970. Old World monkeys: evolution, systematics, and behavior. Acad. Press, New York, xvi + 660 pp.

NAPIER, P. H. 1976. Catalog of primates in the British Museum (Natural History). Part I: Families Callithricidae and Cebidae. Br. Mus. (Nat. Hist.), London.

NICHOLSON, A. J., AND D. W. WARNER. 1953. The rodents of New Caledonia. J. Mammal., 34:168–179.

NIETHAMMER, J. 1974. Zur Verbrietung und Taxonomie grieschischer Saugetiere. Bonn. Zool. Beitr., 25:28–55.

NIETHAMMER, J., AND F. KRAPP, eds. 1978. Handbuch der Saugetiere Europas. Vol. 1, Rodentia I (Sciuridae, Castoridae, Gliridae, Muridae). Akad. Verlag, Wiesbaden, 476 pp.

NORRIS, K. S., ed. 1966. Whales, dolphins, and porpoises. Univ. Calif. Press, Berkeley, xv + 789 pp.

OGNEV, S. I. 1928. Zveri vostochnoi Evropy i severnoi Azii. [Mammals of eastern Europe and northern Asia.] Gosudarst. Izdat., Moscow-Leningrad. Vol. 1, Insectivora and Chiroptera, xv + 631 pp. Engl. translation, Off. Tech. Serv., Washington, D.C., 1962.

OGNEV, S. I. 1931. ———. Vol. 2, Carnivora (Fissipedia), xi + 776 pp. Engl. translation, Off. Tech. Serv., Washington, D.C., 1962.

OGNEV, S. I. 1935. Zveri SSSR i prilezhashchikh stran. [Mammals of the USSR and adjacent countries.] Vol. 3. Carnivora (Fissipedia, Pinnipedia), viii + 752 pp. Engl. translation, Off. Tech. Serv., Washington, D.C., 1962.

OGNEV, S. I. 1940. ———. Akad. Nauk, Vol. 4. Rodentia (Duplicidentata, Simplicidentata), 615 pp. Engl. translation, Off. Tech. Serv., Washington, D.C., 1966.

OGNEV, S. I. 1947. ———. Vol. 5, Glires, 809 pp. Engl. translation, Off. Tech. Serv., Washington, D.C., 1963.

OGNEV, S. I. 1948. ———. Vol. 6, Glires, 559 pp. + 12 maps. Engl. translation, Off. Tech. Serv., Washington, D.C., 1963.

OGNEV, S. I. 1950. ———. Vol. 7, Glires, 706 pp. + 15 maps. Engl. translation, Off. Tech. Serv., Washington, D.C., 1963.

OGNEV, S. I. 1957. See Tomilin, 1957.

ORLOV, V. N. 1974. Kariosistematika mlekopitayushchikh. [Karyosystematics of mammals.] Nauka, Moscow, 207 pp.

OSBORN, D. J., AND I. HELMY. 1980. The contemporary land mammals of Egypt (including Sinai). Fieldiana Zool., n.s. 5, 1309:xix + 1–579.

OSGOOD, W. H. 1943. The mammals of Chile. Field Mus. Nat. Hist., Zool. Ser., 30:1–268.

PALMER, T. S. 1904. Index generum mammalium: a list of the genera and families of mammals. N. Am. Fauna, 23:1–984.

PANOUSE, J. B. 1957. Les mammiferes du Maroc. Trav. l'Institut Sci. Cherifien, Ser. Zool., 5:1–210.

PANTELEEV, P. A., ed. 1980. Gryzyny. Materialy v Vsesoyunovo Soveshchaniya. [Rodents. Materials for the All-Union Conference.] Nauka, Moscow, 471 pp.

PEARSON, O. P. 1951. Mammals in the highlands of southern Peru. Bull. Mus. Comp. Zool., 106:117–174.

PEARSON, O. P., AND J. L. PATTON. 1976. Relationships among South American phyllotine rodents based on chromosome analysis. J. Mammal., 57:339–350.

PETERSON, R. L. 1955. North American moose. Univ. Toronto Press and R. Ontario Mus. Zool. Paleontol., xi + 280 pp.

PETERSON, R. L. 1966. The mammals of eastern Canada. Oxford Univ. Press, Toronto, xxxii + 465 pp.

PETTER, A., AND J. J. PETTER. 1977. Part 3.1, Infraorder Lemuriformes, in J. Meester and H. W. Setzer, eds., 1971–1977.

PETTER, F. 1972. The rodents of Madagascar: the seven genera of Malagasy rodents. Pp. 661–665 in Biogeography and ecology in Madagascar, R. Battistini and G. Richard-Vindard, eds. W. Junk, The Hague.

PETTER, F. 1972. Part 5, Order Lagomorpha, in J. Meester and H. W. Setzer, eds., 1971–1977.

PETTER, F. 1975. Part 6.2, Family Nesomyinae, in J. Meester and H. W. Setzer, eds., 1971–1977.

PETTER, F. 1975. Part 6.3, Subfamily Gerbillinae, in J. Meester and H. W. Setzer, eds., 1971–1977.

PETTER, F., AND R. MATTHEY. 1975. Part 6.7, Genus Mus, in J. Meester and H. W. Setzer, eds., 1971–1977.

PETTER, J. J. 1965. The lemurs of Madagascar. Pp. 292–319 in Primate behavior, I. DeVore, ed., Holt, Rinehart, and Winston, New York.

PILROVSKII, A. V., AND I. A. MAKARANETS. 1980. Pp. 259–260 in P. A. Panteleev, ed. Nauka, Moscow, 471 pp.

POCOCK, R. I. 1939. The fauna of British India, including Ceylon and Burma. Mammalia. Vol. I. Primates and Carnivora (in part). Families Felidae and Viverridae. Taylor and Francis, Ltd., London, xxxiii + 463 pp., 31 pls., 106 figs.

POCOCK, R. I. 1941. ———. Vol. II. Carnivora (cont. . .), suborders Aeluroidea (part) and Arctoidea. Taylor and Francis Ltd., London, xii + 503, 12 pls., 115 figs.

POOLE, A. L. 1973. Wild animals in New Zealand, 2nd ed. A. H. and A. W. Reed, Wellington.

PRAKASH, I. 1965. A list of the mammals of the Rajasthan Desert. J. Bengal Nat. Hist. Soc., 28:1–7.

PRAKASH, I., AND P. K. GHOSH, eds. 1975. Rodents in desert environments. W. Junk, The Hague, xxi + 624 pp.

PRATER, S. H. 1965. The book of Indian animals, 2nd (rev.) ed. Bombay Nat. Hist. Soc., Bombay, xxii + 323 pp.

QINGHAI-GANSU RESEARCH TEAM. 1964. Report on the mammals of Qinghai and Gansu provinces. Science Press, Peking, 80 pp. (in Chinese).

RAHM, U. H. 1970. Ecology, zoogeography and systematics of some African forest monkeys. Pp. 589–626 in Old World Monkeys: evolution, systematics, and behavior, J. R. Napier and P. H. Napier, eds. Acad. Press, London.

RANCK, G. I. 1968. The rodents of Libya. Taxonomy, ecology and zoogeographical relationships. Bull. U.S. Nat. Mus., 275:1–264.

RAUSCH, R. L. 1977. O zoogeografii nekotorykh Beringiiskikh mlekopitayushchikh. [On the zoogeography of some Beringian mammals.] Pp. 162–177 in Uspekhi Sovremennoi Teriologii. [Advances in modern theriology], V. E. Sokolov, ed. Nauka, Moscow (in Russian with English summary).

REIG, O. A. 1978. Roedores Cricetidos del Plioceno de la provincia de Buenos Aires (Argentina). Publ. Mus. Munic. Cienc. Nat. Mar del Plata "Lorenzo Scaglia," 2(8):164–190.

REPENNING, C. A. 1967. Subfamilies and genera of the Soricidae. Geol. Surv. Prof. Pap., 565:iv + 74 pp.

RICE, D. W. 1977. A list of the marine mammals of the world, 3rd ed. Nat. Oc. At. Admin., Tech. Rep., Nat. Marine Fish. Ser. SSRF-711, 15 pp.

RIDE, W. D. L. 1970. A guide to the native mammals of Australia. Oxford University Press, Melbourne, 249 pp.

ROBERTS, A. 1951. The mammals of South Africa. Central News Agency, South Africa, xlvii + 700 pp.

ROBERTS, T. J. 1977. The mammals of Pakistan. Ernest Benn, London, xxvi + 361 pp.

ROSEVEAR, D. R. 1953. Checklist and atlas of Nigerian mammals, with a foreword on vegetation. Nigerian Government, Lagos.

ROSEVEAR, D. R. 1965. The bats of West Africa. Br. Mus. (Nat. Hist.), London, xvii + 418 pp.

ROSEVEAR, D. R. 1969. The rodents of West Africa. Br. Mus. (Nat. Hist.), London, xii + 604 pp.

ROSEVEAR, D. R. 1974. The carnivores of West Africa. Br. Mus. (Nat. Hist.), London.

ROURE, G. 1962. Animaux Sauvages de la Cote d'Ivoire et du versant atlantique de l'Afrique intertropicale. Imprim. Nationale de la Cote d'Ivoire, Abidjen.

RUMBAUGH, D. M., ed. 1972. Gibbon and siamang. Vol. 1, Evolution, ecology, behavior, and captive maintenance. Karger, Basel, x + 263 pp.

RUMPLER, Y. 1975. The significance of chromosomal studies in the systematics of the Malagasy lemurs. Pp. 25–40 in Lemur Biology, I. Tattersal and R. W. Sussman, eds. Plenum Publ. Co., New York.

SAEMUNDSSON, B., AND M. DEGERBOL. 1939. Mammalia, in The zoology of Iceland, 4(76):1–52. E. Munksgaard, Copenhagen and Reykjavik.

SAINT GIRONS, M.-C. 1973. Les mammiferes de France et du Benelux (faune marine exceptee). Doin, Paris, ii + 481 pp.

SAVAGE, R. J. G. 1977. Evolution in carnivorous mammals. J. Paleont., 20(2):237–271.

SCHANPP, B. 1966. The mammals of Romania. Trav. Mus. d'Hist. Nat. "Grigore Antipa," 4:473–496.

SCHEFFER, V. B. 1958. Seals, sea lions, and walruses, a review of the Pinnipedia. Stanford Univ. Press, Stanford, Calif., x + 179 pp.

SCHEFFER, V. B., AND D. W. RICE. 1963. A list of the marine mammals of the world. U.S. Fish and Wildlife Serv., Spec. Sci. Rept. Fisheries, 431:1–12.

SCHMIDT-KITTLER, N. 1981. Zur Stammesgeschichte der marderverwandten Raubtiergruppen (Musteloidea, Carnivora). Eclogae Geol. Helv., 74(3):753–801.

SEAL, U. S., AND D. G. MAKEY. 1974. ISIS mammalian taxonomic directory. International Species Inventory System, Minnesota Zool. Garden, St. Paul, Minn., viii + 645 pp.

SETZER, H. W. 1956. Mammals of the Anglo-Egyptian Sudan. Proc. U.S. Nat. Mus., 106:447–587.

SETZER, H. W. 1957. A review of the Libyan mammals. J. Egyptian Publ. Health Assoc., 32:41–82.

SETZER, H. W. 1975. Part 6.5, Genus *Acomys, in* J. Meester and H. W. Setzer, eds., 1971–1977.

SHEN, S. 1963. Faunal characteristics of Tibetan mammals and the history of their organization. Acta Zool. Sin., 15:139–150 (in Chinese).

SHOU, C.-H. [=SHAW, T. H.]. 1964. Records of economic mammals of China. Sci. Publ. Off., Peking, xiii + 554 pp., 72 pl. (in Chinese).

SIDNEY, J. 1965. The past and present distribution of some African ungulates. Trans. Zool. Soc. London, 30:1–397.

SIIVONEN, L. 1967. Pohjolan nisakkaat. [Mammals of northern Europe.] Helsingissa Kustannusosakeyhtio Otavo, 194 pp. (in Finnish). (Swedish translation, Nordeuropas daggdjier, Stockholm, 1968.)

SILVA-TABOADA, G. 1979. Los Murcielagos de Cuba. Editorial Academia, Havana, 423 pp.

SIMPSON, G. G. 1945. The principles of classification and a classification of the mammals. Bull. Am. Mus. Nat. Hist., 85:i–xvi + 1–350.

SLAUGHTER, B. H., AND D. W. WALTON, eds. 1970. About bats: a chiropteran symposium. South. Methodist Univ. Press, Dallas, vii + 339 pp.

SLUDSKII, A. A., ed. 1969. Mlekopitayushchie Kazakhstana. [Mammals of Kazakhstan.] Vol. 1, Gryzuny (surki i susliki). [Rodents (marmots and ground squirrels).] Nauka, Alma-Ata, 455 pp.

SLUDSKII, A. A., AND E. I. STRAUTMAN, eds. 1980. Mlekopitayushchie Kazakhstana. [Mammals of Kazakhstan.] Vol. 2, Zaitseobraznye. [Lagomorphs.] Nauka, Alma-Ata, 236 pp.

SMITHERS, R. H. N. 1966. The mammals of Rhodesia, Zambia and Malawi. Collins, London, 159 pp.

SMITHERS, R. H. N. 1971. Mammals of Botswana. Nat. Mus. Rhodesia, Mus. Mem., 4:1–340.

SMITHERS, R. H. N., AND J. L. LOBAO TELLO. 1976. Checklist and atlas of the mammals of Mocambique. Nat. Mus. Monum., Salisbury, Mus. Mem., 8:1–184.

SMITHERS, R. H. N., AND V. J. WILSON. 1979. Checklist and atlas of the mammals of Zimbabwe Rhodesia. Nat. Mus. Monum., Salisbury, Mus. Mem., 9:1–193.

SOKOLOV, I. I., ed. 1963. See Gromov *et al.*, 1963.

SOKOLOV, V. E., ed. 1977. Uspekhi sovremennoi teriologii. [Advances in modern theriology.] Nauka, Moscow, 296 pp.

SOKOLOV, V. E. 1973. Sistematika mlekopitayushchikh [Systematics of mammals]. Vol. 1 [Monotremes, marsupials, insectivores, dermopterans, chiropterans, primates, edentates, pangolins], Vysshaya Shkola, Moscow, 430 pp.

SOKOLOV, V. E. 1977. ———. Vol. 2 [Lagomorphs, rodents], 494 pp.

SOKOLOV, V. E. 1979. ———. Vol. 3 [Cetaceans, carnivores, pinnipeds, tubulidentates, proboscideans, hyracoids, sirenians, artiodactyls, tylopods, perissodactyls], 528 pp.

SOKOLOV, V. E., YU. M. KOVALSKAYA, AND M. I. BASKEVICH. 1980. Gryzuny severnovo Kavkaza. [Rodents of the northern Caucasus.] Nauka, Moscow.

SOKOLOV, V. E., AND V. N. ORLOV. 1980. Opredelitel' mlekopitayushchikh Mongol'skoi Narodnoi Respubliki. [Guide to the mammals of the Mongolian People's Republic.] Nauka, Moscow, 351 pp.

SOUKOP, J. 1960–1961. Materiales para el catalogo de los mamiferos Peruanos. Biota, 3(26):240–276; 3(27):277–324; 3(28):325–331.

STARCK, D. 1974. Die Saugetiere Madagaskars, ihrs Lebenstraume und ihre Geschichte. Sitzungb. Ges. J. G. Goethe-Univ., Frankfort am Main, 11:67–124.

STONEHOUSE, B., AND D. GILMORE, eds. 1977. The biology of marsupials. Macmillan Press, Baltimore, viii + 486 pp.

STROGANOV, S. YU. 1957. Zveri Sibiri. Nasekomoyadnie. [Mammals of Siberia. Insectivores.] Acad. Sci., Moscow, 267 pp.

STROGANOV, S. YU. 1962. Zveri Sibiri. Khishchye. [Mammals of Siberia. Carnivores.] Acad. Sci., Moscow, 458 pp. Engl. translation, Jerusalem, 1969.

STRUHSAKER, T. T. 1970. Phylogenetic implications of some vocalizations of *Cercopithecus* monkeys. Pp. 365–444 *in* Old World monkeys: evolution, systematics, and behavior, J. R. Napier and P. H. Napier, eds. Acad. Press, New York.

STUBBE, M., AND N. CHOTOLCHU. 1968. Zur Saugetierfauna der Mongolei. Mitt. Zool. Mus. Berlin, 44:5-21.

STUBBE, M., AND N. CHOTOLCHU. 1971. Zur Saugetiere der Mongolei—III. Taiga Pfeifhasen, *Ochotona alpina hyperborea* (Pallas, 1811), aus dem Chentej. Mitt. Zool. Mus. Berlin, 47:349-356.

SWANEPOEL, P., R. H. N. SMITHERS, AND I. L. RAUTENBACH. 1980. A checklist and numbering system of the extant mammals of the Southern African subregion. Ann. Transvaal Mus., 32(7):155-196.

SWEENEY, R. C. H. 1959. A check list of the mammals of Nyasaland. Nyasaland Society, Blantyre.

SWYNNERTON, G. H., AND R. W. HAYMAN. 1951. A checklist of the land mammals of the Tanganyika territory and the Zanzibar Protectorate. East Afr. Nat. Hist. Soc., 20:274-392.

SZALAY, F. 1977. Phylogenetic relationships and a classification of eutherian Mammalia. Pp. 315-374 *in* Major patterns in vertebrate evolution, M. K. Hecht, P. C. Goody, and B. M. Hecht, eds. Plenum Press, New York.

SZALAY, F. S., AND E. DELSON. 1979. Evolutionary history of the primates. Acad. Press, New York, xiv + 580 pp.

TATE, G. H. H. 1939. The mammals of the Guiana region. Bull. Am. Mus. Nat. Hist., 76:151-229.

TATE, G. H. H. 1951. Results of the Archbold Expeditions. No. 65. The rodents of Australia and New Guinea. Bull. Am. Mus. Nat. Hist., 97:185-430.

TATTERSAL, I., AND R. W. SUSSMAN, eds. 1975. Lemur biology. Plenum Publ. Co., New York, ix + 365 pp.

TAVROVSKII, V. A. 1971. Mlekopitayushchie Yakutii. [Mammals of Yakutiya.] Nauka, Moscow, 660 pp.

TAYLOR, E. H. 1934. Philippine land mammals. Monogr. Bur. Sci., Manila, 30:1-548 pp., 25 pls.

TEDFORD, R. H. 1977. Relationship of pinnipeds to other carnivores (Mammalia). Syst. Zool. (for 1976), 25:363-374.

THENIUS, E. 1979. Zur systematischen und phylogenetischen Stellung des Bambusbaren: Ailuropoda melanoleuca David (Carnivora, Mammalia). Z. Saugetierk., 44:286-305.

THORINGTON, R. W., AND C. P. GROVES. 1970. An annotated classification of the Cercopithecoidea. Pp. 631-647 *in* Old World monkeys: evolution, systematics, and behavior, J. R. Napier and P. H. Napier, eds. Acad. Press, New York.

TOMICH, P. W. 1969. Mammals of Hawaii—a synopsis and notational bibliography. Bernice P. Bishop Mus. Spec. Publ. 57, Bishop Mus. Press, Honolulu, Hawaii, 238 pp.

TOMILIN, A. G. 1957. Zveri SSSR i prilezhashchikh stran. [Mammals of the USSR and adjacent countries.] Vol. 9, Cetacea. Engl. translation, Clearinghouse Fed. Sci. Tech. Inf., Springfield, VA, 1967.

TOPACHEVSKII, V. A. 1969. Slepyshovye [Blind mole rats] (Spalacidae). Fauna SSSR, Mlekopitayushchie [Mammals], 3(3):1-248. Engl. translation, Amerind Publ. Co., New Delhi, 1976.

TOSCHI, A. 1965. Faune d'Italia, Mammalia: Lagomorpha, Rodentia, Carnivora, Artiodactyla, Cetacea. Edizioni Calderini, Bologna.

TOSCHI, A., AND B. LANZA. 1959. Fauna d'Italia, Mammalia: Generalita, Insectivora, Chiroptera. Edizioni Calderini, Bologna.

TOVAR, A. 1971. Catalogo de mamiferos Peruanos. An. Cient., 9:18-37.

TROUESSART, E. L. 1898-1899. Catalogus mammalium tam viventium quam fossilium. Friedlander, Berolini, Vol. 1, vi + 1-664 pp.

TROUESSART, E. L. 1898-1899. ———. Vol. 2, v + 665-1469 pp.

TROUESSART, E. L. 1904-1905. ———. Suppl., vii + 929 pp.

TROUGHTON, E. 1947. Furred animals of Australia (1st Amer. ed.). Charles Scribner's Sons, New York, xxvii + 374 pp.

TROUGHTON, E. 1967. Furred animals of Australia, 9th ed. Angus and Robertson, Sydney.

TYLER, – –, ed. 1979. The status of endangered Australasian wildlife. Royal Zoological Society of South Australia, Adelaide, 220 pp.

U TUN YIN. 1967. Wild animals of Burma. Ukhin Pe Gyi, Rangoon, 301 pp.

VAN DEN BRINK, F. H. 1957. Die Saugetiere Europas. Paul Parey, Berlin. 225 pp.

VAN DEN BRINK, F. H. 1972. ———, 2nd ed., 217 pp.

VAN DEN BRINK, F. H. 1967. A field guide to the mammals of Britain and Europe. Collins, London, 221 pp.

VAN DEN BRINK, F. H. 1968. A field guide to the mammals of Britain and Europe. Houghton Mifflin Co., Boston, 221 pp.

VAN GELDER, R. G. 1977. Mammalian hybrids and generic limits. Am. Mus. Novit., 2635:1–25.

VAN GELDER, R. G. 1978. A review of canid classification. Am. Mus. Novit., 2646:1–10.

VAN PEENEN, P. F. D., P. F. RYAN, AND R. H. LIGHT. 1969. Preliminary identification manual for mammals of South Vietnam. Smithson. Inst., Washington, D.C., vi + 310 pp.

VAN ZYLL DE JONG, C. G. 1972. A systematic review of the Nearctic and Neotropical river otters (genus Lutra, Mustelidae, Carnivora). Life Sci. Contr., R. Ont. Mus., No. 80, iv + 104 pp.

VARONA, L. S. 1974. Catalogo de los Mamiferos Vivientes y Extinguidos de las Antillas. Acad. Cienc. Cuba (Havana), vii + 139 pp.

VAUGHAN, T. A. 1978. Mammalogy. 2nd ed. W. B. Saunders Co., Philadelphia, 522 pp.

VERESHCHAGIN, N. K. 1959. Mlekopitayushchie Kavkaza. [Mammals of the Caucasus.] Akad. Nauk, Moscow-Leningrad, 703 pp. Engl. translation, Clearinghouse Fed. Sci. Tech. Inf., Springfield, VA, 1967.

VERSCHUREN, J. 1967. Les grand mammiferes du Burundi. Mammalia, 42:209–224.

VILLA-R., B. 1967. Los murcielagos de Mexico. Anal. Inst. Biol., Univ. Nac. Auto. Mexico, xvi + 491.

VINOGRADOV, B. S., AND A. I. ARGYROPULO. 1968. Fauna of the USSR. Mammals. Key to rodents. Jerusalem, 1–241 (Translation of 1941 Russian publ.).

VON BLOEKER, J. D., JR. 1967. The land mammals of the southern California islands. Pp. 245–263 in Proc. Symp. Biol. Calif. Islands, Santa Barbara Botanic Garden.

VORONTSOV, N. N., ed. 1969. Mlekopitayushchie (evolyutsiya, kariologiya, sistematika, faunistika). [Mammals (evolution, karyology, systematics, faunistics).] Acad. Sci. U.S.S.R., Sib. Branch, Novosibirsk, 167 pp.

VORONTSOV, N. N., AND J. M. VAN BRINK, eds. 1980. Animal genetics and evolution. W. Junk B.V., The Hague, iv + 383 pp.

WALKER, E. P. et al. 1968. Mammals of the world, 2nd ed., Johns Hopkins Press, Baltimore, 2 vols., xlviii + 646; viii + 647–1500 pp.

WALKER, E. P. et al. 1975. Mammals of the world, 3rd ed., revised by J. L. Paradiso. Johns Hopkins Press, Baltimore, 2 vols., xlviii + 1–644; viii + 645–1500 pp.

WANG, S. 1965. Species accounts of mammals from southern Sinkiang. Pp. 158–212 in The Birds and Mammals of Southern Sinkiang, Mammalia. Y. W. Chien, C. Chang, P. L. Cheng, S. Wang, G. S. Guan, and S. Z. Shen, eds. Science Publishing House, Peking (in Chinese).

WATSON, L., AND T. RITCHIE. 1981. Sea guide to whales of the world. E. P. Dutton, New York, 302 pp.

WEBB, S. D., ed. 1974. Pleistocene mammals of Florida. Univ. Presses of Florida, Gainesville, x + 270 pp.

WESTON, M. 1979. Craniometric variation in the pika (Ochotona sp., O. Lagomorpha, F. Ochotonidae) and its evolutionary implications. Abstracts, World Lagomorph Conf., Univ. Guelph, 12–17 Aug. 1979. mimeo.

WETTSTEIN-WESTERSHEIMB, O. 1955. Mammalia. Catalogus Faunae Austriae, Teil 21c:1–16.

WETZEL, R. M., AND E. MONDOLFI. 1979. The subgenera and species of long-nosed armadillos, genus Dasypus L. Pp. 43–63 in Vertebrate ecology in the northern Neotropics, J. F. Eisenberg, ed. Smithson. Inst. Press, Washington, D.C.

WHITEHEAD, G. K. 1972. Deer of the world. Constable, London, xii + 194 pp.

WINGE, H. 1941. The interrelationships of the mammalian genera. Vol. 1. C. A. Reitzels Forlag, Copenhagen, 418 pp. (Danish translated by E. Deichmann and G. M. Allen.)

WITSTRUK, K.-G. 1970. Zur Wirbeltier Fauna Nordkoreas. Wissensch. Z. Univ. Halle, 19:123–139.

WODZICKI, K. A. 1950. Introduced mammals of New Zealand. Bull. Dept. Sci. Ind. Res., Wellington, 98:x + 255 pp.

WON, P. O., AND H. C. WOO. 1958. A distributional list of the Korean birds and mammals. Forest Exper. Sta. Inst. Agric. Korea, pp. 71–96.

WOOD, A. E. 1974. The evolution of the Old World and New World hystricomorphs. Pp. 21–60 in The biology of hystricomorph rodents, I. W. Rowlands and B. Weir, eds. Symp. Zool. Soc. London. Acad. Press, New York.

WOZENCRAFT, W. C. 1980. The phylogenetic status of civets and mongooses. Abstracts of papers . . . , 60th Ann. Meet. Am. Soc. Mammal., Univ. Rhode Island, Kingston, no. 119.

XIMENEZ, A., A. LANGGUTH, AND R. PRADERI. 1972. Lista sistematica de los mamiferos del Uruguay. An. Mus. Nat. Hist. Nat. Montevideo, 7:1–49.

YALDEN, D. W., M. J. LARGEN, AND D. KOCK. 1976. Catalogue of the mammals of Ethiopia. 2. Insectivora and Rodentia. Monitore Zool. Ital., n.s. suppl., 8:1–118.

YALDEN, D. W., M. J. LARGEN, AND D. KOCK. 1977. ———. 3. Primates. Monitore Zool. Ital., n.s. suppl., 9:1–52.

YALDEN, D. W., M. J. LARGEN, AND D. KOCK. 1980. ———. 4. Carnivora. Monitore Zool. Ital., n.s. suppl., 13:169–172. (1. Chiroptera, see Largen et al.)

YOUNGMAN, P. M. 1975. Mammals of the Yukon Territory. National Museum of Natural Sciences (Canada), Publ. Zool., No. 10:1–192.

YUDIN, B. S. 1971. Nasekomoyadnye mlekopitayushchie Sibiri. [Insectivorous mammals of Siberia.] Nauka, Novosibirsk, 169 pp.

YUDIN, B. S., L. I. GALKINA, AND A. F. POTAPKINA. 1979. Mlekopitayushchie Altae-Sayanskoi gornoi strany. [Mammals of the Altai-Sayan mountain region.] Nauka, Novosibirsk, 296 pp.

YUDIN, B. S., V. G. KRIVOSHEEV, AND V. G. BELYAEV. 1976. Melkie mlekopitayushchie severa Dal'nevo Vostoka. [Small mammals of the northern Far East.] Nauka, Sibir. Otd., Novosibirsk, 270 pp.

ZIEGLER, A. C., AND W. Z. LIDICKER. 1968. Keys to the genera of New Guinea Recent land mammals. Proc. Calif. Acad. Sci., 36:33–71.

ZIMMERMANN, K. 1964. Zur Saugetier-Fauna Chinas. Ergebnisse der Chinesisch-Deutschen Sammelreise durch Nord-und-Nordost-China. 1956. Mitt. Zool. Mus. Berlin, 40:87–140.

ZIMMERMANN, K., O. VON WETTSTEIN, H. SIEWERT, AND H. POHLE. 1953. Die wildsauger von Kreta. Z. Saugetierk., 17:1–72.

Index

gratus, Mus 532
gravesi, Sorex 93
gravida, Crocidura 74
gravipes, Dipodomys 383
grayi, Crocidura 75
  Mesoplodon 301
graysoni, Sylvilagus 603
greenhalli, Molossops 208
greenwoodi, Crocidura 75
gregalis, Microtus 492, 494
gregorianus, Thryonomys 592
grevyi, Equus 308
greyi, Macropus 46
  Rattus 549
greyii, Nycticeius 193
grimmia, Sylvicapra 342
grisea, Marmosa 21
  Murina 183
griseicauda, Stenocephalemys 556
griselda, Blarinella 68
  Lemniscomys 520, 521
griseoflavus, Graomys 418
  Phyllotis 410, 418
  Sciurus 363
griseoventris, Cryptotis 85
grisescens, Crocidura 74
  Myotis 187
griseus, Bradypus 53
  Cricetulus 405, 406
  Dusicyon 246
  Grampus 292
  Hapalemur 217
  Pteropus 121, 122
  Sciurus 364
  Zygodontomys 476
Grison 257
Grisonella 257
grobbeni, Gerbillus 416
groenlandica, Phoca 288
groenlandicus, Dicrostonyx 482, 483, 484
grootensis, Hydromys 518
gruberi, Soriculus 98
grunniens, Bos 328
  mutus, Bos 328
grypus, Halichoerus 286
guadeloupensis, Eptesicus 174
guairae, Proechimys 589, 590
guamara, Dasyprocta 576
guanicoe, Lama 318
guardafuensis, Genetta 267
guardia, Peromyscus 450
guariba, Alouatta 226
guatemalensis, Microtus 500
  Peromyscus 450, 452
  Pitymys 500
guaycuru, Myotis 191
guba, Xenuromys 559
gud, Microtus 492, 495, 497
gueldenstaedti, Crocidura 72, 75
guentheri, Madoqua 336
  Microtus 490, 492, 497
guereza, Colobus 233
Guerlinguetus 362, 364, 365
  aestuans 362
  cuscinus 364
  flammifer 364
  ignitus 364
  igniventris 364
  urucumus 365
guerrerensis, Cryptotis 85
  Liomys 386
  Sigmodon 466
guhai, Mus 551

guianae, Cavia 573
  Holochilus 419, 420
  Neacomys 428
guianensis, Sotalia 295
guigna, Felis 279
guillarmodi, Chlorotalpa 63
guineae, Tatera 469
guineensis, Crocidura doriana 74
  Crocidura odorata 74
  Eptesicus 174
  Rhinolophus 144
Gulo 257
  gulo 257
  luscus 257
gulo, Gulo 257
gundi, Ctenodactylus 594
gunnii, Perameles 35
gunningi, Amblysomus 62
gunnisoni, Cynomys 348
gurkha, Apodemus 507
gutturosa, Procapra 340
guyannensis, Proechimys 590, 591
gwatkinsi, Martes 259
gyas, Taterillus 470
Gymnobelideus 40
  leadbeateri 40
gymnocercus, Dusicyon 246
Gymnomys 558
gymnonotus, Pteronotus 149
Gymnopyga 235
gymnotis, Peromyscus 450, 452
  Phalanger 37
Gymnura 65
gymnura, Saccopteryx 132
Gymnuromys 418
  roberti 418
gymnurus, Cabassous 54
  Cratogeomys 380
  Echinosorex 65
  Hoplomys 588
  Pappogeomys 380
  Tylomys 475
Gyomys 545

H

habbema, Hydromys 518
habessinica, Procavia 313
habessinicus, Lepus 599
Habromys 418, 419, 448
  chinanteco 419
  ixtlani 419
  lepturus 419
  lophurus 419
  simulatus 419
Hadromys 517
  humei 517
Hadrosciurus 362, 364, 365
Haeromys 517
  margarettae 517
  minahassae 517
  pusillus 517
hageni, Dorcopsis 44
  Petinomys 359
haggardi, Ourebia 338
  Phyllotis 457
hainana, Petaurista 359
hainanensis, Neohylomys 66
haitiensis, Phyllops 162
halconus, Crocidura 76

Halichoerus 286
  grypus 286
halicoetes, Reithrodontomys 461
halli, Lestodelphys 19
hallucatus, Dasyurus 29
Haltomys 567
hamadryas, Papio 237
hamatus, Rattus 549
hamiltoni, Taphozous 132
hamlyni, Cercopithecus 232
hammondi, Oryzomys 440
handleyi, Lonchophylla 158
  Marmosa 21
hanglu, Cervus 320
  Cervus elaphus 320
hansae, Eumops 207
hantu, Chimarrogale 69
Hapalemur 217
  griseus 217
  simus 217
Hapalomys 517
  delacouri 517, 518
  longicaudatus 517
Haplomylomys 448
Haplonycteris 116
  fischeri 116
hardwickei, Kerivoula 178
  Rhinopoma 127
hargravei, Saccolaimus 131
harpax, Cynopterus 112
harpia, Harpiocephalus 176
Harpiocephalus 176
  harpia 176
Harpiola 182, 183
Harpyionycteris 116
  celebensis 116
  whiteheadi 116
harquahalae, Thomomys 381
harringtoni, Pelomys 540
  Taterillus 470, 471
harrisi, Scotinomys 465
harrisii, Ammospermophilus 345
  Sarcophilus 31
harrisoni, Hylopetes 353
  Kerivoula 178
  Musonycteris 161
  Thryonomys 592
harterti, Massoutiera 594
hartii, Artibeus 152
hartmannae, Equus zebra 309
hartmanni, Rheomys 462
hartwigi, Praomys 543
harveyi, Cephalophus 332
harwoodi, Gerbillus 413
hasseltii, Myotis 187
hastatus, Phyllostomus 163
hatti, Otonyctomys 446
haussa, Mus 532
hawashensis, Acomys 504, 505
hawkeri, Sorex 94
haydeni, Sorex 91, 93
haymani, Gerbillus 413
heathi, Scotophilus 203
heavisidii, Cephalorhynchus 291
hebes, Vulpes velox 250
hecki, Macaca 237
hectori, Cephalorhynchus 291
  Mesoplodon 301
hedenborgiana, Crocidura 74
heermanni, Dipodomys 383, 385
Heimyscus 518, 519
heinrichi, Bunomys 511
  Hyosciurus 353

M